D0212709

MODERN MANUFACTURING PROCESS ENGINEERING

McGraw-Hill Series in Industrial Engineering and Management Science

Consulting Editor: *James L. Riggs, Department of Industrial Engineering, Oregon State University*

Barish and Kaplan: *Economic Analysis: For Engineering and Managerial Decision Making*
Blank: *Statistical Procedures for Engineering, Management, and Science*
Cleland and Kocaoglu: *Engineering Management*
Denton: *Safety Management: Improving Performance*
Dervitsiotis: *Operations Management*
Gillet: *Introduction to Operations Research: A Computer-oriented Algorithmic Approach*
Hicks: *Introduction to Industrial Engineering and Management Science*
Huchingson: *New Horizons for Human Factors in Design*
Law and Kelton: *Simulation Modeling and Analysis*
Leherer: *White-Collar Productivity*
Love: *Inventory Control*
Niebel, Draper, and Wysk: *Modern Manufacturing Process Engineering*
Polk: *Methods Analysis and Work Measurement*
Riggs and West: *Engineering Economics*
Taguchi, Elsayed, and Hsiang: *Quality Engineering in Production Systems*
Riggs and West: *Essentials of Engineering Economics*
Wu and Coppins: *Linear Programming and Extensions*

Also available from McGraw-Hill

Schaum's Outline Series in Mechanical and Industrial Engineering

Each outline includes basic theory, definitions and hundreds of solved problems and supplementary problems with answers.

Current List Includes:

Acoustics
Basic Equations of Engineering
Continuum Mechanics
Engineering Economics
Engineering Mechanics, 4th edition
Fluid Dynamics
Fluid Mechanics & Hydraulics
Heat Transfer
Introduction to Engineering Calculations
Lagrangian Dynamics
Machine Design
Mechanical Vibrations
Operations Research
Strength of Materials, 2d edition
Theoretical Mechanics
Thermodynamics

Available at Your College Bookstore

MODERN MANUFACTURING PROCESS ENGINEERING

Benjamin W. Niebel
Professor of Industrial Engineering
Pennsylvania State University

Alan B. Draper
Late Professor of Industrial Engineering
Pennsylvania State University

Richard A. Wysk
Professor of Industrial Engineering
Pennsylvania State University

McGraw-Hill Publishing Company

New York St. Louis San Francisco Auckland Bogotá Caracas
Hamburg Lisbon London Madrid Mexico Milan Montreal
New Delhi Oklahoma City Paris San Juan São Paulo
Singapore Sydney Tokyo Toronto

This book was set in Times Roman.
The editors were John Corrigan and John M. Morriss;
the production supervisor was Denise L. Puryear.
The cover was designed by Joseph Gillians.
Project supervision was done by The Universities Press.
Arcata Graphics/Halliday was printer and binder.

MODERN MANUFACTURING PROCESS ENGINEERING

2 3 4 5 6 7 8 9 0 HAL HAL 8 9 4 3 2 1 0 9

ISBN 0-07-046563-0

Library of Congress Cataloging-in-Publication Data

Niebel, Benjamin W.
Modern manufacturing process engineering/Benjamin W. Niebel, Alan B. Draper, Richard A. Wysk.
p. cm. —— (McGraw-Hill series in industrial engineering and management science)
Includes index.
ISBN 0-07-046563-0
1. Manufacturing processes. I. Draper, Alan B. II. Wysk, Richard A., (date).
III. Title. IV. Series.
TS183.N54 1989
670.42——dc19 88-18770

ABOUT THE AUTHORS

Benjamin W. Niebel is Professor Emeritus of Industrial Engineering at The Pennsylvania State University. He was Professor and Chairman of the Department of Industrial and Management Systems Engineering at The Pennsylvania State University from 1955 until his retirement in 1978. Previous to entering education, he was Chief Industrial Engineer at the Lord Manufacturing Company. He has been consultant to more than 100 companies in the areas of process engineering, operations analysis, standards, methods, and engineering maintenance. He has authored more than 70 publications, including 12 books. The majority of this work is in the areas of process engineering, standards, methods, work simplification, wage payment, and maintenance. He is the holder of five patents.

A Fellow in the Institute of Industrial Engineers, he is the holder of the Phil Carroll Award for outstanding achievement in methods engineering and work measurement. He is also a holder of the Frank and Lillian Gilbreth Industrial Engineering Award for distinguishing himself by contributing to humanity through the use of Industrial Engineering. In addition, he holds an American Die Casting Institute, Inc., Doehler Award.

Alan B. Draper, deceased, was Professor of Industrial Engineering at The Pennsylvania State University. With more than 75 publications in the cast metal and metal processing field, Professor Draper was internationally known for his research relative to the use of the cupola as a production melting facility of ferrous materials, porosity of die castings, and die casting die materials for molding ferrous metals. Dr. Draper supervised some 35 Industrial Engineering theses and dissertations in the metal processing area. Included in these were seven dissertations related to metal casting.

Dr. Draper was the key Professor at Penn State representing the Foundry

Educational Foundation. More foundry scholars participated in this program at Penn State than at any other of the more than 40 colleges that are or have been involved in the FEF program. He is listed in both *Who's Who in Engineering* and *Who's Who in Technology*.

Richard A. Wysk is a Professor of Industrial Engineering and Director of the Manufacturing Research Center at The Pennsylvania State University. His current research focuses on the design, control, and analysis of automated manufacturing systems and the economic implementation of these systems. Dr. Wysk is the recipient of the 1981 SME Outstanding Young Manufacturing Engineer Award, the 1982 IIE Region III Award of Excellence, and the 1986 College of Engineering Research Award at The Pennsylvania State University. He is listed in *Who's Who in the East* (*1985*) and *Who's Who in Science and Technology* (*1985*).

Prior to joining the faculty at Penn State, Dr. Wysk was on the faculty at Virginia Polytechnic Institute and State University. Dr. Wysk has also held positions as Production Control Manager at General Electric and as a Research Analyst at Caterpillar Tractor Company. Dr. Wysk is a Senior Member of SME and IIE. He currently serves as a member of the SME ABET Accreditation Committee and the SME Manufacturing Engineering Education Foundation Capital Equipment Committee.

Dr. Wysk has published more than 50 articles in the open literature. He is also the coauthor of five textbooks.

CONTENTS

PREFACE

Manufacturing processes touch our lives every day. Manufacturing methods effect our industrial productivity which affects our very standard of living. This book is intended to provide an in depth introduction of manufacturing processes to the engineering and technology student. A constant focus of the book is to view manufacturing as a total system which begins with the functional design of the product and continues through the manufacturing process. The book provides a background of how to design so as to help ensure a successful product. In so doing, the book provides fundamental information on engineering materials, manufacturing processes, and automation so that the product designer can capitalize on the most favorable materials, processes, and methods available to transform his or her ideas into specifications for the product and process. The principal constraints associated with product design are developed so as to determine their influence on the design itself. These constraints include: quantity to be produced, delivery requirements, product reliability, the price the potential customer will pay, the value of appearance, human engineering, product function, and preventative maintenance.

This volume is a suitable text (at the undergraduate level) for students in various engineering and technology curricula. It should be of significant value to manufacturing, industrial, mechanical, and electrical engineering students concerned with product and process design. The book is also a valuable reference text for product design and manufacturing departments in industry. Design, manufacturing, and tool engineers will find the book a valuable addition to help meet increasing competition in the United States.

As is the case with any book the authors are indebted to a number of people that helped in the preparation of the book. Special thanks are due Paul H. Cohen, for his efforts to revise the chapter on Metal Cutting, Dr. Robert Lindsay, Professor Emeritus of Metallurgy; Dr. Samuel Zamrick, Professor of Engineering Science and Mechanics; and Paul C. Roche, Chairman of the Board, Corry Plastics, Inc., for their criticisms and contributions. Special thanks are also due to the following reviewers: William E. Biles, University of Louisville; David Bourell, University of Texas; Stephen Dickerson, Georgia Institute of Technology; David Dornfeld, University of California at Berkeley; Joseph ElGomayel, Purdue University; R. E. Goforth, Texas A & M University; George E. Kane, Lehigh University; John K. Schueller, University of Florida; and Klaus J. Weinmann, Michigan Technological University.

Benjamin W. Niebel
Alan B. Draper
Richard A. Wysk

MODERN MANUFACTURING PROCESS ENGINEERING

CHAPTER 1

THE ECONOMIC IMPACT OF MODERN PROCESS ENGINEERING

Modern process engineering, working in concert with the design and development of new products, is the basis for any advanced economy. The life-blood of any individual product-producing enterprise is the continual introduction of products or product improvements at competitive prices. Without this effort, the enterprise will stagnate and ultimately fail. With this effort, productivity will improve, the firm will become more competitive, and it will capture an increasing proportion of the market.

In order for industries to compete in an integrated world economy, it is necessary that output per hour be high at competitive wages and at an acceptable quality level. Science and technology are continually changing the physical limits to economic growth. The object of this book is to present fundamentals for the economic manufacture of a marketable commodity. This effort may include specifying and planning the sequence of operations and inspections to be performed; the design of the jigs, fixtures, and gauges; and special equipment needed to produce the work. Thus, process engineering will include the design, specification, or creation of tools, equipment, methods, and manufacturing information for the economic manufacture of a product.

There are three broad areas of study involved with effective process engineering:

1. Properties and behavior of materials

2. Basic, secondary, finishing, and assembly processes
3. Integration and quality control of manufacturing processes

The better the engineer understands the nature of materials and the reasons for their physical and mechanical properties, the more quickly and wisely he/she will be able to specify the most favorable shaping, forming, cutting, finishing, and test processes to produce the product in accordance with quality requirements and at the least cost.

The materials used to produce the majority of "hard" goods can be classified as ferrous, nonferrous, plastics, ceramics, powdered metals, and composites. It is to the manufacture of products made with these materials that this text is directed. Chapters on these materials families will be discussed in detail in the first third of this book.

It is fundamental that process engineers be knowledgeable regarding competitive process that can be specified to transform raw materials into the finished products being produced in their plants. This knowledge should include details of the dimensional capability of the various processes, rates of output, cost of tooling, and skills required. In this text processing has been divided into basic processes of the solid state and the plastic state, secondary material-removing processes, secondary forming processes, joining processes, decorative and coating process. The second third of the text covers this material.

Finally, today's process engineers need to be familiar with the interfacing of those related manufacturing systems that complement one another. They should be able to interface the process planning system with industrial engineering and work measurement systems, including automated time standards. Also, the interfacing of material requirements planning systems, using automated process planning and group technology to achieve a just-in-time manufacturing environment will be an area in which they should be competent. Process engineers will be able to implement the interfacing of quality assurance, traceability issues, and coordination with statistical process control efforts. These subject areas, including computer-integrated manufacturing, programmable controllers, numerical control and robotics, flexible manufacturing systems, reliability, and quality control are discussed in the final third of this text.

PROCESS ENGINEERING IS DYNAMIC

Improvement in the ways materials are processed is taking place continually. Industries that do not keep up to date and utilize innovative procedures will soon lose out to more progressive enterprises—witness the United States steel industry during the past twenty years. By not keeping up to date with the latest techniques in steel making, including continuous casting, electrogalvanizing, and thermomechanical control processing, the US steel industry lost considerable volume to the Japanese, Korean, and Brazilian producers. For example, controlled rolling and cooling adopted by the Japanese gives a weldable, tougher steel with no alloy additives. Previously, a rolled steel plate was left to cool by itself. But in this way, its grains expanded during the long cooling period that allowed for recrystallization. Japanese shipbuilders desired stronger steel for ship hulls. Strength could be improved readily by adding appropriate alloys. However, the alloyed steel was not as weldable and ship hulls made of alloyed steel plates were more likely to crack at the welds. Also, the alloyed steel plates were more costly to produce. The continuous on-line controlled process transforms the metal's microstructure from ferrite-pearlite to a ferrite-bainitic structure studded with martensitic islands that bolster its strength without a serious loss of toughness. Previously, slab coming out of a continuous caster was rolled and then left to cool by itself. Subsequently, it was reheated for rerolling, consuming more energy. The new controlled rolling and cooling process needs no reheating.

Steel users are quick to take advantage of improved quality brought about by better production methods. Steel companies that do not take advantage of the latest techniques will soon find no demand for their product. For example, world automobile makers in recent years have made more stringent specifications for flat-rolled steel, and consumers of products such as bar and rail have boosted their surface-quality requirements. In order to acquire orders for the highest-quality steel, producers need to increase their efforts to meet the new expectations for clean metal. Steelmakers recognize that the first step in making zero-defect steel is to manufacture clean steel. In this effort, some progressive mills have installed refining equipment on their basic oxygen furnaces in order to obtain cleaner steel. The lance bubbling equilibrium process injects gases at the bottom of the vessel to reduce the sulfur content and to lower the amount of impurities in the molten metal. Other modern techniques being introduced include ladle metallurgy, which further refines molten steel after it has been poured from oxygen furnaces into a ladle; air-mist cooling as opposed to spray cooling; nickel–chrome plating on copper caster molds; four-point straightening rather than single-point unbending; modifications in the mold oscillator; changes in the pitch of rolls, etc. Perhaps the most important ladle process that results in a higher quality product that is better suited for many of today's exacting specifications is vacuum degassing, which employs a purging vacuum action to remove dissolved gases such as hydrogen, oxygen, and nitrogen from molten steel.

Similar improvements have been introduced in almost all basic, secondary, and finishing processes. Today, the typical injection-molded thermoplastic part is produced complete in the molding process—no secondary operations are required. Modern machining centers will produce complex parts much faster and at closer tolerances, and with more refined surface finishes, than their counterpart individual machine tools of a decade ago.

PROCESSING COSTS

Costs are the basis of action within an organization. When the cost of processing a part becomes high in comparison to competitive ways of producing a component, then consideration will be given to making a change. Invariably, there are several alternatives for producing a given functional design that will compete on a cost basis. For example, casting will compete with forging, reaming with broaching, die casting with plastic molding, powdered metal with automatic screw machine, etc.

When determining the cost of a given process, engineers will need to consider both the direct and indirect costs. Under direct cost, they consider the cost of the direct labor and material. For example, in estimating the direct cost to produce an ABS component that is injection-molded, the analyst will first multiply the cost per pound of the ABS resin by the weight in pounds of the product, with due allowance for the sprue, runners, and normal shrinkage (typically 3 to 7 percent on complex thermoplastic parts). To this figure, the analyst will add the direct labor cost. For example, if the injection press operator services five machines and has an hourly rate (including the cost of fringe benefits) of $18 per hour, the direct labor cost per injection press would be $3.60 per hour or $0.06 per minute. If the cycle time for molding the part in question were 30 seconds, then the direct labor cost per piece would be $0.03.

Let us assume the cost of the resin to be $1.20 per pound and that the weight of each piece is 1 ounce. Also, let us assume the weight of the runners and sprue is 0.1 ounce and 5 percent shrinkage is characteristic of this particular molding room. Then the estimated material cost would be

$$\left(\frac{(\$1.20)(1.1)(1.05)}{16}\right) = \$0.087$$

In this example, the direct cost per piece would be $0.03 + $0.087 = $0.117.

Indirect costs, overhead or burden, and tool expense may have more influence on the selection of a particular process than material and direct labor costs. For example, referring to the above illustration, the single cavity mold may have cost $30,000 and the machine rate (cost of operating the injection press exclusive of the press operator) may be $20.00 an hour.

The reader should understand that most progressive companies today, through their accounting departments, carry a machine rate on the various facilities within the plant. This machine rate is developed by considering the following indirect cost increments: power consumed by the facility, depreciation of the facility, maintainence of the facility, floor space or cubic footage occupied by the facility, proportion of line supervision allocated to the facility, and small tool and/or perishable tools and supplies regularly consumed by the facility. Machine rates are often as low as a few dollars per hour and can range to as much as 50 dollars or more per hour in a complex machining center.

Of course, the allocation of tool costs has a very significant relationship to the quantity to be produced. Going back to our example, let us assume that 10,000 pieces were to be produced. This would give a unit tool cost of $30,000/10,000 or $3.00 per piece. This is, of course, much greater than the combined material and direct labor cost (more than 10 times). If 1,000,000 pieces are to be produced, the unit cost for the mold is only \$0.03 (about $\frac{1}{3}$ the cost of the direct labor and material). Let us assume the machine rate of this equipment (not including the mold cost) is \$20.00 per hour or \$0.333/minute and that a million pieces are desired. Then the total factory cost (direct material + direct labor + factory expense) would equal:

$$\$0.087 + \$0.03 + \$0.03 + \$0.333/2 = \$0.3135$$

In this text, we will be concerned primarily with factory cost, since this is the cost that impacts upon the choice of alternative ways of producing a given design. Other expenses (treated as overhead or burden) going into the total cost of a product include general expense (often referred to as administrative expense or burden) and sales expense or burden. Figure 1.1 illustrates the elements of cost and profit entering into the development of selling price.

An understanding of the basis of cost will enable the engineer to use good judgment in the selection of materials, processes, and functions to create the best product. An increase in perfection from 90 to 95 percent may result in a 50 percent increase in development and product cost and destroy the sales value of the product. Cost, quality, and degree of perfection should be considered carefully to obtain the greatest profit over a period of time. Cost is usually the deciding factor. The relationship between cost, sales, profit or loss, and volume is best revealed on what is known as a break-even (cross-over) chart. Figure 1.2 illustrates a typical break-even chart. Note, in this chart, expense or overhead is referred to as variable and fixed burden.

The reader should recognize that the distribution of cost factors will vary dramatically with the number of units to be produced. This has been demonstrated in connection with fixed burden shown in Fig. 1.2 and the

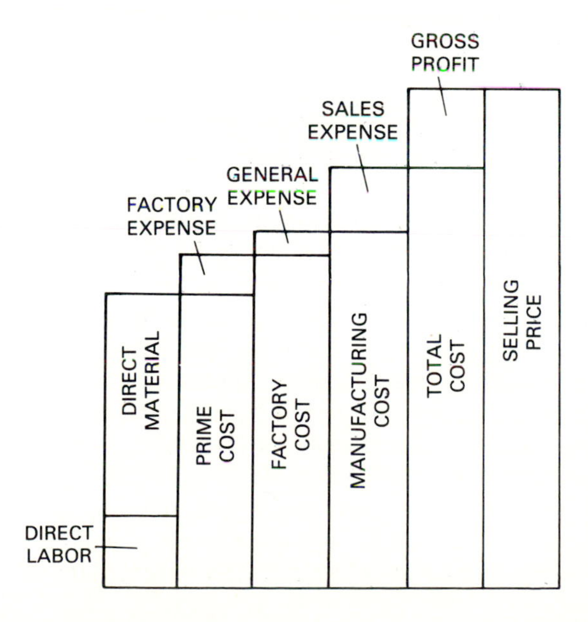

FIGURE 1.1
Elements of cost and gross profit (profit before taxes) entering into the development of selling price. Note that, in this particular product, material cost is approximately 53 percent of total cost and 17.5 percent of total cost is anticipated as gross profit.

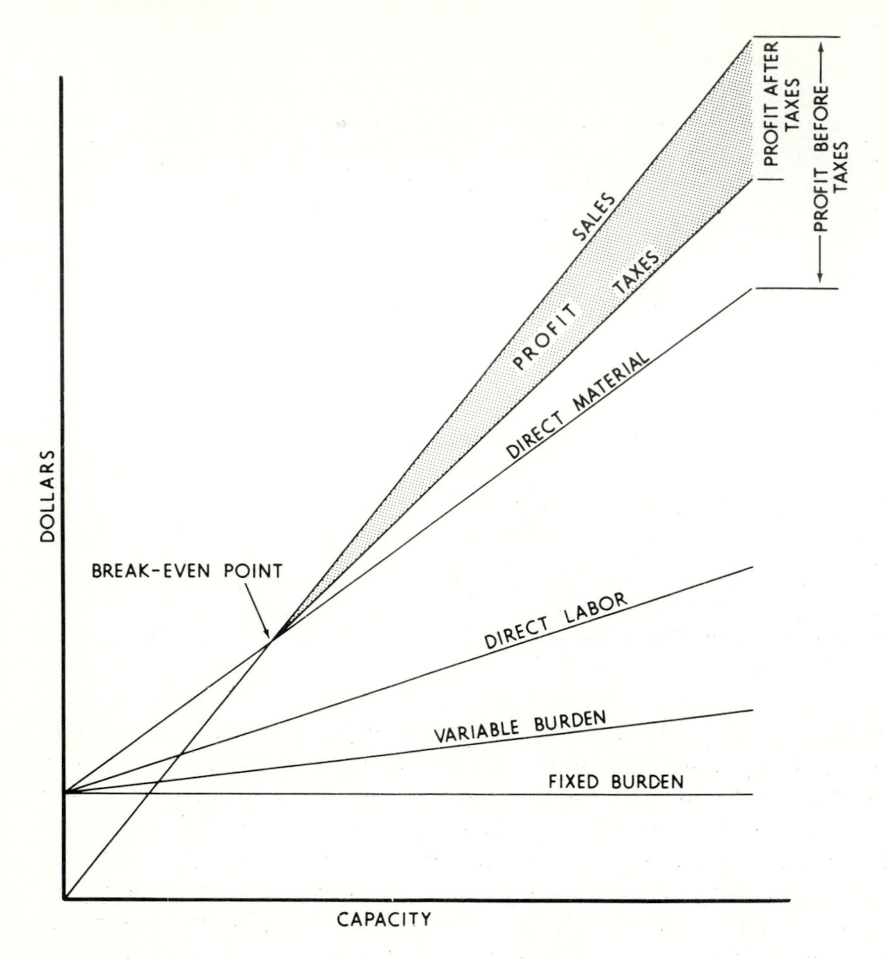

FIGURE 1.2
Break-even chart indicating relationship between cost, sales, profit or loss, and volume.

distribution of the mold cost referred to in the previous example. Figure 1.3 illustrates the emphasis on the various factors of cost. When quantities are low, the proportion of development cost is high compared to the cost of manufacturing overhead, direct labor, raw material, and purchased parts. The development cost includes design, preparation of drawings, writing manufacturing information, designing and building tools, testing, inspection, and many other items incidental to placing the first parts into production. As the number of units increases, the emphasis centers on reduction in factory overhead, direct labor, and material costs through advanced process engineering and manufacturing methods. When quantities to be produced are low, expenditures on tooling, automation, use of robotics, and engineering refinements will net less return in cost reduction. When quantities are high, an expenditure of engineering effort will result in reduced labor, material, and overhead costs per unit of output and give large returns for even small savings per unit. For example, if an automobile manufacturer produces 2 million cars per year of which 1 million have four cylinders per engine and the other 1 million have six cylinders per engine, and if there are four piston rings per cylinder, it is necessary to produce 40 million piston rings per year. A saving of $0.01 on each ring equals $400,000 per year. Therefore, to obtain the minimum cost, considerable engineering effort can be profitably used beginning with raw material and continuing up to the installation of the finished product.

There is constant competition between materials and processes based on costs that are influenced by the number of pieces made during a period of time. The activity of parts affects the amount of time the activity is operated compared with the hours available. The ratio of hours operated to hours available has considerable effect upon cost. Consider the following representative example.

A large hydraulic extrusion press, including the hydraulic pumps and building to house it, costs $3 million. Depreciation, maintenance, and interest on investment amount to 20 percent ($600,000) per year. There are normally 2000 working hours a year in one shift (8 hours/day × 5 = 40 hours/week × 50 weeks/year = 2000 hours/year). Three shifts would represent 6000 hours/yr available: 600,000/6000 = $100/h, the minimum of the cost of the facility during a 24-hour day. Actually, the sales department can sell only enough to keep the equipment busy 8 hours a day. Therefore, the machine costs 600,000/2000 = $300/hour. If sales decrease, the cost of the machine per hour will increase and make it difficult to operate the business at a profit.

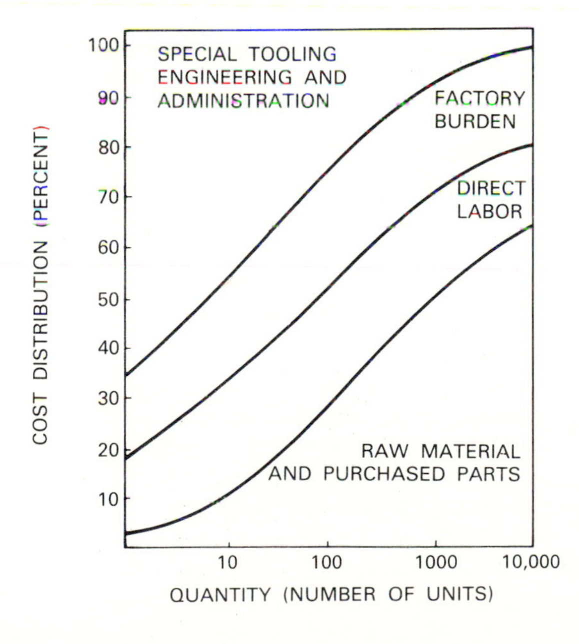

FIGURE 1.3
Cost-distribution factors charted in relation to quantity to indicate changing emphasis as production increases.

DISTRIBUTION OF FACTORY OVERHEAD

It is important that the engineer engaged in cost reduction activities understand the company's accounting system. For example (see Table 1.1), when budgets and costing ratios are based on direct labor hours, a line supervisor who is progressive and introduces new and/or improved methods may be penalized because of the system. He may have a part that he is boring and turning on a vertical turret lathe in 5 hours. By the use of coated carbides, he finds he can reduce the machining time to 2.5 hours. This is 50 percent less labor, over which he must spread the additional cost of maintenance (since the machine is working harder) and the additional material handling, inspection, supplies, production clerks, etc., which vary according to the actual number of parts handled per unit of time rather than direct labor hours. When the line supervisor's budgets go into the red, the supervisor can only point out that the department is turning parts out 50 percent faster than before and that therefore the departmental expenses are higher in proportion to the direct labor. Suppose, by introducing coated carbides and other improvements throughout the department, that the line supervisor reduced all his direct labor costs by 50 percent and still produced the same number of parts. His budgets are based on direct labor. Therefore, adjustments in budgets should be made

TABLE 1.1
Reduction in direct labor on a part may cause hourly overhead to exceed hourly budgeted overhead

		New method: Coated carbide tools	
	Old method	Overhead based on established hourly rate	Overhead based on true costs
Allowed time for turning sheave in NC vertical boring mill	5.00 h	2.5 h	2.5 h
Direct labor rate per hour	$ 20.00	$ 20.00	$ 20.00
Direct labor costs	$100.00	$ 50.00	$ 50.00
Overhead costs for inspection, repairs, material handling, depreciation, etc. per hour	$ 40.00	$ 40.00	$ 80.00*
Cost of overhead	$200.00	$100.00	$200.00
Cost of direct labor overhead	$300.00	$150.00	$250.00

Apparent savings per piece = $300.00 − $150.00 = $150.00
Actual savings per peice = $300.00 − $250.00 = $50.00

* Since overhead costs per piece will tend to remain constant, the actual overhead costs for the new method will be about the same as for the old method.

as improvements are made. When the pressure is to increase budgets and costing rates (which has been the characteristic situation during the past twenty years) many times, methods engineering can be effective by decreasing indirect costs. For example, better material handling facilities and improved layout can reduce overhead expense. The efficient shop usually has a low direct labor cost and a high overhead rate. A low costing rate may not indicate an efficient shop. When making cost reduction, the manufacturing engineer should study the efficiency of the overhead departments (maintenance, shipping, receiving, material handling, stock, and stores) as well as the direct labor areas and thus obtain overall reductions in cost.

COST REDUCTION—AN ECONOMIC NECESSITY

In the progressive manufacturing firm, cost reduction or cost improvement should be taking place continually. This effort is frequently referred to as "operations analysis" or "methods engineering." The procedure used by the engineer is to analyze all productive and nonproductive elements of an operation with the thought of improvement. Certainly, the major part of the analyst's effort is directed toward studying the current design, the materials being used, and the processes that are transforming the raw material into the finished components. In these observations, the analyst will endeavor to:

1. Reduce the number of parts, thus simplifying the design
2. Reduce the number of operations by making the machining, joining, and assembly easier
3. Liberalize the tolerances and specifications without deteriorating the quality of the part
4. Utilize a material that enhances the quality and/or reduces the cost and/or improves the appearance
5. Utilize material more economically
6. Mechanize manual operations
7. Automate mechanized operations
8. Utilize robotic instead of manual control
9. Utilize a machining center rather than several independent machining, joining, and assembly work stations
10. Operate all work centers and/or stations more efficiently through the application of sensors and/or programmable controllers
11. Design tooling to utilize the full capacity of the facility
12. Introduce more productive and/or efficient tooling

These twelve points, when carefully pursued by a competent engineer with a background in manufacturing systems, will result in cost reduction ideas. The engineer must recognize that all people (those representative of both management and labor) inherently dislike any change from the current way of operation. Thus, they must be sold the idea that the improved design, material, or process will result in a better product that can be sold at a more attractive price. This effort then will result in greater volume, providing more work for more workers and greater company profit.

JOB COST SYSTEMS

It is important that engineers associated with processing engineering have an understanding of the flow of cost data in a typical job cost system. With this understanding, they are in a much better position to talk with the accountants about cost analysis of a given product design. They will also be better able to plan processes so as to minimize costs by routing to areas that do not carry heavy burden charges in the form of fixed charges and indirect labor. They will be better able to take steps to reduce costs in areas where they appear disproportionate.

In any job cost system, debit and credit are names given to left-side and right-side entries in the general ledger. For every left-side entry there must be an equal and offsetting right-side entry.

Assets reflect debit balances and include such items as cash, receivables, and inventories. Liabilities are shown by credit balances and include payables

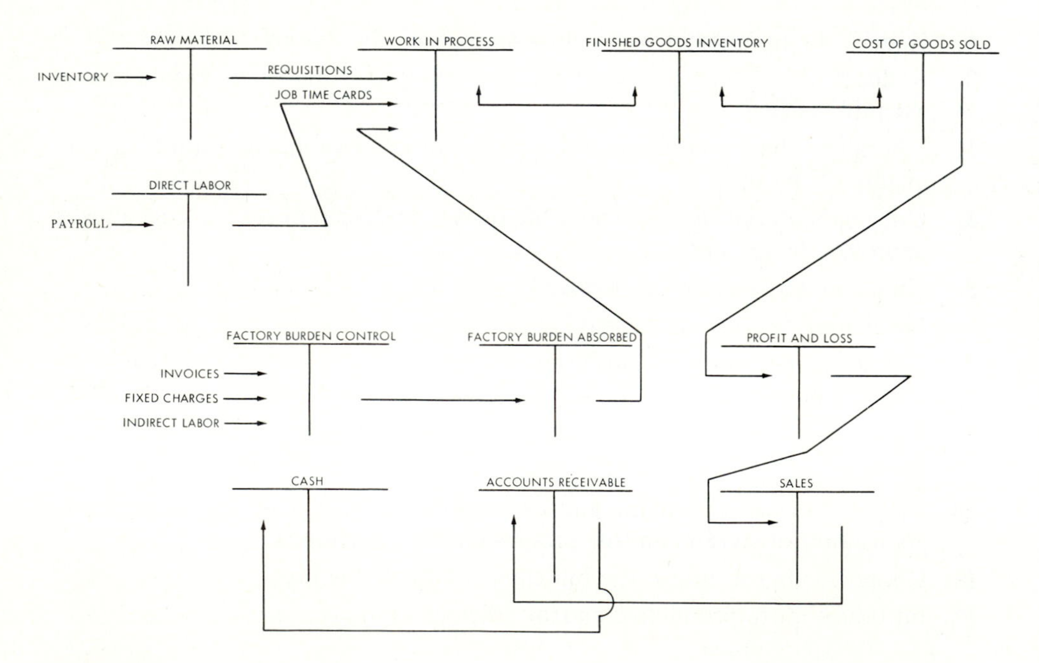

FIGURE 1.4
Flow of entries in a typical job cost system.

as well as accrued items. Net worth is shown as a credit balance on common stock, preferred stock, and surplus. The relationship between assets, liabilities, and net worth is

$$\text{Assets} = \text{liabilities} + \text{net worth}$$

The principal accounts maintained in a job cost system include raw material, direct labor, factory burden control, factory burden absorbed, work in process, finished goods inventory, cost of goods sold, cash, accounts receivable, sales, and profit and loss. The flow of entries in these accounts is illustrated in Fig. 1.4. The process engineer should note that a sale is always a credit to the sales account and a debit to the accounts receivable account. The engineer should carefully study the entries made in the various accounts and understand the relationship between debit and credit balances and their significance to the status of the enterprise.

PURCHASED PARTS VERSUS MANUFACTURED PARTS

One of the most difficult decisions for the process planner is whether to make a part in the shop or purchase it. Frequently, the supplier's cost is less than the factory cost of the part; in this case, it is generally economical to purchase from the supplier. However, the department in which the part would be produced may have an overhead charge that has no relation to the part to be manufactured. If this expense is disregarded and a correct overhead charge is assigned, the factory cost may be favorable. Given the same equipment and materials and an overhead charge applicable to the product, the plant should be able to meet outside competition.

Progressive companies encourage competition between divisions and encourage suppliers to bid in order to improve the products' quality and reduce costs. Good accounting practice will avoid unrealistic accounting procedures that distort costs and cause unfair competition.

Many components that are produced from special-purpose high-volume tools and equipment and are manufactured in large quantities—bearings, bolts, nuts, washers, screws, cotters, etc.—can usually be purchased at a significant savings over producing these parts on general-purpose equipment in the shop. Of course, if the required volume of any of these components reaches such a size that it would be cost-effective to procure the necessary special-purpose machine and fabrication tools and produce them internally, then this procedure should be followed. When parts are made internally, quality as well as delivery can be more readily controlled. Also, the cost of transportation from the vendor's plant is saved. It is good practice to indicate on all vendor-procured component cards that quantity above which it may be economical to produce the components internally. When demand approximates the quantity shown, a detailed analysis should be made to determine whether it would be desirable to manufacture the part or parts internally.

SUMMARY

Process engineering, the engineering of determining how a part may best be produced with due regard for quality, quantity, and cost requirements, is the principal basis of competitive production. New materials and processes of manufacture are being developed continually. Therefore, it is fundamental that manufacturing firms keep abreast of modern process technology if they are to be competitive. Industries ripe for foreign competition are not selected by random drawing—they tend to be those that have let costs get too high or quality too low.

Costs are the ultimate criterion of the value of a product. Direct costs are a good basis for judgment; but indirect costs, overhead or burden or tool expense, may have more influence than direct material and direct labor costs. Errors can creep into calculations made by clerks and persons unfamiliar with the product; therefore, the engineer should have sufficient data to check costs in order to detect costing errors. The many variables encountered in cost considerations make it necessary to consider each case on its individual merit.

QUESTIONS

1.1. Explain why modern process engineering is the basis for an advanced economy.

1.2. What broad areas of study are involved in effective process engineering?

1.3. What four classifications of materials are used in the manufacture of hard products?

1.4. What recent improvements have the Japaneses introduced in the making of steel?

1.5. What costs are included in indirect costs?

1.6. Why is it that efficient shops often have a high indirect cost?

1.7. Outline the structure of cost that goes into the establishment of a selling price.

1.8. Explain the significance of the break-even point.

1.9. Explain why the costs of raw materials and purchase parts become more significant as the number of units to be produced increases.

1.10. What 12 considerations should the engineer consider when conducting an operations analysis for the purpose of cost reduction?

1.11. Explain how the distribution of overhead as a percentage of direct labor can introduce problems with the typical line supervisor.

1.12. Explain the relationship between assets, liabilities, and net worth.

1.13. A sale is a —— to the sales account and a —— to the accounts receivable account in a typical job cost system.

1.14. What type of components is it usually advantageous to purchase rather than produce in the shop?

1.15. Why is process engineering the principal basis of competitive production?

1.16. Explain why selling price is frequently not determined by cost.

PROBLEMS

1.1. How large an original investment could you justify on a new product having an anticipated sales life of 3 years, if the estimated profit at end of the first year is

$100,000, the estimated profit at the end of the second year is $150,000, and the estimated profit at the end of the third year is $20,000? Management expects a return of at least 20 percent on capital for investments lasting from 2 to 5 years. (*Hint* : $PW = R/(1+i)^n$, where PW is the present worth of return, R is return on profit, i is interest or return on capital, and n = number of interest periods.)

1.2. The development cost of a new product is estimated to be $10,000 with a tooling cost of $26,500 to produce the jigs, dies, molds, and fixtures needed to produce the design in the quantities desired. In view of the anticipated short life of the design, all tools are written off during the first year. Sales of the product have been estimated to be between 3500 and 4000 units in the first year. The estimated manufacturing cost is $16.75 per unit. If the selling price is established at $27.00 per unit, what is the break-even point? How much profit will the plant make if the selling cost is taken as 50 percent of the manufacturing cost? If the manufacturing cost rose to $21.35 per unit because of a change in the union contract and an inflation of 10 percent in raw materials, what would the new selling price have to be to make the same profit margin as originally planned?

1.3. A manufacturer of small electric hand-tools such as drills, sanders, hedge clippers, and buffers had sales of $2,000,000 last year and made a gross profit of only $200,000. The main cause of the problem appeared to be failure to increase the sales price, although labor cost had increased to $750,000 and raw material had risen to $800,000 per year. At a staff meeting, the product manager suggested that an increase of 25 percent in the selling price of their product line would place the company in a good profit position again. But the vice president of sales disagreed strongly, claiming that such a large increase in price would drive the company out of the market. After a discussion, the two finally agreed that the net result of such a price increase would be a 10 percent decrease in sales and production. The chief engineer proposed another alternative—operations analysis directed toward cost reduction. By making some improvements in the processing equipment and the purchase of more efficient tooling at an estimated cost of $200,000 for both items, he estimated that costs could be reduced by 25 percent. Analyze the data presented and make a report to the president showing him which proposal is better and explain why.

SELECTED REFERENCES

1. Dopuch, N., Birnberg, J., and Dempski, J. S.: *Cost Accounting—Accounting data for Management's Decisions,* 3rd ed., Harcourt Brace Jovanovich, New York 1982.
2. Green, J. H.: *Operations Management Productivity and Profit,* Reston Publishing Company, Reston, Va., 1984.
3. Horngren, C. T.: *Cost Accounting a Managerial Emphasis,* 5th ed., Prentice-Hall, Englewood Cliffs, N.J., 1982.

CHAPTER 2

PROPERTIES AND BEHAVIOR OF MATERIALS

INTRODUCTION

Designers and engineers are usually more interested in the behavior of materials under load or when in a magnetic field than in why they behave as they do. Yet the better one understands the nature of materials and the reasons for their physical and mechanical properties the more quickly and wisely will he/she be able to choose the proper material for a given design. Generally, a material property is the measured magnitude of its response to a standard test performed according to a standard procedure in a given environment. In engineering materials the loads are mechanical or physical in nature and the properties are recorded in handbooks or, for new materials, are made available by the supplier. Frequently such information is tabulated for room-temperature conditions only, so when the actual service conditions are at subfreezing or elevated temperatures, more information is needed.

The properties of materials are sometimes referred to as structure-sensitive, as compared to structure-insensitive properties. In this case structure-insensitive properties include the traditional physical properties: electrical and thermal conductivity, specific heat, density, and magnetic and optical properties. The structure-sensitive properties include the tensile and yield strength, hardness, and impact, creep, and fatigue resistance. It is recognized that some sources maintain that hardness is not a true mechanical property, because it varies somewhat with the characteristics of the indentor and therefore is a technological test. It is well known that other mechanical properties vary significantly with rate of loading, temperature, geometry of notch in impact testing, and the size and geometry of the test specimen. In that sense all mechanical tests of material properties are technological tests. Furthermore, since reported test values of materials properties are statistical averages, a commercial material frequently has a tolerance band of ±5 percent or more deviation from a given published value.

In the solid state, materials can be classified as metals, polymers, ceramics, and composites. Any particular material can be described by its behavior when subjected to external conditions. Thus, when it is loaded under known conditions of direction, magnitude, rate, and environment, the resulting responses are called mechanical properties. There are many possible complex interrelationships among the internal structure of a material and its service performance (Fig. 2.1). Mechanical properties such as yield strength, impact strength, hardness, creep, and fatigue resistance are strongly structure-sensitive, i.e., they depend upon the arrangement of the atoms in the crystal lattice and on any imperfections in that arrangement, whereas the physical properties are less strucure-sensitive. These include electrical, thermal, magnetic, and optical properties and do depend in part upon structure; for example, the resistivity of a metal increases with the amount of cold work. Physical properties depend primarily upon the relative excess or deficiency of the electrons that establish structural bonds and upon their availability and mobility. Between the conductors with high electron mobility and the insulators with no free electrons, precise control of the atomic architecture has created semiconductors that can have a planned modification of their electron mobility. Similarly, advances in solid-state optics have led to the development of the stimulated emission of electromagnetic energy in the microwave spectrum (masers) and in the visible spectrum (lasers).

In studying the general structure of materials, one may consider three groupings: first, atomic structure, electronic configuration, bonding forces, and the arrangement of the aggregations of atoms; second, the physical aspect of materials, including properties such as electrical and thermal conductivity, specific heat, and magnetism; and third, their macroscopic properties, such as their mechanical behavior under load, which can be explained in terms of impurities and imperfections in the lattice structure and the procedures used to modify that behavior.

PHYSICAL PROPERTIES OF MATERIALS

In the selection of materials for industrial applications, many engineers normally refer to their average macroscopic properties, as determined by engineering tests, and are seldom concerned with microscopic considerations.

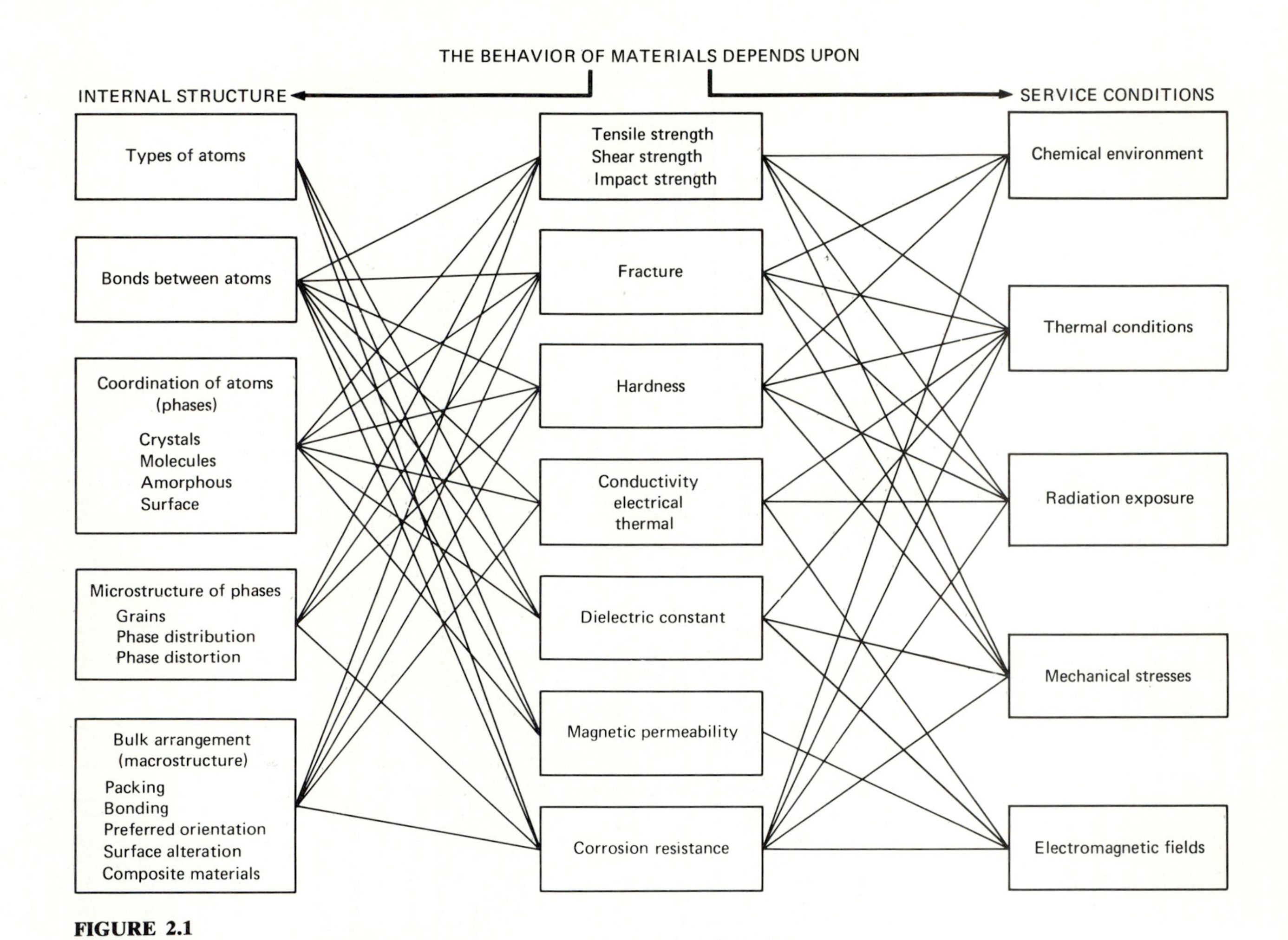

FIGURE 2.1
Interrelationships between material behavior, internal structure, and environment.

Others, because of their specialty or the nature of their positions, have to deal with microscopic properties.

The *average* properties of materials are those involving matter in bulk with its flaws, variations in composition, and variations in density that are caused by manufacturing fluctuations. *Microscopic* properties pertain to atoms, molecules, and their interactions. These aspects of materials are studied for their direct applicability to industrial problems and also so that possible properties in the development of new materials can be estimated.

In order not to become confused by apparently contradictory concepts when dealing with the relationships between the microscopic aspects of matter and the average properties of materials, it is wise to consider the principles that account for the nature of matter at the different levels of our awareness. These levels are the commonplace, the extremely small, and the extremely large. The commonplace level deals with the average properties already mentioned, and the principles involved are those set forth by classical physics. The realm of the extremely small is largely explained by means of quantum mechanics, whereas that of the extremely large is dealt with by relativity.

Relativity is concerned with very large masses, such as planets or stars, and large velocities that may approach the velocity of light. It is also applicable to smaller masses, ranging down to subatomic particles, when they move at high velocities. Relativity has a definite place in the tool boxes of nuclear engineers and electrical engineers who deal with particle accelerators. For production engineers, relativity is of only academic interest and is mentioned here for the sake of completeness.

Application of Concepts from Quantum Mechanics

Quantum mechanics, once restricted to the academic halls, has now become a bread-and-butter topic. It generally deals with particles of atomic or subatomic sizes. But from the understanding of the behavior of these particles comes a better understanding of such phenomena as thermal conductivity, heat capacity, electric conductivity . . . and the very existence of transistors and thermistors.

Quantum mechanics is a complex subject, largely outside the scope of this text, yet a brief mention of two of its most important concepts may aid the production-design engineer in the study of the basic characteristics of materials. One of these is Planck's quantum hypothesis, later extended by Einstein and others. The other is Pauli's exclusion principle.

The quantum hypothesis postulates that the bound energy state of particles of very small size cannot be represented by a continuous function but is discrete in nature. Thus, between any two permissible energy states of a bound particle there exists a region of states that are forbidden.

Each of the electrons of any atom has a particular energy level, E_1, E_2, E_3, etc. For an electron to go from one energy state to a higher one, it must

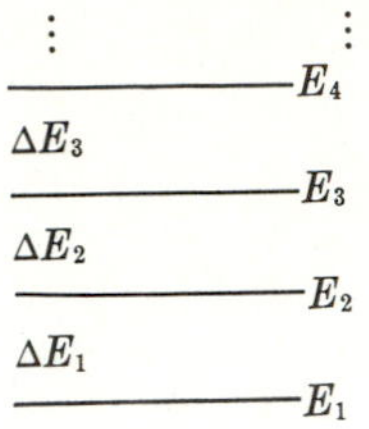

⋮ ⋮

———————— E_4

ΔE_3

———————— E_3

ΔE_2

———————— E_2

ΔE_1

———————— E_1

FIGURE 2.2
Concept of energy levels of planetary electrons.

receive sufficient energy to jump through the forbidden-energy regions. Thus if an electron at energy level E_2 is to go to the higher-energy level E_3, it must receive an energy ΔE_2 (Fig. 2.2).

For it to go to energy level E_4 it must receive an energy of $\Delta E_2 + \Delta E_3$. If an electron receives insufficient energy to jump a forbidden energy region, it will get rid of its extra energy in the form of electromagnetic radiation and remain in its original energy state. By the same token, if an electron at, say, energy level E_1 receives energy greater than ΔE_1 but smaller than $\Delta E_1 + \Delta E_2$, it will be able to go to E_2 and the excess energy over ΔE_1 will be emitted as electromagnetic radiation.

All matter has a spontaneous tendency to be in the lowest possible energy state. Consequently, the electrons of any atom fill the lowest permissible energy levels and are only excited to upper empty levels when they are given energy by means of interaction with electromagnetic radiation, particle bombardment (by electrons, protons, etc.), or an electric field, or by thermal excitation (collision with neighboring atoms brought about by an increased amplitude of atomic vibration caused by an increase in temperature).

Thus, if an electron is dislodged from its initial energy level it can only go to a higher unoccupied level or entirely out of the atom. On doing this, it leaves an empty state that will be immediately occupied either by the original dweller or by any of the electrons at a higher energy; in any case, the electron that drops down will release its excess energy in the form of electromagnetic radiation. This in turn will leave its own original level vacant and there will be more transitions until all the normally filled levels are filled again.

It may be recalled that the number of electrons in any atom is equal to the number of protons in the nucleus. If for any reason an atom is stripped of some or all of its electrons, free electrons in the vicinity of the ion will rapidly fall into the empty energy levels, emitting their excess energy in the form of radiation.

The preceding discussion states that each electron in an atom is associated with a particular energy state—but there can be only one electron at each particular energy level. This is explained by Pauli's exclusion principle, which states that no two bound particles in an interacting system can exist at exactly the same energy state. Thus, when there are two electrons in one of the electron "shells" of an atom (principal quantum numbers), that normally could be considered to be at the same energy, their actual individual energy levels

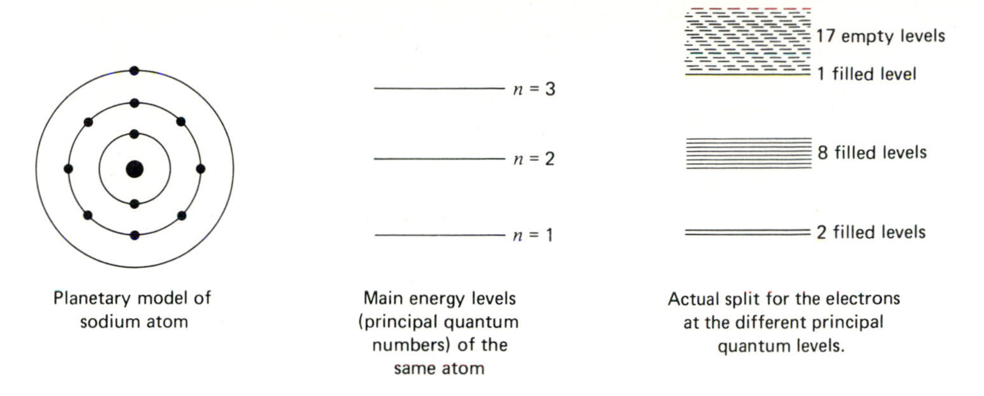

FIGURE 2.3
Relationships between the orbital electrons of the sodium atom and their energy levels.

are different although rather close to one another. By the same token, when there are 8 electrons in a "shell," they have eight separate energy levels. When two or more atoms are brought together, there is a further shift and splitup of energy levels.

The maximum permissible number of electrons in the first "shell" (principal quantum number) is 2, in the second 8, in the third 18, in the fourth 32, etc. Then for an isolated atom there will be 2 individual levels at the first principal level, 8 at the second, and so on (Fig. 2.3). If two atoms are brought together (Fig. 2.4), then the first principal level will have 4 energy states, the second 16, and so forth. When many atoms are brought together, there will be many electrons at any principal quantum level and the spacing between individual energy levels will be so small that there will in effect be energy bands rather than individual levels. Note that there are 6.023×10^{23} atoms in one gram-atom, which is Avogadro's number.

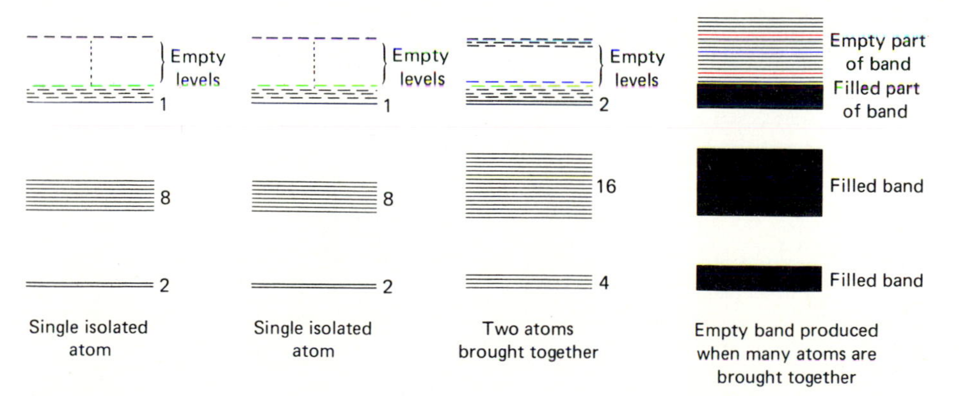

FIGURE 2.4
Formation of energy bands in a solid.

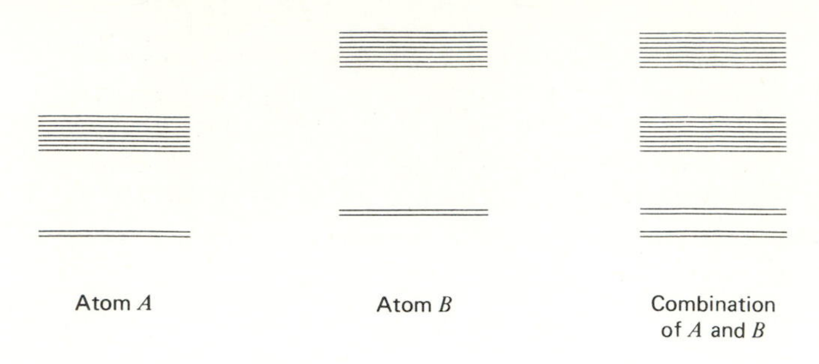

FIGURE 2.5
Superposition of energy levels.

It must be understood that each kind of atom has its own particular energy-level arrangement and that energy splits and band formations only take place where there would be a coincidence of energy levels, such as is illustrated in Fig. 2.4. But if there are two different atoms, A and B, and their energy levels do not coincide, the energy-level arrangement of their combination will be a superposition rather than a level shifting (Fig. 2.5).

In the same way, if there is a large group of atoms of the same kind and one different atom (as an impurity atom) is brought into the group, there will be a shift in the levels that coincide and a superposition where there is no coincidence (Fig. 2.6). It is to be noted that this type of superposition is the same as introducing permissible steps in the forbidden energy regions. By means of these steps the electrons may jump to higher-energy levels without requiring as much energy at any one time.

If the bound arrangement of a given solid is such that all the energy levels in a certain band are filled, whereas the energy levels in the next band higher

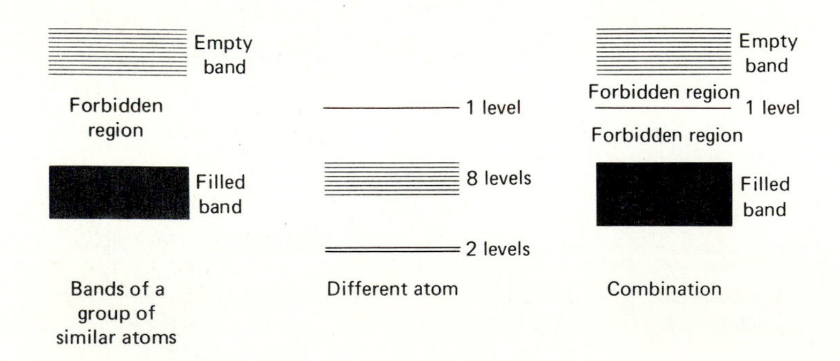

FIGURE 2.6
Superposition of an impurity atom on the band structure of similar atoms.

up are entirely empty, the material is an insulator. For this material to conduct, the valence electrons would require acceleration to gain some energy above the ground state—but in such a case there are no immediately adjacent empty higher-energy levels, only a wide forbidden energy gap. To jump that energy gap would require considerably more energy than is provided in the normal mode of excitation, and therefore the material is an insulator—but it could conduct if a sufficient input of energy were supplied.

If the valence electrons of a solid occupy a partially empty band, these electrons can be accelerated with very little energy and the material is classed as a conductor.

When the valence electrons of a solid occupy a filled band but the forbidden energy gap is small, of the order of kT (where k is Boltzmann's constant, 1.38×10^{-9} J/K and T is absolute temperature in Kelvins (K)) at ordinary temperatures, then it is not difficult for the valence electrons to jump to the empty band. Such a material is known as an intrinsic semiconductor.

In the case of insulators and semiconductors, the top filled band is known as the valence band and the higher adjacent empty band is known as the conduction band.

In conductors, the higher-energy electrons are already in the conduction band and can move freely at the least excitation. On the other hand, insulators require considerable energy for an electron in the valence band to go to the conduction band. Any insulator may be made to conduct electricity if its electrons are sufficiently energized. The electrons that can move about freely in the conduction band are viewed as "free electrons" that are not associated with any atom in particular. These electrons are considered principally responsible for the electrical and thermal conductivity of matter.

Before entering into a detailed discussion of the physical properties of materials, it is convenient to review another mechanism that also intervenes in the evaluation of these properties. This is the oscillatory motion of atoms or molecules that are bound in a solid. In gases and liquids the atoms are free to move about, to a greater or lesser extent, respectively, but in a solid the atoms or molecules are restricted to oscillation about fixed points.

The atoms or molecules of a solid above absolute zero always vibrate about fixed centers. The higher the temperature, the greater the amplitude of the oscillation, and this amplitude of oscillation has a direct effect on the conductivity of solids. The oscillation stores energy and this energy is directly related to the heat capacity of solid materials.

De Broglie's relationship between moving particles and wave motion is not only applicable to the theory of relativity and quantum physics but also to modern physics as a whole. De Broglie conceived that a moving particle can be associated with a wave motion and that propagation of a wave can be considered as a particle. This viewpoint can account for the diffraction of particles such as electrons and neutrons—a phenomenon that was formerly considered an exclusive property of wave motion. It also accounts for the particle-like collision of photons.

Heat Capacity

Heat capacity is the amount of heat that is transferred in raising the temperature of a unit mass of material by one degree. The heat capacity of a solid depends upon the amplitude of oscillation of the particles about their centers of equilibrium. The greater the temperature, the greater the amplitude, and consequently the greater will be the heat capacity. It is measured in terms of energy per unit mass. (Recall that the specific heat of a substance is the ratio of its heat capacity to the heat capacity of water; it is a dimensionless quantity.) This variation is represented as a continuous function in different ranges; however, it is actually a discrete function, since the particles cannot vibrate in a continuous mode but in quantized ones. For the majority of engineering applications, heat capacities may be considered to be numerically equal to specific heats.

Starting from a very low temperature the heat capacity increases proportionally to the third power of the temperature and tapers off to a constant value of 6 C_v [about 6 cal/(mol · °K)] for any substance after passing a critical temperature called the Debye temperature, θ (Fig. 2.7 and Table 2.1).

As the gram-molar values for different substances vary, their heat

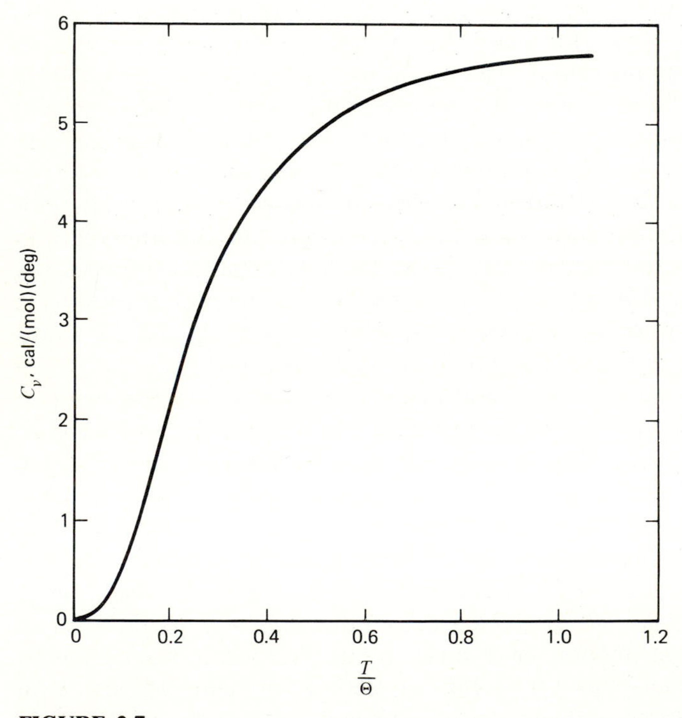

FIGURE 2.7
Heat capacity of a solid (in three dimensions), according to the Debye approximation. (*Redrawn from Charles Kittel,* Introduction to Solid State Physics, *2d ed., Wiley, New York, 1956.*)

TABLE 2.1
Debye temperature for some elements

Substance	θ, K
Beryllium	1,160
Magnesium	406
Molybdenum	425
Tungsten	379
Iron	467
Nickel	456
Copper	338
Zinc	308
Aluminum	418
Lead	94.5

Source: From Charles Kittel, *Introduction to Solid State Physics,*" 2d ed., Wiley, New York, 1960.

capacity per pound will vary even though their heat capacities per gram-mole are the same.

Thermal Conductivity

The thermal conductivity of a solid may be viewed from the standpoint of the free-electron theory. As these electrons drift because of a difference of thermal potential, they may travel along until they interact with an atom. This travel is denoted as the mean free path. When the atoms are lined up with low amplitude of oscillation (at low temperatures), the electrons may travel a fairly long distance without interference (a large mean free path); but as the temperature of the solid increases so does the amplitude of oscillation and the electrons interact with the atoms more often (a small mean free path). The larger the mean free path, the larger the thermal conductivity of the material. Thus, at low temperatures the thermal conductivity of conductors is usually larger than at higher temperatures (Fig. 2.8). The mean thermal conductivity is defined by

$$k_m = \frac{1}{t_0 - t_1} \int_{t_1}^{t_0} k\,dt$$

where k is thermal conductivity, Btu/h/ft^2/unit temperature gradient °F/ft.

In the case of a thermal insulator, the electrons have to be given considerable energy before they can jump into the conduction band and act as conductors. Therefore, insulators gain in conductivity as the temperature increases, despite the interference caused by the increased oscillation of the atoms.

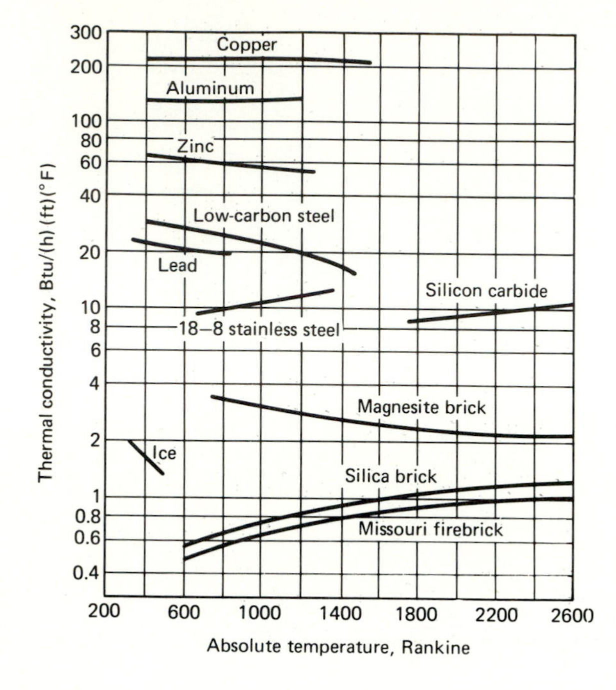

FIGURE 2.8
Variation of the thermal conductivity of solids with absolute temperature of the environment.

Another way to look at thermal conductivity is on the basis of wave propagation. It has been determined that the transport of heat through a body can be considered as the propagation of waves. From De Broglie's relationship these waves may be treated as particles and the whole phenomenon can be studied as the scattering of these particles, which are called *phonons*. Using this model of phonon scattering, it has been determined that the isotope distribution in a material has a bearing on the scattering of the phonons, for the minority isotopes act as scattering centers thus reducing the thermal conductivity of the material.

Electrical Resistivity

The phenomenon of electrical conductivity is similar to that of thermal conductivity. They are both related to the drift of free electrons. The electrical conductivity is often replaced by its inverse, resistivity for practical engineering calculations. Thus the resistance R in ohms of a conductor at a given temperature is calculated by

$$R = \rho \frac{l}{A}$$

TABLE 2.2
Resistivity and density of various elements

Element	Resistivity ρ μΩ · cm	Density	
		lb/in.^{3}	g/cm^3
Silver	1.59	0.379	10.60
Copper	1.673	0.324	8.89
Aluminum	2.655	0.098	2.70
Magnesium	4.46	0.063	1.74
Tungsten	5.5	0.697	18.60
Zinc	5.92	0.258	7.04
Nickel	6.84	0.322	8.60
Iron	9.71	0.284	7.85
Titanium	47.8	0.164	4.50
Selenium	12.0	0.179	4.30
Mercury	94.1	0.490	13.55
Carbon	1.35×10^3	0.080	2.25–3.52
Silicon	1×10^5	0.084	2.35–2.92
Germanium	6.0×10^7	0.192	5.46
Boron	1.8×10^{12}	0.083	2.45–2.54

Source: Theodore Baumeister, *Standard handbook for Mechanical Engineers,* 8th ed., McGraw-Hill, New York, 1978.

where ρ is the resistivity of conductor material, $\Omega \cdot$ cm, l is the length of conductor in cm, A is the cross-sectional area in cm^2.

Table 2.2 gives the resistivity of various metallic elements and semiconductors at room temperature arranged in order of increasing resistivity. Note that semiconductors have high resistivity in the unaltered state.

The resistance of pure metals increases with temperature. It may be calculated at any temperature below the melting point by means of the relationship

$$R = R_1[1 + \alpha_1(t - t_1)]$$

where α is the temperature coefficient of resistance and t is the temperature. The subscript 1 refers to the reference-temperature values. Thus, if the resistance R_1 and the temperature coefficient α_1 have been evaluated at temperature t_1, the resistance R at temperature t can be calculated (Table 2.3).

SEMICONDUCTORS. In the case of conductors, the electrons are already in the conduction band and can drift at the least excitation. In contrast, the electrons of insulators have to receive considerable energy before their electrons can jump from the valence band to the conduction band. When this happens it is referred to as a breakdown of the insulator.

When some impurity atoms are included in some materials, this introduces intermediate permissible steps in the forbidden band and then the

TABLE 2.3
Temperature coefficients of resistance

Initial temperature, °C	Increase in resistance/°C	
	Copper	Aluminium
0	0.00427	0.00439
5	0.00418	0.00429
10	0.00409	0.00420
15	0.00401	0.00411
20	0.00393	0.00403
25	0.00385	0.00396
30	0.00378	0.00388
40	0.00364	0.00373
50	0.00352	0.00360

valence electrons can jump out of the valence band with much less excitation than in the pure insulator. Materials arranged in this manner are known as extrinsic semiconductors

The motion of the positive hole that remains in the valence band contributes to the conductivity of the material as well as the motion of the electrons that have reached the conduction band. When the impurity atoms introduce vacant levels close to the valence band, electrons can jump from the valence band to these levels, leaving behind their positive holes. The free levels are called acceptor levels. Materials arranged in this way are called *p*-type semiconductors for the *p*ositive hole responsible for the electrical conductivity (Fig. 2.9).

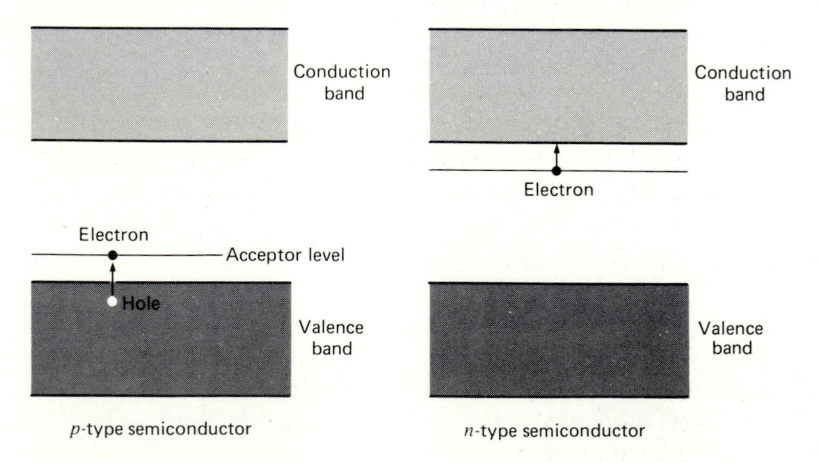

FIGURE 2.9
Band representation of *p*-type and *n*-type semiconductors.

The n-type semiconductors are those in which the impurity atoms introduce filled energy levels close to the lower edge of the conduction band. In this case, the electrons from these levels can be easily expelled into the conduction band. The interposed levels are called donor levels. The n designation results from the *n*egative charge introduced into the conduction band.

Impurities with loosely bonded electrons introduce donor levels in semiconductors; that is, they can easily contribute negative charges to the empty conduction band of the parent material, and thus produce n-type semiconductors. Common "impurities" used to produce n-type semiconductors are arsenic, phosphorus, and antimony. The typical parent material for n-type semiconductors is germanium. The p-type impurities are those that introduce acceptor levels in a semiconductor—and permit the production of positive "holes" in the valence band. Examples of these elements are gallium, boron, and indium. Typical parent materials are silicon and germanium.

The calculation of resistances or conductivities of semiconductors is outside the scope of this book. It is suggested that interested students consult standard works concerning the physics of electrical engineering.

Magnetic Properties of Materials

A moving electric charge produces a magnetic field that is concentric with its direction of motion. If the charge moves in a circle, the magnetic field concentrates in the center of the circle and produces a magnetic dipole (Fig. 2.10). A moving charge within an atom contributes to the magnetic behavior of a material. The magnetic flux B induced in a material depends on how well the tiny dipoles line up when an external magnetic field H is applied. The ratio of B to H is known as the magnetic permeability M of the material:

$$M = \frac{B}{H}$$

Solids exhibit three major types of magnetic behavior: diamagnetism, paramagnetism, and ferromagnetism. Diamagnetism is observed in materials

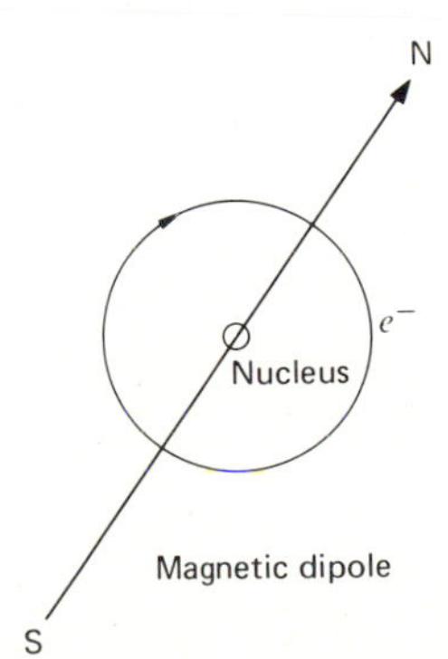

FIGURE 2.10
Schematic representation of a magnetic dipole.

that have complete outer electron rings, such as the elements in the eighth column of the periodic table. Such elements normally have zero *net* magnetic moment. The presence of an external field induces an opposite weak magnetism. The magnetic permeability of diamagnetic materials is less than unity.

Paramagnetic materials have permanent microscopic magnetic moments caused by the spin and orbital motions of the electrons. Paramagnetism is found in all atoms with partially filled inner shells. It is also produced from the spin of the free electrons that are present in metals. The microscopic magnets readily line up when the material is placed in a magnetic field. The permeability of paramagnetic substances is greater than unity.

Paramagnetic and diamagnetic behavior are briefly mentioned only for the sake of completeness, since the magnitude of the effects is so small when compared with its magnitude for ferromagnetic materials. Ferromagnetism is the important magnetic property of an engineering material. It stems from the spontaneous alignment of the small atomic magnets into magnetic domains. The permeability of ferromagnetic material is much greater than unity. This magnetic behavior is not obvious in soft magnetic materials, such as pure iron, because all their magnetic domains are oriented at random. In permanent magnets there is a greater number of domains with a particular orientation than there is without a domain. The magnetic domains can be aligned in the presence of a magnetic field H, thus producing a magnetic flux B. The rotation of the magnetic domains must overcome internal friction within the magnetic field, therefore B and H are not linearly proportional. If H is cycled and B and H are plotted, a hysteresis curve whose internal area is proportional to the energy consumed per cycle is obtained. Engineers considering materials for use in electromagnetic circuits should select ferromagnetic materials having a narrow hysteresis loop.

Most materials are nonmagnetic. Only iron, nickel, cobalt, ferric oxide, and alloys containing chromium, nickel, manganese, or copper are magnetic to any extent. High-silicon iron is the most permeable material and is widely used in magnetic circuits such as those in transformers.

MECHANICAL PROPERTIES OF MATERIALS

Once the important physical properties of a material have been established, mechanical properties such as yield strength and hardness must be considered. Mechanical properties are structure-sensitive in the sense that they depend upon the type of crystal structure and its bonding forces, and especially upon the nature and behavior of the imperfections that exist within the crystal itself or at the grain boundaries.

An important characteristic that distinguishes metals from other materials is their ductility and ability to be deformed plastically without loss in strength. In design, 5 to 15 percent elongation provides the capacity to withstand sudden dynamic overloads. In order to accommodate such loads without failure, materials need dynamic toughness, high moduli of elasticity, and the ability to dissipate energy by substantial plastic deformation prior to fracture.

To predict the behavior of a material under load, engineers require reliable data on the mechanical properties of materials. Handbook data is available for the average properties of common alloys at 68°F. In design, the most frequently needed data are tensile yield strength, hardness, modulus of elasticity, and yield strengths at temperatures other than 68°F. Designers less frequently use resistance to creep, notch sensitivity, impact strength, and fatigue strength. Suppliers' catalogs frequently give more recent or complete data.

Production-engineering data that is seldom found in handbooks include strength-to-weight ratios, cost per unit volume, and resistance to specific service environments.

A brief review of the major mechanical properties and their significance to design is included to ensure that the reader is familiar with the important aspects of each test.

Tensile Properties of Materials

When a material is subjected to a tensile or compressive load of sufficient magnitude, it will deform, at first elastically and then plastically. The deformation is *elastic* if after the load is removed the material returns to its original shape and dimensions; if it does not, the material has undergone *plastic* in addition to elastic deformation. The *engineering* stress–strain diagram presents the elastic data for a material, whereas the *natural* stress–strain diagram can present both elastic and plastic properties in a form that is useful to the designer and the engineer.

1. ENGINEERING STRESS–STRAIN RELATIONSHIPS. In the elastic range, deformation is proportional to the load that causes it. The proportionality is stated by Hooke's law in terms of the stress S and strain e:

$$S = Ee \tag{2.1}$$

where E is Young's modulus, the constant of proportionality between stress and strain, lb/in.^{2}, S is the stress $= P/A =$ load/initial cross-sectional area, lb/in.^{2}, e is the engineering strain $= \Delta l/l_0 =$ change in length/original length, in./in.

If, in the process of pulling a tensile specimen to failure, the stress is plotted as a function of the strain at selected strains, the engineering stress–strain curve results (Fig. 2.11). Note that the stress is plotted as a function of the strain, contrary to the usual procedure of plotting the dependent variable as a function of the independent variable.

The straight part of the stress–strain curve is the proportional range

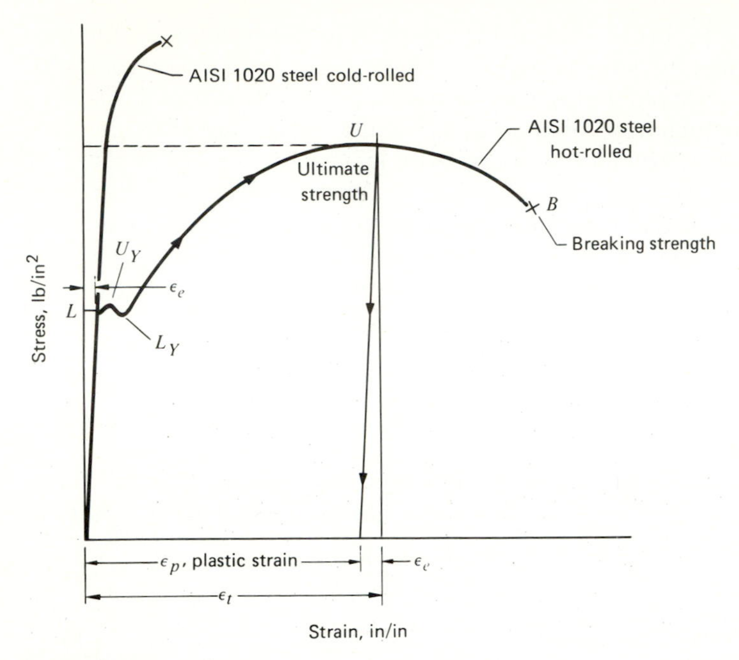

FIGURE 2.11
Engineering stress–strain diagram for AISI 1020 hot-rolled and cold-rolled steels.

described by Hooke's law. The upper limit L of this range is known as the proportional limit. In the case of annealed low-carbon steel there follows a region, the yield point, where the strain increases without any significant increase in stress. As indicated in the diagram, there can be an upper yield point, U_Y, and a lower yield point, L_Y. The maximum stress to which a material may be subjected is labeled U and the breaking point is denoted by B.

When a material is stressed beyond the elastic limit a permanent deformation results. However, this permanent deformation is associated with an elastic one. Notice that if the material of Fig. 2.11 is stressed to point P, upon the removal of the load so that the stress returns to zero, the plastic (permanent) strain is obtained by following downward from point P along a line parallel to the proportional line. Here we have at point P the total strain ε_t, composed of an elastic component ε_e and a plastic one ε_p.

The area under the curve in Fig. 2.11, up to the proportional limit, represents the potential energy per unit volume (resilience) stored in an elastically deformed body; on the other hand, the total area under the curve up to point B represents the total energy per unit volume (modulus of toughness) required to break the specimen.

The engineering properties of commercial plastics vary considerably. For any given plastic, the stress–strain curve depends upon both the temperature and the rate of strain. Slow straining and higher temperatures result in larger

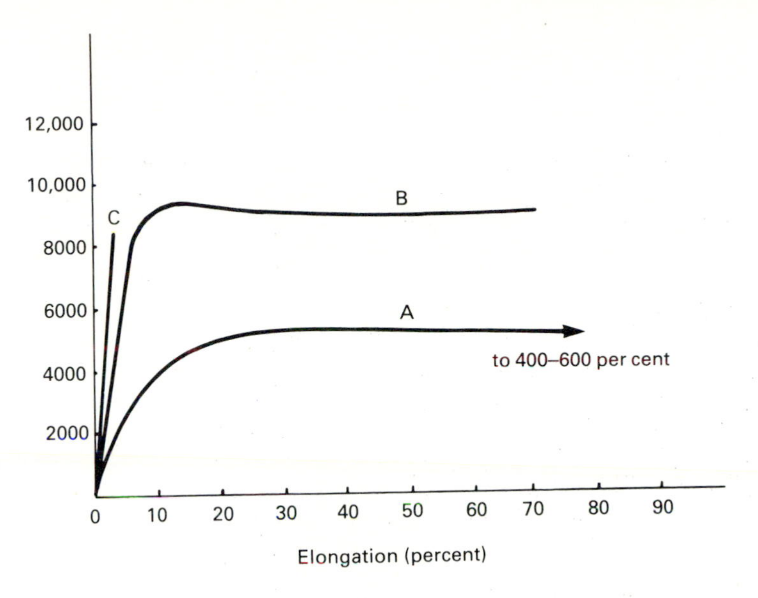

FIGURE 2.12
Representative load–elongation curves for plastics. Curve A is representative of flexible and tough plastics; curve B of rigid and tough plastics; curve C of rigid and brittle plastics. (*Redrawn from Stephen L. Rosen,* Fundamental Principles of Polymeric Materials for Practicing Engineers, *Barnes & Noble, Inc., 1971.*)

ultimate elongations and lower moduli (slopes). Figure 2.12 illustrates typical load–elongation curves for three classes of plastics. Those plastics characterized by curve C have a high initial modulus and a small area under the stress–strain curve (they are *brittle*). These plastics characteristically fail by catastrophic crack propagation at strains of about 2 percent.

Those plastics following curve B are usually referred to as *engineering thermoplastics*. These plastics have good tensile strength and hardness and will deform or draw beyond the yield point, which is evidence of much molecular orientation prior to failure. Thus, these plastics have impact resistance or toughness as well as good strength. They will withstand shock without brittle failure.

Curve A illustrates those plastics that are both tough and flexible. In these plastics, ductile deformation results from converting a material of low crystallinity to one of high crystallinity. These plastics have low tensile strength and moduli. The reader should understand that polymers are much more sensitive to temperature than either metals or ceramics in connection with their response to an applied stress. Figure 2.13 illustrates the relationship between temperature and elastic modulus of a typical thermoplastic material.

2. NATURAL STRESS–STRAIN RELATIONSHIPS. In many engineering designs, only the elastic behavior of a material is significant, so the engineering

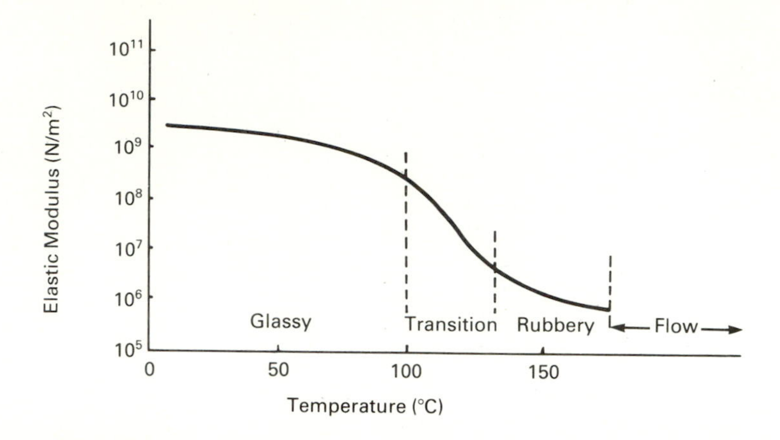

FIGURE 2.13
Relationship between temperature and elastic modulus of a representative thermoplastic material. (*Redrawn from Edward Miller (ed.),* Plastics Products Design, *Dekker, Inc., 1981.*)

stress–strain curve was commonly used to depict the behavior of a material up to its yield point. The yield strength of a material is calculated by dividing the load at the yield point by the *original* cross-sectional area of the test specimen. However, when dealing with a material that is stressed beyond the elastic range, as in plasticity or metal forming, it is apparent that a way should be found to define the relationships in terms of the changing cross-sectional area. Thus, the *natural* or *true stress,* σ, was defined as the load at any instant divided by the cross-sectional area *at that instant.* Thus,

$$\sigma = \frac{P_i}{A_i} \tag{2.2}$$

Similarly we may define the true strain as the change in linear dimension divided by the instantaneous value of the length. Thus:

$$\varepsilon = \int_{l_0}^{l_i} \frac{dl}{l} = \ln \frac{l_i}{l_0} \tag{2.3}$$

Typical natural stress–strain curves for steel and brass are given in Fig. 2.14.

The plastic region of the true stress–strain diagram may be expressed by

$$\sigma = \sigma_o \varepsilon_t^m = \sigma_0(\varepsilon_e + \varepsilon_p)^m \tag{2.4}$$

where σ is the true stress, σ_0 is the work-hardening constant (a constant of the material), ε_t is the total strain = elastic strain ε_e + plastic strain ε_p, m is a constant of the material.

Equation (2.4) is known as the strain-hardening equation. If both Eqs. (2.1) and (2.4) are expressed in logarithmic form, the result is:

$$\ln S = \ln E + \ln \varepsilon_e \tag{2.5}$$

$$\ln \sigma = m \ln \varepsilon_t + \ln \sigma_0 = m \ln(\varepsilon_e + \varepsilon_p) + \ln \sigma_0 \tag{2.6}$$

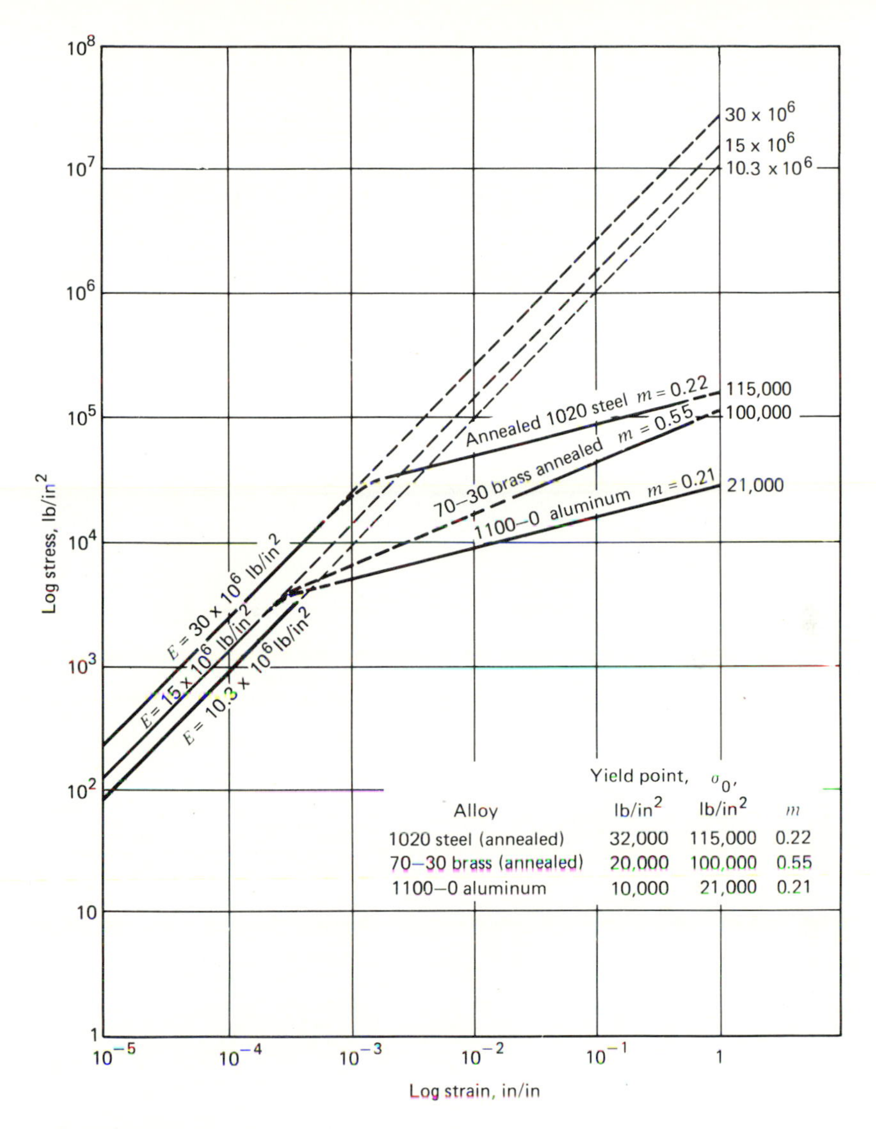

Alloy	Yield point, lb/in^2	σ_0, lb/in^2	m
1020 steel (annealed)	32,000	115,000	0.22
70–30 brass (annealed)	20,000	100,000	0.55
1100–0 aluminum	10,000	21,000	0.21

FIGURE 2.14
True stress–true strain curves for AISI 1020 steel, 70–30 brass, and 1100–0 aluminum.

These two equations plotted on log–log paper yield two straight lines with a discontinuity about the region of the yield point (Fig. 2.14). It will be noted that Eq. (2.5), for the elastic zone, is the equation of a straight line with a slope of unity (45°) and an intercept of ln E. Equation (2.6) is also a straight line but with intercept ln σ_0 and slope m.

3. CORRELATION OF ENGINEERING AND NATURAL STRESSES AND STRAINS. Hooke's law applies for the elastic range of the true stress–strain diagram as well. The relationship is expressed in this case by

$$S = Ee \tag{2.7}$$

where E is Young's modulus, as for Eq. (2.1),

$$S = \text{engineering stress} = P/A_0 = \text{loading force/original cross-sectional area} \tag{2.8}$$

$$e = \text{engineering strain} = (l_i - l_0)/l_0 = \text{change of length/original length} \tag{2.9}$$

It is convenient to derive a relationship between the true stress–strain curve and the engineering stress–strain curve. For true stress, there is the relationship

$$\sigma = \frac{P}{A_i} \qquad \text{[see Eq. (2.4)]}$$

As the volume of the material remains constant during the loading, $A_i l_i = A_0 l_0$; thus

$$A_i = A_0 \frac{l_0}{l_i} \tag{2.10}$$

Consequently,

$$\sigma = \frac{P}{A_0} \frac{l_i}{l_0} \tag{2.11}$$

Substituting Eq. (2.6) into Eq. (2.9) results in

$$\sigma = S \frac{l_i}{l_0} \tag{2.12}$$

which is the relationship between true stress and engineering stress.

On the other hand, true strain is defined as

$$\varepsilon = \int_{l_0}^{l_i} \frac{dl}{l} = [\ln l]_{l_0}^{l_i} = \ln \frac{l_i}{l_0} = \ln\left(\frac{A_0}{A_i}\right) \tag{2.13}$$

Recalling from Eq. (2.7) that the engineering strain is

$$e = \frac{l_i - l_0}{l_0} = \frac{l_i}{l_0} - 1 \tag{2.14}$$

there results

$$\frac{l_i}{l_0} = e + 1 \tag{2.15}$$

Substituting Eq. (2.15) into Eq. (2.13) gives

$$\varepsilon = \ln(e + 1) \tag{2.16}$$

the relationship between the true strain and the engineering strain.

It is to be noted that the strain corresponding to the elastic limit of most engineering materials is very small, usually smaller than 0.005; consequently,

$$\frac{l_i}{l_0} \approx 1 \qquad \text{and} \qquad \ln(e+1) \approx e \tag{2.17}$$

Therefore,

$$E = \frac{\sigma}{\varepsilon} = \frac{S(l_i/l_0)}{\ln(e+1)} \approx \frac{S}{e} \tag{2.18}$$

It is worthwhile for the practical engineer to be conscious of the differences between the basic stress–strain diagrams for different materials, and also of how the stress–strain diagram for one material varies according to its mechanical or thermal treatment. Also, it must be emphasized again that the strain at the elastic limit for most engineering materials is very small. Many textbook presentations represent the elastic part of the stress–strain curve tilted completely out of proportion for ease of illustration.

Combined Stresses

Stresses in solids can be described in several ways. The simplest method, using principal stresses, is the most practical procedure for the manufacturing engineer. Principal stresses are a set of three direct stresses, mutually at right angles, acting on planes of zero shear stress. The directions along which principal stresses act are principal axes, and the planes normal to the principal directions are principal planes.

The technique for finding principal stresses and principal directions is to sum components of force on an arbitrary free body (of the solid under stress) in three noncoplaner directions, to set the shear stresses to zero on a plane whose direction is to be determined, and to solve for the three unknowns: the magnitude of the principal stress and two of the direction cosines of the principal plane. The process is repeated in two dimensions to obtain a second principal plane. The third principal plane is at right angles to each of the two planes so determined. Usually, for simplicity, the arbitrary free body is a tetrahedron containing the given stresses.

In practice, the principal planes are often apparent. Not only are the principal planes free from shear stress, but the principal stresses include both the maximum and the minimum normal stress. Figure 2.15 provides examples in which the direction of principal stress is obvious. A ram pressing on metal in a closed die would have the direction of ram motion as a principal axis. The reader should note, however, that if very high pressures are involved, so that the metal tends to flow laterally beneath the ram face, the ram axis would not necessarily be a principal axis. A cylinder pressurized by a fluid would have principal axes normal to the cylinder wall, parallel to the cylinder axis, and circumferentially. The reader should understand that the directions of the

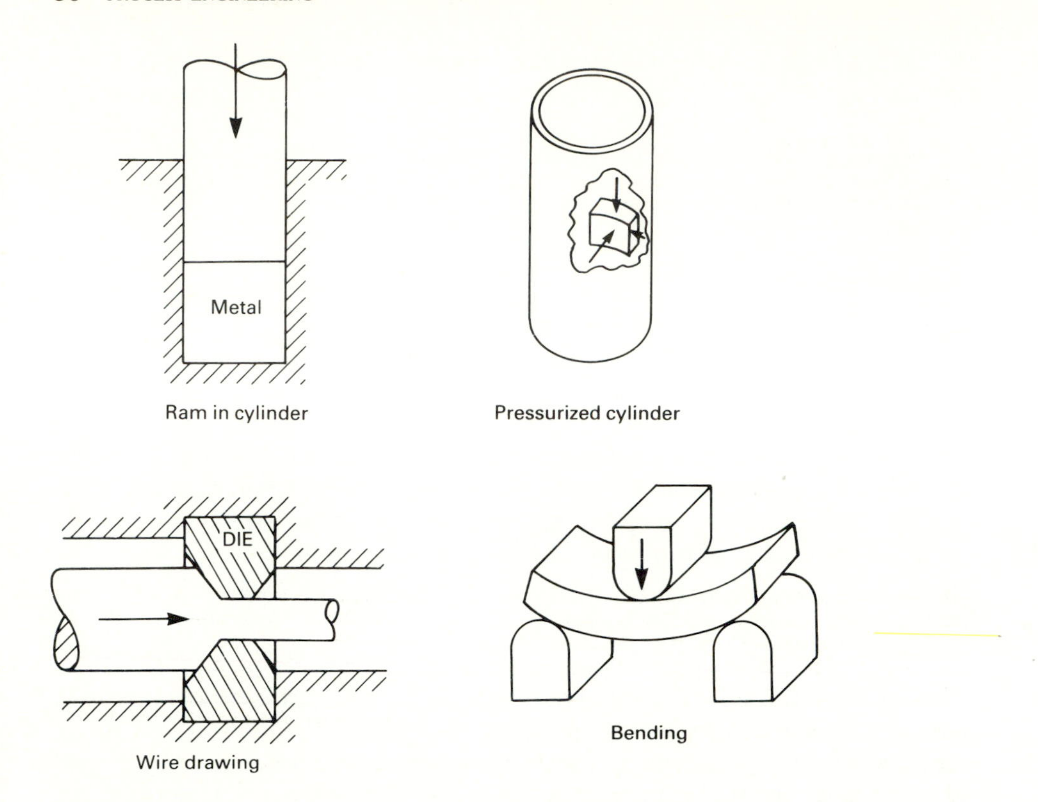

FIGURE 2.15
Familiar examples in which the direction of principal stress is obvious.

principal axes can change from point to point in a metal that is not under uniform stress. Note in Figure 2.15 that the axis of wire drawing is a principal axis. In three-point bending, a principal axis (in the absence of friction) at the plane of the sheet is normal to the sheet. This is not true at the center of the sheet, since there is a parabolic distribution of horizontal shear, which is a maximum at the center and zero at the edges.

It can be shown that the maximum shear stress in a body under stress is equal in magnitude to half the algebraic difference of the maximum and minimum shear stresses, and lies in planes at 45° angles to these principal shear stresses. Figure 2.16 shows the location of planes of maximum shear stress from the position of planes of maximum and minimum principal stresses.

The average normal stress S is of importance in metal working under pressure. It is defined as the algebraic average of the three principal stresses. See Figure 2.17. Thus

$$S = \frac{S_1 + S_2 + S_3}{3}$$

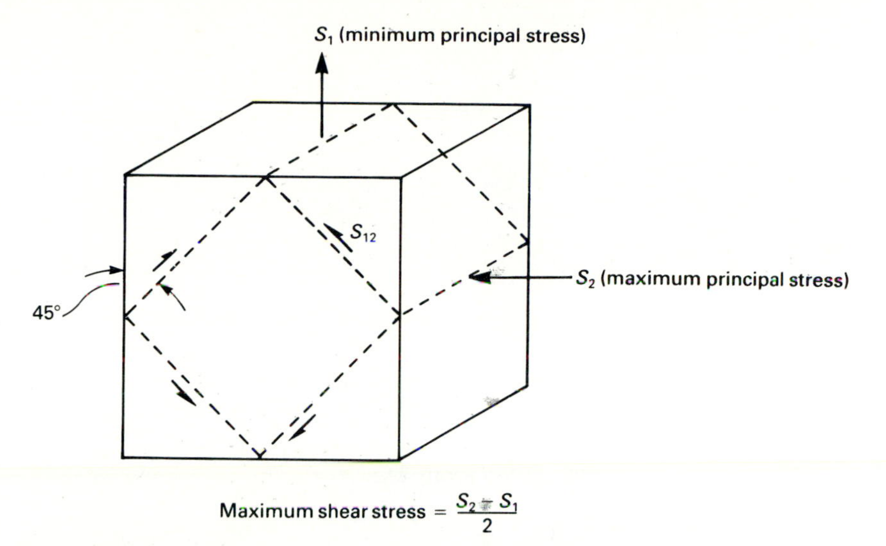

FIGURE 2.16
Maximum shear stress from principal stresses.

Compression here is taken as positive and tension as negative. For examples, let the three principal stresses be 16,000 psi compression, 6000 psi tension and zero. The maximum shear stress is $(16{,}000 + 6000)/2 = 11{,}000$ psi, and the average normal stress is $(16{,}000 - 6000 + 0)/3 = 3333$ psi.

The average normal stress S is related to the ductility or brittleness of a material. It does not indicate whether plastic flow will take place. Two

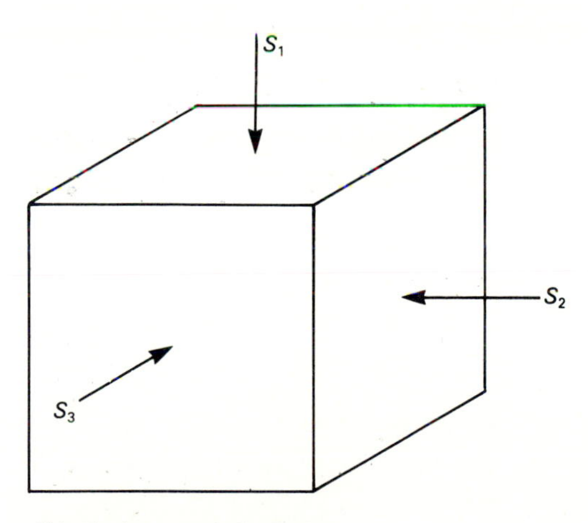

Principal stresses: S_1, S_2, S_3
Average normal stress: S

$$S = \frac{S_1 + S_2 + S_3}{3}$$

FIGURE 2.17
Stress nomenclature.

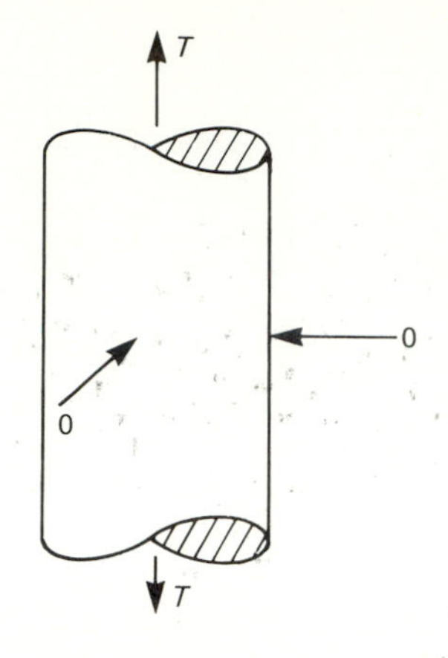

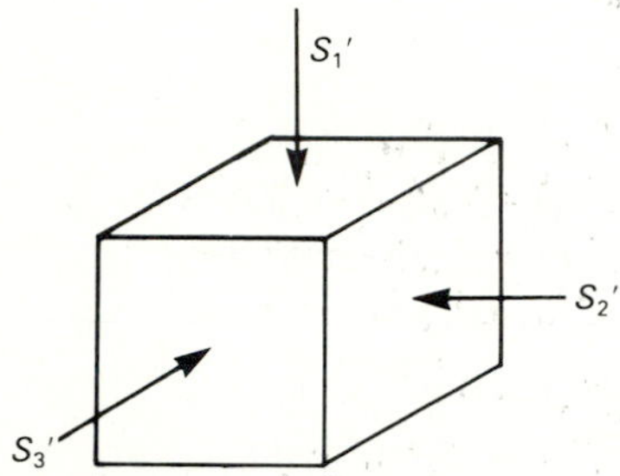

FIGURE 2.18
Use of distortion energy theory for onset of plastic flow.

σ_{YP} = yield stress

$S_1 = T, \quad S_2 = 0, \quad S_3 = 0$

$$S = \frac{T+0+0}{3} = \frac{T}{3}$$

$$S_1' = \tfrac{2}{3}T, \quad S_2' = -\frac{T}{3}, \quad S_3' = -\frac{T}{3}$$

$J = \tfrac{2}{3}T^2$

Plastic flow occurs when

$(S_1')^2 + (S_2')^2 + (S_3')^2 = \tfrac{2}{3}\sigma_{yp}^2$

where

$S_1' = S - 1 - S$
$S_1' = S_1 - S$
$S_2' = S_2 - S$
$S_3' = S_3 - S$

commonly used criteria for determining the beginning of yielding or plastic flow are:

1. Distortion energy theory, sometimes referred to as octahedral shear stress theory or the Huber–von Mises–Hencky theory
2. Maximum shearing stress, or Tresca, theory

Both of these criteria are intended for use only with materials that are isotropic (having the same physical properties in all directions) and homogeneous (having the same physical properties at every point).

The distortion energy theory states that plastic flow occurs in a material when a quantity J equals the value of J at onset of plastic flow in a uniaxial tensile test. Here we have the relationship:

$$(S_1 - S_2)^2 + (S_2 - S_3)^2 + (S_2 - S_1)^2 = 2(S_{ys})^2$$

where S_1, S_2, and S_3 are the principal normal stresses (see Fig. 2.18).

The maximum shear stress theory states that yielding occurs in a specimen when the uniaxial tensile yield strength of the material equals the largest principal stress minus the smallest principal stress. Thus

$$S_{\max} - S_{\min} = S_{ys}$$

Failure is predicted by this theory if $S_1 - S_2 \geq S_f$ (fracture stress).

The question arises of which criterion of plastic flow is the more exact, the distortion energy criterion or the maximum shear stress criterion. It has been found that one criterion may be better for one material while another critierion is better for a different material.

It is thus apparent that for some materials a simple tension test suffices to determine the onset of plastic flow; such materials possess a known criterion. As a rule of thumb, the distortion energy criterion is suitable for heat-treated (normalized or hardened) ferrous alloys, the maximum shear stress criterion for annealed ferrous alloys, and neither for highly hardened ferrous alloys and for magnesium alloys. In all cases, one can assess the onset of plastic flow from scale experiments.

Hardness

Hardness is an engineering property that is related to the wear resistance of a material, its ability to abrade or indent another material, or its resistance to permanent or plastic deformation. The selection of the appropriate hardness test is dependent upon the relative hardness of the material being tested and the amount of damage that can be tolerated on the surface of the test specimen. Table 2.4 provides information relative to the most used hardness standards, each having application for certain types of materials. Hardness can be measured by three types of tests: (1) indentation—Brinell, Rockwell, Knoop, Durometer; (2) rebound or dynamic—Scleroscope; (3) scratch—Moh. These standards together with the materials to which they apply are as follows.

1. Brinell—ferrous and nonferrous metals, carbon and graphite
2. Rockwell B—nonferrous metals or sheet metals
3. Rockwell C—ferrous metals
4. Rockwell M—thermoplastic and thermosetting polymers
5. Rockwell R—thermoplastics and thermosetting polymers
6. Knoop—hard materials produced in thin sections or small parts
7. Vickers diamond pyramid hardness—all metals
8. Durometer—rubber and rubber-like materials
9. Scleroscope—primarily used for ferrous alloys
10. Moh—minerals

Characteristics of the various hardness tests are compared in Fig. 2.19.

The Brinell hardness test is applicable to most metals and their alloys. This test provides a number related to the area of the permanent impression made by a ball indentor, usually 10 mm in diameter, pressed into the surface of the material under a specified load. The larger the Brinell number, the harder the material. The Brinell number (Bhn) is computed as follows:

$$\text{Bhn} = \frac{P}{(\pi D/2)(D - \sqrt{D^2 - d^2})}$$

TABLE 2.4
Comparison of major hardness scales commonly used in engineering

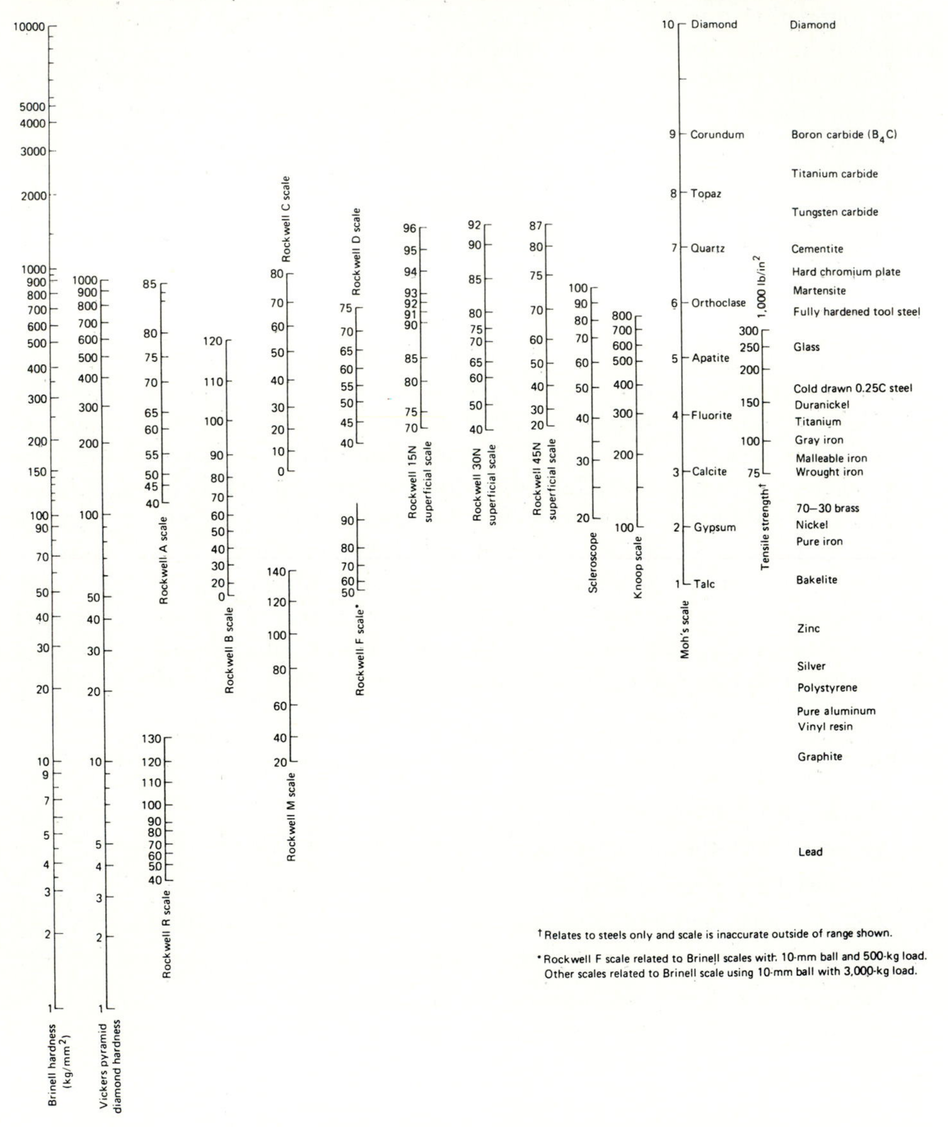

where P is the applied load in kg, D is the diameter of ball in mm, d is the diameter of impression in mm and Bhn is the Brinell hardness in kg/mm^2.

Rockwell hardness is also an indentation test used on both metals and plastics. The Rockwell method makes a smaller indentation compared with the Brinell test, is more applicable to thin materials and is more rapid. The Rockwell number is derived from the net increase in depth of an impression of

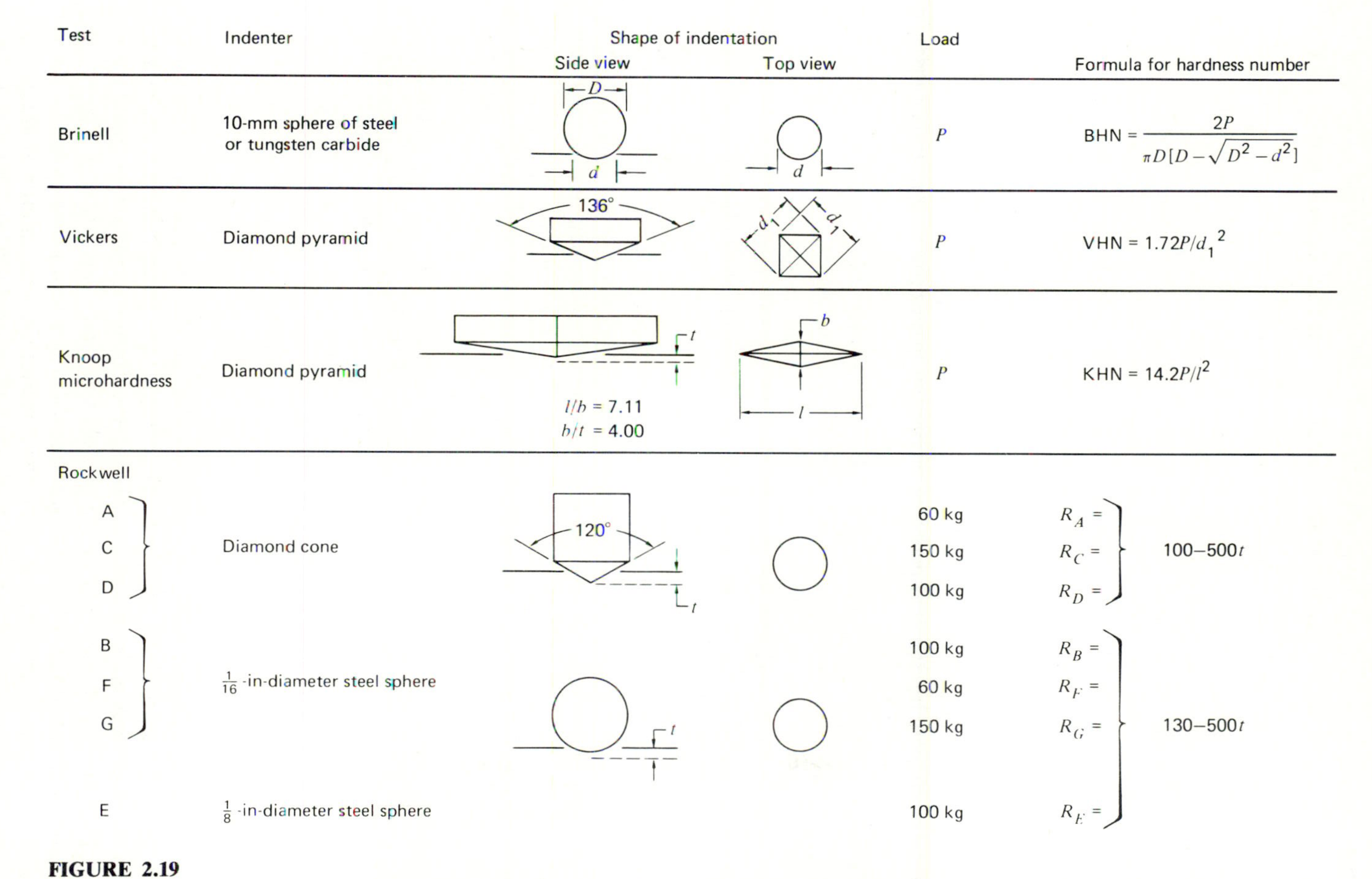

FIGURE 2.19
Comparison of hardness tests. (*Redrawn from H. W. Hayden, W. G. Moffat, and V. Wulff*, The Structure and Properties of Materials, *Wiley, New York, 1965.*)

a standard indentor as the load is increased from a fixed minor load to a major load and then returned to the minor load. The Rockwell C test (using a brale as indentor and a major load of 150 kg) is used primarily for ferrous alloys, whereas the Rockwell B procedure (using as indentor a $\frac{1}{16}$ in. ball and a major load of 100 kg) is used for softer or thinner alloys for which a minimum indentation is preferred. Rockwell M numbers are applied to hard plastics.

The Knoop and Vickers hardness tests measure the microhardness of small areas of a specimen. They are especially suited for measuring the hardness of very small parts, thin sections, and individual grains of a material, using a microscope equipped with a measuring eyepiece. Both tests impose a known load on a small region of the surface of a material for a specified time. In the case of the Knoop test, the indentor is a diamond with a length-to-width ratio of about 3 : 1, whereas the Vickers indentor is a square diamond pyramid with an apical angle of 136°. Both values of hardness are calculated by dividing the applied load by the projected area of the indentation. For the Vickers method, $V = P/0.5393d^2$, where V is the Vickers hardness number, P is the imposed load (1 to 120 kg), and d the diagonal of indentation in mm.

The Shore Scleroscope is a rebound device that drops a ball of standard mass and dimensions through a given distance. The height of the rebound is measured. As the hardness of the test surface increases, the rebound height increases because less energy is lost in plastic deformation of the test surface.

The Durometer hardness test is used in conjunction with elastomers (rubber and rubberlike materials). Unlike other hardness tests, which measure plastic deformation, this test measures elastic deformation. Here an indentor of hardened steel is extended into the material being tested. A hardness value of 100 represents zero extension of the indentor; at zero reading, the indentor extends $0.100^{+0.000}_{-0.003}$ in. beyond the presser foot that surrounds the area being measured.

Moh hardness values are used principally in the designation of hardness of minerals. The Moh scale is so arranged that each mineral will scratch the mineral of the next lower number. Ten selected minerals have been used in the development of this scale. The scale, along with the values for these ten minerals are: talc, 1; rock salt or gypsum, 2; calcite, 3; fluorite, 4; apatite, 5; feldspar, 6; quartz, 7; topaz, 8; corundum, 9; diamond, 10. According to the Moh hardness scale, a human fingernail has a hardness of 2, annealed copper 3, and martensite 7. The Moh scale has too few values to be of practical use in the metal-working field, since four values cover the range of the softest to the hardest metals.

In view of the wide variation in the characteristics of engineering materials, one hardness test for all materials is not practical, but conversion from one hardness scale to another can be done in some cases. The Brinell and Rockwell B and S scales can be converted from one to the other. To approximate equivalent hardness numbers, the alignment chart shown in Table 2.4 can be used. This chart has not included a Durometer scale, since correlation between Durometer and Brinell values is inconsistent. The

Durometer test measures the elastic properties of a material, whereas all other indentation hardness tests measure the plastic properties of a material.

Impact Properties

In some designs, dynamic forces are likely to cause failure. For example an alloy may be hard and have high compressive strength and yet be unable to withstand a sharp blow. In particular, low-carbon steels are susceptible to brittle failure at certain temperatures (Fig. 2.20). Experience has demonstrated that the impact test is sensitive to the brittle behavior of such alloys. Most impact tests use a calibrated hammer to strike a notched or unnotched test specimen. In the former, the test result is strongly dependent on the base of the notch, where there is a large concentration of triaxial stresses that produce a fracture with little plastic flow. The impact test is particularly sensitive to internal stress producers such as inclusions, flake graphite, second phases, and internal cracks.

The results from an impact test are not easily expressed in terms of design requirements because it is not possible to determine the triaxial stress conditions at the notch. There also seems to be no general agreement on the interpretation or significance of the result. Nonetheless the impact test has proved especially useful in defining the temperature at which steel changes from brittle to ductile behavior. Low-carbon steels are particularly susceptible to brittle failure in a cold environment such as the North Atlantic. There were

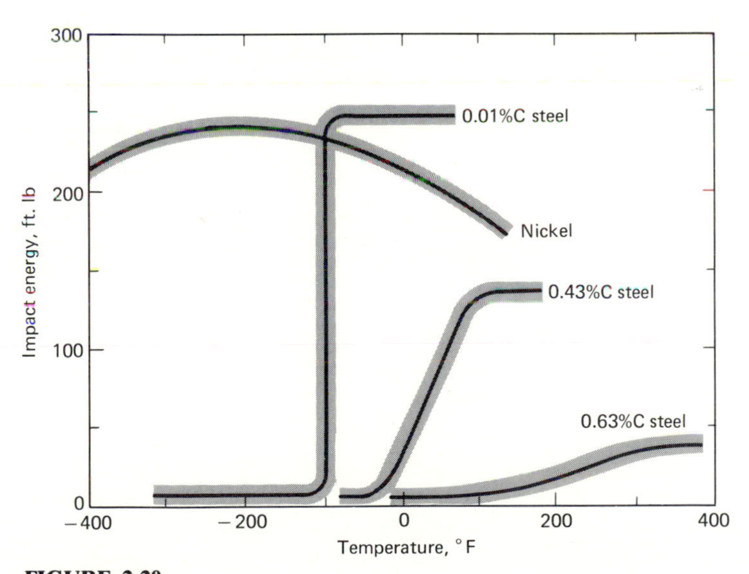

FIGURE 2.20
Impact-test results for several alloys over a range of testing temperatures. (*Redrawn from H. W. Harden, W. G. Moffatt, and V. Wulff,* The Structure and Properties of Materials, *Wiley, New York, 1965.*)

cases of Liberty ships of World War II vintage splitting in two as a result of brittle behavior when traveling in heavy seas during the winter.

In a particular design having a notch or any abrupt change in cross section, the maximum stress occurs at this location and may exceed the stress computed by typical formulas based upon simplified assumptions in connection with stress distribution. The ratio of this maximum stress to the nominal stress is known as a *stress concentration factor,* usually denoted by K. Stress concentration factors may be determined experimentally or by calculations based on the theory of elasticity. Figure 2.23 illustrates stress concentration factors for fillets of various radius divided by the thickness of castings subjected to torsion, tension, and bending stresses.

Fatigue Properties and Endurance Limit

Although yield strength is a suitable criterion for designing components that are to be subjected to static loads, for cyclic loading the behavior of a material must be evaluated under dynamic conditions. The fatigue strength or endurance limit of a material should be used for the design of parts subjected to repeated alternating stresses over an extended period of time. As would be expected, the strength of a material under cyclic loading is considerably less than it would be under a static load (Fig. 2.21). The plot of stress as a function of the number of cycles to failure is commonly called an S–N curve. It is interesting to note that for specimens of SAE 1047 steel there is a stress called the *endurance or fatigue* limit below which the material has an infinite life

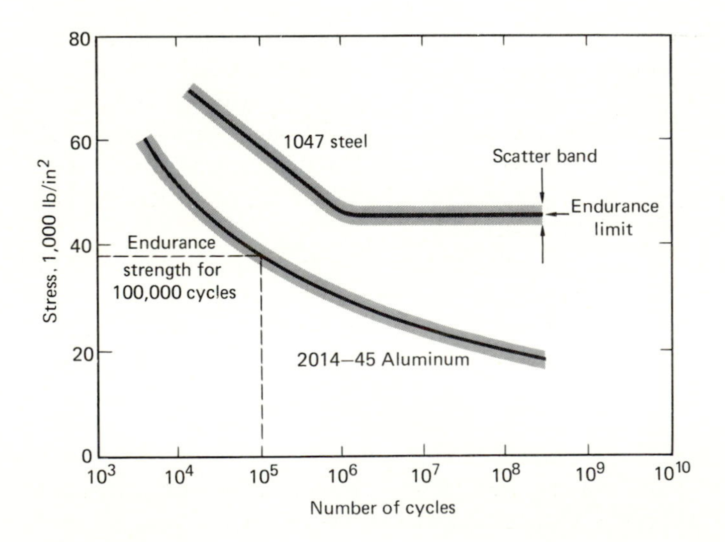

FIGURE 2.21
Typical fatigue-failure curves for Al and 1047 steel. (*Redrawn from H. W. Hayden, W. G. Moffatt, and V. Wulff,* The Structure and Properties of Materials, *Wiley, New York, 1965.*)

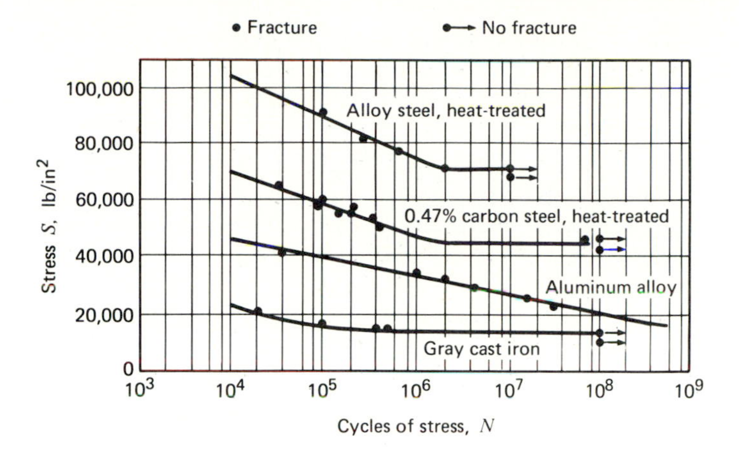

FIGURE 2.22
Fatigue (S–N) curves and endurance limits for various alloys. (*Redrawn from A. G. Guy,* Elements of Physical Metallurgy, *Addison-Wesley, Reading, Mass., 1960.*)

($>10^8$ cycles), i.e., the steel would not fail regardless of exposure time (Fig. 2.22). In contrast, the 2014–T6 aluminum alloy has no limiting stress value.

Fatigue data are inherently more variable than tensile test data. In part the scatter is caused by variation in surface finish and environment (Fig. 2.24). Polished specimens of the same material give significantly better life than machined or scaly surfaces. Since most fatigue failures initiate at surface notches, fatigue behavior and notch sensitivity are closely related. Thus mechanical or other treatments that improve the integrity of the surface, or add residual compressive stresses to it, improve the endurance limit of the specimen.

Overstressing above the fatigue limit for cycles fewer than necessary to produce failure at that stress reduces the fatigue limit. Also, understressing below the fatigue limit may increase the fatigue limit. Thus, cold working and shot peening usually improve fatigue properties.

Several standard types of fatigue-testing machines are commercially available. The results of extensive fatigue-life testing programs are now available in the form of S–N curves that are invaluable for comparing the performance of a material that is expected to be subjected to dynamic loads.

Creep and Stress Rupture

As metals are exposed to temperatures within 40 percent of their absolute melting point, they begin to elongate continuously at low constant load (stresses beyond the proportional limit). A typical creep curve is a plot of the elongation against time of a wire subjected to a tensile load at a given temperature (Fig. 2.25). Creep is explained in terms of the interplay between

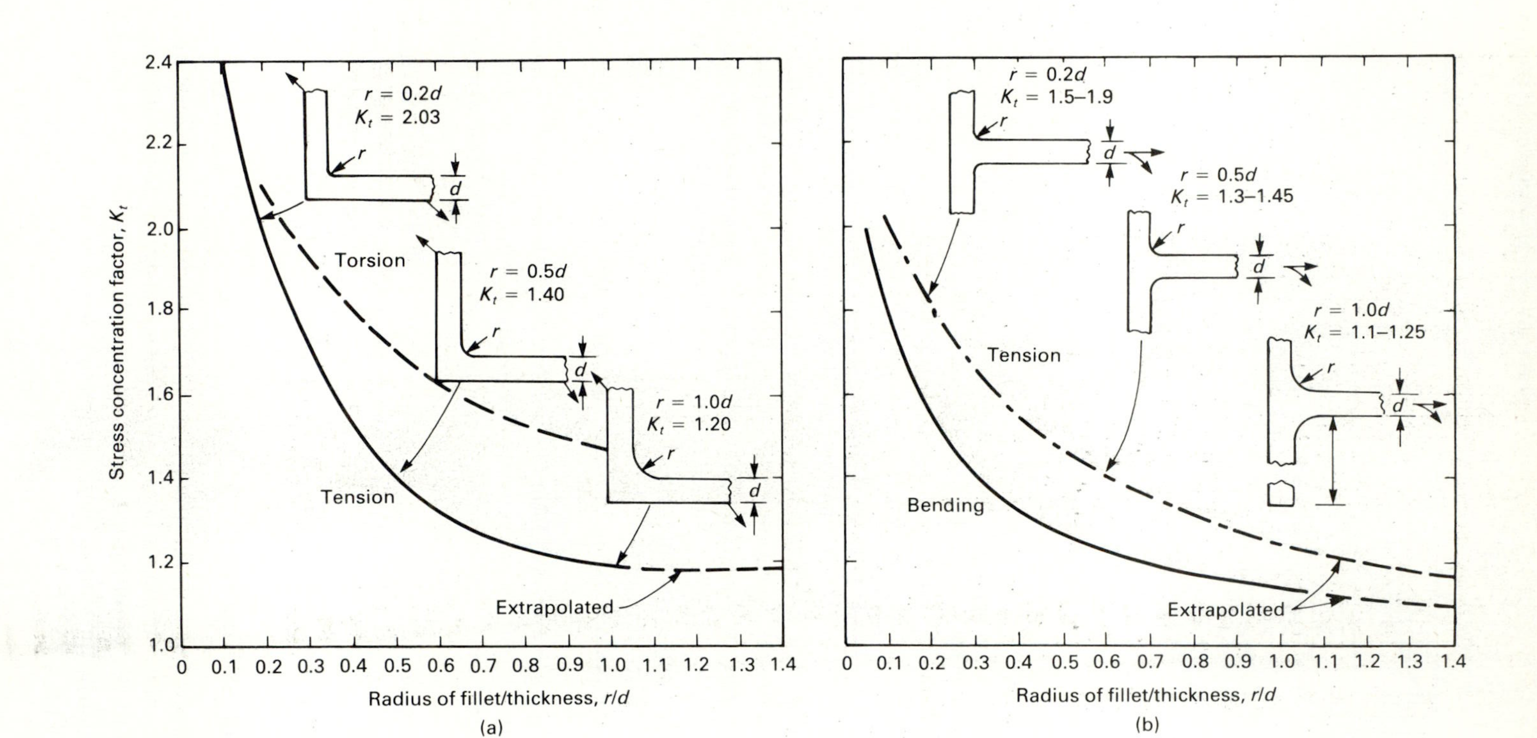

Stress concentration factor, K_t
2.4
2.2
2.0
1.8
1.6
1.4
1.2
1.0
$r = 0.2d$
$K_t = 2.03$
r
d
Torsion
$r = 0.5d$
$K_t = 1.40$
r
d
$r = 1.0d$
$K_t = 1.20$
r
d
Tension
Extrapolated
0 0.1 0.2 0.3 0.4 0.5 0.6 0.7 0.8 0.9 1.0 1.1 1.2 1.3 1.4
Radius of fillet/thickness, r/d
(a)
$r = 0.2d$
$K_t = 1.5–1.9$
r
d
$r = 0.5d$
$K_t = 1.3–1.45$
r
d
$r = 1.0d$
$K_t = 1.1–1.25$
r
d
Tension
Bending
Extrapolated
0 0.1 0.2 0.3 0.4 0.5 0.6 0.7 0.8 0.9 1.0 1.1 1.2 1.3 1.4
Radius of fillet/thickness, r/d
(b)

FIGURE 2.23

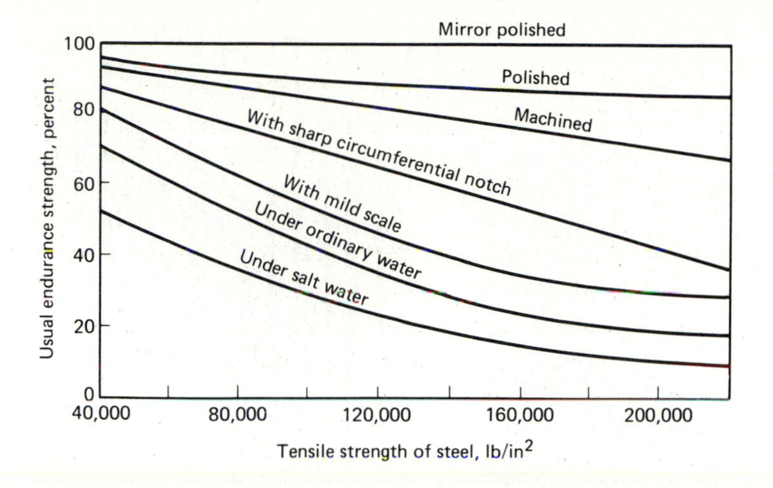

FIGURE 2.24
Effect of surface condition and environment on the endurance strength of steel (Karpoo and Ballens). (*Redrawn from A. G. Guy,* Elements of Physical Metallurgy, *Addison-Wesley, Reading, Mass., 1960.*)

work-hardening and softening from recovery processes such as dislocation climb, thermally activated cross slip, and vacancy diffusion. During primary creep the rate of work-hardening decreases because the recovery processes are slow, but during secondary creep both rates are equal. In the third stage of creep, grain-boundary cracks or necking may occur that reduce the net cross-sectional area of the test specimen. Creep is sensitive to both the test

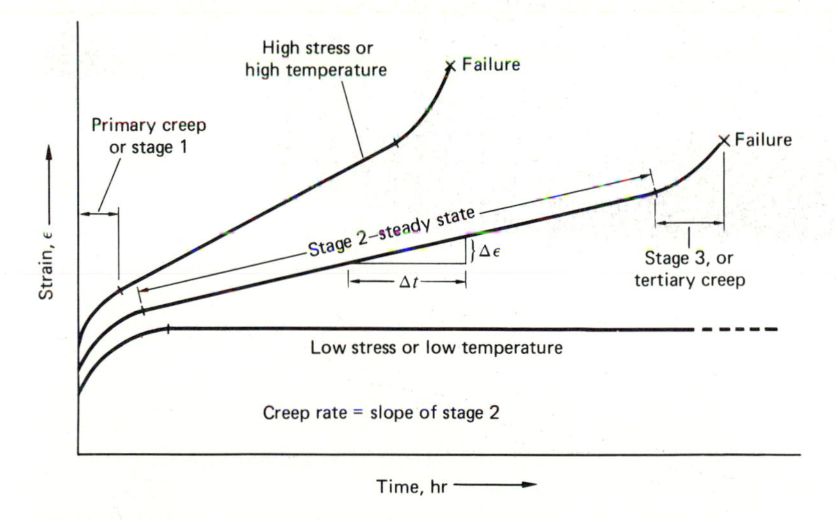

FIGURE 2.25
Typical creep curve showing the stages of creep at constant temperature. Stage 2 determines the useful life of the material.

temperature and the applied load. Increases in temperature speed up recovery processes and an increased load raises the whole curve.

The region of constant strain rate or secondary creep (stage 2—steady state in Fig. 2.25) is usually used to estimate the probable deformation throughout the life of the component. The working stress is then chosen so that the total predicted deformation is within tolerance.

In the stress rupture test, the specimen is held under an applied load at a given temperature until it fails. A series of curves are plotted that can be helpful to the designer who must consider high-temperature applications.

COMPARISON OF THE PROPERTIES OF MATERIALS

The materials commonly used in manufacturing can be divided into four groups for the purpose of contrasting their properties: ferrous, nonferrous, thermosetting, and thermoplastic. In some cases these four groups are not particularly appropriate, since, for example, dielectric strength is not applicable to metals. On the other hand, polymers have such low yield strengths that they fare poorly when compared with even the weakest metal on the basis of strength.

The bar charts in Fig. 2.26 compare selected physical properties of the four classes of materials, including dielectric strength, resistivity, coefficient of thermal expansion, thermal conductivity, and specific weight. It can be readily seen that mica and glass have up to four times the breakdown strength of polymers or rubber. Some thermoplastics exceed all other materials in electrical insulation, with most other materials fairly close to second place. The coefficient of thermal expansion of polymers is an order of magnitude greater than that of metals, whereas the thermal conductivity of metals far exceeds that of all plastics. The specific weight of ferrous alloys is four to five times that of polymers and three times that of aluminum alloys.

The same groups of materials were compared with respect to their mechanical properties, including ultimate strength, yield strength, modulus of elasticity, hardness, and useful temperature range (Fig. 2.27). The ultimate tensile strength of metals far exceeds that of polymers—by a factor of 25 for unreinforced materials and a factor of 8 for the woven glass-reinforced resins. In the case of the modulus of elasticity, the best thermosets are below the lowest nonferrous materials with a maximum of 5×10^6 lb/in.^{2} (Fig. 2.28). The ferrous materials range from 12×10^6 for class 25 gray iron to 30×10^6 for most steels. Tungsten carbides are off the graph with values up to 90 to 95×10^6 lb/in.^{2} Tungsten carbide, therefore, despite its cost and weight, is useful for long, cantilever-beam, boring-tool holders because of its low deflection under load. In the comparison of yield strengths, metals are far superior to polymers. Nonferrous materials range from a few hundred to 80,000 lb/in.^{2} for aluminum and 175,000 lb/in.^{2} for certain copper- and nickel-based alloys. Ferrous alloys

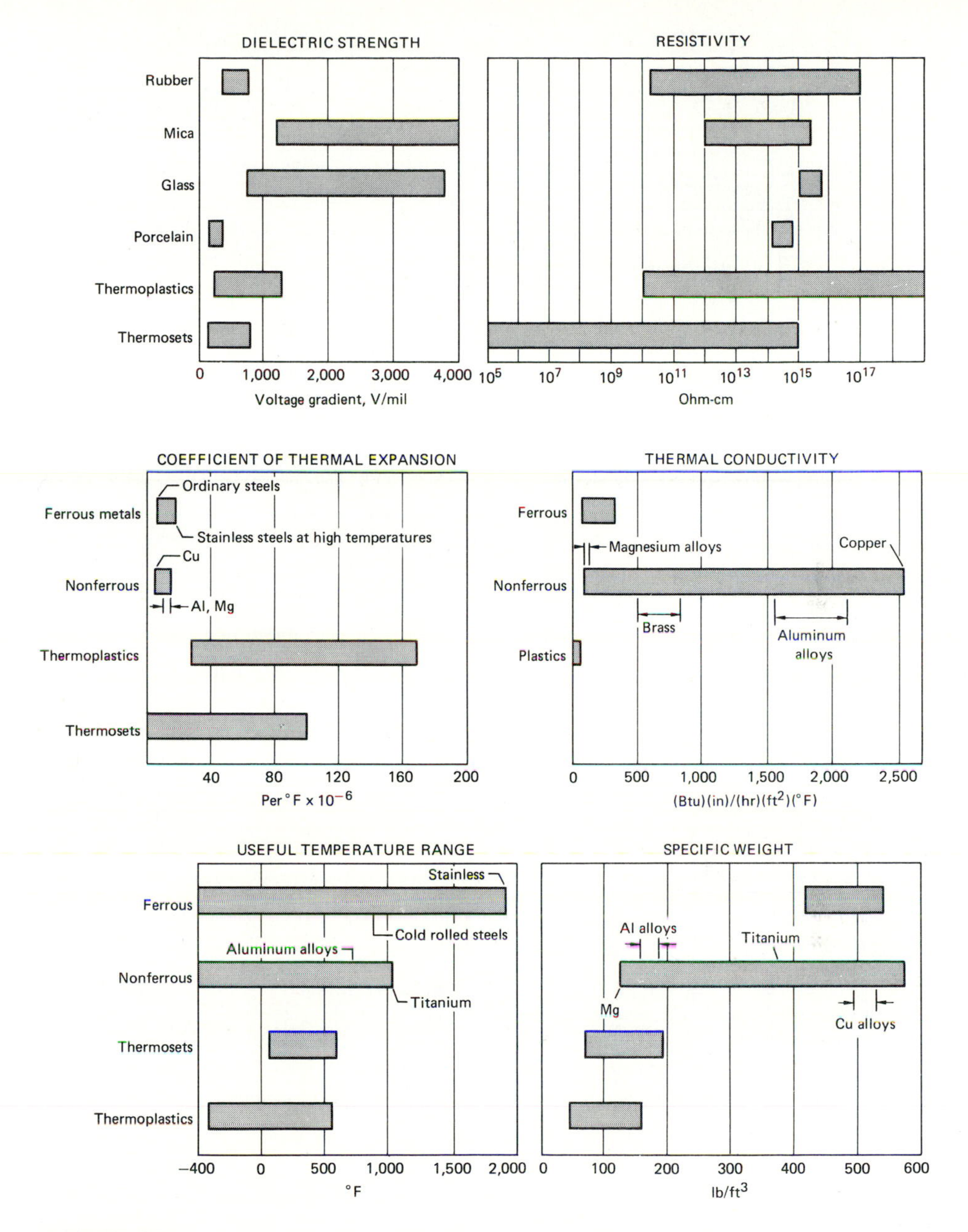

FIGURE 2.26
Comparison of the physical properties of materials.

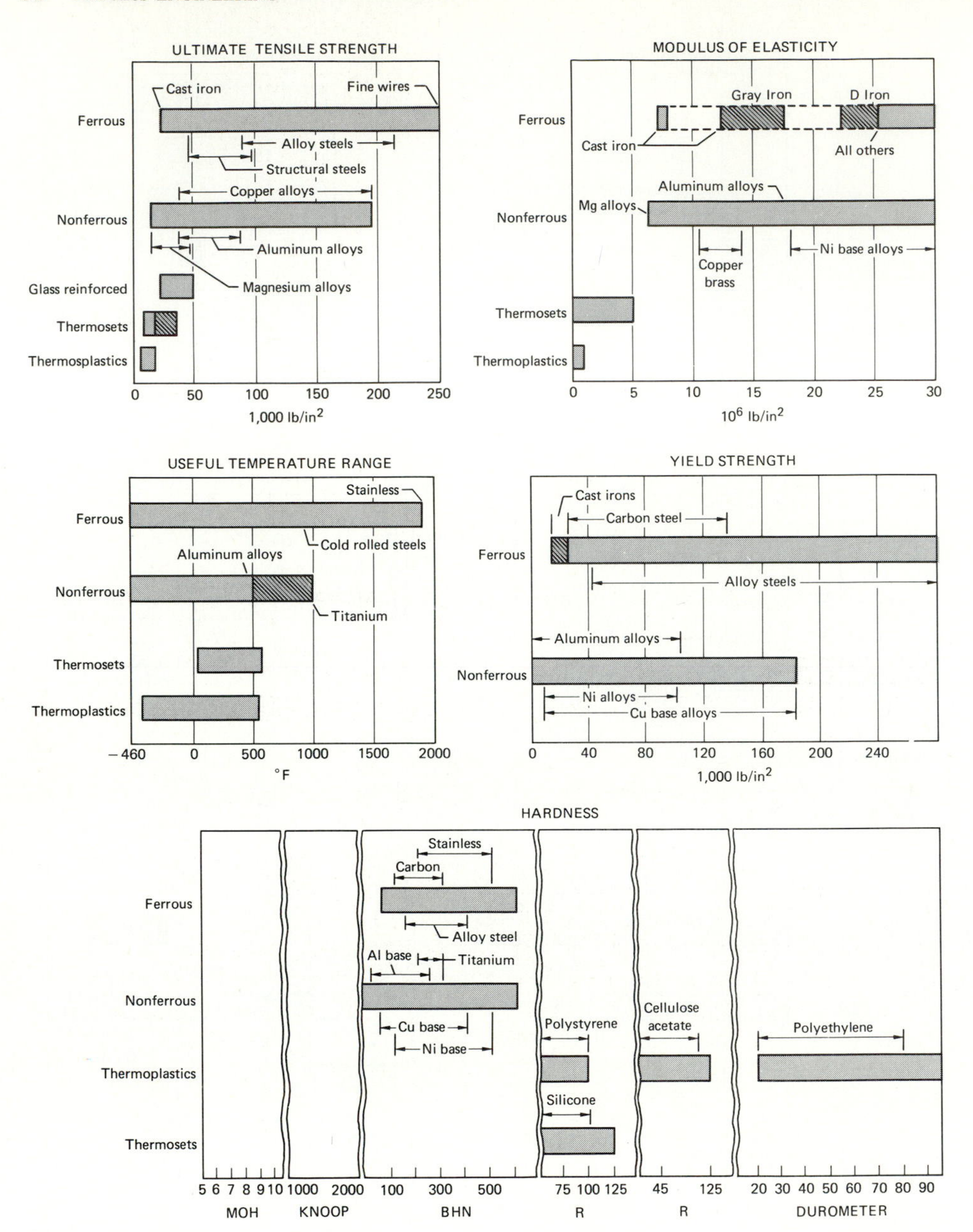

FIGURE 2.27
Comparison of the mechanical properties of materials.

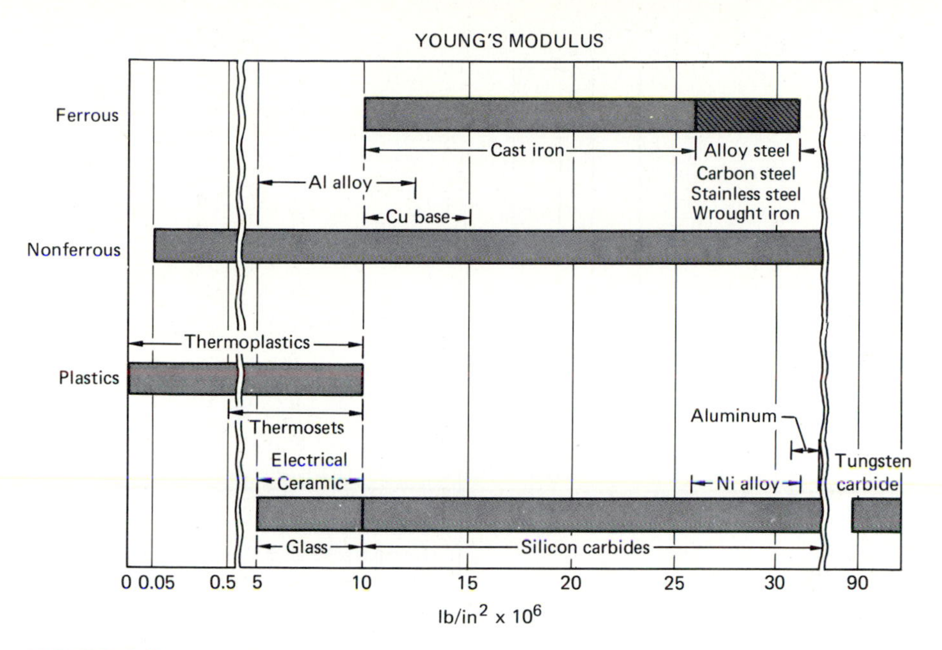

FIGURE 2.28
Comparison of the elastic moduli of materials.

range from 20,000 lb/in.^{2} for certain gray irons to 280,000 lb/in.^{2} or more for particular alloy steels.

The usable range of temperature for steel ranges from −460°F to almost 2000°F for specific stainless steels. Aluminum alloys can withstand temperatures from 300 to 500°F, and some titanium-reinforced polymers are useful up to 400 to 900°F, but the vast majority are good only to 200°F. Hardness is the most difficult property to use for making valid comparisons, because the deformation of plastics and elastomers under an indentor is different from that of metals. As a group, polymers are far softer than metals. Ferrous and nickel-base alloys range from 100 to 600 Bhn, which is a tremendous range of values.

RELATIVE COST OF ENGINEERING MATERIALS

The design engineer is responsible for the initial specification of a material, yet in many cases he spends little time checking alternatives. Many designers feel that their obligation ends with the functional design, so they tend to minimize the importance of material selection. The reasons are many, but two are likely to be (1) lack of proper education in their engineering courses and (2) inadequate time because of the need to meet a deadline.

In over 90 percent of all designs a material is selected primarily on the

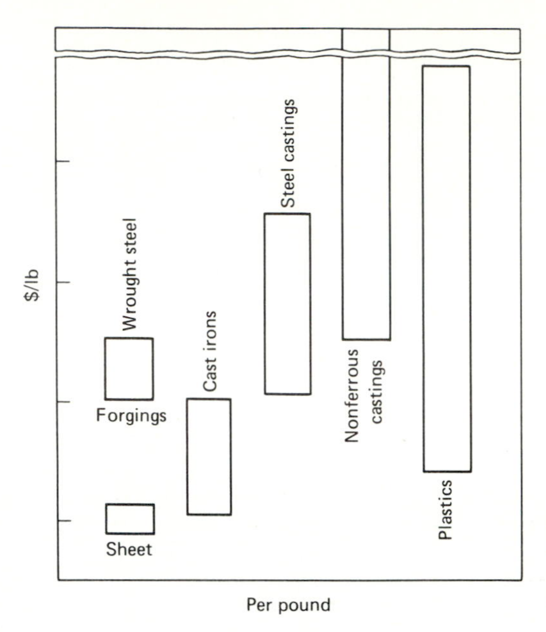

FIGURE 2.29
Comparative cost/unit volume for several materials. (*Redrawn from Foundry magazine.*)

basis of yield strength and ability to fill space at the lowest cost. In some cases, other criteria, depending primarily on physical properties such as electrical conductivity or chemical properties such as corrosion resistance, are dictated by the end-use of the design and form the basis for the material specification.

It is the designer's responsibility to match the functional and esthetic requirements of his design with the behavior of a particular material so that the total cost after manufacture is minimized. In a large organization, this undertaking is a team effort in which design engineers, production engineers, and materials specialists all play an important role. In a small company, the designer may handle the entire assignment.

The costs of engineering materials are in a constant state of flux. Three components may be recognized: first, the general price changes that follow the ebb and flow of the national economy; second, the supply-and-demand effect on the price of some metals, specifically copper and nickel; and finally, the effect of rising production capacity and strong competition, as found in the polymer industry since the 1950s. In general the polymers have dropped in price in the past decade, whereas the cost of all metals but aluminum has increased. Inflation had led to increasing prices.

Materials cannot be compared on the basis of cost alone because they have widely varying densities and strengths. If the cost per pound is converted to cost per unit volume, a more meaningful comparison can be made. When the same materials are compared on the basis of cost per unit of strength per unit of volume it is evident that not only do most materials become strongly competitive, but that in the future, if stronger polymers are created, the 4000-year dominance of metals may be severely challenged (Fig. 2.29).

QUESTIONS

2.1. Explain clearly the difference between the physical properties and mechanical properties of a material. Give several examples.

2.2. Why is a designer or production engineer interested in the properties of materials?

2.3. Would a designer and a production engineer be equally interested in the same properties of a material to be used, for example, in an electric motor frame? Explain your answer.

2.4. What is the difference between the structure-sensitive and structure-insensitive properties of a material?

2.5. Give several distinguishing characteristics of the four major classes of solid materials.

2.6. What are three general groupings that can be used in the study of the structure of materials?

2.7. What is the major feature of Planck's quantum hypothesis? What physical property of a material does it help us to understand?

2.8. What is the basic factor that affects the value of the heat capacity of a solid?

2.9. How would the heat capacity of two metals vary, in general?

2.10. What is the specific heat of a substance?

2.11. What property of two materials would lead to a difference in the magnitude of their heat capacities?

2.12. How does the thermal conductivity of any material vary with temperature?

2.13. Why does the thermal conductivity of a metal exceed that of a plastic?

2.14. Make a sketch of the stress–strain diagram of a typical "engineering thermoplastic."

2.15. What is the relationship between true strain and engineering strain?

2.16. What is the advantage of the true stress–true strain relationship for a design engineer?

2.17. Why does the elastic region of the true stress–true strain graph plot as a 45° slope?

2.18. At what point will the compression stress-strain curve diverge from the tensile stress–strain curve?

2.19. How is hardness determined? Give the nature of the various hardness tests.

2.20. Why is the Moh hardness test seldom used in connection with checking the hardness of metals?

2.21 What is the relationship between hardness and tensile strength for steel? Is this relationship also found for other alloys?

2.22 How do fatigue tests aid the design engineer?

2.23. In a casting having a radius of fillet to thickness ratio of 1, what stress concentration factor would apply to tensile stresses?

2.24. What is the value of a creep test in a production design?

2.25. In a production design, what are the principal advantages of thermosets and thermoplastics compared to metals?

2.26. What is the value of "shot peening" a particular part?

2.27. What are the major advantages of ferrous alloys compared with plastics and other metallic alloys?

2.28. What material is strongest on a strength-to-weight basis?

2.29. How does cost affect the choice which material to use for a given situation?

PROBLEMS

2.1. Explain the behavior of electrical and thermal insulators and conductors as a function of temperature in terms of the band theory of solids and free electrons. Show by an appropriate sketch how a semiconductor can become a conductor or an insulator.

2.2. A standard tensile test on a specimen 0.505 in. diameter with a 2-in. gauge length yielded the following data: yield load, 11,700 lb; maximum load, 65,000 lb; gauge length at the maximum load, 3.018 in.; reduction in area at fracture, 35 percent. Determine the yield strength, the engineering fracture stress, the natural fracture stress, the strain-hardening equation, and its constants.

2.3. Given the strain-hardening equation for an annealed material as $\sigma = 95{,}000\varepsilon^{0.25}$, estimate its yield and tensile strengths as annealed and after 60 percent cold work.

2.4. If the value for the Bhn of AISI 1040 steel as water-quenched and tempered at 400°F is 560, what would be its probable tensile strength?

2.5. From a long time-creep test of Inconel X at 1500°F carried out in a nonoxidizing atmosphere, the following data were obtained:

(*a*) Elongation after 10 h, 1.2 percent
(*b*) Elongation after 200 h, 2.3 percent
(*c*) Elongation after 2000 h, 4.5 percent
(*d*) Elongation after 4000 h, 6.8 percent
(*e*) After 5000 h, neckdown began
(*f*) At 5500 h, rupture occurred

Determine the creep rate of Inconel X at 1500 h.

2.6. A specimen of 60–40 brass was cold-rolled so that it received a 63 percent reduction, what is the magnitude of the true strain in the material?

2.7. A cast iron bearing plate $10 \times 10 \times 2.5$ in. is subjected to a uniformly compressive load. The compressive strength of gray iron is 100,000 lb/in.^{2} and the strain at fracture was found to be 0.002 in./in. What was

(*a*) The compressive fracture load?
(*b*) The total contraction at fracture?
(*c*) The total strain energy corresponding to fracture?

SELECTED REFERENCES

1. American Society for Metals: *Metals Handbook,* Metals Park, Ohio, 1985.
2. Brophy, J. H., Rose, R. M., and Wulff, J.: "The Structure and Properties of Materials," Vol. II. Thermodynamics of Structure, Wiley, New York, 1964.
3. Cahners, Inc.: *Ceramic Data Book,* Cahners, Chicago [published annually].
4. Chapman–Reinhold: *Materials Selector Issue,* Materials Engineering Series, Chapman–Reinhold, New York [published annually].

5. Leslie, W. C.: *The Physical Metallurgy of Steels,* McGraw-Hill, New York, 1981.
6. Maki, T., Tsuzaki, K., and Tamura, I.: "The Morphology of Microstructure Composed of Lath Martensite in Steels," *Transactions of the Iron and Steel Institute of Japan,* vol. XX, 1980.
7. McClintock, F. A., and Argon, A. S.: *Mechanical Behavior of Materials,* Addison-Wesley, Reading, Mass., 1966.
8. McLean, D.: *Mechanical Properties of Metals,* Wiley, New York, 1962.
9. Meshil, M.: *Mechanical Properties of BBC Metals,* TMS-AIME, Warrendale, Pa., 1982.
10. Parker, E. R.: *Materials Data Book for Engineers and Scientists,* McGraw-Hill, New York, 1967.
11. Rosenthal, D.: *Introduction to Properties of Materials,* Van Nostrand, Princeton, N.J., 1964.
12. Samaus, C. H.: *Metallic Materials in Engineering,* Macmillan, New York, 1963.
13. Smithells, C. J.: *Metals Reference Book,* 3d ed., Butterworths, London, 1962.
14. Winchell Symposium on Tempering of Steel: *Metallurgical Transactions,* vol. 14A, pp. 985–1146, 1983.

CHAPTER 3

ENHANCEMENT OF THE PROPERTIES OF MATERIALS

Heat treatment is commonly used to enhance the mechanical properties of materials in the solid state. Although the process is usually thermal and modifies only the structure, there are thermomechanical treatments that also alter both the structure and shape, and thermochemical treatments that may modify both the structure and the surface chemistry. All three of these processes for the enhancement of properties can be classified as heat treating.

Many alloys are heat-treated, the more important being ferrous and aluminum alloys. The mechanical properties of metallic materials can be increased by strain-hardening. Hardening mechanisms include:

1. Strain-hardening
2. Control of grain size
3. Eutectoid decomposition
    (*a*) Equilibrium
    (*b*) Nonequilibrium
    (*c*) Hardening operations
4. Alloy hardening
    (*a*) Solid solution
    (*b*) Dispersion hardening

(*c*) Precipitation hardening
(*d*) Diffusion reaction

STRAIN HARDENING

When a metallic alloy is plastically deformed, its yield strength increases with increase in strain as long as the recrystallization temperature is not exceeded. Thus, controlled amounts of cold working may be used to increase the mechanical properties of a metallic material (Fig. 3.1). The true stress–true strain curves from Chapter 2 show that on a log–log plot, the strain-hardening equation is indeed a straight line and its slope is defined as the coefficient of strain-hardening. Through the strain-hardening equation, an engineer can predict the improvement in properties that a given operation will impart to a material.

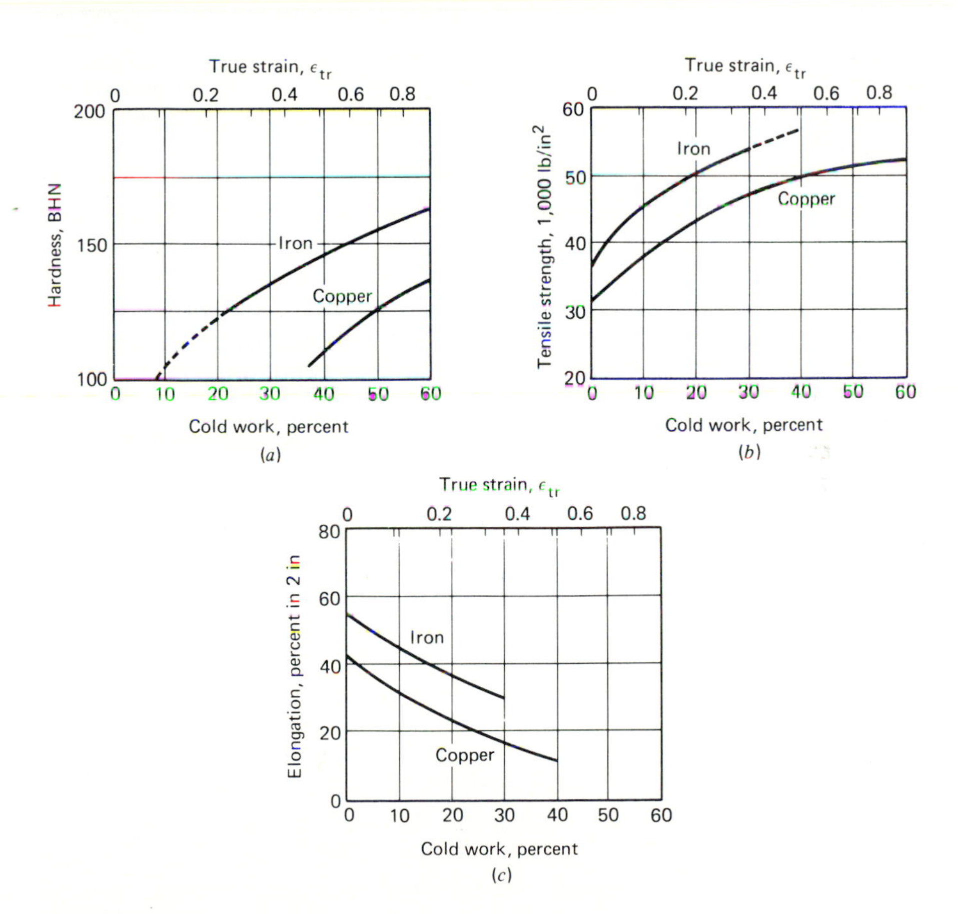

FIGURE 3.1
The effect of cold work on the mechanical properties of iron and copper. (*Redrawn from L. H. Van Vleck,* Materials Science for Engineers, *Addison-Wesley, Reading, Mass., 1970.*)

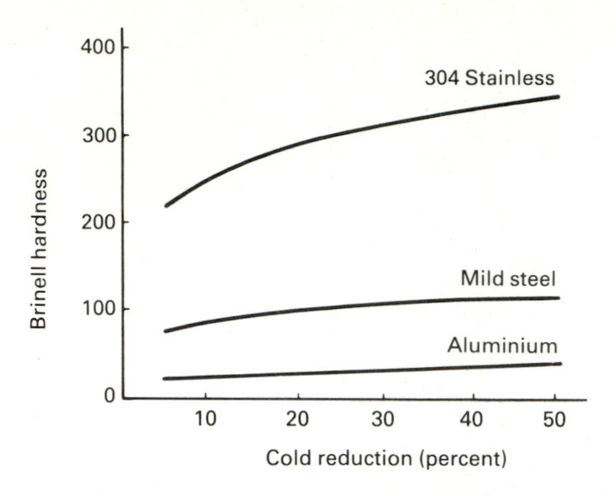

FIGURE 3.2
The rate of strain-hardening in three metallic materials.

In polycrystalline alloys, the mutual interference of adjacent grains causes slip to occur on many intersecting slip planes with accompanying strain-hardening. This progressive strengthening with increasing deformation stems from the interaction of dislocations on intersecting slip planes.

The strain-hardening behavior of a metal depends on its lattice structure. In face-centered cubic crystallographic structures, the rate of strain-hardening is affected by the stacking fault energy through its influence on mechanical properties. Copper, nickel, and austenitic stainless steel strain-harden more rapidly than aluminum. Hexagonal close-packed metals are subject to twinning and strain-harden at a much higher rate than do other metals because there is only one plane of easy glide available in the close-packed hexagonal structure. Strain-hardening is also effected by grain size, impurity atoms, and the presence of a second phase. The strain-hardening rate increases with the complexity of the structure of the alloy.

Severely strained metals may have elongated grains with distorted and twisted lattices and a strong anisotropy that can be used by an astute designer if he takes advantage of directional strength in his design. On the other hand, because of the increased strength and hardness brought about by strain-hardening, the number of reductions in a forming sequence will be limited before annealing is required; see Fig. 3.2.

Recovery

During a cold-working process most of the energy used is dissipated as heat but a small percentage is stored within the distorted lattice structure of the alloy. That energy is the thermodynamic driving force that tends to return the metal to its original state provided there is sufficient thermal energy to enable the reactions to occur. Recovery is a gradual change in the mechanical properties of an alloy, i.e., loss in brittleness, or marked increase in toughness brought about by controlling the heat-treating times and temperatures so that there are no appreciable changes in the microstructure. However, recovery does significantly reduce the residual stresses within the distorted lattice structure.

Recrystallization

All the properties of a cold-worked metal are affected to some degree by a yield behavior heat treatment, but the yield strength and ductility can only be restored by recrystallization. This may be defined as the nucleation and growth of strain-free grains out of the matrix of cold-worked metal. In general the properties of the recrystallized alloy are those of the metal before the cold-working operation. This is of commercial importance because if an alloy has become work-hardened in a drawing operation, it can be recrystallized and the drawing operation can be continued. Nucleation is encouraged by a highly cold-worked initial structure and a high annealing temperature. The presence of alloying elements in solid solution decreases the rate of nucleation.

SOLID-STATE REACTIONS THAT IMPROVE MECHANICAL PROPERTIES

Strain hardening (Fig. 3.1) and alloy hardening (Fig. 3.3) are widely used in industry, especially in aluminum- and copper-based alloys. It is possible to augment the mechanical properties of certain alloys by other solid-state reactions that can produce much greater hardness than is possible by alloying alone, and plastic deformation is not required. However, solid-state reactions are usually limited to certain alloys for several reasons.

1. Relatively few alloys can be affected by a given solid-state reaction. Euctectoid decomposition, for instance, is rare in alloy systems other than the iron–carbon system.

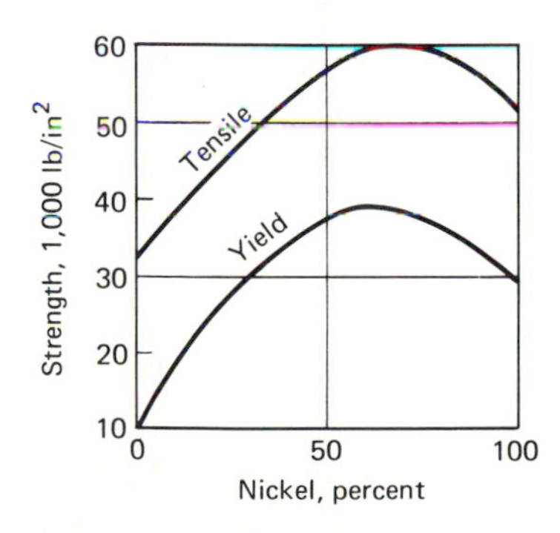

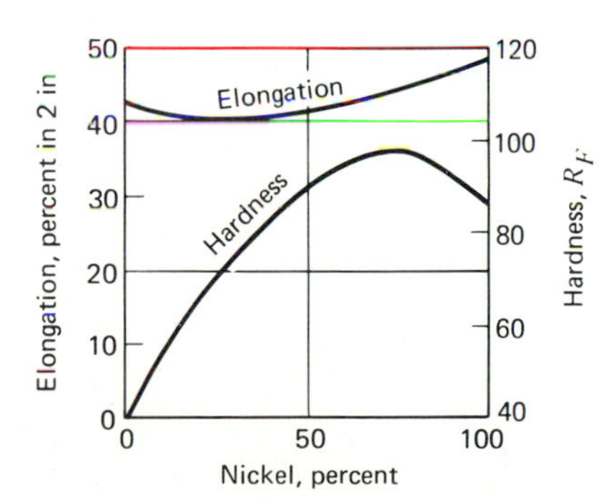

FIGURE 3.3
Effect of increasing solute on the strength of a substitutional solid solution alloy of copper and nickel. (*Redrawn from L. H. Van Vlack,* Materials Science for Engineers, *Addison-Wesley, Reading, Mass., 1970.*)

2. To achieve significant hardening, a solid-state reaction must form a nonequilibrium or metastable structure such as martensite.
3. Even though an energetically favorable solid-state reaction is possible according to the equilibrium phase diagram, it may produce little or no hardening. Therefore the occurrence of a given reaction is a necessary, but not a sufficient, condition for strengthening.

Several solid-state reactions produce gains in mechanical properties that are important to engineers. Typical phase diagrams indicate the combination of phases that is needed before a solid-state reaction can occur, but that alone does not mean that a significant improvement will occur. A typical eutectoid reaction is a prerequisite to eutectoid decomposition (Fig. 3.4*a*). However the iron–carbon system is the only one of commercial value. When the solvus line which separates one- and two-phase regions slopes to indicate decreasing solid solubility of B in A, there is a chance that precipitation-hardening may occur (Fig. 3.4*b*). Several aluminum alloys have this type of phase diagram. Although many binary alloys have similar solvus lines, aluminum alloys are the best example of precipitation hardening. The requirements of the diffusion reaction are indicated in Fig. 3.4*c*. In this case the diffusion of metal C into the alloy causes the composition of the hardenable alloy (metal B in metal A) to shift from a single-phase region to a two-phase region. As metal C diffuses into the solid solution, the overall composition gradually shifts into the α plus β region, so that the β-phase (B_zC_y) begins to precipitate. The nitriding process works according to this reaction because aluminum, chromium, and vanadium form nitrides. Thus, when nitrogen gas diffuses into the steel surface that contains those elements, their nitrides form within the surface of the metal to produce an extremely hard surface. The unusual surface hardness is thought to be a result of the fine dispersion of nitride particles rather than of the inherent hardness of the nitrides alone.

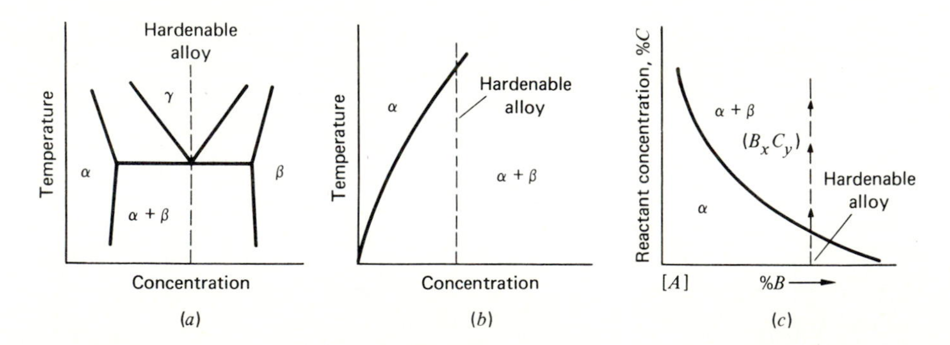

FIGURE 3.4
The form of equilibrium diagram required for three solid-state reactions: (*a*) eutectoid decomposition; (*b*) precipitation from the solid; (*c*) diffusion reaction. (*Redrawn from A. G. Guy,* Elements of Physical Metallurgy, *2d ed., Addison-Wesley, Reading, Mass., 1959.*)

SOLID-STATE TRANSFORMATIONS IN FERROUS ALLOYS

Steel is unique in its ability to exist as a soft ductile material that can easily be formed or machined and then, as a result of a heat treatment, assume the role of a hard, tough material that resists changing shape. There are two reasons for this behavior. The first is the fact that iron undergoes an allotropic change at 1330°F. Carbon has little solubility in the body-centered cubic lattice that is characteristic of iron at room temperature, but up to 2 percent carbon is soluble in the face-centered cubic lattice which is stable above 1330°F. The second is the solid-state eutectoid reaction in which a solid solution at a certain temperature can react to form two new solid phases. In the case of carbon steels, the γ-phase, austenite, transforms to α-iron (ferrite) and cementite, (Fe_3C) by the eutectoid reaction. The eutectoid reaction is common in many materials, but only steel exhibits such a marked change in properties.

Thus, when iron with 0.8 percent carbon is heated above 1330°F, the carbon dissolves and the resulting solid solution is called austenite or γ-iron. If the austenite is suddenly quenched in water, the carbon cannot escape and thus is trapped within the lattice structure as an interstitial atom that strains the lattice because of the increased volume it must occupy in the new body-centered tetragonal structure called martensite, which is characterized by a needlelike microstructure (Fig. 3.5). Martensite is a hard, brittle, metastable structure, i.e., a nonequilibrium structure, that is a supersaturated solid solution of cementite, Fe_3C, in a body-centered tetragonal iron. In the presence of a moderate temperature rise (300 to 400°F), tempering or recovery occurs and the brittle structure becomes tougher, while still retaining its strength and hardness.

Grain Size

An important factor in the heat treatment of steel is the grain size, by which is meant the size of the microscopic grains that are established at the last

FIGURE 3.5
The microstructure of martensite.

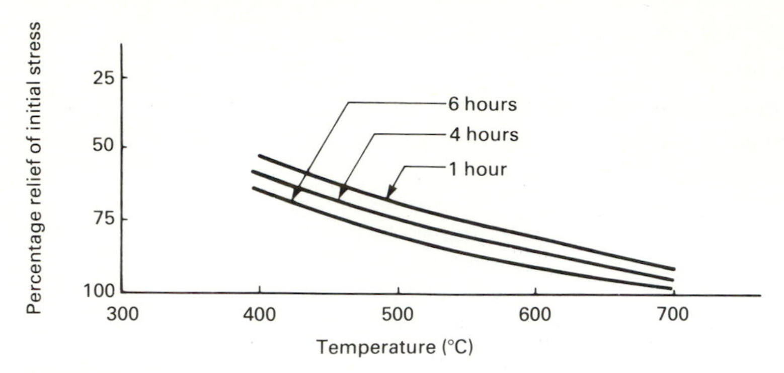

FIGURE 3.6
The relation between temperature vs. soak time and the amount of stress relief.

temperature above the critical range to which the piece of steel has been treated.

1. Fine-grained steels show better toughness at high hardness.
2. Grain-size-controlled steels will show less warpage.
3. Coarse-grained steels harden better.

Grain-size-controlled steel can be secured from the steel mill within certain limits. A high percentage of carbon steels used for heat treating is grain-size-controlled and sold to grain-size limits. Almost all the alloy steels are grain-size-controlled. If a given grain size is specified, it must be obtained by suitable control of the recrystallization process and by prevention of excessive grain growth.

Austenitizing

Most ferrous heat treatments require that austenite be produced as the first step in a heat-treating operation. The iron–carbon phase diagram of Figs 3.7 and 3.8 shows the minimum temperature at which austenite can form, but a temperature about 100°F higher is required if a reasonable austenitizing time is desired. The phase diagram assumes equilibrium, that is, the carbon and iron have adequate time to distribute themselves in the phases as shown. The engineer should recognize that at times equilibrium is difficult to achieve, especially in steel that contains elements which diffuse slowly. Certain heat treatments, such as hardening, are designed to prevent formation of equilibrium structures. As shown in Fig. 3.7, the austenitizing temperature varies with the carbon content, decreasing along line A_3 to A_1 at 0.8 percent carbon and increasing along the A_{Cm} line. Austenitizing is a function of both time and temperature (Fig. 3.8). In practice, a soaking time of 1 hour/in. of cross

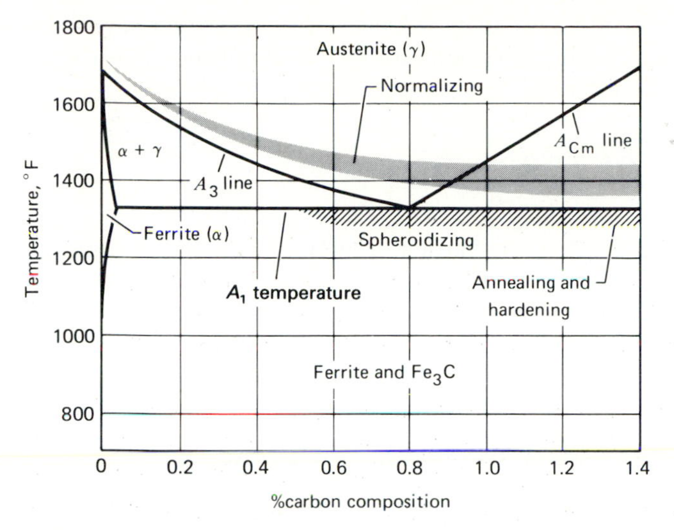

FIGURE 3.7
A portion of the iron–carbon equilibrium diagram showing the temperature ranges usually employed in various heat treatments.

section is considered to be adequate for austenitizing a carbon steel, although temperature and initial carbide particle size are both important factors.

Steels and cast irons contain, in addition to iron and carbon, other elements that shift the boundaries of the phase fields in the Fe–C diagram. Some alloying elements such as manganese and nickel are austenite stabilizers

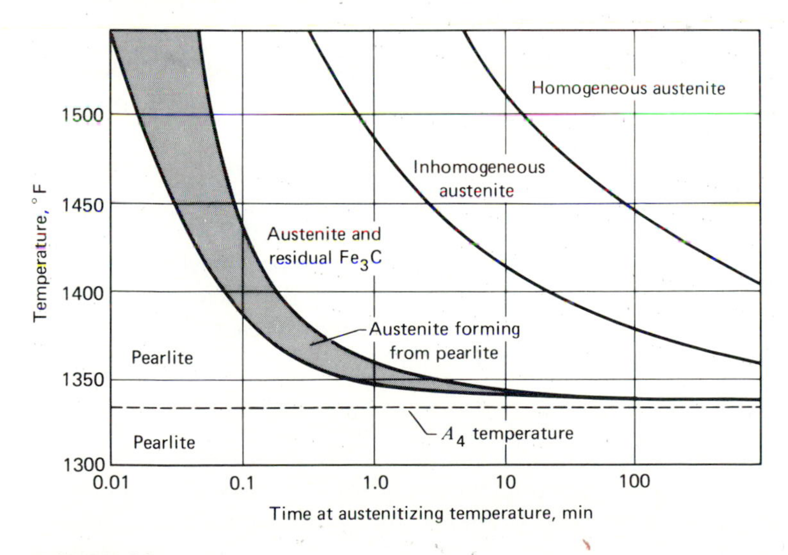

FIGURE 3.8
Approximate times necessary for the isothermal formation of austenite in a normalized eutectoid steel at various austenitizing temperatures. (*After Roberts and Mehl.*)

and extend the temperature range over which austenite is stable, while elements such as chromium and molybdenum are ferritic stabilizers and restrict the range of austenitic stability.

HEAT TREATMENT OF FERROUS MATERIALS

All heat treatments either soften or harden a metal. The most common treatments for softening are stress relieving, annealing, and normalizing; while the major hardening processes are case-hardening or surface-hardening, and through-hardening. The softening heat treatments for steel will be presented first.

Stress Relieving

Stress relieving is used to relieve stresses that remain locked in a structure, usually as a consequence of some manufacturing process. It involves heating to a temperature below the critical temperature (1100 to 1200°F for low-carbon steel) and cooling slowly. This process is particularly applicable after straightening and cold-working operations and either prior to or after heat treatments to reduce distortion. In stress relieving, no change in microstructure is involved, but residual stresses are markedly reduced and toughness is improved. The relief of locked-in stresses is a time–temperature phenomenon; see Fig. 3.6. The thermal effect for the relief of residual stresses has been correlated by the Larson–Miller equation:

$$\text{Thermal effect} = T(\log t + 20) \times 10^{-3}$$

where T is the temperature (Rankine) and t is the time in hours. Thus, holding a part at 575°C (1067°F) for 5 hours will provide about the same relief of residual stresses as heating to 600°C (1112°F) and holding for 2 hours. Typical stress-relief temperatures for low-alloy ferritic steels are between 595 and 675°C (1100 and 1250°F).

When consideration is given to the simplicity of the stress-relieving operation and the equipment required, this important process, which can save considerable amounts of straightening and reworking time, cannot be overlooked. It may be applied to relieve stresses induced by casting, quenching, normalizing, machining, cold working, or welding.

Annealing

Annealing involves heating to and holding at a suitable temperature, followed by cooling at an appropriate rate, primarily for softening of metals. When annealing iron-based alloys, the work is 50 to 100°F above the critical temperature range. The alloy is held for a period of time to insure uniform temperature throughout the part, and then allowed to cool slowly by keeping the parts in the furnace and allowing both to cool. See Fig. 3.6. The cooling method should not permit one portion of the part to cool more rapidly than another portion. Its purpose is to remove stresses; induce softness; alter ductility, toughness, and magnetic properties; change grain size; remove gases; and produce a definitie microstructure.

Steel is annealed for one or more of the following reasons:

1. To soften it for machining or fabrication operations
2. To relieve stresses in the material. (often necessary after casting or welding operations)
3. To alter its properties
4. To condition the steel for subsequent heat treatments or cold work
5. To refine grain size, improve ductility and promote dimensional stability.

Perhaps the principal reason for annealing is to improve machinability characteristics of the steel. It is not possible to present a general rule for annealing that can be used on all types of steel subjected to diversified metal-removal operations. For heavy roughing cuts, the material should be as soft as possible (Rockwell B80). This is especially true when accuracy and finish are not important; however, if a close tolerance is required, as on a broaching operation, then a finer lamellar microstructure with a hardness of approximately Rockwell B100 is more desirable. The principal point for the engineer to keep in mind when using medium-carbon steel is to specify the condition of the steel as purchased.

As the hardness of steel increases during cold working, ductility decreases and additional cold working becomes so difficult that the steel must be annealed to restore its ductility.

Localized or spot annealing can be accomplished with an oxyacetylene torch and is a valuable technique when salvaging scrap work, but tool steels should not be torch-annealed—they will crack.

Normalizing

Normalizing is heating iron-based alloys to 100 to 200°F above the upper critical line of the iron–carbon phase diagram, that is above A_3 for hypoeutectoid steels and above A_{Cm} for hypereutectoid steels (see Fig. 3.7) and then cooling in still air at ordinary temperatures. Normalizing is performed in order to relieve internal stresses resulting from previous operations and to improve the mechanical properties of the steel. For example, the hammer stresses developed in a forging need to be relieved prior to machining.

Normalizing may be done to improve machinability, to refine the grain structure, and to homogenize the part. For example, homogenization of steel castings may be done in order to break up or refine the dendritic cast structure and facilitate a more uniform response to subsequent heat treatment. In the case of wrought products, normalization can reduce large grain size where large and small grains exist, thus developing a more uniform grain structure. Toughness can be increased. This is frequently desirable in the use of "as-rolled" medium carbon steels.

A broad range of ferrous products can be normalized. All of the standard, low-, medium-, and high-carbon wrought steels can be normalized. Stainless steels, austenitic, and maraging steels are not usually normalized. The reader should recognize that normalizing may increase or decrease the strength and hardness of a given steel depending on the carbon content and the thermal and mechanical history of the part.

Gears, bolts, nuts, washers, and other parts in which low distortion is an important criterion should be made in the following general operation sequence: rough machine, normalize, finish machine, carburize, heat treat, grind.

Spheroidizing

Spheroidizing produces globular carbides in a ferritic matrix (Fig. 3.9). The iron-base alloy is held for a prolonged period of time (10 to 12 hours) at a temperature near but slightly below the A_1 temperature, and then slowly cooled (furnace-cooled). A spheroidized steel has minimum hardness and maximum ductility. The structure improves machinability markedly. Normalized steels are one of the better starting spheroidized materials because their fine initial carbide size accelerates spheroidization.

Hardness and Hardenability

The maximum hardness of a steel is a function of its carbon content (Fig. 3.10). Although alloying elements such as chromium increase the rate at which

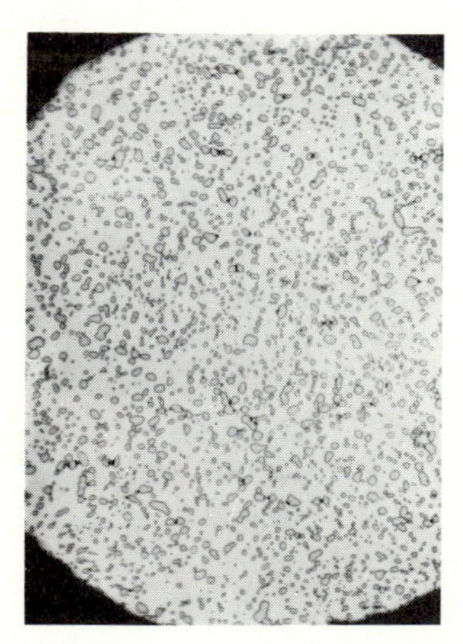

FIGURE 3.9
The microscopic structure called spheroidite.

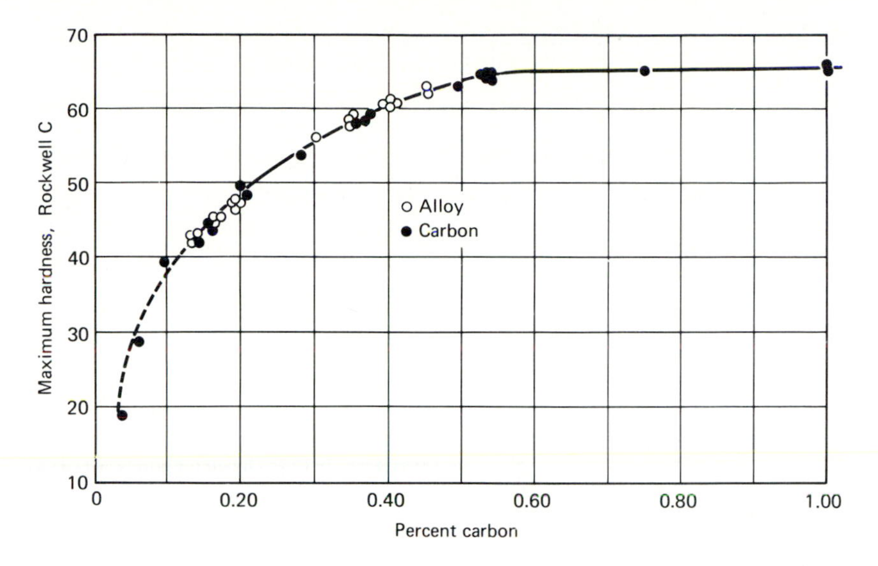

FIGURE 3.10
Maximum hardness versus carbon content. (*From J. L. Burns, T. L. Moore, and R. S. Archer, "Quantitative Hardening,"* Transactions ASM, *vol. XXVI, 1938. Courtesy American Society for Metals.*)

the martensite transformation occurs and thus the depth to which full hardness can be achieved, no alloy steel can exceed the hardness of SAE 1055 steel.

Hardenability refers to the distance within a specimen normal to its surface that appreciable hardness can be developed. Since the quenching rate of a steel is limited by its heat diffusivity and the rate at which austenite can transform, the hardenability of steels depends on the alloy content and the grain size of the original austenite. Hot-work steel die blocks, which must be hardened throughout their mass, are alloyed to such an extent that they can transform to a bainitic structure with an air quench. The hardenability of a typical alloy steel and a plain carbon steel are compared in Fig. 3.11. Note that for all sections above $\frac{1}{2}$-in. diameter the plain carbon steel could not be quenched sufficiently rapidly to achieve full hardness even at the outside diameter.

A standard test for checking hardenability is the Jominy test. It consists of heating in a 1-in. diameter bar to its austenitizing temperature and then setting it over a jet of water that hits only the bottom face. Consequently, there are various cooling rates along the length of the specimen; later, hardness values can be measured along the side of the bar, representing cooling rates that vary from that of a full water spray down to an air cool. This is a wide-range test, yet its results can be correlated with tests on fully quenched bars, as well as with work on actual parts.

Another useful fact may be mentioned. A gear or a shaft made from a

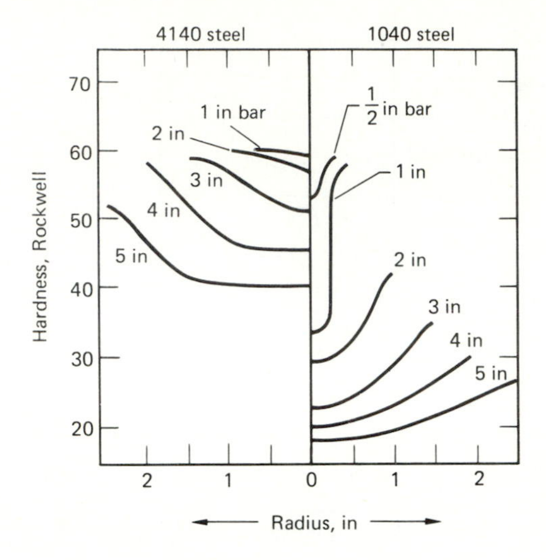

FIGURE 3.11
Comparison of the hardenabilities of 4140 and 1040 steels.

particular steel may be found to have at some point a Rockwell hardness of C50. A Jominy hardenability test is then run, and it is found that C50 is $1\frac{9}{16}$ in. from the water-cooled end. Any other steel that will show C50 at $1\frac{9}{16}$ in. from the quenched end will produce a similar hardness distribution when it is used to make the same part.

Table 3.1 collected from various sources, shows the sizes of bars that are fully hardenable, that is, that will develop a hardness of at least Rockwell C50 at the center when quenched as indicated. These are interesting figures, particularly when correlated with steel prices. You will see that SAE 2340 (more expensive than 4150) is much less hardenable. It is also interesting to note that the mechanical properties obtained for 4150 are just as good. The 1300 steels are practically as good as 2300 (3.5 percent nickel) or as 5100 chromium steels in the matter of depth of hardening; they are almost as good as nickel–chromium steels.

TABLE 3.1
Size of bars fully hardenable (Rockwell C50 minimum at center)

SAE	Water Quench, in.	Oil quench, in.	SAE	Water quench, in.	Oil quench, in.
1050	$\frac{3}{4}$–1	—	3130	$1\frac{1}{8}$	$\frac{9}{16}$
1330	$1\frac{1}{4}$	$\frac{11}{16}$	3140	$1\frac{5}{8}$	1
1340	$1\frac{1}{4}$	$\frac{11}{16}$	X-3140	$2\frac{1}{2}$	$1\frac{3}{4}$
2330	1	$\frac{1}{2}$	4150	$3\frac{1}{2}$	$2\frac{1}{2}$
2340	$1\frac{3}{8}$	$\frac{7}{8}$	X-4340	6	4
5130	1	$\frac{1}{2}$	6150	$1\frac{3}{8}$	$\frac{7}{8}$
5145	$1\frac{1}{2}$	$\frac{7}{8}$	3240	4	3

Microstructure of Heat-treated Steel

Today there is a broad variety of heat-treating equipment providing various ways of heating the steel, quenching it, and tempering it. Temperature control of the various stages can be assured; thus, the production engineer can depend upon accurate and reproducible control of heat-treated parts after he has established a sound heat-treating cycle.

The purpose of heat treatment is to change the form in which the carbon is distributed in the steel. Alloying elements present in a steel will affect the rate at which the reactions of heat treatment occur, but have little effect on the tensile properties of the steel. In ordinary carbon steel that is not hardened, the carbon is present (as carbides) in either globules or rodlike particles that can be readily discerned with the microscope. The structure having the parallel-plate appearance is known as pearlite (Fig. 3.12), while the globular structure is referred to as spheroidite (Fig. 3.9). When steel such as this is heated to a relatively high temperature (1500°F), the carbide particles become dissolved in the surrounding ferrous structure and a solution of carbon in iron known as austenite is formed (Fig. 3.13).

As austenite is cooled, the carbon tends to separate from the solution and return to its original form. However, by controlling the rate of cooling, the return to pearlite or spheroidite may be avoided. For example, if the austenite of a 0.80 percent carbon steel is cooled from 1500°F down to 1200°F and is then allowed to remain at 1200°F until it transforms, it will take the form of pearlite, as illustrated in Fig. 3.12. This steel would be quite soft, having a Brinell hardness of about 200. However, if the austenite is cooled quickly to a lower temperature, say 600°F, it will escape the 1200°F transformation stage, and will change to a structure known as bainite, which has a hardness of about 550 Brinell. If the austenite is cooled to a still lower temperature, for example

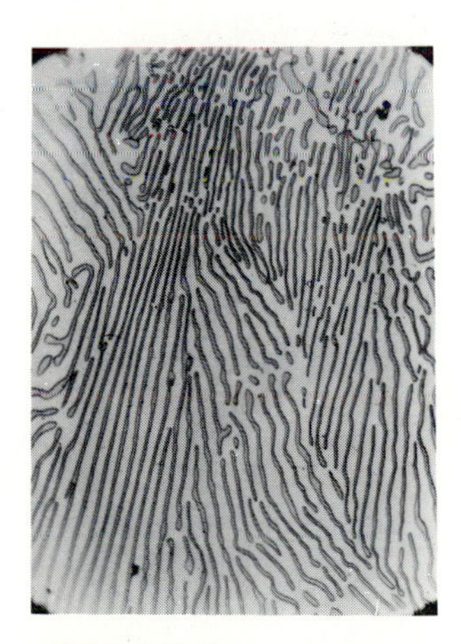

FIGURE 3.12
The microstructures of pearlite.

FIGURE 3.13
The microstructure of austenite.

250°F, before transformation takes place, the very hard structure known as martensite will be formed. Martensite gives a Brinell of about 650 (Fig. 3.5).

Figure 3.14 gives information on the hardness of the transformation products of coarse pearlite to martensite. An important rule that the

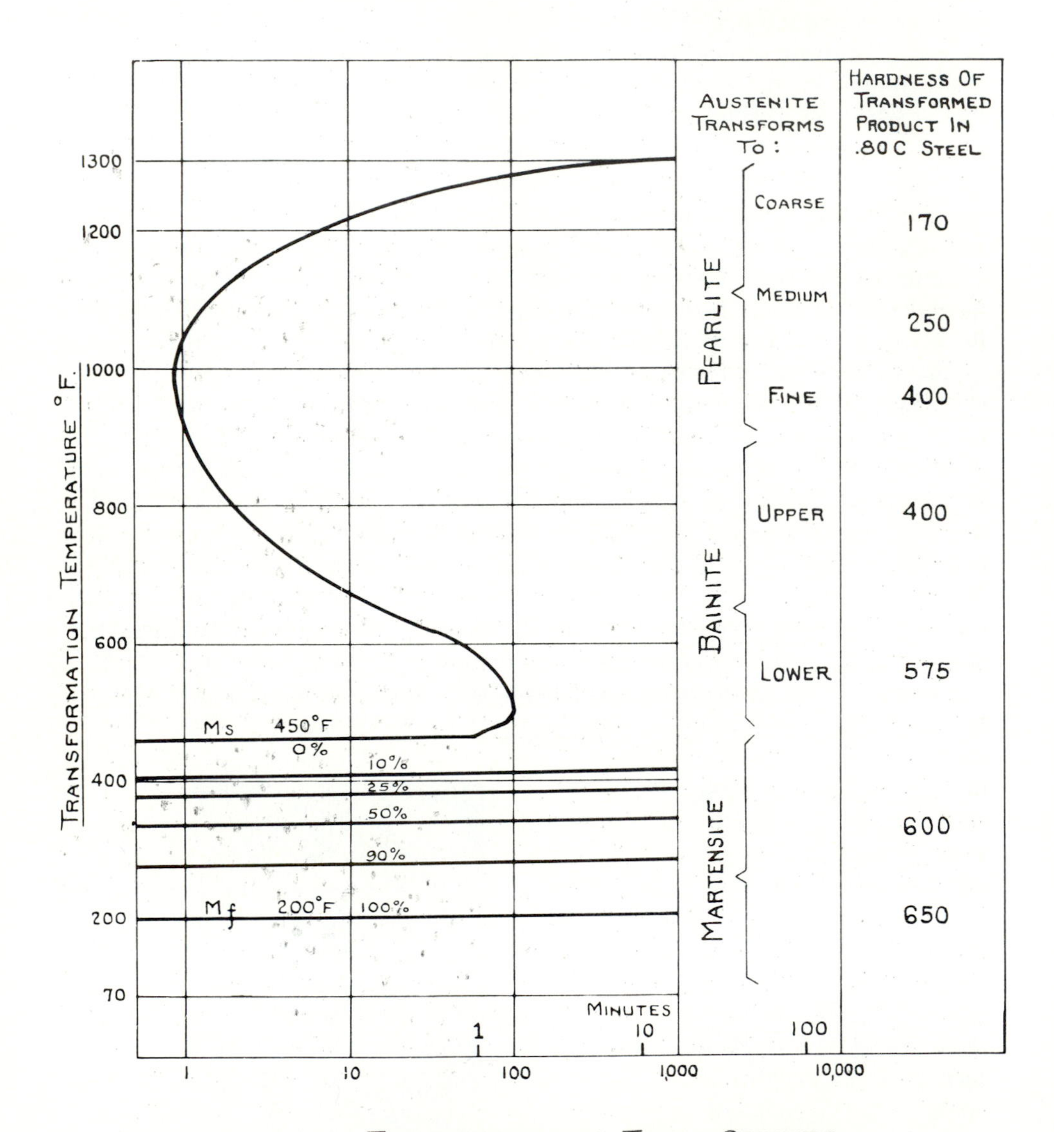

FIGURE 3.14
Time required by austenite to transform at different temperature (TTT curve or isothermal transformation diagram). (*Courtesy USX Corp.*)

production engineer should keep in mind is that, when forming any particular transformation product, the steel must escape transformation at a higher temperature. If this is not observed, then the desired transformation product will not be achieved. From the foregoing, it can be seen that a hardened piece of steel takes the form of martensite or bainite. The formation of pearlite must be avoided. Thus, the process of heat treating involves first heating and soaking at the correct temperature, then cooling the steel rapidly enough to avoid the formation of pearlite, and finally holding at the desired temperature to form either bainite or martensite. The amount of martensite formed is a function of the amount of undercooling below the M_s temperature, the temperature at which martensite starts to form on cooling a given steel. Figure 3.16 shows the M_s temperature as a function of carbon content and Fig. 3.14 illustrates the transformation time in seconds required by austenite at different temperatures. This chart is known as a TTT curve (time, temperature, transformation). This isothermal transformation diagram is very useful as it shows the various structures formed by a steel during its cooling period or formed while the steel is held at a temperature to which it was cooled.

A study of this diagram will indicate the importance of cooling time in order to avoid the formation of the soft pearlite. Small parts that cool rapidly (i.e., dwell in the 1000°F zone for less than 1 to 3 s) are readily transformed to either bainite or martensite and escape the pearlite formation. However, in large sections, where the cooling period is longer, it would be necessary to use a steel that had the nose of its TTT curve moved to the right so there would be a longer period of cooling time in order to avoid the formation of pearlite. When steel is alloyed with certain element such as chromium or molybdenum, thick sections can be hardened even in the center. Such steels are referred to as having greater hardenability.

From this, it can be seen that the effect of reducing the carbon content is to induce more rapid transformation rates. The effect of alloying is to shift the entire isothermal transformation diagram to the right; that is, transformation at all temperature levels starts later and is slower to go to completion. It should be recognized that although this is a general characteristic of alloying, various alloys differ substantially in both the magnitude and the nature of their effect. That is, the shape of the isothermal curve will vary considerably as well as its location with respect to the time–temperature axes. Several heat treatments are shown schematically on TTT diagrams in Fig. 3.16.

Figure 3.15 indicates two forms of martensite in carbon steels: lath and plate. These forms are based on the morphology and microstructural characteristics of the martensite. The lath morphology forms in low- and medium-carbon steels and consists of regions or packets where many fine laths or board-shape crystals are arranged parallel to one another. The plate morphology forms in high-carbon steels and consists of martensite plates that form at angles with respect to each other.

As-quenched martensite is supersaturated with carbon; it has a very high interfacial energy per unit volume associated with the fine laths and/or plates

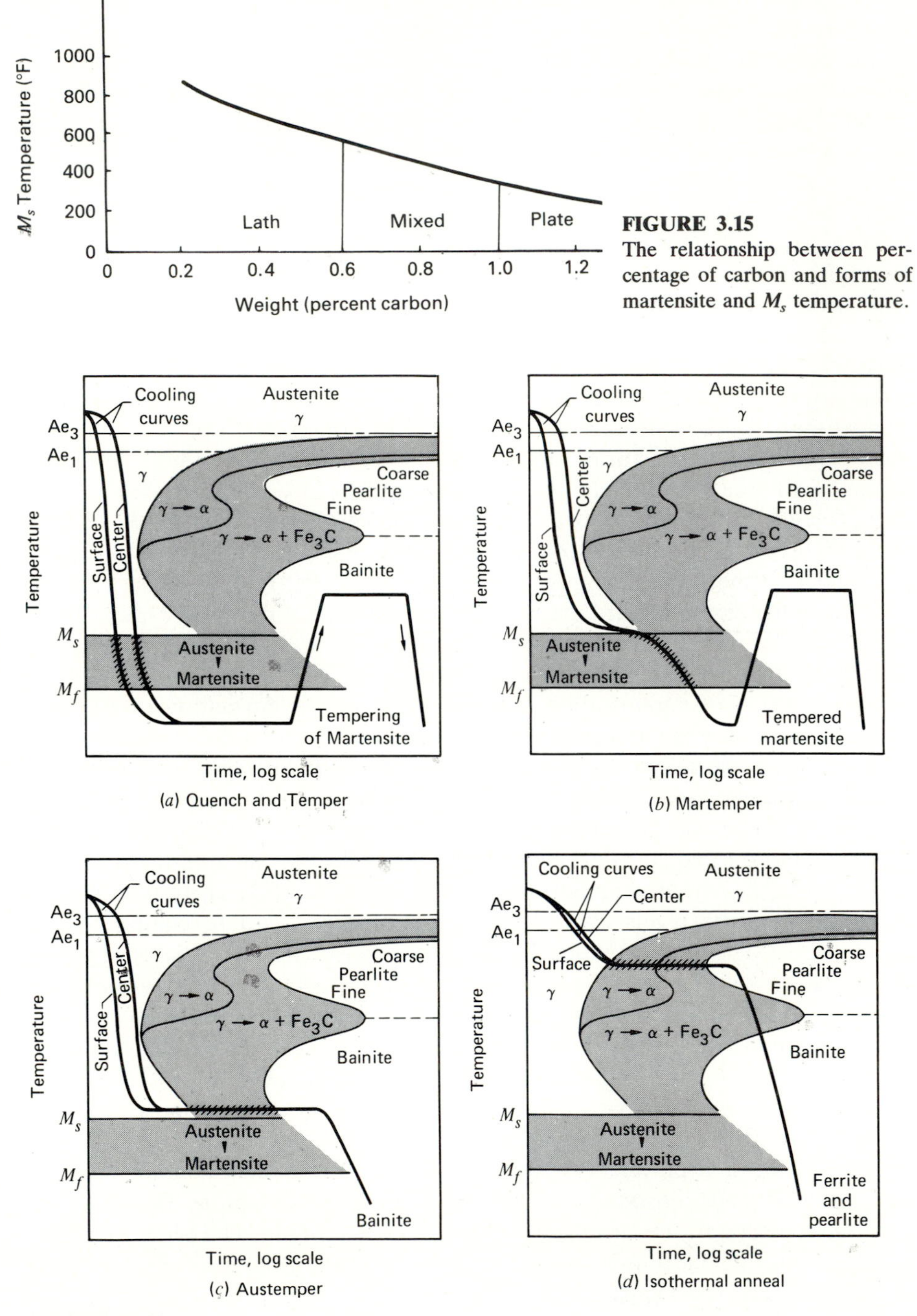

FIGURE 3.15
The relationship between percentage of carbon and forms of martensite and M_s temperature.

(*a*) Quench and Temper

(*b*) Martemper

(*c*) Austemper

(*d*) Isothermal anneal

FIGURE 3.16
Various heat treatments related to the TTT diagram.

of the martensite microstructure. It contains numerous dislocations that store considerable strain energy. In view of these characteristics, martensitic microstructures are unstable and will decompose when heated. Upon decomposition, the steel increases in toughness and for this reasons most hardened steels are heated to some temperature below the A_1 temperature (Fig. 3.7) in order to temper the steel. Tempering between 100°C and A_1 produces various types of carbide-particle dispersions as well as major changes in the matrix martensite. The reactions that produce the carbide are classified as stages of tempering: T_1, T_2, etc. (see Fig. 3.17.)

Toughness, or the ability to deform slightly under load prior to fracture, is the principal reason for tempering. Toughness is brought about by changing the shape of the individual carbide particles from plates to spheroidal shape. Tempering of hardened steel results in this change, which gives it toughness characteristics. The degree of toughness varies with the steel treated. A high-carbon steel given the same heat treatment and tempering will result in a stronger but less tough steel than a low-carbon steel that has undergone the same treatment. In treating steel the engineer must realize that a gain in one characteristic such as hardness results in a loss in another characteristic such as toughness.

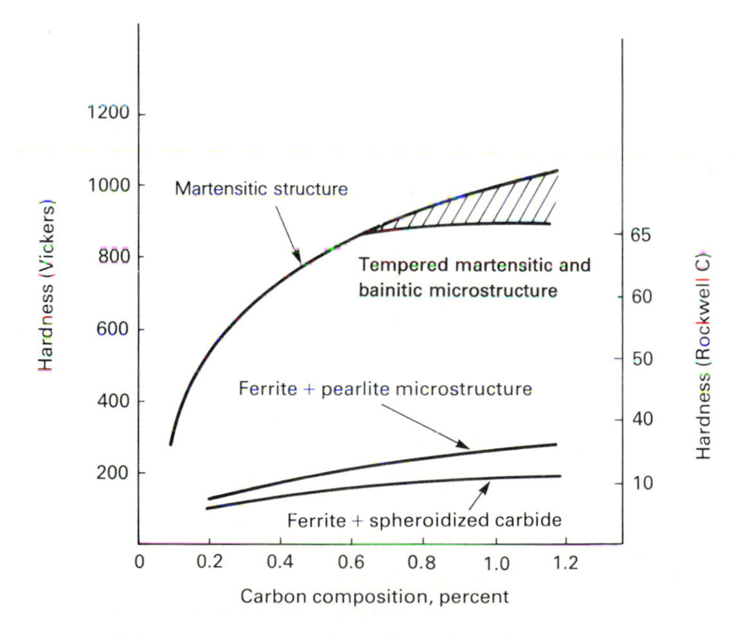

FIGURE 3.17
Relation between percentage carbon composition and hardness for various microstructures. Note cross-hatched area shows effect of retained austenite. (*Redrawn from George Krauss,* Physical Metallurgy and the Heat Treatment of Steel.)

Effects of Grain Size

When steel is heated through the critical range (approximately 1350 to 1600°F) transformation to austenite takes place as mentioned. When first formed, the austenite grains are small, but they grow in size as the temperature above the critical range is increased and, to a limited extent, as the time is increased.

When temperatures are raised significantly above the critical range, different steels show wide variations in grain size, depending upon the chemical composition of the steel and the deoxidation practice used in making the heat. Heats are usually deoxidized with aluminum, ferrosilicon, or a combination of other deoxidizing elements.

Fine-grain steels do not harden as deeply as coarse-grain steels, and they are more resistant to cracking upon quenching after heat treatment; they tend to be tougher and have better application under impact load conditions. However, coarse-grain steels produce a discontinuous chip that permits a better machined finish and less tendency toward a built-up edge, thereby making them more applicable in designs requiring intricate machining.

Increasing the grain size of austenite lengthens the time for both the beginning and completion of transformation.

Through-Hardening

Through-hardening consists of heating the part to some point above the critical temperature and then quenching. This is frequently followed by tempering in order to instill toughness to the part (Fig. 3.16a). Through-hardening is generally applied to medium- and high-carbon steels in order to develop a particular combination of both hardness and toughness.

Oven furnaces, pit furnaces, and liquid baths are used for the heating stage for through-hardening, with the oven type being the most popular. The furnaces may be heated by gas, oil, or electrical resistance elements. On smaller parts, induction heating and flame heating may be used.

In order to ensure that the correct heat treatment will be achieved, the production engineer should verify that the "soaking" period in both the heat and the draw are adequate. In the heating period, it is essential that the part be at the correct hardening temperature so that full hardness throughout the section is achieved. Also it is important that the atmosphere of the furnace be controlled so that the decarburization of the surface will not be greater than anticipated.

Quenching is an important factor in the through-hardening of a piece of steel. The most common quenching media are water, brine solutions, caustic solutions, oils, polymer solutions, molten salts, molten metals, gases (still or moving), and fog. Water usually is used on the plain carbon steels that require fast cooling in order to achieve the highest hardness. Water quenching is quite severe and should not be used on intricate shapes and contours or cracking at the time of quench may be induced. Oil quenching is considerably milder than

TABLE 3.2
Effect of agitation on quenching

Amount of agitation	Oil	Water	Caustic soda or brine
None	0.25–0.30	0.9–1.0	s
Mild	0.30–0.35	1.0–1.1	2–2.2
Moderate	0.35–0.4	1.2–1.3	—
Good	0.4–0.5	1.4–1.5	—
Strong	0.5–0.8	1.6–2.0	—
Violent	0.8–1.1	4	5

Source: Adapted from *Metals Handbook,* American Society of Metals 1985.

water quenching and is usually used with alloy steels and on intricate shapes. Cooling in an air blast is used on the higher-alloy steels such as some of the tool steels.

Agitation of the quenching media has a pronounced effect on the severity of the quench. Table 3.2 provides data on the effect of agitation on quenching. For any amount of agitation, the effectiveness of the quench in hardening increases from left to right and for a quenching media from top to bottom. Thus, a violent agitation of water would be twice as effective as a quenching medium as brine with no agitation.

To achieve the required toughness in a through-hardened piece it is usually necessary to temper. In tempering of steel, previously hardened or normalized steel is heated to a temperature below the transformation range and then cooled. Tempering usually follows quenching from above the critical temperature. It is accomplished by reheating the part to a temperature between 300 and 1000°F (depending upon the alloy and toughness required) and then cooling. Tempering steel composition will not only improve toughness but will relieve quenching stresses and help insure dimensional stability.

Tempering furnaces include baths of oil, molten lead, molten salt media, and forced air circulation. The last type can be made to cover a wide range of tempering temperatures than other tempering media and are therefore quite popular in view of the wider variety of work that can be handled. Tempering oils are used from 300 to 600°F. Oils have the disadvantage that the parts require cleaning once they are extracted from the tempering tank. Tempering salts usually are used when drawing at higher temperatures (500 to 1000°F). Salts have the advantage of holding the oxidation of high-temperature draws to a minimum. The principal disadvantage of tempering salts lies in the fact that the salt must be boiled or washed off the work once it is removed from the furnace. Lead baths are also frequently used. When using a lead bath, care must be exercised in placing the part in the bath so as to avoid cracking. Lead is a rapid-heating medium and the through-hardened piece may not be able to withstand the quick immersion in the molten lead. Also, lead tends to adhere to the parts when they are removed from the bath, especially if they have deep recesses. This is frequently objectionable.

The length of time the part should be held at tempering heat is also important. Generally speaking, the higher the alloy content, the longer should be the draw. For example, carburized SAE 1020 steel should be drawn at 350°F for 1 hour; parts of 2315 or 4615 steel should be drawn at 350°F for a period of $1\frac{1}{2}$ to 2 hours.

Interrupted-quench treatments are used when it is necessary to keep distortion to a minimum and guard against cracking. The most common interrupted-quench treatments are martempering, austempering, and isothermal anneal (Fig. 3.16*b, c,* and *d*).

Martempering is an elevated-temperature quenching procedure for the purpose of reducing cracks, distortion, and residual stresses. Typically it consists of quenching a piece of steel from the austenitizing temperature into a hot fluid medium (hot oil, molten salt, or molten metal) at a temperature usually above the martensitic range (300 to 500°F) and then holding the part in the bath until the temperature is equalized throughout the piece (Fig. 3.16*b*). The part is then removed from the salt bath and air-cooled. Following this, it is tempered to the desired hardness. Being less severe than quenching in cool oil, this process permits the structure of the part to change more readily with fewer internal strains. This treatment is most effective on the oil-quenching steels, although water-hardening steels are handled also in this manner. In martempering of water-hardening steels, the parts are first quenched in brine to a point somewhere between the transformation point and the lowest temperature of the critical-cooling curve. The parts are then rapidly transferred to a bath the temperature of which is slightly above the transformation point.

Austempering is similar to martempering except that no draw is performed after quenching in the salt bath (Fig. 3.16*c*). In austempering, the salt bath is maintained between 400 and 850°F, depending upon the section thickness. This process is used on sections of alloy and tool steel not over 1 in. thick.

Isothermal transformation is a method that gives similar characteristics to the austempering technique on larger sizes (up to 2 in.). Here the part is quenched in an agitated salt bath between 450 and 600°F, and held to permit isothermal transformation. The part is then transferred to another bath at a higher temperature in order to draw the piece (Fig. 3.16*d*).

SURFACE HARDENING

Case Hardening

Case hardening is the hardening of a surface layer of a part made from a low-carbon steel so that this outside layer is substantially harder than the interior. This is accomplished by one of several processes that call for addition of either carbon or nitrogen to the part being hardened. The case-hardening processes include carburizing, cyaniding, carbonitriding, and nitriding. Flame-hardening and induction-hardening may be used to austenitize the surface carbon prior to quenching.

Carburizing involves the heating of the steel part to between 1600 and 1850°F in the presence of either a solid carbonaceous material such as charcoal, a carbon-rich gaseous atmosphere, or a liquid salt. This process is usually applied to steels with low carbon content. Open-fired or semimuffle gas furnaces of the continuous type or electric and gas-fired bell-type furnaces have been designed for controlled-atmosphere carburizing. In all carburizing processes, the amount and depth of case is determined by temperature and time. The greatest amount of control is obtained with the gas-carburizing technique. This method also permits the greatest depth of case. Cases between 0.30 and 0.40 in. are readily obtained by the gas-diffusion method. Case depths up to about 0.070 in. are obtained by pack carburizing.

F. E. Harris developed the following expression for the effect of time and temperature on case depth for normal carburizing:

$$\text{Case depth (in.)} = \frac{31.6t}{10^{(6700/T)}}$$

where t is the time in hours at temperature T in degrees Rankine.

The salt-bath method of carburizing is particularly adapted to small parts where many components may be immersed and treated simultaneously. A depth of case up to 0.030 in. is obtained by this method.

Cyaniding is performed at temperatures between 1500 and 1650°F in salt baths, electric or radiant-tube furnaces, or in electric or gas-fired bell-type furnaces. The cyaniding technique provides a case of high hardness and good resistance to wear. The process is the result of the formation of iron nitrides in the case brought about by the release of nitrogen from the bath. Cyaniding is used effectively for case depths up to 0.025 in. It is used especially on smaller parts that can be handled in salt-bath equipment without mechanical handling equipment.

Nitriding involves subjecting the parts to be hardened to the action of ammonia gas at temperatures of 920 to 980°F, so as to introduce nitrogen to the surface of the parts. Since that temperature is below the normal or critical hardening temperature of the steels being nitrided, it is said to be "subcritical" hardening operation. Sealed retort furnaces with close temperature control are required for the process. The time of exposure is relatively long, 72 hours being required for about 0.025 in. case depth. No quenching is necessary in nitriding, since the hard case is formed by the nitrides developed from the combination of nitrogen, iron, and the various alloying elements in the steel. This fact, combined with the low temperature used, minimizes distortion. This case-hardening process is usually used on special Nitralloy steels. Other steels, however, are also nitrided successfully. A typical nitrided steel is a chromium–aluminum–molybdenum steel in which a 50-hour cycle gives a 0.015-in. case. The hardness of a nitrided surface ranges from 60 to 72 Rockwell C and the hard, tough nitrided case is from 0.002 to 0.035 in. deep.

Care must be exercised in using hardness-testing devices on case-hardened parts. Tests may be misleading if they puncture the hardened surface.

LIGHT VERSUS HEAVY CASES. Under most conditions there is a better wear resistance from the first few thousandths of an inch of the heavy case than there is for the light case. The reason is that when a heavy case is produced, the maximum carbon at the surface is greater; in a shallow case, the maximum carbon is less because the migration of the carbon away from the surface is so fast that the surface carbon does not reach the maximum value. Better wear resistance is generally obtained from a carbon steel having 1.25 percent carbon than from another with 0.90 percent carbon; 1.25 percent surface carbon is to be expected after 8 hours, but after 1 hour not more than 0.80 or 0.90 percent is to be expected.

To summarize, the principal situations that warrant a heavy case are:

1. To withstand a heavy load, so that the case will not collapse. The danger of case collapse is a function of the load and also of the hardness of the material under the case. If the case is thin relative to the load, and if the core is soft relative to the case depth, trouble may develop.
2. To withstand severe wear. This applies only if the part can lose dimension without losing its usefulness. It is useless to put a 0.075-in. case on a gauge, for example, if it is discarded after wearing 0.0005 in. However, if the part is a link pin, which will wear under sandy conditions and still be good until the day it breaks, the more case that is put on it, the more life it will have.
3. Because of grinding to be done after heat treatment. Any case that is taken off by grinding not only represents wasted heat treatment, but requires extra expense to remove; therefore, excessive finish grinding should be avoided. The part ground off is the most valuable part of the case in terms of hardness and wear resistance. Not infrequently, a heavy case is specified and then too much stock is allowed for grinding. What is left is a very poor case.
4. To build up the mechanical properties of the surface. It should be emphasized that hardness means tensile strength; the reverse is also true—tensile strength in steel means hardness. If a high tensile strength is needed to give a high endurance limit at the surface, a deep case is required.

PROPERTIES OF THE CORE. There is still another factor in the selection of case depth—the balance between core properties and case depth. If wear resistance of the outer part of the steel is to be increased, and resistance to crushing is also required, there are two ways to do it: (1) by putting a heavy case on a poorly hardened steel; (2) by putting a light case on steel that has already hardened fairly well without a case.

The automotive industry has arrived at a compromise between the older plans of carburizing a low-carbon steel or merely oil quenching an alloy steel containing 0.50 percent carbon. The new procedure is to cyanide or lightly case-carburize a steel with perhaps 0.40 percent carbon. The plan is to get from 0.006 to 0.010 in. of carburized case, and yet to have the core quench out to Rockwell C50. A gear is economically produced in this manner because the case depth is shallow, the steel blank is relatively easy to machine because of its intermediate carbon, and the gear offers excellent resistance to crushing because the core is stronger, yet the gear teeth have desirable surface hardness. This compromise is an indication of the current trend toward use of intermediate steels case-hardened to shallow depths rather than use of extremely heavy cases.

Where localized case hardening is required, the piece may be copperplated on all sections except where the hard case is required, provided that a carburizing grade of base metal was specified. The copperplating will prevent the part from absorbing any carbon in the plated areas. Another method is to carburize the entire part and then machine off the unhardened case in the sections where it is not desired. In this method, allowance has to be made for the finish-machining operations.

WOODVINE DIAGRAM. Strength of the metal below the case is important, as indicated in Fig. 3.18 known as the *Woodvine diagram.* This shows the stresses on a piece that withstands bending. Line OA indicates that from the neutral axis (center) of a bar that is being stressed by bending toward the surface, the imposed stress increases from zero at the center to a maximum at the surface.

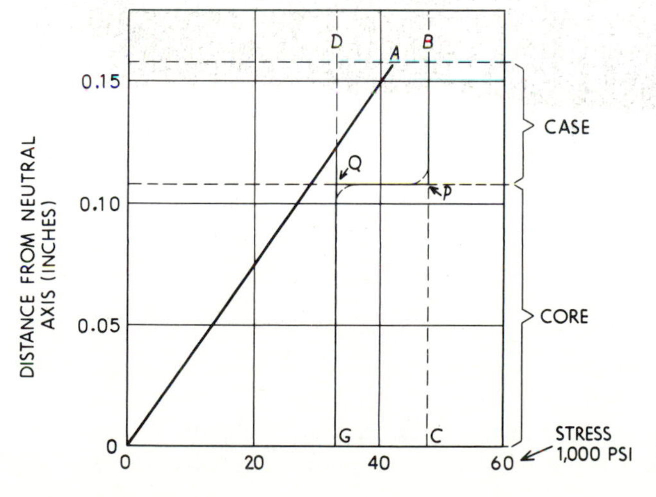

FIGURE 3.18
Woodvine diagram.

In this particular example, a diameter of 0.3 in. was selected and the stress on the outermost fiber was assumed to be about 45,000 lb/in.^{2} If the case is on a steel part that has a stress resistance of 47,000 lb/in.^{2} (plotted as line CB), it is obviously safe at the surface. If the case extends inward to a distance equivalent to 0.045 in., the strength of this outer region is indicated as BP, still on the safe side of the requirements. At the bottom of the case a fairly rapid transition can be assumed to the unaffected core with its stress resistance of only 32,500 lb/in.^{2} (distance OG on Fig. 3.18). The metal here is barely strong enough to stand the stress, yet it would probably "get by."

Now suppose a light case were used. The effect would be to move line QP upward, since the stress resistance of the core is approached more rapidly. Point Q would then be on the wrong side, i.e., the upper side of line OA: there would be less resistance at the junction between the case and core than the magnitude of the applied stress, and a subsurface fatigue failure could result if the bar were subjected to alternating stresses.

Figure 3.18 might also be used to explain the use of a shallow case on a toughened core. Line QP would be closer to DB, but line GQD would move over to the right because the stress resistance of the core would have a higher value. If a shallow case is used on a poor-hardening core, a subsurface failure can be expected, whereas with a stress-resistant core such a failure will be avoided.

A higher core hardness produces a more impact-resistant gear tooth. In this connection, nickel–chromium–molybdenum alloy steels have proven to be superior to all others. In some cases it is possible to harden the case and to toughen the core in a single heat treatment. If there is sufficient carbon in the alloy, the gear teeth can be flame- or induction-heated and quenched, leaving the core with the properties produced at the steel mill.

One precaution should be noted: the harder a steel, the more notch-sensitive it becomes. A case-hardened surface has a maximum hardness and as a result a maximum sensitivity to nicks and tool marks. Every effort must be made to provide ground or lapped surfaces and blended radii at all surface intersections to ensure maximum impact strength.

Induction Hardening and Tempering

The engineer should be familiar with induction heating because it has so many applications. The induction-heating circuit is essentially a transformer in which the conductor carrying a supply of alternating current is the primary, and the material to be heated is the secondary (Fig. 3.19). The current flowing through the inductor (copper tubing in the form of coils) sets up magnetic lines of force in a circular pattern that pass through the surface of the work. A flow of energy is therefore induced and internal molecular friction is developed in the work that dissipates itself in the form of heat. This heating effect is known as hysteresis. Heat is also generated in the work by eddy currents set up by the I^2R loss. The work is heated electrically by the alternating current, although

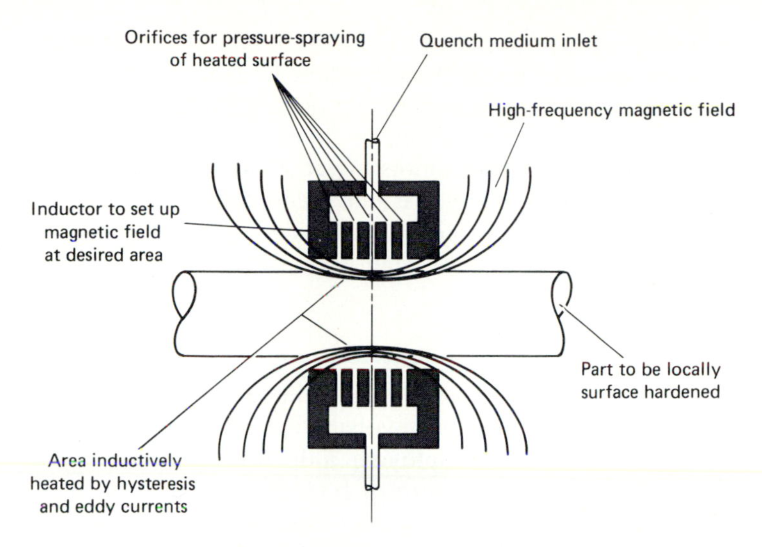

FIGURE 3.19
Schematic diagram of an induction heat-treatment device.

the work itself does not come in contact with the power source. Thus, it may be seen that any conductor will become heated when placed within the immediate area of a conductor that is carrying an alternating current.

In induction-hardening, electrical frequencies from 1000 to 500,000 cycles per second (Hz) are employed. Very high frequencies, 3000 to 500,000 Hz, cause the current to remain on the surface of the part; therefore, only the surface is heated. Lower frequencies, from 60 to 2000 Hz, are used for heating the part to greater depth. When a variety of parts are to be hardened, the heat-inducing coils must be changed to accommodate the particular part being hardened. The induction-heating machines on the market readily permit coil replacement. Induction hardening is a fast, reliable method for surface hardening; however, since specially designed coils are necessary to accommodate different parts, it is usually economical only for use on production runs. The pattern and amount of heating obtained by induction is a function of the shape of the induction coil, operating frequency, the alternating-current power input, the time that the part is subject to the power input, and the nature of the work part.

The equipment may be automatic or it may be operated manually. When heated to the proper temperature, the part can be quenched by a spray or stream of liquid, or it can be dropped into a tank. Little distortion results, and a uniformly hardened surface is achieved without affecting the properties of the core. To ensure good control, a temperature-controlled quenching liquid is often forced directly through the inductor elements to quench the surface as soon as it is austenized.

Induction-tempering has application in a variety of production lines. It can be used for selective tempering in an automated line or progressive tempering of bar stock that has previously been hardened.

Very high frequencies are used to obtain dielectric heating of nonmetallic parts. Plywood adhesives are cured by heating in a dielectric field; thus, a thermosetting plastic can be used to give a moisture-resisting bond. Induction heating is applied to soldering and brazing. It is quick and clean. The heat can be applied where desired, and distortion or excessive losses of heat are avoided. This process is especially useful on small parts.

Flame-Hardening

Flame-hardening is used on small lots because of its adaptability and freedom from special tooling and setup. The method involves the rapid heating of the part by means of a torch or torches employing a high-temperature flame obtained through the combustion of a mixture of fuel gas with oxygen or air (Fig. 3.20). This is followed by a quench. The part may be moved under the nozzles or the nozzles may move over the part. The shape and position of the nozzles are modified to suit the shape of parts, such as gear teeth or flat or V-shaped lathe bedways.

The speed of travel is 6 to 8 in./min and the penetration depth is $\frac{1}{32}$ to $\frac{1}{4}$ in. Machines are made fully automatic or semiautomatic. SAE 1040 to 1070

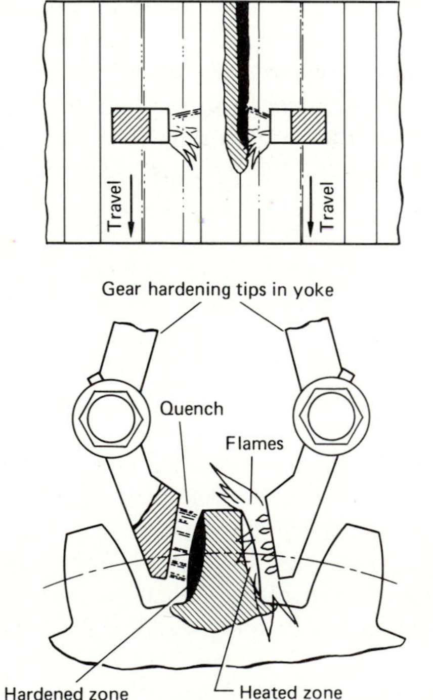

FIGURE 3.20
Flame-hardening setup for gear teeth. (*Redrawn from Doyle* et al., Manufacturing Processes and Materials for Engineers, *2d ed., Wiley, New York, 1969.*)

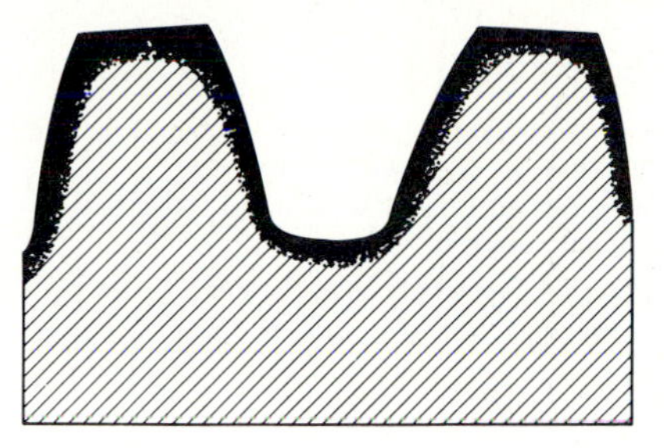

FIGURE 3.21
Section of a carburized gear tooth after hardening.

steels are suitable for flame hardening. Surface cracks may occur on high-carbon steels and the quenching rate must be reduced. Cast-iron lathe and machine toolways are flame-hardened to give a hard-wearing surface. Little distortion occurs, consequently little machine finishing is required. Wearing surfaces can be hardened without affecting the rest of the part. Crankshaft bearing surfaces, cams, gear teeth, and machine parts can be hardened (Fig. 3.21). Large gears, 20 or more feet in diameter, and large shafts, 20 or more feet in length, can be flame-hardened; thus, the danger of cracking is avoided and the use of large heating and quenching equipment is unnecessary.

Once a decision has been made to flame-harden a part and the most suitable process has been selected, the next step should be the determination of the depth of case required. In the interest of economy, the case should be no deeper than necessary. The amount of finish grinding needed after heat treatment must be considered. A part that is subject to considerable abrasion, pressure, or pounding should have a relatively deep case (0.050 to 0.075 in.). It should be remembered that too deep a case will result in cracking or checking on parts that are subjected to elevated temperatures. This is due to the lack of flexibility of a hard, deep case.

HEAT TREATMENT OF ALUMINUM

Aluminum is the only nonferrous material of structural importance that can be effectively heat-treated to enhance its mechanical properties. The mechanism of this heat treatment is known as precipitation-hardening. The process of precipitation, and the hardening that accompanies it also has application in some magnesium-, copper-, and nickel-based alloys.

Precipitation-hardening is possible wherever the phase diagram of the alloy shows a sloping solvus line, as shown in Fig. 3.22, for aluminum and copper. If a particular aluminum alloy, as depicted by the vertical line, is cooled slowly from its molten state to room temperature, first the liquid completely solidifies to a solid phase, the α-phase, then upon further cooling, a different phase, the β-phase, evolves, to some extent from the α-phase, so that at room temperature there will be a mixture of phases α and β—the resulting material being rather soft and without internal stresses.

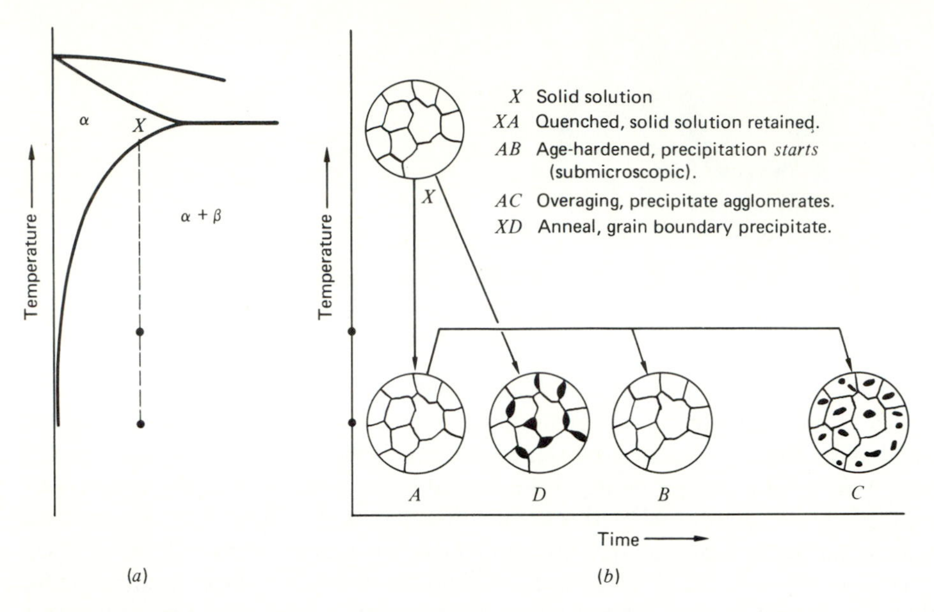

FIGURE 3.22
Phase diagram for aluminum and copper.

On the other hand, if the alloy is cooled very fast from a high-temperature zone within the α-phase and across the solvus line, then the β-phase does not appear, the solid solution being supersaturated at low temperature. The hardness of the material in this condition is slightly higher than in the annealed condition. However, given enough time, some of the β-phase will precipitate out of the supersaturated solution at room temperature and the hardness of the alloy will considerably increase. The alloy is then said to be naturally aged. This precipitation can be accelerated by a slight heating to some 200°F; the alloy is then artifically aged (see Fig. 3.23).

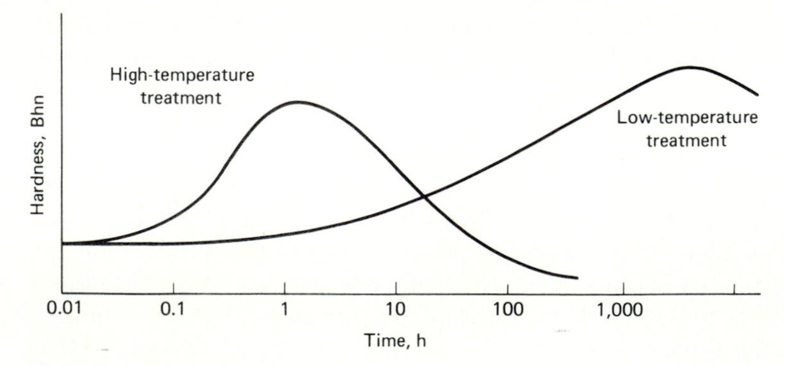

FIGURE 3.23
Variation of hardness with aging time.

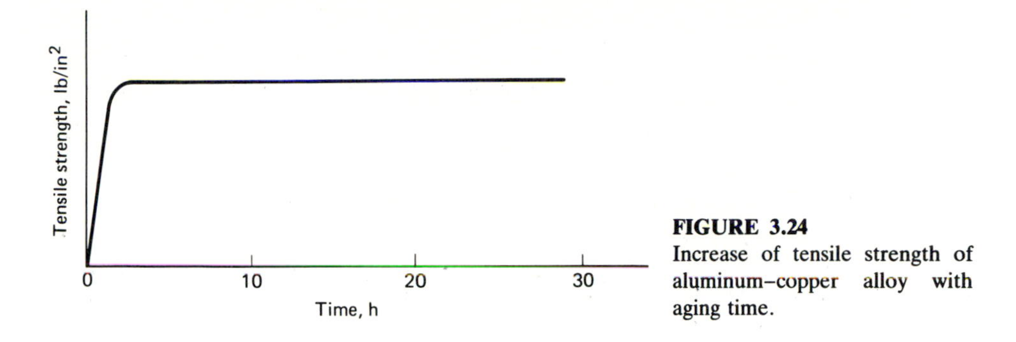

FIGURE 3.24
Increase of tensile strength of aluminum–copper alloy with aging time.

As the cooling rates of aluminum alloy castings can vary considerably, yielding anything from a completely annealed material to a supersaturated one, it is necessary to solution heat-treat aluminum castings before their precipitation hardening. This is done in order to establish a standard procedure. In solution heat treatment, the material is heated into the α-phase region, close under the solidus line, and kept there long enough to ensure that all the material is in the α-phase. After this, the material is quenched to obtain the supersaturated solid solution. Then the aluminum alloy is heated to a moderate temperature (about 200°F) and kept there as long as necessary to obtain the precipitation-hardening desired.

The tensile strength of the alloy is also enhanced during precipitation heat treatment (Fig. 3.24). Strength, as well as hardness, depends on the aging temperature and time.

Designation of Alloy Treatment

As the properties of an alloy very often depend on its processing history, it is convenient to specifiy that history by means of code letters. The condition or temper of an alloy is specified by adding a letter to the alloy designation:

Letter	Condition
F	As fabricated
O	Annealed
H	(Cold)-work-hardened
T	Heat-treated

The degree of strain hardening is designated by a number affixed to the initial letter H; e.g. H1, H2, H3, . . . , H9. The various tempers produced by heat treatment are denoted by adding an additional number after the T designation:

T2 Annealed (cast products only)

T3 Solution heat-treated and cold-worked

T4 Solution heat-treated and naturally aged
T5 Artificially aged only
T6 Solution heat-treated and artificially aged
T7 Solution heat-treated and stabilized (by overaging)
T8 solution heat-treated, cold-worked, and artifically aged
T9 solution heat-treated, artificially aged, and then strain-hardened.

HEAT-TREATMENT PROCEDURES FOR ALUMINUM ALLOYS

The heat-treatable aluminum alloys are those that contain elements that have solid solubility at higher temperatures. Included in the heat-treatable group are the copper-bearing alloys, the magnesium silicide type alloys, the magnesium alloys, and the zinc-bearing alloys.

If foreign atoms are introduced into the space lattice, resistance to slip will be pronounced. This increased resistance to slip along the slip planes will result in an increase in the inherent strength of the aluminum. This strengthening effect is enhanced by the precipitation of intermetallic compounds along the slip planes as the alloy is cooled. For example, when copper is alloyed with aluminum, it reprecipitates to form the compound $CUAl_2$. The particles of $CuAl_2$ precipitating out at higher temperatures (above 350°F) coalesce into large masses and, consequently, interfere little with slip. However, those particles that precipitate at lower temperatures will remain at sizes comparable to space-lattice distances and cause dislocation pileup, thus retarding slip.

In the heat treatment of aluminum alloys, rapid cooling should take place from the solution temperature in order to maximize the precipitate that most retards slip. This is handled by subjecting the alloy to a severe quench (usually by water or oil held at 100°F).

Precipitation begins as soon as the alloy is quenched. The time required for complete precipitation at room temperature may be several days. When precipitation takes place at room temperature, the alloy is said to be naturally aged and is given the designation T4.

By elevating the temperature, it is possible to shorten the period for complete precipitation (that point at which the mechanical properties stabilize). This process is referred to as artificial aging. Aluminum alloys that are hardened by thermal precipitation techniques are heated to a temperature ranging from 250 to 375°F, and soaked from 6 to 24 hours, depending upon the alloy. After adequate soaking, the alloy is air-cooled.

Annealing

Aluminum alloys can be purchased in the annealed state (O temper) directly from the manufacturer; however, the production engineer will frequently find it necessary to anneal aluminum that has been strain-hardened during forming and requires still further forming. The cold working of the metal may be considered to result from using up the available "slip planes." Thus, an aluminum that is soft is thought of as having a fresh, undisturbed bank of slip planes that allow adjoining lattice planes to slide against each other. When the metal is formed by the application of force, the metal's dimensions are changed as slip occurs. The planes available for subsequent slippage will not be located for the most favorable movement; consequently, greater force must be applied in order to bring about an equal amount of cold-forming.

The original workability of the aluminum can be restored by heating it above a certain temperature, known as the recrystallization temperature. This temperature is usually between 600 and 750°F. The application of heat creates fresh slip planes along which the material can move, permitting easy deformation.

Homogenizing

Homogenizing is a high-temperature soaking treatment to eliminate or reduce segregation by diffusion. It facilitates a proper distribution of the alloying elements throughout a material so that a homogeneous structure is obtained. This is accomplished by heating the metal to a temperature just under its melting point, and following this by a slow cooling process.

Heat-treating Equipment

The salt bath and the air furnace are the two types of furnaces used in the heat-treatment of aluminum alloys.

Salts baths are heated by electricity, gas, or oil; air furnaces are usually heated by either gas or electricity. Both types of furnaces should include automatic temperature regulators so that control of temperature can be maintained. Accurate temperature control is a primary requisite for a furnace intended to be used for the heat-treatment of aluminum alloys. The temperature should be held to a 20°F spread between the high and low points of control.

Gas-fired air furnaces should be of the muffle type so that the parts being treated will not come in contact with the products of combustion. These air-conversion-type furnaces permit accurate temperature control in low-temperature ranges and consequently are quite popular.

Salt baths are usually made up of a mixture of molten sodium nitrate and potassium nitrate. This combination provides the flexibility for both heat treating and annealing. For heat treating only, a salt bath of just sodium nitrate is suitable. Because of its high melting point, it is not sastisfactory for annealing operations. It might also be mentioned that sodium nitrate alone is not as stable as the mixture of the two salts, and tends to corrode the tank as well as the parts being treated.

Procedure for Heat Treating Aluminum

As with ferrous materials, the heat treatment of aluminum alloys involves three distinct steps: heating to a predetermined temperature, soaking at the required temperature for a predetermined interval, and quenching.

Depending on the alloy, the heat-treating temperature for aluminum varies between 900 and 1000°F. It is important that the heat-treating temperature of the specific alloy be accurately maintained in order to arrive at the desired physical properties.

The soaking period will run from a few minutes to more than an hour, depending on the heating medium, the alloy being treated, and the shape and size of the part. The soaking time increases with the thickness of the part. When different-sized pieces are soaked, the cycle should be governed by the part having the heaviest section.

Quenching of aluminum alloys is usually done by immersion in oil or water. A fast quench is considered necessary in order to prevent the precipitating $CuAl_2$ from coalescing into larger masses. This would lead to corrosion because of electrolytic reaction when the alloy is exposed to salt water. In addition to keeping the time interval of the quench as short as possible, the parts should be thoroughly washed so that all the salt is removed from the surfaces. This step is necessary as a further guard against corrosion.

If the parts being treated are of intricate design, it may not be advisable to subject them to a rapid quench because of the danger of warpage. An alternate method of cooling such parts is to use spray or fog quenching. In both of these methods, a fine spray of water is thrown on the parts from all directions.

Some warpage can be corrected after hardening, and this straightening should be done immediately after the quench before age-hardening takes place. If it is impractical to straighten the parts immediately after quenching, they can be kept under refrigeration, thus delaying age-hardening.

SUMMARY

Heat treatment of parts is often avoided because the engineer is unfamiliar with the possibilities of heat-treated materials. Tool steels, stainless steels, alloys of iron, and aluminum acquire many useful properties with proper heat treatment. The processes must be controlled carefully in the shop; proper equipment, temperature controls, quenching media, procedures, materials, and skills are necessary for uniform and satisfactory performance. By visualizing the process and what occurs to the part during treatment, the engineer quickly realizes that a quenched part can have internal stresses locked within. This is revealed by cracks occurring during treatment or breakage after a slight load has been applied. Therefore, stress concentrations caused by sharp corners, uneven sections, rapid and uneven cooling, gas pockets, and complicated designs must be avoided by the designer. Many of the principles applied to casting apply to heat-treated parts. Consider the problem of a cube. Which portion of the cube cools the slowest? The temperature gradient is affected by the shape, location, and extent of surface area. This is governed by the design of the part. The heat-treater can help alleviate some of the poor conditions of high-temperature gradients by masking or by using progressive quenching, but the problem of internal strains is present with the best possible form of part.

Typical difficulties that arise during heat-treating processes are

1. Uneven support during heating causing warpage
2. Parts having variable surface characteristics may warp when quenched
3. Warpage frequently resulting from quenching at too high a temperature, even though the parts are within the allowable temperature range
4. Low overall hardness occurring because of poor circulation in the quenching medium
5. An even penetration of case being retarded by soot deposits during gas carburizing

Some general rules for quenching, tempering, and austempering, as developed by the U.S.X. Corporation, will provide helpful guideposts to the engineer.

1. The useful structures in hardened pieces are bainite and tempered martensite. To obtain either of these structures, the piece is heated to secure adequate solution of carbides, and then quenched so rapidly that the formation of pearlite or upper bainite in the 1000°F zone is avoided, bainite or martensite being formed at lower temperatures.
2. When quenching a specific size of piece, the avoidance of pearlite is made easier by lengthening the time of possible formation. The slower rate of pearlite formation is obtained by adding alloys. If the piece is fully hardened (martensitic), the hardness is affected very little by the alloys, being governed almost exclusively by the carbon content.
3. For a given size of piece, avoidance of pearlite or upper bainite is also aided by employing a greater severity of quench. This is obtained by avoiding excess scale on the piece, by employing quenching liquids that generally show a greater severity of quench, and by agitating either the quenching liquid or the piece being quenched. It is well to remember that the old method of hand agitation during quenching was very effective, and it may be desirable to duplicate it in modern mechanized equipment.
4. When measuring toughness, remember that the results are affected profoundly, not only by the magnitude of the stress imposed, but also by its nature and direction and, consequently, also by the shape of the piece. A laboratory test should always simulate a service test as closely as possible.

5. Toughness in martensitic structures is obtained by tempering the quenched piece.
6. Bainite structures, obtained in austempering, are inherently tough and, when produced with Rockwell C 48 or above, offer a combination of strength and toughness superior even to that of tempered martensite.

QUESTIONS

3.1. What are the most common forms of heat treating?

3.2. Define the following terms: stress relieving, annealing, normalizing, spheroidizing, quenching, hardening, tempering, and carburizing.

3.3. For what reasons is annealing performed?

2.4. In finishing materials requiring a close tolerance, to what hardness would you recommend bringing the material?

3.5. Give the operation sequence on products requiring low distortion, close tolerance, and fine finishes.

3.6. To what processes may stress relieving be advantageously applied?

3.7. What is the relative effect of violent agitation quenching in oil and no quenching in water?

3.8. What is the effect of normalizing on grain size?

3.9. What steels usually are not normalized?

3.10. Explain the difference between "lath" and "plate" forms of martensite.

3.11. In the stress relieving of a part at 500°C, what would be the approximate soak time?

3.12. What alloying elements are austenite stabilizers? Ferrite stabilizers?

3.13. Where would you recommend the use of Nitralloy steels?

3.14. Explain the process of induction heating.

3.15. What advantages are offered by the flame-hardening process when compared to induction hardening?

3.16. What methods may be used to localize case hardening?

3.17. What is meant by "soaking time"?

3.18. For what type of steels is water preferred as a quenching medium?

3.19. Explain the martempering process.

3.20. Distinguish between pearlite and spheroidite.

3.21. Explain the importance of time in the cooling stage of heat treatment.

3.22. What is meant by "slip planes?"

3.23. How can the workability of aluminum be restored after a deep-drawing operation?

3.24. What is meant by precipitation when referred to aluminum alloys?

3.25. What is homogenizing?

3.26. What six general rules for quenching, tempering, and austempering of ferrous materials have been established by the U.S.X. corporation?

PROBLEMS

3.1. To provide a tough core and a hardenable surface in a carbon steel member, it is possible to diffuse carbon into a low-carbon steel by a carburizing process. Since only carbon solid solution can take part in diffusion, the maximum effective carbon concentration at the surface of a steel is determined by the solubility of carbon in austenite. At 1700°F this value is approximately 1.3 percent by weight. It is desired to carburize 18-in. lengths of 1 in. round SAE 1010 steel. Plot the carbon penetration curve (carbon content versus distance from the surface of the work) that would be produced by carburizing for 3 hours at 1700°F. Determine the time required to obtain a carbon content of 0.6 percent at a depth of 0.05 in. The diffusion constant for carbon in austenite may be estimated from

$$D = (0.07 + 0.06)(\text{weight percent C}) \exp \frac{-3200}{RT} \quad \text{cm}^2\text{s}$$

where R (gas constant) = 1.987 cal/(K · gmol) and T is the temperature, in K. Make whatever assumptions you feel necessary.

3.2. A casting of plain carbon steel is specified for a wear-resistant application. Outline the proper heat treatment for the optimum carbon content to give the best microstructure for a tough core and maximum wear resistance on the surface, after the casting has been cooled to room temperature and has passed through the cleaning room. Give the temperatures, quenching media cooling times, and cooling rates.

3.3. In a certain corrosive atmosphere only a copper–nickel alloy has been considered suitable. Such alloys form substitutional solid solutions in all proportions. Choose the least cost alloy and its proper diameter to support a 5000-lb tensile load without yielding.

3.4. An annealed copper wire that was 0.010 in. OD was cold-drawn to 0.008 in. OD: what is its tensile strength after that amount of cold work?

3.5. A severely cold-worked metal is 50 percent recrystallized after each of the following heat treatments: 1 min at 162°C, 10 min at 138°C, and 100 min at 97°C. Estimate the temperature which would be required for 50 percent recrystallization in 10^4 min by treating recrystallization as a self-diffusion process.

3.6. Compute the case depth for the normal carburizing of a steel component that is soaked at a temperature of 1800°F for a period of 4 hours.

3.7. On the basis of Figs. 3.1 and 3.2, what would you estimate the tensile strength of a piece of mild steel to be after it was exposed to a cold reduction of 50 percent?

3.8. What would be the approximate austenizing temperature of a piece of steel having 0.3 percent carbon?

3.9. Using the Larson–Miller equation, what would be the holding time in hours at 580°C to provide the same relief of residual stresses when the work is heated to 600°C and soaked for 4 hours?

SELECTED REFERENCES

1. American Society for Metals: *Metals Handbook,* eds. Howard E. Boyer and Timothy L. Gall, Metals Park, Ohio, 1985.

2. Krauss, G.: *Deformation, Processing, and Structure,* American Society of Metals, Metals Park, Ohio, 1984.
3. Krauss, G.: *Principles of the Heat Treatment of Steel,* American Society of Metals, Metals Park Ohio, 1980.
4. Leslie, W. C.: *The Physical Metallurgy of Steels,* McGraw-Hill, New York, 1981.
5. Maki, T., Tsuzaki, K., and Tamura, I.: "The Morphology of Microstructure Composed of Lath Martensite in Steels", *Transactions of the Iron and Steel Institute of Japan,* vol. 20, pp. 207–214, 1980.
6. Meshii, M.: *Mechanical Properties of BCC Metals,* TMS–AIME, Warrendale, Pa., 1982.
7. Winchell Symposium on Tempering of Steel: *Metallurgical Transactions,* vol. 14A pp. 985–1146, 1983.

CHAPTER 4

FERROUS METALS

Ferrous metals are widely used because they are abundant, economic (on a basis of cost per unit of strength), and have unique magnetic and other physical properties. Ferrous alloys can be easily formed in the annealed state and can then be heat-treated to be as hard as a file, as tough as the hook of a 50-ton crane, or tempered properly to drill holes in other metals.

This chapter presents the types of ferrous materials available to design engineers, the properties of ferrous alloys, and how design objectives can be developed in ferrous alloys.

Ferrous metals are furnished to industry in the form of shapes and plates, bars and tool steels, sheets and strips, castings of iron and steel, and forgings (Fig. 4.1). Over the past 35 years the production of cast iron, weldments, and sheet products has increased markedly, whereas in most other categories there has been relatively little change and the production of forgings and semifinished steel shapes and plates has decreased. Competition from the casting process and from plastics as well as foreign imports of both steel and automotive components has been responsible for a large part of the decrease.

Ferrous materials can be broadly divided into cast irons, which contain relatively high carbon content, and steels, which usually contain 1 percent carbon or less. The latter can be further subdivided into plain carbon steels, with low, medium, and high amounts of carbon; low- and high-alloy steels; and

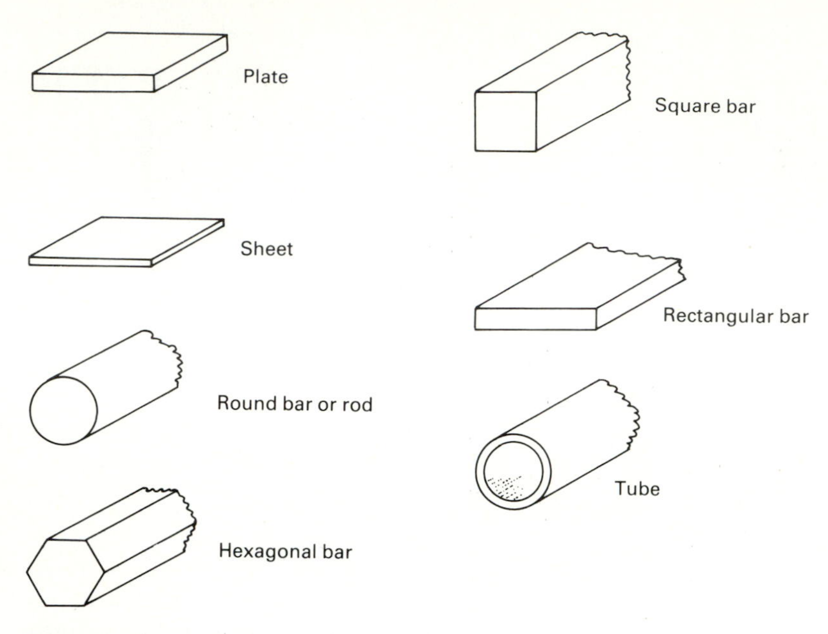

FIGURE 4.1
Standard material shapes.

tool steels. Each type of steel has unique properties, but low-carbon steel far exceeds all others in tonnage produced.

IRON–CARBON EQUILIBRIUM DIAGRAM

Before beginning a discussion of ferrous alloys, it is well to review briefly the general relationships shown in the iron–carbon equilibrium diagram. This phase diagram shows the types of alloy structures that can be formed in a two-component system, in this case iron and carbon, under equilibrium conditions. For any binary alloy system a similar plot of change of phase from liquid or liquid plus solid to solid could be constructed from carefully measured freezing-point data. Ferrous alloys have greatly differing properties and microstructures with changes in carbon content. These can be related to one another in terms of the iron–carbon diagram.

Figure 4.2 illustrates this diagram for carbon contents up to 5 percent. It will be noted that two allotropic forms of iron exist. At temperatures below 1670°F (910°C) and at temperatures ranging from 2552°F (1400°C) to the melting point, the stable form of iron has a body-centered cubic structure. (For convenience, the low temperature form is designated α, while the high-temperature form is designated δ.) At temperatures varying between 1670°F and 2552°F, the stable form of iron has a face-centered cubic structure that is

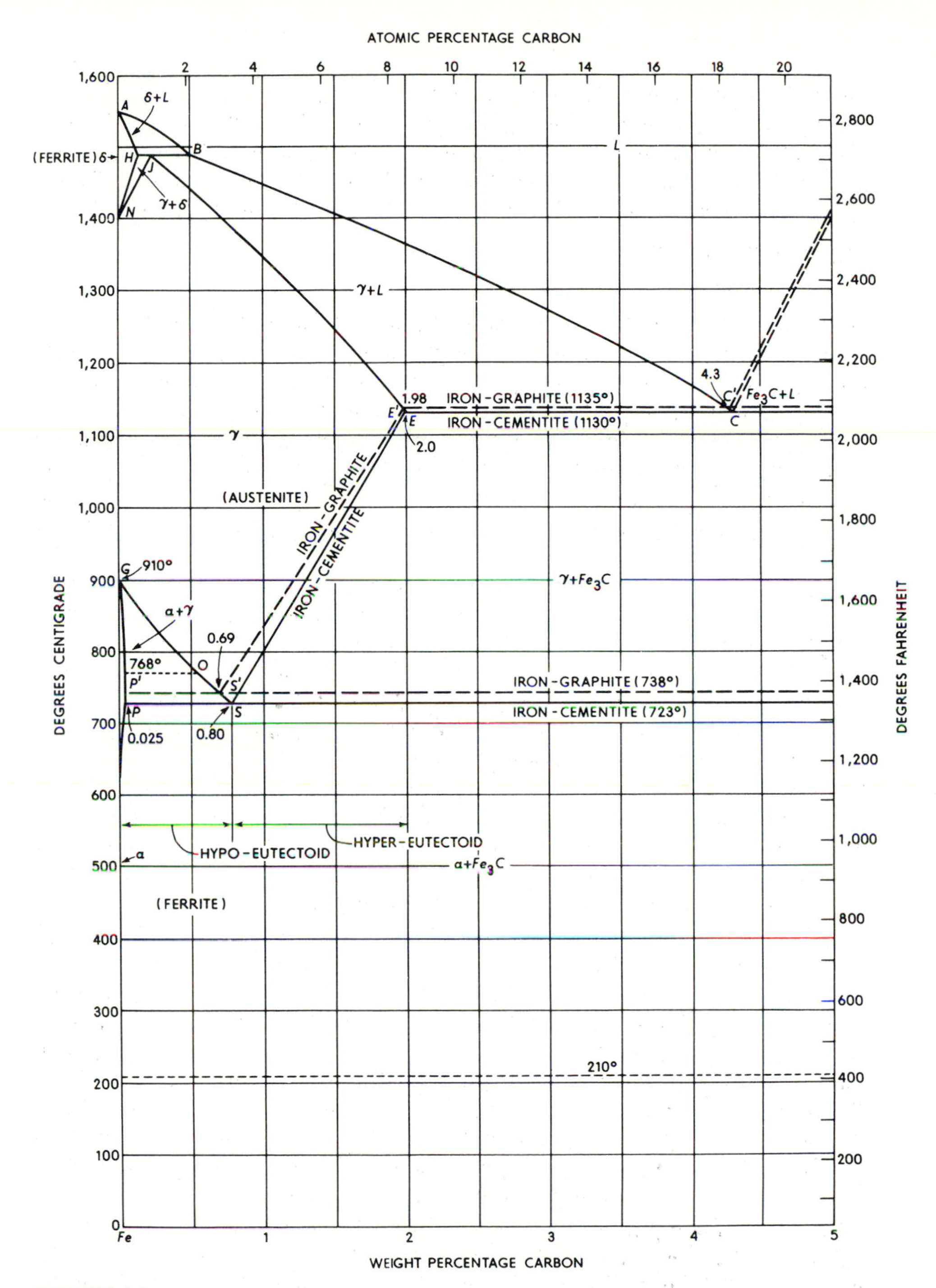

FIGURE 4.2
The iron–carbon equilibrium diagram for carbon contents up to 5 percent.

designated as γ. An equally common practice is to refer to the body-centered forms of iron as ferrite and the face-centered form as austenite. The high-temperature body-centered cubic form is designated δ-ferrite.

On heating pure iron from room temperatures, at 1670°F a change from body-centered α-iron to face-centered γ-iron occurs. Further heating to 2552°F will cause the γ-iron to revert back to the body-centered form, δ. Since these changes are reversible, on cooling from above 2552°F, the reverse reactions will occur. At 1670°F and 2552°F,

$$Fe_{bcc} \underset{\Delta T}{\rightleftharpoons} Fe_{fcc}$$

The bcc–fcc reactions occur at critical temperatures depending upon the carbon content. The iron–carbon equilibrium diagram shows, among other things, the variation of transformation temperatures with the addition of carbon, and the regions where the phases are stable. It also shows that when carbon is added to iron a third phase may be stable under certain conditions. This phase is the iron–carbon phase, composed of Fe_3C, commonly referred to as cementite or as iron carbide.

The fact that iron–carbon alloys undergo various phase transformations largely accounts for the importance of steel as an engineering material. It is these transformations that make the heat treatment (and therefore the variation and control of mechanical properties) possible. However, it must be remembered that the conditions shown by the equilibrium diagram are not necessarily those that exist in a steel after heat treatment. The phase diagram shows those phases that are thermodynamically stable (i.e., of lowest energy), whereas heat treatment takes advantage of the variation in the kinetics of the reactions in order to obtain structures that are thermodynamically metastable.

The dominance of ferrous alloys in manufacturing stems from their wide range of properties with changes in carbon content. As little as one part in one thousand (0.1 percent) changes pure iron to steel. This is soft sheet steel used in drawing and forming. Mild steels have carbon contents of 2 to 3 parts per thousand (ppt). Rail and tool steels have from 6 to 9 ppt of carbon. With somewhat above 1 percent carbon, the ability to hold a sharp edge is greatest, so such steels are used for razor blades and wood chisels.

The cast-iron range extends from 2 to 4.8 percent carbon. Actually as much as 6 percent carbon can be dissolved in molten iron, but less than 2 percent can remain in solution in the solidified alloy. It is important to recognize that cast irons also contain significant amounts of silicon. The silicon adds a third dimension to the cast irons and creates ternary alloys that are more complex than the binary alloys of steel.

Ferrous metals are produced in a variety of ways in order to obtain different shapes and material properties. Cast ferrous materials are usually specified when the shape of the part is reasonably complex. Rolled or extruded irons are usually specified when the part being produced has regular polyhedral or symmetric characteristics. Although the shape of the part reflects the basic manufacturing process required for raw material selection, the physical, magnetic and electrical properties must all be considered in the selection of raw material.

STEELS

Steel is one of the most valuable metals known to man; approximately 200 million tons can be produced in the United States annually. In 1900, US capacity was but 21 million tons. Although the process of steelmaking is familiar to most engineers, a review of this process would be appropriate at this time.

Iron ore, limestone, and coal are the principal raw materials used in making iron and steel. Coke is produced by heating bituminous coal in special ovens. Skip cars go up the skip hoist with loads of iron ore, coke, and limestone and dump them into the top of the blast furnace. Hot air from the stove is blown into the furnace near the bottom. This causes the coke to burn at temperatures up to 3000°F. The ore is changed into drops of molten iron that settle to the bottom of the blast furnace. The limestone that has been added joins with impurities to form a slag that floats on top of the pool of liquid iron. Periodically (approximately every 6 hours), the molten iron is drained into a ladle for transporting to either the Bessemer converter, electric furnace or open-hearth furnace. The slag is removed separately so as not to contaminate the iron.

The making of steel from iron involves a further removal of impurities. Regardless of which process is used for making steel—open-hearth, Bessemer-converter, or electric-furnace—steel scrap is added along with desired alloying elements and the impurities are burned out.

Liquid steel removed from the furnace is poured into ingot molds. The ingots are then removed to "soaking pits" where they are brought to a uniform rolling temperature.

At the rolling mill, the white-hot steel passes through rolls that form the plastic steel into the desired shape: blooms, slabs, or billets. These three semifinished shapes then go to the finishing mills where they are rolled into finished forms as structural steel, plates and sheets, rods, and pipes (see Fig. 4.3).

Steel is the basic and most valuable material used in apparatus manufactured today. Its application is based on years of engineering experience, which serves as a guide in choosing a particular type of steel. Each variable, such as alloy, heat treatment, and processes of fabrication (casting, forging, and welding) has its influence on the strength, ductility, machinability, and other mechanical properties, and affects the type of steel selected. The following basic concepts also assist in determining which steel should be used:

1. The modulus of elasticity in tension falls within the range of 28×10^6 to 30×10^6 lb/in.^{2}, regardless of composition or form; therefore, sizes as determined by deflection remain the same regardless of the steel chosen.

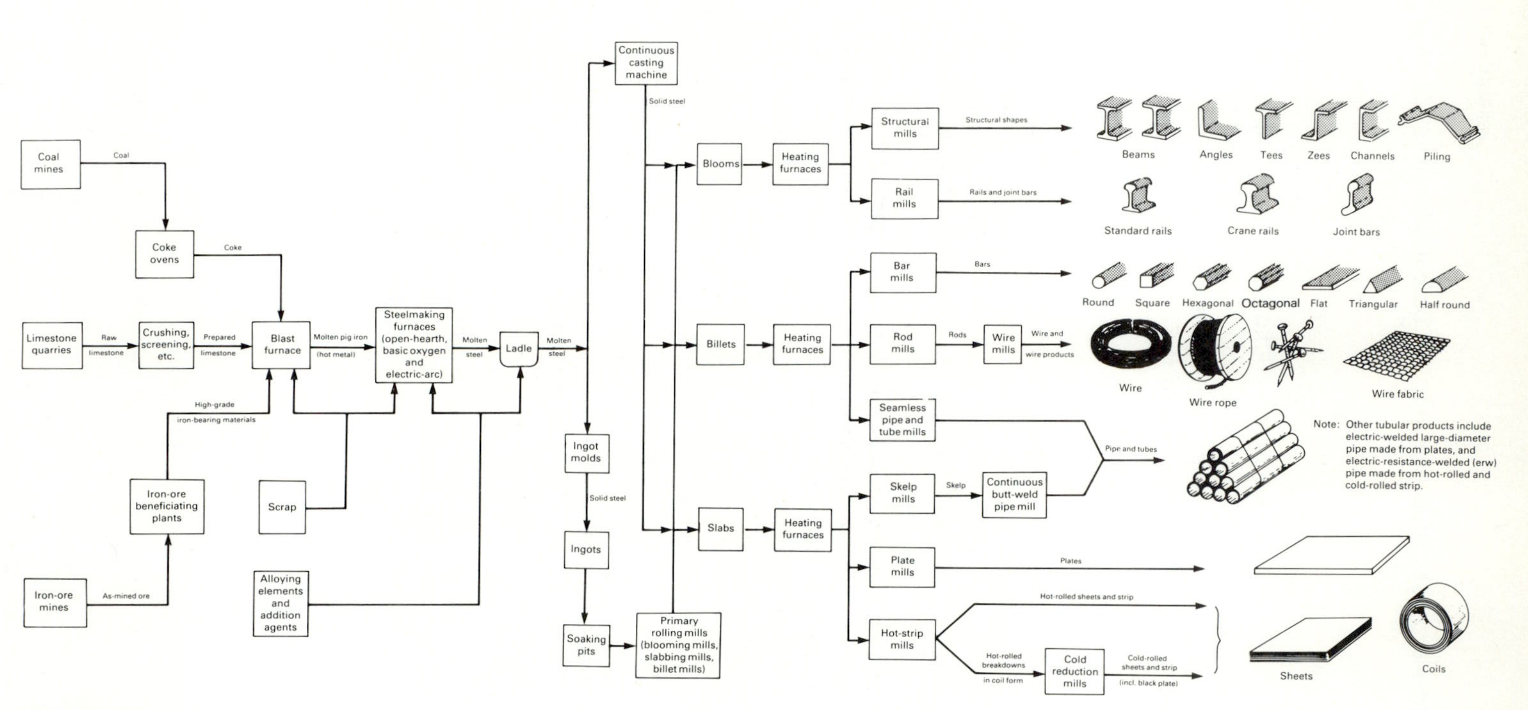

Coal mines
Coal
Coke ovens
Coke
Limestone quarries
Raw limestone
Crushing, screening, etc.
Prepared limestone
Blast furnace
Molten pig iron (hot metal)
Steelmaking furnaces (open-hearth, basic oxygen and electric-arc)
Molten steel
Ladle
Molten steel
High-grade iron-bearing materials
Iron-ore beneficiating plants
Iron-ore mines
As-mined ore
Scrap
Alloying elements and addition agents
Continuous casting machine
Solid steel
Ingot molds
Solid steel
Ingots
Soaking pits
Primary rolling mills (blooming mills, slabbing mills, billet mills)
Blooms
Billets
Slabs
Heating furnaces
Heating furnaces
Heating furnaces
Structural mills
Structural shapes
Rail mills
Rails and joint bars
Bar mills
Bars
Rod mills
Rods
Wire mills
Wire and wire products
Seamless pipe and tube mills
Pipe and tubes
Skelp mills
Skelp
Continuous butt-weld pipe mill
Plate mills
Plates
Hot-strip mills
Hot-rolled sheets and strip
Hot-rolled breakdowns in coil form
Cold reduction mills
Cold-rolled sheets and strip (incl. black plate)
Beams
Angles
Tees
Zees
Channels
Piling
Standard rails
Crane rails
Joint bars
Round
Square
Hexagonal
Octagonal
Flat
Triangular
Half round
Wire
Wire rope
Wire fabric
Note: Other tubular products include electric-welded large-diameter pipe made from plates, and electric-resistance-welded (erw) pipe made from hot-rolled and cold-rolled strip.
Sheets
Coils

FIGURE 4.3

2. Carbon content determines the maximum hardness of steel regardless of alloy content. Therefore, the strength desired, which is proportional to hardness, can determine the carbon content.
3. The ability of the steel to be uniformly hardened throughout its volume depends on the amount and kind of alloy. This is more complex, but does not necessarily change the calculation of the size of the part.
4. Ductility decreases as hardness increases.

The preliminary choice of steel for a part as well as for other factors, such as notch sensitivity, shrinkage, blowholes, corrosion, and wear, is simplified when based on the above principles. The final selection is made by matching the material with the process of manufacture used in order to obtain the shape, surface, and physical requirements of the part. The selection may be made from among low-carbon steels, low-alloy steels, high-carbon steels, and high-alloy steels.

Steel is one of the few common metals that has an endurance limit. You will recall that fatigue is the failure of a material due to repeated loading. Most metals become tired as they are subjected to stress over and over again. The stress a material can withstand under constant loading is much less than under static loading. As steel is continually loaded, it will reach a lower limit of strength. This property is quite pronounced in wire shapes. Common copper and aluminum wire can easily be broken by flexing the wire in a local spot.

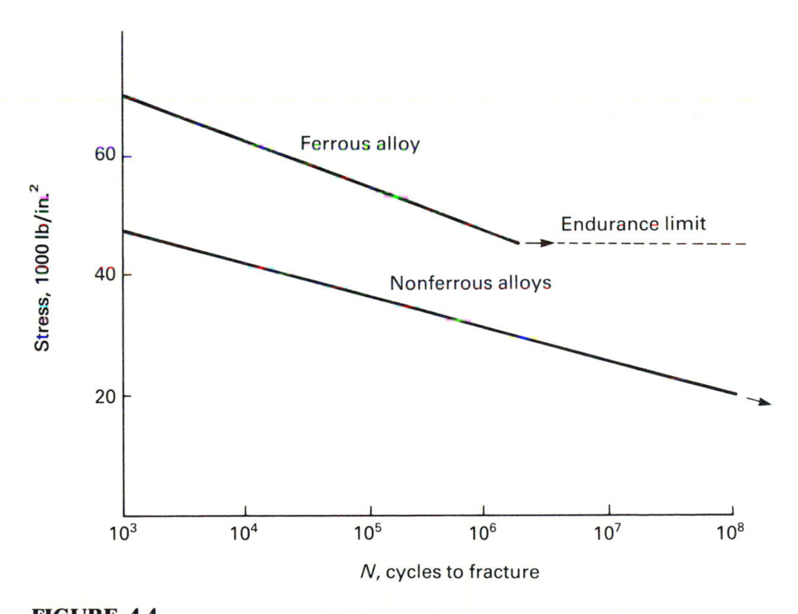

FIGURE 4.4
Idealized fatigue curves for ferrous and nonferrous alloys. For ferrous alloys an endurance limit, below which stress failure does not occur, is reached near 10^6–10^7 cycles. For nonferrous alloys a fatigue limit is not reached, even beyond 10^8 cycles.

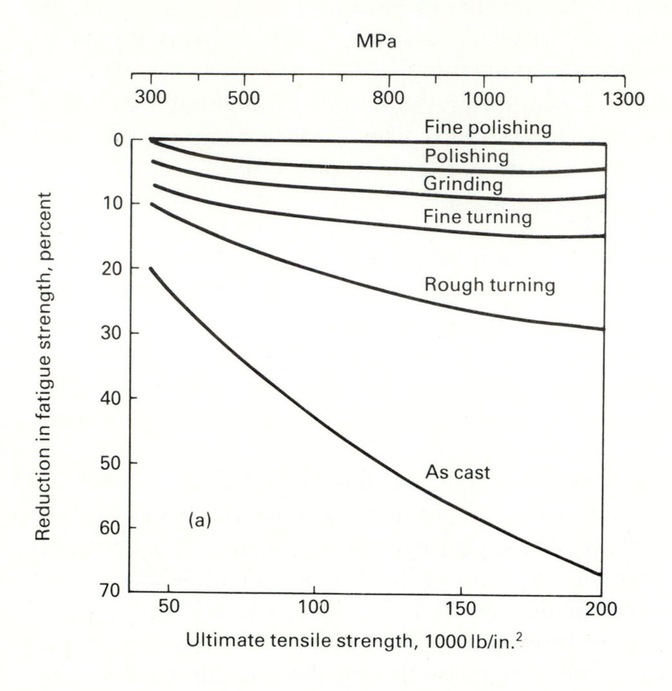

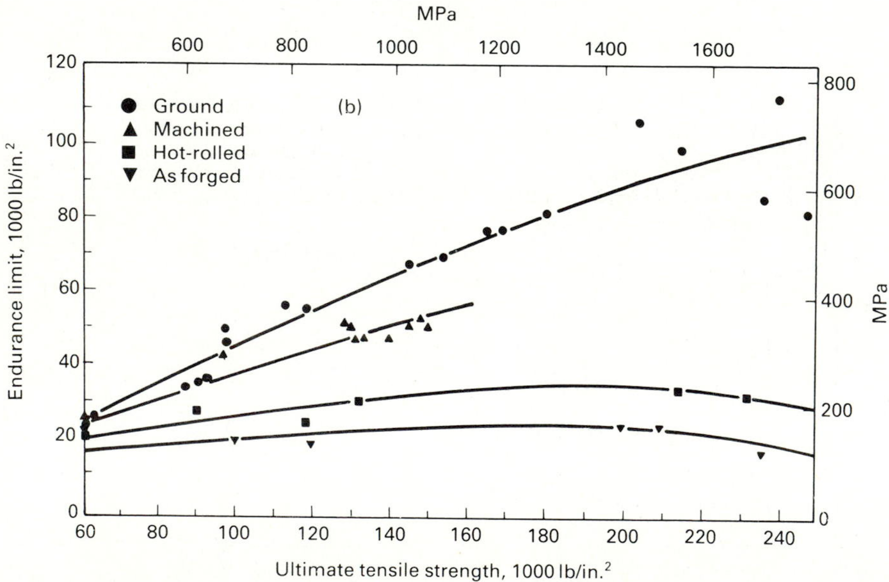

FIGURE 4.5
Effect of surface quality on endurance limit and fatigue strength (*Serope Kalpakjian, Manufacturing Processes for Engineering Materials, Addison Wesley,* 1984.)

Normally after a few dozen flexes, the wire breaks. Steel wire, however, is very tough and flexing the wire simply cold works the material making the process futile for the unknowing person trying to break a steel wire. At some point steel will resist weakening due to repeated loading. This is known as an "endurance limit". The endurance limit of steel is around 60% of its original strength.

This property of having an endurance limit makes steel invaluable for use in structural applications like bridges, springs, struts, beams, etc. Figure 4.4 shows a typical effect of fatigue on common steel. Of course, there are many factors that effect the endurance limit of a material. A primary factor is the surface quality of the material and/or the manufacturing process used to produce the specimen. Figure 4.5 shows the basic relationship of surface quality on ultimate tensile strength.

Fatigue is attributable to the initial material not being an ideal homogeneous solid. In each half cycle, irreversible minute strains are produced. Fatigue failure usually develops from:

1. Repeated cyclic stresses that cause incremental slip and cold working locally in the material
2. Gradual reduction of ductility of the strain hardened areas that develop into cracks
3. A notching effect from submicroscopic cracks

The endurance limits of steels create some very desirable physical properties. These properties can be detrimental to the manufacturability of the material. For instance, in the cold rolling of steel the endurance limit creates a limitation on the amount of cold working that can be input to any part. After this limit has been reached the material must be heated above its critical temperature to permit further cold working.

Carbon Steel

Plain carbon steels represent the major proportion of steel production. Carbon steels have a wide diversity of application, including castings, forgings, tubular products, plates, sheets and strips, wire and wire products, structural shapes, bars, and tools. Plain carbon steels, generally, are classified in accordance with their method of manufacture as basic open hearth, acid open hearth, or acid Bessemer steels, and by carbon content.

The principal factors affecting the properties of the plain carbon steels are the carbon content and the microstructure. The microstructure is determined by the composition of the steel (carbon, manganese, silicon, phosphorus,and sulfur, which are always present, and residual elements including oxygen, hydrogen, and nitrogen) and by the final rolling, forging, or heat-treating operation. However, most of the plain carbon steels are used without a final heat treatment and, consequently, the rolling and forging

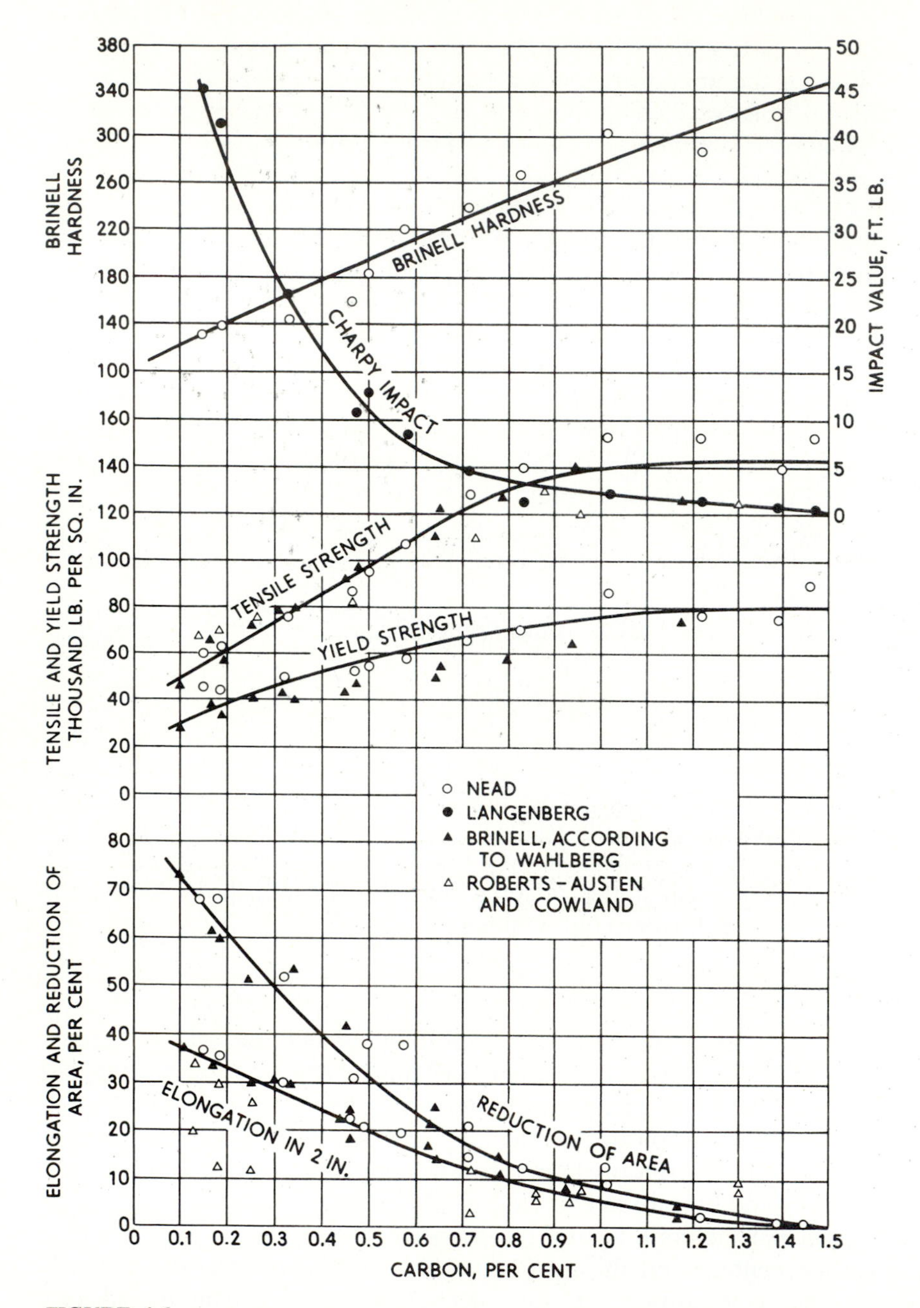

FIGURE 4.6
Variations in average mechanical properties of as-rolled 1-in.-diameter bars of plain carbon steels, as a function of carbon content.

operations influence the microstructure. The average mechanical properties of as-rolled 1-in. bars of carbon steel as a function of carbon content are shown in Fig. 4.6.

Carbon steels are predominantly peralitic in the cast, rolled, or forged conditions. The constituents of the hypoeutectoid steels are therefore ferrite and pearlite, and of the hypereutectoid steels are cementite and pearlite.

Alloy Steel

Alloy steel is an alloy of iron and carbon containing alloying elements, one or more of which exceeds the following: manganese, 1.65 percent; silicon, 0.60 percent; copper, 0.60 percent; and/or specified amounts of other alloying elements, including aluminum, boron, and chromium up 3.99 percent; cobalt, niobium, molybdenum, nickel, tungsten, vanadium, zirconium, or other elements added in sufficient quantity to give the desired properties of the steel.

Since there are more elements, some expensive, to be kept within the specified ranges in alloy steel than are required in carbon steel, alloy steel requires more involved techniques of quality control and, consequently, is more expensive.

Alloy steel can give better strength, ductility, and toughness properties than can be obtained in carbon steel. Consequently, the engineer should consider alloy steels in designs subject to high stresses and/or impact loading.

Almost all alloy steels are produced with fine-grain structures. A steel is considered to be fine-grained if its grain size is rated 5, 6, 7, or 8. Number 1 grain size shows $1\frac{1}{2}$ grains/in.^{2} of steel area examined at 100 diameters magnification. Fine-grain steels have less tendency to crack during heat treatment and have better toughness and shock-resistance properties. Coarse-grained steels exhibit better machining properties and may be hardened more deeply than fine-grained steels.

To select the alloy steel that is best suited for a given design, the effects of the principal alloying elements must be taken into account. They are as follows.

1. *Nickel* provides toughness, corrosion resistance, and deep hardening.
2. *Chromium* improves corrosion resistance, toughness, and hardenability.
3. *Manganese* deoxidizes, contributes to strength and hardness, decreases the critical-cooling rate.
4. *Silicon* deoxidizes, promotes resistance to high-temperature oxidation, raises the critical temperature for heat treatment, increases the susceptivity of steel to decarburization and graphitization.
5. *Molybdenum* promotes hardenability, increases tensile and creep strengths at high temperatures.
6. *Vanadium* deoxidizes, promotes fine-grained structure.
7. *Copper* provides resistance to corrosion and acts as strengthening agent.

8. *Aluminum* deoxidizes, promotes fine-grained structure, and aids nitriding.
9. *Boron* increases hardenability.

Tool Steel

Tool steels constitute a class of high-carbon alloy steels that have properties such as shock and wear resistance, hardness, strength, and toughness produced by heat treatment and fabrication. Tool steels are essentially combinations with iron of one or more of the following elements: carbon (0.80 to 1.30 percent), manganese (0.20 to 1.60 percent), silicon (0.50 to 2.00 percent), chromium (0.25 to 14.00 percent), tungsten (1.5 to 20.00 percent), vanadium (0.15 to 3.00 percent), molybdenum (0.80 to 5.00 percent), and cobalt (0.75 to 12 percent).

Tool steels are characterized by hardenability and timbre. Hardenability refers to the ability of steel to harden upon heat treatment, and timbre is the property that refers to grain size and toughness after being subjected to a given heat treatment.

The hardenability of a tool steel may be expressed as the depth to which a certain standard size will harden when quenched in a certain medium. Hardness is influenced by the speed of cooling from above the critical temperature to the necessary transformation temperature (about 1160°F) and the critical quenching speed of the steel, which is influenced by the alloy additions. Plain carbon steels have a high critical quenching speed, while the addition of most alloys decreases this speed. Thus, it may be said that the critical quenching speed determines the depth to which a steel will harden in any given cooling medium.

Timbre of a tool steel is determined by inspection of its grain size. If the grain size is fine, the steel is referred to as having *tough timbre*; if the grain size is coarse, the timbre is referred to as being of *brittle timbre.*

In the selection of a tool steel, the engineer will encounter four requirements that will vary to a degree depending on the use of the tool. Each of these requirements may be realized in greater or lesser degrees by selecting the proper tool steel. However, improvement of one requirement will result in the diminishing of at least one of the remaining three requirements. These requirements as outlined by the Carpenter Steel Company are:

1. Wear resistance for cutting or abrasion
2. Toughness or strength
3. Hardening accuracy and increased hardenability
4. Ability to do its work at elevated temperatures

Stainless Steel

The term "stainless steel" denotes a large family of steels containing at least 11.5 percent chromium. They are not resistant to all corroding media.

Stainless steel competes with nonferrous alloys of copper and nickel on a corrosion-resistance and cost basis and with light metals such as aluminum and magnesium on the basis of cost and strength-weight ratio. Stainless steel has a number of alloy compositions and there are many suppliers. Information on its properties and fabrication can be obtained readily. Sound techniques have been evolved for casting, heat treating, forming, machining, welding, assembling, and finishing stainless steel. It will be found that this material usually work-hardens (which makes machining, forming, and piercing more difficult), and it must be welded under controlled conditions and under inert gas. It has desirable high strength, corrosion resistance, and decorative properties.

A bright, clean surface is essential for best corrosion resistance. Traces of scale and foreign matter should be removed by machining, pickling, or polishing. Dipping in nitric acid will ensure the formation of a good oxide film on new pieces. Stainless steels may be electroplated and electropolished, anodically etched, covered with porcelain enamel, or given colored coatings through the dying of surface oxides. Highly polished sheets may be purchased directly from stainless-steel producers. A coating of plastic may be used to protect the surface during fabrication.

Stainless steel can be made very hard and its strength can be more than doubled by cooling to −300°F and simultaneously rolling under high pressure, then heating to 750°F for 24 hours.

Corrosion resistance is the most important single characteristic of the stainless steels. This quality is due to a thin transparent film of chromium oxide that forms on the surface. It will withstand oxidizing agents such as nitric acid, but will be attacked by reducing agents such as hydrochloric acid or any of the halogen salts. Scaling and corrosion are accelerated in applications in which the oxide layer is constantly being broken. Repeated heating and cooling, with the accompanying expansion and contraction, cracks off the oxide layers. Since the straight-chromium grades of stainless steel have lower thermal expansion than the chromium-nickel grades, they serve best where constant heating and cooling is involved. Most stainless steels show good short-time strength at 1500°F and a few special types are good at 2000°F. Compare this with ordinary carbon steels, which lose their usefulness above 900 to 950°F. The heat-conducting properties of stainless steel are poor, so copper cladding is often used in cooking utensils to distribute heat. See Table 4.1 for physical properties of rust-resisting steels. Table 4.2 is a summary of some of the more common characteristics of carbon steel. (SAE numbers refer to the Society of Automotive Engineers.)

Groups of Stainless Steel

Although there are many types of stainless steel available to the engineer today, they may be classified in one of three groups according to their microstructures: austenitic, ferritic, and martensitic.

The austenitic group is often referred to as chromium-nickel alloy, having

TABLE 4.1
Properties of stainless and rust-resisting steels[a]

Carbon %	Cr %	Ni %	1,000 Lbs/Sq. In. Ult. Str.	1,000 Lbs/Sq. In. Yield Str.	% Elong. 2	% Red.	Brin.	Machining (2)	Weldability	Rel. Corr. Res.	Scaling Temp. °F (3)	Elec. Res. (4)	Shape of Mat.	DESCRIPTION AND COST (5)	
.15	5		65	25	35	70	150	--	Fair (10)	Poor	1100	260	Sheets	H.R. annealed. Range ovens.	
.30	13		235	(6) 150	--	--	477	--	(11)	Fair	--	284	Springs	Rust resisting springs.	
.12	13		90	60	30	45	180	--			1200	286	Bars	Cold drawn, low sulphur, electrodes.	
.12	13		80	48	25	50	190	F.M.	(12)	Good	1200	293	Bars	Cold drawn, high sulphur. Screw-mach. wk.	
.15	13		65				131	--		(14)	1200	287	Strips	C.R. for electrical resistance.	
.10	13		65	35	30		140	--			1200	270	Strips	H.R. & C.R. General use.	
1.00	18		95	55	6	10	190	--	(11)		1200	306	Bars	H.R. Rollers, cam for brush arm.	
.12	17		70	40	25	--	159	--	(12)	Good	1550	307	Sheets	C.R. annealed. General use.	Non-Hardenable.
.15	17		70	40	25	50	175	--		(15)	1550	279	Bars	H.R. & C.R. annealed.	
.20	18	8	80	25	35	45	210	F.M.			1650	345	Bars	H.R. General use.	Non-Magnetic.
.20	18	8	90 (7)	(7) 40	35 (7)	45 (7)	230 (8)	F.M.	(13)		1650	345	Bars	Cold finish. General use.	
.16	18	8	90	40	35	45	230	--			1650	345	Bars	Cold finish. General use.	
.20	18	8	90	35	35	--	--	--		Very	1650	345	Sheets	H.R. & C.R. General use.	
.08	18	8	85	30	50	--	--	--	Good	Good	1650	345	Sheets	H.R. & C.R. for deep drawing.	
.10	18	8	182	--	--	--	375	--	--	(16)	1650	345	Springs	Cold finished. Flat spring steel.	
.15	18	8	246 (9)	--	--	--	--	--	--		1650	345	Springs Wire	Cold drawn spring wire.	
.25	18	8	200 (7)	(6) 135 (7) 170	--	--	--	--	--		1650	345	Wire	Cold drawn. Tinning banding wire.	
.07	18	8	85	30	50	55	--	--	Good		1650	345	Tubes	El. welded, annealed.	
.07	18	8	85	30	50	55	--	--	Good		1650	345	Tubes	Seamless, annealed.	
.20	24	12	100	45	35	45	185	--	Good	Very	2100	378	Bars	Cold drawn.	
.20	24	12	90	40	40	50	170	--	Good	Good	2100	369	Sheets	Annealed. Furnace lining.	
.10	24	12	90	40	40	50	170	--	Good	(17)	2100	369	Tubes	Seamless, annealed.	

(1) The data given in this table are only relative for comparative purpose except the data marked (7), (8), (9).
(2) F.M.--Free machining, all other material is difficult to machine.
(3) Scale resistance is not merely a function of temperature. The atmosphere encountered is quite important.
(4) Microhms per square inch foot.
(5) Cost increases approx. with total percentage of chromium and nickel.
(6) Eleastic Limit. (7) Minimum. (8) Maximum.
(9) Min. for .0625 diam. Strength varies inversely with diam.

(10) Preheating from 300 to 500°F. essential. Anneal after welding.
(11) Not recommended for welding.
(12) Limited, preheating and annealing essential.
(13) Good; anneal after welding for corrosion resistance.
(14) Improved if heat treated.
(15) Better than the preceding specifications.
(16) Improved if annealed.
(17) Better than the preceding specifications.

TABLE 4.2

Steel bar, carbon—minimum properties[a]

SAE no.	Description	Tensile strength lb/in^2. Yield	Tensile strength lb/in^2. Ultimate	Percent elong. in 2 in.	Pecent red. in area	Brinell hardness	Chemistry percent, and SAE equivalent	Bending recommendation	Application
1070	Hot-rolled, annealed spring steel **1.** 46563 gr. B type II **2.** QQ-S-663, FS1070 **3.** 51-107-WD-1070 **5.** A107, gr. 1070 **6.** 1070					184 max.	0.65–0.75 C, 0.70–1.0 Mn, 0.35 Si max; SAE 1070	180° over pin = 2 × thk.	Springs. Requires heat treatment following forming to develop spring properties
1015	Hot-rolled, unpickled, unoiled **1.** 47S11, CLC, Fin **2.** QQ-S-636, as-rolled **3.** 57-136-WS 1010, as-rolled **5.** A 107, G 1010 **6.** 1010 or 1015	26,000	48,000	30	55	100	0.08–0.18 C, 0.30–0.60 Mn SAE 1015	180° over pin. Size $\frac{1}{2}$ × size; $\frac{3}{4}$ or less 1 × size; $\frac{3}{4}$ to 1 $1\frac{1}{2}$ × size; 1 to $1\frac{1}{2}$ $2\frac{1}{2}$ × size; $1\frac{1}{2}$ to 2 3 × size; 2	General use where a soft ductile steel is required and where fit and finish are not important. Suitable for welding but contains scale undesirable for resistance welding. Threads may be chased or cut with special dies, but this steel cannot be used to advantage on automatic screw machines. Much cheaper than cold-rolled steel
1015	Hot-rolled, pickled and oiled **1.** 47S11, CLC, fin. 3 **2.** QQ-S-636, pickled and oiled **3.** 57-136-WS 1010, pickled and oiled **6.** 1010 or 1015	26,000	48,000	30	55	100	0.08–0.13 C, 0.30–0.60 Mn SAE 1015	Same as 1015 above	Same as 1015 except superior for resistance welding due to absence of scale

Table 4.2 contd.

SAE no.	Description	Tensile strength lb/in^2. Yield	Tensile strength lb/in^2. Ultimate	Percent elong. in 2 in.	Pecent red. in area	Brinell hardness	Chemistry percent, and SAE equivalent	Bending recommendation	Application
1045	Hot-rolled, forging quality **1.** 46S4, Gr. P **2.** QQ-S-663, FS1045 **3.** 57-107-WD 1045 **4.** A-107, Gr. 1045 **6.** 1045	40,000	70,000	18	30	140	Comp. A / Comp. B 0.42–0.50 C / 0.42–0.50 C 0.40–0.60 Mn / 0.55–0.75 Mn 0.15–0.30 Si / 0.15–0.30 Si SAE 1045	180° over pin = 4 × size	Shafts, gears, cams and pinions subject to drastic heat treatment. Suitable for upsetting and forging.
1035	Hot-rolled straightened Size: 3 in. dia. or less ⟶	40,000	75,000	18	30	150	0.30–0.40 C Approx., 0.60–0.90 Mn, 0.15–0.30 Si SAE 1035	180° over pin = 4 × dia.	Miscellaneous shafting, bolts, studs
	2. QQ-S-663, FS1035 Size: Over 3 to 6 dia. ⟶	40,000	to	15	25	150			
	Size: Over 6 dia. ⟶ **3.** 57-107-WD 1035 **4.** AN-S-4, cond. B **5.** A 306, Gr. 75 **6.** 1035 **7.** 5080	36,000	95,000	15	25	150			
1111	Cold-finished, Free machining **2.** QQ-S-663, FSB 1111 **3.** 57-107-WD 1111 **5.** A 108, Gr. 1111 **6.** B 111	45,000	60,000	8	30	120	0.13 C Max, 0.60–0.90 Mn, 0.07–0.12 P, 0.08–0.15 S SAE 1111	Not recommended	Smooth-finish high-speed automatic screw machine work. Do not use for application requiring shock resistance such as for transportation apparatus. Low temperatures reduce the impact resistance still further. Not recommended for pack carburizing

1111	Centerless ground, round, free-machining, close plus and minus tolerances, bright smooth finish	45,000	60,000	8	30	140	0.13 C Max, 0.60–0.90 Mn, 0.07–0.12 P, 0.08–0.15 S SAE 1111	Not recommended	Close plus and minus tolerance applications. Same restrictions for use as SAE 1111
1016	Cold-drawn, carburizing quality **2.** QQ-S-663, FS1016 **3.** 37-107-WD 1016 **4.** MIL-S-866, CIA **5.** A108, Gr. 1016 **6.** 1016 **7.** 5060	50,000	70,000	15	45	140	0.13–0.18 C, 0.60–0.90 Mn SAE 1016	Not recommended	Same as 2084-1 except for parts to be carburized such as circuit-breaker triggers, latches, and links
1137	Cold-drawn, relatively free-machining, stress-relief annealed **5.** A311, Gr. 1137 **6.** 1137 **7.** 5024	80,000	95,000	16	45	200	0.32–0.45 C, 1.35–1.65 Mn, 0.08–0.13 S SAE 1137	Not recommended	Shafts. Where moderate strength, negligible warping during machining and relatively free machining characteristics are required. Automatic screw machine work

[a] Bold specification numbers refer to the following: 1, Navy, 2, federal; 3, Army; military or Air Force-Navy; 5, ASTM; 6, SAE or AISI number. 7, SAE-AMS specification.

18 percent chromium and 8 percent nickel, and is known as 18–8 stainless. The 8 percent nickel is generally sufficient to stabilize the high-temperature, austenitic face-centered cubic structure so that it does not transform upon cooling to room temperature. This gives the alloy high-temperature strength. There are many modifications (over 20) based on this grade in which the chromium-nickel ratio is modified, carbon content decreased, or other elements added for stabilization or increased resistance to oxidation.

In general, the austenitic stainless family may be hardened only by cold working. As a family, they are nonmagnetic and have good resistance to atmospheric corrosion. However, after heating in the critical range, 800 to 1300°F, the steel may become prone to intergranular corrosion in a corrosive environment. This intergranular failure may develop in areas that have been welded. Intergranular attack may be avoided or greatly reduced by using a grade that is stabilized with certain alloying elements (titanium, niobium), by using a grade with extra-low-carbon content (0.3 percent carbon maximum—known as the ELC grades), or by heat treating the steel in the range 1800 to 2100°F followed by quenching after it is exposed in the critical range.

The austenitic stainless grades are generally stronger than the other grades at temperatures above 1000°F. Since the austenitic steels are more ductile than the ferritic and martensitic, they are considered to have better fabrication properties. However, all the wrought stainless-steel grades can be fabricated by the methods applicable to carbon steel.

The ferritic grades differ from the austenitic in that they contain 18 to 30 percent chromium and no nickel. Ferrite is magnetic and has a body-centered cubic crystal structure. This group can be hardened to some extent by cold working, but cannot be hardened by heat treatment. These steels are restricted to a somewhat narrower range of corrosive conditions than the austenitic grades but may be considered to be interchangeable with austenitic types under strongly oxidizing conditions.

The martensitic grades contain chromium from 12 to 17 percent, and usually no nickel. There are exceptions where martensitic grades do contain some nickel. The martensitic stainless steels differ from the ferritic and austenitic in that they can be hardened by heat treatment. Martensitic stainless steel is magnetic and has a body-centered tetragonal crystal structure characterized by a needlelike pattern. It has excellent corrosion-resistant properties under mild conditions such as weak acids, fresh water, and the atmosphere. Severely corrosive environments will attack martensitic stainless steels.

Other Types of Steel

There are many other types of steel classed according to their alloying elements, such as boron, copper-bearing, manganese, nickel, nickel–chromium, molybdenum, chromium, chromium–vanadium, silicon–manganese, and silicon steels. There are also groups classed according to molding and forming properties and surface conditions, which include free-cutting, deep-drawing, texturized, precision-ground, polished, cold-drawn, and cold-rolled steels. Although it is not practical to describe them in detail, much of the information in the data sheets and in the chapter as a whole is applicable to all of these types of steels.

DATA SHEETS FOR STEEL

Useful reference information that can help to apply steel to a design project is included in Tables 4.2 and 4.3. These are the type of data supplied to designers in large corporations. The same type of information exists for other materials, such as aluminum, magnesium, copper, brass, and bronze. The information listed is based upon the following general data.

Mechanical properties. The values listed indicate minimum mechanical properties (except where noted otherwise) that may be expected from 0.505-in. diameter tension-test specimens at room temperature. The properties listed are obtainable only in the section sizes specified. If no section size is specified, it may be assumed that the properties apply to all sizes commercially obtainable.

Section sizes. Where size ranges are indicated to which certain properties apply, the term "dia." refers to solid round bars; the term "thk." refers to the minimum dimension of a rectangular section; and the term "size" refers either to the diameter of round bars or the minimum dimension of a rectangular section.

Modulus of elasticity in tension. The modulus of elasticity in tension for all specifications listed falls within the range of 28×10^6 to 30×10^6 lb/in.^{2}

Endurance limit in tension. In the absence of actual data, the endurance limit in tension is approximately 0.45 times ultimate tensile strength where ultimate tensile strength does not exceed 125,000 lb/in.^{2}

Torsion. In the absence of actual data, the yield point in torsion may be taken as one-half that for tension, the endurance limit in torsion as one-half that for tension where ultimate tensile strength does not exceed 125,000 lb/in.^{2}, and the modulus of elasticity in torsion as 11.5×10^6 lb/in.^{2}

Machinability. Steels having a hardness below 225 Brinell are generally machined easily. For average applications, the cost of machining rises rapidly above 250 Brinell. Steels may be machined at hardness over 300 Brinell, but with considerable difficulty and generally at excessive cost. Although possible, machining of steels at hardnesses exceeding 400 Brinell is rarely attempted, but it is possible using aluminum oxide inserts.

Chemistry. Where phosphorus and sulfur contents are not given, the maximum phosphorus content is in the range of 0.03 to 0.05 percent and the maximum sulfur content is in the range of 0.03 to 0.055 percent. The approximate SAE equivalent specification is listed immediately following the chemistry of each specification.

TABLE 4.3

Steel sheet, carbon, cold-rolled—thickness tolerances (plus or minus)

	Thickness, in.[a]														
	0.2299	0.1874	0.1799	0.1419	0.0971	0.0821	0.0709	0.0567	0.0508	0.0388	0.0343	0.0313	0.0254	0.0194	0.0141
Width, in.	0.1875	0.1800	0.1420	0.0972	0.0822	0.0710	0.0568	0.0509	0.0389	0.0344	0.0314	0.0255	0.0195	0.0142	and less
To $3\frac{1}{2}$ incl.													0.003	0.002	0.002
$3\frac{1}{2}$–6											0.004	0.003	0.003	0.002	0.002
6–12								0.005	0.005	0.004	0.004	0.003	0.003	0.002	0.002
12–15	0.008	0.007	0.007	0.007	0.006	0.006	0.006	0.005	0.005	0.004	0.004	0.003	0.003	0.002	
15–20	0.008	0.008	0.008	0.008	0.007	0.007	0.006	0.006	0.005	0.004	0.004	0.003	0.003	0.002	
20–32	0.009	0.009	0.009	0.008	0.007	0.007	0.006	0.006	0.005	0.004	0.004	0.003	0.003	0.002	
32–40	0.009	0.009	0.009	0.009	0.008	0.007	0.006	0.006	0.005	0.004	0.004	0.003	0.003	0.002	0.002
40–48	0.010	0.010	0.010	0.010	0.008	0.007	0.006	0.006	0.005	0.004	0.004	0.003	0.003	0.002	0.002
48–60			0.010	0.010	0.008	0.007	0.007	0.006	0.005	0.004	0.004				
60–70			0.011	0.011	0.009	0.008	0.007	0.007	0.006	0.005	0.005				
70–80			0.012	0.012	0.009	0.008									
80–90			0.012	0.012	0.010										
90			0.012	0.012											

[a] Thickness is measured at any point on the sheet not less than $\frac{3}{8}$ in. from an edge.

TABLE 4.4
Steel sheet, carbon—thickness. (Preferred for uncoated, thin, flat metals under 0.250 in. thick)[a]

0.004	0.014	0.040	0.112
0.005	0.016	0.045	0.125
0.006	0.018	0.050	0.140
0.007	0.020	0.056	0.160
0.008	0.022	0.063	0.180
0.009	0.025	0.071	0.200
0.010	0.028	0.080	0.224
0.011	0.032	0.090	
0.012	0.036	0.100	

[a] The use of the American Standard preferred thicknesses eliminates the confusion caused by the various gauge number systems.

Note: The same composition and finish as indicated for bars in Table 4.2, plus many more types of sheets, can be found on the market.

Bending recommendations. Bending recommendations show maximum bendability and are based on specification values or are estimated. They are not based on actual shop practice.

Corossion resistance. The corrosion resistance of steels is complicated by many factors. In general, carbon and alloy steels do not differ greatly in their resistance to air atmospheres in rural, urban, marine, and industrial locations; they rust readily if moisture is present. Where resistance to rust is required, it is necessary to use a protective finish on carbon and alloy steel or to specify a special steel which develops an adherent oxide coat such as Corten steel.

Application

In choosing between carbon, alloy, and stainless steels, the most important considerations are usually mechanical properties, corrosion resistance, magnetic properties, and cost. Wherever applicable, carbon steel should be used because of its lower cost. Use of alloy steel is usually restricted to applications requiring high mechanical properties involving heat treatment in section sizes larger than can be hardened throughout with carbon steel. Stainless steels are used primarily to obtain corrosion resistance, although the nonmagnetic characteristics of the austenitic chromium–nickel steels are frequently of major importance. The magnetic chromium stainless steels are used generally for parts requiring corrosion resistance (significantly greater than carbon or alloy steels, but inferior to chromium–nickel stainless steel) together with high mechanical properties obtainable by heat treatment. Another group of magnetic chromium stainless steels, which are nonhardenable, are occasionally used for their magnetic and corrosion-resisting characteristics.

FACTORS IN THE SELECTION OF STEEL

Once engineers have decided upon the properties they need, they can then select the material that best fits the criteria established. Some factors that will assist in the selection of steel follow.

Useful and important properties are added to steel by heat treatment (see Chapter 3). Strength and hardness increase when steel is cold-worked. Rolled and forged steel has directional properties. Properties of steels are better with the grain than across the grain. Grain direction can be used to an advantage in forgings and should be incorporated in design. All steels have practically the same stiffness (modulus of elasticity). It is impossible to stiffen a piece of steel by heat treating. The modulus of elasticity in alloy steels ranges from 27×10^6 to 32×10^6 lb/in.^{2} Deflection is often more important in design than strength.

Temperatures less than room temperature increase tensile strength and reduce ductility.

Designs are not based on tension alone. Failures may occur because of many unknown factors, such as the overloading of equipment by an operator. The absence of failures in the field indicates that the parts are designed too conservatively, except in cases where life would be endangered by failure. It is sometimes best to choose the optimum size and conditions first and modify as

TABLE 4.5
Strength–weight factors

Material	Ultimate tensile strength, lb/in.^{2}	Average specific gravity	Strength–weight ratio
Aluminum, commercial 2S	13,000	2.71	4.8
Iron, ingot (wrought)	40,000	7.87	5.1
Steel, cold-rolled	60,000	7.84	7.6
Aluminum, cold-rolled, 2S-H	24,000	2.71	8.9
Steel, SAE low-alloy, low-carbon	160,000	7.85	20.4
Aluminum alloy, 17S-T (duralumin)	58,000	2.79	20.8
Spruce for aircraft	10,000	0.435	23.0
Aluminum alloy, C17S-T	65,000	2.8	23.2
Steel, SAE medium-alloy, medium-carbon, heat-treated	150,000	7.85	24.2
Aluminum alloy, 24S-RT	68,000	2.77	24.5
Magnesium alloy, AM58S	46,000	1.85	24.9
Steel, 18–8 stainless, heavily cold-rolled	200,000	7.93	25.2
Steel, high-alloy, high-carbon, heat treated	250,000	7.85	31.8
Steel, piano-wire, cold-drawn, very fine	400,000	7.84	51.0

experience dictates. Table 4.5 gives the tensile strength, specific gravity, and strength–weight ratio of representative ferrous and nonferrous materials.

Failure of Steels

Very few machines fail through tension or shear. They usually wear out or fail because of fatigue. *Fatigue failure* occurs after repeated application of a load that can be borne safely in a single application. It seems surprising that a hardened steel specimen that will carry 215,000 $lb/in.^2$ in a tension test without showing any plastic deformation will break by the application of 120,000 $lb/in.^2$ if such a stress is repeated often enough. The endurance limit is ordinarily, but not necessarily, below the yield point and is close to 50 percent of the ultimate tensile strength. The effect of the shape and size of notches on fatigue strength is considerable and should be checked by actual tests. The higher the tensile strength of the material, the greater the effect of surface corrosion. In design, notches and anything that weakens the surface, such as decarburization, should be avoided. Fatigue is affected by grooves on the surface. Corrosion severely diminishes the endurance limit of a material. Corrosion resistance can be improved by surface burnishing.

"Crystallization in service" is a myth. Innumerable mechanical experts have looked at a piece of metal, broken off sharply without any apparent sign of ductility, and said, "crystallization." They believe that the metal was fibrous at the start of its life, but that with the stresses and strains of its existence the material had changed from tough fibers to large brittle crystals. They prove it by pointing to the coarse crystalline fracture (when there is one; when it is a fine porcelainic fracture, they still say "crystallization").

If the part after failure exhibits a coarse crystalline fracture, it had a coarse crystalline structure when it left the hands of the craftsman who fashioned it. At least 90 percent of the failures called crystallization are fatigue failures. Service can not change the grain size of metallic alloys.

Scoring and galling. *Scoring* is similar to abrasion, but the abrasive, instead of being dirt, is one or more hard particles or projections of metal, embedded in one of the surfaces. As the pieces slide past each other, a high spot on one may interlock with a slightly high spot on another piece; a fragment is torn out, and scoring results. Scoring may occur soon after a part is placed in service, and, once having made suitable tracks for the high spots on the mating part, may not go any further. All too frequently, it does go further, and then it is known by another name—*galling*. High finish is usually helpful in reducing scoring: of course, lubrication is an important factor. If galling is severe enough to stop the sliding motion of two parts, it is known as *seizure*. Some high-alloy steels are notoriously bad in the matter of galling and seizure as, for instance, 18–8 stainless steel, which is difficult to form in dies.

Pitting. Pitting is another important type of failure. A true case of pitting involves compression fatigue of the surface layers; the fatigue failures first occur below the surface in the weaker, lower-carbon material found there.

Flow and corrosion. Flow under pressure, vibration, or shock may cause a soft bearing to yield or a ball socket to expand, permitting greater movement and pounding during operation. This causes loss of dimension.

Corrosion can cause loss of dimension such as the rusting of a frame or rod until it breaks under load.

The loss of dimension due to abrasion, scoring, galling, pitting, flow, and corrosion may determine the life of a product. The loss of 5 lb of metal may make the difference between a brand new and a completely worn-out 5-ton truck; a 10,000-lb machine becomes completely worn out by the loss of 5 lb in critical places. Indefinite life can be obtained by smooth surfaces and the use of noncorrosive lubricants.

Metallurgical Factors in the Selection of Steels

Metallurgical factors that determine the useful properties of steel are not very numerous. Details of heat treatment, critical points, critical cooling rates, and similar matters will not be discussed here (information can be obtained from books on metallurgy and metals), but these things must be borne in mind in selecting steel.

General data on representative properties of dozens of much-used steels are available from any steel company. The typical chart contains the recommended heat treatment; forging temperature; quench, normalizing, annealing, or quenching temperature; size of piece treated to give the indicated results; the critical points; chemical specifications; and often the analysis of particular samples used. Curves show the Brinell hardness, tensile strength, yield point, elongation in 2 in., reduction of area, and Izod impact, all as they vary with the drawing temperature.

A group of actual test data would not fall neatly on these curves, but would occupy a band across the diagram, whose lower region is marked by the published curves. Manufacturers usually publish minimum values in order that they can be reasonably sure of meeting their own specifications.

Properties of Steels As Purchased

What can the engineer normally expect of forging billets, hot-rolled bars, normalized bars, annealed bars, and cold-drawn bars? One alloy steel will not be praised or belittled; the primary aim is to show the similarity between steels—features that they have in common—not the differences between steels that have been talked of so much by proponents of single alloying elements. Those differences are sometimes exaggerated. As far as rolling-mill practice goes, the finishing temperature as the steel emerges from the last roll is of most significance to the buyer. The steelmaker can "roll hot," using a high temperature to the benefit of the buyer if he wants a coarse grain in the bar, as rolled. Low finishing temperatures usually induce a fine grain and an associated high degree of toughness, even after subsequent heat treatment.

Grain-size differences cause a noticeable difference in machinability and hardenability. Thus, finishing temperature can be considered another variable in quality steelmaking practice that must be kept under control.

Steel can be purchased to many size tolerances. Tolerances are relatively large for hot-rolled steels; likewise, such bars may be out-of round 0.010 or even 0.020 in. Since they have scaled surfaces, there will be handicaps for some uses, particularly in automatic screw machines where the work is held in collets. If the engineer wants to use the OD (outside diameter) of a bar as a portion of a completed part without any finishing, then he will want closer tolerances. He can buy such accurate hot-rolled bars by paying extra for them, but ordinarily he would want to choose cold-finished bars.

What could fairly be expected in a steel bar of a certain carbon content—as-rolled, normalized, or annealed (any of these conditions being purchasable from the steel mill)?

Figure 4.7 summarizes the tensile properties as affected by carbon content and final mill treatment. Note the solid lines, which show the as-rolled properties of these steels. The tensile strength goes up as the carbon content

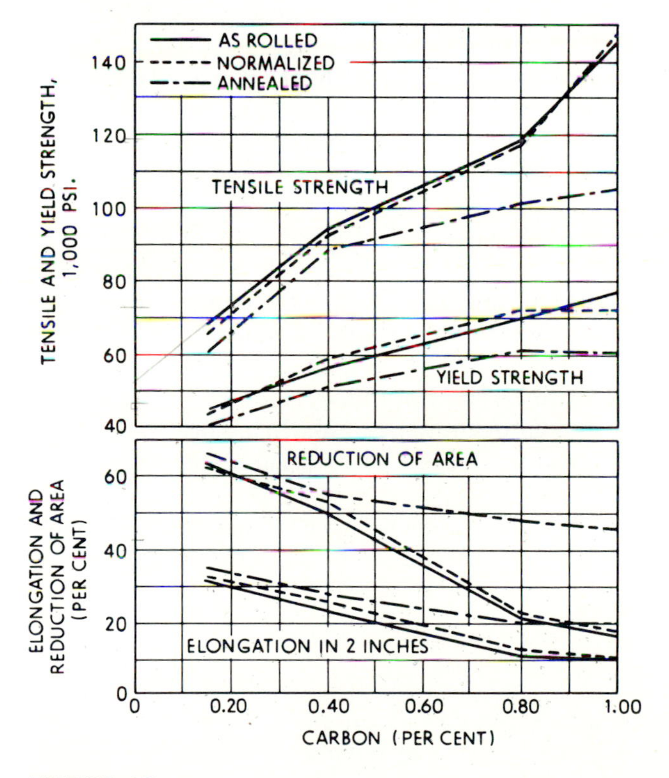

FIGURE 4.7

Average properties of 1-in. commercial carbon-steel bars, as they vary with carbon content and with treatment subsequent to final rolling.

goes up. With 0.20 percent carbon steel, there may be 70,000 lb/in.^{2} in an as-rolled bar; when the carbon content goes up to 0.80 percent, 120,000 lb/in.^{2} is obtained. The yield strength also goes up, but not strictly in proportion. The reduction of area and the ductility go down, as might be expected. Between 0.20 and 0.80 percent carbon, there is a gain of perhaps 70 percent in tensile strength, but a loss of about two-thirds of the reduction of area, and almost as much of the elongation. In other words, as the carbon goes up the tensile strength increases, but the ductility goes down just as fast.

When the tensile strength passes beyond 100,000 lb/in.^{2}, machining becomes difficult. It is, therefore, frequently necessary to buy the higher-carbon steels in the annealed or normalized condition to counteract the hardening and strengthening effect of the rather rapid cooling on the runout table from the last stand in the hot-rolling mill. This is particularly true in small sizes. In such cases, better machinability is associated with a lower hardness number. For lower-carbon steels where normalizing does not affect the hardness much, the as-rolled bar may be expected to machine a little better, because its less-uniform microstructure would probably produce a slightly more brittle, free-breaking chip.

Note that the tensile curves for normalized steel approximately coincide with the ones for as-rolled steels. Since the difference in tensile properties between normalized and as-rolled bars is very slight, if the final rolling temperature is low and cooling on the hot bed is at a moderate rate, normalizing would be done to ensure a more uniform microstructure and to release internal strains that would cause warpage on machining or subsequent heat treatment.

The broken lines in Fig. 4.7 show the properties of annealed bars. Annealing removes all hardening that may have taken place during the more rapid cooling after rolling. It lowers the tensile and yield strengths, more so in the higher-carbon than in the lower-carbon steels. Reduction of area and elongation are likewise improved greatly.

Above 0.20 percent carbon the HR (hot-rolled) alloy steels are usually purchased in the annealed condition for the sake of machinability. Alloy steels are too expensive not to be heat-treated so as to secure full use of their properties.

Bars may be purchased with the following special surfaces: (1) rough-turned; (2) turned and polished or centerless-ground; (3) cold-drawn; (4) cold-rolled.

Effects of Cold Work

Cold work, even as little as 12.5 percent ($\frac{1}{16}$ in. on 1-in. bar) affects the properties of the bar to the very center, as anyone knows who has used cold-drawn steel for its good machining characteristics. This important property is often not understood; the gain is more than surface gain. Hot-rolled bars have Brinell hardness of about 125 all the way across the

TABLE 4.6
Effect of cold-drawing carbon steels and screw stock

SAE steel	Draft on 1-in. bar	Tensile strength	Yield point	Elongation	Reduction of area	Charpy impact
1020	None[a]	64,000	48,000	36	68	53
	$\frac{1}{16}$ *in.*	83,000	74,000	19	61	44
	$\frac{1}{8}$ *in.*	91,000	82,000	15	58	34
1112	None[a]	71,000	54,000	34	59	30
	$\frac{1}{16}$ in.	97,000	89,000	16	48	16
1045	$\frac{1}{8}$ in.	105,000	98,000	13	44	12
	None[a]	109,000	58,000	25	52	—
	$\frac{1}{16}$ in.	118,000	79,000	15	44	19
	$\frac{1}{8}$ in.	130,000	85,000	12	38	14

[a] Hot-rolled 1-in. rod.

section. After a $\frac{1}{32}$-in. draft, the surface hardness has increased to 185 and the center to 165. After a $\frac{1}{16}$-in. draft, the surface increases further and the center catches up more nearly to the surface; and with a $\frac{1}{8}$-in. draft, the hardness is approximately uniform from center to surface (see Table 4.6).

The proportional limit of steels (carbon or alloy) is not raised by cold-drawing; in fact, the stress–strain curve in a tension test departs from a straight line at a rather low figure. However, the load that causes definite yield (permanent set) is raised. This is shown in part by Fig. 4.8.

Figure 4.8 shows two stress–strain curves for the same steel, plotted one on top of the other; one curve was taken from a cold-drawn sample, the other from a piece of the same bar after stress relief by annealing at 900°F. The curve for cold-drawn steel deviates from true proportionality at about 42,500 lb/in.^{2} (It might have been even lower if the test had been performed with extensometers of higher sensitivity.) The actual yield point (beginning of plastic deformation) is much higher, at a point that cannot be noted on a simple test to fracture, but the test piece is acting in an elastic manner. The process known as stress relieving (heating the steel up to a temperature of 630 to 900°F) produces a proportional limit of 84,000 lb/in.^{2} in the cold-drawn steel bar.

The residual stresses within cold-worked materials cause unpredictable size changes during machining. Proper stress relief is necessary for such alloys if distortion is to be avoided and proper tolerances are to be achieved. For example, a plant made a series of large shafts from welded tubes that were swaged shut on each end for a distance of 8 to 10 in. The swaging operation was often finished below the recrystallization temperature of the steel. However, the company engineers did not realize that and spent many thousands of dollars trying to solve the machining problem by calling in consultants and machining experts. In the end, proper stress relief turned out to be the answer to the problem.

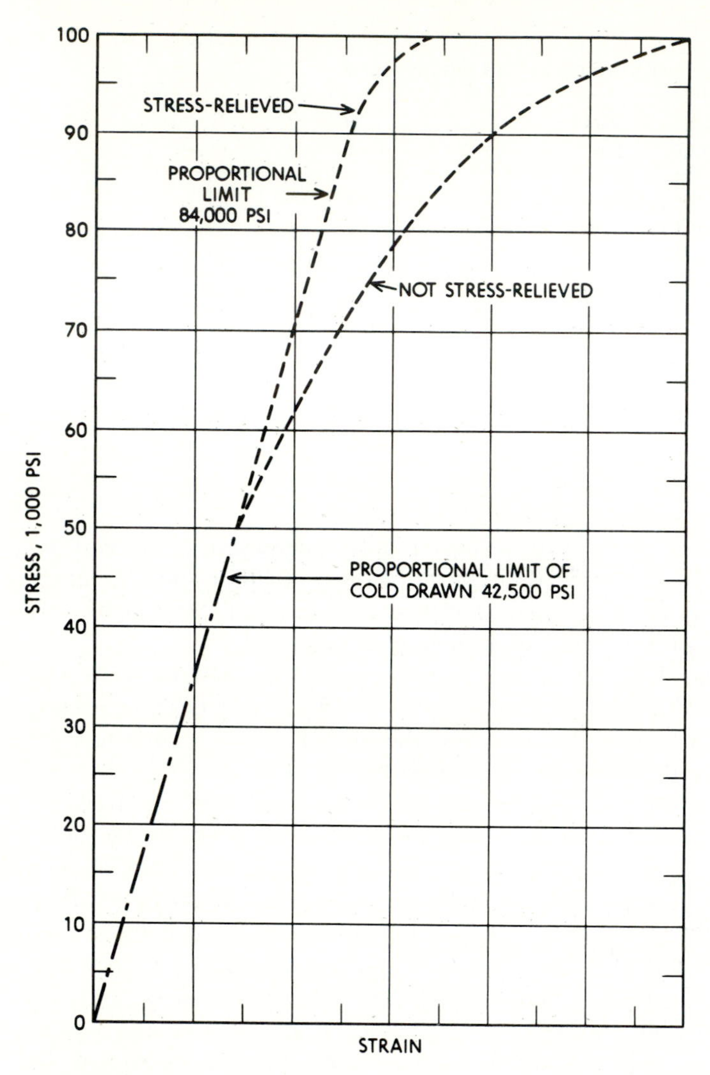

FIGURE 4.8
True proportional limit is reached by cold drawing, although the steel can be more heavily loaded without taking a permanent set. Stress relief at 900°F will restore and increase the proportional limit to a high figure.

Selection of Suitable Steel

There is not a great difference between "this" steel and "that" steel; all are very similar in mechanical properties. Selection must be made on factors such as hardenability, price, and availability, and not with the idea that "this" steel can do something no other can do because it contains 2 percent instead of 1 percent of a certain alloying element, or because it has a mysterious name. A tremendous range of properties is available in any steel after heat treatment; this is particularly true of alloy steels.

Ease of hardening (hardenability). The correct selection for a given part will often depend on hardenability, which can be defined as the "depth to which steels can be hardened in quenching." Various alloy steels will differ considerably in this respect. When they are properly hardened, all of these alloy steels with a given carbon content have similar properties. The difficulty is in hardening them. The effectiveness of heat treatment is what is bought in alloy steels. The maximum hardness attainable is a function of the carbon content and practically nothing else, but the quench necessary to get the hardness may be difficult to obtain. The effect of the alloy is to raise the hardness not at the surface, but at the center of the bar when it is quenched.

"The size of bar fully hardenable," as listed, is regarded as that bar that will show, when quenched, a hardness of not less than C50 at the center. It is well known that the quenching rate in oil is not as fast as in water, and consequently does not have as much effect at the center of bar. Therefore, the size that can be hardened to Rockwell C50 minimum is smaller when quenched in oil than when quenched in water. One thing to be emphasized is that 1.5 percent manganese steels 1330 and 1340 will harden in water to Rockwell C50 at the center of a $1\frac{1}{4}$-in. bar. These are the cheapest alloy steels shown on the chart. More expensive steels have scarcely any difference: 2340, $1\frac{3}{8}$ in.; 5140, $1\frac{1}{2}$ in.; 3130, $1\frac{1}{8}$ in.; 6150, $1\frac{3}{8}$ in.

Considerations in Fabrication

The properties of the final part (hardness, strength, and machinability), rather than properties required by forging, govern the selection of material. The properties required for forging have very little relation to the final properties of the material; therefore, not much can be done to improve its forgeability. Higher-carbon steels are more difficult to forge. Large grain size is best if subsequent heat treatment will refine the grain size.

Low-carbon, nickel–chromium steels are just about as plastic at high temperatures under a single 520-ft · lb blow as plain steels of similar carbon content. Nickel decreases forgability of medium-carbon steels, but has little effect on low-carbon steels. Chromium seems to harden steel at forging temperatures, but vanadium has no discernible effect; neither has the method of manufacture any effect on high-carbon steel.

Formability. The cold-formability of steel is a function of its tensile strength combined with ductility. The tensile strength and yield point must not be high or too much work will be required in bending; likewise, the steel must have sufficient ductility to flow to the required shape without cracking. The force required depends on the yield point, because deformation starts in the plastic range above the yield point of the steel. Work-hardening also occurs here, progressively stiffening the metal and causing difficulty, particularly in the low-carbon steels.

It is quite interesting in this connection to discover that deep draws can sometimes be made in one rapid operation that could not possibly be done leisurely in two or three. If a draw is half made and then stopped, it may be necessary to anneal before proceeding, that is, if the piece is given time to work-harden. This may not be a scientific statement, but it is actually what seems to happen.

Surface. A good surface on the steel is very important for any drawing operation for two reasons: (1) to get slippage through the dies, and (2) because there is a large amount of stress and of deformation at single points, and stress raisers will cause trouble there. For example, if a deep draw is attempted around a punched hole that is rough inside, the part may split from the punch marks around the original hole. (Better results will be obtained from reaming, followed by deep drawing.) The stress raisers combined with the work-hardening that has taken place locally will interfere with satisfactory drawing. Of course, the finish of the dies or forming tools is equally important from the standpoint of easy slippage.

Internal stresses. Cold forming is done above the yield point in the work-hardening range, so internal stresses can be built up easily. Evidence of this is the springback as the work leaves the forming operation, and the warpage in any subsequent heat treatment. Even a simple washer might, by virtue of the internal stresses resulting from punching and then flattening, warp severely during heat treating.

When doubt exists as to whether internal stresses will cause warpage, a piece can be checked by heating it to about 1100°F and then letting it cool. If there are internal stresses, the piece is likely to deform. Pieces that will warp severely while being heated have been seen, yet the heat-treater was expected to put them through and bring them out better than they were in the first place.

Welding. The maximum carbon content of plain carbon steel safe for welding without preheating or subsequent heat treatment is 0.30 percent. Higher-carbon steels are welded every day, but only with proper preheating. There are two important factors: (1) the amount of heat that is put in, and (2) the rate at which it is removed.

Welding at a slower rate puts in more heat and heats a large volume of metal, so the cooling rate due to loss of heat to the base metal is decreased. A preheat will do the same thing. For example, SAE 4150 steel, preheated to 600 or 800°F, can be welded readily. When the flame or arc is taken away from the weld, the cooling rate is not so great, owing to the higher temperature of the surrounding metal, and slower cooling results. Even the most rapid air-hardening steels are weldable if preheated and welded at a slow rate.

Machinability. Machinability means several things. To production men it generally means being able to remove metal at the fastest rate, leave the best possible finish, and obtain the longest possible tool life. Machinability applies to the tool–work combination.

It is not determined by hardness alone, but by the toughness, microstructure, chemical composition, and tendency of a metal to harden under cold work. In the misleading expression "too hard to machine," the work "hard" is usually meant to be synonymous with "difficult." Many times a material is actually too soft to machine readily. Softness and toughness may cause the metal to tear and flow ahead of the cutting tool rather than cut cleanly. Metals that are inherently soft and tough are sometimes alloyed to improve their machinability at some sacrifice in ductility. Examples are the use of lead in brass and of sulfur in steel.

Machinability is a term used to indicate the relative ease with which a material can be machined by sharp cutting tools in operations such as turning, drilling, milling, broaching, and reaming.

In the machining of metals, the metal being cut, the cutting tool, the coolant, the process and type of machine tool, and the cutting conditions all influence the results. By changing any one of these factors, different results will be obtained. The criterion upon which the ratings listed are based is the relative volume of various materials that may be removed by turning under fixed conditions to produce an arbitrary fixed amount of tool wear.

Although machinability ratings are relative it must not be assumed, for example, that a steel with a rating of 40 would simply require twice as much time and cost to machine as another steel with a rating of 80. Actually, the steel with a rating of 40 may not even be machinable by ordinary methods (see Table 4.7).

The machinability of materials is a major factor in the cost of the product. The size, strength, composition, and heat treatment of the material can be selected to give the best machining conditions and the lowest costs.

TABLE 4.7
Relative machinability rating as related to machining characteristics

Relative machinability rating	Machining characteristics
85 and above	Free machining
70 to 80	Easily machined
55 to 65	Difficult machining
40 to 50	Very difficult machining. May be unmachinable in thin sections or with operations requiring extreme pressure or tools of weak section. Carbide tools generally recommended
35 and under	Unmachinable by ordinary shop methods. Such materials require grinding or special techniques

TABLE 4.8
Machinability ratings for certain steels

Cold-drawn		Annealed	
SAE no.	Value	SAE no.	Value
1113	135	3140	55
1112	100	4130	65
1315	85	4150	45
1335	70	5140	45
1020	75	4640	55
1020	75	6140	40
1040	60	1330	50
3115	50	1340	40
4615	60	2350	35
2515	30	52100	30

Machinability is an important item to be considered by the designer, as well as by the engineer in the shop.

In general practice, a Brinell hardness in the neighborhood of 180 is acceptable to the machine shop, especially if this hardness is combined with poor ductility, although, with the will and the tools to do it, materials that have been heat-treated to a surprisingly high hardness can be machined. Frequently the machinability of a given steel can be helped by heat treatment; that is, it can be normalized, if it is a low-carbon material, in order to bring up the hardness to something approximating 180 and likewise reduce the ductility. SAE 1015 or 1020 steels may even be quenched to bring up the hardness.

Table 4.8 summarizes the machinability ratings for a number of steels in the cold-drawn and annealed condition. Table 4.9 lists ratings for different materials. These figures give some indication of the relative machinability, but so many things influence machining that the use of a table is not completely satisfactory. For the purpose of the table, the machinability of cold-drawn screw stock is taken as 100 percent. The improved 1112 will rate 135 percent. The implications of that can be realized by comparing it with the value for SAE 1020, which is similar in carbon and manganese to the 1112, but has lower sulfur and low phosphorus, and will have a machinability somewhere in the neighborhood of 75 percent. The considerably higher percentage of sulfur and phosphorus makes 1112 less ductile in the transverse direction.

Lead has also been successfully added to steel to facilitate the cutting action. It apparently acts in essentially the same manner as sulfur, producing discontinuities in the chip transversely without wearing the tools at a high rate. The actual lubricating value of the lead is probably negligible.

Improving the machinability of a given material means that the mechanical properties must be changed, either by reducing the tensile strength or

TABLE 4.9
Machinability ratings for various metals

Material	Machinability rating	Brinell Hardness
B1112 screw-machine steel, high-sulfur	100	179–229
Standard melleable	120	110–145
Cast iron		
Soft	80	160–193
Medium	65	193–220
Hard	50	240–250
Cast steel (0.35 C)	70	170–212
Wrought iron	50	101–131
Carbon steel (C1010)	50	131–170
Stainless steel (18–8 FM)	45	179–212
Tool steel	30	200–218
Free-cutting brass (35.0% Zn, 3.0% Pb)	100	77
Naval brass (39.25% Zn, 0.75% Sn)	30	87
10 to 20 percent silver	20	88–94
12 percent leaded nickel–silver	60	88
5 percent aluminum–bronze	20	92

damaging the ductility. Cold drawing will help the low-carbon materials because it will increase the hardness.

From the strictly metallurgical standpoint, the material in process must be homogeneous, developing strength uniformly. If one day a piece of 1045 machines with a given setup at a satisfactory rate, and the next day a new batch of material can no longer achieve the standard output, it may be because of nonhomogeneity of the steel or its treatment.

CAST IRONS

There are five types of cast iron: gray, ductile, malleable, high-alloy, and white: by far the most common of these are ductile and gray iron. These cast irons cannot be specified or identified by chemical analysis alone. It is the form of the excess carbon that determines the type of cast iron. The mechanical properties depend on both the form of the free graphite and the matrix that surrounds it.

In *gray iron* the silicon content is high enough to cause the iron carbides to break down so that flake graphite precipitates during solidification. The simultaneous solidification of the iron and precipitation of the graphite flakes greatly reduces its volumetric shrinkage during solidification. Since gray iron is a complex alloy, it freezes in a mushy manner and therefore has little tendency to develop internal shrinkage as would a skin-forming alloy. This makes it

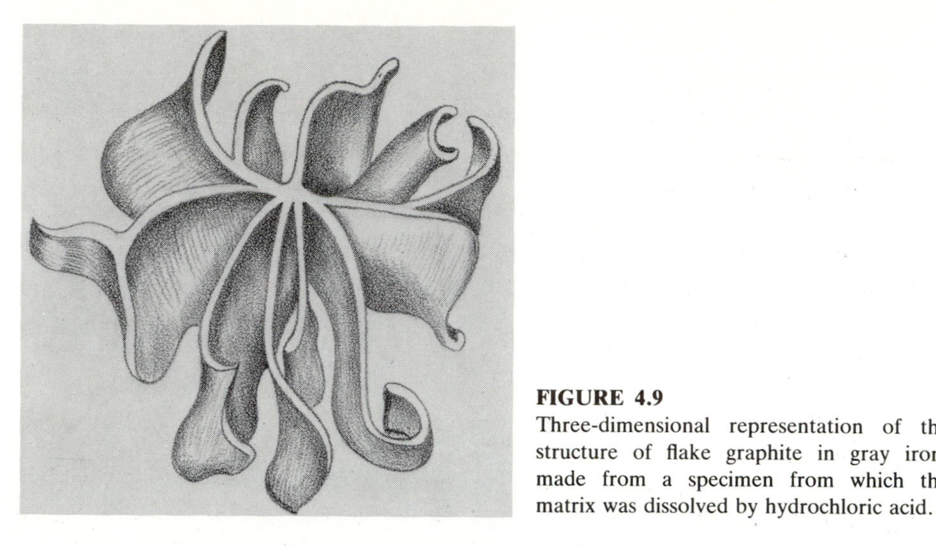

FIGURE 4.9
Three-dimensional representation of the structure of flake graphite in gray iron, made from a specimen from which the matrix was dissolved by hydrochloric acid.

relatively easy to obtain a sound casting even in complex shapes such as motor blocks, air brakes, or control valves.

The excess carbon is the basis for many of the good properties of gray iron, such as high fluidity, high damping capacity, low notch sensitivity, and good machinability. The amount of free carbon in a cast iron depends on the chemical composition, the rate of freezing, and the amount of silicon present. The flake graphite is in the form of a three-dimensional structure similar to the petals of a newly opened rose bud on a microscopic scale (Fig. 4.9). The space between is filled with the matrix of either α-iron or pearlitic iron or a mixture of both.

Alpha or ferritic iron is pure iron with a small amount of dissolved carbon appearing as white grains in the photomicrograph. Pearlite is the eutectoid composition that occurs at 0.8 percent carbon. It has grains that are composed of alternate plates of α-iron and combined carbon as Fe_3C (Fig. 4.10).

Gray cast iron is comparatively soft, of low tensile strength, and easily machined. There are at least eight engineering grades of gray iron (Table 4.10) with tensile strengths ranging from 30,000 to 80,000 lb/in.^{2} One of the outstanding characteristics of gray iron is its compressive strength. Cast iron is stronger than many steels in compression, having strengths ranging from 105,000 lb/in.^{2} for ASTM class 30 to 225,000 lb/in.^{2} for ASTM class 80.

The strongest gray irons have a pearlitic matrix; those with a ferritic matrix are softer and more machinable. The presence of free carbides in a cast iron reduces its machinability markedly, so foundrymen take great care to produce irons with no carbide inclusions.

The form and size of the flake graphite also has a bearing on the properties of a cast iron. A finer, uniformly distributed type A graphite is

FIGURE 4.10
Pearlitic microstructures found in the matrix of pearlitic gray, malleable, and ductile cast irons. Nital etch; 1000 ×. (*Courtesy of John Hoke, Pennsylvania State University.*)

usually specified, because gray irons with that type of graphite have the best mechanical properties.

Ductile iron is a relatively new alloy. It is the fastest growing ferrous alloy because it has such a wide range of properties; it can be stronger than mild steel, yet poured from a low-cost melting furnace such as a cupola. It is frequently also called nodular iron because its free graphite is in the form of spheres rather than flakes.

Ductile iron is produced by adding trace amounts of elements such as magnesium to the molten alloy. The trace elements alter the surface tension of the graphite in the molten iron and cause it to condense into spheroids (Fig. 4.11). In the form of tiny balls, the graphite has no detrimental effect upon the mechanical properties of the matrix, so the strength of ductile iron depends upon the type of metallic matrix. With a perlitic matrix ductile iron can have strengths up to 120,000 lb/in.^{2}, which is equivalent to the strength of high-carbon steel but with superior castability and machinability. The major types of ductile iron are listed in Table 4.11.

Malleable iron starts as a white-iron casting that is then heat-treated. In this case the influence of silicon is most evident, because white iron, which has all its carbon combined in Fe_3C, can only be produced at low silicon and carbon levels. As the white iron is soaked at temperatures above 1600°F, the silicon causes the iron carbide to break down into iron and carbon in the form of irregular-temper carbon nodules (Fig. 4.12). The resultant cast structure is very ductile and is easily machined. For structural designs, malleable iron is limited to relatively thin walls (less than 1 in.) because of large shrinkage values and the need for rapid chilling to produce a white iron as cast structure.

In *high-alloy irons,* those with over 3 percent alloy, there is such a radical change in the basic microstructure that the material can no longer be regarded as gray, ductile, or malleable iron. High-alloy irons are more costly to produce and are used for special applications, such as those requiring extreme wear or corrosion resistance.

TABLE 4.10

General engineering grades of gray iron—mechanical and physical properties

Class	30	35	40	45	50	60	70	80
Tensile strength, lb/in.^{2} (min)	30,000	35,000	40,000	45,000	50,000	60,000	70,000	80,000
Compressive strength, lb/in.^{2}	105,000	115,000	125,000	135,000	150,000	175,000	200,000	225,000
Torsional strength, lb/in.^{2}	40,000	45,000	54,000	60,000	67,000	75,000	85,000	90,000
Modulus of elasticity,[a] lb/in.$^2 \times 10^{-6}$	14	15	16	17	18	19	20	21
Torsional modulus, lb/in.$^2 \times 10^{-6}$	...	...	5.5	6.6	7.0	8.0	8.1	
Impact strength, Izod AB (1.2-in.-diam. unnotched), ft · lb	23	25	31	36	65	75	120+	120+
Brinell hardness	180	200	220	240	240	260	280	300
Endurance limit, lb/in.^{2}								
Smooth	15,500	17,500	19,500	21,500	25,500	27,500	29,500	31,500
Notched	(13,500)	(15,500)	17,500	(19,500)	21,500	23,500	25,500	27,500
Damping capacity	Excellent	Excellent	Excellent	Good	Good	Good	Fair	Fair
Machinability	Excellent	Excellent	Excellent	Good	Good	Fair	Fair	Fair
Wear resistance	Fair	Fair	Good	Good	Excellent	Excellent	Excellent	Excellent
Pressure tightness	Fair	Fair	Good	Good	Excellent	Excellent	Excellent	Excellent
Specific gravity:								
g/cm^3	7.02	7.13	7.25	7.37	7.43	7.48	7.51	7.54
lb/in.^{3}	0.254	0.258	0.262	0.267	0.269	0.270	0.272	0.273
Thermal coefficient of linear expansion: (in./in., °F) $\times 10^{-6}$								
(50–200°F)	6.5–6.7	...	6.6–6.8	6.4–6.4				
(50–500°F)	6.9–7.2	...	7.1–7.3	6.8–7.0				
(50–800°F)	7.4–7.6	...	7.4–7.6	7.0–7.2				
Magnetic properties	Mag	Mag	Mag	Mag	Mag	Mag	Mag	Mag
Pattern shrinkage, in./ft	$\frac{1}{10}-\frac{1}{8}$	$\frac{1}{8}$	$\frac{1}{8}$	$\frac{1}{8}$	$\frac{1}{8}$	$\frac{1}{8}-\frac{3}{16}$	$\frac{1}{8}-\frac{3}{16}$	$\frac{1}{8}-\frac{3}{16}$
Coefficient of friction (against steel)	...	...	...	(0.19)	(0.195)	(0.20)		

[a] At 25% of tensile strength.

Note: Values in parentheses are estimated.

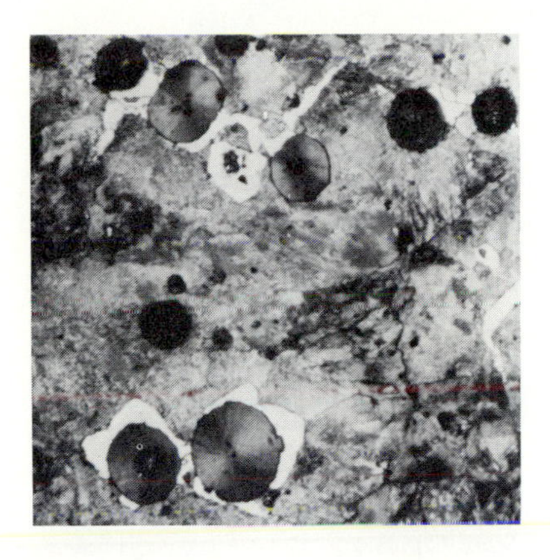

FIGURE 4.11
Microstructure of ductile iron showing the spheroidal graphite in a pearlitic matrix. Nital etch; 230×. (*Courtesy of John Hoke, Pennsylvania State University.*)

TABLE 4.11
Principal types of ductile iron

Type no.[a]	Brinell hardness no.	Characteristics	Applications
80–60–03	200–270	Essentially pearlitic matrix, high-strength as cast. Responds readily to flame or induction hardening	Heavy-duty machinery, gears, dies, rolls for wear resistance, and strength
60–45–10	140–200	Essentially ferritic matrix, excellent machinability and good ductility	Pressure castings, valve and pump bodies, shock-resisting parts
60–40–15	140–190	Fully ferritic matrix, maximum ductility and low transition temperature (has analysis limitations)	Navy shipboard and other uses requiring shock resistance
100–70–03	240–300	Uniformly fine pearlitic matrix, normalized and tempered or alloyed. Excellent combination of strength, wear resistance, and ductility	Pinions, gears, crankshafts, cams, guides, track rollers
120–90–02	270–350	Matrix of tempered martensite. May be alloyed to provide hardenability. Maximum strength and wear resistance	Same as 100–70–03

[a] The type numbers indicate the minimum tensile strength, yield strength, and percent of elongation. The 80–60–03 type has a minimum of 80,000 lb/in.^{2} tensile, 60,000 lb/in.^{2} yield, and 3 percent elongation in 2 in.

Source: Courtesy Gray and Ductile Iron Founder's Society.

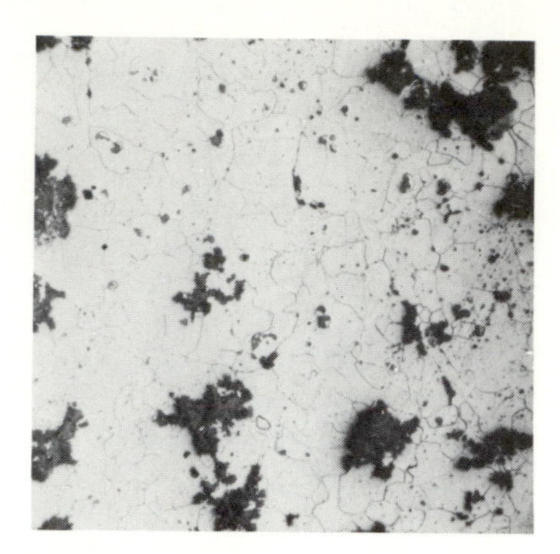
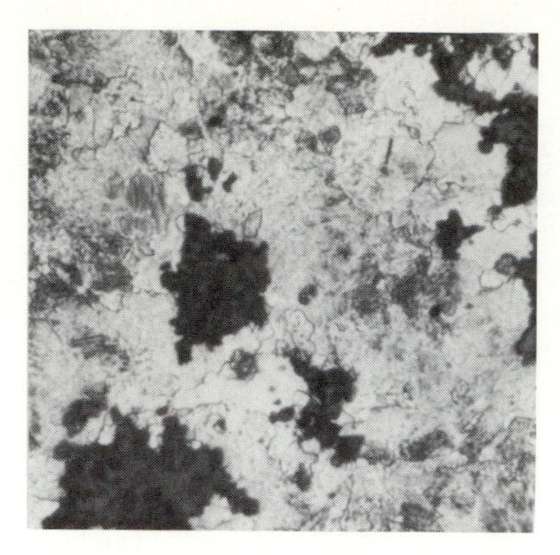

FIGURE 4.12
Microstructure of malleable iron showing the typical carbon with (left) a ferritic matrix and (right), a pearlitic matrix. (*Courtesy of John Hoke, Pennsylvania State University.*)

Mechanical Properties of Cast Irons[1]

Tensile strength. Tensile strength is the most frequently specified property of cast iron. Although, as mentioned, ASTM specifications list classes of gray iron by tensile strength, a given class of gray iron will have its properties signficantly affected by the cooling rate in the mold. Thus, thickness of the casting will influence both the tensile strength and hardness. Figure 4.13 illustrates these relationships. Typical production strengths of several cast irons for both the as-cast and the heat-treated condition are also shown in Fig. 4.14.

Since there is no definite yield point for cast iron, it is possible to use the material at stresses approaching its maximum tensile strength under static loading. However, care must be exercised, since a slight overload results in fracture.

Compressive strength. The compressive strength of gray iron is usually 3 to 5 times its tensile strength and the shear strength is approximately equal to its tensile strength. Thus, where components are stressed in compression, gray iron is comparable in strength to the higher-strength steels. Consequently, gray iron has had wide application for use in machine-tool bases, die blocks, etc., and where reinforcing ribs stressed in compression may be introduced into the design.

Modulus of elasticity. Figure 4.15 illustrates the stress–strain relationships of the principal cast irons in tension. In determining the modulus of elasticity of gray cast iron, it is common to use the slope of the load–deflection curve at 25

[1] Much of this material was taken with permission from *Gray Iron Castings.*

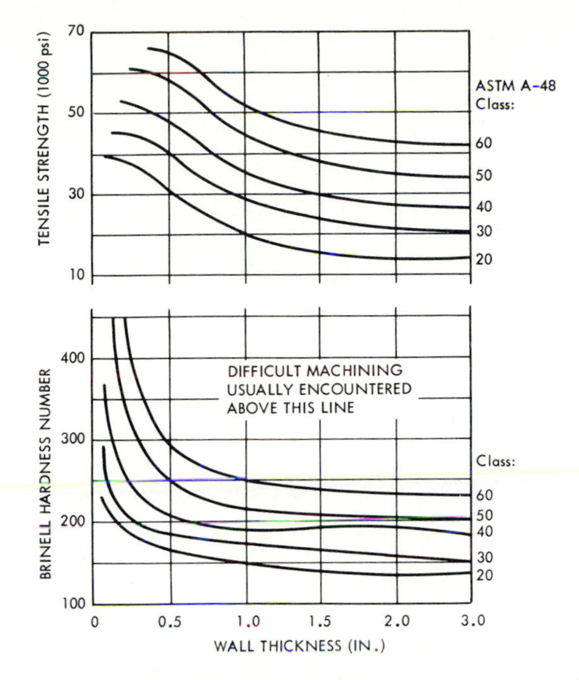

FIGURE 4.13
Tensile strength and hardness as a function of section thickness for various grades of cast iron.

percent of the tensile strength. The designer should select cast irons with a low modulus of elasticity in applications requiring resistance to heat shock. It will be noted that high-yield nodular and white cast iron have moduli of elasticity approaching that of steel (see Fig. 4.16).

Yield strength. Since a clearly defined yield point is not apparent during the typical tensile test, the value for yield strength of gray iron is taken at the point

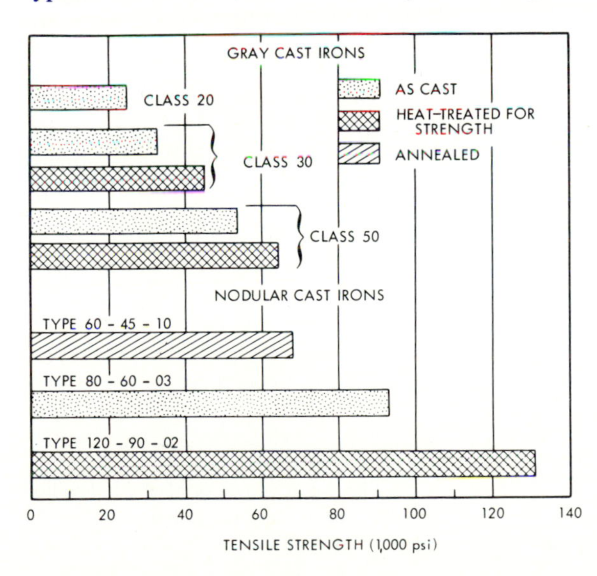

FIGURE 4.14
Tensile-strength properties of cast irons.

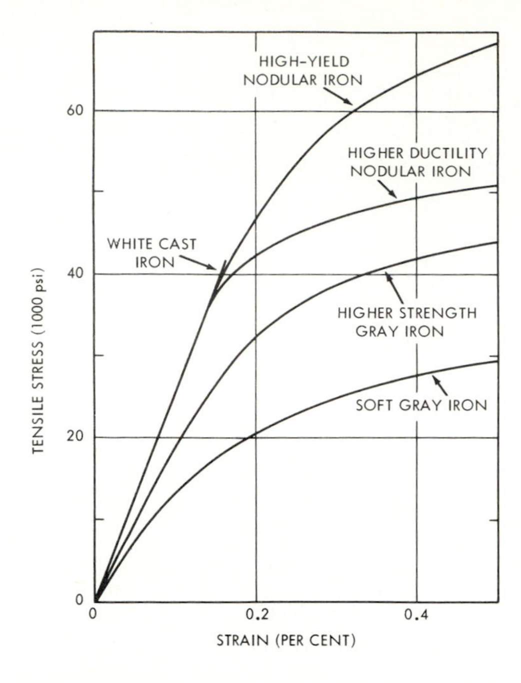

FIGURE 4.15
Stress–strain relationships of the principal cast irons in tension.

of 0.2 percent elongation. Figure 4.17 illustrates yield strengths of the principal cast irons.

Endurance limit. The engineer is frequently interested in the endurance limit, which is a measure of the resistance to fatigue of a material. The values shown in Fig. 4.18 were obtained on a rotating-beam machine, where the stresses in the surfaces of the samples were alternated between tension and compression. For gray cast iron, the endurance limit usually is computed as being between 35 and 50 percent of the tensile strength.

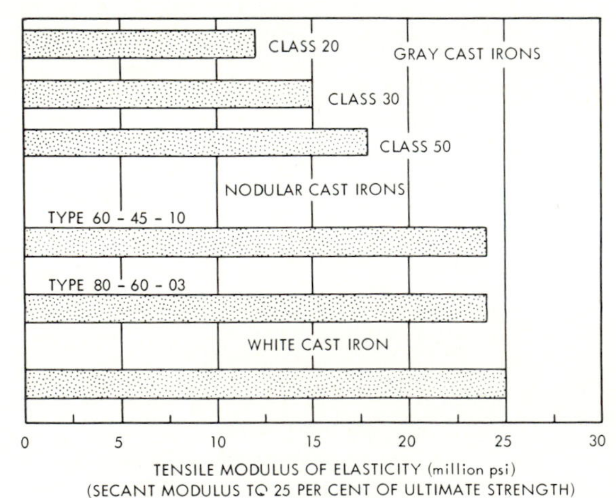

FIGURE 4.16
Tensile modulus of elasticity (million lb/in.^{2}) (secant modulus to 25 percent of ultimate strength).

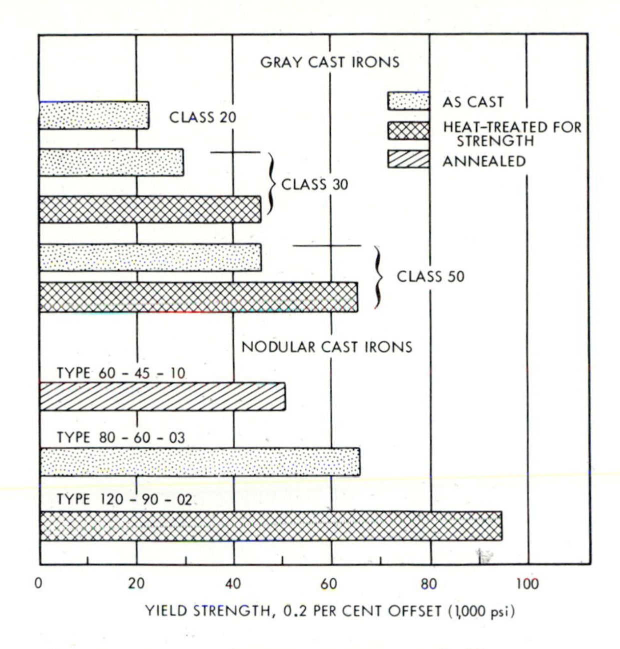

FIGURE 4.17
Yield strength, 0.2 percent offset (1000 lb/in.^{2})

Ductility. Cast irons have low ductility. Gray iron will give elongation ranging from 0.2 to approximately 1 percent; while the high-strength heat-treated irons show elongations of less than 0.2 percent. Nodular irons test in the range of 3 to 20 percent depending upon type. Values enumerated are based on tensile tests.

Gray cast iron is much more brittle than steel and also has lower impact value. This characteristic has given cast iron a performance edge over steel in

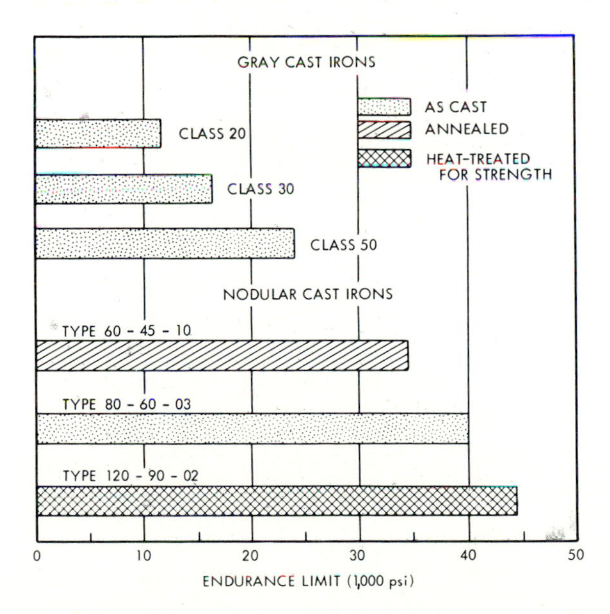

FIGURE 4.18
Endurance limit (1000 lb/in.^{2}).

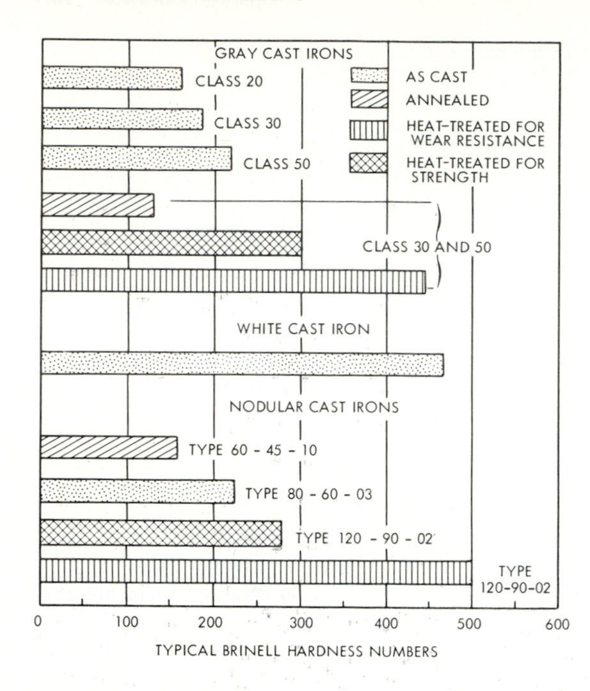

FIGURE 4.19
Typical Brinell hardness numbers.

the design of jigs and fixtures. An impact blow that will break cast iron will often deform steel to an extent that will destroy the necessary relationship between drill bushings, mastering surfaces, and so forth.

Hardness. The hardness of gray iron varies with its tensile strength. Typical Brinell hardness numbers are illustrated in Fig. 4.19. The Brinell test is preferred in measuring the hardness of cast iron because its ball indentor covers sufficient area to give a reading that is representative of the overall hardness.

Production and Design Characteristics of Cast Iron

Castability. To the engineer, castability refers to the ease with which the material can be cast in thin and complex sections. Cast iron is a fluid casting alloy and can be considered as having good castability characteristics. Its castability is better than that of steel or aluminum. For example, automobile engine blocks and heads are produced readily in cast iron and would be difficult to cast with other ferrous materials.

Machinability. Machinability refers to the ease of cutting material with due regard to surface finish and tool life as well as rate of metal removal. Cast irons generally are machined easily and good finishes at low total costs are obtained.

Those cast irons having higher strengths and higher hardness values are obviously not as easily machined as those with lower hardness/strength values.

Corrosion resistance. Cast iron will corrode readily under atmospheric conditions. This corrosion, however, forms a protective surface that offers resistance to further atmospheric corrosion as well as soil corrosion. Widespread use of cast iron for water–main and soil pipe is a good indication of its generally favorable corrosion resistance. Gray-cast-iron water mains and gas lines have been in service in some areas for more than 100 years.

Vibration damping. Damping capacity is an inverse function of the modulus of elasticity. Gray cast iron has good damping qualities, making it a valuable engineering material in the design of parts subject to vibration due to dynamic forces. Figure 4.20 illustrates the relative damping capacities of several engineering materials.

One way of demonstrating damping capacity is to strike the casting: steel will ring, while cast iron will thud. The energy absorbed per cycle in gray iron is about 10 times that of steel, but damping capacity decreases with applied load.

Wear resistance. In sliding friction, gray iron is outstanding in its wear resistance. This is amply demonstrated by the fact that practically all engine cylinders or liners are made of gray iron. The ways of many machine tools are made of gray iron for the same reason.

Weldability. Two methods lend themselves most readily to the welding of gray-iron castings. These are oxyacetylene welding with cast-iron filler and metal arc with nickel or copper–nickel welding electrodes. Oxyacetylene welding is the fastest method of depositing metal and offers sound deposits and

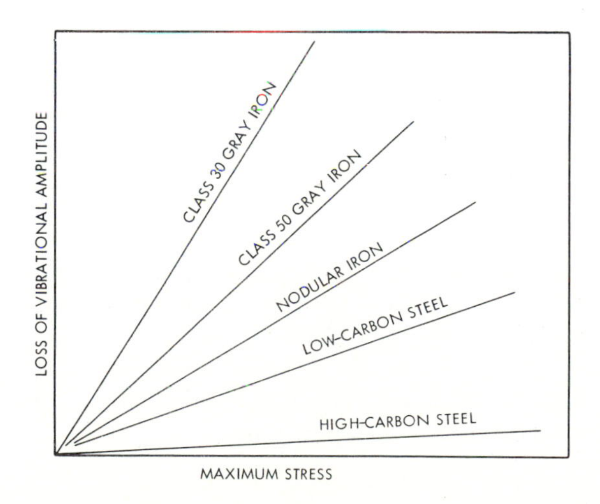

FIGURE 4.20
Relative damping capacities of several engineering materials.

TABLE 4.12
Machine-finish allowances for gray cast iron

Casting dimension, in.	Expected tolerances for as-cast dimension, in.
Up to 8	$\pm\frac{1}{10}$
Up to 14	$\pm\frac{3}{32}$
Up to 18	$\pm\frac{1}{8}$
Up to 24	$\pm\frac{5}{32}$
Up to 30	$\pm\frac{3}{16}$
Up to 36	$\pm\frac{1}{4}$

identical color match. When properly executed, the welds are readily machinable and free from defects. The items to be joined must be preheated.

Shrinkage rules and machining allowances. Unlike most metals, gray iron shrinks very little when it solidifies. Depending upon the grade, 0.100 to 0.130 in./ft must be provided by the patternmaker to allow for solid shrinkage as the casting cools from 2100°F to room temperature.

The amount of material necessary to provide for machine finish will vary with the casting size and the materials' tendency to warp, as well as the analysis of cast iron itself. As a guide to the designer, Table 4.12 illustrates typical machine-finish allowances.

Ductile Iron

Various alloys of gray cast iron are in use today. Iron, with sufficient additions of magnesium (in the form of Mg Fe Si or Mg Ni) or cerium can form what we refer to as ductile or nodular cast iron. These alloys have greater strength and ductility and compete well with forged steel.

Having several times greater tensile strength than ordinary cast iron plus greatly increased ductility and shock resistance, ductile iron combines such advantages of cast iron as availability, ease of founding, and machinability with many of the product advantages of steel. The presence of small amounts of magnesium in ductile iron produces graphite in spheroids. The elimination of the weakening effect of flake graphite gives the magnesium-containing iron excellent engineering properties; it has partcularly high tensile strength, elastic modulus, yield strength, toughness, and ductility. Under stress, ductile iron behaves elastically like steel, having proportionality of stress to strain up to high loads. It has a modulus of elasticity of about 25 million lb/in.^{2} Table 4.11 presents the principal types of ductile iron and its applications.

DESIGN AND PROCESS SELECTION

The basic part geometry for ferrous materials can usually be obtained by using more than one possible basic process. The question of when a casting (and what type of casting) should be used versus a plate (or perhaps a weld fabrication) is a recurrent problem for the product engineer as well as the process engineer. The resolution of this question often determines the profitability of a company.

Figures 4.21 and 4.22 show the relationship of component size and surface finish as a function of the various processing methods. The two ends of

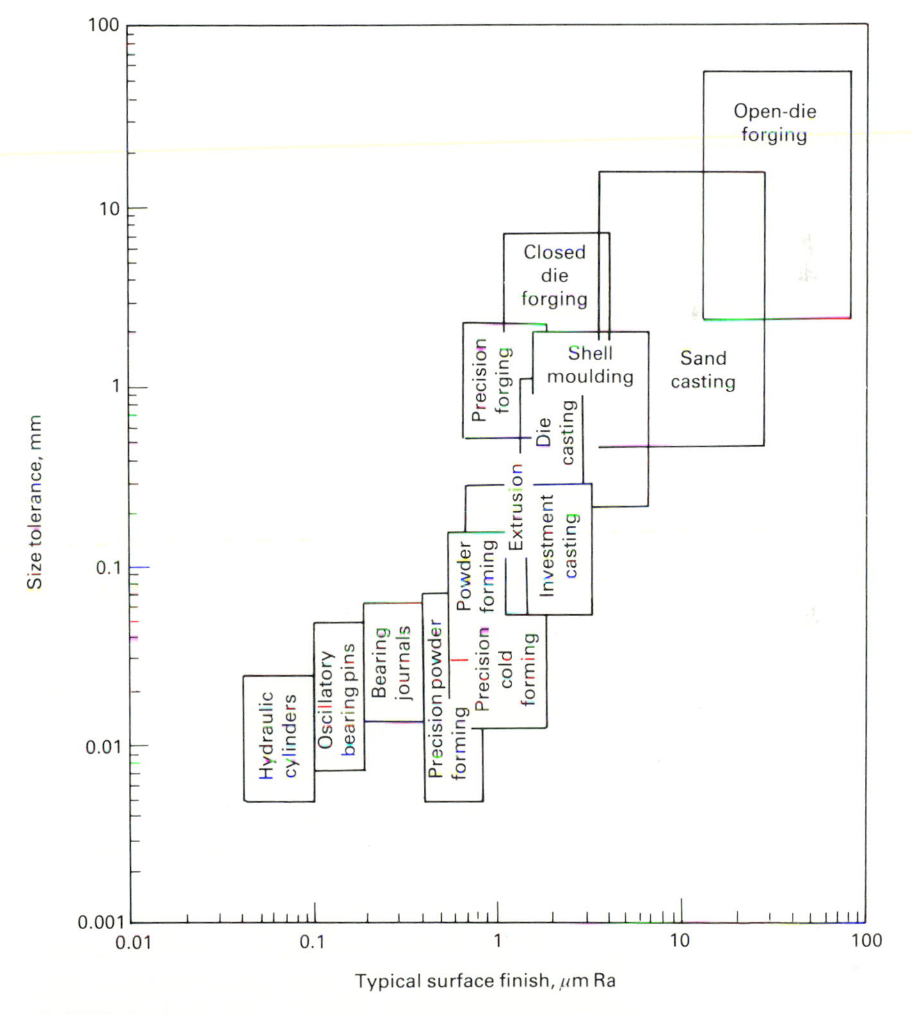

FIGURE 4.21
Interaction of tolerance, component size, and various manufacturing process. (Redrawn from Manufacturing Engineering [7]).

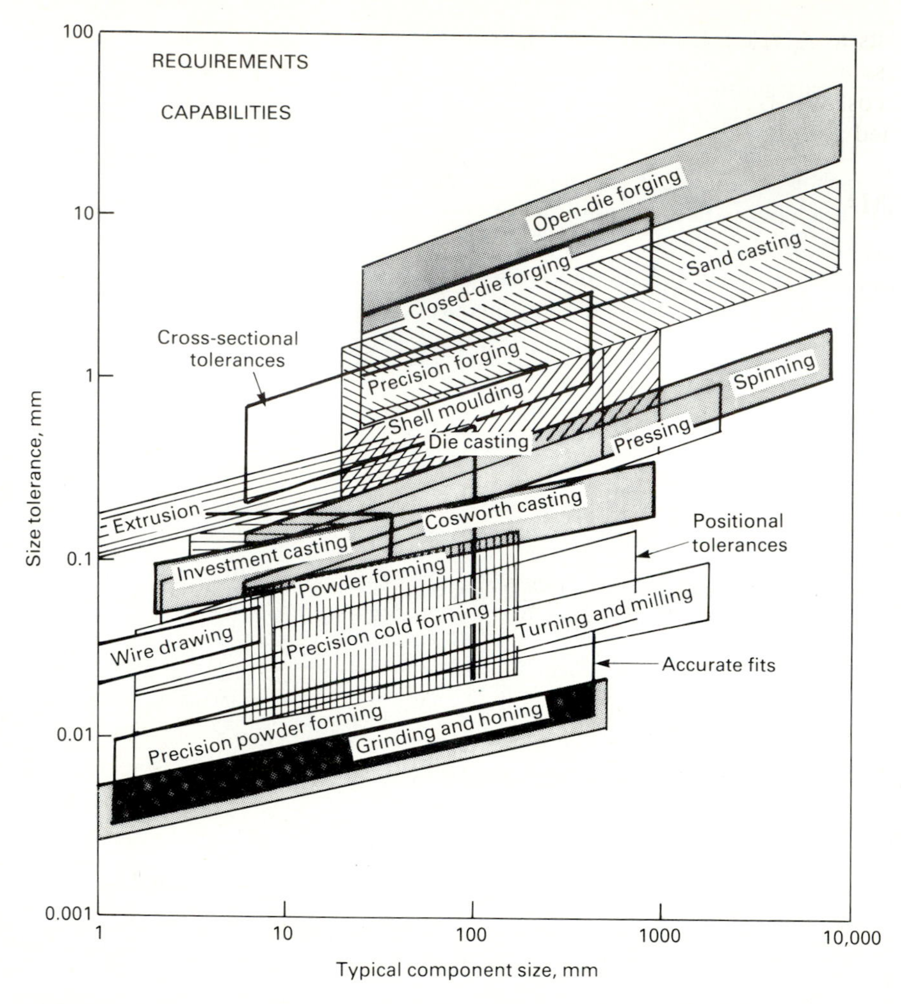

FIGURE 4.22
Surface finish and tolerances obtainable from various processes. These are compared to requirements for some sample components. (Redrawn from Manufacturing Engineering [7]).

the spectrum would be "open die forging" and "precision powder forming." While open die forging produces large bulky items that usually require additional processing, precision powder forming is usually used to produce small reasonably accurate items requiring little or no additional processing. In forming and casting, the accuracy of the process does not always dictate the cost of the item. If few parts are to be made, then one could generally conclude that increased accuracy and increased cost would go hand and hand. However, the geometry of the part and the quantity of parts required usually contribute more significantly. Thin web parts with complex geometric features are normally difficult to engineer, making the fixed costs very high. Incremental costs are usually lower. One general conclusion that can be drawn is that the less accurate casting and forming methods require the smaller set up or fixed cost. Additional processing may, however, be required to obtain the specified accuracy.

SUMMARY

Ferrous materials are certainly the most widely used and most important to the majority of product design engineers. It is essential that the engineer understand the usefulness of ferrous materials, their potentialities and available properties, and how they can be used to the best advantage. He/she should have the knowledge to get the most out of one specific type of cast iron or steel and be certain what that "most" is. He/she should also understand the significant factors that interfere with the attainment of these useful properties.

Mechanical properties, as ordinarily revealed by the tensile, notched-bar, fatigue, hardness, and wear tests, are important in correct material selection. Likewise, metallurgical characteristics, such as grain size and hardenability, are important criteria in material selection. Furthermore, shop factors, such as welding and machinability, have a direct bearing on steel selection as well as the cost and availability of the ideal analysis.

The use of statistics can prove quite helpful in the selection of a material or a material treatment. For example, Student's t distribution provides information on whether a sample whose mean value is $\bar{x}$ could have come from a population whose mean value is μ and whose standard deviation is σ.

QUESTIONS

4.1. Why is cast iron frequently used in the construction of large jigs and fixtures?

4.2. What properties of cast iron make it a much-used material for frames and beds of machine tools?

4.3. What is ductile iron? How is it produced?

4.4. What basic concepts assist the production-design engineer in the selection of a steel?

4.5. What are the principal characteristics of carbon steel? Of alloy steel? Of tool steel?

4.6. What does the term "stainless steel" denote?

4.7. What is meant by "camber" tolerances?

4.8. How would you recommend the finishing of the outside diameter of a piece of steel of 300 Brinell?

4.9. What would be the approximate tensile strength of a piece of steel with a Brinell reading of 180?

4.10. Differentiate between the Rockwell, Brinell, and scleroscope hardness test. What would be the equivalent value of Brinell 200 on the Rockwell and scleroscope scales?

4.11. What does "toughness" imply in a specimen of steel?

4.12. What is the most important factor in the heat treatment of steel?

4.13. What would be the maximum hardness of a piece of steel with 0.50 percent carbon?

4.14. Describe the Jominy test.

4.15. What would be the approximate hardness of a $1\frac{5}{8}$-in.-diameter chromium–nickel bar $\frac{7}{8}$ in. from the outside surface? Assume a 0.75 percent carbon content and a quench into a warm brine after heat treatment.

4.16. Why is water usually preferred as a quenching medium?

4.17. How may case-hardening be accomplished?

4.18. How is machinability determined?

4.19. About what Brinell hardness is thought of as being desired for general machining work?

4.20. How can a piece of steel be checked to determine whether internal stresses will cause warpage?

4.21. Describe the two allotropic forms of iron.

4.22. Explain why the iron–carbon equilibrium diagram is important to the production-design engineer.

4.23. A class 50 gray-iron casting has a wall thickness of $\frac{1}{2}$ in. It is to be machined to a 100 root-mean-square surface finish. What recommendations can you make?

PROBLEMS

4.1. Compare the structure and mechanical properties of AISI 1020, 1050, 1080, and 1.2 percent carbon steels in the fully annealed state and as-quenched.

4.2. Clearly differentiate between the following

(*a*) Substitutional and interstitial solid solutions
(*b*) Eutectic and eutectoid compositions
(*c*) Peritectic and peritectoid compositions
(*d*) Ferrite and austenite
(*e*) Pearlite and cementite
(*f*) Martensite and bainite

4.3. A farm-machinery company received field complaints that an SAE 1020 steel lever deflected excessively when a celery picker was operated at high speed. A quick check showed that the operating load on the lever was well within the yield strength of the material, but the dynamic force imposed by inertia caused by the mass and acceleration of the lever in operation created a problem. The design engineer decided to change to an aluminum alloy for the lever to reduce its mass and thereby to decrease the inertial forces and the deflection.

(*a*) Derive an analytical relationship to test the validity of the proposed solution.
(*b*) Explain what properties of a material affect its resistance to deflection.
(*c*) Would changing from a low-carbon steel to a high-strength, heat-treated alloy steel solve the problem? Explain your answer.
(*d*) Compare the deflection of levers of the same design but made from (1) aluminum, (2) gray iron, (3) malleable iron, (4) magnesium, (5) steel, (6) bronze.
(*e*) In what ways, other than changing the material, could the stiffness of the lever be increased?

(*f*) What material(s) could be used to provide an inherently stiffer lever of the original design? Would such materials be economically feasible?

4.4. Traditionally, only nonferrous alloys have been die cast, but a number of steel castings weighing up to a pound or more have been successfully produced by the die-casting process. Several potential die materials have been investigated, three of which are TZM, a molybdenum alloy; SiN, silicon nitride; and W–Cu, a powdered metal compact of tungsten and copper.

(*a*) Compare the heat absorption by each of those die materials when a flat plate $3 \times 3 \times \frac{1}{4}$ in. thick is die cast from 18–8 stainless steel that is injected from the shot cylinder at 2780°F (about 100°F superheat), and from aluminum alloy when injected into an identical cavity in an H-13 alloy die at 100°F superheat. The cavity is cut $\frac{1}{8}$ in. into each die half.

(*b*) Find the steady-state die temperature of a plane $\frac{1}{4}$ in. from the flat bottom of each die cavity. Of course, that temperature would be only a rough comparison between materials and not the temperature reached in practical die-casting operations. Base your calculations on a $\frac{1}{4} \times \frac{1}{4}$ in. element at the center of the die cavity and consider the latent heat of fusion.

(*c*) Determine the solidification time for each alloy in each die material assuming quasi-steady-state conditions.

(*d*) On the basis of your calculations, discuss the relative merits of each die material from the standpoint of heat transfer compared with that of aluminum alloy in H-13 dies.

Problem 4.4: Physical properties of the materials

Material	Conductivity, Btu/(hr · ft · °F)	Density, lb/ft³	Specific heat, Btu/(ft³ · °F)	Thermal diffusivity, ft²/h	Latent heat, Btu/lb	Injection temperature, °F
TZM	101.8	633.4	0.07	2.34		
W–Cu	81.4	1076	0.036	2.09		
H-13	30.2	490.8	0.12	0.52		
SiN	13.1	208.2	0.24	0.27		
304 Stain St.	20.0	501.1	0.12	0.33	117.0	2780
43 Aluminum	82.2	171.6	0.23	2.09	149.5	1250

SELECTED REFERENCES

1. American Society for Metals: *Metals Handbook*, vol. 7, "Atlas of Microstructure of Industrial Alloys," 8th ed., Metals Park, Ohio, 1972.
2. American Society for Metals: *Tool Steels*, 3d ed., Metals Park, Ohio, 1962.
3. Briggs, Charles W.: *Steel Castings Handbook*, Steel Founders' Society of America, 1960.
4. Dumond, T. C. (ed.): *Engineering Materials Manual*, Reinhold, New York, 1954.
5. Walton, Charles F.: *Gray Iron Castings*, Gray Iron Founders' Society, Inc., 1968.
6. U.S. Steel Corp.: *The Making, Shaping and Treatment of Steel*, Pittsburgh, 1987.
7. Ludema, K. C., Caddell, R. M., and Atkins, A. G.: *Manufacturing Engineering: Economics and Processes*, Prentice-Hall, Englewood Cliffs, N.J., 1987.
8. Doyle, L. E., Keyser, C. A., Leach, J., Schrader, G. F., and Singer, M.: *Manufacturing Processes and Materials for Engineers*, Prentice-Hall, Englewood Cliffs, N.J., 1986.
9. Van Vlack, L. E.: *Elements of Material Science*, Addison-Wesley, Reading, Mass., 1977.

CHAPTER 5

NONFERROUS METALS

Although nonferrous metals comprise only approximately 15 percent by volume of all metals used in commercial production, some are of major importance, such as aluminum to the aircraft industry, lead in storage batteries, or chromium for plating. A comparison based on the dollar value of the finished product would show that nonferrous metals hold a considerably higher place because they are used for components that have high finished value.

The unique properties of aluminum, copper, zinc, lead, magnesium, nickel, cobalt, chromium, and titanium make them useful for components of a large number of designs. In fact about 40 nonferrous metals are of commercial value today. Nonferrous alloys are most often used for parts in which there is considerable labor added in the form of secondary operations or in applications where their unique properties, such as high resistance to corrosion, justify their higher cost, Usually, a combination of several properties is a sufficient basis for selecting a nonferrous alloy; these include ease of fabrication, light weight, corrosion resistance, good machinability, electrical or thermal conductivity, color, ability to absorb energy, or good strength–weight ratio.

ALUMINUM[1]

Aluminum alloys are the most important nonferrous alloys. The art and science of fabricating and applying aluminum has evolved a vast store of information with which the engineer must be familiar. Aluminum today serves as a basic raw material for more than 20,000 businesses throughout the country. The probability that aluminum will emerge as the most viable source of material used in automobile manufacturing by the year 2000 is significant. Since 1982, the aluminum and automobile industry have been working together to develop a strong, lightweight automobile for eventual high-volume production. In 1985, a prototype of an aluminum automobile body was displayed. This prototype includes an aluminum alloy structure with aluminum exterior body panels that reduce the weight of the car by 288 lb when compared to similar designs in steel. By the time the 1991 cars are produced, it is estimated there will be a 35 percent increase over the 1988 models. It is anticipated that cast, forged, and extruded components, including those used in power trains, drive lines, wheels, and suspension systems will be converted to aluminum alloys.

The selection of an aluminum-based alloy for use in a particular product or structure is usually based on one or more of the following design characteristics: (1) when weight–strength ratio is important, that is, in applications where lightness of the final product is desirable (aluminum has a specific gravity of 2.70); (2) when ease of machining, fabrication, or forming is of major importance; (3) where resistance to atmospheric corrosion or attack by certain chemicals is required; (4) where low electrical resistance is a requisite; (5) when high heat and/or light reflectivity or low emissivity is needed; (6) where such properties as acoustical deadness, nontoxicity (concerning foods), or nonsparking or nonmagnetic properties are desirable.

Aluminum alloys are produced in practically all of the forms in which metals are used: plate, sheet, rod, wire, tube, forgings, castings, and ingots, as well as rivets and screw-machine products. Shapes may be the standard structural shapes or they may be of special design that can be produced only by extrusion. Forgings may be made by pressing or by the drop-forging method. Castings may be poured into sand, in cast-iron molds, or in special plaster or other refractory material, or may be made by forcing the molten alloy into a steel die under pressure. The various forms are fabricated into finished shapes and structures by drawing, stamping, spinning, hammering, machining, welding, brazing, riveting, and, in some cases, soldering. Ease of fabrication is one of the reasons for the choice of aluminum alloys.

Notice that the reference is to aluminum alloys. High-purity aluminum, while it has many desirable characteristics, is soft and ductile: it has a tensile

[1] Much of the material on aluminum has been provided by the Aluminum Association from *Aluminum Standards and Data,* 1984.

strength of about 13,000 lb/in.^{2} Even though this strength can be approximately doubled by cold working, the resulting strength is still not high and aluminum is not heat-treatable.

Pure aluminum finds little use in the production of castings, not only because of its low strength, but also because of its inferior casting qualities. The only castings practically produced from commercially pure aluminum are those requiring the higher electrical conductivity of the pure metal. Two examples are induction-motor rotors and cable clamps.

Greater strength in aluminum is achieved by the addition of other elements to produce alloys and then either heat treating or work-hardening the material depending upon the particular alloy. Today, aluminum alloys having tensile strengths approaching 100,000 lb/in.^{2} are available.

Designation System for Aluminium Alloys and Wrought Aluminum

The wrought aluminum designations follow a four-digit index system in which the first digit indicates the alloy type, the second digit indicates the alloy modifications or impurity limits and the last two digits identify the aluminum alloy or indicate the aluminum purity. The table below gives the various types of aluminum alloys and indicates the corresponding first digit in the aluminum identification.

Aluminum—99.00 percent mininum	1xxx
Copper	2xxx
Manganese	3xxx
Silicon	4xxx
Magnesium	5xxx
Magnesium and silicon	6xxx
Zinc	7xxx
Other element	8xxx
Unused series	9xxx

The effects of the alloying elements are as follows:

1000 series. Aluminum of 99 percent or higher purity has many applications, especially in the electrical and chemical fields. These compositions are characterized by excellent corrosion resistance, high thermal and electrical conductivity, low mechanical properties, and excellent workability. Moderate increases in strength may be obtained by strain-hardening. Iron and silicon are the major impurities.

2000 series. Copper is the principal alloying element in this group. These alloys require solution heat treatment to obtain optimum properties; in the heat-treated condition, mechanical properties are similar to, and sometimes exceed, those of mild steel. In some instances, artificial aging is employed to further increase the mechanical properties. This treatment materially increases yield strength with attendant loss in elongation; its effect on tensile (ultimate) strength is not as great. The alloys in the 2000 series do not have as good corrosion resistance as most other aluminum alloys, and under certain conditions they may be subject to intergranular corrosion. Therefore, these alloys in the form of sheet are usually clad with a high-purity alloy or a magnesium–silicon alloy of the 6000 series that provides galvanic protection to the core material and thus greatly increases resistance to corrosion. Alloy 2024 is perhaps the best known and most widely used aircraft alloy.

3000 series. Manganese is the major alloying element of alloys in this group, which are generally not heat-treatable. Because only a limited percentage of manganese, up to about 1.5 percent, can be effectively added to aluminum, it is used as a major element in only a few instances. One of these, however, is the popular 3003, which is widely used as a general-purpose alloy for moderate strength applications requiring good workability.

4000 series. The major alloying element of this group is silicon, which can be added in sufficient quantities to cause substantial lowering of the melting point without producing brittleness in the resulting alloys. For these reasons, aluminum–silicon alloys are used in welding wire and as brazing alloys where a lower melting point than that of the parent metal is required. Most alloys in this series are not heat-treatable, but when used in welding heat-treatable alloys they will pick up some of the alloying constituents of the latter and so respond to heat treatment to a limited extent. The alloys containing appreciable amounts of silicon become dark gray when anodic oxide finishes are applied, and hence are in demand for architectural applications.

5000 series. Magnesium is one of the most effective and widely used alloying elements for aluminum. When it is used as the major alloying element or with manganese, the result is a moderate-to high-strength alloy that is not heat-treatable. Magnesium is considerably more effective than manganese as a hardener, about 0.8 percent magnesium being equal to 1.25 percent manganese, and it can be added in considerably higher quantities. Alloys in this series possess good welding characteristics and good resistance to corrosion in marine atmospheres. However, certain limitations should be placed on the amount of cold work and the safe operating temperatures permissible for the alloys of higher magnesium content (over about 3.5 percent for operating temperatures above about 150°F) to avoid susceptibility to stress corrosion.

6000 series. Alloys in this group contain silicon and magnesium in approximate proportions to form magnesium silicide, thus making them heat-treatable. Major alloys in this series is 6061, one of the most versatile of the heat-treatable alloys. Though less strong than most of the 2000 or 7000 alloys, the magnesium–silicon (or magnesium-silicide) alloys possess good formability and corrosion resistance, with medium strength. Alloys in this heat-treatable group may be formed in the T4 temper (solution heat-treated but not artificially aged) and then reach full T6 properties by artificial aging.

7000 series. Zinc is the major alloying element in this group, and when coupled with a smaller percentage of magnesium results in heat-treatable alloys of very high strength. Usually, other elements such as copper and chromium are also added in small quantities. The outstanding member of this group is 7075, which is among the highest-strength alloys available and is used in airframe structures and for highly stressed parts.

In the aluminum 99.00 percent minimum and greater group, the last two of the four digits indicate the minimum aluminum percentage. These two digits are the same as the two digits to the right of the decimal point in the minimum aluminum percentage when it is expressed to the nearest 0.01 percent. For example, if the minimum aluminum were 99.2 percent, the last two digits would be "20". The second digit uses zero to indicate aluminum that is not alloyed and has natural impurity limits. The integers 1 through 9 are used to indicate the special control of one or more impurities or alloying elements.

In the 2xxx to 9xxx alloy groups, the last two digits serve to identify the different alloys in the group. The second digit is used to indicate alloy modifications.

A tempered designation system is used for all forms of wrought and cast aluminum and aluminum alloys except ingot. The tempered designation follows the alloy designation; a hyphen being used to separate the two. The basic designations are:

F *As fabricated.* Applies to the products of shaping processes in which no special control over thermal conditions or strainhardening is employed. For wrought products, there are no mechanical property limits.

O *Annealed.* Applies to wrought products that are annealed to obtain the lowest-strength temper, and to cast products that are annealed to improve ductility and dimensional stability. The O may be followed by a digit other than zero to indicate a product in the annealed condition having special characteristics.

H *Strain-hardened* (wrought products only). Applies to products that have their strength increased by strain-hardening, with or without supplementary thermal treatments to produce some reduction in strength. The H is always followed by two or more digits.

W *Solution heat-treated.* An unstable temper applicable only to alloys that spontaneously age at room temperature after solution heat-treatment. This designation is specific only when the period of natural aging is indicated; for example W $\frac{1}{2}$ h.

T *Thermally treated to produce stable tempers other than F, O, or H.* Applies to products that are thermally treated, with or without supplementary strain-hardening, to produce stable tempers. The T is always followed by one or more digits.

The following subdivisions of "H temper: Strain-hardened" are used.

H1 *Strain-hardened only.* Applies to products that are strain-hardened to obtain the desired strength without supplementary thermal treatment.

The number following this designation indicates the degree of strain-hardening.

H2 *Strain-hardened and partially annealed.* Applies to products that are strain-hardened more than the desired final amount and then reduced in strength to the desired level by partial annealing. For alloys that age-soften at room temperature, the H2 tempers have the same minimum ultimate tensile strength as the corresponding H3 tempers. For other alloys, the H2 tempers have the same minimum ultimate tensile strength as the corresponding H1 tempers and slightly higher elongation. The number following this designation indicates the degree of strain-hardening remaining after the product has been partially annealed.

H3 *Strain-hardened and stabilized.* Applies to products that are strain-hardened and whose mechanical properties are stabilized either by a low-temperature thermal treatment or as a result of heat introduced during fabrication. Stabilization usually improves ductility. This designation is applicable only to those alloys that, unless stabilized, gradually age-soften at room temperature. The number following this designation indicates the degree of strain-hardening remaining after the stabilization treatment.

Temper designation of an aluminum alloy employs the letter O, the letter F, and letters H and T followed by one or more numbers. The letter follows the alloy designation and is separated from it by a hyphen. The letter O indicates the annealed temper of wrought materials and T2 indicates the annealed temper of casting materials. Temper designation F, in the case of a wrought alloy, indicates the as-fabrication condition. (This indicates no control has been exercised over the temper of the alloy.)

Temper designation T followed by one or more numbers indicates a heat-treated alloy. The first digit following the T signifies the type of heat treatment given the alloy. The heat treatments available, and their symbols, are:

T3 Solution heat treatment followed by strain-hardening. Different amounts of strain-hardening of the heat-treated alloy are indicated by a second digit.

T4 Solution heat treatment followed by natural aging at room temperature to a substantial stable condition.

T5 Artificial aging after an elevated-temperature, rapid-cool fabrication process such as casting or extrusion.

T6 Solution heat treatment followed by artificial aging.

T7 Solution heat treatment and then stabilization to control growth and distortion.

T8 Solution heat treatment, strain-hardening, and then artificial aging.

T9 Solution heat treatment, artificial aging, and then strain-hardening.

Castings

As in the wrought aluminum and aluminum alloy designation system, a four digit designation is used to identify aluminum alloys in the form of castings and foundry ingot. Here, the alloy groups identified by the first digit are as tabulated below.

Aluminum, 99.0% and greater	1xx.x
Copper	2xx.x
Silicon, with added copper and/or magnesium	3xx.x
Silicon	4xx.x
Magnesium	5xx.x
Zinc	7xx.x
Tin	8xx.x
Other elements	9xx.x
Unused series	6xx.x

The next two digits are used to identify the aluminum alloy or indicate the aluminum purity. The fourth digit (which is separated from the other three by a decimal point) indicates whether the form is an ingot or a casting (0 indicates castings and 1 indicates ingot).

The improvements in the casting qualities of the aluminum alloys as compared with the pure metal are perhaps even greater than the improvements in their mechanical properties. The commercial casting alloys, containing varying percentages of one or more alloying elements, have been developed to combine good qualities from the standpoint of foundry operations with desirable mechanical properties in the finished castings.

Sand castings are most generally used when the quantity requirement is small, or if the casting requires intricate coring, or if the castings are very large. The minimum thickness of sections that can be cast in sand depends on the size and intricacy of the casting, the pressure tightness, and the casting alloy. The minimum thickness for small medium-sized castings is $\frac{3}{16}$ in., although small castings with $\frac{1}{8}$-in. wall thickness have been made. Tolerances in the order of $\frac{1}{32}$ in. are practical in small castings but should increase with increased size. If machined, a casting is normally given $\frac{1}{8}$-in. allowance, but for large castings, $\frac{1}{4}$ in. or more is frequently provided.

The minimum wall thickness of castings poured in permanent molds and semipermanent molds is substantially the same as sand castings ($\frac{3}{16}$ in.). However, tolerances and machining allowances need be only one-fourth as large. Molds should have a minimum draft of 1 to 5°; 3 to 5° is preferable.

Relatively thin and uniform sections are desirable for die casting. These castings may weigh a fraction of an ounce or as much as 80 lb. For castings having dimensions up to about 6 in., 0.045-in. sections are practical; for 15-in. castings, 0.080-in. sections are practical; for larger dimensions, 0.150-in. sections are usually used. Tolerances of 0.002-in./in. can readily be maintained.

The principal casting alloys of aluminum are copper, silicon, magnesium, zinc, and nickel. The specific gravity of the various alloys does not vary significantly and, for practical purposes, all alloys may be considered to weigh about 0.10 lb/in.^{3}. All alloys have a modulus of elasticity of approximately 10,300,000 lb/in.^{2}, a modulus of rigidity of 3,900,000 lb/in.^{2} and Poisson's ratio of 0.33.

It is apparent, in view of the low modulus of elasticity as compared with ferrous materials, that deeper sections must be used when a design indicates loading as a beam. However, the lower modulus of elasticity is an asset when greater deflections are desired under a given load, such as in a relay contact flexible support and when greater resilience is desired. Resilience is the amount of energy absorbed per unit weight of material upon loading within the elastic limit. Resilience is directly proportional to the yield strength and indirectly proportional to the modulus of elasticity. Thus, aluminum with a lower modulus of elasticity can absorb more energy than steel with a higher modulus of elasticity at the same stress.

Creep in aluminum castings will take place at or near the yield strength but will not usually occur at stresses lower than half the yield-strength value.

The majority of the aluminum casting alloys are welded easily; thus, castings may be joined readily with other aluminum components.

Alloys

Since the desirable qualities associated with aluminum depend on the alloy, the next question is, "What are these alloying elements?" While a variety of alloying elements are used in the production of aluminum alloys, copper, magnesium, silicon, manganese and, more recently, zinc are the more common. In some special-purpose alloys, nickel, tin, lead, and bismuth are added, while in many alloys, both cast and wrought, titanium or chromium comprise an important part of the composition. Generally speaking, the total alloy content is greater in casting alloys than wrought alloys.

The aluminum–silicon alloys have the best castability qualities but their mechanical properties are somewhat inferior to those aluminum alloys in which copper or magnesium are the principal alloying agents.

Nickel is used mainly to aid in maintaining mechanical properties at elevated temperatures.

Heat-treatable alloys contain at least one element that is more soluble in aluminum at higher temperatures than at room temperature. Copper is the element most frequently used in the heat-treatable alloys, although magnesium, zinc and silicon are also used extensively.

The initial strength of an alloy that is not heat-treatable is dependent upon the hardening effect of elements such as manganese, silicon, iron, and magnesium. These alloys are work-hardenable and further strengthening is accomplished by various degrees of cold working, as indicated by the "H" series of tempers.

Tables 5.1 and 5.2 provide some of the important properties of both heat-treatable and non-heat-treatable alloys. It should be recognized that aluminum and its alloys lose strength at elevated temperatures although some alloys maintain good strength to temperatures up to 500°F. At subzero temperatures, their strength increases without loss of ductility.

Heat Treatment

An important feature of some aluminum alloys is that their strength can be varied by heat treating. The metal can be annealed to make it extremely soft and easily worked during drawing and forming. After it has been shaped the metal can be given another (and different heat) treatment to make it extremely strong and hard, so that the surface is more resistant to wear. By properly selecting the alloying elements and combining heat treatments with small amounts of cold work, strengths in excess of 80,000 lb/in.^{2} can be obtained. Heat treatment of aluminum alloys is covered in Chapter 3.

Plate, Sheet, and Foil

Aluminum that has been rolled to a thickness of more than 0.25 in. is classified as plate, while thicknesses of 0.006 to 0.25 in. are known as sheet. Foil refers to aluminum that has been rolled to thicknesses less than 0.006 in. Today, foil is being produced to thicknesses of less than 0.0002 in.

Sheet of 0.102 in. or less is available in both flat and coiled form. Usually there is a slight price advantage to purchasing sheet in coil form. If required, sheet can be purchased (at extra cost) from the mill with a bright finish. The degree of brightness will, of course, depend upon the grade and the temper specified. The standard mill finish will vary from a dull gray to a relatively bright finish, depending on the process conditions. It is possible to procure sheet with a standard mill finish on one side and a bright finish on the other.

Properties

Aluminum is not affected by magnetic fields; it is nonsparking, and is considered acoustically dead (desirable for such uses as in ducts where transmission of noise is objectionable). Aluminum is about three times as flexible as steel.

The information given on the accompanying Tables (5.1–5.5) is for the guidance of the engineer when choosing materials. Such information is limited and should be supplemented by current catalogs and price lists.

Toxicity. Pure aluminum is nontoxic and will not change the flavor, purity, or color of food. It is thus suitable for use in processing, shipping, and cooking foods and beverages. Long and careful research has proved that there is no such thing as aluminum poisoning. As a matter of fact, traces of aluminum are

TABLE 5.1

Aluminum non-heat-treatable—properties

TYPE	COMMERCIALLY PURE ALUMINUM												
P.D. SPEC. NO.	7601-1 2 S	7601-2 2 S-O	7601-3 2 S	7601-4 2 S-O	7601-5 2S-1/2 H	7601-6 2 S-H	7601-12 2 S-O	7601-13 2S-1/2 H	7601-14 2 S-H	7601-7 2 S-O	7601-15 2S-1/4 H	7601-8 2S-1/2 H	7601-9 2 S-H
	BAR			FLAT STRIP			SHEET AND PLATE			TUBING			
GRADE	AS FABRICATED COLD FIN.	SOFT-COLD FIN. OR ROLLED	AS FABRICATED ROLLED	SOFT	HALF HARD	HARD	SOFT	HALF HARD	HARD	SOFT	QUARTER HARD	HALF HARD	HARD
TENSILE STRENGTH ① (LB/SQ. IN. IN 1,000)	17	15.5 Max.	15	15.5 Max.	16 Min.	22 Min.	15.5 Max.	16 Min.	22 Min.	15.5 Max.	14 Min.	16 Min.	22 Min.
YIELD STRENGTH ② (LB/SQ. IN. IN 1,000)	14	4	13	4	14	20	4	14	20	4	12	14	20
SHEAR STRENGTH ③ (LB/SQ. IN. IN 1,000)	11	9.5	10	9.5	11	13	9.5	11	13	9.5	10	11	13
ENDURANCE LIMIT ④ (LB/SQ. IN. IN 1,000)	7	5	6	5	7	8.5	5	7	8.5	5	6	7	8.5
AVERAGE ELONGATION % IN 2 IN. ⑤	15 to 25	25 Min.	20 to 30	15 to 30	2 to 6	1 to 4	15 to 30	3 to 7	1 to 4	15 to 30	4 to 10	2 to 6	1 to 4
BRINELL HARDNESS	32	23	28	23	32	44	23	32	44	23	28	32	44
BEND TEST SPECIMEN MIN. DIAM. ON PIN	180° Flat	180° Flat	180° Flat	180° Flat	180° Flat	180° 3 X Tks.	180° Flat	180° Flat	180° 3 X Tks.				
ELEC. COND. % OF COPPER	58	59	57	59	58	57	59	58	57	59	58	58	57
THERMAL COND. AT 100° C. C.G.S. UNITS	.52	.53	.51	.53	.52	.51	.53	.52	.51	.53	.52	.52	.51

TYPE	MANGANESE ALUMINUM ALLOY										
P.D. SPEC. NO.	7602-1 3 S	7602-2 3 S-O	7602-3 3 S	7602-4 3 S-O	7602-12 3S-1/4 H	7602-5 3S-1/2 H	7602-13 3S-3/4 H	7602-6 3 S-H	7602-7 3 S-O	7602-8 3S-1/2 H	7602-9 3 S-H
	BAR			FLAT STRIP SHEET AND PLATE					TUBING		
GRADE	AS FABRICATED COLD FIN.	SOFT-COLD FIN. OR ROLLED	AS FABRICATED ROLLED	SOFT	QUARTER HARD	HALF HARD	THREE QUARTER HARD	HARD	SOFT	HALF HARD	HARD
TENSILE STRENGTH ① (LB/SQ. IN. IN 1,000)	21	19 Max.	18	14.5 to 19	17 Min.	19.5 Min.	24 Min.	27 Min.	19 Max.	19.5 Min.	27 Min.
YIELD STRENGTH ② (LB/SQ. IN. IN 1,000)	18	5	15	5	14	18	20	25	5	18	25
SHEAR STRENGTH ③ (LB/SQ. IN. IN 1,000)	14	11	12	11	12	14	15	16	11	14	16
ENDURANCE LIMIT ④ (LB/SQ. IN. IN 1,000)	9	7	8	7	8	9	9.5	10	7	9	10
AVERAGE ELONGATION % IN 2 IN. ⑤	10 to 20	25 Min.	15 to 25	20 to 25	4 to 9	3 to 8	1 to 4	1 to 4	20 to 25	3 to 8	1 to 4
BRINELL HARDNESS	40	28	35	28	35	40	47	55	28	40	55
BEND TEST SPECIMEN MIN. DIAM. ON PIN	180° 1 X Tks.	180° Flat	180° Flat	180° Flat	180° Flat	180° Flat	180° 4 X Tks.	180° 3 X Tks.			
ELEC. COND. % OF COPPER	41	50	42	50	42	41	41	40	50	41	40
THERMAL COND. AT 100° C. C.G.S. UNITS	.37	.45	.38	.45	.39	.37	.38	.36	.45	.37	.36

TABLE 5.1 (Contd.)

DESIGN CHARACTER-ISTICS	Maximum corrosion resistance. Can be torch, arc, and spot welded without appreciable loss in properties. Used for equipment to be left with bright finish. If painting is required, prepare surface by sand blasting, scratch brushing or etching. Can be anodized for additional corrosion resistance or decoration. Electrical characteristics good. Machining characteristics fair. Application: General use bright finish aluminum.	Corrosion resistance comparable to aluminum #7601. Can be torch, arc and spot welded without appreciable loss in properties. Can be painted without preparation of surface because of gray finish. Can be anodized for additional corrosion resistance. Cost same as aluminum #7601. Machining characteristics fair. Application: General use gray finish aluminum. Chemical Properties: 1.0 to 1.5% Manganese.
The above materials are produced in the following sizes: 7601-1: Rounds 1-1/2 diam. or less Squares & Hexagons 1/32 to 1-1/2 across flats Rectangles 1/16 x 1/8 to 1-1/2 x 4 7601-2: Cold Finish--Same as 7601-1 Rolled--Same as 7601-3 7601-3: Rounds 1-9/16 to 8 diam. Hexagons 1-9/16 to 2 across flats Squares 1 to 4 square Rectangles 3/32 x 1-1/8 to 3 x 7 7601-4: Thicknesses .006 to .102 for Flat Strip 7601-5: Thicknesses .010 to .085 for Flat Strip 7601-6: Thicknesses .006 to .102 7601-12: Thicknesses .010 to .249 for Sheet .250 and over for Plate 7601-13: Thicknesses .010 to .249 for Sheet .250 and over for Plate 7601-14: Thicknesses .010 to .128 for Sheet .250 and over for Plate		7602-1: Same as 7601-1 7602-2: Same as 7601-2 7602-3: Same as 7601-3 7602-4: Thicknesses .006 to .102 for Flat Strip .010 to .249 for Sheet .250 and over for Plate 7602-12: Thicknesses .017 to .102 for Flat Strip .017 to .249 for Sheet .250 and over for Plate 7602-5 Thicknesses .010 to .085 for Flat Strip .010 to .249 for Sheet .250 and over for Plate 7602-13: Thicknesses .006 to .053 for Flat Strip .010 to .162 for Sheet .250 and over for Plate 7602-6: Thicknesses .004 to .102 for Flat Strip .006 to .128 for Sheet .250 and over for Plate

Values specified for tensile, yield and shear strength and endurance limit are averages, unless otherwise specified.

① Young's Modulus of Elasticity is approximately 10,300,000 lb/sq.in.

② Stress which produces a permanent set of 0.2% of initial gage length (A.S.T.M. Specification E8-40T).

③ Single shear strength values obtained by double shear test.

④ Based on withstanding 500,000,000 cycles of completely reversed stress, using the R.R. Moore type of machine and specimen.

⑤ Elongation values vary with size and shape.

TABLE 5.2
Aluminum, heat-treatable—properties

TYPE	COPPER-MAGNESIUM-MANGANESE-ALUMINUM ALLOYS														
	7603-1 17 S-O	7603-2 17 S-T	7603-7 17 S-T	7603-3 17 S-O	7603-4 17 S-T	7603-5 17 S-O	7603-6 17 S-T	8490-1 24 S-O	8490-2 24 S-T		8490-4 24 S-T	8490-8 24 S-RT	8490-5 24 S-O	8490-6 24 S-T	8490-7 24 S-RT
	BAR			SHEET AND PLATE		TUBING		BAR		FLAT STRIP, SHEET & PLATE					
GRADE AND TEMPER	SOFT ANNEALED	HEAT TREATED	HEAT TREATED ROLLED	SOFT ANNEALED	HEAT TREATED	SOFT ANNEALED	HEAT TREATED	SOFT ANNEALED	HEAT TREATED	SOFT ANNEALED	HEAT TREATED	ROLLED AFTER HEAT TREATING	SOFT ANNEALED	HEAT TREATED	ROLLED AFTER HEAT TREATING
TENSILE STRENGTH ① (LB/SQ IN. IN 1,000)	35 Max.	50 to 55	50 to 53	35 Max.	55 Min.	35 Max.	55 Min.	35 Max.	57 to 70	35 Max.	58 to 62	68	35 Max.	64 Min.	70
YIELD STRENGTH ② (LB/SQ IN. IN 1,000)	10	28 to 30	28 to 30	10	32 Min.	10	40 Min.	10	40 to 52	10	40 Min.	50 Min.	10	42 Min.	58 Min.
SHEAR STRENGTH ③ (LB/SQ IN. IN 1,000)	18	36	36	18	35	18	35	18	41	18	41	42	18	41	42
ENDURANCE LIMIT ④ (LB/SQ IN. IN 1,000)	11	15	15	11	15	11	15	12	18	12	18		12	18	
AVERAGE ELONGATION % IN 2 IN. ⑤	16 Min.	16 to 18	16 Min.	12 Min.	10 to 18	10 to 12	12 to 16	16 Min.	10 to 14	12 Min.	6 to 17	10 to 12	12 Min.	10 to 16	10 Min.
BRINELL HARDNESS	45	100	100	45	100	45	100	42	105	42	105	116	42	105	116
BEND TEST SPECIMEN MIN. DIAM. ON PIN	180° 1 X Tks.	180° 4 X Tks.	180° 4 X Tks.	180° Flat to 6 X Tks.	180° 3 to 8 X Tks.			180° Flat to 1 X Tks.	180° 6 X Tks.	180° Flat to 6 X Tks.	4 to 10 X Tks.	4 to 9 X Tks.			
ELEC. COND. % OF COPPER	45	30	30	45	30	45	30	50	30	50	30	30	50	30	30
THERMAL COND. AT 100° C C.G.S. UNITS	.41	.28	.28	.41	.28	.41	.28	.45	.28	.45	.28	.28	.28	.28	.28

	7603 alloys	8490 alloys
DESIGN CHARACTERISTICS	Corrosion resistance inferior to aluminum #7601. Can be spot welded without appreciable loss of properties. If spot welding is not practical, riveted construction is recommended. For high efficiency joints hot steel rivets are preferred. Can be painted without special surface preparation. Annealed material can be formed and then heat treated to obtain maximum properties. More expensive than #7601. Machines freely. Application: High strength construction, frames, rotor wedges, screw machine products. 3.5 to 4.5% Cu, .20 to .75 Mg, .40 to 1.0 Mn	Corrosion resistance comparable to aluminum alloy #7603. Can be spot welded without appreciable loss of properties. If spot welding is not practical, riveted construction is recommended. For high efficiency joints hot steel rivets are preferred. Can be painted without special surface preparation. Annealed material can be formed and then heat treated to obtain maximum properties. Slightly more expensive than #7603. Machines freely. Application: Similar to #7603. 3.8 to 4.9% Cu, 1.2 to 1.8 Mg, .3 to .9 Mn

The above materials are produced in the following sizes:

- 7603-1: Rounds 8 diam. or less; Squares & Hexagons 1/32 to 4 across flats; Rectangles 1/16 to 1/8 to 3 x 7
- 7603-2: Same as 7603-1
- 7603-7: Rounds 8 diam. or less; Squares & Hexagons 1/32 to 4 across flats; Rectangles (round edge) 1/8 x 5/8 to 1/2 x 6
- 7603-3 and 7603-4: Thicknesses .010 to .249 for Flat Strip & Sheet; .250 and over for Plate
- 8490-1: Same as 7603-1
- 8490-2: Same as 7603-2
- 8490-3 and 8490-4: Same as 7603-3
- 8490-8: .019 to .249 for Flat Strip and Sheet; .250 and over for Plate

Values specified for tensile, yield and shear strength and endurance limit are averages, unless otherwise specified.

① Young's Modulus of Elasticity is approximately 10,300,000 lb/sq.in.

② Stress which produces a permanent set of 0.2% of initial gage length (A.S.T.M Specification E-8-40T).

③ Single shear strength values obtained by double shear test.

④ Based on withstanding 500,000,000 cycles of completely reversed stress, using the R.R. Moore type of machine and specimen.

⑤ Elongation values vary with size and shape.

TABLE 5.3
Tolerances of rolled aluminum structural shapes

Dimensions	Tolerance
Thickness of section	Plus or minus $2\frac{1}{2}$ percent of nominal thickness—minimum tolerance: ± 0.010 in
Overall dimensions; length of leg of angles or zees	Plus or minus $2\frac{1}{2}$ percent of nominal—minimum tolerance: $\pm\frac{1}{16}$ in
Length	
Up to 20 ft not incl.	Minus 0, plus $\frac{1}{4}$ in
20 ft to 30 ft incl.	Minus 0, plus $\frac{3}{8}$ in
Over 30 ft	Minus 0, plus $\frac{1}{2}$ in
Channels, depth	Plus $\frac{3}{32}$ in, minus $\frac{1}{16}$ in
Channels, width of flange	Plus or minus 4 percent of nominal width
Weight of a lot or shipment of sizes 3 in or larger*	Plus or minus $2\frac{1}{2}$ percent of nominal weight

* For sizes under 3 in, dimension tolerances only apply.

found naturally in many foods, including yellow string beans, beets, lettuce, carrots, celery, onions, milk, and calves' liver. The use of aluminum in cooking utensils and for packaging foods is a major market for this material.

Heat and light reflectance, emissivity, thermal conductivity. Aluminum is a good reflector of light and heat. When brightly polished, it will reflect up to 95 percent of the light and up to 98 percent of the infrared energy falling upon it. Aluminum, even when weathered, will reflect 85 to 95 percent of the radiant heat striking it. These properties, desirable in roofing materials and other similar applications, make aluminum ideal for use in heat reflectors. The emissivity of aluminum is extremely low (4 to 5 percent of a perfect blackbody radiator); for this reason it is used in hot-air ducts and similar applications where low heat loss is desirable. Aluminum has high thermal conductivity, thus making it an effective material for heat exchangers in the aircraft, food, petroleum and other industries.

Weight. The advantages of lightness can be readily appreciated in a qualitative sense, by engineer and layman alike; the design engineer must also learn to place a quantitative valuation on lightness. There are a few applications in which the value of "lightness" can be computed directly. An aircraft is an example of this type of product. The commercial airline operators have calculated the cost of carrying each pound of airframe. The exact cost depends of course, on the type of service in which the particular aircraft is used. In most products the value of lightness cannot be measured directly in terms of

TABLE 5.4
Aluminum and aluminum alloy sheet and plate—some thickness tolerances (Plus or Minus)[a]

	Width, in				
Thickness, in	18 or less	Over 18 to 36 (incl.)	Over 36 to 48 (incl.)	Over 48 to 54 (incl.)	Over 54 to 60 (incl.)
0.007 to 0.010	0.001	0.0015			
0.011 to 0.017	0.0015	0.0015			
0.018 to 0.028	0.0015	0.002	0.0025		
0.029 to 0.036	0.002	0.002	0.0025		
0.037 to 0.045	0.002	0.0025	0.003	0.004	0.005
0.046 to 0.068	0.0025	0.003	0.004	0.005	0.006
0.069 to 0.076	0.003	0.003	0.004	0.005	0.006
0.077 to 0.096	0.0035	0.0035	0.004	0.005	0.006
0.097 to 0.108	0.004	0.004	0.005	0.005	0.007
0.109 to 0.140	0.0045	0.0045	0.005	0.005	0.007
0.141 to 0.172	0.006	0.006	0.008	0.008	0.009
0.173 to 0.203	0.007	0.007	0.010	0.010	0.011
0.204 to 0.249	0.009	0.009	0.011	0.011	0.013
0.250 to 0.320	0.013	0.013	0.013	0.013	0.015
0.321 to 0.438	0.019	0.019	0.019	0.019	0.020
0.439 to 0.625	0.025	0.025	0.025	0.025	0.025
0.626 to 0.875	0.030	0.030	0.030	0.030	0.030
0.876 to 1.125	0.035	0.035	0.035	0.035	0.035
1.126 to 1.375	0.040	0.040	0.040	0.040	0.040
1.376 to 1.625	0.045	0.045	0.045	0.045	0.045
1.626 to 1.875	0.052	0.052	0.052	0.052	0.052
1.876 to 2.250	0.060	0.060	0.060	0.060	0.060
2.251 to 2.750	0.075	0.075	0.075	0.075	0.075
2.751 to 3.000	0.090	0.090	0.090	0.090	0.090

* Grades: Alclad 145; Alclad 245, 525, 615, and 755.

dollars, but nonetheless the engineer must evaluate the intangible benefits in cases where weight of product can be reduced at a cost. Weight is related to wear on bearings and surfaces in sliding contact in machines, for the forces resulting from the motion of the parts take the form of added pressure on these surfaces.

In bridge spans of extreme length, the truss weight is greater than the

TABLE 5.5
OD and wall tolerances of aluminum and aluminum alloy extruded tubing (Plus or Minus)

OD, in	Allowable deviation of mean dia. from specified dia.	Allowable deviation of dia. at any point from specified dia.	Specified wall thk., in	Allowable deviation of mean wall thk. from specified wall thk.			Allowable deviation of wall thk. at any point from mean wall thk. (eccentricity)
				Under 3 OD	3 to 5 OD	5 OD and over	
0.50–0.99	0.010	0.020	0.062 and less	0.007	0.008	0.010	Plus or minus 10% of mean wall thickness; max. ± 0.060 min. ± 0.010
1.00–1.99	0.012	0.025	0.063–0.124	0.008	0.010	0.015	
2.00–3.99	0.015	0.030	0.125–0.249	0.009	0.013	0.020	
4.00–5.99	0.025	0.050	0.250–0.374	0.011	0.016	0.025	
6.00–7.99	0.035	0.075	0.375–0.499	0.015	0.021	0.035	
8.00–9.99	0.045	0.100	0.500–0.749	0.020	0.028	0.045	
10.00–11.99	0.055	0.125	0.750–0.999		0.035	0.055	
12.00–12.25	0.065	0.150	1.000–1.490		0.045	0.065	

weight of useful load supported by the bridge; the structural members are, in effect, mainly supporting themselves. Here reduced weight of members would permit reduced loads on other members and the reduction would have a multiplying effect. Aluminum structural members offer interesting possibilities in this area, but little work has been done in this field so far.

The engineer should recognize two types of weight problems in design, and should use two sets of criteria in selecting light alloys. First, there is a so-called *strength–weight criterion*. In this type of problem, the strength requirements are set, and the problem is to find the material that will meet these requirements without adding any more weight than necessary. In this case the designer will consult a strength-to-weight ratio table and select the material with the highest strength-to-weight ratio that meets all requirements in other respects.

The other type of design problem is one in which the interest is in weight per unit volume. For many parts, almost any of the common aluminum alloys will meet the strength requirements because other design considerations necessitate the use of a large volume of metal. Cylinder blocks and aircraft wheels are examples of this type of product. In this case, cost and other design considerations will determine the choice of material.

This distinction is quite important to the designer. High-strength aluminum alloys are generally more expensive, and harder to machine and weld, and are often less corrosion-resistant than the low-strength alloys. The high-strength alloys will also require heat treatment if extensive forming is to be done.

Density and specific gravity values are, of course, dependent upon composition. The density (lb/in.^{3}) of aluminum alloys varies from 0.095 for 5056 to 0.099 for 3003. The respective specific gravities for these two alloys are 2.64 and 2.73. Thus the mass (weight) of aluminum is about 35 percent that of iron and 30 percent that of copper.

Machining. Aluminum is about the cheapest of all common metals to machine (magnesium can be machined slightly faster). The table below gives a comparison of relative cutting speeds for aluminum and various other metals.

Material	Relative cutting speed (drilling and turning)
Aluminum alloys	0.5–1.0
Steel, mild (up to 0.3 percent)	0.25–0.35
Brass or bronze	0.6–1.0
Magnesium and its alloys	0.9–1.3

Actually, in most instances, the equipment controls the cutting speed, as the material can be machined faster than the speed at which the facility is capable of running. The process designer for general machining of aluminum should employ high machining speeds and moderate feeds and depths of cut.

Where extensive forming and machining operations are required to produce a part, manufacturing costs can be substantially lowered if aluminum can be substituted for a ferrous metal. The reduction in manufacturing costs may more than offset the higher cost of the raw material.

Forming. Most aluminum alloys can be drawn, forged, or extruded, but the high-strength, heat-treatable alloys must be annealed before extensive cold working is done. This will in turn necessitate heat treating the alloy after forming to restore strength properties.

A substantial savings in both cost and weight that may result from the use of extruded or forged parts instead of machined or fabricated members should not be overlooked by the designer. Aluminum extrusions in a wide variety of shapes are stocked by distributors; for volume production, specially made extrusions and forgings can be ordered from aluminum producers. Extruded parts are almost always stronger and lighter than comparable machined or fabricated members; forgings are likewise stronger than castings or sections machined from solid stock.

Fabricating. Aluminum can be joined in numerous ways, but riveting is probably the most common process. It is usually preferable to use rivets made of the same alloy as the material being joined. When designing for riveted joints, the holes that are drilled, pierced, or subpunched and reamed to size to take the rivets should provide a clearance between the inside diameter and the rivet. The best clearance (approximately 0.010 in.) is the smallest that will allow the rivet to be inserted readily. Hot rivets require more clearance than cold so that they can be inserted without delay.

All of the aluminum alloys can be joined by some form of welding. Spot welding is widely used to join aluminum sheet. When this process is selected it is necessary to remove any protective oxide from the surfaces of the metal being joined so as to prevent electrode pickup and produce sound, consistent welds. Since spot welding is a form of resistance welding and aluminum has high electrical conductivity, extremely high welding amperage is required to produce ample heat for fusion. Current requirements are between 13,000 and 36,000 A (amperes), depending upon the alloy and the thickness of the material. Electrode pressures of 58,000 lb/in.^{2} immediately before and after and 23,000 lb/in.^{2} during the welding process are typical.

Aluminum is also satisfactorily joined by the gas, arc, and inert-gas-shielded arc processes; however, these techniques should not be selected for sections less than 0.032 in. in thickness. When welding thick sections, it is wise to preheat the material to around 600°F so as to prevent cracking. This procedure will also simplify bringing the base material to a molten condition at the time the molten filler material is introduced.

Today, adhesive bonding of aluminum parts is widely used. This joining procedure is particularly evident in both the aircraft and automotive industries.

Corrosion resistance. Aluminum is highly resistant to progressive atmospheric oxidation. Pure aluminum readily forms an oxide, but this oxide is tough and protects the base metal from further oxidation. For this reason, it is not necessary to give aluminum or the near-pure aluminum alloys any protective coating for service under ordinary atmospheric conditions. If the service conditions of the design result in erosion or abrasion, the protective oxide will be removed and corrosion will be accelerated.

Several aluminum (mainly, high-strength alloys in which copper or zinc are major alloying constituents) will corrode even under normal conditions. To overcome this difficulty, such alloys are sometimes coated with pure aluminum. The alloys sold under the trade name Alclad fall in this category. The cladding is usually between 3 and 5 percent of the total thickness on each side.

Substantial savings in painting costs may be realized from the use of aluminum instead of the more easily corroded ferrous metals. Aluminum roofing has been in use for some time and prefabricated aluminum wall panels have recently come into limited use. In this latter application, speed and ease of erection have been the chief selling points.

Aluminum is far from being chemically inert, but there are numerous compounds that will attack steel and not aluminum. Aluminum thus has found many applications in the chemical industry; piping in chemical plants, containers for shipping chemicals, and tank cars for chemical shipment serve as common examples of this use.

Care must be exercised to avoid direct contact of aluminum with other metals in the presence of moisture so as to avoid galvanic corrosion. Cadmium or zinc plating on ferrous materials will reduce the corrosive effect.

Electrical properties. Aluminum is one of the best electrical conductors known. It is inferior in this respect only to gold, silver and copper. The relative conductivity of electric conductivity grade (1350) aluminum is about 62 percent that of copper. Conductivity is measured on a unit-area basis, and copper weighs 3.3 times as much as aluminum. An aluminum conductor with the same resistance as a comparable copper conductor has only about 40 percent of the weight of the copper conductor. Both metals are sold by the pound and aluminum is the cheaper of the two metals. For power-line wire, aluminum conductors are wound on steel wire, the latter providing the strength required for long unsupported spans. The long spans that are made possible by this combination of a light conductor and strong steel wire permit the use of approximately 30 percent fewer supporting poles and towers on transmission lines.

Aluminum conductors should also be considered for use on aircraft and other applications where weight is important. It must be remembered, though, that aluminum is more difficult to solder than copper, consequently its use to form good electrical connections can be a problem.

MAGNESIUM

Magnesium, with a specific gravity of only 1.74 is the lighest metal available that is stable under ordinary conditions. It is 64 percent the weight of aluminum and 23 percent the weight of iron. The chief source is sea water, which contains about 0.13 percent magnesium.

Like many other pure metals, magnesium must be alloyed to give maximum strength and usefulness. Although lightness is its outstanding characteristic, it has other properties that are equally significant. Magnesium is extremely easy to machine and is adaptable to practically all the usual methods of metal working. It can be joined by gas, arc, and electrc-resistance welding and by riveting, it can be cast by the sand, permanent-mold, and die-casting methods, it can be painted, plated, or anodized. Other properties include good stability to inland atmospheric exposure and resistance to attack by alkalies and most oils. The alloys of magnesium have relatively low electrical resistivity and high thermal conductivity. Magnesium is also nonmagnetic.

Magnesium as an alloy is more expensive than aluminum. It is more difficult to produce both in wrought and cast forms. Its tensile strength is less and it must be heat treated to get good strength characteristics. If weight represents a critical item, or if much machining is required, then the selection of magnesium should be considered. Standard ASTM alloys are identified by the following to show the percentage of the principal alloying elements:

A Aluminum
M Manganese
S Silicon
K Zirconium
H Thorium
Q Silver
Z Zinc

The alloy AZ92A would contain 9 percent aluminum and 2 percent zinc. Table 5.6 provides nominal composition of magnesium sheet and plate and extrusion alloys.

Uses

Sand castings of magnesium find widespread use in aircraft engines and aircraft landing wheels, where light weight, high strength, and shock resistance are of prime importance. In high-speed rotating and reciprocating parts, its light weight is used to effect smoother operations. Every principal American aircraft power plant uses numerous magnesium sand castings for parts such as the

TABLE 5.6

Nominal composition and physical properties of magnesium alloys

Alloy and temper	Nominal composition—%						Density lb/in.^{3} 70°F	Melting point °F	Thermal conductivity cgs units, 70°F	Electrical resistivity μΩ·cm 70°F
	Al	Mn	Th	Zn	Zr	Mg				
(TCH) Sheet and plate alloys										
AZ31B-H24									0.18	9.2
-H26	3.0	—	—	1.0	—	Bal.	0.0639	1160	0.18	9.2
-0									0.18	9.2
HK31A-H24	—	—	3.0	—	0.7	Bal.	0.0647	1202	0.27	6.1
-0									0.25	6.6
HM21A-T8 -T81	—	0.8	2.0	—	—	Bal.	0.0640	1202	0.33	5.0
(TCH) Extrusion alloys[a]										
AZ31B, or C-F	3.0	—	—	1.0	—	Bal.	0.0639	1160	0.18	9.2
AZ61A-F	6.5	—	—	1.0	—	Bal.	0.0647	1140	0.14	12.5
AZ80A-F	8.5	—	—	0.5	—	Bal.	0.0649	1130	0.12	14.5
-T5	8.5	—	—	0.5	—	Bal.	0.0649	1130	0.12	14.5
HM31A-T5	—	1.2	3.0	—	—	Bal.	0.0651	1202	0.25	6.6
ZK21A 1	—	—	—	2.3	0.6	Bal.	0.0645	—	0.30	5.4
ZK60A 1	—	—	—	5.5	0.5	Bal.	0.0659	1175	0.28	6.0
-T5	—	—	—	5.5	0.5	Bal.	0.0659	1175	0.29	5.7

[a] For extrusion alloys, Mn values are min.

blower and supercharger housings, gear cases, carburetor bodies, ignition harnesses, oil sumps, air-induction systems, and numerous small parts and covers. Of particular current interest is the extensive use of magnesium parts in turbojet engines; parts of the air-compression system are built almost entirely of magnesium castings. An example of the use of magnesium castings is the cast magnesium aircraft wing section made by the Northrop Aircraft Corporation. The casting is 16 ft long and adheres to exacting tolerance.

In 1976, there were no magnesium diecastings used in the typical American car. Ten years later (1986), the average American-built motor car used three pounds of magnesium diecastings. It is estimated that the typical 1991 car will use about six pounds of magnesium in the form of diecastings.

The excellent surface finish, reduction in machining, and low cost have caused magnesium die castings to be widely used in such applications as instrument parts, various kinds of housings, vacuum cleaner parts, and innumerable other parts where light weight is essential.

Characteristics

The outstanding characteristic of magnesium is, of course, its light weight, which is only one-quarter that of iron and two-thirds that of aluminum. Electric conductivity on a volume basis is about 38 percent that of standard copper and 60 percent that of pure aluminum. Like aluminum, magnesium can be fabricated by any of the industrial methods, although its strong tendency to oxidize rapidly makes precautions necessary in foundry, welding, and machining processes. Magnesium and its alloys are the most readily machinable of all industrial metals, and deep cuts can be made at high speeds. The amount of cold work that can be done on magnesium alloys is limited by its close-packed hexagonal structure. Because of its rather narrow plastic range, close temperature control is required. There is grain growth with repeated heating, thus facilitating hot working. It is readily deep drawn at temperatures of approximately 600°F. Commercially pure magnesium is of little value as a construction material, having a tensile strength of only 14,000 $lb/in.^2$ when cast and 25,000 $lb/in.^2$ when rolled. However, the tensile strength of some of its alloys is in excess of 50,000 $lb/in.^2$ (see Table 5.7).

Tensile strengths and yield strengths decrease at elevated temperatures to the extent that magnesium alloys are usually unsatisfactory for use at temperatures exceeding 400°F. The strength of magnesium alloys is not changed significantly at subnormal temperatures. The modulus of elasticity of magnesium alloys is 6,500,000 $lb/in.^2$ In view of this low modulus of elasticity, magnesium alloys have a good capacity for energy absorption.

Magnesium alloys have thermal conductivities lower than those of aluminum alloys but greater than those of ferrous alloys. The coefficient of thermal expansion is approximately twice that of ferrous alloys, being about 0.0000145/°F. Thus allowances must be made in designs for the thermal-expansion differential between magnesium and other members if an assembly involves dissimilar metals and temperature variations will be encountered.

Magnesium and its alloys have satisfactory resistance to corrosion by solutions of most alkalies and by many organic chemicals. They are attacked by practically all common acids except chromic and hydrofluoric. They are also attacked by salt solutions. Protective coatings are essential for any magnesium alloy used in the vicinity of salt water and are recommended in all cases for protection against atmospheric corrosion when the installation is not inland. Small amounts of iron or nickel in magnesium–aluminum alloys increase susceptibility to corrosion, so these impurities are eradicated as far as possible in the extraction processes. Because magnesium is higher in the electrochemical series than copper and iron, magnesium in contact with either of those metals or their alloys, especially in the presence of moisture, is subject to severe galvanic corrosion. Magnesium's good sound-damping characteristics make it a useful shielding material.

Available Alloys

Magnesium is alloyed readily with most elements except iron and chromium. However, aluminum is the most commonly used alloying element and is used in amounts varying from 3 to 10 percent. Alloying magnesium with aluminum results in alloys with increased strength, hardness, rigidity, and refined grains. The greater the percentage of aluminum added, the harder the alloy.

Zinc and manganese are used as alloying materials in quantities to 3 and 0.3 percent respectively in order to provide greater resistance to corrosion.

Diecasting alloys typically contain aluminum, manganese, and silicon, while wrought alloys contain only aluminum and manganese. Forgings usually contain aluminum, manganese, and zinc.

Forms

Magnesium alloys are available as forgings, extruded shapes, sheets, plates, strips, and castings, including sand, permanent-mold, and die casting. When design requirements specify lightness, but the components are subjected to high or repeated impact stresses, a magnesium forging may be the most appropriate fabricated form. As with other metals, forging results in a fine-grained, homogeneous structure. In the forging of magnesium, hydraulic presses are usually used because more dependable forgings are obtained when forging is done at slow speeds.

Magnesium can be extruded easily and is available in standard shapes, including round, square, rectangular, and hexagonal. The maximum width of extrusions is approximately 12 in. Extruded tubing is available with ODs from $\frac{3}{16}$ to 12 in. and with walls ranging from 0.022 to 1 in. Sheet, plate, or strip are available in the annealed, as-rolled, and hard conditions. Sheet magnesium alloy is available in thickness ranging from 0.016 to 0.219 in. and in widths from 24 to 48 in. Magnesium sheets can be welded (arc, gas, or spot welded), formed, or riveted. Table 5.8 provides a summary of forming methods used on magnesium.

TABLE 5.7
Room-temperature mechanical properties of magnesium bars, rods, and shapes

Alloy and temper[a]	Least dimension, in.	Area, $in.^2$	TS 1000 $lb/in.^2$		TYS 1000 $lb/in.^2$		E in 2 in.,%		CYS 1000 $lb/in.^2$		Shear, 1000 $lb/in.^2$	Bearing		Approx. Brinell hardness
			Typ.	Min.	Typ.	Min.	Typ.	Min.	Typ.	Min.	typ.	Ult. 1000 $lb/in.^2$ typ.	Yield 1000 $lb/in.^2$ typ.	
AZ31B-F	Under 0.250	—	38	35	28	21	14	7	15	—	19	56	34	—
	0.250–1.499	—	38	35	29	22	15	7	14	12	19	56	33	49
	1.500–2.499	—	38	34	28	22	14	7	14	12	19	56	33	—
	2.500–5.000	—	38	32	28	20	15	7	14	10	19	56	33	—
AZCOML[b]	Under 0.250	—	37	—	27	—	13	—	14	—	—	—	—	—
	0.250–1.499	—	37	—	28	—	13	—	13	—	—	—	—	—
	1.500–2.499	—	36	—	27	—	13	—	13	—	—	—	—	—
	2.500–5.000	—	36	—	27	—	14	—	13	—	—	—	—	—
AZ61A-F	Under 0.250	—	46	38	33	21	17	8	—	—	23	65	38	—
	0.250–2.499	—	45	40	33	24	16	9	19	14	22	68	40	60
	2.500–5.000	—	45	40	31	22	15	7	21	14	22	68	42	—
AZ80A-F	Under 0.250	—	49	43	36	28	12	9	—	—	22	68	48	60
	0.250–1.499	—	49	43	36	28	11	8	—	17	22	68	48	60

	1.500–2.499	—	49	43	35	28	11	6	—	17	22	68	48	60
	2.500–5.000	—	48	42	36	27	9	4	—	17	22	68	48	60
AZ80A-T5	Under 0.250	—	55	47	38	30	8	4	34	—	24	60	57	82
	0.250–1.499	—	55	48	40	33	7	4	35	28	24	60	58	82
	1.500–2.499	—	53	48	39	33	6	4	32	27	24	66	54	82
	2.500–5.000	—	50	45	38	30	6	2	31	26	24	68	53	82
HM31A-T5	—	Under 1.000	44	37	38	26	8	4	27	19	22	62	49	—
	—	1.000–3.999	43	37	34	26	13	4	23	15	21	62	44	—
ZK21A-F	—	Under 5.000	42	38	33	28	10	4	25	20	21	60	46	—
ZK60A-F	—	Under 2.000	49	43	38	31	14	5	33	27	24	76	56	75
	—	2.000–2.999	49	43	37	31	14	5	28	26	24	76	50	75
	—	3.000–4.999	49	43	36	31	14	5	27	25	24	76	49	75
	—	5.000–39.999	48	43	37	31	9	6	23	20	24	75	44	75
ZK60A-T5	—	Under 2.000	53	45	44	36	11	4	36	30	26	79	59	82
	—	2.000–2.999	52	45	43	36	12	4	31	28	26	78	53	82
	—	3.000–4.999	51	45	42	36	14	4	30	25	25	77	52	82
	—	5.000–9.999	—	45	—	34	—	6	—	23	—	—	—	—
	—	10.000–24.999	—	45	—	34	—	6	—	22	—	—	—	—
	—	25.000–39.999	—	43	—	31	—	6	—	20	—	—	—	—

[a] Temper, F = as extruded, T5 = artificially aged.

[b] Typical properties, no guaranteed minimum values for this alloy.

TABLE 5.8
Summary of forming methods used on magnesium

Type of forming	Basic equipment	Accessory equipment	Heating methods	Dies
Bending	Leaf or press brake bending presses; bending rolls; tube benders	Flexible mandrel, sand, lead shot	Ovens; electrically heated press brake blade holder	Insulated top, bottom dies; metal dies with polished edges
Press drawing	Hydraulic presses; mechanical presses	Draw and blank holder rings; graphitic and dry soap lubricants	Gas or electric heaters in dies; ovens, platen heaters for sheet	Heated dies—open type punch and die sets, mating dies
Rubber forming	60–70 Durometer main rubber pads for el. temp. use	Rubber throw pads	Blank and form block oven heated, or platen is heated	Mg and Al form blocks
Dimpling	Die sets	Automatic temperature and speed controls	Sheet preheated in oven	Electrically heated dies
Stretch forming	Various types of stretch presses	Gripper jaws, wiper shoes, synthetic rubber or glass wool blankets; graphitic lubricant	Electric resistance or infrared heating	Form die and mandrel
Spinning	Hardwood or metal chucks, sometimes segmented	Standard tools, mandrels; graphitic, soap or paraffin-tallow lubricant	Hand torches, or burner on spinning lathe	Form blocks
Impact extrusion	Vertical or horizontal mechanical or hydraulic presses	Automatic feeding equipment, lubricant applied by tumbling	Slug heated on automatic heated feed track	Tool steel punches: 1-pc. tool steel dies or 3-pc. carbide die
Drop hammer forming	Pneumatic hammers; gravity type hammers	Rubber pads; vegetable-lecithin oil lubricant	Oven heating of stock and large dies; ring burners or blow torches	Zinc alloy or cast iron
Forging, coining, embossing	Hydraulic and slow acting mechanical presses; standard forging hammers	Trimming press or bandsaw; graphitic lubricant	Stock, heated, in ovens; dies by gas, electric resistance, or heat transfer materials	Typical die steels for 400 to 900°F
Hand forming	Soft hammers, wood or metal form blocks	Torches; soft jawed vises; clamps	Torch, oven or electric heaters	Wood or metal form blocks

Speeds	Pressures	Formability	Temperature	Uses	Remarks
36.3-in. strokes per minute maximum			From room temperature to 325–550°F		High-speed process; little springback
Up to 50 feer per minute	50 to 100 psi cushion pressure	Up to 70% 65% is a practical limit	300 to 600°F	High production of shallow and deep-drawn parts	Very deep single stage draws possible
	900 psi average	Stretch flange limits about 25%	325°F typical on AZ31B-H24	Small number of shallow parts of intricate shape	For room and el. temp. forming lub. seldom needed
Average dwell time 2–3 seconds			Dies at 550–600°F	For flush riveting	Heat control critical on -H24
			Sheet at 270 to 340°F; die below 450°F	For low curvature asymmetrical parts	Uniform properties in large sections; heat control critical on -H24
	Manual and machine		600°F (nominal)	Symmetrical shapes, e.g., cones	Cheap tooling; high worker skill
Hand fed 10 to 20 pcs/min automatic 75 to 100	25 to 35 tons/in.^{2} to reduce area 85%	95% reduction in area has been made	350–700°F	High production of deep 1-pc. parts, e.g., battery cans	Variable wall thickness, sharp corner radii
		Nominal 10% reduction	Dies up to 450°F. Parts made at 450–500°F	Shallow draws and asymmetrical shapes	For multistage parts
Slow speeds	15 tons/in.^{2} is min. pressure on hydraulic press	Up to 80% reduction in height	550 to 750°F; forged colder	High strength, long life, or reproducible quality	High production, sound structure
				Intricate contours	Few parts, prototypes

The characteristics of good design of magnesium-alloy castings are similar to those of aluminum castings. To prevent stress concentration leading to fatigue failure, sharp internal corners, gouges in surfaces, and other surface irregularities in stressed areas should be avoided. Ample fillets should be used around bosses, but not to the extent that the mass is such as to cause draws and cracks. Bosses should be located so that they can be risered or chilled to secure maximum soundness if this is desired. The thin and heavy sections must be so blended as to avoid a sharp transition line. Wall thickness in sand and permanent-mold castings preferably should not be less than $\frac{5}{16}$ in. but may be $\frac{3}{16}$ in. in limited areas. In die castings, walls may go down to 0.050 in.

Machining

Magnesium alloys may usually be machined at very high speeds. In view of its low flash point, a fire hazard exists in the grinding and machining of magnesium alloys when a finely divided magnesium dust is produced. In order to guard against this danger, coolants of neutral, high-flash-point mineral oil are recommended.

Cutting tools should be ground with large clearances (10 to 12°) and extra-smooth tool faces. It is not difficult to maintain a good smooth finish while machining magnesium alloys. Magnesium machines easily, even though it has a hexagonal close-packed structure, because the temperature in the cutting zone is above its recrystallization temperature.

Joining

Magnesium alloys can be joined by most of the common joining techniques including welding, riveting, adhesives, and soldering. Of these, riveting is the most commonly used method. Aluminum-alloy rivets are used, since magnesium rivets would become exceedingly hard after being driven because of the work-hardening characteristics of magnesium. In this connection, it is advisable to drill the rivet holes in the magnesium rather than pierce then so as to avoid work-hardening of the sheet adjacent to the rivet. If it is desirable to pierce the rivet holes, the sheet should be heated to at least 400°F so as to minimize cold work-hardening.

Certain alloys are suitable for gas welding and can be joined by means of oxyacetylene, oxyhydrogen, or oxycarbohydrogen gas. Butt joints are recommended so as to avoid trapping of fluxes that might enhance subsequent corrosion. It should be mentioned that magnesium alloys are not satisfactorily welded to other metals and should not be cut by torch methods.

The inert-gas-shielded arc is the most widely used welding method for joining magnesium alloys. No flux is used in this method, so any type of joint is satisfactory.

Spot, butt, seam, or flash welding may be employed on magnesium alloys. However, these resistance welding methods are only satisfactory when joining magnesium to magnesium.

COPPER AND COPPER-BASED ALLOYS

Copper, one of the first metals used by man, has become one of the most useful metals to the engineer. The major reasons for its importance are: (1) its high electrical conductivity, surpassed only by silver; and (2) the alloys it forms, the most important of which are brasses and bronzes. Brass is the second most commonly used nonferrous alloy. (Table 5.9 shows some of the common coppers and their alloys.)

Properties

Copper is a yellow–red metal having a face-centered cubic crystal structure of specific gravity 8.91 (0.321 lb/in.^{3}) and a melting point of 1083°C. The metal is very malleable; it can be rolled to sheets $\frac{1}{16}''$ thick. It is very ductile and this property improves with purification. Its heat conductivity is next to those of gold and silver and its electrical conductivity is second only to silver (97 percent of silver).

The principal reasons for the widespread industrial use of metallic copper are its ease of working, its resistance to corrosion, the pleasing color of the metal itself and of its alloys, the wide variety of alloys possible, the ability for the metal to be hardened by alloying or by cold working, and its high electrical and thermal conductivity.

Copper is not used extensively in the pure state because of its mechanical properties: pure copper is soft and relatively weak having an as-cast tensile strength of about 19,000 lb/in.^{2} The majority of all pure copper used today is in the form of electrical conductors. Annealed wrought copper has a strength of 32,000 lb/in.^{2} while cold-drawn copper has a tensile strength of approximately 56,000 lb/in.^{2} Pure copper is not easy to cast because it readily absorbs oxygen, thus forming oxides.

Uses

Copper bus bars or buses are used for carrying heavy currents over short distances, while copper wire is used principally for conductivity purposes over longer distances. Copper rod and wire is not restricted to electrical work but is used for bars for locomotive staybolts, woven wire screen or cloth, and welding electrodes. Aluminum is the principal competitor of copper for certain types of electrical conductors.

Copper sheet is used for producing fabricated objects by such processes as cold working, stamping, spinning, welding, soldering, and brazing. Sheet may be used directly for roofing, sheathing, flashing drains, and numerous other purposes.

Copper pipe can be made by rolling copper sheet into a cylinder and welding the seam; or seamless tubing can be made by piercing billets and rolling them over a mandrel. This tube is widely used in radiators, refrigerators, air-conditioning equipment, and similar equipment where maximum heat transfer is desired.

TABLE 5.9
General data for copper-base rod alloys

ALLOY GROUP	ALLOY	NOMINAL COMPOSITION Percent					DENSITY Pounds per Cubic Inch at 68° F	SPECIFIC GRAVITY
		COPPER	LEAD	ZINC	TIN	OTHER		
1	ELECTROLYTIC TOUGH PITCH COPPER	99.90 Min				O 0.04	0.321-.323	8.89-8.94
1	TELLURIUM COPPER	99.5				Te 0.5	0.323	8.94
1	SELENIUM COPPER	99.4				Se 0.6	0.322	8.91
1	LEADED COPPER	99.0	1.0				0.323	8.94
1	TELLURIUM-NICKEL COPPER	98.2				Te 0.5 Ni 1.1 P 0.2	0.323	8.94
2	COMMERCIAL BRONZE, 90%	90.0		10.0			0.318	8.80
2	CARTRIDGE BRASS, 70%	70.0		30.0			0.308	8.53
3	LEADED COMMERCIAL BRONZE	89.0	1.75	9.25			0.319	8.83
3	MEDIUM-LEADED BRASS	64.5	1.0	34.5			0.306	8.47
3	HIGH-LEADED BRASS	62.5	1.75	35.75			0.306	8.47
3	**FREE-CUTTING BRASS**	**61.5**	**3.0**	**35.5**			**0.307**	**8.50**
3	FORGING BRASS	60.0	2.0	38.0			0.305	8.44
3	ARCHITECTURAL BRONZE	57.0	3.0	40.0			0.306	8.47
4	LOW-SILICON BRONZE	96.0 Min	0.05 Max	1.50 Max	1.60 Max	Si 1.5 Max Fe 0.80 Mn 0.75	0.316	8.75
4	PHOSPHOR BRONZE, 5%	95.0			5.0		0.320	8.86
4	LEADED PHOSPHOR BRONZE, 5%	94.0	1.0		5.0		0.322	8.92
4	ALUMINUM-SILICON BRONZE	91.0				Al 7.0 Si 2.0	0.278	7.69
4	FREE-CUTTING PHOSPHOR BRONZE	88.0	4.0	4.0	4.0		0.320	8.86
4	ALUMINUM BRONZE, 9%	87.0 Min				Al 7.0-10.0 Fe 1.5 Max Mn, Ni, Sn 2.0 Max	0.274	7.58
4	ALUMINUM BRONZE, 10%	82.0				Al 9.5 Fe 2.5 Ni 5.0 Mn 1.0	0.274	7.58
4	NAVAL BRASS	60.0		39.25	0.75		0.304	8.41
4	LEADED NAVAL BRASS	60.0	1.75	37.5	0.75		0.305	8.44
4	MANGANESE BRONZE, (A)	58.5		39.2	1.0	Fe 1.0 Mn 0.3	0.302	8.36
5	NICKEL SILVER, 65 - 18	65.0		17.0		Ni 18.0	0.316	8.73
5	LEADED NICKEL SILVER, 61.5 - 10	61.5	1.0	27.5		Ni 10.0	0.314	8.70
5	LEADED NICKEL SILVER, 61.5 - 12	61.5	1.0	25.5		Ni 12.0	0.315	8.73
5	LEADED NICKEL SILVER, 61.5 - 15	61.5	1.0	22.5		Ni 15.0	0.316	8.75
5	LEADED NICKEL SILVER, 61.5 - 18	61.5	1.0	19.5		Ni 18.0	0.317	8.78
5	EXTRUDED LEADED NICKEL SILVER	46.5	2.75	40.75		Ni 10.0	0.306	8.47

Chemical symbols used are as follows: Al Aluminum, Fe Iron, Mn Manganese, Ni Nickel, O Oxygen, P Phosphorus, Se Selenium, Si Silicon, Sn Tin, Te Tellurium

TABLE 5.9 (Contd)

PHYSICAL PROPERTIES				FABRICATION PROPERTIES					
COEFFICIENT OF THERMAL EXPANSION Per 0°F from 68°F to 572°F	ELECTRICAL CONDUCTIVITY (Volume Basis) Percent IACS at 68°F (Annealed)	MELTING POINT (Liquidus) °F	MODULUS OF ELASTICITY (Tension) psi	CAPACITY FOR BEING COLD WORKED	CAPACITY FOR BEING HOT WORKED	RELATIVE POWER REQUIRED FOR BEING HOT WORKED	HOT WORKING TEMPERATURE °F	ANNEALING TEMPERATURE °F	MACHINABILITY RATING Free-Cutting Brass = 100
0.0000098	101.	1981	17,000,000	Excellent	Excellent	Moderate	1400-1600	700-1200	20
0.0000099	90.0	1980	17,000,000	Good	Excellent	Moderate	1400-1600	700-1200	90
0.0000098	99.0	1958	17,000,000	Good	Excellent	Moderate	1400-1600	700-1200	90
0.0000098	99.0	1974	15,000,000	—	—	—	—	—	80
0.0000098	50.0[1]	1980	17,000,000	Good	Good	Moderate	1400-1600	1150-1450	80
0.0000102	44.0	1910	17,000,000	Excellent	Good	Moderate	1400-1600	800-1450	20
0.0000111	28.0	1750	16,000,000	Excellent	Fair	Moderate	1350-1550	800-1400	30
0.0000102	42.0	1900	17,000,000	Good	Poor	—	—	800-1200	80
0.0000113	26.0	1700	15,000,000	Good	Poor	—	—	800-1200	70
0.0000113	26.0	1670	15,000,000	Fair	Poor	—	—	800-1100	90
0.0000114	**26.0**	**1650**	**14,000,000**	**Poor**	**Fair**	**Low**	**1300-1450**	**800-1100**	**100**
0.0000115	27.0	1640	15,000,000	Fair	Excellent	Low	1200-1500	800-1100	80
0.0000116	28.0	1630	14,000,000	Poor	Excellent	Low	1150-1350	800-1100	90
0.0000099	12.0	1940	17,000,000	Excellent	Excellent	Moderate	1300-1600	900-1250	30
0.0000099	18.0	1920	16,000,000	Excellent	Poor	—	—	900-1250	20
0.0000099	16.0	1920	15,000,000	Fair	Poor	—	—	900-1250	50
0.0000100	9.2	1841		Poor	Excellent	—	1300-1600	—	60
—	12.0	1830	15,000,000	Fair	Poor	—	—	900-1250	90
0.0000094	12.6	1908		—	Excellent	Low	1650-1750	1450-1650	20
0.0000094	7.5	1931		Poor	Excellent	Low	1650-1750	1450-1650	20
0.0000118	26.0	1650	15,000,000	Fair	Excellent	Low	1200-1500	800-1100	30
0.0000118	26.0	1650	15,000,000	Poor	Good	Low	1200-1400	800-1100	70
0.0000118	24.0	1630	15,000,000	Poor	Excellent	Low	1150-1450	800-1100	30
0.0000090	6.0	2030	18,000,000	Excellent	Poor	—	—	1100-1500	20
—	—	1830	17,500,000	Fair	Poor	—	—	1100-1300	50
—	7.0	1880	18,000,000	Fair	Poor	—	—	1100-1300	50
—	6.0	1970	18,000,000	Fair	Poor	—	—	1100-1300	50
—	6.0	2010	18,000,000	Fair	Poor	—	—	1100-1300	50
—	8.5	1715	-	Poor		—	—	-	80

[1]Age Hardened.

NOTE: The values listed above represent reasonable approximations suitable for general engineering use. Due to commercial variations in composition and manufacturing limitations, they should not be used for specification purposes.

Copper Alloys

Copper can be alloyed with many elements, including zinc, tin, lead, iron, silver, phosphorus, silicon, tellurium, and arsenic (see Table 5.10). Copper and most of its alloys act like a pure metal. Its alloys are not heat-treatable and cannot be hardened by application of heat followed by a quench. Hardness is produced by cold working, and softening is produced by heating above the recrystallization temperature. Copper-base alloys have good ductility, elongation, electrical conductivity, thermal conductivity, and corrosion resistance. Most alloys are readily drawn, stamped, coined, forged, formed, spun, soldered, brazed, and welded. Also, many of the alloys can be obtained in various tempers and finishes. Copper and its alloys are nonmagnetic and nonsparking. In some instances, these two properties are quite important.

The tensile strength is also good, approaching 80,000 $lb/in.^2$ for several of the alloys (aluminum–bronze, manganese–bronze, and hard bronzes).

BRASS

When copper is alloyed with zinc in various proportions, the resulting material is known as *brass,* which is stronger than either of the materials from which it is made. If the copper crystal structure is face-centered cubic, there will be up to 36 percent of zinc and the solid solution is known as α-brass. When more than 36 percent of the alloy is zinc, a body-centered cubic structure is formed and the solution is known as β-brass. γ-Brass has more than 45 percent zinc. α-Brass combines strength with ductility, while β-brass is relatively high in hardness and is brittle. γ-Brass is difficult to work in both the hot and cold states.

In general, the physical properties of brass depend largely upon manipulation, that is, whether it has been cast, rolled, extruded, drawn, or annealed, to all of which it lends itself readily. Properties such as ductility, springiness, toughness, strength, and stiffness are controlled either by annealing or by the amount of reduction by cold rolling or drawing after the last anneal. The variation of these properties relative to the amount of manipulation can be read from charts issued by the various brass companies.

Machinability, corrosion resistance, hot and cold workability, and other similar properties are controlled largely by modifying the proportions of copper and zinc, and by the addition of small amounts of elements such as lead, tin, silicon, aluminum, nickel, phosphorus, and arsenic. Thus, a wide range of desired physical characteristics can be imparted to brass by variations in composition, heat treatment, and manipulation. The largest measure of satisfaction can be secured when the brass maker works hand in hand with the designer and is cognizant of the exact purpose for which the material is to be employed.

On a basis of overall or final cost, brasses offer a combination of unusual properties for the production of a wide variety of cast, forged, machined, stamped, drawn, or spun articles. The comparatively high initial per-pound price should not influence the choice of material because finishing costs may be lower, especially where buffed or plated surfaces are required. Some of the unique properties that should be taken advantage of are:

1. High ductility
2. Rapid drawing speeds in draw presses and dies
3. Lower maintenance cost on dies
4. Ease of plating or use for plating
5. Ability to be alloyed with many metals to give desirable characteristics
6. Ease of machining, forming, and casting
7. High value of scrap

Brass is one oldest alloys used by man. Every handbook and textbook on materials and alloys covers the essential information. New materials have displaced brass in many fields, but it still has a broad economic use.

The information in Table 5.10 is for guidance of the engineer when choosing materials. Such information is limited and should be supplemented by current catalogs.

Machining

In general, the machinability of most brasses is good. With the free-cutting alloys and with some of the moderately machinable alloys, the chip breaks up at a rapid rate and consequently is only in momentary contact with the cutting tool. Then the principal function of the cutting medium is to serve as a coolant and not a lubricant.

Table 5.10 gives the approximate machineability and tool-wear ratings of the principal brasses and bronzes. Excellent finishes can be obtained by keeping the cutting tools sharp and by using high speeds, lower feeds, and minimum rake angles. Addition of a few percent (0.5 to 3.25) of lead improves the machineability and related operations. Lead also improves the antifriction and bearing properties.

BRONZE

Alloys of copper with material other than zinc or nickel are known as bronze. The major alloying element in bronze is tin, although other elements, such as silicon, aluminum, manganese, phosphorus, and nickel may be included. Similar to the brasses, the various bronzes can be hardened by cold working. The most widely used bronzes are beryllium bronze, aluminum bronze, phosphor bronze, and silicon bronze.

Beryllium bronze (beryllium copper) is an alloy of copper and beryllium that came into active use during World War II. It is used where a nonferrous,

TABLE 5.10
Mechanical properties of rod alloys

TEMPER	DIAMETER Inches	TENSILE STRENGTH psi		YIELD STRENGTH psi		ELONGATION Percent		ROCKWELL HARDNESS ON SURFACE Average Value		CONTRACTION OF AREA Percent Average	SHEAR STRENGTH psi Average	ASTM SPECIFICATION	ALLOY	ALLOY GROUP
		Specification Minimum Value	Average	Specification Minimum Value	Average	in 4 x D Gage Length Specification Minimum Value	In 2" Gage Length Average	F Scale	B Scale					
As Hot Rolled	1		32000		10000		55	40		70	22000	B133	ELECTROLYTIC TOUGH PITCH COPPER	1
Soft	1/8 to 1/2 incl.	37000*	32000		10000	25	45	40			22000			
	Over 1/2	37000*	32000		10000	20	55	40		70	22000			
Hard	1/8 to 1/4 incl.	50000	55000		50000		10	94	60		29000			
	Over 1/4 to 3/8 incl.	50000	55000		50000	8	12	94	60		29000			
	Over 3/8 to 1 incl.	45000	48000		44000	10	16	87	47	55	27000			
	Over 1 to 2 incl.	40000	45000		40000	12	20	85	45		26000			
	Over 2	35000	38000		36000	15	22	85	45		25000			
Half Hard	1/8 to 1/4 incl.	38000	42000	30000	39000	15	20		40		26000		TELLURIUM COPPER	
	Over 1/4 to 1/2 incl.	38000	42000	30000	39000	15	22		40		26000			
	Over 1/2 to 1 incl.	38000	42000	30000	39000	15	25		45		26000			
	Over 1 to 2 incl.	38000	42000	30000	39000	15	35		45		26000			
Hard	1/8 to 1/4 incl.	48000	54000	40000	50000	10	11		50		27000			
	Over 1/4 to 1/2 incl.	44000	48000	38000	46000	10	18		45		27000			
	Over 1/2 to 2 incl.	44000	48000	38000	46000	10	20		50		27000			
Soft Anneal			32000		10000		45	35			22000		SELENIUM COPPER and LEADED COPPER	
Hard			45000		40000		12		50		26000			
Hard	1/8 to 1/2 incl.	70000	80000	60000	70000	12	20		85		43000		TELLURIUM NICKEL COPPER	
	Over 1/2 to 1 incl.	68000	77000	57000	67000	12	27		84		40000			
	Over 1 to 2 incl.	68000	73000	57000	62000	15	38		84		39000			
Soft Anneal	1/2 to 1 incl.	35000	40000	10000			50	55			32000	B134 Alloy 2	COMMERCIAL BRONZE, 90%	2
Half Hard	1/2 and under	45000	52000	27000	47000		25		60		35000			
	Over 1/2 to 1 incl.	40000	50000	25000	45000		20		58		34000			
Hard	1/4 to 1 incl.	48000		30000							—			
Quarter Hard	1 and under		58000		43000		40		65	70	36000	B134 Alloy 6	CARTRIDGE BRASS, 70%	
Half Hard	1 and under	55000	65000	25000	48000	20	30		75	68	42000			
Soft	All sizes	35000	37000	10000	12000	15	45	55			24000	B140 Alloy B	LEADED COMMERCIAL BRONZE	
Half Hard	1/2 and under	50000	55000	30000	50000	10	14		61		31000			
	Over 1/2 to 1 incl.	45000	52000	27000	47000	10	18		58		30000			
	Over 1	40000	50000	25000	45000	12	25		58		30000			

TABLE 5.10 (Contd)

0.25 mm.	1		50000		19000		60	70			34000			
Quarter Hard	1/4 to 1 incl.	50000	55000	20000	42000		35		60		36000	B121 Alloy 3	MEDIUM LEADED BRASS	3
Half Hard	1/4 to 1 incl.	55000	65000	25000	48000		25		75		42000			
Half Hard	1/2 to 1 incl.	55000	60000	25000	42000	10	23		70	45	34000		HIGH-LEADED BRASS	
Soft	**1 and under**	**48000**	**50000**	**20000**	**22000**	**15**	**36**	**68**	**48**	**54**	**30000**			
	Over 1 to 2 incl.	**44000**	**48000**	**18000**	**21000**	**20**	**32**		**45**	**40**	**29000**			
	Over 2	**40000**	**45000**	**15000**	**19000**	**25**	**36**		**40**	**35**	**28000**			
Half Hard	**1/2 and under**	**57000**	**65000**	**25000**	**42000**	**7**	**15**		**73**	**44**	**38000**			
	Over 1/2 to 1 incl.	**55000**	**60000**	**25000**	**42000**	**10**	**23**		**70**	**45**	**34000**	**B16**	**FREE-CUTTING BRASS**	
	Over 1 to 2 incl.	**50000**	**54000**	**20000**	**37000**	**15**	**31**		**66**	**45**	**32000**			
	Over 2	**45000**	**50000**	**15000**	**28000**	**20**	**35**		**62**	**35**	**30000**			
Hard	**1/8 to 3/16 incl.**	**80000**	**85000**	**45000**	**58000**		**7**		**85**	**50**	**47000**			
	Over 3/16 to 5/16 incl.	**70000**	**75000**	**35000**	**52000**	**4**	**9**		**81**	**45**	**42000**			
As Extruded	1		52000		20000		45	78		65	--			
As Forged			42-65000	—			25-60					B124 Alloy 2	FORGING BRASS	
Half Hard	1/2		75000				14		85					
As Extruded			60000		20000		30		65		35000		ARCHITECTURAL BRONZE	
Soft	All sizes	40000	45000	12000	16000	30	40	55			35000			
Half Hard	Up to 1/2 incl.	55000	60000	20000	48000	11	16		—	—	41000			
	Over 1/2 to 1 incl.	55000	60000	20000	48000	12	19		—	—	41000			
	Over 1 to 2 incl.	55000	60000	20000	48000	12	22		—	—	41000			
Hard	Up to 1/2 incl.	65000	70000	35000	55000	8	13			—	45000			
	Over 1/2 to 1 incl.	65000	70000	35000	55000	10	15		80	—	45000	B98 Alloy B	LOW-SILICON BRONZE	4
	Over 1 to 2 incl.	65000	70000	35000	55000	10	17			—	45000			
Extra Hard (Bolt)	Up to 1/2 incl.	85000	90000	55000	67000	6	11		90	—	50000			
	Over 1/2 to 1 incl.	75000	80000	45000	62000	8	13		—	—	48000			
	Over 1 to 1-1/2 incl.	75000	80000	40000	62000	8	15				48000			
Soft	Under 1/4	40000	45000				—		—	—	—			
Half Hard	1/2 to 1 incl.		75-70000		65-58000		25		80-78					
Hard	Under 1/4	80000	105000						—	—	—			
	1/4 to 1/2 incl.	70000	92000			13	23		—	—	—			
	Over 1/2 to 1 incl.	60000	68000		61000	15	32		—	—	—	B139 Alloy A	PHOSPHOR BRONZE, 5%	
	1 and over	55000	60000		57000	18	34		—	—	—			
Spring	Under .026	125000							—	—	—			
	.026 to 1/16 incl.	115000	127000						—	—	—			
	Over 1/16 to 1/8 incl.	110000	123000		—		—		—	—	—			
	Over 1/8 to 1/4 incl.	105000	122000	—	—	3.5	—		—	—	—			
	Over 1/4 to 3/8 incl.	100000	121000		—	5.0	—		—	—	—			
	Over 3/8 to 1/2 incl.	90000	100000		—	9.0	—		—	—	—			

nonsparking, or good electrical conductor (45 percent when hardened; see Table 5.9) with high strength (100,000 to 190,000 lb/in.^{2}) is required. It has a modulus of elasticity of approximately 19,000,000 lb/in.^{2} It is one of the best corrosion-resistant spring materials. Springs of all forms, X-ray windows, molds for plastics, and diaphragms illustrate its versatile use. This copper alloy can be heat-treated and shaped at the same time by holding in steel molds.

Aluminum bronze has wide applications because its physical properties can be controlled over a broad range by varying the amount of aluminum in the alloy and also by heat treating the bronze. This alloy has excellent resistance to corrosion and has superior strength. The family consists of α aluminum bronzes (less than 8 percent aluminum) and α–β bronzes (8 to 12 percent aluminum). The strengthening effect of aluminum results in the highest-strength copper alloys (up to 100,000 lb/in.^{2} in tension). These strength and corrosion-resistant characteristics make this family ideal structural materials.

Phosphor bronze is an alloy of copper and tin deoxidized with phosphorus. This alloy is known for its strength and hardness (which increases as the percentage of tin increases) and low coefficient of friction. These bronzes are produced with varying amounts of tin, although usually the tin percentage is somewhere between 2 and 10 percent. This family finds application as a spring and bearing plate material.

Silicon bronze includes between 2 and 4 percent silicon and usually 1 percent zinc or manganese. Silicon bronze can be worked readily by the common techniques. It is easily cast, forged, welded, stamped, rolled, and spun. In addition to raising strength, silicon increases the electrical resistivity. Silicon bronze finds its principal uses where resistance to corrosion plus high strength are the criteria to be met.

Cast bronzes frequently have lead additions to improve their pressure tightness. Bronze alloys freeze over a long temperature range so leakage can be a problem.

NICKEL AND NICKEL ALLOYS

Nickel's principal use is as an alloying element of both steel and nonferrous metals (there are over 3000 active alloys of nickel), although large amounts of nickel are used in electrodeposition and there are many uses for practically pure nickel in both cast and wrought form.

It is magnetic up to 680°F, is resistant to corrosion and the majority of the nonoxidizing acids (with the exception of nitric acids), melts at 2646°F and has a specific gravity of 8.84. Pure nickel is a silvery-white, tough metal with approximately the same density as copper, but it costs about three times as much.

Grade A nickel, the most commonly used pure nickel (99 percent pure) is used in nickel–cadmium batteries. Here the nickel is deposited as a powder in the manufacture of sintered battery plates.

Duranickel, as produced by the International Nickel Company, is an alloy of 93.7 percent nickel, 4.4 percent aluminum, 0.5 percent silicon, 0.35 percent iron, 0.30 percent manganese, 0.17 percent carbon, and 0.05 percent copper. When age-hardened, the tensile strength is over 175,000 lb/in.^{2}, with elongation of 65 percent and Brinell hardness of 375. It is a useful material for springs, bellows, etc., since it has high resistance to fatigue and withstands heat to 550°F.

The engineer will find nickel and its alloys useful for meeting the requirements of many designs. Often when a cast-iron part has blowholes or chills and is difficult to machine owing to hard spots, the addition of a small percentage of nickel will improve the quality.

The properties of several alloys containing more than 50 percent nickel are summarized in this text. These alloys are stronger, tougher, and harder than copper and aluminum alloys, and are as strong although more costly than steel alloys. Nickel alloys are highly resistant to corrosion and exhibit good heat resistance. In general, they can be cold-worked, although frequent annealing may be required in order to remove the rapid work-hardening effect characteristic of these alloys.

In view of their high heat resistance, ability to maintain strength at elevated temperatures, and high corrosion-resisting characteristics, these alloys are finding increased uses in furthering the space age.

The kind and quality of elements added to the nickel form a convenient method of grouping the alloys. The following summary gives the principal classifications according to composition.

Group 1 Nickel: 93.5 to 99.5 percent nickel (and a maximum of 4.5 percent manganese)
Group 2 Nickel–copper: 63 to 70 percent nickel, 29 to 30 percent copper
Group 3 Nickel–silicon: 83 percent nickel; 10 percent silicon
Group 4 Nickel–chromium–iron: 54 to 78.5 percent nickel, 12 to 18 percent chromium, 6 to 28 percent iron
Group 5 Nickel–molybdenum–iron: 55 to 62 percent nickel, 17 to 32 percent molybdenum, 6 to 22 percent iron
Group 6 Nickel–chromium–molybdenum–iron: 51 to 62 percent nickel, 15 to 22 percent chromium, 5 to 19 percent molybdenum, and 6 to 8 percent iron

In the nickel group of alloys, there are five grades of commercial nickel: A, D, E, L, and Z. Grade A nickel contains an average of 99.4 percent nickel and is distinguished from electrolytic nickel, which is 99.95 percent pure. Grade A nickel is in the wrought form and combines high mechanical properties with excellent resistance to corrosion. It can be fabricated readily and joined by welding, brazing, or soldering.

In composition, grades D and E nickel are similar to A, the principal difference being the inclusion of small percentages of manganese (usually between 2 and 5 percent), which replaces a like percentage of nickel. The manganese increases the resistance of the alloy to atmospheric corrosion and to sulfur compounds at elevated temperatures.

In L nickel, the carbon content is kept low (between 0.01 and 0.02 percent); this alloy is softer than the others, having a lower yield strength and elastic limit. This alloy is fabricated easily and for this reason frequently is specified where drawing and forming operations are required.

Z nickel contains about 93.5 percent nickel. It is harder and stronger than the other group 1 alloys. This alloy can be heat-treated, and tensile strengths up to 240,000 $lb/in.^2$ can be obtained.

The group 2 alloys, referred to as the nickel–copper alloys, are perhaps the best known of all the nickel alloys and are commonly known as the Monel type. Monel is a trade name of the International Nickel Company, and it contains approximately 67 percent nickel and 33 percent copper. Six grades of Monel metal are available. These are known as Monel, K Monel, R Monel, H Monel, S Monel, and KR Monel.

Monel. Monel metal is the most important nickel alloy of this group. It has an average composition of 67 percent nickel, 28 percent copper, and 5 percent iron, manganese, and silicon combined. Monel resists corrosion of distilled water, salt water, foods, and acids, except hydrochloric acid above 120°F. Its strength and its thermal expansion are about the same as steel; therefore, it can be used in machinery parts that must resist corrosion. It is used extensively in the chemical, marine, power-equipment, laundry, food-service, pickling, roofing, petroleum, household-equipment, and pulp and paper industries. Steam valves, pipes, turbine blades, and shafting are normal applications. At 750°F, it retains 75 percent of its strength at room temperature. It resists oxidation up to 1350°F.

The color of Monel is close to that of nickel. This has brought about its general use in kitchen equipment and utensils. It is produced in several alloys giving flexibility in strength and machining qualities without reducing its corrosion-resistance properties. Although its cost per pound is relatively high, it has an equally high scrap or chip value. This factor makes it competitive with stainless steel in many installations. The alloy may be cast, rolled, or forged and can be annealed after cold working.

K Monel. K Monel is a nonmagnetic, age-hardening alloy containing approximately 3 percent aluminum. Its strength and hardness, particularly in large sections, are comparable with those of heat-treated alloy steels. It has a hardness above 300 Brinell when heat-treated and a tensile strength of more than 160,000 $lb/in.^2$ It will retain its hardness up to 700°F.

R Monel. R Monel is a free-machining alloy that is particularly adapted for use in automatic screw machines. Its relatively free-machining properties are the result of a 0.025 to 0.060 percent sulfur content. Cold drawn rods have a tensile strength of 90,000 $lb/in.^2$ and Brinell hardness of 180.

H Monel. H Monel includes 2.75 to 3.25 percent silicon. The silicon provides increased hardness with good ductility. This is a sand-casting alloy having a tensile strength of 100,000 lb/in.^{2} and Brinell hardness of 210.

S Monel. S Monel includes approximately 4.0 percent silicon, which induces age-hardening characteristics. This alloy in the cast state will give a hardness reading of about 320 Brinell, which can be increased to 350 with suitable heat treatment.

KR Monel. KR Monel is similar in composition to K Monel but it has a higher carbon content. This alloy has better machinability characteristics than its companion K.

Group 3 alloys are extremely hard (Brinell of approximately 360) and consequently are quite difficult to machine. These alloys are resistant to corrosion in sulfuric acid, acetic acid, formic acid, and phosphoric acid.

The best known of this group are manufactured by the Union Carbide Corp. under the trade name Hastelloy O.

Group 4 alloys are characterized by large percentages of chromium and iron. In this group, the alloy Inconel as produced by the International Nickel Company is one of the best known. These alloys have excellent resistance to corrosion, good strength and toughness properties, and can withstand repeated heating and cooling in the temperature range of 0 to 1700°F.

Group 5 alloys represent the nickel–molybdenum–iron alloys. This group, in addition to having good corrosion-resistance properties (in particular to hydrochloric acid), has strength and ductility properties comparable to alloy steel. The best known of this group are Hastelloy A and Hastelloy B as produced by the Union Carbide Corp. These alloys can be work-hardened but do not respond to age-hardening heat treatment.

The sixth group of alloys are those containing large amounts of chromium, molybdenum, and iron with nickel. These alloys find application where high mechanical properties combined with resistance to oxidizing acids and agents are required. Hastelloy C is one of the better known of this group.

MOLYBDENUM

Molybdenum has unusual properties that make it useful in alloys of steel, cast iron, and high-temperature alloys. The metal has a specific gravity of 10.2 and a meting point of 4750°F. Rolled molybdenum has a tensile strength of approximately 250,000 lb/in.^{2} with Brinell hardness of about 170. Its electric conductivity is 34 percent that of copper.

Pure molybdenum (99.95 percent) finds application in such products as heating elements for electric furnaces, arc-resistant electric contacts and structural parts in jet engines and missiles. It is used as a flame-resistant and corrosion-resistant coating for other metals. It may be arc-deposited.

Machining and welding of high-purity molybdenum is difficult owing to its high melting temperature of 4750°F and hardness of 150 to 250 Brinell at high temperatures. The high melting temperature makes it a basic material in high-temperature alloys and an alloy agent for steel, cast iron, cast steel, nickel steel, chromium–nickel steel, and tool steel. The general physical properties of these alloys are increased toughness, tensile strength, uniformity, corrosion resistance, and hardenability.

When used in the foundry, molybdenum does not oxidize and permits close control of the alloy. It can be added at any time during melting or casting. During heat treatment and die forging, it promotes free scaling. When the part is cast, increased uniformity and decreased section sensitivity are obtained and a thicker and more complex casting may be designed.

The steel and cast-iron alloys at elevated temperatures have increased hardness and tensile strength.

Molybdenum is sold in the form of pure molybdenum, molybdenum oxide, and molybdenum sulfide. Economy is possible because of high recovery from scrap steel and because it does not oxidize in foundary practice and permits close control of the alloy.

COBALT

Cobalt was first introduced as an alloy in tool steel because it has the property of adding red-hardness to cutting alloys. It also hardens alloys especially in the presence of carbon. Cobalt cutting tools were superior to high-speed steel tools. Later, cobalt was introduced as a binder in cemented carbide tools and it improved their toughness.

Cobalt has a specific gravity of 8.76, melting point of 2719°F, hardness 86 Brinell and electric conductivity of about 16 percent that of copper. The tensile strength of cast cobalt with 0.25 percent carbon is approximately 65,000 $lb/in.^2$

Cobalt is an important constituent (along with other alloys) of high-temperature, high-strength steels having dampening characteristics. These steels are used in turbine blades subject to vibration.

Cobalt and its alloys can be spot-welded, brazed, and fusion-welded. Cobalt and cobalt alloys have the properties and applications shown in Table 5.11. An interesting cobalt alloy is marketed under the trade name Invar. This alloy, containing 54 percent cobalt and 36 percent nickel, has a thermal coefficient of expansion of about zero over a small temperature range, making it a useful material in the design of control mechanisms.

TITANIUM

Titanium is classed as a light metal (0.16 $lb/in.^3$). It is 60 percent heavier than aluminium but only 56 percent as heavy as alloy steel. Titanium-based alloys are extremely strong, far more so than aluminum alloys. Because of its high strength–weight ratio, which is superior to both steel and aluminum, and its resistance to corrosion, titanium competes primarily with aluminum and to a lesser extent with various alloy steels.

TABLE 5.11
Properties and applications of cobalt and cobalt alloys

Properties	Alloy name	Application
Low expansion	Invar Kovar	Thermostat controls Glass to metal seals Ceramic to metal seals
Constant modulus (E) during temperature change (has low elastic limit)	Elinvar	Hair springs in watches Electric-current-conducting springs Springs working in high temperatures
Good castability, casts to high dimensional accuracy and surface smoothness Resistance to tarnish and abrasion	Vitallium	Replaces precious metals in dental work
Compatible with mouth tissues		Bone surgery
High strength and stiffness, lower weight and low cost in comparison with precious metals		Prosthetic devices
High-temperature strength		Electronic-tube parts
Corrosion resistant		Springs Cemented carbide tools
Alloys with and hardens copper, gold, silver, platinum		Jewelry
Colorant and decolorizer		Glass Ceramics
Reduced resistance to electric current at increased temperatures		
Catalysts		Oil refineries
Radioactive		Radioisotopes
Nutrition element		Animal nutrition (Vitamin B12)

Because of the difficulty of removing the metal from its contaminants and because the metal is so very active chemically when molten (absorbing oxygen or nitrogen from the air with ruinous rapidity), it has been difficult and costly to produce.

Commercially pure titanium has the following approximate mechanical properties:

1. Hardness, 85–95 Rockwell B
2. Ultimate strength, 80,000 lb/in.^{2}
3. Yield strength, 70,000 lb/in.^{2}
4. Elongation, 20 percent
5. Sp.ht.: 0.13 Btu/(lb · °F)
6. Thermal conductivity: 105 Btu/(ft^2 · h · in.)
7. Coefficient of expansion: 5×10^{-6}/°F
8. Stress rupture: @750°F; 35,000 lb/in.^{2} for 1000 hours

Commercially pure titanium does not respond to heat treatment, but may be cold-worked to well above 120,000 lb/in.^{2} tensile and 100,000 lb/in.^{2} yield, with a resultant drop in elongation to about 10 percent.

The good strength and weight characteristics result in a higher strength–weight ratio for titanium alloys than can be obtained from alloy steel at temperatures up to 800°F. Above 800°F its strength–weight ratio falls below that of alloy steels.

Many titanium-based alloys that have superior physical properties are available. For example, the Titanium Metals Corporation type T1-155A is a high-strength forging alloy, β-stabilized by iron, chromium, and molybdenum, with sufficient aluminum to maintain strength at temperaures to 1000°F. It has the following mechanical properties:

1. Hardness, 300–320 Brinell
2. Tensile strength, 155,000 lb/in.^{2}
3. Yield strength, 140,000 lb/in.^{2}
4. Elongation, 12%

Standard aircraft-quality titanium alloys contain 6 percent aluminum and 4 percent vanadium. They are known as "6/4" types. These and other alloys of titanium are commercially available and are used as flat products, bars, and forgings. They contain combinations of iron, aluminum, chromium, manganese, molybdenum, tin, and vanadium. The elements iron, chromium, vanadium, molybdenum, and manganese stabilize the high-temperature modification and produce alloys that can be hardened by heat treatment. Aluminum, tin, nitrogen, and oxygen facilitate the strengthening of the low-temperature modification.

The melting point of titanium (3135°F) is higher than those of the majority of metals. This property led to the belief that the alloy would be suitable for high-temperature application. However, because of its affinity for both oxygen and nitrogen at elevated temperatures it has been found impractical for sustained use at temperatures greater than 1000°F.

Titanium is very resistant to corrosion by salts and to oxidizing acids. However, it is not resistant to hydrofluoric, sulfuric, oxalic, and formic acids. Although it is attacked by concentrated alkalies, it is resistant to dilute alkalies. Its adherent oxide film resists moist air and oxygen up to 600°F.

Today, titanium is fabricated by sheet-metal forming, forging, casting, and machining. In these processing methods, titanium behaves much as stainless steel. A word of caution should be stated, however, relating to titanium castings. In producing castings, both the mold and furnace must be housed where either a vacuum or inert gas atmosphere is maintained since liquid titanium reacts with all elements except the inert gases. Titanium alloys forge well at 1600 to 1800°F in inert atmosphere. Strip, wire, and rod may be cold-worked up to 50 percent reduction. The tensile strength increases to 135,000 lb/in.^{2} and elongation drops to 10 percent.

Titanium is being successfully clad onto carbon steel by rolling; and investigation continues on the use of titanium as an electrolytic plating material.

On the basis of response to heat treatment, titanium alloys have for the most part not been hardened by heat treatment. However, both the martensitic alloys and the metastable β-alloys are potentially capable of hardening by heat treatment. Surface hardening does take place when titanium-based materials are heated in air, because of the absorption of oxygen and nitrogen. Commercial titanium can be annealed and this is usually done in the range 1100 to 1350°F.

Titanium can be joined by inert-gas-shielded arc-welding procedures as well as brazing. Adhesion and riveting find application as joining methods in addition to welding and brazing.

LOW-MELTING POINT ALLOYS

The field of low-melting-point alloys includes all combinations of metals that have melting temperatures below 1000°F. This classification can be further broken down into three categories: alloys that melt between 300 and 1000°F, those that melt between 100 and 300°F, and finally the ultra-low-melting alloys that melt below 200°F.

Alloys Melting between 300 and 1000°F

The first of these three categories embraces, primarily, the zinc-, tin-, and lead-based alloys. These are by far the most important of all of the low-melting alloys.

Zinc-based alloys. Zinc-based alloys are primarily used for die castings. Here they are used widely at the present in all types of products. The auto industry uses many zinc-based alloys in the manufacture of carburetors, fuel pumps, door handles, and numerous other items. They are also used in the construction of household appliances, business machine parts, blower rotors, and other functional parts for which the load-bearing and shock requirements are relatively low.

Zinc-based alloys can be cast easily; in fact, more die castings are made of zinc-based alloys than of any other metal. The advantages of using this type of alloy are low production costs of complicated shapes produced to close tolerances, low die-maintenance costs due to the low temperatures inherent in the process, smooth surface of product, and relatively good strength of the cast product when correctly designed.

A widely used zinc alloy is ASTM XXI, which corresponds to SAE 921 and to New Jersey Zinc Company's alloy Zamak 2. This alloy consists of 4.1 percent aluminum, 2.7 percent copper, 0.03 percent magnesium, and 93.17 percent pure zinc. The melting point is 733.6°F and the shrinkage is

0.0124 in./in. Average tensile strengths of 47,300 lb/in.^{2} are obtained from 6-month-old cast specimens. Another important alloy, ASTM XXIII (SAE 903 or Zamak 3), contains 4.1 percent aluminum and 0.04 percent magnesium, and the remainder is pure zinc. This alloy has slightly lower tensile strength and hardness but it has excellent retention of impact strength and dimensions.

Tin-based alloys. Although tin is used in brasses and bronzes, it will be briefly discussed here in connection with Babbitt metals that are used principally for bearings. The main alloying elements are copper, antimony, and lead. The melting point of pure tin is 450°F; however, for complete liquefaction of the alloys, a temperature of 700 to 800°F must be used. The effect of the antimony is to harden these alloys and increase their antifriction properties.

The alloy ASTM B23-36, grade 1 (SAE 10), contains 91 percent tin, 4.5 percent copper, and 4.5 percent antimony. Grade 2 is composed of 89 percent tin, 3.5 percent copper, and 7.5 percent antimony. Both of these alloys are used extensively in automotive and aircraft bearings. The latter grade is slightly harder and stronger. Alloy SAE 12, which has 60 percent tin, 3.5 percent copper, 10.5 percent antimony, and 26 percent lead, is much cheaper than the other two alloys because of the high lead content. This alloy is used chiefly for light-duty bearings.

Tin-based alloys are corrosion-resistant and can consequently be used in food-handling equipment and for soda-fountain hardware.

Lead-based alloys. Lead-based alloys are composed principally of lead and antimony. Pure lead melts at 621°F; however, the addition of 17 percent antimony reduces the melting point to 570°F. The antimony in these alloys hardens the lead and reduces shrinkage. The chief lead-based alloy used in bearings is ASTM B23-26, grades 12 and 17, which have respectively 10 and 15 percent antimony, copper held to 0.50 percent maximum, and arsenic held to 0.25 percent maximum.

These alloys have low-strength properties, but are cheap and easy to cast. They are used for light-duty bearings, weights, battery parts, X-ray shields, and for noncorrosive applications.

Shrinkage of both the lead- and tin-based alloys is exceptionally low, ranging from 0.002 to 0.003 in./in. This characteristic facilitates the casting of parts to accurate size and also lessens the danger of shrinkage cracks in casting.

Solder. The final alloys of importance in this heat range are the tin–lead alloys, more commonly called *solder,* in which the metals are present as a mechanical mixture of two separate phases. The solders also contain small percentages of antimony and sometimes silver. The physical properties are affected most significantly by varying the lead–tin ratio. Solders have a wide possible range of composition from 100 percent tin to 100 percent lead. The most widely used solders are those that contain 38.1 percent lead and 61.9 percent tin, and those that are composed of 68 percent lead and 32 percent tin.

The first of these is known as *tin solder.* It is completely molten at 362°F and, after losing its latent heat, is completely solid at 362°F. From this it can be seen that it is quick setting. The latter composition is termed *plumbers' solder.* it is completely molten at 488°F and totally solid at 362°F. This range of temperature between the molten and solid state makes possible the plumbers' wiped joint. Today lead free solders are primarily used in connection with supplying drinking water. These alloys are composed of silver, copper and tin.

Alloys Becoming Fluid between 200 and 300°F

The second category of low-melting alloys, those alloys that melt between 200 and 300°F, are among the most interesting and the most misunderstood of the low-melting alloys. There are more than 2000 commercial and experimental types, many of which are relatively new, and as yet their full potentialities have not been appreciated.

Bismuth is a common constituent of all of the alloys in this and the ultra-low-melting groups. It has a specific gravity of 9.75, melting point at 520°F, and hardness of 73 Brinell. Its thermal conductivity is less than any other metal except mercury. In its pure state, it has few uses. The usual alloying elements are tin, lead, and cadmium. Certain peculiar characteristics of bismuth and bismuth alloys are responsible for their wide range of usefulness.

Bismuth is one of the few elements that does not shrink when it solidifies. Water and antimony are two other substances that expand on solidification, but bismuth expands more than water, namely, 3.3 percent of its volume. This expansion is somewhat modified when bismuth is alloyed. Some of the alloys shrink slightly upon solidification but expand again after a few hours of aging at room temperature.

This characteristic of bismuth alloys makes them excellent for casting because, when they expand into the mold, they pick up every detail. This expansion characteristic is also the basis of the alloys' widest use, anchoring. The growth enables them to grip keyed, notched, or bossed surfaces on which they are cast, and to fill annular spaces tightly, despite the absence of fusion or bond with the materials with which they come in contact. Examples of this are the anchoring of bearings, bushings, and stationary parts in machinery. This is done by providing oversize holes to receive the bearings or bushings and then accurately aligning the parts in jigs. Babbitting clay dams are formed around the part and the molten alloy is poured in. Upon solidification, the alloy holds the parts rigidly. This method of locating bearings and parts eliminates all costly machining to press-fit parts, makes possible easy field servicing, and safeguards against overheating through failure of oil supply, since the inexpensive alloy melts before the bearing is damaged.

Dies and punches for blanking, piercing, and trimming are secured in a steel shell by use of these alloys. This method of construction eliminates the three-stage dies formerly required. Chucks are also made of low-temperature alloys to hold odd-shaped parts while they are being machined, ground, and inspected.

Cores for electroforming are also made from these alloys. The core is easily molded into intricate shapes and is then placed in the electrolytic bath. The plating metal is then deposited upon the core, and as the final step, the low-melting alloy is melted out.

Alloys Becoming Fluid at Less Then 200°F

In alloys melting in the 100 to 200°F range, tensile strengths up to 1500 lb/in.^{2} and compressive strengths up to 20,000 lb/in.^{2} can be obtained. Bismuth is the principal metal used in this class of alloys. These permit the use of low-melt alloy casts as form blocks or dies for the fabrication of sheet-metal products with hydropress and drop-hammer equipment.

The ultra-low-melting alloys have only limited application in present-day industry. As materials for finished products, they have been used for such purposes as fire-extinguishing systems, the production of special molds and patterns for dentures, and molds for plaster casting. Their greatest and most important use is in the bending of thin-walled tubing. This tubing requires internal support during bending to prevent wrinkling, buckling, and flattening. The tubes are filled with alloy and the operations are completed. The alloy is then melted out by immersing in boiling water.

Low-melting alloys, although relatively new, are finding increasingly important roles in industry. It would be wise for every engineer to keep abreast of all new developments in this field for the many moneysaving applications they provide.

SUMMARY

Only nonferrous metals that are principally used by the engineer have been discussed in this chapter. Many other metals, such as gold, silver, zirconium, vanadium, thorium, thallium, sodium, tantalum, rhodium, ruthenium, osmium, palladium, platinum, rhenium, manganese, mercury, indium, iridium, lithium, germanium, gallium, niobium, cerium, and calcium, have application, but to a lesser degree; consequently, they will not be discussed in this text.

When the engineer selects a nonferrous material, it is usually because he is looking for certian characteristics discussed in this chapter. Familiarity with the metal's physical properties and ability to be worked will facilitate the selection of the material that will meet design requirements.

QUESTIONS

5.1. For what reasons is aluminum being considered as a body material by the automotive industry?

5.2. Describe the four-digit index system for aluminum alloys and wrought aluminum.

5.3. What are the principal advantages of selecting a "2000 series" aluminum?

5.4. What is the major alloying element in the 7000 series?

5.5. Describe several products that might advantageously specify aluminum because of its thermal conductivity properties.

5.6. What are some of the uses of aluminum foil?

5.7. Why has aluminum been widely used as a roofing material? Why is it advisable to use aluminum nails with aluminum roofing?

5.8. Describe the characteristics of the single-point cutting tool when used to cut aluminum.

5.9. Why is it not advisable to join thin sections (less than 0.032 in.) of aluminum by arc-welding methods?

5.10. What is the realtionship of the density of aluminum to that of copper? Of iron?

5.11. How can the corrosion of aluminum be retarded?

5.12. What is the conductivity of aluminum relative to that of copper?

5.13. For what reasons is the amount of cold work that can be done on magnesium limited?

5.14. What advantages does magnesium have over aluminum in product design?

5.15. What precautions should be observed when grinding magnesium?

5.16. What are the principal reasons for the widespread use of metallic copper?

5.17. What metal has better electrical conductivity than copper?

5.18. Why is pure copper difficult to cast?

5.19. How is hardness produced in copper alloys?

5.20. What is the principal difference between α- and β-brass?

5.21. What metal is sometimes added to brass in order to improve its machining characteristics?

5.22. What are the relative machineabilities among free-cutting brass, manganese bronze, and aluminum bronze?

5.23. Make a sketch of the recommended tool geometry using high-speed and carbide single-point tools for turning brass.

5.24. What is the relationship in ductility between free-cutting brass and a leaded commercial bronze?

5.25. When would the engineer specify bronze in preference to brass?

5.26. What are the characteristics of Duranickel?

5.27. What is Monel metal? What is its strongest competitor? Why?

5.28. Briefly describe the application of the low-melting alloys.

5.29. Why has titanium received so much publicity in recent years?

5.30. What are the 6/4 types of titanium alloys? Where are they mainly used?

5.31. Would you specify a zinc-based alloy die-casting to produce the piston of a $\frac{1}{2}$-H.P. "home use" air compressor? Why?

5.32. What metal is a common constituent of all of the alloys in the ultra-low-melting group?

PROBLEMS

5.1. The plant engineer at the XYZ Company is planning the installation of a new press in the plastic molding department, where a corrosive environment exists. He decides to mount each of the four feet of the press on a square cast-aluminum bearing plate. The bearing plate is 8 in. square and 2 in. thick. The compression strength of the aluminum is 30,000 lb/in.^{2} and the strain at fracture is 0.006 in./in. Calculate:

(*a*) The compression load at fracture
(*b*) The total contraction at fracture
(*c*) The total strain energy corresponding to fracture

5.2. A solid aluminum cylinder with a diameter of 4 in. and a length of 20 in. is subjected to a tensile load of 120,000 lbf. Young's modulus of elasticity for this material is 10,300,000 lb/in.^{2} What would be its increase in length?

5.3. A special structural shaped-aluminum channel was rolled with a nominal thickness of 0.30 in. What tolerance can be assigned to this dimension? Explain.

5.4. Based upon the comparison of equal strength (bending moments are equivalent) for bars having the same width and using steel as 100 for comparison, what would be the stiffness value of titanium? Base your answer on the following:

Young's modulus for steel = 29×10^6 lb/in.^{2}
Young's modulus for titanium = 15×10^6 lb/in.^{2}
Yield strength of structural steel = 36,000 lb/in.^{2}
Yield strength of titanium = 72,000 lb/in.^{2}

5.5. A rolled aluminum channel is 1 in. deep with $\frac{3}{4}$-in. flanges. The section thickness is $\frac{5}{32}$ in. and lengths are 20 ft. What tolerances can the engineer expect on this structural shape as it is received from the supplier?

5.6. A Zamak 2 diecasting is specified in connection with the design of instrument panel knobs. The outside diameter of these knobs is 0.875 in. What diameter should the mold be made in order to accommodate shrinkage? If a lead-based alloy were substituted, would you change the mold dimensions? If so, by how much?

5.7. You have designed a fire-extinguishing system that makes use of bismuth to actuate the sprinklers. The overhead system carries a static load of 120 lb in tension. Allowing a 30 percent factor of safety, what section size would you recommend the supporting bismuth component be?

SELECTED REFERENCES

1. Aluminum Association Inc.: *Aluminum Standards and Data,* 8th ed., Washington, D.C., 1984.
2. Aluminum Company of America: *Structural Handbook: A Design Manual for Aluminum,* Pittsburgh, Pa., 1958.
3. American Society for Metals: *Metals Handbook,* 8th ed., vol. 1, "Properties and Selection of Metals," Metals Park, Ohio, 1972.
4. Brady, George S., and Clauser, Henry R.: *Materials Handbook,* 12th ed., McGraw-Hill, New York, 1985.
5. Dow Chemical Co.: *Fabricating Magnesium,* Midland, Mich., 1984.

CHAPTER 6

PLASTICS

INTRODUCTION

The term *plastics* has been applied to those synthetic nonmetallic materials that can be made sufficiently fluid to be shaped readily by casting, molding, or extruding, and that may be hardened subsequently to preserve the desired shape. Synthetic rubber, ceramics, and glass, which may seem to be in this category, are often classified separately. Both natural and synthetic rubber are included in this chapter, while ceramics, including glass, are discussed in Chapter 7.

Plastics as engineering materials are increasingly useful in present day designs. The use of plastics is following the same course of development and application that was characteristic of the introduction of stainless steel and light metals. Costs have been reduced, uniformity and reliability have been improved, and new applications have been determined. Plastics are attractive materials and offer advantages in weight, cost, moisture and chemical resistance, toughness, abrasive resistance, strength, appearance, insulation (both thermal and electrical), formability, and machinability. However, the structural design of plastic parts requires a somewhat different design approach from that to metals. Plastics, unlike metals, do not respond elastically to stress and undergo permanent deformation under sustained loading. The term *viscoelastic* describes the behavior of plastics when subject to stress. After the application of an initial load, the plastic structure will move in response to the applied stress. The relation between the load taken to cause flow and the load taken by elastic deformation will change. Also, the load (stress) and the deformation (strain) will change with time. When compared with metals, plastics lack stiffness and their physical properties are temperature-dependent.

The earliest and largest user of plastics was the electrical industry, which is a leader in the use of thermosetting and laminated plastics. Electrical products spurred, in turn, the development and application of plastics for many parts that had no electrical function. For example, high-density polyethylene milk containers are predominantly used today in the dairy industry. In fact, the majority of dairy products are packaged in plastic containers, whereas only 6 percent used plastic about 20 years ago. Today, most major stadiums in the United States are covered by synthetic turf. New homes utilizing ABS (acrylonitrile butadiene styrene) drainage plumbing are continuing to increase. On a pounds-per-car basis, plastics increased approximately 35 percent between the 1976 and 1986 models. In 1983, General Motors put its second all-plastic car, the Pontiac Fiero, on the market and in 1990 GM intends to have three more lines with plastic/composite bodies on the road. The Buick-Oldsmobile-Cadillac group is currently considering substituting composite materials for steel in some major underbody and chassis applications.

In considering the materials available, especially for high technology components, the engineer should consider nonmetallic composites. These include materials with the following matrix resins and reinforcements:

Matrix resins	Epoxy, polyester, polyimide, thermoplastics
Reinforcements	Carbon fiber, glass, quartz

Composites, even though many utilize a plastic matrix, will not be discussed in this chapter but will be covered in Chapter 7.

By virtue of their thermal characteristics, plastics are usually divided into two groups: thermoplastic or thermosetting. Those that undergo no chemical change in the molding operation may be softened again by heating to the temperature at which they originally became plastic, and are therefore termed *thermoplastic*. Since they become increasingly softer with increase in temperature, certain members of the thermoplastic family are liable to permanent distortion under mechanical strain at relatively low temperature (140°F). They may flow to an appreciable extent under load at room temperatures.

The basic structural units of thermoplastics are referred to as monomers. Monomers are molecules consisting of carbon atoms with attached ribs of other atoms such as hydrogen, chlorine, and fluorine. These ribs determine to a large extent the intrinsic properties of the plastic. For example, thermoplastics such as polyethylene, composed of the ethylene monomer, C_2H_4

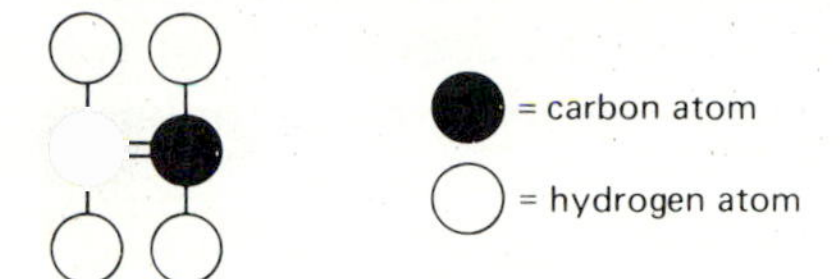

have both good flexibility and toughness. By replacing one of the hydrogen atoms with chlorine, the monomer vinyl chloride is formed:

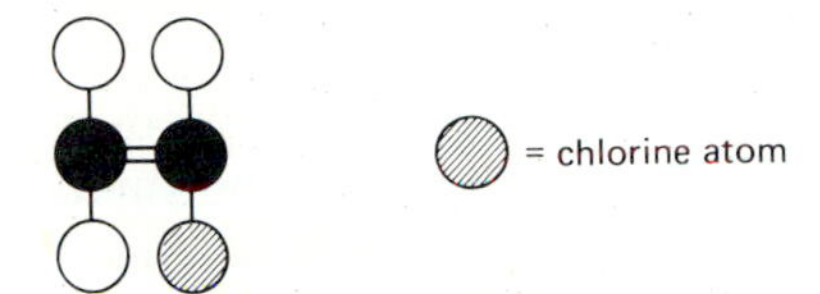

This provides a thermoplastic with more rigid characteristics than ethylene. The graphic representation illustrates the presence of a double bond in the monomer, which is necessary in all addition polymerization processes. Under certain conditions, the two unused valences present in the double bond can become available for reaction:

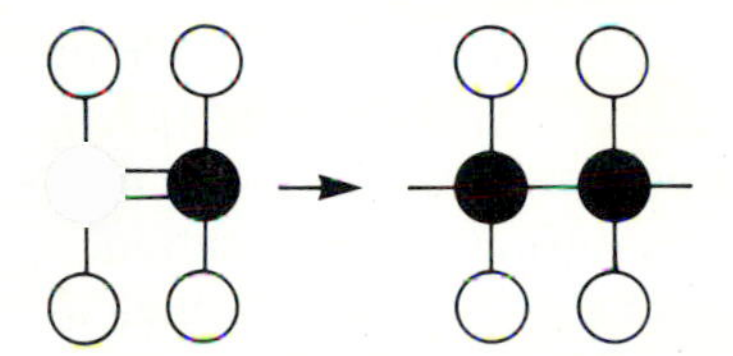

These valences will attach to other atoms. By providing energy via heat, X-rays, high-energy electrons, etc., the molecule (in this case ethylene) is activated and the double bond is sprung, making it ready for reaction. This takes place during what is referred to as the initiation phase. Initiation may also be caused by adding a reactive chemical such as peroxide, for example benzoyl peroxide (C_6H_5CO). Peroxides decompose at the bond between the oxygen atoms. Thus, for hydrogen peroxide: $H—O—O—H \rightarrow 2(H—O—)$. Each of the two molecules contains an unsatisfied valence, referred to as a free radical, that attacks neighboring ethylene molecules. The ethylene double bond is broken and the free radical attaches to one end, causing the other end to become active. Thus:

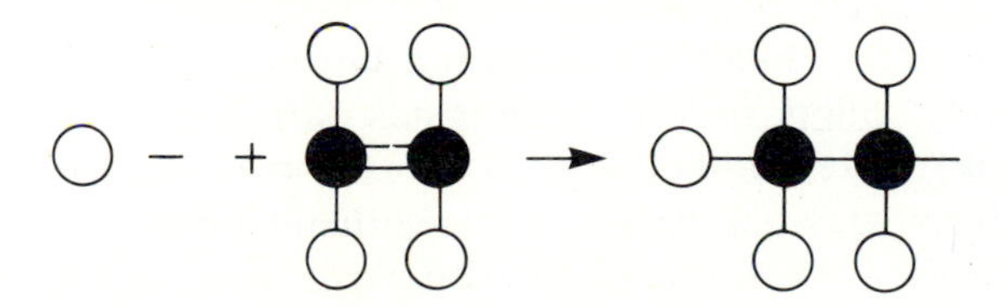

This growth phase continues in a very short interval of time. Many chains are growing simultaneously during this growth phase when a termination phase takes place. Chain termination can take place when the active ends of two growing chains meet and the two free valences join to form an ordinary chemical bond. Thus:

$$\mathrm{H} \sim\sim \mathrm{FF} \sim\sim \mathrm{H} \rightarrow \mathrm{H} \sim\sim \mathrm{H}$$

The F at the end of a chain represents a free radical. Termination can also take place as the result of combining a newly formed free radical with a growing chain. For example:

$$\mathrm{H} \sim\sim \mathrm{F} + \mathrm{FH} \rightarrow \mathrm{H} \sim\sim \mathrm{H}$$

Thus, these structural units, monomers, are joined end to end to produce long chainlike molecules known as polymers. The process by which polymers are formed is known as *polymerization,* which may be defined as "the chemical reaction by which single molecules are linked to form large molecules."

Another technique, using the addition method, for varying the kinds of polymers that can be produced is copolymerization. Here, a mixture of two monomers is used. Depending upon the properties of each, a range of characteristics may be obtained. The copolymer of vinyl chloride and vinyl acetate to produce PVC is one of the most important industrial thermoplastics. The synthetic styrene butadiene rubber, SBR, is a familiar copolymer.

Plastics that are capable of being changed into a substantially infusible or insoluble product when cured by application of heat or chemical means are called *thermosetting* plastics. Here, reactive portions of the molecules may form cross-links that are attached to other molecules, during polymerization.

These materials, once molded, will distort under stress at approximately 250 to 500°F, but they will not become soft or fusible. Thermosetting materials will char and burn at high temperatures. They are inclined to have greater tensile strength and hardness, and, in some cases, are lower in raw-material cost than thermoplastics. Thermoplastics, on the other hand, have generally higher impact strength and pleasing appearance, and can be converted into a finished product at lower manufacturing cost.

All plastic articles are initially derived from molding compounds. These molding compounds consist of a resin or binder and one or more of the following components: fillers, plasticizers, dyes and pigments, and lubricants. The "resin," as the principal component, gives the compound its name and classification, and imparts the primary properties to it. It is the cohesive and adhesive agent that provides rigidity and binds together the filler particles. The "filler" is usually an inert, fibrous material that modifies the properties of the resin or imparts special properties to it. Fillers often form a large part of the bulk of a compound. Structurally, fillers may be aggregates with round or polyhedral shapes such as chalk; plates or flakes such as clay or mica, or lamellar glass fibers such as fiber glass, asbestos, and synthetic fibers; or cellular material such as glass beads, vermiculite, or foamed glass. Some fillers can be catalytic to the structure of the plastics. They can also cause cross-linking of the structure, as in the case of carbon black. Some fillers are completely inert and remain as a void filler. Or again, fillers may absorb the polymer phase because of an ability to wet and/or react chemically to form a bond with the polymer. Wood flour is the filler that is used to produce general-purpose molding powders, while cotton fibers are chosen for improved toughness and mica powder for certain electrical applications.

Plasticizers are added to the compound if flow or softness must be regulated to improve its processability. At times, plasticizers may be added to change the properties of the polymer, thus extending its application. Plasticizers are usually liquids with high boiling points.

Dyes and pigments are added to impart color to the molded part. The pigment added must not decompose at molding temperature.

Lubricants of wax or stearates are added occasionally to a molded compound to facilitate the filling of intricate mold cavities as well as to aid removal of parts from the mold.

Carbon reinforcements are finding wide application in high-performance installations such as aircraft and other products requiring high strength and stiffness. Recent developments in fibrous reinforcements include biodegradability in living tissue fiber (calcium sodium metaphosphate). This fiber, in addition to having improved isotropic properties, may prove to be less of an airborne hazard at the workplace. This fiber has been claimed to be stiffer, stronger, and less brittle than milled glass.

Natural and synthetic rubber are plastics that are very important to industry. These incompressible elastic plastics present problems very similar to those encountered in the manufacture and application of other types of plastics. Some specialized equipment, such as callenders, is required for the preparation of raw material. Rubberlike plastics may be extruded into shapes, filaments, or sheets, and formed by the extrusion, transfer, and compression types of molds.

Some thermoplastic and thermosetting plastics are used as adhesives. They are tough, strong, and reliable, and can be applied between almost any combination of materials. One of the first and most successful applications was the use of transparent plastic sheets between two plates of glass to form our present-day safety glass. The tough plastic adheres to the glass and prevents splinters from flying. Plywood made with plastic adhesives withstands weathering and water and can now be used for concrete forms and outside sheathing for homes. Large wooden columns and thick panels can be built by curing the adhesive by dielectric heating. The combination of plastic and wood makes a strong structural member. The "C" process of sand molding, as described in Chapter 8, uses a phenolic resin for binding the sand in cores and molds. Wooden furniture and metal cabinets are held together by adhesives that simplify their design and reduce cost of manufacture.

Fabrics are made of many plastics in pleasing colors. They are durable, tough, and easy to clean. Natural fabrics such as cotton and wool are facing stiff competition from these new plastic filaments.

Reinforcing of plastics by metal and glass fibers has produced strong, flexible, and light materials, such as that used in bullet-proof vests for the armed services.

Plastics are manufactured under controlled conditions to give uniform raw material and finished products. Color, surface, strength, and size variations are minor. Failures are usually due to misapplication by the engineer. There is no such thing as a bad plastic; all plastics are good if compounded properly for a particular use. Sufficient information for making a propert choice is available from materials suppliers and plastic-molding companies.

The relative cost of plastics is being reduced continually as improvements are made and demand increases. In general, they are more expensive than metals on a per-pound basis. By coating metals with plastics, the benefits of both materials can be obtained. Plastics must do the same job as another material at less cost, or a better job for the same money, or be in the position that a plastic is the only material that can do the job, before they are used more extensively.

The cost of metal molds for plastics is about the same as that of die-casting dies. A superior polish in a mold for plastic is transferred directly to the molded piece. The molding operation in some cases may be slower than die casting and permanent-mold casting; however, the trimming of flash associated with die casting is normally not required with molded plastics.

In the development of new functional designs where plastics are being considered, the engineer should make his decisions on engineering properties and prototyping. The design procedure recommended should approximate the following.

1. Conception of part with reference to size, geometry, and quantity to be produced.
2. Candidate materials should be screened on the basis of properties, ease of processing, secondary operations to be performed and cost.
3. Analysis of design. Here calculation of all dimensions including wall thickness will be made. Design information based upon engineering properties and manufacturing experience will be utilized in this analysis.
4. Prototype models will be developed and tested under either actual or simulated use conditions.
5. Redesign and retooling will take place for the second generation of prototype models.
6. Design is finalized. Material selections and methods of processing are identified.

Today the functional design engineer is handicapped because of the dearth of engineering property data on most commercial plastics.

GENERAL PROPERTIES

The problem of selecting plastic materials is that of finding the material with suitable properties from the standpoint of intended service, methods of forming and fabricating, and cost. New and improved plastic materials possessing progressively broadening characteristics are being introduced continually. There are plastics that do not require plasticizers, that have greater flexibility under lower temperatures, and are stable under higher temperatures. Some resist water, acids, oils, and other destructive agents. The wide use of plastics testifies to their value; however, fundamental limitations should be considered when applying a new material or adapting an old material to new applications.

Thus, the engineer will be concerned not only with those properties such as static strength, rigidity, impact strength, fatigue, but high and low temperature capability, flammability, chemical resistance, and arc resistance. Today, there is a stimulus to replace metals with lighter weight plastics to conserve energy.

Effects of Temperature

Plastics are inclined toward rigidity and brittleness at low temperatures, and softness and flexibility at high temperatures. They are fundamentally unstable dimensionally with respect to temperature, and are susceptible to distortion and flow when subjected to elevated temperatures. The thermoplastics are particularly susceptible, while the thermosetting plastics are much more resistant, though they differ only in degree. The distinction between the thermal stability of the thermosetting and thermoplastic resins is not well defined. A true distinction can be drawn only between individual plastics, rather than between classes of plastics.

Impact resistance is improved at elevated temperatures, while these temperatures not only seriously reduce the mechanical properties of plastics, but accelerate the destructive action of external agents to which they are sensitive. Continuous heating also may induce brittleness and shrinkage in heavily plasticized materials by volatilization of plasticizers. The use of one plastic in contact with a dissimilar plastic in a proposed application should be checked first in the light of possible "migration of plasticizer," sometimes resulting in discoloration or hardening of one of the plastics.

In general, moderate temperatures are required for storage of plastics over long periods; low temperatures are to be avoided because of the low-temperature brittleness of most of the plastics, and high temperatures should be avoided because of the rapid loss of mechanical properties, volatilization of plasticizers, and the susceptibility of a large number of plastics to distortion.

Effects of Humidity

Plastics, with only a few exceptions, are extremely sensitive to the effects of water. High-humidity atmospheres induce water absorption and varied resulting effects, depending upon the composition and formulation of the plastics. Increased water content plasticizes some materials, and there is a general lowering of the mechanical properties. Water absorption is responsible for swelling in certain plastics and the ultimate decomposition of a few. Moist or wet atmospheres may extract plasticizers from heavily plasticized materials and also provide conditions favorable to fungal growth. In recent years, however, new plastics have come into use that have high moisture resistance and may contain water indefinitely and still maintain their normal properties. It is a common practice to de-humidify some thermoplastics immediately prior to molding; in the unique case of nylon (polyamide), moisture is occasionally introduced into the molded product to enhance toughness.

Extremely dry environments may cause brittleness in certain plastics as a result of loss of water that normally contributes to their plasticity. Cyclic wet and dry atmospheres are more destructive to plastics than continuous exposure at constant humidity because of the mechanical stresses induced in the plastics by swelling and shrinking with moisture absorption and moisture emission. Relatively constant, moderate to low humidities are preferred for plastic storage because of the adverse effects of water on the structure and properties of these materials, and the possibility of plasticizer loss by extraction and fungal attack in moist atmospheres.

Effects of Light

Prolonged exposure to sunlight will affect adversely all plastics with the exception of tetrafluoroethylene (Teflon). The change induced by the ultraviolet components may vary in kind and severity from color loss to complete disintegration as a result of chemical degradation of the polymeric compound or plasticizer, which results in loss of strength, reduced ductility, and increased fragility. Many plastics are offered in special formulations containing "ultraviolet inhibitors", which should be used when the product is to be exposed to sunlight for long periods. Therefore, exposure of plastics to sunlight during storage should be avoided especially when the transparency of clear materials is to be preserved.

Weight

As a family, plastics are lighter than metals. Most plastics have a specific gravity slightly over 1.0, while certain polyolefins actually float on water.

Electrical Resistivity

Plastics have excellent electrical resistivity, giving them wide application as insulating materials. In the high-frequency applications, plastics are particularly advantageous and, consequently, are being used to a large extent in the fields of radar and television.

Heat Insulation

Plastics have low heat conductivity and, consequently, have application as insulating materials. In particular, thermosetting plastic handles are used for appliances and tools subjected to or generating heat.

Fabrication

The principal characteristic of plastics from the fabrication standpoint is ease of molding. Both thermosetting and thermoplastic materials lend themselves to molding irregular and complex shapes with relatively short curing cycles.

Plastics may be joined by using various cements, chemical solvents, and mechanical fasteners. Heat sealing, which parallels somewhat the welding of metals, is used extensively in joining light thermoplastic films by means of dielectric heating. Friction adhesion has had moderate application in the joining of small thermoplastic parts.

Plastics can be machined with conventional machine tools. However, certain precautions should be exercised. In order to maintain a good finish, a heavy flow of coolant should be used so as to avoid temperatures that will distort the work. In some thermosetting laminates (incorporating glass, for example), the customary high-speed steel tool will not stand up in view of the abrasive action of the laminating material. Here, either tungsten carbide or ceramic cutting tools must be used.

Effects of Oxygen

Organic plastics are nearly all subject to oxidation when exposed to the atmosphere. The process is accelerated by high temperatures and light; but, over long periods of time, oxidative deterioration may take place at room temperature. Oxidation susceptibility depends largely upon the chemical nature of the plastic and its compounding. Materials with the greatest number of double bonds in their molecular structure will generally be the most sensitive to oxidation. Yellowing and a gradual loss of strength and ductility are the principal results of oxidative processes.

Effects of Loading

Under moderate conditions the common thermoplastic materials are subject to distortion and flow when significantly loaded. Such plastics cannot be expected to maintain a high degree of mechanical stability over extended periods when subjected to stress; this is especially true when they are also exposed to relatively high temperatures. The thermoplastics should, however, maintain themselves fairly well when not subjected to load or when subjected to only moderate load. Recently, fillers, such as glass wool, have been added to thermoplastics to further improve this property.

The thermosetting plastics are much more load-stable than the thermoplastics because of their structure and the inclusion of fillers in their formulation. In the laminated form they provide a rather high order of distortion and creep resistance. When not subjected to mechanical stress they may be considered to be highly stable. These materials, however, may suffer creep over long periods, especially when maintained at elevated temperatures.

Thermoplastics should not be subjected to load when stored; and, whenever possible, the loading of stress-bearing thermosetting moldings or laminates should be removed or reduced. Thermoplastics are sometimes annealed for stress relief, in order to enhance their resistance to otherwise detrimental conditions.

Chemical Stability

Plastics, in general, possess a high degree of inherent stability with respect to chemical deterioration. In many instances, this stability may be fortified by the addition of the proper stabilizers during compounding. While there is a vast difference from one plastic to another, generally there is a plastic available to resist virtually any commercial chemical.

THERMOPLASTICS

Thermoplastics can be reshaped upon being reheated, or more significantly—can be reground and remolded. The reader should understand that success in thermoplastic waste recycling requires practical techniques as well as markets for the reprocessed resin. For example, polyethylene terephthalate (PET) used in cans cost about $0.65/lb in 1988 and the reprocessed clear about $0.45/lb. Although there is a saving of about 31 percent in the reprocessed material, there may not be a market for it. It is estimated that plastics, by the year 2000 will be responsible for about 10 percent of our municipal solid waste.

Many uses have been found for the various types of thermoplastic materials, and processes have been developed to produce economical finished and semifinished products. Thermoplastics can be molded, rolled into sheets, used for strip coatings, and extruded into shapes. New materials that offer additional advantages are constantly being developed. Most thermoplastics can be obtained in a variety of desired shades or mixtures of colors and many thermoplastics are available in any degree of transparency from crystal-clear to opaque. This is a definite advantage in product development and often eliminates costly painting or finishing operations. Some of the principal thermoplastic materials available to the engineer today are discussed briefly. After the type of polymer is shown, its recurrent unit of chemical structure is illustrated.

Poly(methyl methacrylate)

```
        CH3
        |
CH2—C—
        |
        COOCH3
```

Poly(methyl methacrylate) is primarily used in products in which color stability and high light transmittance are important in addition to good chemical and environmental resistance. Poly(methyl methacrylate) types (Plexiglas, Lucite) are almost perfectly clear and transparent. Sheets and shapes are stable under normal temperature (140 to 190°F) and light loading (4000 to 10,000 lb/in.^{2}), but are subject to creep under heavy loads. They have low water absorption and are affected very little by weather conditions. They have the best optical properties of the transparent plastics and are known for their ability to pipe light, their use in windows and optical lenses, and their beauty of color. They are also widely used in producing automotive lenses, wine glasses, canopies, machine cover guards, and table tops.

Cellulose Acetates

```
H       C——C        O_
 \     /|   |\     /
  \   / OR  H \   /
    C   H       C
   /  \ |      / \
_O     C——O      H
       |
       CH2OR  (R is the acetate or nitrate group)
```

Cellulose acetates are tough, transparent materials with high impact strength and are pleasant to touch because they are poor conductors of heat. They can be varied from semirigid types to elastic types. Good molding qualities make them suitable for compression and injection molding. All colors can be provided. Water absorption is high. Tensile strength is 5000 to 11,000 lb/in.^{2}, and flexural strength is 4000 to 8000 lb/in.^{2} Their physical properties are quite sensitive to temperature. Strength, weather, and chemical resistance are not good because of crazing, embrittlement, and loss of transparency. They are used for general-purpose parts.

Cellulose Acetate Butyrate

Cellulose acetate butyrate (Tenite II) is tougher and has higher impact strength and weather resistance than cellulose acetates. Otherwise, it is similar in many respects. It is vacuum-formed easily and has proven especially useful for laminating with thin-gage aluminum foil. It can also be vacuum metalized. Metalized butyrate sheet is used for printed and formed signs and for interior decoration. Cellulose acetate butyrate is also used for fuel lines, tool handles, and general purpose parts.

Cellulose Nitrate

Cellulose nitrate (Pyralin, Nitron) is tough, water-resistant, and the most stable of the cellulosic plastics. Otherwise, the general characteristics listed previously apply. It is commonly known as "celluloid" and is extremely combustible. Today, the greatest use of cellulose nitrate is as a base for lacquers and cements. In its molded form, it has been used for fashion accessories and decorative inlays.

Ethyl Cellulose

Ethyl cellulose (Ethocel, Colon, Nixon) has the general characteristics of the other cellulose types and is used for parts requiring good strength and electrical characteristics at low temperatures, such as for camera cases, flashlights, and some military equipment. This thermoplastic is light amber and slightly hazy in color. It cannot be produced in crystal-clear sections, although it can be provided in many transparent, translucent, and opaque colors. It is available both in pellet form for molding and extrusion and in sheet form for fabrication. It has good processability and can be produced in heat-resistant formulations, high-impact formulations, and formulations for use in contact with foods.

Polyamide Resins

Polyamide resins $(CH_2)_nCONH(CH_2)_mHNOC$ (nylons) are a family of thermoplastics that have physical stability at high temperatures and a high, rather sharp melting point (450 to 505°F). These resins excel in toughness and strength (see table below), chemical resistance, and low friction value.

Temperature, °F	(TSM) Strength, lb/in.^{2}	
	Type FM1,	Type FM3,
70	15,700	12,900
77	10,530	7,600
170	7,600	6,760

Today, there are two general classes of nylons: those polymerized by condensation of a dibasic acid and a diamine and hexamethylene diamine to form polyhexamethylene adipamide, such as nylon 6/6, and those polymerized by addition reactions of ring compounds that contain both acid and amine groups on the monomer, such as nylon 6. These two nylons (6 and 6/6) comprise the vast majority of the sales, since they offer the most favorable combination of properties, ease of processing, and price.

Molded nylon materials have a high tensile strength, large elongation, and comparatively good impact strength at normal temperatures, but, like other thermoplastics, their tensile strength decreases with rising temperature and the elongation increases. Because of its crystalline structure, nylon preserves its physical properties at high temperatures to a much greater extent than most thermoplastics.

Nylon is water-sensitive and will absorb large amounts of water when immersed or exposed to atmospheres of high humidity. Variation in the equilibrium moisture content causes a corresponding variation in the mechanical properties of nylon. With an increase in moisture content, the stiffness and tensile and flexural strengths decrease, and the impact strength increases as does the flexibility. When exposed to elevated temperatures, unmodified nylons undergo dehydration and resulting loss in mechanical properties. For practical purposes, unstabilized nylons can withstand continuous exposure to 65°C (150°F), which is the rating that Underwriters Laboratories apply to all standard nylon materials.

Fibers and fabrics of nylon possess the same general characteristics as the molded materials. They are tough, extremely strong, stable over wide temperature variation, and more resistant to weathering and fungal attack than natural materials. Nylon fabrics are not completely weather resistant, and they may support fungal growth when fabricated with susceptible lubricants or sized.

Nylon gears tested in a kitchen mixer showed practically no wear after a 24-hour run with the beaters turning in a bowl of sand. Bronze gears on a similar test were worn out at the end of that time. Three months later, after 2400 hours of continuous operation, the nylon gears were still running and showed very little wear. Another application of nylon is for bearings. A thin, $\frac{9}{64}$-in.-thick section of nylon was molded to a steel shell to form the bearing. A $\frac{1}{2}$-in.-diameter shaft of SAE 1112 steel operating at 350 rpm was loaded at 200 lb/in.^{2} with a clearance of 0.005 in between shaft and bearing. The nylon showed no visible wear after 2,500,000 revolutions without lubrication.

Since molding requires special experience, the supplier should be selected with care. Many parts are made by machining because nylon works like metal. Guides for the best way to machine can be obtained from the E. I. DuPont de Nemours Co.

Polyamide-imide

This high-performance opaque engineering thermoplastic, characterized by good dimensional stability, creep resistance, impact resistance, and excellent mechanical properties to temperatures approaching 500°F is available under the commercial name Torlon.[1] Characteristic mechanical properties of unfilled polymide-imide at room temperature include: tensile strength approximately 27,000 lb/in.^{2}, flexural strength about 31,000 lb/in.^{2}, compressive strength around 40,000 lb/in.^{2}, and elastic modulus of about 750,000 lb/in.^{2} Impact strength is about 2.5 ft · lb/in., with a tensile elongation of approximately 15 percent. At temperature of approximately 500°F the tensile strength will deteriorate to about 7500 lb/in.^{2} The creep resistance of this thermoplastic is excellent and it has superior thermal stability.

It is available in several grades. The most common being unfilled, 30 percent glass fiber-filled, 30 percent graphite fiber-filled and 40 percent glass fiber-filled. The unfilled grade provides the best impact resistance while the graphite fiber-filled gives the stiffest grade. The 40 percent glass fiber-filled is the least expensive.

Products made from this engineering plastic include mechanical parts for valve trains, pump and compressor parts, seals, vanes, flow control parts, generator parts, and electrical devices.

Polycarbonate

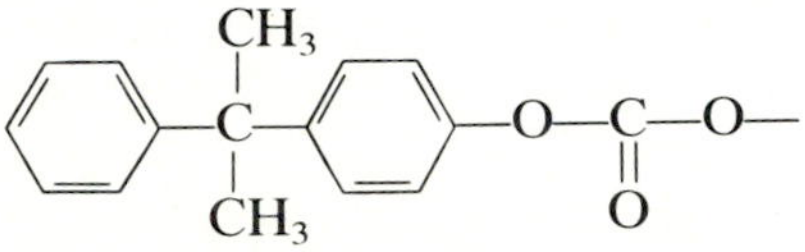

CH_3

—C—O—C—O—

CH_3 O

Polycarbonate is a polyester of carbonic acid produced from dihydric phenols with a suitable carbonate precursor. Today, it is sold by the General Electric Company by the trade name Lexan and by Mobay Chemical Company under the name Merlon.

This thermoplastic has high impact strength and stiffness, good dimensional stability and creep resistance, low water absorption, good electrical properties, and glass-like transparency. It can be blended with other polymers, thus improving specific properties such as chemical resistance and impact resistance.

Polycarbonate is available in a variety of grades, including blow molding and extrusion grades. Flame-retardant grades have been developed to meet a variety of thermal environment conditions. It is commonly used in such products as automotive headlamps and tail lights, instrument panels, traffic light housings and lenses, signal lenses, returnable bottles, office water coolers, etc. Housings for hand-tools, internal components in television sets, and fixtures in the marine and boat industry are broadening the application of this material.

[1] Trademark of Amoco Chemicals Corp.

Polystyrenes

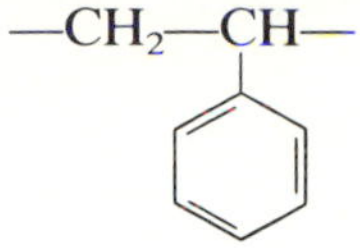

Polystyrene (PS) is a transparent, amorphous resin produced in both high-heat and general-purpose grades. It is usually molded without a plasticizer as it has very good molding qualities. The resins have good electrical properties, clarity, and low sensitivity to water (5 percent in 48 hours). They are rigid to approximately 150 to 205°F (in various formulations), resistant to distortion, and do not creep or flow below their load-bearing capacities. They have limited impact resistance. Films of the material offer good protection as they resist both water and weathering. Representative physical properties include: specific gravity of 1.05; flexural strength of approximately 12,000 lb/in.^{2}; tensile strength 7000 lb/in.^{2} They are very little affected by many chamicals, including alcohol and gasoline. They have broad colorability advantages.

Today, packaging is the largest application for both crystal and impact grades. Flame-retardant grades find application in such products as housing for motors, television cabinets, and appliance parts. An interesting application of low-density polystyrene is in the cast-metal industry. Duplicates of metal castings are first mass-produced in polystyrene. These polystyrene patterns with polysyrene gates and risers attached are coated with a refractory wash and immersed in unbonded silica sand, which is poured in the flask around the pattern and jolted to insure proper compaction. Molten metal is then poured into the gating system, which evaporates along wth the pattern under the heat of the metal. The pattern is thus replaced with an exact duplicate in metal.

Polyvinyl and Vinyl Copolymers

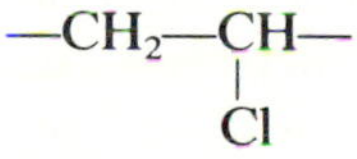

Polyvinyl chloride (PVC) is the principal member of this family. Others that will simply be mentioned include:

1. Polyvinyl acetate (PVAC), used mainly in latex paints, adhesives, and surface coatings
2. Polyvinyl alcohol (PVAL), mainly used in adhesives, paper coating, and textile finishing
3. Polyvinyl fornal, used in enamels to provide heat resistant insulation on wire
4. Polyvinyl butyral (PUB), used as an interlayer in the manufacture of safety glass

5. Polyvinyl fluoride (PVF), used as a weather-resistant coating.
6. Polyvinylidene chloride (PVDC), used mainly as a food packaging film

PVC is a thermoplastic produced by the addition of the following compounding ingredients: heat stabilizers, lubricants, plasticizers, impact modifiers, fillers, and pigments. PVC is processed by extrusion, compression molding, blow molding, calendering, injection molding, powder coating, and liquid processing. Depending on the use of plasticizers, PVC ranges from strong, rigid products to flexible or elastomeric ones. In general, they are insensitive to water at low temperature. They are tough, have high impact strength and abrasion resistance, and have unlimited colorability. PVC has good dielectric characteristics, does not support flame, is not toxic, and resists moisture.

PVC products include pump parts, handles, plumbing pipe and elbows, building siding, window frames, gutters, interior molding and trim, shower curtains, refrigerator gaskets, appliance components, and auto tops. PVC is extruded into tubing, tape, and wire coating. It is formed into sheets used to enclose and seal packages for storage under dehydrated conditions. It is made into filaments that are woven into warm fabrics, window screen, and shoe fabrics. The resins are used to coat and finish paper and as components of varnishes and paint.

Polyethylene

Polyethylene, —CH_2—CH_2—, is produced in three types: low-, intermediate- and high-density. All three types, formed through the polymerization of ethylene gas, have in common the properties of resistivity to solvents, alkalies, and mold acids; excellent dielectric properties; good colorability; very low moisture absorption; and relatively low cost. The three types vary in physical properties from hard to soft, rigid to flexible, and tough to weak.

Type I (low-density polyethylene) is quite flexible, having a stiffness modulus of approximately 20,000 $lb/in.^2$ It has high impact strength but low heat resistance: 175°F represents the upper temperature limit for sustained use. Film and sheeting is the major market for low-density polyethylene.

Type II (intermediate-density) is less flexible than type I. It has a stiffness modulus of about 45,000 $lb/in.^2$ and can withstand temperatures up to 200°F.

The high-density polyethylene materials known as type III are fairly rigid, having a stiffness in flexure up to 140,000 $lb/in.^2$ Type III materials are able to withstand temperatures similar to those sustained by type II. The principal outlets for high-density polyethylene is in the injection molding of such items as pails, tubs, closures, bowls, plates, paint-brush handles, and hardware. Polyethylene is easily colored with either dry pigments or liquid colorants. Since it is an inert material, it is suitable for containing food products and various chemicals and reagents for which nonreactivity is important. Although polyethylene is processed by all of the techniques used in connection with thermoplastics, extrusion is the principal method used in the production of film, sheeting, coating, and profiles, and coating on wire and cable.

Polypropylene

$$-\underset{\displaystyle CH_3}{\underset{|}{CH}}-CH_2-$$

Polypropylene is similar in properties to type III polyethylene. It is slightly lighter (specific gravity approximately 0.9) somewhat more rigid, and capable of withstanding higher temperatures (up to 230°F). Polypropylene has good resistance to creep and it also has good chemical resistance, electrical insulation, stiffness, impact strength, abrasion and stress crack resistance. It has low water absorption and can be extruded, blow-molded and injection-molded. Its principal application is as a packaging material when high strength and heat resistance are necessary. For example, it is used extensively for food and household chemical containers. It is also used as a material for appliance components such as small gears, cans, housings, etc. Continuous fiber-reinforced PP sheet is finding application as a stampable and compression moldable material for bumper beams in the automotive industry. This has reduced the weight of cars by approximately 20 pounds.

ABS

$$-CH_2-CH{=}CH-CH_2-\underset{\displaystyle C{\equiv}N}{\underset{|}{CH}}-CH_2-$$

ABS is named from the three monomers used to produce the polymer: acrylonitrile, butadiene, and styrene. Acrylonitrile contributes to heat stability, chemical resistance, and aging resistance; butadiene assists in providing low-temperature property retention, toughness and impact strength; styrene helps provide gloss, rigidity and facilitates molding. The most important property of the ABS family is toughness. Tensile strength is about average for thermoplastics at about 7000 lb/in.^{2} at 65°F. Figure 6.1 illustrates representative stress–strain data for general-purpose ABS. These materials can be injection-molded and calendered into sheets.

ABS can be purchased in both unpigmented and colored pellets. It is also available in powder form for alloying with other polymers such as PVC and polycarbonate. Since ABS is slightly hygroscopic, it should be dried for a few hours at about 200°F just prior to processing. This polymer can be molded, hot stamped, painted, vacuum metallized, printed, electroplated, laminated, and embossed. Typical applications include pipe and fittings, housings for household appliances, electrical conduit, football helmets, utensil handles, telephones, refrigerator tanks and door liners, luggage, power tools, faucets, etc.

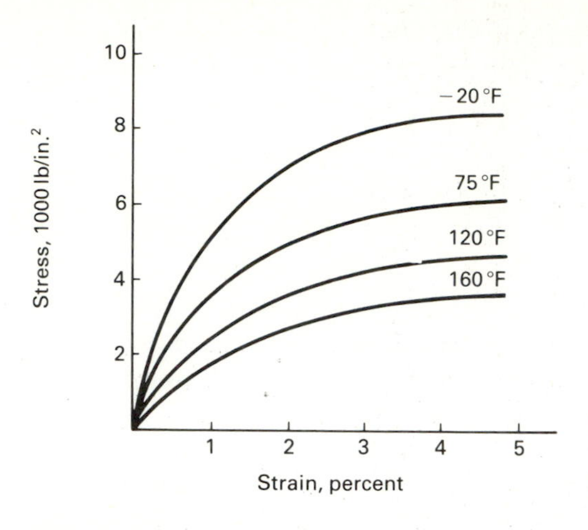

FIGURE 6.1
Typical stress–strain data for general-purpose, high-impact ABS.

Fluorocarbons

This family includes tetrafluoroethylene, commonly known as Teflon and chlorotrifluoroethylene, usually referred to as Kel-F. Both of these fluorocarbons have the disadvantage of comparatively low strength and high cost. The principal advantages of this family of thermoplastics include high temperature resistivity (500°F for Teflon, 390°F for Kel-F) and inertness to most chemicals. They have low coefficients of friction and good dielectric properties.

Acetal

Acetal copolymer resins provide strong, hard, highly crystalline components. This thermoplastic in its natural color is a translucent white, although it is available in a variety of colors. It is one of the most creep-resistant crystalline thermoplastics and has good tensile strength (approximately 10,000 lb/in.^{2}), resistant to temperature (240°F), low friction characteristics, and resistance to most solvents. The acetal copolymer family has wide application in connection with plumbing (sprinkler nozzles, valves, pumps, sinks, fittings, faucets), automotive (hardware, seatbelt components, handles, gears, housings), pen barrels, aerosol bottles, and containers. This copolymer is competitive with ABS, especially the elastomer-modified grades that provide good impact resistance.

Polyurethane

Polyurethanes, —R_1—NHCOO—R_2—OOCHN—, can be formulated as thermoplastics for extrusion and injection molding applications or as thermosets in connection with liquid casting. The major advantage of polyurethane plastics is high durability. They are also excellent thermal and electrical insulators. The major use of polyurethanes is to protect surfaces from wear, although foamed polyurethanes find application in automobile padding and insulation. Other applications of polyurethane elastomers include pipe lining and pipe, flexible tooling molds, fork lift truck tires, shoe soles, and electrical insulation.

Aromatic Polyesters

Aromatic polyester is a relatively new thermoplastic resin that is being used today in such applications as solar energy collectors, electrical connectors, fasteners, hinges, tinted glazing, and appliances. Polyarylate (trade name for aromatic polyester) can be injection, transfer, and compression molded. This engineering plastic provides excellent thermal resistance and resistance to ignition and flame spreading. This family maintains good retention of physical properties under ultraviolet irradiation and has low haze development when subjected to outside weather conditions. Its typical tensile strength is 10,000 $lb/in.^2$ at room temperature and it has about 8.5 percent yield elongation.

Thermoplastic Polyester

The principal application of this thermoplastic resin, commonly known as PET, is as a packaging material for carbonated beverages, frozen vegetables, and the production of fiber for the manufacture of clothing. It has recently been approved as a packaging material for distilled spirits, thus it will compete with glass and other ceramics in a variety of applications. Its success as a container material is due primarily to its shatter resistance, lightness in weight, ability to be recycled, clarity, toughness, and low cost. Blow-molding grades of PET can be purchased in either clear, green, and amber colors.

Shellac Compounds

Shellac resin is a natural product of the lac bug *Tacchardia lacca*. The resin is thermoplastic and is generally employed as an impregnant and binder in heavily filled moldings or in laminates. Shellac compounds are characterized by their good electrical properties, scratch hardness, and resilience.

The mechanical properties of shellac compounds depend upon the type and characteristics of the filler or lamina. Wood-flour-filled moldings have low tensile strength, while that of the laminates may be quite high. Shellac resins soften at rather low temperatures, and the compounds are not suitable for use much above 150°C.

THERMOSETTING PLASTICS

Thermosetting resins are used with or without modifying agents or fillers. The base materials are compounded and then formed in molds. Heating the compound changes the resin to a binder. The molded product usually is hard and dense as a result of forming under pressure.

Thermosetting Resins

The resin imparts certain properties to the finished product and functions principally as a binder for the filler. Thermosetting plastics are classified according to the resin used. The main classes are: (1) phenol formaldehyde, (2) urea formaldehyde, (3) melamine formaldehyde, (4) polyesters, (5) epoxies, (6) silicones.

Modifying Agents and Fillers

Modifying agents include dyes, pigments, lubricants, plasticizers, accelerators, graphite, and so forth.

Fillers, together with the resin, affect the properties of the final product. Wood flour, for example, reduces the cost and improves the mechanical properties. Cotton flock improves impact resistance still further. Fabric cuttings, cord, and string are used for high impact, and paper pulp for improved impact of special shapes. Mica filler will improve the dielectric strength and reduce electric losses. Asbestos improves the heat and moisture resistance. Slate, silica, soapstone, or clay may be selected as filler for improved appearance, for electrical resistance, and for thin sections of the final product.

The mineral fillers increase the wear of the mold and the wear of such tools as cutters, saws, drills, and taps when further machining is required after molding. Short-fiber fillers keep the cost down; the long-fiber materials, having improved impact, are more expensive. It is not economic to select an impact material better than that required for the application under consideration.

Phenolic Resins

OH OH

—⌬—CH_2—⌬—CH_2—⌬—CH_2—

CH_2 OH OH CH_2

—⌬—CH_2—⌬—CH_2—⌬—CH_2

OH OH

Phenolic resins have been a popular engineering material since the early part of this century because their heat resistance, chemical resistance, good dielectric properties, surface hardness, dimensional and thermal stability, and ease of molding. Single-stage phenolic resins are produced with an alkaline catalyst and the formaldehyde necessary to complete the reaction with phenol. The ingredients are reacted under controlled conditions to produce chains that are thermoplastic. The process is stopped before cross-linking occurs, so that the resin has enough shelf-life for subsequent processing. In the high-temperature molding process, the reaction between the phenol and formaldehyde is then completed, producing an infusible cross-linked thermoset material. Today, there is a wide variety of parts produced from this important plastic. Perhaps the principal uses are in electrical equipment and automotive components.

The main advantages of the phenolic family are heat resistance, rigidity, dimensional stability, wear resistance, good dielectric properties, resistance to many chemicals, and low cost. The main disadvantages are its impact strength and colorability. The high-impact category (there are three types: general-purpose, impact resistant, and heat resistant) of phenolic resins are capable of impact strength of only 10 to 33 ft · lb/in. in the Izod notched test. The natural color of phenolics is black. If black or the darker colors of brown and blue are not satisfactory, the only alternative is to paint the phenolic part, which will of course increase the cost.

The impact- and shock-resistant grades incorporate either glass fibers, cotton flock or cellulose as filler material, while the heat resistant phenolics usually use hydrated alumina, mica, clay, and other minerals. These mineral fillers also contribute to the dielectric properties and the dimensional stability.

Phenolic resins can be molded successfully by the screw-injection process in addition to transfer and compression molding methods. The more rapid injection process allows phenolics to compete economically with many thermoplastic materials. In the screw injection process, the temperature relationships between the barrel and the mold when molding phenolics is just the reverse of what is characteristic when molding thermoplastics. With phenolics, barrel temperatures are kept at 200 to 230°F to plasticize the resin, whereas thermoplastics need much higher temperatures, usually in the range of 500 to 600°F. With phenolics, the mold temperatures are raised to 320 to 370°F to cure the material, while with thermoplastics the mold temperature is reduced to around 200°F to cure the molded part.

Laminated-type Phenolics

Another group of plastic products, known as laminated phenolic plastics, uses sheets of cloth or paper to give greater mechanical and electrical strength. The term "laminations" applies to the layers of plastic-treated paper, cloth, and wood veneer that are piled up to form (by pressure and temperature) the thickness of the molded sheet or plate. The use of this material requires another set of instructions, a portion of which will be given so they may be compared with other molded materials. The greater proportion of laminated materials is used in the form of sheets, plates, channels, angles, bars, rounds, and tubes (round and rectangular). These materials are molded under greater limitations than are regular molding materials.

Plates are furnished in sizes of 36 × 36 in., 36 × 72 in., 48 × 48 in., and 48 × 96 in. The thickness starts with $\frac{1}{32}$ in. and varies as follows

From $\frac{1}{32}$ to $\frac{1}{16}$:	in $\frac{1}{64}$-in. steps
From $\frac{1}{16}$ to $\frac{3}{16}$:	in $\frac{1}{32}$-in. steps
From $\frac{1}{4}$ to $\frac{3}{8}$:	in $\frac{1}{16}$-in. steps
From $\frac{1}{2}$ to $\frac{3}{4}$:	in $\frac{1}{8}$-in. steps
From 1 to $\frac{1}{2}$:	in $\frac{1}{4}$-in. steps

Thicker plates can be molded to order. Thickness tolerances vary with types of material, thickness, and size (0.030 to 0.060 per inch of thickness).

Information on the use of rods, bars, angles, Z bars, and channels is omitted. The tubes and rods are made by wrapping the continuous sheet around a mandrel of the desired shape, and baking. Angles and channels are made by molding laminated sheets with an inside radius at least equal to the thickness of material and a corresponding outside radius at the corners.

The dielectric strength of the laminated type is best at right angles to the laminations and, therefore, in applying the laminated type electrically, advantage should be taken of this characteristic wherever possible.

Thermosetting phenolic resin should be applied mechanically with the thought in mind that it is not a ductile material like common metals. The laminated-type products, plates, angles, channels, tubing (also bars to some extent), and other laminated-molded-type shapes have the inherent characteristic of splitting relatively easily (as compared with metals) between the laminations as a result of incorrect machining or application of mechanical loads that place the bond or binder of these laminations under adverse stresses. The higher the bond-strength value, the more resistant to splitting.

The laminated type (also the laminated-molded type generally) has better mechanical properties (except resistance to splitting) than the chopped-molded types when manufactured using the same paper, cloth, or other material in relatively simple shapes or molds. The chopped-molded type may be produced in fairly intricate molds similar to those of certain nonlaminated compositions. The chopped-molded type generally has better resistance to splitting as compared with the previously mentioned laminated type.

This material, especially in plate form, is subject to a small but indeterminate warping, and is therefore not adaptable to parts requiring refined accuracy of alignment. Wherever one surface of laminated plate is subjected to a different treatment (such as machining, sanding, or varnishing) from the other, warping is almost certain to result. Heat treatments of laminated plate are also likely to cause warping.

Subjecting the laminated type to temperatures above 140°C may result in splitting or blistering. Exposure of this material to temperatures above 115°C for more than 4 hours will also cause blistering.

Urea Formaldehyde

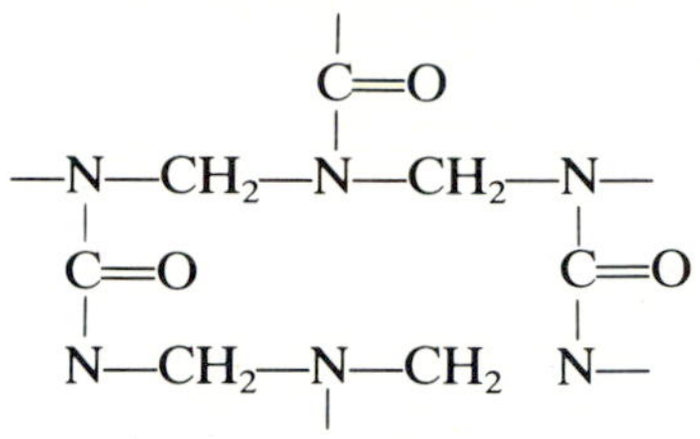

```
              |
              C=O
              |
 —N—CH2—N—CH2—N—
  |               |
  C=O             C=O
  |               |
  N—CH2—N—CH2     N—
        |
```

Urea formaldehyde (Beetle, Plaskon) is a thermosetting polymer formed by the controlled reaction of formaldehyde with various compounds that contain the amino group. They are valuable to the engineer primarily because pastel shades and translucent colors can be produced readily. Although comparatively low in impact strength, they are remarkable for their hardness, color stability, and color fastness. They have excellent electrical properties, notably high arc resistance. Their resistance to organic commercial solvents, weak acids, and weak alkalies is excellent. The ureas are odorless and tasteless, and have high heat resistance that makes them suitable for use in contact with foodstuffs. Urea formaldehydes are primarily either compression-molded or transfer-molded at temperatures varying from 260 to 340°F.

As with phenolics, the properties of the urea resins can be modified by varying the composition. The resins are supplied as liquids, spray-dried solids, and filled molding compounds. The liquid and dried resins find extensive use in laminates and chemically resistant coatings. The resin is used with water or in solvent solution for adhesives of several kinds. As an adhesive, it is used for the bonding of plywoods and fine furniture veneers, for which use it combines good bonding properties with ease of application. It is also used for paper-backed adhesives (for belt sanding). Urea resins are compounded with various alkyd derivatives as a low-cost surface on metal and wood.

Melamine Formaldehyde

Melamine resins (Melmac) belong to the amino resin family, as does urea formaldehyde. However, the melamine resins maintain their properties over a substantially greater range of temperatures and offer higher electrical resistance than do either the phenol or urea resins. The melamines have all the advantages of the ureas and phenolics, with the added feature of better water, acid, and alkali resistance.

Because of their special properties, melamine resins can be used with all fillers, including glass fibers. When melamine is compounded with α-cellulose fiber, a hard, tough plastic results. For general-purpose molding, a wood-filled material shows good flexural strength. High-heat-resistance plastics are made with asbestos as the filler and these combine high arc resistance with their elevated temperature characteristics, making them valuable materials in the electrical field. Melamine plastics have gained extensive use as commercial and domestic tableware in view of their durability, pleasing appearance, and resistance to sterilizing temperatures.

Polyesters

The polyester resins are copolymers of a polyester and usually styrene. They have good strength, toughness, and resistance to chemical attack; low water absorption; and ability to cure at low pressures and temperatures.

These resins have had wide application in low-pressure laminates. Here, the unpolymerized resin can be ladled onto the laminating fabric and subsequent cure can be completed at temperatures as low as 160°F for the majority of formulations, while some will cure at room temperature. Curing pressures range from ordinary contact pressure to 30 lb/in.^{2} One widespread application of this family of thermosetting plastics is the manufacture of boat hulls, using glass fibers as the laminating material and a polyester as the resin.

Alkyl is formulated from unsaturated polyester resins, cross-linking monomers, mineral fillers, catalysts, lubricants, colorants, and fibrous reinforcements. Alkyd polyester is useful in applications requiring high-temperature electrical properties, arc-track resistance, and dimensional stability. Depending upon its fibrous reinforcement, this plastic has flexural strength from 8000 to 24,000 lb/in.^{2} and tensile strength from 5000 to 15,000 lb/in.^{2} This family has poor chemical resistance and low resistance to hydrolytic degradation. These compounds may be molded by compression, injection, and transfer methods. The material is available in both granular and pelletized condition, allowing easy feeding in automatic equipment. The principal use of this thermosetting plastic is in connection with such products as circuit breakers, transformer housings, distributor caps, rotors, etc.

PBT (polybutylene terephthalate or polytetramethylene terephthalate)

This polyester-type resin provides a range of properties characterized by high heat resistance, good mechanical strength and toughness, general chemical resistance, good lubricity and wear resistance, low moisture absorption, and eye appeal.

It is mainly fabricated using the injection process. When reinforced with glass, its tensile strength approaches 20,000 lb/in. Some of the more significant applications of PBT include automotive body components, ignition components, automotive window and door hardware, transmission components, switches, relays, motor brush holders, fuse holders, terminal blocks, food processor blades, fans, gears, and frame and bracket parts.

Epoxies

Epoxies are a broad family of polymer materials characterized by the presence of epoxy groups in their molecular structure. Although high-molecular-weight linear epoxies are often used as thermoplastics, particularly in coatings, they are most often used as thermosetting materials that cross-link to form a three-dimensional nonmelting matrix.

The epoxies are some of the most versatile plastics, since their toughness, chemical resistance, and adhesion make them a popular resin for protective coatings. Their laminates have been used extensively as tool materials because they can reproduce contours and dimensions readily and because they have good physical and mechanical properties. Dies for stamping and forming have made use of this material at considerable savings. For example, one manufacturer experienced a $16,000 saving in the cost of dies used in producing station-wagon roofs by converting to epoxy from steel. Because of epoxy's lighter weight, it also has considerable application as a jig and fixture material where much material handling is involved.

As a casting resin, it has been used extensively for encapsulating capacitors, resistors, and other electrical equipment.

Epoxies require two separate components that react chemically and harden soon after mixing. The hardening is accelerated by application of moderate heat.

The epoxy resins have good chemical resistance, excellent bonding properties, and exceptionally high strength when reinforced with fiberglass or similar laminates, and the resins are resistant to temperatures up to 350°F. Their principal drawback is the relatively high resin cost.

Silicone Resins

Silicone resins have excellent resistance to heat, water, and certain chemicals. The resins are usually filled or used in laminated parts. Good physical properties are maintained from 450 to 500°F.

This family of synthetic polymers is partly organic and partly inorganic. Silicones are classified as fluids, elastomers, and resins. Thus, silicone polymers may be fluid, gel, elastomeric, or rigid in their final form. The properties that make silicones attractive to engineers include: low surface tension, high lubricity with rubber and plastic surfaces, excellent water repellency, good electrical properties, thermal stability, chemical inertness, and resistance to weather.

Silicone fluids are used mainly in the plastics and rubber industry as mold-release agents. When added to organic plastics in molding, they provide water repellency, abrasion resistance, lubricity, and flexibility.

Silicone rubbers are used to make seals, gaskets, tubing, hoses, wire insulation, coated fabrics, contact lenses, and artificial toe and finger joints. Silicone resins and composites provide thermal stability to temperatures to about 300°C (572°F).

DESIGN SUGGESTIONS

Selecting Materials

The following step-by-step procedure will help in selecting the most suitable material for an application, that is, the material giving the best service at the lowest cost.

1. Analyze carefully the requirements of a new application and the conditions under which the apparatus has to operate in service. Determine whether such factors as the mechanical (impact-resistance), electrical, or heat-resistance properties are of primary importance for selecting a specific material.
2. With the above data on hand, check the properties of several plastics. Compare the specific values of those materials selected in order to make the choice of the material that comes closest to the requirement.
3. Estimate the working stresses within the material which may be prevalent in the assembly under consideration.
4. Determine the required minimum sizes of the important cross sections of the part to be designed, using the property values of the material selected. Calculate the strength of the cross sections, using a minimum safety factor of 4 for the mechanical properties and a minimum safety factor of 6 for the electrical properties listed on the property sheet. These minimum safety factors must be increased by the designer, depending upon:
   a. The importance of the molded part in the functioning of the assembly
   b. The accuracy of the estimates of the working stresses involved
   c. The deterioration or decrease of the test values given due to conditions prevalent at the place of use or service
5. Before releasing the design, consult the molding department.

The following information is given as a guide for designing molded parts of plastic.

Minimum wall thickness. The minimum thickness of any wall or rib should be between $\frac{3}{32}$ and $\frac{5}{32}$ in. depending upon the moldability of the plastic employed.

The above ruling will not affect the minimum wall thicknesses around holes and inserts. Deep draws ($\frac{1}{2}$ in. and above) require separate consideration. Consult the molding department.

Maximum wall thickness. It is advantageous not to exceed the above minimum wall thicknesses. Heavier walls waste material and increase the time for molding and curing. Any wall thickness of urea resin should not exceed $\frac{1}{8}$ in. Design suggestions for the size of holes apply to urea resins only as long as a $\frac{1}{8}$-in. thick wall is not exceeded. Heavier walls will be porous or undercured. Consult the molding department in case deeper holes or inserts are required.

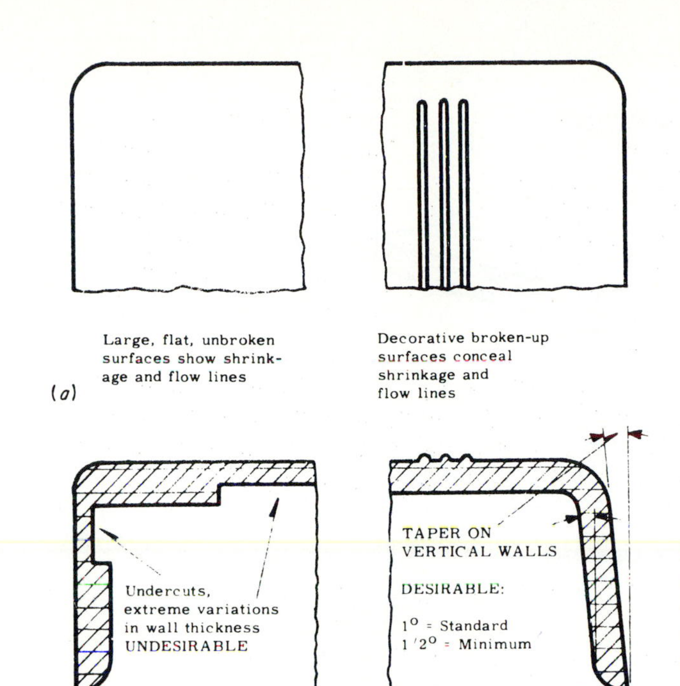

FIGURE 6.2
(*a*) Using decorative design to conceal shrinkage and flow lines. (*b*) Illustration of undesirable undercuts. (*c*) Application of taper on vertical walls.

Draft on side walls. A draft or taper of 1 to 2° is desirable (see Fig. 6.2b) on the vertical surfaces or walls parallel with the direction of mold pressure. A minimum draft of 0.5° may be permissible in order to facilitate removal of molded parts from the mold cavity. However, this amount of draft may necessitate the use of a more substantial knockout mechanism. Short outside surfaces approximately $\frac{1}{2}$-in. high formed vertically by the matrix of a compression mold in some cases may have the taper omitted. Thin walls or barriers require a minimum taper of 1° on both sides (Fig. 6.2*c*). A minimum surface of $\frac{3}{16}$ in. should be provided at the end for the stripper or ejector pin (Fig. 6.3).

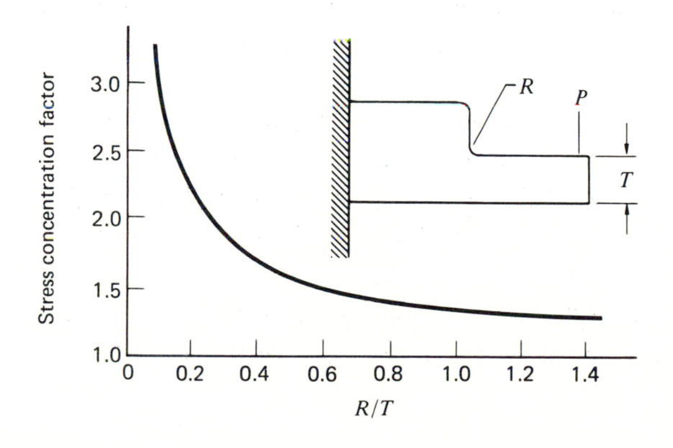

FIGURE 6.3
Fillet, rib, and surface for stripper pin design.

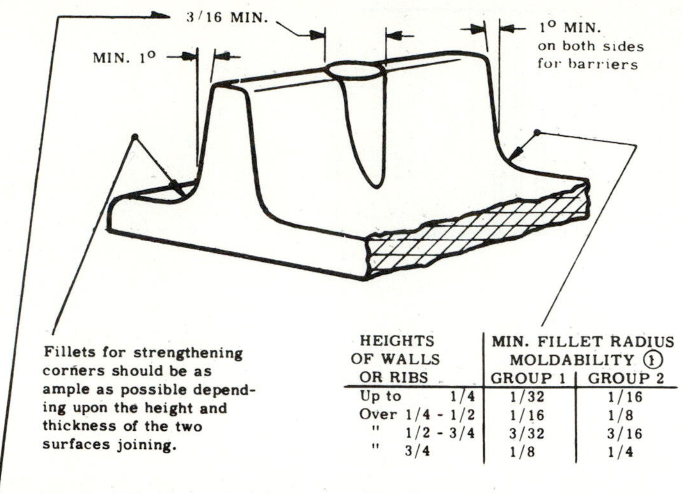

HEIGHTS OF WALLS OR RIBS	MIN. FILLET RADIUS MOLDABILITY (1) GROUP 1	GROUP 2
Up to 1/4	1/32	1/16
Over 1/4 - 1/2	1/16	1/8
" 1/2 - 3/4	3/32	3/16
" 3/4	1/8	1/4

FIGURE 6.4
Effect of fillet radii on stress concentration.

Fillets. Fillets (Fig. 6.4) must be added to facilitate molding with minimum distortion and breakage. The use of fillets and generous radii facilitates the flow of plastic material into the mold cavity, minimizes stress concentrations, and lessens material warpage after molding. Figure 6.4 shows the effect of fillet radius for a given section thickness.

Ribs and bosses. Ribs (Fig. 6.5) can be used to increase part strength without increasing wall thickness and bosses can provide reinforcement around holes. Ribs and bosses must have 5° taper and adequate fillets.

Pronounced markings (flow lines or shadows) may appear opposite ribbed surfaces. Specify decorations (Fig. 6.2*a*) to conceal this undesirable

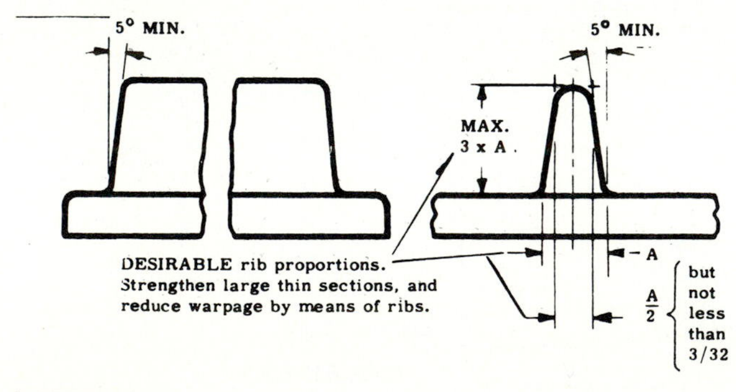

FIGURE 6.5
Design of rib proportions.

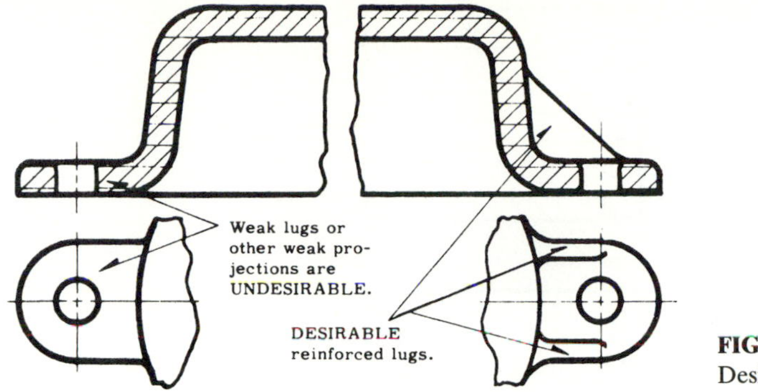

FIGURE 6.6
Design of lugs.

effect. Ribs should not be higher than three times the thickness. Bosses should not be higher than twice the diameter.

Lugs and projections. Lugs (Fig. 6.6) and projections on molded parts should be reinforced by ribs and fillets to reduce warpage and distortion and to increase the strength without increasing the section of lugs or projections.

Irregular parting lines, surfaces, and projections. Irregular surfaces at the parting line for the mold and projections requiring featheredge members of the mold should be avoided as they increase the cost of the mold, of the molded piece, and of the mold maintenance (Figs. 6.7 to 6.9).

Location of cavity number, parting line, and ejector pins. Cavity numbers on parts are required for identification purposes during manufacture. Such numerals will be engraved in the mold 0.094-in. high and 0.006 to 0.008 in. deep, which will make them raised on the molded piece. The location of these numerals should be indicated on the drawing at a place on the finished piece where it will not affect the appearance, design, or function of the assembly.

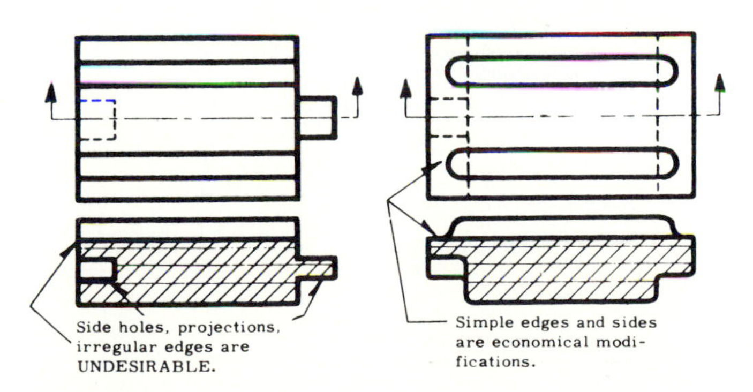

FIGURE 6.7
Design should avoid side holes and side projections.

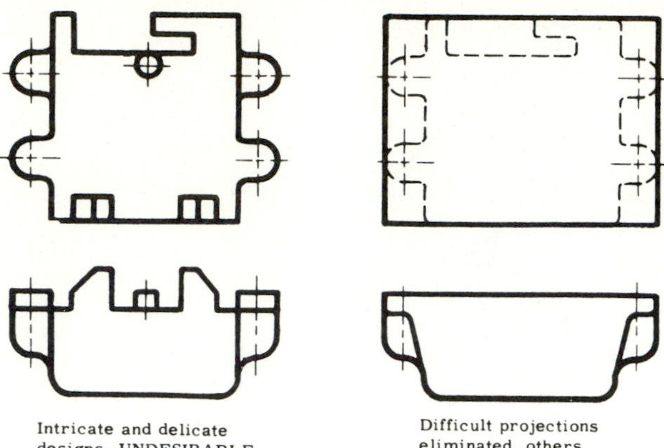

FIGURE 6.8
Design should avoid intricate and delicate sections.

The location of the parting line and ejector (stripper) pins should also be discussed with the molding department because they leave a mark on the molded surface of the part and thus affect the appearance.

Threads. Threading (Fig. 6.10) below $\frac{5}{16}$ in. diameter should be cut after molding. Threads of $\frac{5}{16}$ in. diameter and above should be molded. Threaded through holes, tapped or molded, should start with a countersink in order to eliminate chipping. Owing to shrinkage, long-molded thread will not assemble with standard metal thread, but will fit mating molded thread or mating short-metal thread approximately $1\frac{1}{2}$ times its diameter. A suitable grip or straight knurls should be provided for unscrewing mating thermoplastic parts.

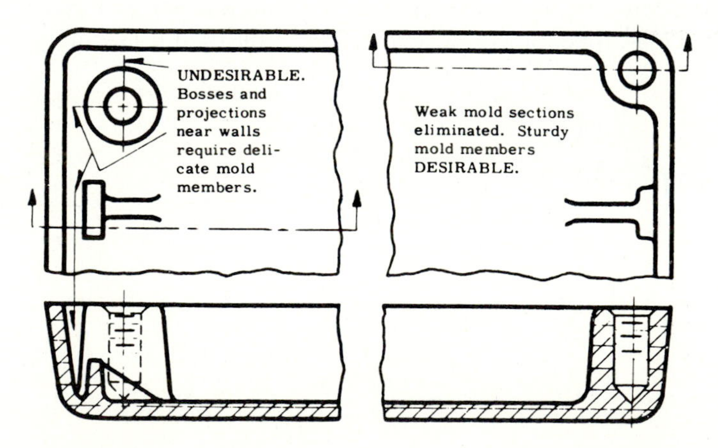

FIGURE 6.9
Design for sturdy mold members.

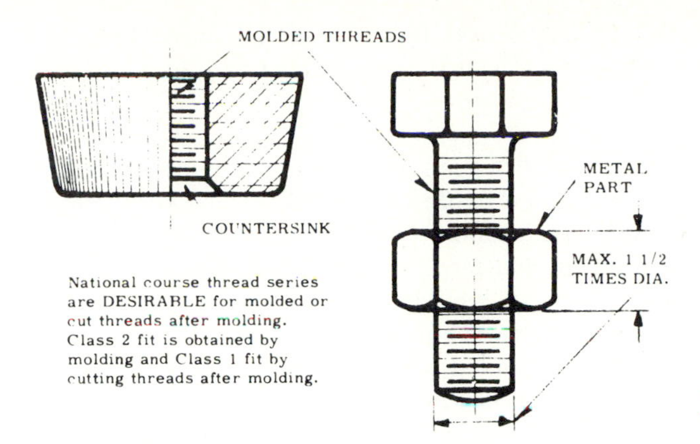

FIGURE 6.10
Thread design.

Letters. Molded letters (Fig. 6.11), raised 0.006 to 0.008 in. above the basic surface of the molded part, are reproductions of engraved characters in the mold. These characters may be engraved on pins or blocks inserted in deep molds or cavities if the chosen location is inaccessible for the engraving tool. These engraving pins or blocks permit raised letters on depressed surfaces of the molded piece by inserting the pins or blocks to 0.010 in. above the level of the mold cavity if molded raised letters are objectionable.

Standard characters are recommended in order to avoid costly engraving. Filled characters engraved in the surface of the molded piece are expensive and therefore should be avoided. They may be obtained by embossing or branding characters into the surface of dark-colored pieces. A heated die is pressed into the surface of the finished part and is coated by means of a colored foil for better contrast.

Labels and printings do not adhere sufficiently to a glossy surface although the adhesion may be improved by sand blasting the mold or the molded surface of the finished piece. However, special paints and inks that etch and adhere very satisfactorily to glossy surfaces are now available for most plastics.

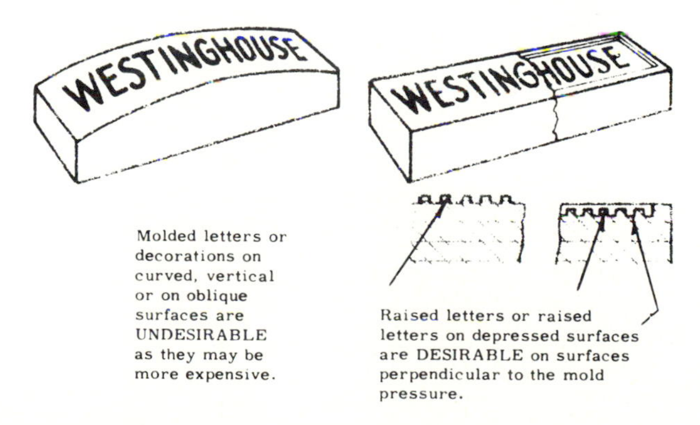

FIGURE 6.11
Letter design.

Baselines for tolerances and oblong holes. In order to eliminate loose or careless fits, it is well to establish baselines for dimensioning and for specifying tolerances.

Holes—slots and recesses. Holes smaller than $\frac{1}{16}$ in. diameter must be drilled or formed after molding. The maximum depth of molded vertical holes, slots, or recesses depends upon the type of hole, method of molding, and the moldability of the material used, as indicated in Table 6.1. The problem of varying diameters due to draft must be considered. This is, of course, especially prevalent on deep holes. The minimum thickness of walls or solid material required around or between molded vertical holes or recesses depends principally upon the depth of hole as shown in Table 6.1.

The molded recessed through hole in Fig. 6.12 should be given preference, because this type will result in the sturdiest mold members—and slots. Therefore, ample fillets on holes are desired for economical production and mold maintenance.

If any of the previously mentioned rounded edges or corners are critical for the proper functioning of an assembly and if the rounding must not exceed a permissible radius, the maximum allowable radius should be specified on the drawing. A recommended minimum radius of 0.020 applies to most designs. The elimination of sharp corners reduces the stress concentration at these points and consequently the molded parts will have greater structural strength.

Straight-wall through holes are formed by molding pins, mounted usually in the bottom half of the cavity and entering into holes on the top half. The vertical holes formed by this type of pin have a radius at the lower end only.

The four extreme edges of a narrow slot may be improved as shown in Fig. 6.13. By increasing the center section of the mold pin, it is strengthened to assure greater service life.

TABLE 6.1
Hole specifications

Type of detail	Molding method	Max. depth of holes
Blind hole and slot with straight wall and recess (Fig. 6.12), taper 5° min.	Compression Transfer	$H \leq 2D_1$ $H \leq 3D_1$
Through hole with recessed wall (Fig. 6.12)	Compression Transfer	$T \leq 2(D_2 + D_3)$ $T \leq 3(D_2 + D_3)$
Through hole with offset wall (Fig. 6.12)	Compression Transfer	$T \leq 4D_2$ $T \leq 6D_2$
Through hole and slot with straight wall (Fig. 6.15)	Compression Transfer	$T \leq 3D_6$ $T \leq 3D_6$

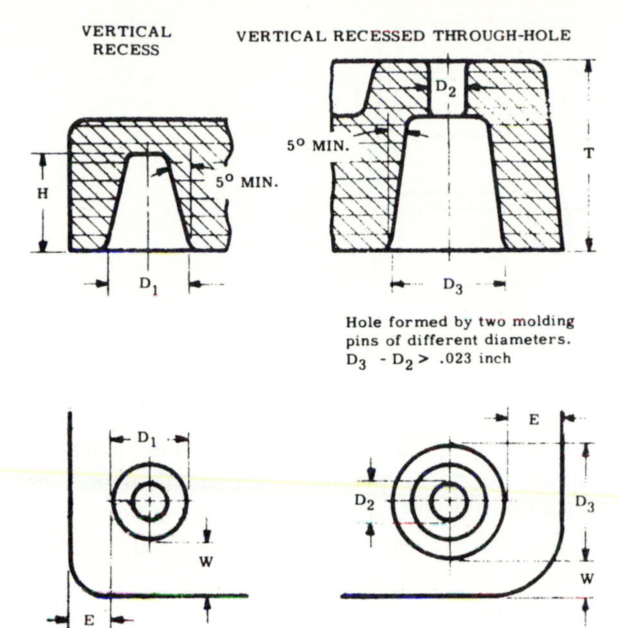

FIGURE 6.12 Design for vertical recesses and recessed through holes.

Molded horizontal holes. Molded horizontal holes are frequently uneconomic, as they require removable mold members. Short side openings or slots may be formed by vertical mating mold members as indicated in Fig. 6.14.

Long molded horizontal holes require a support at the free end of their molding pins and a removable mold section, increasing the cost of the mold

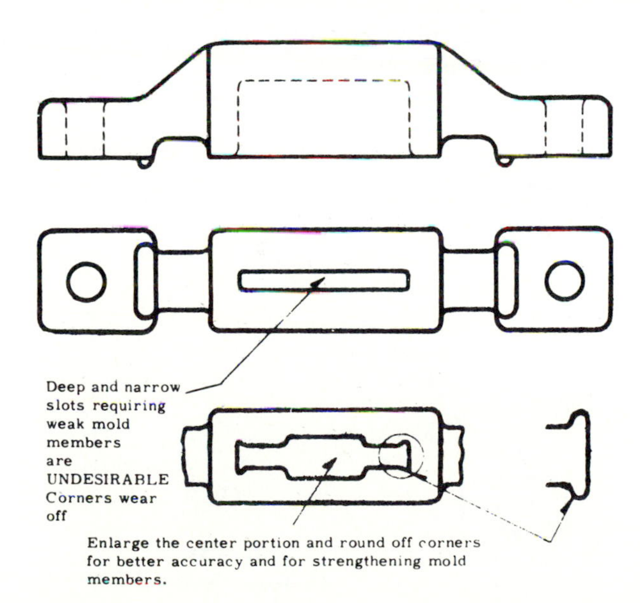

FIGURE 6.13 Design for slots.

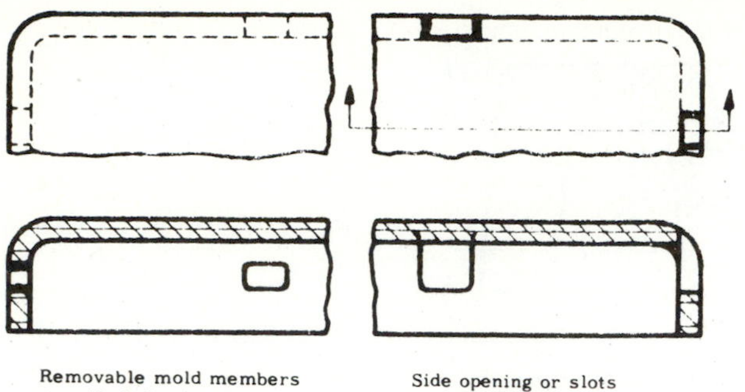

FIGURE 6.14
Design for side openings.

and the cost of production. An alternative is to use cam-activated cores in conjunction with the mold. If quantities to be produced are small, it may be more economical to drill such round horizontal holes or other holes that do not have their long longitudinal axes parallel with the direction of the molding pressure.

Inserts—general. Avoid inserts if a satisfactory assembly may be made with speed nuts, speed clips, spring clips, threads, self-tapping screws, drive screws, or tubular rivets. Inserts in urea resins are impractical and are in general more suitable in plastics with higher "elongation" property. When it is necessary to use inserts, it is important that the insert be surrounded with enough plastic to absorb shock, so that the insert does not break loose when the part is in service.

Molded-in inserts. Use standard inserts of brass. The metal insert surface must be cleaned for good contact or alignment after molding.

Use round-end inserts instead of square or other shapes (Fig. 6.15).

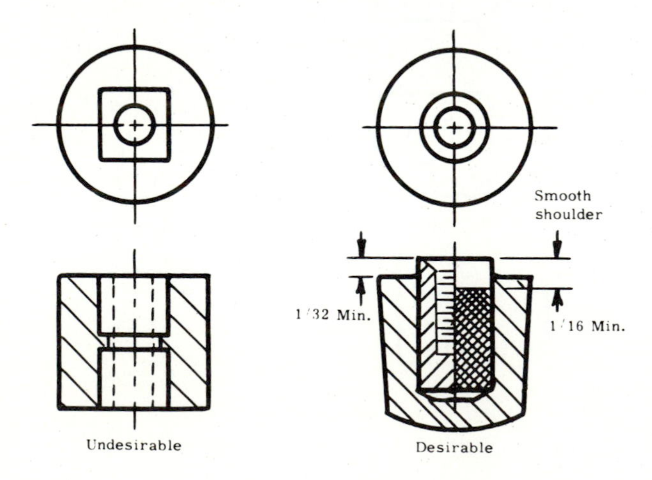

FIGURE 6.15
Design for inserts.

Shapes other than round will be expensive to anchor and seal in recesses required in the mold. Use closed-end inserts to prevent flow of molding material inside them.

Standard molded-in inserts are held by the diamond knurl on the imbedded metal surfaces. A straight knurl would not prevent the inserts being pulled out. Providing a smooth shoulder of $\frac{1}{16}$-in. minimum width on the projecting end of the insert will allow a better seal in the recess of the mold to eliminate flow of material inside.

The imbedded end of the insert should not extend close to the surface. The recommended minimum thickness of material around molded-in inserts should be $\frac{3}{32}$ in. for inserts up to $\frac{1}{4}$-in. diameter, $\frac{5}{32}$ in. for inserts above $\frac{1}{4}$- to $\frac{1}{2}$-in. diameter, and $\frac{1}{4}$ in. for inserts above $\frac{1}{2}$- to 1-in. diameter.

Molded-in inserts extending through the part should be avoided for compression-molded parts.

Standard inserts in a horizontal position while molding must be supported on both ends. A short insert having an imbedded length shorter than 120 percent of its diameter may be drilled deeper and retapped. Whenever possible, avoid inserts in a horizontal position while molding, because they require special mold construction and molding procedure that increase the cost. Use pressed-in inserts if the pull-out load is light.

Avoid inserts if they weaken the maximum cross section permissible for the design. Molded or tapped threads, drilled holes together with self-tapping screws, or drive screws may be used for light loads and permanently assembled parts.

Special inserts anchored in holes of the mold, instead of on pins, should be designed with the imbedded section (diameter and length) conforming with standard inserts; that is, the imbedded length should be limited to 150 percent of the diameter. They should have round smooth shoulders for good seal in the recesses of the mold. The threads should end $\frac{1}{16}$-in. minimum from the shoulder, as shown in Fig. 6.16.

Long, slender inserts may be bent during molding. Inserts having a long,

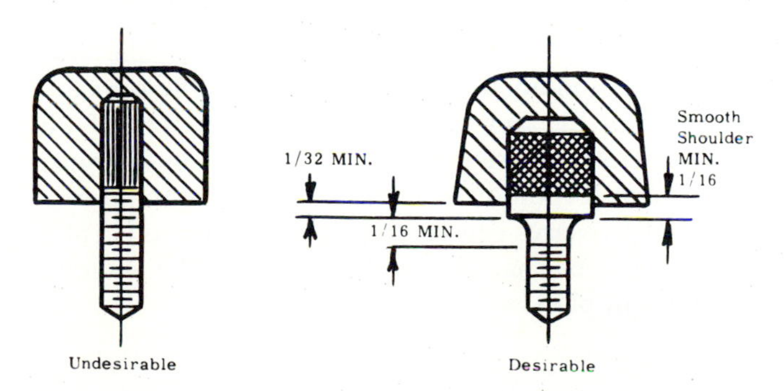

FIGURE 6.16
Do not extend threaded portion of the insert into the molding.

TABLE 6.2
Press fit for pressed-in inserts

Hole diameter, in.	Wall thickness, in.	Recommended interference, in.
$\frac{1}{8}$	$\frac{3}{16}$ to $\frac{1}{4}$	0.0005–0.0035
$\frac{1}{8}$	Over	0.001–0.004
$\frac{3}{16}$	$\frac{3}{16}$ and over	0.002–0.005
$\frac{1}{4}$	$\frac{1}{4}$ and over	0.002–0.005
$\frac{3}{8}$	$\frac{3}{8}$ and over	0.002–0.005
$\frac{1}{2}$	$\frac{3}{8}$ and over	0.003–0.006

knurled surface or anchoring shoulders far apart ($\frac{5}{8}$ in. or further, depending upon the shrinkage) may cause cracking. Specify short knurls and short distance between outer shoulders to permit the shrinkage, or specify compounds with small differences of shrinkage between metal and molding material, in critical cases.

Inserts of other materials, like asbestos, are feasible for such purposes as improving the arc resistance or the mechanical properties.

Pressed-in metal inserts. These inserts should be considered only if the previously mentioned assembly methods and the molded-in metal inserts are not practical or are not economical. These inserts are advisable only in cases where the pull-out load is light, as for terminals. The resulting press fits of the pressed-in inserts are variable owing to the necessary working tolerances in the diameters of the holes and across the knurled inserts, as shown in Table 6.2. Heavy walls and impact-resisting materials are most suitable for pressed-in inserts. Specify 30 straight knurl on this type of insert when it has an OD of $\frac{3}{16}$ in. and below, and 46 straight knurl above $\frac{3}{16}$-in. OD.

Tolerance Specifications

Several factors must be taken into account in establishing the tolerance specifications that can be expected in molded plastic compounds. The principal factor is the tolerance capability of the toolmaker in the construction of the mold. The magnitude of the material shrinkage also plays a factor in tolerance capabilities of molded parts, and distortion of the molded piece due to internal stresses resulting from high mold pressures causes dimensional variation. Table 6.3 provides a summary of tolerances that can be held during the molding process for several of the important plastics.

Coating of Plastics

Both painting and metal-vapor coating are practical and economical ways of coating part or all of plastic components to give the product certain

TABLE 6.3

Plastic compound	Typical molding tolerance, in./in.
Methyl methacrylate	0.0015
Polystyrene	0.0020
Cellulose acetate butyrate	0.0025
Vinyl	0.0100
Polyethylene	0.0150

characteristics. Painting includes the use of either lacquers or enamels. Enamels are more frequently used with thermosetting plastics in view of the high baking temperatures required.

Most metal-vapor coating involves the deposition of a thin film of aluminum on the plastic surface to be coated. The thickness of this coat is only two to three millionths of an inch and is applied primarily for reasons of appearance. One pound of aluminum will cover over 20,000 ft^2. Aluminum is usually used because it is inexpensive and is easily evaporated in the vacuum chamber.

Copper alloys, silver, and gold may also be used for vacuum metallizing.

IDENTIFYING PLASTICS

The engineer often needs to identity an unknown plastic. Perhaps the easiest way is by flame testing. Hold any flame (burner, cigarette lighter, match) at an edge of the unknown plastic resin and observe the following points.

1. Does the material burn?
2. If it burns, what color is the flame?
3. Does the material smoke?
4. If it smokes, what color is the smoke?
5. Does the material drip?
6. Does it continue to burn after removal of the flame?
7. What odor is given off?

Table 6.4, developed from information published by E. I. DuPont de Nemours, Inc. provides characteristic information that helps in the identification of several of the more commonly used commercial plastics.

Electroplating is also an important coating process applied to plastics. There are several distinct advantages in the use of plated plastics over plated metals. Some of the more important of these are the following.

1. Plated plastics are lighter, making them easier to handle and less expensive to ship. For example, a recent model of Chevrolet used a plated ABS grille

TABLE 6.4
Polymer burning identification chart

Polymer	Ease of lighting	Self extinguishing	Odor	Nature of flame	Behavior of material
Acetal	Moderate	No	Formaldehyde	Clean blue flame, no smoke	Melts, drips; drippings may burn
ABS	Readily	No	Use known sample to compare	Yellow with black smoke	Softens, drips, chars
Cellulose acetate	Readily	No	Acetic acid burned sugar	Dark yellow flame, some sooty black smoke	Melts, drips; drippings continue to burn
Epoxy	Readily	No	Use known sample to compare	Yellow, spurts black smoke	Chars
Nylon	Moderate	Yes	Burned wool	blue flame, yellow top	Melts, drips, froths
Phenolic, molded	Very difficult	Yes	Burned fabric, phenolic	Yellow, a little black smoke, sparks	Cracks badly, chars, swells
Polycarbonate	Difficult	Yes	Sweet carbon odor	Yellow flame, dense black smoke, carbon in air	Softens, spurts, chars, decomposes
Polyester	Moderate	No	Burning coal	Yellow, black smoke, burns steadily	Softens, does not drips; continues to burn
Polyethylene	Readily	No	Burning paraffin	blue flame, yellow top	Melts, drips; drippings may continue to burn or swell
Polypropylene	Readily	No	Burning paraffin	Blue flame, yellow top, some white smoke	Melts, swells, drips
Polystyrene	Readily	No	Illuminating gas	Orange-yellow flame, dense black smoke, carbon in air	Softens, bubbles
Polyvinyl chloride	Difficult	Yes	Hydrochloric acid, chlorine	Yellow flame, green on edges, spurts green and yellow, white smoke	Softens

weighing slightly over 6 lb. This is about one-third the weight of a comparable zinc die-cast grille.

2. Plated plastics can be handled in higher-temperature environments because of their lower thermal conductivity.
3. Complex shapes may be more readily achieved with plastics.
4. Cost per unit may be less in view of a lower cost of resin on a volume basis.

To electroplate plastics, a conductive substrate must be applied to the plastic. This is accomplished by chemical means and is referred to as *electroless metal plating*.

Although all plastics can be plated, there are only three types today that provide good adhesion supplemented with a high-gloss surface and can be plated economically on a production basis. These are ABS, polysulfone, and polypropylene.

Interior applications for plated plastics have proved quite successful during the past 10 years. However, when plated plastics are subjected to severe extremes of outdoor weathering, blistering, cracking, and peeling are more likely to occur. For this reason the automotive industry has been cautious in converting zinc die-cast exterior parts such as grilles, trim, headlamp bezels, etc.

NATURAL AND SYNTHETIC RUBBER

Rubber is the term commonly applied to substances, either natural or synthetic, that are characterized by exceptional elastic deformability. Properly prepared rubbers may be stretched without rupture far beyond the limits of any other engineering material, and upon release of the deforming stress they will return almost to their original shape. Thus, in addition to high elasticity, rubbers may possess resilience to a useful though not perfect degree. Even if rubbers possessed no other useful properties, these two would make them unique among engineering materials.

As its name implies, natural rubber occurs in nature as a latex, a milky fluid contained in specialized tissues that grow between the bark or protective sheath and the main body of certain plants. A much larger number of plants than is generally known yield rubber latex. Several, such as milkweed and dandelions, grow in temperate climates. The rubbery component known as rubber hydrocarbon is identical in all these plants. The nonrubber components of the latex, however, vary widely from plant to plant, and are often injurious to the quality of the finished rubber. It has not been economically practicable to remove these imputities. As a result, commercial natural rubber is produced from less than a dozen species of rubber-bearing plants. Of these, by far the largest source is the tree *Hevea brasiliensis,* a native of the Amazon valley, which has been transplanted to other tropical regions.

Rubber hydrocarbon is more technically termed polyisoprene. Isoprene is

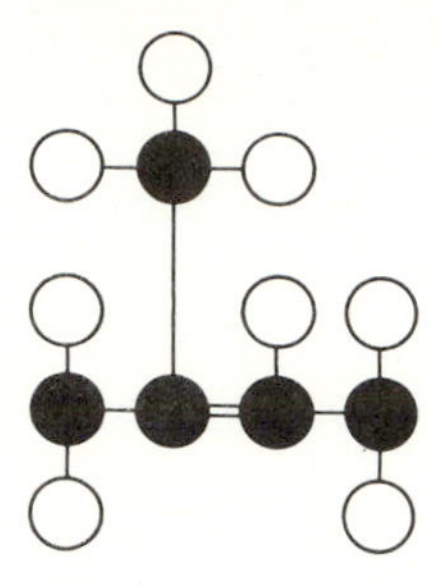

FIGURE 6.17
Isoprene monomer composed of five carbon atoms and eight hydrogen atoms. This is the monomer that makes up natural rubber.

a volatile organic liquid with the chemical formula C_5H_8 (see Fig. 6.17). Like many organic substances, isoprene molecules have the ability to combine with other isoprene molecules to form very large molecules called polymers. Rubberlike polymers are unique in that the individual molecules are believed to join together to form long chains that coil around somewhat like a spring. This is believed to account for their unusual elasticity, and has led to the creation of the special term "elastomer" to describe rubbery polymers.

Because of its unique properties, rubber is vital to the operation of many types of machinery, both military and civilian. It thus has an importance as a strategic material that far surpasses its annual dollar volume. The so-called "synthetic rubbers" that are now commercially available have done much to free the rest of the world, and the United States in particular, from the grip of a monopoly that at times has seriously threatened the whole economy.

During the search for synthetic rubber, a number of polymers have been developed that possess to some degree the properties of natural rubber. These have become generally known as synthetic rubbers. It should be noted, however, that this is a misnomer. They are synthetic elastomers, but none of them is real rubber. Some of them possess one or more properties that far surpass natural rubber. For example, polysulfide elastomers, which are marketed under the trade name of Thiokol, greatly exceed natural rubber in resistance to petroleum oils, greases, and fuels. Silicone elastomers retain useful rubberlike properties at temperatures both higher and lower than those rubber can withstand. By the proper choice of a synthetic elastomer the design engineer can achieve many results that would not be possible through the use of natural rubber. But one or more desirable properties will often be obtained at the sacrifice of others.

The search for improved elastomers has been made more difficult by the dual nature of the rubbery state. All elastomers act in part like solids and in part like fluids. As a consequence, elastomers follow completely neither the laws of solids nor the laws of fluids. This dual nature accounts for many of the virtues as well as the defects of rubberlike materials. It has also made a systematic and rigorous study of the rubbery state very difficult. As a pragmatic art, rubber chemistry has grown to be a highly developed and most useful industrial tool. But the essential nature of an elastomer is still only imperfectly understood. As a result, most of the research and development in this field is by the "cut and try" method, and relatively few fundamental laws have yet been discovered.

In the raw state, none of the elastomers possesses much engineering or practical utility. Raw natural rubber, for example, is quite temperature-sensitive, being hard and hornlike in the winter, and soft and tacky in the summer. In general, the fluid properties of raw elastomers are objectionably prominent.

The first great technical advance, and the one that really established the rubber industry, was the discovery of vulcanization by Charles Goodyear. Vulcanization is the establishment of chemical cross-links between the elastomer molecules. This acts in much the same way as a reinforcing strut in a mechanical system, and in effect ties the whole mass of rubber together into a single molecule. By this means the slippage between molecules is restrained, which minimizes the fluid properties and reinforces the solid properties. Sulfur is the principal vulcanizing agent for natural rubber, and will cross-link several of the synthetic elastomers also. Subsequent advances in the art of vulcanization have been largely the development of catalysts or accelerators to speed up and control the reaction.

If very finely subdivided solids, ranging roughly from 20 to 100 nanometers (nm) in diameter, are mixed into elastomers, they appear to wedge into the molecular interstices. Again this interferes with plastic or fluid properties and reinforces the solid properties. For this reason, such materials are sometimes known as reinforcing fillers. Carbon has been relatively easy to prepare in this small particle size, and has long been the principal reinforcing filler. More recently, finely divided silica, magnesia, and other minerals have become available. Light-colored stocks of good physical properties have thus become possible.

The sulfur-accelerator system and the reinforcing fillers contribute most to the improvement of elastomers as engineering materials. In addition, ingredients are added to minimize attack by oxygen, ozone, and sunlight. All in all, a dozen or more ingredients may be incorporated in a modern rubber compound, and these may be widely varied to meet specific needs.

PROPERTIES OF VULCANIZED RUBBER

In speaking of rubber as an engineering material, we thus do not refer to a singles species, but to a complex family of compounds. Each individual elastomer possesses its own specific virtues, and each may be widely modified by compounding to meet a myriad of specific engineering needs. No single compound excels in all virtues. By wise selection, elastomers may be used by engineers to do things that no other engineering materials can do. Elastomers are useful engineering materials by virtue of their mechanical, physical, and chemical properties. The properties most useful to the engineer will be discussed.

Mechanical Properties

Elasticity. The ability of elastomers to undergo large deformations and yet to return to substantially their original shape makes them very useful for connecting structures that undergo large relative motions. No other engineering material will accommodate as much relative motion in such a small space. Several elastomeric materials can be elongated as much as five to ten times their original length.

Resilience. Because of their ability to restore the energy by which they are deformed, elastomers may be used as springs. Natural rubber can be compounded to have a resilience approaching 95 percent. Synthetic elastomers, in general, will have lower resilience. Because of their large elasticity, elastomeric springs can be made very compact. Moreover, when rubber is used in systems apt to become resonant, high resilience is undesirable. Metal springs for such uses require dash pots or other dampers. These are bulky and often expensive. Elastomers, on the other hand, can be compounded for low resilience. With no increase in size, elastomer springs can be built with inherent damping.

Tensile strength. Natural rubber can be compounded to a tensile strength in excess of 4000 lb/in.^{2} Synthetic elastomers will, in general, have tensile strengths of the order of 3000 lb/in.^{2}, while silicone rubbers have much lower tensile strengths, of the order of 800 to 1000 lb/in.^{2} Although high tensile strength is a measure of a high-quality rubber compound, it should be noted that a rubber stock loaded in tension to this value would be extended to seven or more times its original length. Hence, tensile strength does not have the significance to the design engineer for rubber that it has for steel.

Fatigue resistance. Because elastomers serve some of their most useful functions interposed between structures in relative motion, ability to withstand repeated and often severe flexure is of great engineering importance. The nature of the principal function to be served may determine the type of compound to be used. Elastomers are relatively poor conductors of heat. Therefore the heat generated through hysteresis loss (the inverse function of resilience) may be significant. If ability to withstand flexure is the major consideration, as in a flexible coupling that must accommodate shaft misalignment, low-hysteresis stocks would be desirable. In a vibration isolator under steady-state vibration, some hysteresis might be desirable to damp out high resonant amplitudes. In a shock absorber, it may be desirable to absorb the energy of large-magnitude impacts. In such cases high-hysteresis stocks may be desirable, provided the impacts are sufficiently infrequent to permit the energy absorbed to be dissipated as heat.

Cut-growth resistance. As with most structural materials, cuts, sharp notches, and other surface defects can cause local stress concentrations, and thus serve as focal points for fatigue failure. The designer of rubber parts for dynamic service should provide fillets or radii at sharp corners to minimize this effect. This situation illustrates a point frequently encountered in designing elastomeric compounds for specific uses: strengthening of one property of a compound will be at the sacrifice of another. High-resilience stocks, which may be desirable for resistance to fatigue, are often not especially resistant to cut growth. This is quite comparable to spring steels, which are often poor in notch-impact resistance.

Abrasion resistance. Perhaps because of their ability to yield in the path of a sharp object that would otherwise shear off some of their surfaces, elastomers in general possess unusual resistance to abrasion. This property has been very important in the development of tires, the largest segment of the rubber industry. High abrasion resistance is dependent in large measure upon skillful compounding, although proper design, especially to provide for the proper dissipation of heat, is also of great importance.

Friction properties. It is well known that elastomers possess very high coefficients of friction when in contact with dry surfaces. This property has been of great importance in the tire industry, for shoe soles, and in many mechanical applications. In the presence of films of water or of other substances, the coefficient of friction is greatly reduced. Because of this property, propeller-shaft bearings for ships are frequently lined with rubber. Not only are the friction properties adequate for this service, but the high abrasion resistance and ability to yield to stresses helps prevent scratching of the shaft in the presence of sand or other abrasive substances. Friction properties may be modified through choice of elastomer, through compounding, and through design. The process for making the butadiene–styrene elastomer known as Buna-S or GR-S had to be modified before this elastomer could be safely used for tire construction. Many motorists will remember the poor performance on wet highways of the first synthetic tires brought out during the early days of World War II. By proper choice of compounding ingredients, the coefficient of friction can be modified over a fairly wide range. In particular, it is often possible to include lubricants in the compound to reduce the coefficient of friction. Finally, design may be very important. In tire design, for example, proper tread design will permit the wiping away of the water film on a wet pavement through squeegee action. Again, the design engineer must provide for the dissipation of heat, since the elastomer surface may otherwise melt and provide a fluid film at the friction interface.

Moldability. With varying degrees of ease, depending on the elastomer chosen and the type of compounding employed, elastomers may be molded under heat and pressure during the vulcanizing or curing operations. Molded articles can be made to conform accurately to very intricate mold configurations, as is demonstrated in many tire treads. If the mold surfaces are polished, high-gloss surfaces may be imparted to the molded object. Since the coefficient of thermal expansion of most elastomers is about 10 times as great as for steel, proper allowance must be made in the mold design for the shrinkage that will occur as the article cools to room temperature from the curing temperature.

Extrudability. By pressure applied either by a hydraulic piston or by a worm screw, elastomers can be forced through orifices to form extruded shapes of considerable complexity. Hose and gaskets are examples of items conveniently and economically formed by this means. The unvulcanized stocks must be fairly loaded with fillers to give them enough stiffness to prevent their collapsing under their own weight; therefore, high-quality mechanical goods of complex shape will not ordinarily be formed by extrusion. However, simple, solid shapes of high mechanical quality, such as tire-tread camelback, are often extruded. Because of the extreme elastic deformation that occurs as elastomers pass through the extrusion orifice, the design of these orifices for complex shapes is a highly skilled and expensive operation, applicable, in general, only to large-volume mass production.

Bondability. Most of the elastomers can be made to adhere to metals, glass, plastics, and fabrics by a process known as bonding. By proper attention to detail and by using high-quality stocks, the bond so attained is usually stronger than the elastomer itself, so that rupture will occur within the body of the elastomer. The first great advance in the art of bonding was the discovery that excellent bonds can be formed to brass plate of closely controlled composition. While brass plating for bond formation is still used extensively, bonding adhesives have been developed that will form excellent bonds to the thoroughly cleaned surfaces of most commercial metals or other structural materials without brass plating. Because of the great difficulty of controlling brass plating, and the expense of the installation, this process is now becoming obsolete. A variety of bonding adhesives are available to meet the individual needs of the engineer.

Electrical Properties

Electrical resistance. Uncompounded elastomers are, in general, good nonconductors of electric currents. Most of the compounding ingredients used are also nonconductors; hence, the majority of elastomeric products are excellent insulators and are widely used for that purpose. Carbon, however, is a moderately good conductor of electricity, and its use as a filler in rubber compounding can be taken advantage of to make the stock itself an electric conductor. A number of types of carbon are available for rubber compounding, some of which are better conductors than others. In order for the stock to conduct electric current, a large proportion of carbon must be added so that the particles are in contact with each other. In effect, therefore, the rubber remains nonconductive, and merely acts as the matrix for a continuous chain of conducting particles embedded within it. Because of the large volumes of filler required, only relatively inelastic compounds are conductive. Fairly low ohmic resistance can be achieved, but because of the poor heat conductivity of elastomers it is difficult to dissipate the heat generated as I^2R loss. Conducting stocks are thus used primarily to conduct static electricity or where the power is low.

Electrodeposition. In addition to natural rubber, several of the synthetic elastomers either occur as latices at some step in their synthesis or can be otherwise prepared as latices. These are colloidal suspensions in an aqueous medium. The particles are electrically charged. If a direct-current field is created in the latex, the charged particles of elastomer will migrate to the pole of oposite charge and be deposited upon it. Thereupon their charge is neutralized and a layer of the elastomer will be built up. By the use of compounding ingredients, also in colloidal suspension, it is possible to deposit compounded elastomers requiring only drying and vulcanization to complete. This process is used to line vessels with rubber or to coat wire mesh for abrasion-resistant screens. Because of the problem of removing residual water from massive sections, the process is usually restricted to the deposition of relatively thin coatings for chemical or abrasion resistance.

Chemical Properties

An examination of Table 6.5 will disclose that natural rubber, in general, will surpass the synthetic elastomers in useful physical and mechanical properties. In the area of useful chemical properties, the synthetic elastomers will often surpass natural rubber. It should be noted that the pattern varies, and that each elastomer excels in a limited area but may be relatively inadequate in other areas. Even the useful physical properties may, in time, be impaired through undesirable chemical reactivity of the elastomer. Therefore the design engineer must carefully balance physical and chemical properties to select the elastomer most useful for his purposes. Again, it will frequently be necessary to accept a compromise. For this reason it will be of great importance to recognize and provide for those properties that can least afford to be impaired and to make sacrifices in other areas of relatively less importance. From the point of view of the engineer, the useful chemical property of an elastomer may be defined as its resistance to serious and irreversible damage through chemical interaction with atmospheric or other agents to which it must be exposed throughout its proposed service life. It should be remembered that wherever a system is created in which materials of different chemical nature are associated, each can react in a manner different from the way it might behave if its neighbor were not present. This is well known in the field of metals, where the importance of galvanic interaction of dissimilar metals in contact in the presence of a corrosive atmosphere is recognized. It is less well

TABLE 6.5
Comparison of various properties of types of rubber[a]

Property	Natural rubber NR	Synthetic rubber: Gr-S butadiene-styrene copolymers (Buna S)	Synthetic rubber: GR-M chloro-butadiene polymers (Neoprene)	Synthetic rubber: GR-N buradiene-acrylonitrile (Buna N)	Synthetic rubber: GR-1 Isobutylene copolymers (Butyl)	Reclaimed rubber RR
Cost	3	2	4	6	5	1
Hardness, durometer "A"	20–100	30–95	25–100	25–100	—	40–90
Tensile strength	1	5	2	3	4	6
Elongation	1	3	2	4	5	6
Low compression set	1	4	3	2	5	6
Resilience	1	4	2	3	6	5
Color adaptability	1	2	3	4	—	5
Flexibility	1	4	2	3	—	5
Insulation	1	2	5	4	3	—
Tear resistance	2	5	4	3	1	6
Low staining effect on enamel	2 (Pale crepe)	1	4	5	3	6
Minimum oder	1 (Pale crepe)	2	3	4	5	6
Adherence to metals	1	4	2	3	5	6
Adherence to fabric	1	2	3	4	6	5
Resistance to:						
Abrasion	1	4	2	3	5	6
Aging	5	4	1	3	2	6
Alcohol	2	4	3	1	—	5
Animal fats	3	4	2	1	—	—
Acids (weak)	1	5	2	4	3	6
Benzene	—	—	2	1	—	—
Cold (to 40°F)	1	3	5	4	2	6
Cold (to 70°F)	1	3	5	4	2	—
Carbon dioxide	3	4	2	1	—	—
Carbon tetrachloride	—	—	—	1	—	—
Ethylene glycal	—	—	2	1	—	—
Flame	—	—	1	—	—	—
Freon	—	—	2	1	—	—
Gas (natural)	3	4	2	5	1	6
Gasoline	—	3	2	1	—	—
Heat (to 180°F)	1	4	3	2	—	5
Heat (to 250°F)	4	3	2	1	—	—
Kerosene	—	3	2	1	—	—
Methyl chloride	—	3	2	1	—	—
Moisture	2	4	5	3	1	—
Naphtha (coal ter)	—	—	—	1	—	—
Naphtha (petroleum)	—	3	2	1	—	—
Oil (Vegetable)	1	4	3	2	—	—
Oil (Lubricating)	4	3	2	1	—	—
Prestone	—	3	2	1	—	—
Sunlight	5	4	1	3	2	6
Turpentine	—	—	—	1	—	—
Weather aging	5	4	1	3	2	6

[a] *Note:* No. 1 indicates the compound most favorable for each property, etc.

Source: Adapted from data furnished by Lavelle Rubber Co., Chicago.

known, but equally true, that a similar situation can also exist when nonmetallic materials are involved in the system. The design engineer must therefore keep in mind all materials that can enter into a proposed system and allow for their possible interaction. If a bonded-rubber system is to be employed, the bond interface should not be ignored.

Atmospheric effects. Except in very special cases, almost all elastomer structures are exposed to the air. The oxygen in the air slowly attacks natural rubber at room temperatures, and more rapidly at elevated temperatures. This reaction constitutes part, at least, of the phenomenon known as *aging*. The first manifestation is usually a hardening of the surface owing to the formation of a resinlike film somewhat similar to hard rubber. At this stage rubber may crack at the surface when flexed, thus accelerating fatigue failure. Later the surface may become gummy and tacky. Because aging is practically universal, antioxidants are regularly used in compounding practically all natural rubber, and to the degree justified by the expected service life of the compound.

Ozone, which is normally present in air at sea level in only microscopic proportions, is extremely active in its attack. Other oxidizing substances, such as the oxides of nitrogen, will have a similar effect. Certain ground-level areas appear to be particularly bad in this respect, the Los Angeles area being a notorious offender. Electric sparks, corona discharges, and ultraviolet radiation will all generate ozone. Synthetic elastomers should be chosen for use in close proximity to such ozone generators whenever their other properties will permit. If not, excellent antiozonant materials may be used in compounding, or coatings may be applied to protect the rubber. It should be noted, however, that most resistant coatings tend to crack or flake off if the elastomer is subject to much flexure. Elastomers exposed to sunlight are also attacked in a similar fashion. The effect appears to be indirect, since those surfaces subject to direct exposure may be less attacked than shaded surfaces in a generally sunlit area. Normal compounding of high-quality mechanical rubber stocks will include ingredients to protect against this "sunchecking."

Temperature effects. Two types of temperature effect must be considered. The first is irreversible but indirect. In general, the rate at which any chemical reaction proceeds increases rapidly with increasing temperature; therefore, it may be expected that any of the chemical effects here discussed will take place at a faster rate if the system is at elevated temperatures. This does not necessarily mean that the effect will proceed to a greater extent. However, the service life of any elastomer subject to degradation through chemical means may be expected to be shorter at elevated temperatures.

The second effect is purely physical and reversible. As the temperature of any elastomer is lowered, the elastomer becomes gradually less resilient. After going through a stage where it is still pliable but leathery, it eventually becomes brittle and will fracture easily upon impact. By the use of plasticizers, which are, in effect, internal lubricants, the useful low-temperature ranges of elastomers can be extended somewhat below their inherent value. The silicone elastomers have the broadest low-temperature range, and are useful below −100°F. Specially compounded natural rubber is useful to −70°F, but the low limit for most of the synthetic elastomers will not exceed −40°F.

Raw rubber undergoes what is known as a second-order transition as it is cooled below about 40°F. It loses elasticity, becomes more opaque, and undergoes a volume change. Vulcanized rubber either loses this transition or has it shifted below the temperature at which it will impair its useful properties. Some of the synthetic elastomers, notably some of the neoprenes, undergo this transition even in the cured state. It is important that although it is a reversible phenomenon, a higher temperature is required for reversal; it may be necessary to heat above 100°F. If much work is done on the elastomer in a dynamic system, the heat generated by hysteresis may bring about reversal. Otherwise care may be required in selection of elastomers for service or storage in cold climates.

With increase in temperature above room temperature, both reversible and irreversible changes may take place. Reversibly, as the temperature rises, all elastomers lose in tensile strength. Relatively, the silicone rubbers suffer least in this respect. The poor initial tensile strength must be remembered, however. At a given temperature, other elastomers that have lost a much greater percentage of their initial tensile strength may still have higher absolute tensile strength.

The acceleration of aging reactions at elevated temperature has been discussed previously. This effect is irreversible. Most reports in the literature to date have dealt only with this phase of high-temperature effects. Very little has been published to date concerning tests run at the actual temperature of operation. Obviously, both phases should be carefully considered by the design engineer before an actual installation is made.

Solvent effects. The degrading effect of mineral oils, gasoline, benzene, and similar lubricants, fuels, and solvents, is well known. This is a *solvent effect.* In unvulcanized natural rubber, the molecules are actually separated to form a colloidal suspension. This is not a true solution, but is similar to one. After the molecules have been cross-bridged by vulcanization, complete separation does not occur; however, a more or less pronounced swelling takes place, and the elastomer becomes weak and flabby. The degree of swelling will depend on the nature of the solvent. Aromatic substances of the benzene family will in general be more severe than the straight-chain compounds of the aliphatic family. The size of the solvent molecules may affect the rate at which swelling occurs, but may not greatly affect the degree of swelling that will ultimately occur. Heavy lubricating oil, for example, may swell a rubber stock quite slowly. If only slight or occasional exposure is to be met, and if the elastomeric part can be fairly easily replaced, advantage may be taken of other desirable properties of that elastomer, even though it may not have the highest oil swell resistance. On the other hand, if constant exposure is expected, or if the part is inaccessible for maintenance, another elastomer may be necessary, even at the sacrifice of other desirable properties.

Swelling due to organic solvents is to some degree reversible. Volatile fluids may evaporate in time, leaving little permanent swelling or damage. The process is slow, however, since the solvent must first diffuse to the surface. Moreover, the work done in swelling shows that some active forces of attraction are at work. It is doubtful, therefore, that swelling is ever completely reversed, so it is preferable to choose an elastomer more resistant to swelling wherever possible.

Corrosive chemical effects. Many elastomers possess excellent resistance to inorganic acids, bases, and salts. Most of them are not highly resistant to strong oxidizing chemicals such as nitric acid, concentrated sulfuric acid, and peroxides. With these exceptions, elastomer containers or linings are widely used in the chemical industry. Since the compounding ingredients may be less resistant, however, special compounding is necessary for such uses.

Although the elastomers themselves may be resistant, bonded assemblies or other combinations of elastomers with nonelastomeric materials must often be considered for such service. In such cases consideration must be given not only to each of the materials involved, but also to the interaction each may have upon the other. A bonded-rubber assembly, for example, contains at least three major components: an elastomer, usually (though not always) a metal, and the thin but very real bonding system. Such an assembly is usually attached to another structure by screws or rivets or by other mechanical means. The entire system now consists of many parts, all assumed to be exposed to the corroding medium. In such cases, for example, galvanic corrosions may occur between the bolt and the supporting structure. Even though the entire bonded assembly has been proved to be satisfactorily resistant to this corroding medium, the electrochemical effects of the bolt-supporting structure may be in some degree related to it, with consequent early failure of the bond. The design engineer should bear in mind that all marine atmospheres contain appreciable amounts of salt that can deposit on any exposed equipment. Moreover, inorganic salts are frequently used on highways either for dust or ice control. Finally, rains wash appreciable amounts of acids and other corroding materials out of smoke-filled or industrial atmospheres. There are, therefore, few places where elastomers can be designed into equipment where some problem of corrosion may not potentially exist.

SYNTHETIC RUBBERS

None of the synthetic rubbers is equal to natural rubber in resilience, yet each synthetic has some physical or chemical property that is superior to the same property in natural rubber. Since World War II, the synthetics have opened up new fields in the design of such items as oil seals, gasoline hose, gasoline-pump

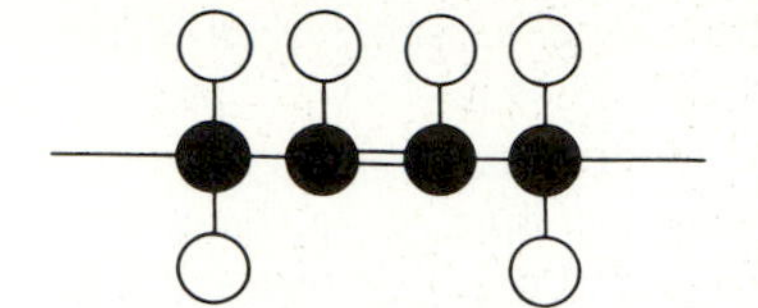

FIGURE 6.18
The monomer butadiene consists of four carbon atoms linked with six hydrogen atoms.

diaphragms, couplings used under oil exposure conditions, inner tubes, and linings impermeable to air, gases, and chemical solutions.

SBR

SBR, frequently referred to as Buna-S, is a synthetic copolymer of styrene and butadiene (see Figs. 6.18 and 6.19). It is one of the earliest synthetics and even today represents the largest variety being made. It is about 70 percent as resilient as natural rubber at room temperature, and has a lower tensile strength than natural rubber. At 40 Durometer hardness, SBR is approximately only 10 percent as strong as natural rubber. Natural rubber has a tensile strength of 2000 lb/in.^{2} and SBR will run only about 200 lb/in.^{2}, with compounds resulting in 40 Durometer or softer.

The tear resistance of SBR is poorer than that of natural rubber. As with natural rubbers, SBR should not be used in products where resistance to oil and solvents is needed. SBR compounds have fair resistance to set, and resistance to abrasion is also satisfactory. Certain types give good wear resistance and consequently, SBR has had application as a tire material. It has superior water resistance as compared with natural rubber. Resistance to sunlight and ozone is about the same as natural rubber. It is less expensive than natural rubber.

NBR

NBR is a copolymer of acrylonitrile and butadiene (see Fig. 6.20). NBR has low tensile strength in comparison with natural rubber and is approximately

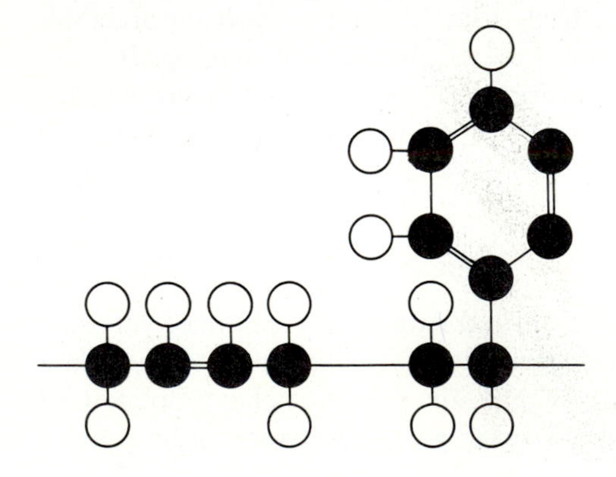

FIGURE 6.19
One butadiene monomer linked with a styrene monomer to form a chain produces the styrene-butadiene copolymer.

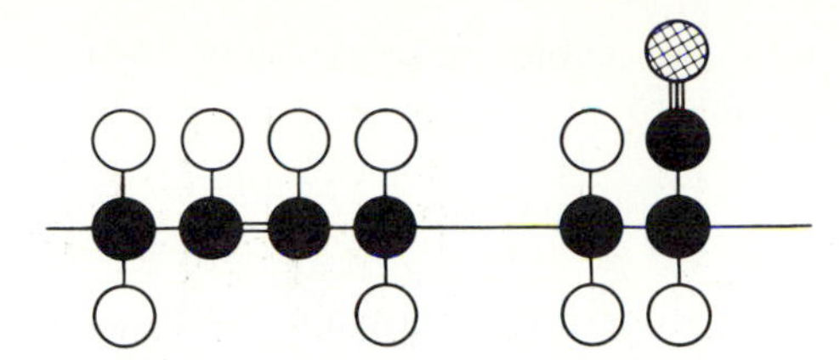

FIGURE 6.20
One butadiene monomer linked with an acrylonitrile monomer to form a chain produces the acrylonitrile butadiene copolymer.

two-thirds as resilient as natural rubber; however, it can be compounded to give tensile strengths approaching that of natural rubber. It loses its resiliency rapidly as the temperature declines, and at −40°F it becomes brittle. Its resiliency improves at higher temperatures. For this reason, it was used extensively during World War II years as a substitute for natural rubber on installations where temperatures were higher than normal room temperatures.

NBR has good compression set resistance and consequently has had wide use as a gasket or seal material. The important quality of NBR is its resistance to swelling and deterioration when exposed to gasoline and oils. For this reason, it is being used in quantity for such purposes as gasoline hose, oil-pump seals, and carburetor and fuel-pump diaphragms.

Neoprene

Neoprene is a polymer of chloroprene (see Fig. 6.21). It approximates natural rubber in resiliency at room and elevated temperatures. In this respect, it is superior to the other synthetic materials. The general physical properties of Neoprene, including resistance to set and tensile strength are good; however, because of its cost and the fact that it stiffens appreciably at lower temperatures, it has not been seriously considered as a replacement for rubber.

Because of certain chemical properties that Neoprene possesses, it does have wide application. Although Neoprene swells when subject to gasoline and oils, it will not disintegrate as does natural rubber. Also, at elevated temperatures (150 to 250°F), it will not soften as does natural rubber. In general, Neoprene resists the effect of oxygen, ozone, and aging better than the other rubberlike materials. Neoprene tends to be impermeable to air and many gases, although in this respect it is inferior to butyl rubber. It is flame-resistant and will not support combustion. It has good resistance to the corrosion action of chemicals and to water.

Because of its desirable characteristics, Neoprene is being used for such

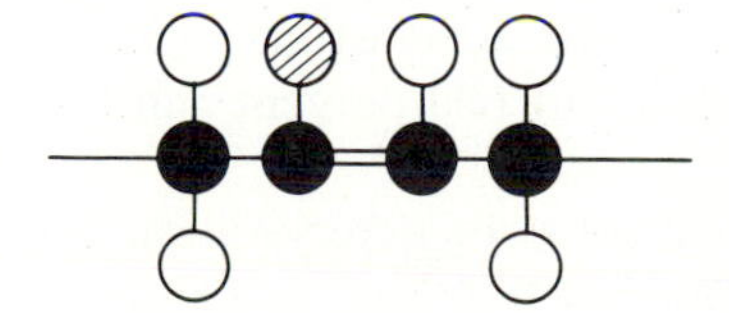

FIGURE 6.21
Replacing one of the inner hydrogen atoms of the butadiene monomer with a chlorine atom gives the chloroprene monomer characateristic of Neoprene.

items as garden hose, insulation for wire and cable, gasoline-pump hose, packing rings, motor mountings, and oil seals.

Butyl Rubber

Isobutyl-isoprene rubber is produced by copolymerizing isobutylene with small amounts of isoprene. The general physical properties of butyl rubber, including tensile strength, resistance to tear, and resiliency, are somewhat poorer than those of natural rubber. Butyl rubber is not resistant to gasoline and oils, and has only fair resistance to set under compression and tension. Also, its compounds stiffen considerably when subject to low temperatures.

Butyl rubber is highly resistant to ozone and oxidation and thus is used when resistance to aging is a requisite. It has excellent dielectric properties, and is resistant to sunlight and all forms of weathering. It is also practically impermeable to air and many gases; thus it is useful for products such as inner tubes, tubeless tires, dairy hose, and gas masks. In addition, since it is resistant to acids, it is useful for many applications in the chemical industry.

Polysulfide

Linear polysulfide's mechanical properties, including tensile strength and resistance to tear and abrasion, are poor compared to natural rubber's. Abrasion resistance is about one-half that of natural rubber and tensile strengths average about 1300 lb/in.^{2}

Polysulfide polymers are resistant to swelling and deterioration in the presence of gasoline and oils. They are also quite resistant to oxidation, ozone, and aging. They are not permeable to liquids and are more impermeable than natural rubber to most of the gases.

Polysulfide rubber has been used for solvent-resistant molded parts and coated fabrics. It is used also for such items as gasoline, paint and lacquer hose, printers' rolls, and newspaper blankets. It has not proved very useful for mechanical goods.

Silicone Rubber

Silicone rubber has a different type of structure from other elastomers. It is not made up of a chain of carbon and hydrogen atoms but of an arrangement of silicon, oxygen, carbon, and hydrogen atoms (see Fig. 6.22).

The mechanical properties of the silicone rubbers, including tensile strength, tear and abrasion resistance, resistance to set, and resiliency, are inferior to those of natural rubber. Today, silica-reinforced elastomers have tensile strengths of about 2000 lb/in.^{2}, and elongations of 600 percent can be realized.

This rubber's most important property is resistance to deterioration at elevated temperatures. Intermittent temperatures as high as 550°F and

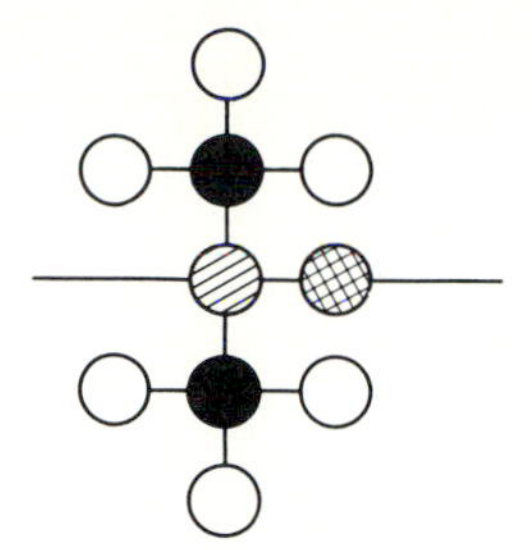

FIGURE 6.22
The silicone elastomer is based on arrangement of silicon, oxygen, carbon, and hydrogen atoms known as the dimethyl siloxane polymer.

continuous temperature of 450°F produce negligible changes in flexibility and surface hardness of silicone rubber, whereas natural rubber and the other rubberlike materials would quickly deteriorate or harden.

Silicone rubber also has excellent resistance to oxidation, ozone, and aging. It is fairly resistant to oils, although it does deteriorate in gasoline.

Because of the high cost of silicone rubber (about 20 times the cost of natural and most synthetic rubbers), it has had limited industrial application.

SUMMARY

The number of plastic materials that have become available to the engineer has greatly increased during the past decade. Today there are more than 30 chemically distinct families of plastics. The selection and design of elastomeric, thermoplastic, and thermosetting plastics is complicated.

The alert engineer should become familiar with the fundamentals of good design as related to plastics in general and should have an understanding of their mechanical and physical properties as well as methods of fabrication and manufacturing limitations of the principal thermosetting, thermoplastic, and elastomeric materials. To meet both static and dynamic forces, chemical, electrical, or mechanical needs, plastics are widely used in many structures and potentially could be used to advantage in many more. For example, reaction-injection-molded (RIM) polyurea polymers, RIM nylon block copolymers, and injection-molded nylon polymer blends are being seriously considered for use in the bodies of many modern automobiles. These plastics are especially attractive to car producers because they can often be used on conventional electrocoat priming, bake oven, and top-coating equipment owing to their high temperature resistance, dimensional stability, and good surface characteristics. There is a good chance that the cars of the future will have plastic or largely plastic skins. By the use of properly selected plastic systems, longer useful life, more comfort, more convenience, and utility can be built into many engineering structures.

Although plastics cost more than aluminum, magnesium, and steel on a per-pound basis, it must be remembered that for a given section thickness plastics weigh approximately only 15 percent as much as steel and 45 percent as much as aluminum; so on a strength to weight basis, they are reasonably competitive.

QUESTIONS

6.1. How are plastics, as engineering materials, proving themselves useful in present-day designs?

6.2. Explain how the structural design of plastic parts requires a different design approach from that of metals.

6.3. Why was the electrical industry one of the earliest users of plastics?

6.4. What is meant by polymerization?

6.5. Of what do molding compounds consist?

6.6. What is the effect of plasticizers in the molding compound?

6.7. When would it be advisable to add a lubricant to a molding compound?

6.8. What materials are used as fillers?

6.9. What are the effects of temperature on thermoplastic materials?

6.10. Why is "migration of plasticizers" an important consideration in designing assemblies?

6.11. Outline the effects of humidity on plastics.

6.12. Outline the design procedure to be followed when designing a new part to be made from a plastic.

6.13. What is the reason for introducing moisture into molded nylon?

6.14. How does prolonged exposure to sunlight affect plastics?

6.15. Is oxidation a serious objection to designing components to be made of plastics? Why?

6.16. Why are thermoplastics sometimes annealed?

6.17. What are the effects of loading on plastic materials?

6.18. In what type of parts would it be advisable to call for the use of cellulose-acetate plastics?

6.19. What are the principal applications of polycarbonate resins?

6.20. What are the desirable characteristics of nylon?

6.21. When would it be desirable to specify the use of phenol formaldehyde resins? Of urea formaldehyde resins? Of melamine formaldehyde?

6.22. What five-step procedure will help in selecting the most suitable plastic material?

6.23. When would you recommend the engineering plastic Torlon?

6.24. What change in strain would take place when using a general-purpose high-impact ABS resin under a stress of $2000\ lb/in^2$. and the temperature was raised from 75°F to 160°F?

6.25. Why will a long male thread cut in metal not be readily assembled in a female threaded plastic member?

6.26. What thought should be given to molded horizontal holes?

6.27. What are the relative costs of plastics, ferrous metals, and nonferrous materials? Explain the relationship of these costs to their specific gravities.

6.28. Design a plastic part with inserts and the mold for making the part according to good molding practice. Select materials and give reasons for selection.

6.29. What is the chemical constitution of isoprene?

6.30. How are rubberlike polymers unique?

6.31. What does the vulcanization process entail?

6.32. Define the mechanical property "resilience." Where would materials of high resilience be used?

6.33. What is the approximate tensile strength of natural rubber? Why is the tensile strength of rubber of little concern to the design engineer?

6.34. How can a compounded elastomer be made an electrical conductor?

6.35. In the design of what kind of products would you recommend the use of Buna-S?

6.36. Why is Neoprene widely used as a material in the manufacture of oil seals?

6.37. Why has Thiokol not proved very useful for mechanical goods?

6.38. What is silicone rubber's most important property?

PROBLEMS

6.1. Design a plastic cover for an oil reservoir for an automatic forming machine. There is a filter element inside the reservoir that may become clogged in use. In that event the cover will be subjected to an internal pressure of up to 200 lb/in.^{2} The cover must have eight threaded bolt holes with the heads of the bolts on the other side of the flange to which the cover is attached. The cover must also have the company name on the outside in raised or indented letters—whichever you recommend. Specify the type of plastic, the size of the bolts, the size of the enlargements for the bolt holes, the minimum wall thickness, the type and size of the lettering and how the holes are to be threaded or whether inserts are to be used. Also specify the molding process to be used if the production quantity is expected to be 100,000 per year.

6.2. After reading this chapter and reviewing the tables you should understand that most plastics have a coefficient of expansion from five to ten times greater than the majority of metals. Obtain a glass bottle that is fitted with a plastic cap. Remove the plastic cap and place it in a cup of hot water for a few seconds and then screw it back on the glass bottle (which has a lower coefficient of expansion than most metals). Note how much more difficult it is to remove the cap after it has shrunk tightly around the glass as it cools to room temperature. Use a torque meter and measure the increase in torque required to loosen the plastic cap. Give some other design applications where the high coefficient of expansion can be useful.

6.3. A type FM3 polyamide resin was used in a design subject to a tensile load of 200 lb at room temperature. A design was developed so that the smallest section that accommodated the load in tension was 0.025 in.^{2} The design engineer specified a factor of safety of 0.3. Temperatures during usage ranged from 30 to 100°F. Is the design satisfactory? If not, what recommendations would you make?

6.4. A general-purpose high-impact ABS resin is subject to a stress of 7000 lb/in.^{2} while installed at a temperature of 50°F. What would be the estimated strain in percent?

6.5. Obtain a small piece of thermoplastic sheet approximately 70 mils in thickness and about 6 in. square. Clamp the plastic sheet in a frame so that all four edges are held firmly. Now place the clamped sheet inside an oven and heat until the sheet is softened. Quickly remove the rubbery sheet of plastic and form it over a tapered mandrel, being sure to push down on the end of the mandrel and draw the material over the mandral, thus forming a cone. Remove the mandrel and then squeeze the cone with one hand, causing a split along the axis of the cone. Now carefully cut a strip of plastic about $\frac{1}{2}$ in. wide and 6 in. long from the cone shape with the length in the direction of the cone axis. Stress the strip in tension and bend it in the direction of the cone axis and note that it seems very strong. Now bend the strip in a direction transverse to the cone axis and note that the strip splits when bent this way. Why did the sheet lose strength in the direction transverse to the stretch? Why was the sheet strengthened in the direction of the stretch? Explain how this phenomenon can be used in the design of products.

SELECTED REFERENCES

1. Hertzberg, R. W. and Manson, J. A.: *Fatigue of Engineering Plastics,* Academic Press, New York, 1980.
2. Kaufman, M.: *Giant Molecules, the Technology of Plastics, Fibers, and Rubber,* Doubleday Science, New York, 1968.
3. Legge, N. R., Holden, G., and Schroeder, H.: *Thermoplastic Elastomers,* Macmillan, New York, 1987.
4. Rosen, S. L.: *Fundamental Principles of Polymeric Materials for Practicing Engineers,* Barnes and Noble, New York, 1971.
5. Saechtling, H.: *International Plastics Handbook,* Macmillan, New York 1987.
6. The Society of the Plastics Industry, Inc.: *Plastics Engineering Handbook,* Reinhold, New York, 1985.

CHAPTER 7

CERAMICS, POWDERED METALS, AND COMPOSITES

CERAMICS

The American Ceramics Society has defined ceramic products as those manufactured "by the action of heat on raw materials, most of which are of an earthy nature (as distinct from metallic, organic, etc.), while of the constituents of these raw materials, the chemical element silicon, together with its oxide and the compounds thereof (the silicates), occupies a predominant position." Ceramics, like plastics and metals, have been driven to new levels of sophistication. With the introduction of such materials as carbide, fused cordierite, pure alumina titania, and others, the ceramic industry, which is one of the oldest, has made rapid strides in the development of new materials. These new ceramics are frequently referred to as high-performance ceramics, high-tech ceramics, or advanced ceramics. The distinguishing feature of these new ceramics include: higher performance levels, more exacting control in composition, and closer tolerance on specific properties. Almost every plant has a chance to garner extra profits by taking full advantage of recent developments in this field. High-performance ceramics frequently have application in a wide variety of machinery, processing equipment and end-use products since they are frequently able to perform some functions better than competing polymers and metals. The principal properties for which this superiority may exist include electromagnetic properties, chemical inertness, temperature capabilities, hardness, and strength. Often a combination of one or more of these properties results in the desirability of selecting a ceramic. For example, ceramic cutting tools show promise in the fine machining of many materials because the ceramic tool tip allows less friction loss between the tool face and the cut chips and is more resistant to cratering than conventional carbide and high-speed tools. Again, the ceramic coating of molds used in the diecasting and plastics industries to increase service life and reduce maintenance shows promise. The coatings can improve the temperature uniformity of the molds while reducing sticking problems of the parts in the mold.

The engineer should recognize that the basic difference when working with ceramics as opposed to metals is that ceramics are brittle while metals are ductile. Because of their brittleness, when ceramic parts do fail, the results often can be more disastrous than failure of other materials.

Ceramics may generally be classified into six groups: (1) high-performance ceramics, (2) whitewares, (3) glass, (4) refractories, (5) structural clay products, and (6) enamels. Of these, high-performance ceramics, whitewares, glass, refractories, as well as carbon, will be discussed briefly.

High-Performance Ceramics

The unique thermal, wear, and corrosion properties characteristic of high-performance ceramics are achieved by the careful control of the compositions and microstructures entering into the material and during the processing. The steps in the processing include the manufacture of the ceramic powder, compaction of the powder into a "green" ceramic form, sintering, and secondary operations to bring the fired blank to the dimensional configuration of the end-use component.

The unique properties of high-performance ceramics result in their application in high-temperature, wear, and corrosive environments—such as heat engines. Because of light weight and heat tolerances of more than 1832°F, high-performance ceramics may well be the material most used in the aircraft engines of the future, since they will permit engines to reach higher speeds with less energy than metal components. Other applications of high-performance ceramics include shrouds, bearings, insulation parts, cases, and rings.

Perhaps the best known use of high-performance ceramics is for cutting tool inserts. This is anticipated to be a $160 million industry by the year 2000. Today, silicon carbide is being used in military rocket nozzles, mechanical seals for the shafts of pumps; zirconia is being used for dies in powder metallurgy compacting, extrusion dies, bearing blocks, etc. The reader should understand that monolithic ceramics (those that have the same ceramic material throughout) have limited value and are used only in simple nonstressed components. Today, the emphasis is on developing composite ceramics, those in which fibers from a second material have been introduced in the primary ceramic to add toughness and crack resistance.

High strength and fine surface finish require the use of fine grain-size powders (average particle size less than 1 μm). When the desired grain size is not achievable through chemical synthesis processes, reduction of grain size can be achieved through attrition and/or ball milling and grinding. Today, colloidal chemistry techniques are permitting the production of fine oxide particles and controlled nucleation and growth in gas phase reactions for the development of fine carbide and nitride powders.

Powders need to be compacted into a geometrical configuration representative of the final part. Since this compaction necessitates some flow, the powder must be converted into a viscous fluid that will take the desired form with a minimum of porosity. This is done through the addition of materials that, upon the exposure to heat (sintering), will decompose and escape as gases.

In producing the green ceramic form, it is important that the design engineer provide for material shrinkage during sintering as well as stock for finish-machining operations. At times, final dimension size can be achieved if the mold is designed properly, thus giving products requiring no final machining.

Pressing of the ceramic powder into the mold is done either by uniaxial methods (applying pressure along a single direction) or isostatic methods (applying of pressure on all sides). Complex parts are produced by isostatic pressing (both hot and cold) as well as simple designs, since more uniform green densities are achievable by this more costly process. In isostatic pressing, the powder is frequently initially formed uniaxially in a mold and is subsequently placed in a rubber mold that is immersed in a pressurized liquid. Thus, pressure is applied on all sides of the part.

Sintering follows the pressing of the ceramic powder. At times, pressing is carried out at elevated temperatures so that the sintering process is performed simultaneously with the shaping process. During sintering, two phenomena take place: densification resulting from the forces of surface tension, and grain growth caused by differential grain boundary energies. Since grain growth needs to be minimized in order to control geometry of the part, high-temperature materials such as oxides are often added in order to reduce grain growth.

The more important families of advanced ceramics include silicon nitride, silicon carbide, zirconia, aluminum oxide and ceramic–ceramic composites. These high-performance materials are finding broad applications in electronic, gas sensor, optic, bioceramic, and structural applications.

Both silicon nitride and silicon carbide possess properties making them candidates for heat engine and industrial applications requiring good strength retention at elevated temperatures at low thermal expansion.

Zirconia is a ceramic that has significant application as a structural material. However, its use is limited to moderate temperature applications because its strength decreases to approximately 25 percent of its room-temperature strength when the temperature reaches 1832°F (1000°C). However, its strength is good at moderate temperatures. Also, its thermal expansion

TABLE 7.1
Properties of advanced ceramic compounds

Compound	Density, g/cm^3	Hardness, kg/mm^2	Melting point, °C	Thermal conductivity, cal/(cm s °C)
Aluminum oxide	3.98	2100	2050	0.07
Zirconium oxide	6.27	1200	2715	0.005
Silicon carbide	3.22	2500	2220[a]	0.16
Silicon nitride	3.17	2400	1900	0.04
Silica glass	2.20			0.002

[a] Decomposes

Source: Courtesy of Dr. J. B. Wachtman, Jr., Rutgers University Center for Ceramics Research, taken from *Manufacturing Engineering,* February 1985, p. 60.

parallels that of iron–based alloys, making it a desirable material when used in concert with ferrous materials.

Aluminum oxide is available with almost 100 percent purity. Its average particle size is around 0.5 μm. As well as in the cutting tool industry, it has wide application for sodium vapor lamp tubes and electronic ceramics. Table 7.1 provides useful properties for some of the principal high-performance ceramics.

Whitewares

Ceramic whiteware is a ceramic body that is usually white and may be glazed. It includes such families of products as earthenwares, china, and porcelain. The typical ceramic whiteware body is composed of a nonplastic, a plastic, and a flux. The nonplastic comprises the structural skeleton, which is held together by the bond developed from the flux during the forming process. The plastic gives the ceramic the workability required in the forming process.

Production methods. Once the body has been prepared, there are four possible ways to form the required shape: jiggering, casting, extruding, and pressing.

Jiggering. Jiggering, which is usually automatic, is a forming process patterned after the old potter's wheel. Here blanks of the material are placed in a heavy mold made of plaster of paris. The inside of the mold has the desired shape of the outside of the ceramic piece being made. The inner surface of the blank is formed by having a shaped tool forced into the plastic material while the blank is rotated.

Casting. In casting, a suspension of the body composition is poured in a gypsum mold. The mold, which absorbs moisture from the suspension, causes the body to cast against the plaster face of the mold. If the part is to be hollow, the excess body may be poured off once the correct wall thickness has been realized. This is known as *drain casting*.

Extruding. In extruding, the raw material is mixed to a plastic state in the extrusion press. Frequently the plastic material, prior to the actual extrusion, is de-aired in what is known as a *pug mill*. The nozzle of the extrusion press allows mandrels to be placed in position so that the extruded rod can have a variety of internal openings as well as shapes.

Pressing. In pressing, the base aggregate (which has a reduced amount of water as compared with extrusion) is placed in a steel die and subjected to pressure. Pressing, which is done anywhere from several hundred to several thousand pounds per square inch with hydraulic or automatic presses, has gained extended use where it is adaptable. However, since the granular material is limited in its plasticity, complex shapes in the lateral plane cannot be formed successfully.

The formed ceramic is in the *green* state; after drying it is said to be in the *leather-hard* state.

If a glaze is to be applied, it is sprayed on and the body and the glaze are matured in a single firing. Sometimes the body is matured first with a bisque fire; then the glaze is applied, which is matured with a lower-temperature firing known as a ghost fire. Figure 7.1 illustrates the manufacturing processes commonly used in making porcelain insulators.

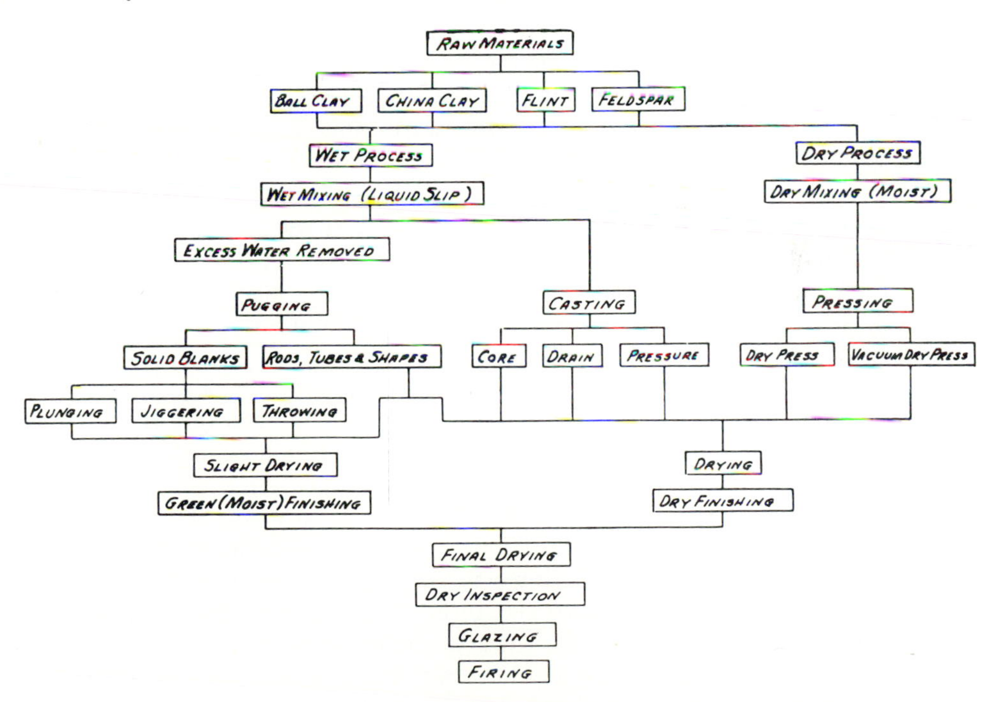

FIGURE 7.1
Manufacturing process commonly used in making porcelain insulators.

Joining. In joining ceramic whiteware to metals, five techniques are employed: mechanical seals, brazed seals, soldered seals, matched seals, and unmatched seals. Mechanical seals are the most widely used; although with the expanding field of electronics, soldered and brazed seals are popular in the miniaturization of components.

Properties of whitewares. Almost all whiteware ceramics are resistant to all chemicals except hydrofluoric acid and, to some extent, caustic solutions. They will withstand prolonged heating to temperatures over 1800°F. Indeed, many will bear temperatures considerably higher. Ceramics, in general, are weak in tensile strength and are much stronger under compressive loading. There is a wide range of strength in the different ceramics depending upon their composition and vitrification.

Whitewares have good thermal endurance and high dielectric strength, making them valuable materials for many electrical applications.

The required characteristics of the spark-plug insulator illustrate those properties attainable with high-tension procelain. Here we must have: (1) good dielectric properties at elevated temperatures, (2) high dielectric strength, (3) resistance to heat shock, (4) good mechanical strength, (5) resistance to lead compounds, and (6) resistance to attack by carbon. A typical composition of such a vitrified porcelain would be 30 percent kaolin, 20 percent ball clay, 30 percent feldspar, and 20 percent silica.

Types of whitewares. The production-design engineer today will inevitably find application for a ceramic whiteware at some time. The principal types of these ceramics today, exclusive of earthenwares and chinas, include: alumina-type bodies, beryllia-type bodies, steatite-type bodies, zirconium bodies, and titania ceramics.

The alumina-type bodies are typified by the spark-plug insulator. These ceramics may be defined as those that contain at least 80 percent of alumina in the α form. Most of the remaining composition is silica plus the alkaline earths. The alumina, silica, and alkaline earths when fired form a stable, hard, dense, chemically inert, gas-tight mass.

The typical characteristic of a high-alumina ceramic is its relatively good strength. Transverse strengths of 15,000 lb/in.^{2} can be achieved at 2500°F in some alumina ceramic materials. Dimensional stability and chemical inertness are other important attributes of alumina ceramics.

High-alumina ceramics have made possible construction of high-powered klystrons that have been used to bounce radio signals from the Moon. The very low dielectric loss-factor property of alumina ceramics has been instrumental in allowing these klystrons to be built. High-alumina ceramics are being used to a great extent in microwave communications.

Beryllia-type ceramics usually have beryllia contents that range from 95 to 100 percent BeO. As with alumina ceramics, the beryllia ceramics are hard, dense, gas-tight materials but they have a much higher thermal conductivity.

Thus, with the beryllia ceramics, we are able to incorporate the thermal conductivity of a metal like aluminum with electrical resistivity. This ceramic promises to be one of the important nuclear materials because it is capable not only of shielding a reactor but of withstanding temperatures that are approximately double those experienced in reactors today.

Steatitic porcelain is produced largely from talc ($3MgO \cdot 4SiO_2H_2O$). This family of ceramics has its greatest use as an insulating material, and has such applications as crystal cases, coaxial cable insulators, coupling insulators, heater-wire supports in appliances, etc. These ceramics have high mechanical strength (20,000 lb/in.^{2} flexural strength and 85,000 lb/in.^{2} compressive strength), high hardness values (7.5 Moh hardness), and good resistance to abrasion and dielectric strength (240 V/mil).

Where large temperature gradients are encountered, an improvement in the thermal endurance of steatite may be realized by adding appreciable (up to 20 percent) barium zirconium silicate in place of an equivalent amount of talc for zirconium types of ceramics. Other ceramics recently developed for improvement of thermal-shock characteristics include: zircon–calcium, zirconium, and zircon–alumina.

Cordierite-type ceramics have application where low thermal expansion and/or thermal endurance is a requisite. The theoretical composition of this type is $2MgO \cdot 2Al_2O_3 \cdot 5SiO_2$. In general, the cordierite ceramics are characterized by having fair mechanical properties (8000 lb/in.^{2} flexural strength and 40,000 lb/in.^{2} compressive strength), high dielectric strength (220 V/mil), wide firing range with low thermal expansion, and excellent thermal endurance.

The most important characteristic of titania ceramics is the high dielectric constant. Representative dielectric-constant values of the more important titanias are:

Barium-strontium titanate	10,000
Barium titanate	1,200–1,500
Strontium titanate	225–250
Calcium titanate	150–175

High dielectric values have given the titania ceramics much application in resistors, high-capacity capacitors, waveguides, photoelectric cells, and subsitutes for mica capacitors.

Design considerations. Since most whiteware parts shrink during the firing process, which takes place at about 2500°F, shrinkage must be taken into consideration in the design. In order to finish-machine the formed material in the dried state, a certain amount of material must be allowed; however, it is usually not necessary to finish-machine formed ceramics, as fairly close tolerances can be achieved when the material is formed. These are ±0.012 in. for casting and ±0.007 in. for dry pressing.

Where finishing speed operations are required, they are usually achieved by wet grinding at high speeds. In cylindrical grinding, ±0.004 in. can be held, and parallel grinding will give an accuracy of ±0.0012 per 4-in. length.

The minimum thickness of flat surfaces should be $\frac{1}{8}$ to $\frac{1}{4}$ in., depending on the method of production and the size of the finished part. As the surface area increases, the minimum thickness will of course increase.

Glass

About 95 percent of the glass produced today is made from a mixture of oxides of silicon and certain metals. Modern refinements in the method of manufacture, together with the general properties of this material, have made possible the development of one of the most versatile of all materials. Today, glass is woven into cloth; made into doors, cookware, and self-defrosting windshields; used as a glazing material for buildings; made into filters, prisms, and other light-separating devices; and made into bottles, jars, and many other products.

Types of glass. The principal types of glass, classified according to composition, are silica, borosilicate, lead, and lime. Table 7.2 shows a comparison of these types of glass.

Silica glass. Silicon oxide, when fused to a clear glass, produces properties that have considerable application under thermal conditions. This type of glass, frequently referred to as *quartz glass,* has very low thermal expansion. It can be heated red hot and then immersed in cold water without

TABLE 7.2
Comparison of principal types of glasses

	Silica	96% Silica	Borosilicate	Lead	Lime
Cost	Highest	High	Moderate	Low	Lowest
Electrical Resistance	High	High	High	Highest	Moderate
Thermal Shock Resistance	Highest	High	Good	Low	Low
Strength	Highest	High	Good	Low	Low
Hot Workability	Poorest	Poor	Fair	Best	Good
Chemical Resistance	Highest	High	Good	Fair	Poor
Impact Abrasion Resistance	Best	Good	Good	Poor	Fair
Heat Strengthening Possibilities	None	None	Poor	Good	Good
Ultraviolet Light Transmission	Good	Good	Fair	Poor	Poor
Weight	Lightest	Light	Medium	Heaviest	Heavy

cracking or failure. Silica glass also has high strength characteristics, chemical resistance, electrical resistance, and good ultraviolet light transmission. If it were not for the difficulty in producing it, and its consequent high cost, it would be an excellent general-purpose glass.

Borosilicate glass. The borosilicate glasses represent a class of glasses to which oxides of boron, sodium, and sometimes potassium and aluminum are added. Boron oxide is the principal additive, usually comprising 14 percent of the glass. A representative borosilicate glass would be made up as follows: silica, 80 percent; boron oxide, 14 percent; sodium oxide, 4 percent; aluminum oxide, 2 percent. Borosilicate glasses are readily worked and retain high strength, high chemical resistance, high electrical resistance, and low thermal expansion. Because of these characteristics and a lower cost than the silica group, the borosilicate glasses have wide industrial usage. Typical applications include sight glasses, gage glasses, electrical insulators, laboratory glassware, and glass cookware.

Lead glass. Lead glasses are those in which lead oxide has been added to the sodium oxide–silicon oxide combination. The amount of lead oxide added varies considerably with the purpose of the final product. For example, some glasses used for protective purpose against X-rays have as much as 90 percent lead oxide. A representative lead glass for utility purposes could be: silica, 70 percent; lead oxide, 15 percent; sodium oxide, 10 percent; potassium oxide, 5 percent.

Lead glasses have excellent electrical-resistance properties and ability to be readily hot-worked. Consequently, they are used for such products as thermometer tubing, fluorescent lamps, lamp tubing, and television tubes. Lead glasses also have a high refractive index, ranging from 1.50 to 2.2, making them ideal for optical glasses. Since these glasses have a high dispersion of light, they are especially desirable for cut glassware and jewelry.

Lead glasses, the heaviest of the glasses, are soft and, consequently, easily scratched. These inferior characteristics should be considered before selecting lead glass as an engineering material.

Forms of glass. Glass is available in numerous forms including sheet, rod, tube, and various finished forms. When special finished forms are desired, the production-design engineer should consult the glass manufacturer, who usually produces the glass in the finished form required. Costs can frequently be lowered by following the suggestions of the glass manufacturer.

Window glass. Window glass is usually a soda-lime sheet glass produced by extruding a thin strip of the material vertically from the melting tank. It cools as it descends and is cut to length from the bottom of the extruded sheet. Window glass is produced in thicknesses up to $\frac{1}{4}$ in., in standard size sheets of 76×120 in. Window glass is not acceptable for automotive or aircraft glazing but is widely used for domestic and commercial buildings. Sheet window glass also finds wide application in such items as mirrors, table tops, and photographic plates.

Sheet window glass can be heat-treated so as to increase its tensile strength from two to five times. With increased strength, the sheet glass can be converted to many additional uses, including fire screens, safety mirrors, office-building glazing, and gage shields. Wire mesh can be imbedded in sheet glass in the molten state so as to give a glass that is stronger from the standpoint of penetration of missiles and less vulnerable to the danger of fragmentation.

Plate glass. Plate glass refers to sheet glass that has been ground. It is produced by rolling the plastic glass to thicknesses ranging from $\frac{3}{16}$ to $1\frac{1}{4}$ in. This rough-rolled stock is then ground and polished with special-purpose equipment to obtain a surface that is optically flat. This quality glass has been found useful for automotive glazing, storefront windows, tracing tables, and surface plates. Plate glass can be heat-treated to give extra strength; when this is done, it is sold under the name of tempered plate glass.

Laminated glass. Laminated glass, frequently referred to as safety glass, is produced by placing a transparent vinyl plastic between two layers of plate glass. The plastic layer prevents splintering of the outside glass layers if broken. It is used for such purposes as automotive and aircarft glazing, protection shields, and storefront windows.

Formed glassware. Formed glassware may be classified as pressed ware, blown ware, and drawn ware. Pressed ware is produced by forcing molten glass into a mold. This is usually done automatically, although it may be done by hand. Typical products by this method are household applicances, eyeglasses, glass gages, and decorative and ornamental pieces.

Blown ware includes those hollow forms produced by blowing a jet of air into a glob of the molten material so that it takes the form of the closed mold upon solidification. Typical blown ware includes bottles, jars, vases, and bulbs.

Drawn ware includes rod and tubing and is used in gage glasses, chemical pipe, and insulation.

Properties of glass. When designing for the use of glass, structurally a maximum load of 1000 lb/in.^{2} is considered suitable for annealed glass and a load of 2000 to 4000 lb/in.^{2} (depending upon its composition) for heat-treated glass. Glass usually fails in tension, as it is about 10 times as strong in compression as it is in tension. Glass is subject to fatigue: under load, it will lose strength in time. Temperature affects the strength of glass. At low temperatures, glass is the strongest; it tends to reach its lowest strength at around 350°F; thereafter, its strength increases with additional temperature rise. Generally speaking, glass is a hard material, being harder than soft steel, aluminum, and brass. The thermal conductivity of glasses at room temperature ranges from 0.0016 to 0.0029 cal/(cm · s · °C). Glass makes an excellent material for cookware, since about 96 percent of radiated heat is absorbed by

TABLE 7.3
Glass as an industrial material

Type	Preparation	Composition, Sizes, Etc.	Uses
		Sheet Glass	
Window	Drawn vertically from the molten bath as a continuous sheet, firepolished and annealed. Shows only slight wave or distortion.	Soda-lime glass. Thicknesses from very thin to ¼ in. max. standard sheets 76 in. x 120 in.	For general purpose glazing of building. Also for table tops, induction heating jigs, shields, electrical insulators, small mirrors, etc. Thin sheets for photographic and microscopical uses.
Tempered	Glass is quickly heated to about 1150 F. and chilled quickly. Strength increased 2-5 times.	Same as above—must be cut to size, drilled, etc., before tempering.	For gage glasses, safety mirrors, business machine windows, fire screens, hospital glazing, etc.
Wire	Wire mesh is embedded in the molten glass.	Sometimes uses heat-absorbing types of glass.	For skylights, roofing, air raid precautionary glazing, etc.
Heat-Resisting	Special compositions, usually of borosilicate type.		For oven sight glasses, heat protection shields, electrical insulators, furnace door glasses, etc.
Ultraviolet	Glass of a special composition to transmit about 50% of the ultraviolet rays in solar radiation.		For hospital sun porches, greenhouses, poultry houses, etc.
Water White	An exceptionally "white" glass with higher light transmission.		For greenhouses, picture glazing, photographic uses, etc.
Light Reducing	Blue-tinted glass for cutting down amount of sunlight transmitted.		For glazing in sunny climates.
Colored	Obtainable in a wide range of colors, transparent and translucent. Plain and varigated colors.		For ornamental glazing, decorative panels, modern furniture.
Figured	A figure is rolled into the glass to diffuse or otherwise reduce lighting.		For obscured windows, skylights, glare-reducing windows, etc.
Polarizing	Glass is coated with a chemical that polarizes the light transmitted.		For anti-glare applications, as sunglasses.
		Plate Glass	
Rough Rolled	Rolled to sheet form with a knurled surface in a variety of patterns.	Soda-lime glass. Standard thicknesses 3/16 in. to 1¼ in. in usual grades.	The rough stock for manufacture of plate glass. Also used for ornamental glazing, obscured windows, skylights, etc.
Polished	Rough rolled stock ground and polished to substantially optical flatness of surface.	Standard thicknesses 7/64 in. to 1¼ in.	For tracing tables, surface plates, blueprint machines, and show windows and counters, illuminated signs, mirrors, windows, doors, stairways, walls, etc.
Tempered	Plate glass is rapidly heated to about 1150 F. and cooled suddenly in a blast of air. Strength is increased to 4 or 5 times that of ordinary plate.	Same as above—must be cut to size before tempering.	For pickling tanks, gage glasses, pressure tanks, oven sight glasses, store front panels, advertising signs, etc.
Special Purpose: X-Ray	Specially prepared with lead content of about 61%, and lead equivalent of about 0.32. About ¼ in. thickness. About twice as heavy as ordinary plate glass.		For X-ray protection.
Special Purpose: Water White	Special composition to provide higher transmission of all light.		For photographic and blueprint purposes, refrigerated showcases, display windows, etc.
Special Purpose: Document	Designed to cut down the amount of ultra-violet light transmitted. About ¼ in. thickness.		For protection of old documents and collections in museums.
Special Purpose: Heat Absorbing	A blue-tinted glass transmitting 70% of solar light but only 45% of the heat.		For double-glaze units on trains, skylights, airport control towers.
Special Purpose: Colored	Blue, flesh-tinted, and other colors are produced in transparent and opaque glasses.		For decorative mirrors, panels, chalkboards, bulletin boards, etc.
		Laminated Glass	
Safety Plate	Two or more lights of glass bonded by interlayers of transparent plastic.	Plate glass—composite is usually ¼ in. thick.	For automobile windshields and windows, aircraft and bus glazing, railroad car windows, protective shields, jewelry store windows, safety lenses, etc.
Safety Sheet		Sizes from about 3/32 in. to ½ in.	

SOURCE: From "Materials and Methods Portfolio of Engineering File Fact," 5th ed., no. 81. May, 1945. pp. 65–68.

TABLE 7.3 (*continued*)

Type	Preparation	Composition, Sizes, Etc.	Uses
Bullet Resisting Plate / **Bullet Resisting Sheet**	Several lights of glass, usually three, laminated as for safety plate.	Standard thicknesses from ½ in. to 1½ in.	Primarily for protection against firearms, as in banks, money collecting trucks, police cars. Also for protective shields in laboratories.
Double Glass	Not strictly a laminated glass—2 lights are separated by cemented glass spacer around edges, leaving air space between.	Must be made to order as to sizes—usual thickness is two ⅛-in. lights with ⅛-in. space.	For glazing air conditioned buildings, railway cars, etc.
Formed Glassware			
Pressed Ware	Molten glass is pressed into a mold, by hand or automatically. Metal inserts or attachments may be used. Parts may be tempered.	Compositions to suit requirements. Borosilicate glasses where thermal resistance or extra strength is required. Lead glass or soda-lime-magnesia-alumina for household ware.	For electrical insulators, centrifugal pump parts, glass gages, godet wheels, rotary valves, sight glasses, etc. for industry; household utensils, tableware, decorative glass articles, etc., for commercial fields.
Blown Ware	Molten glass is blown into a mold to produce hollow forms. Work may be done by hand or automatically. Parts may be tempered.	Frequently, a soda-lime-magnesia-alumina glass is used because of its easy workability. Heat resisting glasses or other special compositions where desired.	For bottles and similar containers, battery jars, floats, laboratory glassware, reaction columns, lantern globes, domestic utensils, etc.
Drawn Ware	Glass is drawn directly from the melting furnace in some standard form.	A variety of compositions can be used, usually soda-lime-magnesia-alumina or a heat-resisting variety.	For tubing, glass pipe, rods, gage glasses, and similar forms.
Multiformed	Glass is cold molded to close tolerances and sintered. Wide range of shapes.	A variety of compositions can be used. The glass usually is white and translucent.	For electrical insulators, pump seal rings, etc.
Glass Fiber Materials			
Batts	Very fine glass filaments drawn out from molten glass and collected as a woolly mass—may be treated with a resin binder and compacted to any desired degree, or faced with wire mesh. Forms a non-absorptive, non-combustible, chemically resistant material having good sound-, heat-, and electrical-insulating properties.		For insulation of military aircraft, domestic stoves, roofs of industrial buildings, railroad tank cars; as tower packing; insulation of furnaces; sound deadening in test rooms, as a dust filter in air conditioning.
Blocks	Mats of glass fiber compacted to blocks or boards under pressure and held with a binder. Available in thicknesses to 2 in. for use to 600 F. for one grade, to 1200 F. for another.		For insulation of boilers, ovens, breeching, pipes, valves, tanks, low temperature grade for refrigerated spaces.
Textiles / **Cord** / **Sleeving** / **Tape or Cloth**	Glass fiber filaments may be spun to threads and woven into fabric on standard textile equipment. The textile materials so produced have exceptionally high strength, exceeding that of the natural and of most other synthetic fibers, weight for weight.		As heat insulating material—for lagging pipe, aircraft engine exhausts, etc. As electrical insulation—for winding wire, for covering motor coils, ignition cable. As a decorative or fireproof, or rotproof fabric—for draperies, military fabrics.
Composites	*Glass Fiber-Plastic Laminates*—an extremely high-strength plastic laminate. *Glass Fiber-Asbestos*—for high-temperature insulation. *Glass Fiber-Neoprene*—for conveyor belts to operate at elevated temperatures. *Glass Fiber-Mica*—slot insulation in motors and generators, etc. *Glass Fiber-Rubber*—for military and naval tarpaulins, etc.		
Glass Structural Forms			
Block	Made by fusing together two halves of pressed glass to form a hollow, partially evacuated block.	In 3 standard sizes—5¾ in. sq., 7¾ in. sq., 11¾ in. sq., all 3⅞ in. thick. Standard radial and corner blocks available.	For light-transmitting masonry walls, interior partitions, permanent windows or panels in air-conditioned structures, etc.
Cellulated	Gas is trapped in molten glass to give a lightweight material containing numerous closed cells.	As usually made, is lighter than water. Available as blocks in sizes to 12 in. x 18 in. x 6 in.	For building blocks, thermal insulation; also for life rafts, fishing floats, etc.
Cast Ornamental Panels	Sculptured designs cast in glass, and back face frosted, polished, mirrored, etc.	In several standard patterns and modeled designs. Strip or square panels of various sizes.	For door or fireplace trim, indirect lighted panels, screens and partitions, strip decoration, fountains, etc.

glass. When in loosely compacted batts of glass fibers, it is a poor conductor of heat and so an excellent thermal insulator. Because of its electrical resistivity, glass is widely used in the electrical industry as a standard resistance material.

Design considerations. Since glass is weak in tension, designs should avoid stressing it in this manner. Large radii should be provided on molded shapes,

TABLE 7.4

Type	Fused Silica	96% Silica	Soda-Lime Glasses: Plate	Soda-Lime Glasses: General Purpose	Alumino-Silicate
PHYSICAL PROPERTIES					
Density, Lb/Cu In.	0.079	0.078	0.09	0.089	0.091
Softening Point, F (Approx)	3050	2800	1330	1285	1675
Thermal Cond, Btu/Hr/Sq Ft/Ft/F @ 212 F	0.80	0.80	0.53	0.53	—
Coeff of Exp per ° F, 32–570 F	3.0×10^{-7}	4.5×10^{-7}	48×10^{-7}	51×10^{-7}	23×10^{-7}
Spec Ht, Btu/Lb/F	0.185	0.185	0.20	0.20	—
Elect Res, Ohm-Cm @ 212 F	$>10^{15}$	$>10^{15}$	—	4×10^{9}	$>10^{15}$
Power Factor, 68 F 1 MC, %	0.025	0.02–0.04	0.80	0.90	0.37
Dielectric Constant, 68 F 1 MC	3.75	3.8	7.4	7.2	6.3
Thermal Stress Resistance[1]	—	390	—	65	85
Thermal Shock Resistance[2]	Very high	Very high	135	125	240
MECHANICAL PROPERTIES					
Mod of Elasticity, Psi	10×10^{6}	9.7×10^{6}	$9\text{–}10 \times 10^{6}$	$9\text{–}10 \times 10^{6}$	12.7×10^{6}
Normal Work Stress (Annealed) Psi	1000	1000	1000	1000	1000
THERMAL TREATMENT					
Annealing Temp, F (Stress Relieving)	2080	1670	1010	950	1315
FABRICATING PROPERTIES	Glass can be ground or polished without great difficulty. Machining is limited to sawing or drilling. Sawing is readily accomplished with an impregnated wheel. Drilling is difficult, but is done with special carbide-tipped drills using kerosene as a lubricant.				
Joining	Glass can be joined by heat-sealing using a blast-lamp, gas torch or by electric heating. Parts must have closely similar coefficients of expansion or intermediate pieces having intermediate coefficients must be used. Glass-to-metal seals are made by this latter process, also.				
Highest Service Temp F, Annealed	1650	1500	900	840	1200
Tempered	—	—	550	480	840
CORROSION RESISTANCE	Glass is attacked by hydrofluoric acid, hot concentrated phosphoric acid and strong alkalies. It is resistant to most other chemicals.				
USES	Ultra-violet energy transmission, chemical apparatus, thermocouple protection tubes, laboratory apparatus.	Ultra-violet energy transmission, chemical reaction vessels, thermocouple protection tubes, laboratory apparatus.	Sheet and plate glass, molded glassware, bulbs for electric lamps, bottles, vials, fluorescent lamp tubing.		

as glass is quite notch-sensitive. Designs that require heat-treated glass should be symmetrical and have constant section thickness. With formed glass, sharp changes in section, long thin necks, and heavy sections near the top of the mold should be avoided.

Pressed glass necessitates sections heavy enough to permit adequate fluidity to fill the mold. Ample draft and liberal radii should be provided so that the molded part may readily be removed from the mold. Block molding is the lowest in cost of the pressing methods and should be utilized when possible. Tables 7.3 to 7.5 provide the engineer with pertinent information as to the various types of glass.

TABLE 7.5

	Borosilicate Glasses			
Type	Low Expansion Chem. Resistant	Baking Ware	"Kovar" Sealing	Low Electrical Loss
PHYSICAL PROPERTIES				
Density, Lb/Cu In.	0.080	0.081	0.082	0.77
Softening Point, F (Approx)	1500	1425	1300	—
Thermal Cond, Btu/Hr/Sq Ft/Ft/F @ 212 F	0.67	—	—	0.67
Coeff of Exp per ° F, 32–570 F	18.5×10^{-7}	20×10^{-7}	25×10^{-7}	18×10^{-7}
Spec Ht, Btu/Lb/F	0.195	0.195	—	—
Elect Res, Ohm-Cm @ 212 F	10^{12}	—	3×10^{13}	$>10^{15}$
Power Factor, 68 F 1 MC, %	0.46	0.28	0.26	0.06
Dielectric Constant, 68 F 1 MC	4.6	4.7	5.1	4.0
Thermal Stress Resistance[1]	—	—	—	—
Thermal Shock Resistance[2]	320	—	—	—
MECHANICAL PROPERTIES				
Mod of Elasticity, Psi	9.8×10^6	—	—	6.8×10^6
Normal Work Stress (Annealed), Psi	1000	1000	1000	1000
THERMAL TREATMENT				
Annealing Temp, F (Stress Relieving)	1020	975	890	910
FABRICATING PROPERTIES	Glass can be ground or polished without great difficulty. Machining is limited to sawing or drilling. Sawing is readily accomplished with an impregnated wheel. Drilling is difficult, but is done with special carbide-tipped drills using kerosene as a lubricant.			
Joining	Glass can be joined by heat-sealing using a blast-lamp, gas torch or by electric heating. Parts must have closely similar coefficients of expansion or intermediate pieces having intermediate coefficients must be used. Glass-to-metal seals also are made by this latter process.			
Highest Service Temp F, Annealed	900	840	800	810
Tempered	550	500	420	450
CORROSION RESISTANCE	Glass is attacked by hydrofluoric acid, hot concentrated phosphoric acid and strong alkalies. It is resistant to most other chemicals.			
USES	Heat exchanger tubes, chemical apparatus, cooking-ware, electrical insulators, sight and gage glasses, containers for chemicals and medicinals, metal sealing, industrial piping, industrial glassware requiring thermal resistance, special lighting ware, heat resisting lenses.			

Refractory Ceramics

A refractory ceramic is a nonmetallic material resistant to a state of fluidity or breakdown due to heat. Refractories are primarily used to provide linings for furnaces, where they must resist high temperatures, slag corrosion, and abrasive action of the charge. In the form of slurries, refractories are forced through a hose and nozzle and sprayed over the inside of the cupola and open-hearth furnace or the fire-box of a boiler. In this manner, spalled areas can be repaired and a sealed furnace can be obtained. Almost all refractory materials are available as brick in a wide range of standard sizes and shapes. Most refractories are also available as mortars for forming joints between the brick or for patching or coating brickwork.

Refractory types. Refractories are classified according to chemical composition into four major groups: alumina–silica, silica, basic, and special. The alumina–silica group includes the refractories composed practically entirely of silica and alumina. Fire-clay brick is the most widely used type of refractory, comprising about 75 percent of all refractories produced in the United States.

High-temperature ceramics are being used in paints and coatings. In paints, they are used to coat metal surfaces exposed to temperatures of 1800°F. Ceramic paints may be applied without firing. They provide protection against oxidation and radiant-energy transmission. Ceramic coatings, including graphites, oxides, and endothermic materials, are fired on the base metal, which they protect within the plastic range of the metal.

Design of refractory and high-temperature ceramics. Refractory ceramics have little ductility and points of stress concentration, such as holes, notches, and sharp corners, should be avoided. Tolerances should be generous so as to avoid machining, which is often difficult. On coated ceramics, the thickness of coat will thin out on sharp corners, so they should be avoided if uniform thickness is required.

Carbon

Carbon graphite is coming into more and more prominence as an engineering material. Its value as an automotive water-pump seal nose is widely recognized and it is employed considerably throughout the industry. In the aeronautical field, especially for fuel pumps, carbon in the form of seals and bearings is one of the few successful materials. Similarly, general industry, textile bleacheries, oil refineries, chemical processing plants, and oven-conveyor users find that carbon is frequently a solution in cases where bearings cannot be lubricated owing to high temperatures or inaccessibility, or because they are immersed in corrosive mixtures.

The most important properties of carbon and graphite materials are inertness to chemical action, good high-temperature strength, excellent resistance to thermal shock, high sublimating or boiling points, good heat and electrical conductivity, high heat of vaporization, low friction losses, and freedom from swelling or warping.

Fabrication of carbon. Carbon powder is fabricated by combining the powder with a tar that acts as a binder material and shaping either by extrusion or pressure molding. The green parts are then baked at around 1800°F in order to remove the volatile compounds in the binder and give a stronger finished product.

Designing carbon parts. In the baking process, the material shrinks and distorts somewhat, so that shrinkage should be provided for when designing the mold. Relatively close tolerances can be achieved in the molding process.

Small parts are frequently held to dimensional tolerances as close as 0.0005 in. On most dimensions, a tolerance of 2 percent can readily be maintained; this amount of variation represents standard molding tolerance for carbon parts.

Some general design suggestions for molded parts are:

1. Avoid close tolerances (0.001 in./in. or closer).
2. Avoid intricate external and internal shapes.
3. Avoid thin-sectioned flanged parts. Better to use two parts.
4. Avoid press-fit of carbon sleeve over shafts.
5. Do not make seals (washers) from extruded tubing. Better to let manufacturer produce finished seals to specifications.
6. Do not specify wall thicknesses of less than $\frac{3}{16}$ in.
7. Do not specify threads in graphite.
8. Specify chamfered edges rather than radius edges.

For special application, the pores of molded carbon parts can be impregnated with thermosetting resins to make the product impervious to liquids and gases. Greases, oils, and metals such as copper and babbitt can also be used as impregnants.

Carbon parts can be vulcanized to rubber and can be attached to metal parts through the use of adhesives. Carbon parts are readily machined after molding and are usually processed dry. Parts can be lapped and polished to accuracies of a few millionths of an inch.

POWDER METALLURGY

Powder metallurgy is the production of metallic components by feeding metal powder into a die of the desired shape, pressing the loose metal powder into a compact briquette of the approximate geometry of the desired finished product, and then heating the briquette in a furnace so as to bond the powder particles together to produce parts with the desired physical and mechanical properties. If necessary, final sizing or coining of the product to meet specified dimensional tolerances may take place.

Today, parts weighing as much as 30 lb are being produced in quantity from iron powders. Powdered-metal components are competing advantageously with machined castings, forgings, and stampings. They are finding increasing usage in the automotive and appliance industries for producing gears, bearings, splined parts, piston rings, valve seats, valve guides, tappets, rockers and pedestals, valve retainers, valve caps, camshaft and crankshaft sprockets and pulleys, clutch hubs and plates, hydraulic pistons, and the like. A recent U.S. Army contract permits the use of 4600 series powder-forged steel in the manufacture of 20 small-caliber weapons parts. This contract permits the substitution of powder-forged for ingot-metallurgy material. Powder-forged materials frequently compare favorably, and in some cases the tensile strengths are greater than the ingot-metallurgy materials. In general, the impact properties are lower. However, at higher hardness (greater than Rockwell 45) powder-forged material is usually competitive with some of the AISI steels.

This process is required to produce many components than cannot be manufactured economically by any other method. For example, the fabrication of metals with high melting points, such as incandescent lamp filaments, nonconsumable welding electrodes, and resistance wire for high-temperature furnaces. It also has application in connection with certain components of a composite that must retain properties that would be lost if melted. For example, in the manufacture of cemented carbides. Another group of products that is only achievable by this method is those that have properties that would be impossible to obtain by other methods, such as porous electrodes for alkaline batteries, and porous bearings.

Advantages of Designing for Powder-Metallurgy Components

The two main advantages of powder-metallurgy components are that (1) it allows production of complex parts such as gears, eccentrics, splines, and irregularly shaped holes with a minimum of machining and scrap, and (2) it can produce parts, such as porous bearings, cemented carbide cutting tools, and refractory metals, with special properties not otherwise attainable.

Today, not only are iron, carbon steel, alloy steel, copper, brass, bronze, and nickel available in powder form, but also stainless steels, refractory metals, and the precious metals. Thus, parts of complex geometry can be produced with a choice of a variety of materials giving unique properties of the specific material employed. Where high-density parts are needed with increased physical and mechanical properties, a re-pressing and second sintering operation will provide beneficial results. Tensile strengths up to 180,000 lb/in.^{2} are achieved in heat-treated high-density, low-porosity, powdered-metal components.

Special properties such as self-lubrication in porous bearings are an important advantage of powder metallurgy. The porosity characteristic of powdered-metallurgy construction allows between 10 to 40 percent of oil by volume to be contained within the component. As the part heats, because of friction, the contained oil will expand and emerge on the bearing surface providing a film of lubricant. Subsequent cooling results in reabsorption of the oil into the pores of the part.

Powder-metallurgy components have damping characteristics that result in quieter operation of mating parts.

Powder metallurgy can be used to produce high-melting (refractory) metal such as tungsten, molybdenum, and tantalum. An example of the type of product is heavy metal, a tungsten alloy containing 6 percent nickel and 4 percent copper. It can be used for radium containers, balancing weights, and similar applications where high density is desirable.

Copper and tungsten can be joined together by pressing and sintering a mixture of the metal powders to produce special electrical contacts and resistance-welding electrodes. With such a mixture it is possible to retain much of the relatively high electric conductivity of the copper while realizing high strength from the tungsten.

One of the important applications of the powder-metallurgy process is the production of cemented carbide products. Mixtures of hard metal carbides, tungsten carbide, molybdenum carbide, and tantalum carbide are made with a small amount of cobalt. After pressing, sintering is carried out at a temperature above the eutectic and the cobalt cements the hard carbide particles into a strong solid mass.

Combinations of metals and nonmetals are used for special purposes. Examples are clutch plates and brake linings, in which nonmetallic abrasive powder is included. Copper alloys containing graphite are widely used as bearings.

With controllable density, the engineer can select predetermined mass–weight ratios. Density can be distributed within a structure to achieve high strength in one area and self-lubrication in another. Metals can be oil-impregnated for self-lubrication or made to be completely porous for use as filters.

Limitations of Powder Metallurgy

The principal limitations are those imposed on the size and the shape that can be produced economically. Relatively high pressure is required to compact the powder into the desired shape, and the size of a part is consequently limited by the available press capacity. Seldom is a part weighing more than 30 lb produced by this process.

Since the compacted part must be ejected from the die without fracture, the shapes that may be made by the powder method are limited. Designs with undercuts or re-entrant angles cannot be made. Thin sections must be avoided, as well as featheredges and narrow or deep splines.

Another limitation is the die cost. Precision dies for forming the powder are expensive, and unless the high cost can be distributed over large production runs, the die cost per piece produced may be excessive. Usually 10,000 pieces represent a minimum quantity before this method of fabrication is economically practical.

A final limitation is that protection against corrosion during the processing cycle must be provided.

Designing for Powder Metallurgy

Before specifying a powder-metallurgy component, the engineer should consider four constraints:

1. Is the quantity large enough to justify the tooling? As mentioned earlier, the minimum figure in most cases is approximately 10,000 pieces of the same part.

2. Can the shape of the part be produced to the required tolerances by powder metallurgy? The expected tolerance that is achieved on diameters and dimensions at right angles to the direction of compaction is ±0.001 in./in. Tolerances of ±0.004 in./in. are characteristic of tolerances held in the direction of compacting.

   Secondary sizing operations such as shaving can, of course, be performed at added cost to improve tolerance accuracy.

   The engineer should recognize that undercuts cannot be molded, since it would not be possible to eject the briquette from the die. Likewise, complete spheres cannot be molded since the point of contact of the mating punches would involve a sharp edge that would fracture as the compressing pressure built up.

   Re-entrant angles cannot be produced. Thin walls (less than 0.030 in.) should be avoided as well as abrupt changes in cross section, and large thin forms.

   Five design rules relating to shape should be borne in mind. First and most important is that the shape must be such that the briquette can be ejected from the die. Second, you should not design a part that requires the metal powder to flow into a thin wall, or flange, or sharp corner. Third, the geometry of the part must permit the design of strong die sections. Fourth, the length–diameter ratio of 2.5:1 should not be exceeded for most thin-walled parts. Fifth, to assure uniform density and consequent high strength, changes in section thickness should be minimized.

   These design rules have the following important corollaries.

   a. Threads cannot be molded. They must be machined on sintered parts.
   b. Gear hub diameters should be less than the root diameter of the gear.
   c. Avoid narrow deep splines.
   d. Round corners give better powder flow in the die and result in increased strength of the part.
   e. Round holes simplify pressing and reduce the cost of the part.
   f. Reverse tapers cannot be molded.

3. Are the required physical and mechanical properties within the capabilities of powdered metals? There is a wide range of physical properties that can be achieved. Low-density iron-product parts have tensile strengths as low as 10,000 lb/in.^{2}, whereas high-density high-strength steel powders have tensile strengths over 70,000 lb/in.^{2} Heat-treated high-density alloy powders have tensile strengths of 180,000 lb/in.^{2} or more. Figure 7.2 provides a graphic illustration of the relationship between density and yield strength of three ferrous powders.

   Other properties, such as hardness, yield strength, etc., should be carefully checked to ensure that the attainment of the required specifications is within the realm of powder metallurgy.
4. Economic considerations enter into every design decision. It should be established that powder-metallurgy design is the most economical possible,

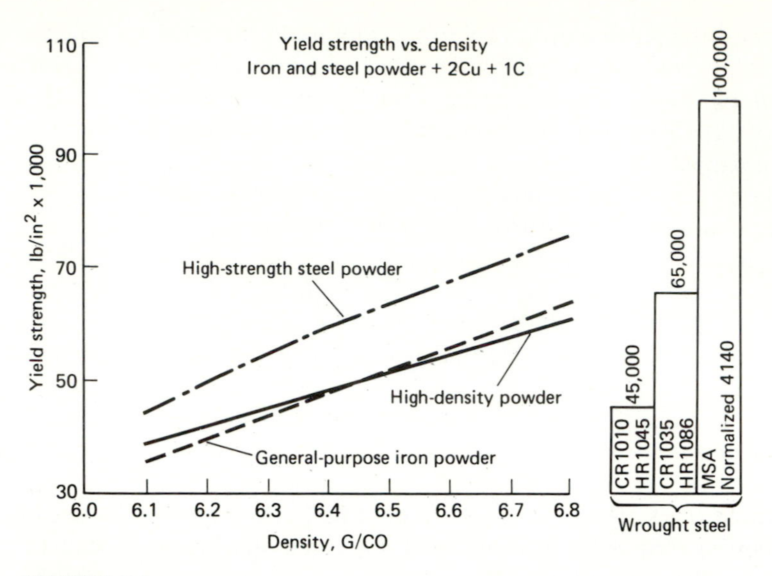

FIGURE 7.2
Yield strength versus density—iron and steel powder with 2 percent copper and 1 percent carbon added.

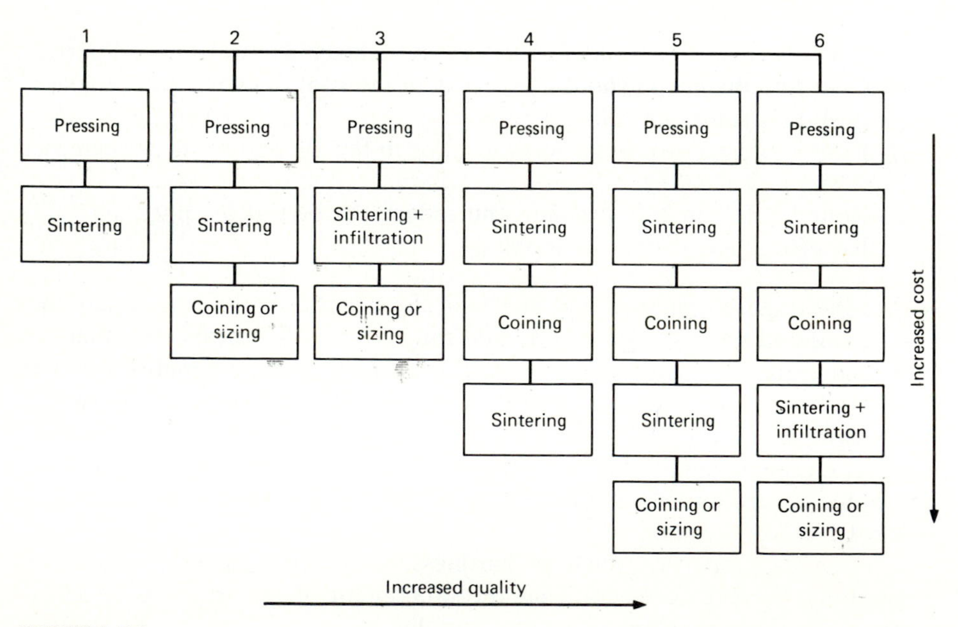

FIGURE 7.3
Processes for the manufacture of sintered structural parts and their relation to increased cost.

assuming other factors are at least equal to alternative methods. Increased physical properties can be obtained by additional sintering and infiltration. However, this increases cost. The various common processes for the manufacture of sintered structural parts and their relation to increased cost are illustrated in Fig. 7.3.

Steps in Producing Powder-Metallurgy Parts

Three steps are involved in producing powder-metallurgy parts: mixing and blending; compacting; and sintering.

Mixing and blending. This involves the mixing of the metal powder with any alloying elements and lubricants required. Precise control of the volume of the raw powders and lubricants dispensed is essential. The best mixing is obtained when all particles have similar size, shape, and density.

Compacting Compacting the powder is an important step in the powder-metallurgy process. In addition to producing the required shape, the compacting operation also influences the subsequent sintering operation and governs, to a large extent, the properties of the final product. The most important effects of compacting include:

1. Reduction of voids between powder particles and increased density of the compact
2. Adhesion and cold welding of the powder and sufficient green strength
3. Plastic deformation of the powder to alloy recrystallization during subsequent sintering
4. Plastic deformation of the powder to increase the contact areas between the powder particles, increasing green strength and facilitating subsequent sintering

Compacting methods such as isostatic pressing and roll compacting, which were performed only in the laboratory a few years ago, are today in use on a production basis. The principal methods of powder compacting are:

1. Unidirectional
   a. Single-action pressing
   b. Double-action pressing
2. Isostatic pressing
3. Rocking-die compacting
4. Explosive compacting
5. Powder rolling
6. Powder extrusion
   a. Unidirectional
   b. Isostatic
7. Powder swaging
8. Powder forging

TABLE 7.6
Typical tonnage requirements and compression ratios for various powdered-metal products

Type of compact	t/in.^{2}	t/cm^2	Compression ratio
Brass parts	30–50	4.7–7.8	2.4:1 to 2.6:1
Bronze parts	15–20	2.4–3.1	2.5:1 to 2.7:1
Carbides	10–30	1.5–4.7	2.0:1 to 3.1:1
Alumina	8–10	1.2–1.5	2.5:1
Iron parts			
Low-density	25–35	3.8–5.5	2.0:1 to 2.4:1
Medium-density	35–40	5.5–6.2	2.1:1 to 2.5:1
High-density	35–60	5.5–9.3	2.4:1 to 2.8:1
Tungsten	5–10	0.8–1.5	2.5:1
Tantalum	5–10	0.8–1.5	2.5:1

Source: Courtesy of Henry H. Hausner, and Mal M. Kumar, *Handbook of Powder Metallurgy*, Chemical Publishing Co. Inc., New York, 1982.

Compacting is performed by pouring a measured amount of the appropriate powder into the die cavity and then introducing one or more plungers to press the metal powder into a coherent mass. Pressures up to 100,000 lb/in.^{2} are used. Table 7.6 provides tonnage requirements and compression ratios for various powder products. A measured volume of powder is usually used, although in some cases a definite weight of powder is more suitable.

An ideal pressed compact should have the desired shape and uniform density. However, metal powders do not flow uniformly, nor do they uniformly transmit pressure. Friction along the die walls and between powder particles, and the inability of powder to transmit uniform pressure at right angles to the direction of the applied force, result in nonuniform compacting force within the die. The powder is not compressed to uniform density throughout the compact, and does not react uniformly to subsequent sintering. However, this inequality may be minimized in several ways. Thin compacts are more evenly pressed throughout their thickness than are thick compacts; consequently, the design thickness of powder-metallurgy parts should not be excessive. Double-end pressing, in which the part is simultaneously pressed from both top and bottom, helps enhance the pressure gradient in the compact and provides more even density. When irregularly shaped parts are made, multiple punches are often necessary. Isostatic pressing is also applicable when producing complex parts. Figure 7.4 illustrates the added density achieved by isostatic pressing and increasing the compacting pressure. Temperature has an influence on both the density and strength of pressed powder compacts. Figure 7.5 illustrates the typical relationships between compacting temperature and the parameters of density and strength.

Lubrication of the die walls and between the powder particles is helpful

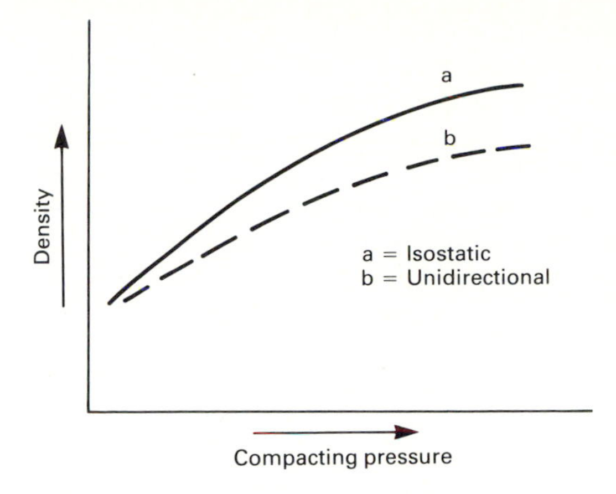

FIGURE 7.4
Method of compacting and its effect on the density of the pressed powder.

in reducing friction and pressure gradients within the compact. Stearic acid, precipitated zinc stearate, precipitated lithium stearate, amide wax, polyethylene, paraffin, graphite, melamine and other lubricants may be mixed with the powder or coated on the die walls to provide lubrication. The lubricants are burned off during the early stages of sintering. Typically, the lubricant added ranges between 0.2 and 1 percent of the total powder weight.

Binder is a cementing medium that is either added to the powder to increase the green strength of a compact or added to a powder mixture for the purpose of cementing together the powder particles that would otherwise not sinter into a strong body. Typically, on a weight basis, binder content ranges from 1 to 5 percent. Binders find application when porous compacts, or very fine powders, or powders with spherical particles are being used.

For production runs, the punches and dies are rigidly mounted in a suitable press and the die cavity is filled with a measured amount of properly mixed metal powder. The volume of the loose powder is approximately 2 to 10 times the volume of the pressed compact.

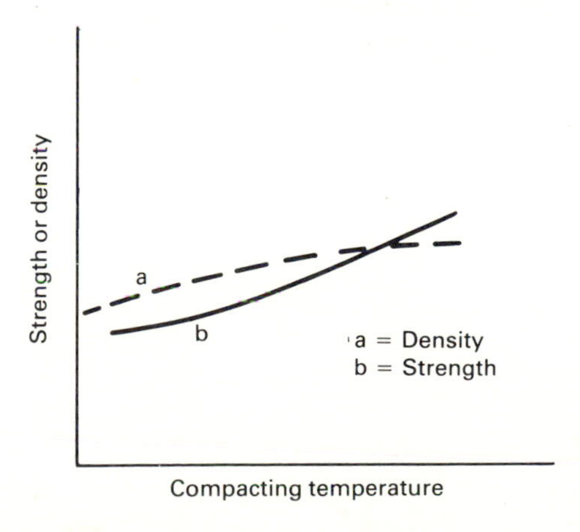

FIGURE 7.5
The effect of compacting temperature on the density and strength of pressed powder compacts.

The speed of compacting, which is proportional to the rate of punch travel, is relatively unimportant. Compacting pressure may be applied rapidly or slowly, the only limitation on speed being due to entrapment of air within the pressed compact. When the pressing speed is excessive, air between the loose powder particles does not have sufficient time to escape, and is trapped within the compact. Subsequent sintering may then produce considerable expansion of the compact, rather than the usual shrinkage.

Following the compacting operation, the relatively fragile but coherent mass of metal powder must be ejected from the die without injury, and transferred to a furnace for sintering.

Sintering. Sintering entails heating the pressed compact to below the melting temperature of any constituent of the compact, or at least below the melting temperature of all principal constituents of the compact. The purpose of such heating is to facilitate a bonding action between the individual powder particles that increases the strength of the compact.

Sintering is generally carried out in a controlled-atmosphere furnace, because oxidation of individual metal particles weakens the sinter bond or even entirely prevents sintering.

The mechanism by which individual metal particles are joined into a coherent mass having increased density and strength has been the subject of many investigations and much discussion. The mechanisms involved in sintering appear to be principally the following: (1) diffusion, (2) recrystallization, (3) grain growth, and (4) densification.

Increase of strength developed in the powder compact during sintering is principally due to a disappearance of the individual particle boundaries through diffusion and recrystallization. The powder particles are usually deformed by greater than critical strain during pressing, so that recrystallization and coalescence are common during sintering.

The important factors controlling sintering are type of reaction during sintering, sintering temperature, rate of heating and cooling, sintering time,

TABLE 7.7
Representative sintering times and temperatures

	Temperature		
Material	**°F**	**°C**	**Time (min)**
Bronze	1400–1600	760–870	10–20
Copper	1550–1650	840–900	12–45
Brass	1550–1650	840–900	10–45
Iron	1850–2100	1010–1150	8–45
Nickel	1850–2100	1010–1150	30–45
Stainless steel	2000–2350	1095–1285	30–60
Tungsten	4250	2345	480

and sintering atmosphere. An increase in sintering temperature and time increases the sintering effect. Strength, hardness, and elongation are increased by longer sintering time or higher sintering temperature. Table 7.7 provides information as to the typical sintering times and temperatures for the more important metal powders.

Ordinarily, the density of the compact is increased by sintering, and shrinkage will occur. Shrinkage entails the filling of holes within the compact, a process that necessitates closer contact between individual particles, increase of areas of contact, and reduction of the size of holes in the compact.

In some instances sintering is done at a temperature at which a liquid phase exists. When a liquid phase exists, the compacting conditions are less critical and high density can be attained quickly. In products of this type the nature of the alloy formed between the solid and liquid metals, the ability-of the liquid metal to wet the solid particles, the capillary action of the liquid in the solid compact, and other factors play important roles in sintering.

Preparation of Metal Powders

A great variety of mechanical, physical, and chemical methods are employed for the production of metal powders; and powders widely differing from each other in particle size and shape, chemical purity, and microstructure are on the market. The powders for use in powder metallurgy should be nearly spherical in shape, with a minimum of fine dust and no large particles. The properties of the final product and the techniques used for its production depend upon the characteristics of the powder, which are principally influenced by the method of manufacturing the powder.

Most powders intended for use in powder-metallurgical processes are produced by one of two methods—reduction of metal oxides or electrolysis—although atomization, mechanical comminution, thermal decomposition of

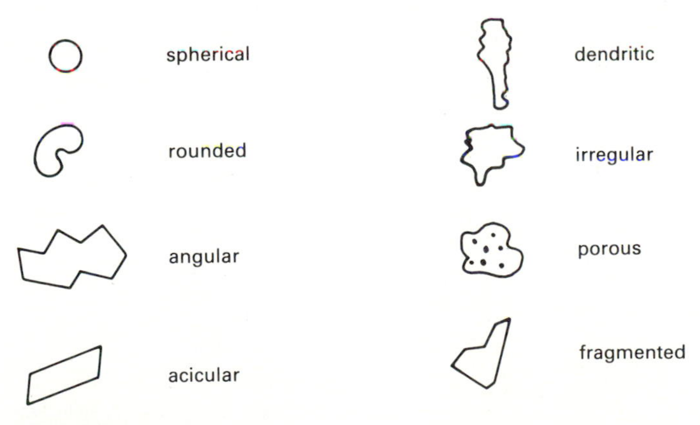

FIGURE 7.6
Shapes of metal powders.

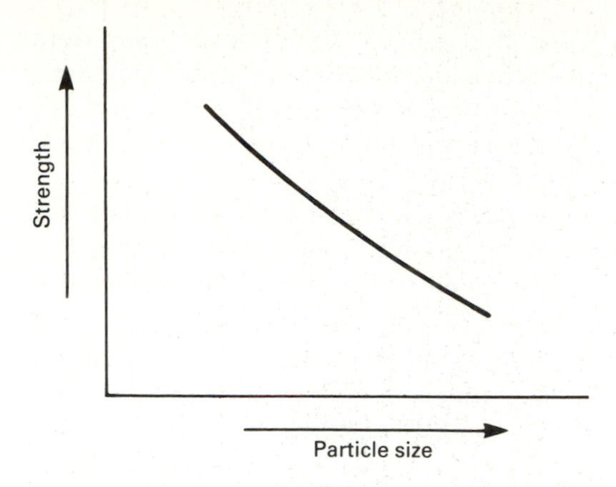

FIGURE 7.7
Effect of particle size on the strength of the pressed powder compact.

carbonyls, and intergranular corrosion are employed sometimes. There are eight shapes of metal powders. These are illustrated in Fig. 7.6. Figure 7.7 illustrates the effect of particle size on the strength of the pressed powder compact and Fig. 7.8 shows the effect of particle size on the pore size of compacted powders.

Reduction of metal oxides. Very fine metallic oxides are reduced in a gaseous medium to the metallic state, producing a spongy metallic mass that is later pulverized to the desired particle size. Gases such as hydrogen, carbon monoxide, and coal gas, and also carbon are employed as reducing agents. The powders produced by this process are in most cases spongy and of angular and granular shape.

For the carbide and nitride ceramics, high-temperature reactions are used. For example, the Acheson process, used to produce silicon carbide crystals, involves heating, by electric current from carbon electrodes, a mixture of silica sand and coke. The resulting silicon carbide crystals are subsequently broken up and sized. Today, silicon nitride particles are produced by utilizing a laser-driven reaction between ammonia and silane.

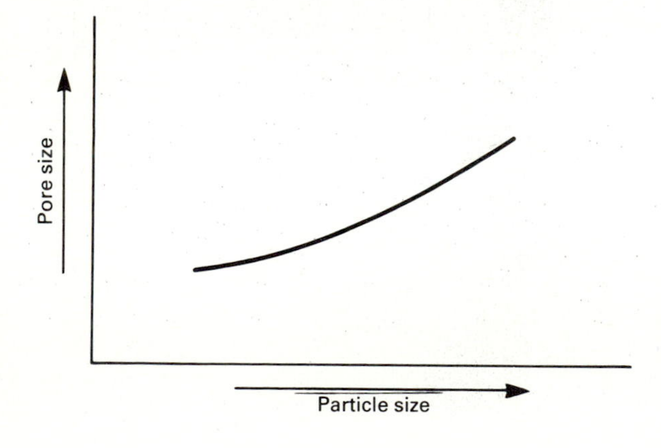

FIGURE 7.8
The effect of powder particle size on the pore size of compacted powders.

Electrolysis. Metal powders can be produced by electrodeposition from solutions, as well as from fused salts. The electrolysis of solutions is employed for the commercial production of metals such as iron, copper, nickel, cadmium, tin, antimony, silver, and lead. The electrolytic process allows a control of powder characteristics, particularly particle size. Means of control are the regulation of current density, temperature, composition and circulation of the bath, and size and arrangement of the electrodes.

Atomization. This process consists of forcing molten metal through a small orifice and breaking up the thin stream of liquid metal by a jet of compressed air, steam, or inert gases. The cooling effect produced by the sudden expansion of the metals leaving the nozzle, combined with the cooling effect of the gases, instantaneously solidifies the metal. The controlling factors for particle size are nozzle design, temperature and rate of flow of the molten metal, and temperature and pressure of the gas.

Mechanical comminution. Crushing and milling of the brittle metals generally produce angular particles that are suitable for powder metallurgy. Ductile metals may also be milled to powder form, but the resulting powder is flakelike and not desirable for most powder-metallurgy work.

Thermal decomposition of carbonyls. Iron and nickel powders are often made by the decomposition of iron carbonyl and nickel carbonyl. The metallic carbonyls $Fe(CO)_5$ and $Ni(CO)_4$, which are gaseous at the operating temperature maintained during the process, are decomposed to iron or nickel. These powders are usually spherical.

Intergranular corrosion. This method is of particular interest because it permits the preparation of powders of 18:8 stainless steel that cannot be prepared by other methods. In this process, the steel is sensitized and after cooling is disintegrated at the grain boundaries by a corrosive solution—for example, by a boiling aqueous solution of 11% $CuSO_4$ and 10% H_2SO_4. The particle size of the resulting powder is determined by grain size of the corroded material after sensitization.

Forgings from Powder Preforms

By combining preform sintering with subsequent forging it is possible to produce parts made of powdered metal that have little flash and are of dimensional accuracy sufficient to minimize secondary operations while providing parts with strength and toughness properties much superior to the typical sintered powdered-metal component.

Furthermore, the preform is usually forged in only one die cavity and is normally given only one blow to attain the desired component. In those cases where the composition of the preform is similar to a conventional forged grade of material and the density is approximately 100 percent of its theoretical value, the powder-metallurgy forgings will have about the same properties as the conventional material. For example, die-forged preforms of 1050 carbon steel have shown yield strengths of 76,000 lb/in.^{2} when the density is 7.86 g/cm^3. Tensile strength, yield strength, and ductility fall off rapidly as the percentage porosity increases. For example, yield strengths of 70,000 lb/in.^{2} at 0.005 percent porosity fall off to 40,000 lb/in.^{2} at 0.05 percent porosity when using low-alloy steel powders. Mechanical properties appear to be a function of final pore shape rather than particle size prior to sintering.

Cost

The cost of metal powders for ferrous materials typically approximates 15 to 25 percent of the total unit part cost. For example, the distribution of costs of several parts made of different iron powders and weighing approximately 2 ounces is as follows:

Material	15–25%
Compacting	20–30%
Sintering	20–30%
Secondary operations	10–20%
Unit tool cost	10–20%
	100%

It is interesting to note that the prices of ferrous powdered metals have deviated very little in the past eight years. For example, the following table illustrates the relative stability in the cost of metal powders.

	Cost, $/lb		
Material	**1981**	**1985**	**1989**
Aluminum	1.05	1.20	1.82
Brass	1.30	1.27	1.53
Iron	0.27	0.31	0.33

COMPOSITES

The use of composites for such products as tennis rackets, skis, sailboats, football equipment, fishing rods, medical implants, and automobile components is well known. Perhaps the earliest composites were produced by the Egyptians more than 2000 years ago when they improved the strength and durability of bricks by adding compressed straw to mud. In modern times, the development in the 1960s of fiberglass, in which special glass fibers were lined up in a resin matrix, represented the first broad use of composites. Du Pont later introduced Kevlar, which was a patented fiber placed in an organic matrix. This was a strong but light engineering material developed to replace ferrous materials. Graphite fibers were subsequently developed by Union Carbide; these had not only unusual strength but the ability to withstand high temperatures. Then, in the early 1970s, to improve temperature capabilities, a graphite epoxy was developed to add to the carbon matrix. These carbon–carbon composites were produced for designs such as rocket nozzles and re-entry nosetips that needed to withstand very high temperatures.

Then, in the mid-1970s and into the 1980s, metal matrix composites were introduced to improve on the lightness and temperature resistance of the light structural metals (aluminum and titanium). For example, aluminum, which can tolerate temperatures up to around 300°F under load, was now able to withstand temperatures of more than 500°F. Metal matrix composites are primarily used for missile and aircraft components (see Fig. 7.9). These composites use graphite or silicon carbide as the major reinforcements, while the metal matrix is usually aluminum or titanium. Other matrixes include copper, magnesium, and lead.

Organic matrix resins are typically epoxy, polyester, polimide, and thermoplastics. Reinforcements are either glass (mainly "E" and "D"), quartz (pure silica) fiber, and Kevlar. Recent interest in advanced thermoplastics has led to several materials, among which thermoplastic polyimide, polysulphones, polyphonylen sulphide, polyethersulphone and polyetheretherketone are notable.

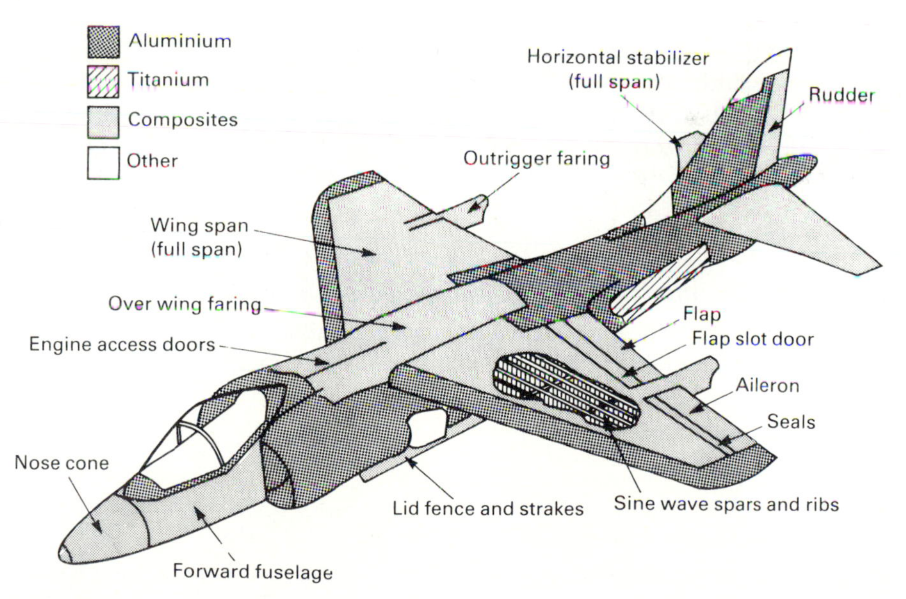

FIGURE 7.9
Typical use of composites, titanium, and aluminum in the manufacture of a modern fighter plane.

In the next decade, reinforced ceramics will be introduced in much large quantities for airplane engine components for which there is a need to withstand high temperatures. It has been estimated that by the year 2000 fighter planes will be 50 percent composites.

Auto makers are considering fiber-reinforced plastic composites for such stressed items as control arms and cross-members in order to take advantage of potential weight reductions of 25 to 85 percent, better corrosion resistance, and improved noise and vibration damping.

A technique referred to as *squeeze casting* is a method of producing ceramic–metal and metal matrix composites for a variety of applications, including connecting rods, pistons, and similar components. This process involves metering liquid metal into a preheated lubricated die and forging the metal while it solidifies. The load is applied immediately after the metal begins freezing and is maintained until the casting has solidified. Casting ejection from the die is analogous to the procedure used in closed-die forging. In this process, a preheated, porous ceramic is placed into a die cavity, which is subsequently filled with the liquid melt. Upon the application of pressure, liquid melt will infiltrate the pores of the ceramic preform, resulting in a metal–ceramic composite.

Advantages of Composites

Engineering composites can be defined as those materials in which nonwoven fibers are oriented in a matrix in such a manner so as to improve their structural capability. Composites utilize combinations of unusually strong, high-modulus fibers and organic, ceramic, or metal matrices. In view of the large variety of both combinations and arrangements of fibers and matrices combined in concert with lamination, engineers have increased opportunities to meet challenging new product designs for use in diverse environments. The principal advantages of those composites that are of structural interest include the following.

1. They have high stiffness-to-weight and strength-to-weight ratios
2. Corrosion and stress corrosion problems are eliminated. Thus, there is no need for costly anticorrosion treatments during manufacture or time consuming corrosion control efforts.
3. Problems of fatigue are significantly reduced—especially with carbon fiber components.
4. The number of parts going into a product is usually reduced, thus saving in inventory control, purchasing, and paper work.
5. With the use of composites, there is usually a reduction in structural mass, resulting in many direct and indirect cost savings (for example cost of shipping, increase in miles per gallon of fuel, etc.).

6. Control of surface contour and smoothness is improved, which can yield better performance (for example, higher aerodynamic efficiency).
7. Improved appearance results in greater sales appeal.

Processing of Composites

In the processing of composites, the basic process cycle can be a resin injection process using matched metal molds (see Fig. 7.10) with woven fabric shapes as reinforcement. The successive steps in such a process would be as follows.

1. Prepare mold by cleaning and applying a release agent.
2. Build-up reinforcement on the male mold.
3. Close mold and attach ancillary equipment.
4. Prepare resin by melting (200°F approximately) to remove any entrapped air.
5. Inject resin from a heated pressure pot into the closed mold.
6. Cure resin (typically about 400°F).
7. Remove part from mold.
8. Post-cure (age) at approximately 450°F.

Figure 7.11 illustrates a typical time–temperature relationship of an injection molding cycle using an organic matrix with fiber reinforcement.

Carbon Fiber Reinforcement

Carbon fibers are available in a range of compounds, supplied as granules, with different levels of fiber reinforcement. Typically, granules of 20, 30, or 40 percent by mass are available. The design engineer can select the level of

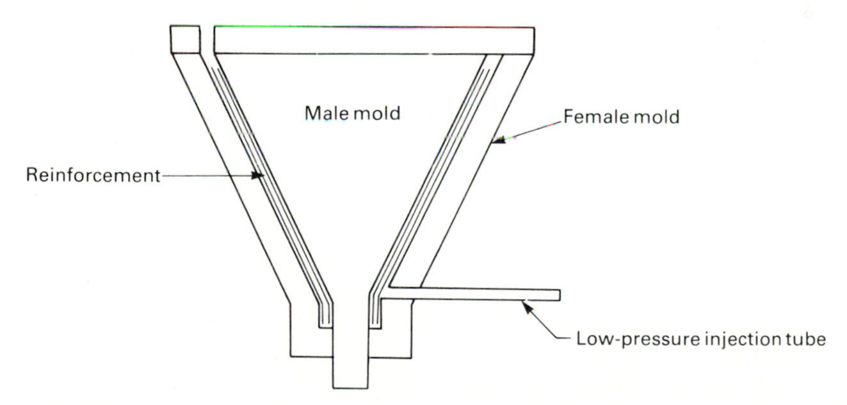

FIGURE 7.10
Representative injection mold for the molding of organic matrix resin with fibre reinforcement.

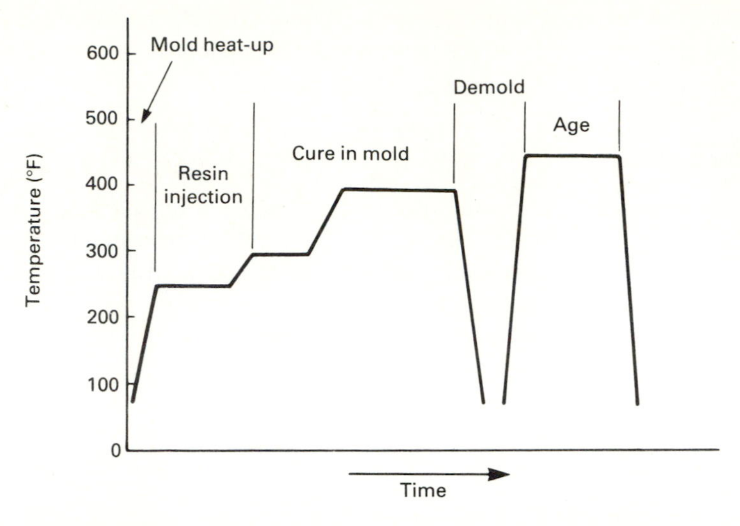

FIGURE 7.11
Time–temperature relationship for typical injection molding of organic matrix with fibre reinforcement.

reinforcement on the basis of properties and cost. These short carbon fiber reinforcement thermoplastics are high-performance materials often referred to as CFRTP. In load-bearing applications, carbon fibers are used frequently with the matrix nylon 66, which gives good performance in both strength and modulus over a broad temperature range. Compounds based on polyethersulphone (PES) and polyetheretherketone (PEEK) are used frequently in applications where performance at higher temperatures is required (see Table 7.8).

CF/nylon 66 service characteristics include the ability to withstand temperatures in the range −140°F to 190°F, relative humidity up to 100

TABLE 7.8

Property	40% CF/PA66	40% CF/PES	40% CF/PEEK
Density, g/cm^3	1.34	1.52	1.45
Tensile strength, MPa			
24°C	246	176	227
100°C	108	140	165
140°C	75	112	129
Tensile elongation, %	1.65	1.1	1.35
Flexural strength, MPa	413	244	338
Compressive strength, MPa	240	216	—

Source: Adapted from *Engineering with Composites,* the Third Technology Conference of the European Chapter, March, 1983.

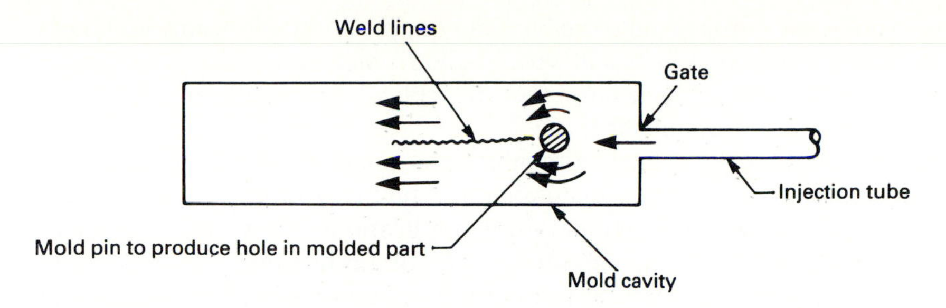

FIGURE 7.12
The formation of weld lines caused by two fronts converging.

percent, and a shelf life or more than 10 years. The material is resistant to gasoline, oils, solar radiation, and fire. It does not support combustion, is self-extinguishing, and does not emit toxic fumes upon being extinguished.

Design Considerations

Components molded from short fiber reinforced thermoplastics have anisotropic properties because the fibers are not randomly oriented. Fiber orientation is caused by melt flow during the filling of the mold cavity and the position of the mold gate (cavity feed point). Defects due to melt flow are referred to as "weld lines", which are caused by the flow around obstacles such as in the formation of holes or at corners and changes in section thickness (see Fig. 7.12). Weld lines are formed in two ways: when two melt fronts meet head on, or when two melt fronts converge, as illustrated in Fig. 7.12.

When the melt flow has induced a high degree of orientation, in the flow direction, then both strength and modulus are higher in this direction compared with direction transverse to flow. This can be put to design advantage by the engineer through the selection of the position of the gate in the mold. The ratio between the properties in these two directions is termed mechanical anisotropy and is typically 2:1 for CFRTP.

In designs, a radius of 0.060 in. (1.5 mm) will double the fracture toughness compared with a sharp corner. The strength and modulus properties for CFRTP are reduced by approximately 25 percent at weld lines. Components having thick wall sections and/or complicated geometry are likely to have more isotropic properties.

SUMMARY

New ceramic materials are being developed continually. Advanced ceramics with precisely controlled compositions and crystal structures are in increasing demand for such products as manufacturing dies, cutting tools, and high-performance engine components. We will see greater use of these engineering materials in structural, load-bearing applications as well as electrical and electronic uses, and furnace and crucible lining applications. Ceramic materials that have been developed in recent years include high-alumina and magnesia ceramics with density approaching the theoretical recrystallized graphites; foamed ceramics; and ceramic fibers.

Ceramic refractory coatings (zirconium oxide and zircon–alumina–silica) and ceramic metal coatings (silica–alumina and ball clay–silica) are now being used successfully in elevated temperature applications, providing significant energy savings and extending the service life of both metal and refractory substrates. Thus, ceramic coatings are cutting energy costs in the steel, heat treating, forging, aluminum, and diecasting industries. Currently, they are being evaluated for heat shields in space shuttles.

Today, many metals are challenged by the ceramics, glass, and carbon. Thus we have fiberglass furniture, automobile bodies, fishing rods, tennis rackets, and many other products that for a long time were produced exclusively from metallic or wood materials. The alert engineer will continually follow the development of ceramic materials. The material that seems to have little application in a given design today may be the ideal material in that same design tomorrow.

Today, powder metallurgy is one of the lowest-cost mass production methods of manufacturing small to medium-size complex parts to close tolerances. Ferrous and nonferrous-based powders are finding an ever widening market in the design of metal parts in almost every consumer industry. As a result of improved control and tooling, the powder-metallurgy process is being advantageously chosen for producing complex and critical application parts in the automotive, appliance, farm equipment, electric and electronic, and other industries.

Nonmetallic composites, particularly those with carbon as the main load-carrying element, are being specified in more and more products in order to improve performance and reduce operating costs. Metal matrix and reinforced ceramics will be utilized much more extensively during the period of 1990 to 2000. The alert engineer will keep abreast of composite technology, as this family of materials will be used extensively not only in the aircraft industry but in many designs subject to fatigue and stress corrosion problems.

QUESTIONS

7.1. What is meant by "high-performance ceramics?"

7.2. Explain how the design engineer establishes dimensional values on green ceramic forms.

7.3. Differentiate between isostatic and uniaxial pressing.

7.4. What unique properties do silicon nitride and silicon carbide possess that make them candidates for heat engine components?

7.5. What four processes are available to the production design engineer for forming ceramic whitewares? Describe each.

7.6. Where does the property of high dielectric strength find application?

7.7. What type of glass is best suited to high electrical resistance?

7.8. What is the difference between laminated and plate glass?

7.9. Would you recommend glass as a suitable material for the manufacture of Go–No Go plug gages? Why?

7.10. What are the properties of carbon that make it a valuable engineering material?

7.11. Would you recommend a threaded graphite bushing? Why?

7.12. What are the three main steps involved in the powder-metallurgy process?

7.13. Why are powder-metallurgy components frequently used in the manufacture of bearings?

7.14. What are the principal limitations of powdered metallurgy parts?

7.15. What finishing operations, if any, would be required on a 2-in. powdered metal gear having a tolerance of plus or minus 0.008 in. on the major diameter?

7.16. What yield strength can be expected from a general-purpose iron powder having a density of 6.3 to 6.5 g/cm^3?

7.17. What six advantages are characteristic of composites?

7.18. What elastomer is used in the compression molding of graphite–epoxy composites? Explain why this material can be used successfully in this process.

7.19. Explain why the location of the injection gate is important in connection with the design of a mold for producing themoplastic composites.

7.20. Explain how the formation of weld lines can be retarded.

7.21. In the injection molding of a composite component having a $\frac{1}{4}$ in. hole central to the part, explain why it might be cost effective to drill the hole as a secondary operation rather than mold it.

PROBLEMS

7.1. A general-purpose iron powder is compressed to a density of 6.5 g/cm^3. The expected yield strength of the resulting sintered product is 50,000 $lb/in.^2$ with a standard deviation of 2,000 $lb/in.^2$ With further compaction the production-design engineer can expect a density of 6.8 g/cm^3 at an added cost of \$0.11 per product unit. This added compaction will give an expected yield strength of 65,000 $lb/in.^2$ with a standard deviation of 4,000 $lb/in.^2$ The product is stressed at an average tensile load of 42,000 $lb/in.^2$ with a standard deviation of 5,000 $lb/in.^2$ If the company computes the cost of product failure at \$10 each and they expect to produce 500,000 parts, what level of compaction would you recommend?

7.2. Calculate the volume of metal required to compact a T-shaped body to 85 percent of its theoretical density. Give the size of the dies and height of the compact if the sintering operation results is 5 percent linear shrinkage in all directions. The finished parts need to be 1 in. $\pm^{0.005}_{0.000}$ OD on the body, $1\frac{1}{4}$ in. $\pm^{0.005}_{0.000}$ on the head diameter. The cored hole wants to be 0.625 in. $\pm^{0.003}_{0.000}$. The head is $\frac{3}{8} \pm 0.015$ in. thick and the overall length is $1\frac{1}{4} \pm 0.015$ in.

7.3. A tantalum ball joint for hip replacement has a projected area of 4.28 square inches. A four cavity die has been designed for a unidirectional single action press. The press is calculated to be 80% efficient. What tonnage press is required to produce pressed powder compacts of this ball joint?

7.4. The design engineer of the XYZ Company has two alternatives for the design of a powdered metal valve component: brass and stainless steel. He estimates the cost of the brass material as one unit, compacting brass one unit, sintering brass one unit, secondary operations with brass 0.75 units, and tool costs of brass one unit. If the cost of the stainless steel powder is 0.8 units, the secondary operations for stainless steel one unit, what would be the total cost in units of the stainless steel component? (Note sintering cost is directly proportional to sintering time and temperaure).

7.5. A ceramic material was being considered as a possible mold material in conjunction with the die casting of brass faucet handles. A cylinder of material 8 in. in diameter and 10 in. in length was used to obtain a stress–strain relationship in compression. The material failed under a load of 265,000 lb and a total strain of 0.012 in. What was the material's fracture strength? Percent contraction at failure? Modulus of toughness?

7.6. Two structural aluminum plates 2 in. wide by $\frac{1}{2}$ in. thick are connected by a lap joint using a single rivet 1 in. in diameter and subjected to a tensile load of 10,000 lb. Determine the factors of safety with respect to tension, shear, and bearing if the tensile, shear, and bearing ultimate strengths are respectively: 20,000, 13,000, and 27,000 lb/in.^{2}

7.7. In the XYZ Company, the process engineer is asked to improve the grain structure of an aluminum–bronze casting. He wishes to obtain 80 to 86 grains/in.^{2} at $\times 100$ magnification. A sample of 10 castings (using larger chills at the suggestion of the metallurgist) gives an average of 82 grains/in.^{2} at $\times 100$ magnification, with a standard deviation of 4. A previous test using smaller chills involving five castings provided an average of 76 grains/in.^{2} with a standard deviation of 3.2. Can we conclude that the use of chills is refining the grain structure? Explain.

7.8. An aluminum rod is cold-rolled so that it encounters a 70 percent reduction in area. What would be the magnitude of the true strain? Would this material be suitable for deep drawing? Why?

7.9. You have designed a part that will be subject to a 3000 lb/in.^{2} stress in tension while in service. At times the part will be exposed to temperatures approaching 250°F. Sound design suggests a factor of safety of 20 percent. What carbon-fiber reinforcement would you recommend? What minimum section thickness would you specify? Identify the radius you would specify at all junctures.

REFERENCES

1. Haloin, J. C.: *Primer on Composite Materials: Analysis,* Air force Materials Center, Technomic Publishing Co., Lancaster, Pa., 1984.
2. Harris, Bryan: *Engineering Composite Materials,* The Institute of Metals, North American Publications Center, Brookfield, Vt., 1986.
3. Hausner, Henry H.: *Powder Metal Processes,* Plenum Press, New York, 1967.
4. Hausner, Henry H., and Kumar, Mal M.: *Handbook of Powder Metallurgy,* Chemical Publishing Co. Inc., New York, 1982.
5. Howard, P., and Koczak, M. J.: *International Journal of Powdered Metallurgy and Powdered Technology,* Vol. 17, 1981.
6. Tendokar, Sebastian: *International Journal of Powdered Metallurgy and Powdered Technology,* Vol. 15, 1979.

7. Third Technology Conference of the European Chapter of Composites: *Engineering with Composites,* Vol. 1 and 2, London Tara Hotel, March 14–16, 1983.
8. Tsai, Stephen W.: *Composites Design,* The United States Air Force Materials Laboratory, Dayton, Ohio, 1986.
9. Wachtman, John B., and McLaren, J. L.: "Advanced Ceramics: Structural Materials With a Hot Future", *Manufacturing Engineering,* February, 1985.

CHAPTER 8

CASTING PROCESSES

Casting is the fastest way to go from a raw material to a simple or complex shape that is hollow or nonuniform in cross section. More than 85 percent of all metal castings are poured into sand molds, the balance are made in ceramic shell or metal molds. In a sand foundry, up to 90 percent of the molding sand can be reprocessed to make new molds. This is fortunate, because disposal of spent sand presents a significant solid waste problem, since there is a good possibility that the organic core binders will leach into the ground water beneath the disposal site.

The casting process is basically accomplished by pouring a liquid material into a mold cavity of the shape of the desired item and allowing it to solidify. For example, one might cast a bookend in the shape of Rodin's sculpture of the Thinker, or a chocolate rabbit. The former might be poured with a copper-based alloy such as bronze, the latter with molten chocolate. In each case the cooled shape would be a casting.

Castings are frequently hollow, as might be the case with the rabbit. The hole may be made by a core that the molten metal would surround, or one could allow a skin to freeze about $\frac{1}{4}$ in. thick and the liquid center would then be poured out. The latter would be called a slush casting. Cores are discussed more fully later.

HISTORICAL REVIEW

Bronze arrowheads were cast some 6000 years ago in open-faced clay molds: small statues and religious items were poured not much later (Fig. 8.1). The art of founding is fundamental to the production of tools for all civilizations. Casting was practiced throughout the ancient world: in Europe, Central and South America, India, the Orient, and North Africa. The bronze age appeared at different times in each area, but it is not known whether the knowledge was transferred from people to people or whether it was, like language, discovered independently by the different civilizations. According to the Bible, Tubal Caine was the world's first historical foundryman.

Certainly the Church nurtured the art of founding and preserved much of the science of casting and practical metallurgy through the publication of Biringuccio's *Pirotechnia* in the 16th century. In that work he gave detailed instructions for casting large church bells using a pit mold (Fig. 8.2) built at the base of the steeple in order to minimize the handling of the large castings. Pit molding is still used today for castings such as a large turbine for a hydroelectric plant.

In later times, Paul Revere was a foundryman. His enterprise grew to be the forefather of the Revere Copper and Brass Company of today.

The Saugus Iron Works, built just north of Boston in 1640, was the first American foundry. Although it had only a short life, it trained many of the foundrymen who started similar enterprises from New Jersey and eastern Pennsylvania down to the Birmingham area of Alabama. The charcoal blast

FIGURE 8.1
Cast bronze cat (made with core that had been removed). Cast in Egypt, probably in Sakkarah, in the seventh century B.C.

FIGURE 8.2
Sweeping cope of bell mold in pit. Note complete core and cope, at left. Vertical sweeping coupled with pit molding, was usually employed in molding bells. (*From B. L. Simpson.*)

furnaces for smelting iron were significant establishments in the early days of the United States and were a major factor in establishing the steel industry in Pennsylvania. The Hopewell Iron Plantation, south of Reading, Pa., was typical of a way of life for the many people required to produce iron in the 17th and 18th centuries. Later, Juniata Iron was produced in dozens of charcoal furnaces, which dotted the streams of central Pennsylvania. The local forests furnished the fuel, the swampy areas yielded bog iron ore that was smelted to produce blast-furnace pig or cast iron ore; this was smelted to produce blast-furnace pig or cast iron items such as fire backs, stove parts, and cooking utensils. Cupolas were developed in the early 1700s. A cupola is a vertical-shaft re-melt furnace that could take pig iron from the blast furnace and re-melt it for pouring commercial gray iron castings.

Today the huge American scrap-metal industry produces some 80 percent of the ferrous metals that are needed for cupola operation. Thus, the foundry industry is now a major recycler of metals from discarded automobiles, scrap machinery, plumbing installations, etc. In contrast, the People's Republic of China and the USSR do not have such an abundance of scrap iron and steel, so they must use more costly blast-furnace pig as the source of raw material for their gray iron foundry operations. In the United States, recycling plays a significant role in increasing the useful life of most of our primary metals reserves and in saving energy that would otherwise be needed to smelt metals from their ores (Fig. 8.3).

As the use of aluminum increases, so does the secondary metal market. It is much more energy efficient to use recycled scrap aluminum than to reduce aluminum from its bauxite ore. It is in the national interest to keep iron scrap

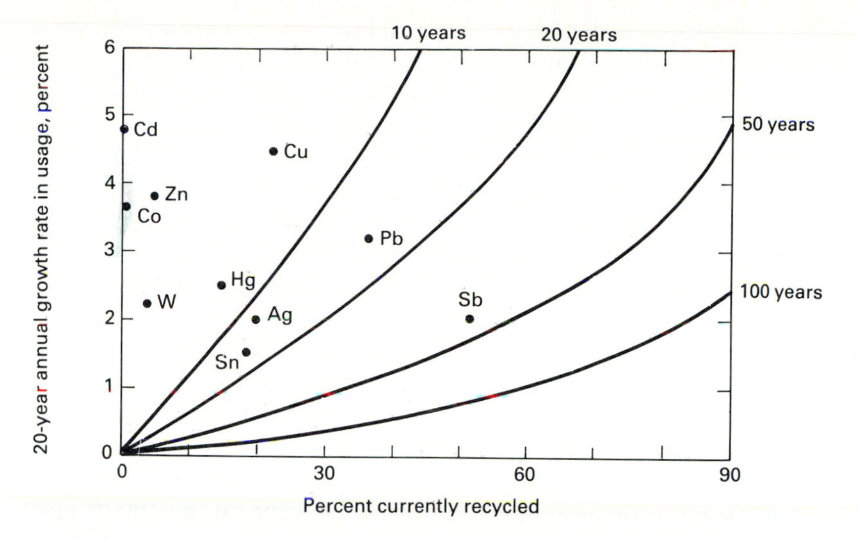

FIGURE 8.3
Effectiveness of recycling. Numbers 10, 20, 50, and 100 indicate the number of years that metal exhaustion is postponed by recycling. (*From A. Hurlich,* Metal Progress, *Oct. 1977.*)

and especially aluminum scrap within the U.S. borders rather than export scrap and its contained energy overseas.

SAND CHARACTERISTICS AND BONDING

Since the majority of castings are poured in green sand molds, it is appropriate to examine the characteristics of molding sands and clays as well as the nature of the clay bond. Foundry sands have particle sizes between 53 and 4000 μm in size, but the term is applied to the particles without regard to chemical composition or shape. Foundry sands are also defined in terms of refractoriness and chemical composition (Table 8.1). Silica is the least expensive and by far the most abundant sand in the U.S. In Norway, olivine sand is common.

TABLE 8.1
Major foundry sands

Sand	Source	Principal mineral constituents
Quartz	U.S. and World	Silica SiO_2
Olivine	Norway, U.S.	Forsterite $2MgO \cdot SiO_2$ ~10 percent
		Fayalite $2FeO \cdot SiO_2$ ~90 percent
Chromite	Southern Africa	Chromite, $FeO \cdot Cr_2O_3$
Zircon	Australia	Zirconium silicate, $ZrO_2 \cdot SiO_2$

Sand is further identified by its geographic origin, such as Ottawa, IL silica sand, Michigan Lake sand, or Providence River naturally bonded sand, Today, most molding sands are mulled using graded, washed, and dried silica sand that is then bonded with a bentonite clay and various additives.

Molding sands consist of sand, clay, water, and additives. At least four minerals are the base materials for foundry sand: silica, olivine, chromite and zircon, as in Table 8.1. The characteristics of each mineral are discussed briefly below. The nature of green sand bonding is included together with the common laboratory tests used to obtain the proper quality assurance for production molding.

Foundry sands also are classified by average grain size, shape (rounded, subangular, angular, or compound) and grain-size distribution. The grain size is defined as the mesh size of a U.S. Standard Sieve Set that the average grain would pass through if all the grains were the same size. Coarse sands with a grain fineness number (GFN) of 35 to 45 would be used for large ferrous castings; a 60 to 70 grain fineness sand would be used for 10- to 100-lb gray iron castings; and 120 GFN would be used for aluminum sand castings. Generally, a broad distribution of sieve sizes is preferable for clay-bonded sands; thus, a four-sieve sand has a broader distribution than a three-sieve sand. In a four -sieve sand at least 80 percent of the sand mass would be found on four adjacent sieves in the grading stack. The total surface area of a sand is important because it must be covered by the binder. A round-grain sand has less surface area per unit weight than does an angular grain sand for the same grain-fineness class. This is particularly important for core sands that are coated with organic resins, because a narrow distribution of round-grain sand has the least contact area for a given core volume and would therefore minimize binder cost for a given strength. Olivine sand is used in two types of applications: (1) for small jobbing foundries, because the olivine suffers less breakdown from thermal and mechanical stress than silica does; (2) for manganese-steel castings, because of the mold–metal interfacial reactions between silica and manganese. Since steel has already reacted with iron to produce fayalite ($FeSiO_4$), olivine sand, which is about 90 percent fayalite, cannot react with iron again. Thus, it is a fine sand for casting manganese steel because it is not subject to mold–metal interfacial reactions. Olivine sand is heavier than quartz and tends to have angular, needlelike particles because it must be crushed to size from large boulders or chunks of rocks.

Chromite sand is a heavy expensive subangular jet black sand from South Africa. It is chromium oxide, which bonds easily with bentonite clay to produce a mold with a relatively high thermal conductivity. However, its use can be justified for large steel castings where metal penetration and heat resistance are problems.

Zirconium silicate has the best refractory properties of any sand. It is not wetted by liquid metals and is about 1.8 times as dense as silica. It also is scarce, being commercially available only in Australia and Florida. Because of its high cost, zircon sand is usually ground to a fine powder and is extensively

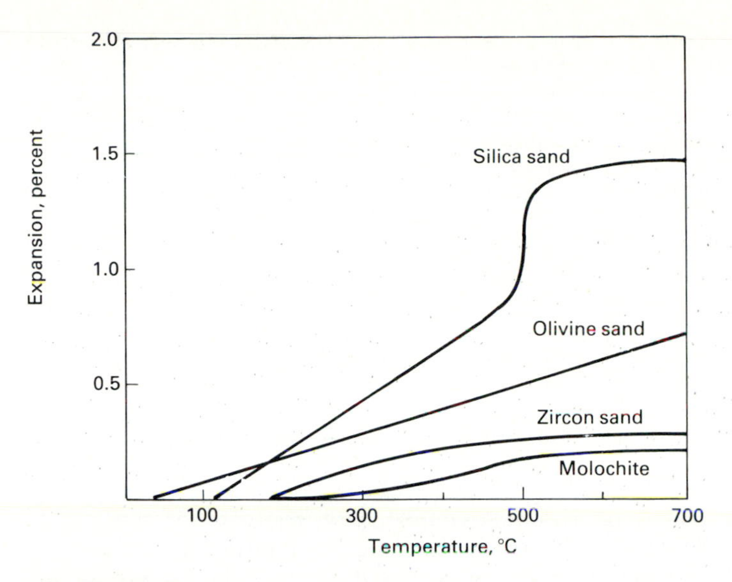

FIGURE 8.4
Thermal expansion of mold refractories. (From Middleton), courtesy of Institute of British Foundrymen.)

used as a slurry for coating molds and cores, especially to prevent metal penetration into the cores and molds for large ferrous castings. Penetration can easily occur in such cases because of the relatively large expansion of silica (Fig. 8.4) and the fact that iron oxide easily wets and dissolves silica to produce fayalite.

Bonding Clays

Clay bonding has been known since prehistoric times, but the widespread use of bentonite clays in foundry sands has been common in the U.S. since World War II. Clays consist of thin colloidal-sized platelets, similar to snowflakes (see Table 8.2). Western (Wyoming) bentonite contains sodium ions, and southern Alabama bentonite has calcium ions primarily. Various deposits have some of both types of ions. Sodium ions permit a great amount of swelling to occur between the platelets, whereas with calcium bentonite there is little swelling. Both bentonites have about equal strengths if thoroughly mulled (mixed), but western bentonite requires more intensive mulling and has high hot strength, whereas calcium bentonite has little hot strength. Therefore, for large machine-compacted molds, with clay in the order of 10 percent, a mixture of western and southern bentonites must be used to assure proper mold shake-out. Such high-clay sands are needed for flaskless molding.

Water is the most important variable in a molding sand. It plasticizes the clay and it forms a stiff gel with the clay when the mix reaches the right temper or water content (this is at a water–clay ratio of about 0.3 for well-mulled

TABLE 8.2
Composition and sizes of foundry clays

Type of clay	Foundry name	Bonding mineral	Platelet size[a] thickness nm	width nm
Sodium montmorillonite	Western bentonite	$(Al_{1.67}Mg_{0.33})Si_4O_{10}(OH)_2$ with $Na_{0.33}$ ↑	1	100–300
Calcium montmorillonite	Southern bentonite	$(Al_{1.67}Mg_{0.33})Si_4O_{10}(OH)_2$ with $Ca_{0.33}$ ↑	1	100–300
Illite	Natural clay	$Al_4K_2(Si_4Al_2)O_{20}(OH)_4$	20	100–250
Kaolinite	Fire clay	$Al_2(Si_2O_4)(OH)_4$	20	100–250

[a] $1\ nm = 10^{-9}\ m = 10^{-3}\ \mu m$.

western bentonite. Because water is dipolar and asymmetric in nature, each molecule has a positive and negative charge, with the far side of the oxygen atom being negative and the two hydrogen atoms being positive on the other side. This structure permits layers of water to build up surrounding the positive areas around the adsorbed sodium or calcium atoms. This results in the orientation of water molecules in a hexagonal array on the surfaces of the clay platelets. The electrostatically bonded water has the same configuration as that of ice. Experimentally, it has been found that the maximum strength of the clay bond occurs with three adsorbed molecular layers of water in western bentonite and with four adsorbed layers in southern bentonite. In either case, it is interesting to note that the maximum strength of a sand compact occurs when the sand is at its minimum density and close to its maximum permeability. This is unexpected and opposed to the behaviour of homogeneous materials!

Organic Resin Bonding of Foundry Sands

Historically, foundry sands were bonded by nonclay materials as early as 1898. But only after World War II did foundrymen in the United States become aware of the Croning and the CO_2 processes of bonding founding sands. Moreover, the two decades from 1955 to 1975 saw more changes in foundry binder technology than occurred in the previous 5000 years of foundry evolution. In fact, the family of nonclay binders has increased to such an extent that there are now organic and inorganic types of vapor-cured binders (Table 8.3). The inorganic sodium silicate bonding system is by far the oldest, dating back before the turn of the century. In the post World War II era, synthetic resin-bonded cores began to be accepted in the United States. The Croning Shell Mold Process used a hot pattern onto which was applied a mixture of sand and finely divided phenolic resin particles; later a binder layer

TABLE 8.3
Classification of nonclay binders

Organic	Inorganic
Thermocured	**Thermocured**
1. *Oven cured*	1. *War box process*
a. Core Oil	a. Sodium silicate
2. *Shell processes*	
a. Phenolic resin	
b. Sand coating or dry mix	
(i) Hot coat	
(ii) Warm coat (solvent)	
3. *Hot box processes*	
a. Furan resin	
b. Phenolic resin	
Self-cured	**Self-cured**
1. *No-bake processes*	1. *Polymization processes*
a. Furan resin	*a. Sodium silicates*
(i) High nitrogen: 5 to 11 percent (acid)	(i) Ester-cured
(ii) Medium nitrogen: 2 to 8 percent (acid)	(ii) Ferro-silicon-cured
(iii) Low nitrogen: 0 to 2 percent (acid)	(iii) $2CaO \cdot SiO_2$-cured
	b. Cements
	(i)Hydraulic bond
b. Phenolic resin	(a) Portland cement
(i) Phenolic (acid)	(b) Portland cement (fluid sand)
c. Urethane resins	*c. Phosphates*
(i) Alkyd (organometallic)	(i) Phosphate (oxide cured)
(ii) Phenolic (pyridine)	
Vapor-cured	**Vapor-cured**
1. *Cold box processes*	1. *CO_2 process*
a. Furan–peroxide (SO_2 gas-cured)	*a.* Sodium silicate (CO_2 gas-cured)
b. Phenolic urethane (amine gas-cured)	

was precoated on the sand. After a short contact time with the hot pattern, a layer of partially cured sand about $\frac{1}{8}$ to $\frac{1}{4}$ in. thick adhered to the pattern. The balance of the uncured sand was poured back and the shell was allowed to cure on the plate. This process achieved immediate popularity and it is still one of the most widely used bonding systems. But it is now steadily losing favor because of its relatively high cost, except for large hollow cores, the need for thermal energy to cure the shell, and the pervasive odor of phenolformaldehyde that is difficult to eliminate. In particular, small cores tend to be filled with trapped sand that will not pour out, which adds to the total cost.

The most universally used, and one of the oldest, nonclay binder systems in the world is the inorganic sodium silicate system. It is also the least expensive, yet it is not particularly popular in the United States or West

Germany. Both Japan and England have championed the silicate bonds for many years. There is no doubt that the use of this system will become more widespread in the United States as foundrymen discover that it is truly low-cost and that various additives make it moisture-resistant and give it good shake-out and adequate reclaimability. At the present time, sodium silicate bonds are most commonly used in steel and aluminum foundries, but gray iron foundymen are wary of silicates because of the shake-out problems they encountered in the 1950s.

Hot box cores based on furan technology are certainly one of the most widely used core-making processes, particularly in the high-production automotive foundries in the U.S. However, this is an area where the cold box process will probably be more widely used. The heyday of hot box cores may be over in the United States because of total cost and the greater energy required for core-making.

On the other hand, no-bake technology has improved markedly and many foundries have installed no-bake loops with continuous mixers to handle large cores or molds for low-volume production. The waiting time required before stripping has been greatly reduced and is easily controlled by the amount and type of catalyst added to the sand, but the curing cycle is strongly temperature-dependent. Sand heaters and temperature control of the core room are important. These no-bake systems are gaining in acceptance because of the availability of commercial systems that provide simultaneous mechanical shake-out of the sand and reclamation. Thus, the castings are cleaned and the sand is reclaimed in one process. This automates a formerly messy procedure and results in less manual handling, good casting finish, better dimensional accuracy—particularly in gray iron castings—and higher yield because the more rigid molds require fewer risers. These factors result in the increasing use of the no-bake process with its easy sand reclamation in more and more foundries. It saves up to 80 percent on sand transportation cost, requires no thermal energy, and greatly reduces solid waste disposal problems. However, the possibility of long-term leaching into the ground water makes disposal of discarded sand difficult.

In spite of these recent developments, the use of core oil as a binder is still widespread in the United States. Apparently, if the high-capital-cost core ovens are already installed and depreciated, the relatively low binder cost keeps the oil-based, oven-bake binders attractive on a total cost basis. For larger cores, such as are used on railroad side-frames, the process is good because the savings in binder cost are large.

There are great differences in the relative use of the various types of sand binders in the U.S. It is anticipated that the economic forces are such that silicates will probably increase their percentage of the market in the decade of the 1990s, and that use of phenolic shells will dwindle gradually. A decline undoubtedly will occur in the hot box process and the oven-baked core oil process because these require thermal energy to effect their cure. On the other hand, the sum of the no-bakes and the vapor-cured cores will increase because of the fast production time and ease of mold making and reclamation of the sand.

Organic Bonding of Sands

Linseed oil was the major core sand binder prior to World War II. Since then it has had to compete with synthetic resins such as phenolic, urethane, alkyd, and furan as well as inorganic glassy types of binders. The characteristics of these bonds is presented in this section.

OIL BINDERS. Drying oils are natural resins that have been used in sand bonding for many years. These oils includes linseed, perilla, tung, and dehydrated caster oils, which can combine with oxygen when spread as a thin film on a surface such as a wall or a sand grain. The reaction can be speeded by heating or by adding a catalyst.

Such oils are naturally occurring esters of glycerol and fatty acids. Their ability to cure into a strong film depends on their degree of unsaturation or the number of cross links that can be formed in the polymer chain during curing. Linolenic acid has a high degree of unsaturation, whereas stearic acid has none.

Foundry core oils can also be made economically from tall oil, which is a by-product of the paper industry. In addition, some mineral oils that have a high degree of unsaturation can be blended with natural or synthetic oils to reduce the total binder cost.

During baking, bond strength is developed by a blend of several different processes: (1) a rearrangement of the double bonds of the fatty acid chain into a system of alternate double and single bonds (—CH═CH—), which is called addition polymerization, to form a conjugate system (Fig. 8.5a); (2) linking of the fatty acid chains in adjacent molecules by condensation polymerization, in which two different types of molecules interact to give the polymer plus a condensation product such as water or alcohol. Water is the product of the phenol-formaldehyde or urea-formaldehyde reaction that occurs in the curing of shell molds (Fig. 8.5b); (3) cross-linking, in which two molecules join at double-bond points when a foreign atom such as oxygen becomes part of the system, as in the curing of drying oils (Fig. 8.5c) and in evaporation and autoxidative polymerization. The latter reaction can be speeded up by the addition of a catalyst (dryer such as a napthanate of cobalt, vanadium, lead, molybdenum, or zirconium to the binder.

SYNTHETIC-RESIN BINDERS. To produce a resin, its chemical components are weighed, blended, and cooked in a reaction kettle outfitted to provide heating and cooling at controlled rates and times. Vacuum is used to remove vapors and water; water is either added with the charges or may be a condensation product of the resinification process. The finished resin can have a number of different properties depending on the amount and composition of

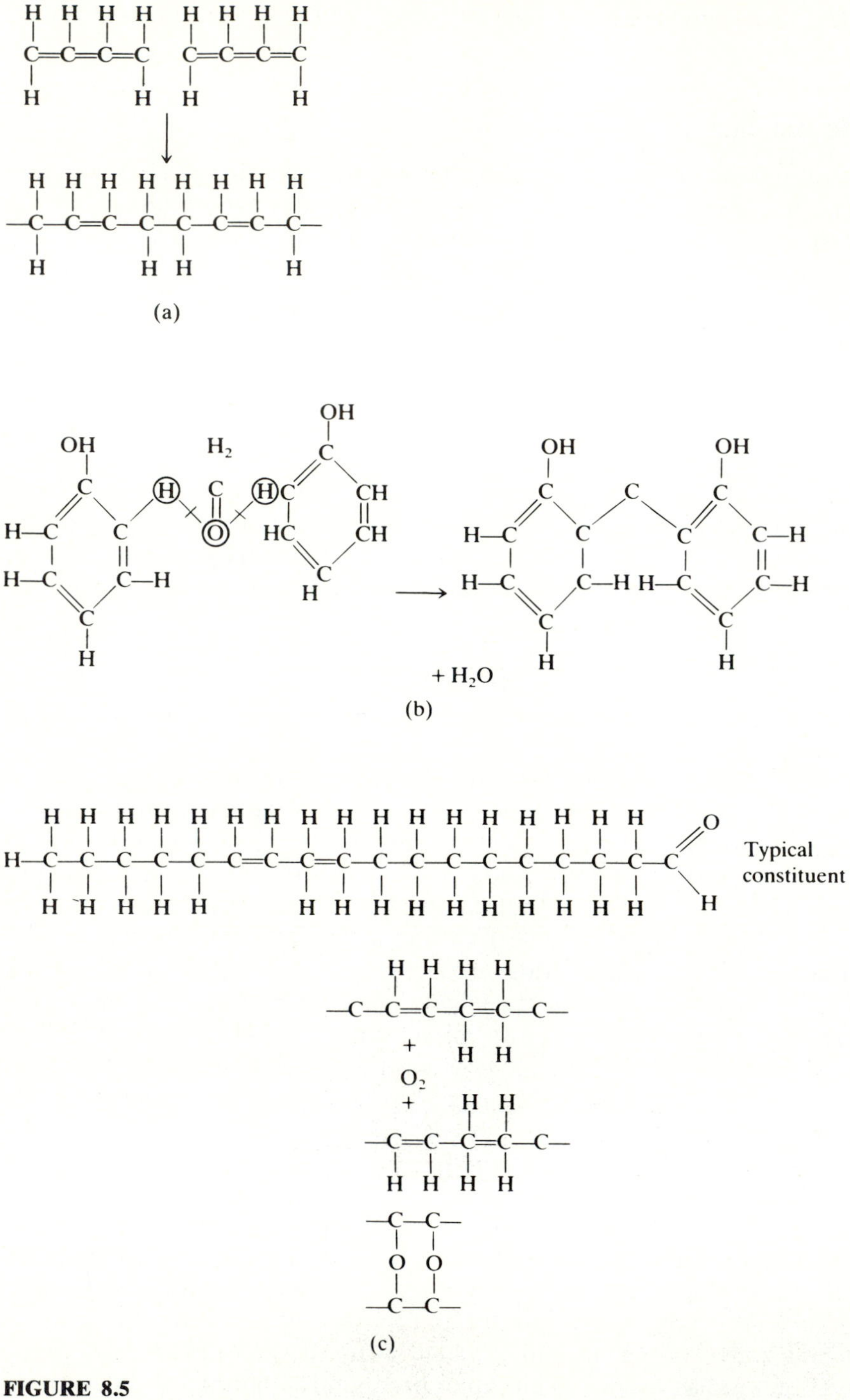

(a)

$+ H_2O$

(b)

Typical constituent

O_2

(c)

FIGURE 8.5
(*a*) Addition polymerization. (*b*) Condensation polymerization. (*c*) Cross-linking.

the charge make-up, the kettling time, the temperature cycle, and the addition of solvents such as water, ethanol, or isopropyl alcohol. The final resin is shipped as viscous fluid, or it can be solidified in the form of granules or flakes or prepared as powder.

Four types of resins are commonly used in the production of cast metals: phenolics, furans, ureas, and polyols. The latter are unique because of the importance of their hydroxyl groups in producing a polyurethane bond with the inorganic groups.

Phenolic resins. Phenol and formaldehyde have been found to produce a wide range of resins formulations, which have the generic name of phenolics. There are two classes, the novalak or two-stage resins, which are only used for the shell process, and resole or single-stage resins. These materials have limited shelf-life because they polymerize in the container. The resins cure by condensation polymerization as a result of heating, or heat and an acid salt solution, or heat and dilute acid, or the use of a concentrated acid in the case of a self-setting process. The rate of polymerization is approximately proportional to the acid concentration (the hydrogen ion concentration). In the cold-setting process, phosphoric acid and *para*toluene sulphonic acid may be used to initiate the curing process. The acid is added to the sand, followed by the resin addition. A set time of 3 to 20 minutes can be achieved, depending on the resin chosen and the acid concentration used.

Furan resins. This resin is produced from furfuryl alcohol, which is the product of furan and formaldehyde. Originally the furfuryl alcohol was obtained directly from agricultural by-products such as corn husks, oat hulls, or bran, but it can now be produced synthetically.

Furfuryl alcohol is cured by acidification, which is accomplished by acid salts in the hot box process or by more concentrated acids in the cold-setting processes. The curing mechanisms are complex but they involve linear condensation and cross-linking by methylene bridges and hydrogen ions via double bonds. The result is a complex three-dimensional structure. Furfuryl alcohol is rarely used alone; it is blended with urea or phenol-formaldehyde to provide improved hot strength to the cured sand.

Urea resins. Foundry resins of this class are derived from the condensation polymerization of urea and formaldehyde under slightly alkaline conditions.

Curing of the resin occurs under acidic conditions that may be produced either by heat plus an acid salt or by the addition of a more concentrated acid. In practice, the resin is a mixture of dimethylol and monomethylol ureas.

Polyol resins. The hydroxyl group is the most important factor in polyol resins. These resins are used in the phenol/isocyanate binders, in which the polyol is a phenolic resin, and in the cold box process, in which a similar resin reacts with the isocyanate aided by exposure to amine gas that acts as a catalyst and which is either triethylamine (TEA) or diethylamine (DMEA).

The hydroxyl groups of the polyol resin react with the NCO groups of the isocyanate to produce a polyurethane bond. The isocyanate hardener produces the cross-links between the molecules of the phenolic resin.

SYNTHETIC RESIN BONDING OF NO-BAKE MOLDS. Furan, urethane, and polyol no-bake resins have gained wide acceptance by the cast-metals industry in the last decade. Their major advantages are based on the improved control of the bench-life and setting times that can be achieved by the addition of specified amounts of the catalyst. Furthermore, ribbon mixing of the raw materials just before use and the automated cleaning and recycling of the sand make this a simple process to control with a minimum of labor. All these advantages tend to offset the high binder costs. However, they do not keep out foreign competition from Third World nations that have much lower costs of both labor and management.

There are four types of no-bake resins.

1. Phenol-formaldehyde resins use a sulphonic acid catalyst.
2. Urea-formaldehyde resins may use ferric chloride as a catalyst.
3. Furan resins are derived from furfuryl alcohol that has been reacted with phenol-formaldehyde or urea resin. The former resin is used with a sulphonic acid catalyst and the latter is catalyzed with phosphoric acid.
4. Polyols use either phosphoric acid or sulphonic acid as the catalyst. These binders are nearly nitrogen-free with through-hardening.

All of the above binder systems form tough, resinous films at low binder concentration without needing additional heat. In fact, the reactions are exothermic according to the following general scheme: liquid resin plus a catalyst makes a solid binder film plus water and releases heat. The speed of the reaction is a function of the sand characteristics and temperature, the thoroughness of mixing, the binder formulation, and the amount and type of catalyst used.

Furan-urea no-bake binders are free-flowing and are thus easily compacted with a minimum of ramming. They are available in several grades to meet metallurgical needs as well as economic competition. Unfortunately, the least-cost binders produce the most nitrogen. For ferrous castings, especially for low-carbon steel, it is important to use a nitrogen-free binder, because nitrogen can cause surface nitride formation and subsurface porosity.

Polyurethane binders are nearly nitrogen-free. They cure completely at room temperature by means of an isocyanate accelerator. This binder is based on a two-part system that consists of an alkyd resin with a cobalt naphthanate dryer and a methylene diphenyl diisocyanate (MDI) hardener. This binder has a two-stage curing reaction: the first stage is cross-linking between the base resin and the isocyanate; the second stage is the slower oxidation and polymerization of the alkyd resin into a solid gel similar to the film of an alkyd resin house-paint.

In low-production mold making with synthetic resin, binders, a facing of 3 to 4 in. of self-curing sand is placed around the pattern. This may be backed up with normally compacted green sand to reduce total molding cost, or for higher production the entire flask is often made from the synthetic resin-bonded sand so that it can be reclaimed without mixing with clay from the green sand backup.

When using furan-bonded sands, care must be taken to keep the incoming sand at a uniform temperature both in summer and winter. Also, there must be a sufficient volume of sand in the system so that it can cool to room temperature between cycles. The reaction time doubles for every 10°C rise in the mix temperatures. If hot return sands are used, process reliability and control are lost. In addition, a continuous mixer must be used; conventional mullers are poor substitutes and do not insure proper control of the reaction time.

A comparison of the advantages of polyurethane and furan bonded sands is given in Table 8.4.

TABLE 8.4
Comparison of polyurethane and furan bonded molds

Polyurethane	Furan
1. Mixes are not sticky, therefore clean-up of mixer is easier.	**1.** Furan binders cost less per pound of mixed sand.
2. Stripping of molds or cores is easier because of the more rubbery consistency of the mix after the initial set. Pattern equipment can be less than perfect to be usable.	**2.** Spalling of cope sand is less likely because the heat resistance of the furan-bonded sand is better.
3. The polyurethane system is less sensitive to variations in pH. Thus, the cure speed and tensile strength are not significantly affected by normal fluctuations in pH or acid demand values of foundry sands.	**3.** Furan-bonded sands are less sensitive to variations in moisture content. In polyurethane-bonded sands the isocyanate portion has low moisture resistance.
4. Polyurethane bonds keep a constant cure speed at low temperatures more consistently with only a slight increase in the amount of catalyst required.	**4.** Furan binders yield less carbon for low carbon steels to pick up.
5. Molds with polyurethane binder *cannot* be poured for 12 to 24 hours.	**5.** Furan bonding speeds production. Molds and cores with furan binder can be poured in 2 to 4 hours, whereeas polyurethane bonded sands require up to 16 hours.
6. Lower nitrogen content makes these binders best when surface nitrides or subsurface pin holes are a problem.	**6.** Odor may cause a problem, but not if adequate ventilation is provided.
7. Better collapsibility and shake-out are available because the residual carbon bond is weaker in the polyurethane bonded sands.	
8. Urethane systems are more tolerant to sand additions such as cereal or iron oxide.	

In summary, no-bake system sands provide the following advantages.

1. Excellent flowability, which results in a minimum of ramming and tucking
2. High strength from thorough curing, which results in elimination of most core rods and strengthening wires
3. Greater inherent strength, which results in improved dimensional control and greater quality assurance

There are several disadvantages.

1. The bench-life of the sand mix is limited. Therefore, there is waste sand whenever the mixer is idle for a period of time. The sticky mix of furan-bonded sands is more difficult to clean, whereas polyurethane-bonded sands are not sticky.
2. Hot return sand cannot be tolerated in reclaimed no-bake sand systems. Temperatures over 95°F cannot be tolerated. The curing action is exothermic in polyurethane-bonded sands and there is little moisture to produce evaporative cooling.
3. Polyurethane-bonded sands require 12 to 24 hours before they can be poured, whereas furan-bonded sands can be poured in 2 to 4 hours.

Inorganic Bonding of Molds and Cores

Historically, sodium silicate has been used as a foundry binder since the latter part of the 19th century. But the process produces a glass type of bond that results in difficult shake-out and sand reclamation. However, considerable progress has been made in reducing the shake-out difficulty by using up to 10 to 15 percent of various organic additives.

The increasing use of aluminum alloys has focused attention on the shake-out of cores used with low-melting-point metals. Again, proper breakdown of the core binder is essential to avoid hot tears, which occur because the core binder fails to disintegrate from the heat of the liquid metal and thus certain areas of the casting are put in tension. This leads to tensile failure of the casting wall as the metal tries to shrink. Phosphate bonding materials have been found to be suitable for cores for aluminum alloys.

SODIUM SILICATE BINDERS FOR FOUNDRY SAND. The molding aggregate is bonded with 3 to 5 percent sodium silicate (waterglass). After the sand is packed around the pattern or into the core box, it is cured by passing carbon dioxide gas at about 10 lb/in.^{2} through the mix for about 10 seconds using a probe or a gassing head. The sand immediately becomes extremely strongly bonded as the sodium silicate (Na_2SiO_3) becomes a stiff gel:

$$Na_2O{\cdot}xSiO_2 + nH_2O + CO_2 + Na_2CO_3 + xSiO_2{\cdot}n(H_2O)$$

where x is 1.6 to 4; most often $x = 2$.

The ratio of silica to soda can vary quite widely, which results in many colloidal silica gels that produce a wide range of basic pH values and great differences in viscosity. However, for use in foundry sand binders, the practical ratios of silica to soda usually lie between 1.5:1 and 3:1.

Silicone parting agents should be sprayed on the patterns and core boxes before each use to prevent sticking, because the cured molds and cores are very strong and have little yield.

The process is more expensive than molding with clay-bonded sand because the binder is more costly and CO_2-bonded cores have poor collapsibility, yet the process offers a number of advantages that are more widely accepted abroad than in the United States.

1. It uses conventional equipment for molding and sand mixing.
2. It eliminates the need for internal support for cores and for mold jackets.
3. Molds and cores can be used immediately after processing.
4. It eliminates baking ovens and core dryers.
5. Improved dimensional accuracy is available because of greater fidelity of pattern or core box details in the cured mold.

Sodium silicate binders can also be solidified by various acid salts that react with the sodium hydroxide to form a colloidal silica gel. A Japanese process uses about 2 percent powdered ferrosilicon, and a Russian fluid sand process achieves gel formation by adding powdered dicalcium silicate. However, removal from the core box is difficult in the latter case. The primary action of the powders is the dehydration of the binder. These processes are limited in use and the ester–silicate system appears to be more effective.

Organic esters generally react with the sodium silicate to produce an organic sodium salt and alcohol. In practice, from 0.2 to 0.4 percent of the organic ester is added to the sand first, then 3 to 4 percent of the silicate binder is blended into the sand. The mold can be poured as soon as it is removed from the pattern plate, although several hours are needed to develop full strength. Self-setting silicate binders have easy shake-out characteristics for both molds and cores. Although the costs are now comparable with other synthetic resin binders, the sodium silicofluoride catalyst appears to permit a significant cost reduction.

The organic ester-modified sodium silicate bond gives excellent dimensional control, complete through-curing, ease of control, excellent flowability, good shake-out, and ecological acceptability because it produces no toxic by-products. The system still suffers from a somewhat brittle, glass-like bond structure and it is sensitive to variations in temperature and moisture.

PHOSPHATE BONDING. With the advent of high-production automotive casting poured from aluminum alloys, a high-production method of core making was needed. Oil-bonded cores were used originally, but the modern trend away from core ovens did not permit such a solution. Most core binders used for ferrous casting are too strong to be used for aluminum casting, because they do not get hot enough to break down sufficiently for proper cleaning.

It was discovered that phosphate bonding materials would work well for use as core binders for cast aluminum alloys. They did not cause hot tearing and disintegrated sufficiently for easy cleaning after exposure to molten aluminum.

Gas Curing of Synthetic Resins

One of the most significant advances in core making has been the ability to cure synthetic resin binders in a few seconds by a gassing operation. There were a number of difficulties encountered in making the process safe, but new technology solved the problems.

The Ashland Chemical Co. has patented the use of amine gas to cure synthetic resins that have isocyanate components. The gassing process takes only a few seconds, but purging the gassing chamber to eliminate any environmental contamination can take from 30 to 60 additional seconds. Proper equipment is readily available to handle this task.

In the middle of the 1970s, a process to cure furan resin formulations by passing SO_2 gas through the core box was developed in France. Both processes take about the same amount of time to produce a core, but the latter has not developed as wide a usage as the amine gas process. As far as can be ascertained, the SO_2 process has not yet become the major competitor of the amine gas process.

In either process, the rapid curing of foundry cores without external heat has proved to be a major factor in saving energy, reducing production costs, and increasing casting quality.

SAND CASTING

A sand casting is produced when molten metal is poured at atmospheric pressure into a cavity formed in a compacted molding sand. In typical ferrous casting, the mold would be bonded with 6 to 8 percent western bentonite mulled with 2 to 3 percent water. Patterns would be reusable, but the sand molds would be broken from the casting after the solidified casting had cooled sufficiently. The sand would be returned to the muller for recycling by mulling with some new clay and water to reactivate the old clay. Sand molds define the external geometry of a casting, whereas cores define the interior or undercuts.

Sand casting accounts for more than 90 percent of all metal poured, on a tonnage basis. It can be an extremely high-production process—for instance, 125-lb gray-iron bathtubs are molded and poured at the rate of 72 an hour using a crew of five men. Automotive engine blocks are poured at the rate of one every 15 seconds! For comparison, a die-cast engine block requires at least 2 minutes floor-to-floor time for one block.

On the other hand, the process requires a high investment in plant facilities and sand-handling equipment to make such production rates possible. Sand casting is versatile and is capable of making parts ranging from small knobs to huge radiotelescope bases weighing more than 200,000 lb.

Sand Mold

Sand molds are made by ramming sand around a pattern in a flask (Fig. 8.6). The sand is either shoveled by the molder, thrown by a sand slinger, or dropped in from an overhead chute. The molding sand is compacted by jolting several times and squeezing, or by jolting alone for larger, deep flasks. The pattern is withdrawn, cores are added, and the mold is reassembled, weighted, and sent on to the pouring station. Types of sand mold include green sand, skin dried, carbon dioxide, shell molded, and cement bonded.

The texture of the mold surface is imparted to the surface of the casting. If the pouring temperature is too high, sand grains may adhere to the cast metal surface. Such sand inclusions cause problems in the machine shop and are to be avoided. Proper selection of sand grain size and distribution, and proper clay content, additives, and compaction, can result in excellent surface appearance and detail in sand molds for most foundry alloys.

In high-pressure molding, the compaction force must exceed 100 lb/in.^{2} at the parting plane, which is a 50 percent increase over conventional practice. The greater compaction pressures yield harder, more uniform molds that permit much less mold wall movement and therefore result in closer dimensional control of the final casting.

Sand molds are made in many sizes, and machines have been developed to mechanize molding of all but the largest sizes. Thus, sand molds are produced by bench molding, jolt–squeeze molding, automatic molding, flaskless molding, high-pressure molding, floor molding, and pit molding (Fig. 8.7).

Green-sand molds must be strong enough to withstand the poured weight of the molten metal, must resist erosion by the flowing metal, must be sufficiently permeable to liberate mold gases, and must be refractory and yet break out easily after cooling.

Pit Molding

Castings that are too large to be made as bench molds or floor molds are made by the pit-molding process (Fig. 8.2). If possible, the pattern is set in the pit in the position in which the casting is to be poured; sand is then rammed and tucked around and under the pattern. In former times, equestrian statues and church bells were poured in pit molds. If the design of the casting prevented the drawing of the pattern, the mold would be constructed of assembled cores.

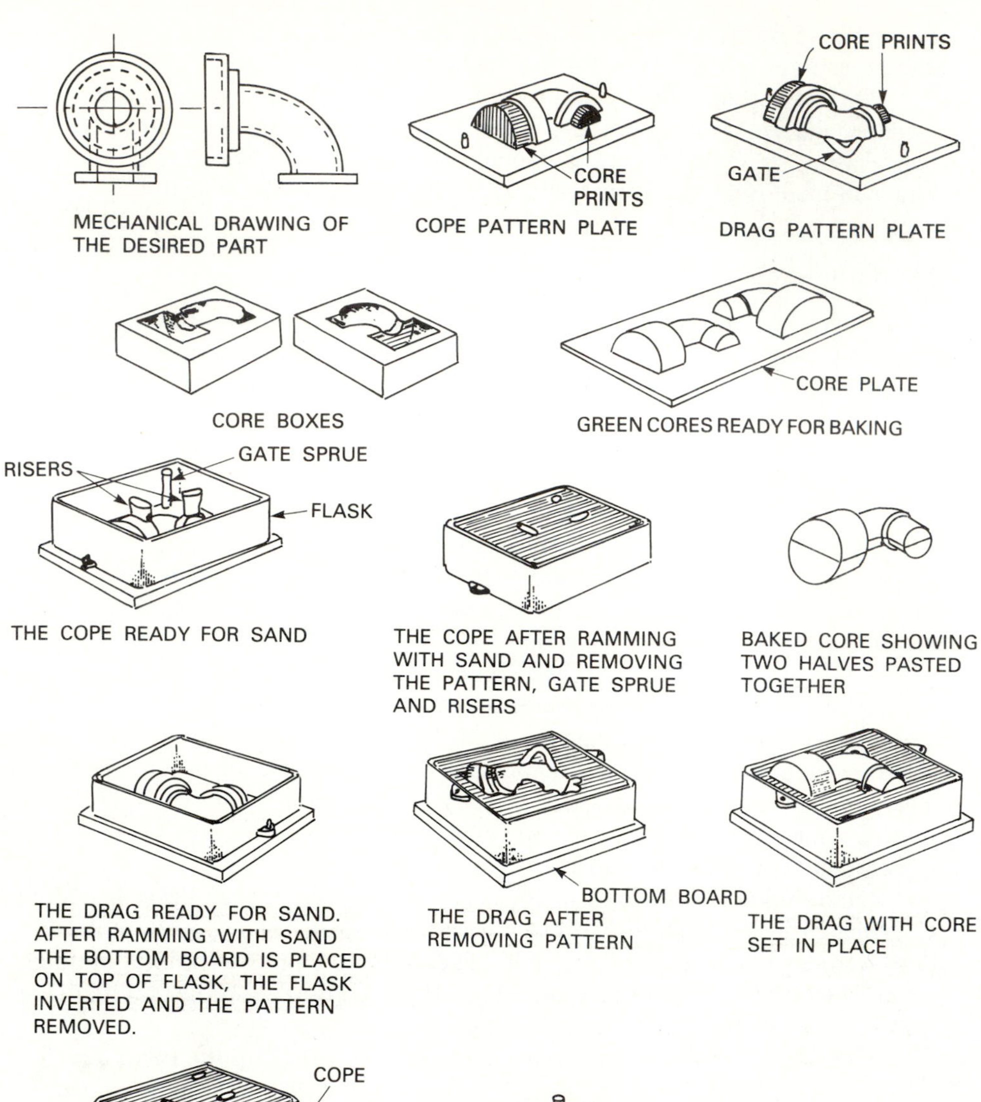

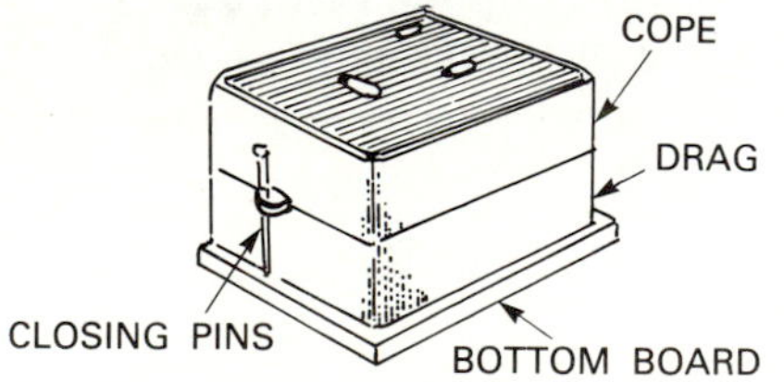

FIGURE 8.6
Series of sketches showing steps and equipment used in making a simple steel (*Courtesy of Adirondack Foundries and Steel, Inc.*)

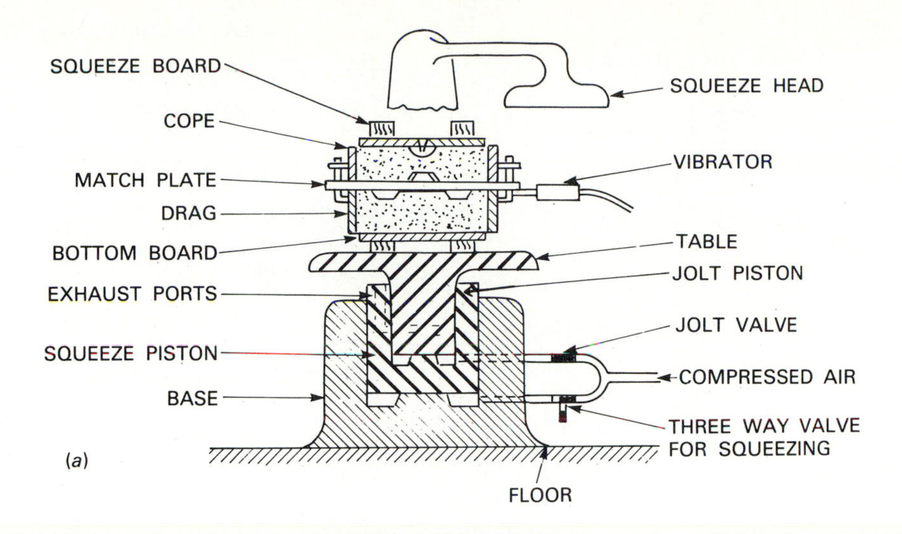

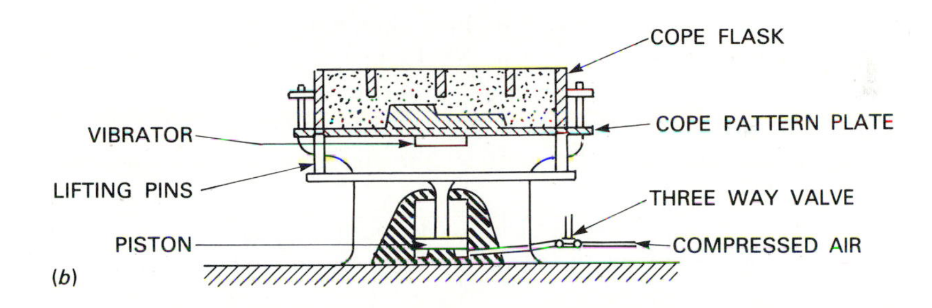

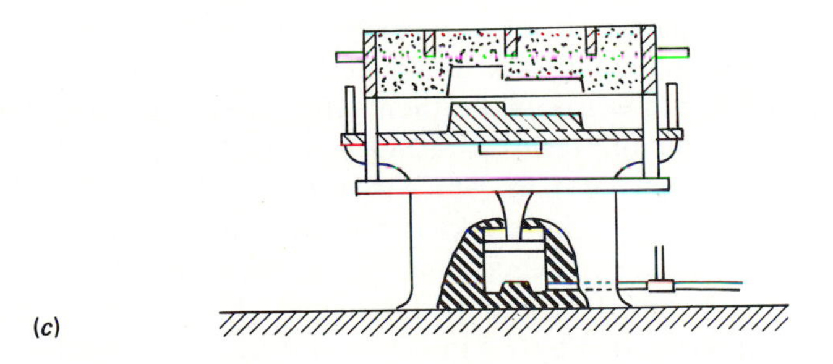

FIGURE 8.7
Various types of molding machines. (*a*) A schematic sketch showing the operation of a jolt squeeze molding machine. This is called matchplate molding. (*b*) A pattern-draw molding machine ready to draw the pattern after ramming a cope. (*c*) Drawing the pattern by lifting the mold. (*d*) Side view of a jolt rockover pattern-draw molding machine shown in the jolting position. (*e*) Shown after rocking over the mold ready to draw the pattern by unclamping and lowering the mold. The rockover arm cannot rotate further to the left.

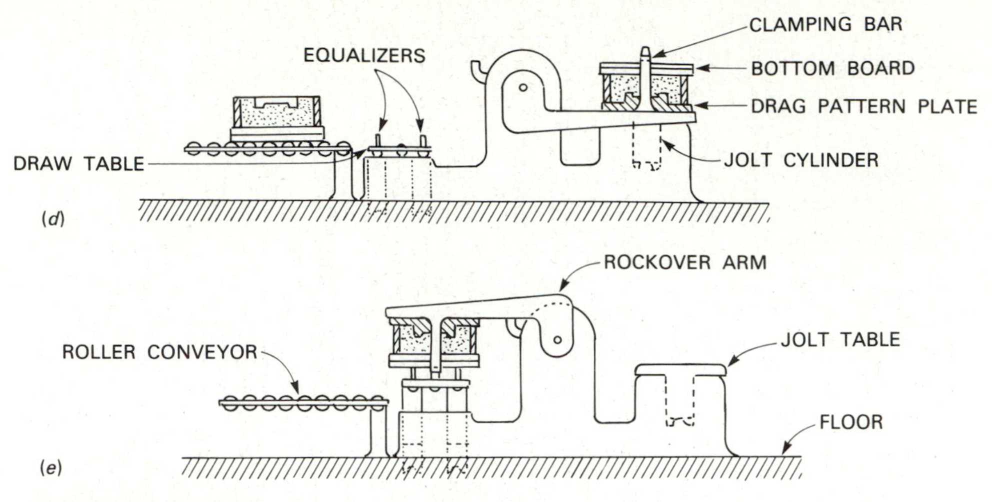

FIGURE 8.7 *(continued)*

Large nonferrous ship propellors weighing 300,000 lb, or the base for a radiotelescope, would be made in a pit mold. In fact, the gates for the Saint Lawrence Seaway locks were cast in pit molds, but the sections had to be made small enough to be carried to northern New York State by railroad. The sections were welded on site to produce the large gates by cast–weld construction.

Such large castings require several hours to pour and up to six weeks to cool. The long cooling times lead to extremely large-grained structures.

Shell-mold Casting

Shell molding uses a phenolic resin binder mixed with sand for casting steel, iron, or nonferrous alloys. The pattern is made of metal, preferably cast iron. It should be machined, because the process duplicates parts to close tolerances. The pattern is heated to 400 to 500°F; then, after a silicone parting agent is sprayed onto the surface, the resin-and-sand mixture is deposited on the pattern by blowing or dumping (Fig. 8.8). The excess material is shaken off for re-use and the shell or crust ($\frac{1}{8}$ to $\frac{1}{4}$ in. thick) is removed from the pattern after curing is complete. Halves are matched and located by integral bosses and matching recesses and are glued or clamped together. Finally they are placed in a metal case, and surrounded by 1.5 in. of steel shot, sand, or other backup material to support them during pouring. The gates, sprues, and risers are usually a part of the mold.

The shell-molding process illustrates how one industry (plastics) affects another industry (sand casting). The process today has proved practical for parts ranging from malleable iron chain hooks ($\frac{1}{2} \times 1 \times \frac{1}{2}$ in.) to automotive crankshafts.

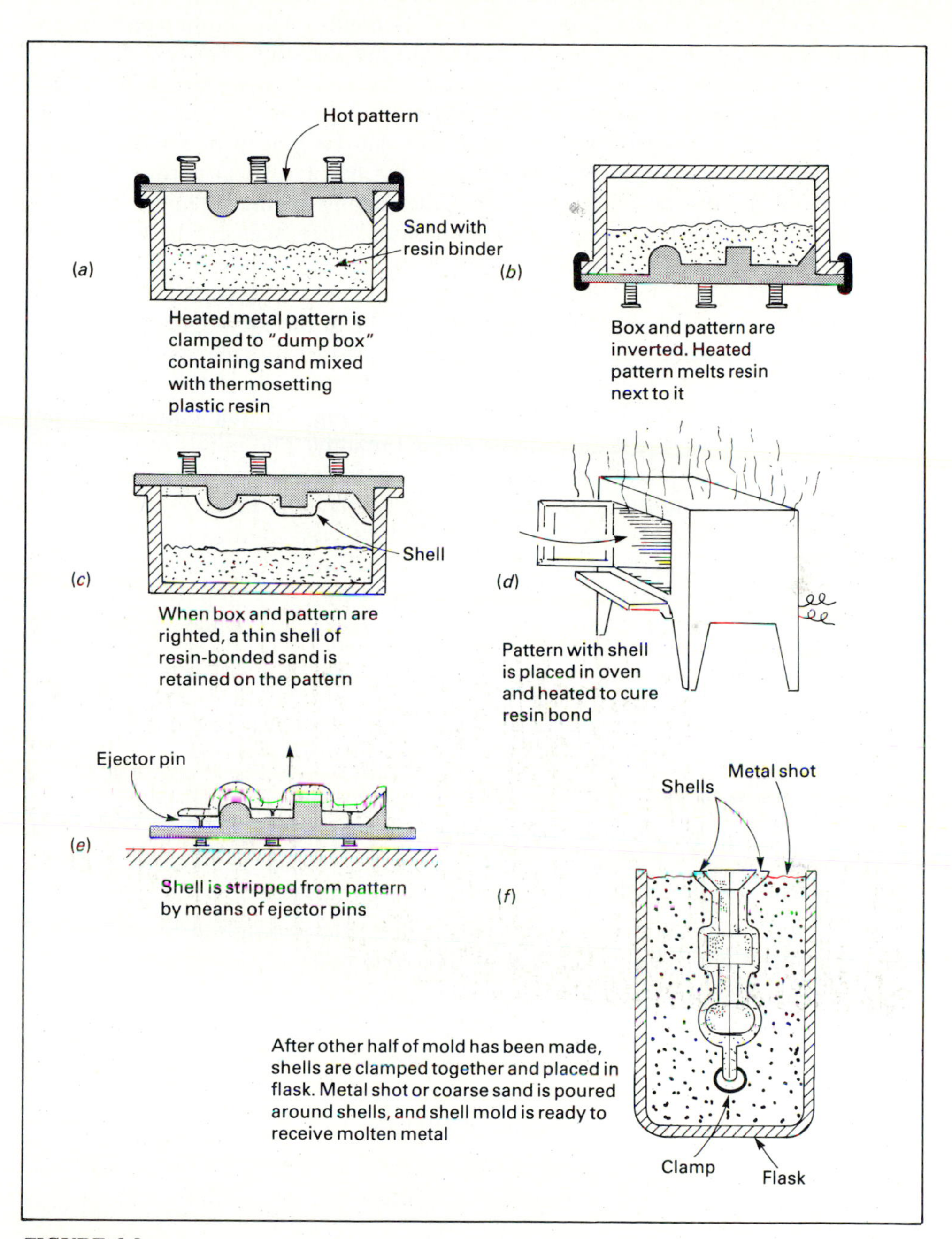

FIGURE 8.8
Shell-molding process.

Fully automatic shell-molding machines are available that can produce a thin shell of resin-bonded sand every 30 seconds. This equipment, in the production shop, is coupled with a shell-closing machine that seals the cope and drag halves of the mold together. Close tolerance between the cope and drag is maintained, thus minimizing objectional fins in the finished casting. After the shell-closing operation, the molds can be placed in a mold-storage area until needed. The mold shown in Fig. 8.8 has a vertical parting line, but most molds have a horizontal parting if the casting is small. High production rates, good surface finish, and close dimensional tolerances can be achieved in gray iron and small steel castings.

No backing is used, just a bed of dry sand and a weight. When the metal is poured, the smooth shell promotes easy flow and the metal is not held back by gas pressure; the porous mold permits gases to escape easily. Heat is removed quickly through the thin shell and backup material (Fig. 8.8). Some resin formulations are slightly flexible after setting, so that patterns do not need as much draft and molds can be ejected readily. The molds can be stored and do not contain moisture. Thus, they are suitable for pouring at any time.

This process reduces the dust present in the typical sand foundry and reduces the amount of sand to be handled. Since the sand cannot be reused, without pneumatic or thermal reclamation, shell molding is relatively expensive, which is the major disadvantage of the process.

The shell-molding process has the following advantages over typical green sand molding.

1. Productivity can exceed that of conventional sand-casting practice.
2. Thin sections (down to 0.010 in.) can be cast.
3. Machining of the castings is reduced and, in some cases, eliminated.
4. Cleaning is considerably reduced and, in some cases eliminated.
5. Saving of metal through the use of smaller gates, sprues, and risers in the casting process results in a higher yield from the metal.
6. Savings in work space, material handling, and storage are also realized through use of 90 percent less molding materials.
7. The cured resins are not hygroscopic, thus permitting prolonged storage of the molds.
8. Closer dimensional tolerances can be obtained (0.002 in./in. in one-half of a mold, but 0.010 in. across the parting line).
9. Better surface finishes are realized (100 μ in.)
10. Shell molding is particularly useful for small steel castings (up to 20 lb).
11. Shell core production is particularly attractive because it reduces core weight and eliminates core ovens.

As mentioned, the principal disadvantage of the shell-molding process is that, because of the cost of patterns and curing equipment, the process is not economically advantageous in small quantities. A representative phenolic resin will cost about $0.40/lb and cannot be re-used. The typical clay binder in green sand molds costs only about $0.015/lb and may be used many times. Secondly, the strong phenolic odor bothers some people and may be enough to cause disagreeable working conditions.

"V" Process

The principle of vacuum forming of transparent plastic film around merchanidise prior to sale is widely used throughout the world. In 1970, an obscure Japanese foundryman wondered what would happen if plastic was vacuum-formed around a gated pattern mounted on a plate and then the flask was filled with loose sand and jolted before the top surface was covered with a plastic film of ethylene vinyl acetate about 4 to 7 mil thick. A separate vacuum would then be applied to the interior of the flask. Thus, after the vacuum on the pattern was released the fine sand (about 180 GFN) would be held in a stiff block by 2100 lb/ft^2 of air pressure on the plastic films at the parting line and the top of the flask. (Fig. 8.9). The pattern impression is particularly distinct because of the fine unbonded sand and the compacted mold having a hardness of 90 Dietert. Of course, a separate cope is made in the same way. The process now uses a heater to soften the plastic film just before it falls on the pattern.

In the past decade, great strides have been made in adapting the process to a higher level of production of larger parts. In the beginning it was best suited to parts, such as grille work, that had relatively small mold depths from the parting line and that needed fine textures and surface details. Now, however, parts as large as a bath tub or a milling machine stand can be made at rates of 10 to 20 molds per hour. In addition the process can be used for iron pipe fittings, stainless-steel valve bodies, ship anchors, railroad car bolsters and side frames, engine parts, and agricultural castings. The process has been used to cast items ranging from those weighing less than a pound to those exceeding 10 tons and can accept conventional cores so that complex internal configurations can be produced.

In the gating and risering of "V" process castings, provision must be made for venting the risers, and the sprue should be perforated to permit mold gases from the initial pouring to escape. A nonpressurized gating system with ingates into the drag should be used. It is important that unfilled portions of the mold are always at atmospheric pressure. As the mold is poured, the molten plastic film vaporizes and diffuses back into the sand, where it condenses and glazes the surfaces of the sand grains. Shake-out occurs by releasing the vacuum and permitting the loose sand grains to drop through a grille into the funnel-shaped enclosure for cooling prior to reuse in the subsequent mold. Since there are no binders, there is a minimum of odor and smoke except from the core binders.

It should be noted that the patterns for the "V" process must have small vent holes connected to a large runner to permit the vacuum to be transferred

1

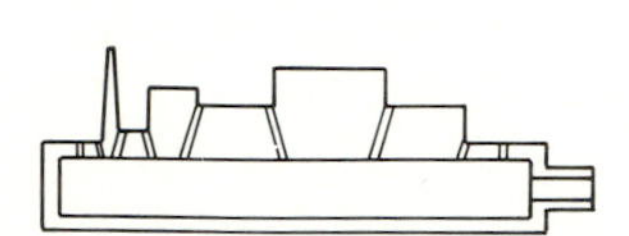

Pattern (with vent holes) is placed on hollow carrier plate.

2

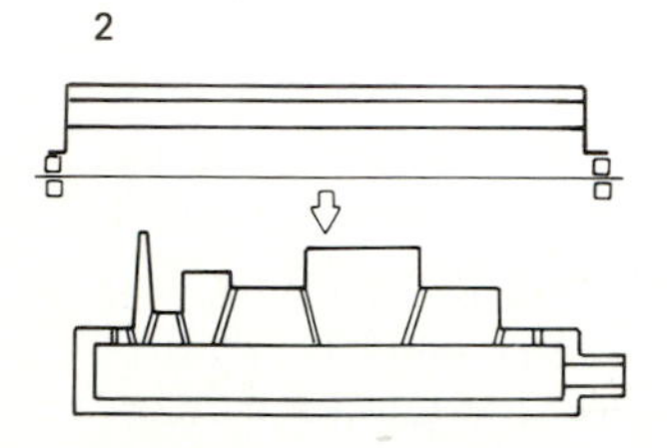

A heater softens the .002″ or .005″ plastic film. Plastic has good elasticity and high plastic deformation ratio.

3

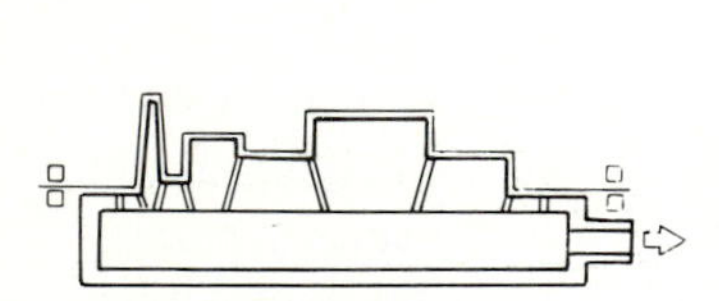

Softened film drapes over the pattern with 200 to 400 mm Hg vacuum acting through the pattern vents to draw it tightly around pattern.

4

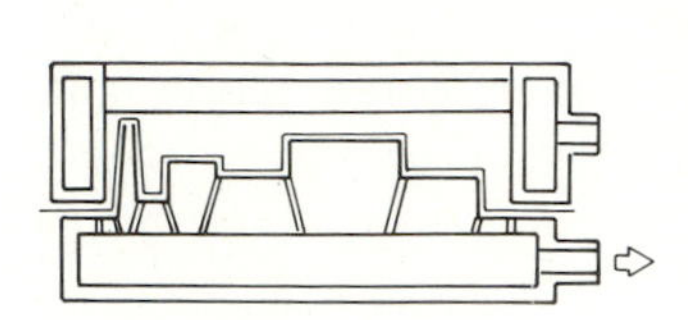

Flask is placed on the film-coated pattern. Flask walls are also a vacuum chamber with outlet shown at right.

5

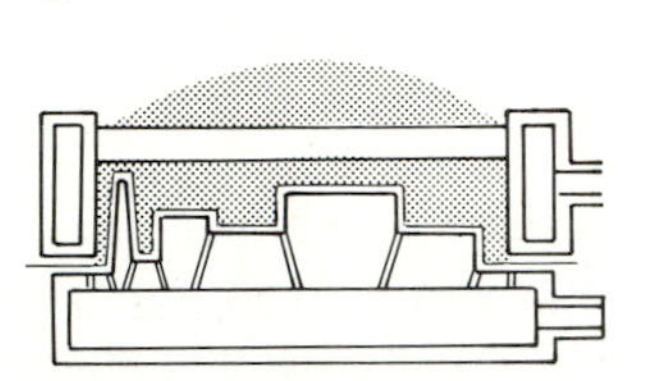

Flask is filled with dry unbonded sand. Slight vibration compacts sand to maximum bulk density.

6

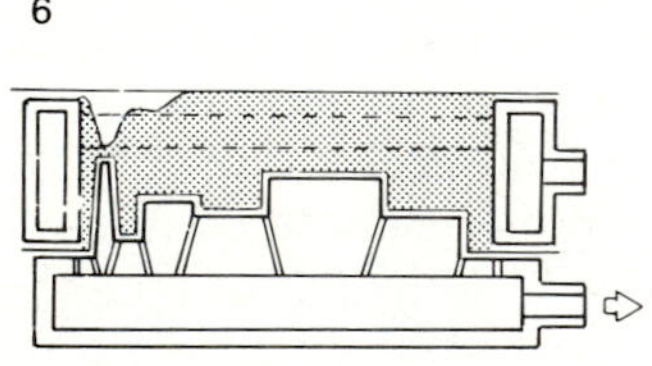

Sprue cup is formed and the mold surface leveled. The back of the mold is covered with unheated plastic film.

7

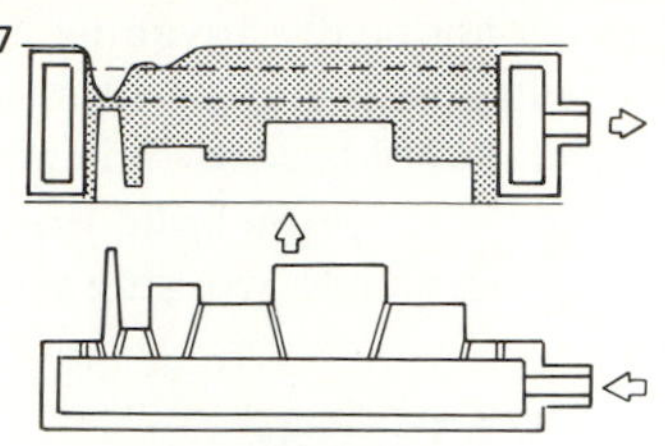

Vacuum is applied to flask. Atmospheric pressure then hardens the sand. When the vacuum is released on the pattern carrier plate, the mold strips easily.

8

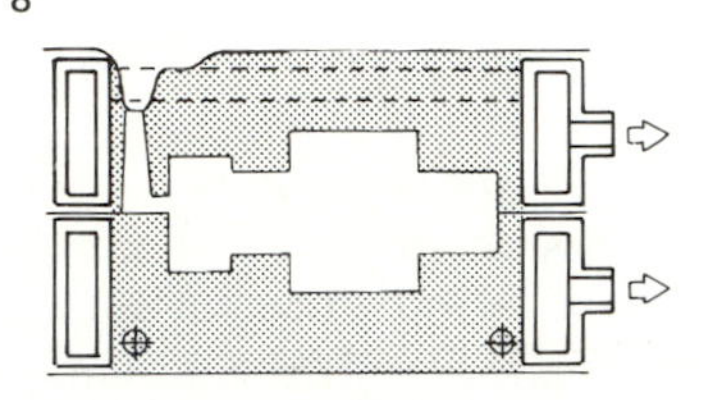

Cope and drag assembly form a plastic-lined cavity. During pouring, molds are kept under vacuum.

9

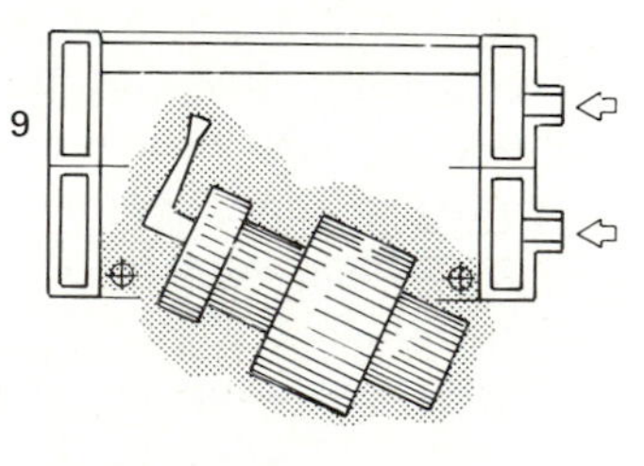

After cooling, the vacuum is released and free-flowing sand drops away leaving a clean casting, with no sand lumps. Sand is cooled for re-use.

FIGURE 8.9
The vacuum-molding process.

to the heated film as soon as it touches the pattern otherwise wrinkling may occur. Proper venting of the pattern requires experience; the number of holes, in particular, depends on changes in the pattern surface contours.

The surface chill of "V" process castings is reduced, with a consequent 5 to 10 percent reduction in surface hardness; but metal fluidity is improved, so that thinner sections in rangey castings can be easily poured without cold shuts.

The machinability of cast irons is also improved because of the coarser graphite structure and the reduction in surface carbides.

Flaskless Molding

Flaskless molds with a vertical parting line are made at a rate of up to 750 molds per hour by high compaction of sand into blocks that are about 14×18 in. in cross section and up to 8 in. thick. The sand is blown into the machine shuttle at 90 to 100 lb/in.^2, then the entrance is stopped off and a squeeze ram moves forward and the block is compacted at up to 2000 lb/in.^2 The shuttle moves to one side, bringing a second shuttle to the molding station, (Figs. 8.10, 8.11). While the second mold is compacted, a hydraulic ram strips the first compact and adds it to the previous blocks that are already on the pouring conveyor. Note that the conventional cope and drag patterns are on opposite sides of the compacted block. Only when two blocks have been completed and pushed together can a mold be poured.

To meet the high pouring rates that are required by the fast molding cycle, a multilipped pouring ladle can be used to pour as many as five molds at one time. The shake-out and sand-return details are indicated also. Of course, all elements of the operation must be fully balanced, and the length of the cooling conveyor is dependent upon the section thickness of the casting and its cooling rate.

A specially blended high-clay, synthetically bonded molding sand is required to make the process feasible in production. With the proper sand and a balanced cooling rate, high production rates can be achieved on such diverse parts as gas stove grilles, brass valve bodies, and malleable iron pipe fittings. The process is competitive with die casting on small brass and ferrous parts. In fact, it may make the latter process desirable only when a superior surface finish is needed as cast. Flaskless molding can achieve a tolerance as small as 0.005 in. across the parting line on a routine basis.

Expanded Polystyrene Process

This is also called the foam vaporization process. The polystyrene foam pattern, complete with sprues, bottom gates, runners, and risers is placed in a suitable flask, which is then filled with loose sand and rammed up. The pattern is expendable and no parting line is needed (see Fig. 8.12). During pouring, the heat from the molten metal vaporizes the pattern, which then moves through the capillary-porous molding media to a point where it condenses. Cores are not needed if holes are properly located and sized, because the molding medium itself can be used as green-sand core, or an air-setting sand may be rammed into the cored areas. Thus, the pattern itself can serve as a core box! Suitably formed polystyrene foam can also be used as an extension of or modification of an existing pattern or as loose pieces that remain in the mold. In addition, spherical risers can be rammed up in a conventional cope

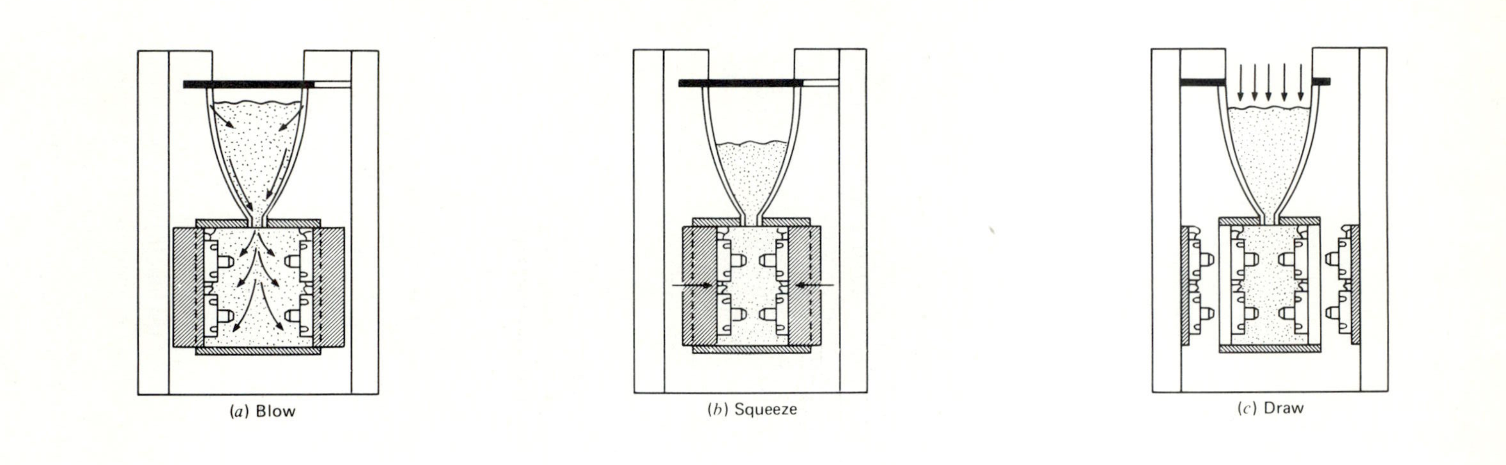

(*a*) Blow (*b*) Squeeze (*c*) Draw

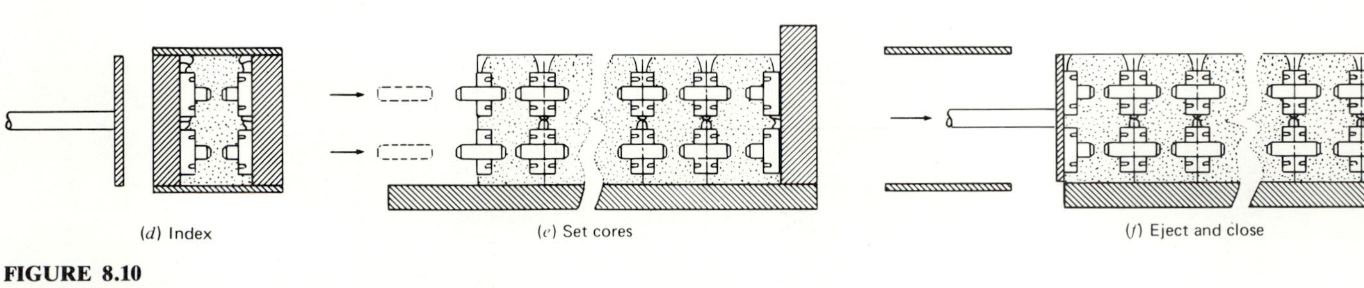

(*d*) Index (*e*) Set cores (*f*) Eject and close

FIGURE 8.10
Diagram of operation of flaskless molding machine.

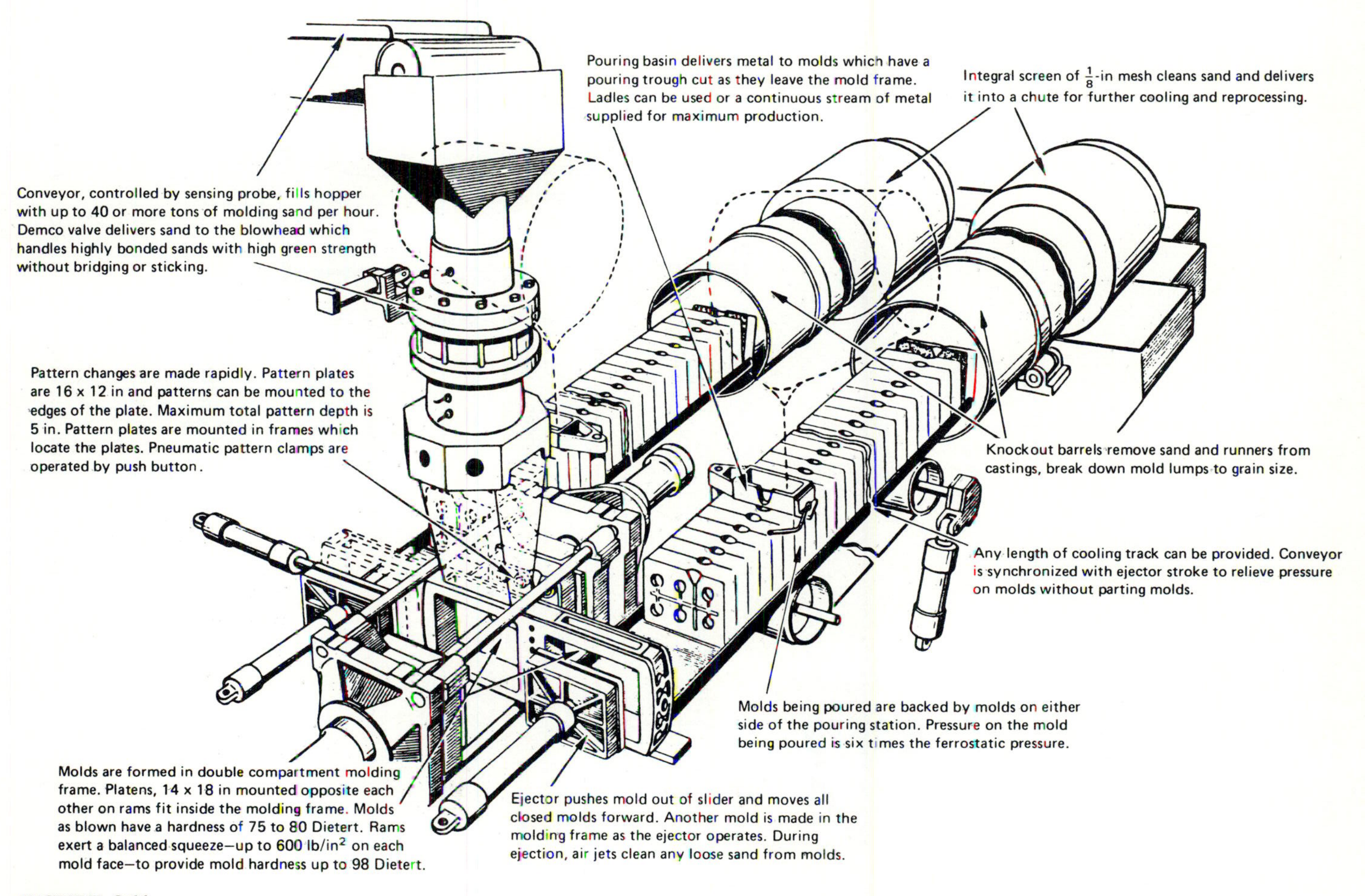

FIGURE 8.11
Flaskless molding machine. (*Courtesy of The Herman Corp; Zelienople, Pa.*)

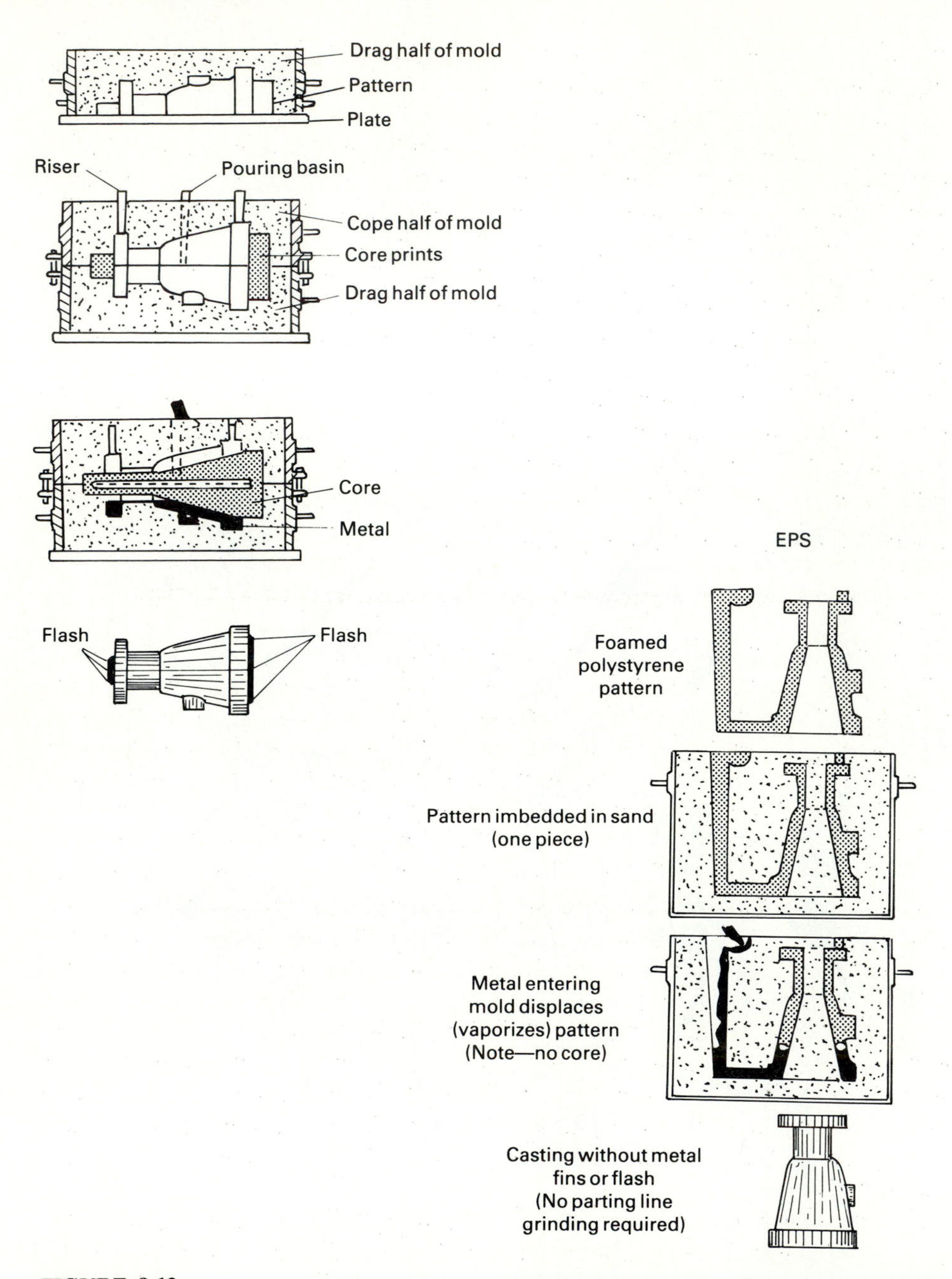

FIGURE 8.12
A schematic comparing the sand molding process and the expanded polystyrene (EPS) molding process.

and drag mold wherever desired for best feeding without regard to a parting line, because they will stay in the mold when the pattern is drawn.

The process is especially useful for large castings of which only one or two pieces are required. Most automotive forming dies for quarter panels, roofs, or hoods are made by the expanded polystyrene process, because the foam can be formed so easily. Pattern shops have determined that they can make three polystyrene patterns before they reach the break-even point with a wood pattern.

The future of the process is particularly bright if one considers that a foam of the proper density could be made right in the foundry in the proper steel pattern box. A nonbonded molding media could be poured in and vibrated around the pattern and core assembly. The mold could be poured, the casting shaken out, and the sand passed through a pneumatic reclaimer. Then the process would be repeated.

Evaporative pattern casting is most advanced for aluminum materials. With current interest in introducing silicon carbide or alumina fibers into aluminum in order to provide greater wear resistance, the application of evaporative pattern casting can only expand. Metal matrix composites will be one of the primary materials used in engineering designs in the next decade.

The application of evaporative casting for gray and ductile iron, steel, stainless-steel, brass, and copper-based alloys is expanding. Users of evaporative casting point out that one of the more important advantages of the process over permanent mold casting is the savings in time in changing tools when producing a different casting. Again, the process results in improved quality of parts because of reduced porosity, and the process will reduce costs of production.

Precision Casting in Ceramic-Shell Molds

Historically, investment casting dates back at least to 1766–1122 BC, when the Egyptians made religious items from precious metals. Until World War II it was still used for art castings and jewelry, and in dentistry for producing gold inlays. Such things as superchargers and the turbine blades for jet engines became the backbone of a new industry.

Ceramic-shell casting has proved to be a boon to investment casters because it has permitted the automation of the ancient art of *cira perdue* or *lost-wax casting*. Ceramic-shell molds may cost less than 50 percent as much as conventional investment molds for the same pattern. The patterns are wax-welded to a central sprue (Fig. 8.13). The assembly is carefully cleaned in a suitable solvent, which makes the wax surface wettable by water; it is dipped in a colloidal ethyl silicate gel and then drained for several minutes. When dripping stops, the assembly is stuccoed by dipping it into a fluidized bed of a fine-grained fused silica. The process is repeated. On the third and subsequent stuccoing operations, a coarser grade of fused silica is used; usually five coats

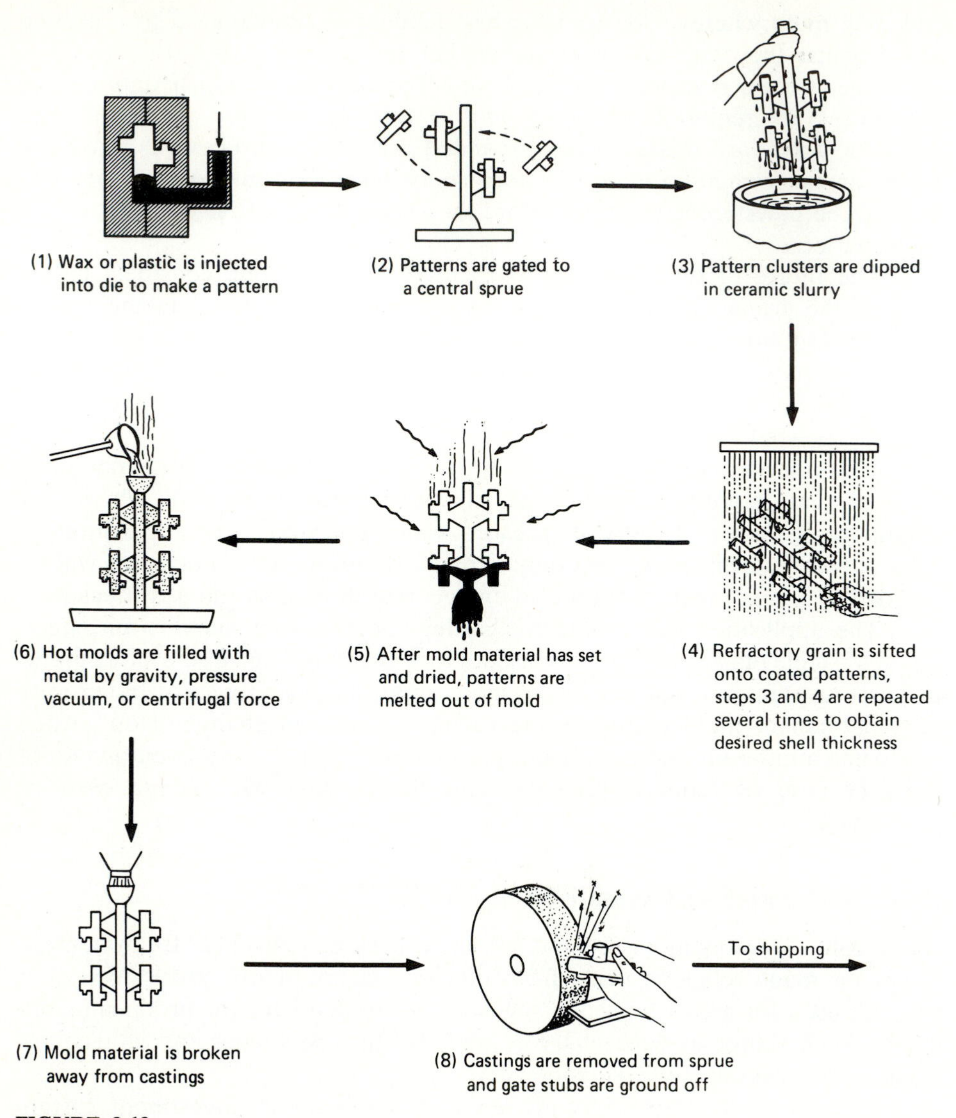

FIGURE 8.13
Sequence of operations in the ceramic-shell process.

are sufficient. Over 85 percent of all precision casting is carried out in ceramic-shell molds.

Investment casting is of interest to the production-design engineer because this process offers greater freedom of design than any other metal-forming operation. Accurate and intricate castings can be made from alloys that melt at high temperatures. Parts such as gas-turbine blades and

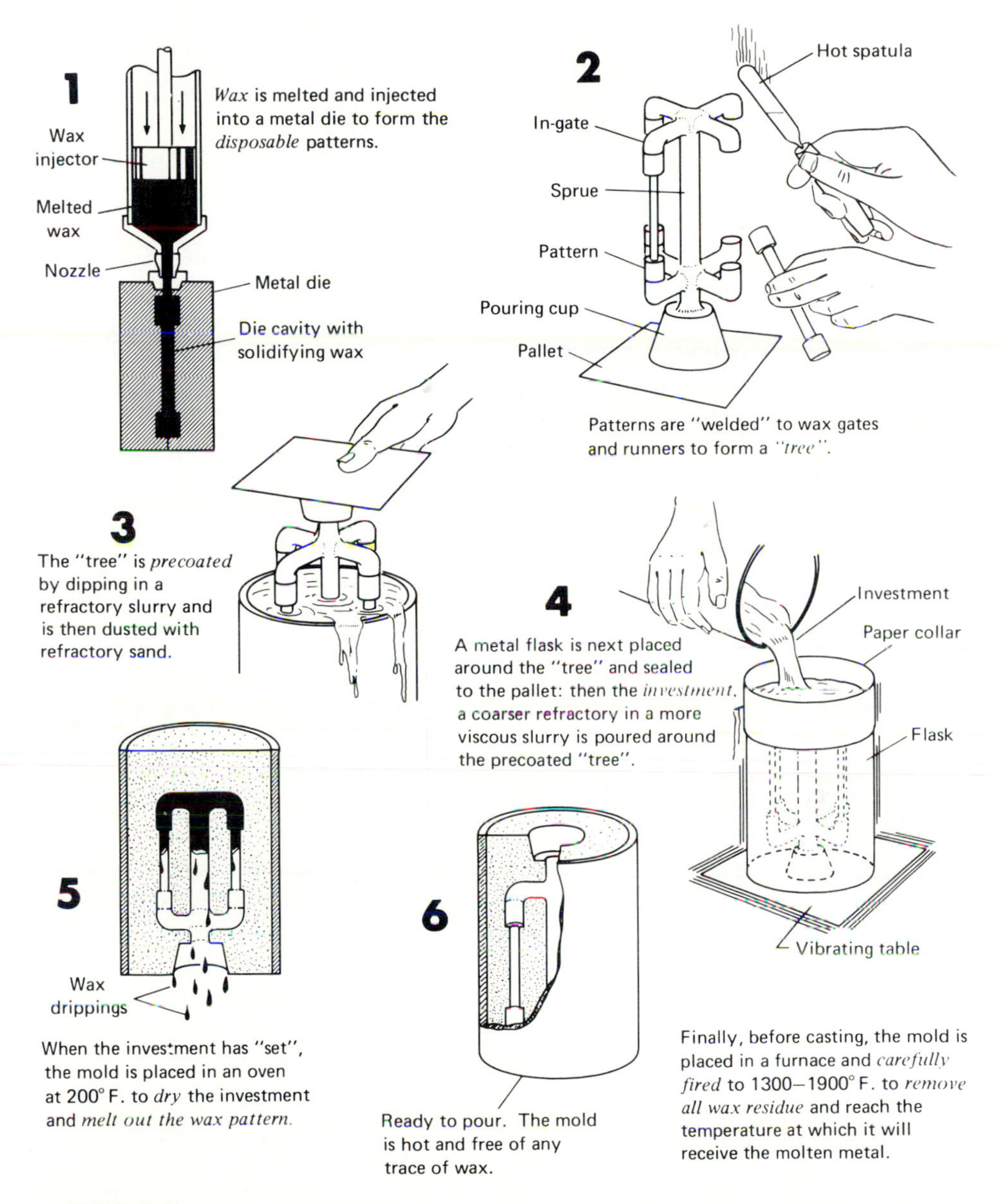

FIGURE 8.14
Preparing a mold for investment casting.

latches can be cast to such close tolerance that little or no machining is required.

Investment casting is like sand casting in that the mold is broken to release the casting; it is like permanent molding and plaster casting in that good surfaces, accuracy, and consistency can be obtained; and it is like plaster molding in that the mold is made from a slurry and cured. Investment casting, precision casting, and lost-wax processes are all essentially the same.

Full investment casting is unlike other coating processes in that an expendable pattern of wax, plastic, or frozen mercury is used. (Fig. 8.14). Intricate castings can be made by forming expendable patterns separately in dies and assembling them to make up the final and more complicated pattern unit. Gates and risers are attached for molding. The whole expendable pattern unit is coated with a fine colloidal silica wash and is dusted with a refractory sand. A slurry is then poured around it and allowed to set. By this means no parting-line inaccuracies exist, so that close tolerances can be maintained in all directions. The flask and its contents are vibrated to pack the slurry around the pattern and to separate the solid material from the water, which can be poured or withdrawn from the top. After the slurry has set, the flask is then placed upside down in a furnace and cured. The pattern material melts and runs out, leaving the mold ready for final curing at high temperatures. The mold is in one piece and cannot be inspected before pouring. After it is baked at high temperature, it is ready for casting a high-melting-point alloy while the mold is still hot. When sufficient cooling time has elapsed, the mold is broken and the casting is removed. This process can be used for any type of material. Low-temperature alloys such as brass, silver, gold, and bronze were cast in this manner in ancient times. It is particularly useful for extremely thin castings.

The various controls used for materials, temperatures, curing time, and pouring ensure uniform castings. The wax is of a known composition and has definite shrinkage factors. The ingredients can be adjusted to vary the shrinkage; thus compensation can be made for mold shrinkage, and the difficulty of adjusting mold sizes for shrinkage is reduced.

Dies for the wax pattern can be made of rubber, plaster, soft metals, or steel. Their composition is determined by the accuracy requirements, quantity, and ease of removing the wax. Plastic and mercury require metal molds. Units made in one mold are joined to gates to make multiple patterns or trees, so that many parts can be cast at the same time. The accuracy of the original mold determines the final accuracy of the part.

Castings are usually poured by gravity, with parts arranged in a vertical fashion. Centrifugal force, vacuum, or air pressure are sometimes used to pour metal into the hot mold and to obtain more accurate, homogeneous castings.

Ceramic-mold Casting

This process, which is also known as cope and drag investment casting has two proprietary variations: (1) Shaw process and (2) Unicast process. The major distinction between ceramic-mold casting and investment casting is that the former relies on precision-machined metal patterns rather than the expendable patterns used in conventional investment casting. Ceramic-mold casting is similar to plaster molding except that the mold materials are more refractory, require higher preheat, and are suitable for most castable alloys, particularly

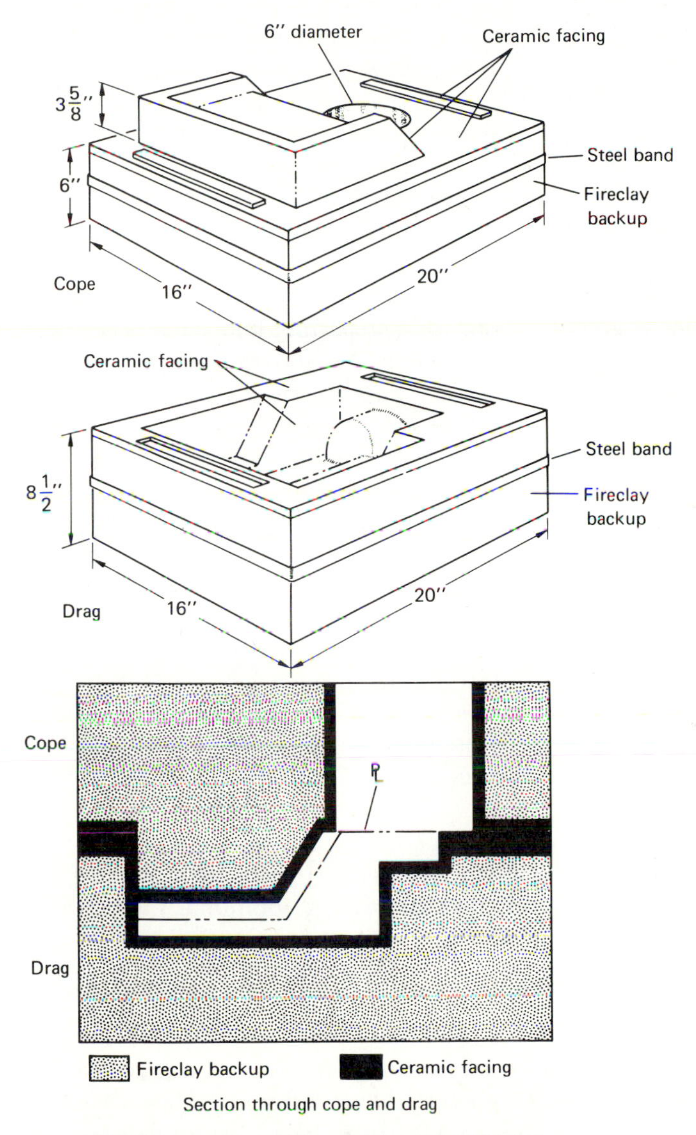

FIGURE 8.15
Typical cope and drag setup for ceramic molding. (*Redrawn from American Society for Metals,* Metals Handbook, *vol. 5, Metals Park, Ohio, 1970.*)

ferrous alloys. The refractory slurry consists of fine-grained zircon and calcined high-alumina mullites, or in some cases fused silica.

In ceramic molding, a thick slurry of the mold material is poured over the reusable split and gated metal pattern, which is usually mounted on a match plate. A flask on the match plate contains the slurry, which gels before setting completely. The mold is removed during the time when the gel is firm in order to prevent bonding to the pattern (Fig. 8.15). The basic difference between the Shaw and Unicast processes is that in the Shaw process stabilization results from the burn-off of an alcohol binder, whereas in the Unicast process the mold is stabilized by immersion in a liquid bath or a gaseous atmosphere that cures the gelled slurry. Before pouring, the molds are usually preheated in a furnace to reduce the temperature difference between the mold and the molten metal and to maximize the permeability available through the microcrazed mold structure.

Ceramic-mold casting is preferred when (1) the parts are too large for conventional precision-investment casting or (2) parting-line defects are not objectionable. The process replaces sand casting if a superior surface finish is required, if improved dimensional accuracy is needed, or if sand inclusions or hot tears cannot be tolerated in a casting of complex geometry. Ceramic molding can handle parts weighting 1500 lb or more. The process has excellent accuracy and reproducibility even with such large castings.

It is impractical to control dimensional tolerance across the parting line to the same tolerance as within one half of the mold. Also, mechanical properties suffer the loss of the chill effect, because the mold materials are such good insulators that a coarse-grained structure results. The designer can compensate for such a reduction in strength by making critically stressed sections somewhat heavier. The process is expensive because the mold materials are high in cost and are expendable. Ferrous alloys are the most commonly cast; they include die casting dies, large trim dies, components for food machinery, milling cutters, structural components for aircraft, and hardware for aerospace vehicles and atomic reactors.

An as-cast surface finish of 125 μin. or better can be achieved readily. Dimensional tolerances are +0.003 in./in. for the first inch, with incremental tolerances of +0.002 in./in. for larger dimensions. An additional tolerance of +0.001 in./in. should be provided across the parting line.

Plaster-mold Casting

Plaster-mold casting as an art has been known for years. Today, it is competing successfully for castings because engineers have improved the accuracy, quality, range of applications, and cost. Costs have been reduced by standardization of methods, flasks, materials, and material-handling facilities. The process is similar to sand casting except that the mold and cores are made of plaster (or a combination of plaster and sand) instead of packed sand. The mold is made usually in a wood or metal frame containing the pattern. After

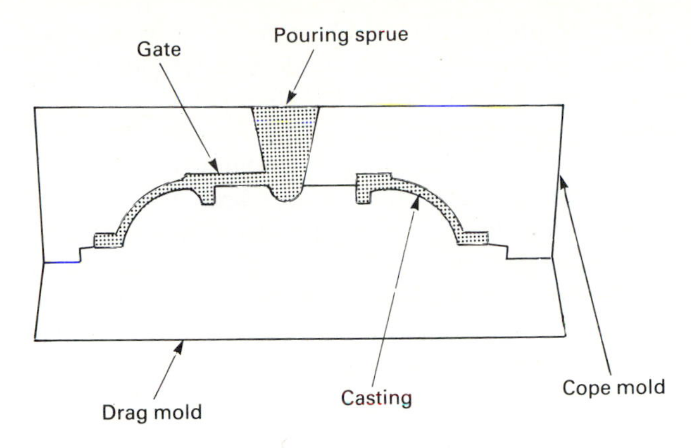

FIGURE 8.16
Plaster mold after pouring.

the plaster sets, the frame and pattern are removed, and the mold halves and cores, if any, are assembled and are normally baked (see Fig. 8.16). Plaster molds cannot be used for ferrous castings because the plaster ($CaSO_4 \cdot \frac{1}{2}H_2O$) is not sufficiently refractory. However, nonferrous castings can be made with smooth, accurate surfaces and fine details. Their cost is about three times that of sand castings, but the elimination of machining and finishing operations frequently compensates for the additional cost. This process, because of the recent improvements in foundry mechanization and techniques, should receive consideration when the design and process are selected.

Mold material. Gypsum or calcium sulfate (plaster of paris), talc, asbestos, silica flour, and others, and a controlled amount of water, are used to form a slurry.

The steps in processing are as follows.

1. Parting compound is sprayed on flask and pattern.
2. Slurry is poured from hose around pattern and flask is filled.
3. Pattern is removed after setting.
4. Mold is dried and baked in conveyor oven.
5. Mold is separated from flask and flask is returned to step 1.
6. Inserts and cores are placed and guide pins are placed in holes in four corners of mold.
7. Cope and drag are matched or assembled and guided by pins.
8. Metal is poured.
9. Casting is cooled on conveyor in mold.
10. Casting is shaken out and the mold is destroyed. (The mold material is not salvaged.)
11. Castings are trimmed of gates, sprues and flash, and are inspected.

Antioch process. In the Antioch process, a mold material is used that combines the advantages of sand and gypsum plaster. Cast surfaces and details are comparable to those produced by plaster molds, but, like plaster molding, the Antioch process is confined to nonferrous casting. Silica sand is the bulk material and gypsum is the binder. Talc, terra alba, sodium silicate, and asbestos are used to control some characteristics. Water and dry material produce a slurry that is piped to the flask and pattern setup. The material develops an initial set in 5 to 7 minutes, is air-dried for about 5 hours, and is then placed in an autoclave at about 2 atmospheres of steam pressure. The molds are again air-dried for about 12 hours and then reheated in an oven at 450°F. This autoclave operation and drying procedure helps develop excellent permeability. Tolerances of +0.005 in. can be kept on small castings, and +0.015 in. on automotive tire molds.

Permanent-mold Casting

When sand castings are made, the sand mold is destroyed. Therefore, it has been an ambition of engineers to develop a mold that can be reused thousands of times. Permanent molds that are filled by gravity pouring have come into general use because they are economical, and because they are particularly useful in casting low-melting-point alloys (Fig. 8.17). Metal-mold castings have some distinct advantages over the typical sand-mold casting. These include closer dimensional tolerances, better surface finish, greater strength, and more economical production in larger quantities. Some disadvantages of metal molds are their lack of permeability, the high cost of the mold, the inability of the metallic mold to yield to the contraction forces of the solidifying metal, and difficulty in removing the casting from the mold because the mold cannot be broken up. Recently, bronze, cast iron, and even steel have been cast in metal molds and have extended the use of permanent molds.

In general, castings to be produced by permanent-mold methods should be relatively simple in design with fairly uniform wall sections and without undercuts or complicated coring. Undercuts on the exterior of the casting complicate the mold design, resulting in additional mold parts and increased cost.

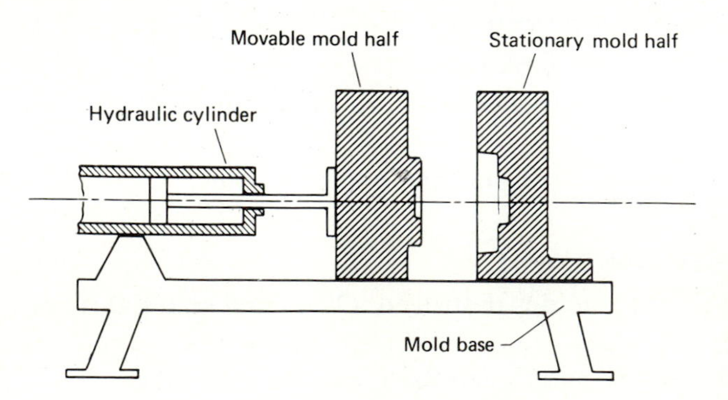

FIGURE 8.17
Schematic drawing of straight-line permanent-mold machine equipped with two-piece deep-cavity die.

If the design requires undercuts or relatively complicated coring, semipermanent molds should be considered. The semipermanent molds are like permanent molds, except that they use sand cores. The metal is poured into the permanent mold under the force of gravity and therefore there are no high pressures.

It should be emphasized that all permanent molds have a thick (0.030-in.) coating of sodium silicate and clay or other insulating materials over the cast-iron surface. The coating causes a poorer surface finish and wider tolerance than is found in a typical die casting. Generally speaking, permanent-mold castings are sounder than equivalent die castings. Automotive pistons are cast in permanent molds rather than die-cast for this reason.

Permanent-mold dies are preheated to 300 to 400°F before pouring and are given a graphite dusting every three or four shots. Thermal balance is important, and auxiliary water cooling or radiation pins are used to cool heavy sections. Proper venting of the cavity is important to avoid misruns.

Low-pressure permanent-mold casting. This casting method is gaining acceptance as a production process. Low-pressure permanent molding depends on directional solidification and proper die design. Other factors include the position of the casting within the dies, the location of the ejection pins, and die coating and venting. Finally, heat balance in the die, ingate location, and gate removal are important factors in the economical production of sound castings.

In this type of casting, molten aluminum is forced by low air pressure from a silicon carbide crucible up the delivery tube and into the die cavity. The solidified casting is removed from the upper die by ejector pins, which are activated when the upper platen is raised (Fig. 8.18). The casting is located so

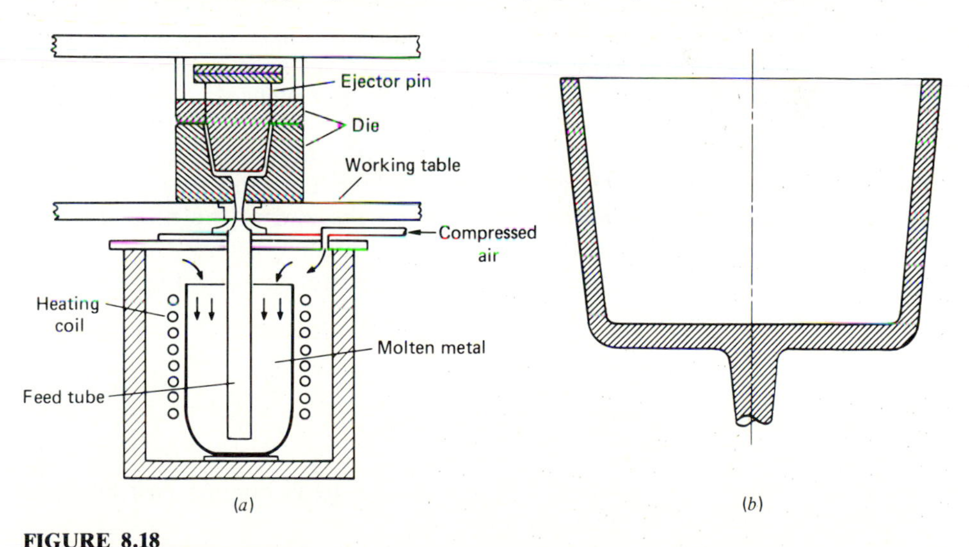

FIGURE 8.18
Low-pressure permanent-mold casting. (*a*) General equipment; (*b*) enlarged view of casting showing typical "carrot" ingate.

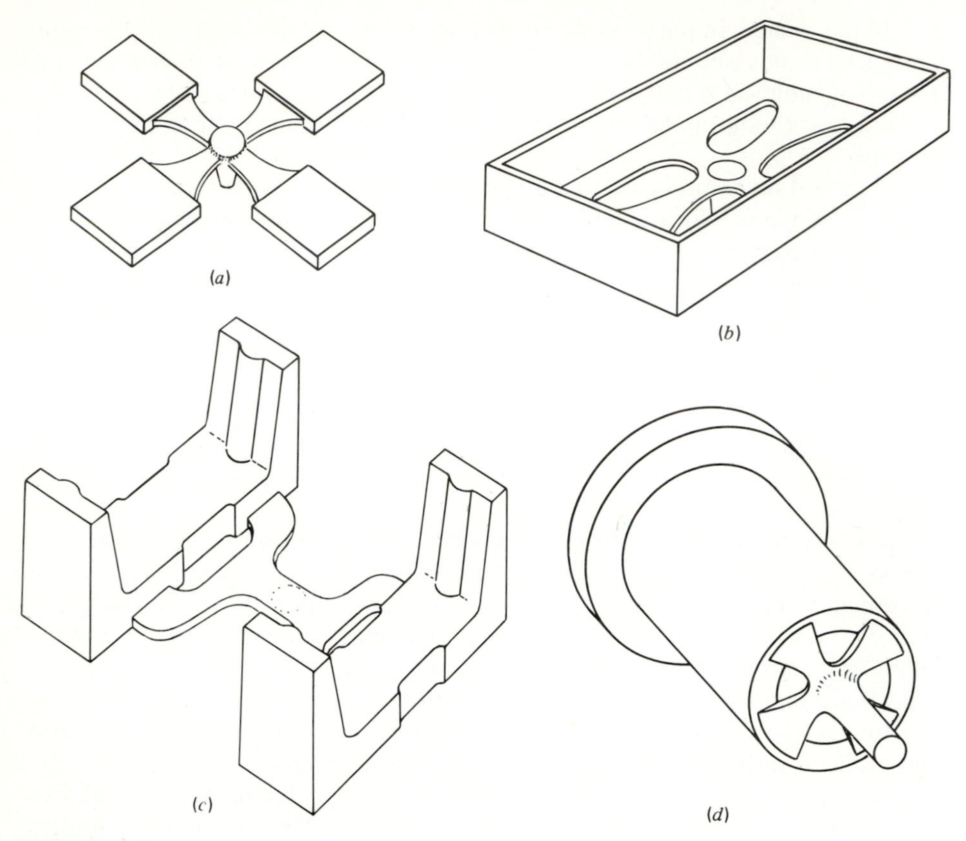

FIGURE 8.19
Typical parts cast of aluminum alloys by the low-pressure permanent-mold process. (*a*) Four pieces on a central sprue; (*b*) box-shaped part with a gate system which is sheared all around; (*c*) twin pieces to keep heat flow in balance; (*d*) cylinder with chill at far end.

that directional solidification takes place toward the mouth of the tube. If a proper heat balance is maintained, risers are eliminated and there is a high casting yield. As soon as solidification occurs at the end of the tube, the air pressure is released and the metal returns to the crucible, where it is reheated for the next shot. Typical parts cast by this process are shown in Fig. 8.19.

The process has a number of useful features.

1. It makes possible thin-walled castings.
2. Aluminum alloys can be cast by this process.
3. The process can use expendable cores.
4. Heavy sections and castings of large projected areas may be cast because there is no need for high lock-up forces.
5. Low-pressure permanent-mold castings usually require no risers to ensure adequate feeding, so the average yield approaches 90 percent.

6. Die materials and die coatings are similar to those used for permanent-mold castings.
7. Accurate control and reproducible results are virtually assured.

Die Casting

Die casting is the technique of rapidly producing accurately dimensioned parts by forcing molten metal under pressure into metal dies. The term also applies to the resultant casting. Die castings can be used economically in designs having moderate to large activity because the completed piece has a good surface, requires relatively little machining, and can be held to close tolerances. The principles of die casting follow those of good practice in any casting operation. The steel dies are permanent and should not be affected by the metal introduced into them, except for normal abrasion or wear. Die-casting dies are usually more expensive than those used in plastic or permanent molding of a part of similar size and shape.

The rapidity of operation depends upon the speed with which the metal can be forced into the die, cooled, and ejected; the casting removed; and the die prepared for the next shot (Fig. 8.20).

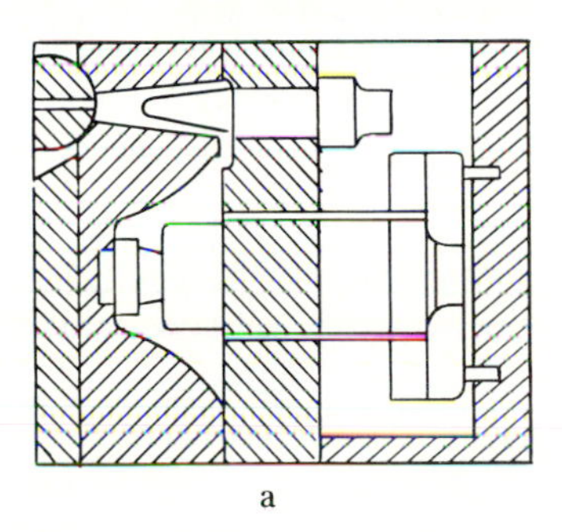

a

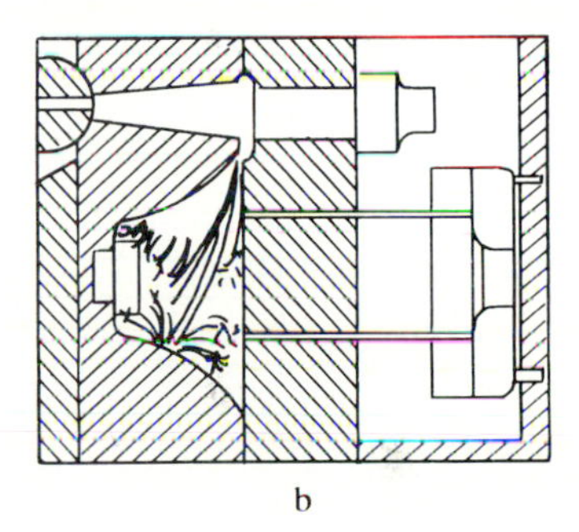

b

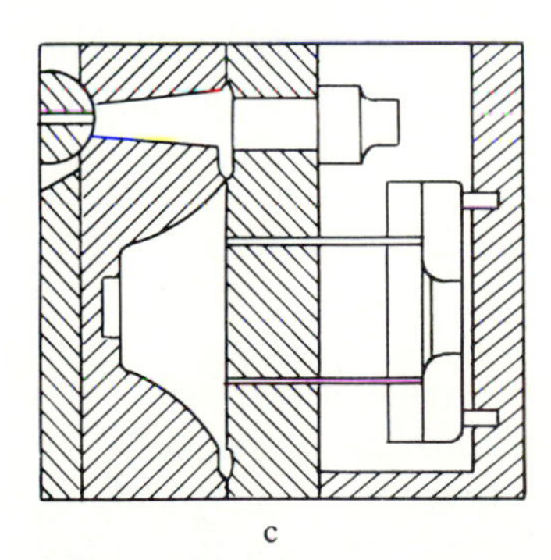

c

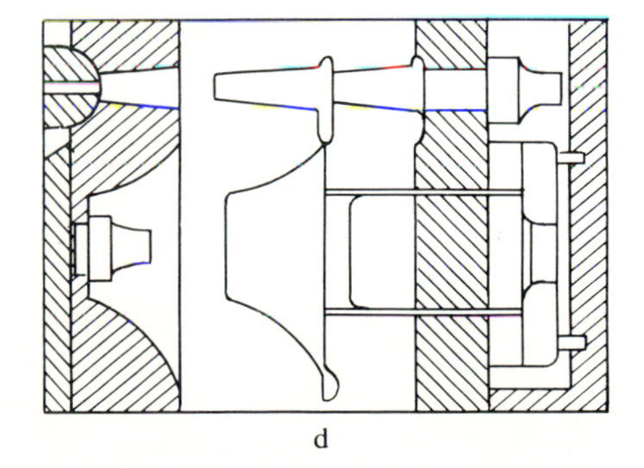

d

FIGURE 8.20
Filling sequence in hot-chamber die casting.

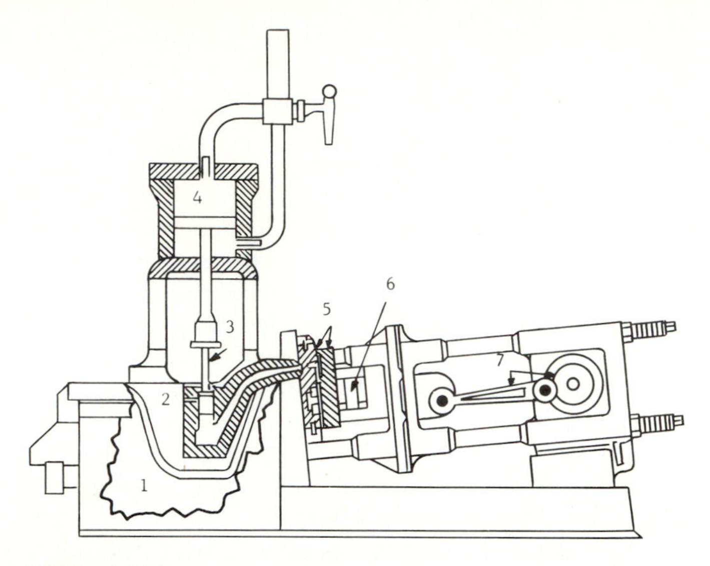

FIGURE 8.21
A typical hot-chamber die-casting machine on which are indicated the seven basic elements. (1) Furnace for keeping the alloy molten. (2) Alloy-holding pot. (3) Plunger that forces the alloy from the cylinder through the passage into the die cavity. (4) Air cylinder that operates the plunger. (5) Die with its cavity. (6) Mechanisms for ejecting the casting. (7) Connecting rod and crank for opening and closing the dies. The transition from molten alloy to die-cast part is accomplished in a fraction of a minute with each cycle of the machine.

Hot-chamber type (Fig. 8.21). The hot chamber refers to the pot in which the metal is melted and from which it is led to the die by what is known as a gooseneck. The hot metal is forced into the die by one of two methods.

1. By maximum air pressure of 600 $lb/in.^2$ over the molten metal. With this method, 100 shots per hour can be made. This is obsolete for zinc and aluminum alloys, but can be used for ferrous alloys.
2. By a cylinder and plunger that are able to exert a pressure of 1500 to 6000 $lb/in.^2$ The cylinder is submerged in the pot from which the molten metal flows; up to 700 shots per hour can be made.

These machines are used for metals having a maximum melting point of 800°F.

Cold-chamber type (Fig. 8.22). The second group of machines is called the cold-chamber type because the metal is heated in a separate pot suitable for high temperatures. The metal is poured into the cylinder by a hand ladle or automatic pour device, and the plunger forces the metal through the orifice into the die. Pressures of 2000 to 6000 $lb/in.^2$ are normal, but some machines offer much higher pressures for castings with a small projected area. The metal

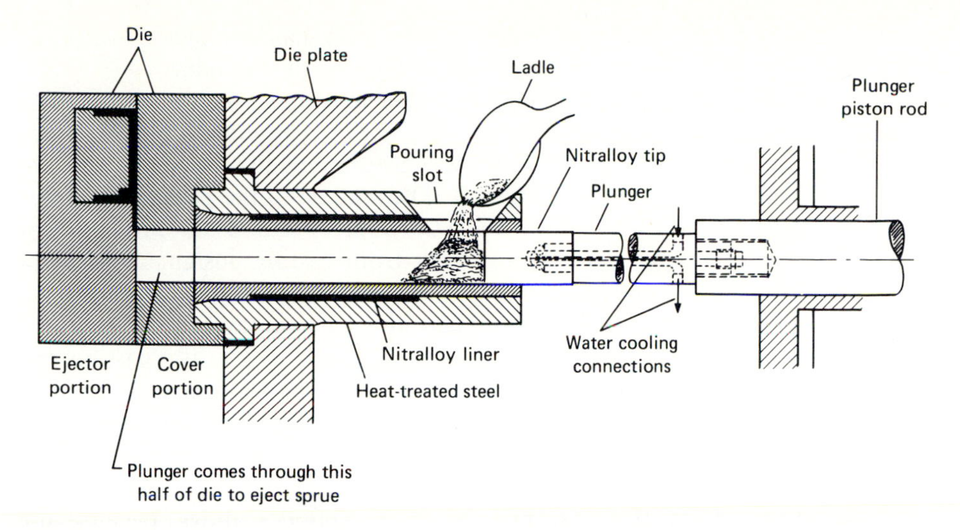

FIGURE 8.22
Schematic view of the cold-chamber die-casting process.

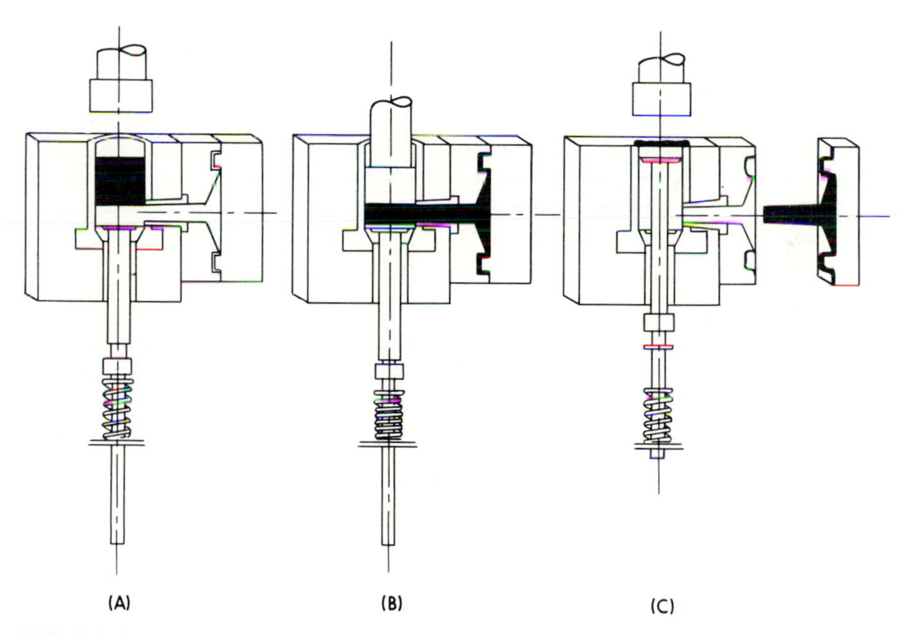

FIGURE 8.23
Cycle in a vertical cold-chamber die-casting machine. (*a*) Cold chamber filled with metal for injection. (*b*) Upper piston descends, lower piston is pushed down, and metal flows into die under pressure. (*c*) Ascending lower piston shears off excess metal, die opens, and casting is ejected.

is usually hand-ladled into the cylinder, but automatic ladles have been developed to deliver the correct amount directly into the cylinder.

In both the hot-chamber and cold-chamber machines, the metal may pass through the orifice and gates into the die as a spray. The metal quickly covers the surface and fills in the voids. Thus all corners and shapes of the die are completely filled, giving remarkable detail. The surface of the part has no folds and reproduces the surface in the die. Of course, the usual problems of trapped air and shrinkage defects can result from casting design, die design, variations in operation, and metal conditions. Generally, all die castings have trapped mold gases, but a solid-front fill through larger gates results in better-quality castings. Locking pressure is the limiting factor. If a casting had a total projected area of 400 in.^{2} including the runners and biscuit, then at 3000 lb/in.^{2} injection pressure, $400 \times 3000 = 12 \times 10^5$ lb locking force would be required to contain the metal after the shot. An 800-ton machine would be needed to provide about a 30 percent safety factor.

Cold-chamber machines can be used to cast copper alloys, because the die can receive metals at high temperatures and pressures. The shot cylinder is placed close to the die, thus cutting down the length of travel of the molten metal through gates and orifices. The injection cylinder may be horizontal or vertical (Figs. 8.22 and 8.23).

Centrifugal Casting

Centrifugal-casting techniques have been used in routine production operations in this country for more than a quarter of a century. Up to the advent of World War II, centrifugal casting was employed mostly in the production of cast iron pipe in sand and metal molds. Necessity developed its use as a method of producing tubular steel shapes, which until then, had been produced by rolling and welding.

Theory of centrifugal casting. Figure 8.24 indicates the chilling action of sand castings compared with that of centrifugal castings. The sketch at the upper left shows the direction of grain flow during the cooling of a static or sand casting. Note that the chilling action originates at both the outside and inside surfaces and progresses to the center of the casting, thus developing an area of weakness in the center of the wall. This is caused by the meeting of the grain boundaries at final solidification and the entrapment of impurities in this central section.

The three lower diagrams indicate the progressive formation of the grain structure in a centrifugal casting. The sketch at the lower left illustrates the layering effect present immediately after pouring, with minute impurities distributed throughout the mass or casting.

The center sketch shows the casting in the state of partial solidification from the outside surface, inward. Note that the impurities are now concentrated in the still molten metal near the inner diameter.

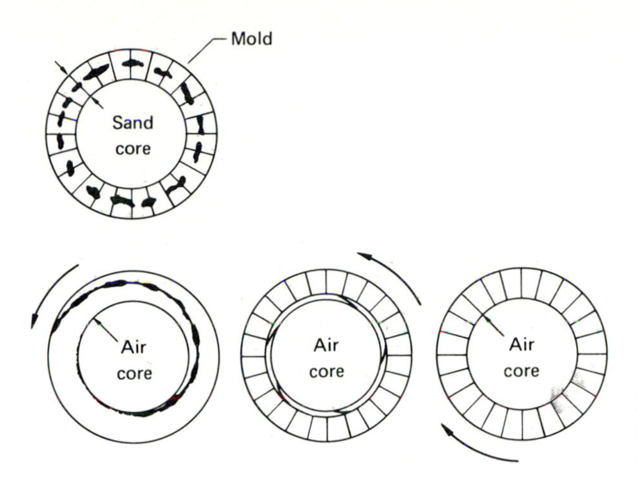

FIGURE 8.24
Sand-cast versus centrifugally cast cylinder.

The lower right sketch shows the casting in the state of complete solidification—dense, sound, and free of impurities, with no weakened sections caused by grain growth from outer and inner surfaces at the same time. All impurities, being lighter than the molten metal, have been forced to the inner diameter of the casting by centrifugal force and lie deposited in the inner bore, from which they may be removed by a rough machining operation if required.

Centrifugal casting can be broken down into three general types: centrifuged, semicentrifugal, and true centrifugal (Fig. 8.25).

True centrifugal casting. True centrifugal casting not only involves feeding the metal by centrifugal force, but also involves using no core or riser. An example of this is the production of steel tubing by centrifugal methods.

A tubular flask with a baked- or green-sand lining is placed horizontally on a casting machine, where it is rotated by a system of rollers actuated by an electric motor. Molten metal is fed into one end of the mold cavity and is carried to the walls of the cavity by centrifugal force. Speeds of rotation sufficient to produce a force of 75*g* are feasible. The wall thicknesses of the tubes are controlled by carefully weighing each charge of metal and knowing the volume, and hence the weight, of metal needed to produce a given wall thickness. Tolerances of $\frac{1}{64}$ in. have been adhered to in this type of casting.

Semicentrifugal casting. Semicentrifugal casting is a means for forming symmetrical shapes about the axis of rotation, which is usually placed vertically for small parts. The molten metal is introduced through a gate, which is placed on the axis, and flows outward to the extremities of the mold cavity. During World War II, nozzles for aircraft turbo-superchargers were made by this method. Wheels are often cast by this method with the gate placed in the hub, but risers and cores are needed.

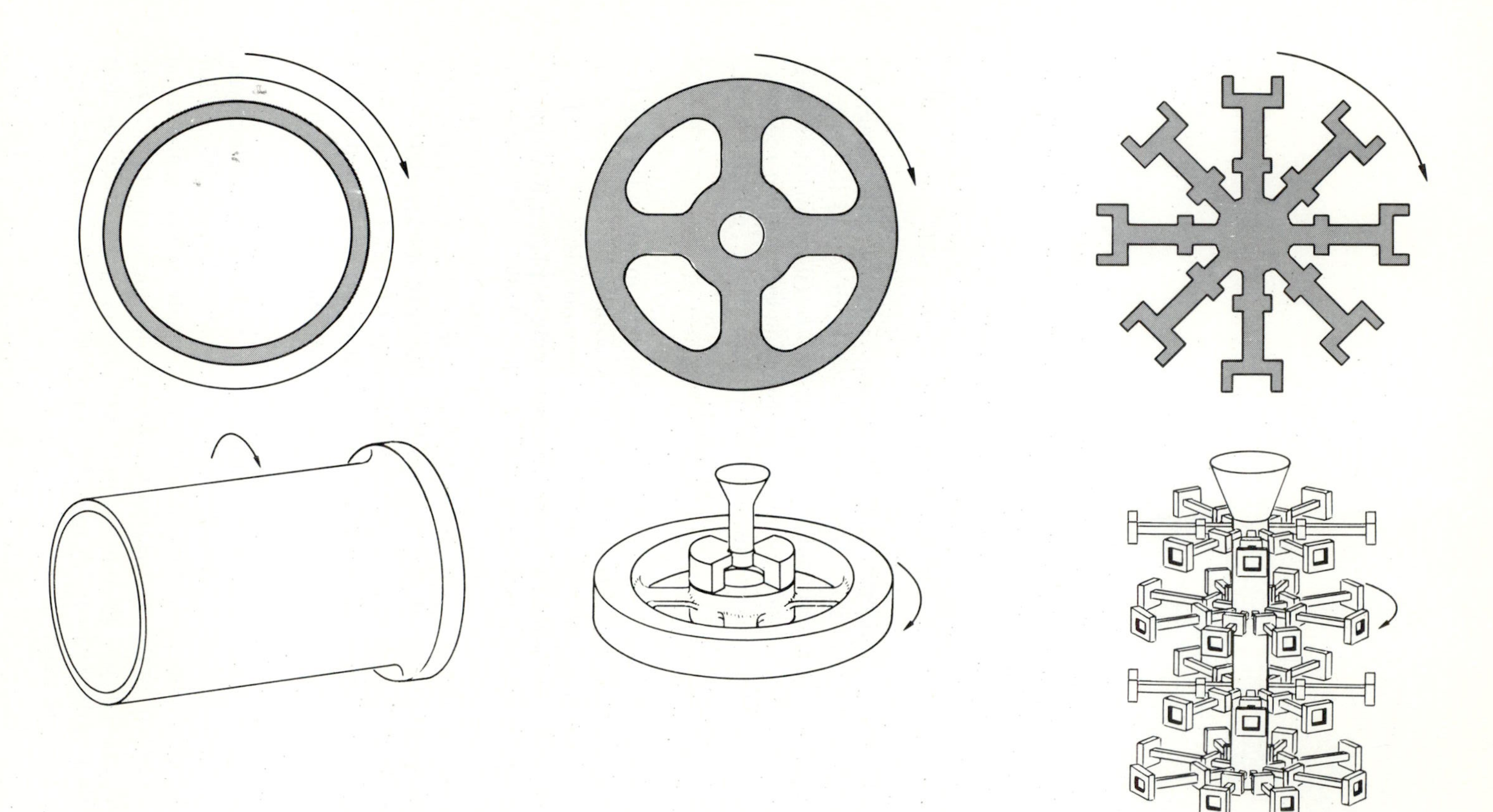

FIGURE 8.25
Three types of centrifugal castings. (*From Ekey and Winter, Introduction to Foundry Technology, McGraw-Hill, New York, 1958.*)

Centrifuged casting. Centrifuging provides a means for obtaining greater pouring pressures in casting. In this process, several molds are located radially about a vertically arranged central riser or sprue, and the entire mold is rotated with the central sprue acting as the axis of rotation. The centrifugal force provides a pressure, at every point within the mold cavity. This pressure is directly proportional to the distance from the axis of rotation and the square of the speed of rotation. This type of centrifugal casting is best used for small, intricate parts where feeding problems are encountered. This method can be used to advantage for the stack molding of six or more molds mounted one above the other (Fig. 8.25).

Applications and problems. Centrifugally cast steel tubes in small diameters, and made of the ordinary carbon steels, are not competitive with welded or rolled tubes. Economy will result when thick-walled tubes are cast, and when mechanical properties are a major consideration and high-alloy grades of steel are required.

Iron and steel centrifugally cast tubes and cylinders are produced commercially with diameters ranging from $1\frac{1}{8}$ to 50 in., wall thicknesses of $\frac{1}{4}$ to 4 in., and lengths up to 50 ft. It is impractical to produce castings with the outside diameter to inside diameter ratio greater than about 4:1.

The as-cast tolerances for centrifugal castings are about the same as those for static castings. On cast gray-iron pipe, for example, tolerances of 0.06 in. for 3-in. diameter and 0.12 in. for 48-in. diameter are typical.

Centrifugally cast tubes can be produced in various external shapes, such as square, elliptical, hexagonal, or fluted, simply by constructing the proper mold. In large quantities, production of these shapes will prove to be particularly economical.

A designer or other user may secure long-, centrifugally cast steel tubes in a large range of diameters and of almost any analysis he may require. The foundry can produce centrifugally cast steel tubes from minimum heats of 2000 lb. The designer may experience difficulty in securing the correct sizes and alloys from the high-production rolling mills, which only produce and stock standard sizes of the most popular alloys.

Centrifugally cast steel tubes are used as propeller shafts for naval and coast guard vessels. Hollow shafts, which are stronger than solid shafts of the same weight, have been used successfully and have started a new trend in propeller-shaft design.

In the De Lavaud method, gray and ductile iron is centrifugally cast in water-cooled metal molds to produce high-grade cast-iron pipe; 3- to 48-in. diameter pipe in standard lengths up to 20 ft is easily cast. Sand cores are used to form the bell end of the pipe. The pipe is poured from a long-lipped ladle that moves along the axis of rotation at a given rate while the pipe spins. The pouring temperature and fluidity of the iron are very critical.

Several different problems have arisen in the production of centrifugally cast shapes, but solutions have been found. One of these problems was shrinking and cracking from pouring directly into bare metal molds. The resultant chilling and contraction of the outer surface caused cracks and tears to appear. This was cured by spraying the mold cavity with a refractory coating that retarded the heat transfer to the mold and allowed a uniform outer wall to form. Heat transfer to the inner air space was also encountered. This caused a thin wall to form on the inside diameter, with the result that feeding of the outer section caused shrinkage within the casting itself. To overcome this it was necessary to minimize the heat transfer to the inner air pocket. This was accomplished by using an exothermic and insulating material of a lower specific gravity than the metal, which stopped the heat transfer and permitted the metal to solidify at 100 percent density.

One of the most interesting and significant applications of centrifugal casting is the production of dual-metal tubes that combine the properties of two metals in one application. One example of this is the production of dual-metal cylinder liners for diesel engines, with a mild steel outer shell for strength and gray iron on the inside for wear resistance.

In producing these dual-metal cylinders, the workman prepares each metal in a separate furnace. The metal that is to form the outer wall is poured first and allowed to freeze partially; then a flux is added; and finally the inside metal is poured. This process must be properly timed if the metals are to fuse properly at the line between the two metals. With proper timing, this bond is as strong as the metals themselves because all of the basic requirements of a weld are met.

The following data are a summary of process-design considerations regarding centrifugal casting.

Dimensions handled. Length, up to 347 in.; diameters, 2 to 50 in.

Tolerances. $\frac{1}{64}$-in. minimum—about the same as other casting techniques.

Surfaces. Same as static casting processes.

Materials handled. Steel (carbon steel, alloy steel, stainless steel), copper and its alloys, aluminum alloys, aluminum and manganese bronzes, Monel, Everdur, and plain and alloy cast irons.

Economical quantities. Foundry heats as low as 2000 lb can be economically cast.

Supplementary operation. Metal molds sprayed with refractory material.

Design factors. Odd sizes and uncommon alloys can be economically used.

Process control. Heat transfer to walls of metal molds and to the center air space must be controlled.

Slush and Pressed Castings

Slush casting, a form of permanent-mold casting, is limited to some tin-, zinc-, or lead-based alloys. The process involves filling the metal molds with molten metal. Then, after a brief solidification period, the mold is turned over and the liquid metal in the center is allowed to run out. The casting is in effect a shell, the outside having the same appearance as the mold contour and the inside being an irregular cavity. The thickness of the casting can be controlled by the length of the solidifying period before dump-out.

Another minor variation of permanent-mold castings is Corthias or pressed casting. In this procedure, a core is inserted into the partially filled permanent mold causing the fluid metal to fill the remaining space. Corthias castings are also shell-like but with a controlled inside cavity.

These types of castings find application in the production of such products as toys, ornaments, and lighting fixtures, where strength is not of prime importance and good appearance is an absolute necessity. Considerable quantities of slush and pressed castings are made in Britannia metal, pewter, and zinc.

Continuous Castings

Savings in equipment, fuel, and processing time have attended the development of continuous casting for metals. The process is different from extrusion in that the raw materials are melted and transferred to equipment that transforms the molten material directly into semifinished mill products. This eliminates soaking pits, blooming mills, and reheating cycles before the finishing operations for mill products. The process eliminates the casting of ingots that go through a steel mill and reduces floor space and maintenance costs.

Material cast continuously has uniform grain structure and composition, and little slag, impurities, or porosity; is more suitable for future processing operations than noncontinuously cast material. On nonferrous castings, close tolerances can be met, and different shapes and sizes can be made with moderate changeover expense.

The process (Fig. 8.26) involves a source of molten alloy, which is prepared in batches, poured into a holding reservoir, kept at the proper temperature, and controlled to remove impurities and prevent contamination. From the reservoir, the metal is continuously fed through launders into the water-cooled mold, which removes heat rapidly and solidifies the molten metal. The casting exit is regulated by rollers, which travel at the proper speed and contact the cooled alloy. The alloy is water-cooled in the mold for only a short distance (on the order of 1 ft) because contraction prevents contact. The internal portion is still molten and, as the material contracts, the molten metal feeds down from the top. After passing through the die, the metal is cooled by direct-contact sprays or muffled, as the requirements dictate. The material is then cut to suitable lengths by saws or burning equipment that travel at the speed of the casting. These lengths are processed in mills for various applications. Alternatively, large-diameter castings may be made 12 or 20 ft long. As soon as the proper length is reached, pouring stops, the hot casting is removed, and another pour is begun.

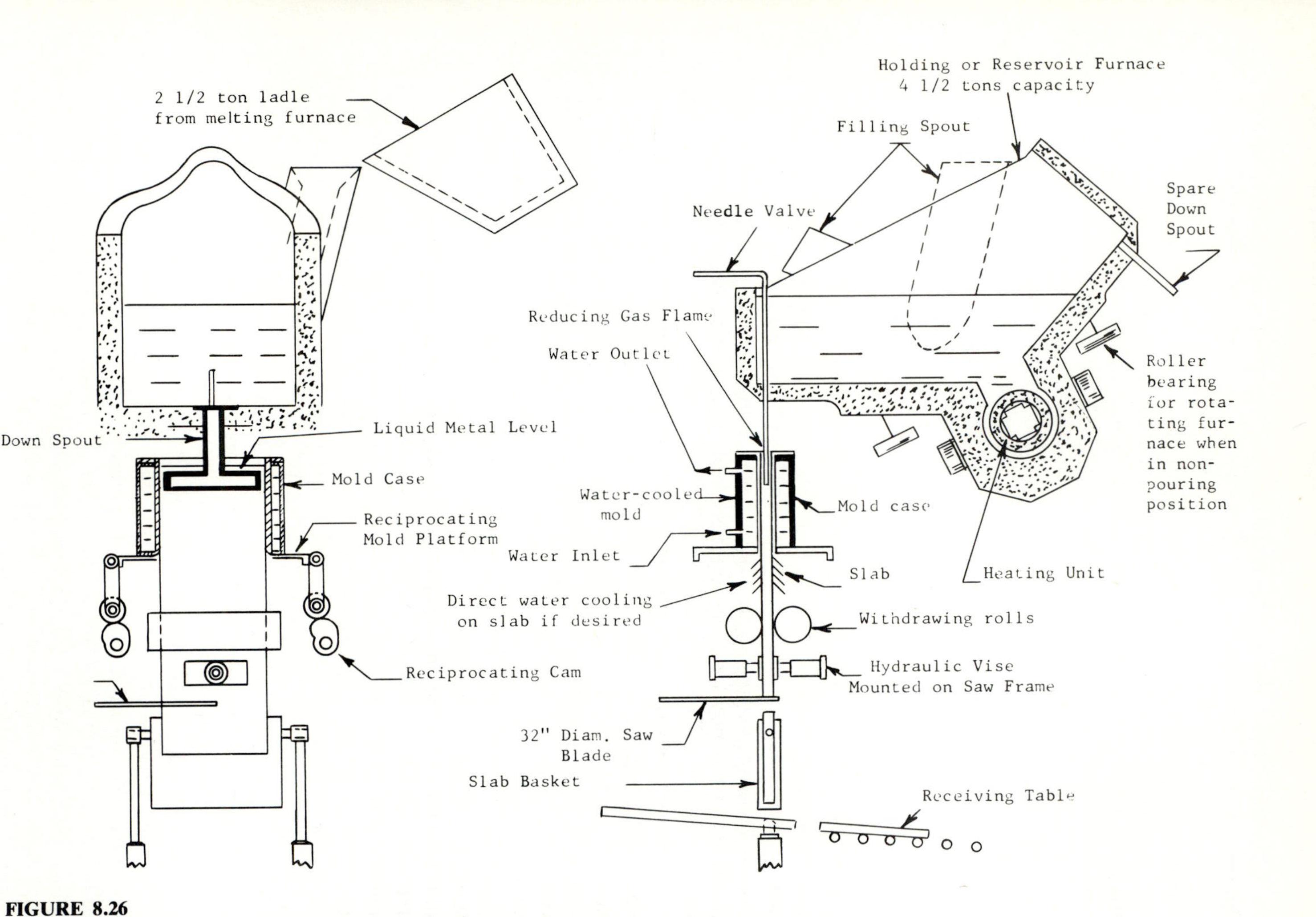

FIGURE 8.26
Sketch of a continuous-slab casting unit, illustrating particularly the twin-spout arrangement of the holding furnace.

Nonferrous equipment can make tubes and other shapes in a manner similar to the extrusion process.

Steel is much more difficult to handle because of its high melting temperatures and the resulting problem of heat transfer at the mold. Mold sections are very thin and are made of steel, copper, and brass. Large quantities of water are forced around the mold to cool it and the molten material.

Since this is a basic process, the production-design engineer would not use this method in normal manufacture unless it was in connection with designing equipment for the continuous-casting processes. A description has been given in order to point out how ideas of long standing are being developed into practical processes by engineers.

Vacuum Casting of Metals

The need for metals and alloys that are free from contaminants has led to the development of furnaces for melting, reducing, and pouring metals under a vacuum at low pressures. Typical vacuum-melting equipment operates at temperatures up to 3000°F and under 10 μm pressure. Undesirable gases are abstracted from the molten metal, and little contamination can take place in view of the controlled environment.

Vacuum casting has progressed from the small electric furnace in the laboratory to equipment that can pour the largest ingots made. In the typical vacuum-molding installation, an electric furnace melts the alloys (usually through induction heating) in a vacuum furnace. The ingot mold is poured within the vacuum furnace, and the vacuum is maintained until the ingot is completely solid.

Turbine spindles are produced today from vacuum-poured steel ingots. The steel ingots are forged to shape and the resulting forgings are more uniform and free from occlusions. Metals such as titanium, aluminum, and steel are affected by the oxygen, hydrogen, and nitrogen of the air, and titanium, for example, can be cast only in a vacuum in view of its great affinity for nitrogen and other elements.

SUMMARY

The casting processes provide a versatility and flexibility that is second to no other broad process classification. In small and medium-sized production runs, sand casting with expendable patterns represents the most economical method for producing a vast variety of parts. Intricate designs can frequently be cast, at an overall savings over other methods, using the investment-casting technique. Likewise, each of the other casting processes offers the production-design engineer definite advantages under certain conditions.

Table 8.5 indicates some of the principal characteristics of the various casting processes.

TABLE 8.5
Common factors in casting processes

			Feed type						Surface		Cores				Castings								
Process	**Permanent molds**	**Expendable molds**	**Gravity**	**Vacuum**	**Air pressure**	**Plunger**	**Centrifugal**	**Squeeze of mold**	**Rough**	**Smooth**	**Expendable**	**Metal**	**Mechanical removal**	**Mechanical removal**	**Annealing**	**Shake-out**	**Expendable inserts**	**Permanent inserts**	**Chills**	**Risers**	**Draft**	**Expendable patterns**	**Permanent patterns**
Investment (high-temperature alloys)		×	×	×			×			×	×				×	×	×		×	×	×	×	
Steel sand casting		×	×						×		×				×	×	×		×	×	×	×	×
Carbon-iron sand casting		×	×						×		×				×	×	×		×	×	×		×
Nonferrous sand casting		×	×						×		×					×	×	×	×	×	×		×
Plaster casting (nonferrous only)		×	×							×	×					×	×	×	×	×	×		×
Shell mold		×	×							×	×				×	×	×		×	×	×		×
Permanent molding	×		×				×			×	×	×						×		×	×		
Slush and Corthias	×		×							×											×		
Dies casting	×									×		×	×	×				×			×		
Bronze	×					×				×		×	×	×				×			×		
Aluminum	×				×					×		×	×	×				×			×		
Zinc alloys	×				×					×		×	×	×				×			×		
Plastic casting	×																						
Compression	×							×		×		×	×	×				×			×		
Transfer	×					×		×		×		×	×	×				×			×		
Rubber	×							×		×		×	×	×				×			×		
Ceramic	×							×		×		×	×								×		
Continuous	×		×							×					×								
Centrifugal	×	×	×				×		×						×	×							×

In spite of competition from welded structures, sheet-metal parts, and forgings, the foundry and molding industries have expanded. Larger equipment has been developed. The frame of the automobile door has been die-cast. These same industries cast many smaller items in large quantities—items as small as the die-cast lugs used in zippers for clothing. All of the casting industries have become highly mechanized, especially the foundries. It is difficult to get skilled labor because of relatively poor working conditions; therefore, dust removal, material handling, and reduction of heavy labor receive considerable attention from engineers and management.

Each new material (alloy, plastic, ceramic) opens new opportunities for developing processes and making parts at lower cost. As accuracy and quality of castings are improved, the amount of machining is reduced. Castings can join several parts into one, and as mechanization of processes improves, more parts will be cast.

QUESTIONS

8.1. How do casting and molding processes differ from forging or extrusion processes? (See Chapter 10.)

8.2. How long have metallic castings been produced? Give a brief history of the art of metal casting.

8.3. What is the impact of the American scrap metal industry on the ferrous cupola operation?

8.4. Today what approximate percentage of copper is recycled?

8.5. What is the range in size of foundry sands?

8.6. When would it be advisable to specify green-sand cores instead of dry-sand cores?

8.7. What are the principal mineral constituents of Olivine sand?

8.8. What sand has the best refractory properties?

8.9. Describe the characteristics of western bentonite?

8.10. What type of sand binders will be used extensively in the year 2000?

8.11. What was the major core-sand binder up until 1950? What are the major sand binders in use today?

8.12. Differentiate between the mechanisms of addition polymerization and condensation polymerization.

8.13. Upon what does the accuracy of any casting depend?

8.14. What are the principal metal casting processes?

8.15. In what quantity are sand castings usually advantageous?

8.16. What are the four types of no-bake resins? Make a comparison of polyurethane- and furan-bonded molds? What are the major advantages of phosphate bonding in connection with core binders for cast aluminum alloys?

8.17. What advantages are characteristic of high-pressure molding? What is the magnitude of compaction force at the parting line of this method?

8.18. Give some examples of parts made by the pit-molding process.

8.19. For what type of castings and in what quantities would you recommend shell molding?

8.20. Explain how the vacuum-molding process works.

8.21. Why should the sprue be perforated in connection with the "V" process?

8.22. What tolerances can be achieved in flaskless molding?

8.23. What is the future of the expanded-polystyrene process?

8.24. Why is the plaster-mold process not adaptable to ferrous materials?

8.25. Why can investment casting be produced to close tolerances?

8.26. When would you advocate the investment-casting process?

8.27. What economies are gained by the shell-molding casting process?

8.28. Explain the theory of centrifugal casting.

8.29. Why is slush casting limited to the alloys of lower melting temperature.

PROBLEMS

8.1. A foundry is producing gray-iron blocks $12 \times 6 \times 4$ in. The parting line is at the midpoint of the height of the block, so that there the mold cavity extends 2 in. into both the cope and the drag. The flasks are $20 \times 12 \times 5$ in. over 5 in., so the combined height of the cope and drag is 10 in. The mold is poured with molten gray iron that weighs 0.22 lb/in.^{3} and the compacted sand weighs 100 lb/ft^3. What would be the total lifting force in pounds tending to separate the cope from the drag as a result of the metallostatic pressure within the mold?

8.2. The volume of a sand core is 100 in^3. Find the buoyant force on the core if it is poured with the following alloys: (a) cast iron, (b) cast steel, (c) aluminum. Note that the density of molten metal must be obtained by taking the density at room temperature and correcting for expansion by using the volumetric coefficient of expansion applied to the solid. The error in neglecting the liquid expansion is not large because most foundry alloys are poured at only 100 to 200°F superheat.

REFERENCES

1. American Foundrymen's Society: *Metal Casting and Molding Processes,* DesPlaines, Ill., 1981.
2. American Society for Metals: *Metals Handbook,* 8th ed., vol. 5, Forging and Casting, Metals Park, Ohio, 1970.
3. Dantzig, J. A., and Berry, J. T. (eds.): *Modeling of Casting and Welding Processes,* II, Metallurgical Society of AIME, Warrendale, Pa., 1984.
4. Flemings, M. C.: *Solidification Processing,* McGraw-Hill, New York, 1974.
5. Gray and Ductile Iron Founders' Society: *Gray and Ductile Iron Castings Handbook,* Rocky River, Ohio, 1971.
6. Heine, R. W., Loper, C. R., and Rosenthal, C.: *Principles of Metal Casting,* 2d ed., McGraw-Hill, New York, 1967.
7. Malleable Founders' Society: *Malleable Iron Castings,* Rocky River, Ohio, 1960.
8. Minkoff, I.: *The Physical Metallurgy of Cast Iron,* Wiley, New York, 1983.
9. Steel Founders' Society of America: *Steel Castings Handbook,* 5th ed., Rocky River, Ohio, 1980.
10. Walton, C. F., and Opar, T. J. (eds.): *Iron Castings Handbook,* 3d ed., Iron Casting Society, DesPlaines, Ill., 1981.

CHAPTER 9

DESIGN FOR CASTING

The present-day engineer is indeed fortunate, for, according to Dr. M. C. Flemings of MIT, "As the secrets of alloy solidification unfold, our most versatile metal-working process comes of age. The fruits are vital to engineer and artist alike—castings of unparallleled quality and economy." Premium-quality castings are in routine production in aluminum foundries and the development of ferrous die casting may herald cast steels of untold strength and economy. But major breakthroughs in die materials, die coatings, and metal input are still needed to make the process a competitive success.

Yet, to be effective and economic, a cast component must be especially designed for the casting process or converted to a casting design. In several cases, quality products made by other processes have been successfully converted to the casting process. Some said that it couldn't be done! But engineers from General Motors did it. Cast crankshafts for automobiles and trucks are here to stay, as are steering knuckles and universal joints. In the late 1940s, no engineer would consider casting such parts even with a 4:1 factor of safety. Today, engineers often think of castings first, even for dynamically stressed parts. Careful engineering design and testing are required for successful conversion to the casting process.

Designers must realize that cost estimates are based on the normal production of good castings. It a casting design leads to greater than normal scrap rates, for example, the original estimate will be invalid. To avoid that pitfall, the engineer must be aware of casting theory, design, and practice.

Since all castings begin with patterns and cores, we should logically look at them first.

PATTERNS

A pattern forms a mold cavity such that after the cast metal reaches room temperature the product will be the expected size. Shrinkage allowances (Tables 9.1 and 9.2) are at best approximations; the size and shape of a casting and the mold wall's resistance to deformation by the liquid metal affect the shrinkage allowance significantly. For example, a round steel bar shrank $\frac{9}{32}$ in./ft; with a large knob on each end, it shrank $\frac{3}{16}$ in./ft; when the knobs were replaced by flanges, it shrank $\frac{7}{64}$ in./ft. Thus, if tolerances are important, an engineer must work closely with qualified foundrymen. In addition, the geometry of a casting may cause it to shrink differently from the allowances in the tables. For example, in a large bull gear the shrinkage across the diameter will be very different from the shrinkage across the rim.

TABLE 9.1
Shrinkage, machining, outside-dimension allowance, minimum section thickness, and typical tolerances for sand-cast ferrous alloys

Pattern dimension, in.	Shrinkage, in.		Machining allowance,[a] in.			Minimum section size, in	Typical tolerance, in
	Solid	Cored	Bore	Outside dimension	Cope side		
			Gray iron				
Up to 6	$\frac{1}{8}$	$\frac{1}{8}$	$\frac{1}{8}$	$\frac{3}{32}$	$\frac{3}{16}$	$\frac{1}{8}$	$\pm\frac{1}{32}$
6–12	$\frac{1}{8}$	$\frac{1}{8}$	$\frac{1}{8}$	$\frac{1}{8}$	$\frac{1}{4}$	—	$\pm\frac{1}{16}$
13–24	$\frac{1}{8}$	$\frac{1}{8}$	$\frac{3}{16}$	$\frac{5}{32}$	$\frac{1}{4}$		
25–36	$\frac{1}{10}$	$\frac{1}{10}$	$\frac{1}{4}$	$\frac{3}{16}$	$\frac{1}{4}$		
37–48	$\frac{1}{10}$	$\frac{1}{12}$	$\frac{5}{16}$	$\frac{1}{4}$	$\frac{5}{16}$		
49–60	$\frac{1}{12}$	$\frac{1}{12}$	$\frac{5}{16}$	$\frac{1}{4}$	$\frac{5}{16}$		
61–80	$\frac{1}{12}$	$\frac{1}{12}$	$\frac{3}{8}$	$\frac{5}{16}$	$\frac{3}{8}$		
81–120	$\frac{1}{12}$	$\frac{1}{12}$	$\frac{7}{16}$	$\frac{3}{8}$	$\frac{7}{16}$		
			Cast steel				
Up to 1	—	—	Cast solid	—	—	$\frac{3}{16}$	
Up to 6	$\frac{1}{4}$	$\frac{1}{4}$	$\frac{1}{4}$	$\frac{1}{8}$	$\frac{1}{4}$	—	$\frac{1}{4}$
6–12	$\frac{1}{4}$	$\frac{1}{4}$	$\frac{1}{4}$	$\frac{3}{16}$	$\frac{1}{4}$	—	$\frac{1}{4}$
13–18	$\frac{1}{4}$	$\frac{1}{4}$	$\frac{9}{32}$	$\frac{1}{4}$	$\frac{5}{16}$	—	$\frac{1}{4}$
19–24	$\frac{1}{4}$	$\frac{3}{16}$	$\frac{9}{32}$	$\frac{5}{16}$	$\frac{3}{8}$	—	$\frac{5}{16}$
25–48	$\frac{3}{16}$	$\frac{3}{16}$	$\frac{5}{16}$	$\frac{3}{8}$	$\frac{1}{2}$	—	$\frac{3}{8}$
49–66	$\frac{3}{16}$	$\frac{5}{32}$	$\frac{3}{8}$	$\frac{3}{8}$	$\frac{1}{2}$	—	$\frac{1}{2}$
67–72	$\frac{3}{16}$	$\frac{1}{8}$	$\frac{1}{2}$	$\frac{7}{16}$	$\frac{9}{16}$	—	$\frac{5}{8}$
Over 72	$\frac{5}{32}$	$\frac{1}{8}$	$\frac{5}{8}$	$\frac{1}{2}$	$\frac{5}{8}$	—	$\frac{3}{4}$

TABLE 9.1 (*continued*)

			Ductile iron		
Up to 24	$\frac{1}{10}-\frac{1}{8}$	$\frac{1}{10}$	$\frac{3}{16}$	$\frac{5}{32}$	$\frac{1}{4}$
			Malleable iron		
Section thickness, in.	**Shrinkage,[a] in.**	**Section thickness, in.**	**Shrinkage,[b] in.**	**Minimum section size, in.**	**Typical tolerance, in.**
$\frac{1}{16}$	$\frac{11}{64}$	$\frac{1}{2}$	$\frac{7}{64}$	$\frac{1}{16}$	Up to 5 $\pm\frac{1}{32}$
$\frac{1}{8}$	$\frac{3}{32}$	$\frac{5}{8}$	$\frac{3}{32}$	—	5–8 $\pm\frac{3}{64}$
$\frac{3}{16}$	$\frac{19}{128}$	$\frac{3}{4}$	$\frac{5}{64}$	—	9–12 $\pm\frac{1}{16}$
$\frac{1}{4}$	$\frac{9}{64}$	$\frac{7}{8}$	$\frac{3}{64}$	—	13–24 $\pm\frac{1}{8}$
$\frac{3}{8}$	$\frac{1}{8}$	1	$\frac{1}{32}$		

[a] Allowance on bore given for radius.

[b] No core.

Material for patterns must resist the moisture and abrasion of green sand. Which material is chosen depends on the quantity of castings to be made and on the casting process to be used. The most common pattern materials are wood and metals, but epoxy resins and plaster are used on occasion.

Softwood (white pine) patterns are suitable for short runs (1 to 50 pieces) of medium castings (up to 6 ft long) but they wear rapidly; hardwoods (mahogany, cherry, birch) are used for runs of 50 to 200 pieces. At the latter

TABLE 9.2
Shrinkage and machining allowances for nonferrous castings poured in sand molds

	Pattern dimensions, in.	Section thickness in.	Shrinkage allowances, in.	Machining allowances, in.	Minimum section size, in.	Typical tolerance, in.
Aluminum	Up to 24	$\frac{5}{32}$	$\frac{5}{32}$	$\frac{3}{32}$	$\frac{3}{16}$	$\pm\frac{1}{32}$
	25–48	$\frac{5}{32}$	$\frac{9}{64}-\frac{1}{8}$	$\frac{1}{8}$		
	49–72	$\frac{9}{64}$	$\frac{1}{8}-\frac{3}{16}$	$\frac{1}{8}$		
	Over 72	$\frac{1}{8}$	$\frac{1}{8}-\frac{3}{16}$			
Magnesium	Up to 24	$\frac{11}{32}$	$\frac{5}{32}$	$\frac{3}{32}$	$\frac{5}{32}$	$\pm\frac{1}{32}$
	25–48	$\frac{11}{32}$	$\frac{5}{32}-\frac{3}{16}$	$\frac{1}{8}$	—	$\pm\frac{1}{32}$
	Over 48	$\frac{5}{32}$	$\frac{5}{32}-\frac{3}{16}$	$\frac{1}{4}-\frac{3}{8}$	—	$\pm\frac{1}{32}$
Admiralty metal	Up to 24	—	$\frac{1}{8}$			
Copper	—	—	$\frac{3}{16}-\frac{7}{32}$	—		
Brass	—	—	$\frac{3}{16}$	—	$\frac{3}{32}$	$\pm\frac{3}{32}$
Bronze	—	—	$\frac{1}{8}-\frac{1}{4}$	—	$\frac{3}{32}$	$\pm\frac{3}{32}$
Beryllium–copper	—	—	$\frac{1}{8}$		—	$\pm\frac{1}{16}$
Everdur	—	—	$\frac{3}{8}-\frac{1}{4}$			
Harteloy	—	—	$\frac{1}{4}$			
Nickel and nickel alloys	—	—	$\frac{1}{4}$			

production requirement, gates should be attached to the pattern; flasks and core boxes should have metal wear strips. In some cases, additional equipment may be needed (e.g., core setting and pasting fixtures, dryer patterns, or core dryers). At still higher production requirements (200 to 5000 pieces), patterns and accessories must be metal with hardened-steel wear plates, and core boxes should be constructed with venting devices to permit blowing.

To decide what pattern material to use, it is necessary to make an engineeering cost estimate considering the average annual investment in tooling, the annual maintenance cost, the production usage, and the risk of obsolescence. Wood is inexpensive and easily worked; therefore, it is commonly used for low-production needs and large patterns. But even when the wood is painted, humidity causes wood patterns to vary in size and shape, and they are easily damaged on the molding floor or on their way to and from storage. Metal patterns are more accurate and durable. They are used for all match plates (aluminum for manual operations and gray iron or brass for machine-lifted plates).

The following factors must be considered in designing a pattern.

1. Have gating and risering been provided?
2. Can the pattern be removed from the mold?
3. Has the minimum castable section thickness been considered?
4. Is there proper draft to permit the removal of the pattern?
5. Will draft be permissible on the final casting?
6. Can cores be anchored?
7. Can loose pieces, multiple partings, and irregular partings be eliminated?
8. Is size of pattern adjusted to allow for shrinkage of casting and expansion of mold?
9. Has machining allowance been provided?
10. Has a warping or distortion allowance been provided?
11. Has the foundry been consulted?

Classification of Patterns

Single patterns. Single patterns are usually solid with no gates attached. They are single copies of the casting to be produced with adequate allowances for making the casting. A skilled molder is required to gate them properly and determine a parting line (Fig. 9.1*b*). They are used mostly in casting limited quantities of a part.

Gated patterns. Gated patterns are one or more single patterns fastened to a gate (Fig. 9.1*c*). They require the same skill as the single pattern in providing a parting line; however, the hand cutting of the gates is eliminated. They are used when the production requirements are medium.

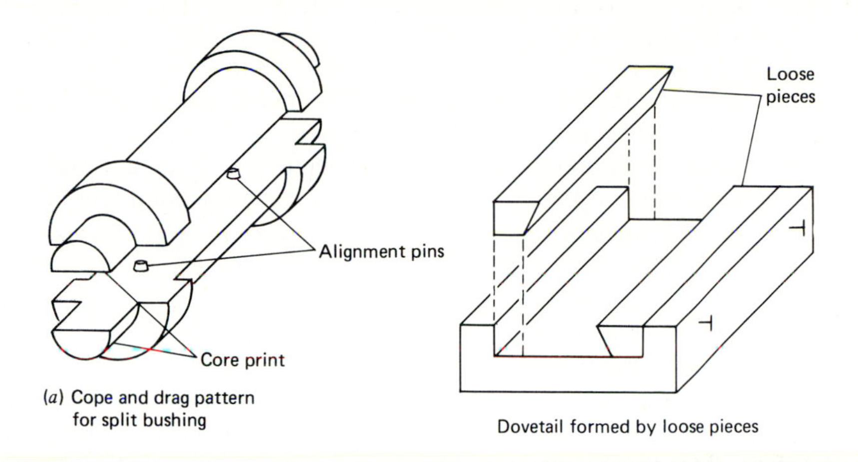

(a) Cope and drag pattern for split bushing

Dovetail formed by loose pieces

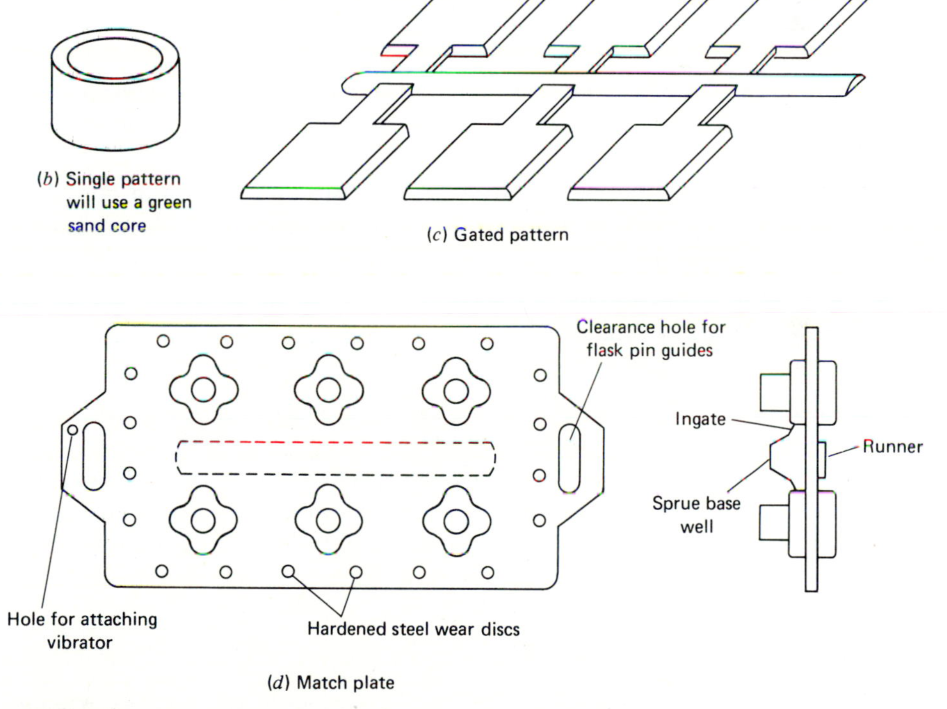

(b) Single pattern will use a green sand core

(c) Gated pattern

(d) Match plate

FIGURE 9.1
Typical small pattern equipment.

Match-plate patterns. Match-plate patterns are mounted on a board or plate with gates and runners provided (Fig. 9.1*d*). The pattern may be split along the parting line formed by the board. The cope pattern is on one side and the drag pattern is on the opposite side. Match-plate patterns are economical only for quantity production, handled by unskilled operators on rollover, jolt, and squeeze-type molding equipment. In very high production or for large castings there may be two plates—one for the cope and one for the drag—which are made at two separate places on the production line.

Cope and drag patterns. Cope and drag patterns are of great variety (Fig. 9.1*a*). They are made so they will split and can be drawn from the cope and drag. Most patterns are of this type and are used for the largest castings. At times the cope or drag may be separated into one or more parts and the patterns are divided accordingly.

Temporary and permanent patterns. Patterns or core boxes are called "temporary" or "permanent," depending upon the material used in their construction. Temporary patterns, of soft wood, are easily made. They soon wear, warp, or crack, and have a short life. Hardwood patterns and core boxes are used more than any other type of pattern. The portions that wear may be protected by sheet metal. Permanent patterns are made of metals—usually aluminum or brass—or plastics that are easily cast and machined.

Pattern colors. In 1958, a tentative pattern color code for new patterns was adopted by the Pattern Division of the American Foundrymen's Society:

1. Unfinished casting surfaces, the face of core boxes and pattern or core-box parting faces: clear coating.
2. Machined surfaces: red.
3. Seats of and for loose pieces: aluminum.
4. Core points: black. In the case of split patterns and where cores are used, paint the core area black.
5. Stop-offs:[1] green.

The clear coating on most surfaces will disclose the quality of material and workmanship of the pattern. Likewise, important construction details and layout or centerlines used by the patternmaker will not be obliterated and will thus be available for use during repair or modification of the equipment.

[1] Stop-offs are portions of a pattern that form a cavity in a mold that is refilled with sand before pouring. This might be desirable to prevent breakage of a frail pattern member. The stop-off may be filled by a core later. This procedure may simplify the location of the pattern parting line by making it unnecessary to carry it out of a plane position (to provide for lugs or certain cored holes).

There are numerous types of pattern-coating materials on the market. Nitrocellulose lacquers cure rapidly but have relatively poor moisture resistance. Present-day modified lacquers have much better moisture resistance. Shellac modified with various synthetic resins is superior to pure shellac. Wood patterns coated with synthetic resin (plastic) have been used under the most adverse conditions, such as being rammed up with hot sand or left in the mold for many hours, without noticeable damage. Plastic coatings require considerable time and care for application, but prolong the pattern life up to several times. In addition, there is less tendency for the sand to cling to the pattern surface, and so such patterns draw more easily.

DRAFT ALLOWANCE

Once a mold has been rammed around a pattern, the latter must be removed from the mold cavity. This is aided by rapping or vibrating the pattern and by a taper called *draft* on all vertical surfaces of the pattern. Usually, draft will vary between $\frac{1}{16}$ to $\frac{1}{8}$ in./ft (1 to $1\frac{1}{2}^\circ$). Inside surfaces such as cored holes require greater draft than outside surfaces, and manual molding equipment needs greater draft than that required if mechanical drawing equipment is used.

SHRINKAGE AND MACHINING ALLOWANCE

Solidification shrinkage must be compensated by the riser, but shrinkage that occurs during the cooling of a casting from the solidification temperature to room temperature is compensated for by making the pattern oversize. The allowances given in Tables 9.1 and 9.2 are general averages. Closer dimensional tolerances can be achieved in green-sand molding provided that there is sufficient production to warrant pattern modification and that a controlled foundry practice is followed. Machining allowances also can be reduced if, for instance, high-pressure molding is used with good sand control or if shell molding is used in the foundry production. For instance, automotive foundries consistently use smaller-allowance figures than those published here. This is where a detailed knowledge of the practice of a particular plant is important because the saving in machining can be significant.

VENTING OF MOLDS

Molds have other functions besides forming the cast material. They provide passages or gates through which the metal flows into the casting cavity. In steel castings the mold must be designed with parting line, gates, and vents, so that the molten metal can enter at the bottom, without turbulence and without creating hot spots in the mold by impinging on certain points. In sand molds aluminum is directed through traps or strainer cores or glass-fiber screens that help eliminate sand and other foreign material that may have been picked up by the molten metal.

The molten metal generates steam and gases as it comes into contact with the molding material. These gases must escape and cannot be confined in pockets. In synthetic sand molds there is usually sufficient permeability in the rammed mold to vent any trapped gases. However, if the mold has been compacted by a high pressure molding machine, it may require the drilling of additional vents. Vents should be so arranged that gas in the mold, as it is being pushed ahead of the gradually rising level of molten metal, is not trapped in crevices or other branches not directly in line with risers or overflows. If these gases did not escape, the casting would be defective and the mold might explode. The molds are vented by small holes through the mold material. Definite vent passages are provided in metal molds and, when required, in large sand molds. Overflow of excess material is provided for by passages to pockets.

CORES

When it is necessary to leave an opening in a casting that cannot be made by the external mold, a core made to the shape of the opening is provided and located in position by anchoring it to the external mold or supporting it by chaplets. Chaplets are fusible metal supports placed between the cores and the mold wall for the purpose of separation. They melt and are absorbed by the liquid metal. If the chaplet is not absorbed, it may cause porosity.

Cores can be made by any of the sand processes described under sand molding, i.e., the traditional oil-bonded cores (usually a linseed-oil derivative), the carbon dioxide, synthetic resin, or inorganic bonded ceramic cores. Die casting and permanent-mold casting dies usually use metallic cores of H13 die steel or a molybdenum alloy such as TZM. Metal cores are actually a part of the die and will not be discussed further.

Sand cores are usually made from specially bonded and cured silica sands rammed into a core box. Although hand-ramming methods are used, blowing of the bonded mix is by far the most common production method. Curing may be by baking, but that method is fast losing out to curing in the core box by gassing or by stripping from the box in a semicured condition and allowing the core to set through the use of catalysts.

Components of cores may be made separately and then pasted together to make complicated cores. The cores used in an automobile engine block for water jackets and gas passages are an example of intricate coring. The cores for a railroad airbrake cylinder and valve are complex and are held to very close tolerances. Each individual core part is molded in a separate core box, baked, put in place, and held there by the core paste.

Cores are used externally to form flat and perpendicular surfaces that will not permit draft, on vertical surfaces to form undercuts or bosses that cannot be drawn, to form surfaces in pit molding where there is no cope to mold the top part of the casting, and to prevent mold erosion at critical points. An example of a completely cored part is the crankshaft casting of an automobile, which is made of pancakelike layers stacked one on top of the other and keyed in place.

Metals are heavy in comparison with nonmetallic mold materials; if they flow too rapidly, they will wash the molds and pick up foreign material. The heavy liquid can float the cores, as well as the cope, as the metal enters the mold. This causes shifts in cores and distortion of the mold. Weights on the molds and clamps are used to counteract this liquid pressure. A feeder 1 ft high for a typical ferrous material will exert 4 $lb/in.^2$ on a core. If the core has a face area of 5×8 in., the force tending to shift it out of place is 160 lb. The heavy, hot fluid material must be guided to its place without damage to the mold or contamination of the material. This mass of hot molten or mushy material has no strength at temperatures above and near the melting point. Therefore, the mold and cores must support the material until it cools to the point where it is strong enough to carry its own weight.

Cores are made on automatic machines, on production lines, where a combination of machine and hand operations produce them in great quantities, and at benchwork stations. In considering cores, it must be remembered that the labor for making a core is almost equivalent to the labor required to make a casting of the same size. They must be formed in a core box, baked or cured (unless a green sand core is used), placed in the mold, and removed after the casting is made. The removal of a core should be provided for in the design of the part. A green-sand core, formed in a similar manner but not baked, is less expensive. Green-sand cores are often not practicable, however, owing to their low strength.

Core-sand bonds disintegrate under heat, and thus the sand can be easily removed after the casting is made.

Cores are often anchored solidly by core prints in the mold so they will not wash out. The correct location of the core depends upon the accurate size of the cores and molds and the care taken by the molder in placing the cores. At times the molder is supplied with measuring gauges to check that the core location is correct.

Cores are expensive and should be kept to the minimum when designing parts. Holes that can be made by a separate core can sometimes be drilled at a cost less than that of the core.

Core-making Equipment

Core boxes contain cavities to form sand cores to the desired shape. Patternmakers construct core boxes from wood, metal, synthetic resin, and other suitable materials. Generally, the core box material would be the same as that chosen for the pattern. There are several types of core boxes and accessories.

1. Two- or three-piece split core boxes (hand- or machine-rammed)
2. Dump boxes for making a half-core (later the two halves are pasted together)

3. Blow boxes for the high production of relatively small cores (a few ounces to 400 lb).
4. Multiple-piece, loose-piece, or special-purpose core boxes
5. Auxiliary equipment such as core dryers and pasting jigs
6. Shell-core boxes, equipped with integral heating devices
7. Hot-box core equipment for curing furan resin-bonded sand

The design and construction of most core boxes and patterns are often left largely to the discretion of the patternmaker. But a tool engineer should understand the principles of core-box design, vent location, vent areas, and the general break-even point at which it is better to move from wooden core boxes to metal core boxes or from hand ramming to core blowing.

Core Blowing

The core blower rapidly produces small and medium-sized cores. It clamps the core box shut, seals the sand reservoir tube to the box, then fills and rams the core by the kinetic energy of a sand-laden air stream. Filling can only be achieved if the sand is free to enter the box through the blow-holes and to exit from the box cavity through suitable vents that will impound the sand (Fig. 9.2). Proper orientation of the blow and vent holes will promote uniform filling and ramming of the core in less than 2 seconds—even for the largest core boxes.

Core blowers require large volumes of compressed air—up to 12 to 30 ft^3/min of free (14.7 lb/in.^{2}) air per cycle. Consequently, high-capacity air compressors are required for large core-blower installations. Supply piping must also be of sufficient diameter to permit large flow (2 to 4 in. diameter).

Suitable air cleaners such as centrifugal filters and oil traps must be provided to eliminate oil and water from the air stream. Also, adequate clamping pressure must be developed to assure tight sealing of the box sections in both horizontal and vertical planes. All joined edges must be machined square and parallel to assume correct seating and to eliminate blow-by. Core boxes are usually aluminum with added wear plates of steel at points of high attrition.

Blowholes

A steel blow plate fits the bottom of the sand reservoir and provides holes from $\frac{3}{16}$ to $\frac{1}{2}$ in. diameter to direct the sand to the proper location within the core box (Fig. 9.2). The number and size of the blowholes are at present largely a matter of experimentation. An insufficient number of blowholes prevents the box from filling completely and promotes channeling. On the other hand, too many holes cause the box to fill before the remote corners are rammed sufficiently hard.

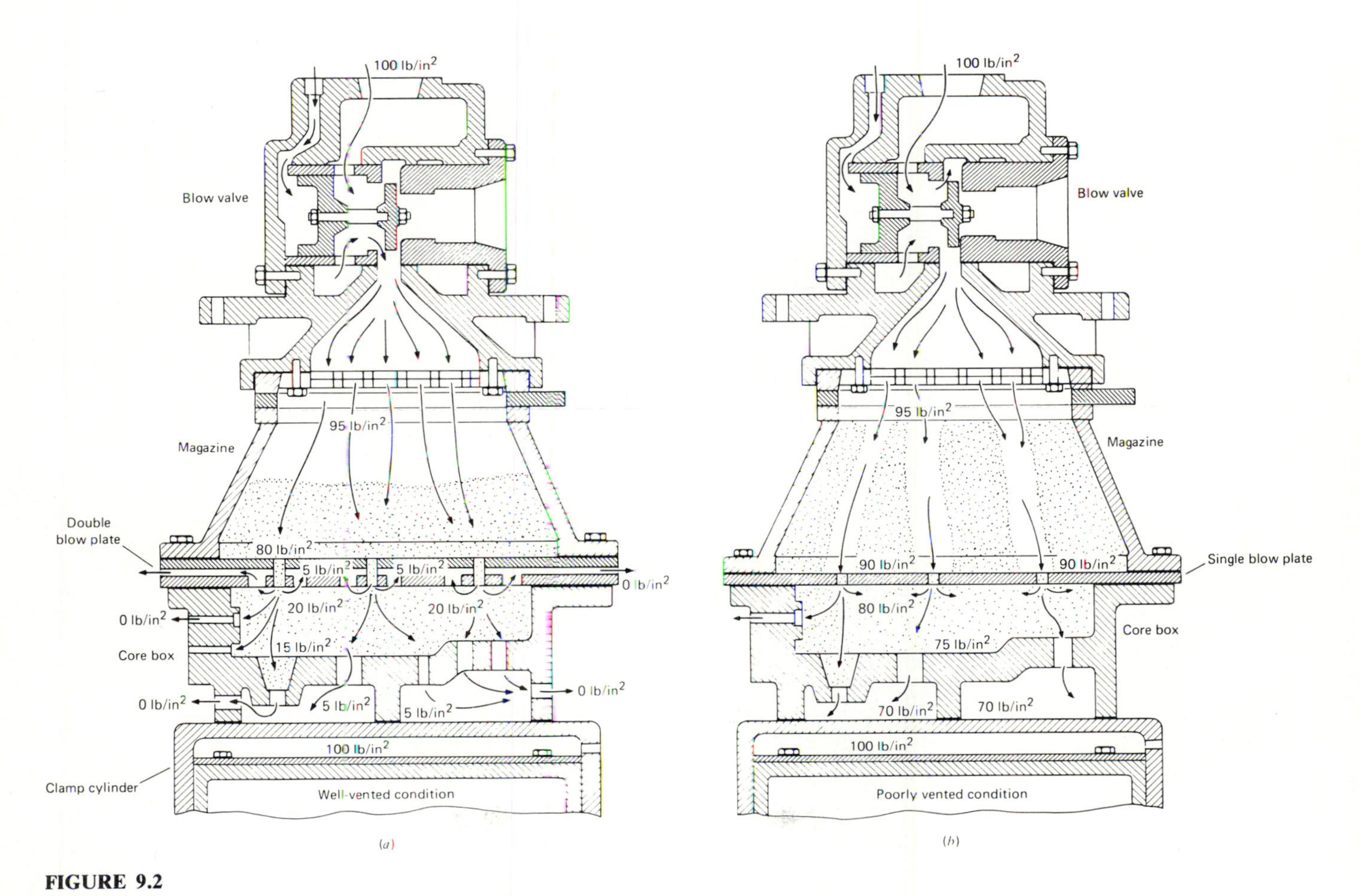

FIGURE 9.2
Pressure distribution in a core-box blowing system. good venting versus poor.

The blow plates and sand reservoirs should be designed to be interchangeable for as many core boxes as possible to obtain maximum returns from the original expense and to minimize storage space.

Core-box Venting

All the air that enters the core box through the blowholes must be vented. The air enters the core box, expands, moves at a lower velocity, deposits the entrained sand, and flows out through the vents. The impact of one grain upon another rams the core.

Venting must be designed in proportion to the blowhole area to achieve adequate ramming. Less venting is required for more flowable sands because a smaller volume of ramming air is required to ram cores to suitable strength.Conversely, more vent area is required for core sands with high green properties because larger volumes of air are needed for equivalent rammed properties.

Venting is obtained by using vent screens or slotted vents (Fig. 9.3). If screens are used, the venting area should be about twice the blowhole area; slotted vent plugs require greater area. Vent plugs do not leave surface imperfections on a core as do screens because the former can be contoured to

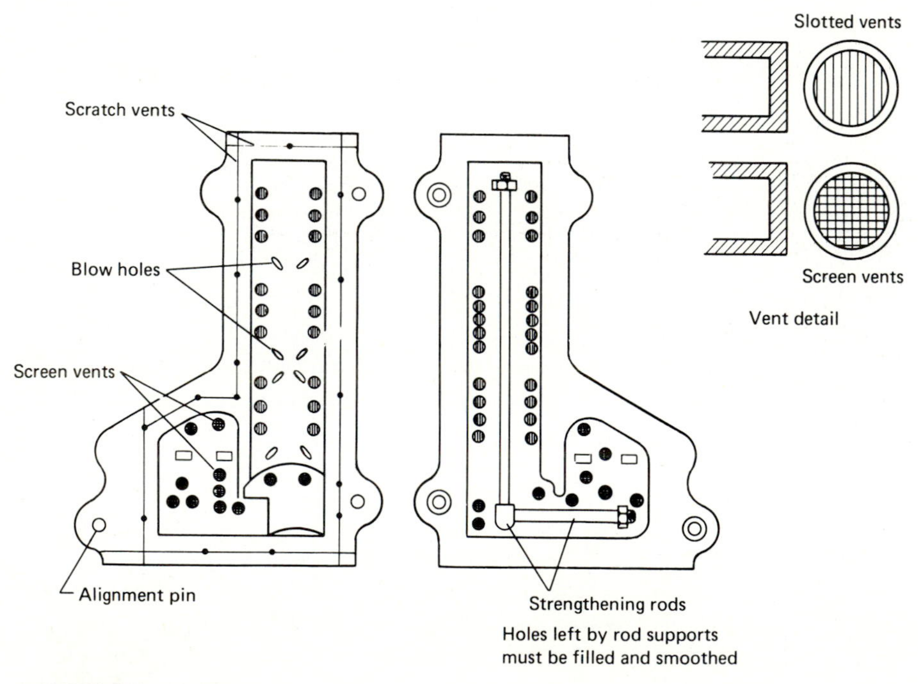

FIGURE 9.3
Vents for core boxes.

the cavity. If surface finish is important, either vent plugs must be used or screens should be placed on core prints and at the other spots where blemishes are not critical.

Venting of Core Gases

Cores must be permeable to permit the hot gases that are generated by the burning of bonding materials to escape to the atmosphere. Such vents are usually aided by placing wires or wax rods in the core box prior to blowing the core. Subsequently, the wires are removed or the wax melts during the core baking. If possible, the vents should pass through the locating prints to permit gases to escape through the back of the core prints.

Core Support

Larger cores require internal reinforcement with embedded wires or rods to prevent breakage caused by handling or premature sagging in the mold after the casting is poured.

Irregularly contoured cores require special core dryers for support during the curing process. In high production the dryer may be the lower half of the core box itself. Such a practice would require a considerable investment in tooling, which would have to be balanced by a cost saving in core production and scrap loss.

INSERTS

Inserts are separate parts made of a metal that is generally different from the metal of the casting, which are "cast in" to provide locally some special properties such as hardness, wear resistance, strength, bearing qualities, electrical characteristics, corrosion resistance, resilience, or special decoration not obtainable from the cast metal. Typical inserts are the heating units cast in aluminum flatirons and hot plates, tubing for the passage of a liquid, bushings, bearings, anchorage for soldering connections, and studs. Inserts are knurled, grooved, or surfaced in such a manner that the material, in freezing around the insert, grips it firmly so that it cannot turn or pull out in tension.

Attention must be paid to the danger of electrolytic or galvanic corrosion when the assembly is exposed to any kind of humidity. The corrosion is caused by the galvanic potential difference between the base metal of the casting and the dissimilar metal of the insert. Aluminum- or magnesium-based alloys in intimate contact with copper-based, tin-based, lead-based, or nickel-based alloys, or steel and iron, are apt to corrode when covered by moisture, especially when the area of the aluminum- or magnesium-based casting surrounding the insert is smaller than that of the inserted metal. The same holds true for aluminum joined to magnesium. When joints of dissimilar metals are exposed to moisture, it is necessary to protect the joints against the entry of moisture by painting, dipping, or plating.

Furthermore, it is advisable in the design of inserts to avoid sharp corners, projections surrounded by thin sections of material, or other factors that might lead to stress concentration with its injurious effect on the mechanical properties of the casting, particularly in fatigue and shock. Yet inserts should have a knurled or grooved surface at the casting-insert interface if a tight mechanical fit is desired.

When inserts are used, the cost of the casting is usually increased because of the cost of the insert and the cost involved in placing it in the mold. Also, allowances must be made for damaged inserts if they are not reclaimable; however, inserts may reduce the overall cost of the product.

GATING SYSTEMS FOR SAND MOLDS

Designs

The gating system for a casting is a series of channels that lead molten metal into the mold cavity. It may include any or all of the following: pouring basin, sprue, sprue base, runners, and ingates. A well designed gating system should:

1. Minimize turbulence within the molten metal as it flows through the gating system. The use of tapered sprues and proper streamlining will reduce excessive erosion and gas entrainment.

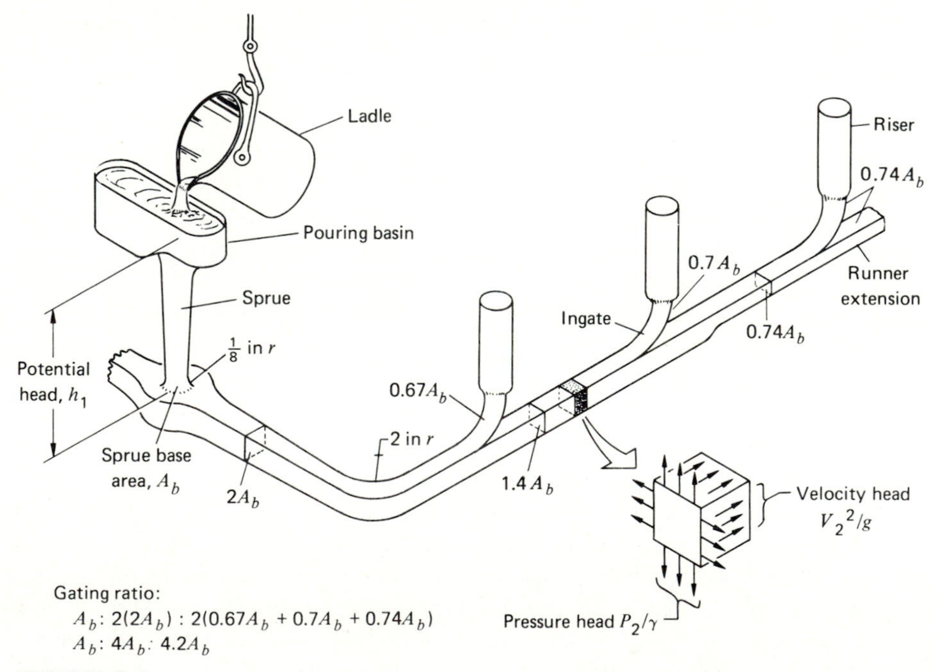

FIGURE 9.4
Design of a nonpressured gating system using a 1:4:4 gating ratio.

2. Reduce the velocity of the molten metal in order to attain minimum turbulence.
3. Deliver the molten metal at the best location to achieve proper directional solidification and optimum feeding of shrinkage cavities.
4. Provide a built-in metering device to permit uniform, standardized pouring times regardless of variations in pouring techniques.

Figure 9.4 is a typical nonpressurized gating system for an aluminum alloy. Note that each gate has a riser.

Turbulence within a Gating System

Extensive research has shown that molten metal and water flow similarly and that gating systems can be designed using the principles of fluid mechanics.

Several limitations are apparent. The high density of metals (up to 10 times that of water) makes it difficult to force molten alloys to turn a corner as from a runner to an ingate. But once Newton's law of inertia is applied to flowing metal, proper gating systems can readily be designed. The density of a metal does not affect its flow characteristics, because the rate and nature of fluid flow depend on the inertia of the fluid and the forces applied to it. Both factors depend on density in the same way, so it has no effect on the fluid flow. But impact depends upon density alone; hence mold erosion increases directly with density.

Although water has a surface tension approximately one-tenth that of molten metals, recent experiments have shown that Wood's metal and mercury have nearly equal to water surface tension and ability to entrain air. However, the greater surface tension and the natural oxide coating that envelopes a stream of molten metal seem to permit it to flow in a nondisruptive fashion at a greater velocity than that suitable for water in a given channel.

Velocities Within a Gating System

The flow of molten metal in a gating system is a function of a number of other variables. Bernoulli's theorem states that the sum of the potential, pressure, kinetic, and friction energies at any point in a flowing liquid is a constant:

$$Wh_1 + \frac{WP_1}{\gamma} + \frac{WV_1^2}{2g} + WF_1 = Wh_2 + \frac{WP_2}{\gamma} + \frac{WV_2^2}{2g} + WF_2 \tag{9.1}$$

Dividing by W the total weight of liquid flow per unit time, one obtains the usual form:

$$h_1 + \frac{P_1}{\gamma} + \frac{V_1^2}{2g} + F_1 = h_2 + \frac{P_2}{\gamma} + \frac{V_2^2}{2g} + F_2 \tag{9.2}$$

where $h_1 h_2$ are the respective heads at stations 1 and 2, in. P_1, P_2 are the respective pressures on liquid, lb/in.^{2} V_1, V_2 are the respective liquid velocities,

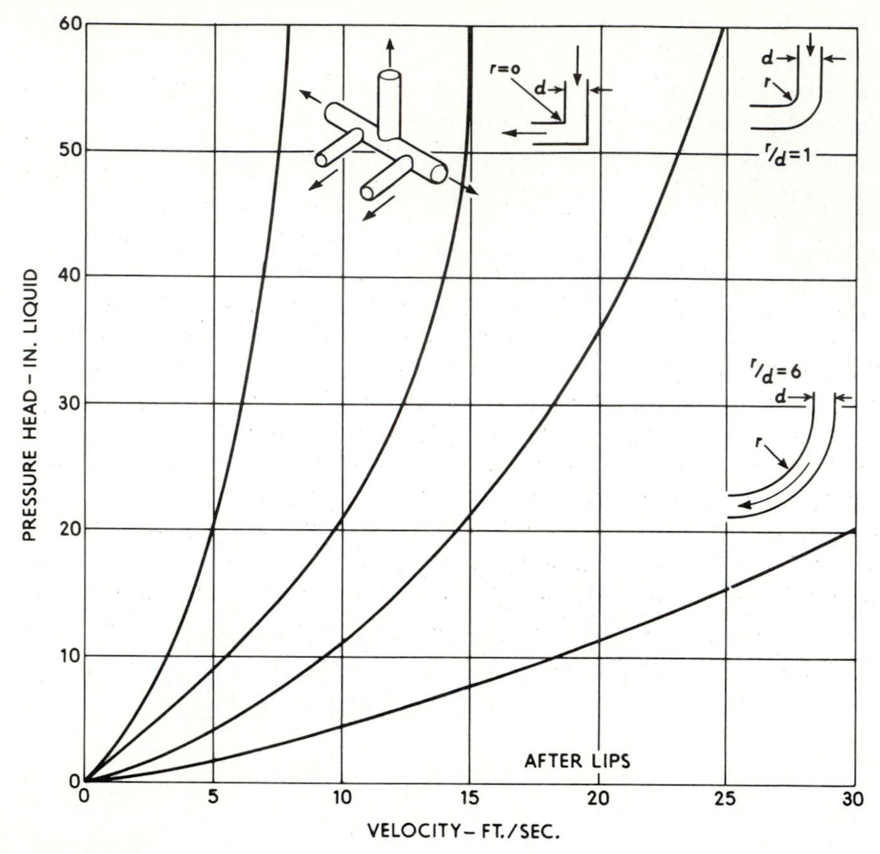

FIGURE 9.5
Effect of streamlining gating on velocity.

in./s. γ is the specific weight of liquid, lb/in.^{3} g is the gravitational constant on Earth, 386 in./s^2 and F_1, F_2 are the respective head losses from friction, in.

The velocity of the molten metal at any point in a gating system can be evaluated by the use of Bernoulli's theorem. Proper streamlining will permit a significant increase in the flow rate of a gating system (Fig. 9.5).

Effect of Streamlining a Gating System

Gates of rounded cross section are more efficient than those of any other shape because they have the smallest surface-area-to-volume ratio and, consequently, can pass a greater volume of metal with the least heat loss. The gating system should be streamlined and of correct magnitude so as to control the velocity of the flowing metal. Too high a velocity will cause disruptive turbulent flow, resulting in sand inclusions and erosion of the mold cavity wall. Streamlining can effectively increase the volumetric capacity of a gating system and thereby allow smaller-size gates and runners that will consequently increase effective melt utilization. The effect of streamlining on metal velocity is shown in Fig. 9.5. A method for improving streamlining at the base of a T section from the sprue to the runner is shown in Fig. 9.4.

The Law of Continuity

A second fundamental relationship in fluid flow is the *law of continuity*, which states that the flow rate of a fluid is a constant at any point in a continuous stream:

$$q = A_1 v_1 = A_2 v_2 \qquad \text{or} \qquad Q = Avt \tag{9.3}$$

where q is the flow rate, in.^{3}/s. Q is the volume of flow in a given time, in.^{3} A_1, A_2 are the respective cross-sectional areas of the flow channel at points 1 and 2, in.^{2} v_1, v_2 are the respective velocities of flow at points 1 and 2, in./s and t is the time, s.

The Vertical Elements of a Gating System

The law of continuity requires that the same quantity (flow rate) of material must exist at all points in a flowing stream. In the vertical part of a gating system (sprue), the acceleration of gravity increases the velocity of flow. If a straight-sided sprue is used, the cross-sectional area of the flowing stream at the sprue base will be less than that of the sprue. Consequently, air will be aspirated from the surrounding mold until the sprue volume is completely filled. However, if a tapered sprue is designed to conform to the dimensions of the descending stream, such a condition can no longer exist and the metal quality will improve.

If we neglect friction and take a horizontal plane through the ingate as a reference, Bernoulli's equation becomes

$$h_t + \frac{P_t}{\gamma} + \frac{v_t^2}{2g} = h_b + \frac{P_b}{\gamma} + \frac{v_b^2}{2g} \tag{9.4}$$

where t refers to the top and b to the base. Then $h_b = 0$ because it is on the reference plane, $v_t = 0$ because there is no velocity at the top, and P_t/γ and $P_b/\gamma = 14.7\ \text{lb/in.}^2$ because the system is at atmospheric pressure at both ends. Then we have

$$h_t = \frac{v_b^2}{2g} \qquad \text{or} \qquad v_b = \sqrt{2gh_t} = 27.8\sqrt{h_t} \tag{9.5}$$

To design a sprue of suitable proportions, let the area at the top of the sprue be A_t and the velocity there be v_t at a flow rate q_t; then $q_t = A_t v_t$ from continuity. Similarly, at the sprue base $q_b = A_b v_b = q_t = A_t v_t$. Then $A_t = A_b(v_b/v_t)$. From Bernoulli's equation, $v_t = \sqrt{2gh_t}$ and $v_b = \sqrt{2gh_b}$, where h_t and h_b are the heads, in inches of metal, at the top and bottom of the sprue.

Then

$$A_t = A_b \frac{\sqrt{2gh_b}}{\sqrt{2gh_t}} \quad \text{or} \quad A_t = A_b \sqrt{\frac{h_b}{h_t}} \tag{9.6}$$

Thus, once the area of the sprue base is known, the vertical gating system can be designed. Equation (9.6) indicates that the sprue should have parabolic sides, but experience shows that a straight-sided sprue having the calculated diameters at the top and bottom is satisfactory (when solved for a series of heights).

When the height h_t of the pouring basin above the sprue base is known, if a tapered sprue is used, and if the pouring basin is kept full throughout the pour, Bernoulli's equation can give an approximate answer, but much of the flow is transient and the cavity must be in the drag and nonpressurized.

To find the diameter of the sprue base, it is convenient to use the result from previous research in which it was found that for an unpressurized system, poured with aluminum alloy, an average of 5.75 lb of alloy per minute passed through each square inch of the sprue area. This is equivalent to 60 in.^{3}/minute per square inch of cross-sectional area.

In the case where $h_t = 9$, the following calculation may be made.

$$A_b = \frac{60\ \text{in.}^3/(\text{min/in.}^2)}{27.8\sqrt{h_t}} = 0.72\ \text{in.}^2$$

or

$$A_b = \pi r^2 = 0.72\ \text{in.}^2 \tag{9.7}$$

$$\therefore r = \sqrt{\frac{0.72}{\pi}} = \sqrt{0.229} = 0.479$$

$$\therefore d = 0.95\ \text{in.}$$

If a bottom gate is used, then the filling time is longer because the metal is subject to a decreasing head during filling. Therefore in an increment of time dt, the height will increase dh and the volume of the metal will increase by the area of the mold $A_m\,dh$, while the flow through the ingate in time dt will be $A_g v\,dt$ and the velocity of the metal through the gate will be $\sqrt{2g(h_t - h)}$ at any instant. If we equate the increase in casting volume in time dt to the flow through the ingate in that same time interval, we find

$$A_m\,dh = A_g\sqrt{2h(h_t - h)}\,dt \quad \text{or} \quad \frac{A_g}{A_m}\,dt = \frac{dh}{\sqrt{2g(h_t - h)}} \tag{9.8}$$

Let t_p be the time to fill the mold and h_m be the height of the mold cavity. Then

$$\frac{1}{2g}\int_0^{h_m} \frac{dh}{\sqrt{h_t - h}} = \frac{A_g}{A_m}\int_0^{t_p} dt \tag{9.9}$$

$$\therefore t_p = \frac{2A_m}{A_g\sqrt{2g}}(\sqrt{h_t} - \sqrt{h_t - h_m}) \tag{9.10}$$

It is found that bottom gating takes twice the pouring time of a top gating system; this is obvious. If parting-line gating is used, the calculation is made in two parts: (1) top gating until the drag is filled, and (2) bottom until the mold is filled. If a top riser is used, a third calculation is required. Of course, we have considered the simple case of a mold of constant cross section without a core. Appropriate corrections must be made for the more complex shapes.

Ratio Gating

There are two types of gating systems: *nonpressurized,* or free flowing like a sewer system (Fig. 9.4), and *pressurized.* The latter has less total cross-sectional area at the ingates to the mold cavity than at the sprue base. The gating ratio relates the cross-section areas of each component of the gating system taking the sprue base areas as unity, followed by the total runner area and finally the total ingate area. Thus a pressurized system would have a ratio of 1:0.75:0.5, whereas a nonpressurized system might be 1:1.5:2 or 1:4:4 as in Fig. 9.4.

Unpressurized gating systems reduce velocity, turbulence, and aspiration but must use tapered sprues, enlarged sprue base wells, and pouring basins to achieve proper flow control. In addition, they can deliver metal uniformly to each ingate only if the runners are in the drag with the ingates in the cope and if the runner area is reduced by the area of each ingate after the junction, in a manner similar to that used for ducts in a heating system (Fig. 9.4).

RISERS FOR SAND MOLDS

Risers serve a dual function: they compensate for solidification shrinkage and they are also a heat source, so that they freeze last and promote directional solidification. Risers provide thermal gradients from a remote chilled area to the riser. Gating systems and risers are closely interrelated; in some cases the ingate is through a riser (Fig. 9.4).

Riser design includes supplying feed metal for shrinkage and any mold enlargement, riser location, spacing of risers for casting soundness, adequate sizing, proper connection to the casting, and the use of chills or insulation. Risers designed according to these concepts can result in improved casting quality and reduced cost.

Solidification Shrinkage

Gray iron with a carbon equivalent of 4.3 percent actually expands up to 2.5 percent because of graphite precipitation, but other ferrous alloys contract 2.5 to 4 percent during freezing. Nonferrous metals contract even more: pure aluminum contracts 6.6 percent and copper 4.9 percent. Their alloys usually shrink somewhat less, with near eutectic compositions contracting least. Lead has 7.7 percent reduction in volume at its phase change.

When an alloy has a short solidification range, as in a eutectic, pure metal, or low-carbon steel, a solid skin freezes at the mold–metal interface. Solidification then proceeds slowly toward the thermal center of the casting. Alloys that solidify over a long freezing range, such as aluminum or bronzes, are subject to dispersed microporosity more or less uniformly distributed throughout the cast structure. Skin-forming alloys are likely to have centerline shrinkage. Eutectic alloys require the least feed metal and therefore most commercial alloys are of near-eutectic composition. The risering concepts given are for low-carbon steel.

Riser Location

No matter how complex, any casting can be reduced to a series of geometrical shapes that consist of two heavier sections joined by a thinner one. Each heavier section needs its own riser. If the thinner section is not tapered to become larger toward the heavier sections, centerline shrinkage is probable. Chills at the thinner section may prevent such shrinkage and may promote directional solidification from the chill to the riser.

Feeding Distance

Past research has shown than an adequate riser can provide soundness for a distance of $4.5t$ for a plate casting; $2t$ is the riser contribution and $2.5t$ is from the edge effect. The maximum distance between risers is $4t$ for plates but is only $1t$ to $4t$ for bars. Chills increase the feeding distance for plates to a total of $4.5t + 2$ in. for a plate and to $6\sqrt{t} + t$ for a bar. Thus, the maximum spacing between risers if chills are used midway between them is $9t + 4$ in. for plates and $12\sqrt{t} + 2t$ for bars. Note that the distances are from the outside edge, not the centerline, of the riser (Fig. 9.6).

Riser Size

The riser size for a given application depends primarily on the alloy poured and the volume-to-surface-area ratio of riser relative to that of the casting section which is to be fed. Obviously, to be effective a riser must freeze more slowly than the casting.

Chvorinov's rule is the basis of most methods now used to calculate the proper size for short-freezing-range alloys such as steel or for pure metals. There is no satisfactory method for calculating the riser size for nonferrous alloys. Chvorinov's rule states that the solidification time for an alloy is

$$t = k\left(\frac{v}{sa}\right)^2$$

where t is solidification time, min. v is the volume of the casting section, in.^{3} sa is the cooling surface of the casting section, in.^{2}

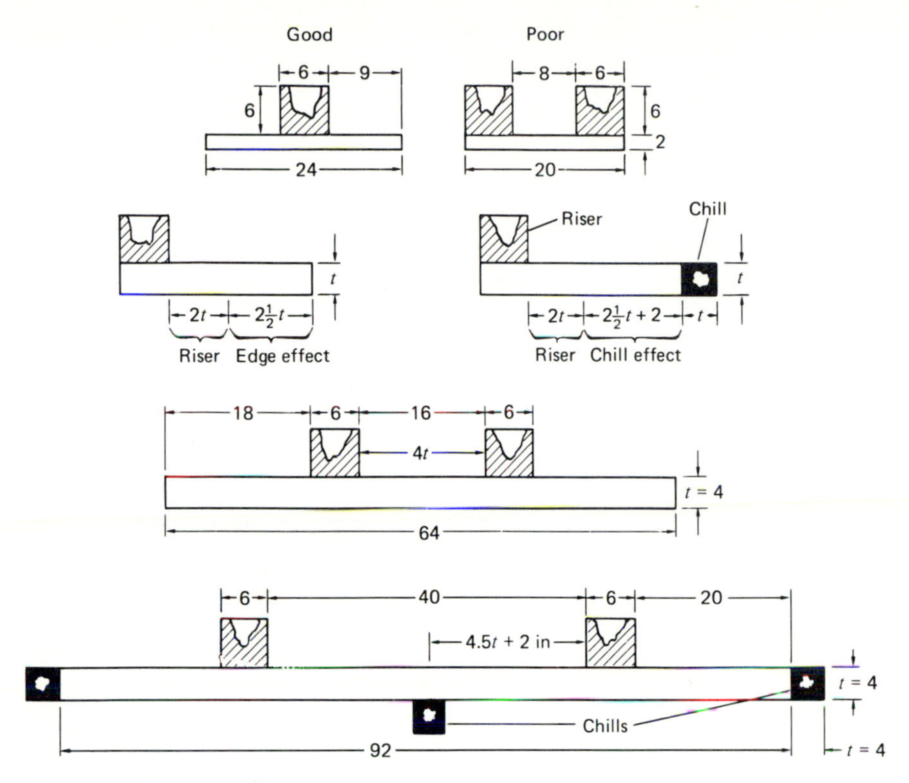

FIGURE 9.6
Feeding distance of risers for steel plates.

There are two types of risers—top risers and side risers. Top risers are placed above the volume to be fed and extend to the top of the cope. Side or blind risers are located at the parting line to feed locally; in the case of skin-forming alloys they can feed many times their own height provided that a cured or green sand core connects the mold to the thermal center of the riser. This permits atmospheric pressure to be exterted on the molten metal to force it to rise to a height corresponding to atmospheric pressure.

There are several alternative procedures for determining an adequate riser size for an alloy that freezes in a skin-forming manner. All give approximately the same result. One of the most direct depends on the *shape factor,* which is defined as the sum of the length and width of the section in question divided by its average thickness. In the 1950s the Naval Research Laboratory devised the shape factor chart (Fig. 9.7*a*) and riser height and volume chart (Fig. 9.7*b*). With these charts the proper risers for steel castings can easily be selected, once the casting has been divided into the proper sections and their volume-to-surface-area ratios have been calculated. This method gives conservative answers, but that is good when only one casting is to be poured and it must be correct the first time. In high-volume production,

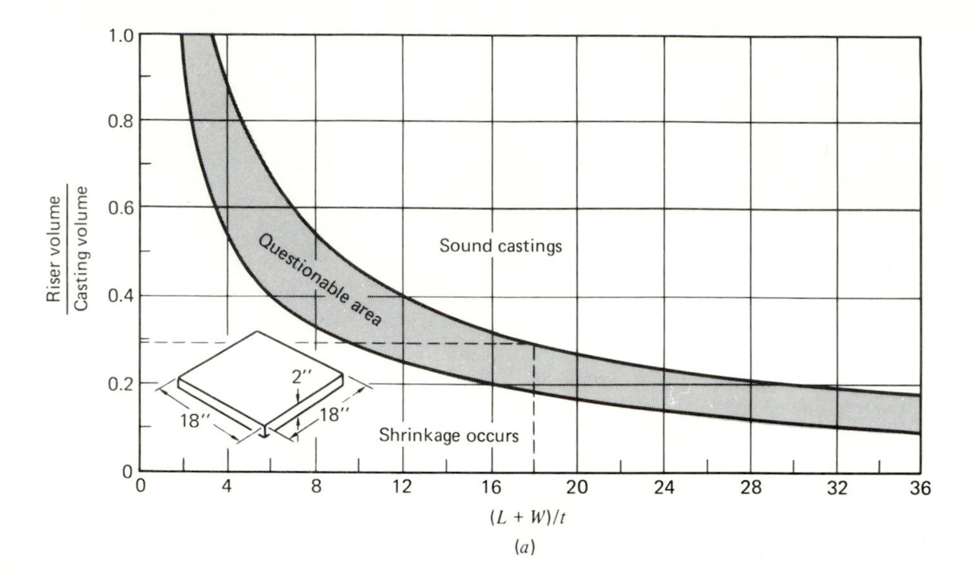

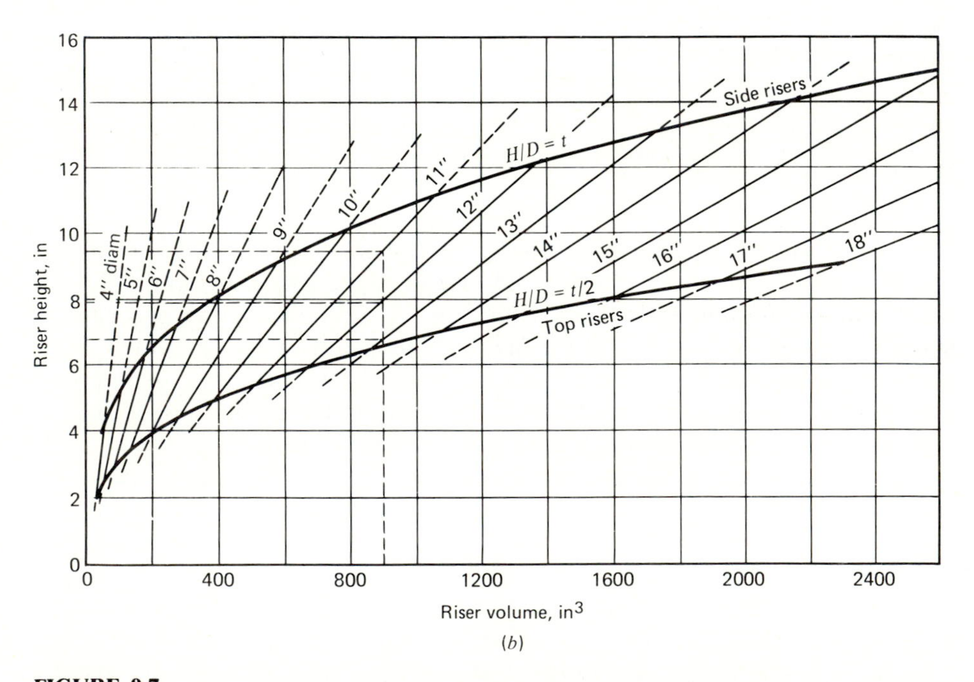

FIGURE 9.7
Riser size by the shape factor method. (*a*) Riser volume to casting volume ratio as a function of the shape factor. (*b*) Chart for determining riser diameter. (*From E. T. Myskowski, H. F. Bishop, and W. S. Pellini.*)

some additional experiments would be in order before final pattern construction.

Riser Connections

The riser connection to the casting warrants considerable care because it determines how well the riser feeds and, secondly, how readily a riser can be removed. To aid in top riser removal, it is wise to consider placing an annular core similar to a large washer in the riser neck. The thin section of core sand soon becomes hot, so that little chilling occurs and feeding takes place through the center hole. The connection length for a riser should not exceed $D/3$ to $D/2$, where D is the diameter of the riser and the inner hole should be about 1.2 times the connection length.

Riser Topping and Insulation

Riser size can be reduced by about 50 percent if insulated sleeves and riser topping or antipiping compound are placed on top of open risers as soon as pouring is complete. Sleeves are most important for aluminum and copper alloys where the radiant energy loss from the riser is a smaller proportion of the total heat loss from the riser. For steel castings, though, the antipiping compounds are most important, because they reduce radiant energy loss and leave a flat-topped riser. The proper selection of these materials and procedures can increase casting yield significantly. Not only is remelt cost reduced but the labor for cleaning and cutting off risers is much reduced.

DESIGNS FOR PLASTER-MOLD CASTING

Flasks. Frames of the flask are made of solid metal with internal drafts so that the plaster mold can be pushed out. They are accurately machined so they fit snugly on locating points of match plates. Conveyors carry flasks from station to station and are used over again after cleaning. Flask sizes are as follows:

$12 \times 18 \times 3\frac{1}{2}$ to 4 in. depth (standard)
$10 \times 18 \times 2\frac{1}{2}$ in.
$12 \times 21 \times 6$ in.
$24 \times 36 \times 12$ in.

Patterns. Patterns are of split type mounted on metal plates. There is one for drag and one for cope molds. They are made of engraver's brass, are accurately machined, and are easily changed. More than one part can be grouped on the standard pattern plates so that they can be poured at one time. There is no wear from slurry. The average cost is $600 to $1200.

Size. Size is determined by standard flasks. Maximum weight cast is usually 200 lb, and the maximum size is approximately 35 in. diameter. (The size capable of being cast is continually increasing, and these limitations will probably soon be exceeded.)

Material. Yellow brass, aluminum bronze, managanese bronze, silicon–aluminum bronze, nickel brass, and aluminum alloy can all be cast in plaster molds. The material of the mold can stand a maximum pouring temperature of 2200°F. Further development of mold materials may raise this temperature.

Quantity. Small or large quantities can be run by merely placing the pattern in the line when proper metal is being poured. One pattern can produce 150 to 250 pieces per week. Multiple patterns or plates will increase production accordingly.

Surface. The surface of the pattern can be reproduced. It is satin in texture, but can be plated easily without special preparation and can be buffed and polished easily (30–70 to 150 μin).

Tolerances. Tolerances are as tabulated below

	Tolerance
Dimensions wholly in one-half of the mold	(0.005 in./in.) 0.05 mm/cm
Dimensions across mold-parting line	(0.010 in./in.) 0.10 mm/cm
Dimensions subject to horizontal shift between cope and drag of mold	(0.010 in./in.) 0.10 mm/cm
Flatness	(0.005 in./in.) 0.05 mm/cm in any direction
Surface finish	(30–50 μin.) 0.8–1.3 μm

Machining allowance. Stock allowances for machining should be 0.8 mm (0.030 in.). On small parts the machining allowance may be less than 0.8 mm for holes and broaching.

Markings. Markings can be clearly reproduced.

Inserts. Inserts can be used with due regard to corrosion problems and ability to locate properly.

Wall thickness. this process can produce thin and sound walls:

Maximum of 2 in.^{2}	0.040-in. minimum thickness
Maximum of 6 in.^{2}	0.062-in. minimum thickness
Maximum of 30 in.^{2}	0.093-in. minimum thickness

Tapers with knife edges can be cast easily.

Stock allowances. Stock allowances are $\frac{1}{32}$ in. for machining. On small parts, they are less than $\frac{1}{32}$ in. for holes and broaching. (Note this can be a great saving over other processes.)

Cored holes. Reasonable holes are cored easily with good accuracy and finish. Holes less than $\frac{1}{2}$-in. diameter are more economically drilled. Drill spots are accurate and can eliminate jigs.

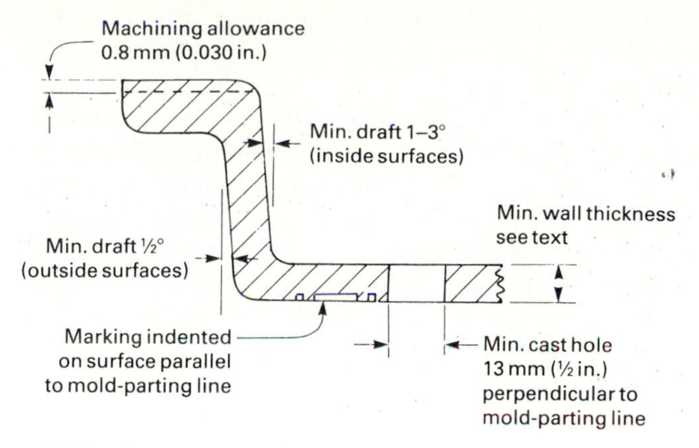

FIGURE 9.8
Design recommendations for plaster-mold castings.

Draft. Draft for outside surface is $\frac{1}{2}°$; for inside surface, 1 to 3°. Figure 9.8 illustrates the principal design recommendations for plaster-mold casting.

Advantages and Disadvantages. Plaster-mold casting has the following advantages:

1. Ability to vary mold material in thermal capacity from a heat insulator to a chill
2. Slow cooling of metal
3. Mold yield is minimal
4. Distortion and warping at a minimum
5. Smooth surfaces
6. Easy venting
7. No metal agitation
8. Uniform density; no gas pockets
9. Close tolerances

This process has the following disadvantages.

1. It can be used only for nonferrous materials.
2. The mold material is expendable and cannot be reused.
3. Large castings can be made, but they are expensive compared to sand castings.

INVESTMENT CASTINGS

Design

The fundamentals of good sand-casting design apply to the production of investment castings. Three of the most important elements of good casting design would include: (1) minimizing nonfunctional mass, (2) providing uniform wall sections, and (3) providing a suitable gating system to ensure complete filling of the mold cavity.

By minimizing nonfunctional mass, material is saved and a sound casting free of shrinkage porosity is more likely. The larger the mass, the greater the difficulty of providing sufficient gating to "feed" the part. By taking advantage of the correct placement of ribs, strengthening webs, etc., it usually is possible to design a part so as to minimize heavy sections.

As with all castings, the wall thickness should be as uniform as possible. If abrupt section changes are necessary, generous radii and fillets should be incorporated to reduce the turbulence of metal flow and to minimize stress concentration and porosity in these areas.

Tolerances and Quality

Investment castings have been called precision castings because they are much more consistent in size and composition than sand castings. The precision does not approach that of precision machining: ±0.002 in./in. for low-temperature alloys and ±0.005 in. for ferrous and high-temperature alloys are far from precision-machined tolerances. Nothing has as profound an effect on the shape and size of the desired casting as the master pattern or pattern die used to make the disposable patterns from which the part will ultimately be formed. Master patterns are required when the disposable pattern is to be made from a castable soft-metal alloy.

The size of a casting depends on many factors.

1. The master pattern must be made to allow for the four shrinkages and other adjustments listed below. If no master pattern is used, the machined mold eliminates the first shrinkage factor.
   a. Shrinkage of the die alloy during solidification and cooling
   b. Shrinkage of the wax or plastic pattern
   c. Shrinkage of the investment during setting and firing
   d. The solidification and cooling contraction of the metal in the casting
2. Experience in the art of investment casting governs the quality of castings. Variations are affected by design details such as:
   a. Presence or absence of cores
   b. Location of cored passages
   c. Ratio of core mass to metal mass
   d. Conjunction of light and heavy sections
   e. Location and design of angular projections
   f. Position and mass of gates
   g. Size and location of risers and vents
   h. Orientation of the pattern in the flask
   i. Position of casting in the flask relative to the gate and the sprue

j. Method of forcing metal into the mold
k. The position of the gates, which may distort the pattern and casting when they solidify
l. The inability of the pattern material to shrink due to restrictions of the die
m. The pouring temperature of the alloy and mold temperature when poured

3. Wax patterns must not be distorted when removed from mold or when stored. Plastic patterns do not require racks for storing in order to maintain size.
4. Parting lines and the location of cores influence the removal of the pattern from the die and the final tolerance of the investment casting.
5. Wax, plastic, and low-melting metals and alloys may be used for disposable pattern materials, but the low-melting alloys need more development because they do not leave a clean mold surface when they are melted out. Frozen mercury has good accuracy, but has other limitations.
6. In producing a pattern die it is usually necessary to hold the deviation from design tolerance to one-tenth of total permissible deviation. For a dimension to be held to ±0.005 in., the toolmaker must work to ±0.0005 in. For ±0.002 in. in the casting, the toolmakers must atempt to work to ±0.0002 in. It is desirable that all nonfunctional dimensions never be specified closer than 0.010 in. to simplify the die or master construction problem and thereby minimize tool costs. It is comparatively easy for a competent toolmaker to work to tolerances of ±0.001 in. in a die or on a master pattern. The fact that dies can be held to tolerances of ±0.0002 in. is no justification for requiring such expensive work unless the function of the casting can justify it.
7. If the refractory material next to the casting is different from that in the rest of the mold, it must be of such composition that it will not crack or spall when it is fired and thus affect the cleanliness and soundness of the casting. The ferrous and refractory metals are usually cast into molds that were formed from two different investments. The first material is a dip coat. This is applied to the gated patterns to impart a fine finish to the castings and to prevent contact between the metal and the coarse, porous, and usually chemically active material used to form the bulk of the mold. The expansion characteristics of the coating and backup investments must be closely matched.
8. The greatest variation in size comes from cooling the metal after pouring, rather than from the steps preceding.

There are many other factors, such as water content of refractory, uniformity of the materials, and pouring temperatures of wax or metals. All have an effect on the size of the final casting. Therefore, the engineer must not demand undue accuracy, but should design for more allowances.

The usual cautions concerning fillets, large flat areas, and threads (internal threads are very difficult to cast) that are considered in other casting processes apply to investment casting.

Cost of Parts

Investment casting has received the attention of engineers because it can be used to cast high-temperature and very hard alloys to a size that can be ground to final dimensions economically, and because it can be used to combine several parts into one cast unit and eliminate machining operations. For example, L. G. Daniels, in *Metal Progress,* discussed an operation that combined 13 parts into one casting and saved 57 machining operations (see Table 9.3).

An example of the potentialities of intelligent application of investment castings is a radar waveguide mixer body cast in B195 aluminum. Part cost is $350 when fabricated; casting cost is $75. Production rate as investment casting is 2 parts per hour. Interior rectangular passages are held within ±0.0002 in. parallel and from center to center.

TABLE 9.3
Operation for shuttle lifter

Number	Part	Operations	Cycle hours	Setup hours
		Old method of machining subassemblies		
1	Lifter body	7	0.255	0.148
1	Upper plate	20	0.390	0.388
1	Lower plate	20	0.300	0.316
1	Plate	9	0.235	0.168
1	Pad	8	0.145	0.124
4	Lock washer			
4	Cap screw			
	Subassembly	1	0.060	
13		65	1.385	1.144
			Grand total	2.529 h
		New method of machining single casting		
1	Lifter	8	0.260	0.164
			Total	0.424 h
		Savings		
12 parts		57 operations		2.105 h

Source: Courtesy of *Metal Progress*.

Size of Parts

The size of parts depends upon the size of equipment available in the foundry. Average flask size is 10 in. in diameter and 20 in. long. Box flasks are $9 \times 12 \times 15$ in. The weight of steel castings made by investment is 5 lb maximum. The sizes can be increased by further development and installation of larger production equipment.

TOOLING FOR CASTING IN METAL MOLDS or Permanent Molds

Casting in metallic molds often successfully challenges highly mechanized match-plate molding or automatic shell molding. A typical gray-iron "permanent" mold yields 50,000 or more pieces of aluminum or magnesium before failure. Such mold life is achieved by coating the interior mold surfaces periodically with a refractory mold wash.

Permanent molding is most likely to be successful if

1. Production volume is high (at least several thousand pours)
2. Strength of a nonferrous product (or gray iron) is sufficient
3. Shape is basically simple and suitable to the process

Compared with casting in sand, tooling and auxiliary equipment is more costly, but the advantages include lower cost and weight per unit of product, less entrained gas, finer grain structure, smoother surfaces, closer dimensional tolerances, and lower machining costs. Tooling for permanent molds is more challenging; the designer must thoroughly understand metal flow, solidification, and shrinkage. Certain casting problems demand more careful attention. Fillet design is more critical; mold-wall thickness becomes significant in terms of the cavity volume enclosed, the alloy cast, proximity to the sprue, and other variables. New problems such as cyclical thermal balance and casting removal appear.

Most of these comparisons are also true for die castings as compared to sand casting but in greater degree. For example, tooling cost is still higher but is often more than offset by less operator time per unit of product, more production relative to die life, smoother surfaces, and closer dimensional fidelity.

Tooling for permanent molding, which utilizes gravity feeding, is treated here. Tooling for die casting, which is distinguished by pressurized injection, is the subject of the following section. Parts can be more economically made by die casting up to a certain production rate; thereafter, permanent-mold casting becomes more attractive (Fig. 9.9).

In practice, a permanent-mold casting usually weighs less than 20 lb and not more than 300 lb. Die castings now weigh less than 100 lb, and sand castings frequently weigh many tons. Shapes too complex for metallic molds

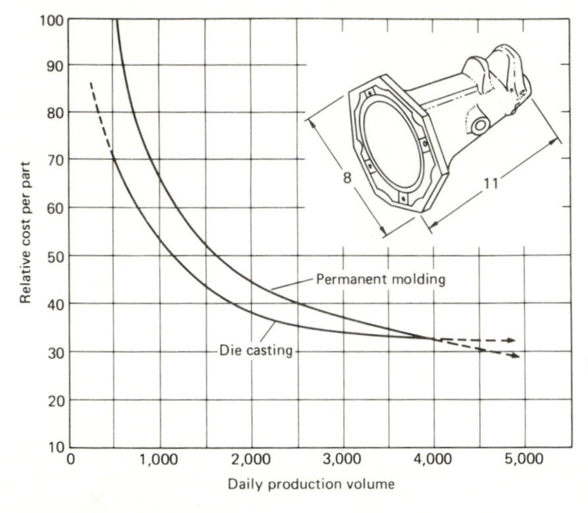

(*a*) Cost comparison of a transmission extension housing showing breakeven point for permanent molding and die casting. For a production rate of 4,000 the permanent mold casting method was chosen.

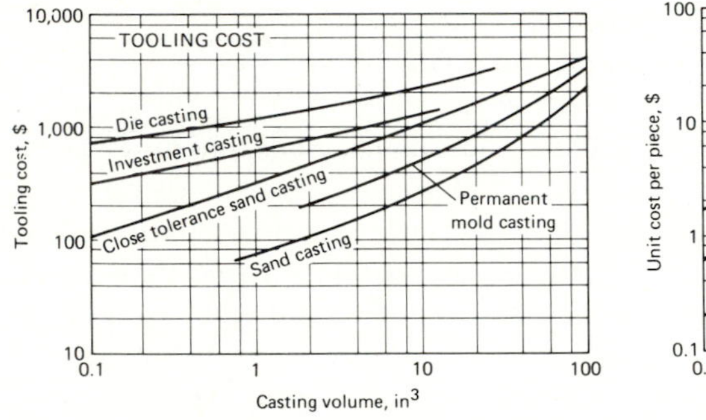

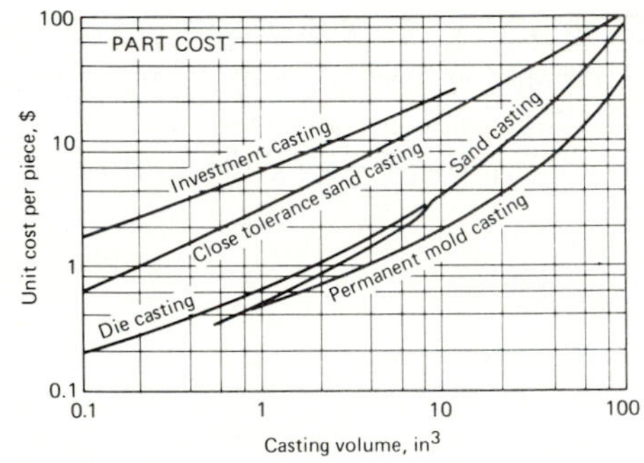

(*b*) Summary of the tooling and parts costs for 49 different aircraft castings

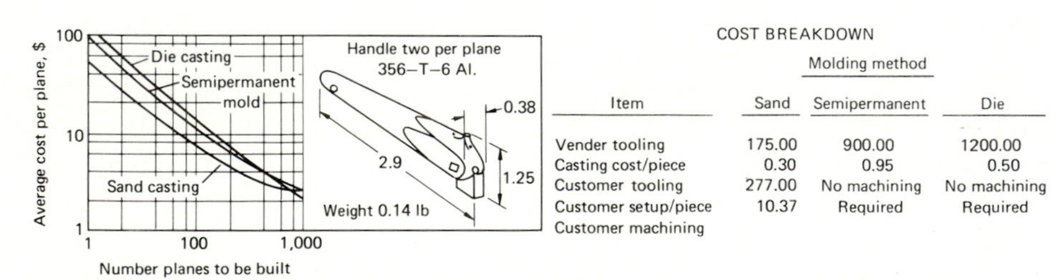

COST BREAKDOWN

Item	Sand	Semipermanent	Die
	Molding method		
Vender tooling	175.00	900.00	1200.00
Casting cost/piece	0.30	0.95	0.50
Customer tooling	277.00	No machining	No machining
Customer setup/piece	10.37	Required	Required
Customer machining			

(*c*) Cost comparison of a valve handle cast by three different methods

FIGURE 9.9
Cost comparisons for aircraft castings. (*Redrawn from American Society of Metals,* Metals Handbook, *Metals Park, Ohio, 1970.*)

can be cast in sand. The strength of permanent-mold castings exceeds that of die castings partly because correctly designed gating systems in the former entrap little or no gas. Tooling costs depend on production volume (Fig. 9.9).

DESIGN FOR PERMANENT MOLDING

A product that is suitable for permanent-mold casting must be carefully studied, and modified if needed, before the permanent mold can be considered in detail. The basic features of such a mold design include:

1. Simplicity—to mimimize the cost of the mold, turbulence, and the need of machining
2. Liberal tolerances
3. Foresight in the choice of the parting plane, which largely establishes the location of the gating and risers
4. Progressive solidification toward the riser from thinner sections remote from it

Dimensional standards and tolerances for permanent molding (Tables 9.4 and 9.5) must also be included in the production design of the casting. Figure 9.10 illustrates the principal design fundamentals.

Gating System Design

Success of the casting once depended largely on the operator's skill; today the designer uses fundamental fluid-flow principles for gating design, as shown in gating for sand casting.

Using a gating ratio of 1:2:2 or 1:2:1.5 with a pouring basin, conical sprue, and sprue base well that are proportioned as in sand casting, will give faster flow with least heat loss and turbulence. Most molds are gated to fill from the bottom to the top (Fig. 9.11). The metal flows from the sprue base

TABLE 9.4
Comparison of sand- and metal-mold casting processes

	Sand	Permanent mold	Die casting
Minimum section thickness, in.	$\frac{3}{16}$	$\frac{1}{8}$	$\frac{1}{16}$
Dimensional tolerance, in.	$\pm\frac{1}{32}$–$\frac{1}{8}$	±0.020–0.050	±0.004–0.020
Machining allowance, in.	$\frac{1}{8}$–$\frac{1}{4}$	$\frac{1}{32}$–$\frac{1}{16}$	0.010–0.020
Minimum core diameter, in.	$\frac{3}{8}$	$\frac{1}{4}$–$\frac{1}{8}$	$\frac{3}{32}$
Maximum weight, lb	Almost unlimited	Up to 300	Up to 100

TABLE 9.5
Typical dimensional standards for permanent molds and cores

Casting wall thicknesses, average

Casting size, in.	Under 3	3–6	Over 6
Minimum wall thickness, in.	1	$\frac{5}{32}$	$\frac{3}{16}$

Fillet radii

Inner radius = t (t = average wall thickness)
Outer radius = $3t$ blend tangent to walls

Unsupported length of cantilever for cores (must be over $\frac{1}{4}$ in. of diameter, d): $10d$ maximum; $8d$ preferable; $4d$ if $d < \frac{5}{8}$ in.

Draft (cores and cavity), degrees

	Short cores	Average cores	Outside surfaces	Recesses
Minimum	2	2	1	2
Preferable	3	3	3	5

Typical tolerances, in.

	Up to 1 in.	Additional over 1 in.
Basic tolerance (in one mold section)	$\pm\frac{1}{64}$	±0.001 in./in.
Between points produced by core or slide and mold	$\pm\frac{1}{64}$	±0.002 in./in.
Dimensions across parting plane	±0.020	±0.002 in./in.

Surface finish

Low-pressure process	20–125 μin.
Conventional process	150–500 μin.

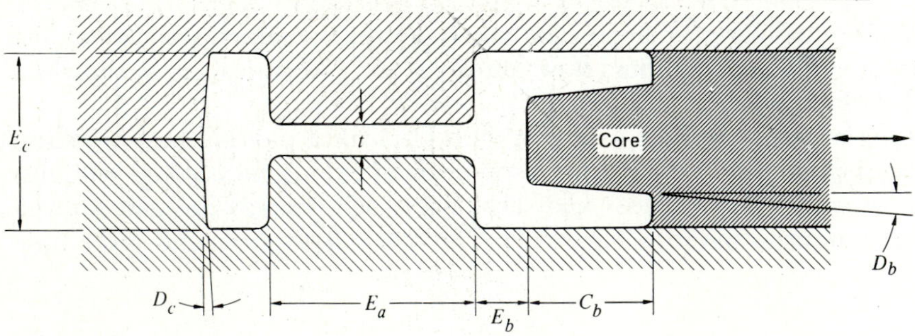

Dimensions produced by one mold half (A)	±0.015 in. (±0.4 mm)	±0.001 in./in.[a] (±0.1 mm/100 mm)
Dimensions between points produced by a core and the mold (B) or across parting plane (C)	±0.020 in. (±0.5 mm)	±0.002 in./in.[a] (±0.2 mm/100 mm)

Machining allowance, in.

	<10 in.	>10 in.	Sand-cored
Minimum	$\frac{1}{32}$	$\frac{3}{64}$	$\frac{1}{16}$
Preferred	$\frac{3}{64}$	$\frac{1}{16}$	$\frac{5}{64}$

[a] For copper alloys, allow 0.005 in./in. (0.5 mm/100 mm) additional.

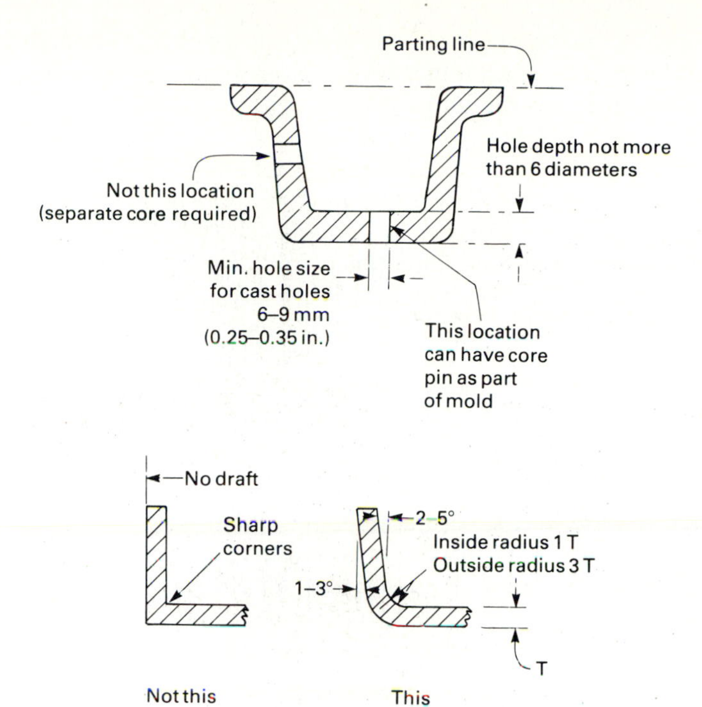

FIGURE 9.10
Design fundamentals for permanent-mold castings.

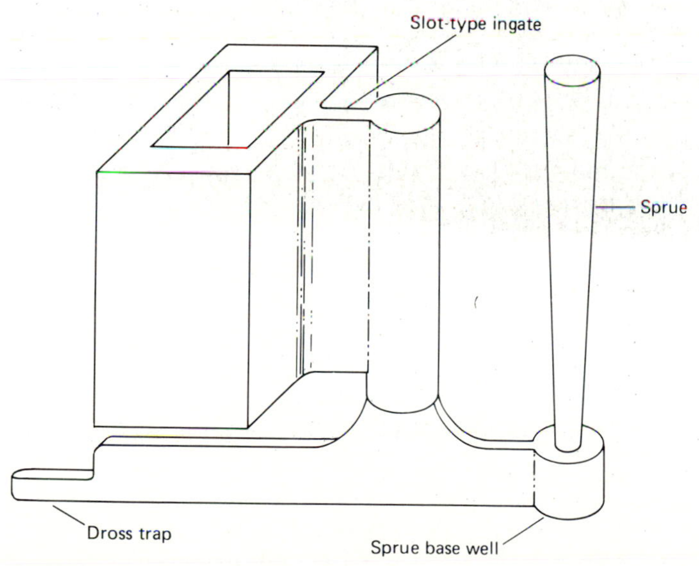

FIGURE 9.11
A typical gating system for a permanent mold.

well along the runner bottom, feeds the riser, and passes through a slot gate along the vertical side of the casting into the cavity. Note that a horizontal extension of the base runner receives the initial dross.

Risers

For proper feeding, the casting must have directional solidification from thin remote sections toward the riser. The riser in turn is fed last with hot metal and has a surface-area-to-volume ratio such that it will freeze more slowly than the casting.

There exists at present no detailed study of risering for aluminum alloys. The concept of casting yield may be used to determine a reasonable estimate of the riser size for permanent molds. Yield refers to the ratio of the casting weight divided by the total weight of metal poured. Metals that exhibit greater shrinkage characteristics require larger risers to make them sound. Low-shrinkage alloys such as a eutectic alloy of aluminum with 12 percent silicon, require little feed metal for a sound casting (Table 9.6). The figures given are at best approximations, because the weight of the gating system must be estimated and the surface-area-to-volume ratio is not considered.

Vents

Vents must discharge the gas in the gating system and mold cavity as fast as the metal enters the mold; natural venting along sliding members and at the parting line is usually inadequate. Additional venting may be added as follows.

1. Cut slots as deep as 0.010 in. and of suitable width across the parting seal surface.
2. Drill small clusters of holes 0.008 to 0.010 in. in diameter in the mold wall at a point where venting is needed.
3. Drill one or more $\frac{1}{4}$-in.-diameter holes into the area requiring additional venting. Drive into them square pins $\frac{1}{4}$ in. across the corners (pin vent).
4. Drill holes and install slotted plugs (plug vent).

TABLE 9.6
Empirical yield data for aluminum alloys

Alloy	Yield, percent
380 (4 percent copper, 7.5 percent silicon)	80
Aluminum–12 percent silicon	80
Aluminum–zinc	50
Alcoa 142 or 750	40

TABLE 9.7
Typical gray-iron composition for permanent molds and cores

		For longer life add:	
Element	**Composition, percent**	**Element**	**Composition, percent**
Carbon	3–3.5 (with 0.4–0.5 percent combined C)		
Silicon	1.6–2	Chromium	0.6
Manganese	0.5–0.8	Molybdenum	0.4
Phosphorus	0.2 max.	Nickel	0.2
Sulfur	0.05 max.		

Mold Material

The mold material is chosen on the basis of three criteria: material cost, the expected total number of pours required, and the casting alloy. Most common and suitable for a permanent mold is high-quality pearlitic gray iron, inoculated at the ladle to achieve uniform grain size and highly dispersed fine graphite. This material, frequently referred to as meehanite, has the composition given in Table 9.7.

Large castings or high pouring temperatures may require cores of alloy cast iron or H11 die steel. Undercuts or complex internal features can be formed with sand cores.

VENTING DETAILS FOR LOW-PRESSURE PERMANENT MOLDING

In low-pressure permanent molding, ejection of the casting and venting are the major problems. The use of ejector pins that bear on overflow wells can reduce the damage to a casting in any metallic-mold process, as is shown in Fig. 9.12*a*. Several means of venting large flat areas include the use of ribbon vents (Fig. 9.12*b*) and plut vents (Fig. 9.12*c*).

DIE CASTING

Design Considerations

Die castings have the advantage of reproducing accurate casting in large quantities and with good die life. Thus, they are able to incorporate features that minimize or eliminate subsequent machining and finishing operations. The extent of incorporating features that would otherwise require machining operations, such as holes and external and internal threads, in the casting

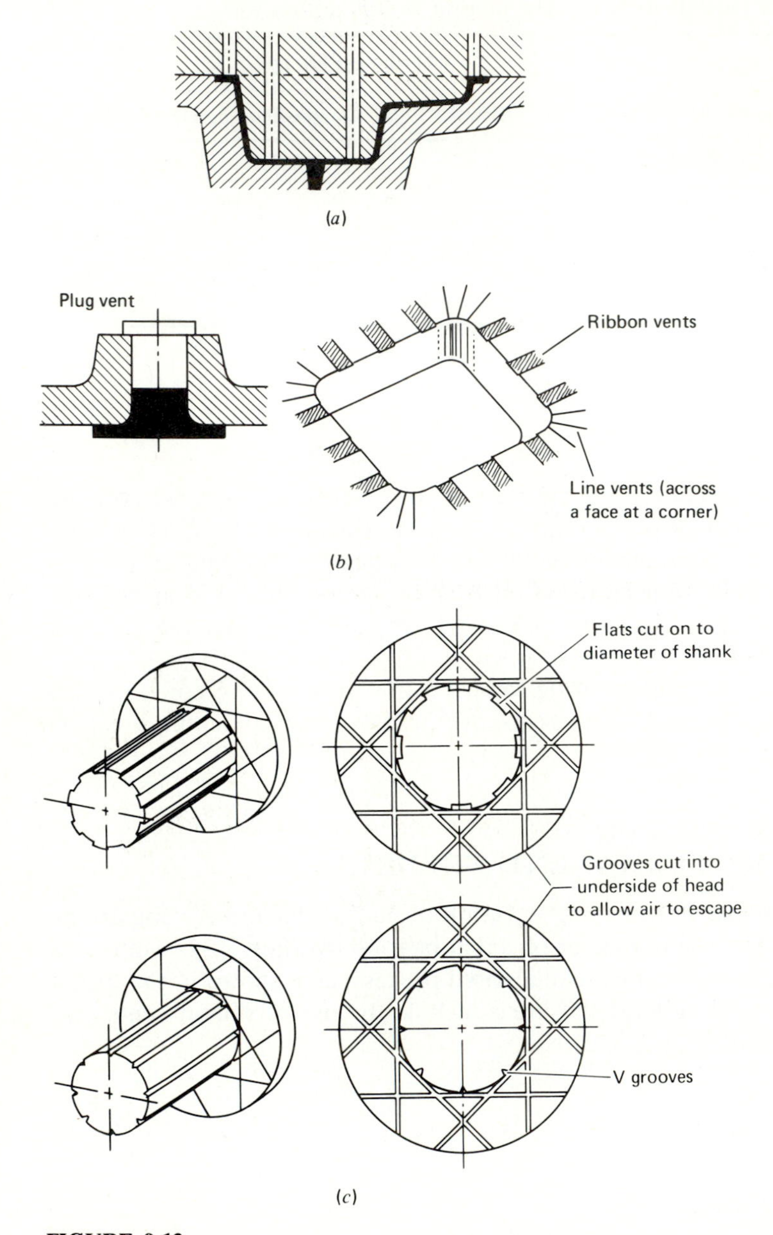

FIGURE 9.12
Ejection and venting details for dies to make low-pressure permanent-molded parts. (*a*) Ejector pins must be balanced to provide for uniform part removal; (*b*) typical plug and ribbon vents; (*c*) plug vent details.

depends on the ingenuity of the designer and the additonal die and operating expense.

If inserts are used, the time for placing them in the die will increase the molding-machine operation time. If it is necessary to use slides and cores operated by separate mechanisms other than those opening the die, these slow down the operation and increase die and machine expense.

Ejection of the casting with distortion and unfavorable ejection pin marks requires the attention of the engineer. When the die-casting operation is observed, it is often discovered that more labor is expended in punching, machining, and cleaning the casting of fins and burrs than in the casting operation itself. Therefore, the engineer must consider all operations for making the part, whether by the toolmaker or in his own shop. For example, a cored hole may be made by a two-part core that leaves a fin in the hole. This fin must then be punched out. Since this punching operation is required, the punching die can be designed to pierce the hole and eliminate the metal core in the casting die, making the die less expensive and possibly speeding up the operation.

The cost of tools can be reduced by using drill prints and pilot holes for drilled holes. This eliminates the need for a drill jig. The cores for small-diameter holes are expensive to maintain in die-casting molds, and often the holes can be drilled at less expense. It is the prerogative of the process-design engineer to determine which is the most economical method for a given design.

Die-cast parts made of ductile material can be formed, bent, spun, or joined to another part after casting. They can be pierced, twisted, embossed, swaged, and staked; and lugs can be used as rivets. Such operations make it practical for the production-design engineer to consider the use of die casting for a broad variety of assembly components.

The cleaning of flash depends largely on the design of the part and the die. The polishing and finishing costs can also be reduced when the designer eliminates re-entrant corners, avoids flat surfaces, and places ejector pins where marks can be removed or hidden.

Cores are an asset in die casting; nevertheless, their use should be carefully questionsed in the light of subsequent operations and die maintenance costs. Cores save weight, keep sections more uniform, and thus secure sounder castings. They help carry the heat away and reduce the casting cycle of the machine. Often the dies can be vented through the core. Cored holes will reduce machining, owing to the accuracy of their size and location.

Minimum Section Thickness

Good casting design dictates the use of as uniform a wall thickness as possible or one which tapers slightly from the thinnest section remote from the gate to the heaviest section at the gate. In addition, alloy fluidity dictates the minimum practical wall thickness for each die-castable metal (Table 9.8). Walls must be

TABLE 9.8
Minimum wall thickness for die castings

	Base casting alloy		
Surface area, in.^{2}	**Tin, lead, zinc**	**Aluminum, magnesium**	**Copper**
Up to 3.9	0.0236–0.0394	0.0315–0.0471	0.0589–0.0787
4.0–15.5	0.0395–0.0589	0.0472–0.0707	0.0788–0.0982
15.6–77.5	0.0590–0.0787	0.0708–0.0982	0.0983–0.118
Over 77.6	0.0788–0.0982	0.0983–0.118	0.119–0.157

thick enough to permit proper filling but sufficiently thin for rapid chilling to obtain maximum mechanical properties.

Undercuts and Inserts

Wherever possible, one should redesign a part to eliminate undercuts. In most cases, such part modification is more economical than the labor required to handle a loose piece each casting cycle. In one case, redesigning to reduce undercuts saved $1.50 per casting on a part that is still in production 15 years later. Over 4.5 million castings have been made during that time.

Inserts such as bearings, wear plates, bushings, and shafts can be incorporated in die castings, but they must be easily and precisely located in the die. If they are not securely placed, they may slip between the dies during the casting cycle and cause great damage. Inserts must be provided with properly knurled, crimped, or grooved surfaces to ensure a good mechanical bond between the insert and casting. If the insert is large relative to the casting, better results will be obtained if the insert is preheated, e.g., cylinder liners for automotive engines.

Cored Holes

Holes can be readily cast in all alloys according to the data given in Table 9.9.

TABLE 9.9
Depths of cored holes in die castings

		Hole diameter, in.								
Alloy	**Minimum diameter castable**	$\frac{1}{8}$	$\frac{5}{32}$	$\frac{3}{16}$	$\frac{1}{4}$	$\frac{3}{8}$	$\frac{1}{2}$	$\frac{5}{8}$	$\frac{3}{4}$	**1**
Zinc	0.039	$\frac{3}{8}$	$\frac{9}{16}$	$\frac{3}{4}$	1	$1\frac{1}{2}$	2	$3\frac{1}{8}$	$4\frac{1}{2}$	6
Aluminum	0.098	$\frac{5}{16}$	$\frac{1}{2}$	$\frac{5}{8}$	1	$1\frac{1}{2}$	2	$3\frac{1}{8}$	$4\frac{1}{2}$	6
Magnesium	0.078	$\frac{5}{16}$	$\frac{1}{2}$	$\frac{5}{8}$	1	$1\frac{1}{2}$	2	$3\frac{1}{8}$	$4\frac{1}{2}$	6
Copper-based	0.118	—	—	—	$\frac{1}{2}$	1	$1\frac{1}{4}$	2	$3\frac{1}{2}$	5

Draft Requirements

The amount and location of draft in a die-cast part depends upon how it is located in the die and whether it is an external surface or cored hole. Draft on the die surfaces normal to the parting line permits the parts to be ejected without galling or excessive wear of the die impression (Fig. 9.13). The values shown represent normal production practice at the most economic level. Greater accuracy involving extra close work or care in production should be specified only when necessary because it may involve extra cost.

Dimensional Tolerance

The dimensional tolerance that can be achieved in die casting depends on several factors.

1. The accuracy to which the die cavity and cores are machined.
2. The thermal expansion of the die during operation

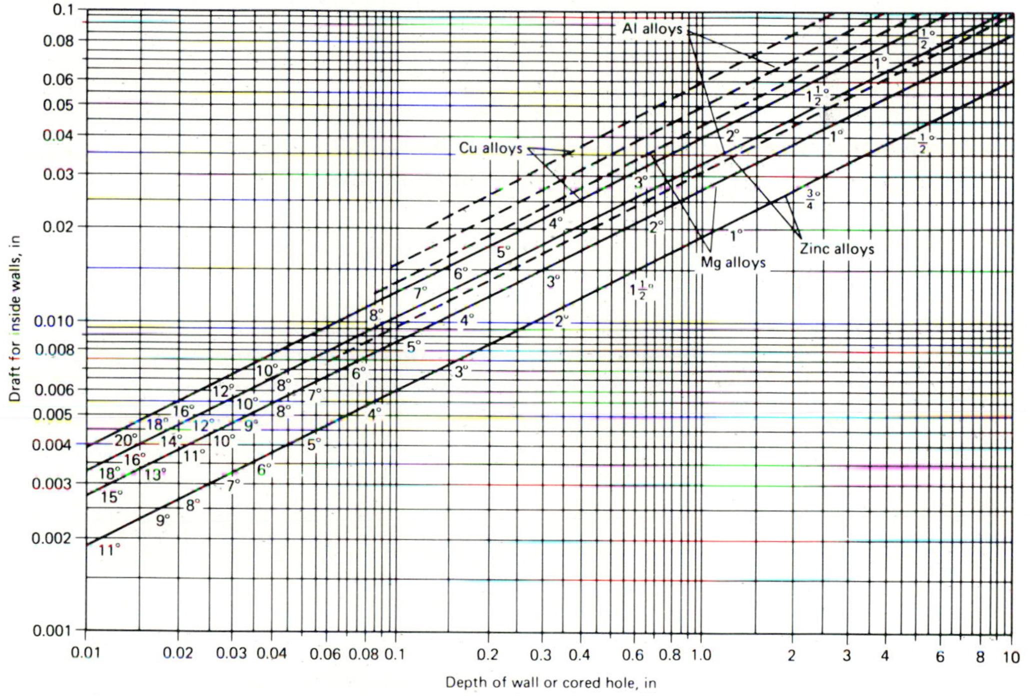

([1]) Draft on cored holes is the total for both sides.

([2]) Draft for outside walls is 50 percent of inside wall value.

([3]) Values given herein represent normal production practice. Greater accuracy should be specified only when absolutely required because it usually involves extra cost.

FIGURE 9.13
Draft allowance for die-casting dies (wall: solid lines; cored holes: dashed lines). Note that (1) draft on cored holes is the total for both sides; (2) draft for outside walls is 50 percent of inside wall value; (3) values given represent normal production practice.

TABLE 9.10
Tolerance for dimensions of die castings

	Zinc	Aluminum	Magnesium	Copper
Basic tolerances (up to 1 in.)	±0.003	±0.004	±0.004	±0.007
Additional tolerance (1–12 in.)	±0.001	±0.0015	±0.0015	±0.003
Additional tolerance (over 12 in.)	±0.001	±0.001	±0.001	
Additional tolerance across movable die section, projected area				
Up to 10 in.^{2}	±0.004	±0.005	±0.005	±0.010
11–20 in.^{2}	±0.006	±0.008	±0.008	
21–50 in.^{2}	±0.008	±0.012	±0.012	
51–100 in.^{2}	±0.012	±0.015	±0.015	
Additional tolerance across parting line, projected area				
Up to 50 in.^{2}	±0.004	±0.005	±0.005	±0.005
51–100 in.^{2}	±0.006	±0.008	±0.008	
101–200 in.^{2}	±0.008	±0.012	±0.012	
201–300 in.^{2}	±0.012	±0.015	±0.015	

3. The injection temperature and shrinkage of the alloy being cast
4. Normal wear and erosion of the die cavity and cores
5. Position of the movable parts relative to each other during casting
6. Surface finish—50 to 125 μin. is common, a better finish can be achieved if needed.

The tolerances for basic linear dimensions in one-half of the die are summarized in Table 9.10. Another tolerance must be added to the basic tolerance if the linear dimension is for a part that extends across a movable die section. More tolerance must be added to dimensions that extend across the parting line.

TABLE 9.11
Recommended materials for die-casting dies

	Total castings produced		
Alloy	**50,000 or less**	**250,000**	**1,000,000**
Zinc (1-in. cavity)	P20	P20	P20 or H13
Zinc (4-in. cavity)	P20	P20 or 4150	4150 or H13
Aluminum or magnesium	H11 or H13	H13 or H11	H13 or H11
Copper	H12, H20, H23		
Copper	Molybdenum		

TABLE 9.12
Normal composition of die-casting die steels

Steel	Carbon	Chromium	Tungsten	Molybdenum	Vanadium	Manganese
H11	0.35	5.00	—	1.50	0.40	
H12	0.35	5.00	1.50	1.50	0.40	
H13	0.35	5.00	—	1.50	1.00	
H20	0.35	2.00	9.00			
H21	0.35	3.50	9.00			
H22	0.35	2.00	11.00			
P4	0.05	5.00				
P20	0.30	0.75	—	0.25		
4150	0.50	0.95	—	0.20	—	0.87

Die Materials

Die-casting die materials must be resistant to thermal shock, softening, and erosion at elevated temperatures. Of lesser importance are heat treatability, machinability, weldability, and resistance to heat checking. Tool steels of increasing alloy content are required as the injection temperature of the molten alloy, the thermal gradients within the die, and the production cycle increase. Dies for use with zinc can be prehardened by the manufacturer in a range of R_c 29 to 34. The higher-melting alloys require hot-work tool steels (Tables 9.11 to 9.13).

Wear of Cores, Slides, and Pins

Cores, slides, and pins must provide abrasion resistance in addition to heat resistance. Wear can be reduced by the following procedures.

1. Use of contacting materials of differing hardness
2. Use of nitride on one or both surfaces in contact
3. Use of a lubricant on the areas of contact (avoid contamination of the molten metal)

TABLE 9.13.
Materials for cores, slides, and ejector pins

Casting alloy	Cores and slides	Ejector pins[a]
Zinc	440B,[b] H11, H12, H13	H12, 7140 nitrided, H11, H13
Aluminum or manganese	TZM, H11, H12, H13	H12, 7140 nitrided, H11, H13
Copper	TZM, H11, H12, H13	H21, H20, H22
Ferrous	TZM	H13

[a] Nitride all steels except for copper.

[b] Use for cores only.

4. Establishing and maintaining proper clearance between mating parts
5. Polishing the wearing surfaces of the mating parts.

Materials for such moving parts are listed in Table 9.13.

OTHER CONSIDERATIONS IN CASTING AND MOLDING

Quality–Accuracy–Uniformity

Castings are fairly accurate and uniform in size in any of the casting processes. Accuracy of casting depends upon the following.

1. The accuracy of the pattern and mold and its construction to prevent warpage and shrinkage
2. The uniform temperatures of molten or molded material and mold, and the moisture content of the mold
3. The kind of surface produced
4. The construction and accuracy of the flasks and molding machines
5. The skill of the operator in placing cores and inserts in the mold, and his skill in withdrawing the casting from the mold
6. The uniformity of the cast material in terms of composition and quantity
7. The wear of the pattern or die

For example, the accuracy of precision casting depends upon the size of the wax pattern, the uniformity of the wax material, wax-expansion characteristics, the care and assembly of wax parts, the mixing and baking of the slurry material for the mold, the wax residue, and the pouring of the metal. When these factors are uniform and under control, the casting will stay within reasonable limits of accuracy and the finish will be uniform (usually ±0.005 in./in., but ±0.0005 in./in. is possible). The dentist casts crowns, dentures, and bridges to exact sizes by this process. Dentists have been pioneers in adapting the lost-wax process and other precision-casting techniques for modern use.

Good casting practice now produces homogeneous castings that are strong and reliable. X-ray, sonic, penetrant, gamma-ray, and holographic inspection techniques detect faults and assist the molder or foundryman in correcting the defects. Castings are not subject to grain flow and are of nearly equal strength in all directions. Castings of the same chemical composition and heat treatment are competitive, with respect to mechanical properties, with rolled and forged products. Cast crankshafts for automobiles are an example of a product that is now improved because of the inherent characteristics of the cast material.

The quality and uniformity of plastic, die-cast, permanent-mold, and sand castings are generally accepted and can be relied upon. These high standards, along with low costs, have kept casting in competition with other processes.

Drawing Information

All the requirements of draft, risers, gating, parting lines, cores, locating points, finish allowance, size of sections, and pressure-tight castings should be placed on the drawing to guide machinists and foundrymen. If possible, draft, finish, and core supports and prints should be indicated on the drawing. All such specifications should have the approval of the foundry. This vital information is supplied by the process designer as he works and collaborates with the product designer.

As an engineer works closely with these casting industry representatives, he learns the limitations and possibilities of the particular foundry or mold shop he is using. If another supplier is used, definite and detailed information about that shop must be obtained, because each shop has distinct characteristics and offers specific equipment.

The design drawing should indicate the locating points and the chucking or holding procedure desired for the machining of the casting, so that the foundry, mold shop, and machine shop will use the same checking routine.

Quantity and Costs

The quantity of castings to be made within a period of time determines to a large extent the cost of the casting. The greater the quantity, the more that can be invested in pattern or die equipment and the less the casting will cost. Multiple patterns or dies produce two or more castings in practically the same time it takes to produce one. Increased quantities resulting in affording better equipment improve the quality of the casting and permit the spreading of the cost of molds or patterns over more castings with less development cost per unit.

Close competition exists between the various casting processes, as well as with other types of processes. Use of each process is determined by the following factors: quantity to be produced; material required to meet strength, corrosion, weight, and appearance conditions; accuracy required; complexity; and cost of subsequent operations such as machining and finishing.

For example, say a cover plate for a piece of equipment is required. As the market for it develops, improvements are made, quantities increase, and functions change. In Figs. 9.14 to 9.18 note the influence of quantity and function on the type of material and operation used. For a brief period, castings lost to rolled steel. Note how considerations of appearance and accuracy played an important part in the choice of process. The metal molds

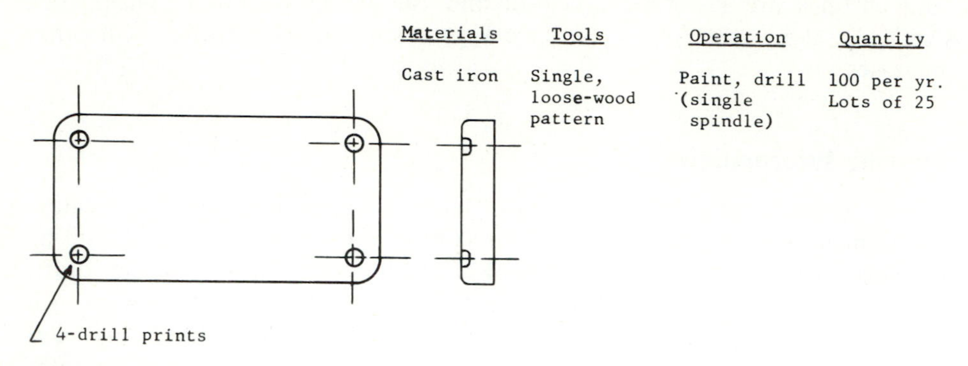

FIGURE 9.14
Cast-iron cover plate.

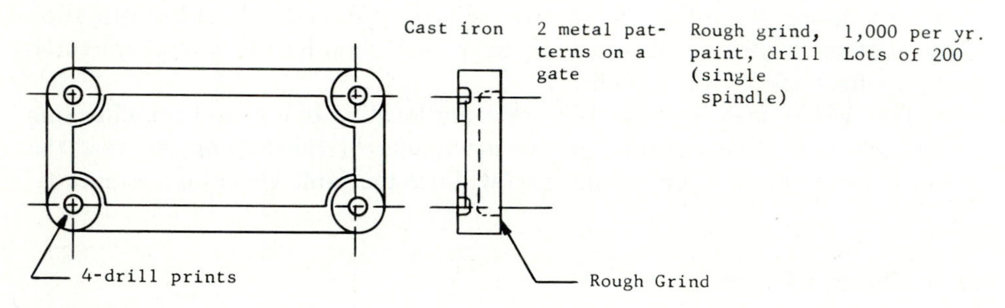

FIGURE 9.15
Cast-iron cover plate molded two at a time. Quantity increases; tools and dies can be paid for in 1 year.

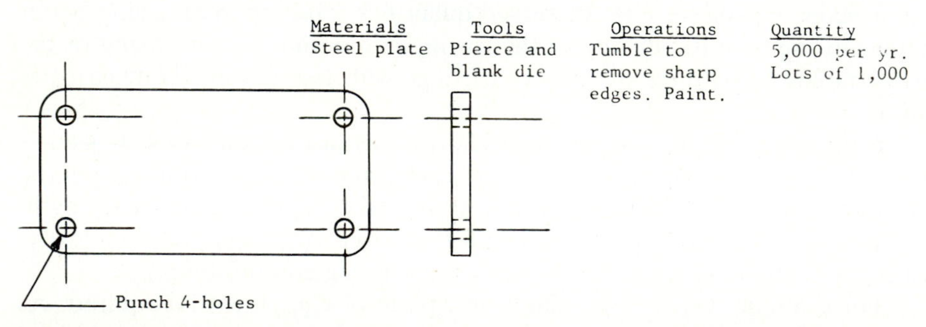

FIGURE 9.16
Stamped-steel cover plate. Quality of apparatus demands better appearance, and aluminum is chosen. No other finish is required.

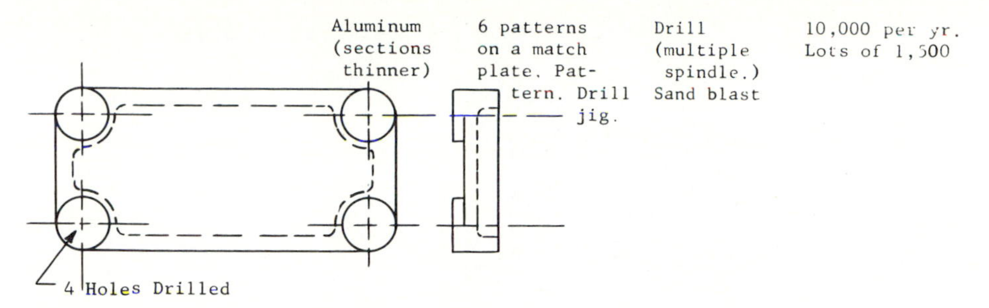

FIGURE 9.17
Sand-cast aluminum cover plate. Cover plate is applied to other equipment that requires a better seal and gasket that must be held in a gasket groove. Better finish and greater accuracy are required. (Bolt holes can be cast, and better finish and accuracy can be had using permanent metal molds.)

(Fig. 9.18) could not be used until the activity of the part justified the expensive mold.

In looking at the figures, assume that a gasket is necessary to prevent oil leakage. Assume also that the part comes from the foundry or molder with all operations—such as removal of fins and marks due to parting lines, gates, pushout pins, and cores—performed to give a piece that is finished except for machining and final finishing operations.

A permanent-mold aluminum cover plate in Fig. 9.18 is being applied to equipment that requires a better seal and a gasket that must be held in a gasket groove. Better finish and greater accuracy are required. Bolt holes can be cast and better finish and accuracy can be had by permanent metal molds. Figure 9.19 shows a die-cast zinc cover plate. A lacquer finish was applied to harmonize with equipment. Countersink head screws were necessary, and an insignia on the cover was added. Here the gasket is cheaper, but sand blasting is required. This cover plate is used when activity has increased and the accuracy of the part must be improved. The quantity required has also increased. By using a zinc die-casting alloy, cost of material is reduced and less material is required in die casting the part. Greater accuracy is obtained. The

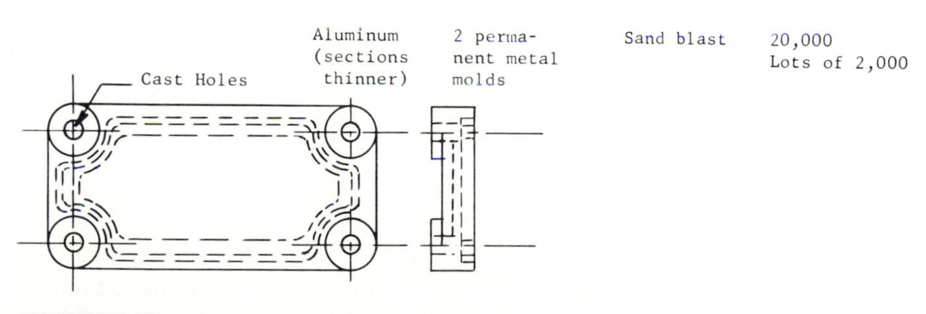

FIGURE 9.18
Die-cast aluminum cover plate.

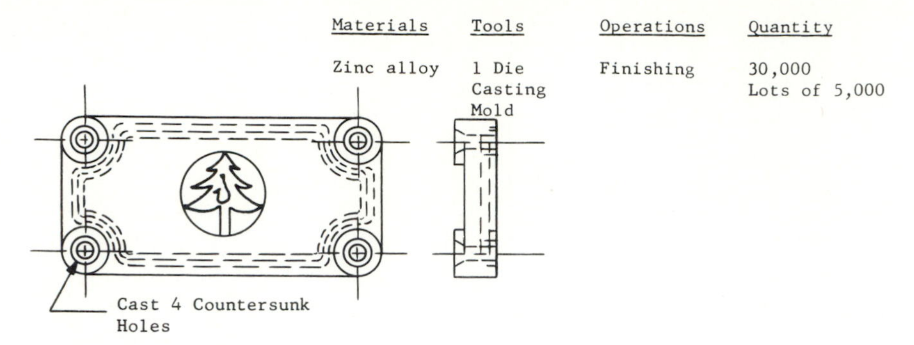

FIGURE 9.19
Die-cast zinc cover plate.

die-cast part requires no finishing operations other than cleaning and application of finish, as the surface is better than with permanent-molded parts.

Material and Overhead Costs

The material cost is usually not the highest cost of casting. Material costs can be reduced by using a standard material that is used frequently in the foundry or molding shop. A special material may not be uniform. It may be expensive to prepare or to adapt equipment to handle it in small quantities.

With the mechanization of foundries to handle sand, and raw and finished material, by conveyers, special equipment, and molding machines, the overhead costing rates have increased considerably over the days when sand was hand-shoved. The maintenance cost of such equipment is high because of the sand and dust hazards. In spite of high overhead costs, the foundries have reduced the cost per pound of castings through advanced mechanization.

Molding shops have high costs of maintaining molding equipment—with air, steam, electric power, and water service. "Permanent" molds are also subject to wear and are replaced after a given service life. This requires expensive machine equipment and high-priced toolmakers to maintain the quality of the product.

An example of competition between materials and processes might be a single automobile company die casting an engine block out of aluminum and at the same time perfecting the sand molding and casting of thin-walled accurate casting for a similar engine block.

Auxiliary Operations

The cost of molding is sometimes less than the auxiliary operations of removing gates, risers, and parting-line material; trimming, cleaning, grinding, and polishing surfaces; and inspection. Costs are reduced by eliminating these auxiliary operations. Heat-treating and straightening operations are expensive and their cost can sometimes be reduced by proper design and specification of material.

Purchasing Castings

Purchasing castings on the straight cost-per-pound basis is simple from the accounting standpoint, but purchasing castings on a classified weight basis is less misleading. The most economical arrangement is to deal with an organization that knows the detailed costs of each operation required to make the casting, pattern, and tools. Then these costs can be compared with detailed costs of other processes and the correct decisions can be made.

For example, company A purchases castings at a flat price of $0.30/lb; the weights of the castings vary from 1 to 2000 lb; and the volume of castings is distributed as shown in Table 9.14. It can be seen that inexpensive small castings are being purchased owing to the excessive price of the heavy castings. If the true costs were allowed for, better pattern equipment would be ordered for the production of the smaller castings. For example, small parts may be more economically made as weldments than as castings. The possible savings resulting from using the best process, on an *overall* cost basis, are shown in Table 9.15.

Design Factors for Castings

The part to be made from a casting may be intricate or simple. Intricate castings can be economical because they may combine many parts into one piece and thus save the cost of the fabrication and joining of several separate pieces. Joints must be machined and bolted together. Misalignment and loosening during vibration may result. Sand castings weighing several tons (such as a locomotive frame) have been made to advantage because of their ability to combine several parts into one piece. The more complicated the casting, the more ingenuity and control required. The simpler the part, the less

TABLE 9.14
Annual cost at $0.30/lb regardless of volume

Weight classification, lb	Volume, lb/yr	Actual labor cost, $/lb	Raw material cost, $/lb	Actual cost $/lb	Selling price per year, @ $0.30/lb	Actual cost per year
1–9	20,000	0.90	0.06	0.96	$ 6,000	$19,200
10–49	20,000	0.60	0.06	0.66	6,000	13,200
50–199	10,000	0.45	0.06	0.51	3,000	5,100
200–499	50,000	0.1575	0.0525	0.21	15,000	10,500
500–2,000	100,000	0.075	0.045	0.12	30,000	12,000
					$60,000	$60,000

TABLE 9.15
Actual cost cast versus fabricated at several production levels

Weight classification, lb	Actual cost of casting	Correct process: Cost per lb	Correct process: Total cost	Savings
1–9	$19,200	Fabricated, $0.75	$15,000	$4,200
10–49	13,200	Fabricated, $0.54	10,800	2,400
50–199	5,100	Fabricated, $0.45	4,500	600
200–499	10,500	Cast, $0.21	10,500	
500–2,000	12,000	Cast, $0.12	12,000	
	$60,000		$52,800	$7,200

the cost of the mold and pattern equipment and hence the less expensive the part. Variations in size and strength may be more difficult to control in more complex parts; thus a more highly skilled molder may be required.

The design of a part to be cast depends upon the behavior of the material as it cools, the construction of the mold, and the functions of the part in service. The art of casting and molding has progressed to such an extent that practically anything can be cast that is within the size range of the equipment available. It may not be economical to cast the part today, but in a few years the process may be improved so that it may be economical to redesign the part from a sheet metal or welded design into a casting. To make a casting simple and easy to cast demands the highest skill and the best judgment on the part of the designer and foundryman.

Design factors to be considered include the following.[1]

1. For maximum strength and stiffness, material should be kept away from the neutral axis. This is important to all designs and is related to the moment of inertia I, which is a measure of the ability of a cross section to resist rotation or bending about the axis passing through its center.

   The modulus of rupture, which designates the load in pounds per square inch imposed on a beam, is given by the well-known equation

$$S = \frac{MC}{I}$$

   where S is the modulus of rupture or maximum fiber stress, lb/in.^{2}, imposed on metal at greatest distance from the neutral axis, M is the bending moment, or load in pounds times distance from concentrated load

[1] Much of this information is taken from *Steel Castings Handbook*, see references.

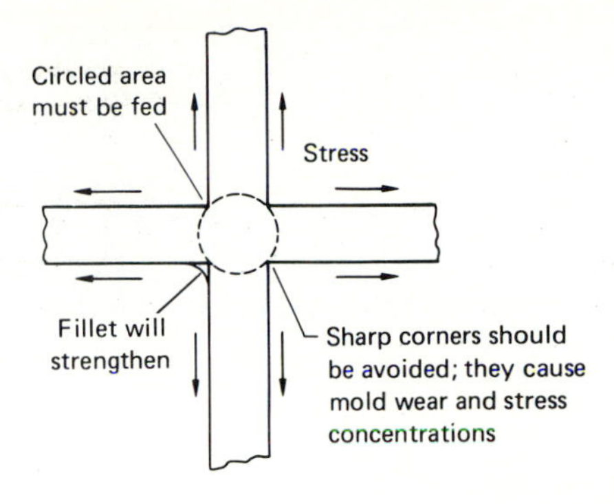

FIGURE 9.20
Stress conditions around hot spot.

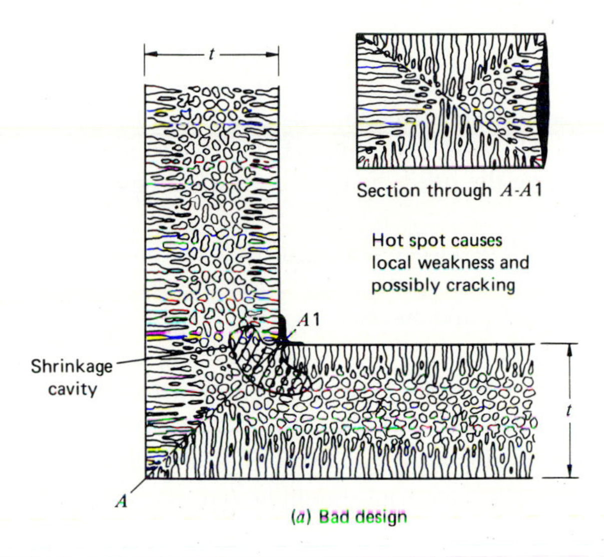

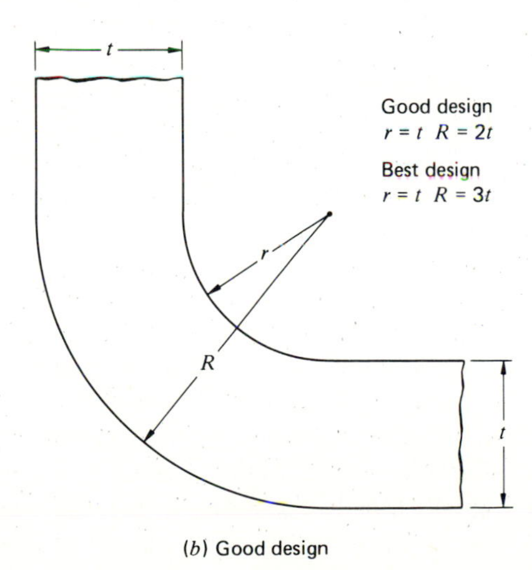

FIGURE 9.21
Design of a corner joint to produce soundness.

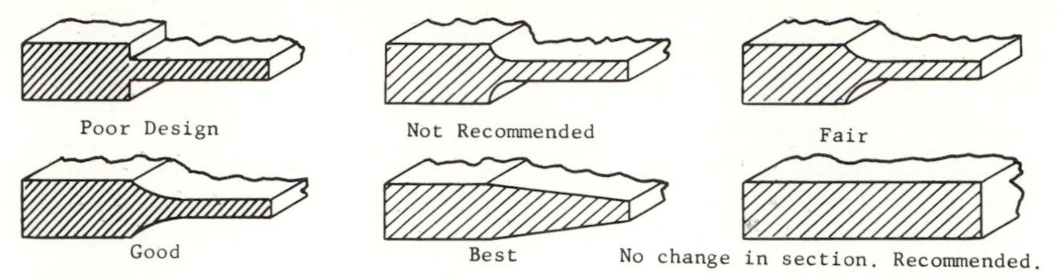

FIGURE 9.22
Design for changes in section thickness.

to point under study and C is the distance from netural axis to outer surface

2. Keep plates in tension and ribs in compression. This is desirable because the compressive strength of some materials, particularly cast iron, is much greater than the same materials tensile strength. The plate distributes the load over its entire surface and is more effective when placed in tension rather than compression. Ribs, on the other hand, are most effective when placed in compression.
3. Insure that the pattern or part can be easily removed from the mold. This factor is probably violated more than any other and causes the use of loose pieces and complicated cores, gates, and risers.
4. Ensure that cores can be removed.
5. Use smoothly tapered sections to eliminate high stress concentrations.
6. Sharp corners and abrupt section changes at adjoining sections should be eliminated by employing fillets and blending radii (see Figs. 9.20 to 9.22).
7. Determine the best location(s) where material should be fed into the part.
8. Does the design avoid the development of tears and spongy sections?
9. Does the necessary draft on the pattern interfere with the part design?
10. Can parts be clamped and located easily for machine operations without interference from fins or excess material at parting lines and junctions of gates and feeders?
11. Are locating points for chucking or holding indicated on the drawing for foundry or molding-shop and machine-shop information?

Effect of Grain Size

A relatively large dendrite grain size may be developed in the process of solidification as a result of the freezing rate or section size of the casting. Large castings and ingots freeze with coarse grains. Thin-sectioned castings, or castings made in metal molds, develop a fine grain size owing to rapid freezing. Normally, a fine grain size is desirable, since higher ductility and impact strength values are obtained at a given tensile strengths level with fine grain size.

Hot Spots

Hot spots are the last portions of the casting to solidify. They usually occur at points where one section joins another, or where a section is heavier than that adjoining it—at a square corner for instance. The outside of the corner should be rounded to reduce the section change (Fig. 9.21).

Hot spots are weak and the metal may tear where they occur. Figure 9.20 shows stress conditions around the hot spot within the circle. In casting steel and other high-temperature metals, the spot enclosed by the circle must be fed to avoid cavities.

The shape of the part should be such as to enable directional solidification of the molten metal in unbroken sequence from the farthest part of the casting to the point of entry. If the casting should "freeze" somewhere along the line ahead of its turn, the sections between it and the farthest end, which are still liquid, would be cut off from the supply of feeding liquid metal; cracks, or at least dangerous internal stresses, would result. In view of this, the casting should be designed to avoid abrupt changes from a heavy to a thin section (Fig. 9.20). Where light and heavy sections join, the thickness of the lighter section should be increased gradually as it approaches the heavier section.

Steel is difficult to cast. Most steel castings should be stress-relieved because of the high shrinkage of the metal and resulting internal stress. The design rules for steel castings can be modified for other ferrous materials. If they are followed in principle, better castings will be obtained regardless of the material being cast. The following six design rules for casting provide an important supplement to the 11 general principles mentioned previously.

1. In designing unfed joining sections in L or V shapes, it is suggested that all sharp corners at the junctions be replaced by larger radius corners, so that these sections become slightly smaller than those of the arms (Fig. 9.23).

   Reason: Outside corners of joining sections are positions of extra mass and result in hot spots. If these positions are not fed from outside sources (by risers) they will be the location of shrinkage cavities. Cutting off the outside corner is similar to producing a uniform section and is a fundamental of good design.

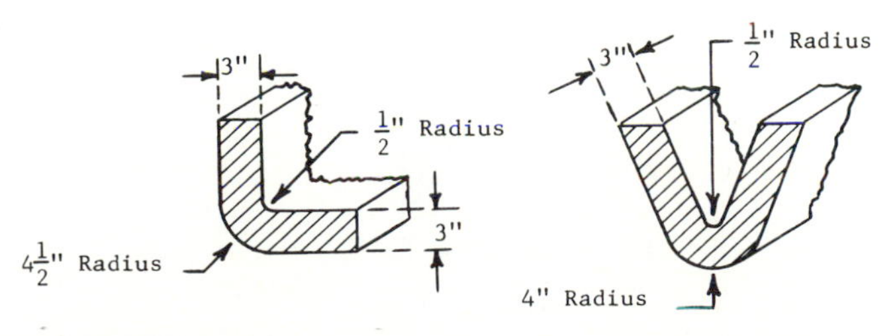

FIGURE 9.23
Unfed joining section.

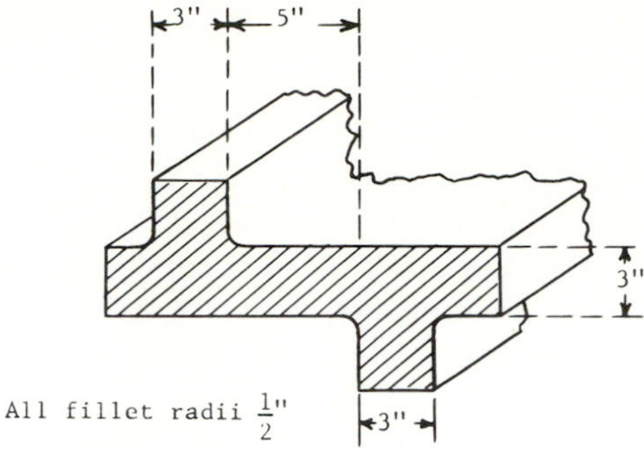

FIGURE 9.24
Offset the arms of an X section to produce two T sections, thereby reducing the hot spot.

2. In designing members that join in an X section, it is suggested that two of the arms be offset considerably (Fig. 9.24).

    Reason: The center of an X section that cannot be fed by a riser is a hot spot and will result in the formation of a shrinkage cavity. Offsetting the arms permits the use of external chills by the foundryman to produce a section free from cavities. If the design does not permit offsetting the arms at the X section, then the possibility of placing a core in the middle of the junction should be considered. This will permit uniform sections with small stresses through the center (Fig. 9.25).

3. Isolated masses not fed directly by risers are details of poor design. They should be hollowed out with a core construction or constructed of a lighter section. If this cannot be done, then internal chills should be used.

    Reason: Areas of heavy metal attached on all sides to members of much smaller thicknesses are considered isolated masses. When these areas are so located that the foundryman has no opportunity to feed the heavy portions properly by means of conveniently placed risers, shrinkage cavities occur within the section.

4. The use of webs, brackets, and ribs at joining sections should be kept to a minimum. If they are required as stiffeners for a casting, the poor design effect can be remedied by the extensive coring of the web, bracket, or rib in

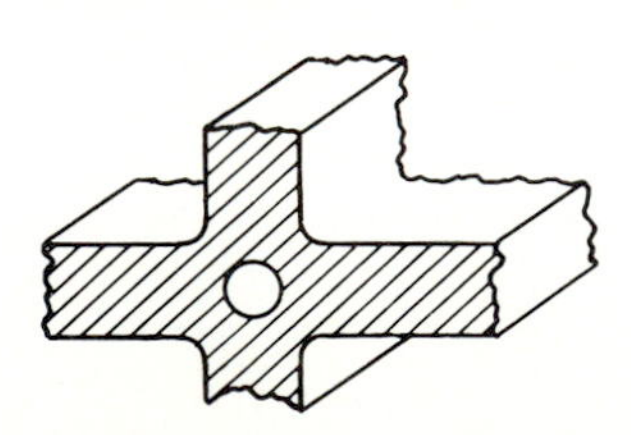

FIGURE 9.25
Reduction of a hot spot by coring.

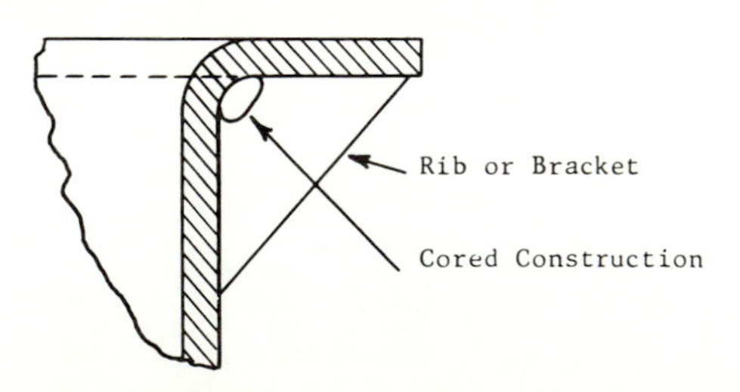

FIGURE 9.26
Cored construction in stiffening members.

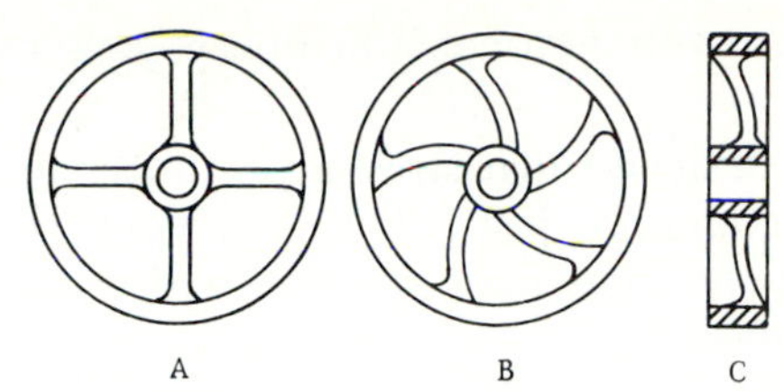

FIGURE 9.27
Wave construction. Design (A) produced cracked spokes. It is corrected by the use of five curved spokes as in (B) or the alternative design as in (C).

the region of the adjoining section (Fig. 9.26). Small brackets will help prevent hot tears and consequently are frequently used.

Reason: Stiffening members at joining sections are sources of considerable trouble with regard to the formation both of hot tears and of shrinkage cavities. They increase the mass at the intersection of the adjoining members. Coring will not impair the stiffening features of these webs, brackets, or ribs.

5. When the design of a one-piece cast steel structure is very complicated or intricate or when section thickness variations are large, the piece can sometimes be broken up into parts so that they may be cast separately and then assembled by welding, riveting, or bolting.

Reason: Very complicated designs often lead to enclosed stress-active systems and any mold relieving that may be employed by the foundry would not be sufficient to produce the casting without cracking or failure. In such cases, appendages, cast-in baffles, and other extraneous members should be removed from the design and cast or prepared separately, and then welded into place.

6. Cast members that are parts of an enclosed stress-active system should be designed slightly waved or curved (Fig. 9.27).

Reason: Curve construction or wave design reduces distortion resulting from contraction of the steel during cooling. A good example of this type of design is the use of curved spokes in many classes of wheels (Fig. 9.27). In such castings the rim, hub, and spokes may cool at different rates, thus subjecting the casting to considerable internal stress, Spokes designed with a wave in them will, under stress, tend to flex, thus preventing tearing or distortion. Residual stresses can be relieved through heat treatment.

Details of Design for Steel Castings

The greater the shrinkage and the higher the temperature of the metal, the more difficult the problem of casting. While iron castings are common, and iron is relatively easy to cast, steel casting is much more difficult, and the principles established for steel castings are described as a review and as an illustration of the more severe conditions. The practices described may be modified for other kinds of materials.

1. To prevent cavities in the casting resulting from shrinkage, thought must be given to the casting design, so that the volumetric contraction is compensated for by a supply of liquid steel.

2. The casting and mold should be so designed that stresses are as small as possible when the casting is at a temperature just below the solidification temperature. It is at this point that steel has poor strength and ductility characteristics so that it is likely to hot-tear.
3. Steel has low fluidity compared with other metallic alloys. Thus, thin sections should be avoided, especially when located remotely from the orifice feeding the casting.
4. A poured steel casting will solidify from the outside of the casting inward and the rate of solidification is about the same regardless of section size; therefore, heavier sections will take proportionally longer to solidify than lighter sections. The principal defects in steel castings that result from poor design are hot tears, cold shuts, shrinkage cavities, misruns, and sand inclusions.

Hot tears. Hot tears are cracks at various points in a casting brought about by internal stresses resulting from restricted contraction. Sharp angles and abrupt changes in cross section contribute to large temperature differences within a casting, which may result in hot tears.

Shrinkage cavities. Shrinkage cavities are voids in the casting brought about by insufficient metal to compensate for the volumetric contraction during the solidification of the casting. Shrinkage cavities are more pronounced in areas fed by thin sections. This results because the thin feeding section will solidify too rapidly to allow the introduction of additional metal to the casting which has diminished in size because of volumetric contraction.

Misruns. When a section of a casting is incompletely filled with metal, it is known as a misrun. This is usually brought about by solidification prior to the complete filling of a mold cavity. Thin casting sections should be avoided.

Sand inclusions. When a portion of the mold breaks away and is wholly or partially enclosed by the molten metal, a sand inclusion occurs. This may be brought about by abrupt turns and complicated passages of flow of the metal as it is poured into the casting.

Comparison of Cast and Forged Connecting Rods

In 1962 the Central Foundry Division of General Motors introduced the cast connecting rod. Since then it has become the standard of the industry because of its economy and reliability. This development is a triumph of engineering perseverance and casting design. The success of this design depends on the use of the inherent strength of *pearlitic malleable iron* and its machinability. The presence of temper carbon in the matrix reduces the cost of cutting operations by acting as an internal lubricant and eliminating deburring.

Why cast a connecting rod? The casting process permitted greater freedom of design because the engineer could put the metal where it was needed. But stress analyses were needed to determine where the greatest stresses occurred. In their original work, General Motors engineers took conventional forged connecting rods and ran exhaustive stress analyses. First static stress loads were calculated as follows:

$$\text{Compressive load} = P_{\max} \times A$$

where $P_{\max}$ is the maximum gas pressure in cylinder, lb/in.^{2}and A is the area of cylinder bore, in.^{2}

$$\text{Tensile load} = KW\left(\frac{S}{2}\right)N^2\left(1 + \frac{S}{2L}\right)$$

where K is a constant $= 2.84 \times 10^{-5}$, W is a reciprocating weight, lb, S is the stroke, in., n is the engine speed, rev/min and L length, pin center to crank center, in.

Analysis of stresses in a forged rod. Brittle-laquer-stress-coat patterns showed the location of highly stressed areas in the connecting rod (Fig. 9.28). Strain

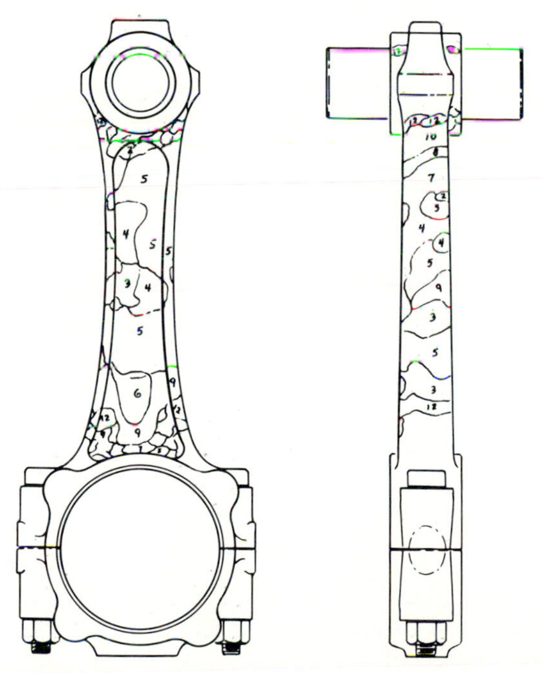

Compression load indicating compressive stresses

FIGURE 9.28
Brittle-lacquer stress analysis.

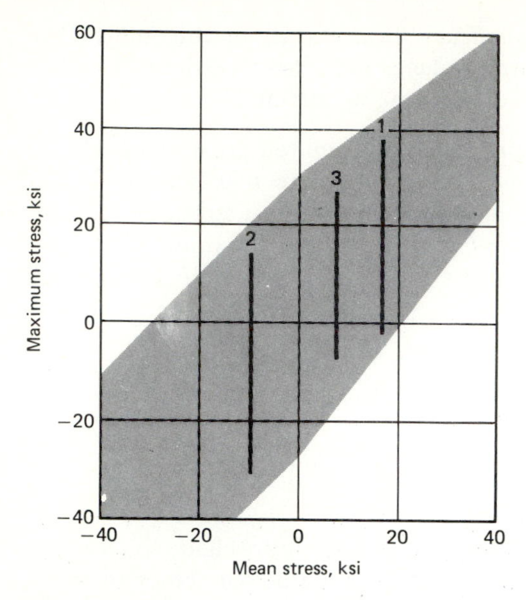

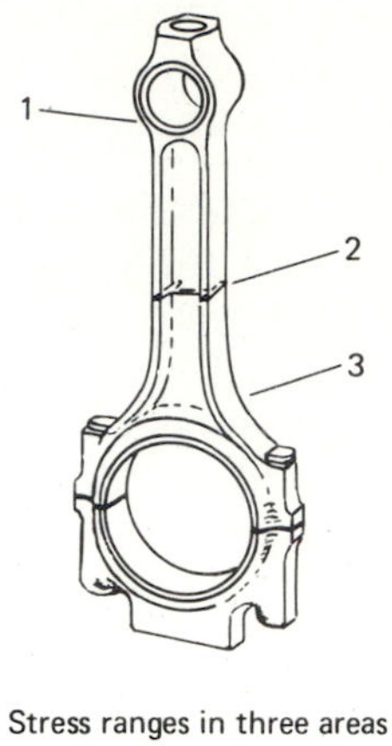

Stress ranges in three areas of a connecting rod

FIGURE 9.29
Goodman diagram of fatigue stress levels at three sections of a forged connecting rod.

gages attached to those areas gave the quantitative values of stresses in the rod. The largest stress was found to be in the center of the column of all rods studied.

Next, fatigue testing was carried out with alternating axial loading to produce a Goodman diagram (Fig. 9.29), which represents fatigue strength under combined alternating loading and a static load. If no failure occurred in 10×10^6 cycles, the rod was considered to be acceptable.

On the basis of the Goodman diagram the stresses in three particular areas were plotted (Fig. 9.29).

Area 1. Compressive loading should result in high tensile stresses normal to the long axis of the rod on the underside of the wrist pin.

Area 2. Compressive and tensile loads each produce moderate stresses in the column in the direction of the long axis, but combined they produce a high alternating stress.

Area 3. The tensile load should induce high tensile stresses in the rail at the crankpin end, but, in fact, relatively low stress was induced there because of the compressive loading.

The validity of the fatigue tests was revealed when actual failure occurred as predicted when the rods were subjected to engine tests. Failures took the form of transverse fractures in the center of the column and longitudinal fractures below the wrist-pin bosses (Fig. 9.30). Although the fatigue tests

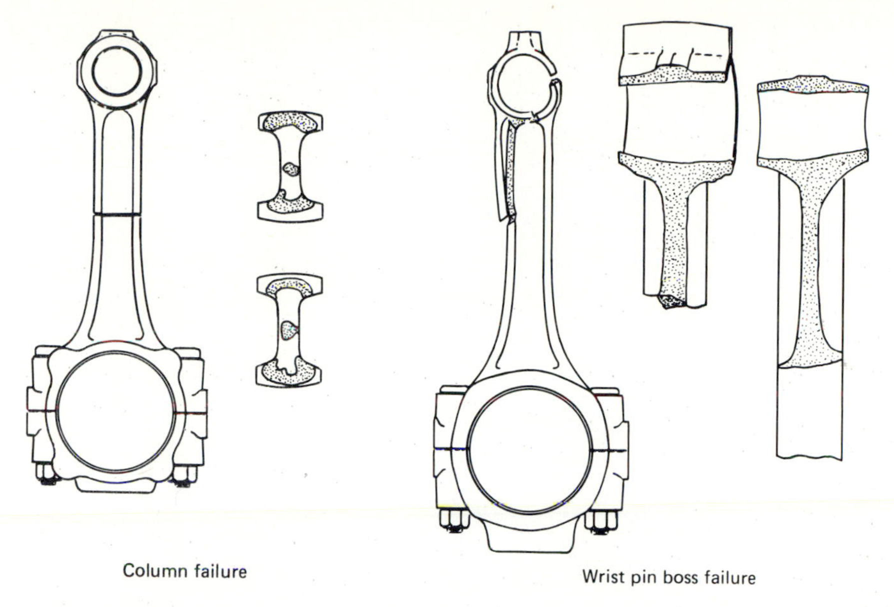

FIGURE 9.30
Typical connecting-rod fatigue-test failures.

were based on the combined maximum compression and tension loads, in actual practice the rod never reached such a load combination at any speed (Table 9.16).

The design must be such that the alternating stress in all areas falls within the limits of the Goodman diagram.

Design of a cast connecting rod of pearlitic malleable iron. After the basic design parameters had been determined, the design of the pearlitic malleable-iron connecting rod was tailored to the casting process. The column was

TABLE 9.16
Rod loading as a function of engine speed

Engine speed, rev/min	Tensile load, lb	Compressive load, lb
500	27	8,156
1,000	109	8,077
2,000	437	7,755
3,000	983	7,224
4,000	1,747	6,477
5,000	2,728	5,519
6,000	3,925	4,372

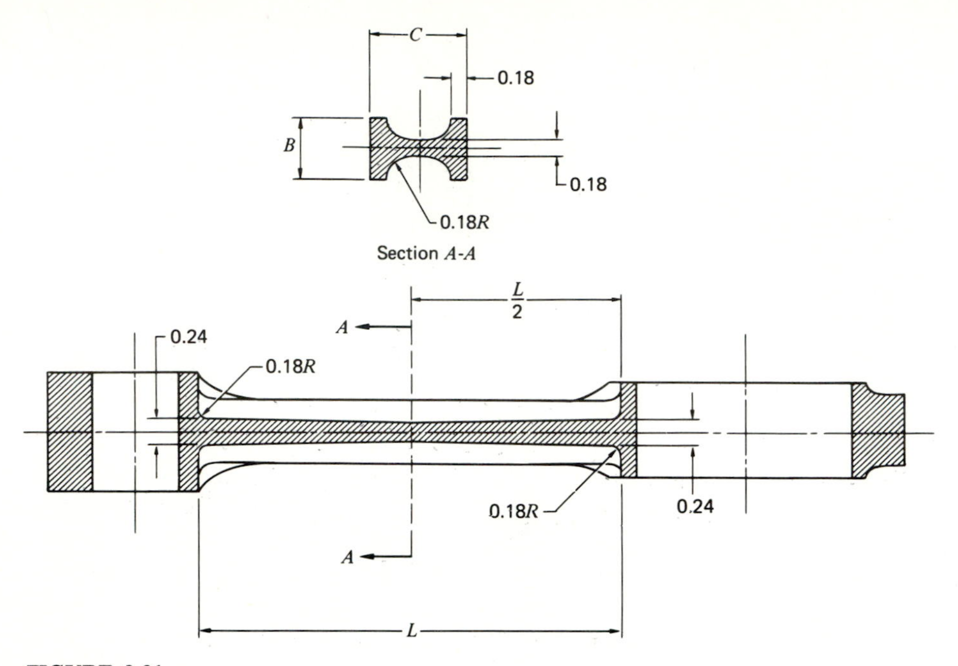

FIGURE 9.31
Typical cross sections of a cast connecting rod.

specified to have a minimum and uniform cross section consistent with good foundry practice (Fig. 9.31). Wherever the stress level was low, metal was removed, and it was added where more stress was found. The ability to redistribute the metal permitted the weights of the cast and forged rods to be equalized (Fig. 9.32).

Design of gates and risers. To determine the best foundry practice, gating studies were made using high speed photography and Lucite models. On the basis of the tests, the casting was gated at only one end because the meeting of the streams from dual end gates could cause defects in the highly stressed center of the rod arm.

Risering was designed to achieve directional solidification by tapering the central web so that solidification began at a point remote from the risers, in this case the center of the thin arm. Thermocouples placed in test castings at critical points indicated the sequence of freezing and helped achieve proper riser sizes for both ends of the connecting rod. Directional solidification was achieved from the center of the arm to each end despite gating from the small end. The *I*-beam section of the arm provided additional chilling, so that solidification began at the arm center where it was critically stressed and traveled to the risers. Cooling fins were added to the boss and column rail to provide better local solidification.

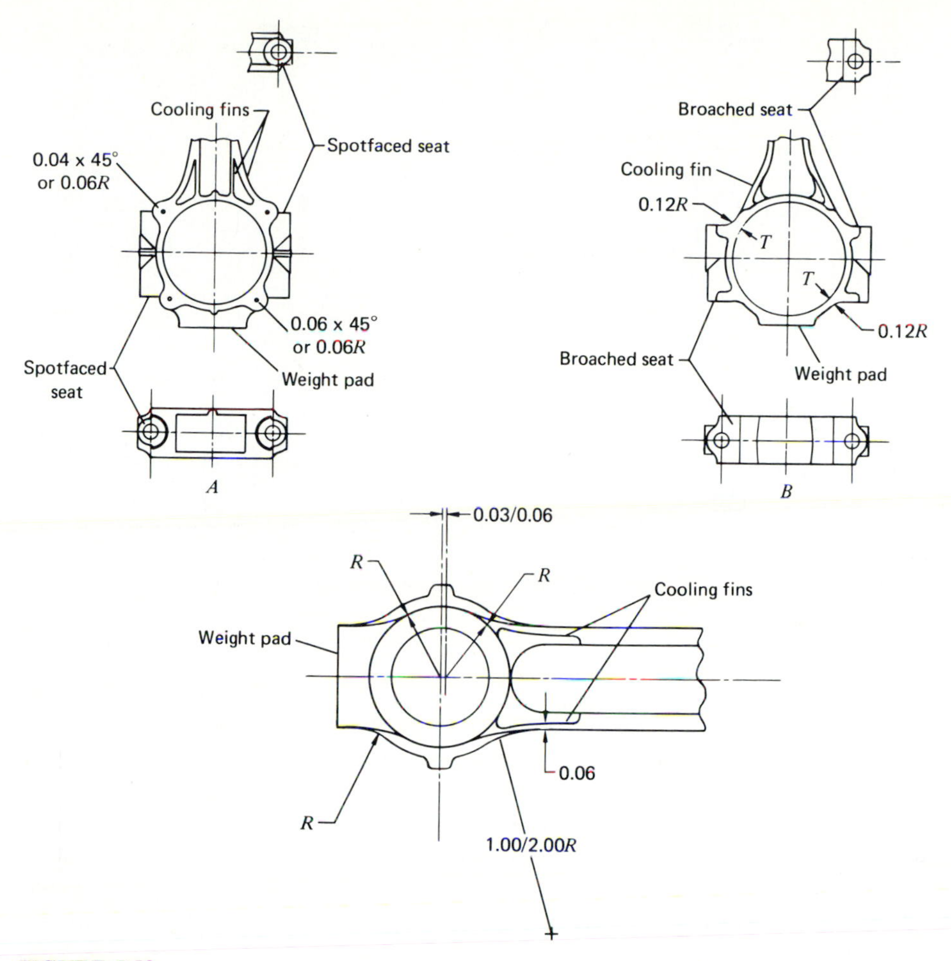

FIGURE 9.32
Typical design details for the crankpin and wrist-pin ends of the cast connecting rod.

A match plate showing the gates, risers, and other details of the arrangement reveals the care that was taken in maximizing the product yield when using the casting process (Fig. 9.33).

Match-plate features. Use of rounded corners, generous fillets, and uniform section thickness with increasing taper toward the direction of solidification provided for sound castings. Other general requirements include:

1. Draft, 4°
2. Stock for machining, 0.060 in.
3. Fillet radius, 0.12 in.; corner radii, 0.06 in.
4. Dimensional tolerances: connecting rod length ±0.01 in.; other dimensions, ±0.02 in.; gate removal, +0.06 to −0.03 in.

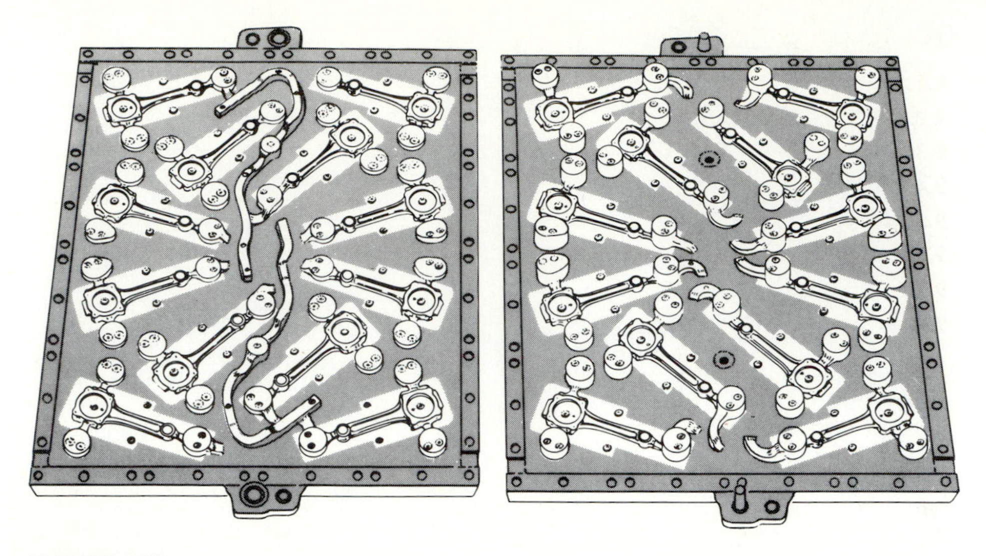

FIGURE 9.33
Match plate for the cast connecting rod. Note that 12 are cast in one mold, gating is from one end, and three risers are used on each rod.

Inspection features. Of course, not only must the proper melting procedures and foundry practice be followed but also hot trimming and mechanized inspection procedures must be provided. All connecting rods pass through conveyorized sonic, ultrasonic, and Magnaglow inspection stations, where defective castings are immediately removed from the line. Provision for such elaborate inspection procedures is needed in order to guarantee to the customer that the quality of cast product is equal to or exceeds the quality of rods produced by any other process.

SUMMARY

In this chapter, the fundamentals of sound casting design have been presented. Castings, in some form, represent the best way to produce many components. If the engineer designs parts that are simple and easy to produce, there is no doubt the end product will be able to meet international competition. The principal cost reductions and cost avoidances are those that result from sound product design. The engineering student is urged to become familiar with good casting design for the various casting processes. Since castings are used to a large extent in practically all of the metal trade industries, it is highly probable that practicing engineers will be confronted with problems relating to casting design on many occasions no matter whether their efforts be centered in the functional design of products or in manufacturing, where the principal concern is the utilization of the optimum process.

QUESTIONS

9.1. What automobile components were converted to castings by General Motors?

9.2. What design factors should the engineer observe when specifying castings?

9.3. Explain why it is difficult to provide exact allowances for shrinkage.

9.4. How can riser size be calculated in order to feed a given casting satisfactorily?

9.5. What is meant by directional solidification?

9.6. Give a sketch of the ideal shape of a casting to obtain directional solidification.

9.7. Why is steel difficult to cast?

9.8. A corded pattern is 42 in. long and is being used for the production of gray iron castings. What shrinkage in inches should the pattern accommodate? What machining allowance in inches should be provided on the bore? On the outside diameter?

9.9. When would you advocate aluminum patterns? Brass patterns? Gray iron patterns?

9.10. What color would you apply to an aluminum pattern that forms a cavity in a mold that is refilled with sand prior to pouring?

9.11. For what reasons are molds vented?

9.12. What is the purpose of core prints?

9.13. For what reason are clamps and/or weights placed on a mold?

9.14. How are patterns classified?

9.15. What factors should be considered in pattern design?

9.16. What is the function of chaplets? Make a sketch of chaplets in use.

9.17. What information is the engineer expected to provide on the engineering drawing?

9.18. What precautions should be observed in the design of a steel casting so that internal stresses are as small as possible?

9.19. Why is it especially important not to have thin sections when designing steel castings?

9.20. What is meant by hot tears? How can they be eliminated?

9.21. What is a misrun? How is it caused?

9.22. What are sand inclusions and how can they be minimized?

9.23. What finish allowance would you specify for a gray-iron casting 5 ft in length?

9.24. What is the strength–weight ratio of aluminum versus bronze permanent-mold castings?

9.25. Why is it difficult to core holes less than $\frac{1}{4}$ in. in diameter in permanent-mold castings?

9.26. Describe the various types of core-boxes and accessories.

9.27. Explain why both the size and number of blowholes are important in connection with the blow plate in a core-box blowing system.

9.28. Identify those important design considerations that the engineer should observe in the design of inserts.

9.29. Explain how Bernoulli's theorem is utilized in the design of a gating system.

9.30. What is the function of the riser?

9.31. Explain why gray iron with a carbon equivalent of 4.3 may actually expand rather than shrink during freezing.

9.32. What would be the maximum distance between two risers on a plate casting that is 3 in. thick?

9.33. Explain the use of Chvorinov's rule in the design of risers.

9.34. What tolerance across the parting line would you assign to a plaster mold casting made in a 12 × 21 × 6-in. flask?

9.35. Would you be able to cast a taper with a knife edge using 60–40 brass and the plaster-mold method?

9.36. What precision is characteristic of investment castings?

9.37. What is the approximate size of the largest casting that is economically die cast today?

9.38. From an economic standpoint, for a precison casting of 1 in.^{3} in volume and daily production requirements of 2000 pieces, would you recommend permanent molding or die casting. Explain your answer.

9.39. How could die life be prolonged for the die casting of zinc-based parts?

9.40. If porosity were causing trouble in the die casting department of your plant, what steps would you take to minimize this complaint?

9.41. Why are the weight and area limitations of zinc-based die casting greater than for aluminum-based die castings?

9.42. When would you recommend incorporating internal threads on die-cast parts?

PROBLEMS

9.1. Sketch a match-plate layout for the gate valve casting shown in Fig. P9.1. Determine the parting line, the number of castings on a plate if the production rate calls for delivery of 800 valves per month excluding scrap that has run at an average of 8 percent of castings for porosity at the leak test that follows the machining operations at the customer's plant. The casting is a globe valve for domestic water systems, so choose the alloy accordingly. Specify the alloy to be used, the casting process, and the type of melting equipment. Would you modify the design of the part to make a better casting? If so, how? What gating ratio would you use? Show the calculations on your drawing.

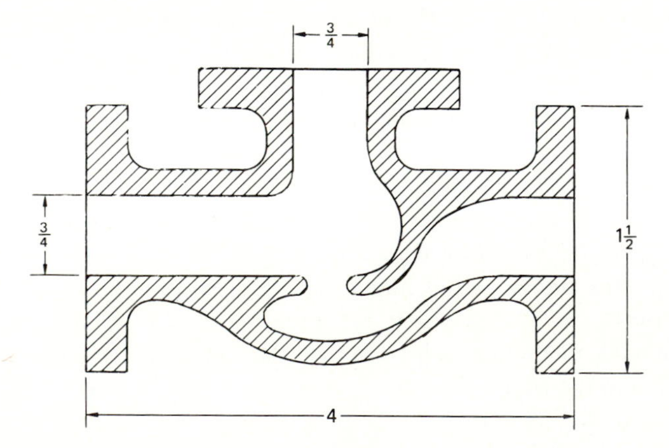

FIGURE P9.1

9.2. Design a cantilever load-carrying member for a machine tool that must withstand 2500 lb at its outer end. Use a flange design with a supporting rib. What is the best design in terms of strength-to-weight ratio. Compare this design to that of a steel weldment using an I beam to carry the same load. Use a class 40 gray iron for the cast beam.

9.3. A small aluminum casting is required to be cast to close dimensional tolerances, including the location of its two cored holes. The casting weighs 5 lb. Specify the tolerances and pattern equipment and materials for production quantities of 5, 100, and 20,000 castings. The part has an average wall thickness of 0.200 in. and is the cover half of an enclosure for electronic equipment that will be exposed to outdoor environments throughout the world.

9.4. The part shown in Fig. P9.4 is to be made as a permanent-mold casting. Suggest what changes in design should be made and explain why. Make a careful sketch to scale to show the details of the part redesigned for permanent molding.

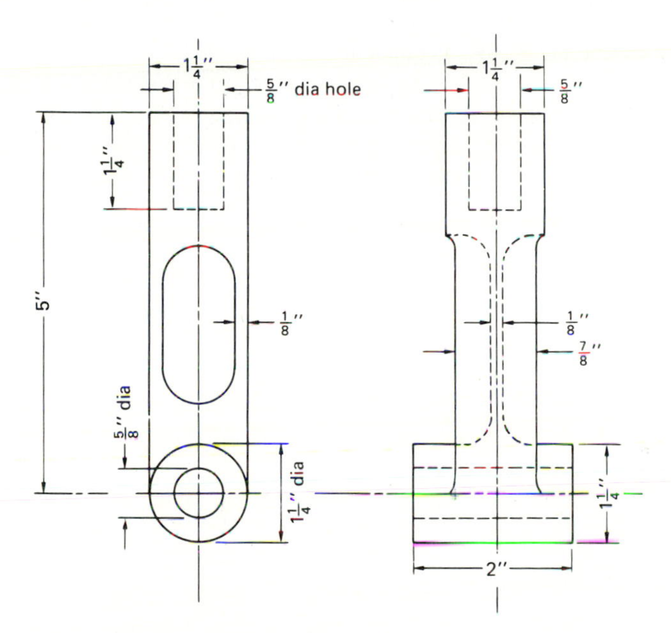

FIGURE P9.4

9.5. Given below are the solidification times for top cylindrical risers that are 4 in. diameter × 4 in. high for steel, copper, and aluminum.

	Solidification times, min			
Metal	**No treatment**	**Insulating sleeve**	**Radiation shield**	**Insulation + shield**
Steel	5	7.5	13.4	43.0
Copper	8.2	15.1	14.0	45.0
Aluminum	12.3	31.1	14.3	45.6

(*a*) Determine the effective constant for Chvorinov's rule from the data for each of the metals listed above.
(*b*) Discuss the relative effect of radiation shielding for each metal listed, in terms of specific heat, latent heat of fusion, and pouring temperature.
(*c*) Repeat item (*b*) in terms of using an insulating sleeve only.
(*d*) Discuss why the solidification times for the steel and the aluminum risers are about the same when they have both insulating sleeves and radiation shielding.

9.6. Explain the use of blind risers and the type of alloy that can be fed by them. Calculate the height in inches to which a blind riser can feed a casting that has a specific weight in the molten state of 0.264 lb/in.^{3} if it freezes like low-carbon steel.

9.7. Derive an equation for the injection velocity of an alloy in a die-casting operation as a function of the maximum plunger pressure p. Assume that the shot chamber is large in cross section compared with the ingate, the cavity is vented through the entire injection process, the energy at any point in the molten metal stream is a constant, and all friction and orifice coefficients are neglected.

9.8. A casting as shown in Fig. P9.8 is produced in a green-sand mold that measures 30 × 30 in. with a 6-in. cope over an 8-in. drag. The casting is specified to be class 40 gray iron. Determine the total weight of metal required to fill the mold, the gating ratio, whether the gating system is pressurized or unpressurized, and the casting yield, i.e., the ratio of good casting to the total weight poured.

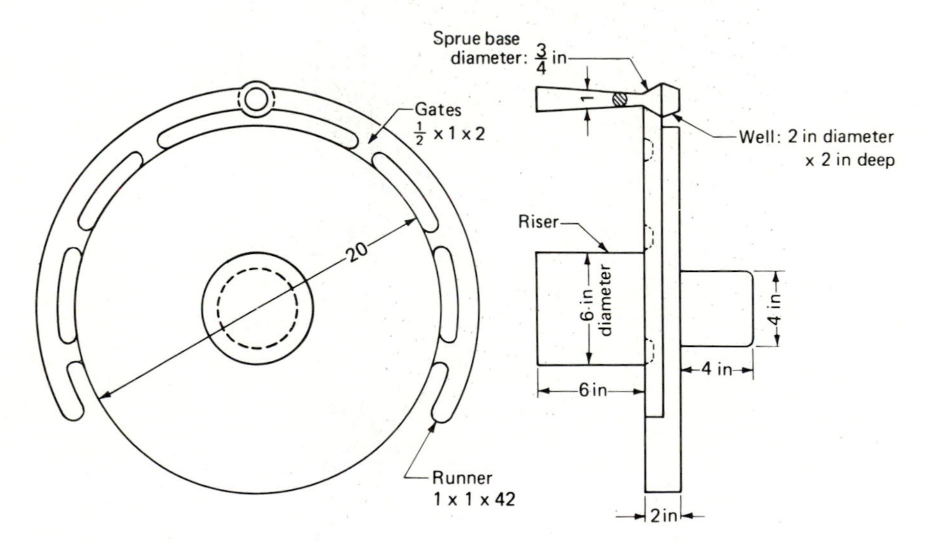

FIGURE P9.8

9.9. Two castings are molded in green sand. They differ in weight by a factor of 3.8 but they are both cubes. An experiment has shown that the lighter casting solidifies in 8.7 min. How much time would you estimate that it would take for the larger casting to solidify?

9.10. In true centrifugal casting, i.e., with spinning about a horizontal axis, too low a speed will permit slipping or raining of the metal and too high a speed will result in hot tears on the periphery of the casting. A force on the metal rim of about 75g works well for sand molds, whereas about 60g is sufficient for metal molds because of their greater chilling power. A force of as much as 100g is needed for molds spun about a vertical axis.

(*a*) Derive a relationship to express the minimum force in terms of g to make a true centrifugal casting D in. in diameter and spinning at n rev/min. Use the centrifugal force on the annular ring and the angular velocity to obtain the relationship.

(*b*) Find the centrifugal force f at any radius r for a molten metal of specific weight d.

(*c*) Use the relationship obtained above to calculate the rotational speed required to cast ductile iron centrifugally in the form of a pipe with a $\frac{3}{8}$-in. wall thickness. What speed would be needed if the casting were produced in a sand mold? What speed if cast in a metal mold?

(*d*) If an unbonded sand layer is first distributed uniformly throughout the inner surface of the centrifugal mold, explain why molten metal can be poured into the mold without dislodging the sand layer.

(*e*) Why is a greater force needed to make a centrifugal casting with a vertical axis than is required to make the same casting about a horizontal axis. Consider the wall thickness and contour in each case.

REFERENCES

1. American Foundrymen's Society: *Metal Casting and Molding Processes,* DesPlaines, Ill., 1981.
2. American Society for Metals: *Metals Handbook,* 8th ed., vol. 5, Forging and Casting, Metals Park, Ohio, 1970.
3. Dantzig, J. A., and Berry, J. T. (eds.): *Modeling of Casting and Welding Processes, II,* Metallurgical Society of AIME, Warrendale, Pa., 1984.
4. Flemings, M. C.: *Solidification Processing,* McGraw-Hill, New York, 1974.
5. Gray and Ductile Iron Founders' Society: *Gray and Ductile Iron Castings Handbook,* Rocky River, Ohio, 1971.
6. Heine, R. W., Loper, C. R., and Rosenthal, C.: *Principles of Metal Casting,* 2d ed., McGraw-Hill, New York, 1967.
7. MacLaren, John L.: "Die Castings," in *Handbook of Product Design for Manufacturing,* edited by James G. Bralla, McGraw-Hill, New York, 1986.
8. Malleable Founders' Society: *Malleable Iron Castings,* Rocky River, Ohio, 1960.
9. Minkoff, L.: *The Physical Metallurgy of Cast Iron,* Wiley, New York, 1983.
10. Spinosa, Robert J.: "Investment Castings", In *Handbook of Product Design for Manufacturing,* edited by James G. Bralla, McGraw-Hill, New York, 1986.
11. Steel Founders' Society of America: *Steel Castings Handbook,* 5th ed., Rocky River, Ohio, 1980.
12. Walton, C. F., and Opar, T. J. (eds.): *Iron Castings Handbook,* 3d ed., Iron Castings Society, DesPlaines, Ill., 1981.
13. Zuppann, Edward C.: "Castings Made in Sand Molds", In *Handbook of Product Design for Manufacturing,* edited by James G. Bralla, McGraw-Hill, New York, 1986.

CHAPTER 10

BASIC MANUFACTURING PROCESSES: SOLID STATE

When metals are formed at temperatures that exceed their recrystallization temperature, which is about 50 percent of their absolute melting temperature (Fig. 10.1), they are being *hot-worked* and behave as perfectly plastic materials (Fig. 10.2). The products of such hot-working operations are called *wrought* metals, and are important engineering materials because they are worked under pressure to:

1. Obtain the desired size and shape from the original ingot, thereby saving time, material, and machining costs
2. Improve the mechanical properties of the metal through: refinement of the grain structure; development of directional "flow lines" (Fig. 10.3); and breakup and distribute unavoidable inclusions, particularly in steel
3. Permit large changes in shape at low power input per unit volume

Examples of wrought metals are structural shapes such as *I* beams, channels, and angles; railroad rails; round, hexagonal, and square bar stocks; tubes and pipes; extrusions and forgings. Most wrought products become the raw material for secondary processes that produce finished items by cutting, forming, joining, and machining.

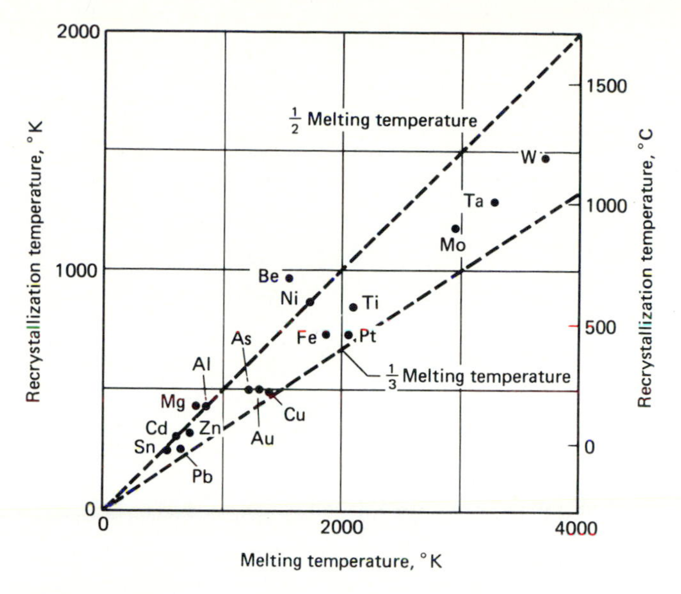

FIGURE 10.1
Recrystallization temperature of a number of metals as a function of their melting temperature.

FUNDAMENTALS OF HOT WORKING

When a metal is hot-worked, as by rolling, it is passed through opposing rollers and reduced in section thickness. The original grain structure of the initial material is elongated and broken up in the deformation zone, and the fragments of crystals become nuclei for new, smaller crystals so that a fine-grained structure is produced at the exit end (Fig. 10.4). The higher the metal temperature and the longer the time at that temperature, the more rapidly and larger the new crystals will grow. The larger the grains, the less the hardness and strength. Hot working gives the following properties to an alloy.

1. Hot working makes little change in the hardness or ductility. Recrystallization is spontaneous and the resultant fine-grained structure is stronger than

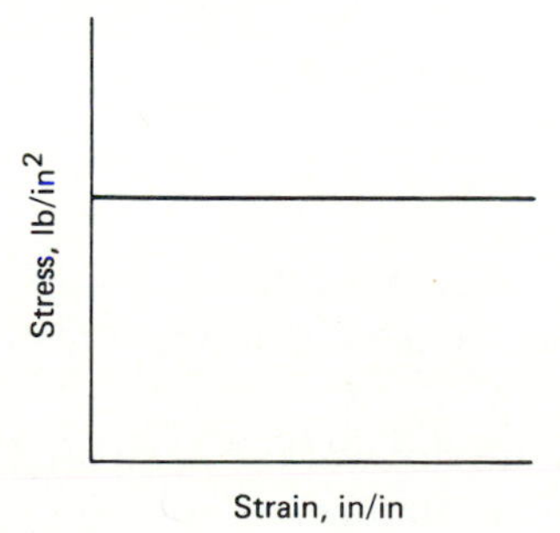

FIGURE 10.2
Typical stress–strain diagram for a hot-worked alloy during working.

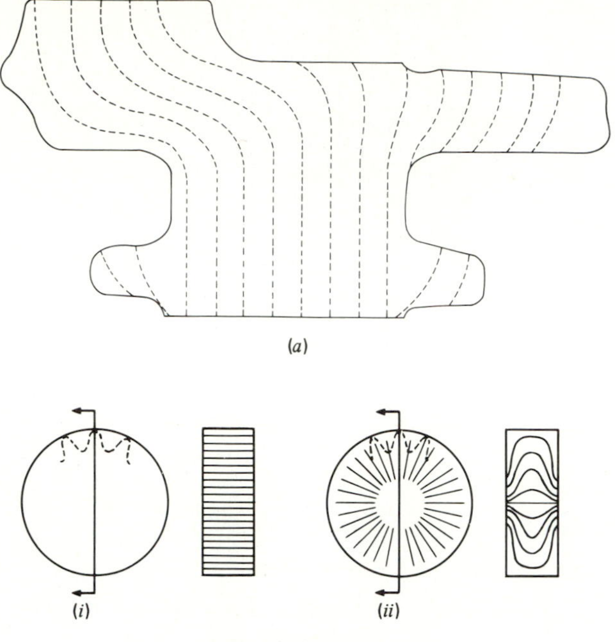

Direction of grain flow in a gear blank;
(i) bar stock and (ii) forged stock.

(b)

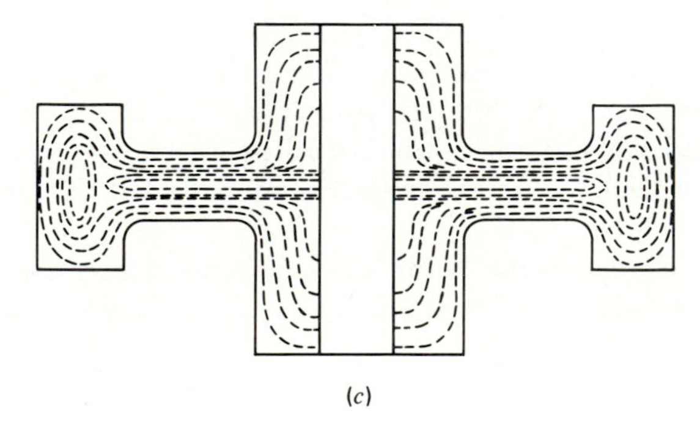

FIGURE 10.3
Typical fiber flow lines of a forging. (*a*) Flow lines in a forged crankshaft; (*b*) flow lines in a gear blank [*i*, bar stock, *ii*, forged blank]; (*c*) flow lines in a typical flanged gear blank.

before deformation, whereas cold working increases the strength and hardness of an alloy (Table 10.1).

2. The worked metal has enhanced properties in the direction of working because impurities segregate into stringers which lie parallel to the direction of metal flow.
3. Metals can receive large reductions when they are hot-worked compared to when they are cold-worked—and there will be no ruptures or hot tears.

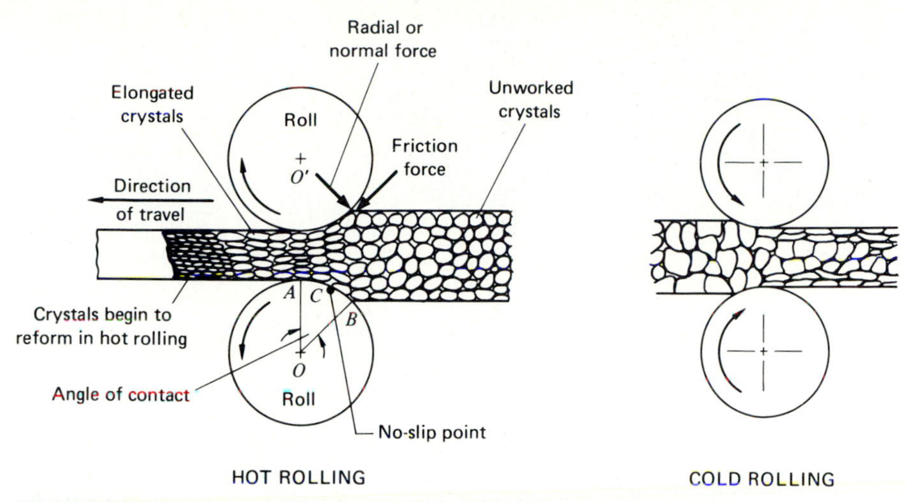

FIGURE 10.4
Hot rolling refines the grain structure, whereas cold rolling distorts it.

There are disadvantages too. Hot working of steels requires expensive heat-resistant tools, though the hot working of aluminum alloys is not so severe on tools because the temperatures are less and there is little scale and oxidation. In the forging of steel, close tolerances cannot be maintained, although with aluminum alloys this is not a problem.

Forging may be defined as shaping a metal under impact or pressure and improving its mechanical properties through controlled plastic deformation. Forged products can be given the optimum strength properties with a minimum variation of properties from piece to piece. Typical structures of a casting, bar stock, and forging are given in Fig. 10.5.

TABLE 10.1
Comparison between properties of hot-worked and cold-worked metals

Alloy	Condition	Ultimate tensile strength, lb/in.^{2}	Yield strength, lb/in.^{2}	Hardness, Brinell No.
Aluminum 1100	O	16,000	6,000	28
	H18	29,000	25,000	55
Electrolytic tough	Hot-rolled	34,000	10,000	42
pitch copper	Extra spring	57,000	53,000	99
Steel, SAE 1010	Hot-rolled	62,000	32,000	94
	Cold-rolled	81,000	50,000	174
Brass, yellow	Annealed	49,000	17,000	57
	Spring hard	91,000	62,000	174

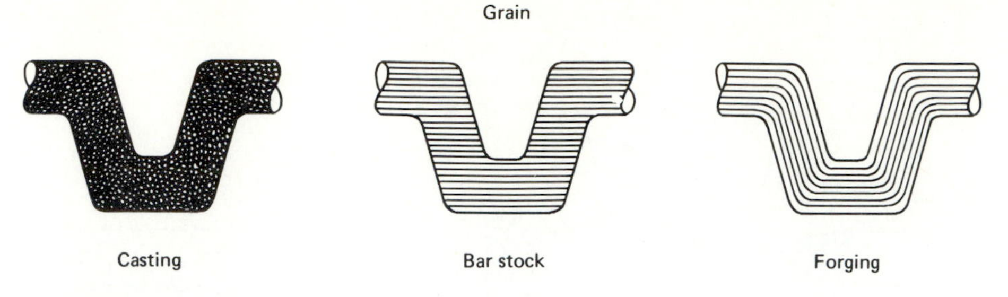

FIGURE 10.5
Schematic diagram of the grain flow in a casting, bar stock, and forging.

Forgings have a history of structural integrity that is not surpassed by any other metalworking process. Internal voids and gas pockets are dispersed and welded shut. Forging is also foremost in uniformity and resistance to impact and fatigue. Yet when all factors are considered, forged connecting rods, crankshafts, and steering knuckles have had to yield to castings in a number of cases, on the basis of economics, despite the fact that these parts are critically stressed. The ability of the casting to put the metal where the stress is located has outweighed the advantage of enhanced strength because of the fiber flow pattern. In other cases, the total cost of the casting is less because of the improved machinability of the equivalent cast alloy; or the ductility, strength, and economy of a properly designed ductile iron casting have made it more economical than a competitive forging. Welding and cast-weld design also give forgings severe competition. But in nonferrous forgings the improvement in properties and the low strength–weight ratio frequently give large aluminum forgings a competitive edge.

Materials

The predominant characteristic of a forging material is its tough, fibrous structure. When a heterogeneous material is formed or rolled to shape in a suitable temperature range, the distortion and subsequent recrystallization of the material results in a fibrous structure. In steel forging, this fibrous structure imparts directional properties that are an advantage when the lines of flow of the fibers are parallel to the applied forces and a disadvantage when forces are applied at an angle to the lines of flow. There is less resistance to shearing forces parallel to the lines of flow than to those applied across the lines of flow.

Forged materials have the following special properties.

1. They have a homogeneous structure, free from voids, blowholes, and porosity
2. They have greater strength per unit of cross-sectional area under static loads, and better resistance to shock.

3. They have superior machining qualities. The uniform structure permits higher machining speeds; the freedom from inbedded impurities allows longer tool life and fewer grindings; and homogeneity reduces machining scrap.

Depending upon the initial quality of the ingot and the processing history of the part, the wrought product may either be superior to the cast product in every respect or be only partly superior in some respects and distinctly inferior in others. The directional properties can be reduced by working the material in all directions. Grain size has very little effect on working properties of material that has high ductility. Grain size is controlled by heat treatment.

Forgeable Metals[1]

An almost unlimited variety of forging metals is available in ferrous and nonferrous alloys. The following are general classifications of forgeable metals. An exhaustive treatment may be found in books on forging and metallurgy.

1. Carbon steels:
   a. Low-carbon (up to 0.25 percent)—forgings for moderate conditions and for carburized parts where resistance to abrasion is important.
   b. Medium-carbon (0.30 to 0.50 percent)—forgings for more severe service. Some heat treatment is generally desirable.
   c. High-carbon (above 0.50 percent)—forgings for hard surfaces and for springs. Heat treatment is essential.
2. Alloy steels (manganese, nickel, nickel–chromium, molybdenum, chromium, vanadium, chromium–vanadium, tungsten, silicon–manganese).
3. Corrosion- and heat-resisting steels and stainless steels: Generally, but not necessarily, forged surfaces should be polished to obtain the full benefit of corrosion-resisting properties.
4. Iron: Either wrought iron or ingot iron is forged for special applications where ductility is required. Wrought iron furnishes a moderate degree of corrosion resistance. The copper-bearing irons and low-carbon steels are in this class.
5. Copper, brasses, and bronzes.
6. Nickel and nickel–copper alloys: Pure nickel is forgeable. The alloy of nickel and copper known as Monel metal offers a desirable combination of strength, toughness, and corrosion resistance.
7. Light alloys (aluminum, magnesium).
8. Titanium alloys.

---

[1] *Standard Practices and Tolerances for Impression Die Forgings,* (1963) Drop Forging Association.

Note that any ductile metal can be forged. The material is selected primarily for its ultimate properties in the part, such as corrosion resistance, strength, durability, and machinability. The forging process is a secondary consideration. It can be used as long as the high tooling cost can be spread over a large number of pieces.

Effect of Temperature on Material

The forming properties of any metal or alloy depend on the temperature of the material.

1. The hot-working range is near the melting point. The working temperature cannot exceed the melting point of any one of the elements of the alloy. Many such alloys can only be cold-worked. Most hot-worked metals do not acquire a permanent hardening.
2. The cold-working range usually is near room temperature. Most metals strain-harden when cold-worked, except zinc, lead, and tin.
3. At low temperatures, steel becomes brittle. Face-centered cubic metals like aluminum, copper, nickel, gold, and platinum remain ductile at all temperatures.

The forging processes work material cold or hot, depending upon the nature of the material and its size. The ability to work cold material depends upon its ductility and malleability, as indicated by its stress–strain curve. The ability to work hot material depends upon its range of plasticity at higher temperatures. The greatest ductility is near the melting point. In general, most materials become more plastic as the temperature increases. The characteristics of each material at higher temperatures should be studied. For example, steel has a blue-brittle range between 450 and 700°F. Also, materials may oxidize rapidly and objectionable scale may form at high temperatures. The region of plasticity before the material becomes liquid may be very short and close temperature control may be required.

Stress–strain curves cannot predict the behavior of materials formed at various forming speeds. A study of equilibrium diagrams and phase changes used in metallurgy indicates little about the forming properties of a material. Experiments are required to determine the best temperatures and speeds of forming. Here again, statistical methods are quite helpful. Let us look at a typical example. A four-impression die involving drawing, edging, prefinal, and final form is designed. The part has intricate geometry and the material being forged is alloy steel. The production design engineer notes that there is considerable variation in the total number of blows required to shape the part. He suspects that the supplier of the steel is permitting undue variation in the material being furnished, since the other parameters, including forging temperature and forging speed, are being closely controlled. He takes four samples of five forgings. Obviously, he would not expect each sample to have the same mean. He would, however, expect the within-sample variation to be not significantly different from the between-sample variation, unless the samples were drawn from different populations.

The results of these four samples are:

	Sample 1	Sample 2	Sample 3	Sample 4
	7 blows	7 blows	4 blows	10 blows
	6 blows	8 blows	4 blows	9 blows
	8 blows	8 blows	5 blows	9 blows
	5 blows	7 blows	4 blows	9 blows
	4 blows	5 blows	3 blows	8 blows
Totals	30	35	20	45
	$\bar{x}=$ 6	7	4	9
	$N=20$	$\bar{\bar{x}}=$ 6.5 blows		

Using simple analysis of variance, we get:

Source of variation	Sum of squares	Degrees of freedom	Variance estimate
Between samples	65	3	$\frac{65}{3}=21.67$
Within samples	20	16	$\frac{20}{16}=1.25$
Total	85	19	

Using an F test, we get a computed value of F equal to:

$$F=\frac{\text{greater estimate of the variance of the population}}{\text{smaller estimate of the variance of the population}}$$

$$=\frac{21.67}{1.25}=17.4$$

Comparing this value of F with the tabulated values for $F_{.99}$ (ninety-ninth percentile of the F distribution), which is 5.29, we can conclude that it is highly probable that the samples were not drawn from the same population. Thus, either a change in the steel furnished, a variation in forging temperature, or some other parameter was altered.

Failures

Failures of material as it is worked or formed depend on a combination of many factors. Material, as it is forced into the various types of tools and

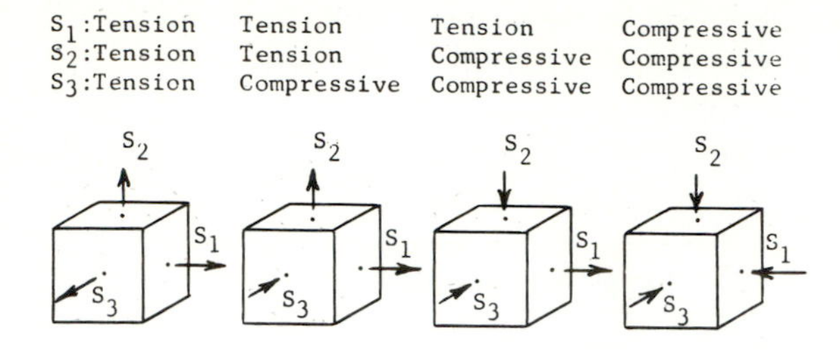

FIGURE 10.6
Four possible combinations of principal stresses.

processes, may fail in shear, compression, buckle, or neck; or it may break under tension. If the friction between material and tool is great and the forces are insufficient, the material may not fill the die and conform to the proper shape. Overlaps and folds may occur because of friction between die and material, or as a result of buckling. The cause of split sheets and cracked or buckled edges can be traced by analyzing the forces applied to the material.

The combination of forces shown in Fig. 10.6 can be applied to a small cube of material. The principle can be illustrated by cutting a cube from a rubber eraser and applying forces according to Fig. 10.6.

External forces always cause other forces in other planes. For example, in Fig. 10.7 a compressive force S_2 on a cube of material will result in tension force on the periphery at S_1. Thus, we can have failure of material in the form of cracks around the middle due to tension forces (S_1) at right angles to the compression forces (S_2), which force the material into the center and outward faster than the outer material can stretch. This same phenomenon occurs when edges crack as sheets are rolled without proper annealing and the right amounts of reduction at each roll. Failures of materials because of splits, tears, and cracks can be understood by the action of the applied forces. A modification of these forces by annealing, reduction of the speed of the material passing through the rolls, or reduction of working pressures will prevent these failures. The engineer learns from these failures to improve his designs and tools.

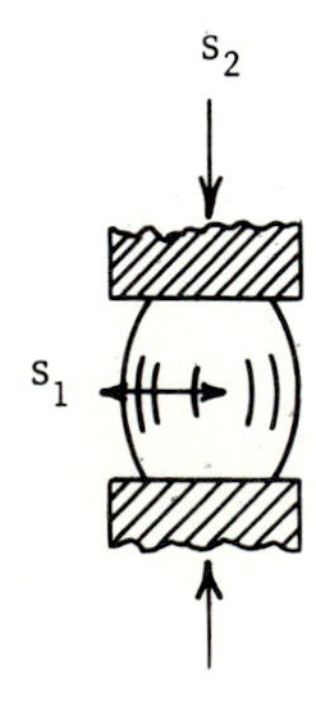

FIGURE 10.7
Cracking occurring in direct-compression-type working processes owing to the presence of secondary tension (S_1).

TABLE 10.2
Relative forgeability of various alloys (in order of decreasing forgeability)

Good	Somewhat difficult	Difficult	Very difficult
Aluminum alloys	Martensitic stainless steel	Titanium alloys	Nickel-based alloys
Magnesium alloys	Maraging steel	Iron-based super alloys	Tungsten alloys
Copper alloys	Austenitic stainless steel	Cobalt-based super alloys	Beryllium
Carbon and alloy steels	Nickel alloys	Columbium alloys	Molybdenum alloys
	Semiaustenitic PH stainless	Tantalum alloys	

Forgeability

The basic lattice structure of metals and their alloys seems to be a good index to their relative forgeability. The face-centered cubic metals are the most forgeable, followed by body-centered cubic and hexagonal close-packed. In alloys, other metallurgical factors may affect their forgeability, such as: (1) the composition and purity of the alloy, (2) the number of phases present, and (3) the grain size. Forgeability increases with billet temperature up to the point at which a second phase appears or where there is incipient melting at grain boundaries or if grain growth becomes excessive. Certain mechanical properties also influence forgeability. Metals that have low ductility have reduced forgeability at higher strain rates, whereas highly ductile metals are not so strongly affected by increasing strain rates.

The relative forgeability of various alloys is presented in Table 10.2.

HOT-WORK PROCESSES

Hot-working processes, i.e., rolling, forging, upsetting, and extruding (Fig. 10.8) change the shape of metals by pressure. The metal is brought to the

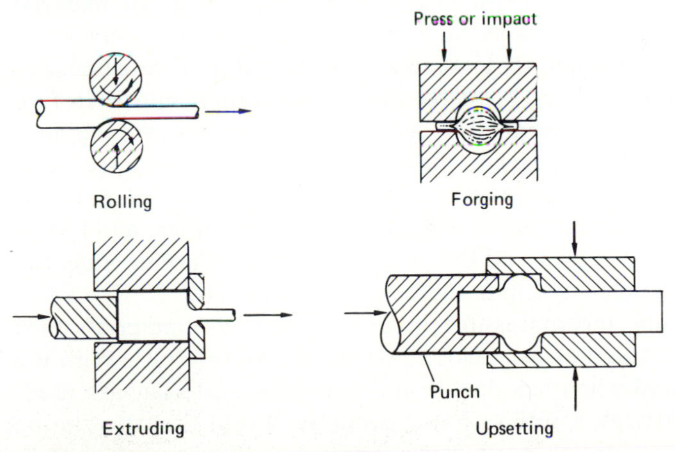

FIGURE 10.8
Schematic diagram of the four major hot-working processes.

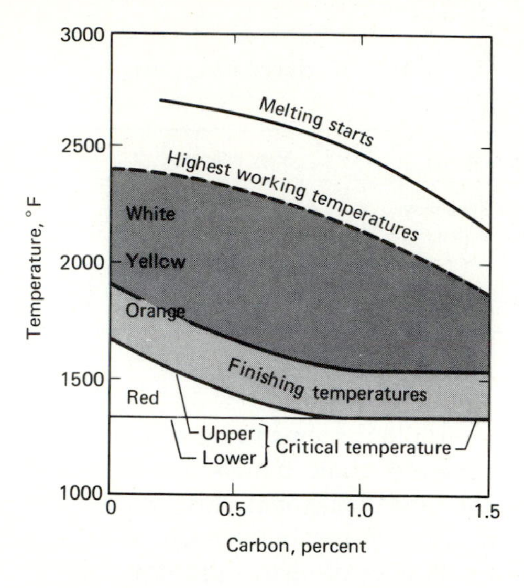

FIGURE 10.9
Range of rolling and forging temperatures for carbon steel.

viscous or plastic state by subjecting it to elevated temperature, and under the influence of pressure it flows without fracture, in contrast to the casting process in which the metal flows as a liquid and is chilled and solidified in the mold or die. In forging and other metalworking processes, the metal may be cold or it may be heated in order to bring it to a more plastic condition. The hot working of metals is accomplished at temperatures from 1700 to 2500°F (Fig. 10.9) for ferrous materials, from 1100 to 1700°F for copper, brasses, and bronzes, and from 650 to 900°F for aluminum and magnesium alloys.

In forging, the die is often heated slowly to working temperature before starting the operation so that stresses due to differential temperatures will be avoided. Also, cold dies in hot-forging operations may chill the blank sufficiently to cause considerable resistance to metal flow. The hot material remains in the die long enough to take the form of the die and is not chilled to any degree. Materials, even when they are in a plastic condition, resist change of shape with considerable force. Most operations are rapid, and therefore large amounts of power must be applied by the machines that operate the dies. This quick application of power heats up the material and aids the plastic flow. The art in forging, pressing, and working this viscous, doughy material is obtaining the final shape with the least number of tools, operations, and loss of material. The skills of the designer and the forger join in the art of creating the shape of the part and the dies to obtain the best flow of metals.

Metal is practically incompressible; therefore, excess material must escape or it will prevent the dies from closing or will break the rolls. With the more complex shapes and when less ductile and malleable materials are used, it is necessary to go through a series of stages before the final shape can be produced from the raw material. Each stage requires a different set of tools and sometimes requires different equipment and skills. Reheating of the material is usually required between these stages.

The forging and metalworking equipment described in this chapter requires hydraulic, steam, compressed-air, and mechanical systems, and all forms of mechanical actuating devices and controls. The heat is supplied by coal, oil, gas, and electrically heated furnaces. Sometimes resistance heating or induction heating is applied directly to the material in the machine.

HAMMER AND PRESS FORGING

The original forging of metals was done with a hammer held in the artisan's hand. The blacksmith with his forge and anvil shaped metal into useful objects; up until 1920 (and even now in a few places) he could be observed by the boys who were to become the future engineers. Today only a few engineering schools teach hand forging in their laboratories, because the hand process has been mechanized by industry.

When steam activates a vertical piston attached to the hammer, we have the steam drop hammer that can be controlled by a skilled operator as easily as the hammer in the hands of a blacksmith. Handling the raw material are mechanical and electrical manipulators that perform the same operations as the arm, the wrist, and the hand holding the tongs. The board-type drop hammer is a drop hammer in the true sense, because the hammer is attached to and lifted up by a board clamped between two counterrotating rollers. When the hammer is at various heights, the pressure between the rollers is released and the hammer drops on the material and anvil. Drop hammers are very noisy and the shock of impact is difficult to isolate. The size of the drop hammer is indicated by pounds of falling weight. Board drop hammers range in size from 200- to approximately 8000-lb falling weight and steam hammers range from 600 to about 50,000 lb.

The Chambersburg impactor has two hammers placed in horizontal position so that the two die parts held by impellers or hammerheads meet in the center. The impellers are actuated by air cylinders that drive them and return them to their resting positions. Special electronic controls regulate the entrance of air into the cylinders and govern the speed of impact and the location of the point of striking on the centerline of the machine (Fig. 10.10).

The impactor is a new application of a long established principle that states that when two inelastic bodies of equal mass traveling at the same but opposite speeds collide, both bodies come to rest with a complete absorption of energy. If the bodies are elastic, the rebound of both is equal and opposite. The work done in both cases in bringing the two bodies to rest is concentrated at the impacting faces of the two bodies. Figure 10.11 compares conventional forging with the forging results of the impactor. No shock is felt when the impactor strikes. The equipment does not require special foundation or buildings. Multiple operations can be performed rapidly by placing additional

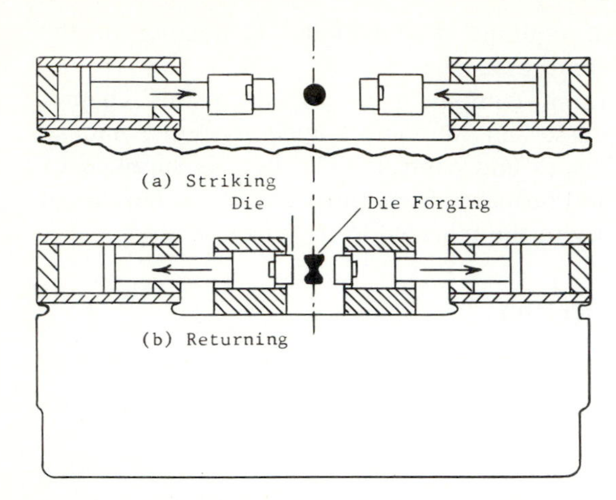

FIGURE 10.10
Operation of the impactor.

units side by side so that no reheating is necessary between operations. The part is passed by a conveyor from one impactor to the other.

Hydraulically Actuated Forging Presses

Hydraulically actuated forging presses are much slower than drop-hammer presses. In press forging, pressure or squeeze is applied to the raw material and the intensity of this pressure increases as the plastic metal resists deformation (Fig. 10.12). Owing to the great pressures available, these presses

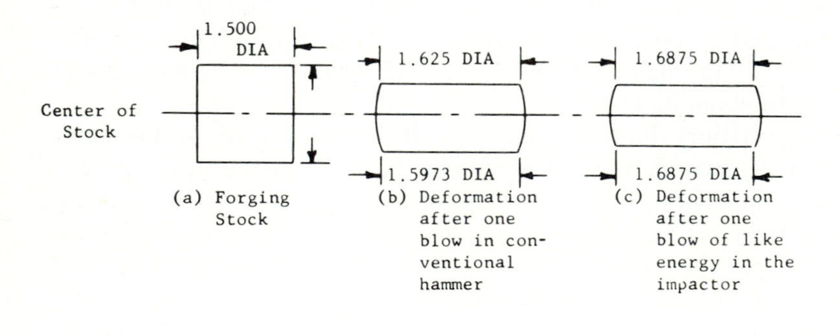

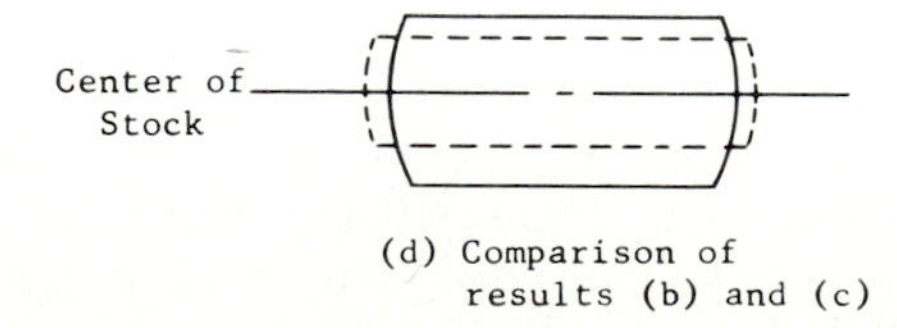

FIGURE 10.11
Comparison of results of a single blow of a given intensity on the same type of forging stock in conventional hammers and in the impactor.

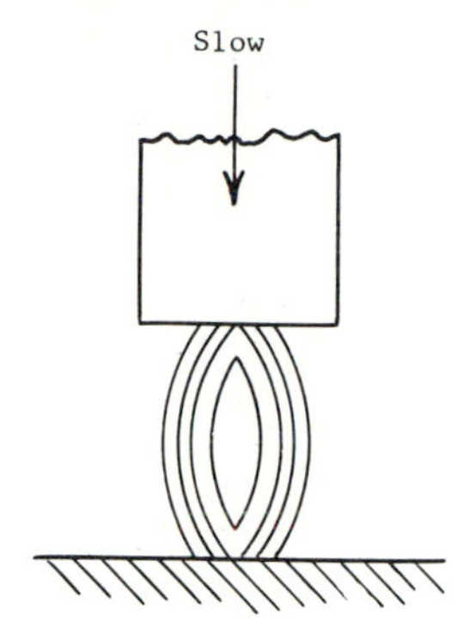

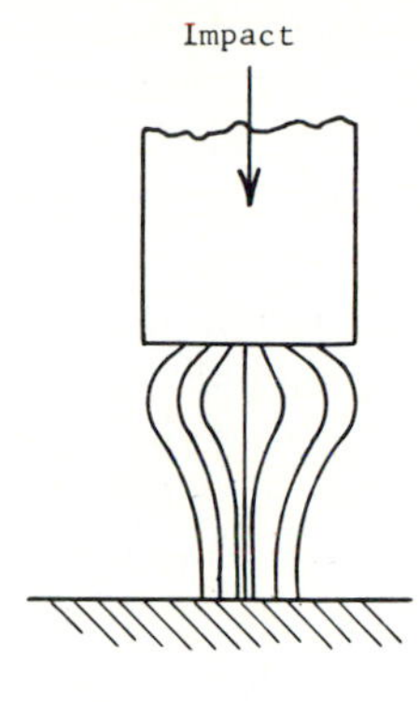

FIGURE 10.12
Flow of material under slow pressure or squeeze and under impact.

can be made to have very large capacities. In World War II, Germany used a 30,000-ton press; presses of more than 75,000 tons are built today. Such presses, the equipment to operate them, and the furnaces to heat the billets or blanks cost several million dollars. The presses are used to develop materials and processes for the cold extrusion of steel and extrusion molding. Of course, hydraulic-actuated presses also are made in small sizes.

Operator skill is an important factor in the production of hammer and press forgings. The press may be entirely operated by one man, although more frequently, depending upon the size and nature of the work, two or more operators may work as a team. Hammer forging has low setup and tool costs and therefore is economical for low-quantity items. Tolerances should be liberal and in line with skills and customary practices; otherwise costs will be excessive. The cost of heating the material is a major item, as well as the loss of material due to scaling and removal of crop ends.

Mechanically Actuated Hammers

For the most part, mechanically actuated hammers are made in small sizes. They impart a series of blows regulated by the speed and power applied to the hammer beam pivoted at one end. The power goes through a spring system, which permits greater amplitude of the beam, and thus a heavier blow is struck when full power is applied.

DIE FORGING

Die forgings are made in steel dies constructed of hard, tough, and wear- and heat-resisting tool steel. As the mating halves of the die close, the material is shaped into the form of the die. When the die halves are separated, the forged part may be removed easily. In forming the part, several successive operations may have to be performed. The material may be shaped by extrusion or rolling and then cut to length. For example, preliminary to making a four-pronged handle for a water faucet, the shape may be cruciform before forging. This material may be partially formed into shape one or more times and then given the final form. The operations may be done in the same press and die, and the operations may be so close together that it is unnecessary to heat between them. As the part takes its final shape, a flash begins to form between the two halves of the die. This will be trimmed off either before or after the final forming operation. The amount of material within the die determines the thickness of the flash. The production-design engineer must exercise care in determining the amount of material to be inserted into the die. If there is not enough material, the part will not fill up the die; if there is too much material, the flash will be excessive, resulting in wasted material and greater die wear.

Parting lines usually are straight, although they may have a contour shape. In order to remove the forging from the die without distortion, the drafts in the die are made larger than in the casting and molding processes.

Die forgings usually are formed hot. The hot material must not stay in contact with the die too long, for the forging then will not be hot enough for the next operation. Also, if the hot material is in contact with the die too long, it will overheat the die and so cause excessive wear, softening, and breakage. The rapidity with which the part can be formed into uniform shapes that have uniform properties permits the die-forging operation to compete with other processes on high-quality parts.

The use of closed-impression dies improves both the strength and toughness of the metal in all directions. The fiber structure characteristic of metals can be formed so as to improve the mechanical properties in areas where they are most needed to meet specific service conditions.

After the forging operation, the part must be trimmed to remove the flash. If the carbon content is low and the forging is small, it is usually trimmed cold. Most medium-sized and large forgings are trimmed hot. Subsequent operations to remove the scale or oxide include shot blasting, tumbling, or pickling.

ROLL DIE FORGING

In roll die forging, often referred to as die rolling, the die configuration is placed on the periphery of the rolls and the material is formed as it passes between the rolls. The die cavities in the upper and lower rolls match and are held in position by a gear train on the ends of the rolls. The rolls are large in diameter and the frames and bearings are strong enough to withstand the forging pressures as the material is squeezed into shape. A series of dies can be placed side by side on one roll, so that the part can be passed through each die until the final shape is achieved. Such multiple operations are usually performed on single pieces. Roll forming is generally used for shaping parts that can be formed in a continuous chain from long bars heated to proper temperature and then cut to length. Such parts as automobile axle blanks, chain-sprocket rim segments, and shaker bolts are made by this method.

In many respects, roll die forging is similar to the regular mill technique of hot rolling, the principal difference being that in hot rolling several passes are made to bring the billet to the desired section, whereas in roll die forging only one pass is made.

The most important use of this process is in the preparation of preformed blanks for forging.

Die rolling is a fast, mass-production process that requires considerable volume to justify the investment in dies and equipment. On light articles, a minimum requirement of about 150 tons would be required, whereas on heavier parts a volume of 600 tons should be available.

It should be recognized that roll die forging has limited application to parts that can be made continuously in chain form. Any material that can be forged can be die-rolled. In fact, even materials that present difficulties when subject to forging operations, such as AISI 6412 steel, can be die rolled at temperatures from 1850 to 2300°F. Another form of rotating die consists of two disks, set at an angle, which rotate while pressed together. Dies are mounted on each disk and the forging is squeezed into one die by the disk that is set at an angle. The pressure is applied at only one point. Gear blanks are made in this manner.

Roll die forming permits close tolerances. Outside diameters may be held from $\pm\frac{1}{32}$ in. on small work to $\pm\frac{1}{16}$ in. on large work.

UPSETTING

Upsetting is similar to die forging in that the shaping of the part is done by dies. The term *upsetting* refers to the process of increasing the size of a portion of the part by forcing material from the rest of the part. The forming of the head of a bolt from a small rod is an example. Upsetting was performed by the blacksmith by heating one end of the rod and then hammering it on the end until the desired amount of material was available to produce the final shape. Today automatic machines clamp the hot rod and a plunger strikes the end, forcing the material into a die that shapes the formed head. The depth of the plunger travel determines the height of the head. Other machines can perform more than one forming operation, as well as cut the material to proper length.

HOT EXTRUSION

Extrusion (Fig. 10.13) is the process of shaping material by forcing it, under pressure, to flow through a die. In hot extrusion, the material is heated to a temperature that will cause it to become plastic without becoming liquid as it is forced through the die. In order to maintain its shape, the material is cooled rapidly on leaving the die. The hot-extrusion process is applied to both ferrous and nonferrous metals, plastics, ceramics, rubber, and other materials that have a plastic or doughy consistency.

Material can be extruded with internal holes running lengthwise. The sizes and shapes extruded are limited by the strength of the die and the

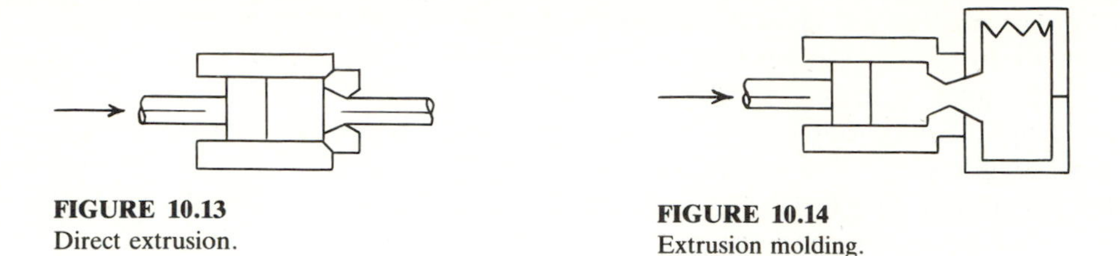

FIGURE 10.13
Direct extrusion.

FIGURE 10.14
Extrusion molding.

capacity of the press that forces the material through the die. Metals can be formed into a final bar shape with good surface and workable limits. Dies are not expensive and they are maintained by the supplier of the extruded material. Owing to the material's plastic condition on leaving the die, the bar may twist or bend and may have to be straightened after it cools. This distortion depends on the symmetry of the section. It is corrected by passing the section through drawing dies to obtain the final shape and straightness.

EXTRUSION MOLDING

Extrusion molding combines the hot-extrusion process with molding (Fig. 10.14). The dies are closed firmly so that the enormous internal pressures will not separate them. The plastic material is then forced in under extreme pressures. The metal remains in a plastic state—no portion of it is liquid—and therefore it does not have the shrinkage characteristics of die-cast materials, which cool from the liquid to the solid state. The raw metal is in billet form, prepared by the usual casting and forging methods. It is heated so that it can be extruded continuously. The heat and pressure change the solid material into a mushy plastic. Great pressures and a shock force are used to push the material into every portion of the die. There are no vents. The air or gas is trapped in overflow cavities so that the material is under pressure in all parts of the mold. Speed of entry and the shock are controlled by a hydraulic accumulator, which is permitted to fall at a given rate. Cores that can be pulled after the material is molded can be used. Bosses and flanges can be produced without setting up internal stresses, slip planes, or concentrated grain flow. The physical and mechanical properties of the resulting shape are comparable to those of the same material when forged, and are better than when the material is cast. The parts are uniform and accurate. The material breaks down or plasticizes as it passes through the restricting orifice, which reduces wear because the material does not touch the orifice except at the entrance. This orifice action is the same as that in the flow of fluids and as that described under impact extrusion (See Chapter 13). Because of the pressure and compression of the metal, heat is generated that causes the metal to be uniformly refined and to have a fine-grained, mushy, plastic consistency. Too great a reduction in the orifice, the wrong orifice contour, or too fast an extrusion speed may cause the metal to overheat and become partly liquid, a most undesirable condition. This causes segregation and actual burning, which results in a cavity full of oxidized or burned metal.

The process is controlled through electrical and hydraulic apparatus, and with properly regulated temperatures and skilled operators, uniform results that exceed those of other processes are obtainable. The extrusion-molded parts are designed similarly to other metal-molded parts in terms of uniform cross sections, radii, and entrance of material. The molding of high-temperature alloys presents the same problem of obtaining suitable materials and producing the molds, cylinders, and plungers that can withstand the high temperatures.

Since material molded by the extrusion-molding process is so uniform and possesses the best characteristics, the designer should base his factor of safety on the high proportional limits and the uniform performance of the actual part under test. In many cases, the weight of airplane parts can be reduced by one-third by using extrusion molding.

HOT ROLLING AND TUBE FORMING FROM SOLID BARS

Unlike the previously described processes, hot rolling and tube forming from solid bars are mill processes usually performed by the supplier of raw material. The end product of the processes is standardized as to size limitations, tolerances, finishes, and materials available. Occasionally, an engineer requires special mill shapes, such as elevator rails and conveyor tracks, for which a mill is willing to make special rolls and run an order of considerable tonnage. Since these processes are seldom used by the production-design engineer, they will not be described in this text. They are important processes used by the raw material supplier who furnishes special shapes for secondary operations.

HIGH VELOCITY FORMING

A number of machines have been designed that can achieve high rates of energy release for forging operations. Their major contribution to conventional forging operations is their high impact velocity which is 2 to 10 times larger than conventional velocities. For example, a large steam hammer with a total die, ram, and piston assembly weight of 50,000 lb can deliver 850,000 ft · lb at an impact velocity of 30 ft/s. In contrast, a high-velocity forming machine with a 2,500-lb ram, piston, and die-block assembly at a velocity of 100 ft/s can deliver 1,2500,00 ft · lb of energy at the moment of impact as determined by

$$KE = \frac{1}{2}\left(\frac{w}{g}\right)v^2$$

where w is the weight of moving ram and die, lb, g is the acceleration of gravity, ft/s^2 and v is the velocity at impact, ft/s.

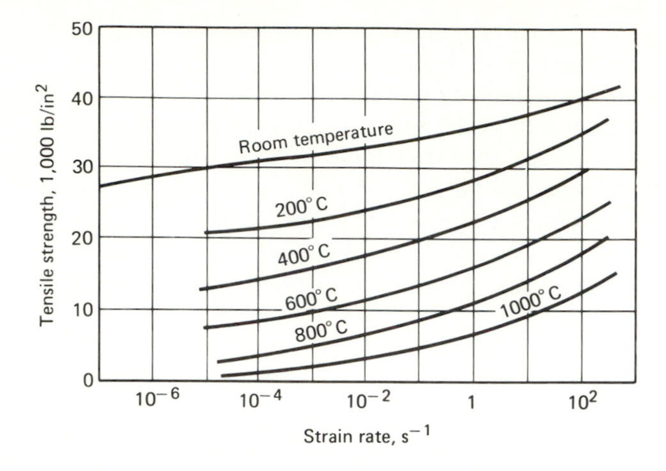

FIGURE 10.15
Effect of strain rate on the tensile strength of copper for tests at various temperatures.

In the conventional practice, the mass term has been made ever larger to obtain heavier forging forces, whereas in high-velocity forming the velocity term has been increased.

Although a great amount of engineering and development effort has been devoted to implementing high-velocity forming, there are two severe limitations. First, most metallic alloys have an inherent resistance to increased strain rates. The faster they are deformed, the greater their flow stress (Fig. 10.15). Unfortunately, that behavior is diametrically opposed to the reduction of the flow stress that is achieved by heating the alloy. Second, during deformation at high velocity, considerable internal energy is released adiabatically. This heat has nowhere to escape, except into the die surfaces where it causes tremendous instantaneous expansion of the surface layers of the die. Thus an extremely steep thermal gradient is developed at the die surface, which in turn causes thermal fatigue at an earlier time than would be the case in conventional forging.

The thermal fatigue is further complicated by the fact that the dies must overcome the augmented flow stress that results from the high rate of energy input to the dies. The two factors lead to a rapid deterioration of the die surface and to reduced die life compared with die life for the same part made by conventional forging practice. Despite the fact that certain more complex parts may be made by high-velocity forming, the total forging cycle time is reduced very little. Short die life has proved to be a major hurdle that has not yet been solved.

DESIGN FACTORS FOR HOT-WORKING PROCESSES

The processes and types of equipment that forge or work metals are based on many factors which apply in some degree to each process and forgeable material.

Some design factors for hot-metalworking processes are listed below.

1. The various sections of the forging should be balanced.
2. Generous fillets and radii should be allowed (see Fig. 10.16).
3. Sufficient draft (preferably 7 degrees) should be allowed.
4. Deep holes and high projections are not desirable.

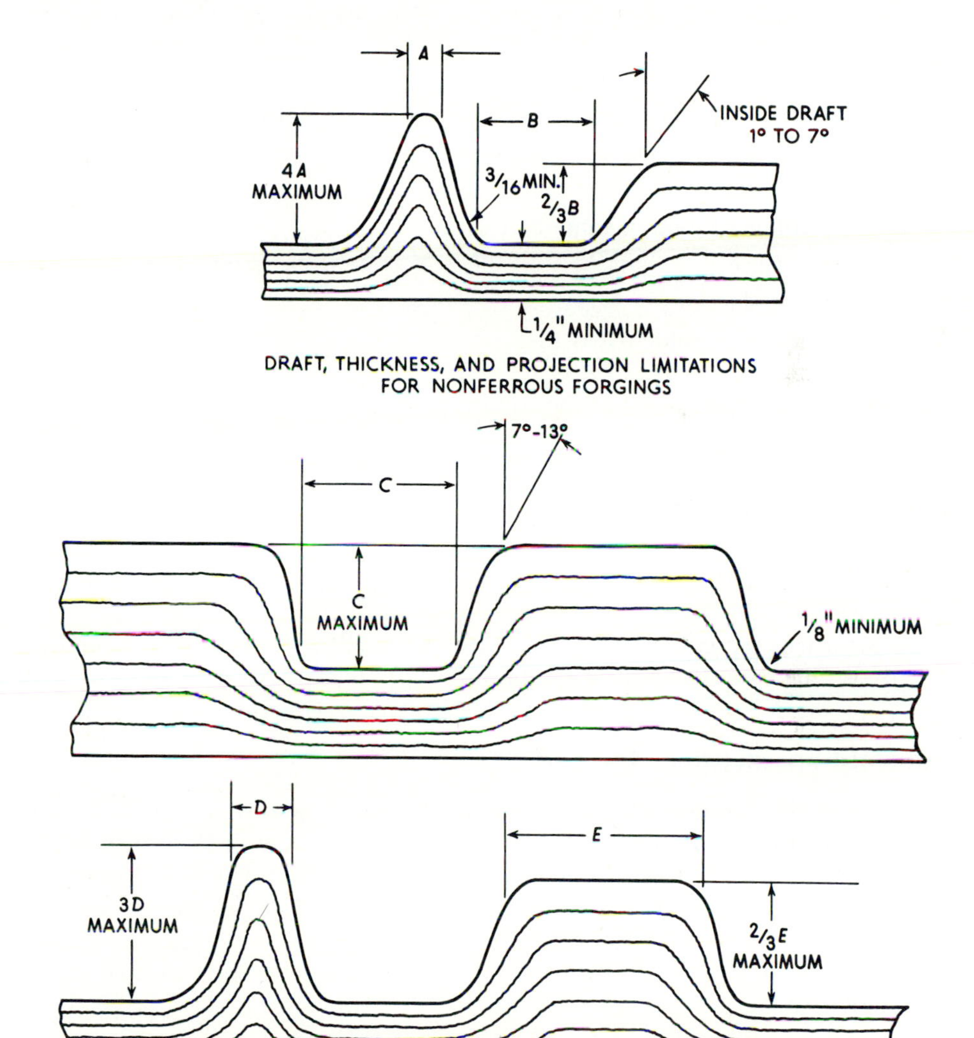

FIGURE 10.16
Forging design factors.

5. Holes in two planes will make removal of the forging from the dies impossible.
6. Flash thickness variation should be specified (not less than $\frac{1}{32}$ in.).
7. Raised letters or numbers for marking should be used.

Advantages and Limitations

Designing parts for the various processes of forging should result in the following advantages.

1. Uniformity of quality for parts subject to high stress or unpredictable loads
2. Weight saving
3. Close tolerances
4. Less machining (none in some cases)
5. Smooth surface
6. Speed of production
7. Incorporation in welded structures

The forging processes also has the following disadvantages.

1. High tool cost
2. High tool maintenance
3. No cored holes
4. Limitations in size and shape

Heat treating is used to increase the physical properties of the material, but it also increases the cost of the part.

Standard Specifications for Forgings

Impression die forgings are sold by the piece and not by the pound. It is understood without specific mention that the excess metal or flash of forgings shall be removed by trimming and that forgings shall be free from injurious defects.

Quantity. The quantity specified permits standard practice limits on overruns and underruns.

Size. Forgings within commercial size limits will be furnished unless closer tolerances are specified.

Coining or sizing. Closer tolerances may be obtained by additional hot- or cold-sizing operations.

Surface conditions. Forgings are generally, but not always, furnished in a cleaned condition by either tumbling, pickling, or blast cleaning.

Special requirements. Any special requirements, such as heat treatment or special tests, should be stated clearly.

Dies. Impression die forgings require special dies and tools for their production. The original charge for dies and tools conveys exclusive use but does not permit removal without additional payment. Dies and tools are maintained without additional charge.

Tolerances. It is important that the production-design engineer be aware of the tolerances that can be held by the forging process. Five tolerance areas should be considered. These are:

1. Thickness (applicable to thickness perpendicular to the plane of the parting line of the die)
2. Widths and lengths
3. Draft angle
4. Quantity
5. Fillets and corner

TABLE 10.3
Thickness tolerances, in[a]

Net weights, max., lb	Minus	Plus
0.2	0.004	0.012
0.4	0.005	0.015
0.6	0.005	0.015
0.8	0.006	0.018
1	0.006	0.018
2	0.008	0.024
3	0.009	0.027
4	0.009	0.027
5	0.010	0.030
10	0.011	0.033
20	0.013	0.039
30	0.015	0.045
40	0.017	0.051
50	0.019	0.057
60	0.021	0.063
70	0.023	0.069
80	0.025	0.075
90	0.027	0.081
100	0.029	0.087

[a] These apply to a thickness perpendicular to the plane of the parting line of the die.

TABLE 10.4
Shrinkage plus die wear

Lengths or widths, max., in.	Plus or minus	New weight, max., lb	Plus or minus
1	0.002	1	0.016
2	0.003	3	0.018
3	0.005	5	0.019
4	0.006	7	0.021
5	0.008	9	0.022
6	0.009	11	0.024
For each additional inch add	0.0015	For each additional 2 lb add	0.0015

Width and length tolerances include shrinkage and die wear, mismatching, and trimmed size.

The tolerances shown in Tables 10.3 to 10.8 may be used as a guide by the production design engineer when designing a part to be produced by the forging process.

Designing for Upset Forging

1. Design for smallest diameter or section of stock.
2. Use the minimum of upset material and shape it into conventional styles—round, hexagon, and square.
3. Avoid square corners. Use as large a radius as possible at inside corners.
4. Avoid using head diameter greater than four times stock diameter. A maximum of $2\frac{1}{2}$ in. diameter can be upset at one time.

TABLE 10.5
Mismatching toletances[a]

Net weight, max., lb	in.
1	0.010
7	0.012
13	0.014
19	0.016
For each additional 6 lb add	0.002

[a] Mismatching is due to the displacement of one die block in relation to the other. This tolerance is independent of, and in addition to, any other tolerances.

TABLE 10.6
Draft-angle tolerances

	Normal angle	Close limits
Outside	$5\frac{1}{2}°$	$0–7\frac{1}{2}°$
Inside holes and depressions	7°	$0–7\frac{1}{2}°$

5. For hot forging, follow these general rules when the part is to be made by one stroke of the press without injurious buckling.

   a. Maximum length of unsupported stock is three times diameter of bar.
   b. When material is gathered in a recess $1\frac{1}{2}$ times (maximum) the diameter of the bar, any length can be gathered that is within the limits of the stroke of the machine without buckling.

Ball bearing blanks, small cups, and nuts, as well as bars with upset portions within their lengths, are made by the upsetting process.

TABLE 10.7
Quantity tolerances

No. of pieces on order	Overrun pieces	Underrun pieces
1–2	1	0
3–5	2	1
6–19	3	1
20–29	4	2
30–39	5	2
40–49	6	3
50–59	7	3
60–69	8	4
70–79	9	4
80–99	10	5
	Percent	**Percent**
100–199	10	5.0
200–299	9	4.5
300–599	8	4.0
600–1,249	7	3.5
1,250–2,999	6	3.0
3,000–9,999	5	2.5
10,000–39,999	4	2.0
40,000–299,999	3	1.5
300,000 and up	2	1.0

TABLE 10.8
Fillet and corner tolerances

Net weight, max., lb	in.
0.3	$\frac{3}{64}$
1	$\frac{1}{16}$
3	$\frac{5}{64}$
10	$\frac{3}{32}$
30	$\frac{7}{64}$
100	$\frac{1}{8}$

Hot electric upsetting consists of passing a high current through the portion to be heated and pressing it into shape as it is heated. Longer lengths of material can be gathered by means of electric upsetting than are possible when the material is upset after furnace heating.

Die Design

The following considerations should be observed so as to minimize forging die cost per piece.

1. Forge two (or more) pieces as one and, subsequently, separate them in either a cutting or a trimming operation.
2. Design forgings to incorporate straight lines and circular arcs so as to minimize toolroom time.
3. If possible, have the lower die a flat surface; thus only one half (upper) will require machining to the required geometric contour.
4. In order to eliminate draft, consider locating the parting line so that the forging geometry provides a natural draft.
5. Consider the possibility of subsequently hot (or cold) twisting or bending of a simply designed forging rather than developing a complex die to forge completely the more complicated forging.

SUMMARY

When the production-design engineer specifies a forging or metalworking process, he is usually trying to design a product with more attractive mechanical characteristics, such as greater strength, or else to economize on the weight of the finished part, thus saving on material cost.

If a forging is selected, the designer must keep in mind the following principles of sound manufacturing design.

1. Plan carefully all draft angles so that the forging can readily be removed from the die and machining costs can be kept low.

2. Forgings should be designed to have a flat plane parting if possible. Equal volumes of metal should be on either side of the parting line.
3. Use generous fillets at all times.
4. Avoid abrupt changes in section thickness.
5. Avoid sections thinner than $\frac{1}{8}$ in.
6. Consider where it will be necessary to hold the forging for machining. Avoid using the parting plane as a location point.
7. Allow sufficient stock on areas that are to be machined. Usually this will vary from $\frac{1}{16}$ to $\frac{1}{8}$ in., depending on the size of the forging.
8. Design forgings so that they are supplied with indentations to spot holes that are to be subsequently drilled.
9. Avoid the design of forgings with deep recesses and pockets. Especially avoid complex contoured deep pockets.
10. Locate identification marks on surfaces that do not require machining.

Today, computer software programs exist that ease the job of the die designer. For example, the Chambersburg Engineering Company have developed software for such impact forging equipment as gravity drop hammers, double-acting hammers, die forgers, and automated model "C" impactor systems. With this software, the user can calculate the forces involved in the development of an end-use forging. It will calculate product cost based on the computer-selected hammer than will do the job most economically.

Input required includes material specifications, geometry of part to be produced, hammer type, and company cost factors. From this basic data, stock size, yield, flash land geometry, flash ratio, maximum force and total energy requirements can be calculated. Also, the software will estimate the number of flash-forming die impressions and the number of forging blows required to produce the part.

To assist in the setup for programmable forging hammers, the computer software provides for estimating the optimum forging requirement for each die station in terms of the energy per blow and the number of blows.

Other output data includes die limitations in terms of length, width, height, and weight based on the equipment in use; minimum die bearing and striking area for the job under study; tonnage requirement for hot or cold trimming and heating and production rates.

QUESTIONS

10.1. What is the temperature range for the hot forging of ferrous metals? Nonferrous metals?

10.2. What is the difference between hammer and press forging?

10.3. Explain how a board hammer operates.

10.4. What is the range of capacity of the steam hammer?

10.5. Upon what basic principle is the Chambersburg impactor based?

10.6. Why is it important that in the final forging operation the dies should close?

10.7. When would you advocate the continuous heating of a forging die during a production run of forgings?

10.8. What is roll die forging? Why is the process limited to parts that can be made continuously in chain form?

10.9. What seven design rules should be observed by the production-design engineer when designing for hot upsetting?

10.10. To what materials is the hot-extrusion process applied?

10.11. When would you advocate extrusion molding?

10.12. What design factors should be considered when specifying a forging process?

10.13. What is meant by the "blue-brittle" range?

10.14. Why are forgings sold by the piece rather than by the pound?

10.15. What are some of the capabilities of modern CAD systems?

10.16. What five tolerance areas should be considered in the forging process?

10.17. What property of metals makes justification of high-velocity forming difficult?

PROBLEMS

10.1. Design the upsetting dies needed to produce a blank for this pinion. Sketch the die features to scale on quadrille paper and include dimensions. Include for each pair of dies the shape at the stage of the preform with dimensions of the pinion form. Give supporting calculations as needed.

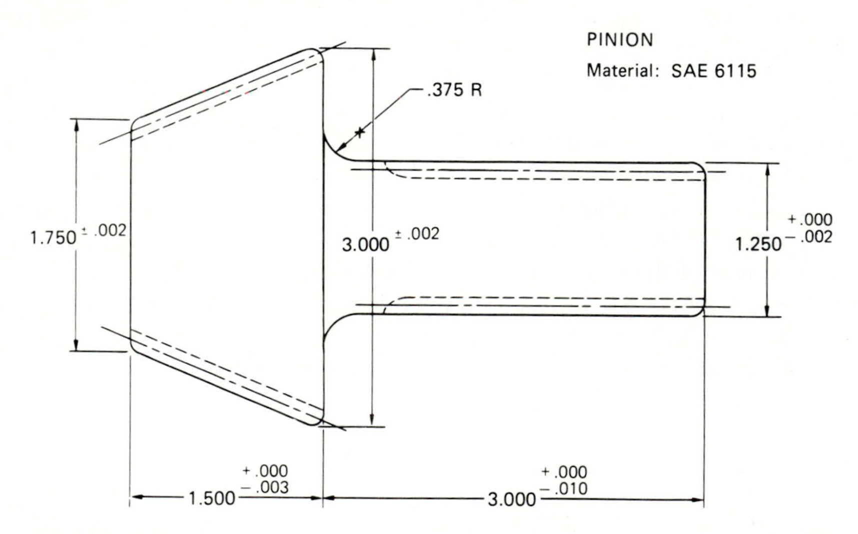

FIGURE P.10.1

10.2. A drop hammer weighing 1200 lb has a free fall of 30 in. What is (a) the energy of the blow and (b) the velocity at the instant of impact? What is the average force exerted by the hammer if it deforms the forging $\frac{3}{32}$ in. in one blow? What would be the projected area of the workpiece if it were made of steel that has an average yield strength of 12,000 lb/in.^{2} at 2000°F?

10.3. A gear blank 4 in. in diameter is hot forged from a height of 3 in. to an average height of 2 in. The engineer in charge of die design has estimated the flow stress at forging temperature of this steel to be 12,000 lb/in.^{2} The forging is being done on a 2000 lb board drop hammer. The height of the vertical drop is 40 in. How much work is accomplished in this forging operation? How many blows would be required to forge the part?

10.4. In the XYZ plant, three electric furnaces are producing steel with a nominal manganese content of 0.36 percent. The engineer in charge of product design wished to determine if the three furnaces are producing steel with essentially the same average manganese content. Consequently he took a random sample of four specimens from each of the three furnaces. The observed sample data found manganese in 0.01 percent increments above and below the nominal of 0.36 percent. The data obtained were as follows.

Manganese content, percent		
Furnace A	**Furnace B**	**Furnace C**
−4	+3	−3
0	+2	−4
−3	0	−3
+3	+3	−6

Do these three furnaces essentially produce the same average manganese content?

10.5. The XYZ Company is developing a new cleaning compound for the cleaning of their stainless steel forgings. The cleaning capability S of the solution is a function of the chlorine content C. The engineer developed the following relationship:

$$S = 3C^2 + C - C^3 + K$$

where K is a constant.

What percentage of chlorine would you recommend be introduced in order to maximize the cleaning capability of the cleaning compound?

10.6. A cold-extruded magnesium alloy rod has a modulus of elasticity of 6.5×10^6 lb/in.^{2} yield strength of 29,000 lb/in.^{2} and an ultimate strength of 44,000 lb/in.^{2} The rod is $\frac{1}{2}$ in. in diameter and 10 feet in length. Determine:

a. The load required to produce an elongation of $\frac{1}{4}$ in.
b. The load required to produce yielding
c. The maximum load
d. The total elastic resilience

REFERENCES

1. Avttzur, B.: *Handbook of Metalforming Processes,* Wiley-Interscience, New York, 1983.
2. Heilman, Paul M.: "Forgings", in *Handbook of Product Design for Manufacturing,* edited by James G. Bralla, McGraw-Hill Book Co., New York, 1986.

3. Hosford, W. F., and Caddell, R. M.: "Metal Forming," in *Mechanics and Metallurgy,* Prentice-Hall, Englewood Cliffs, N.J., 1983.
4. Jenson, J. E.: *Forging Industry Handbook,* Forging Industry Association, Cleveland, Ohio, 1970.
5. Kohser, Ronald A., and van Tyne, Chester J.: "Forming Processes," in *Production Handbook,* edited by John A. White, Wiley, New York, 1987.
6. Lange, K.: *Handbook of Metalworking,* McGraw-Hill, New York, 1985.
7. Parkins, R. N.: *Mechanical Treatment of Metals,* American Elsevier, New York, 1968.
8. Sabroff, A. M., Boulger, F. W., Henning, H. J., and Spretnok, J. W.: *A Manual on Fundamentals of Forging Practice,* Battelle Memorial Institute, Columbus, Ohio, 1964.
9. Wick, C., Benedict, J. T., and Veillux, R. F. (eds.): *Tool and Manufacturing Engineers Handbook,* vol. 2: "Forming," 4th ed., Society of Manufacturing Engineers, Dearborn, Mich., 1984.

CHAPTER 11

BASIC MANUFACTURING PROCESSES: PLASTICS

INTRODUCTION

In the fabrication of plastics, either thermoplastic or thermosetting, only a limited number of basic processes are available. Each basic process is able to impart some geometrical form to the plastic. If we know the geometry and the tolerance requirements of the design, the plastic that is to be used, and the quantity requirements, we can select the most advantageous basic process to produce the design. The selection of this "best" process can be done in advance of design completion if we are able to identify in advance those parameters that affect process selection. By having this information early, we can design for the most appropriate process and thus develop a functional design that is truly designed for production.

The following basic processes represent those that can be used in the fabrication of plastics.

COMPRESSION MOLDING

Compression molding is that basic forming process in which an appropriate amount of material is introduced into a heated mold that is subsequently closed under pressure. The molding material, softened by heat, is formed into

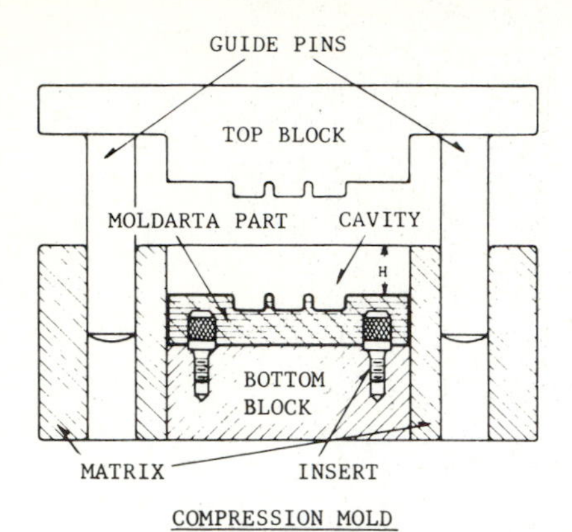

FIGURE 11.1
Typical compression mold.

a continuous mass having the geometric configuration of the mold cavity. For thermosetting plastics further heating results in the hardening of the molding material. If thermoplastics are the molding material, hardening is accomplished by cooling the mold. However, thermosetting resins are generally used so that the continued application of heat completes an irreversible chemical reaction.

Figure 11.1 illustrates a typical compression mold. Here the molding compound is placed in the heated mold with inserts in position. After the plastic compound softens and becomes plastic, the top block moves down and compresses the material to the required density by a pressure of $\frac{1}{2}$ to 3 tons/in.^{2} Some excess material will flow (vertical flash) from the mold as the mold closes to its final position.

Continued heat and pressure produce the chemical reaction that hardens the compound. The time required for polymerization or curing depends principally upon the largest cross section of the product and the type of molding compound. The time may be less than a minute, or it may take several minutes before the part is ejected from the cavity.

The dimension H (in Fig. 11.1) depends upon the nature of the charge. Bulky, nonpreformed materials require more depth than the more dense types of molding materials: a mold made for dense materials (molding compound) frequently cannot be used for the bulky types. The difference in shrinkage of the materials will further reduce the number of cases in which it is feasible to change from one material to another after the mold has been made. The difference in the flow characteristics of the molding compounds is another factor limiting the change from one material to another type if the same mold is to be used without expensive alterations.

Compression molds are of three types: flash, positive, and semipositive. An illustration of a flash compression mold is shown in Fig. 11.2. Here, any excess material is squeezed out between the plunger and cavity across the land area. The land is usually $\frac{1}{8}$ in. in width. Figure 11.3 illustrates a typical positive

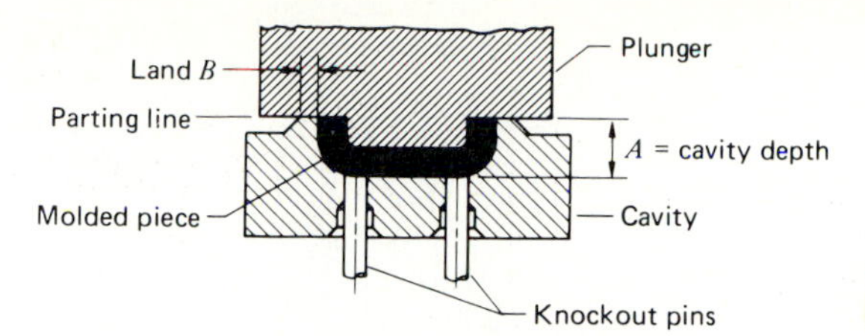

FIGURE 11.2
Flash compression mold.

mold. Here the excess material moves vertically in the clearance between the plunger and cavity into a clearance cavity. The maximum clearance between the plunger and the cavity is 0.006 in. across the entire part or diameter. Positive molds find application for long-draw parts or parts involving fabric-filled or impact-type thermosetting materials. In this type of compression mold it is important that exact weight be used when introducting the molding powder or preform.

In the semipositive mold shown in Fig. 11.4, we have a design that is partly positive and partly flash. When the semipositive mold is partially closed, some of the excess material escapes across the land as shown at y. When the corners at x pass into the mold cavity, the mold becomes positive and the remaining entrapped compound is compressed.

This type of compression mold finds particular application when the design involves sections of various thicknesses. As in the case of positive molds, it yields a vertical flash that is usually easier to remove by belt sanding equipment.

Compression molding of thermosetting plastics (it is seldom used for production quantities of thermoplastic materials) offers the following advantages.

1. It permits molding of thin-walled parts (less than 0.060 in.) with a minimum of warpage and dimensional deviation.

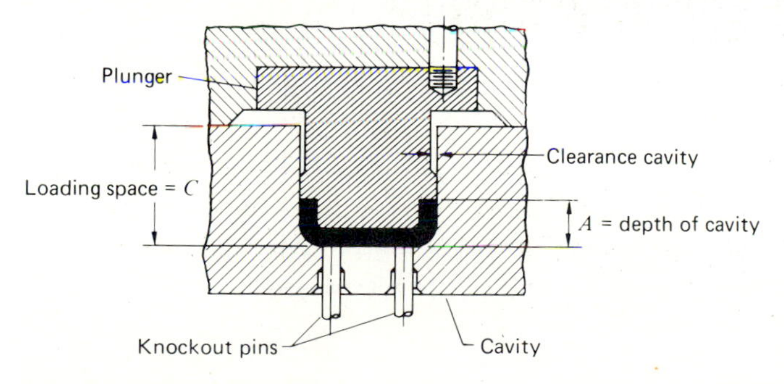

FIGURE 11.3
Positive compression mold.

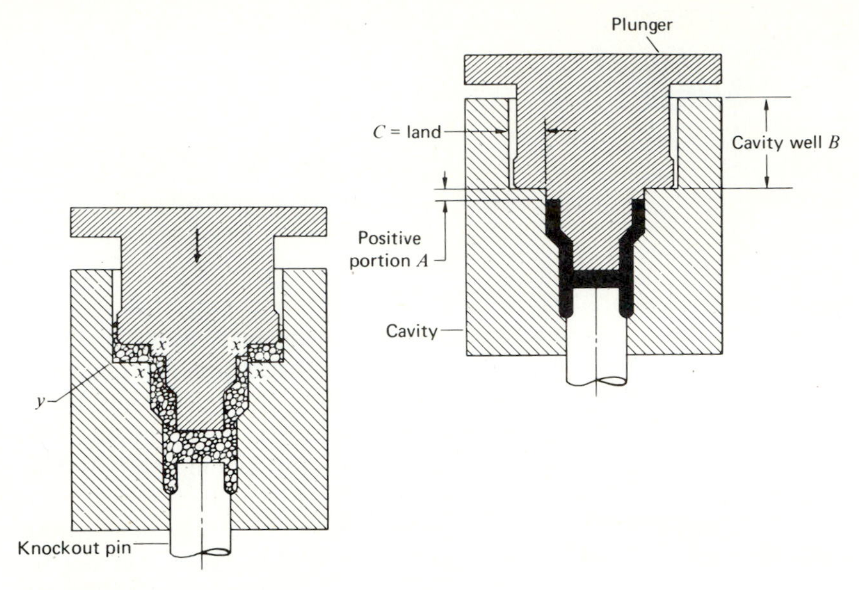

FIGURE 11.4
Semipositive compression mold.

2. The absence of gate markings is advantageous. This is especially true on small parts.
3. Lower and more uniform shrinkage is a characteristic of compression molding.
4. Compression molding is usually more economical for large parts (parts weighing over 3 lb). The process can provide larger parts with greater detail.
5. High-impact materials are difficult to find and consequently may be more easily processed by compression molding.
6. First costs are usually less because compression molds are usually simpler to design and construct.
7. In molding materials with reinforcing fibers, maximum impact strength is permitted, since the reinforcing fibers are not broken up as is the case with closed-mold methods.
8. Little material is wasted since there are no sprues, culls, or runners.
9. Less clamping pressure is required than in transfer and injection molding, thus more parts can be handled in a press of a given tonnage.
10. Mold maintenance is low because the wash action erosion of cavities is minimal.

The principal disadvantages or limitations of the process include the following.

1. Inserts can be damaged by the process as the press closes on cold material.
2. Complex shapes usually can be more easily filled out by the transfer and injection molding processes.
3. Large heavy parts take longer to cure than in transfer or injection molding.
4. Material handling is more time consuming because each cavity is usually loaded individually.
5. Considerable more time must be expended in flash removal than in either the transfer or injection molding processes.

TRANSFER MOLDING

Transfer molding is the process of forming articles in a closed mold to which the thermosetting material is conveyed under pressure from an auxiliary chamber. The molding compound is placed in this hot auxiliary chamber and subsequently forced in a plastic state through an orifice into the cavities by the application of pressure. Channels referred to as sprues and runners direct the flow of material from the auxiliary chamber (pot) to the mold cavities, passing through a gate just prior to entering the cavity. Air in the cavities must be displaced by the incoming plastic material through carefully located vents. The temperature and pressure must be held for a definite time for the chemical reaction (polymerizing or curing) to take place, depending upon the cross sections of the piece and the material used for molding. Curing temperature typically ranges from 280 to 380°F depending on the material used, the mold design, and the configuration of the part. These same conditions are found in compression molding. The molded part and the residue (cull) are ejected upon opening the mold, and there is usually no need for flash trimming, in contrast to the case of conventional compression molding. Figure 11.5 illustrates a transfer mold used to produce a plastic component with inserts.

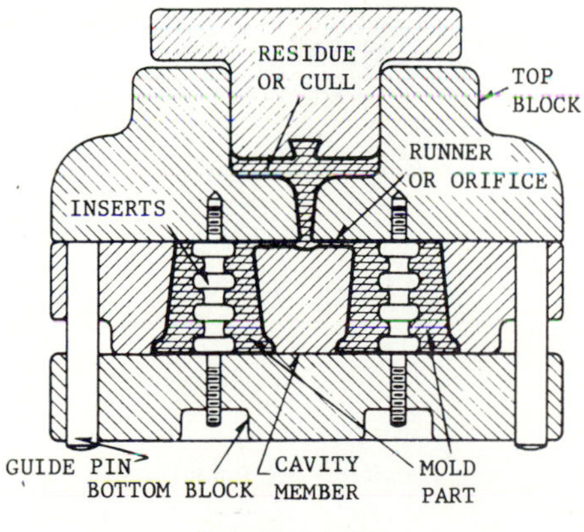

FIGURE 11.5
Transfer mold.

Transfer molds cost more than compression molds. In transfer molding, some sections of the finished part may be weak owing to sparsely filled sections. The pattern of flow may introduce sectionalizing and internal stresses that will result in weak or warped sections. Changing the material without producing a new mold is limited to approximately the same few cases as mentioned under compression molding.

The engineer will frequently need to compute the maximum number of mold cavities that can be placed on a mold plate, giving consideration to pressure limitations. This can be determined through use of the relationship of mold-clamping pressure and plunger pressure. It has been determined[1] that the mold-clamping pressure should be at least 15 percent more than the plunger pressure in order to ensure safe operation and to avoid flashing. Since the mold-clamping pressure also equals the mold-clamping force divided by the area of the mold, we can express the projected area of the mold in terms of the plunger pressure and the mold-clamping force.

For example, let us assume the following conditions applied and we wished to determine the maximum number of cavities that we would be able to attach to the mold plate in order to produce, satisfactory molded parts.

Projected area of one molded article = 4 in.^{2}
Projected runner area per mold cavity = 1 in.^{2}
Total area per cavity = 5 in.^{2}
Diameter of clamping ram = 16 in.
Diameter of injection ram = 6 in.
Plunger diameter = 4 in.
Line pressure = 2,000 lb/in.^{2}
Clamping ram force = $(\pi)(8^2)(2{,}000) = 128{,}000\pi$ lb
Injection ram force (preform) = $(\pi)(3^2)(2000) = 18{,}000\pi$ lb

$$\text{Plunger pressure} = \frac{18{,}000}{(2^2)} = 4500 \text{ lb/in.}^2$$

$$\text{Mold clamping pressure} = (1.15)(4500) = 5175 \text{ lb/in.}^2$$

$$\text{Mold clamping pressure} = \frac{\text{clamping run force}}{\text{mold area}} = \frac{128{,}000\pi}{5N}$$

Then the number N of mold cavities would be

$$N = \frac{128{,}000\pi}{(5)(5175)} = 15.6 \text{ or } 15 \text{ cavities}$$

[1] Society of the Plastics Industry, Inc., *Plastics Engineering Handbook*, Reinhold, New York, 1960.

INJECTION MOLDING

Injection molding is the process of forming articles by placing raw materials (granules, pellets, etc.) into one end of a heated cylinder, heating the material in the heating chamber, and pushing it out the other end of the cylinder through a nozzle into a closed mold, where the molding material hardens to the configuration of the mold cavity. Figure 11.6 illustrates the injection-molding process and Fig. 11.7 shows a multiple-cavity mold used in conjunction with an injection machine for molding thermoplastic materials.

Today, both thermoplastics and thermosetting plastics are injection molded. The injection molding of thermosetting plastics differs from molding thermoplastics because of the difference in the hardening process for each material.

In thermosetting platics also called thermosets, the materials are first heated in order to convert them to a liquid state and then flowed into the cavity/cavities where they are held at an elevated temperature for a controlled time until an irreversible chemical cross-linking or curing takes place. Subsequently, the hot but hardened parts are removed from the mold. In contrast to the injection molding of thermoplastics, in injection molding of thermosets the mold temperature generally is kept in the 300 to 400°F range by thermostatically controlled electric heating cartridges. Frequently, the ejected hot molded parts will not be cured completely. However, they complete the cure and attain full rigidity within one or two minutes because of residual heat in the molded part.

Whereas thermoplastic injection-molded parts usually require no deflashing, thermoset parts often need this operation, because the low viscosity of the material in its fluid state results in a thin film of material flowing across the land area parting line. Deflashing of molded parts can be done in a tumbling barrel or blasting machine, where either ground walnut shells or small plastic pellets are used. Large parts require sanding operations.

To allow adequate time for the heating of successive charges of the

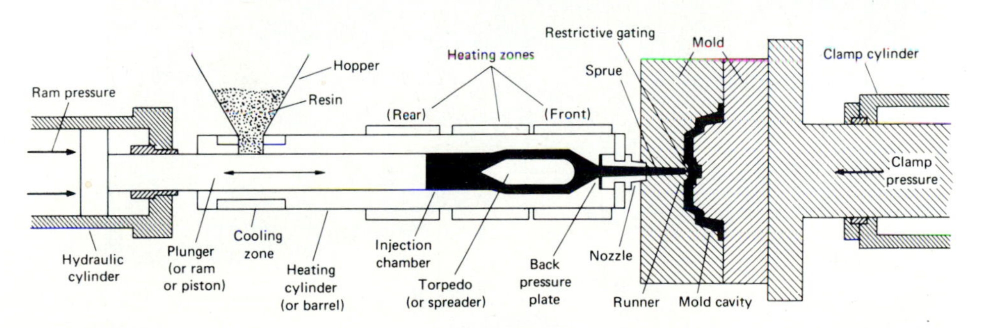

FIGURE 11.6
Schematic cross section of a typical plunger injection-molding machine.

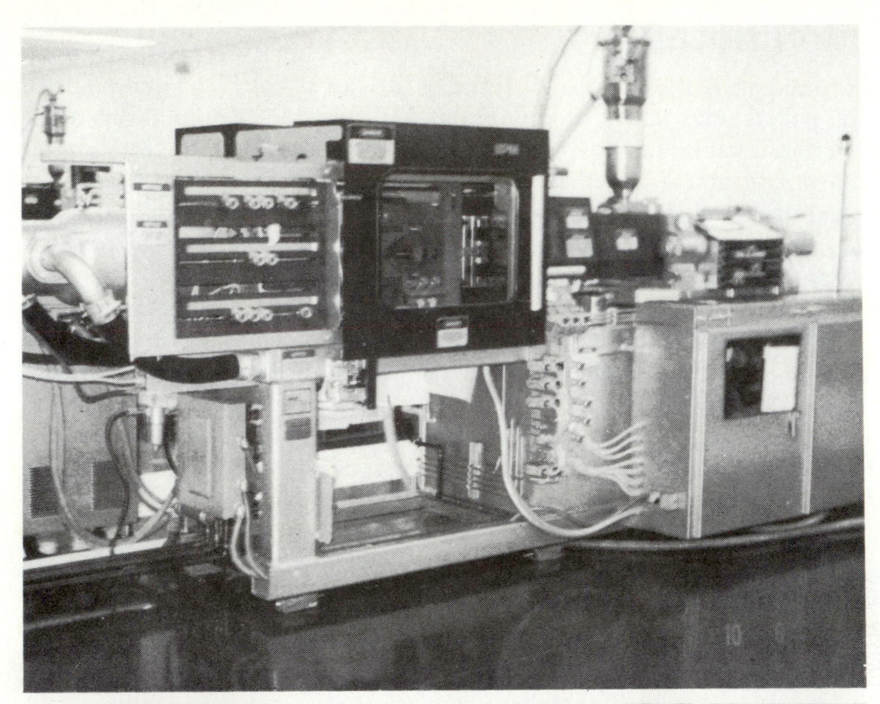

FIGURE 11.7
Multiple-cavity mold (lower) used in conjunction with modern injection-molding machine (above). (*Courtesy of Polestar Plastics.*)

plastic molding material, the heating chamber is designed to accommodate several charges, the number depending upon the size of the molded piece. Unheated compound is metered to the cylinder every cycle to replenish the system, making up for what has been forced into the mold. The pressure required to force the plastic molding compound through the heating cylinder and into the mold cavities varies between 10,000 and 25,000 lb/in.[2] Heating temperatures in the cylinder vary with the plastic being injected, but are usually in the range of 200 to 600°F. Thermoplastics need high barrel temperatures, usually between 350 and 600°F, and those thermosetting plastics that can be injection-molded, such as phenolic resins, use low barrel temperatures, in the range 150 to 250°F.

The number of variables that need be controlled in plastic molding is larger than most industrial processes. Here the equipment should control the material temperature, mold temperature, injection pressure, mold fill rate, clamping force on the mold, machine cycle time, shot volume, and the viscosity of the plastic at injection. Today the most commonly used injection system is the in-line reciprocating screw type (Fig. 11.8). It should be noted that the depth of the screw flights in this equipment varies from a maximum amount at the feed zone to a minimum at the metering zone. Also, the screw drive is designed to cause the screw to reciprocate as an injection plunger. Thus, the screw acts as a combination injection and plasticizing unit. As the plastic material passes from the hopper to the nozzle it encounters three zones: the feed zone, a compression zone, and a metering zone. At the metering zone the melted plastic is conveyed through an antiflowback valve to the nozzle. When the chamber in back of the nozzle becomes full, the resulting pressure forces the rotating screw back to a point where a switch is tripped, causing the screw to be forced forward by hydraulic pressure. This forward thrust of the screw injects plastic into the closed mold. A nozzle shutoff valve is incorporated to prevent plastic from seeping out of the nozzle when the mold is open.

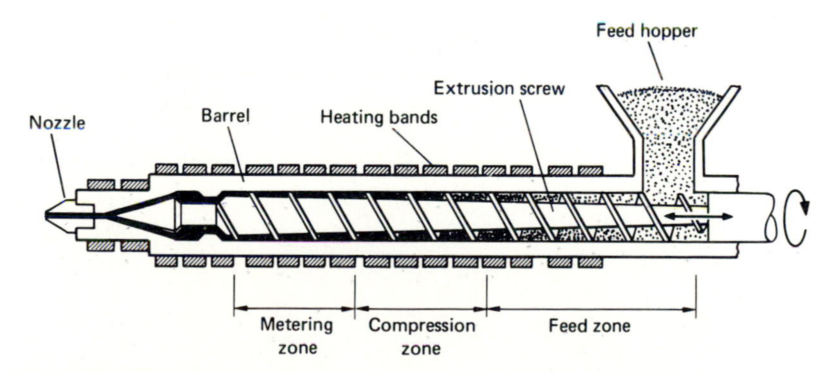

FIGURE 11.8
A typical reciprocating screw as utilized on late model injection molding machines.

It should be noted that the plastic material is melted by frictional heat and by heat externally applied to the barrel (see the heating bands in Fig. 11.8).

Because of the high pressures brought about by the injection process, there must be an adequate clamping force to prevent the mold from flashing. The larger the projected area of the molded part, the greater the clamp force required. The actual clamp tonnage required is a function of the projected area, the part designed, and the material being molded. Typically, 2.5 tons/in.^{2} of projected areas is used in order to estimate the clamp tonnage requirements.

There are three types of clamping systems available on injection equipment: mechanical, hydraulic, and combinations of mechanical and hydraulic. Mechanical clamps are available in sizes ranging from 10 to 3500 tons.

There are several parameters that determine the maximum number of cavities that may be attached to the mold plate. These include:

1. Shot capacity of the machine[1]
2. Clamping capacity of the machine
3. Plastifying capacity of the machine
4. Cost of mold per cavity and cost of the total operation

Usually, it is wise to assign not more than two-thirds the shot capacity of the machine to the mold. Thus, if Q_1 is the number of cavities based on the shot capacity of the machine, W_r is the weight of the sprue and runner in ounces, and W_p is the weight of the molded piece in ounces, we have this relationship:

$$Q_1 = \frac{\frac{2}{3}S - W_r}{W_p}$$

where S is the shot capacity of the machine in ounces.

Shot capacity provides a satisfactory estimate on heavy-sectioned designs. However, when thin flat articles are molded, the clamping capacity of the machine provides a better estimate of the number of cavities. Here we use 2 to 5 tons of clamping force for each square inch of projected cavity area. Thus, if 2.5 tons/in.^{2} was the factor used and Q_2 is the estimated number of cavities based on clamping capacity of machine, C is the clamping capacity of machine, tons, A_r is the projected area of sprue and runner, in.^{2} and A_p is the projected

[1] It is customary to rate machines in ounces of general-purpose polystyrene. It is desirable to use conversion factors to determine the machines capacity for other materials.

area of molded piece, in.^{2}, we have the ralationship

$$Q_2 = \frac{0.4C - A_r}{A_p}$$

It is also desirable to compute the number of cavities (Q_3) based on the plastifying capacity of the heating cylinder.

If we denote the plastifying capacity of the heating cylinder by P (pounds per hour) and the overall cycle in seconds by T, we have

$$Q_3 = \frac{0.00445PT - W_r}{W_p}$$

Of course, final decisions are based ultimately on cost, and the break-even chart can be used to determine the optimum number of cavities from a cost standpoint. Here, fixed cost will be plotted as the cost of the mold (whether it be four, six, eight, etc., cavities) and the variable cost will be the cost of production using that particular mold. Cross-over points of various plottings will reflect the optimum number of cavities for various quantity requirements.

Modern injection machines may be microprocessor-controlled, thus being capable of providing a communication link between machines and subsystems resulting in computer-integrated manufacturing operations. For example, takeout robots that automatically weigh and measure each molded part may be employed. Machines may be tracked and controlled over extended runs, thus minimizing batch variations that cause colors to vary from specifications.

EXTRUSION

Extrusion is the process of continuous forming of plastic articles by softening the plastic material through heat and mechanical working or solvents, and subsequently forcing the soft plastic through a die orifice that has approximately the geometrical contour of the desired profile. The extruded form is hardened by carrying it through cooling media. The operation of an extrusion machine requires exacting heat and feed control accurately regulated for the requirements of the material to be molded, as each plastic has its own individual characteristics. The material, in the form of powder, pellets, flake, beads, or granulated regrinds, often mixed with other ingredients such as colorants, stabilizers, and lubricants are fed to the extruder. These materials pass from the hopper through the machine body and head and are then forced through a die (Fig. 11.9). The extruded material passes on to a conveyor operated at a speed regulated to avoid or control deformation of the extruded shape, and is then wound into coils or cut into desired lengths. Liquid or molten materials may also be fed to the extruder. These polymers may be plasticized from polymerization reactors or continuous mixers.

The continuous extrusion process of plastic fabrication makes possible

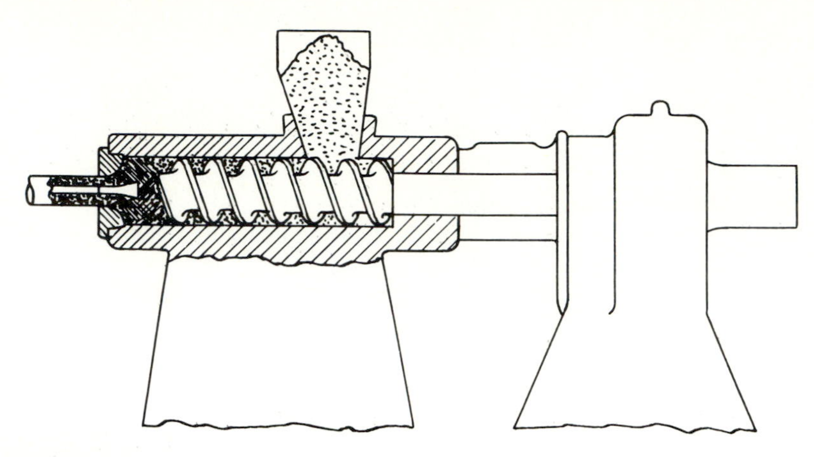

FIGURE 11.9
Continuous extrusion-molding machine.

the use of plastics for many articles that it has previously been impractical or too costly to produce by other methods of plastic fabrication. Rods, tubes, strips, and forms of uniform cross section can be extruded in the wide range of colors offered by thermoplastics, and, in many cases, with the elimination of costly finishing operations. Extrusion molding almost always precedes blow molding.

New extrusion technologies in the production of heavy-duty PVC pipe have resulted in sewer pipe with bores up to 60 inches. This is made by extruding a hollow profile strip on to a conical rotating mandrel that moves it past an infrared unit where it is joined to form a continuous pipe.

Computer-integrated extrusion is characteristic of many plants today. For example, vinyl window profiles represent a product with an expanding market where producing companies have found it economically sound to automate the compounding and extrusion operations.

COLD MOLDING

In cold molding, mixed plastic compounds are introduced into room-temperature molds that are closed under pressure. The formed component is then removed from the mold, transferred to a heating oven, and baked until it becomes hard.

Conventional presses with pressure ranging from 2000 to 12,000 lb/in.^{2} are used in this process. The molding cycle is relatively short. It consists of filling the mold, closing the press, and then removing the formed article. No curing takes place in the press, since the molding is done cold.

The molds are made of abrasive resistant tool steel with wall thicknesses sufficiently heavy to accommodate the pressures required to form the molding material.

THERMOFORMING

Thermoforming is the shaping of hot sheets or strips of thermoplastic materials into a desired geometrical contour through the utilization of either mechanical or pneumatic methods. Mechanical methods involve the use of a solid mold, either moving or stationary. The pneumatic process involves the use of a solid mold and a differential of air pressure, created either by vacuum or compression.

The majority of thermoforming equipment being built today is equipped with microprocessor controls capable of remote multimachine command and monitoring. For example such data as job number, standard cycle time, actual cycle time, production rate, down time, parts made, scrap, etc. can be maintained simultaneously on a battery of machines.

Strips of plastic may also be thermoformed by passing the stock between a sequence of rolls that produce the desired contour. Heat is applied locally to the areas where bending takes place. As the formed shape emerges from the series of rolls, it is cut off to the desired length.

The sheets of plastic used in the thermoforming process are produced by extrusion, calendering, pressing, or casting. Thicknesses of sheets that are processed range from 0.003 in. to as heavy as 0.5 in. or even greater.

The most commonly thermoformed resins include ABS, PBC, PS, PP, Acrylic compounds, and polycarbonate. Such well-known products as auto-headliners, fender wells, auto door panels, shower stalls, bathtubs, refrigerator liners, and freezer panels are commonly made by this process. The largest market for thermoformed products is in the packaging field. Here, such products as egg cartons, disposable cups and containers, and trays for meat packing are in regular use.

The temperature range for forming varies significantly with the material being processed. Most thermoplastic materials become soft enough for thermoforming somewhere between 275 and 425°F. They are brought to these temperatures by infrared radiant heat, electricity, or forced-air ovens heated by gas or oil. The heating of the plastic sheet should be accomplished as rapidly and as evenly as possible over the whole area to be thermoformed.

Those thermoplastics that cannot tolerate intense heat, such as the acrylics, are brought to forming temperature in gas- or electric-heated ovens. The temperatures involved here usually range between 275 and 350°F.

The production-design engineer must realize that, in thermoforming, the wall thickness in the finished product will be reduced proportionally to the increase in area of the formed piece over the area of the original sheet. For example, Fig. 11.10 illustrates a cycle in thermoforming a thermoplastic dish-shaped housing. The wall thickness of the finished part is considerably less

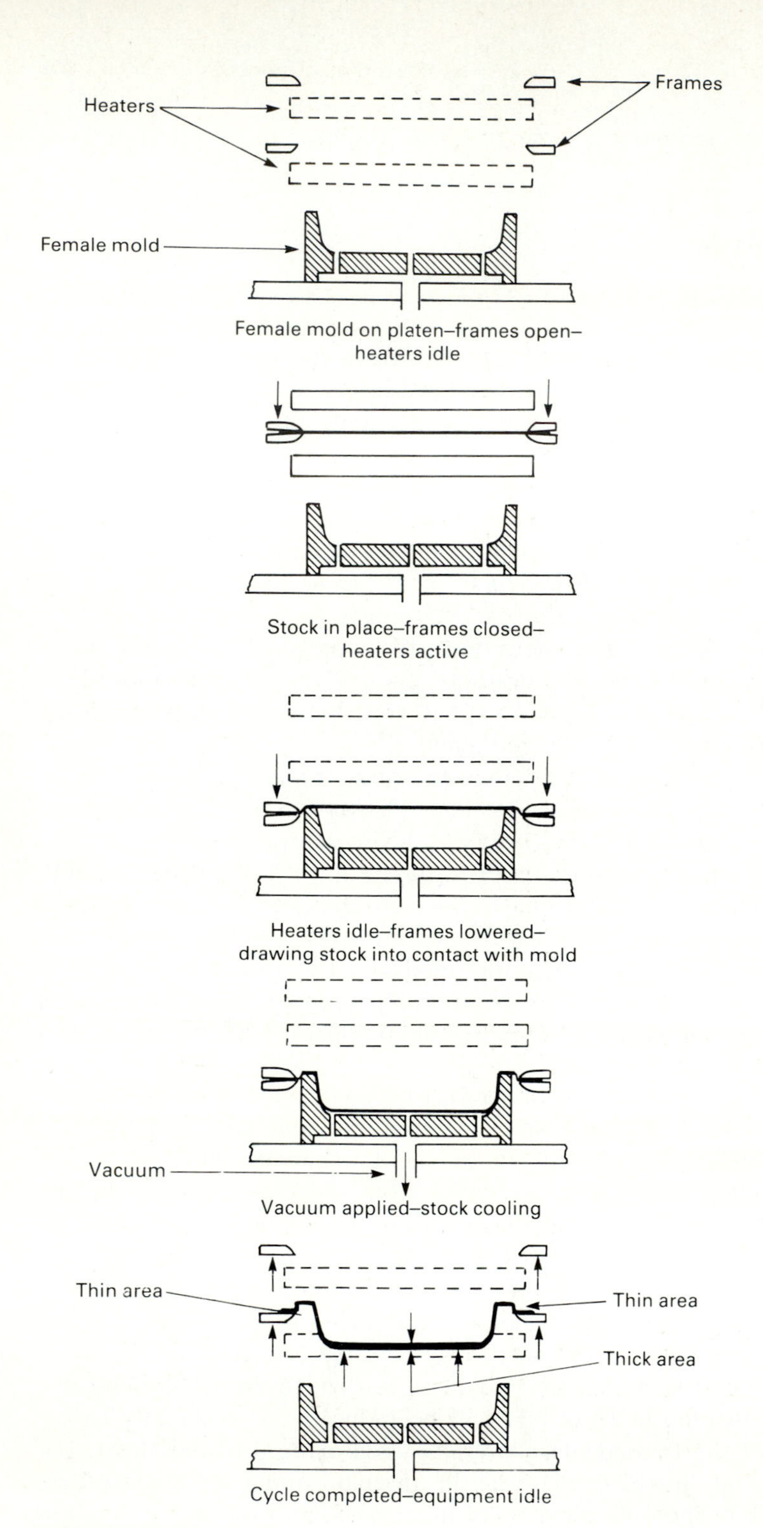

FIGURE 11.10
Cavity forming.

than that of the original stock. Also, it should be noted that areas that reach the mold last are the thinnest.

BLOW MOLDING

In blow molding a tube of molten plastic material is enclosed in a split mold and then this tube is blown out to match the shape of the mold. The tube of plastic material is the end-product of the extruder. Typical materials that are extruded for subsequent blow molding include high- and low-density polyethylene (e.g., for squeeze bottles), nylon, PVC, polypropylene, polystyrene, and polycarbonates.

The following operations are performed in sequence.

1. The plastic (thermoplastic) is melted.
2. The round hollow tube of molten plastic, called the parison, is formed.
3. The mold halves are clamped together around the parison.
4. The parison is expanded into the mold cavity by means of high-pressure air.
5. The part is cooled.
6. Air is exhausted from the part. The mold is opened and the finished part is ejected.

Figure 11.11 illustrates how the blow molding process operates. The use of microprocessor controls for blow molding machines is universal. Thus, modern systems include temperature, pressure, flow and position controllers; sequence control; parison programming control (for up to four heads); machine diagnostics; and computer-linked capabilities for multimachine network control from a central computer.

In order to make this process economical for quality production, the mold must be machined accurately and be made of a material that possesses high heat conductivity and adequate strength. Heat-treated 70/75 aluminum is typically used as a mold material. This material is not only a good heat conductor but it machines easily, and in the heat-treated condition it will provide good strength for the pinch-off cutting edges. Cooling is an important parameter in mold design. It is important that the drilled, interconnected cooling lines are placed so as to accommodate cooling where needed.

Today, blow molding can be integrated with the extrusion process so as to provide a continuous operation. In this arrangement, continuous extrusion of preforms take place by having the extruded stock cut to length and then dropped into open molds mounted on a turntable. After the hot parison enters the open mold, the mold closes so as to seal and pinch off the open ends of the deposited extruded stock. Air is now injected, and the tube of stock is blown out to match the contour of the mold. The part is now cooled and then ejected automatically from the opened mold.

In blow molding, the air pressure required is between 40 and 100 lb/in.^{2}

1

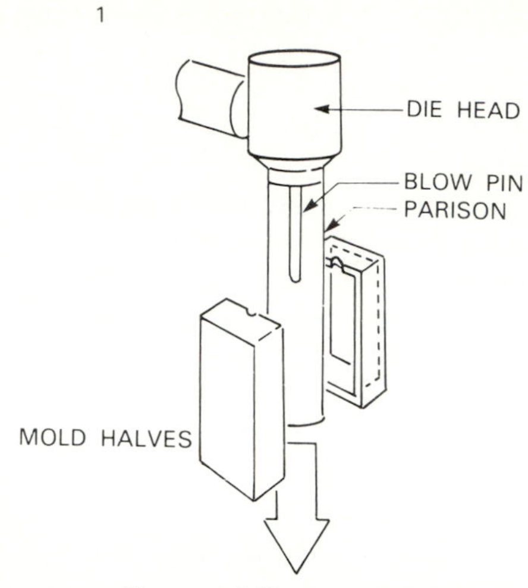

The material is extruded through the die head and forms a hollow tube called a parison. The parison is dropped vertically between the two halves of the open mold

2

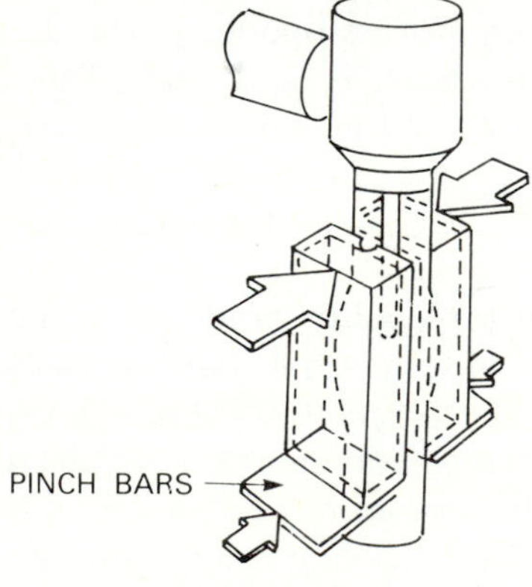

The parison is pinched and partially inflated (preblown) in order to help the parison retain its shape

3

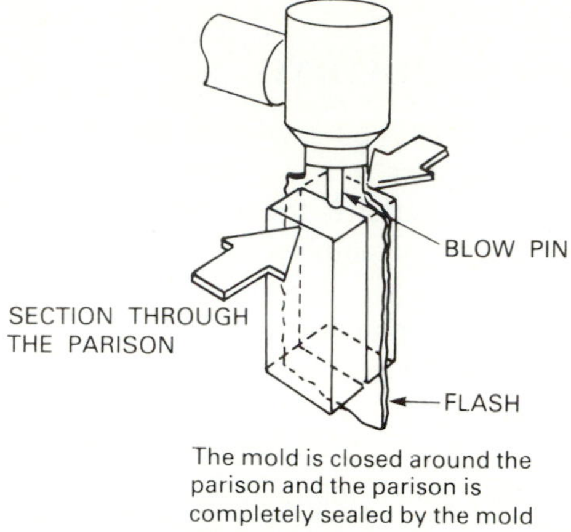

The mold is closed around the parison and the parison is completely sealed by the mold

4

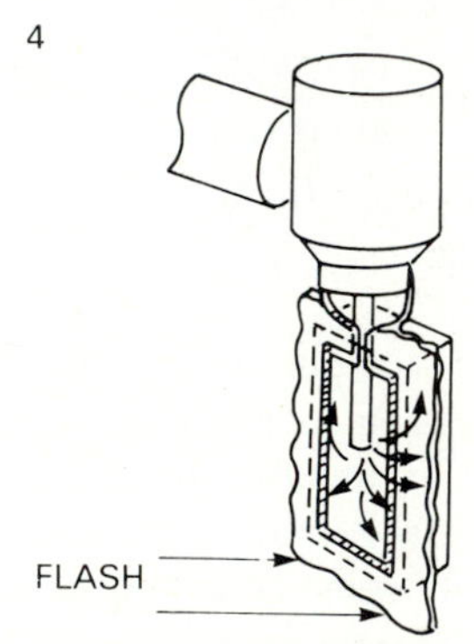

The parison is inflated and expands outward to conform to the surface of the mold

5

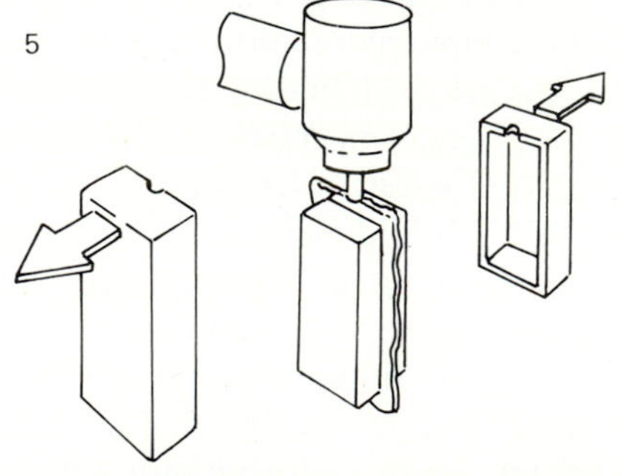

After the part cools, the mold is opened and the part is stripped off. The finished part is obtained once the flash is trimmed

FIGURE 11.11
Principal steps in the blow-molding process.

Usually, the higher the pressure the better the surface finish of the blown item. In order to control the wall thickness of the finished part, a low-pressure (usually between 5 and 20 lb/in.^{2}) is used to blow the molten parison to its initial molded form; subsequently, a higher pressure (usually between 60 and 100 lb/in.^{2}) is used to hold the material firmly against the mold cavity wall while it cures, thus minimizing wrinkling and distortion due to shrinkage.

In very thick sections, carbon dioxide or liquid nitrogen may be used to hasten the internal cooling while the geometry is being formed.

With the use of resins specifically developed for structural parts, blow molding may be a competitive process to produce such large simply designed parts as bumper beams and automobile gas tanks. To produce such products, blow molding uses one of three basic methods: reciprocating screw, ram accumulator, and accumulator head systems. All of these methods can produce large, heavier parts than are able to be produced by continuous extrusion. The reciprocating screw method works best for small parts, while large parts are best handled by the accumulator head system. Wall thicknesses range from about 0.04 in. to approximately 0.35 in. Equipment exists today to produce blow-molded parts up to 36 in. wide and 8 ft long. Such shapes as rectangular, L, U, and S can be molded.

The ability of the plastic to support itself when dropping to form a parison determines the maximum size part that can be produced. The longer the material can "hang," the greater the ability to make large parts.

STRUCTURAL FOAM MOLDING

In the production of structural parts, the structural foam molding process is projected to have the highest growth potential. This process extrudes engineering plastics, most of them glass-reinforced, into a die. Typically a chemical blowing agent (an inert gas such as nitrogen) is used to provide a resin/gas mix that is injected through a manifold system feeding one of more nozzles. The basic steps in the operation are clamping the mold; injecting the resin; de-pressurizing; and ejecting the finished part (see Fig. 11.12). Molds used in the structural molding process are typically produced from cast aluminum and mild steel.

Most people are familiar with the use of structural foams in connection with air application as housings for instruments, business machines, computers, and the like. However, with the current development of reinforced engineering plastics such as ABS, nylon, and glass-reinforced polypropylene, this process may well be used to produce such products as automobile door panels, hoods, trunks, and similar parts.

Combination injection molding/structural foam presses today range in capacity from 500 tons to 3000 tons and sell for between \$$\frac{1}{2}$ million and \$1 million. The typical range of wall thickness for structural foam parts is 0.17 to 0.375 in.

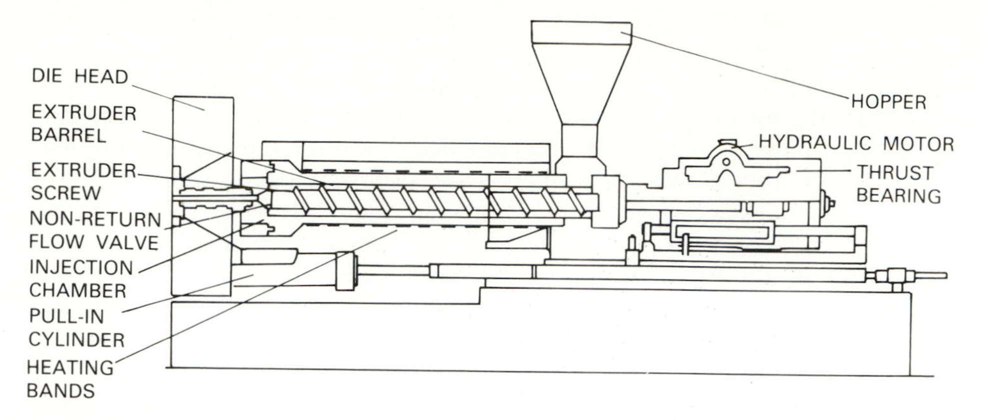

FIGURE 11.12
Fundamental components of a structural foam molding facility.

ROTOMOLDING

Rotomolding is a low-pressure molding process suitable for producing relatively large hollow seamless thermoplastic and some thermosetting parts. The most commonly molded thermoplastics are polyethylene, polyvinylchloride, nylon, and polycarbonate. Unsaturated polyester is the most used thermoset.

The process begins by introducing a premeasured amount of powder or liquid into the mold cavity. The mold is then closed and the molding machine indexes the mold into an oven (see Fig. 11.13) where the plastic is brought to the curing temperature. During the heating cycle, the mold is rotated about its vertical and horizontal axes. This biaxial rotation brings all surfaces of the mold into contact with the plastic material. The mold continues to rotate within the oven until all the plastic material has been picked up by the hot

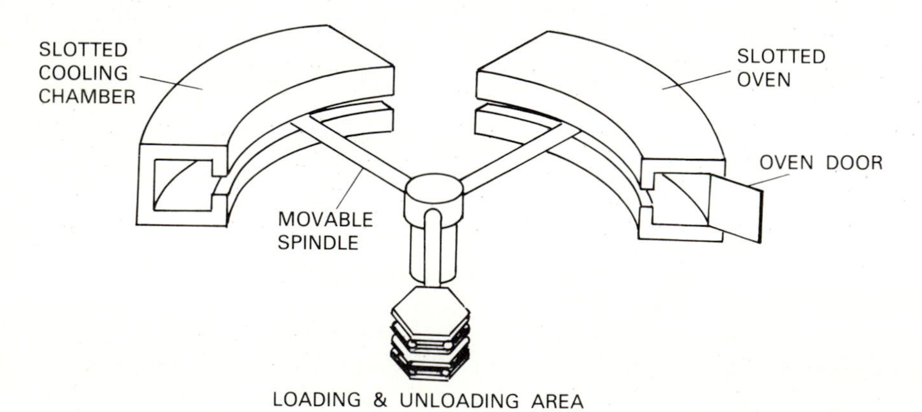

FIGURE 11.13
Rotational-molding process.

inside surfaces of the mold cavity. The machine moves the mold, which is continually being rotated, out of the oven and into a cooling chamber. Here air or a mixture of air and water cools the mold and the layer of plastic. The cooling process continues until the part is sufficiently rigid. The machine then indexes the mold to the loading and unloading station where it is opened, the part is removed, and a new batch of material placed in the mold for repeating of the cycle.

Although there are several different kinds of rotational molding equipment, the three-station layout described is the most common. The biaxial rotation of the mold usually is achieved by a series of gears or chains and sprockets, as illustrated in Fig. 11.14.

Plastic parts of dissimilar shape (such as doll heads and large refuse containers) can be molded simultaneously on the same facility. The limiting factor is that the oven temperature and cycle time must be compatible with the wall thickness of the various parts and the materials being used.

For a given part size, rotational equipment is relatively low in cost compared to other more capital-intensive processes such as injection and blow molding. Molds for rotational molding are generally shell-type molds that define the configuration and outside surface of the part. The molds have no internal core. Thus, the inside surface of the part is formed by the outside shape of the part and its wall thickness.

The engineer who designs the molds needs to give consideration to heat transfer, since the molds are heated and cooled each cycle. Materials of high thermal conductivity such as cast aluminum are commonly used. Sheet metal molds fabricated from aluminum, steel, and stainless steel are also used.

Some new designs of rotomolding molds incorporate up to 16 independently indexed arms that are arranged in modules of four. Microprocessor controls with compatible software are available, permitting full-sequence control of temperature and station time as well as response to operation problems.

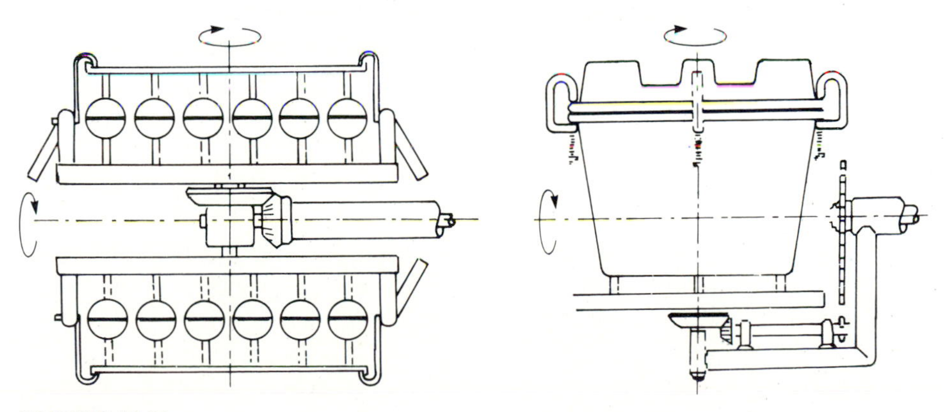

FIGURE 11.14
Methods for achieving biaxial rotation in the rotational-molding process.

MACHINING FROM STOCK

When only a few of the finished parts are needed, it may be more economical to rough- and finish-machine standard stock to the form required rather than provide the tooling to utilize one of the basic processes we have discussed. There are instances where high production runs (10,000 or more) can be produced advantageously by machining from raw plastics stock. Most plastic materials can be purchased in standard sheets, rounds, or flats. These standard shapes can be transformed to the desired geometry by typical machining operations (sawing, drilling, tapping, turning, grinding, etc.).

The machining characteristics of plastics are much different from metals. And there is considerable variation in the machining characteristics between one plastic and another. Since plastics have a low modulus of elasticity compared to metals, the cutting forces as well as those forces required to hold the work must be provided for correctly.

Another point that needs to be considered in the machining of plastics is the amount of elastic recovery that takes place during and after machining. Thus, it is important that the tool geometry provide relief for this recovery through sufficient clearance on the tooling.

Since plastics are poor conductors of heat, the majority of the heat caused by cutting friction between the tool and the work will be absorbed by the cutting tool. This heat needs to be removed or kept to a minimum so as to assure dimensional control and quality surface finish.

Plastics have much greater coefficients of thermal expansion than metals. Thus, adequate cutting tool clearances are important in order to avoid rubbing and the resulting generation of excessive heat. Since the softening temperatures of plastics are relatively low, tolerance control is difficult if heat builds up in the work.

In general, the following guidelines can be helpful in specifying tool geometry for the machining of plastics.

1. Provide polished surfaces on those tool areas coming in contact with the work so as to minimize frictional drag and resulting heat generation.
2. Design tools so that continuous-type chips are produced. This involves the provision of relatively large rake angles. The rake angle will be dependent on the depth of cut, cutting speed, and the type of plastic material.
3. In drill design, the packing of chips should be avoided by providing wide polished flutes and low helix angles.
4. In turning and milling plastics, diamond tools provide the best accuracy, surface finish, and uniformity of finish. Surface speeds of 500 to 600 ft/min with feeds of 0.002 to 0.005 in. are typical.

SUMMARY

With the exception of machining from stock, a mold or a sequence of rolls must be made for all plastic-forming processes; each process involves a different mold type. The design of a part governs whether simple or complex and expensive molds will be used.

Molding may be done in molds mounted in automatic presses, in semiautomatic presses, or in simple hand molds. The proper method of molding is determined by the plastic material being processed; the number of parts to be produced within a given time; the size, geometry, and tolerance requirements of the design; and the inserts to be molded into the piece.

General guides to mold design include the following:

1. Specify how many cavities the mold should contain. Indicate each cavity identification and location in the mold.
2. Specify the make and model of press that will accommodate the mold. Mold length and width and minimum and maximum shut height need be considered.
3. For handling purposes, specify eye bolts on two sides 90 degrees apart. A $\frac{5}{8}$-in.-diameter eye bolt is typical.
4. Consider parting-line location and the location and size of ejector pins in both the mold cavities and the runners.
5. Provide for venting if a gas-trapping problem is predictable.
6. Provide for adequate channeling in the mold for cooling water.
7. In designing the mold, consider the shrinkage rate of the plastic material being processed.
8. Specify runner size. In case of doubt, start small—it can always be enlarged.
9. Specify gate position and gate size.
10. Specify draft.
11. Specify mold finish. Never underspecify finish.
12. Specify the metal, and its hardness, from which mold cavity cores should be made. In the case of molds for vinyl plastic, stainless steel or hard chrome plating is recommended to avoid corrosion.
13. Adopt a working tolerance of plus or minus 0.005 in. in mold construction. Tolerances of plus or minus 0.0002 in. can be achieved readily but at an increase of 15 to 20 percent in cost.

QUESTIONS

11.1. What parameters should be considered in the selection of the most appropriate basic process for producing plastic components?

11.2. Is the compression-molding process capable of producing thermoplastic parts containing inserts? If 1000 pieces of such a design were required, would you recommend compression molding? What other considerations should be investigated? If flash minimization were critical what process would you recommend?

11.3. At what temperature range are most thermoset parts cured in the transfer process?

11.4. Why do injection molded thermoset parts usually require deflashing?

11.5. On what basis is the shot capacity of an injection press rated?

11.6. Where do thermoformed parts find their greatest market?

11.7. If the total projected area of a mold cavity and its runner is 7 in.^{2}, how many cavities can be made on a mold for a transfer press having a clamping ram of 20 in. diameter, an injection ram of 6 in. diameter, a plunger of 5 in. diameter, and a line pressure of 2000 lb./in.^{2}?

11.8. What is the function of the torpedo in the injection molding machine?

11.9. What can you say regarding the cost of molds used in the cold-molding process?

11.10. Would you say the casting process is a high-volume process for producing epoxy components? Explain your answer.

11.11. Describe the blow molding process.

11.12. Explain the relation between vinyl plastics and mold corrosion.

11.13. Why is 70/75 aluminum a good mold material for the blow-molding process?

11.14. Why would it be desirable to provide a polished surface on those tool areas coming into contact with the plastic being machined?

11.15. Why would cast stainless steel seldom be used as a mold material for the rotomolding process?

PROBLEMS

11.1. An injection press has a shot capacity of 18 ounces, a clamping force of 180 tons, and a plastifying capacity of 120 lb/h. The following information has been estimated by the manufacturing engineer in conjunction with a new plastic component that is to be produced on the injection press: weight of molded piece, 3 ounces; weight of sprue, 0.5 ounces; weight of runners 0.5 ounces + $0.1N$, where N is the number of cavities; projected area of piece, 4 in.^{2}; projected area of sprue, 1 in.^{2}; projected area of runners $1.5N$ in.^{2}; estimated cycle time, 1 min; expected shrinkage (losses and rejects), 10 percent; production requirements, 800 pieces per 8-hour shift. How many cavities should the mold design include?

11.2. The case for a volt–ohm meter can be produced from cellulose acetate by injection molding or from a urea formaldehyde compound by compression molding. The case requires 6.8 in.^{3} of material. Cellulose acetate has a cost of \$0.064/in.^{3}, and can be molded in 20 seconds in a die with initial cost of \$13,600. In contrast, urea resin has a cost of \$0.038/in.^{3} It takes 90 seconds to make one piece in a mold that costs \$2600. Determine the break-even point analytically and graphically if the operator rate is \$30/h including the overhead allowance.

11.3. In the transfer molding of a thermosetting resin, the projected area of one molded part is 6 in.^{2} The mold designer has estimated that the projected area of the runner is 0.8 in.^{2} per mold cavity. The diameter of the clamping ram is 20 in. and the injection ram 7 in. The plunger diameter is 4 in. and the line pressure is 2000 lb/in. What would be the clamping ram force? The preform force? The plunger pressure? The mold-clamping pressure? The maximum number of mold cavities?

REFERENCES

1. Association of Rotational Molders: *The Engineer's Guide to Designing Rotationally,* Association of Rotational Molders, Chicago, 1982.
2. Enrenstein, G. W., and Erhard, G.: *Designing with Plastics,* Macmillan, New York, 1984.
3. Gastrow, R.: *Injection Molds 102 Proven Designs,* Macmillan, New York, 1986.
4. Hertzberg, R. W., and Manson, J. A.: *Fatigue of Engineering Plastics,* Academic Press, New York, 1980.
5. Manzione, Louis T.: *Applications of Computer Aided Engineering in Injection Molding,* Macmillan, New York, 1987.
6. Margolis, James M.: *Decorating Plastics,* Macmillan, New York, 1987.
7. Menges, G., and Mohren, P.: *How to Make Injection Molds,* Macmillan, New York, 1986.
8. Rauwendaal, Chris: *Polymer Extrusion,* Macmillan, New York, 1986.
9. Rosen, Stephen L.: *Fundamental Principles of Polymeric Materials for Practicing Engineers,* Barnes and Noble, New York, 1971.
10. The Society of the Plastics Industry, Inc., *Plastics Engineering Handbook,* Reinhold, New York, 1985.
11. Thorne, James L.: *Thermoforming,* Macmillan, New York, 1987.

CHAPTER 12

SECONDARY MANUFACTURING PROCESSES: MATERIAL REMOVAL

INTRODUCTION

In many cases, products from the primary forming processes must undergo further refinements in size and surface finish to meet their design specifications. To meet precise tolerances, the removal of small amounts of material is needed. Machine tools are usually used for such operations, and metals are chosen for the product because of their long life and ease of shaping.

In the United States, material removal is big business—in excess of $36 billion per year, including materials, labor, overhead, and machine-tool shipments, is spent. Since about 60 percent of the mechanical and industrial engineering and technology graduates have some connection with the machining industry either through sales, design, or operation of machine shops, or working in related industries, it is wise for an engineering student to devote some time in his curriculum to studying material removal and machine tools.

It is evident that the advances of our technological civilization have been achieved largely because man has developed measuring devices and machine tools, along with an ability to use them. Without machining, many of America's cherished technological devices would soon disappear and life would revert to a simpler plane. The precise sizing and smooth surfaces that are produced by machining underlie all the technological developments of our day—rapid transportation, nuclear bombs, jet engines, and nuclear power plants, as well as kitchen appliances and sanitary facilities. Machine tools have justly been called the master tools of industry.

Although James Watt designed his steam engine and separate condenser in 1775, it took Wilkinson and Watt 25 more years to build it. They considered it to be a major triumph when their first cylinder was bored so true that "when a piston was tightly fitted at one end, a clearance no greater than a worn shilling was present at the other." Today, machinists routinely bore cylinders to a tolerance of ±0.001 in. and on request can achieve ±0.0001 in. The difference between a new car engine and one that is worn out is only a few ounces of metal in critical areas such as in crankshaft bearings or cylinder walls. Yet this small amount of wear can cause an engine block weighing a hundred or more pounds to be scrapped.

In the early 1800s Eli Whitney developed the concept of mass production in his design of tools and gauges for the manufacture of muskets for the U.S. government. Until that time all guns were custom-made by hand, so that no two muskets were exactly alike. After a number of years of development, under a government grant, Whitney finally traveled to Washington to show the Army and congressional officials his new concept. From a table covered with 10 dismantled muskets, Whitney picked parts at random and quickly assembled 10 muskets, all of which worked perfectly. But even more astonishing was the fact that the jigs, fixtures, and machine tools at Whitney's factory could produce thousands more with relatively little effort—a significant advantage for the United States in the War of 1812. Interchangeable manufacture and mass production are two concepts that foreshadowed America's production supremacy. The final concept emerged in 1914 when Henry Ford introduced assembly-line production. The simultaneous use of these three concepts—mass production, interchangeable manufacture, and assembly-line production—meant an unprecedented increase in the productivity of American labor relative to that of the rest of the world. However, since World War II we have lost much of our advantage in productivity because of our ever-increasing labor cost and the growing use of automation abroad. Other countries have been developing rapidly, and the United States is no longer assured of perennial first place. Japan is vying for leadership in high-quality cameras, radios, televisions, and automobiles.

MECHANICS OF METAL CUTTING

Uniqueness of Metal Cutting

The modeling and analysis of chip formation has been a continuing exercise over the last hundred years. Despite this, our understanding of the process is incomplete owing to its complexity. The chip-forming process in metal cutting is an extreme and unique deformation process distinguished by:

1. Unusually large shear strains on the order of 2 to 5 [1]

2. Exceptionally high shear strain rates of 10^3 to $10^5\,s^{-1}$ with local variations up to $10^7\,s^{-1}$ [2, 3]
3. The rubbing over the tool rake face of a freshly formed surface that is chemically clean and chemically active.
4. The large number of process parameters, like speed and feed, which can have widely varied settings
5. A large number of metallurgical parameters in the workpiece and tool materials that influence the process.

While there is no agreement on the shape of the deformation zone, it is, generally accepted that the magnitude of the shear stress at which the cutting process operates will be constant for a given material at a given temperature. It should be noted that this characteristic shear stress (flow stress) is different from the flow stress for other metal deformation processes, since the material is constrained and deformed in a unique manner.

As mentioned, metal cutting is a chip-forming process. Three types of chips have been traditionally identified [5, 6]. The Type I or discontinuous chip (Fig. 12.1*a*) occurs when a brittle work material is cut and severe strain causes

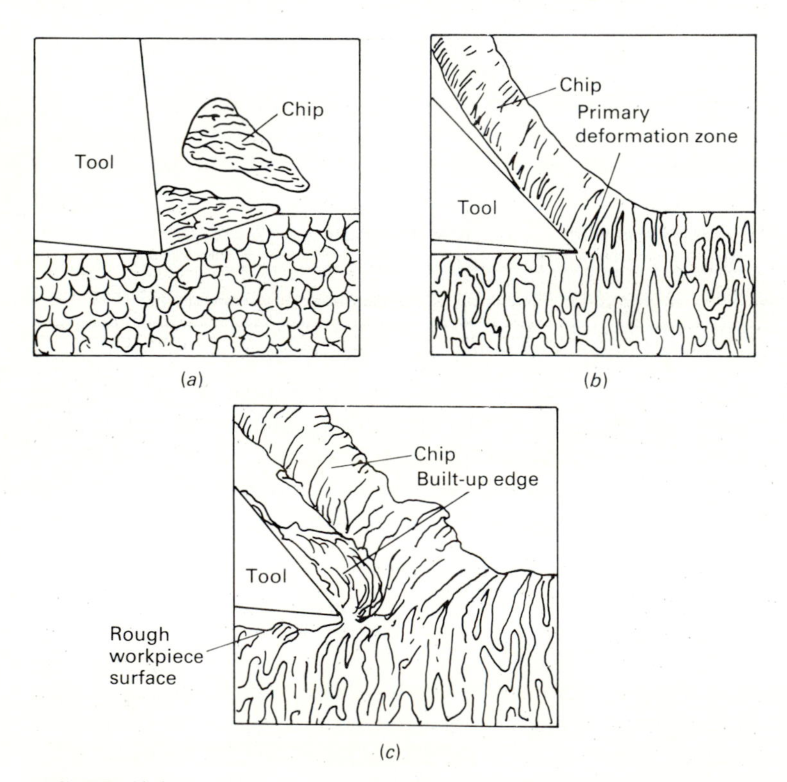

FIGURE 12.1
Chip types in metal cutting: (*a*) discontinuous chip; (*b*) continuous chip, (*c*) continuous chip with built-up edge. (*Redrawn from [4].*)

periodic fractures in the primary (shear) deformation zone. Discontinuous chips may also be produced when ductile materials are machined at very low speeds and high feeds [4]. The Type II or continuous chip (Fig. 12.1*b*) is achieved in the machining of most ductile metals. Most metal-cutting research has dealt with this chip type because its production is a steady-state process. When certain conditions exist between the cutting tool and chip, it is possible for the chip material to actually weld to the rake face of the tool forming a built-up edge (BUE). This BUE material increases friction, which causes layer upon layer of chip material to build up until it becomes unstable, breaks off, and is carried away either by the chip or the work. This is known as a Type III chip—continuous with a built-up edge (Fig. 12.1*c*).

Mechanics of Machining

The classical thin-zone mechanics model was first proposed in this country by Merchant in 1945 [6, 7]. The mechanics were developed for orthogonal cutting with a continuous chip with a planar shear process coupled with the following additional assumptions:

1. The tool tip is sharp and no rubbing or ploughing occurs between the tool and the workpiece.
2. The deformation is two-dimensional or orthogonal, i.e., there is no side spread.
3. The stresses on the shear plane are uniformly distributed.
4. The resultant force R on the chip is equal, opposite, and colinear to the force R' on the chip at the tool–chip interface.

Merchant demonstrated the mechanism of shear with the analogy shown in Fig. 12.2 where the metal is assumed to be displaced forward like a deck of cards. During this process the crystal structure will elongate in a direction defined by ψ that is different from the angle of shear. Each element has a finite thickness ΔS that is displaced through a distance ΔX with respect to its next-closest element. Hence the shearing strain is given by

$$\gamma = \frac{\Delta S}{\Delta X}$$

and from the geometry of Fig. 12.2 it can be shown that

$$\gamma = \cot(\phi) + \tan(\phi - \alpha)$$

where ϕ is the shear angle and α is the back rake angle. Similar geometrical reasoning also yields that the direction of crystal elongation is

$$2\cot(2\psi) = \cot(\phi) + \tan(\phi - \alpha)$$

where ψ represents the direction of elongation. Thus, the elongation direction

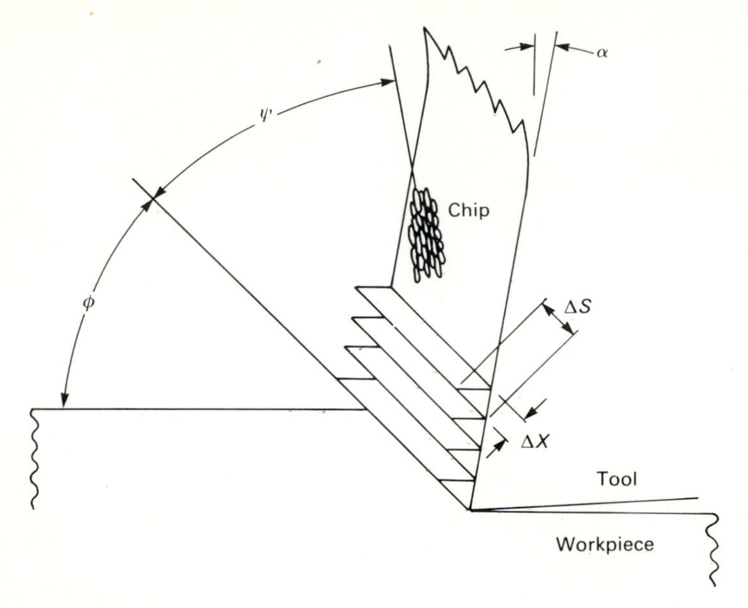

FIGURE 12.2
Schematic representation of shearing strain, γ, in orthogonal cutting.

and shearing strain are related by

$$\gamma = 2\cot(2\psi)$$

As previously stated, one unique characteristic of metal cutting is its very large strain. Strains on the order of 2 to 3 or more are quite common.

Simple velocity relationships can also be derived (see Fig. 12.3), since the

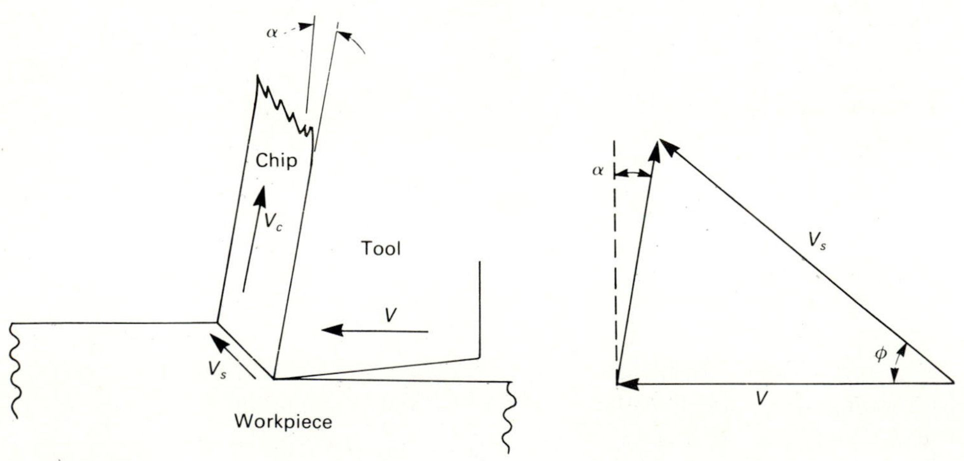

FIGURE 12.3
Velocity relationships in orthogonal cutting.

velocity of the chip relative to the tool (V_c) must equal the vector sum of the velocity of the primary shearing relative to the tool (V_s) and the velocity of the tool relative to the workpiece (V). Thus,

$$V_c = \frac{\sin(\phi)}{\cos(\phi - \alpha)} V, \qquad V_s = \frac{\cos(\alpha)}{\cos(\phi - \alpha)} V$$

where α is the back rake angle of the tool and ϕ is the shear angle. The strain can now be expressed in terms of velocity as

$$\gamma = \frac{V_s}{V \sin(\phi)}$$

and the strain rate of the cutting process by

$$\dot{\gamma} = \frac{\Delta S}{\Delta X \, \Delta t} = \frac{V_s}{\Delta X} = \frac{\cos(\alpha)}{\cos(\phi - \alpha)} \frac{V}{\Delta X}$$

where ΔX is the thickness of the deformation zone, and Δt is the time to obtain the final value of strain. Typical strain rate values in metal cutting are of the order of $10^6 \, s^{-1}$, a high value when compared with strains of $10^{-3} \, s^{-1}$ for ordinary tensile testing and $10^3 \, s^{-1}$ for the most rapid impact tests [8].

The process of metal cutting has three identifiable deformation zones. The primary one involves the periodic shearing of the metal at the shear zone, but this is preceded by compression deformation in the workpiece. A secondary deformation involves the contact region between the chip and tool called the *tool–chip interface.* The force components acting on the shear plane and tool–chip interface are shown in Fig. 12.4 in the free-body diagram of the chip. The shear deformation has a force acting parallel to the shear zone (F_s) and one normal to it (F_N). The chip surface adjacent to the rake face rubs the tool with a velocity V_c. The resulting frictional force F is therefore parallel to the rake face, while N is the force normal to the friction force. All of the force components acting on the deformation zones may be expressed in terms of the measured force components—the horizontal cutting force F_H and its perpendicular force F_V. F_H is sometimes called F_c, the *cutting force,* or F_p, the *power force*; F_v is called F_t or F_q by other authors. On the rake face,

$$F = F_H \sin(\alpha) + F_v \cos(\alpha)$$
$$N = F_H \cos(\alpha) - F_v \sin(\alpha)$$

The coefficient of friction μ on the rake face may be similarly expressed as

$$\mu = \frac{F}{N} = \frac{F_H \sin(\alpha) + F_v \cos(\alpha)}{F_H \cos(\alpha) - F_v \sin(\alpha)} = \tan(\beta)$$

where β is the friction angle. This concept of friction implies that the friction force F and its normal N are uniformly distributed over the tool–chip interface.

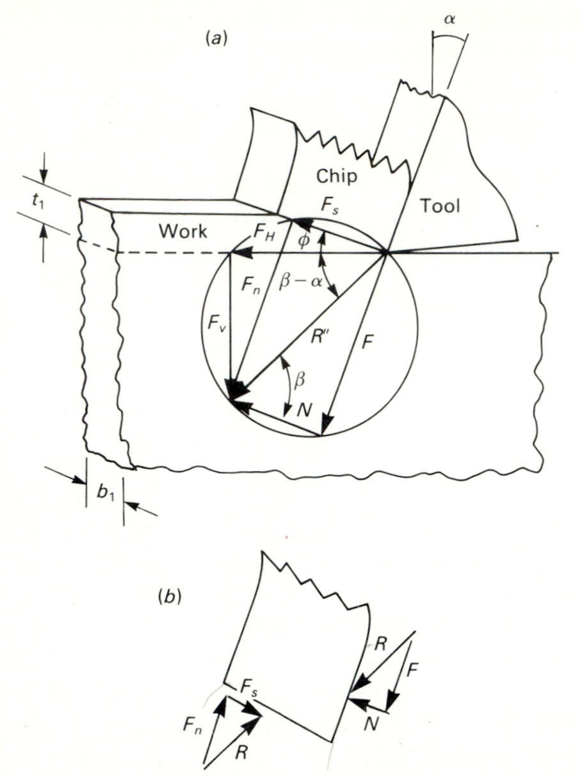

FIGURE 12.4
The geometry and forces in orthogonal cutting.

The forces along the shear zone may be written

$$F_s = F_H \cos(\phi) - F_v \sin(\phi)$$
$$F_N = F_H \sin(\phi) + F_v \cos(\phi)$$

As will be shown, the calculation of the shear (flow) stress τ_s requires knowledge of not only force magnitude but also the direction of flow. In metal cutting, the shear direction defines that direction. The shear angle can be determined in many ways. It can be directly measured through high-speed photographic methods or "quick stop" methods that terminate the deformation with the chip intact, so that the shear angle can be measured. The second group of methods uses the continuity theorem to relate the dimensions and motions of the deformed chip to the undeformed volume. By equating the volume of the deformed and undeformed chip,

$$\rho_1 l_1 b_1 t_1 = \rho_2 l_2 b_2 t_2$$

where ρ_1 is the density of the undeformed chip and l_1, b_1, and t_1 are the length, width, and thickness, respectively of the undeformed chip. The subscript 2 refers to the same quantities for the deformed chip. The material density remains constant despite the severe plastic deformation, and the width is assumed to remain constant (Assumption 2 of the Merchant Analysis). Therefore

$$\frac{l_2}{l_1} = \frac{t_1}{t_2} = r$$

where the quantity r is called the *chip thickness* or *chip length ratio*, depending on which quantity is measured from the chip. Further geometry reveals that r is related to the shear direction by

$$\tan(\phi) = \frac{r\cos(\alpha)}{1 - r\sin(\alpha)}$$

The cross-sectional area A of the undeformed chip is clearly the product of the width of the workpiece and the depth of cut. The inclined shear area is

$$A_s = \frac{A}{\sin(\phi)} = \frac{b_1 t_1}{\sin(\phi)}$$

The shear stress (τ_s) and normal stress (σ_n) on the shear plane may then be found as

$$\tau_s = \frac{F_s}{A_s} = \frac{F_H\cos(\phi) - F_v\sin(\phi)}{b_1 t_1/\sin(\phi)}$$

$$\sigma_n = \frac{F_N}{A_s} = \frac{F_H\sin(\phi) + F_v\cos(\phi)}{b_1 t_1/\sin(\phi)}$$

The shear stress (or dynamic shear stress) is a material property and has been shown to be constant for a given material over a wide range of both cutting and tooling parameters.

Energy in Chip Formation

During a cut, the total energy required per unit time (power) is the product of the horizontal force F_H and the cutting velocity V. If the total energy required per unit time is divided by the metal removal rate, $b_1 t_1 V$, we obtain the energy per unit time, or specific energy, u as

$$u = \frac{F_H V}{b_1 t_1 V} = \frac{F_H}{b_1 t_1}.$$

This specific energy may be partitioned into four portions [9, 10]:

1. The shear energy per unit volume, u_s, required to produce gross deformation in the shear zone:

$$u_s = \frac{F_s V_s}{b_1 t_1 V} = \frac{\tau_s V_s}{V\sin(\phi)}$$

2. The friction energy per unit volume, u_f, expended as the chip slides along the rake face of the tool:

$$u_f = \frac{FV_c}{Vb_1t_1}$$

3. The kinetic energy per unit volume required to accelerate the chip (momentum specific energy), which may be expressed as:

$$u_m = \frac{F_m V_s}{Vbt_1}$$

where F_m is the momentum force $= \rho V^2 bt_0 \gamma \sin \phi$, and ρ is the density of material being cut.

4. The surface energy per unit volume required to produce the new uncut surface area is

$$u_a = \frac{T\,2Vb}{Vbt_1} = \frac{2T}{t_1}$$

where T is the surface energy in inch pounds per square inch of the material being cut. For most metals, T has a value of 0.006 in. · lb/in.^{2}

The kinetic specific energy and surface specific energies are negligible in typical cutting operations, thus

$$u \simeq u_s + u_f$$

where u_s accounts for approximately 75 percent or more of the total energy [10].

While one may wish to describe the energy per unit volume needed to form the chip, machine tools are typically rated in horsepower. Unit energy values (lb/in.^{2}) may be transformed to unit or specific horsepower (hp_u) values (i.e., hp/in.^{3}/min) merely dividing by the appropriate constant

$$\text{hp}_\text{u} = u\,\frac{\text{lb}}{\text{in.}^2} \cdot \frac{\text{hp}}{550\,\dfrac{\text{ft/lb}}{\text{hr}}} \cdot \frac{1}{60\,\dfrac{\text{min}}{\text{hr}}} \cdot \frac{1}{12\,\dfrac{\text{in.}}{\text{ft}}}$$

$$= \frac{u}{396{,}000}\,\text{hp/in.}^3\text{/min}$$

The specific horsepower is a measure of how difficult it is to machine a particular material and may be used to estimate the total cutting horsepower, hp_c, to machine a part.

The total cutting horsepower may be calculated by obtaining the product of the specific horsepower times the rate at which we remove material (Q):

$$hp_c = hp_u \cdot Q$$

The metal removal rate Q may be computed as an uncut area times the rate at which the tool (or table) is moved perpendicular to that area. For the case of a plate, as illustrated in Fig. 12.4, the metal removal rate $Q = 12Vb_1t_1$. Thus, the cutting parameters define the metal removal rate, and in conjunction with the specific horsepower (a material-related property) can be used to predict the total cutting horsepower.

Unfortunately, machine tools are not completely efficient. Losses due to friction, windage, wear, etc., prevent some power from being delivered to perform the machining process. Therefore, the gross horsepower (hp_g) needed may be defined as:

$$hp_g = \frac{hp_c}{\eta}$$

where η is the efficiency of the machine tool.

Heat in Metal cutting

In metal cutting, the mechanical energy used to form the chip is converted into heat. Primary sources for this heat are the energy needed to shear the chip and the energy needed to overcome friction on the rake face (some plastic deformation also occurs on the rake face). Additional energy generated from the rubbing of the tool flank on the newly generated surface, perhaps due to the wearing of the tool, may also occur. The distribution of heat energy generated due to the shearing and friction processes depends on several factors, with cutting speed being the most important. Once generated, the heat may remain in the workpiece, be carried away in the chips, or be transferred to the tool. The distribution between these heat sinks is again largely dependent on cutting speed.

The temperature generated by the plastic deformation of the shearing process may be estimated as

$$\Delta\theta_s = \frac{F_s V_s}{J\rho c}$$

where $\Delta\theta_s$ is the temperature rise due to shearing, J is the mechanical equivalent of work (778 ft · lb/BTU), ρ is the mass density of the workpiece, and c is the mean specific heat. More exact methods allow additional factors, such as the amount of latent heat left in the workpiece and the proportion of heat that stays in the workpiece (versus that being carried away in the chip), to be included. The calculation of the temperature rise due to friction is beyond the scope of this book, but several analyses exist. The total average temperature rise ($\Delta\theta_f$) on the rake face is simply the sum of the temperature rise due to shear deformation ($\Delta\theta_s$) and that due to friction ($\Delta\theta_f$). As shown

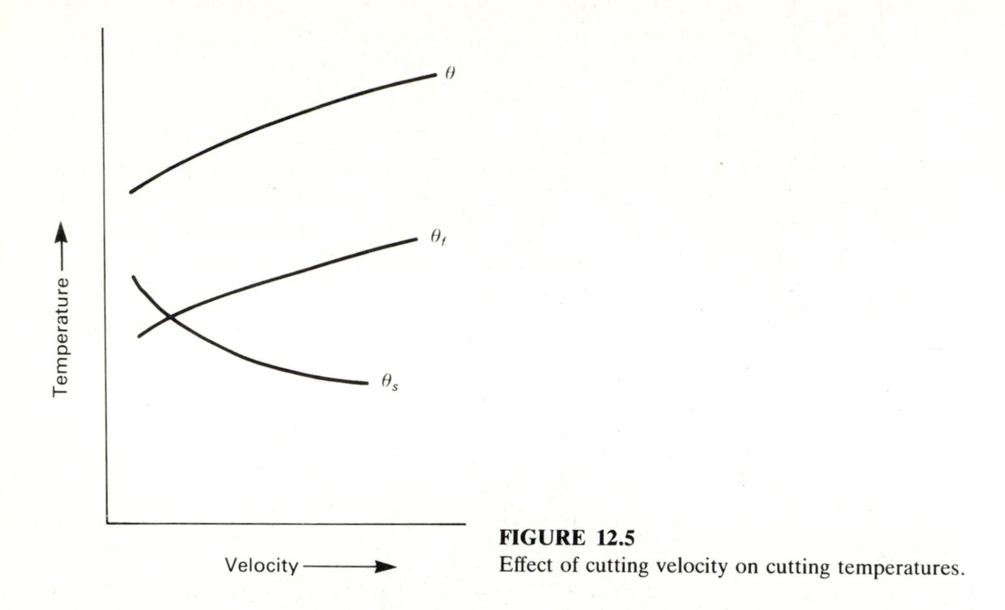

FIGURE 12.5
Effect of cutting velocity on cutting temperatures.

in Fig. 12.5, as speed increases the total temperature increases despite decreasing shear temperature. Decreasing shear temperature is caused by lower friction at higher speed, which yields less strain (i.e., larger shear angles). At higher speeds this is more than compensated for by increased temperature rise due to friction.

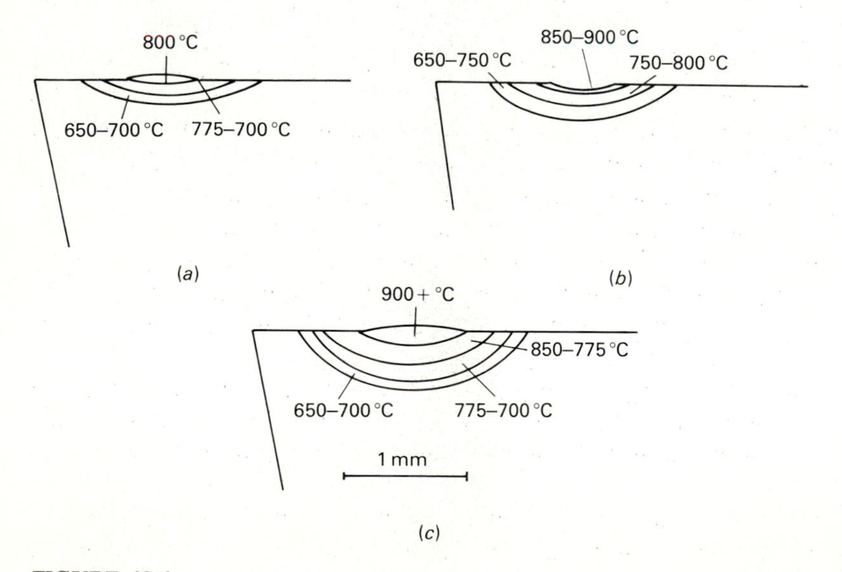

FIGURE 12.6
Influence of feed on temperatures in tools used to cut iron at feeds of (*a*) 0.005 in.; (*b*) 0.010 in.; and (*c*) 0.020 in. per rev. (*Redrawn from [11].*)

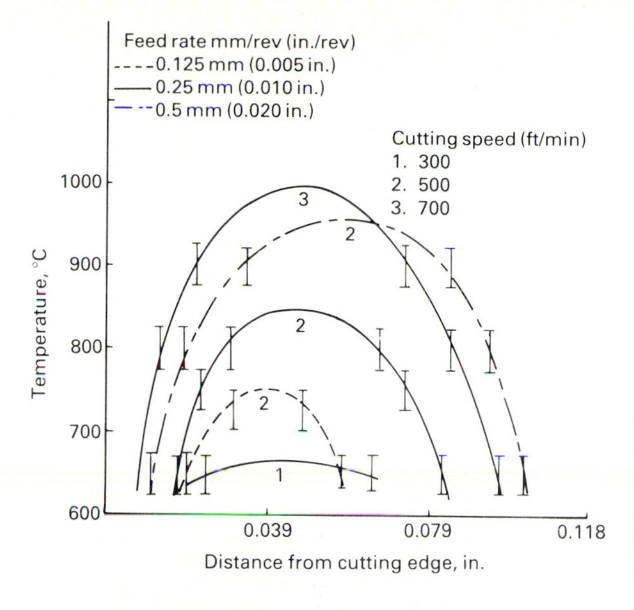

FIGURE 12.7
Temperature distribution on rake face of tools used to cut very low-carbon steel at different speeds and feeds. (*Redrawn from [11].*)

The preceding discussion illustrates the effect of cutting speed on the average temperature and its distribution by source. However, the temperature on the rake face, and consequently in the cutting tool, are not uniformly distributed. The peak temperature can be quite high and can serve to degrade the properties of the tool material, causing a faster wear rate.

Typically, the maximum temperature at the tool–chip interface does not occur at the tip, but at some point further up on the rake face before the chip leaves contact with the tool. Figure 12.6 shows the distribution of temperatures on the rake face of a tool in the machining of iron. Peak temperatures in excess of 1652°F (900°C) can accelerate the wearing of the tool. The speed and feed of the process can greatly affect both the magnitude and location of the peak temperature. As illustrated in Fig. 12.7, for low-carbon steel [11], increased speed and increased feed both cause higher temperatures and shift the location of the peak temperature farther from the tip of the tool.

BASIC MACHINING OPERATIONS

Machine tools have evolved from the early foot-powered lathes of the Egyptians and John Wilkinson's boring mill. They are designed to provide rigid support for both the workpiece and the cutting tool and can precisely control their relative positions and the velocity of the tool with respect to the workpiece. Basically, in metal cutting a sharpened wedge-shaped tool removes a rather narrow strip of metal from the surface of a ductile workpiece in the form of a severely deformed chip. The chip is a waste product that is considerably shorter than the workpiece from which it came but with a corresponding increase in thickness of the uncut chip. The geometrical shape

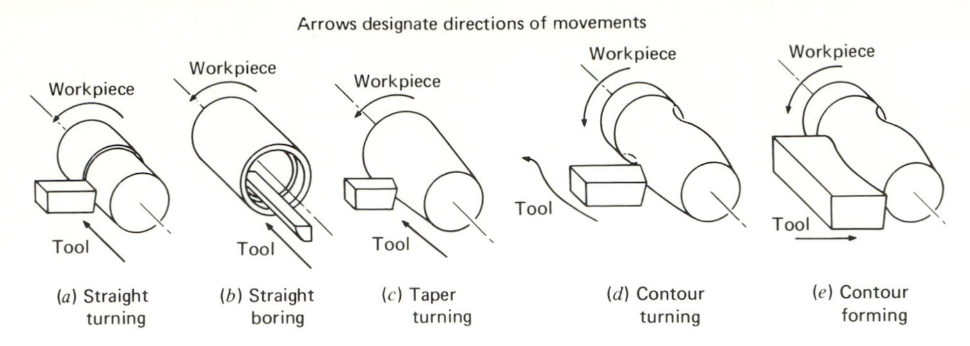

FIGURE 12.8
Diagrams showing how surfaces of revolution are generated and formed. (*Redrawn from [12].*)

of the machine surface depends on the shape of the tool and its path during the machining operation.

Most machining operations produce parts of differing geometry. If a rough cylindrical workpiece revolves about a central axis and the tool penetrates beneath its surface and travels parallel to the center of rotation, a surface of revolution is produced (Fig. 12.8*a*), and the operation is called *turning*. If a hollow tube is machined on the inside in a similar manner, the operation is called *boring* (Fig. 12.8*b*). Producing an external conical surface of uniformly varying diameter is called *taper turning* (Fig. 12.8*c*). If the tool point travels in a path of varying radius, a contoured surface like that of a bowling pin can be produced (Fig. 12.8*d*); or, if the piece is short enough (approximately 1 in.) and the support is sufficiently rigid, a contoured surface could be produced by feeding a shaped tool normal to the axis of rotation (Fig. 12.8*e*). Short tapered or cylindrical surfaces could also be contour formed.

Flat or plane surfaces are frequently required. They can be generated by radial turning or *facing*, in which the tool point moves normal to the axis of rotation (Fig. 12.9*a*). In other cases, it is more convenient to hold the

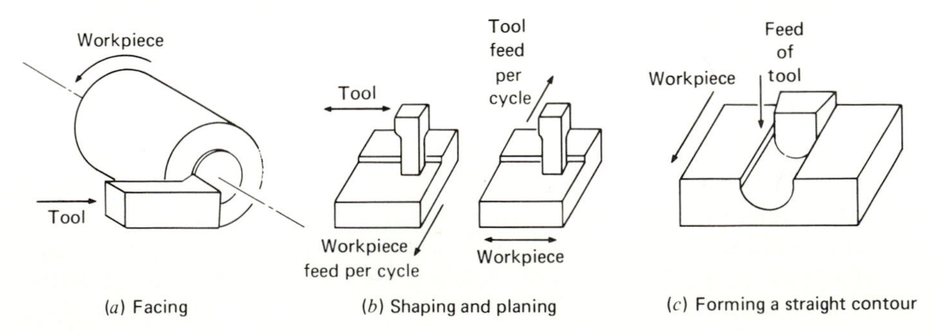

FIGURE 12.9
Diagram showing how plane surfaces are generated and formed. (*Redrawn from [12].*)

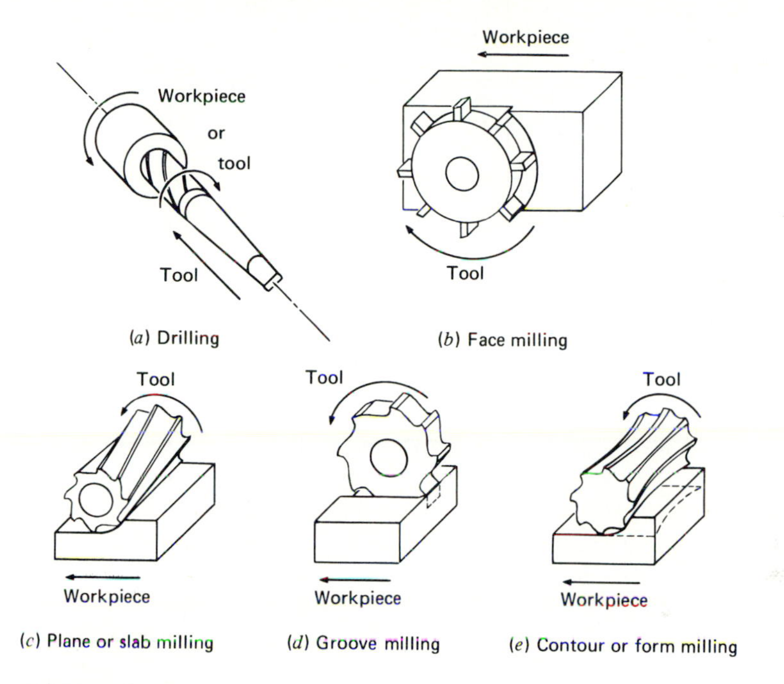

FIGURE 12.10
Typical machining operations performed by multipoint tools. (*Redrawn from [12].*)

workpiece steady and reciprocate the tool across it in a series of straight-line cuts with a crosswise feed increment before each cutting stroke (Fig. 12.9*b*). This operation is called *planing* and is carried out on a *shaper*. For larger pieces it is easier to keep the tool stationary and draw the workpiece under it as in planing. The tool is fed at each reciprocation. Contoured surfaces can be produced by using shaped tools (Fig. 12.9*c*).

Multiple-edged tools can also be used. Drilling uses a twin-edged fluted tool for holes with depths up to 5 to 10 times the drill diameter. Whether the drill turns or the workpiece rotates, relative motion between the cutting edge and the workpiece is the important factor. In milling operations a rotary cutter with a number of cutting edges engages the workpiece, which moves slowly with respect to the cutter. Plane or contoured surfaces may be produced, depending on the geometry of the cutter and the type of feed. Horizontal or vertical axes of rotation may be used, and the feed of the workpiece may be in any of the three coordinate directions (Fig. 12.10).

BASIC MACHINE TOOLS

Machine tools are used to produce a part of a specified geometrical shape and precise size by removing metal from a ductile material in the form of chips.

TABLE 12.1
Comparison of basic machining operations for ductile materials

Operation	Shape produced	Machine tool	Cutting tool	Relative motion		Surface roughness, μin.	Min. prod. tolerance, in.
				Tool	Work		
Turning (external)	Surface of revolution (cylindrical)	Lathe, boring machine	Single point			32–500	±0.001
Boring (internal)	Cylindrical (enlarges holes)	Boring machine	Single point			16–250	±0.0001 ±0.001
Shaping and planing	Flat surfaces or slots	Shaper, planer	Single point			32–500	±0.001
Drilling	Cylindrical (originates holes 0.010 to 4 in. dia.)	Drill press	Drill: twin edges		Fixed	125–250	±0.002
Milling End, form Face, slab	Flat and contoured surfaces and slots	Milling machine	Multiple points (cutter teeth)			32–500	±0.001
Grinding Cylindrical Surface Plunge	Cylindrical and flat	Grinding machine	Multiple points (grind wheel)			8–125	±0.0001

The latter are a waste product and vary from long continuous ribbons of a ductile material such as steel, which are undesirable from a disposal point of view, to easily handled well-broken chips resulting from cast iron. Machine tools perform five basic metal-removal processes: turning, planing, drilling, milling, and grinding (Table 12.1). All other metal-removal processes are modifications of these five basic processes. For example, boring is internal turning; reaming, tapping, and counterboring modify drilled holes and are related to drilling; hobbing and gear cutting are fundamentally milling operations; hack sawing and broaching are a form of planing and honing; lapping, superfinishing, polishing, and buffing are variants of grinding or abrasive removal operations. Therefore, there are only four types of basic machine tools, which use cutting tools of specific controllable geometry: (1) lathes, (2) planers, (3) drilling machines, and (4) milling machines. The grinding process forms chips, but the geometry of the abrasive grain is uncontrollable.

The amount and rate of material removed by the various machining processes may be large, as in heavy turning operations, or extremely small, as in lapping or superfinishing operations where only the high spots of a surface are removed.

A machine tool performs three major functions: (1) it rigidly supports the workpiece or its holder and the cutting tool; (2) it provides relative motion between the workpiece and the cutting tool; (3) it provides a range of feeds and speeds usually ranging from 4 to 32 choices in each case.

Speeds and Feeds in Machining

Speeds, feeds, and depth of cut are the three major variables for economical machining. Other variables are the work and tool materials, coolant and geometry of the cutting tool. The rate of metal removal and power required for machining depend upon these variables.

The depth of cut, feed, and cutting speed are machine settings that must be established in any metal-cutting operation. They all affect the forces, the power, and the rate of metal removal. They can be defined by comparing them to the needle and record of a phonograph. The *cutting speed* (V) is represented by the velocity of the record surface relative to the needle in the tone arm at any instant. *Feed* is represented by the advance of the needle radially inward per revolution, or is the difference in position between two adjacent grooves. The *depth of cut* is the penetration of the needle into the record or the depth of the grooves. A formula for converting cutting speed, usually expressed in feet per minute, to revolutions per minute of the machine spindle is derived as follows:

$$\text{r/min} = \text{V} \cdot \frac{12 \text{ in.}}{\pi D} \qquad \text{or} \qquad \text{r/min} \approx 4 \cdot \frac{\text{V}}{D}$$

where D = diameter of work, in.

Suitable cutting speeds for each workpiece material and tool material have been derived through extensive experimentation, which in turn has evolved into relative machinability ratings for most engineering materials. Such data are available in many machining and engineering handbooks. Whereas such standardized data may vary somewhat from optimum machine settings because of slight differences in material specifications, they are a good basis for original machine-tool setups. Table 12.2 shows recommend tool angles for a variety of materials.

Feed is the advance of the cutting tool (in thousands of an inch per revolution of the spindle) through the workpiece in rotary-motion machine tools. The feed on a milling machine is usally given in inches of table travel per minute, while cutter feed is expressed on the operation sheet as the amount each tooth of the milling cutter advances into the workpiece. Table travel is calculated as feed per tooth × the number of teeth in the cutter × r/min of the spindle. For example,

$$0.002 \text{ in.} \times 20 \text{ cutter teeth} \times 100 = 4 \text{ in./min table feed}$$

On a reciprocating machine tool, the feed is the amount the tool advances

TABLE 12.2
Tool geometry for turning[a]

	High-speed steel and cast-alloy tools					Carbide tools (throwaway)				
Material	**Back rate**	**Side rake**	**End relief**	**Side relief**	**Side and end cutting edge**	**Back rake**	**Side rake**	**End relief**	**Side relief**	**Side and end cutting edge**
Aluminum alloys	20	15	12	10	5	0	5	5	5	15
Magnesium alloys	20	15	12	10	5	0	5	5	5	15
Copper alloys	5	10	8	8	5	0	5	5	5	15
Steels	10	12	5	5	15	−5	−5	5	5	15
Stainless steels, ferritic	5	8	5	5	15	0	5	5	5	15
Stainless steels, austenitic	0	10	5	5	15	0	5	5	5	15
Stainless steels, martensitic	0	10	5	5	15	−5	−5	5	5	15
High-temp. alloys	0	10	5	5	15	5	0	5	5	45
Refractory alloys	0	20	5	5	5	...	...	5	5	15
Titanium alloys	0	5	5	5	15	−5	−5	5	5	5
Cast irons	5	10	5	5	15	−5	−5	5	5	15
Thermoplastics	0	0	20–30	15–20	10	0	0	20–30	15–20	10
Thermosetting plastics	0	0	20–30	15–20	10	0	15	5	5	15

[a] *Source: Machining Data Handbook*, Machinability Data Center, Metcut Research Associates Inc.

TABLE 12.3
Typical feeds and depth of cut for light machining

General feed designation	Typical feeds, in./r, or in./tooth	Depth of cut designation	Depth of cut, 10^{-3} in.
Fine	0.001–0.003	Light	3 × (feed)
Medium	0.005–0.015	Medium	5 × (feed)
Coarse	0.018–0.060	Heavy	10 × (feed)

normal to the direction of cutting (planer) or the amount the workpiece advances normal to the direction of cutting (shaper) at each reciprocation of the table or ram. The proper feed can only be selected after careful consideration of the combined effect of depth of cut, rigidity of the cutting tool and the workpiece, the method by which the workpiece is held in the machine, the microstructure of the material being cut, the geometry of the cutting tool, and the surface finish specified (Table 12.3). The depth of cut is a function of the machine setup, capacity and rigidity of both the machine and the workpiece, and the horsepower of the machine tool.

On rough material or workpieces from which a large amount of material must be removed, three cuts are required. A rough cut should be of sufficient depth to remove all traces of sand, scale, and eccentricity on cast metals and forgings. The semifinish cut should remove from $\frac{1}{32}$ to $\frac{1}{16}$ in. of metal. The purpose of the semifinish cut is to remove any traces of out-of-roundness and also to serve as an index for sizing the finishing cut. The allowance for the finishing cut should be approximately 0.015 to 0.030 in.

Range of Speeds and Feeds

In production tools, the feeds and speeds are frequently provided by a gear-driven transmission in which the various steps are arranged in such a way that no desired cutting speed deviates more than a given percentage from one available through the machine-tool transmission. Thus, if the design criterion was that no more than a 10 percent deviation should be allowed, the successive steps could be no more than ±10 percent apart. The ratio between steps would then be 1.2, and the transmission speeds would vary in a geometric progression from a base of b in the following manner: $b, br, br^2, br^3, \ldots, br^{(n-1)}$. If the last term, $br^{(n-1)}$, is defined as a and n is the number of steps in the progression, the value of r is:

$$r = \sqrt[n-1]{\frac{a}{b}}$$

Many machine-tool drives provide a choice of 8, 12, 16, or 24 speeds with corresponding values of r of 1.58, 1.36, 1.26, or 1.12. As the number of choices rises, the cost of the machine tool also rises; but in most cases a ratio of 1.26 is adequate for a production machine. In a research lathe, an infinitely variable cutting speed is desirable, but for a 30-hp lathe the drive to provide such variability is almost as large as the lathe and is at least as costly as the lathe alone.

CUTTING TOOLS

Cutting-Tool Geometry

Before a systematic study of metal cutting can be undertaken the standard nomenclature for the cutting-tool angles must be understood (Fig. 12.11). There are six single-point tool angles that are important to the machinist and processing engineer. These can be divided into three groups:

1. *Rake angles.* These affect the direction of chip flow, the characteristics of chip formation, and tool life. Rake angles are measured in the plane of the top of the tool bit.
2. *Relief angles.* These avoid excessive friction between the tool and the workpiece and allow better access of coolant to the tool–work interface.
3. *Cutting-edge angles.* The side cutting-edge angle allows the full load of the cut to be built up gradually, thus reducing the initial shock to the cutting tool. The end cutting-edge angle allows sufficient clearance so that the surface of the tool behind the cutting point will not rub over the work surface and cause increased frictional heat. In finish machining, the end cutting edge might be flattened for about 1.5 times the feed increment to permit a burnishing action between the tool and the work surface. The ridges caused by the feed increment are thereby eliminated.

Tool Signature

The elements of a single-point tool are written in the following order, which is the *tool signature*: back rake angle, side rake angle, end relief angle, side relief

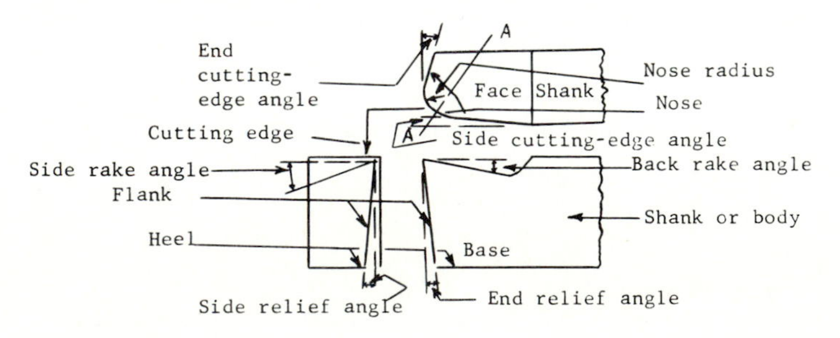

FIGURE 12.11
The geometry of a single-point tool.

angle, end cutting-edge angle, side cutting-edge angle, and nose radius (Fig. 12.11). For a carbide cutting-tool insert in a standard holder the signature might be −5°, −5°, 5°, 5°, 15°, 15°, $\frac{3}{64}$ in. Note that the second angle of each pair refers to the side designation; i.e., first comes the back rake angle and then the side rake angle, and so on.

In general, the cutting force decreases as the rake angle increases. An economic balance must be struck between reduced cutting force and tool life, because larger rake angles mean reduced efficiency of heat removal, less tool rigidity, and shorter tool life.

When using cemented carbide tools and especially cemented oxide (ceramic) tools, the solid angle at the cutting edge must be as large as possible to prevent tool failure by chipping of the cutting edge. In these cases, negative rake angles are commonly used because, despite greater horsepower requirements per unit volume of metal removed, the tool life is much improved and the total cost is significantly reduced.

Note. When a mechanical tool holder is used in conjunction with tool bits of high-speed steel, several tool angles must be corrected by an amount equal to the angle of inclination of the slot for the tool bit. Be sure to determine which angles are involved and make appropriate corrections. The usual angle of inclination of the tool slot is 15° above a horizontal plane. It should always be checked in a specific case.

Drilling and Reaming

Holes are one of the most common features in products manufactured today. Therefore, drilling and other related processes and tools are extremely important. Holes as small as 0.005 in. may be drilled using special techniques. On the other hand, holes larger than 2 to $2\frac{1}{2}$ in. in diameter are seldom drilled, because other processes and techniques are less expensive.

The twist drill (shown in Fig. 12.12) is the most common type of drill. The shank of the drill is held by the machine tool, which in turn imparts a rotary motion. This shank may be straight or tapered. The body of the drill is typically made up of two spiral grooves known as flutes, which are defined by a helix angle that is generally about 30° but can vary depending on the material properties of the workpiece. The point of the drill (see Fig. 12.12) generally form a 118° angle and includes a 10° clearance angle and chisel edge. The chisel edge is flat with a web thickness of approximately $0.015 \times$ drill diameter. This edge can cause problems in hole location owing to its ability to "walk" on a surface before engaging the workpiece. In the case of brittle materials, drill point angles of less than 118° are used, while ductile materials use larger point angles and smaller clearance angles.

Complex hole configurations may often be called for; these include multiple diameters, chamfers, countersinks, and combinations of these, as illustrated in Fig. 12.13. In each of these cases it is possible to make special

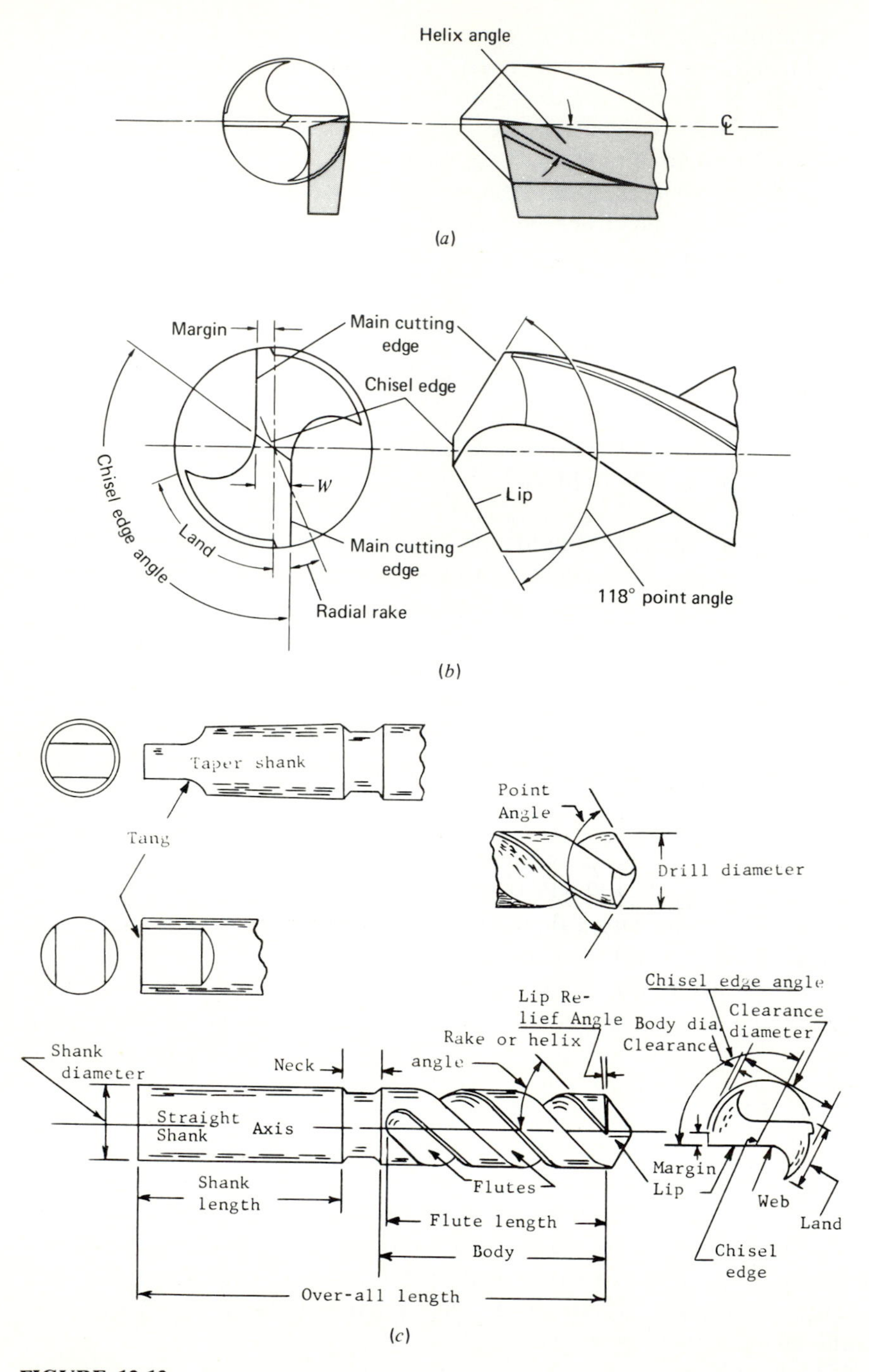

FIGURE 12.12
Geometry of the twist drill. (*a*) Comparison of twist drill and single-point tool; (*b*) standard designation of drill point features; (*c*) standard designation of twist-drill body and shank.

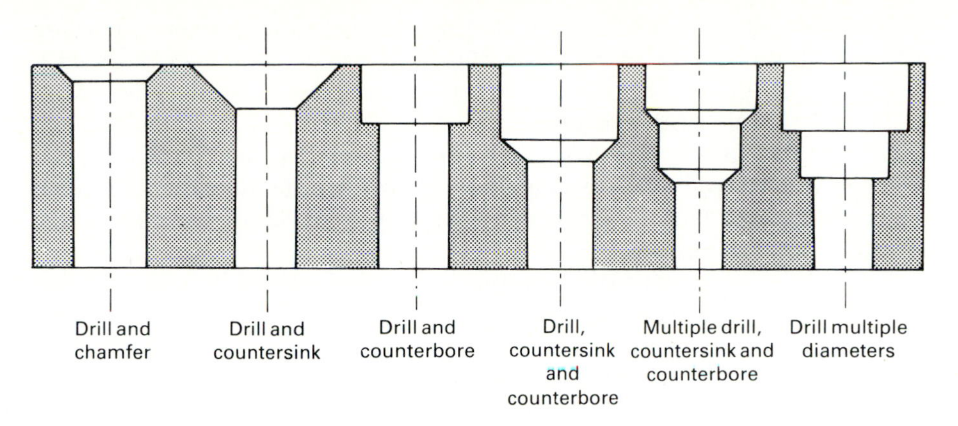

FIGURE 12.13
Internal surfaces produced by special-purpose drills.

combination drills that can produce the configurations shown in a single operation. Although expensive, they can be economically justified for sufficient volume.

The flat chisel edge, which can "walk" on the surface of the workpiece, and the long, slender shaft and body of the twist drill, which can deflect, make it difficult to machine holes to tight tolerances. As shown in Fig. 12.14, a combination center drill and countersink can be used to accurately start a hole, owing to its small web thickness and its tendency to deflect only very small amounts (because of a relatively large diameter-to-length ratio). Truing of the hole to make it straight is accomplished by boring. Reaming the hole provides a better finish as well as more accurate sizing.

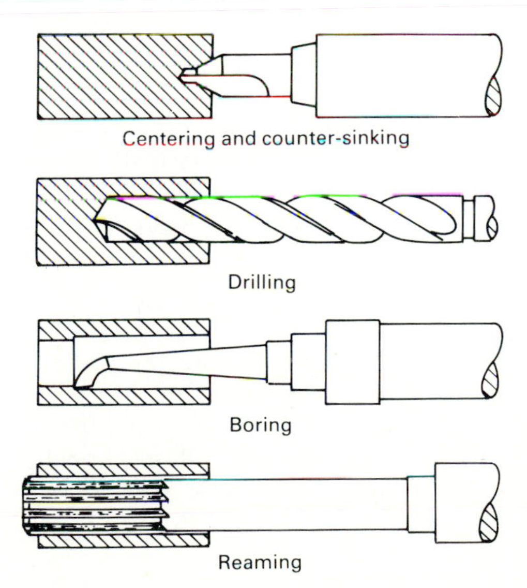

FIGURE 12.14
Tools and sequence of operations needed to produce a hole with accurate size and position, as well as a good finish. (*Redrawn from [20].*)

The feed rate of a drill is normally proportional to its diameter, because it depends on the volume of chips the flutes can handle. However the feed is independent of the cutting speed, which is a function of the tool–work combination. A rule of thumb would give a feed rate as approximately $d/65$, so that a $\frac{3}{4}$-in.-diameter drill would have a feed rate of about 0.012 in./rev. Although the hole wall tends to support the drill when the hole depth exceeds three times the drill diameter, there is a tendency for buckling to occur and the feed rate should be reduced.

Most drills are made from high-speed steel because of its relatively low cost and ease of manufacture. Some types of carbide drills are now available commercially. The demands of numerically controlled machine tools have led to the development of drills that will produce more precise holes and that will originate a hole in line with the centerline of the drill-press spindle. Drills that have heavier webs, less stickout, double margins, and are ground with a spiral point help meet these new demands.

Multipoint Cutting Tools—Milling Cutters

Milling chips are relatively short and, when they are produced by slab milling, the undeformed chip thickness, i.e., original thickness of the layer removed by machining, varies along the length of the chip. There are two major types of milling operations: (1) plain milling and (2) face milling. A plain-milling cutter rotates about a horizontal axis (Fig. 12.15), has cutting teeth only on its periphery, and is designed to produce flat surfaces. A face mill has cutting teeth on the end as well as on the periphery, and is designed to be used with its axis of rotation normal to the surface that is machined.

Each tooth of a milling cutter may be compared to a single-point cutting tool (Fig. 12.15*a*). In fact, a milling cutter may be likened to a series of single-point cutting tools designed to rotate about a common axis (Fig. 12.15*b*). The nomenclature for a typical helical milling cutter is also given in Fig. 12.15.

Cutters may be solid or they may be designed to be supplied with inserted blades of cemented carbide. A solid cutter becomes smaller with repeated sharpening and chip accommodation space becomes reduced. Inserted-blade cutters are usually supplied with cemented carbides because typical cast alloy and ceramic materials have too little shock resistance for interrupted cutting.

Inserted blades permit more economical maintenance of large production cutters, because small inserts can be sharpened individually and the original diameter can be maintained by resetting the insert farther out to compensate for the material loss resulting from the regrinding operation. Step milling or gang milling, which produces two or more surfaces simultaneously, can be economically feasible because a constant differential in diameter or width can be maintained regardless of the number of regrinds.

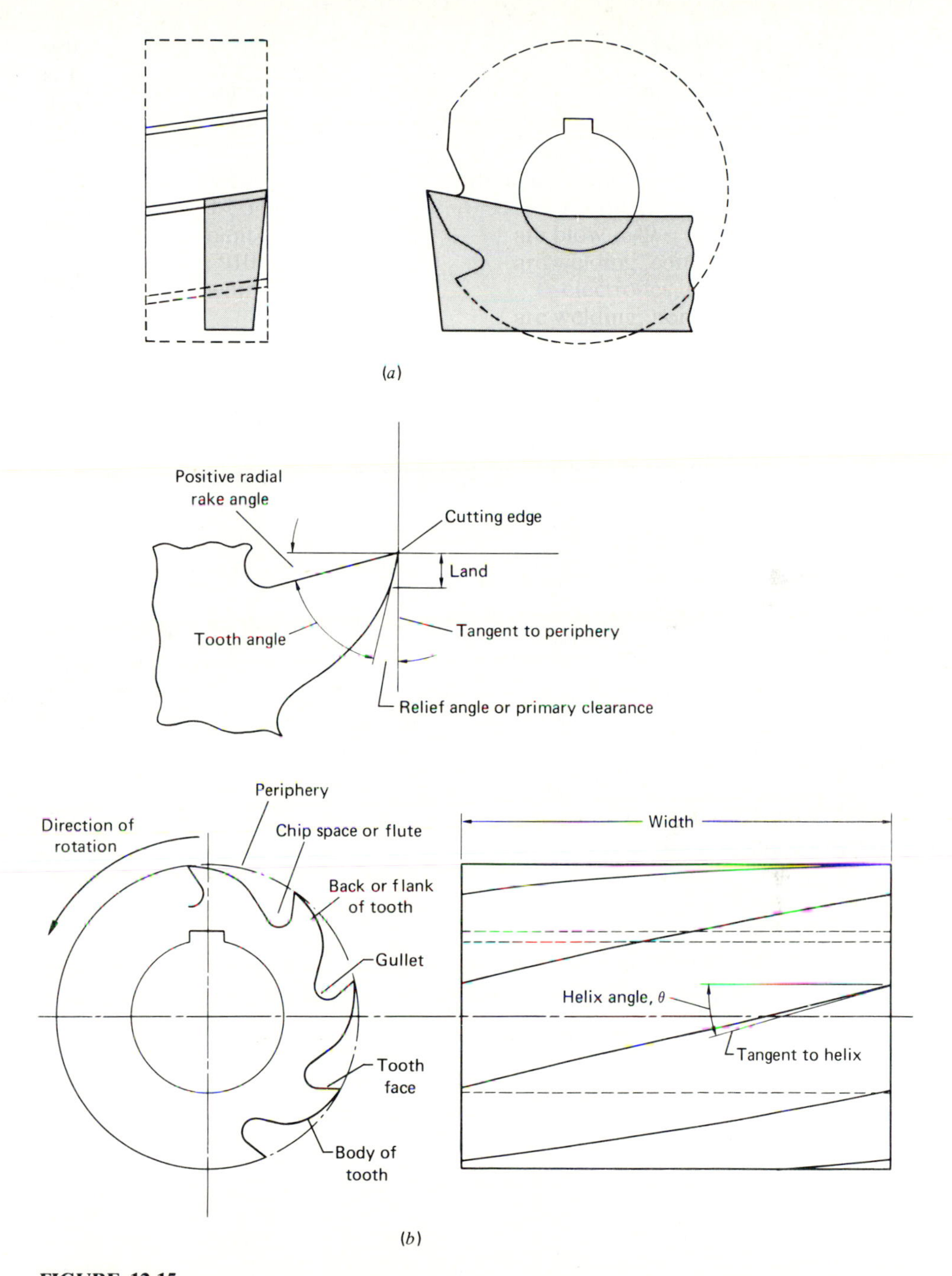

FIGURE 12.15
Principal elements of a helical milling cutter. (*a*) Comparison with a single-point tool; (*b*) nomenclature of a typical milling cutter.

The Grinding Wheel

The grinding wheel approaches a cutting tool with an infinite number of cutting edges. In grinding operations, if a cutting grit is poorly oriented or dull, it pulls away from the wheel and a new grit is exposed. Since the hard, sharp-edged grits are randomly located in the matrix of the grinding wheel, all types of cutting geometry are found. A conceptual view of a grinding wheel and the tiny chips produced is shown in Fig. 12.16. Note that ideally the wheel is much larger than the workpiece so that the effect of wear is negligible when grinding any one piece.

By removing a succession of small chips, the grinding wheel can produce surfaces to very close dimensions and a high degree of smoothness. Hard abrasives can cut hard materials; often grinding is the only way in which some materials, such as tungsten carbides, may be accurately shaped to final size. To obtain various rates of cutting and different kinds of surfaces on the many types of materials, the engineer uses many combinations of abrasive crystals and bonding materials.

DESIGNATION OF GRINDING WHEELS. In the past it was the custom of a manufacturer to accept the recommendations of a supplier for a particular grinding wheel for a special operation. It is good practice to obtain the advice of grinding experts on difficult operations; but now, when wheels are specified, the designations are standardized. Våriations can be made, and more than one supplier can furnish comparable wheels. The standard marking system is given in Fig. 12.17. The markings are in six parts:

1. Abrasive type
2. Grain size
3. Grade
4. Structure

FIGURE 12.16
Relationship between the abrasive grains in a grinding wheel and the work-piece. Note the microchips at *a* and *b*.

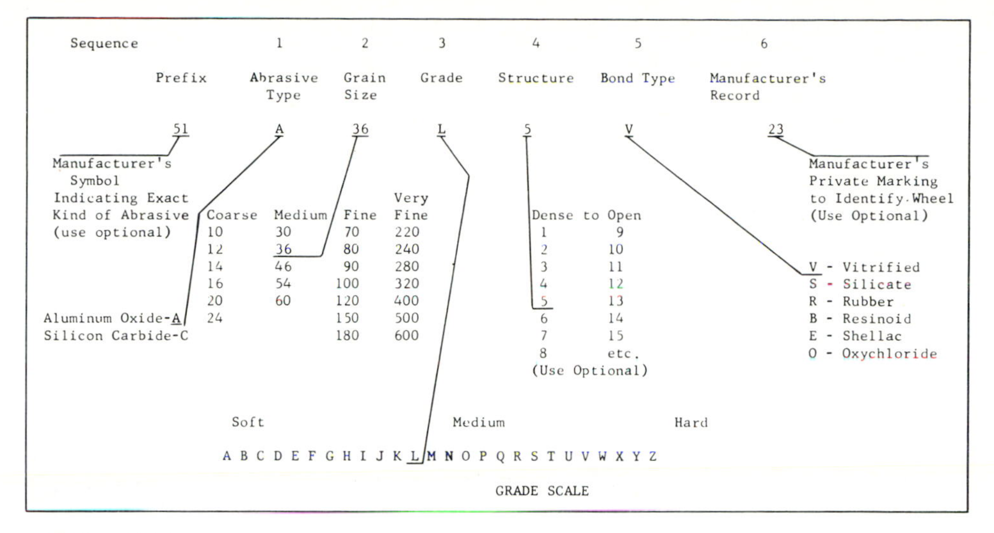

FIGURE 12.17
Standard marking system chart for grinding wheels. (*Approved by the Grinding Wheel Manufacturers Association.*)

5. Bond type
6. Manufacturer's record

Abrasive types. Abrasive types are aluminum oxide and silicon carbide. Aluminum oxide crystals are tough and resist fracture to a high degree and consequently are used principally in grinding ferrous and other materials that have a high tensile strength. Silicon carbide is harder than aluminum oxide, but its crystals are not as tough. The crystals break easily, leaving particles angular in form. Since silicon carbide crystals fracture easily, it is especially adapted to cutting materials with low tensile strength such as brass, aluminum, copper, cast iron, rubber, and plastics. It is also used in grinding hard, brittle materials such as carbide, stone, and ceramics.

Grain size. The number designating size represents the number of openings per linear inch in the screen used to size the abrasive grains. Selection of size of grain will depend on the amount of material to be removed, the finish desired, and the mechanical properties of the material to be ground.

The larger the grains, the faster material will be removed; however, hard, brittle materials will not permit the large grains to penetrate to their full depth, so that any advantage in using grains of this size is lost. In this case, smaller grains are used, as in this way more cutting edges will be brought into contact with the work per revolution and the net result will be the same. Coarse grains are better adapted to grinding soft, ductile materials; fine grains are best when fine finishes and close accuracy are required.

Grade. The letters designating grade indicate the relative strength (holding power) of the bond that holds the abrasive in place. In general, with a given type of bond it is the amount of bond that determines the hardness (grade). When the amount of bond is increased, the size of the bond posts connecting each abrasive grain to its neighbors is increased. The larger bond posts are naturally stronger, thereby increasing the hardness of the wheel. The strength or grade depends upon the forces applied to the abrasive grains. An abrasive grain should remain in a wheel as long as it cuts. When it dulls and does not cut the surface of the material, the increased forces should tear it out of the bond so that fresh, sharp abrasives can cut. Thus:

1. When the area of contact is small (OD of a small cylinder) the forces on each grain are high and a hard wheel is required to hold the abrasives before they dull. The forces are high on each grain because there are fewer grains cutting on the line of contact. When a flat surface or internal surface is ground, the number of grains in contact is greater and the forces are less; therefore, a softer-grade wheel can be used.
2. When material is hard, a soft grade is required because the cutting edges of the abrasive dull quickly and must be removed so that fresh grains can cut.
3. When machine, wheel, or work vibrates, the pounding action causes the grains to break out before they are dull; therefore, a harder wheel is required.
4. When the speed of the cut surface is increased in relation to the speed of the wheel surface, a thicker chip is cut by the abrasive and the forces increase; this tends to break the grains from the wheel. This increased wear of the wheel is called *soft action* because it is similar to the action of a soft wheel. It can also result from increased transverse or infeed.

Structure. The structure number indicates the relative spacing of the abrasive grains. When the grains are close together relative to the grain size, the wheels have low structure numbers such as 1, 2, 3, 4, 5. Wider spacing relative to grain size is designated by higher numbers. Selection of the propert structure, or spacing between the grains, is governed by the finish required, the nature of the operation, and the mechanical properties of the material to be ground.

Wide spacing. A wheel with wide spacing between the grains is best for grinding soft, ductile materials. The grains are thus allowed to penetrate to their maximum depth and clearance is provided between the grains for the large chips. Wide spacing is also best when grinding exceedingly hard materials like cemented carbides, because the grains are released more promptly and new, sharp grains are exposed as needed. In snagging and other operations where a variable application of pressure is involved, as well as in surface grinding, wide spacing is essential so that the pressure on the work will be distributed evenly over all the grains.

Close spacings. A closely spaced structure should be selected when grinding hard, brittle materials in order to utilize the cutting properties of the maximum number of small grains. If heavy pressures are used, there will be a tendency to break down the form of shaped wheels and close grain spacing will be required. From the standpoint of finish, smoothness will vary directly with the closeness of the spacing.

Medium spacing. A medium-spaced structure should usually be selected for center and centerless cylindrical, tool, and cutter grinding. The best policy to follow is to experiment somewhat with work speeds to find the best combination before making any definite selections as to wheel grade. This will give better results than arbitrarily selecting a grade and then fitting the work speed to it.

Effect of grade on wheel life. It might appear that the harder the wheel, the longer its life will be; but this is not always true. If the grade is too hard, the wheel will load and frequent dressing will be necessary. A hard wheel is often worn down more by the dressing than by the grinding it does. This fact merits consideration, especially since dressing-tool cost is high. Another factor to consider in this respect is that while a hard wheel may last longer on a given job, it will cut more slowly, with the result that labor and overhead costs per piece will be higher. If these increased costs are not more than offset by the saving in wheel cost, the use of the hard wheel will be uneconomical.

Bond type. Bond is the medium that holds the grains together in the form of a wheel, or on a belt or disk. The bond functions in the same way as a tool post and holds the grains or cutting tools in position until they become dull and are torn out and fresh grains exposed. The bond itself ordinarily has little or no cutting action. Great advancement has been made in recent years through resinoid and rubber bonds that permit flexibility of the grinding medium, such as thin wheels and belts. They are not affected by water, which can be used to keep them cool. The features of these new bonds have extended the application of abrasives and have reduced costs.

Vitrified bond. Vitrified bond, which is essentially glass, is used in over 75 percent of the grinding wheels manufactured. Wheels made with this type of bond have porosity and strength, and are used when a high rate of stock removal is required. Vitrified bond is unaffected by water, acid, oils, or ordinary temperature conditions.

Silicate bond. Silicate bond is composed principally of silicate of soda or water glass. This bond does not hold the grains as tightly as vitrified bond and, for this reason, less grinding stress is required to tear them out. The principal application of silicate-type wheels is for grinding edged tools that cannot be subjected to excessive heat while grinding.

Shellac bond. Shellac-bonded wheels are elastic and are used to produce high finishes. Other applications are the cool cutting of hardened tool steels and for cutting-off operations. They are affected by coolants and heat.

Resinoid bond. Resinoid-bonded wheels are held together by a synthetic resin or plastic. They can be made in various structures and have the characteristics of cutting cool and removing material rapidly. Resinoid wheels are very strong and can be run at high speeds.

Rubber bond. Rubber-bonded wheels are very strong and tough; for this reason, thin wheels are usually of this type. The principal application of rubber-bonded wheels is in jobs where a good finish is required and in cutting-off operations.

Cutting-Tool Materials

All machine tools must be provided with an easily attached and rigidly supported cutting-tool bit or cutter. It must be able to withstand severe shock, and erosive and abrasive conditions for 10 to 30 min of cutting, with no more than 0.030 in. wear in the flank area. Cutting-tool materials must have *red* (hot) *hardness*; i.e., they must retain their hardness at high temperatures because excessive temperature is the major cause of tool breakdown. Toughness is also a decided asset because cutting tools are subject to shock loading during interrupted cuts or at the beginning of a cutting operation. There are a number of cutting-tool materials available, but high-speed steel and tungsten carbide tools do the bulk of the metal cutting. High-speed steel tools are preferred for use on older machine tools or for interrupted cuts because of their toughness and high tensile strength. Carbide tools have higher hardness at red heat; if properly supported, they will give long life at about $2\frac{1}{2}$ times the cutting speed of high-speed steel. The hardness of the principal cutting-tool materials as a function of operating temperature is given in Table 12.4. The first cutting-tool material was high-carbon steel (SAE 1090), but it could not withstand temperatures above 400°F.

High-speed steels. Although high-carbon steels were the only tool material until 1900, Taylor and White's discovery of high-speed steel and the demands of World War I relegated carbon steel to use in wood chisels and small drills or form tools. The 18% W–4% Cr–1% Va type of high-speed steel (Table 12.4) revolutionized the metal-cutting industry and soon machine tools were redesigned to incorporate heavier bearings, a greater range of feeds and speeds, and individual motor drives that can now provide up to 50 hp or more for a single machine tool.

To save tungsten during World War II, a molybdenum type of high-speed steel was developed. This has proved to be a good, but not necessarily less expensive, substitute. When up to 12 percent cobalt is added to certain types of high-speed steel, super high-speed steel is produced. These alloys have superior red hardness and resistance to abrasion with some loss in toughness, but at 50 percent more cost. However, users find that cobalt high-speed steels are less easily ground and that their heat treatment is considerably more difficult than that used for conventional high-speed steel.

TABLE 12.4
Comparison of cutting-tool materials

Cutting-tool material	Approx. carbide volume, percent	Chemical composition	When first available	Red hardness R_a	Red hardness R_c[a]	Typical cutting speed, ft/min	Max. operating temperature °F	Relative cutting cost, \$/in.^{3}	Cost relative to high-speed steel (h-s steel = 1)
Carbon steel	15	0.9–1.2% C, balance Fe	Only material up to 1900	—	—	40	350–400	0.42	0.3
High-speed steel	10–20	18% W, 4% Cr, 1% Va; 0.6% C, 8% Mo; 4% Cr, 1.5% W, 1% Va (Also many variants)	1900	76	57	90	900–1,000	0.22	1
Cast cobalt alloy	25–35	43–48% Co, 17–19% W 30–35% Cr; 2% C	1915	74	47	150	1,200	0.10	2.5
Sintered tungsten carbide	75–95	Cast iron grade: 94% WC, 6% Co Steel grade: WC, TiC; TaC; 8% Co	~1930 ~1945	82	62	500	1,200–1,400	0.06	10
Ceramic (aluminum oxide)	—	Al_2O_3 plus binders	1955	84	65	800	1,500	0.03	15
Diamond	—	Crystalline C	~1960	—	—	1500	1,100		

[a] R_a = Rockwell A, R_c = Rockwell C.

Infusion of carbides into high-speed steels. A limited amount of work has demonstrated that the tool life of certain materials, such as high-speed steel, could be markedly improved by infusing materials such as tungsten carbide into the surface layers only. Both laser and electron-beam energy sources can be used to produce such carbide-enriched surface layers.

However, typical sputtered coatings of many types seem to spall off the tool materials in erratic fashion. Research along these lines is underway, but it may be a number of years before the results are available for everyday metal cutting.

Cast nonferrous tools. This tool material was first introduced in 1915 just in time for use in the production of material for World War I. A typical cobalt-based alloy, Stellite has a composition of 43–48% Co, 17–19% W, 30–35% Cr, and 1.85–2.15% C. This nonforgeable alloy, which must be cast to shape and ground to give it precise size and shape, cannot be hardened by heat treatment. It is heat-treated after casting to break up the carbide network, although its hardness results from metallic carbides in a softer cobalt matrix. Cobalt alloys are not affected by heat up to 1500°F and are actually tougher at a dull red heat than they are cold. The cobalt alloys take a high polish, which helps to reduce initial edge buildup, and their high abrasion resistance makes them especially good for cutting cast iron at speeds 50 to 100 percent higher than those suitable for 18–4–1 high-speed steel, which places them in competition with carbides at their lower ranges of cutting speed. Cobalt alloys are expensive, so they too are used in the form of inserts.

Cemented carbides. Cemented carbides were first developed in Germany in 1928. Early grades were extremely brittle and could not be used for machining steel because of their tendency to form a built-up edge. In less than 1 min the built up-edge was so large that it put the nose of the tool in tension and the tool tip failed catastrophically. During World War II it was discovered that the addition of up to 30 percent titanium carbide and tantalum carbide to the tungsten carbide, and more cobalt, prevented the formation of the built-up edge by destroying the weldability of the steel to the tool material. The cobalt increased the toughness of the material at the expense of hardness. Although cemented carbides have higher initial cost, they can be used in the form of mechanically clamped inserts. In this way one rectangular tool bit can be used up to eight times before disposal and requires no resharpening cost. Thus cutting cost is in the order of $0.20 to $0.30 per edge. The tool holders have special locating pockets that provide a standard negative-rake angle into which the inserts fit; the base or seat is also carbide for good support and reproducible alignment.

Carbides have now been developed into a wide range of cutting-tool materials varying from hard, brittle, wear-resistant carbides to soft, tough carbides. Their properties depend upon their binder content and the size, distribution, and chemical composition of the carbide particles.

The development of the disposable insert and electrolytic grinding techniques have made cemented carbides the workhorse of industry. However, even though carbides are easily available, each successful installation of carbide tooling must be studied carefully to determine the availability of sufficient power, the tightness of all spindle bearings, and carriage and cross-feed gibs. Only a rigid setup can be used to machine with carbide tools.

Ceramic or metallic oxides. Ceramic inserts were first developed in the late 1930s, but their extreme brittleness made them unsuitable for many of the machines of that day. They did find application for machining cast irons of less than 235 BHN and steels with a hardness of less than 35 Rockwell C. These first ceramic cutting tools were produced by cold pressing. During World War II, the shortage of tungsten made the Germans search for other types of cutting tools, and ceramic tools were intensely investigated. Captured documents led to further development work at Watertown Arsenal in the United States.

Ceramic tools are primarily aluminum oxide compacts with or without additional binders. They must be used in negative-rake tool holders and frequently a 45° chamfer on the cutting edge has been shown to improve tool life. The low coefficient of heat transfer results in most of the heat produced by the machining operation passing to the chips. Thus, the inserts are cool to the touch even after a heavy cut.

In the 1960s, hot-pressed ceramics having a mixture of 70 percent Al_2O_3 powder and 30 percent titanium carbide were introduced. These ceramics had higher transverse rupture strength and greater shock resistance. These ceramics proved successful in the machining of cast irons in excess of 235 BHN and steels with hardness above 35 and up to 68 Rockwell C. They have been successful in machining high-temperature nickel-based alloys such as Inconel 718 at speeds up to six times those recommended with carbide tools. About 90 percent of all steel mill roll turning today is done with hot-pressed ceramics.

It has been found that compressive residual stresses can be induced into the ceramic compact, which in turn may lead to tools that can be used under impact conditions and that can sustain larger rates of stock removal for longer times before failure. Ceramic tools can cut at top speeds, up to 5000 ft/min, but their tool life, as seen from Table 12.4, is less than that of tools made of other cutting tool materials. Their mode of failure is brittle fracture at random locations of point defects. Tool life is less predictable than it is with sintered carbides. One tool may last several times longer than another, even though both appeared to be identical and were subjected to the same cutting conditions.

In the early 1980s and up to the present, significant developments and improvements have taken place in connection with ceramics. The introduction of cermets has improved wear resistance and toughness. Cermets are materials in which the base material is ceramic, and metals and metal carbides are applied as binders and additives. Titanium nitride and titanium carbide serve as principal components of the cermets.

Cermets have higher chemical stability than tungsten carbide, resulting in reducing oxidation, cratering, and built-up edge. Comparison with conventional carbides indicate that cermets have a lower thermal shock factor, so that coolant can be applied on cermet tools during finishing cuts and in threading and grooving, but not on roughing cuts.

In relation to conventional carbides, cermets have up to 25 percent less hot strength but are superior from the standpoint of hot hardness.

Today there are two principal families of cermet cutting tool materials: one is based on aluminum oxide and the other on silicon nitride. Those of the aluminum oxide group contain a small amount of zirconium oxide added as a sintering aid and other additives in order to improve hardness and/or toughness and thermal shock resistance. Among these are zirconium oxide, titanium carbide, and silicon carbide whiskers.

Ceramics based on silicon nitride are known as "sialons" if there is partial substitution of silicon and nitrogen atoms by aluminum and carbon on the crystal lattice. These ceramics are weaker in bending strengths than the aluminum oxide ceramics.

With the new silicon nitride ceramics, which are produced using the same hot-pressing technique, cast irons are being machined at speeds from 350 to 5000 surface feet per min (sfpm) while maintaining good tool life. The hot hardness of ceramics compare favorably with carbides. For example, a C-7 carbide starts losing hardness at 200°C and has lost about 60 percent of its room temperature hardness at 900°C. Silicon nitride ceramics hold almost all their hardness to 750°C and at 900°C lose only about 15 percent of their hardness.

Ceramic tools have tremendous potential because they are composed of materials that are abundant in the Earth's crust. This is one important consideration in view of the limited amount of tungsten available. As our knowledge increases in the areas of fracture mechanics, the role of surface compressive stresses, the control of point defects, and the nature and application of surface films, the use of ceramic tools will increase markedly. Thus there is a need for continuing long-range research in this area.

Diamonds. Diamonds are the hardest known material, but they can be used for cutting only at temperatures up to 1100 to 1200°F because they oxidize readily at high temperatures. They are suitable for cutting very hard materials such as cemented carbides and ceramic tools. They produce a very fine surface finish when used with high speed, low feed, and good point geometry. Diamonds are extremely brittle and are a relatively poor conductor of heat. Consequently, they are limited to depths of cut of only a few thousandths of an inch. Typical applications include the precision boring of holes and the finishing of plastics or other abrasive materials.

Cutting-tool diamonds are off-grade chips of natural diamonds that are not of gem quality, or they may be synthetic diamonds. The present demand is so great that both natural and synthetic diamonds are needed. The growth of inserted carbides has created an increasing market for carbide cutting tools that in turn has caused an increasing demand for industrial diamonds.

TOOL LIFE

Tool Failure

The useful life of a cutting tool may end in a variety of ways, which may be broadly categorized as:

1. Fracture
2. Plastic deformation
3. Gradual wear

Fracture or chipping of the cutting tool may arise when it is overloaded. High forces may be encountered owing to mispecification of cutting parameters (speed, feed, depth of cut), hard particles in the workpiece material, or impact loading when the tool first enters the workpiece. Defects in the cutting tool material may also cause fracture even though the tooling and machining parameter selection is appropriate. Cracking and subsequent chipping or fracture may also occur owing to thermal fatigue during interrupted cutting, because of alternating expansion and contraction of the tool as it goes in and out of cut.

The high temperatures produced during machining, coupled with high compressive stresses that the cutting process imposes on the tool, makes plastic deformation of the tool possible. The maximum deformation is typically near the nose of the tool and can cause higher cutting forces and subsequent fracture.

Tool-wear mechanisms

The wear of materials is a complex phenomenon. Many mechanisms to explain observed wear exist in the literature. Of these, abrasion, adhesion, and diffusion are quite common, especially in conjunction with tools used for machining of metals. To understand the applicability of these mechanisms, special care must be taken to recognize those special factors inherent to the cutting process [13]:

1. The surface against which the tool rubs is always newly cut from the work material and there is little time for oxides or other films to form.
2. The surface on which the tool is rubbing becomes severely hardened owing to the strain developed to form the chip.
3. Pressures and temperatures at the tool–chip interface are high.

In abrasive wear, the cutting tool is worn by hard particles in the workpiece materials. Many steels and cast irons, for example will contain carbides, oxides and nitrides whose hardness will exceed that of the martensitic matrix of a high-speed-steel tool. Sand on the surface of sand-cast parts can also abrade the cutting tool, as can the severely work-hardened built-up edge.

Under conditions of high pressure and temperature it is possible for one material to seize or weld to another. Current theory suggests that as the asperities (high points) of the tool and chip meet, they can deform and weld owing to high localized temperatures and pressures. Subsequently, the relative motion of the chip and tool causes tearing to occur, leaving a wear particle. This type of mechanism is prevalent for soft metals (such as aluminum and copper) that are in an annealed state.

Diffusion is a process of atomic movement caused by an energy input. In the case of machining, the input energy is supplied by the increase in temperature owing to plastic deformation and friction. The diffusion of carbon (in a high-speed-steel tool) or other strengthening constituents can soften the tool and thereby cause a higher wear rate.

Geometry and Models of Tool Wear

The geometry of tool wear is illustrated in Fig. 12.18. Wear on the flank is caused by rubbing of a new surface on the tool flank and is characterized by

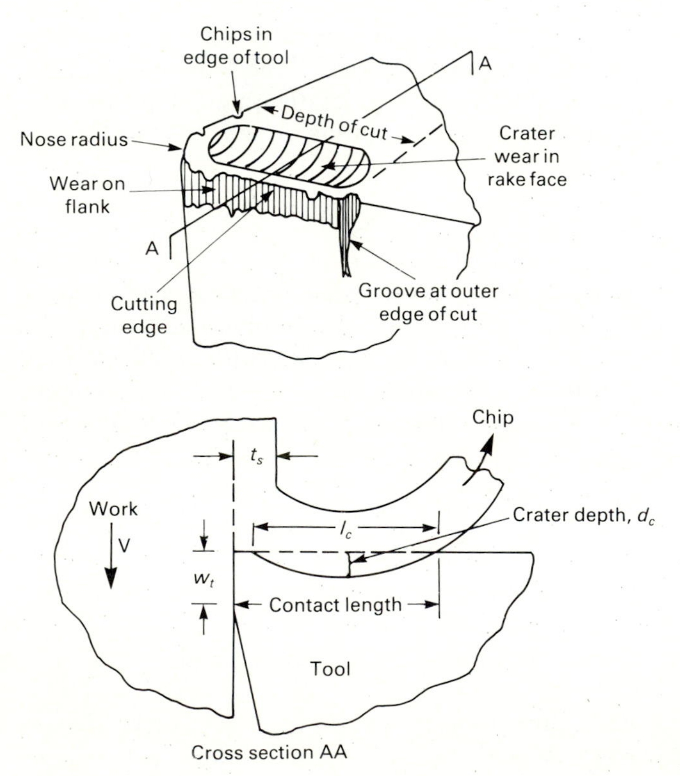

FIGURE 12.18
Geometry of tool wear.

the width of the wear land, w. Crater wear is characterized by either the depth of the crater (denoted d_c) or the length of the crater, l_c, and occurs where the chip rubs on the tool face before it leaves the workpiece.

Models of flank wear that relate it to the primary machining parameters have been studied for many years. Work by F. W. Taylor in 1907 established the following empirical relationship:

$$VT^n = C$$

where V is the cutting speed, ft/min, T is the tool life, min, n is the slope of the best fit line (log–log plot) and C is a constant (the intercept or speed for 1 min of tool life). Typical values of n and C are given in Table 12.5 and apply to only a particular tool–workpiece combination.

To perform a Taylor tool-life test, several speeds are selected to be run with a single feed and depth of cut. The flank wear is measured, periodically recorded, and plotted against time as shown in Fig. 12.19. This is repeated for several speeds and the results are plotted. Each wear curve has three distinct regions. Initially a break-in period with a high wear rate occurs, followed by a region of linear wear. After the tool has worn enough, the cutting temperature and force rise, thereby causing accelerated wear. With the wear curves established, a critical amount of flank wear must be chosen to define tool failure. The value of this cut-off varies, depending on the tool material and type of cut (rough or finished). This then defines a tool life for each speed. This result is then plotted after undergoing a log–log transformation as shown in Fig. 12.20, with n being the slope of the best fit line and C the intercept for 1 min of tool life.

Experimentation and experience has shown cutting speed to be the single most important variable affecting tool life. Feed and depth of cut have a lesser effect and can be included in a tool-life description as follows:

$$TV^A f^B d^C = K$$

where f is the feed, d is the depth, and A, B, C, K are empirical constants. T and V represent the tool life and speed respectively. This form of the model is more accurate but requires much more extensive experimentation.

As well as the machining parameters mentioned, factors dealing with tool geometry, workpiece hardness and the application of cutting fluid can significantly alter tool life. Increases in workpiece material hardness in steels, for example, can radically decrease the useful life of the tool. The effect of cutting fluids is discussed in the next section.

Cutting Fluids

Excess temperature is the most serious limitation to tool life because cutting-tool materials soften markedly at sufficiently high temperatures. Thus, as industry strives for faster metal-removal rates and ever-shorter cycle times, the increased speeds and feeds magnify the thermal problems of the tool and

TABLE 12.5

The relation between cutting speed and tool life for various tool materials and conditions

No.	Material	Shape	Material cut	Size of cut, in.		Cutting fluid	$VT^n = C$		Reference
				Depth	Feed		n	C	
1	High-carbon steel	8, 14, 6, 6, 6, 15, $\frac{3}{64}$	Yellow brass (0.60% Cu, 0.40%	0.050	0.0255	Dry	0.081	242	[14]
2	High-carbon steel	8, 14, 6, 6, 6, 15, $\frac{3}{64}$	Zn, 0.8% Sn, 0.006% Pb)	0.100	0.0127	Dry	0.096	299	[14]
3	High-carbon steel	8, 14, 6, 6, 6, 15, $\frac{3}{64}$	Bronze (0.90% Cu, 0.10% Sn)	0.050	0.0255	Dry	0.086	190	[14]
4	High-carbon steel	8, 14, 6, 6, 6, 15, $\frac{3}{64}$	Bronze (0.90% Cu, 0.10% Sn)	0.100	0.0127	Dry	0.111	232	[14]
5	High-speed steel (18–4–1)	8, 14, 6, 6, 6, 15, $\frac{3}{64}$	Cast iron (160 B)	0.050	0.0255	Dry	0.101	172	[14]
6	High-speed steel (18–4–1)	8, 14, 6, 6, 6, 15, $\frac{3}{64}$	Nickel (164 B)	0.050	0.0255	Dry	0.111	186	[14]
7	High-speed steel (18–4–1)	8, 14, 6, 6, 6, 15, $\frac{3}{64}$	Ni–Cr (207 B)	0.050	0.0255	Dry	0.088	102	[14]
			Steel						
8	High-speed steel (18-4-1)	6, 14, 6, 6, 6, 0, 0	SAE B1113 (C.D.)	0.050	0.0127	Dry	0.08	260	[14]
9	High-speed steel (18-4-1)	6, 14, 6, 6, 6, 0, 0	SAE B1112 (C.D.)	0.050	0.0127	Dry	0.105	225	[14]
10	High-speed steel (18-4-1)	6, 14, 6, 6, 6, 0, 0	SAE B1120 + Pb (C.D.)	0.050	0.0127	Dry	0.060	290	[14]
11	High-speed steel (18-4-1)	6, 14, 6, 6, 6, 0, 0	SAE 1035 (C.D.)	0.050	0.0127	Dry	0.110	130	[14]
12	High-speed steel (18-4-1)	6, 14, 6, 6, 6, 0, 0	SAE 1035 + Pb (C.D.)	0.050	0.0127	Dry	0.110	147	[14]
13	High-speed steel (18-4-1)	8, 14, 6, 6, 6, 15, $\frac{3}{64}$	SAE 1045 (C.D.)	0.100	0.0127	Dry	0.110	192	[14]
14	High-speed steel (18-4-1)	8, 22, 6, 6, 6, 15, $\frac{3}{64}$	SAE 2340 (185 B)	0.100	0.0125	Dry	0.147	143	[14]
15	High-speed steel (18-4-1)	8, 14, 6, 6, 6, 15, $\frac{3}{64}$	SAE 2345 (198 B)	0.050	0.0255	Dry	0.105	126†	[14]
16	High-speed steel (18-4-1)	8, 14, 6, 6, 6, 15, $\frac{3}{64}$	SAE 3140 (190 B)	0.100	0.0125	Dry	0.160	178	[14]
17	High-speed steel (18-4-1)	8, 14, 6, 6, 6, 15, $\frac{3}{64}$	SAE 4350 (363 B)	0.0125	0.0127	Dry	0.080	181	[14]
18	High-speed steel (18-4-1)	8, 14, 6, 6, 6, 15, $\frac{3}{64}$	SAE 4350 (363 B)	0.0125	0.0255	Dry	0.125	146	[14]
19	High-speed steel (18-4-1)	8, 14, 6, 6, 6, 15, $\frac{3}{64}$	SAE 4350 (363 B)	0.025	0.0255	Dry	0.125	95	[14]
20	High-speed steel (18-4-1)	8, 14, 6, 6, 6, 15, $\frac{3}{64}$	SAE 4350 (363 B)	0.100	0.0127	Dry	0.110	78	[14]
21	High-speed steel (18-4-1)	8, 14, 6, 6, 6, 15, $\frac{3}{64}$	SAE 4350 (363 B)	0.100	0.0255	Dry	0.110	46	[14]
22	High-speed steel (18-4-1)	8, 14, 6, 6, 6, 15, $\frac{3}{64}$	SAE 4140 (230 B)	0.050	0.0127	Dry	0.180	190	[14]
23	High-speed steel (18-4-1)	8, 14, 6, 6, 6, 15, $\frac{3}{64}$	SAE 4140 (271 B)	0.050	0.0127	Dry	0.180	159	[14]
24	High-speed steel (18-4-1)	8, 14, 6, 6, 6, 15, $\frac{3}{64}$	SAE 6140 (240 B)	0.050	0.0127	Dry	0.150	197	[14]
25	High-speed steel (18-4-1)	8, 22, 6, 6, 6, 15, $\frac{3}{64}$	Monel metal (215 B)	0.100	0.0127	Dry	0.080	170	[14]
26	High-speed steel (18-4-1)	8, 22, 6, 6, 6, 15, $\frac{3}{64}$	Monel metal (215 B)	0.050	0.0255	Dry	0.074	127	[14]
27	High-speed steel (18-4-1)	8, 22, 6, 6, 6, 15, $\frac{3}{64}$	Monel metal (215 B)	0.100	0.0127	Em	0.080	185	[14]
28	High-speed steel (18-4-1)	8, 22, 6, 6, 6, 15, $\frac{3}{64}$	Monel metal (215 B)	0.100	0.0127	SMO	0.105	189	[14]

29	High-speed steel (MZ)	0, 10, 5, 5, 5, 15, 0.30	AISI 4140		0.0625	0.01	Fluid	0.36	563	[17]
30	High-speed steel (Conrast M2)[a]	0, 10, 5, 5, 5, 15, 0.30	AISI 4140		0.0625	0.01	Fluid	0.32	501	[17]
31	High-speed steel (Conrast M7)	0, 10, 5, 5, 5, 15, 0.30	AISI 4140		0.0625	0.01	Fluid	0.45	750	[17]
32	High-speed steel (Conrast M42)	0, 10, 5, 5, 5, 15, 0.30	AISI 4140		0.0625	0.01	Fluid	0.29	451	[17]
			Steel, annealed							
33	Stellite 2400	0, 0, 6, 6, 6, 0, $\frac{3}{32}$	SAE 3240		0.187	0.031	Dry	0.190	215	[14]
34	Stellite 2400	0, 0, 6, 6, 6, 0, $\frac{3}{32}$	SAE 3240		0.125	0.031	Dry	0.190	240	[14]
35	Stellite 2400	0, 0, 6, 6, 6, 0, $\frac{3}{32}$	SAE 3240		0.062	0.031	Dry	0.190	270	[14]
36	Stellite 2400	0, 0, 6, 6, 6, 0, $\frac{3}{32}$	SAE 3240		0.031	0.031	Dry	0.190	310	[14]
37	Stell, No. 3	0, 0, 6, 6, 6, 0, $\frac{3}{32}$	Cast iron	(200 B)	0.052	0.031	Dry	0.150	205	[14]
			Steel, annealed							
38	WC (T64)	6, 12, 5, 5, 10, 45, 0	SAE 1040		0.052	0.025	Dry	0.156	800	[14]
39	WC (T64)	6, 12, 5, 5, 10, 45, 0	SAE 1060		0.125	0.025	Dry	0.167	660	[14]
40	WC (T64)	6, 12, 5, 5, 10, 45, 0	SAE 1060		0.187	0.025	Dry	0.167	615	[14]
41	WC (T64)	6, 12, 5, 5, 10, 45, 0	SAE 1060		0.250	0.025	Dry	0.167	560	[14]
42	WC (T64)	6, 12, 5, 5, 10, 45, 0	SAE 1060		0.062	0.021	Dry	0.167	880	[14]
43	WC (T64)	6, 12, 5, 5, 10, 45, 0	SAE 1060		0.062	0.042	Dry	0.164	510	[14]
44	WC (T64)	6, 12, 5, 5, 10, 45, 0	SAE 1060		0.062	0.062	Dry	0.162	400	[14]
45	WC (T64)	6, 12, 5, 5, 10, 45, 0	SAE 2340		0.062	0.025	Dry	0.162	630	[14]
46	WC (6Co)		Low-carbon steel		0.012	0.012	Dry	0.23	209	[15]
47	WC[a]		AISI 1045	(190B)		0.010	Dry	0.281	1134	[16]
48	WC	−10, −10, 10, 10, 15, 15, $\frac{1}{16}$	AISI 1045	(170B)	0.062	0.010	Dry	0.30	1670	[19]
49	WC (K20)	0, 5, 5, 5, 15, 15, 0.30	A390 Si-Al alloy (sand cast)		0.005	0.050	Dry	0.21	329	[18]
50	WC (TiC Coating)		Low-carbon steel		0.012	0.012	Dry	0.26	351	[15]
51	WC (TiC Coating)[a]		AISI 1045	(190B)		0.010	Dry	0.312	1883	[16]
52	WC (TiC + Al_2O_3 Coatings)		Low-carbon steel		0.012	0.012	Dry	0.29	384	[15]
53	WC (TiC + Al_2O_3 Coatings)[a]		AISI 1045	(190B)		0.010	Dry	0.429	3037	[16]
54	Polycrystalline diamond	−10, −10, 10, 10, 15, 15, $\frac{1}{16}$	A390 Si–Al alloy (sandcast)		0.005	0.050	Dry	0.32	1906	[18]
55	Polycrystalline diamond	−10, −10, 10, 10, 15, 15, $\frac{1}{16}$	A390 Si–Al alloy (perm. mold)		0.005	0.050	Dry	0.33	1588	[18]
56	Polycrystalline diamond	−10, −10, 10, 10, 15, 15, $\frac{1}{16}$	A390 Si–Al alloy (perm. mold)		0.005	0.050	Fluid	0.23	1211	[18]

[a] Calculated from data supplied in reference.

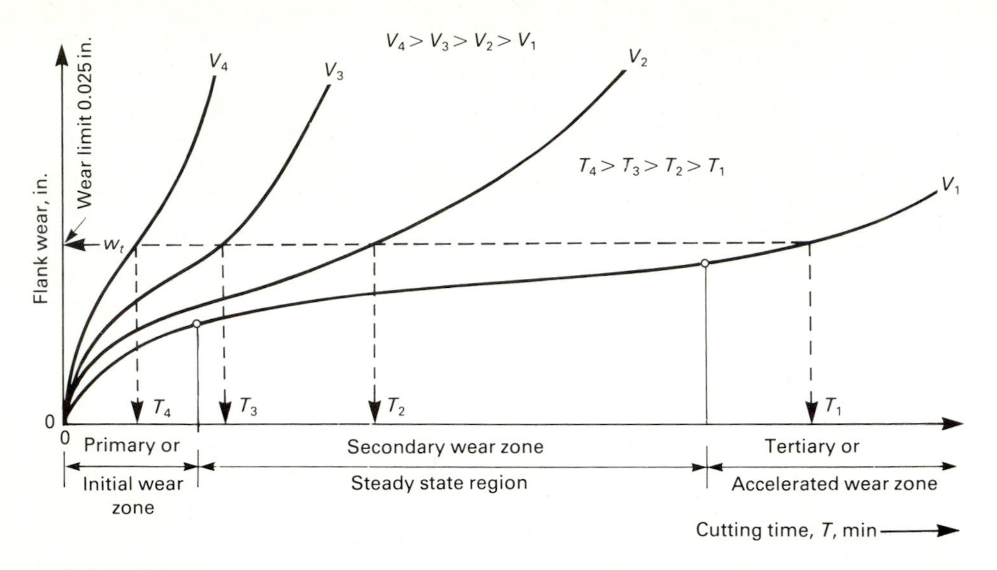

FIGURE 12.19
Wear curves for various cutting speeds.

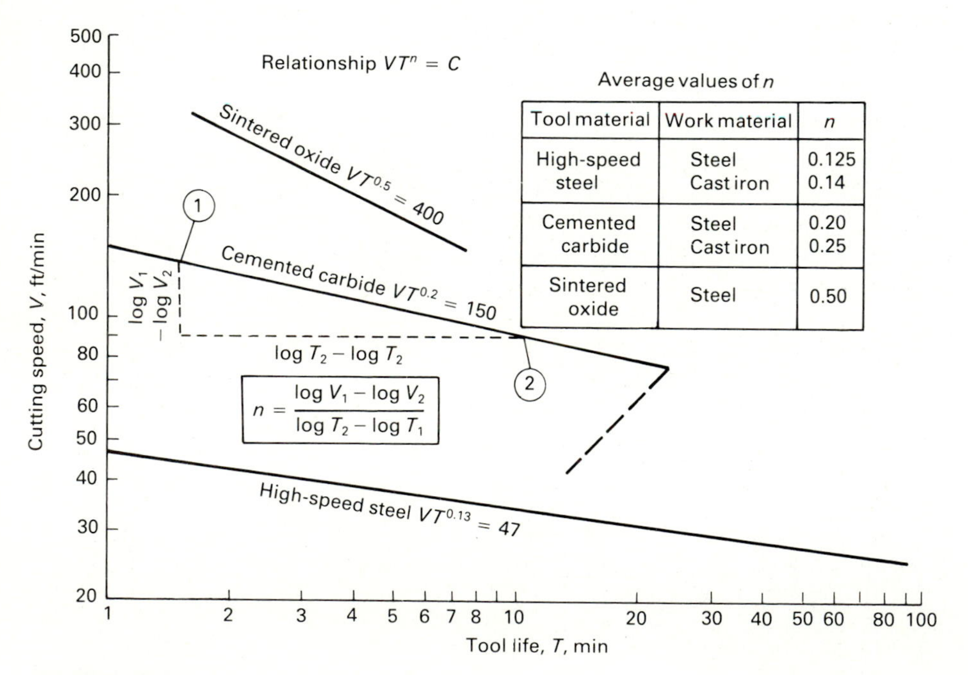

Tool material	Work material	n
High-speed steel	Steel Cast iron	0.125 0.14
Cemented carbide	Steel Cast iron	0.20 0.25
Sintered oxide	Steel	0.50

FIGURE 12.20
Log–log plot of tool life versus cutting speed.

the surface finish of the workpiece. Cutting fluids have long been used to reduce the average temperature within the cutting zone. High specific heat, ready availability, and low cost would seem to make water an ideal cutting fluid, but its low lubricity and tendency to form rust on machine-tool components and ferrous work materials preclude its use.

Several types of fluids are used to reduce the heat from friction and the severe plastic deformation that occurs in metal cutting. Solids may also be used in the form of graphite in gray iron, of manganese sulfide in free-machining grades of ferrous alloys, or of lead in copper-based alloys. Certain low-shear-strength solids may be formed on the tool face by chemical reaction with chlorinated or sulfurized cutting fluids at low cutting speeds.

There are two major types of cutting fluids: water-based and oil-based. Each may have a number of additives to provide germicidal properties, to increase the wettability of the fluids, and to combine with the tool or work material to produce low-shear-strength solids at the tool-work interface. Gaseous cutting fluids include oxygen and water vapor in the air or a foggy mist that may be directed onto the cutting tool.

Cutting fluids perform several functions:

1. Reducing the weldability between the chip and the tool surface, which results in less friction, heat, wear, and built-up edge
2. Cooling the workpiece, chip, and tool and thereby markedly reduce the average temperature in the cutting zone
3. Clearing chips from the cutting region
4. Leaving a residual film on the work surface to reduce corrosion
5. Improving the surface finish of the workpiece, which is especially important in the case of workpieces that are made on screw machines, gear teeth, or the surfaces of tapped holes.

Chemical nature of cutting fluids. There are two general types of actions of cutting fluid. First, there are gaseous boundary reactions that occur during cutting in air. The oxygen in the air reacts instantaneously with the hot nascent surface and produces a very thin boundary-layer film of oxide coating on the work surface and on that of the tool. This layer reduces the tendency for a built-up edge to occur. The second type is an extreme boundary-layer film that is provided by chemical reaction to produce a low-shear-strength solid at the asperities on the interface between the tool and workpiece. In this case, special additives to dipolar lard oils provide an easily adsorbed film of low-shear-strength sulfides or chlorides directly on the tool surface. These materials are adsorbed in the liquid or vapor phase by chemical adsorption. Such boundary-layer lubricants may be physically adsorbed at lower temperatures in multimolecular layers. They are particularly effective at low cutting speeds, i.e., 40 ft/min and below, and they are most useful in tapping and reaming operations or for automatic screw machines where surface finish is most important. When lard oils or less-expensive mineral lard oils are used, a special fungicide or antibacterial additive must also be used to prevent the cutting fluid from deteriorating in the storage tank in the machine base. Also operators must use a special skin cream to protect their arms and hands from rashes and blackheads. Even with the best of care, some people are allergic to cutting fluids with a high oil content. They must be switched to other areas.

Cooling action of cutting fluids. In high-production operations, the type of cutting fluid that uses reactive additives may be of little value, because the cutting speed exceeds the reaction time required to develop a low-shear-strength solid at the tool–work interface. In these cases, the major role of the cutting fluid is that of removal of the heat of friction and deformation. The former is about 25 percent and the latter is about 75 percent of the total heat in metal cutting. On the average, 80 percent of the heat goes to the chip and 10 percent to each of the tool and workpiece. In this case, a cutting fluid with a maximum specific heat is required. Thus, soluble oil emulsions, which have a milky appearance, are commonly used at reduction ratios (i.e., water to oil ratios) of 80:1 to 20:1. Whereas the cutting oils give improved surface finish and reductions in edge buildup and power, soluble oil emulsions give reduced average temperature at the tool--work interface and thereby permit a sufficiently high cutting speed to inhibit the formation of the built-up edge. The latter is not stable at speeds of 150 to 300 ft/min. The higher cutting speed also leads to a higher shear-plane angle and a reduced coefficient of friction, both advantageous with respect to optimum cutting conditions.

Cutting fluids for heat removal should have good thermal conductivity, high specific heat, and a high heat of vaporization. The properties of the three major types of cutting fluids are listed in Table 12.6.

From the table, it is obvious that water would be the most suitable base material, but it must be and is commercially modified to enable it to achieve its potential. Thus, special emulsifiable oils, rust inhibitors, antifoaming compounds, wetting agents, and bactericides are added to water.

TABLE 12.6.
Approximate thermal properties of the three major types of cutting fluids

Cutting fluid	Spec. wt, lb/gal	Spec. ht., Btu/(lb · °F)	Thermal conductivity, Btu/(hr · ft · °F)	Boiling point, °F
Water	8.34	0.998	0.349	212
Lard oil	7.35	0.500	0.075	~350
Air	0.010	0.240	0.015	−318

MACHINABILITY

Machinability has traditionally been defined as the relative ease of cutting a particular material. This definition is admittedly somewhat nebulous and can, as we shall see, be fulfilled by many criteria. The machinability of a material may be defined by (but is not limited to) any combination of the following criteria:

1. Tool life
2. Cutting forces and power
3. Surface finish
4. Chip type and shape
5. Limiting rate of metal removal

Increased cutting velocities for the same tool life indicate superior machinability. One method to quantify this is to assign a machinability rating that is calculated in comparison to a "standard" material, frequently AISI B1112 steel, as follows:

$$\text{Machinability rating} = \frac{V_{\text{material}}\big|_{\text{at } T=60 \text{ min}}}{V_{\text{B1112}}\big|_{\text{at } T=60 \text{ min}}} \times 100$$

Table 12.7 contains machinability ratings for common engineering materials. Care must, however, be taken in employing machinability ratings, since different mechanisms may cause the tool wear.

Lower cutting forces and power consumption also indicate more highly machinable materials. The specific (or unit) horsepower is a good measure to be used for this purpose. Table 12.8 gives typical values of specific horsepower and shows wide variation between materials.

The surface finish imparted to some materials is inherently much worse than that given to others. This may be due to microstructure or prior strain-hardening. For example, a commercially pure aluminum has poor machinability, since it can seize (adhere) to the rake face of the tool thereby causing larger shear angles and higher strains. The same material with some degree of prior strain-hardening will adhere less to the tool, reduce the shear strain, and thereby improve the surface finish.

Chip disposal is a frequent problem for machine shops and automated machine tools. Long, stringy chips that clog the machine tool and cause a potential safety problem are indicative of a far less machinable material than one whose chips fracture.

Thus, qualitative factors and quantitative measures frequently dictate machinability.

TABLE 12.7

Machinability ratings of AISI steels and various ferrous and nonferrous metals to the nearest 5 percent, based on 100 percent rating for Bessemer Screwstock AISI B112, cold-drawn when turned with a suitable cutting fluid at 180 ft/min under normal Screw-machine conditions

AISI no.	Machinability rating, percent	Brinell hardness
C1008	50	126–163
C1010	50	131–170
C1015	50	131–170
C1016	70	137–174
C1020	65	137–174
C1022	70	159–192
C1030	65	170–212
C1035	65	174–217
C1040	60	179–229
C1045	60	179–229
C1050	50	179–229
C1070	45	183–241
C1109	85	137–166
B1111	95	179–229
B1112	100	179–229
B113	135	179–229
C1115	85	143–179
C1117	85	143–179
C1118	80	143–179
C1120	80	143–179
C1132	75	187–229
C1137	70	187–229
C1141	65	183–241
1320	50	170–229
1330*	50	179–235
1335*	50	187–241
1340*	45	187–241
2317	55	174–217
2330*	50	179–229
2340*	45	187–241
2515*	30	179–229
3120	60	163–207
3130*	55	179–217
3140*	55	187–229
3145*	50	187–235
E3310*	40	170–229
4023	70	156–207
4027*	70	166–212
4032*	65	170–229
4037*	65	179–229
4042*	60	183–235
4047*	55	183–235
4130*	65	187–229
4137*	60	187–229
4145*	55	187–229
4150*	50	187–235
4320	55	179–228
4340	45	187–241
6120	50	179–217
6145*	50	179–235
6152†	45	183–241
NE8620 or 8720	60	170–217
NE8630* or 8730*	65	179–229
NE8640* or 8740*	60	179–229
NE8645* or 8745*	55	183–235
NE8650* or 8750*	50	183–241
9260*	45	187–225
E9315	40	179–229
NE9440*	60	179–229

Type	Machinability rating, percent	Brinell hardness
Aluminum 2-S	300–1,500	
Aluminum 11-S	500–2,000	
Aluminum 17-S	300–1,500	
Brass—leaded	300–600	
Brass—red	200	
Brass—yellow	200	
Bronze, lead bearing	200–500	
Bronze, manganese	150	
Copper—cast	70	
Copper—rolled	60	
Dowmetal	500–2,000	
Everdur	120	
Gun metal	60	
Inconel	45	
Iron—cast (hard)	50	220–240
Iron—cast (medium)	65	193–220
Iron—cast (soft)	80	160–193
Iron—ingot	50	101–131
Iron—malleable (standard)	120	110–145
Iron—malleable (pearlitic)	90	180–200
Iron—malleable (pearlitic)	80	200–240
Iron—stainless (12% Cr.F.M.)	70	163–207
Iron—wrought	50	101–131
Magnesium alloys	500–2,000	
Monel metal—cast	35	
Monel Metal—"K"	50	
Monel metal—rolled	45	
Nickel	20	
Ni-Resist*	30	
Steel—cast (0.35 carbon)	70	170–212
Steel—high-speed*	30	

TABLE 12.7 (*continued*)

AISI no.	Machinability rating, percent	Brinell hardness	AISI no.	Machinability rating percent	Brinell hardness
4615	65	174–217	Steel—manganese (oil hardening)†	30	
4640*	55	187–235	Steel—stainless (18–8 austenitic)*	25	150–160
4815	50	187–229	Steel—stainless (18–8 F.M.)	45	179–212
5120	65	170–212	Steel—tool (high-carbon, high-chromium)†	25	
5140*	60	174–229	Steel—tool (low-tungsten, chromium and carbon)†	30	200–218
5150*	55	179–235	Zinc	200	
E52101†	30	183–229			

* † | * = "annealed" and † = "spheroidized anneal." These terms refer specifically to the commercial practice in steel mills, before cold drawing or cold rolling, in the production of the steels specifically mentioned.

Source: E. M. Slaughter and O. W. Boston, January 27, 1947.

MACHINE TOOLS

Block Diagrams of the Basic Machine Tools

The major features of the four basic machine tools can most easily be described by block diagrams. In this case the principal components of a machine tool are represented by blocks and the relative motion of each component is represented by appropriate arrows.

The four basic chip-producing machine tools are: lathes, planers, drilling machines, and milling machines. Grinding machines are specifically excluded from this list because the grinding wheel merely replaces the cutting tool used in the other types. The basic feed motions are the same. For instance, lathes and cylindrical grinders have identical feed motions, but the necessity of providing for a large rotating wheel has modified the machine design.

Engine lathe. The engine lathe produces surfaces of revolution by a combination of a single-point tool moving parallel to the axis of work rotation, and a revolving workpiece (Fig. 12.21). In an engine lathe the workpiece is (1) clampled in a chuck that is mounted on the spindle or is (2) mounted between centers in the spindle and the tailstock and then driven by a dog that connects the workpiece to the work driver. The headstock contains the drive gears for the spindle speeds, and through suitable gearing and the feed rod, drives the carriage and cross-slide assembly. The carriage provides motion parallel to the axis of rotation, and the cross slide provides motion normal to it. The leadscrew is used only for chasing threads and thus is not required on a modern production lathe because thread milling or rolling would be used for threads made in quantity.

The tailstock is adjustable longitudinally along the lathe bed. If the lathe must accommodate long pieces, the bed must be longer, and for slender pieces

TABLE 12.8
Typical values of specific horsepower

Material	Feed, in./rev	Unit power, hp · min/in.^{3}	Specific energy, in · lb/in.^{3}
Steel (120 B[a])	0.001	1.12	443,000
Steel (120 B)	0.003	0.86	347,000
Steel (120 B)	0.005	0.76	301,000
Steel (120 B)	0.010	0.64	254,000
Steel (120 B)	0.020	0.54	214,000
Steel (160 B)	0.001	1.25	495,000
Steel (160 B)	0.020	0.59	234,000
Steel (200 B)	0.001	1.50	594,000
Steel (200 B)	0.020	0.73	290,000
Steel (300 B)	0.001	1.87	740,000
Steel (300 B)	0.020	0.92	364,000
SAE 302	0.003–0.011	0.72	285,000
SAE 350	0.006–0.009	1.20	475,000
SAE 410	0.003–0.013	0.75	297,000
Gray CI (130 B)	0.006–0.012	0.29–0.35	127,000
Meehanite	0.006–0.012	0.55–0.76	262,000
K-Monel	0.004–0.010	0.80	317,000
Inconel 700	0.003–0.007	1.40	554,000
High-temperature alloy A286	0.006–0.011	1.20	475,000
High-temperature alloy S816	0.004–0.009	1.25	495,000
Titanium A55	0.010	0.65–0.76	281,000
Titanium C130	0.010	0.81–0.93	345,000
Aluminum 2014-T6	0.003–0.008	0.24	95,100
Aluminum 2017-T4	0.003–0.008	0.21	83,200
Aluminum 3003-O	0.003–0.008	0.16	63,400
Aluminum 108 (55 B)	0.003–0.009	0.15	59,400
Muntz metal	0.007–0.012	0.55	218,000
Phosphor bronze	0.002–0.006	0.33	131,000
Cartridge brass	0.003–0.009	0.48	190,000

[a] Brinell hardness number.

Source: Courtesy of Monarch Machine Tool Co.

special supports (called back or steady rests) must be provided to prevent workpiece deflection away from the tool during the machining process.

The tool is mounted in the toolpost and must be located on the vertical centerline of the workpiece. The proper height is provided by packing strips and shims in a production lathe. The heavy-duty toolpost is provided with two

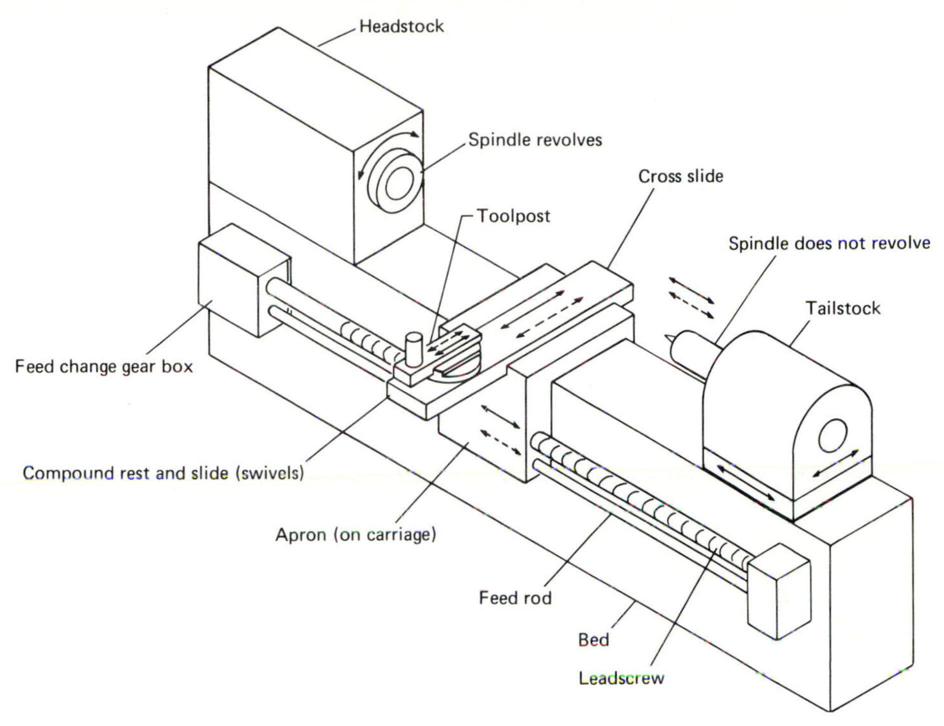

FIGURE 12.21
Block diagram of an engine lathe. (*Redrawn from [12].*)

clamping bolts to support the shank of the tool firmly during the machining operation.

Planing machines. The planing operation generates a plane surface with a single-point tool by a combination of a reciprocating motion along one axis and a feed motion normal to that axis. Slots and limited vertical surfaces can also be produced. Planers handle large, heavy workpieces and shapers handle smaller parts. In the planer, the workpiece reciprocates and the feed increment is provided by moving the tool at each reciprocation (Fig. 12.22). In shaping, the tool is mounted on a reciprocating ram and the knee upon which the workpiece is fastened is fed at each stroke of the ram. To reduce the lost time on the return stroke both planers and shapers are provided with a quick-return mechanism.

The planer drive is furnished by a large 50- to 75-hp ac-dc motor generator set. The large amount of dc power is needed to provide for the huge acceleration and deceleration braking needed because of the inertia of the heavy table and its associated work load. The distinguishing features of a planer are its large size and single-point cutting tool, and the fact that the table reciprocates. The feed is provided by motion of the tool slide along the

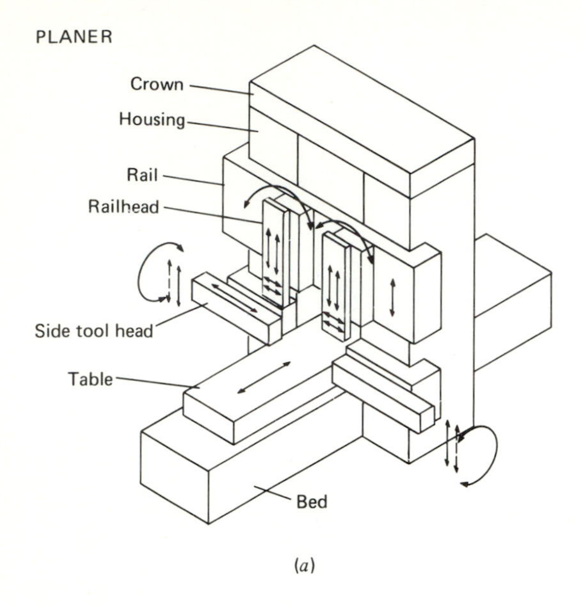

(a)

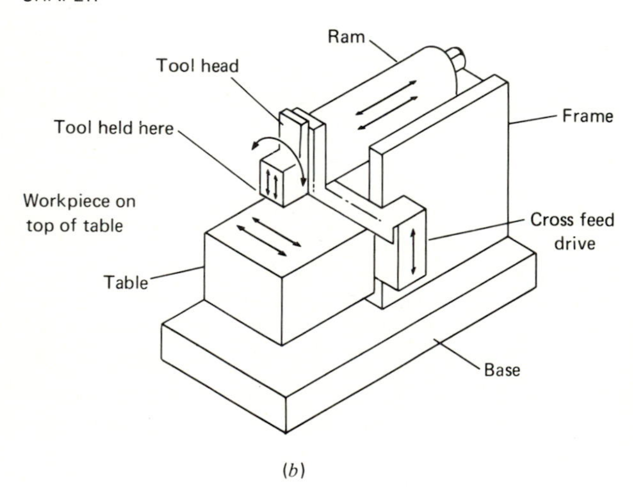

(b)

FIGURE 12.22
Block diagrams for two types of planing machines: (*a*) planer for large work; (*b*) shaper for light work. (*Redrawn from* [12].)

crossrail at each cycle of the table. Two heavy housings are located at the center of the long bed. They support the crossrail that in turn carries one or more tool heads complete with tool slides and clapper boxes for the cutting-tool holders. The clapper box permits the tool to be raised sufficiently to clear the work on the return cycle of the planer table.

Drilling machine. A simple drilling machine consists of a vertical column, on which is located the table for supporting the work, and at a higher level, the

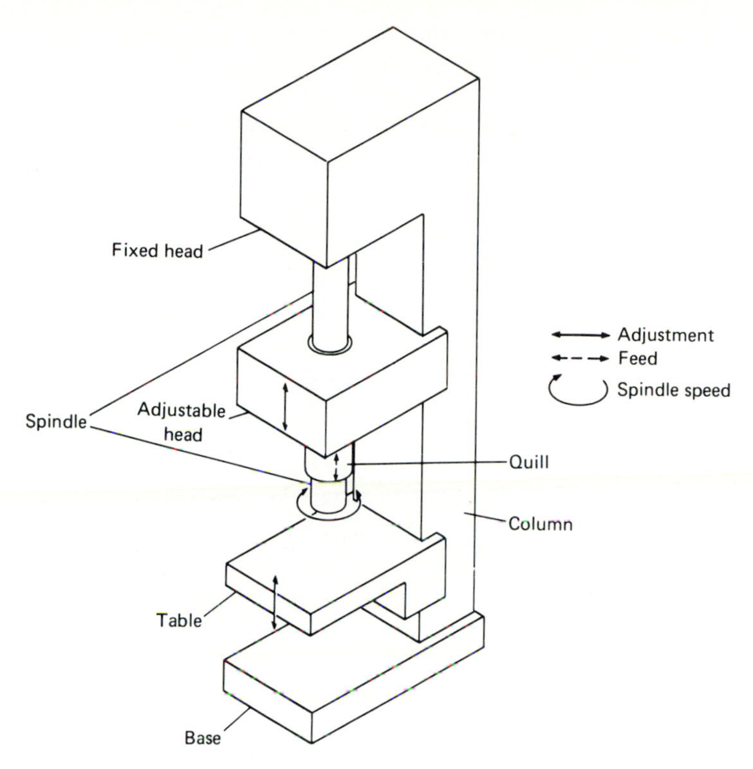

FIGURE 12.23
Block diagram of a typical drilling machine. (*Redrawn from [12].*)

variable-speed drill head (Fig. 12.23). Small, straight-shank drills are mounted in a small three-jaw chuck. Larger drills have tapered shanks that can fit directly in the spindle. Smaller drills frequently use hand feed only; larger-production machines are equipped with a power-feed mechanism. More versatile heavy-duty drilling machines, called radial drills, have a horizontal arm that can be adjusted vertically on a cylindrical column for supporting the drill head. In that case, the arm can be swung radially and the drill head can move along the arm to bring the drill to large cumbersome workpieces.

Milling machine. A milling machine uses a multitoothed rotary cutting tool to remove metal chips from a workpiece. The original cutters used by Eli Whitney were similar to overgrown dentist's burrs. Since that time, milling cutters have been developed to produce a multitude of contours in a finished part. There are two broad classifications of milling operations: (1) plain milling and (2) end or face milling. The basic plain-milling cutter produces a flat surface through the use of cutting teeth on its periphery that are parallel to the axis of rotation.

Plain and form milling are carried out primarily on horizontal milling machines (Fig. 12.24*a*), whereas face milling and end milling operations are carried out on vertical-spindle machines (Fig. 12.24*b*).

An end-milling cutter has its cutting teeth located at the end as well as on its periphery and rotates around an axis that is normal to the surface being cut.

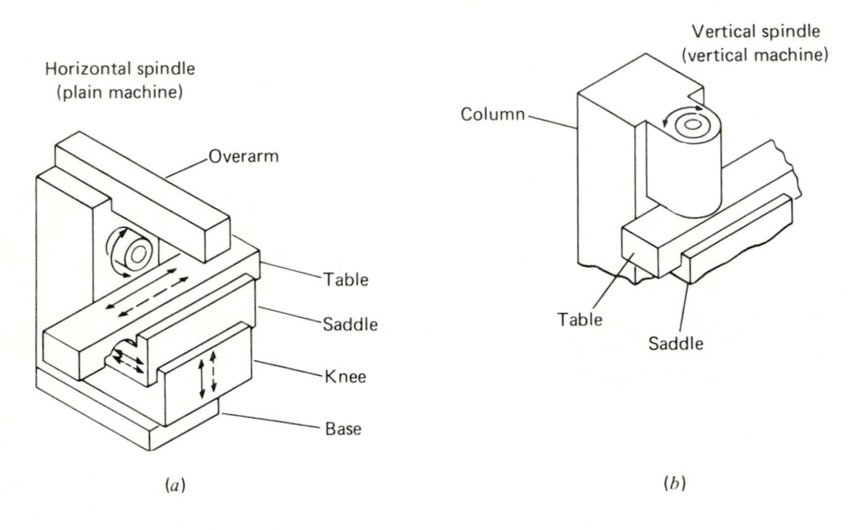

MILLING MACHINE—BED TYPE

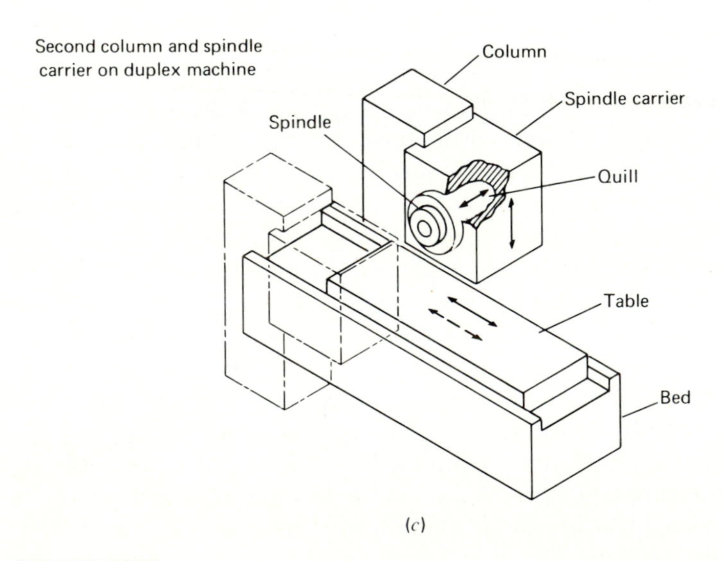

(*c*)

FIGURE 12.24
Block diagrams of typical milling machines: (*a*) horizontal milling machine; (*b*) vertical milling machine; (*c*) bed-type milling machine. (*Redrawn from [12].*)

It also produces a flat surface. A face-milling cutter is large in diameter (>6 in.) and produces flat surfaces such as the top of a six-cylinder block for an internal combustion engine. In face and end milling, the cutting occurs primarily by the peripheral section of the teeth not the end of the cutter itself.

Light- and medium-duty machines are of the column and knee type. That is, the workpiece can be located in space within 0.001 in. in any of the three coordinate directions that lie within the travel range of the machine. This flexibility of location and feed, combined with the diversity of milling-cutter geometry, make the milling machine a tremendously versatile machine tool for toolroom and production use. In the latter case, several operations are frequently performed at once and the versatility of the machine is increased by the use of proper holding fixtures.

High-production milling operations are carried out on bed-type milling machines (Fig. 12.24*c*) which have less versatility but more rigidity. In such cases, machine utilization is increased by placing a work fixture at either end so that one fixture is loaded while the work in the other is being machined. In horizontal milling machines of both types, greater support is provided for the cutter arbor by adding overarms or a ram to provide an outer bearing for the cutter-arbor support. This feature is mandatory to keep the cutter running true and to provide support for heavy loads.

Surface Finish of Machined Parts

Machine processes can be divided into two groups: *primary* machining methods, which use single- or multiple-edged tools and remove large amounts of metal in heavy roughing cuts; and (2) *secondary* methods, which usually follow primary machining and impart greater dimensional accuracy, improved surface qualities, and leave less residual stress in the finished surface. Secondary machining processes are especially useful for removing dimensional inaccuracy caused by heat-treatment procedures (Fig. 12.25), but they add greatly to the total cost of the manufactured item, and should be used only when absolutely necessary to meet engineering requirements. The allowable surface roughness depends upon many factors, such as the functions and size of the piece, the fit and dimensional accuracy required, the load-carrying requirement, the uniformity of motion, or the wear characteristics of sliding or rolling surfaces.

Surface roughness can be evaluated in three ways. First, a profilometer can be used; this utilizes a diamond tracer similar to a phonographic pickup. The tracer reciprocates over the surface at a known speed and generates a current that is proportional to the magnitude of the surface irregularities. Second, evaluation can be done by tactual and visual standards, which are normally arranged with average roughness values of 4, 8, 16, 32, 63, 125, 250, and 500 μin. When a fingernail is moved slowly across the surface of the standard, it can detect a minimum roughness of about 4 μin. Finally, evaluation can be made by direct microscopic examination of the surface, or by

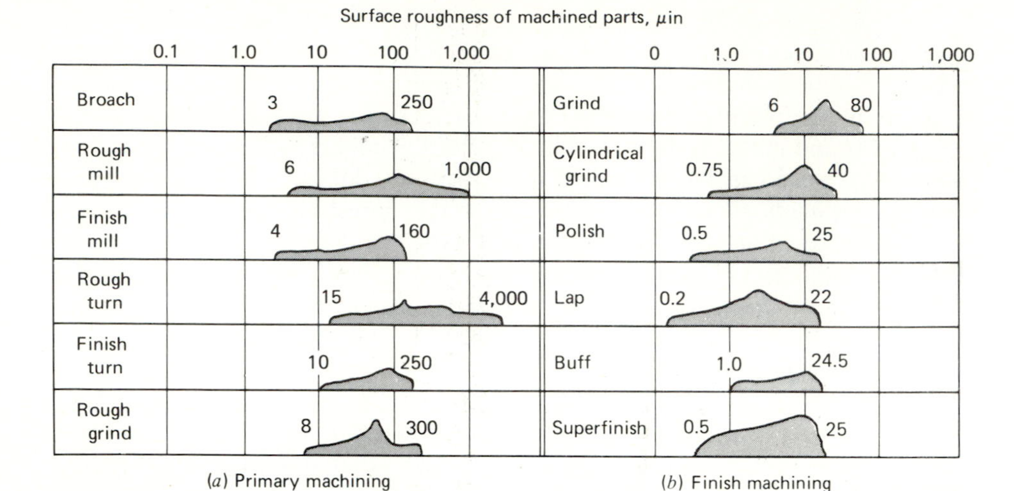

FIGURE 12.25
Machined surface finishes. Height of curve is proportional to occurrence. (*Source. Physicists Research Co., Ann Arbor, Michigan.*)

studying a cellulose (or other plastic) replica produced by pressing the film against a metal surface coated with a thin film of acetone or other solvent. Greater magnification of the surface profile can be obtained by cutting through the surface on a shallow taper (18°), polishing the resultant wedge, and observing the surface profile under magnification.

Formerly, the roughness of a surface was reported as a root-mean-square value; however, this statistical value yields an answer that is consistently approximately 11 percent greater than the arithmetic average based on a true sinusoidal model. The difference is so small that it is negligible for most purposes. Some countries use the maximum peak-to-valley roughness, which gives values considerably greater than the average roughness used in the United States.

In most cases the character of a machined surface depends upon the process used to produce it. For example, there are several sources of roughness when machining with a single-point tool: (1) feed marks left by the cutting tool; (2) built-up edge fragments embedded in the surface during the process of chip formation; (3) chatter marks from vibration of the tool, workpiece, or machine tool itself. When a surface is turned at high speed without chatter, the primary surface roughness lies in an axial direction and may be computed quite accurately from the feed and the tool geometry.[1] The

[1] According to the American National Standards Institute, the average roughness for turned surface is approximately equal to the feed divided by 60.

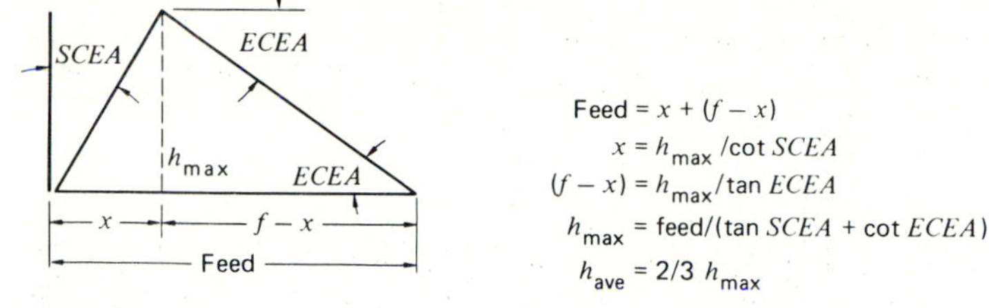

FIGURE 12.26
Analysis of surface profile produced by a pointed tool.

built-up edge causes cyclic gouging and smoothing of the surface as the built-up edge sloughs off, whereas chatter marks are cyclic with the revolving of the workpiece. Both of these latter surface marks are superimposed on the feed marks. An analytical investigation of the characteristics of lathe feed marks as a function of tool geometry and feed is presented in Figs. 12.26 and 12.27.

The surface roughness caused by a built-up edge is primarily affected by cutting speed. As the latter increases, up to a point, the surface finish improves. The critical speed at which the built-up edge becomes insignificant varies from 250 to 500 ft/min, depending on the cutting conditions, tool–workpiece combination, tool geometry, cutting fluid, etc.

For example, from the partial trace of a particular surface (Fig. 12.28) and the tabular data (Table 12.9) from nine particular deviations from the

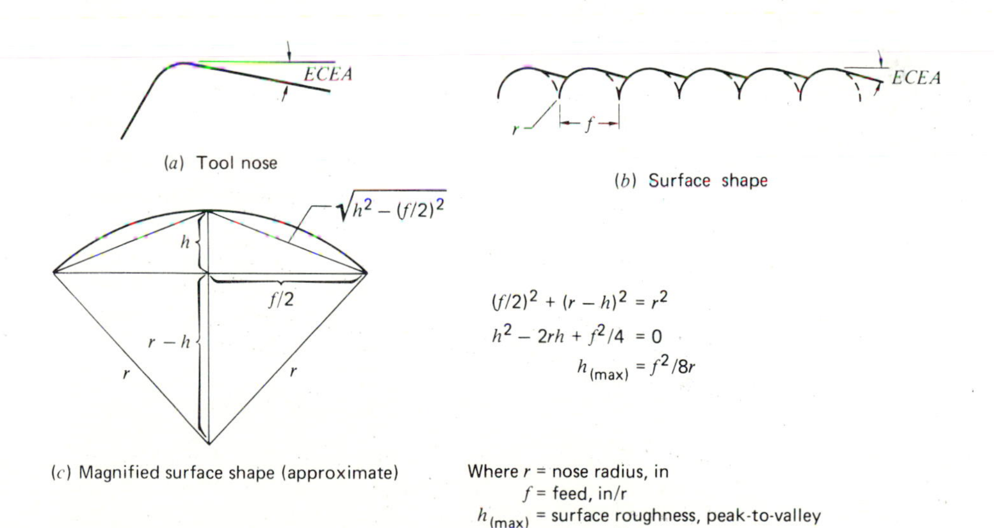

FIGURE 12.27
Analysis of a surface profile produced by nose radius.

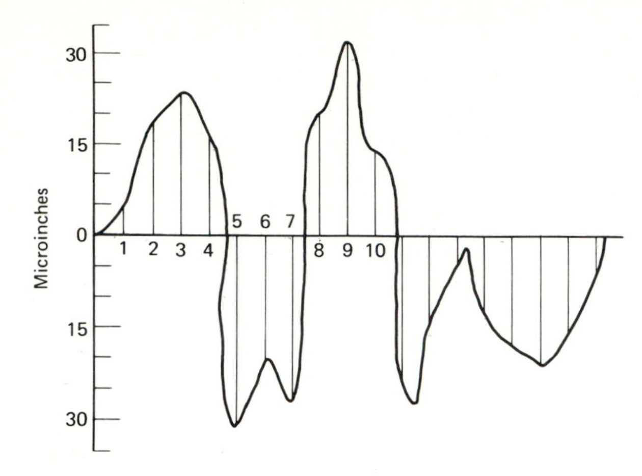

FIGURE 12.28
Partial trace of a surface.

mean value, let us compute the arithmetic and root-mean-square values of surface roughness and compare the answers. The arithmetic average value is

$$\mathrm{AA} = \frac{1}{L}\int_0^L |y|\,dx = \frac{\sum_{i=1}^{N} |y_i|}{N}$$

and the root-mean-square value is

$$\mathrm{rms} = \sqrt{\frac{1}{L}\int_0^L y^2\,dx} = \sqrt{\frac{\sum_{i=1}^{N} y_i^2}{N}}$$

TABLE 12.9
Peak and valley values of the surface trace taken at discrete points (See Fig. 13.28)

Point	Values of y at Δx intervals, μin.	Point	Values of y at Δx intervals μin.
1	4	10	13
2	19	11	23
3	23	12	15
4	16	13	6
5	31	14	12
6	20	15	18
7	27	16	21
8	20	17	17
9	31	18	9

where L is the cutoff length and N is the number of equally spaced discrete points in length L. The rms value is the same as the standard deviation as defined in statistics.

Production Turning Operations

The era of modern machine tools is relatively recent, dating back to the Industrial Revolution in England. The first screw-cutting lathe was invented by Maudsley in 1798. The modern engine lathe takes its name from its gear-driven headstock that typically provides 16 spindle speeds. Although an engine lathe can bring one to four tools mounted in its one toolpost to bear on a workpiece, it can provide only a single feed motion at one time. A typical turned part frequently requires several operations, each of which needs a different tool and feed motion. The replacement and resetting of the tooling and handling of the work in and out of the machine adds a significant cost to the finished part. Moreover, the cost per piece may be three to six times the cost of using a more specialized machine tool. When a number of identical parts are needed, positioning of the tools for each operation not only involves time but may also result in workpieces that will not pass inspection. Thus, an engine lathe is suitable for low production only.

From the simple lathe, more complex machines have evolved for producing duplicate parts efficiently and in quantity. A basic feature of such machines is their ability to apply several tools to the work simultaneously, using different motions. In addition, once the tools are set, they can be used repeatedly in the same sequence without being reset for each operation. Semiautomatic machines like turret lathes (Fig. 12.29) are fitted with two indexing tool holders called *turrets*. A square turret is mounted on a cross slide, and a hexagonal turret replaces the tailstock. They require constant attention by a machine operator, but a high degree of skill is only needed when the machine is being set up for the first piece. Such features as multitool machining, overlapping or simultaneous operations, turning lengths to positive stops, rapid indexing of the square and hexagonal turrets at the end of each cutting operation, and automatic stock feeding and clamping result in increased production. The use of a turret lathe enables each cut to be consistently within normal manufacturing tolerances of +0.002 to −0.003 in. over a production run of several hundred pieces.

The turret lathe can bring at least 11 tools into play in any one sequence of turning operations: four from the square turret, six from the hexagonal turret, and one from the rear toolpost (Fig. 12.30). Special tooling such as a combined stop and center drill or a drill and an external turning tool can provide even more production possibilities through the mounting of two or more cutting tools at a single position.

The ram-type turret lathe (Fig. 12.29, left) is used for light-duty operations, whereas the saddle type (Fig. 12.29, right) with an overhead pilot bar is used for pieces over 2 in. in diameter and for heavy chucking work. The

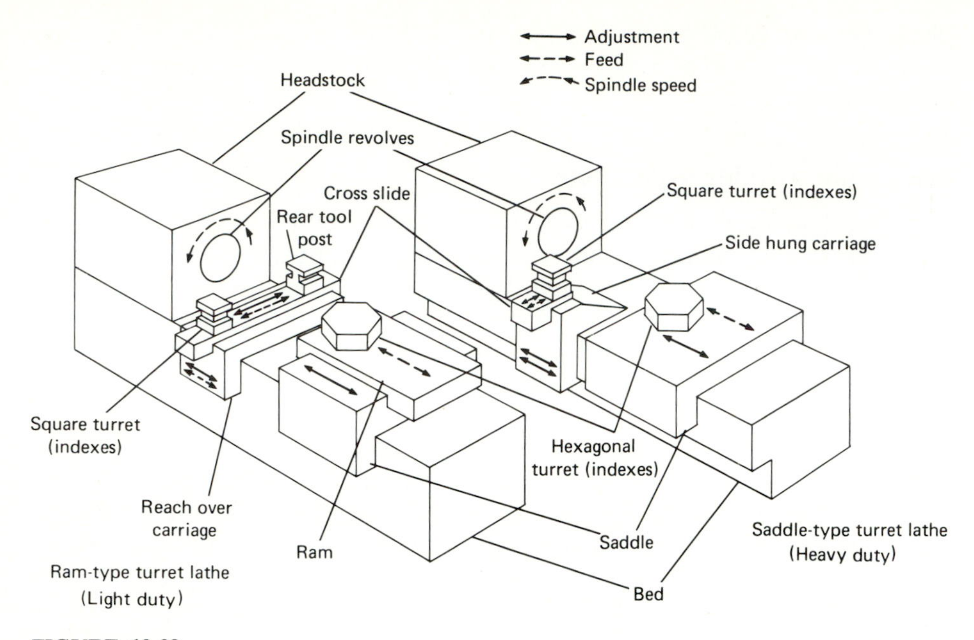

FIGURE 12.29
Block diagrams for turret lathes: ram type for smaller work; saddle type for heavy work. (*Redrawn from [12].*)

saddle type has power assists so that although it is slower, it requires no more manual effort in operation. In fact, this type of turret lathe was one of the first to be numerically controlled. In this case, the operator was replaced by a punched tape programmed with a sequence of instructions that performed all the functions the operator normally carried out.

The automatic screw machine can make small parts from bar stock on a completely automatic basis. In one type of single-spindle automatic screw machine, two cams control the cross slides and a third cam controls the motion of the six-station turret that revolves on a horizontal axis (Fig. 12.31). The three cams must be designed and machined for each new part. However, the tooling is so standardized and the procedure for designing and cutting the cams is so well developed that in less than a day cams can be designed and cut for a standard job. Details of how the cams operate are given in Fig. 12.32.

Multiple-Spindle Automatic Screw Machines

The highest production rates of all automatic screw machines can be achieved by four-, five-, six-, and eight-spindle machines. Bar-type machines are rated by the largest diameter stock that can be passed through the spindles. Chucking machines are rated by the diameter of the part that can clear the tool slides.

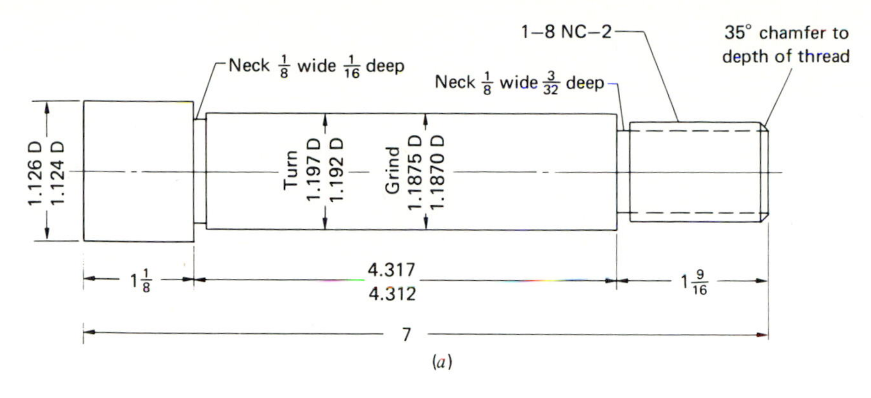

(*a*)

Operation	Hexagon turret	Square turret	Rear toolpost
I	Feed to stock stop, (center withdrawn)		
II	Turn diam (4)		
III	Turn diam (2)	Turn diam (6)	
IV	Face (1)		
V	Center drill (1)		
Double index			
VI	Support on center	Neck diam (3) and (5)	
VII	Thread diam (2)		
VIII	....	....	Cutoff and chamfer (7)

(*b*)

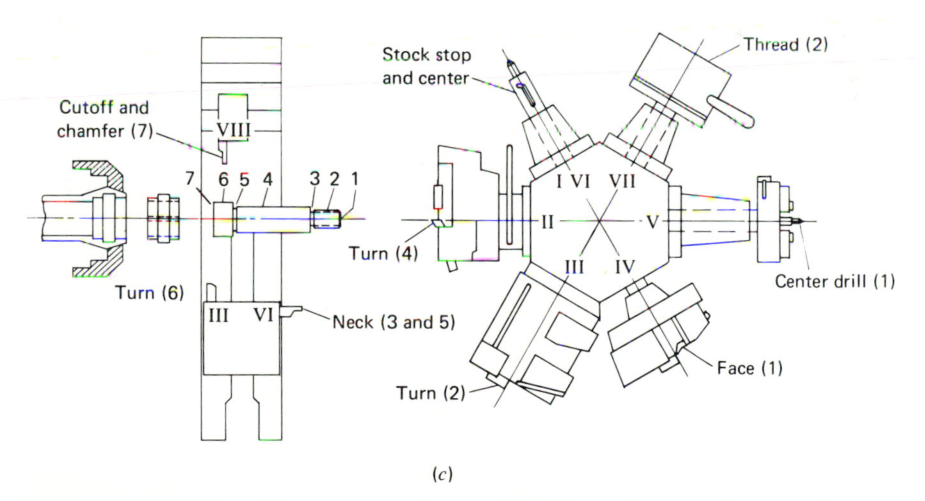

(*c*)

FIGURE 12.30
Turret-lathe production: (*a*) typical piece; (*b*) operation sequence; (*c*) plan view of the required tooling for the square and hexagonal turrets.

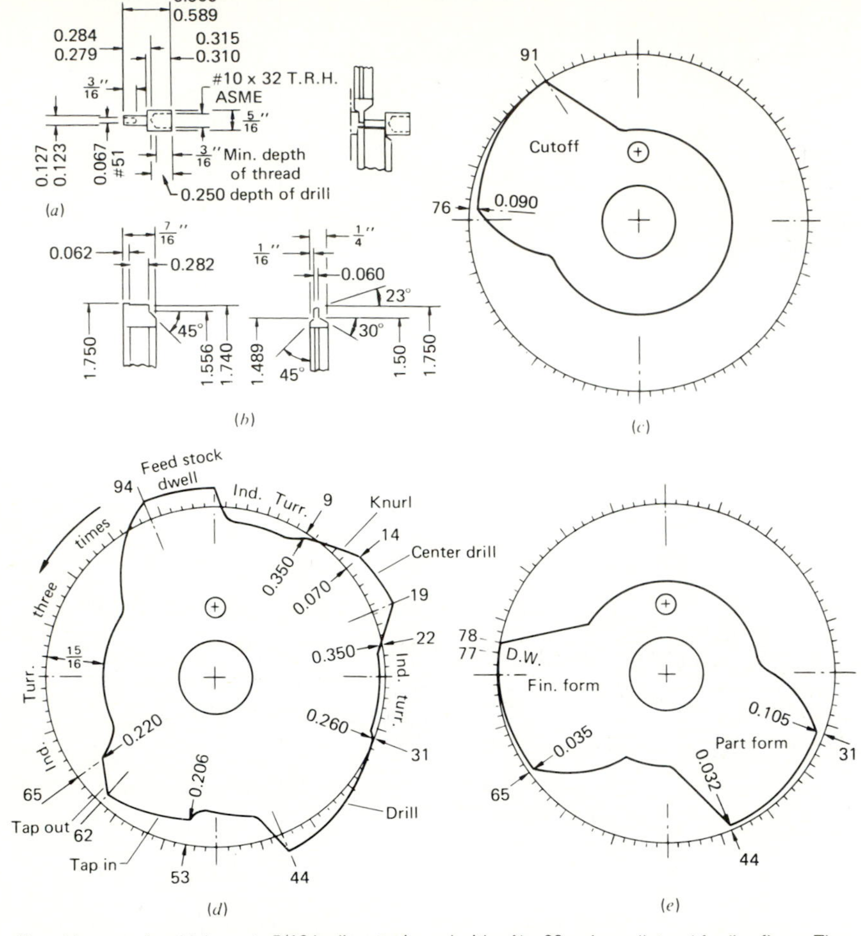

Round brass stock cold-drawn to 5/16-in diameter is used with a No. 00 spring collet and feeding finger. The spindle speed is 1,500 rpm forward and 5,000 rpm reverse. The spindle runs reverse for the carbon-tool-steel cross-slide tools. The driving shaft rotates at 240 rpm with change gears having teeth as follows; on the driving shaft, 40; first on the stud, 60; second on the stud, 52; and on the worm, 30. Production time per piece is 3.25 s, equivalent to 1,108 pieces per hour gross.

Spindle revolutions	*Order of operations and tools*	*Feed per revolution, in*
24	Index turret	0.020
37	Knurl and center drill L.H. 0.250 in diameter	0.040 and 0.005
24	Index turret	
35	Part form with front slide	0.0019
24	Index turret and reverse spindle	
8	Tap in, 10–32 thread	
	Reverse spindle	
8	Tap out, then index turret three times	
35	Finish-form with front slide	0.001
35	Cutoff with back slide	0.0025
24	Feed stock to stop	

(*f*)

FIGURE 12.31
Production tooling for a knurled brass insert made on an OOG Brown and Sharp automatic screw machine. (*a*) Workpiece and position of cross-slide tools; (*b*) high-speed steel cross-slide tools; (*c*) cam for rear-slide cutoff tool; (*d*) cam for turret slide; (*e*) cam for front slide; (*f*) order of operations.

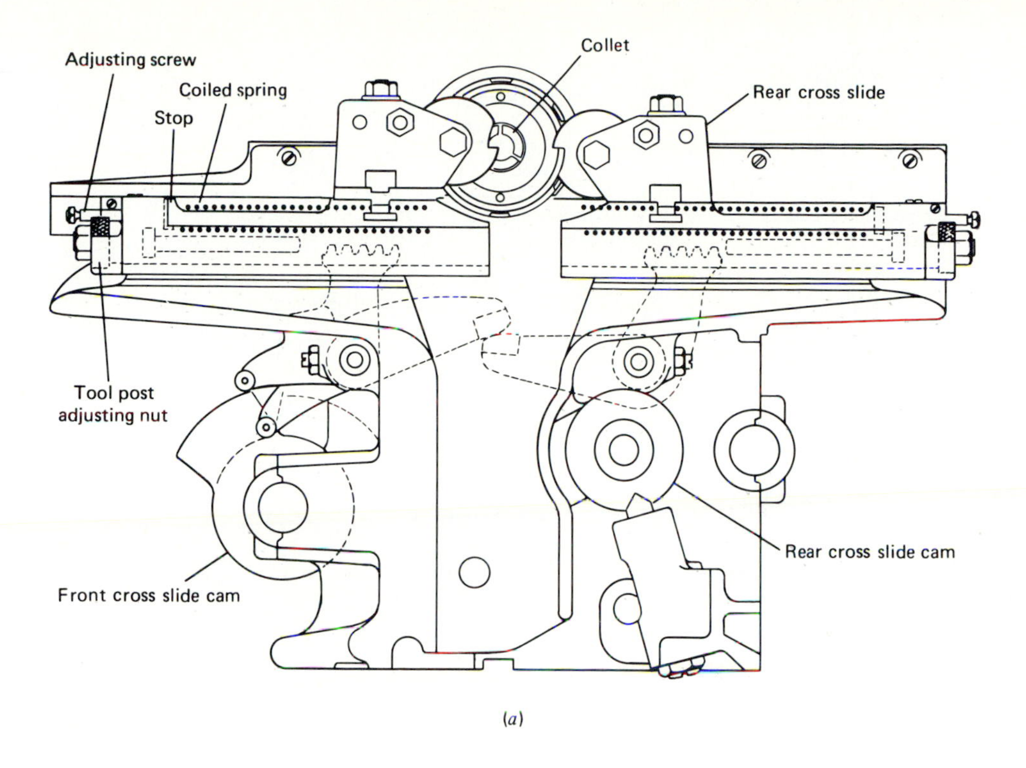

(a)

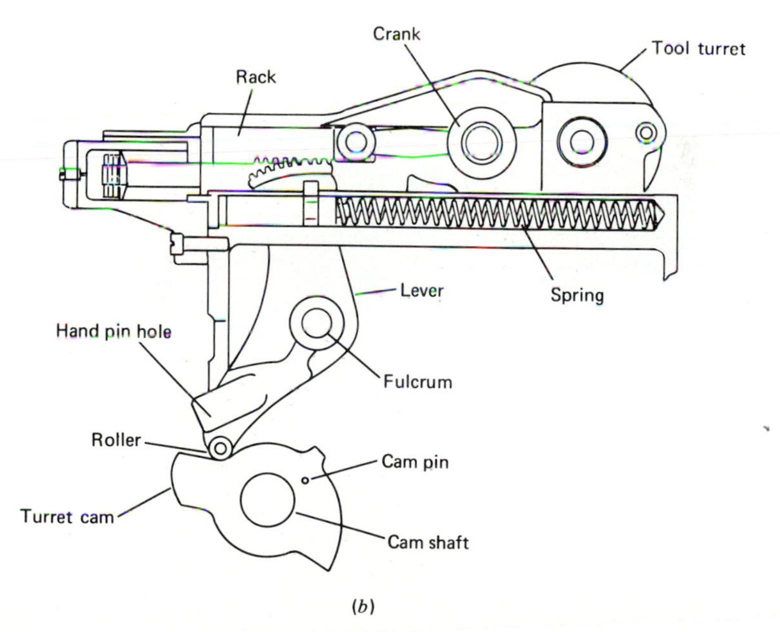

(b)

FIGURE 12.32
Details of cam operation: (*a*) section through cross slides; (*b*) section through turret slide.

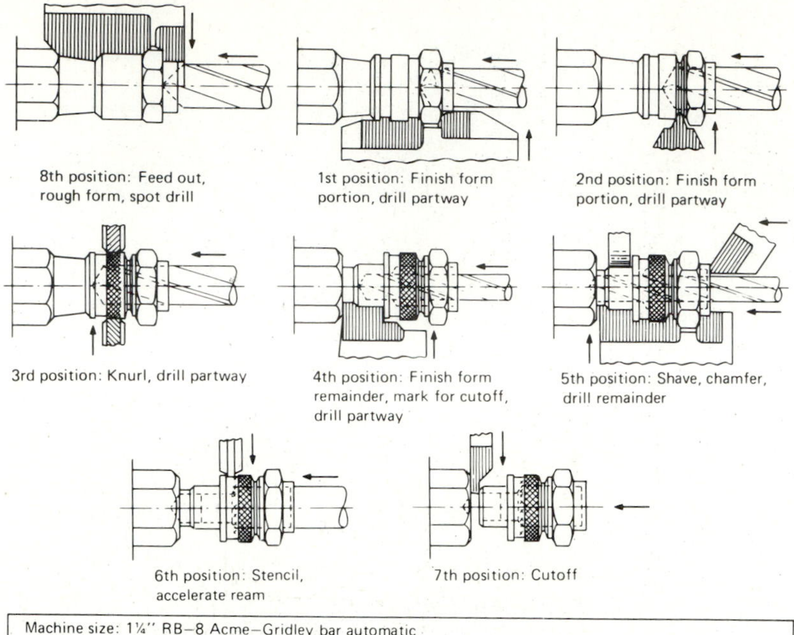

Machine size: 1¼″ RB–8 Acme–Gridley bar automatic
Name of piece: Sparkplug shell; = machine time: min, 4.5 s
Material steel: Open-hearth grade A leaded, (SFM 416); gross production: 800 pieces per hour
Overall dimensions: 13/16 hex x 1¼″ long; spindle speed; 1692 rpm; toolside cam: 11/32 @ 0.0042

(*a*)

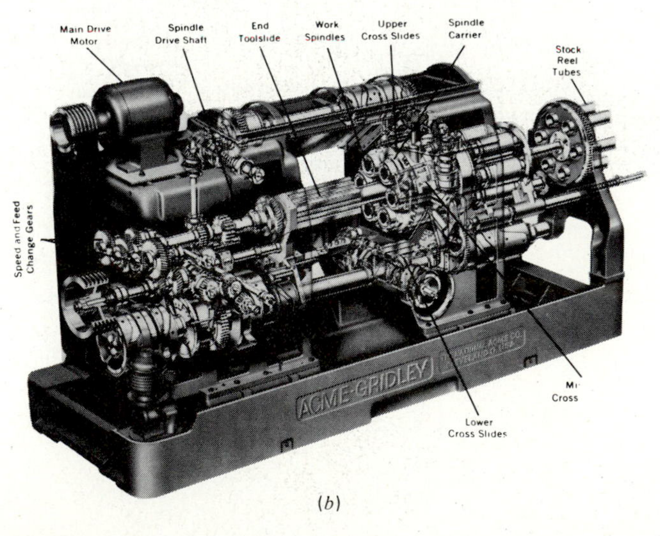

(*b*)

FIGURE 12.33
Multiple-spindle automatic-screw-machine operations for making spark-plug shells at the rate of 800 per hour. (*a*) Details of the operations; (*b*) typical eight-spindle automatic screw machine. (*Courtesy of Acme Gridley, Cleveland, Ohio.*)

End-working tools are mounted on the end-tool slide, which corresponds to the hexagonal turret of a turret lathe. The end tools do not index but slide forward and back on the bedways. The cross slides for forming and the cutoff tools are mounted next to the spindle stations and move radially toward and away from the work.

In operation, the spindle carrier indexes after each operation one station at a time. At each rotation of the spindle carrier, one piece is completed; that is all operations are completed simultaneously and the cycle time for one part is the time for the longest operation plus an indexing time of 1 to 2 seconds. A typical part is the spark-plug shell, which is made at a gross production rate of 800 pieces per hour (Fig. 12.33*a*). Note that the through holes in the body are being drilled in seven of the eight stations. The machine on which the spark-plug bodies are made is shown in Fig. 12.33*b*. Even at the production of 800 parts per hour some 90 machines are required on a three-shift basis, 7 days a week, to meet the daily demand of one manufacturer.

Production Milling Operations

Fixed-bed milling machines are more rigid and have greater accuracy than the column and knee milling machines. The distinctive feature of fixed-bed machines is their automatic cycle. Once a piece is clamped in the fixture, the start button is pushed, the cutter rotates, the table moves at rapid travel to the work, the feed begins, and the piece is cut; immediately the table reverses to the end-of-cycle position at the rapid travel speed and the cycle stops. Thus, the operator merely unloads and loads the fixture and pushes a button to start the cycle. Dual fixtures can be used holding two or more parts at each end of the table. This operation is called continuous reciprocal milling and is described more fully under climb milling.

Special Milling Machines

There are a number of special types of production milling machines. Planer-type milling machines are like planers but have heavy-duty vertical milling heads replacing the planer tools. Another type of heavy-duty mill has large vertical milling heads and a rotary table that can be loaded on one side while cutting along the other side.

Duplicating mills, tracer mills, pantographs, engraving machines, or die sinkers can accurately reproduce one or more cavities by tracing a standard form or master pattern. Larger tracing units are found in die shops where large forging or die-casting dies are produced.

Planning for Milling

Standard parts may be easily supported between the jaws of a standard milling machine vise, but in many cases irregularly shaped castings or forgings require

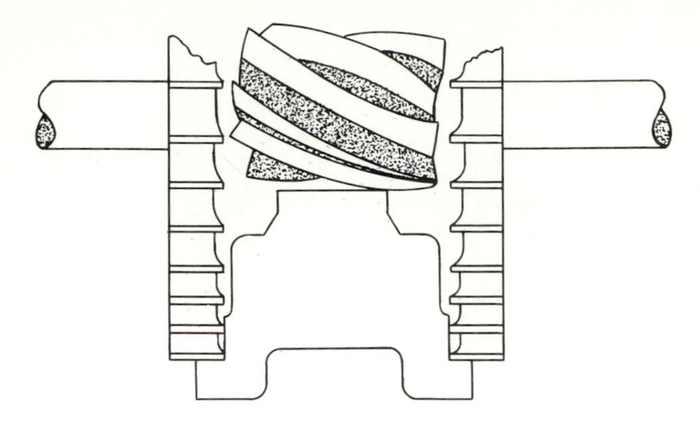

FIGURE 12.34
Gang milling a 20-mm bolt body. Three separate operations are combined. The outer cutters are straddle-milling the body; the helical milling cutter is finishing the top surface.

special holding devices called fixtures to properly support the workpiece during machining. Such fixtures and any special cutters need to be designed and built before any parts can be machined.

Economies in manufacture can be made by milling more than one surface at a time, or by cutting the separate surfaces without removing the part from the machine. More than one surface may be cut by grouping the milling cutters on an arbor, a process called *gang milling* (Fig. 12.34). Different diameters in various positions can mill surfaces parallel or vertical to the axis of the milling arbor. When slots or broad, flat surfaces are milled, interlocking blade cutters can be used to assure a continuously milled surface or the correct width of a slot. The interlocked cutters are spaced apart by washer shims at the hub of the cutter to compensate for the grinding of the sides of the cutter teeth.

Parts are often designed with milled surfaces at an angle to other milled surfaces, or on a plane higher or lower than another. These surfaces can often be located in the same plane by altering the design of the part attached to the milled surface. In this way they can be milled in one operation (Fig. 12.35).

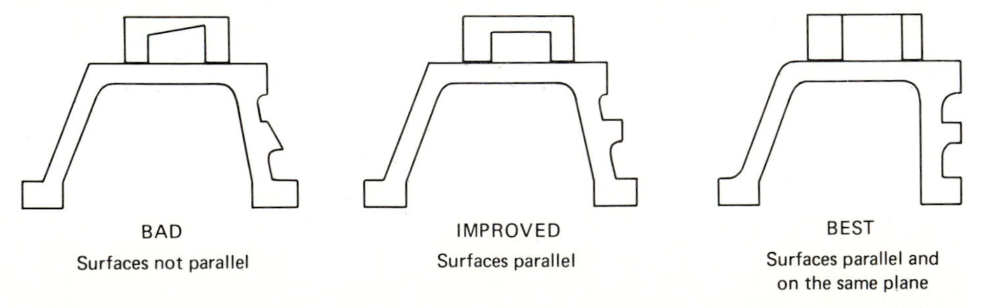

FIGURE 12.35
Design for ease of milling. Keep surfaces parallel and if possible in the same plane.

Climb Milling Versus Conventional Milling

When the cutter enters the material in the direction of feed, it is known as climb milling or down milling; when it enters the material opposite the direction of feed, it is known as conventional milling or up milling (Fig. 12.36). Climb milling requires a well-built and well-maintained machine, with no backlash in the feed screws that will permit the cutter to take too large a bite. On the other hand, it does not have a tendency to lift the part off the table. There are advantages and disadvantages to each method. After some experience, the engineer can determine which is more suitable for a given job.

Climb milling permits simpler fixtures because an end stop is the major requirement for resisting the cutting forces. There is less tendency for chatter so higher speeds and feeds may be used. But the machine must have a backlash eliminator. On the other hand, conventional milling gives less wear and tear on the table feed screw and nut assembly. Conventional milling is used on most column and knee machines because it keeps the screw engaged securely with the nut at all times.

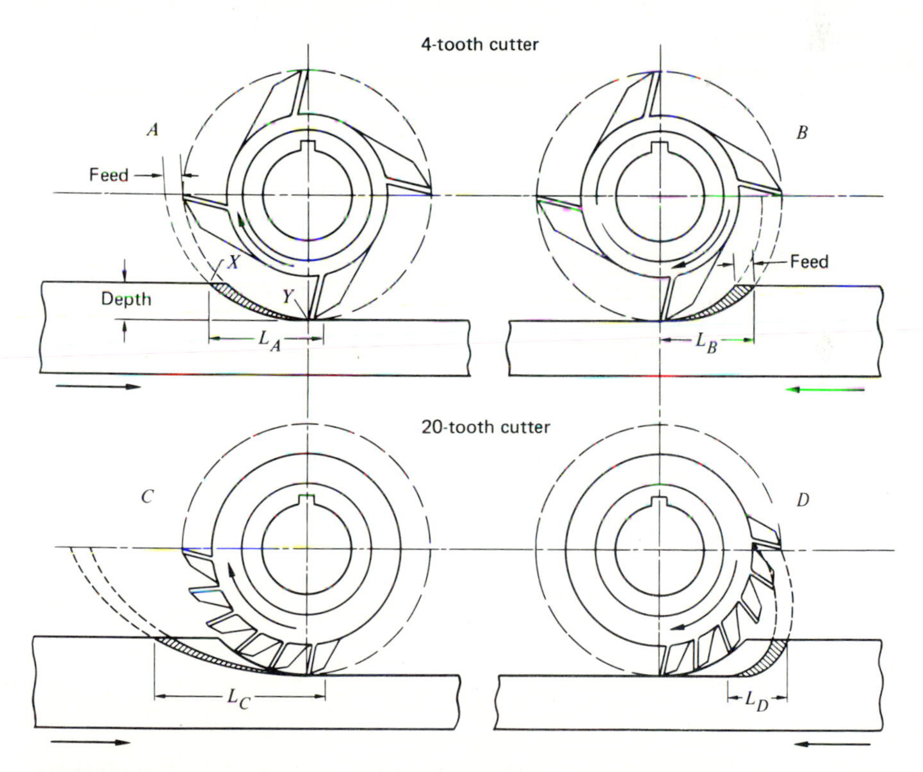

FIGURE 12.36
Comparison of climb milling and conventional milling; *A* and *B* using a 4-tooth cutter, and *C* and *D* using a 20-tooth cutter.

In bed-type machines equipped with a proper backlash eliminator, double-fixture milling is possible. The machine can be set up to run almost continuously; i.e., when a piece is being cut at one end of the table, the operator is loading an identical fixture at the other end. In this case, the machine both climb-mills and conventional-mills alternately with a rapid travel at about 300 in./min between the fixtures. If the milling time is long, an operator can handle two machines.

If the feed per tooth is kept constant and the same tool–work combination is used, the chip made by conventional milling will be longer and thinner than that produced by climb milling (Fig. 12.36*A* and *B*). This phenomenon is even more apparent when a 20-tooth cutter is used because then the feed per tooth is the same but the table feed has increased by a factor of 5 (Fig. 12.36*C* and *D*).

In conventional milling, the tooth comes into contact with the work at point *X* and pushes until the force between the tooth and the work is sufficient to cause the cutting edge to dig in. If the tool is dull, the force reaches a high value before cutting begins. The chip starts at zero and reaches a maximum equal to the feed per tooth at point *Y*. The cross-sectional area of the chip removed can be calculated and is found to be equal to the feed per tooth times the depth of cut.

Gear Cutting

Gear production is really a specialized milling process that can be carried out as a repair job on a milling machine or on special gear-cutting machines. Gears can be produced either by using a series of special form cutters or by using the generation process. In the latter case, a special straight-sided tooth form is used to generate the proper gear-tooth shape. In the milling type of operation the cutter is called a hob (Fig. 12.37). The hob rotates about an axis that is set so that the helix angle of the hob is parallel with the axis of the gear being cut. Then as the gear blank revolves, the hob revolves in mesh with it. The process is called generation because the straight-sided tooth contacts the side of the gear tooth in a series of lines and actually forms a true involute shape as the hob rotates with the gear being cut (Fig. 12.38).

A gear shaper uses a vertically reciprocating cutter that has straight-sided teeth. The cutter slowly revolves while geared to be in mesh with the gear blank being cut. It also generates a true involute shape in much the same manner as the hob does.

Broaching

Broaching may be thought of as a production planing process. It is a machining method wherein one or more cutters with a series of teeth are pushed or pulled across a surface to machine that part to a desired contour. Teeth of the broach increase in height progressively as they approach the finishing end of the tool,

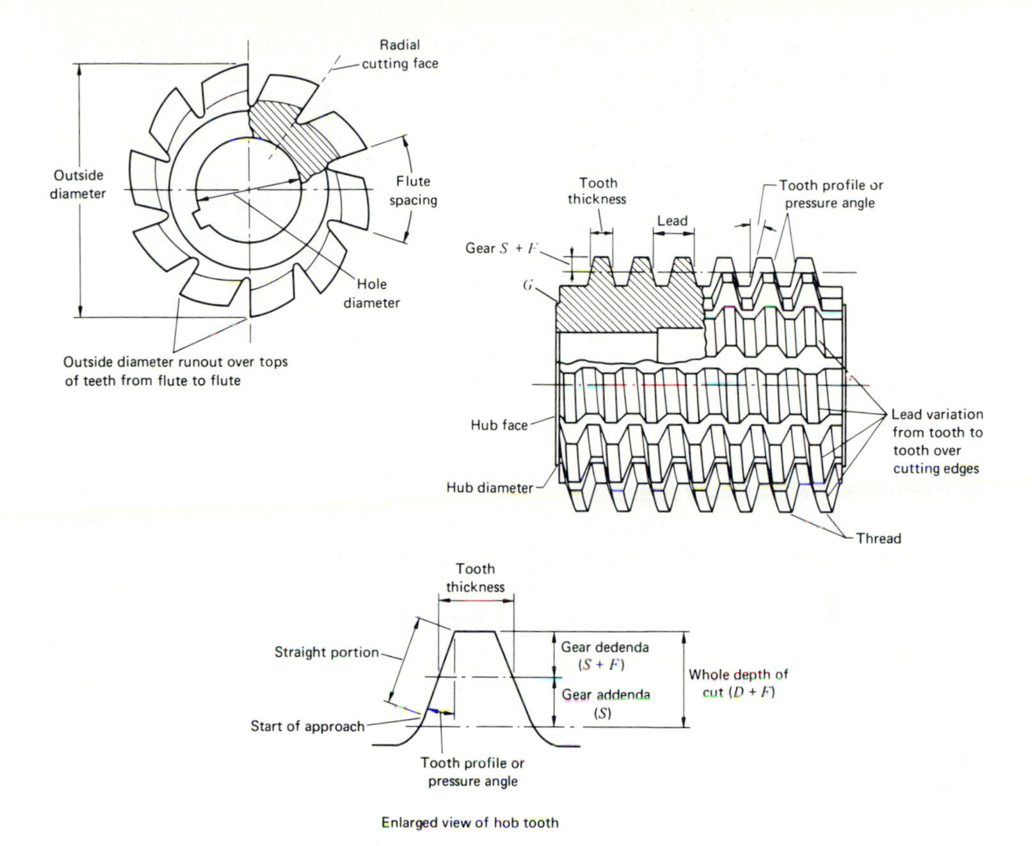

FIGURE 12.37
A hob and its elements.

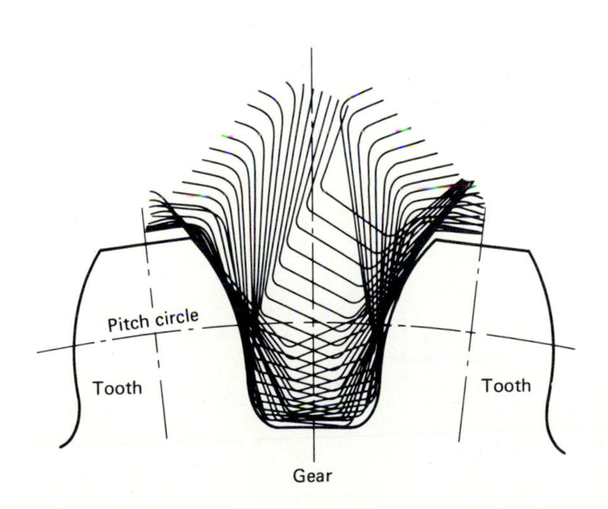

FIGURE 12.38
The action of a hob as its teeth progress through and cut or generate the tooth space in a gear.

so that each tooth removes an equal amount of metal. The last few teeth in the broach are used to bring the work to the desired size and finish.

Broaching has developed from the keyway cutter in a hand press to large machines capable of machining several surfaces simultaneously. For example, the flat surfaces on an internal-combustion engine block are surface-broached in one operation; likewise, a splined internal hole in a gear blank can be sized by one pass of a broach. There is practically no limit to the shape or contour of a broached surface.

Broaching is a fast way to remove metal either externally or internally. It produces a good finish to close tolerances.

Processes similar to broaching include milling, planing, and reaming, although none of these produce the close tolerances and fine finish with the speed of operation of broaching.

Description of process. Figure 12.39 illustrates a typical broach. At the starting end are the roughing teeth, which remove the bulk of the metal. Then follow the finishing teeth, which remove the final increment of stock, bringing the work up to size and finish. When exceptionally fine finishes are required the last few teeth are hemispherical burnishing teeth that remove no metal, but serve only to eliminate surface irregularities. In a single pass, the workpiece is roughed, sized, and burnished, and the finishing operation required with other methods of machining is eliminated. Because little metal is removed by the teeth at the finishing end, there is little wear on these teeth and the broached surfaces can be held to fine tolerances and fine finish. Unlike other machining methods, all of the teeth of the broach cut simultaneously. The rise per tooth

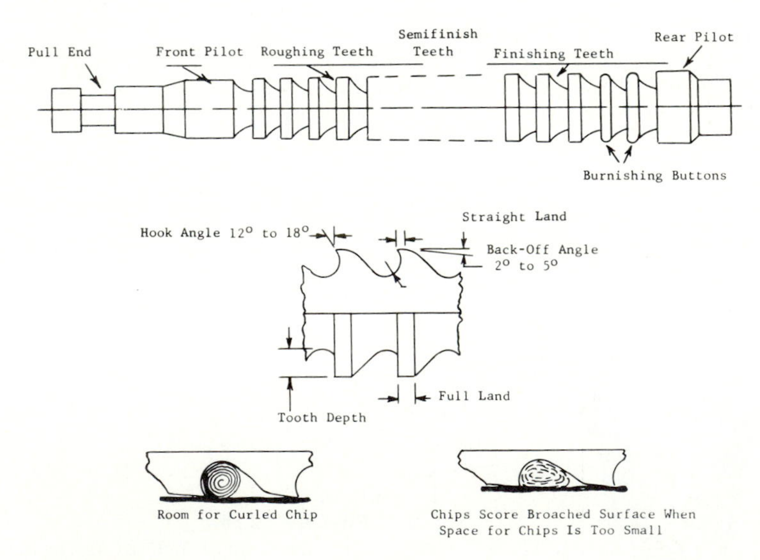

FIGURE 12.39
Broaching tool details.

in a broach is fixed, and the tool pressure cannot be varied. Therefore, the workpiece must either be strong enough to withstand that pressure or must be supported properly in a fixture.

An accurately located drilled hole is satisfactory as a preliminary procedure for an internal broaching operation. Holes may also be prepared for broaching by punching, boring, and coring. The positions of internal broached holes can be adjusted slightly when the holes are broached. The amount of shifting of position to obtain accuracy depends on the strength of the broach and the rigidity of the machine. For external broaching, no preliminary machine operations are required. Properly designed, broaching tools may be used at high rates and are capable of production rates in excess of competitive machining methods.

Depending on the design of the machine, the stroke of the broach is either horizontal or vertical. The broach may be pulled or pushed. Only one pass is made; the broach or piece must be removed before the machine returns to normal operating position. Key slotters or broaches are arranged so that on the back stroke the pressure on the cutting teeth is released. Machines range in size from 2-ton pull, 12-in. stroke to 75-ton pull, 76-in. stroke. Both mechanically and hydraulically actuated broaching machines are on the market and either can be made fully automatic.

Tolerances and finish. The table below gives the tolerances possible through broaching and the recommended design tolerances for minimizing cost when a broaching operation is specified.

Summary	Holes and splines, in.	Gear teeth, in.	External surfaces, in.
Tolerances possible	0.0002	0.0005	0.0002
Tolerances, low cost	0.002	0.002	0.002

When two or more parts are broached simultaneously, high dimensional accuracy (0.002 in.) between the parts can be depended upon.

Surfaces. A fine finish is produced, because burnishing is part of the operation. Usually no further surfacing operations are necessary. The tool marks evidenced in broached holes are axial rather than circumferential as in drilled or reamed holes. This is an advantage in close-fitting reciprocating parts, where circumferential lines may be objectionable because the high spots tend to wear rapidly.

Materials suited for broaching. Steels, cast irons, bronzes, brasses, aluminum, and a broad range of other materials are successfully broached with proper broach design and setup conditions. The best range for broaching of steels lies between 25 and 35 Rockwell C hardness, although steels of higher or lower hardness have been broached successfully. Soft and nonuniform materials are subject to tearing when broached. On surfaces of high hardness, the first tooth of the broach should cut beneath the scale or surface material, thus assuring longer broach life.

Economic quantity. Except when standard broaches may be employed, broaching is economic only for large-quantity production (over 2500 parts). This is because of the relative high cost of the broaching tool rather than because of the cost of setup. Broaching setups are simple, except for fully automatic operations, and can be made relatively quickly.

Broaching tools. As mentioned, broaching tools are expensive and are usually made specially for a given job. The broaching tool is made from a tough, wear-resistant alloy usually containing about 5 percent tungsten and 5 percent chromium.

Broaching tools must be handled carefully in order to prevent nicks in the teeth, which would cause scratches in the work.

Design factors. In designing, the engineer should see to it that the amount of stock to be removed should always be less than $\frac{1}{4}$ in. Good design allows between $\frac{1}{32}$ and $\frac{1}{16}$ in. of stock to be removed from the workpiece. If less than $\frac{1}{64}$ in. is removed, a clean surface cannot be assured.

Since the broach must be able to make an unobstructed pass through or across the workpiece, it is not possible to broach blind holes.

The chip space between successive teeth on the broach must provide sufficient reservoir for the chip. This chip space will limit the axial length of the surface to be broached with a single broach.

Several surfaces can be broached simultaneously. When multiple surfaces on the same workpiece are broached, care must be exercised in designing the fixture to provide adequate strength to withstand the combined cutting-tooth pressure. When gears or splines are to be cut, mechanical, hydraulic, or pneumatic indexing equipment is frequently provided.

Summary. Broaching provides high repetitive accuracy (applicable to production of large numbers of parts of close tolerances and fine finish) and close dimensional relationship of several surfaces broached simultaneously. Broaching is 15 to 25 times faster than other competitive machining methods. The process can be used to accurately produce internal and external surfaces that are difficult to machine by other methods.

The principal disadvantage of broaching is the high cost of special broaching tools. This cost usually does not permit the process to be used when production requirements are low. It cannot be employed economically for the removal of large amounts of stock (more than $\frac{1}{4}$ in.). Lastly, the process has application only on unobstructed surfaces that permit the pass of the broach through the workpiece.

Broaching's principal applications are for the production of almost any desired external or internal contour. This includes flat, round, and irregular external surfaces, round and square holes, splines, keyways, rifling, and gear teeth.

Sawing

Sawing is a multipoint cutting operation that may be related to planing in the case of reciprocating blades and band saws, or to milling as found in circular sawing. In either case, sawing is a parting operation, so the body of the saw is kept as thin as is practical to avoid wasting material in the cutoff operation. Yet it must be rigid enough to support the teeth during the cutting operation. Details of saw teeth and nomenclature are summarized in Fig. 12.40.

Circular saws. Sawing or slitting is performed in milling machines by narrow milling cutters. These cutters are known as circular saws and are also used in sawing machines. The cutters may run slowly and may be of large diameter, approximately 6 ft in some cases. Inserted teeth are used for sawing large sections of metal such as steel forgings and billets. Other types are run at high speeds to cut wood, plastics, and nonferrous metals. The principles of single-point tool cutting apply to saws. Ample room must be provided for chips and side clearance to reduce rubbing friction. The machines may be hand fed or power fed. The material may be fed automatically and cut off. The operator only loads the raw material and removes other cutoff pieces. Sawing provides a fairly smooth and flat surface with a slight burr and does not distort the part as in shearing or slitting. Friction sawing leaves a heavier burr, which can be chipped off easily. Abrasive sawing cuts any type of material with a clean, smooth cut and practically no burr.

Hack saws. Hack saws and hand saws cut the material in the same manner as a circular saw, except that the teeth are in a straight line and act somewhat like a broach. Heavy pressures are required to obtain a cutting action in each tooth throughout the length of cut. Some hack saws are made to feed automatically, and more than one bar can be cut at the same time. Because more than one machine can be tended by an operator, it often costs less to saw than it does to shear, with the advantages of more accurate lengths and no distortion.

Contour sawing. Contour sawing has developed into an important process and has equipment especially designed to obtain maximum results. It is performed on jigsaws, continuous band saws, or long band saws wound on reels. The continuous band saws cut internal holes by separating the saw and inserting it through a pilot hole, and then rapidly welding the ends together on electrical-resistance welding equipment attached to the machine. The band

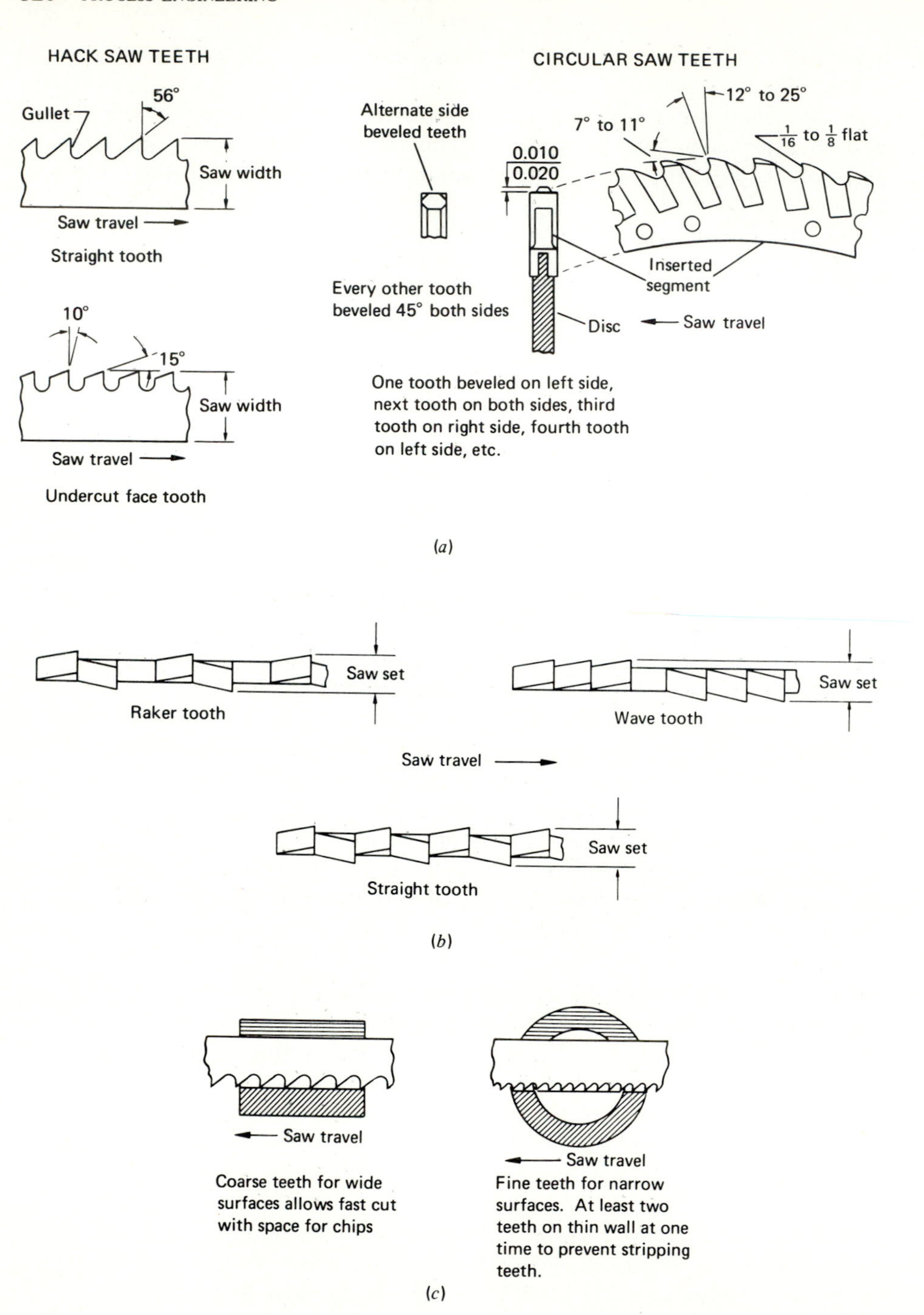

FIGURE 12.40
Details of saw teeth. (*a*) Comparison of hacksaw teeth and circular saw teeth; (*b*) types of set of saw teeth; (*c*) effect of tooth size and pitch.

saws are made in various thicknesses, widths, shapes, and forms of cutting edges. Carbide teeth and abrasive material are fastened to the edge of the band saw in some types. Glass, high-speed steel, armor plate, and all forms of metals and materials are successfully cut on contour band saws.

The engineer can use this process successfully in making model parts and low-quantity parts that are usually blanked, machined, or forged to shape. Dimensions can be held to $\pm\frac{1}{64}$ in. The part is not distorted and surfaces are smooth. Parts up to 6 in. in thickness can be cut. Cut-off pieces can be used for other parts, for example, tool steel used in dies. By using multiple layers of sheets or plates, several parts can be cut out at one time, such as triggers, levers, cams, and cover plates. Speeds as high as 3000 ft/min are used, depending on the type of material and its thickness. Filing, friction sawing, and diamond cutting are performed on these machines in a toolroom or model shop.

Friction sawing. Friction sawing is performed by high-speed circular saws and band saws. The saw has rather dull teeth, which strike the part at high speed. The heat of friction melts the material and the teeth remove the molten material. A spectacular shower of sparks, as well as noise, results. The saw is kept cool by water spray. The teeth gradually wear down and are easily sharpened by enlarging the space between them.

Friction sawing can be applied to any shape of part, angle, channel, I beam, bar, billet, gate, sprue, or casting. A fin forms on the exit edge, but can be chipped off easily. However, hack saws are more economical where they can be applied, because it is not necessary for the operator to tend the machine while sawing and the burr is not usually objectionable.

Summary. Sawing operations are required in many manufacturing processes. The cost of equipment is low; therefore, the cutoff operation should be located near the raw material and the user of the cut-off parts. The shop should not be depended upon to calculate developed lengths; drawings should specify exact lengths so that material can be cut to proper length by the storeroom.

Production Grinding

Traditionally, grinding has been used successfully as a secondary machining operation for high-precision work or to finish hardened workpieces. In fact grinding is still the most economical way to machine external surfaces which are from Rockwell C 45 to 70 hardness. The burden of proof is on the competitive process. Grinding can produce a tolerance of ±0.0001 in., but ±0.0005 in. is more realistically achievable.

Since World War II, the use of high-speed grinding wheels has been made possible by research and improvements in bonding and abrasive materials and in the grinding process itself. In some cases, grinding is economically competitive with primary machining methods. For example, finishing the flat surfaces of forgings or castings can be carried out with less total stock removal if parts are ground from the rough, because the grinding wheel needs only to clean up the surface. There is no need to get under an oxidized or chilled skin by a deep initial cut. Abrasive belt finishing has also grown sufficiently to be recognized as a production finishing operation.

Grinding operations have many points in common. The machines must be particularly rigid because the wheels have a surface speed of at least 5000 ft/min. The high speed makes dynamic balancing particularly important. Second, the wheel bearings must be particularly true, because a cylindrical workpiece has stock removed from both ends of a diameter, so any error in radial advancement is doubled when the work is measured. Third, wheel wear is continuous, and significant compensation must be provided either by having a large ratio of wheel diameter to work diameter, as in external grinding with automatic gauging, or by adaptive control to maintain a constant distance between the wheel surface and the centerline of the workpiece, as is the practice in internal grinding, where the wheel must be smaller than the hole.

There are five major types of grinding operations: cylindrical (center or centerless), internal or hole, surface, tool and cutter, and abrasive belt grinding for precision and surface-finishing operations.

Cylindrical grinding. Cylindrical grinding can be done by grinding from center holes, or it can be centerless. In the former, the machine is similar to a heavy-duty lathe except that both centers are fixed so that the workpiece can not run out at the headstock. To permit easy operator access to the work area, the wheel is located behind the workpiece and is made large in relation to the workpiece diameter so that the wheel wear in one operation is small compared to the stock removed from the workpiece. A typical block diagram of a cylindrical grinding machine is given in Fig. 12.41. Typically, a grinding wheel

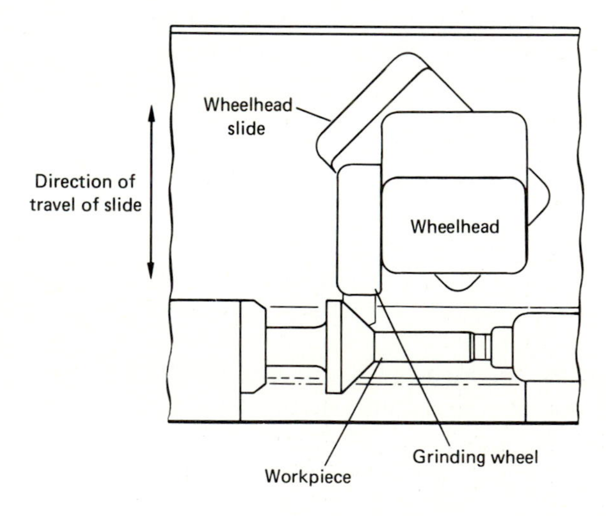

FIGURE 12.41
Block diagram of a typical cylindrical grinder.

for such a machine is 2 to 3 ft in diameter and has a face width of 1 to 3 in. The feed rate is $\frac{1}{4}$ to $\frac{3}{4}$ of the width of the wheel face, depending on the depth of cut, machine rigidity, and the power available at the machine. In plunge grinding, a wheel at least equal in width to the diameter being produced is fed transversely until the proper diameter is reached.

Centerline grinding is a rapid, economical production operation for finishing shafts, pins, ball bearings, bearing races, and similar products to precise tolerances and good finish. When grinding on centers, the work revolves on fixed centers; whereas in centerless grinding the diameter is determined by the periphery of the workpiece. In centerline grinding, time is required to locate and drill center holes, to clean them up after hardening, to clamp the dogs, and to move the piece into and out of the grinder, especially if the piece must be the same diameter from end to end. On the other hand, centerless grinders take more setup time but less operating time and little or no machine loading time.

In centerless grinding, the work is supported by three machine elements: the grinding wheel, the regulating wheel, and the work-rest blade, which is beveled to push the workpiece toward the regulating wheel (Fig. 12.42). The rubber-bonded regulating wheel controls the rotational speed of the workpiece, and its rate of feed is determined by the angle of the horizontal axis of the regulating wheel with respect to that of the grinding wheel. Note that the faces of both wheels are always parallel.

There are two types of centerless grinding: through feed and infeed (Fig. 12.43). In through feed grinding the workpiece passes completely through the grinding zone and exits on the opposite side. Infeed grinding is similar to a plunge cut when grinding on centers. In this case the length of the section to be ground is limited by the width of the grinding wheel.

The infeed method of centerless grinding differs from through feed in the way the workpiece is handled. The infeed method is used for grinding parts with several diameters or heads or shoulders that would prevent complete

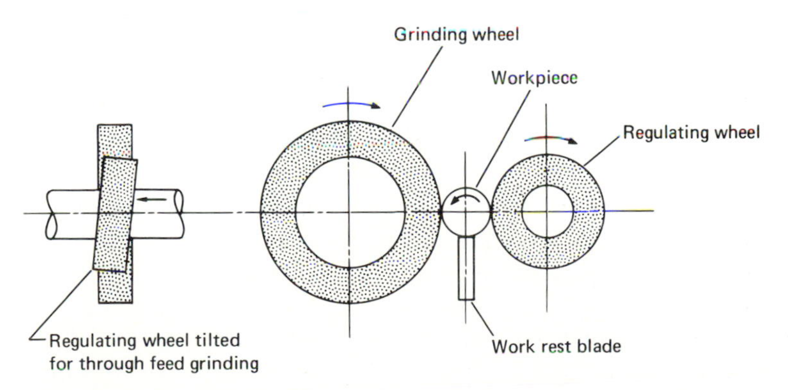

FIGURE 12.42
Schematic sketch showing the principle of centerless grinding.

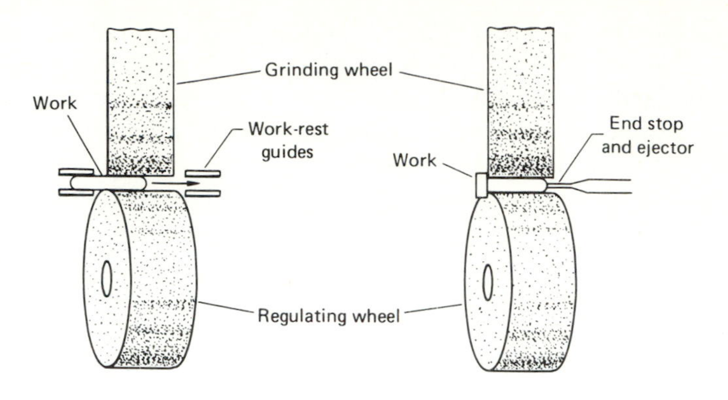

FIGURE 12.43
(*a*) Through feed centerless grinding. (*b*) Infeed centerless grinding.

passage through the machine. The work rest and regulating wheel are withdrawn from the grinding wheel for loading or unloading. They then move toward the wheel to be finished to the final size.

In many cases, special tooling makes centerless grinding a versatile process, but it is often less expensive to design and build new tools. Unless an economic analysis gives a strong indication that centerless grinding will be worthwhile, it is wise to stay with conventional grinding. In some cases it may be economical to centerless-grind as few as 25 or 100 parts, and in others as many as 5000 pieces must be ground before centerless grinding can be justified.

In centerless grinding, the grinding wheel has a surface speed of 5000 to 6500 ft/min, whereas the regulating wheel operates at 50 to 250 ft/min. In through grinding, the rate of feed depends on the speed of the regulating wheel and its angle of inclination; thus the feed in inches per minute is:

$$f = \pi d N \sin \delta$$

where d is the diameter of regulating wheel, in., N is the r/min of regulating wheel and δ is the angle of inclination of the horizontal axis of the regulating wheel (which may vary from 0 to 8° but is usually 3 to 4°. For bent or warped pieces the angle should be 6 to 8°)

The feed force stems from the axis of the regulating wheel, which tilts downward at the entrance to the grinding zone.

The advantages of centerless grinding are as follows.

1. In through-feed grinding the cutting time approaches 100 percent of the operating time, because the grinding action is almost continuous, with little loading and unloading time. Production rates of 150 or more pieces per hour can readily be achieved.

2. Since the workpiece is fully supported by the work-rest blade and regulating wheel, heavy cuts can be taken with minimum danger of distortion or overheating.
3. Centering errors do not exist; therefore, the workpiece is rounded up with a minimum of excess stock allowance.
4. For every adjustment of 0.001 in. of the work-rest blade or regulating wheel there is 0.001 in. off the diameter.
5. Relatively unskilled personnel can be used as machine operators. The setup, specification, and design of the parts for the process requires the most time.

Internal grinding. There are three types of internal grinding machines:

1. Machines in which the workpiece rotates slowly and the wheel spindle rotates and reciprocates the length of the hole
2. Machines in which the work rotates and reciprocates while the wheel spindle rotates only
3. Machines in which the wheel spindle rotates and has a planetary motion while the workpiece reciprocates

Hole-grinding machines require high-speed spindles (up to 100,000 r/min for small holes) because grinding speeds should be in the order of 5000 ft/min and the wheel must be smaller than the hole. Only light feed forces can be used, because otherwise the cantilever-beam effect from the spindle overhang would cause excessive taper. The wheel should not be permitted to leave the hole, in order to minimize bell mouthing.

Wheel wear is rapid because of the small wheel size and the use of a soft grade to permit the wheel to break down under light cutting forces. Thus, accurate hole sizing is difficult to obtain if tolerances are close. To meet this need, production machines can be fitted with automatic gaging and feedback devices that insure accurate control of hole size. The grinding wheel keeps cutting until the hole is 0.001 in. from the proper size and a special cam control is engaged to provide the fine feed that is continued until the hole is finished to the required tolerance.

Surface grinding. Surface grinding is primarily concerned with producing plane and frequently parallel surfaces on steel parts. For this reason, magnetic chucks are standard equipment. As in milling, the grinding wheel may have a vertical or horizontal axis and it may be considered to be equivalent to a face-milling cutter when the axis is vertical and to a plain-milling cutter when the axis is horizontal (Fig. 12.44). Both types of grinders can have either rotary or reciprocating tables. Surface grinders with rotary tables are particularly capable of grinding a number of small pieces simultaneously and at low cost, and they leave a characteristic crisscross pattern on the flat surfaces. The down

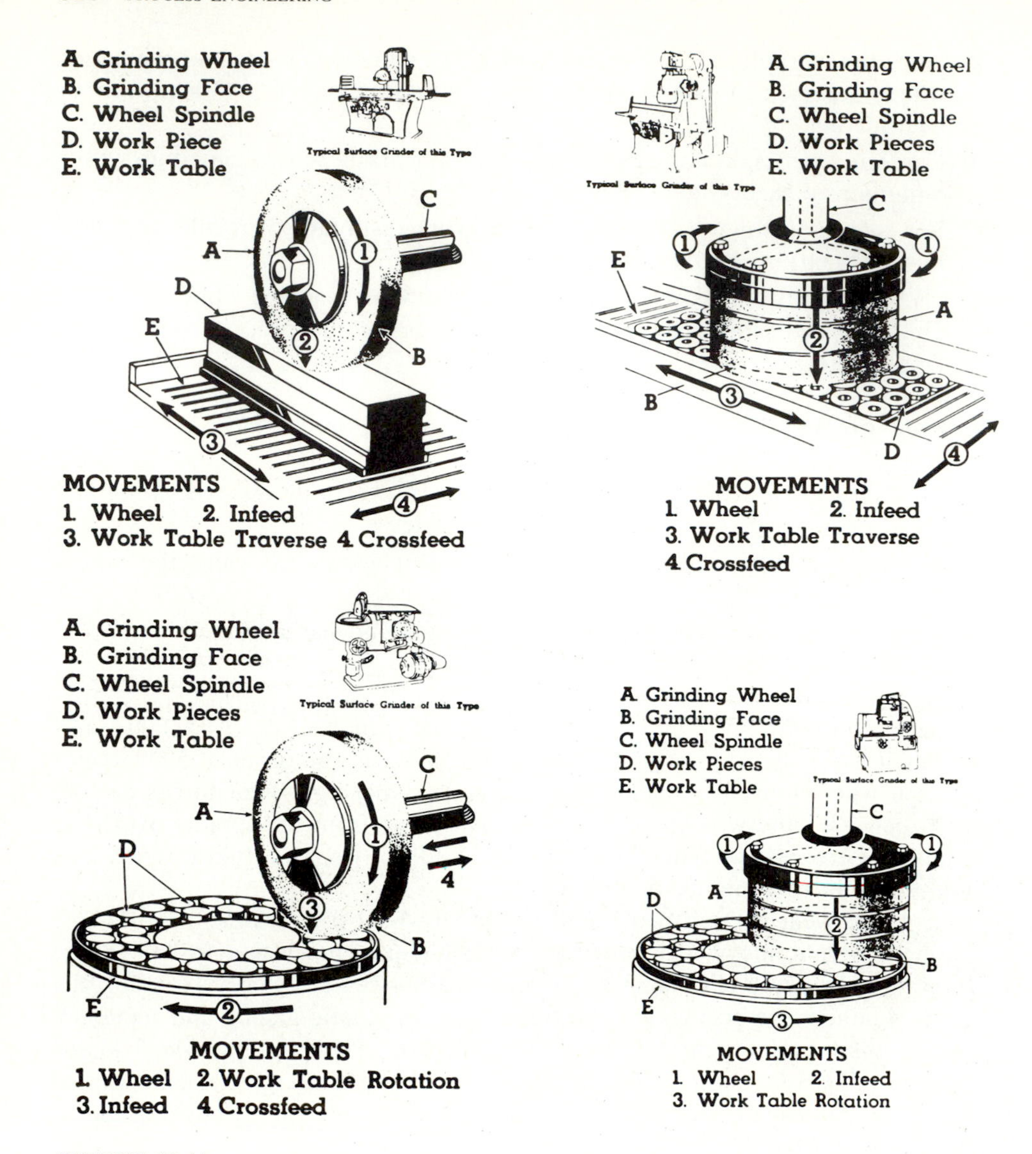

FIGURE 12.44
The principal kinds of surface grinders. Top, reciprocating table grinders; bottom, rotary table grinders; left, horizontal spindles; right, vertical spindles. (*Courtesy of the Carborundum Co.*)

feed on surface grinders can be automatic at 10^{-3} to 10^{-5} in. per revolution of the work table.

Tool-and-cutter grinding. Tool-and-cutter grinders are not production machines and consequently will not be considered here. Any good text on machine tools describes such machines in detail. As a note in passing, the advent of throwaway carbide inserts for turning and milling operations has significantly reduced the work load in the tool-regrinding department, and the importance of these machines has been somewhat curtailed.

Abrasive-belt grinding. With the advent of the flexible plastic bond for applying abrasives to belts, the use of abrasive-belt grinding has increased, because a coolant can be used and the bond is strong. The belt grinder is valuable in the model shop and toolroom, as well as in production. A platen in back of the belt applies pressure to the work and assures a flat or contoured surface as desired. Close dimensions can be held and a minimum of material (0.015 to 0.031 in.) is allowed for machining. Coarse grit will remove material faster, but will not give as smooth a surface. The same general rules described at the beginning of the section on grinding apply to the selection of grain size, grade, and structure. The surface finish produced by belt grinding is better than that by machining. Simple fixtures, or merely holding the part against the belt, are all that are required for cleaning up surfaces, but water must be used to cool the workpiece.

Abrasive cutoff is accomplished by a thin disk-grinding wheel usually made with a rubber bond. The disk cuts principally on its edge, although slightly on the sides too. It gives a smooth surface and very little burr is produced. The process is as fast as sawing, and the part often requires no further machining operations when cut off. It is also accurate as to cutoff length. Water must be used as a coolant.

Any material that can be cut off by sawing, shearing, or flame cutting can also be cut off with an abrasive wheel properly selected for the purpose. This method is applicable to all metals, including hardened tool steels and stainless steel. Other materials that can be cut in this manner, many of which cannot be properly cut with a saw, are plastics in various forms and shapes, ceramic materials, carbon, hard rubber, slate, and casein.

In the foundry, cutoff wheels can be used to remove risers and gates from castings of all kinds, especially nonferrous ones. This can be done without damage to the casting, and the necessity for subsequent grinding may be eliminated. Furthermore, band and hack saws are slower and more easily dulled.

Cutting is performed dry on light, short cuts, and wet on long, heavy cuts. There is little discoloration due to heat when a coolant is used. In submerged cutting, both the work and the lower part of the wheel are completely submerged. At proper wheel speeds, submerged cutting is used to cut glass rods and tubing, plastics, copper tubing, heat-sensitive steels, and other materials that might be affected if the generated heat were not kept as low as possible.

The following are some common faults encountered when operating cutoff wheels; each one suggests its own correction.

1. Movement of work while wheel is in the cut as a result of improper clamping
2. Pinching of wheel caused by work vise high on one side

3. Wheel vibration caused by worn spindle bearings
4. Stalling of wheel in cut because of belt slippage or inadequate power
5. Excessive heating resulting from cuts too heavy for grade of wheel

Precision and surface-finishing operations. After secondary machining, either the precision or the surface finish may require further refinement. In this case, fine abrasive machining is required. If both precision and finish are required, lapping or honing are indicated. If surface finish alone is desired, then fine abrasive blasting, buffing, tumbling, vibratory finishing, or polishing may be sufficient.

Lapping. Lapping is an abrading process that results in a wearing down of the ridges and high spots on a machined surface, leaving the valleys and a random array of fine scratches. Lapping is used to obtain fine dimensional accuracy, to correct minor imperfections in shape, to secure a fine surface finish, and to obtain a fit between two mating surfaces. In the lapping process a fine abrasive, carried in a light oil medium, is supplied to the work surface by a reciprocating type of motion that occurs in an ever-changing path. The resultant abrading action between the work surface and the softer lap such as brass, lead, cast iron, wood, or leather results in the removal of the high spots on the work surface. The cutting action occurs because the soft surface of the lap becomes charged with a layer of abrasive particles that provide a myriad of cutting edges while the lap surface maintains the proper contour.

The service life of lapped surfaces is usually substantially increased over that of the same parts when mated without lapping. Mating gears and worms are lapped to remove imperfections resulting from heat treatment or prior machining. Piston rings and gage blocks are customarily made parallel and to high precision by lapping.

The lapping process can be carried out economically by using the principles of mechanization and automation. If an eccentric spider is used to carry the parts that are to be made parallel in an ever-changing path between two rotating, parallel plates with a recirculating abrasive slurry, rapid and precise lapping will occur. Grooves in a checkerboard pattern in the lapping plate facilitate the delivery of the slurry and help flush away the surface particles. Wet lapping has been shown to be at least six times faster than dry lapping and there is no heat buildup.

Manual lapping is used to bring gage blocks to their final stage of dimensional accuracy and parallelism—a finish of 1 to 2 μin. and a tolerance as small as ± 0.000001 in.

Honing. Honing may be defined as a production method for finishing internal or external surfaces of revolution after machining or grinding. Honing is usually accomplished by reciprocating several spring-loaded sticks of a fine abrasive material over the surface of the rotating workpiece. Honing provides a fine, finished texture in holes and on the exterior surface as well. Bored holes are frequently reamed or honed to obtain final dimensional accuracy and a characteristic figure-of-eight pattern that retains oil on the surface over an extended period of time. This is particularly fortunate in the case of internal-combustion engines, especially during the breaking-in period. Typical honed cylinders are internal-combustion engine cylinders, bearings, gun barrels, ring gages, piston pins, and shafts. Honing is a cutting operation and rarely exceeds a removal of more than 0.001 in. of stock on a side. Boring and reaming should precede honing to assure surface shape and location. Although a 1-μin. surface finish is possible, an 8- to 10-μin. finish is more economical and probably just as good.

Honing machines can be either horizontal or vertical. The former are used for long holes such as cannon or rifle bores or in small models for ring gages. The latter type are more common, ranging from those capable of handling all eight bores of a V-8 engine simultaneously to large machines with an 8-ft stroke capable of handling bores up to 30 in. in diameter.

A typical internal hone consists of a framework for supporting the abrasive stones, which are mounted and internally spring-loaded so that when in the hole they can expand to make contact with the surface of the bore. A cutting fluid is always used to flush away the abraded particles, to improve the finish of the workpiece, and to keep the stones free cutting.

Superfinishing. Superfinishing is a proprietary name, but the process produces the ultimate in refinement of surfaces. Although it is similar to honing, its principles are basically different. A large area of abrasive is used, so that uneven projections and wavy surfaces are removed through a clean cut. Since the abrasive is moved in many directions, it is self-adjusting and becomes automatically a master shape that will correct the work surface. In this respect it is similar to lapping, but the surface is not charged with lapping material. The process does not remove major amounts of material (maximum of 0.0001 in.) in the average production job. The abrasive may be stones, wheels, or belts supported by master platens. Superfinishing is in common use on automobile parts, and automatic machines have been developed to automate this process, which is effective in removing chatter marks and amorphous material until a true surface of the base metal remains. The abrasive is loosely bonded so that each reciprocating stone quickly wears to the contour of the part, but it is large enough so that a representative surface is encountered. Thus, the stone bridges and equalizes a number of defects simultaneously and corrects the surface to an average profile. As the surface becomes smoother the unit pressure decreases until the stone rides on a fluid film and cuts very little. A 1-in. wide ground surface may be refined to a 3-μin. finish in less than 1 min.

Design for grinding. When a part must be ground, the designer should keep the following points in mind.

1. Avoid sharp corners at all shoulders to reduce both tool wear and stress concentrations.
2. Provide recesses if the wheel must grind square with a shoulder.
3. Avoid internal grinding of deep small holes, and be sure to provide an undercut at the base of any ground hole so the wheel can reverse.
4. Avoid narrow slots that require grinding on the sides.
5. All cold-worked metal should be stress-relieved before grinding.

Polishing and buffing. Polishing is the term applied to refining the surface through the cutting action of fine abrasive particles applied to the surface of resilient wheels made of cloth, felt, or wood. Polishing operations may be classified as roughing, dry fining, and finishing. The roughing and dry-fining operations are performed with dry wheels; 20 to 80 abrasives are used in roughing and 90 to 120 in dry fining. In finish polishing, oil, tallow, or beeswax is used with the fine abrasive (150), thus giving a fine finish.

Buffing is the smoothing and brightening of a surface by the rubbing action of fine abrasives applied periodically to a soft wheel. The wheels are spiral sewed buffs of wool, cotton, or fabric. Buffing permits mirrorlike finishes and may be thought of as refined polishing. The abrasive is applied in a lubricating binder commonly known as tripoli.

Barrel finishing. Barrel finishing is of interest because, by its use, many hand operations of burnishing and polishing can be eliminated. The use of tumbling instead of machining, grinding, or hand-finishing operations has resulted in a savings of up to 97 percent in direct labor.

The process differs from conventional tumbling in that, since the parts are always supported by a finishing medium, they are not normally permitted to drop during the rotation of the barrel. The abrading action that takes place is thus concentrated on exposed sharp edges, with material removal varying with the sharpness and relative exposure of these edges. On flat surfaces, while the material removal is practically negligible, there is an action that results in alteration of surface roughness. The degree of action in either case can be controlled accurately, and there is no tendency toward uncontrolled "nicking" of surfaces, which often results from tumbling without a proper abrasive medium.

Barrel finishing can be used to improve the surface of a great variety of small parts made of ferrous, nonferrous, or plastic materials where the problem is to remove burrs or fins, apply a radius to edges, or polish the surface. Since the parts "slide" or "float" in the medium, fragile parts can be tumbled, as well as heavier sections. The process can be applied either wet or dry, depending on the application.

Some typical applications are:

1. Deburring and definning of metal and plastic parts

2. Abrading of desired radii on parts as a substitute for machining, grinding, or hand operations
3. Producing desired surface smoothness for either appearance or performance
4. Producing desired surfaces for subsequent finishing operations, such as plating or painting
5. Removal of scale, rust, and dirt

Almost any metal or plastic part of reasonable size that does not have too great a variation in section size may be tumbled advantageously. A large quantity of parts can be finished at one time, the actual number depending on the size of the part, the size of the barrel, and the surface smoothness desired. Economically and practically, this compares favorably with the individual finishing of pieces by other methods. The only labor involved in production setups is the loading and unloading of the barrel, and occasional screening of the abrasives.

Facilities and materials required. The barrel-finishing operation requires a multisided revolving barrel (preferably rubber-lined), motors and controls, handling facilities, water and drain facilities, and a supply of abrading media.

The selection of the proper abrasive for a given application is important. Experimentation is usually necessary to determine the proper abrasive to use. Aluminum oxide chips, stone chips, steel balls, foundry stars and slugs, molded abrasives and iron slugs, plastic chips, sand, wood balls, and other items are used as abrasive media. In a vast majority of applications, wherever the rigidity of the parts permits, it is possible to use either aluminum oxide chips or granite chips. These materials are furnished in seven or eight sizes, ranging from $\frac{1}{16}$-in. screen to $1\frac{1}{2} \times 2$ in.

The specific size of abrasive chip selected should be such as will not lodge in holes or recesses, or between protruding sections. Where action is required in holes or cavities, the size of abrasive chips used must be such that they will pass through. Where this condition does not exist, the influence of the chip size on surface roughness is usually the deciding factor. Larger chips will cut faster, but will produce a coarser surface. The opposite is true of smaller chips.

On parts that have heavy outside burrs and small burrs around holes, it will be advantageous to use a combination of large and small chips. The large chips will remove the outside burrs faster, while the small chips will pass through the holes, removing the burrs around the hole edges.

Water, alkali compounds, soda ash, trisodium phosphate, burnishing soap, and alkali compounds are used as additives. Water as a coolant is usually added to the mixture of parts and abrasive. It helps prevent the abrasive from loading, provides some lubricating action, and keeps the parts clean. By adding burnishing soap (0.5 percent by weight of the abrasive) to the load, nearly all cutting action is stopped and is replaced by a polishing action. This makes it possible to abrade first and then polish parts without changing the contents of the barrel except for adding the burnishing soap.

Burnishing compounds are often used where high luster and smooth finish are desired. These burnishing soaps and compounds act as lubricants and cushioning agents by preventing abrasion and nicking. In some cases, where unusual conditions cause loading of the chips, an abrasive alkali compound is often used to maintain the stones in a clean, sharp condition.

Barrel speed. The vast majority of parts can be barrel-finished at a speed of 80 to 200 surface feet per minute (with a 32-in. barrel, this varies from 10 to 25 r/min). This wide range indicates that each part should be considered separately when determining the proper speed. Generally speaking, to prevent drastic surface changes, the speed should be such that the parts do not drop when reaching the top of the turn. They should gently turn over and slide with relation to the medium.

Rotation time of barrel. The material from which a part is made determines to a large extent the length of time of rotation. Soft materials, such as copper, brass, aluminum, and magnesium, can be abraded in a fraction of the time required for iron or steel parts, if the shapes are identical. A copper part of the same shape and dimensions as a steel part would be finished in one-half the time of the latter. Aluminum and magnesium require even less time than copper.

The following list gives data on results obtained from a particular operation.

1. Description of work to be done: remove all burrs from a 5-oz steel gear approximately 2 in. in diameter by $\frac{1}{4}$ in. thick, and produce uniform finish.

2. Former method: hand filing. Results: lacked uniformity. Time allowed, 0.033 h; production rate per hour, 30 pieces.

3. Barrel-finishing-method data:
   - *a.* Operator allowed time per piece, 0.00083 h.
   - *b.* Running time per piece (time not allowed), 0.0050 h.
   - *c.* Total cycle time per piece, 0.00583 h.
   - *d.* Production rate per hour (one two-compartment barrel), 171 pieces.
   - *e.* Direct labor cost decrease, 97 percent.
   - *f.* Production rate increase (one barrel), 470 percent.
   - *g.* Barel size, 32 in. diameter by 48 in. length. Lining, wood or rubber. Abrasive, alundum.
   - *h.* Abrasive size, 8*T*. Quantity used, 100 lb for load.
   - *i.* Pieces in barrel, 1200. Speed of barrel, 15 r/min. Time in barrel, 7 h.
   - *j.* Additives: water, level with mass; trisodium phosphate, 2 lb.

Hydrohoning. Hydrohoning is the smoothing of flat and irregular surfaces by the use of a stream of liquid filled with a concentrated mixture of fine abrasive.

This gives a very smooth surface and removes burrs and tool marks. It is used on metal molds.

Sand and grit blasting and hydroblasting. These processes are not classed as machining processes, but are placed here because of their abrasive cutting action. The abrasive is carried by an air stream or water stream and strikes the surface at a high velocity. It removes scale, burrs, light fins, and rust. The surface is marked by the impression of the grit. Steel grit is used in most applications because its rate of recovery is higher. The abrasive is propelled by paddles, which throw the grit onto the work either by centrifugal action or by a blast of air which picks up the abrasive and carries it through pipe or hose on through a nozzle. Nozzles are guided by the operators so the abrasive can reach the proper place.

Blasting equipment ranges from the small bench sizes to rooms large enough to house the largest equipment. Castings and welded steel parts are economically cleaned by these processes before finishing with paint.

Shot peening. Shot peening is similar to shot blasting, except that the surface is peened by the impinging balls from the blast. This hard-works the surface of the material and increases the residual compressive stresses at the surface and resistance to fatigue.

An experiment using shot peening on front-wheel suspension springs resulted in remarkable improvements. Without shot peening, a front-wheel spring made from well-ground steel bars averaged about 170,000 cycles of compression. With the introduction of shot peening, the average life of the springs was increased to 700,000 cycles.

Further research developed a special steel shot, so that the life has been increased to about 4,500,000 cycles. As a matter of fact, the springs wear out the test machine: they simply do not break.

Summary. Grinding has branched into so many fields, and has become so involved, that it seems at times to be an art that is difficult to master. The shop operator soon knows the intricacies of the machine he operates and some of its possibilities, but he seldom knows why the grinding wheel produces the desired results. The production-design engineer who knows why and how the desired finishes are obtained can help make correct decisions.

Chipless Material-Removal Processes

Material removal from ductile materials can be accomplished by using a tool that is harder than the workpiece. During World War II the widespread use of materials that were as hard as or harder than cutting tools created a demand for new material-removal methods. Since then, a number of processes have been developed that, although relatively slow and costly, can effectively remove excess material in a precise and repeatable fashion. There are two

TABLE 12.10

Comparison of chip and chipless machining processes

Machining process	Volts	Power required		Material-removal rate, in.^{3}/h	Average power required, kWh/in.^{3}	Feed rate, in./min	Expected tolerance, in. $\times 10^{-3}$	Expected surface finish, µin.	Depth surface damage, mil
		Amp	Power, kWh						
Turning	220/440	30/15	6.6	360	0.018	3	±0.002	60–300	1
Electro-chemical (ECM)	12–15	10,000 (100–1,500 A/in^2)	100–150	50–80	2	0.01–0.2	±0.002	5–200	0.2
Electric-discharge (EDM)	50	60	3	0.6-7 rough; 0.02–0.2 finish	0.4–20	0.005–0.03	±0.003–±0.001	10–300	5
Electron-beam (EBM)	150,000	0.001	0.15 av.	0.002–0.006	37.5	0.6	±0.001	200–100	10
Laser-beam (LBM)	4.500	0.001	0.0045 av.	0.0004	11,000	0.001	±0.001	20–50	5

types of processes. The first type is based on electrical phenomena and is used primarily for hard materials; the second depends upon chemical dissolution. The first group consists of electric-discharge machining (EDM), electrochemical machining (ECM), electron-beam machining (EBM), laser-beam machining (LBM), and ultrasonic machining (USM). The second includes chemical milling. These processes are useful for machining hard materials because they depend on the physical and chemical properties of a material not its mechanical properties. The designation *chipless machining* has been chosen deliberately because the methods used are in direct contrast to chip-producing methods of machining.

A comparison of the various processes listed above with a typical production turning operation (Table 12.10) shows that there is little likelihood that even the best electrical or chemical machining processes are more than 1 percent as efficient as chip-forming machining processes. Therefore, they are only selected when conditions warrant the extra expense.

The tremendous rise in the use of tungsten carbide for cutting and forming tools has led to a search for methods for shaping it to exact sizes. Either electrical or chemical means can be used, but hydrogen embrittlement and undercutting exclude chemical machining of tools when precise tolerances must be achieved. However, when electrical and chemical methods are combined, as in electrolytic grinding, good results are achieved.

Electric-discharge machining (EDM). The EDM process has become the workhorse of the toolmaking industry for the precise machining of workpieces that can conduct electricity (Fig. 12.45). It can produce holes or cavities of complex cross section and to almost any depth in fully hardened steels or tungsten carbide with relative ease. The high-frequency discharge is developed between the negatively charged tool and the positive workpiece when both are immersed in a dielectric fluid agent that recirculates through a filter and cleans molten droplets and debris from the discharge zone.

At a potential of some 70 V, a critical voltage gradient occurs so that the

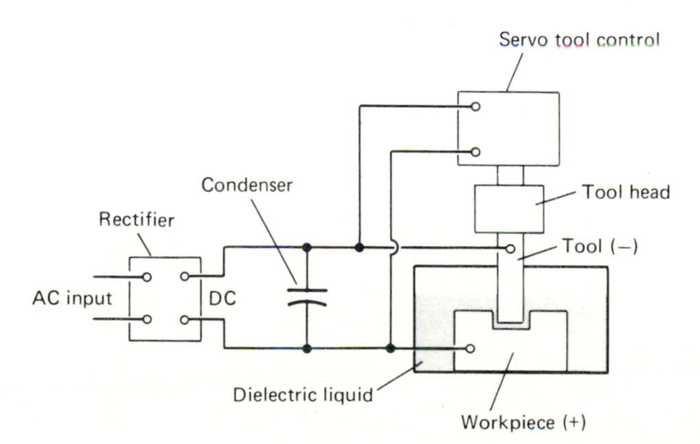

FIGURE 12.45
Electric-discharge machining.

dielectric fluid is locally ionized for a few microseconds, which permits a spark to pass from the work to the electrode with a consequent removal of a small droplet of the work surface, which then is left with a small pockmark. Within 10 to 20 μs, the potential has dropped to 20 V and the capacitor is recharged. When the cycle is repeated at 20,000 to 300,000 Hz at a current density that may approach 10^4 A/in.^{2}, significant metal-removal rates can be achieved. The gap between the tool and workpiece is critical and it must be maintained by a servocontrol device that keeps a constant ratio between the average gap voltage and a suitable reference voltage.

The surface finish in EDM depends on the rate of metal removal. Good finishes require low-energy discharges that leave small craters, but the rate of removal is slow. Good practice indicates that roughing electrodes should be used first. In a typical forging die, finished with two electrodes, a tolerance of ±0.003 in. was achieved; however, when seven electrodes were used, it dropped to ±0.001 in.

EDM is only 1 percent as effective as grinding for removing hardened steel; but it is as good as or better than grinding for use on carbide tools or space-age materials. It is definitely superior for shaping interior configuration in carbide dies. Frail pieces such as honeycomb can be cut without distortion. Again, placing numerous closely spaced holes in a hardened workpiece would be nearly impossible without EDM.

When the EDM process is used for machining dies, it would be advisable to polish or otherwise finish the surface, especially for alloy steels. The surface of the machined contour is carbon-enriched to a depth of 0.1 mil for finish and 5 mils for heavy cuts. There may also be residual-stress cracks from 1 to 30 mils deep, which may result in premature fatigue failure. The residual stresses should be relieved by annealing, and the cracked surface should be removed by lapping to restore the strength of the material at least in part.

Electrochemical machining. In the electrochemical machining (ECM) process, the material is removed by electrolytic deplating rather than by cutting action or solution. This process is being used successfully in drilling of holes from larger than 0.030 in. up to several inches in diameter and in producing irregular slots and contours in solids. Sharp, square corners or sharp-cornered flat bottoms cannot be machined to accuracy.

A tool of conductive material such as lead, tin, or zinc, with the geometry of the inside diameter or surface contour desired in the work, is fed to the workpiece. The tool retains its size and is not affected by the electrolyte. A precise gap is maintained at all times between the tool and the work (see Fig. 12.46). This gap is filled with an electrolyte and, upon the flow of current, metal is electrochemically removed from the work. The maintenance of a precise gap and a uniform feed is the secret to accurate and rapid machining. If the machine deflects, vibrates, or chatters as it moves into the cut, the electrode tool will ground and burn and nonuniform sizes will be obtained.

In accordance with Faraday's law, which states that the mass of any

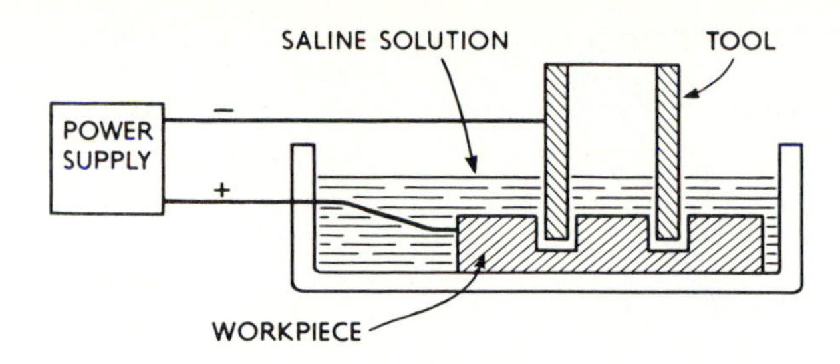

FIGURE 12.46
Electrochemical machining.

substance electrolyzed in a cell is directly proportional to the total quantity of electricity, the rate of metal removal will be proportional to the current flow. Thus, 96,450 C (A · s) are required to remove one equivalent weight of metal, i.e., the atomic weight divided by the valence of the metal. For iron, that would be 28 g ($\frac{56}{2}$), but only part of that energy is available at any one time because of heat and other losses. The amount of current flow also determines the quantity of electrolyte needed. It is common practice to feed the electrolyte through the tool.

As with Chem-milling and EDM, ECM is not affected in any way by the mechanical properties of the work. Irregular holes and contours can be produced in hard materials, such as Stellite turbine blades, free from burrs and with good surface finishes (less than 20 μin.). The metal-removal rate rate is excellent. Holes can be machined at rates more than 0.5 in./min with accuracies of ±0.001 in.

Electrochemical grinding combines the anodic dissolution of a positive workpiece under a conductive rotary abrasive wheel with a recirculating conductive electrolytic. This process has made single-point carbide-insert tooling economically possible. The diamond abrasive in the wheel must remove only about 15 percent of the material, with a consequent saving of up to 50 percent in time and 80 percent for expensive diamond-grinding wheels.

Chem-milling. In the Chem-milling (trade name) process, the material is removed by dissolution rather than cutting action. Three steps are involved in the process. First, the work is masked with a paint or vinyl-type plastic in the areas where no metal removal is desired. Silk screening is often used to apply the paint. the part is then immersed in an etching fluid, which may be acidic or basic, and it is allowed to remain until the required depth has been obtained. Lastly, the maskant is removed from the work.

This process was originally developed for alloys of magnesium and aluminum, but is now being extended to other materials. Dimensional control can be quite accurate. Production tolerances are ±0.005 in., although tolerances as close as ±0.003 in. for aluminum and ±0.002 in. for steel have been reported.

The principal advantages of this process include: low initial tooling costs; burr-free pieces; machining of extremely thin parts (it is possible to machine

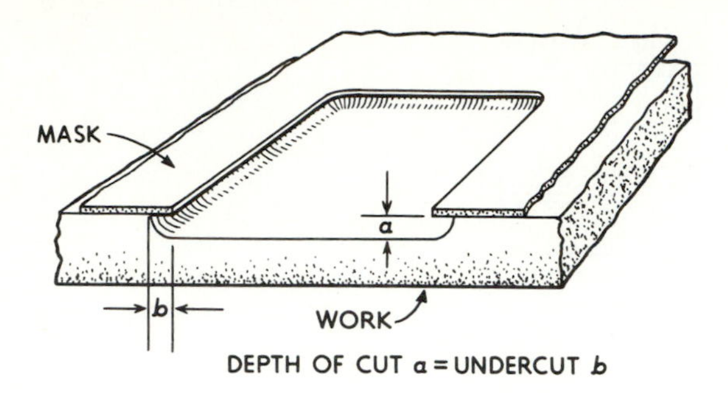

FIGURE 12.47
Undercut in chem-milling.

parts that are only 0.0005 in. thick); and machining of hard and fragile workpieces and those of complex geometry at nominal cost.

In specifying this process, the production-design engineer should be aware of the fact that the template that outlines the position of the work that is to be machined must be smaller than the final size of cut. This is because material is removed under the maskant during the process. Since the metal-removal rate is constant at all areas of the exposed work, the undercut will be equal to the depth of cut and this amount should be allowed for when producing the template (see Fig. 12.47).

Ultrasonic machining. In ultrasonic machining, ultrasonic waves (those having a frequency exceeding the upper threshold of audibility of the human ear, approximately 20,000 Hz) are the means of transmitting energy. Here, a reciprocating tool of the desired shape drives abrasive grains suspended in a liquid that flows between the vibrating tool and the workpiece material. The abrasive granules remove small particles of the material until the desired shape of the workpiece is formed. See Fig. 12.48.

The abrasive particles contained in the liquid are particles of aluminum oxide, boron carbide, or silicon carbide. As would be expected, coarser grits will increase the rate of metal removal, while finer grits are used when surface finish is the principal objective. The liquid containing the abrasive grains is recirculated until the cutting edges become dull and then fresh abrasive-bearing material is introduced to the process. The abrasive grains are driven through movement of the tool, which oscillates linearly at approximately 20,000 times per second with a stroke of only a few thousandths of an inch.

This process is applicable for cutting hard, brittle materials. It can be used to machine holes of any shape and varying depth. The process is equally applicable to conductors or insulators; the process is not particularly suited to machining soft or tough materials. A further limitation is the taper that results when deep holes are drilled.

Tolerances as close as ± 0.0005 in. may be readily attained. Surface finishes to 10 μin. are achieved.

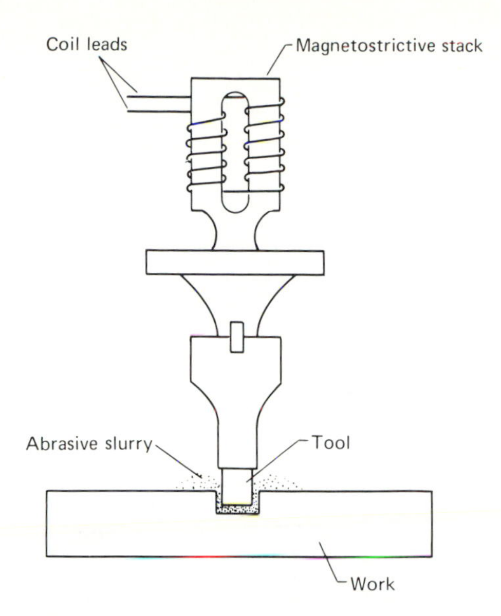

FIGURE 12.48
Typical ultrasonic tooling setup.

Use of Lasers. The use of *l*ight *a*mplification by *s*timulated *e*mission of *r*adiation offers many possibilities in the machining of the "unmachinable." The procedure involves setting the workpiece in a fixture and setting a timer. The timer controls the amount of energy stored in the power supply, which in turn directly affects the power output of the laser gun. Upon the firing of the laser gun, the material upon which it is focused becomes vaporized in 0.002 s or less. The cycle is then repeated in order to machine the next piece or duplicate the previous cycle.

The principle of the laser is the emission of energy at a predetermined level of excitation. Since there does not exist an infinite number of excitation states for any given material, it is necessary for an electron to change its state from one level to a higher one by absorbing energy and thereby becoming excited. The laser beam is a very concentrated monochromatic beam of extremely high intensity.

The application of lasers offers advantages over certain other processes. For example, contact between the laser and the workpiece is not required. The high-power densities available make it possible to vaporize any material, and the small spot permits the removal operation to be performed in microscopic regions. The limits on hole size today are between 0.001 and 0.010 in. in diameter. Laser beams easily drill holes in diamond dies for wire drawing.

PLANNING FOR MACHINE TOOL OPERATIONS

Planning and design of the turning process requires the computation of the time to machine a given component as well as the cutting horsepower to perform the cut.

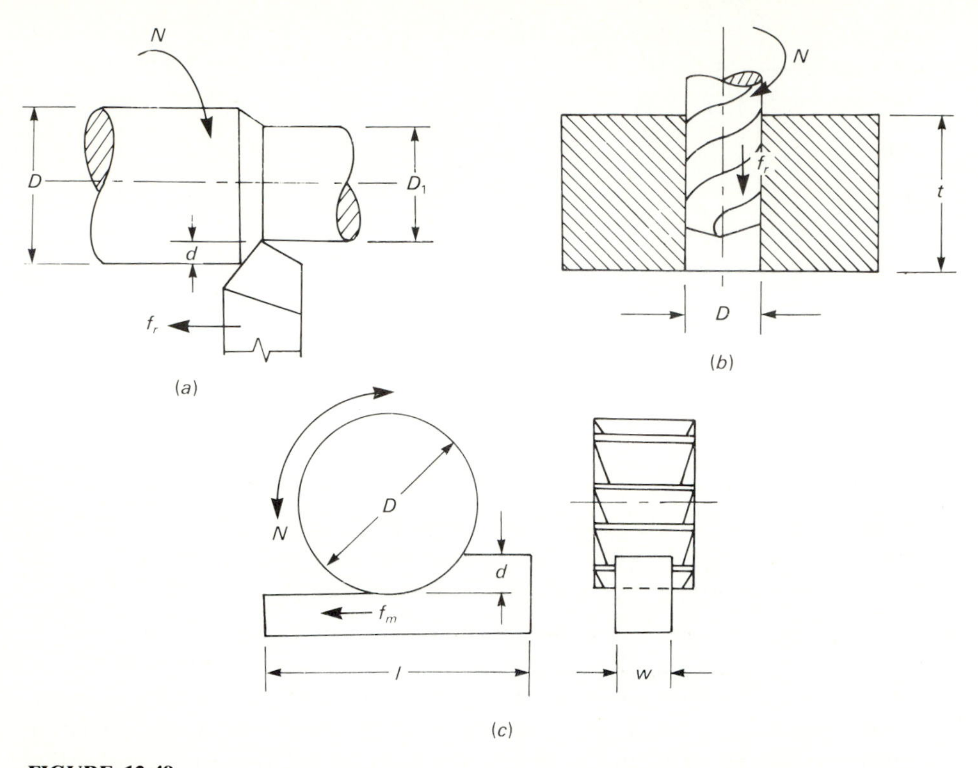

FIGURE 12.49
Notation for machine tool planning. (*a*) Turning operation; (*b*) drilling operations; (*c*) milling operations.

The time to machine a part (t_m) is simply the length to be traversed divided by the rate at which the length is to be traversed. As illustrated in Fig. 12.49*a*, to the length L of the bar there must be added an allowance A, since the tool must start and stop at some distance off the workpiece to account for the cutting tool. The rate at which this distance $L + A$ is traversed is simply the feed rate f_m of the tool in inches per minute. However, the feed set on the machine has the unit of inches per revolution (f_r) since it is actuated from the rotating leadscrew. Hence the feed rate may be calculated as:

$$f_m = f_r \cdot N$$

where N is the revolutions per minute of the spindle. Thus the time to machine may be calculated as

$$t_m = \frac{L + A}{f_m}$$

It is also important to be able to calculate the total cutting horsepower required. As with the power computation for a plate presented previously:

$$\mathrm{hp_c} = \mathrm{hp_u} \cdot Q$$

where the metal removal rate Q must be redefined for turning. The metal removal rate will still be calculated by the product of the area removed times the feed rate (in in./min) perpendicular to area removed. The area removed in turning is an annular ring of area $\frac{\pi}{4}(D^2 - D_1^2)$ as pictured in Fig. 12.49*b*, while the feed rate is $f_r \cdot N$. Hence, the metal removal rate may be computed as

$$Q = \frac{\pi}{4}(D^2 - D_1^2) \cdot f_r \cdot N$$

When the diameter of the part is much larger than the depth of cut, then the following approximation may be used:

$$Q \cong 12Vf_r d$$

where d is the depth of cut.

However, the machine tool inherently contains inefficiency owing to friction and other factors. The gross horsepower required is therefore

$$\text{hp}_\text{g} = \frac{\text{hp}_\text{c}}{\eta}$$

where η represents the efficiency of the machine tool.

The analysis of drilling and related operations is similar to that of lathe operations. The time to machine may be calculated by dividing the distance to be traveled by the drill by the feed rate. The distance traveled is the thickness t of the workpiece plus an allowance owing to the drill angle. An allowance of $\frac{1}{2}D$ is typically used, since a drill angle of 118° is common. The feed rate is the same as in turning. Therefore

$$f_m = f_r \cdot N$$

The cutting horsepower is once again calculated as the product of the specific horsepower and metal removal rate. The metal removal rate is computed as

$$Q = \frac{\pi}{4} D^2 \cdot f_m$$

where $\pi D^2/4$ is the area removed and f_m is the feed rate perpendicular to the area removed.

The cutting horsepower may now be written as $\text{hp}_\text{c} = \text{hp}_\text{u} \cdot Q$ and the gross horsepower as $\text{hp}_\text{g} = \text{hp}_\text{c}/\eta$, just as it was for turning.

In milling (as shown in Fig. 12.49*c*), the cutting time may be calculated by taking the length traversed and dividing by the feed rate. The length traversed will be the length L of the part plus an allowance A. The allowance depends on the type of milling operation. The feed rate in milling in inches per minute is set directly on the machine, unlike lathes and drills. Standard references make their feed recommendations in inches per tooth. Therefore, the feed rate can be calculated from this handbook value as follows.

$$f_m = f_t \cdot n \cdot N$$

where f_t is the feed (in./tooth), n is the number of teeth in the milling cutter, and N is the revolutions per minute of the spindle. The time to machine can therefore be calculated as

$$t_m = \frac{L + A}{f_m}$$

The determination of the cutting and gross horsepower requirements is done precisely as it was for turning and drilling. The only difference is the calculation of the metal removal rate. In milling, the uncut area is simply the product of the depth of cut d and width of cut w. Hence $Q = w \cdot d \cdot f_m$.

It should be clear from the preceding discussion that the general planning methodology is the same regardless of process. The only difference is the calculation of the feed rate (f_m), which depends on the machine tool.

SUMMARY

The production-design engineer should become familiar with the advantages and disadvantages of the five basic metal-cutting processes: drilling, turning, planing, milling, and grinding. Likewise, he should keep abreast of the special metal-removal processes such as Chem-milling, lasers, electrochemical, electric discharge, and ultrasonic machining, so that he will be able to specify and design for the most economical technique.

With the development of ceramic cutting tools and the extended use of carbides, it is possible to remove metal at rates not considered at all feasible a few years ago. Today the machine-tool builder is finding it necessary to redesign so that his machine will be able to withstand the forces developed through the heavy feeds and fast speeds the modern cutting tool is capable of.

The function of production engineering includes the work of tool engineering, which is today a study by itself. Machinability of metals and its relation to the form of the cutting tool is an area that engineers spend a lifetime studying and developing. The engineer who designs for production is vitally interested in all facets of the metal-removal processes in that they affect, to a large extent, the potential of his designs.

Material and machining are expensive; therefore, parts should be preformed, or precast, closer to size in order to reduce the amount of material to be removed.

Finally, the production-design engineer must be cognizant of the limitations of the equipment in his plant and those that service his organization. He should avoid specifications that are impractical because of time and cost. However, he must realize that specifications that are impractical today may be realistic tomorrow. Many poor designs, from the standpoint of machining operations, can be avoided by giving consideration to the following Do Nots.

1. Do not specify tolerances closer than necessary.
2. Do not specify a thread to the bottom of a blind hole.
3. Do not specify a tapped hole unless the tap can cut the entire perimeter of the last thread.

The wide range of equipment necessary to accomplish the many methods of cutting and removing material will occupy the attention of the engineer during his entire career. Improved equipment will require changes in the design of his apparatus in order to take full advantage of the more efficient machinery. This procedure is necessary in order to meet competition satisfactorily.

The production-design engineer should follow the development of ceramic cutting tools so that he will be able to fully utilize the advantages offered. The trend will be to cut materials at faster speeds and heavier feeds. This will require more rigid machine tools. In order for a company to be competitive, it will have to understand and be able to make the most of the latest metal-removal techniques, tools, and equipment.

QUESTIONS

12.1. What are the characteristics of clean-cutting metals?

12.2. How is the property brittleness related to machinability?

12.3. What does the machinist mean when he speaks of a material as being "too ductile?"

12.4. What factors affect the cutting of all materials?

12.5. Make a sketch of a single-point tool and show the back rake angle, end-relief angle, end cutting-edge angle, side cutting-edge angle, and built-up edge.

12.6. What steps can the production-design engineer take to decrease the amount of the built-up edge?

12.7. What would be the cutting speed in feet per minute of a 1-in.-diameter drill running at 300 r/min?

12.8. Explain why large amounts of frictional heat are produced when machining very ductile metals.

12.9. What two pressure areas on cutting tools are subject to wear?

12.10. How does the rake angle affect the life of the cutting tool?

12.11. What is the 18–4–1 type of tool?

12.12. What is the future of the ceramic class of tools?

12.13. What does honing of the cutting and chip-contact surfaces do to the tool life? Explain.

12.14. What are the major functions of the cutting fluid?

12.15. Upon what does the cooling ability of a cutting fluid depend?

12.16. What is the danger of multiple cuts on precision work?

PROBLEMS

12.1. The following measured data were taken from an orthogonal test cutting AISI 1015 steel using a tungsten carbide tool:

Cutting speed	500 ft/min
Feed	0.010 in./rev
Chip width	0.100 in.
Chip thickness	0.022 in.
Cutting force	313 lb
Feed force	140 lb
Tool angle	10°

Compute:

a. Chip-thickness ratio
b. Resultant force
c. Shear angle
d. Length of shear plane
e. Friction angle
f. Friction force
g. Normal compressive force
h. Coefficient of friction
i. Shear force
j. Shear stress on shear plane
k. Shear velocity
l. Net horsepower
m. Specific cutting energy

12.2. In a recent test the following data were collected for 70/30 brass.

Cutting conditions		*Data*	
Speed	200 ft/min	Horizontal force	96.9 lb
Rake angle	10	Vertical force	43.9 lb
Depth of cut	0.0034 in.	Chip thickness	0.0102 in.
Width of cut	0.125 in.		
Contact length	0.040		

a. Determine the shear angle.
b. Calculate the shear and normal stresses on the shear plane.
c. Compute the chip and shear velocities.
d. Calculate the shear and normal stresses on the rake face.
e. Determine the coefficient of friction on the rake face.

12.3. In the experiment in Problem 12.2 a tube was used (4.25 in. outside diameter). The tube was slotted to allow the chip from one revolution to fall into a chip tray. How long would the chip be for the experiment detailed in Problem 12.2?

12.4. A 4.000-in.-diameter brass bar of 10 in. length is to be turned to a new diameter of 3.990 in. A handbook suggests running at 500 sfpm and 0.01 ipr (in./rev).

a. Determine the rpm setting for the lathe.
b. Assuming an allowance of 0.250 in., calculate the cutting time t_c.
c. Determine the metal removal rate Q_t.
d. Calculate the cutting horsepower required if $hp_u = 0.55$ hp/in./min.
e. Calculate the gross horsepower required if the lathe has an efficiency of 0.90.

12.5. A $\frac{3}{4}$ in. diameter hole is to be drilled in a piece of 1020 steel of thickness 1.000 in. Given that we run at 400 rpm and 0.020 ipr:

a. Calculate the suggested speed V.
b. Assuming an allowance of $\frac{1}{2}$ in., compute the time to machine t_c.
c. Determine the metal removal rate Q_t.
d. Calculate the cutting horsepower if $hp_u = 0.80$ hp/in./min.
e. Determine the gross horsepower assuming a machine efficiency of 0.88.

12.6. A slot $\frac{1}{4}$ in. wide and $\frac{1}{4}$ in. deep is to be milled in one pass in a titanium component for usage in a military fighter. The component is 18 in. long and milled with a 6-in.-diameter, 10-tooth cutter. From a handbook:

$$V = 150 \text{ sfpm}, \qquad f_t = 0.005 \text{ ipt (in./tooth)}$$

a. Determine the rpm setting N.
b. Calculate the feed rate f_m.
c. Calculate the time to machine if one cutter diameter is used as the allowance.
d. Determine the metal removal rate Q_t.
e. Calculate hp if the unit horsepower is 1.7 hp/in./min.

12.7. A 4.000-in. diameter aluminum bar of 12-in. length is to be turned to a new diameter with a speed of 525 sfpm, feed of 0.020 ipr and 0.030 in. depth of cut.

a. Determine the rpm to be set on the machine tool.
b. Calculate the cutting time assuming an allowance of $\frac{1}{4}$ in.
c. Determine the productivity of the operation by calculating the metal removal rate.

12.8. In running a tool-life test, you determine that for a cutting speed of 400 sfpm, the tool lasts 1 min and that at a speed of 220 sfpm, the tool lasts 20 min. Determine the Taylor tool life equation (of the form $VT^n = C$) for this high-speed steel tool.

12.9. For the turning of a cast-iron shaft, the following cutting conditions are set:

$V = 100$ sfpm
$f_r = 0.020$ ipr
$d = 0.125$ in. (depth of cut)
$D = 1.5$ in.

a. Assuming $hp_u = 0.8$ hp/in./min, find the cutting horsepower
b. Determine the cutting force F_H and the deflection of the tool, δ, assuming a 1-in. overhang for the $\frac{1}{4} \times \frac{1}{4}$ in. high-speed steel tool ($E = 30 \times 10^6$ lb/in.^{2}, $I = bd^3/12 = 0.0003255$ in.^{4}). Recall that $\delta = F_H L^3/3EI$, where L is the overhang, E is the modulus of elasticity, I is the moment of inertia.

12.10. A titanium component 6 in. wide and 23 in. long is to be milled to final size. A 6 in.-diameter, 10-tooth carbide cutter is to be used. The mill is powered by a 10-hp motor. Recommended machining parameters call for a speed of 400 sfpm and 0.050 in. depth of cut.

a. Assuming a specific horsepower of 1.7 hp/in./min. determine the feed (in./tooth) that will utilize all 10 hp.
b. If the mill has feed rate settings of 2, 5, 7, 10, 12, 15, 18, 22, 30 in./min, which setting would you choose, based on your answer to part (a).
c. Assuming an allowance of one cutter diameter, calculate the cutting time.

12.11. There are several components to the specific energy u; these are shear u_s, friction u_f, chip curl, kinetic energy required to accelerate the chip, and the surface energy required to produce a new surface on the workpiece. The frictional and shear components were considered to comprise most (>95 percent) of the specific energy. Prove that if one neglects the latter three terms and defines u, u_f, u_s by the Merchant mechanics that:

$$u = u_s + u_f$$

(*Hint:* The trigonometric identity $\cos(A - B) = \cos A \cos B + \sin A \sin B$ may come in handy.)

12.12. The metal removal rate is one measure of process productivity and may also be used for horsepower calculations. For turning,

$$Q_t = \frac{\pi}{4}(D^2 - D_1^2)f_r N$$

where D is the initial diameter, D_1 is the new diameter, f_r is the feed, and N is the rpm. It can also be shown that the following approximation for metal-removal rate is valid as long as the depth of cut is small relative to the diameter:

$$Q_t \cong 12Vf_r d$$

where V is the speed and d is the depth of cut. Starting with the exact form, use the condition stated above (i.e., $d \ll D$) to derive the approximation. Units: d, D, D in inches, f_r in in./rev, N in r/mim, V in ft/min.)

REFERENCES

1. Ramalingam, S., and Black, J. T.: "An Electron Microscopy Study of Chip Formation," *Metallurgical Transactions*, Vol. 4, p. 1103, 1973.
2. Black, J. T.: "Flow Stress Model in Metal Cutting," *ASME Journal of Engineering for Industry*, Vol. 101, p. 403, 1979.
3. Von Turkovich, B. F.: "Dislocation Theory of Shear Stress and Strain Rate in Metal Cutting," *Advances in Machine Tool Design and Research*, p. 531, 1967.
4. Boothroyd, G.: *Fundamentals of Metal Machining and Machine Tools*, McGraw-Hill, New York, 1975.
5. Ernst, H.: "Physics of Metal Cutting," in *Machining of Metals*, ASM, p. 38, 1938.
6. Merchant, M. E.: "Mechanics of the Metal Cutting Process—I," *Journal of Applied Physics*, Vol. 16, No. 5, pp. 267–275, 1845.
7. Merchant, M. E.: "Mechanics of the Metal Cutting Process—II", *Journal of Applied Physics*, Vol. 16, No. 6, pp. 318–324, 1945.
8. Shaw, M. C.: *Metal Cutting Principles*, 3d Ed., The M.I.T. Press, Cambridge, Mass., 1957.
9. Vidosic, J. P.: *Metal Machining and Forming Technology*, The Ronald Press Company, New York 1964.
10. Cook, N.: *Manufacturing Analysis*, Addison-Wesley, Reading Mass., 1966.
11. Trent, E. M.: *Metal Cutting*, 2d Ed., Butterworths, London, 1986.
12. Doyle, L. E., Keyser, C. A., Leach, J. L., Schrader, G. F., and Singer, M. B.: *Manufacturing Processes and Materials for Engineers*, Prentice-Hall, Englewood Cliffs, N.J., 1969.
13. Armarego, E. J. A., and Brown, R. H.: *The Machining of Metals*, Prentice Hall, Englewood Cliffs, N.J., 1969.
14. Boston, O. W.: *Metal Processing*, 2d Ed., Wiley, New York, 1951.
15. Kim, N. H., and Chun, J. S.: "A Study of the Tool Life of TiC and TiC plus AlO Chemical

Vapor Deposited Tungsten Carbide Tools," *Journal of Material Science,* Vol. 20, pp. 1285–1290, 1985.
16. Hale, T., and Graham, D.: "How Effective Are the Carbide Coatings," *Modern Machine Shop,* April 1981.
17. Briggs, D. C., and Thomson, R.: "The Properties of High Speed Tool Steel Produced By Horizontal Continuous Casting," *Cutting Tool Materials,* ASM, pp. 93–110, 1981.
18. Wilson, G. F., and Bruschek, D. K.: "Machining of Selected Cast Aluminium Alloys With COMPAX Blank Tools," *Cutting Tool Materials,* ASM, 1981, pp. 297–314.
19. King, A. G. and Wheildon, W. M.: *Ceramics In Machining Processes,* Academic Press, New York, 1966.
20. DeGarmo, E. P., Black, J. T., and Kosher, R.: *Materials and Processes in Manufacturing,* 7th ed., Macmillan, New York, 1988.

CHAPTER 13

SECONDARY MANUFACTURING PROCESSES: FORMING

After a metal has been hot-worked to the desired shape, such as a structural shape or hot-rolled sheet, it is frequently descaled by pickling to make it ready for secondary forming processes that are carried out below the recrystallization temperature. These operations involve cold-rolling into sheet or cold-drawing into finished bars for machining. This chapter is concerned with the secondary forming processes that are common to sheet metal and plate, bar, and tubing. The forming of metals is concerned with bending operations that may or may not be accompanied by stretching. When the metal moves over the edge of a die to change its angle, the operation is drawing. Bulging is the increasing of diameter along a portion of the length of tubing or a shell. Other typical forming operations include edging, folding, cupping and curling. The shaping of metals is the alteration of geometry of massive metal. This alteration in turn can be measured by the change in angle between selected lines on or in the metal caused by the shaping operation. Typical shaping processes include upsetting, extrusion, tube making, and wire drawing. Separation of metal is limited to blanking, piercing, and shaving operations. Here a single piece of material is separated into two or more pieces.

The working of metals is concerned with plastic flow in the workpiece. This plastic flow will alter the mechanical properties such as strength, hardness, ductility, fatigue properties, creep resistance, etc., of the materials so worked. Thus, cold-working operations enhance the strength and hardness of a material at the expense of ductility. The distorted crystal structure uniformly increases in strength in proportion to the degree of deformation, and its yield stress increases in proportion to the amount of reduction. The engineer will take advantage of the increase in yield strength imparted by cold-working when he designs a part to be made by cold-working operations. Examples of such techniques are seen in the design of curved trunk lids, hood designs, and formed patterns used in automobile bodies. The sheet metal gage of the bodies is nowadays considerably reduced from that used in earlier designs.

External forces acting on a solid are either surface forces (pushes or pulls on a surface) or body forces (pushes or pulls throughout the volume of the solid, due to inertia, magnetism, etc.). If a very small area of surface is chosen, then the force acting externally on that area divided by the area is the external stress on that area. In Fig. 13.1, the external stress on area A_1 is F_1/A_1. Any force on a surface may be decomposed into two kinds of components:

1. Components of force perpendicular to the surface, termed direct (normal) force (F_3 in Fig. 13.1)
2. Components of force parallel to the surface, termed shear force (F_2 and F_4 in Fig. 13.1).

Similarly, the external stress can be decomposed into direct stress F_3/A_1 and shear stress F_2/A_2 or F_4/A_1. On any small rectangular parallelepiped as

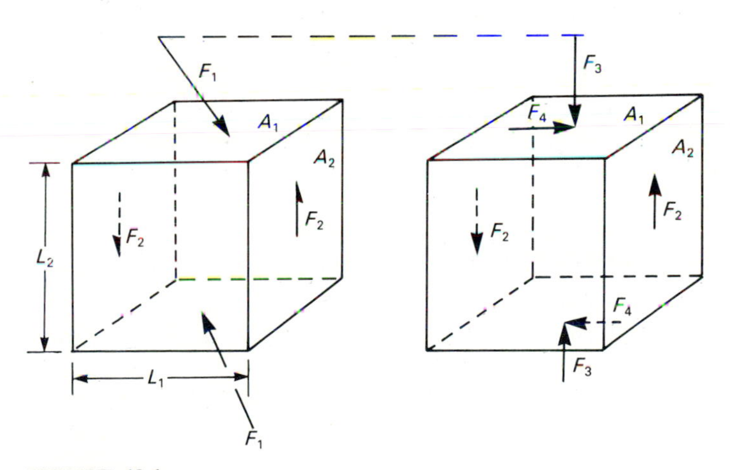

FIGURE 13.1
Decomposition of a composite force into normal and shear components of force.

$$F_4L_2 = F_2L_1$$

$$\frac{F_1}{A_1} = \text{normal stress}$$

$$\frac{F_4}{A_1} = \frac{F_2}{A_2} = \text{shear stress}$$

shown in Fig. 13.1, static equilibrium requires (in the absence of body forces) that the magnitudes of the shear stresses on adjacent faces be equal, that is F_2/A_2 equals F_4/A_1.

The direct stress is termed *compressive* if it tends to move atoms closer together in the direction of the application of the stress. Thus, F_3 in Fig. 13.1 is a compressive stress. The direct stress is termed *tensile* if it tends to move atoms further apart in the direction of application of the stress. Of course, stresses can also exist inside a solid, in which case they are termed internal stresses.

Pressure is the direct compressive external stress at a point on the surface of a solid. If the pressure is the same at each point over an area of a solid, the pressure is *uniform* over that area. If pressure is uniform everywhere over the entire surface of a solid, *hydrostatic pressure* exists. The dimensions of pressure are those of force per unit area. Commonly used dimensions are pounds per square inch (psi), kilograms per square centimeter or per square millimeter, atmospheres, and bars or kilobars. A table of conversion factors follow:

$$1\ \text{kg/cm}^2 = 14.2\ \text{psi}$$
$$1\ \text{kg/mm}^2 = 1422.0\ \text{psi}$$
$$1\ \text{atmosphere} = 14.7\ \text{psi}$$
$$1\ \text{bar} = 14.5\ \text{psi}$$
$$1\ \text{kilobar} = 14{,}500.0\ \text{psi}$$

A convenient approximation is 1 atmosphere = 1 kg/cm^2 = 1 bar = 15 psi.

Ductility is the maximum amount of permanent deformation achieved by a solid material (such as a metal) that has been subjected to increasing loading. Typical tests for ductility include: tension, compression, bending, cupping, etc. Ductility has also been defined as *elongation of a specific tensile specimen* (percent change of a fixed length—the length is frequently expressed as a number of diameters), or as *contraction* of a specific tensile specimen (percent change in a uniform gage section of the cross-sectional area).

COLD-WORKING FUNDAMENTALS

All operations that displace or move metal plastically require forces to move the metal, and the amount of displacement that can be sustained without failure is a function of the metal's ductility and the degree of hardening, strengthening, and embrittlement of the material as a result of the process.

Many metals exhibit an abrupt yielding (the stress in a material at which there occurs a marked increase in strain without an increase in stress[1]). In some materials, the yield point may appear as little more than a "jog" in the stress–strain curve, while in others the yield-point behavior may extend over elongations of several percent.

---

[1] Definition of yield point from ASTM Standards.

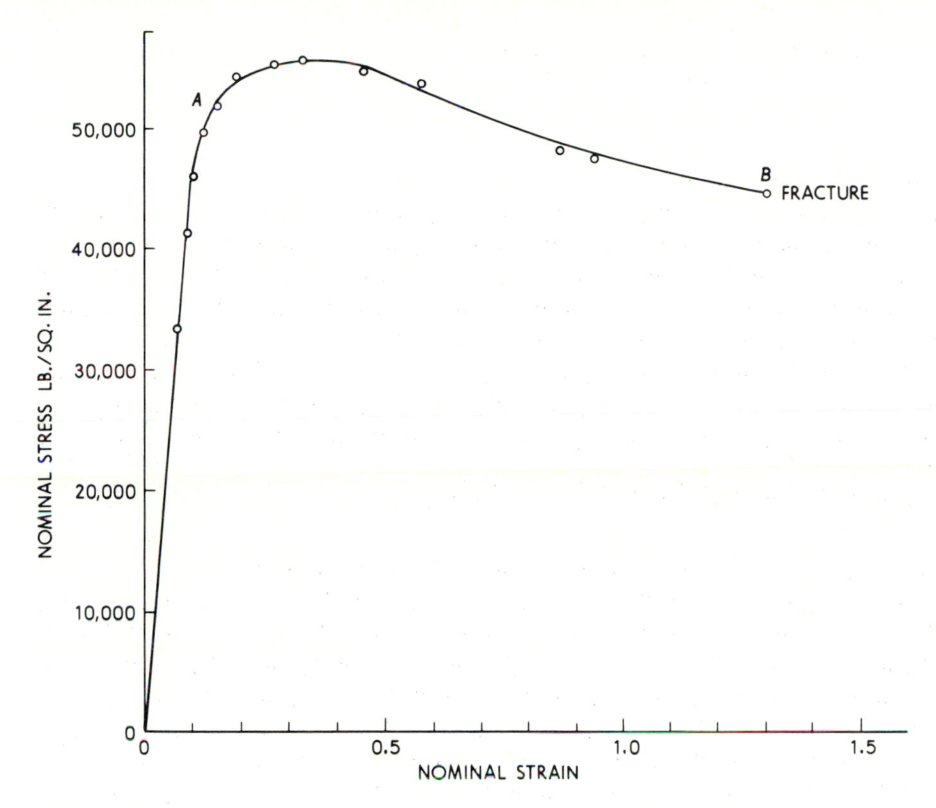

FIGURE 13.2
Conventional stress–strain diagram for cold-worked SAE 1020 semikilled steel.

Figure 13.2 illustrates a conventional stress–strain diagram. The plastic range is shown between the points *A* and *B*. It is within this area that the cold working of metals is effective. Figure 13.3, which was developed from the data used to plot Fig. 13.2, gives a useful straight-line approximation of the true stress versus the reduction of area in the plastic range of SAE 1020 semikilled steel. A study of this relationship provides information as to the amount of reduction of area possible with the material in question and the stress required to obtain such a reduction.

Figure 13.4 illustrates curves showing the commercial cold-working range for representative metals. The common origin of these graphs is at the theoretical limit of annealing at which the yield point is zero. The theoretical yield point of commercial, fully annealed metal is at the left end of the heavy line. The strain-hardened limit (at potential fracture) is at the right end of the heavy line. The rate of strain hardening is indicated by the slope of the graph. Ductility of the annealed metal is shown by comparing the length of the heavy line with the total of the heavy line plus its right-hand extension.

In the process of cold-working (e.g., deep drawing) a large-diameter workpiece is compressed to a smaller diameter. Therefore, we need

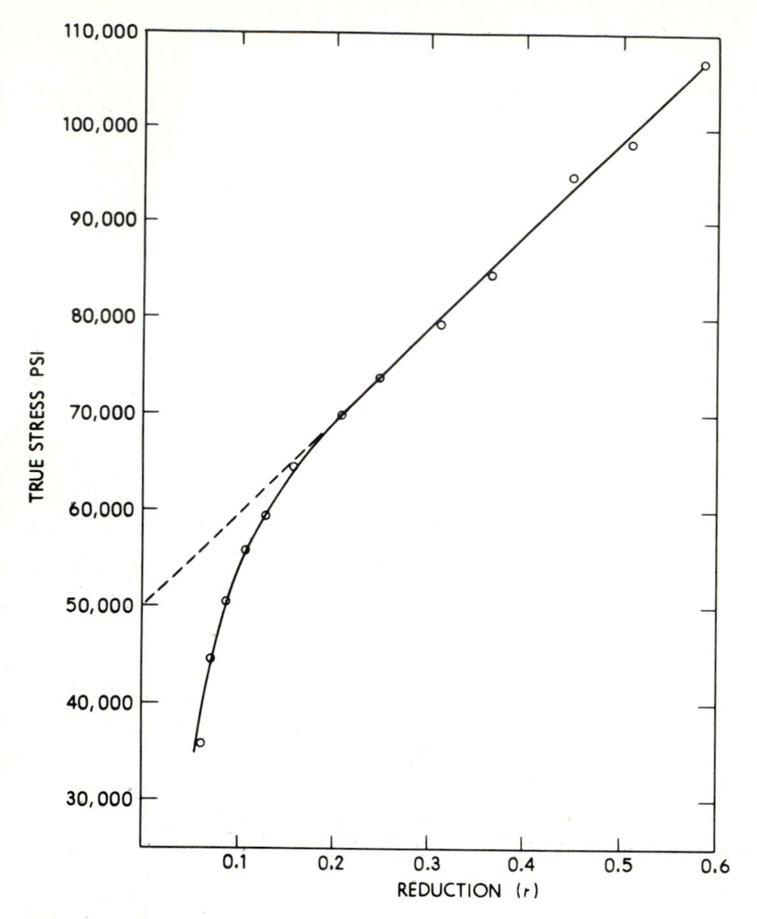

FIGURE 13.3
True stress versus reduction of area in plastic range of SAE 1020 semikilled steel.

compression-test data rather than tensile-test data to predict the behavior of the work material. Steel, at least up to the yield point, has the same properties in both tension and compression, but in deep drawing the deformation takes place beyond the yield point in the plasticity range. Figure 13.5 gives a portion of the stress–strain diagram plotted on logarithmic coordinates.

If a metal is reduced by compression by A percent and the load is removed, yield will not occur when a second load is applied until the stress reaches a new and higher stress σ_A lb/in.^{2} The difference between σ_A and σ_0 is known as the work hardening that occurs as the number of possible slip planes is reduced by the plastic deformation process. The stress σ_0 is known as the flow stress of the annealed alloy, and σ_A is the flow stress of the deformed material.

At the beginning of the reloading operation, we could use A as the

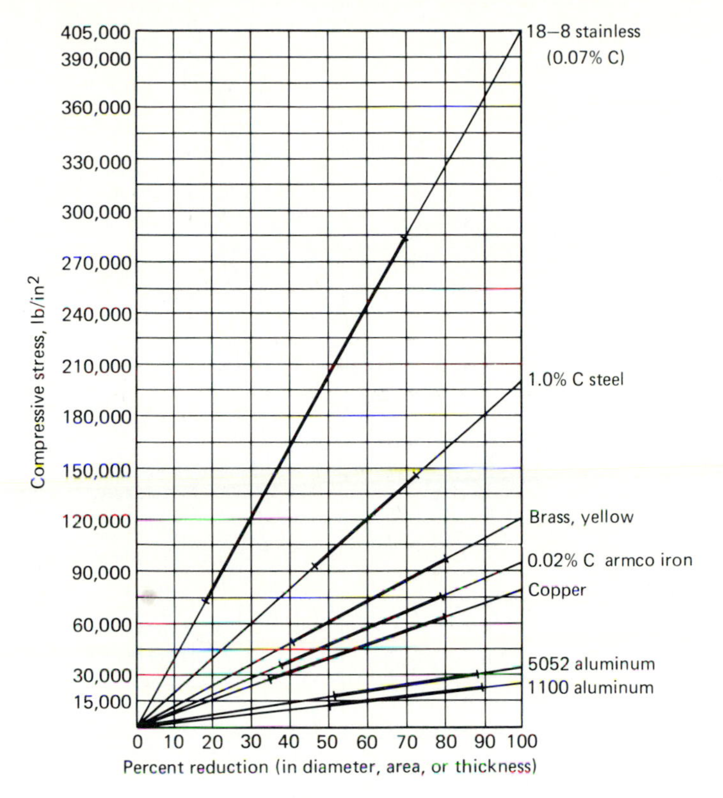

FIGURE 13.4
Tentative curves showing commercial cold-working range for representative metals.

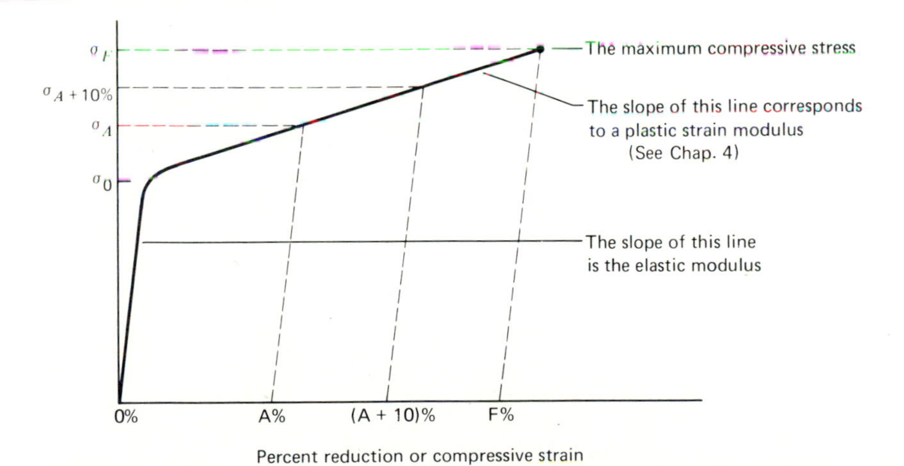

FIGURE 13.5
Compressive stress versus precent reduction curve for a given work-hardening alloy.

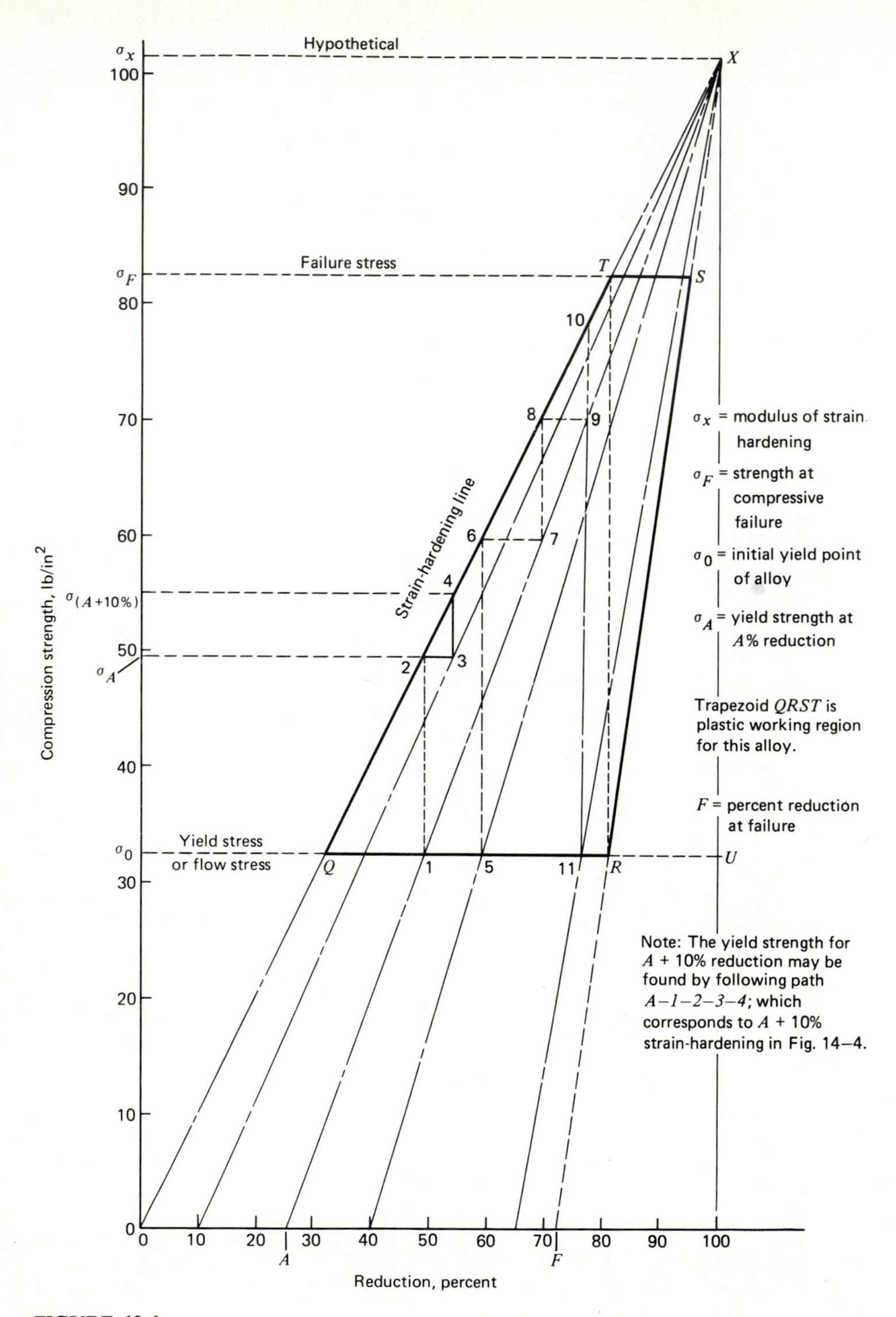

FIGURE 13.6
Typical work-hardenability graph.

starting point and have an additional 10 percent reduction based on A size (not percent). This then would be equivalent to a second draw operation of 10 percent when the first draw operation was A percent. The drawing force for the second draw would be determined by the yield stress at the beginning of the second draw operation, which is the same as the stress σ_A of the stress at the end of the first draw.

The above data are replotted in Fig. 13.6.

MAXIMUM REDUCTION IN DRAWING

The effect of cold-working may be estimated by analysis of the compressive plastic deformation. The percent reduction may be defined as:

$$\text{Reduction} = \frac{\text{initial diameter} - \text{final diameter}}{\text{initial diameter}} \tag{13.1}$$

or

$$R = \frac{D - d}{D}, \qquad R_T = \frac{d_0 - d_n}{d_0} \tag{13.2}$$

Figure 13.6 is a plot of the compressive yield strength of a material as a function of its percentage reduction. Note that trapezoid $QRST$ encloses the area in which plastic deformation can occur without fracture of the material. From the diagram it can be seen that the theoretical yield strength at 100 percent reduction can be found as follows:

$$\frac{\sigma_F - \sigma_0}{F\%\ \text{reduction}} = \frac{\sigma_x - \sigma_0}{100\%\ \text{reduction}} \tag{13.3}$$

$$\sigma_x = \sigma_0 + \frac{100\%\ \text{red.}}{F\%\ \text{red.}}(\sigma_F - \sigma_0) \tag{13.4}$$

Strain-hardening graphs can be constructed for the alloys shown in Fig. 13.4 as follows.

1. Plot the compressive yield stress as a function of the percent reduction from 0 to 100 percent as shown in Fig. 3.6.
2. Draw horizontal lines at σ_0 and σ_F, which are, respectively, the yield and failure strengths in compression.
3. Obtain σ_α by proportion using Eqs. (13.3) and (13.4) from triangles QRT and QUX. Note that any horizontal line within the plastic working region is divided similarly to the original X axis so that its length represents 0 to 100 percent reduction when it is extended to the 100 percent reduction line.
4. Construct the complete strain-hardening graph $QRST$ as in Fig. 13.6.

The strain-hardenability graph is used as follows. If an annealed material is reduced A percent, to find its new yield point start at the σ_0 line and make a horizontal line to the right until it hits the construction line from σ_x to A percent reduction, then rise vertically to the $0 - \sigma_x$ line; that intersection will be σ_A, the new yield stress of the material. If a second draw of $A + 10$ percent is to be made, repeat the former procedure, as in Fig. 13.6. After the second draw the yield stress will have risen to $\sigma_{A+10\%}$ as shown. To compute the value of σ_n, the unit stress at the end of any draw, use Eq. (13.4).

$$\sigma_n = \sigma_0 + R_T(\sigma_F - \sigma_0) \tag{13.5}$$

where

$$R_T = 1 - (1 - r_1)(1 - r_2) \cdots (1 - r_n) \tag{13.6}$$

and

$$r_i = \frac{d_{i-1} - d_i}{d_{i-1}}$$

PROPERTIES OF MATERIALS FOR SHEET-METAL WORKING

The operations described in this chapter are based on the assumption that the material has sufficient ductility that it will not break or crack during the operation. If failures occur, intermediate annealing operations and additional steps in the process may be necessary. To avoid rupture, the material is usually bent across grain between 45 and 90° (Fig. 13.7). When it is necessary to bend with the grain, a larger radius of bend or a softer material is used. A smooth surface facilitates the bending of material to a small radius. The harder the material, the greater the inside radius required. Annealed stock can usually be bent satisfactorily with a radius equal to the thickness of material. Material with rough edges, such as sheared materials, may crack at the edge; this cracking will not occur if the edges are smoothed or rounded.

Work-hardening steels such as stainless steels limit the amount of drawing and bending that may be performed in one operation on conventional presses. High-energy forming and high-speed presses are permitting the forming of work-hardening materials over a larger range of sizes.

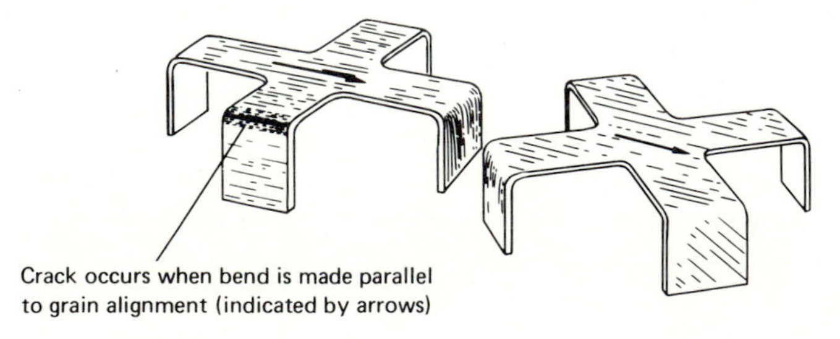

FIGURE 13.7
Mutually perpendicular bends should be made 45° to the grain orientation.

TABLE 13.1
Typical materials for press working

	Theoretical yield points		Reduction in area (maximum annealed), %	Shearing stress and percent penetration to fracture			
	Minimum* (commercially annealed), lb/in.^{2}	Maximum (severely cold-worked), lb/in.^{2}		Annealed to soft temper, lb/in.^{2}	%	Cold-worked to temper noted, lb/in.^{2}	%
Aluminum[a]							
No. 25, commercially pure	(4-)[b] 8,000	H 21,000	80	8,000	60	H 13,000	30
No. 35, Mn alloy	11,500	H 25,000	80	11,000		H 16,000	
No. 175 for heat treatment	HT 30,000						
No. 525, Cr alloy	21,000	H 37,000	80	14,000		H 36,000	
Brass							
Yellow, for cold working	(10-) 40,000	95,000	75	32,000	50	? 52,000	20
Forging, beta							
Bronze, Tobin alloy	25,000	120,000	53	36,000	25	$\frac{1}{4}$H 42,000	17
Copper	(10-) 25,000	62,000	65	22,000	55		
Gold							
Iron							
Cast, gray	8,000 tension 30,000 compr.						
cast, high test	15,000 tension 40,000 compr.						
Lead	(1-) 3,000	4,000		3,500	50		
Monel metal	28,100	H 112,000	63				
Nickel	21,000	H 115,000	75	35,000	55		
Silver							
Steel							
0.03 C Armco	(20-) 35,000	70,000	76	34,000	60		
0.15 C	(32-) 55,000	110,000	60	48,000	38	61,000	25
0.50 C	(55-) 70,000		45	71,000	24	90,000	14
1.00c	90,000	145,000	20–40	115,000	10	150,000	2
Electrical, high-silicon				65,000	30		
Stainless, low-carbon for drawing							
Endure AA	50,000	120,000	60				
18-8	(35-) 50,000	165,000	71	57,000	39		
High-carbon for cutlery, etc.			32				
Tin	(2-) 4,000	4,500		5,000	40		
Zinc	(12-) 28,000		43	14,000	50	19,000	25

[a] Old and new designations of wrought aluminum alloys: 2S = 1100; 3S = 3003; 17S = 2107; 52S = 5052.

[b] Figures in parenthesis represent laboratory test minimum, which might prove misleading for estimating loads. All physical values are subject to some variations with testing methods and analysis of material.

Source: E. V. Crane, *Plastic Working of Metals,* 3d ed., Wiley, New York, 1941.

Three factors govern the choice of a material for press-working (stamping) operations: strength, wear resistance, and corrosion resistance. Typical materials for press working are listed in Table 13.1. Once the alloy has been selected to meet the anticipated service demands, the production engineer must specify the commercial stock dimensions and the proper initial hardness to meet the forming requirements. Finally, he must determine the correct slit or sheared width to obtain maximum material utilization in the blanking and forming operations. Suppliers' handbooks and specification sheets can be consulted to determine the available sheet sizes, temper designations, and standard gages (thicknesses).

Low-carbon steel is by far the most commonly used material for press-working operations because of its good formability, economy, and weldability. For deep drawing, steels with 0.05 to 0.08 percent carbon and 0.25 to 0.50 percent manganese are suitable. In practice, the type of drawing operation is specified and a suitable steel is supplied by the mill.

For large-volume production, automatic presses with coil stock and roll feeders are needed. This automated setup eliminates the press operator and can be used for materials up to 0.044 in. thick on a routine basis. If roll straighteners are furnished, even thicker coil stock can be specified.

LUBRICATION

Lubrication is of two principal types: fluid and boundary (Fig. 13.8). Fluid lubrication is characterized by metal surfaces separated by a continuous film of lubricant having a thickness greater than the height of the surface variation of the two parallel metals. In boundary lubrication, there is some metal-to-metal contact and the lubricant film is a few molecules in thickness in those areas between contact of the two surfaces. In view of the high pressures and low speeds used, in most metal-forming operations lubrication is of the boundary type.

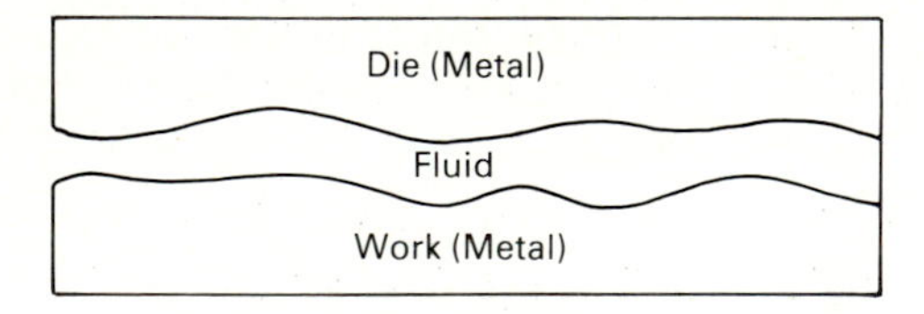

Fluid lubrication

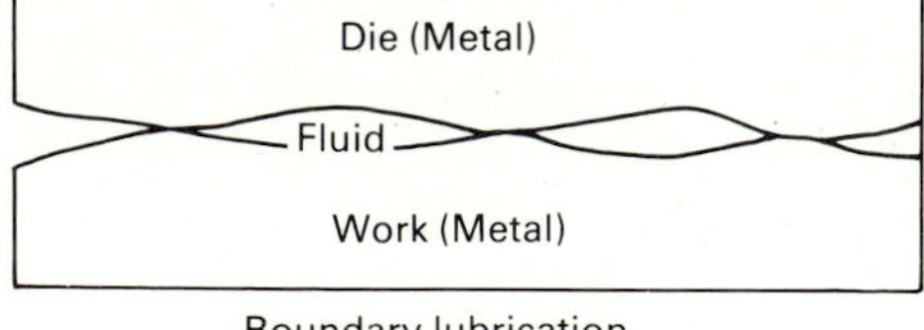

Boundary lubrication

FIGURE 13.8
Fluid and boundary lubrication.

It is important that lubrication take place in press-working operations in order to reduce the friction between the work material and the working surfaces of the dies. A correctly lubricated setup reduces the press tonnage required, increases die life, and gives more uniform work that is free from abrasions and scratches.

When specifying a lubricant, the engineer should select one that has sufficient surface tension and viscosity to adhere well and will spread evenly over the die surfaces. The lubricant may be either oil-soluble or water-soluble but should be nontoxic. It should not have a corroding or abrasive effect on the dies. It should be able to maintain a protective film during the highest pressure to which the work and dies will be subjected. Lubricants must be removable in commercial washing equipment; if not, cleaning costs may outweigh savings made in the press-working operations. Table 13.2 provides recommended lubricants for the principal materials used in metal forming.

In deep drawing, the selection of the proper lubricant is especially important in order to ensure uniform work and minimize rejects. Generally speaking, deep draws and draws entailing heavy gages of steel require a viscous lubricant, such as equal parts of white lead and animal or mineral oil, while shallow draws may be lubricated with light grease or mineral oil. If annealing is required after drawing, then in order to avoid burning of an oil residue into the metal surface, water-soluble lubricants are often preferred. These may include mixtures of water and chalk, soap, or zinc oxides.

Copper and its alloys are usually lubricated with water-soluble mixtures

TABLE 13.2
Commonly used lubricants in blanking and forming

Operation	Aluminum and aluminum alloys	Carbon and low alloy steels	Stainless steels	Copper and copper alloy
Blanking and piercing	Light mineral oils Soap–fat or mineral oil emulsions Mineral-fatty oil blends	Residual fatty oil from light mineral oil Mineral or fatty oil emulsions	Fatty oil emulsions Chlorinated mineral oils Soap–fat emulsions Dry soap films	Soap solutions Straight emulsions Mineral oils Soap–fat emulsions
Light forming	Kerosene Light mineral oils Petroleum jelly Mineral oil plus 10 to 20 percent fatty oil	Residual fatty oil Soap–fat emulsions Chlorinated oils Mineral oils with extreme pressure agents	Fatty oil emulsions Chlorinated mineral oils Soap–fat emulsions Dry soap films	Soap solutions Straight emulsions Mineral oils Soap–fat emulsions
Deep drawing	Mineral–fatty-oil blends. Undiluted fatty oils Undiluted chlorinated oils Pigmented soap–fat compounds Dry soap or wax films	Undiluted chlorinated oil Pitmented soap–fat compounds Sulfochlorinated oils Dry soap films	Chlorinated oil Pigmented soap–fat compounds Dry soap films Dry polymer coatings	Soap–fat emulsions Undiluted fatty oils Noncorrosive extreme pressure oils

such as soap chips and hot water. The deeper or more severe the draw, the greater the concentration of soap required.

GENERAL PRINCIPLES OF SHEARING

Shearing operations, including blanking, notching, parting, piercing, nibbling, trimming, and perforating, involve the shear strength of a material, which is related to but is less than its tensile strength. When a punch penetrates a sheet-metal workpiece and passes into the die (Fig. 13.9), a section through the workpiece shows a smoothly cut area and an angular fractured area (Fig. 13.10). The slug has four features: a plastically deformed and rounded edge, the burnished cut area, a fractured area that corresponds to that of the sheet stock, and a burr at the periphery of the side toward the punch (Fig. 13.10).

Clearance

There is a clearance c between the punch and die at any radial point, so that the difference between the diameter of the punch and die is $2c$. The magnitude of the clearance varies with the hardness, thickness, and properties of the alloy sheet. With too little clearance the initial fractures from the corners of the punch and die fail to meet (this is called *secondary shear*) and a greater shearing force is required. Too much clearance results in excessive plastic deformation. The ideal clearance minimizes plastic deformation and creates no secondary shear.

Since there is a difference in the diameter of the punch and die as a result of the clearance, the size of the slug is determined by the die and the hole is sized by the punch. Thicker and softer sheet metals require a clearance of 10 to 18 percent of the material thickness t; harder metals need less, that is, 6 to 10 percent of t. Harder metals have less plastic deformation and a smaller burnished area than soft metals. For a given metal thickness, the angle of the fracture area decreases with the increasing hardness of the workpiece (Fig. 13.11).

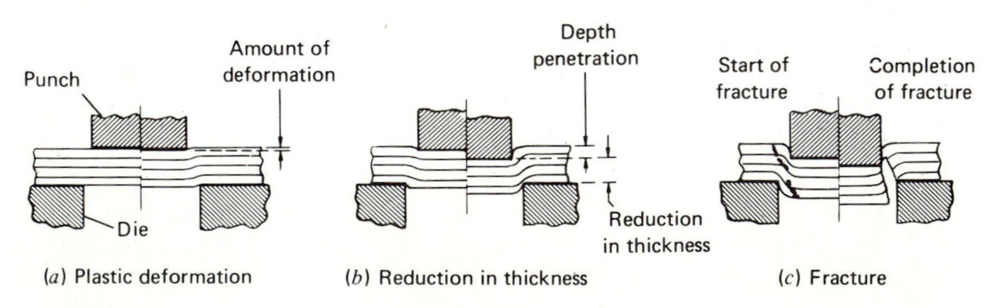

FIGURE 13.9
Several steps in a sheet-metal punching operation. (*From Society of Manufacturing Engineers,* Fundamental of Tool Design, *Prentice-Hall, Englewood Cliffs, N.J., 1962.*)

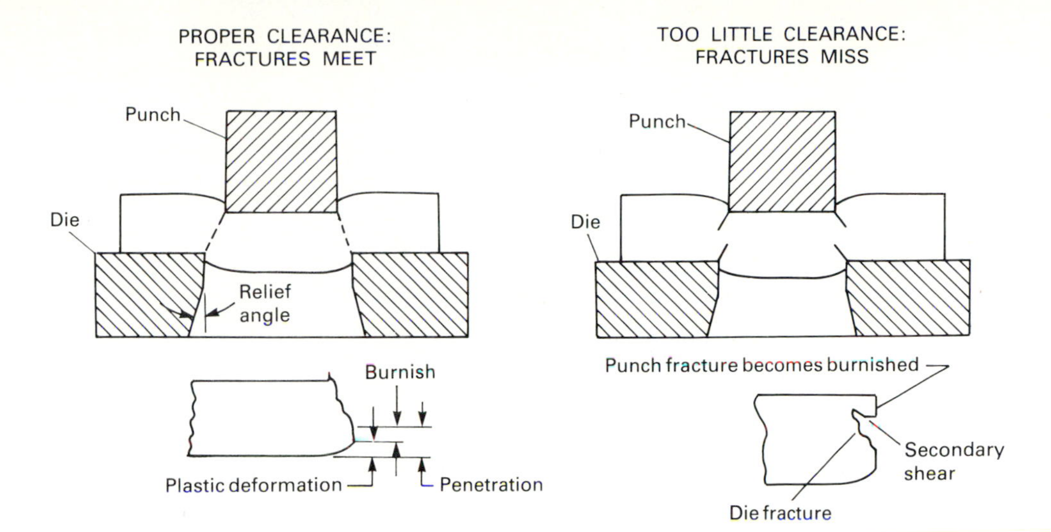

FIGURE 13.10
Cross sections of a typical punching operation showing the effect of clearance and the characteristics of the sheared slug.

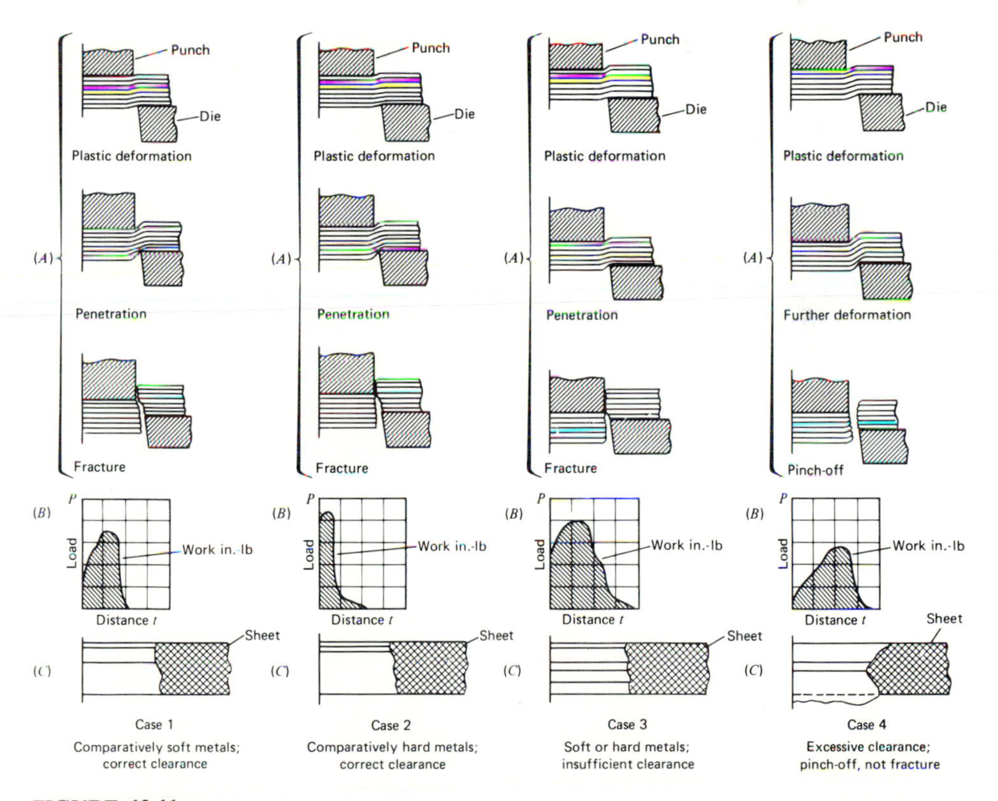

FIGURE 13.11
Effect of different clearances when punching hard and soft alloys. (*From Society of Manufacturing Engineers,* Fundamentals of Tool Design, *Prentice-Hall, Englewood Cliffs, N.J., 1962.*)

Shearing Force

The shear strength of sheet metal is generally about 60 to 80 percent of its tensile strength. In plain carbon steels the shearing strength is directly proportional to the carbon content (Table 13.3).

The force P required to shear sheet metal is

$$P = \pi DSt \qquad \text{(for round holes)}$$

$$P = \pi SLt \qquad \text{(for other contours)}$$

where S is the shear strength of material, lb/in.^{2}, D is the die diameter, in., L is the length of shear (i.e., circumference in a round punch and die set) and t is the thickness of material, inches.

Thus, the force required to punch a 2-in.-diameter hole in annealed SAE 1020 steel plate $\frac{1}{4}$ in. thick is:

$$P = 2\pi(44{,}000)(\tfrac{1}{4})$$
$$= 69{,}000 \text{ lb}$$

Penetration

The penetration of the punch into the workpiece before fracture occurs is the sum of the plastic deformation and the burnished thickness (Fig. 13.10). It is usually expressed as a percentage of the original sheet-metal thickness.

Reduction in Press Load

In many cases the calculated punch load will exceed the capacity of the available press, especially if several punches are shearing simultaneously. Since the press loading is characterized by high force demand in a short time span,

TABLE 13.3
Typical shearing strength for annealed and cold-worked steel

Carbon content	Annealed, lb/in.^{2}	Penetration, percent	Cold-worked, lb/in.^{2}	Penetration, percent
1003	34,000	60		
1010	35,000	50	43,000	38
1015	39,000	45	51,000	33
1020	44,000	40	55,000	28
1030	52,000	33	67,000	22
1050	71,000	24	90,000	14
1% C	110,000	10	150,000	2
Fe 3% Si	65,000	30		
18–8 St. Steel	57,000	39		
SAE 4130	55,000			

the cutting force may be reduced by spreading the energy required over a longer time span, i.e., a longer portion of the total ram stroke. A reduction in the energy demand may be accomplished in two ways.

1. Stagger the length of small punches by increments of the sheet thickness so that only a few are punching at any one time.
2. Add shear (Fig. 13.12) to the punch or die, thereby reducing the area in shear at any one time. The punch load may be cut in half. The change in punch face is about $1\frac{1}{2}$ times the stock thickness. If a punch is large, a double shear is preferable to a single shear because the lateral forces are balanced. Apply the shear to either the punch or the die in such a way that any distortion will appear in the scrap from the punching operation; i.e., the shear will be on the punch. In piercing, the direction of the shear angles should be such that cutting is initiated at the outer extremities of the

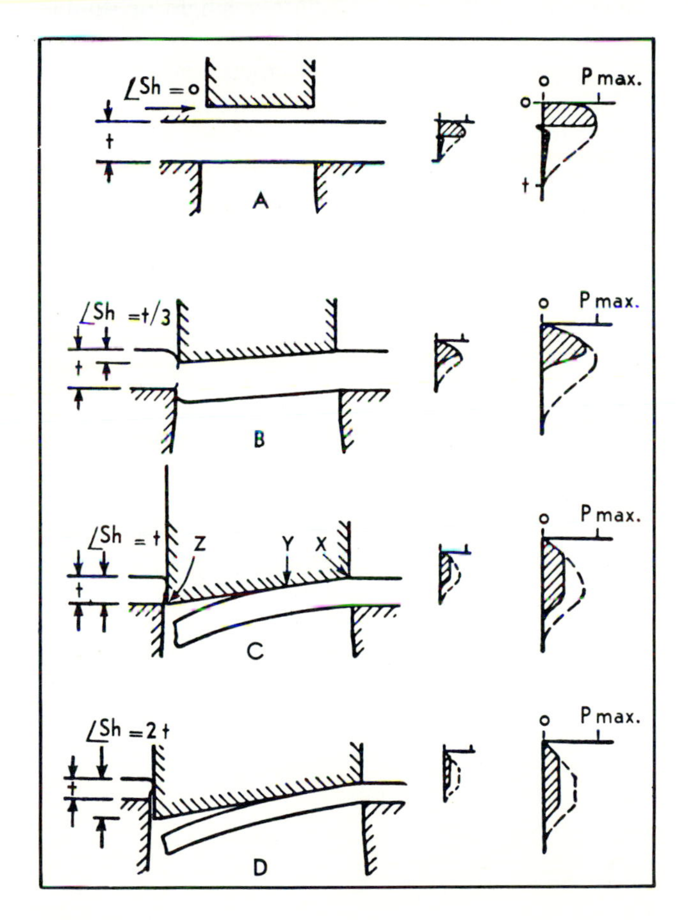

FIGURE 13.12
Effect of shear on the maximum force required for punching sheet metal.

contour and move toward the center in order to avoid stretching the material before it is cut free.

The blanking force is in practice reduced by the penetration allowance for various alloys and states of cold work as given in Table 13.1. It can be reduced even more by grinding an angle called shear on the face of the punch or die. When shear is used, the magnitude of the maximum force is reduced up to 50 percent (Fig. 13.12), but the cutting operation requires a longer time, in view of the progressive cutting.

Shear on the die transfers distortion to the scrap and the blank is undeformed, whereas shear on the punch results in distortion of the slug.

Shear, when defined as kt, is determined by the thickness of the sheet metal, where k usually ranges from 1.5 to 2. If the thickness of the sheet metal is more than $\frac{3}{32}$ in., shear is usually applied in two directions to balance the forces on the workpiece and then should be equal to $2t$. Analysis shows that the magnitude of the punching force is the same for either single or double shear. Figure 13.13 provides examples of shear.

Consider the previous example in which a 2-in.-diameter hole is to be punched in annealed SAE 1020 steel, $\frac{1}{4}$ in. thick. Determine the punching force if penetration is considered. From Table 12.3, the shearing strength is 44,000 lb/in.^{2} and the penetration is listed as 40 percent. Thus the punching force is:

$$
\begin{aligned}
P &= \pi DStp \\
&= (3.14)(2)(40{,}000)(\tfrac{1}{4})(0.40) \\
&= 25{,}120 \text{ lb}
\end{aligned}
$$

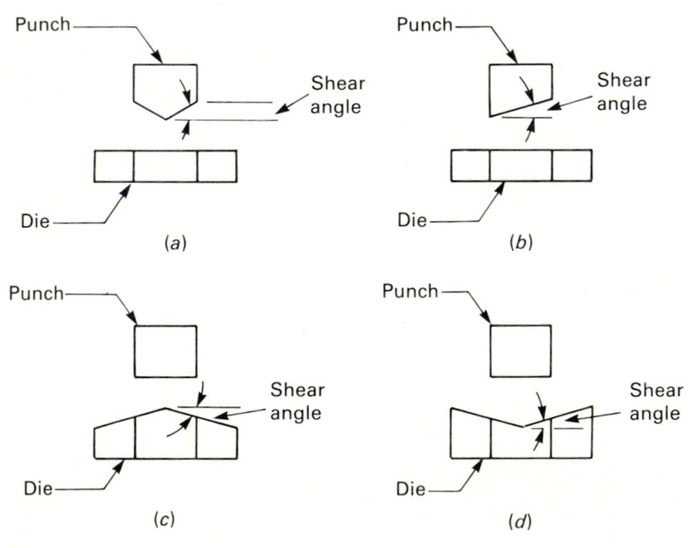

FIGURE 13.13
Representative examples of single and double shear.

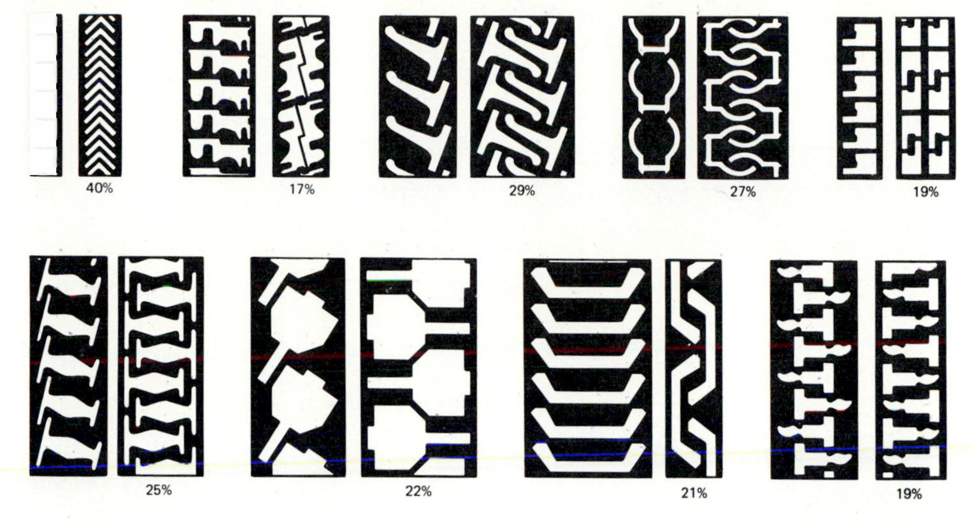

FIGURE 13.14
Several examples of nesting for material saving.

Location of Blanks on the Stock for Economy

The die designer and production engineer must consider the location and spacing of the blanks on the stock. Three factors must be considered: (1) blank location for economy (best use of material), (2) proper location for bending, if required, and (3) location, so that the summation of the shearing forces is symmetrical about the centerline of the ram stroke.

Cardboard templates have proved to be helpful in laying out the strip stock to obtain the maximum economy, but blanks cannot be closer than $\frac{1}{32}$ in. or the stock thickness, whichever is greater. A number of cost-saving layouts are possible (Fig. 13.14). The waste skeleton should be stiff enough to hold together; otherwise waste disposal will be difficult.

The location of the center of pressure may be carried out mathematically as would be done for calculating the center of gravity, but calculations are usually long and tedious. It is adequate to locate the center of pressure within $\frac{1}{2}$ in. of the true center of a punch configuration. A simple procedure may be used to determine the center of pressure experimentally. Balance a wire loop that has been bent to the configuration of the blank or blanks that are made in one stroke, across a pencil in both the x and y coordinate directions. The intersection of the two axes is approximately the center of pressure.

PIERCING AND BLANKING OPERATIONS

The reader should understand that piercing is analogous to blanking, except that piercing involves the cutting of holes in sheet metal so that the metal that

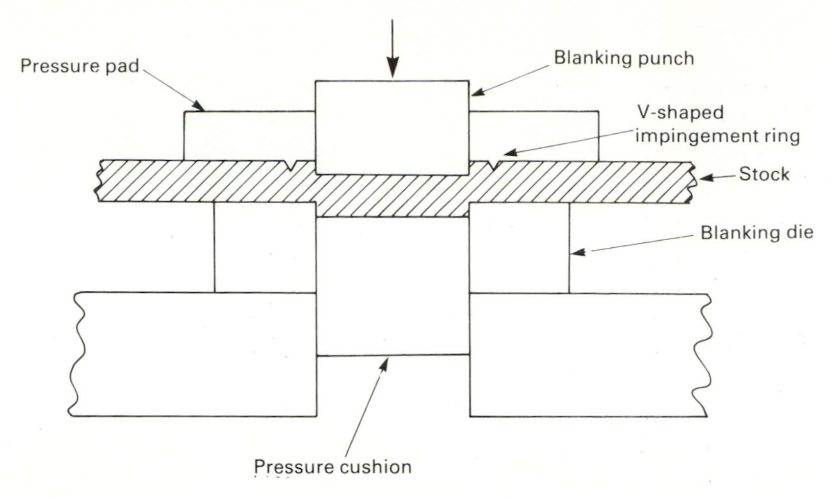

FIGURE 13.15
Fine blanking resulting from extrusion of finished part out of metal strip.

surrounds the piercing punch is the workpiece and the punched-out slug is scrap, while in blanking it is the workpiece that is punched out. As explained in the various shearing operations, there is a fractured edge resulting from the breaking and tearing away of the metal that was not cut during the initial part of the blanking or piercing operation.

In order to produce precise blanks in a single operation without fractured-edge characteristics, the process known as fine-edge blanking or piercing can be used. Here, a V-shaped impingement ring is forced into the stop to lock it tightly against the die and to force the work metal to flow toward the punch so that the part can be extruded out of the strip without fracture. Little die clearance is used and the punch velocity is much slower than in conventional blanking or piercing (see Fig. 13.15).

Production Shearing

Power shears. In shearing material by means of shear blades, the edge of the movable blade is usually at a slight angle to the edge of the stationary blade.

The movable blade usually strikes the material at one edge and advances progressively to the opposite edge. Thus, only a portion of the piece is being sheared at one time. In this way the blade does not carry the load of shearing the entire cross section at once. If this were done, the operation would be called *blanking* or *straight shearing*. In all shearing operations the part under the movable blade is distorted, twisted, or bent. The portion lying on the stationary blade remains straight and may have a burr on the lower edge. When the tool designer says that shear is being placed on a die or punch, it means that the material is being cut progressively along the edge of the die and punch. This lightens the load on the punch and press but distorts the blank or part, depending upon where the shear is placed.

In sheet-metal work and in the making of formed stampings, it is often necessary to cut sheets into blanks having irregular contours or circular shapes. To avoid expensive blanking dies, blanks can be made by slitting, uni-shearing, nibbling, dinking, contour sawing, turning, routing, and milling.

Slitting. Slitting is shearing by circular shears, 4 to 6 in. in diameter, which draw the material through the shears as they rotate. The rollers overlap just enough to shear the material. Excessive overlap distorts the material. Shearing rolls are spaced any desired distance apart on rotating shafts and can shear a sheet or a strip into as many widths as desired. With continuous strip mills furnishing sheet and strip in long lengths, many parts (such as transformer punchings) can be made continuously from slit material. Slitting is usually performed on metal less than $\frac{1}{8}$ in. thick. Circular blanks are sheared by circular shears mounted in a heavy frame. The plate is pivoted at the center and the circular shear blades are forced into the material. The edge is sheared or trimmed to form a circular blank. Depending on the capacity of the equipment, the thickness of material sheared may be as great as $\frac{3}{8}$ in. The material trimmed off is distorted.

Job-Shop Methods of Shearing

Uni-shearing. Uni-shearing is a trade name given to a machine that has a short shear mounted on a portable hand tool or in a frame having a deep throat. The Uni-shear can shear light metal to any shape by guiding the material through the shear or by guiding the portable shear through the material. The material touched by the movable blade is distorted. This is a handy machine for sheet-metal work of gages less than $\frac{1}{16}$ in. thick.

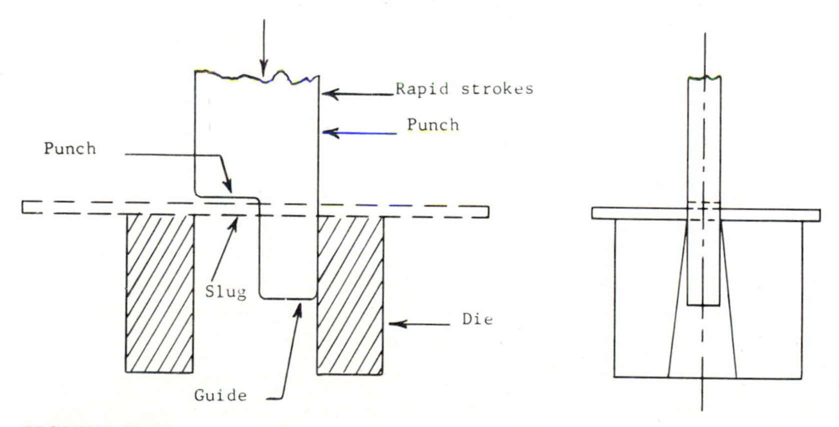

FIGURE 13.16
Typical step punch and die for use in nibbling.

Nibbling. Nibbling is separating material by continuously punching a slot in any direction desired. The nibbling punch and die (Fig. 13.16) are mounted in a press with a deep throat. A support for a circular blank center is usually provided. Any shape hole may be punched. Some nibblers require a starting hole when the hole to be cut is within the sheet. The nibbler consists of a step punch and die. The tip of the punch remains in the die and the rest of the punch cuts a slug. As the punch moves up and down, the tip, which remains in the die, guides the material. Thus a fairly smooth edge is produced and there is no distortion of either part.

Job-Shop Punching and Blanking

Dinking. Dinking or ruler dies are made with steel-cutting flexible-rule material. The die functions like a cookie cutter (Fig. 13.17). Thin aluminum, stainless steel up to 0.040 in. thick, leather, celluloid, linoleum, and cardboard are blanked by this method. The steel rule used to blank metal is hardened after it is bent to shape and then clamped firmly by plywood pieces previously sawed with a jigsaw having the thickness of the cutting rule. The rule is supported its full width except for a 0.02-in. extension. A layer of cork is glued on the inside of the cutter to strip the material from the blade. The assembled die is placed face-up on the bed of the press. The material to be cut is placed on top, and the steel platen of the press comes down and forces the material onto the cutter. Proper spacers are placed between the bed of the press and the platen to prevent the platen from striking the cutters. Material is cut all the way through except for a slight amount which is sheared at the end of the stroke. Wherever these dies can be used, they have proved to be the least expensive method of blanking material when quantity is limited.

Quick-process or continental method. The production-design engineer can have tool-made samples furnished by the continental or quick process in his own toolroom, or he can purchase them from companies that specialize in making small quantities of tool-made parts. Their operators are trained in

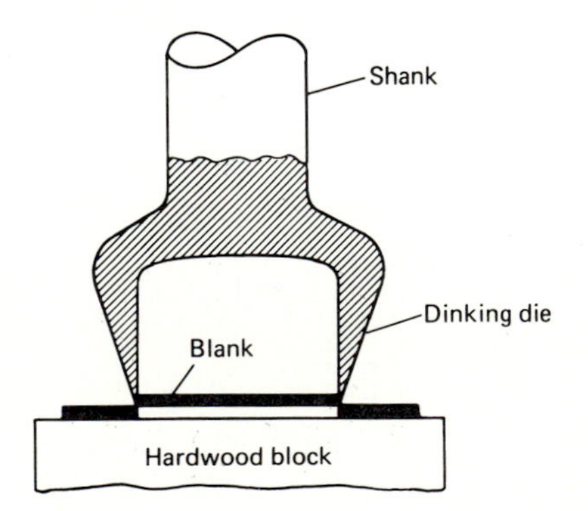

FIGURE 13.17
Dinking die and hardwood die block.

devising ingenious ways of producing blanks that can be punched and formed to the desired shape. Holes are pierced singly by setting gages at the proper distance and punching each piece separately. Standard notching dies for notching 90, 45, and 30° corners are available as are oblong hole punches and dies, so that a wide range of holes can be made in blanks. It is not difficult to make a single odd-shaped punch and die for a hole or blank. The blanking die is made by sawing the desired shape in a steel plate used as a die. The punch is made from the portion sawed out. A guide plate for the punch is sawed to the same shape. It is sometimes sawed together with the die plate. The die is relieved and brought to final shape by peening and by filing. The punch blank is peened out and filed to fit the die. Both may be hardened by applying a slight case. The guide plate is mounted above the die so that the material may be placed between them. The punch is placed in the guide plate on top of the material and the whole unit placed in a press with a flat bed and platen. The flat platen strikes the punch and drives it through the material. Both drop down into the die. The whole set is removed from the press, the blank and punch are removed, and the process is repeated (Figs. 13.18 and 13.19).

Punching and blanking structural steel and plate. There are two general classes of punching, blanking, and stamping work: structural and sheet metal. The first refers to work done on rough, hot-rolled bars, structural members, and thick plates. Extreme accuracy is usually not required or obtainable without excessive costs. Round or oblong holes can be punched in flanges of

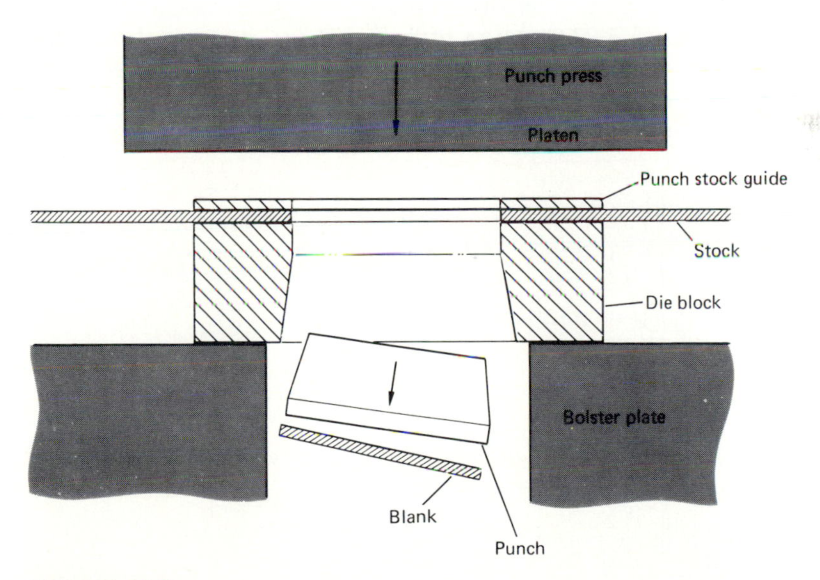

FIGURE 13.18
Simple punch-and-die tooling for the continental method of low-cost hole production in sheet metal.

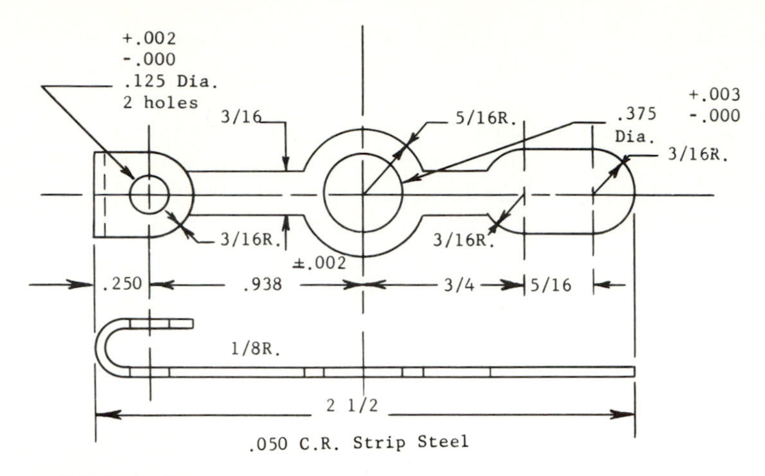

FIGURE 13.19
Typical part made by the continental process.

channels and *I* beams, the bevel portion resting on a bevel die, as shown in Fig. 13.20. Presses for these operations are called structural or plate punch presses. The dies and punches can be quickly changed. Some presses have more than one punch in the press. The desired punch is actuated by moving a gag block between the ram and punch. In plate or girder punching the machines may have 50 or more punches that are set for actuation by an operator according to a template or drawing. The material is moved under and through the punch by a carriage that draws the plates or structural members the proper distance for each hole or series of holes. These gang punches make riveted structures economical to fabricate and place them in competition with welding.

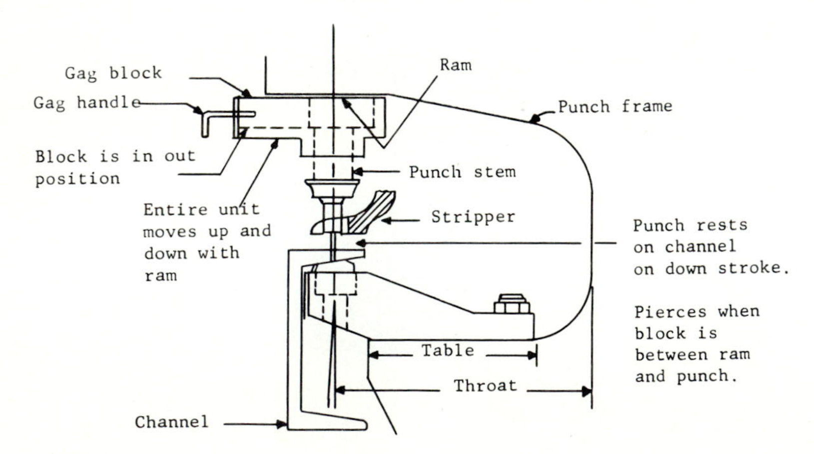

FIGURE 13.20
Heavy-duty press setup for punching holes in structural steel.

Progressive Dies

When the production volume warrants the cost, a single progressive die should be designed and constructed to carry out all the operations required to complete a part automatically in one press. Production of the aluminum cooling fins as used in the evaporator and condenser coils of an air-conditioning unit is a good example of the use of progressive dies. A cross section through a progressive die and the strip for making a notched blank (Fig. 13.21) shows the sequence of operations as a progressive die is used to carry out four cutting operations in sequence to make a notched blank at each stroke of the press.

The cost of a progressive-die setup is high, because it must assume a share of the cost of a fully automatic press with cutoff, feed rolls, straightening

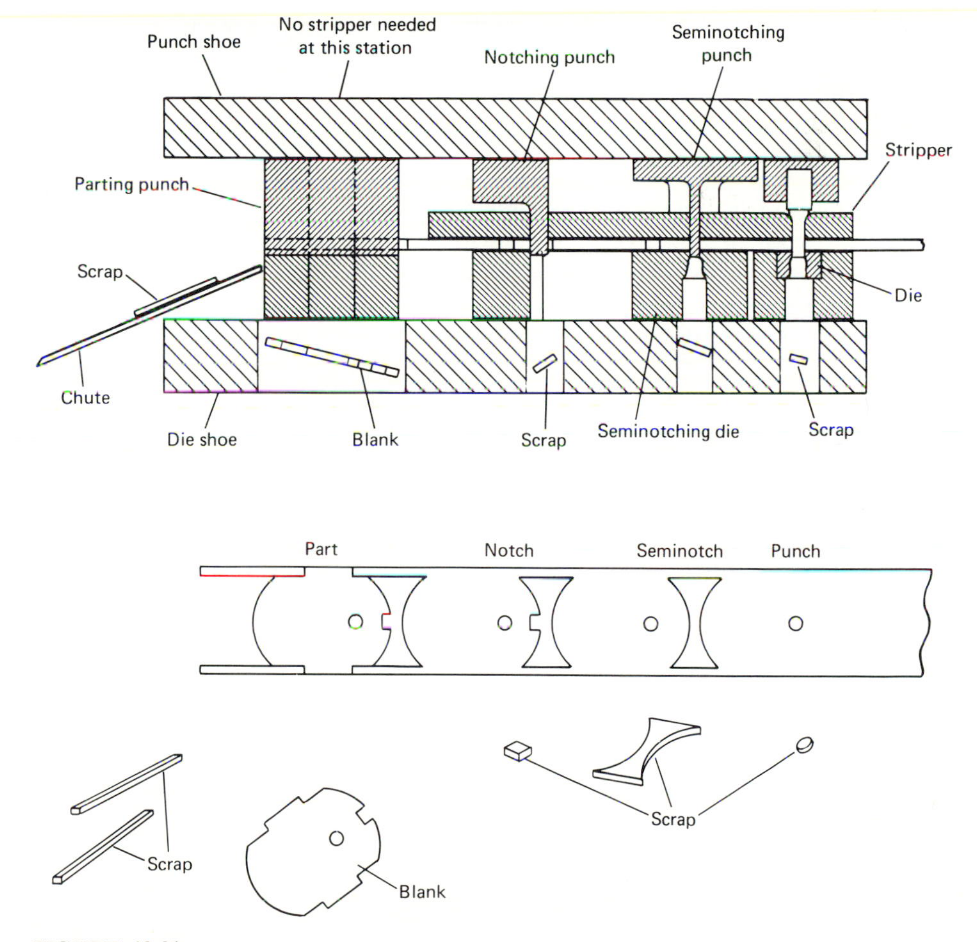

FIGURE 13.21
Typical four-station progressive die for producing a notched blank.

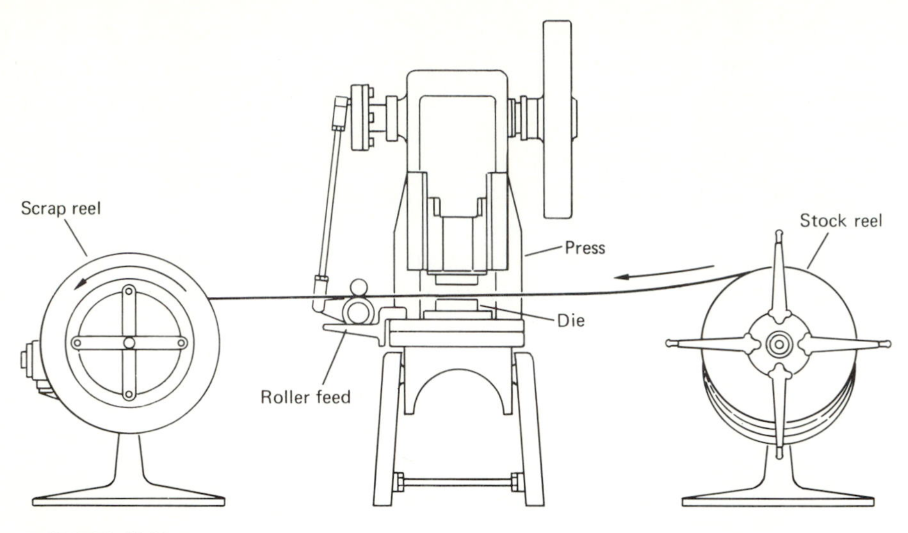

FIGURE 13.22
Roller-feed device for pulling strip off a stock reel and through a progressive die.

rolls (if required), and a scrap reel if needed (Fig. 13.22). In addition, the die design and construction costs are higher than the cost of a series of single-operation dies for making the same part. Provision must be made for obtaining the best sequence of operations at the design stage because once a die is built it is rather difficult to make changes.

On the other hand a number of savings are possible. An operator is required only when a coil of strip stock must be changed on the feed and scrap reels or when there a malfunction such as when a punch breaks or the burrs become excessively large. The latter means that the dies need to be resharpened. But the major saving in addition to direct labor is in the reduction in material handling and the amount of production control required. Since all four operations are combined, the material is handled only once and in bulk as the coiled raw material or in the tub of blanked parts to in-process stores. If made in four operations, three additional paperwork orders, three more setups, and three more material handlings would be required to make the same number of parts in the in-process storage area. There would be additional costs in inspection and supervision. Thus, it is often prudent to consider the use of progressive dies for parts that are to be made in lots of even 10,000 provided there are anticipated to be several years of production before the part becomes obsolete.

Dies for Piercing and Blanking Operations

Considerable ingenuity and engineering development in tools and dies have made possible the high-quality and economical sheet-metal products of today.

The quality and the form of the part determine whether an expensive or simple tool will be required. This is under the control of the production-design engineer. The proper use of the following factors will aid the tool designer in producing an effective tool and quality part.

1. The punch determines the size of the hole and the die determines the size of the blank.
2. Punches should always be as short as practical. Long production runs require punches long enough to grind from $\frac{1}{2}$ to 1 in.
3. All delicate projections or weak parts in a die should, if possible, be designed as inserts.
4. To minimize breaking of slender punches, taper the working end of the punch 0.001 to 0.002 in. This will facilitate stripping, and most breakage occurs during stripping.
5. A die will usually have a life of approximately two to three times the life of the punch or punches.
6. On a blanking die, the burr is toward the punch. Burrs from piercing are on the die side.
7. There is always clearance between the die and punch. Clearance increases with increase in thickness of material. The clearance should be subtracted from the die when designing the punch for blanking and should be added to the die opening when designing the punch for piercing. A good general clearance for blanking dies is 6 percent of the stock thickness on a side between punch and die.
8. The thickness of material should not exceed the diameter of the punch. Greater thicknesses can be punched by using good tool design, but at a sacrifice to tool life.
9. Distortion of holes occurs if they are punched too close to the edge of the sheet, too close to each other, too close to a bend, or in a drawn portion. Holes should be punched after bending and drawing to ensure accurate size and location.
10. Material should be bent across the grain at least up to 45°, especially for sharp bends, so as to avoid cracking of the material. The "grain" is the direction the stock was drawn in passing through the rolls in manufacture.
11. The size of the sheet or strip material should be considered in designing the blank so that a minimum of loss in material is incurred.
12. The quantity of a part to be made determines the quality and complexity of the tools that can be built. A very active part—100,000 to 1,000,000 parts a year—may warrant elaborate progressive dies in which many operations, such as piercing, embossing, forming, blanking, and shearing, may be performed progressively in the die without the part being touched by the operator. The cost of picking up a part, placing it in the die, and removing it is much greater than the actual punching operation by the

TABLE 13.4
Comparison of die productivity and relative cost

Ranking	Type of die construction	Expected production (no. of pieces)	Relative cost (tool steel = 1)
1	Carbide construction	6×10^6	3.00
2	Tool steel (heavy duty, 2 in. thick)	1×10^6	1.00
3	Tool steel (light duty, 1 in. thick)	1×10^6	0.50
4	Flame hardened ($\frac{3}{8}$ in. thick)	0.5×10^6	0.33
5	Template dies ($\frac{1}{16}$ in. thick)	0.2×10^6	0.13
6	Rubber dies	0.1×10^6	0.06
7	Continental dies	1–1,000	0.01
8	Stock dies	1–1,000	0.01

machine. However, as activity of the part decreases so does the economy of making progressive dies; operations are performed at a lower total cost by providing simpler tools or standard tools.

The cost of dies for punching and blanking can vary by as much as 300 to 1 (Table 13.4) when the simplest stock die setup is compared to a complex tungsten carbide die for blanking transformer laminations from 3 percent silicon iron.

FORMING PROCESSES

Forming processes are primarily bending or stretching operations that do not materially change the thickness of the metal. The original material remains about the same except that the thickness or diameter may be reduced by drawing or ironing. This lack of change in thickness is in contrast to forging operations which change the original shape of the plastic material by squeezing, pressing, or rolling.

Forming is based on two priciples.

1. *Stretching and compressing the material beyond the elastic limit on the outside and inside of the bend.* When this condition of compression and tension exists, there is always some springback that frequently varies from piece to piece or from one lot of material to another, unless the shape of the part locks the stresses and does not permit movement after forming.

2. *Stretching the material beyond the elastic limit without compression, or compressing the material beyond the elastic limit without stretching.* When this is done, the springback is reduced to a negligible amount regardless of shape.

The first principle is used in pan- and press-brake forming, punch-press forming, and roll forming. The second principle is used in spinning, stretcher leveling, stretch forming, contour forming, and deep drawing. The processes described may use a combination of these two principles.

Simple Bends

When a wire is bent by hand, it takes a natural bend that is governed by the shape of the thumb and finger and the forces applied to the wire. Usually, when a bend is made its radius is governed principally by the tools making the bend. One leg may be clamped in stationary jaws and a form tool bends the material, as in tube bending, wire bending, pan-brake bending (Fig. 13.23), and dies with clamping pads.

Bending Allowance

The minimum bend radius varies with the alloy and its temper; most annealed sheet metals can be subjected to a bend that has a radius equal to the stock thickness without cracking. The more ductile metals can easily be bent back through a 180° bend, as in a hemming operation.

However, when sheet metal is bent, the total length including the bend is greater than the original stock. This change in length must be considered by

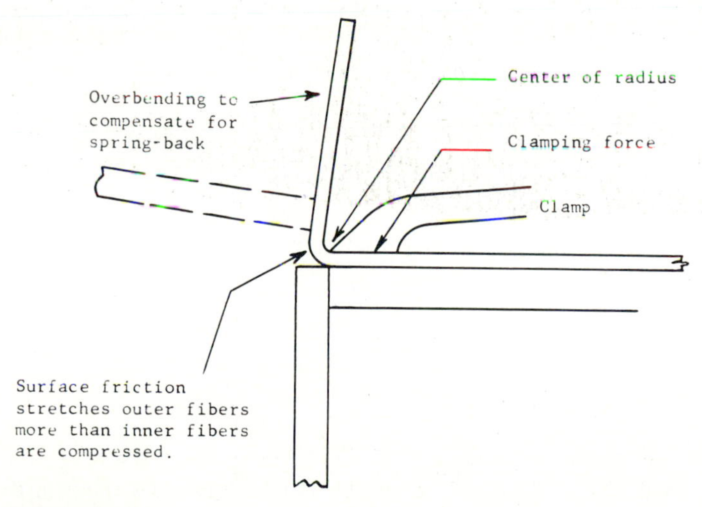

FIGURE 13.23
Pan-brake bender.

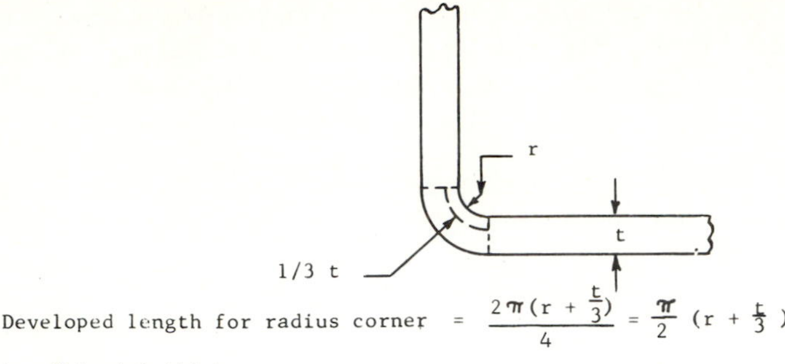

t = Material thickness
r = Inside radius

FIGURE 13.24
Calculation of the length of blank for a part with a 90° bend.

the production engineer and die designer because the length of the sheared stock must be known in order to shear the stock.

In the flat position, the neutral axis of a piece of sheet metal coincides with its centerline. But in a bent position, the neutral axis has shifted to a position 0.33 to $0.40t$ from the inner radius (Fig. 13.24). A relationship for determining the length of the developed blank is given below in Fig. 13.24, where the length of the neutral axis is calculated from the circumference of the quadrant of a circle with a radius of $r + t/3$.

When developing the length of a complex part, first divide it into a series of straight sections, bends, and arcs. Trigonometry can be used to calculate unknown dimensions, but keep the legs of the triangle parallel to the dimension lines, because the hypotenuse is then the bend angle and the length of the legs can easily be added to or subtracted from the blueprint dimensions.

The length of the bent metal can be calculated from the following empirical relationship:

$$B = \frac{A}{360} 2\pi(IR + kt)$$

where B is the bend allowance, in. (along the neutral axis), A is the bend angle, degrees, IR is the inside radius of bend, in., t is the metal thickness, in., and

$$k = 0.33, \qquad IR < 2t$$
$$0.50, \qquad IR > 2t$$

The Glenn L. Martin Company has published a bend allowance chart (Table 13.5) that was developed primarily for the aluminum sheet used in the aircraft industry, but it should be applicable to other alloys. The notes at the bottom make the chart self-explanatory. The chart is based on the empirical

TABLE 13.5

Table for finding the developed length of 90° bends[a]

Birmingham wire gage	Radius $\frac{1}{32}$	$\frac{3}{64}$	$\frac{1}{16}$	$\frac{5}{64}$	$\frac{3}{32}$	$\frac{7}{64}$	$\frac{1}{8}$	$\frac{9}{64}$	$\frac{5}{32}$	$\frac{11}{64}$	$\frac{3}{16}$	$\frac{13}{64}$	$\frac{7}{32}$	$\frac{15}{64}$	$\frac{1}{4}$	$\frac{9}{32}$	$\frac{5}{16}$	$\frac{11}{32}$	$\frac{3}{8}$	$\frac{7}{16}$	$\frac{1}{2}$
0.330																	0.499	0.512	0.525	0.549	0.579
0.284																0.466	0.479	0.492	0.506	0.529	0.557
0.239															0.422	0.435	0.449	0.462	0.476	0.499	0.529
0.238														0.390	0.396	0.410	0.423	0.437	0.450	0.473	0.504
0.220													0.361	0.368	0.374	0.388	0.401	0.415	0.428	0.451	0.482
0.203												0.333	0.340	0.347	0.354	0.367	0.381	0.394	0.408	0.431	0.461
0.180											0.299	0.306	0.313	0.319	0.326	0.339	0.353	0.366	0.379	0.403	0.433
0.165										0.274	0.281	0.288	0.294	0.301	0.308	0.312	0.335	0.348	0.361	0.385	0.415
0.148									0.247	0.254	0.260	0.267	0.273	0.280	0.287	0.300	0.314	0.327	0.340	0.364	0.394
0.134								0.223	0.230	0.237	0.243	0.250	0.257	0.263	0.270	0.283	0.297	0.310	0.324	0.347	0.377
0.120							0.199	0.206	0.213	0.220	0.226	0.233	0.240	0.246	0.253	0.266	0.280	0.293	0.307	0.330	0.360
0.109						0.179	0.186	0.193	0.199	0.206	0.213	0.220	0.226	0.233	0.240	0.253	0.263	0.276	0.243	0.317	0.347
0.095					0.156	0.162	0.169	0.176	0.182	0.189	0.200	0.203	0.204	0.216	0.223	0.236	0.250	0.263	0.276	0.300	0.330
0.083				0.135	0.141	0.148	0.154	0.161	0.168	0.175	0.181	0.188	0.195	0.201	0.208	0.222	0.235	0.248	0.262	0.285	0.315
0.072			0.144	0.121	0.128	0.134	0.141	0.148	0.155	0.161	0.172	0.175	0.181	0.188	0.195	0.208	0.222	0.235	0.248	0.272	0.302
0.065		0.100	0.106	0.113	0.119	0.126	0.133	0.139	0.146	0.153	0.164	0.166	0.173	0.180	0.186	0.200	0.213	0.226	0.240	0.263	0.293
0.058	0.084	0.092	0.098	0.104	0.111	0.117	0.124	0.131	0.138	0.144	0.155	0.158	0.164	0.171	0.178	0.191	0.205	0.218	0.231	0.255	0.285
0.049	0.073	0.081	0.087	0.093	0.100	0.107	0.113	0.120	0.127	0.133	0.144	0.147	0.153	0.160	0.167	0.180	0.194	0.207	0.220	0.244	0.274
0.042	0.065	0.072	0.078	0.084	0.091	0.098	0.105	0.111	0.118	0.125	0.136	0.138	0.145	0.152	0.158	0.172	0.185	0.198	0.212	0.235	0.266
0.035	0.056	0.064	0.070	0.076	0.083	0.090	0.096	0.103	0.110	0.116	0.127	0.130	0.136	0.143	0.150	0.163	0.177	0.190	0.203	0.227	0.257
0.032	0.053	0.060	0.066	0.072	0.079	0.086	0.093	0.099	0.106	0.113	0.119	0.126	0.133	0.139	0.146	0.160	0.173	0.186	0.200	0.233	0.253
0.028	0.048	0.055	0.061	0.068	0.074	0.081	0.085	0.094	0.101	0.108	0.114	0.121	0.128	0.135	0.141	0.155	0.168	0.182	0.195	0.218	0.249
0.025	0.044	0.052	0.057	0.054	0.071	0.077	0.084	0.091	0.097	0.104	0.111	0.118	0.124	0.131	0.138	0.151	0.164	0.178	0.191	0.214	0.245
0.022	0.040	0.048	0.054	0.060	0.067	0.074	0.080	0.087	0.094	0.100	0.107	0.114	0.121	0.127	0.134	0.147	0.161	0.174	0.188	0.211	0.241
0.020	0.038	0.045	0.051	0.058	0.065	0.071	0.078	0.085	0.091	0.098	0.104	0.111	0.118	0.125	0.132	0.145	0.158	0.172	0.185	0.208	0.239
0.018	0.035	0.043	0.049	0.055	0.062	0.069	0.076	0.082	0.089	0.096	0.102	0.109	0.116	0.122	0.129	0.143	0.156	0.169	0.183	0.206	0.236

[a] Subtract the correct figure in the table from the sum of the length of legs.

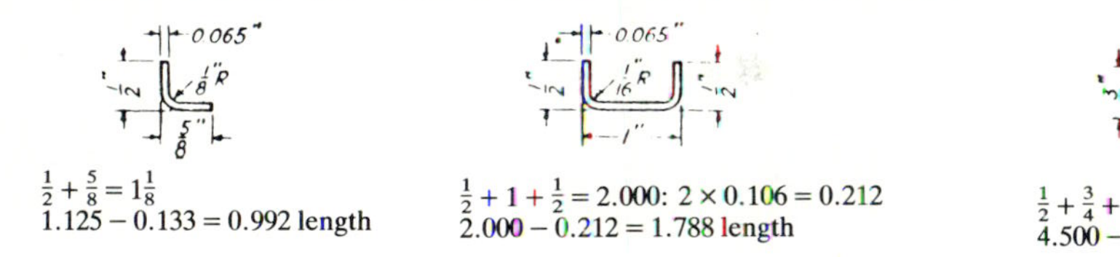

$\frac{1}{2} + \frac{5}{8} = 1\frac{1}{8}$
1.125 − 0.133 = 0.992 length

$\frac{1}{2} + 1 + \frac{1}{2} = 2.000$: 2 × 0.106 = 0.212
2.000 − 0.212 = 1.788 length

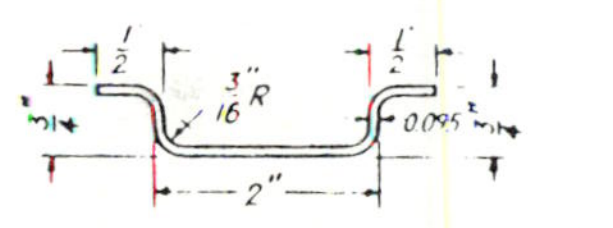

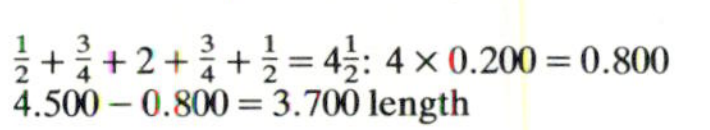

$\frac{1}{2} + \frac{3}{4} + 2 + \frac{3}{4} + \frac{1}{2} = 4\frac{1}{2}$: 4 × 0.200 = 0.800
4.500 − 0.800 = 3.700 length

relationship

$$B = (0.01743R + 0.0078T)90$$

Bending Force

The bending stress in a simply supported beam can be calculated from relationships developed in the field of strength of materials:

$$S = \frac{Mc}{I}$$

where S is the stress, lb/in.^{2}, M is the bending moment, in. · lb, I is the moment of inertia at section involved, in.^{4} and c is the distance from netural axis to outermost fiber in., For the workpiece shown in Fig. 13.25, the bending force P would be calculated as follows:

$$S = \frac{(Pd/2)(t/2)}{wt^3/12}$$

$$= \frac{3Pd}{wt^2}$$

$$P = \frac{0.33Swt^2}{d} \quad \text{lb}$$

where S is the nominal tensile strength, lb/in.^{2}, w is the width of bend, in., t is the thickness of metal, in., and d is the span of bend (approximately $8t$ *for* V dies), in.

Although this is the stress to initiate bending, to make a good 90° V bend requires a certain amount of plastic deformation to eliminate undesirable spring-back. From experiment it has been found that an appropriate value of the total bending force at the bottom of the stroke should be four times greater than the calculated value:

$$P = \frac{1.33Swt^2}{d} \quad \text{lb}$$

As the distance d decreases, the force P increases. The formula for the developed length of bend given in Fig. 13.24 is usually also applied to V dies.

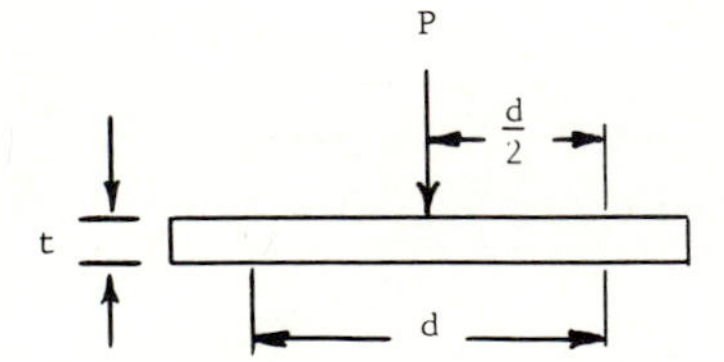

FIGURE 13.25
Bending force for sheet metal treated as a simply supported beam.

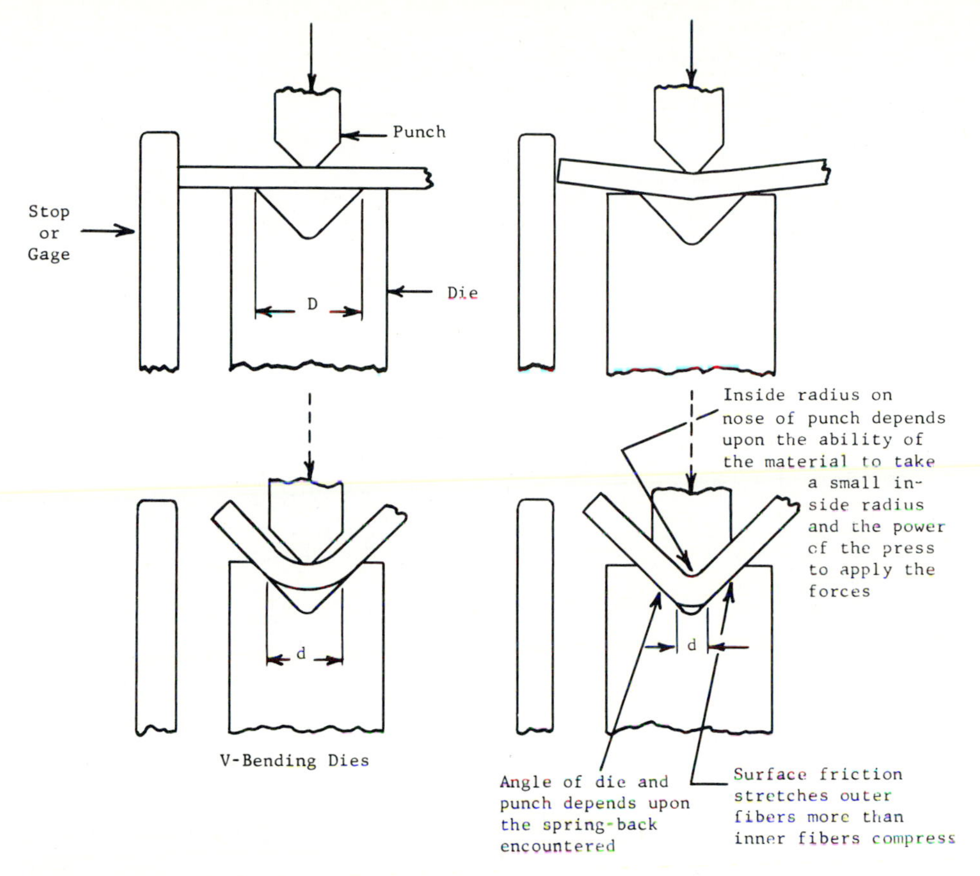

FIGURE 13.26
Details of making 90° bend in a V-die in a press break.

V-Die Bending

V-die bending (Fig. 13.26) is performed easily in hydraulic or mechanical presses. The forces applied to the beginning of the bend are less than those required to complete the bend. V-die bending is suitable to the action of a mechanical crank press.

Punch-Press Bends

In a punch-press die, the bend is usually made with a pressure pad holding the material while a punch forms it (Fig. 13.27).

The conditions that prevail in punch-press bending are similar to those in pan-brake and V-die bending, except that it is more difficult to obtain a right-angle bend owing to the ironing action on the material. Material thickness is important in punch-press bending as far as angle size goes. A few

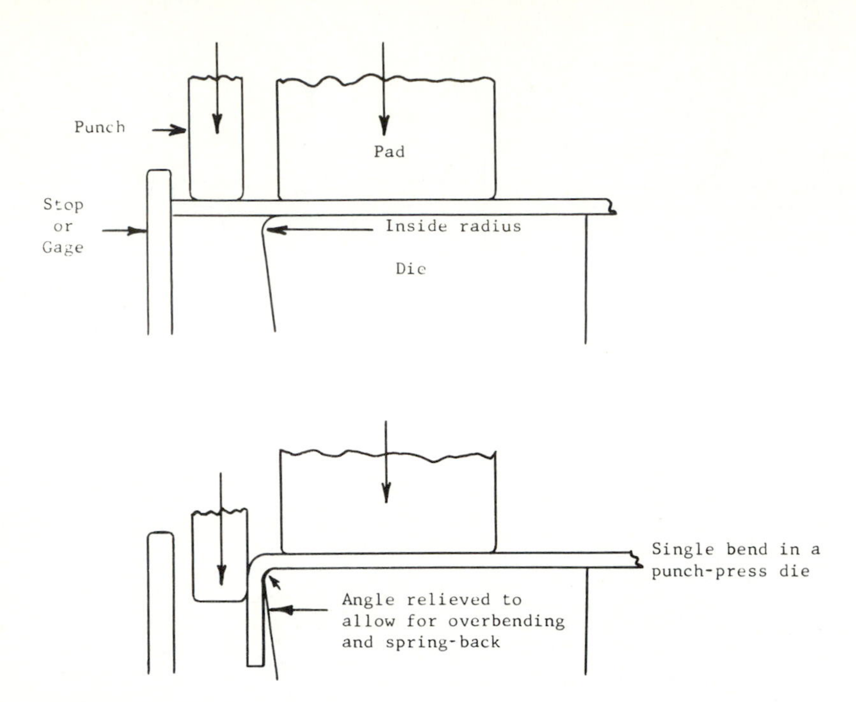

FIGURE 13.27
Bending a cantilevered workpiece in a punch press.

thousandths variation in material thickness will result in appreciable variation of the angle size in the piece being bent.

Although developed lengths may be computed rather closely (Fig. 13.24), it is wise to verify the developed lengths by experiment. For example, when a double bend is made on a punch press, the material may stretch between bends, thus giving a smaller developed length than calculated (Fig. 13.28).

In pan- and press-brake bending of a wide sheet, the radius of bend may be larger in the center owing to the greater deflection of the machine parts at the center (Fig. 13.29).

Limitations to Consider in Design for Bending

The production-design engineer should be aware of and understand the limitations of equipment and design so that interferences are avoided and the usual variations will be acceptable. The pan brake in Fig. 13.23 is subject to deflection in the center when bending wide sheets similar to the deflections of the press brake in Fig. 13.26. A craftsman can vary the radius of the bend by adjusting the machine and by using special tools. Although some pan-type brakes are made to form steel plates 150 in. long and $\frac{3}{4}$ in. thick, the most common sizes are smaller and are used for bending thin sheet metal (12 gage

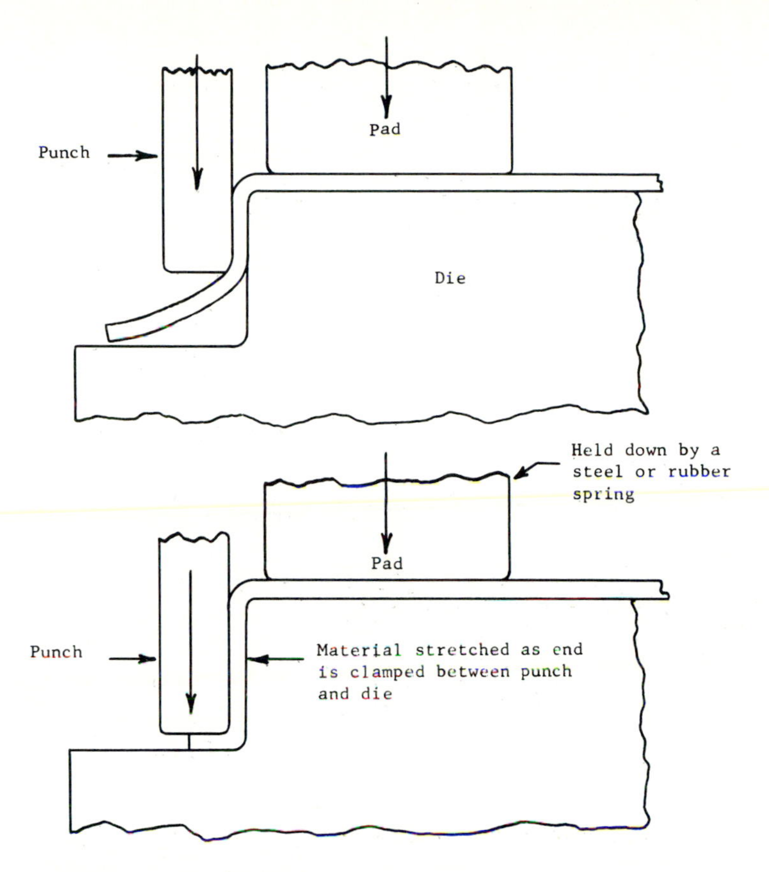

FIGURE 13.28
Case in which material stretches between bends, giving less developed length than expected.

and less). The pan brake can bend the material nearly 180°. Clamps are then used to complete the bend to 180° and give a smooth edge or form a lock-joint as shown in Fig. 13.30.

On second and succeeding bends, the limitations indicated in Fig. 13.31*a* to *d* must be considered. Thus Fig. 13.31*b* illustrates the minimum length between bends, and Fig. 13.31*c* shows the relationship between adjacent lengths and the punch when a flange is "formed up" in the press. A layout of the bent part with the forming tools or a paper representation of the bent part fitted into carboard models of the dies helps to avoid gross errors in forming sheet metal. Dies are not expensive and can be modified into all forms and shapes (Fig. 13.32). Considerable time in the shop would be saved if a progressive picture of each bending operation were shown on the drawing when multiple bends are incorporated in the sheet-metal design. (See Fig. 13.29 and note that the gage stops either on the front or back may interfere and must be removed.)

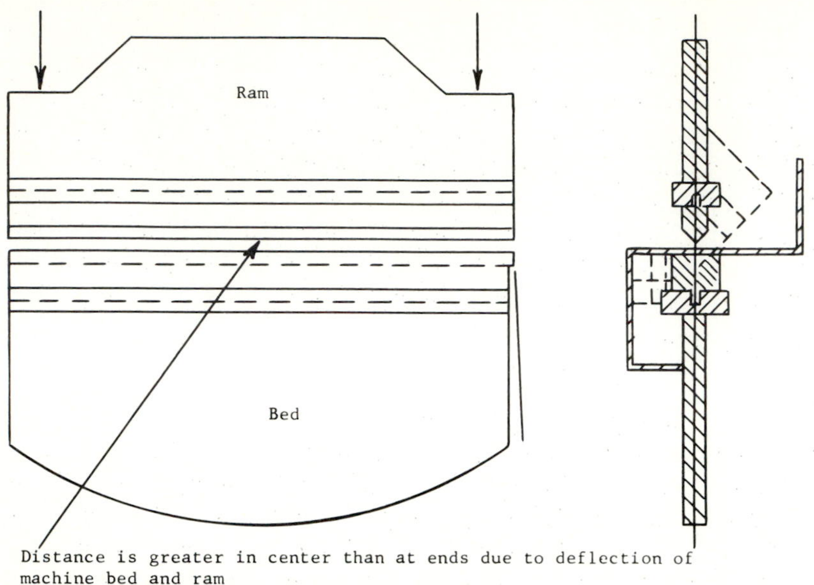

FIGURE 13.29
Typical press brake showing arrangement of ram and bed and possibility of deflection at center.

Many ingenious tools and procedures are used in the trade, such as box-forming brakes (Fig. 13.33) and horn dies on press brakes (Fig. 13.34) that form closed sections such as boxes and channels, and segmented punches that will permit forming the four edges of a sheet (Fig. 13.35).

Tube Bending

A bend in a tube is difficult to make without buckling the walls on the inside of the bend and causing the tube to collapse and become oval-shaped. In order to obtain a uniform cross section, the tube, prior to bending, may be filled with a low-temperature ductile alloy such as lead, or it may be filled with sand and the ends may be sealed. After the bend is made with regular bar-bending equipment, the sand is removed or the tube is heated and the metal is poured out. Another technique is to insert a coiled spring or a snake in the tube at the point of bend in order to support the walls of the tube as it is bent. The snake is a series of balls or lugs connected by flexible joints. Still another method is to insert a mandrel into the tube, which is held at the point of bend as the tube bender rotates (Fig. 13.36). The end of the mandrel is hardened and shaped to

FIGURE 13.30
Lock-type joints made on press-brake or pan-brake machines.

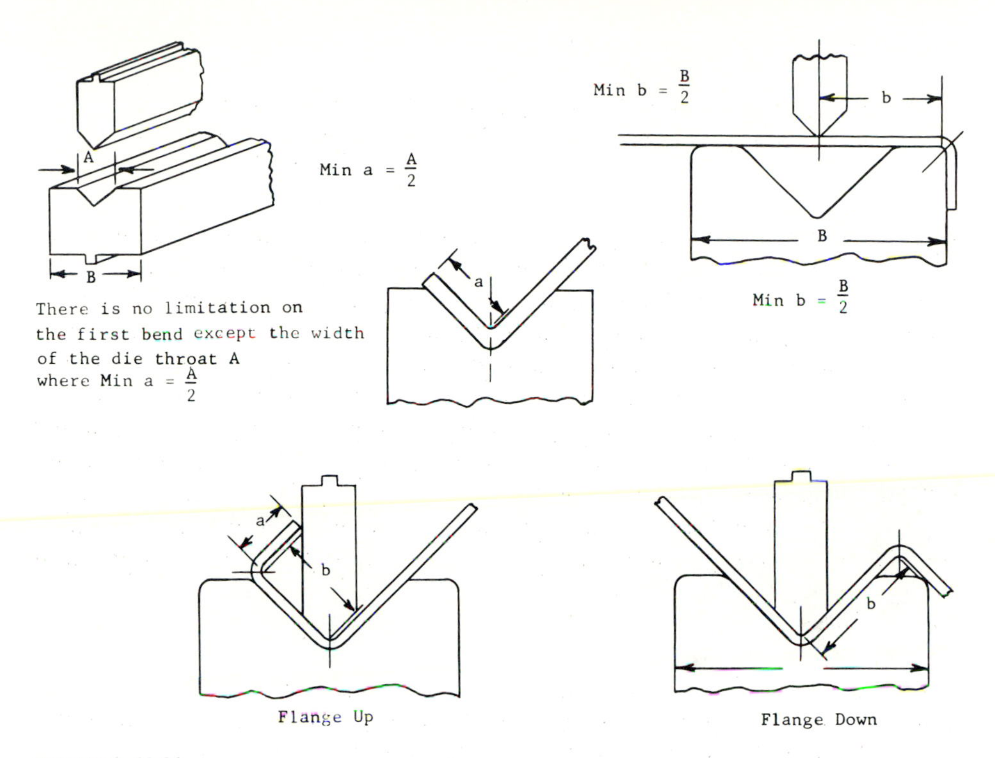

FIGURE 13.31
Limitations on the length between bends. (*a*) No limitation on the first bend; (*b*) width of die limits distance; (*c*) flange up length of leg limits distance; (*d*) flange down width of die limits distance.

the inside of the tube, which it supports and prevents from collapsing. The tube is slightly stretched to prevent collapsing on the inside. Automatic machines are able to make several bends in different directions and at different angles. Such machines are used in the manufacture of furniture and automobile parts. The tools and accessories for bending and forming pipes, tubes, and structural shapes are shown in Fig. 13.37.

Flexible-Die Forming

Rubber forming was used in only a few special applications until the aerospace industry applied it extensively on aluminum sheet metal parts. Rubber, when confined, is incompressible. This property permits the use of high pressure, advantageous in forming parts that would require a collapsible punch. For example, the threads on the inside of an electric light socket are formed by a rubber plunger that forces the metal to the internal form of the die. As the punch descends, the rubber expands; and, since it is confined, the pressure builds up. When the punch ascends, the rubber returns to the original shape

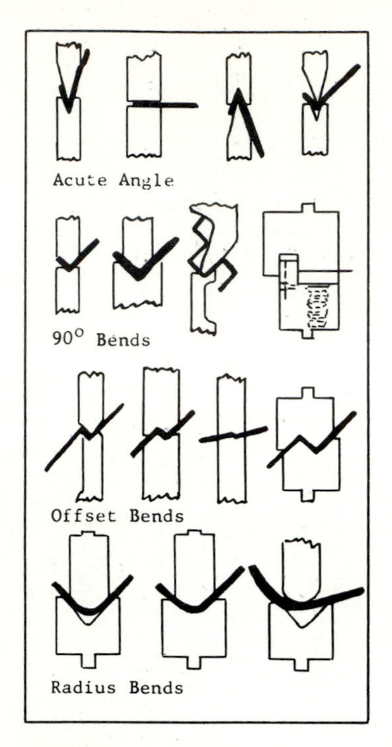

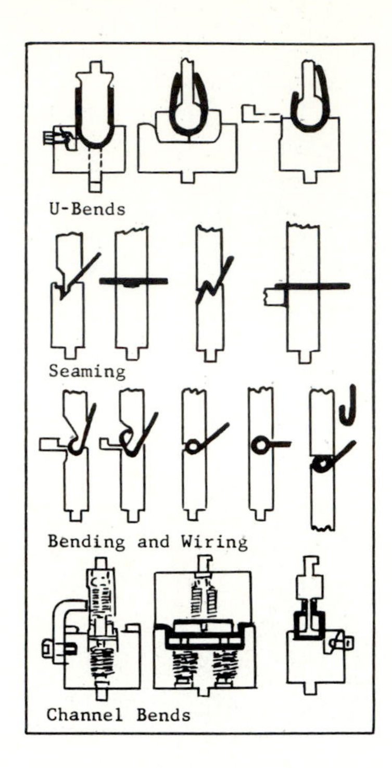

FIGURE 13.32
A group of representative press-brake forming die sets that illustrate some of the contours that can be produced.

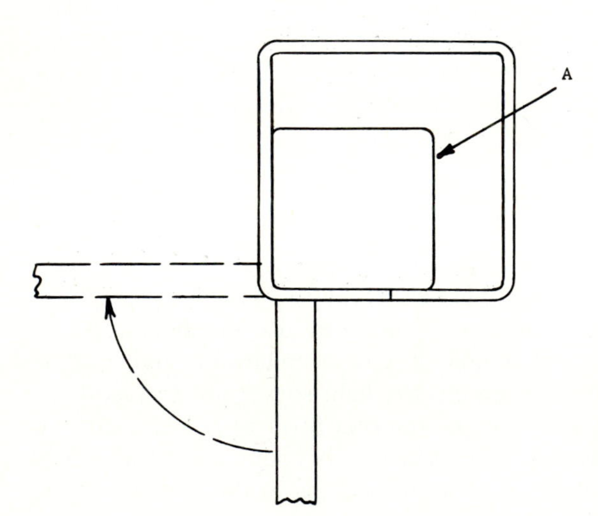

FIGURE 13.33
Box-forming brake. The outside support of bar A swings away so that the part can be slipped off. Otherwise the action is the same as a pan brake.

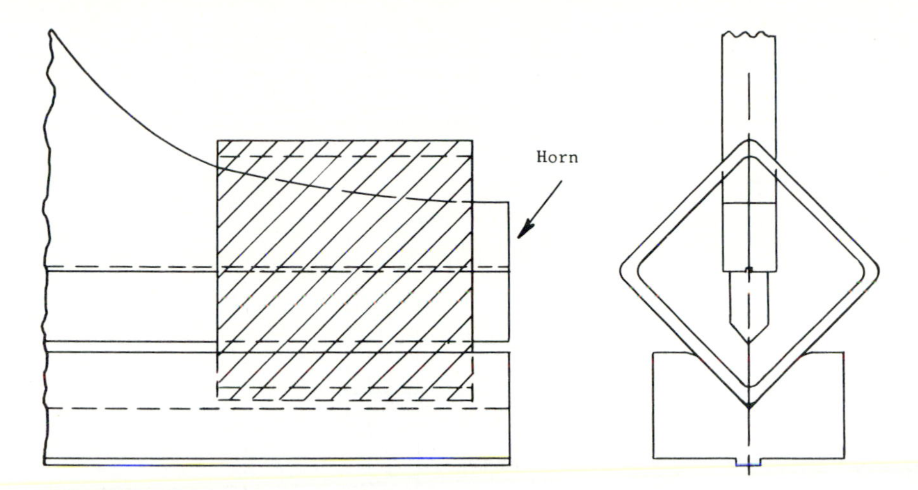

FIGURE 13.34
A horn die on the end of a press brake.

and is easily withdrawn. The die then opens up and releases the socket. In this operation the punch is rubber and only a die is required.

The aerospace industry developed a large hydraulic press with a sponge-rubber cushion on the plunger coming down from above. A typical size is 6 to 8 in. deep, 5 ft wide, and 10 ft long. A platen is slid under this cushion from either side. One platen is prepared while the other is moved under the press and the parts are formed. The material is placed upon the die mounted

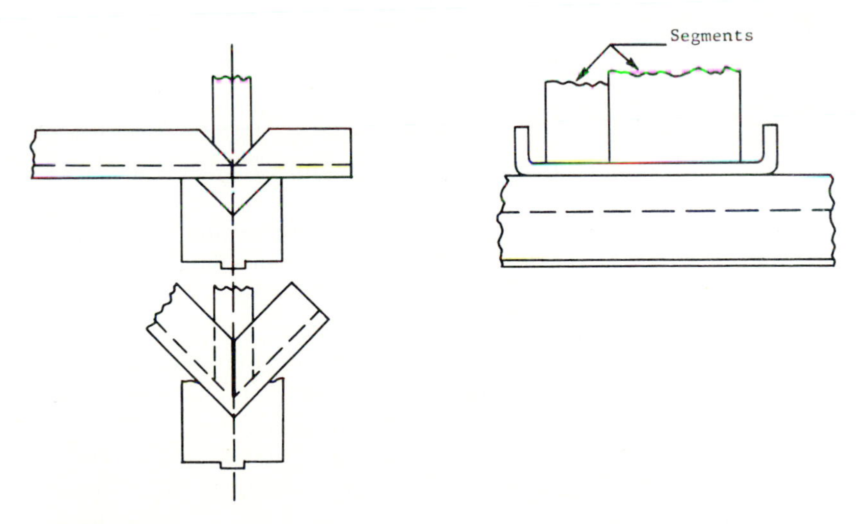

FIGURE 13.35
Segmented punches enabling bending sheets on one or two sides.

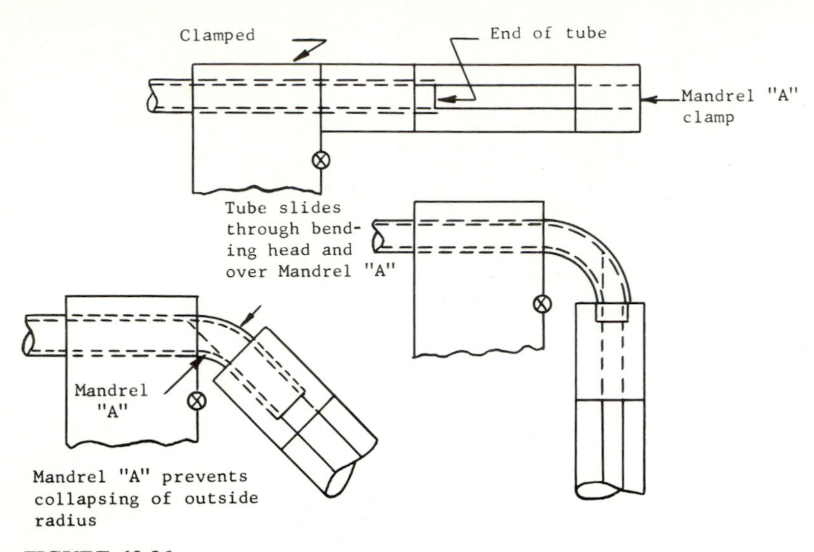

FIGURE 13.36
Tube bending by use of a mandrel.

on the platform and the rubber cushion enfolds the die parts. A part with multiple bends can be formed completely in one operation. A box with four sides and flanges, making eight bends in all, can be formed with one stroke. This process eliminates one half of the forming die. Either the punch or die can be replaced by rubber. (See Fig. 13.38 for a simple example.)

Another use of rubber is for blanking and piercing. The sheet metal is placed over a die and a rubber pad is pressed down on the sheet. The pressure is great enough to shear the material and form a blank. The presses consist of a flat lower platen on which the thin blanking die is placed and an upper and moving platen on which the rubber pad is fastened. After the blank is made, the die and blank are moved out by sliding on the lower platen when the press is open. The blank is removed and the die and sheet are placed in position for the next operation. Holes can be pierced at the same time the sheet is blanked by placing punches in the required positions.

There are a number of flexible die-forming processes available, including the Guerin Process, Marforming, Hydroforming, and Hydrodynamic Drawing (Fig. 13.39). Note that only the first two processes use a rubber die, but all use the principle of having one die flexible. The flexible-die processes are slower than the steel-die processes when the latter are used on a suitable mechanical press and the production quantity exceeds several thousand pieces. At low production levels their short lead time and low tooling cost make them economically justifiable. However the original equipment work load is required to amortize the high equipment cost over a reasonable time period. Production rates vary from 50 to 250 parts per hour and draws as deep as 5 to 12 in. can be achieved by Marforming or Hydroforming.

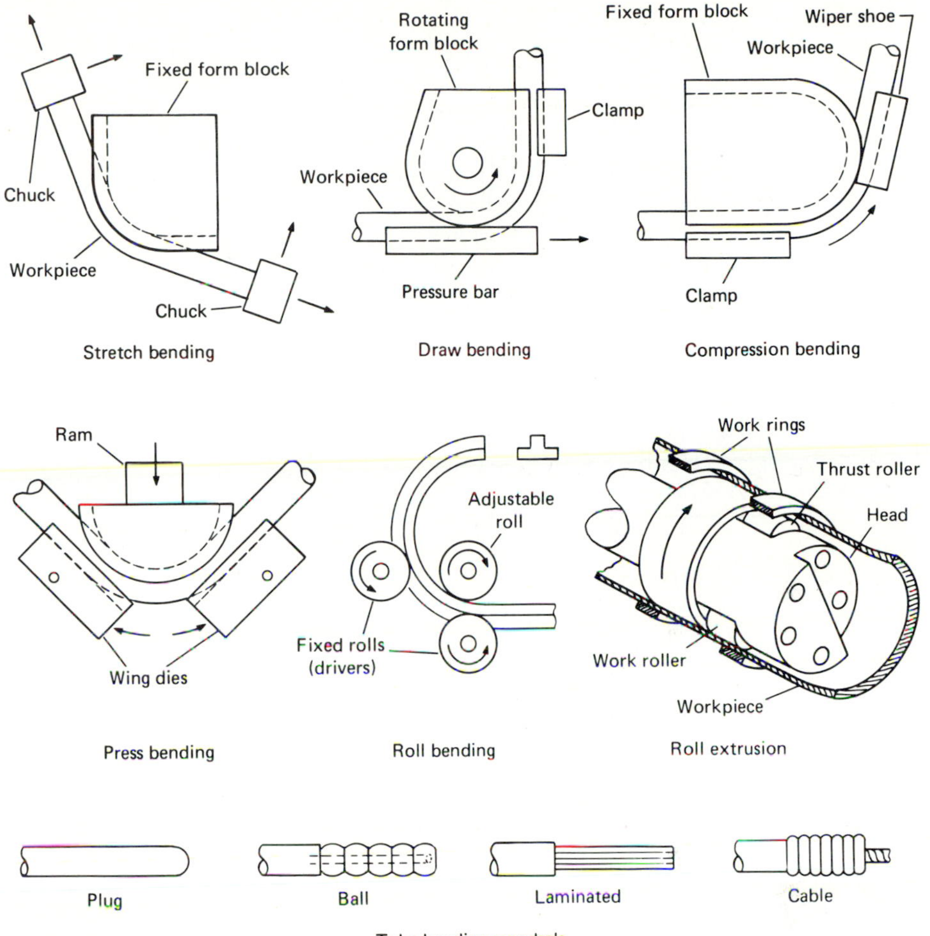

FIGURE 13.37
Operations and accessories for bending pipes, tubes, and structural shapes. (*Redrawn from Doyle et al.,* Manufacturing Processes and Materials for Engineers, *Prentice-Hall, Englewood Cliffs, N.J., 1969.*)

Roll Forming

Bends of large radii, such as circle bends for hoops, angle rings, or tanks, are made on a roll bender. The material is fed into the roll and bent continuously into its final shape. One end is always flat when the inside roll is smaller than the desired radius. Here the work may have to be preformed to the radius desired (Fig. 13.40). Where there are a large number of rolled parts to be made, such as steel-plate motor frames, a roll having a diameter equal to the inside of the frame is used as the center roll. Thus, circular frames can be formed with no flat ends.

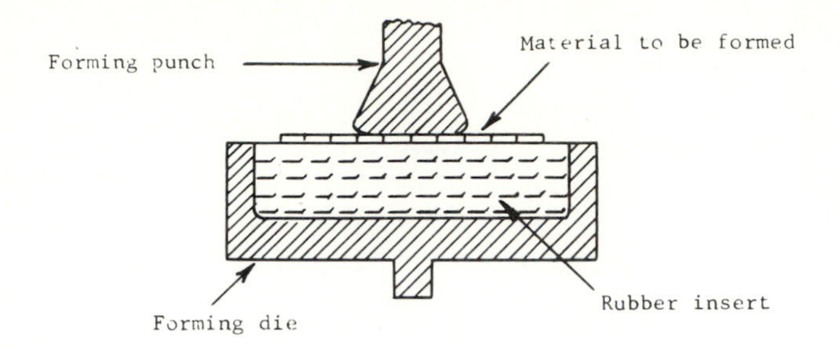

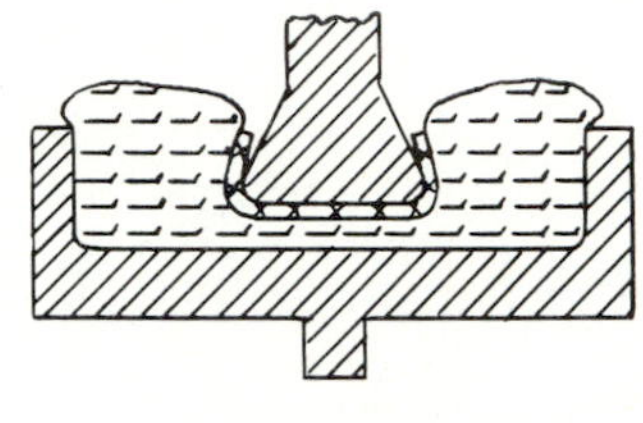

FIGURE 13.38
Use of rubber-forming dies to produce a channel.

Roll-forming strip is the process in which a continuous strip passes through a series of roll stands in order to obtain the final form. The rolls on each roll stand are shaped and positioned to form the contours for each of the required steps (Fig. 13.41). The process is rapid, and the tool expense is moderate. Auxiliary operations such as punching, cutoff, and welding can be added. For example, tubing is roll-formed from strip and butt-welded by gas flame or electric current in the same machine. In roll forming, the bends allow for springback. Roll-formed parts compete with the extruded and molded parts used in building trim, metal window frames, doors, and metal furniture.

High-Velocity Forming

High-velocity forming, frequently called explosive forming, employs the principle of rapid release of energy in order to form or punch a metal part. It has been reported that pressures running as high as 2 million lb/in.^{2} and speeds as high as 3000 ft/s are used. The tooling required includes a female die of the correct geometry into which the flat material is forced by the impact of the explosion.

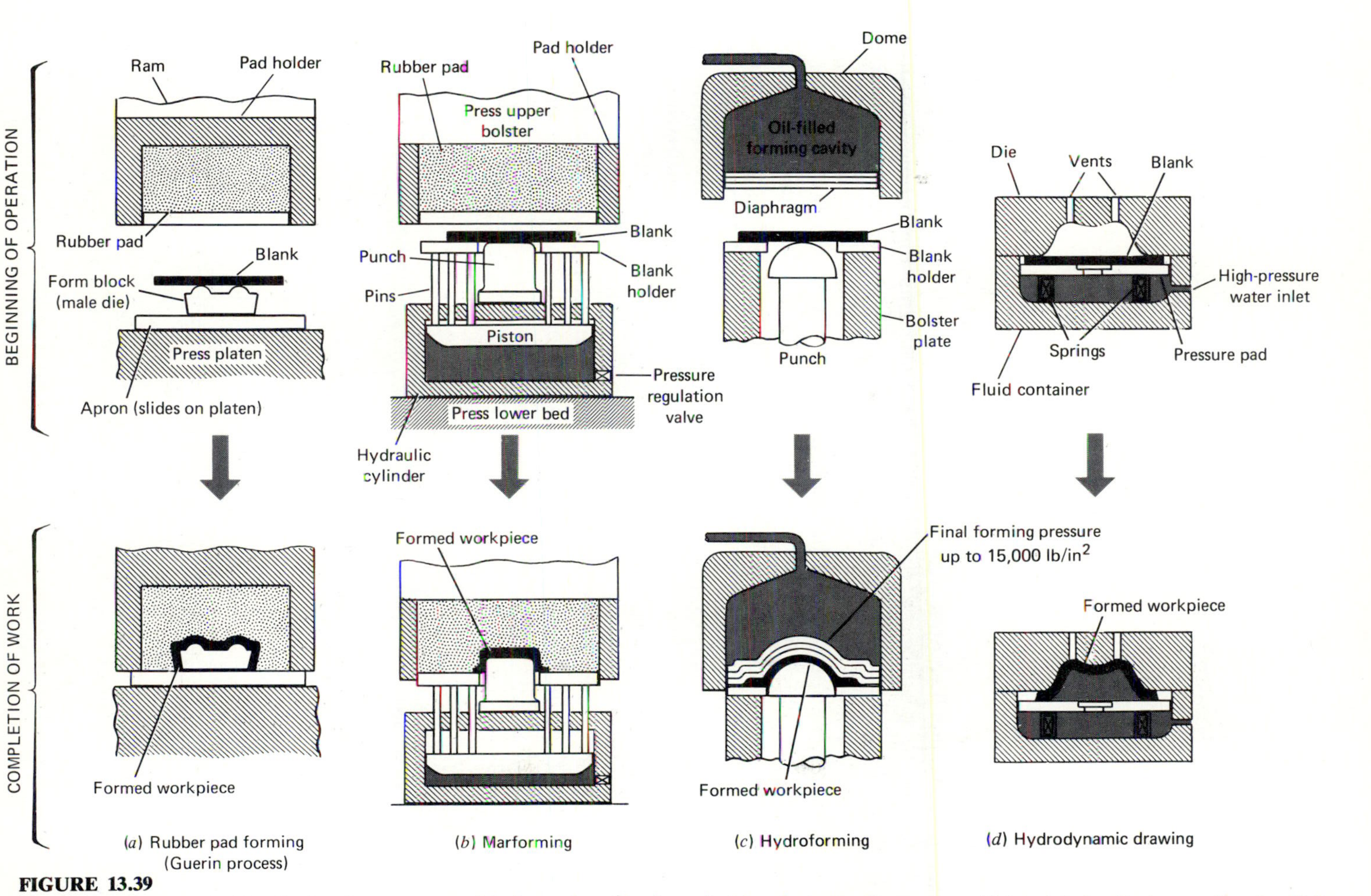

FIGURE 13.39
Diagram of common flexible die-forming processes. (*Redrawn from Doyle et al.*, Manufacturing Processes and Materials for Engineers, *Prentice-Hall, Englewood Cliffs, N.J., 1969.*)

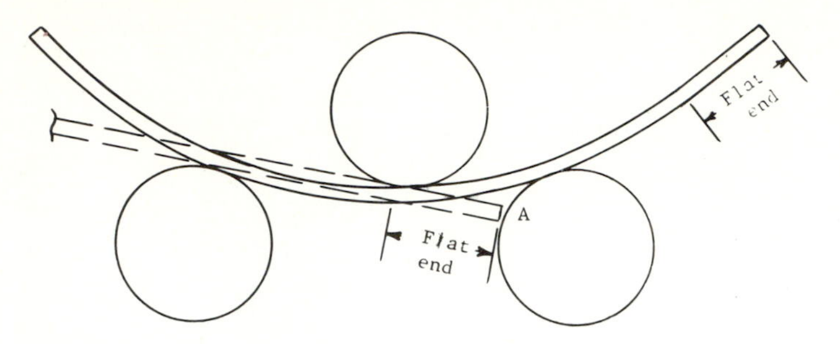

FIGURE 13.40
Use of pyramid rolls to bend plate.

High-velocity forming has application in producing large parts where the cost of conventional equipment would be excessive, for forming materials that are difficult to produce by the usual techniques, and for forming small quantities where the cost of tooling would be high in proportion to the number of pieces to be produced.

High-velocity forming can be applied successfully to the production of such items as expanded tubes, ducts, drawing tubes, forming of sheet products, extruding and forging of steel, piercing of holes, and the compressing of metal and ceramic powders into shapes.

Both low and high explosives are being used in high-velocity forming. Low explosives include black and smokeless powder and other slow-burning, deflagrating explosives. High explosives include dynamite, TNT, RDX, and similar explosives that have a detonation speed of more than 5500 ft/s.

As of this writing, high-velocity forming is not a mass-production process.

Figure 13.42 illustrates the manner of transfer of energy that takes place under high-velocity forming.

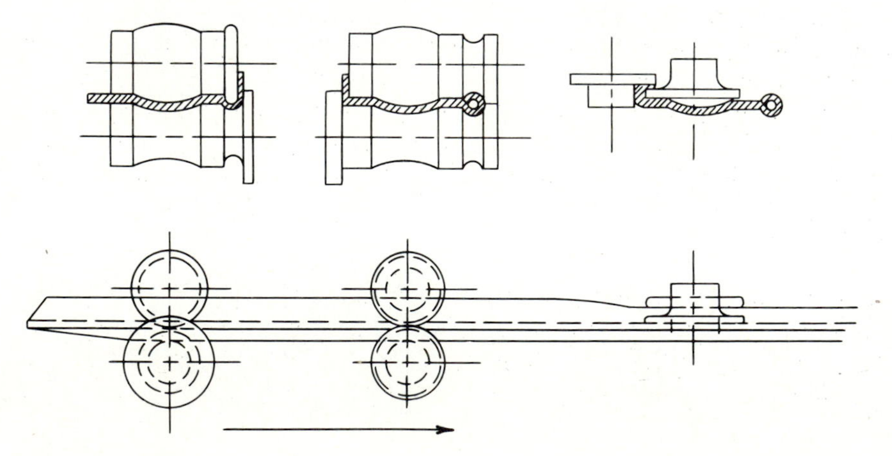

FIGURE 13.41
Continuous-roll forming of strip steel.

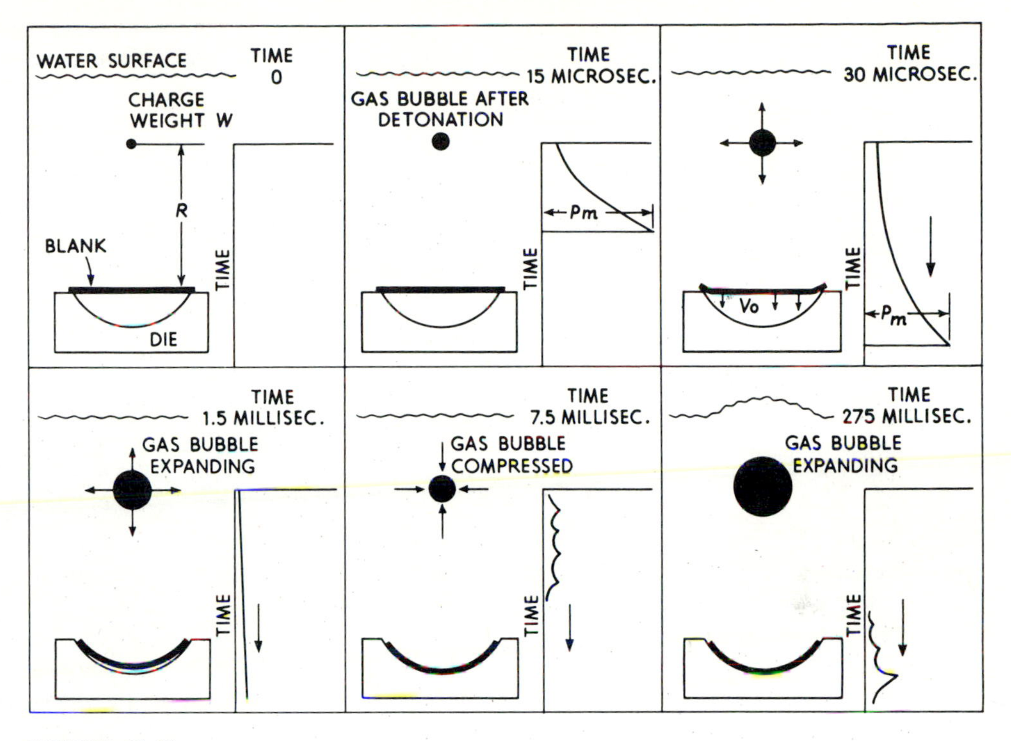

FIGURE 13.42
The development and transmission of explosive energy through water.

COMPRESSION PROCESSES

Cold Rolling

Cold rolling is a mill process, carried out below the recrystallization temperature, which is usually performed by the supplier of the sheet or strip material. Cold rolling entails the passing of hot-rolled metals such as bars, sheets, or strips of cleaned stock through a set of rolls. It is passed through the rolls many times, using light reductions each time, until the correct size is obtained. Cold rolling increases the tensile and yield strength and imparts a bright finish and close tolerances to the finished stock.

Cold and Warm Extrusion of Steel

When steel is cold-extruded, larger-tonnage presses are used and the blank must be homogenous and ductile. Also, the surface must be treated to reduce the friction and permit the flow of the material in the die (see Fig. 13.43). Cold-steel extrusions for making deep shell-shaped parts can be made with less heating and annealing than required in deep drawing. The Ford Motor Company is making steel-extruded axle spindles and other shapes. Here

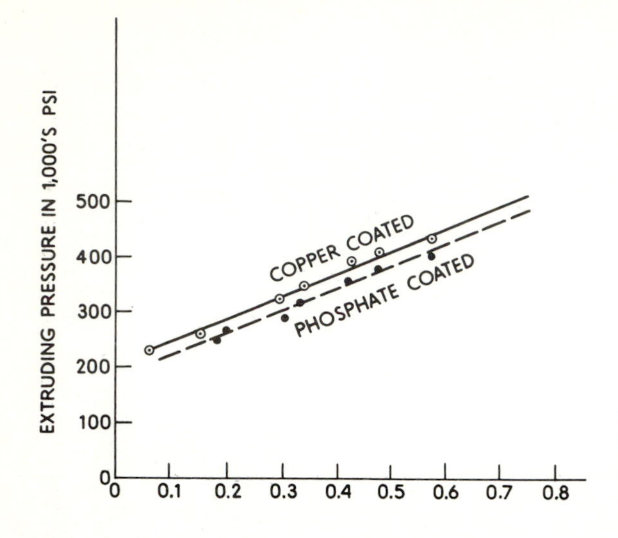

FIGURE 13.43
Relation between carbon content and extrusion pressure in the cold extrusion of steel.

reductions of 30 percent in material, 65 percent in labor, 80 percent in floor space, and 40 percent in die cost have been obtained.

Successful cold extrusion requires properly designed presses, dies, preparation of steel, and lubrication of the billet. Zinc phosphate coatings have met with much success as a suitable lubricant when reacting with other materials, such as soap, to form an integral continuous film. These may be applied either by immersion or by spray.

The sequence of operations in treating the billet in preparation for extruding include: alkaline cleaning, rinse, pickling, rinse, phosphate coating, rinse, lubricate with reactive soap, dry.

Cold extrusion is particularly successful for:

1. A wide range of steels that are suitable for cold extrusion. A5120 and C1010 aluminum-killed in particular give high ultimate tensile strength to extruded products, with little loss of ductility.
2. The heat treatment of steel prior to extrusion, which has proved helpful. A normalizing procedure, followed by quenching in water and annealing, is suggested.
3. A zinc-phosphate coating of controlled formulation and application, which is the best surface preparation for extruding cold steel.
4. The processes of phosphatizing and the addition of a dilute emulsion type of compatible lubricant to increase the efficiency of the extrusion process.

Warm forming refers to forming performed below the recrystallization temperature and above room temperature. This method is often used in connection with precision closed-die flashless forming. Here precision blanks are placed in the die and, upon closing, the die cavity is completely filled and no flash is obtained during the single press stroke. The obvious advantages include no material waste, minimum, if any, subsequent machining, good surface finish, good strength, and high production rate.

Impact Extrusion

In impact extrusion, cold material with a low recrystallization temperature, such as aluminum, copper, or lead is used to make the extruded parts (Fig. 13.44). The slug or disk is placed in the die, where it is struck by a plunger. The rapid travel of the plunger and resulting impact cause the material to heat up and become plastic. The material is thus quickly extruded through the space between the die and the plunger. The fact that the material passes through the opening between the plunger and the die reduces the abrasive wear on the tools and provides a slight clearance between the formed part and the plunger, so that the part can be slipped off easily. Thin-walled items such as toothpaste containers, radio shields, food containers, boxes for condensers, and cigarette lighter cases are made easily when they are symmetrical in shape. The process is very rapid; parts can be made as fast as material can be placed in the die and the cycle of the press completed. Large presses now make parts with thick walls for items such as paint spray-gun containers and grease-gun containers.

A second type of impact extrusion or cold extrusion is performed by forcing the material and plunger through the die. The material flows in the same direction as the plunger travels (Fig. 13.45). The proposed blank for making the part may have a hole with a solid bottom. The solid bottom

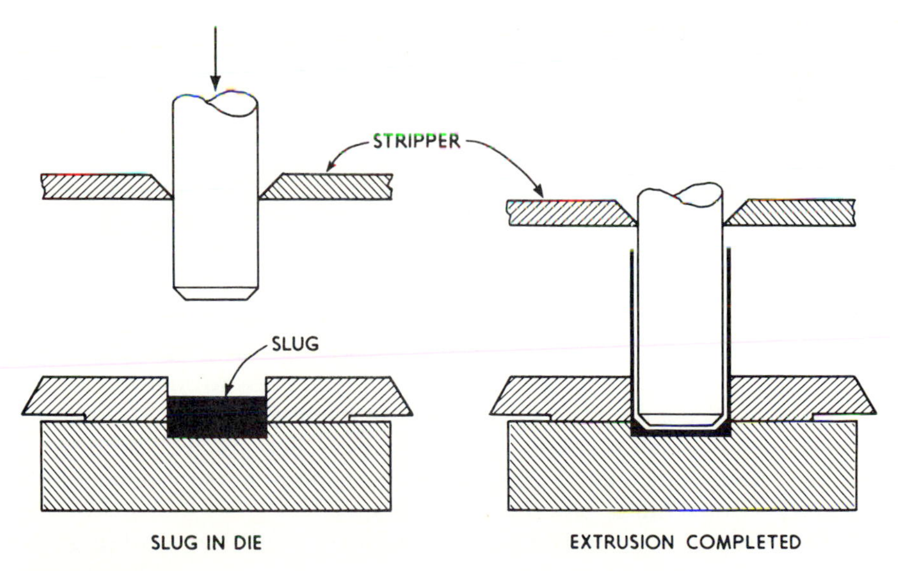

FIGURE 13.44
Backward impact extrusion.

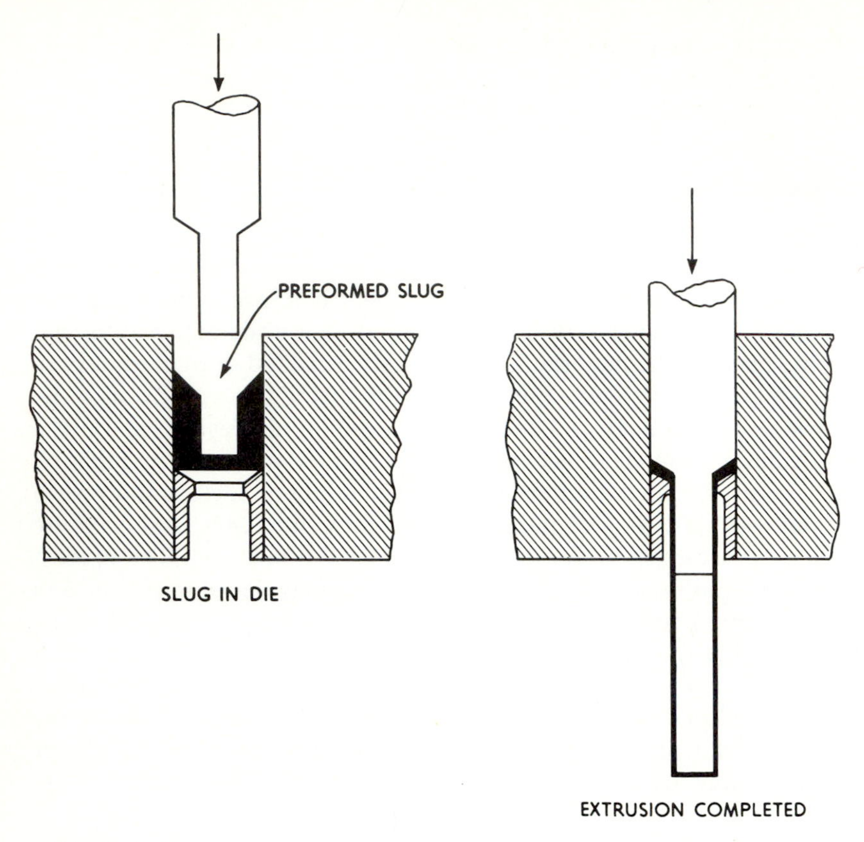

FIGURE 13.45
Forward (Hooker) impact extrusion.

remains untouched by the process. The plunger enters the hole in the blank, presses the walls of the blank between the plunger and the die, and forces the material from the bottom. This process is known as the Hooker process.

Coining

Coining is a cold-forging process in which the material inserted into the die is at room temperature. In this process, a blank is placed in a die and then forced under heavy pressure on a knuckle-joint or toggle-type press to conform to the design and shape of the die. The operation is frequently used for striking up jewelry and medals of all types for which clear-cut designs are required in relief on the surface of the blank. Coining gets its name from the production of coins, as in the United States mint (see Fig. 13.46*c*).

Coining operations are performed with the same general types of press or hammer equipment as used in the production of die forgings. In coining operations, the blank is approximately the same size as the finished piece.

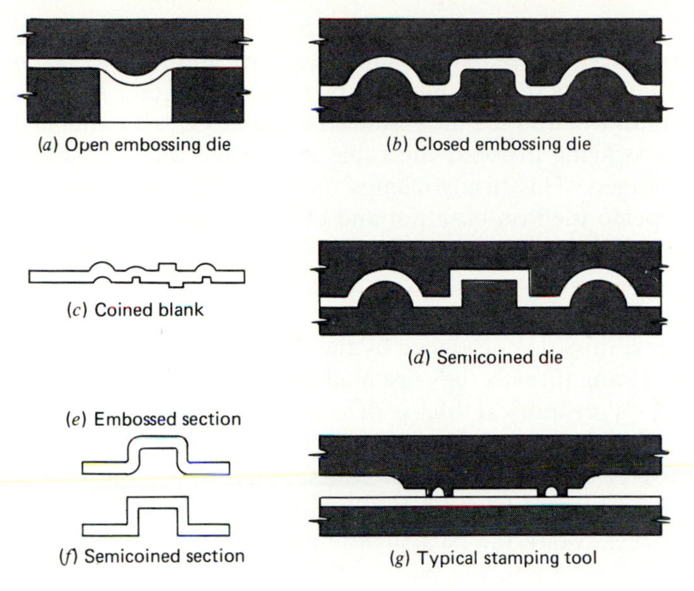

FIGURE 13.46
Typical embossing dies.

Since large pressures are required in order to effect flow of cold materials, it is important that the amount of flow be kept as small as possible. Government tests record the following pressures required for different coins: silver dollar, 160 tons; half dollar, 98 tons; dime, 35 tons; penny, 40 tons.

When accurate size is required, a subsequent shearing or sizing operation may be necessary. For example, in the production of type bars used in book-keeping and adding machines, the following operations are required: (1) blank oversize, (2) coin configurations on type bar, and (3) trim flash.

Embossing Embossing is the placing of configurations on the surface of a part by means of dies that draw the material into position with very little effect on the thicknesses (Fig. 13.46*a* and *b*).

Swaging

Swaging is the shaping of material by a series of blows from a die made to the desired shape. It is usually used for making round shapes such as rod and wire. The outstanding application of swaging is the processing of tungsten into rods and wire by rotary-impact swaging equipment. Tungsten is a high strength material and can be shaped by continuous blows. The swaging equipment consists of two or four dies distributed around the periphery of the material to be formed. Power driven rollers mounted in a ring rotate around the outside of the die and cause the die to strike the material. The constant hammer blows gradually reduce the material in size. Then the material is taken to another machine and reduced further until the desired size is obtained. Swaging equipment can quickly reduce or taper the diameter of wire in order to make possible its entrance into drawing dies. Hollow parts, such as oxygen cylinders, can be reduced on a mandrel that holds the inside size. Cables can be attached to hollow shanks by the swaging process; the cable is inserted into the shank and the shank is then swaged. This firmly clamps the cable within the shank. Swaging operations are performed on both hot and cold material.

Thread Rolling

In thread rolling, the screw thread is produced by the displacement of metal; in all other methods of producing threads they are made by cutting. To produce a screw thread by rolling, a cylindrical blank of a predetermined diameter, approximately the pitch diameter of the screw being produced, is placed between two hardened-steel dies. The dies, of course, have the same thread pattern that is to be formed on the blank. The threading dies may be either flat or circular. The circular dies rotate the part slightly more than one revolution and then release; the flat dies rotate the part to the same degree by moving parallel in opposite directions. As the blank rolls between the dies, the pressure applied to the blank through the dies causes the metal to be displaced and to follow the pattern of the dies. The resulting screw thread will have the same form, pitch, and helix angle as the thread on the dies.

It can be seen that no cutting action takes place in this method, and consequently no material is removed from the original blank. The ridges on the threads of the dies are fed into the blank approximately 50 percent of their height. This action forms the roots of the thread being rolled, while at the same time this displaced metal flows upward into the root of the die thread to form the crest of the screw thread (see Fig. 13.47).

Owing to the cold working of the material and the absence of rough surfaces caused by cutting, rolled threads are stronger and tougher than cut threads. No difficulty is experienced in maintaining class 2 thread tolerances.

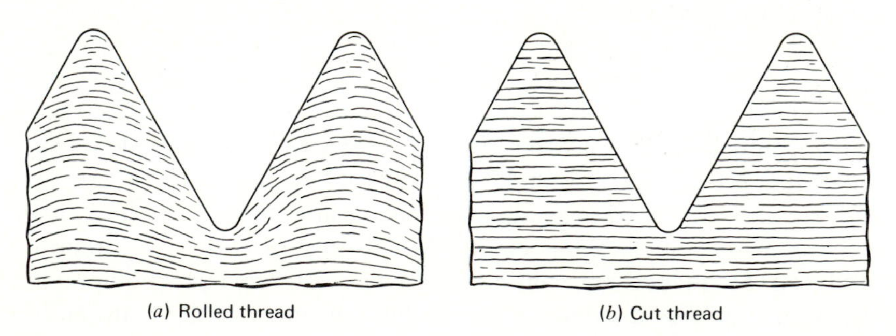

FIGURE 13.47
Comparison of sections through a rolled (*a*) and a cut (*b*) thread. Note the fiber flow lines in the former.

Another advantage of the rolled thread is a savings in material. Since the blank that is used approximates the pitch diameter of the desired thread, a smaller-diameter stock is used than if the same thread were to be cut.

Although the life of thread rolls, when properly cared for, is quite long, the cost is relatively high. Care must be exercised to avoid getting a metallic chip between the roll and the blank, as this may result in chipping or cracking the thread rolls.

Thread-rolling machines today, utilizing the same principle of producing threads, are performing knurling, grooving, burnishing, and spinning operations. Pipe and tubing are formed to circular shapes; worm threads are formed on shafts in worm-gear units; grooves with long leads and grooves parallel to the length of the part can be made easily.

High-Pressure Extrusion of Metals

When a metal rod is drawn through a die from a container in which the rod is subjected to high hydrostatic pressures ranging from 100,000 to 600,000 lb/in.^{2}, it is being cold-extruded under high pressure. The operation is called *high-pressure forming.* A true fluid flows when any shear force is applied to it. Metals with high ductility, i.e., face-centered cubic metals, are prime candidates for use in the high-pressure extrusion of wire.

The principal stresses in a solid are a set of three mutually perpendicular direct stresses that act on a plane of zero shear stress. The direction in which a principal stress acts is a principal axis; a plane normal to a principal axis is called a principal plane (Fig. 13.48). The average normal stress σ is the algebraic average of the principal stresses $\sigma_{av} = (\sigma_1 + \sigma_2 + \sigma_3)/3$.

A hydrostatic state of stress is achieved when all three principal stresses are equal. Then the average normal stress is equal to the principal stress and there are no shear stresses. It is well known that high hydrostatic pressure increases the ductility, whereas hydrostatic tension reduces ductility. For example, TZM molybdenum alloy has a 63 to 65 percent reduction in area in the stress-relieved condition, but under 225,000 lb/in.^{2} its ductility improved to 97 percent reduction in area. Especially impressive is the case of beryllium,

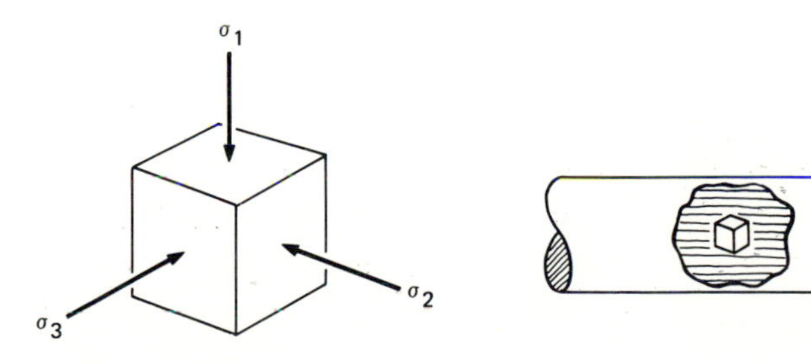

FIGURE 14.48
Principal stresses in a rod.

which metal has been successfully extruded under high pressure despite its usual rating as being nondeformable.

In normal wire-drawing practice using a small die angle, the wire is subjected to an environmental pressure of $(2C - T)/3$ rather than pure tensile stress where C is uniaxial compression stress and T is the uniaxial tension stress. In hydrostatic extrusion, the ductility of the rod increases with the environmental pressure, so that there is no necking failure even at large reduction ratios, i.e., at 2:1 or 5:1 change in diameter.

The Western Electric Company has developed two wire extrusion machines that use the principles of hydrostatic extrusion to achieve large reduction ratios when forming aluminum and copper rods into correctly sized wire in a single die. The secret of their success lies in the simultaneous solution of two major problems:

1. Designing a high-production device that is capable of consistently maintaining pressures up to 150,000 lb/in.^{2}
2. Designing a way to apply a film of lubricant between the die and the rod during the reduction process

The lubrication problem was solved by building a series of viscous drag-feed tube cells that are placed end to end along the rod and within the hydrostatic pressure chamber. By experimentation, it was found that 50 of the viscous drag-reversing cells were needed to increase the pressure from atmospheric to 150,000 lb/in.^{2} Lubricants and the pressure media included heated beeswax and silicone plus polyethylene mixtures. It was found that the pressure media had to have extremely high viscosity to be successful.

The conventional wire-drawing operation at Western Electric Company required from 9 to 12 carbide dies (Fig. 13.49).

Machine no. 1 produced 14-gage (0.065-in.-diameter) wire using 9 to 12 carbide dies and 400 hp, starting with $\frac{5}{16}$-in.-diameter annealed copper rod. End speed was 1000 ft/min.

Machine no. 2 produced 17-gage (0.045-in.-diameter) wire using 10 diamond dies and 150 hp, starting with $\frac{3}{8}$-in.-diameter annealed aluminum rod. The end speed was 3000 to 4000 ft/min.

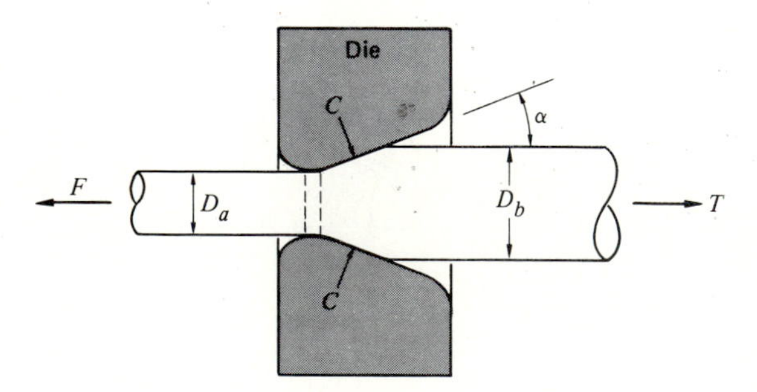

FIGURE 13.49
Conventional wire-drawing die.

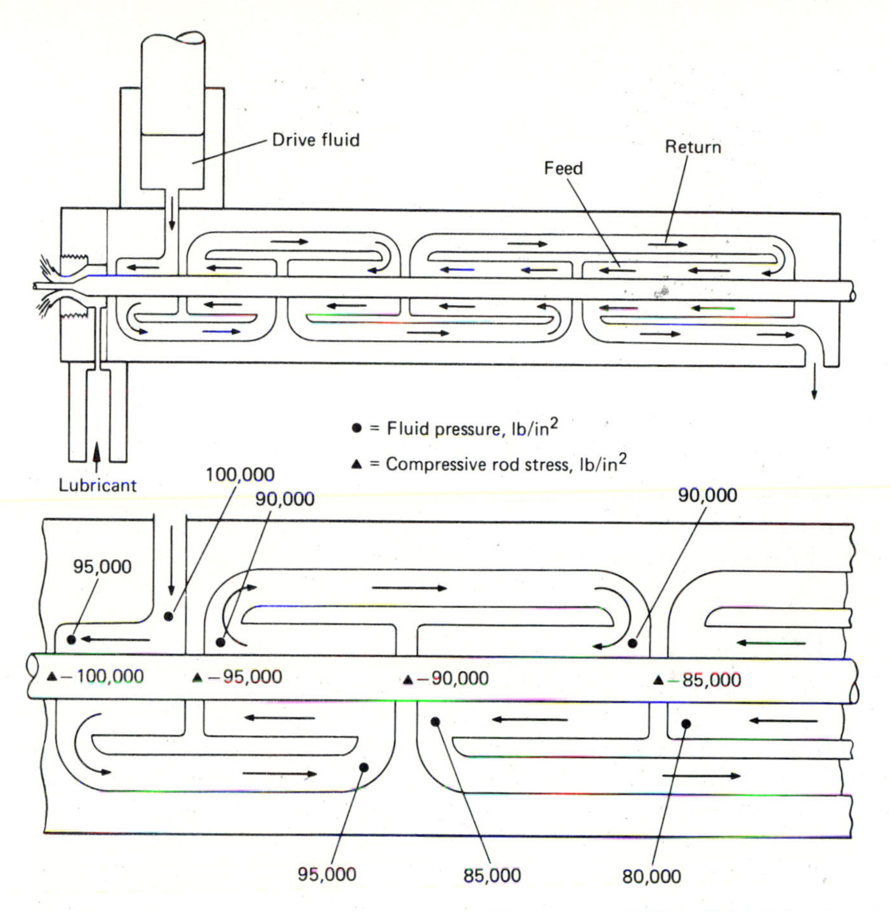

FIGURE 13.50
Schematic diagram of the hydrostatic extrusion of aluminum or copper rod into wire at a reduction ratio of 360:1.

With the prototype models of the hydrostatic extrusion machine (Fig. 13.50), the Western Electric Company engineers have attained a 360:1 reduction of area in a single die and one machine. The resultant saving of 65 percent more than justifies the considerable expense of the engineering and toolmaking man-hours that were required to make the venture a practical success.

DRAWING PROCESSES

Cold Drawing

Cold drawing is the process of pulling material through a die in order to reduce it in size and to form the desired shape. Rounds, hexagons, squares, small angles, channels, and miscellaneous shapes, including tubing, are cold-drawn.

The material must be clean and the surface must be free of scale and large defects before drawing. The resulting material is cold-worked and has a smooth finish. The size and tolerances of the finished part depend upon the ability of the dies to maintain their size. There is considerable abrasion between the material and die, and various lubricants are used to lengthen the life of the die. The dies are made of cemented carbide materials on high-activity sizes. Otherwise, tool steel is used. The material is pulled through by means of a windlass or drawbench. Drawbench work usually straightens the material, thus eliminating straightening operations. Cold-drawn material is in demand because of its accuracy of size, good surface, and adaptability to many shapes and forms. Most bar stock used in screw machines is cold-drawn. In order to obtain the preliminary shape and size before drawing, the raw material may be rolled or extruded to shape. The drawing operations may be done in series. When this is done, the wire passes from one die to the other until it is necessary to anneal or until it is the correct size. Sometimes a back pull is applied to the bar or wire as it is passed through the die in order to stretch the wire as it is being drawn.

Deep Drawing

Deep drawing primarily uses the principle of stretching the material beyond the elastic limit without compression, but may have a combination of compressive and tensile stresses locked in the part. The thickness of the drawn cup is slightly changed; the flange (the last part to be drawn through the die) is reduced in diameter and therefore will be slightly thicker than any other section. Representative sections of deep drawing from disks of 0.040 in. stock are shown in Fig. 13.51. It should be noted that the final cup is slightly conical since the punch that contacts the cup bottom is slightly smaller in diameter than the die diameter. When the cup is drawn through the die, the top portion of the cup will spring outward a small distance. In deep drawing shells, the problem is to obtain the final shape with the minimum number of draws and anneals. When too great a draw is made, the material may buckle, wrinkle, split, or tear. Buckling and wrinkling are sometimes the result of incorrect

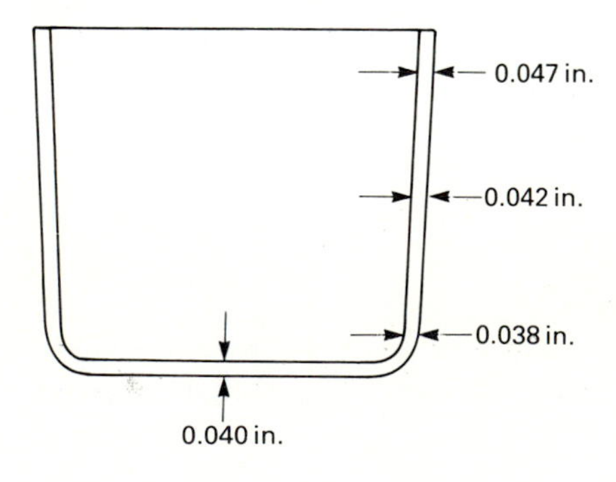

FIGURE 13.51
Typical variation in wall thickness under deep drawing. (*Redrawn from Befzalel Avitzur,* Handbook of Metal-Forming Processes, *Wiley, New York, 1983.*)

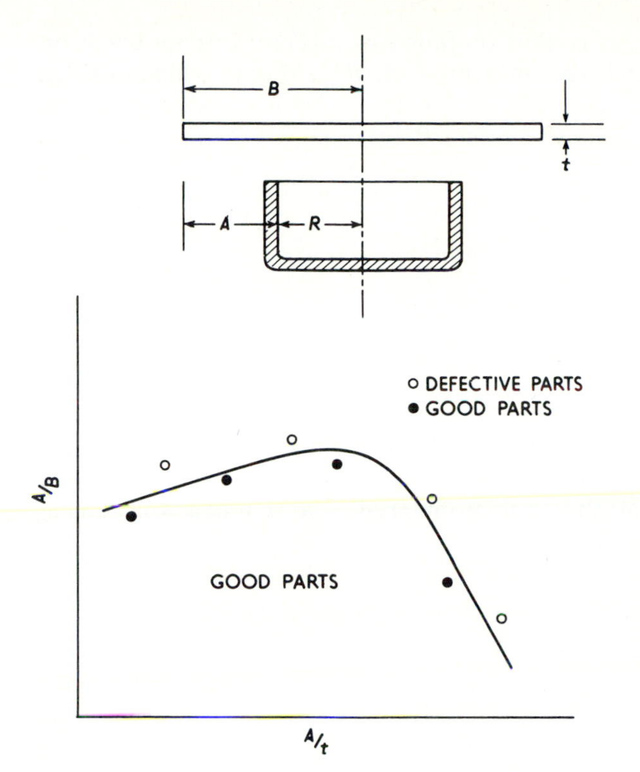

FIGURE 13.52
Curve for a given material whose upper boundary is characterized by plastic buckling and boundary to right by elastic buckling.

design of the blank and too little hold-down pressure on the outer slide (which holds the blank). Incorrect lubrication and die design may also result in faulty work.

The parameters for deep drawing using mechanical dies include the material, blank diameter, material thickness, and the flange overhang. The ratio of the flange overhang to the material thickness is called the buckling ratio. When this ratio for sheet buckling is plotted against the ratio of overhang flange and blank radius, a theoretical curve may be obtained for a given material for predicting plastic and elastic buckling limits in the rim.[1] For circular sheet buckling (Fig. 13.52), we have the relationship:

$$S_b = K\frac{E}{(B/2t)^2}$$

where S_b is the critical buckling stress

$$K = \frac{k}{12(1-\mu^2)}$$

[1] Taken from W. W. Wood, R. E. Goforth, and R. A. Ford, "Theoretical Formability," *Product Engineering,* October, 1961.

E is the Young's modulus, μ is the Poisson ratio and k is the constant based on geometry, edge restraints, and type of loading. B, R, and t are defined in Fig. 13.52.

Values of k for various ratios of B and R are:

$\frac{B}{2R}$	k
1	2.4
2	3
3	3.2

The final piece may have portions that are thicker than the original material where sections were compressed. Other areas where the material was stretched will be thinner than the original material. The thinning will be near the bottom as shown in Fig. 13.51.

Drawing and hold-down forces. An empirical estimate of the drawing force F for a cupping operation can be computed from

$$F = \pi dtS\frac{B}{d} - C$$

where S is the tensile strength of sheet metal, lb/in.^{2}, d is the average diameter of cup, in., B is the blank diameter, in. and C is the constant to account for friction and other losses, with a value of 0.7 appropriate for average conditions.

The hold-down pressure on the rim of the blank should be from $\frac{1}{4}$ to $\frac{1}{3}$ of the drawing force. Excessive hold-down forces will result in rupture at the bottom radius or in the tube wall.

Drawing reduction and number of drawing operations required. Before the proper tooling can be designed for the drawing operations, the production engineer must estimate the number of draws required. Because the metal flow in one operation has a practical limit, on the first draw the area of the blank should be limited to no more than $3\frac{1}{2}$ to 4 times the cross-sectional area of the punch. Also, it is usual practice to limit the reduction in the first draw to 40 percent, with reductions of 20 to 25 percent or less on the next two draws.

The strain-hardening graph (Fig. 13.6) can be used to estimate the number of reductions possible in a series of drawing operations. During a 40–25–25 series of reductions, the mechanical properties of a cup will follow the dashed lines from 40 to point 5 to point 6 during the first draw. The second draw will start at the new yield point of about 60,000 lb/in.^{2} and proceed through points 7 and 8. Likewise, the third draw will follow points 9 and 10 to end with a material with about 79,000 lb/in.^{2}, but almost full hard. By moving vertically downward to point 11 on the σ_0 line, we find that the total reduction is about 65 percent.

Note that two successive draws, at 40 percent reduction each, would have resulted in about the same total reduction, but in practice the second draw would probably have ruptured the bottom. An annealing operation between the two 40 percent reductions would have been a viable alternative. Thus, we must make a judgment with regard to the tradeoff between adding an annealing operation and designing and building the tooling for a third drawing operation. This decision can only be resolved by estimating the probable activity of the part in the future and the cost and lead time of the tooling compared with the cost and delays involved when using the annealing operation.

Design details for drawing tools. The magnitude of the corner radii for drawing punches and dies usually ranges from four to eight times the metal

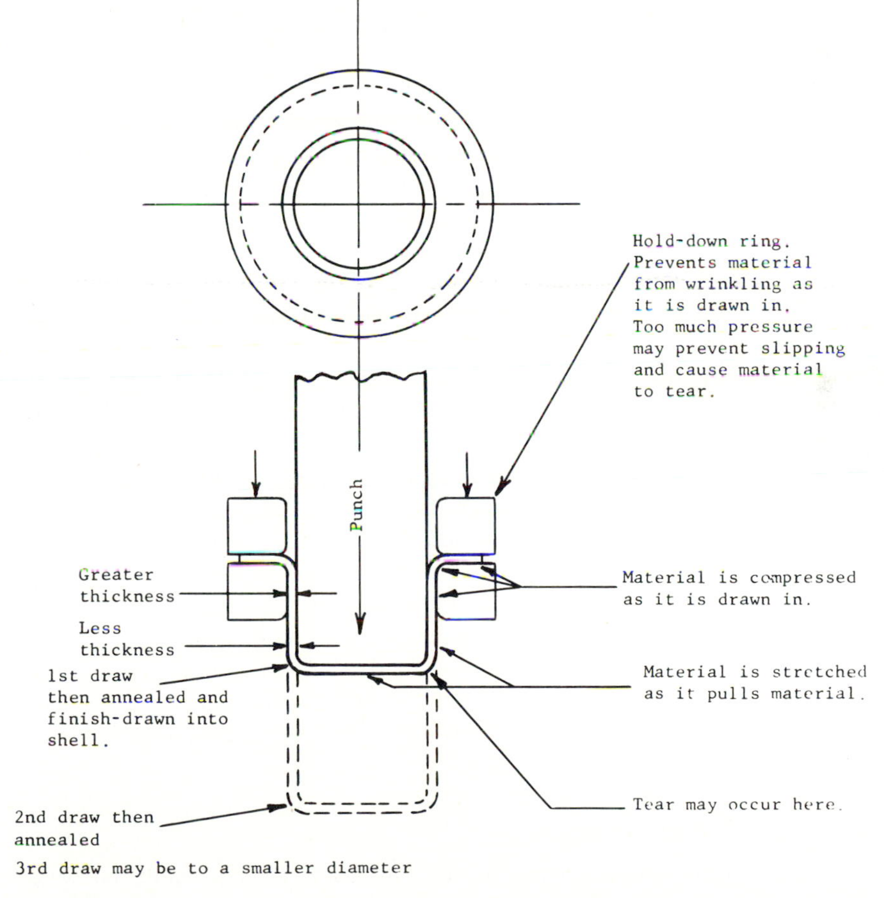

FIGURE 13.53
Typical cupping operation showing hold-down ring, vent hole, and drawing punch.

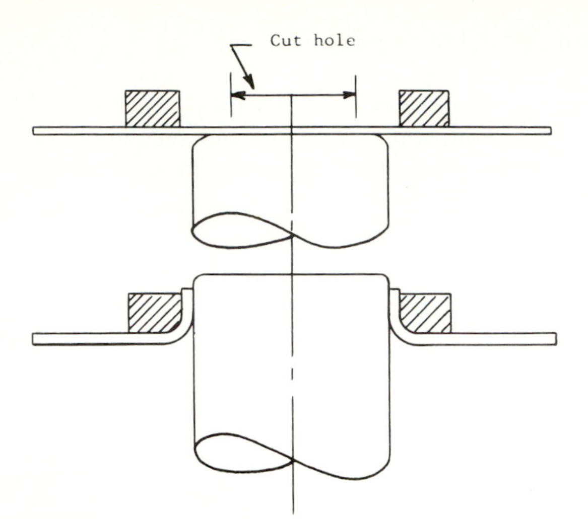

FIGURE 13.54
Holes can be flanged to give material for a threading operation or to stiffen the material near the hole.

thickness. Radii should be as large as possible, but if they are too large, wrinkling will become a problem.

The normal clearance between the punch and die should be about $1.1t$, that is, the die opening would be the punch diameter plus $2.2t$. Ironing occurs when the cup wall thickness exceeds the clearance between the punch and die. This can happen by design if a specified wall thickness is required or it can occur at the top of the cup where the metal has been thickened by the drawing operation or wrinkling.

Air vents are needed to avoid collapse of the cup walls during the stripping operation. The drawing lubricant tends to seal the space between the punch and cup so that atmospheric pressure is applied to only the outside of the hollow cup during the stripping operation. Air vents should be about $\frac{3}{16}$ in. diameter for punches up to 2 in. diameter. Larger punches need several vents.

The production-design engineer can visualize the stresses encountered, and in many cases they can be calculated empirically with reasonable accuracy. The forces depend on unknown friction between material and die, which can be estimated, plus the forces of compression (Fig. 13.53). Drawing operations performed on large parts are done in heavy duty presses. Successive draws may require annealing and cost is increased proportionally. However, small cups and cylinders that can be drawn cold in one operation cost approximately the same as most stamping operations. Often cups, or shells, need to be trimmed. This may be done in a lathe although it can also be done in a press. If the part has a flange, the edge can be accurately trimmed in a blanking die. Holes can be flanged by drawing (Fig. 13.54).

Determination of Blank Size

Graphical methods may be employed in order to compute the blank diameter of the cup (Fig. 13.55). Thus, for a cup of height H and diameter D, we need a

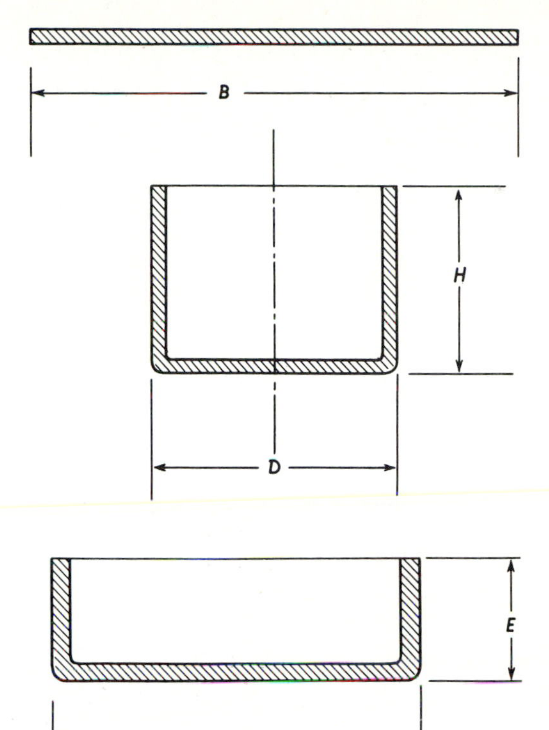

FIGURE 13.55
Standard designations for the dimensions referred to in Fig. 13.56.

blank diameter B which will equal $\sqrt{D^2 + 4DH}$ as found from Fig. 13.56. Thus the dimensions of the first draw are as follows: $C = 0.60B$ and $E = 0.267B$. The estimated percentage reduction of diameter C in the first redraw approximates the values listed in Table 13.6.

The geometrical relationships illustrated in Fig. 13.56 will assist the production-design engineer in estimating the material usage required for producing different designs of drawn shells.

Draw and trimming dies are expensive tools and the operations are costly. By the proper choice of material and intermediate drawn shapes, the number of drawing operations frequently can be reduced. Likewise, by careful design, it is often possible to eliminate the necessity of blanking operations and final trimming.

Considerable interest and development activity are now evident in the hydraulic methods of drawing material. The material is formed by applying hydraulic pressure directly on the material either over a punch or within a die without its companion die or punch. The liquid is sealed in the cavity by clamping the material around the edge tightly enough to prevent the escape of the liquid. The fluid exerts a steady, uniform pressure on every part of the blank. This makes possible the formation of involved contours (as in embossing) and unusual draws (such as cone-shaped or tapered sections). For

	AREA OF BODY A=	DIA. OF BLANK D=
	$\frac{\pi d^2}{4} + \pi d h$	$\sqrt{d^2 + 4dh}$
	$\frac{\pi d_1^2}{4} + \pi d_1 h + \pi f \frac{d_1 + d_2}{2}$	$\sqrt{d_1^2 + 4d_1 h + 2f(d_1 + d_2)}$
	$\frac{\pi d_1^2}{4} + \pi d_1 h + \frac{\pi}{4}(d_2^2 - d_1^2)$	$\sqrt{d_1^2 + 4d_1 h + d_2^2 - d_1^2} = \sqrt{d_2^2 + 4d_1 h}$
	$\frac{\pi d_1^2}{4} + \pi d_1 h_1 + \frac{\pi}{4}(d_2^2 - d_1^2) + \pi d_2 h_2$	$\sqrt{d_2^2 + 4(d_1 h_1 + d_2 h_2)}$
	$\frac{\pi d_1^2}{4} + \pi d_1 h_1 + \frac{\pi}{4}(d_2^2 - d_1^2) + \pi d_2 h_2 + \pi f \frac{d_2 + d_3}{2}$	$\sqrt{d_2^2 + 4(d_1 h_1 + d_2 h_2) + 2f(d_2 + d_3)}$
	$\frac{\pi d_1^2}{4} + \pi d_1 h_1 + \frac{\pi}{4}(d_2^2 - d_1^2) + \pi d_2 h_2 + \frac{\pi}{4}(d_3^2 - d_2^2)$	$\sqrt{d_3^2 + 4(d_1 h_1 + d_2 h_2)}$
	$\frac{\pi d^2}{2}$	$\sqrt{2d^2} = 1.414 d$

	AREA OF BODY 'A' =	DIA. OF BLANK 'D' =
	$\frac{\pi}{4}(d_1^2 + 4h^2) + \pi f \frac{d_1 + d_2}{2}$	$\sqrt{d_1^2 + 4h^2 + 2f(d_1 + d_2)}$
	$\frac{\pi}{4}(d^2 + 4h_1^2) + \pi d h_2$	$\sqrt{d^2 + 4(h_1^2 + d h_2)}$
	$\frac{\pi}{4}(d_1^2 + 4h_1^2) + \pi d_1 h_2 + \pi f \frac{d_1 + d_2}{2}$	$\sqrt{d_1^2 + 4\left[h_1^2 + d_1 h_2 + \frac{f}{2}(d_1 + d_2)\right]}$
	$\frac{\pi}{4}(d_1^2 + 4h_1^2) + \pi d_1 h_2 + \frac{\pi}{4}(d_2^2 - d_1^2)$	$\sqrt{d_2^2 + 4(h_1^2 + d_1 h_2)}$
	$\frac{\pi d_1^2}{4} + \pi S \frac{d_1 + d_2}{2}$	$\sqrt{d_1^2 + 2S(d_1 + d_2)}$
	$\frac{\pi d_1^2}{4} + \pi S \frac{d_1 + d_2}{2} + \pi f \frac{d_2 + d_3}{2}$	$\sqrt{d_1^2 + 2[S(d_1 + d_2) + f(d_2 + d_3)]}$
	$\frac{\pi d_1^2}{4} + \pi S \frac{d_1 + d_2}{2} + \frac{\pi}{4}(d_3^2 - d_2^2)$	$\sqrt{d_1^2 + 2S(d_1 + d_2) + d_3^2 - d_2^2}$

	AREA OF BODY 'A' =	DIA. OF BLANK 'D' =
	$\frac{\pi d_1^2}{2} + \frac{\pi}{4}(d_2^2 - d_1^2)$	$\sqrt{d_1^2 + d_2^2}$
	$\frac{\pi d_1^2}{2} + \pi f \frac{d_2 + d_1}{2}$	$1.414\sqrt{d_1^2 + f(d_2 + d_1)}$
	$\frac{\pi d^2}{2} + \pi d h$	$1.414\sqrt{d^2 + 2dh}$
	$\frac{\pi d_1^2}{2} + \pi d_1 h + \frac{\pi}{4}(d_2^2 - d_1^2)$	$\sqrt{d_1^2 + d_2^2 + 4d_1 h}$
	$\frac{\pi d_1^2}{2} + \pi d_1 h + \pi f \frac{d_1 + d_2}{2}$	$1.414\sqrt{d_1^2 + 2d_1 h + f(d_1 + d_2)}$
	$\frac{\pi}{4}(d^2 + 4h^2)$	$\sqrt{d^2 + 4h^2}$
	$\frac{\pi}{4}(d_1^2 + 4h^2) + \frac{\pi}{4}(d_2^2 - d_1^2)$	$\sqrt{d_2^2 + 4h^2}$

	AREA OF BODY 'A' =	DIA. OF BLANK 'D' =
	$\frac{\pi d_1^2}{4} + \pi S \frac{d_1 + d_2}{2} + \pi d_2 h$	$\sqrt{d_1^2 + 2[S(d_1 + d_2) + 2d_2 h]}$
	$\frac{\pi d_1^2}{4} + \frac{\pi^2 r}{2}(d_1 + 1.274r)$ OR $\frac{\pi}{4}(d_2 - 2r)^2 + \frac{\pi^2 r}{2}(d_2 - 0.726r)$	$\sqrt{d_1^2 + 6.28rd_1 + 8r^2}$ OR $\sqrt{d_2^2 + 2.28rd_2 - 0.56r^2}$
	$\frac{\pi d_1^2}{4} + \frac{\pi^2 r}{2}(d_1 + 1.274r) + \pi h d_2$ OR $\frac{\pi}{4}(d_2 - 2r)^2 + \frac{\pi^2 r}{2}(d_2 - 0.726r) + \pi h d_2$	$\sqrt{d_1^2 + 4(1.57rd_1 + 2r^2 + hd_2)}$ OR $\sqrt{d_2^2 + 4d_2(h + 0.57r) - 0.56r^2}$
	$\frac{\pi d_1^2}{4} + \frac{\pi^2 r}{2}(d_1 + 1.274r) + \pi f \frac{d_2 + d_3}{2}$ OR $\frac{\pi}{4}(d_2 - 2r)^2 + \frac{\pi^2 r}{2}(d_2 - 0.726r) + \pi f\left(\frac{d_2 + d_3}{2}\right)$	$\sqrt{d_1^2 + 6.28rd_1 + 8r^2 + 2f(d_2 + d_3)}$ OR $\sqrt{d_2^2 + 2.28rd_2 + 2f(d_2 + d_3) - 0.56r^2}$
	$\frac{\pi d_1^2}{4} + \frac{\pi^2 r}{2}(d_1 + 1.274r) + \frac{\pi}{4}(d_3^2 - d_2^2)$ OR $\frac{\pi}{4}(d_2 - 2r)^2 + \frac{\pi^2 r}{2}(d_2 - 0.726r) + \frac{\pi}{4}(d_3^2 - d_2^2)$	$\sqrt{d_1^2 + 6.28rd_1 + 8r^2 + d_3^2 - d_2^2}$ OR $\sqrt{d_3^2 + 2.28rd_2 - 0.56r^2}$
	$\frac{\pi d_1^2}{4} + \frac{\pi^2 r}{2}(d_1 + 1.274r) + \pi d_2 h + \frac{\pi}{4}(d_3^2 - d_2^2)$ OR $\frac{\pi}{4}(d_2 - 2r)^2 + \frac{\pi^2 r}{2}(d_2 - 0.726r) + \pi d_2 h + \frac{\pi}{4}(d_3^2 - d_2^2)$	$\sqrt{d_1^2 + 6.28rd_1 + 8r^2 + 4d_2 h + d_3^2 - d_2^2}$ OR $\sqrt{d_3^2 + 4d_2(0.57r + h) - 0.56r^2}$
	$\frac{\pi d_1^2}{4} + \frac{\pi^2 r}{2}(d_1 + 1.274r) + \pi d_2 h + \pi f \frac{d_3 + d_2}{2}$ OR $\frac{\pi}{4}(d_2 - 2r)^2 + \frac{\pi^2 r}{2}(d_2 - 0.726r) + \pi d_2 h + \pi f \frac{d_2 + d_3}{2}$	$\sqrt{d_1^2 + 6.28rd_1 + 8r^2 + 4d_2 h + 2f(d_2 + d_3)}$ OR $\sqrt{d_2^2 + 4d_2(0.57r + h + \frac{f}{2}) + 2d_3 f - 0.56r^2}$ [illegible]

FIGURE 13.56
Formulas for area and blank diameters of drawn shells.

TABLE 13.6
Typical reduction percentages for the first redraw operation

Thickness of stock in.	Reduction percentage		
	Steel	Aluminum	Brass
$\frac{1}{16}$	20	25	30
$\frac{1}{8}$	15	20	24
$\frac{3}{16}$	12	15	18
$\frac{1}{4}$	10	12	15
$\frac{5}{16}$	8	10	12

the ordinary mechanically drawn sections, the mechanical press and tools are less expensive.

The Hydroform press uses a thin flexible diaphragm sealed against high pressure and backed by hydraulic fluid. The material blank is held in position and the punch is moved up by hydraulic pressure into the die cavity, forcing the diaphragm pad and the blank to take the shape of the punch.

When compared with conventional tooling, Hydroforming tooling is relatively simple since only a punch and nest ring is required. The punch should have a good finish since the part being drawn will reproduce the punch. For short runs, kirksite, brass, and wood punches have proved successful.

Stretch Forming

The simplest use of the principle of stretching material beyond the elastic limit without compression is in straightening wire or bars by stretching. The process of stretcher-leveling sheets is slightly more involved, but uses the same concept. The sheet is clamped firmly on each end and stretched well beyond the elastic limit. This results in a flat or level sheet that is used for furniture and cabinet manufacture. In forming contours, the same principle is applied, with the addition of a form on which the sheet or part is stretched. The material takes the shape of the form without any springback. Several machines use this process to greater or lesser degrees. Some machines cannot form reverse bends or complete circles; others can do so to a limited extent. In all cases, allowance must be made for extra material to be clamped by jaws when it is stretched. This portion cannot have a contour and may be removed or used as a part of the piece. For example, electric-stairway track angles that were made on forming rolls (Fig. 13.57) had to be straightened on a surface plate and had one tangential end. With contour forming, the exact radius was obtained without straightening, and it was possible to make both ends tangential. The time save by stretch bending was from $\frac{1}{2}$ to $1\frac{1}{2}$ hours per piece, and better quality was obtained.

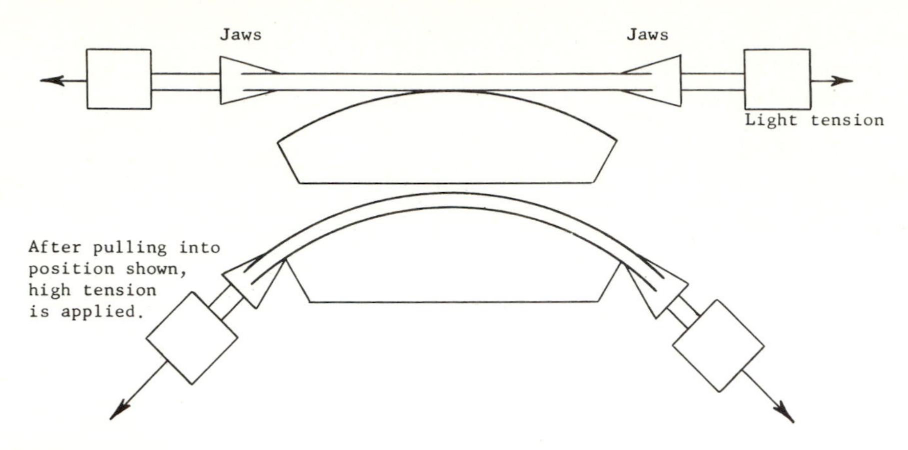

FIGURE 13.57
Schematic representation of a stretch-bending operation.

A STRETCH-BENDING PROBLEM. It is desirable to purchase a stretch bender for forming irregular curves in odd-shaped members. A young engineer is asked to give an engineering explanation of why the parts do not spring back, but take the exact form of the die. An analysis of the stresses incurred during the process and the residual stresses is requested. This information will determine the extent of the application of the equipment and process to materials and parts.

The process consists of shaping a part over a form and pulling at the ends to stretch the metal. No springback occurs. The finished piece has the radius of the die and the part may be a strap or an angle with its flange in or out, or any odd shape.

Discussion of Fig. 13.57, Case 1. Consider a beam supported at each end and loaded in the center. If the load is released, springback will occur (Fig. 13.58*a*).

If tension is applied (Fig. 13.58*b*), the fiber stress will be affected as shown. If the tension applied is great enough, there are no compression stresses. If the load is released, springback will occur.

If further tension is applied (Fig. 13.58*c*), the elastic limit will be exceeded on the top side; when the load is released, some permanent set will occur, but there will be springback owing to the unequal forces on each side of the neutral axis.

If further tension is applied, as in stretch bending, the maximum strength of the fibers will be exceeded and the top fibers will not carry as much as those in the center. When the stress is fairly uniform across the section, the maximum stress will be in the center. When the load is released, there will be no springback, as the forces are balanced about the neutral axis (Fig. 13.59).

The foregoing explanation comes from analyzing what occurs within the

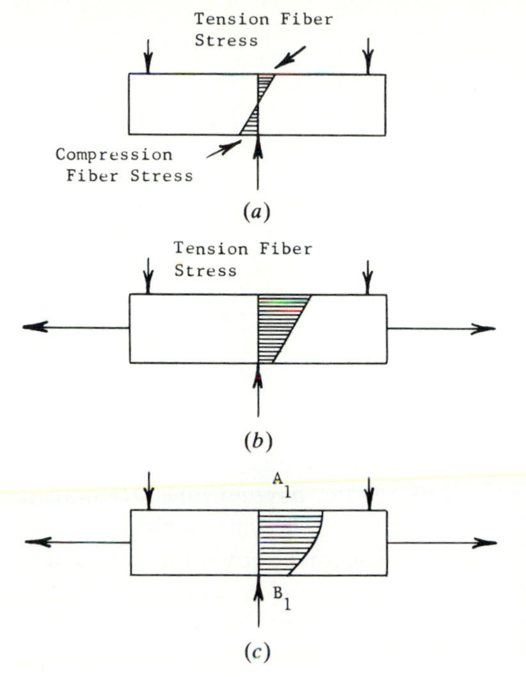

FIGURE 13.58
Schematic representation of the internal stress distribution within a typical work period. (*a*) springback can occur; (*b*) stress when tension is applied; (*c*) sufficient stress is applied to cause permanent set and to eliminate springback.

part and is based on the stress–strain diagram information obtained in the testing laboratory.

Discussion of Fig. 13.57, Case 2. The stress–strain diagram (Fig. 13.60) can also be used to explain the reason for no springback after stretch bending.

1. As long as the stress in the material remains on line OD, the stresses will return to zero and the part returns to its original shape when the load is released.

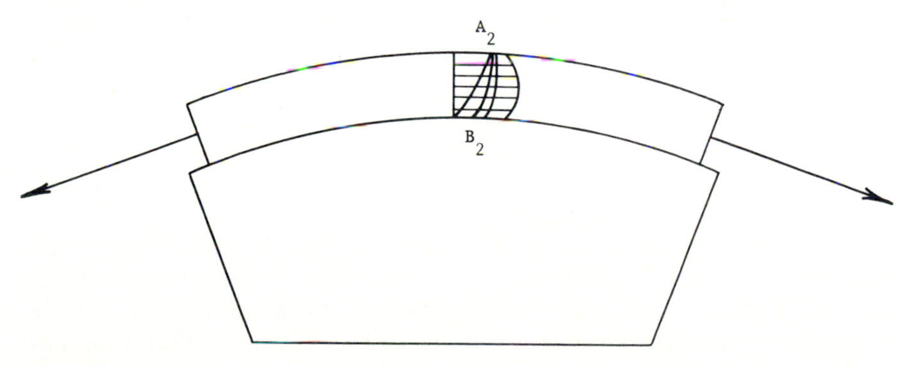

FIGURE 13.59
Typical stretch-forming workpiece and die.

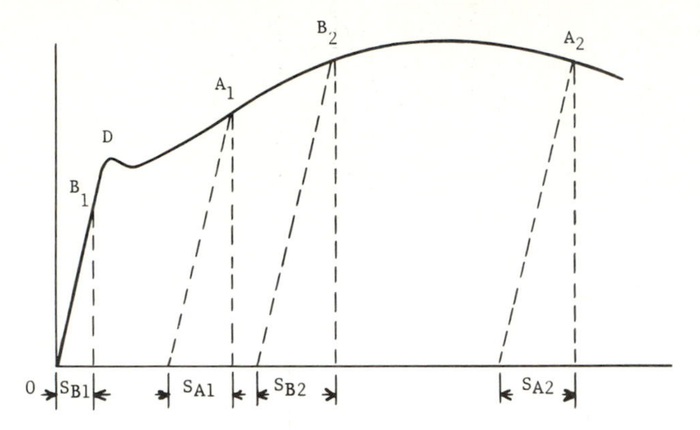

FIGURE 13.60
Typical stress–strain diagram for hot-rolled steel.

2. When the stress in the outer or top fibers A_1 in Fig. 13.60 exceeds the elastic limit, the stress is indicated as progressing beyond line OD until it reaches point A_1. Stresses in the bottom fiber are shown reaching point B_1 on line OD. When the load is released, A_1 will travel down the dotted line to zero stress and indicate a strain S_{A_1} that is greater than strain S_{B_1} which is zero because B_1 was on the elastic portion of the stress–strain curve.
3. When the stress is increased, the stress in fiber A_1 progresses along the line over the maximum stress to point A_2. At the same time, fiber stresses in B_1 progress along to point B_2, both at about the same value of stress. When the load is released, the strains S_{A_2} equal strains S_{B_2}, and therefore there is no springback. It can also be seen that the material is cold-drawn and will have the characteristics of cold-drawn material.

The shape of the stress–strain curve of the material indicates whether it is suitable for stretch bending. Hot-rolled angles bend to shape easily. Thus, unsatisfactory results can be avoided by giving thought to the shape of the stress–strain curves of the material in relation to the fundamental principles of stretch bending.

An analysis of the residual stresses remaining in the stretch-bent bar or sheet can be made to determine other limitations of the process.

Spinning

Spinning usually is performed on a lathe. The mold is fastened to the lathe head and the material fastened to the mold. As the mold and material are spun, a tool (resting on a steady rest) is forced against the material until the material contacts the mold. There is considerable friction between the tool and the material. The skilled operator applies his weight to the overhanging handle of the tool and presses with the proper force so that the tool will not dig into the material. The tools are hardened and round-nosed so they will slide easily and will not wear. Lubricants are used to reduce friction. The material is stretched as it is formed until it is snug against the mold. Brass, aluminum, and copper are the usual materials used for spinnings. Heated steel is spun by mechanical means in large diameters (up to 240 in.) and in thicknesses as great as $\frac{1}{4}$ in. on large vertical boring mills.

A variety of designs can be made. The material can be trimmed and the edge can be beaded. Parts with narrow openings, such as pitchers, vases, and lamp bases, are made on collapsible molds or by mechanical spinning.

Mechanical spinning operations in which rollers are used to form the material are utilized extensively in making tin cans, galvanized tanks, and large steel tanks. Spinnings are economical for experimental models and preliminary production runs before the final design is established. Later, more expensive tools can be made for more economical operation costs. Spinnings are also frequently used in medium-volume production parts, becasue they are less expensive than drawn parts. Typical spinnings are lamp reflectors, mixing bowls, and cooking utensils.

Roll flowing is similar to spinning in that the operation is usually performed on a lathe (Fig. 13.61). It also is known by the trade names Floturning, Shear Spinning, and Hydrospinning. This process of plastic working of metal produces parts to a predetermined geometrical shape by displacing metal in advance of a roller that is fed along a mandrel machined to the desired inside diameter of the part. Usually, hydraulic pressure is used to force the roller against the part and for clamping the blank to the tailstock. In this process, the metal is displaced in the direction of the feed of the roller along the length of the mandrel.

From the production designer's viewpoint, the principal problems associated with this process are blank development and proper feed pressure. Blank development requires determination of wall thickness and diameter of

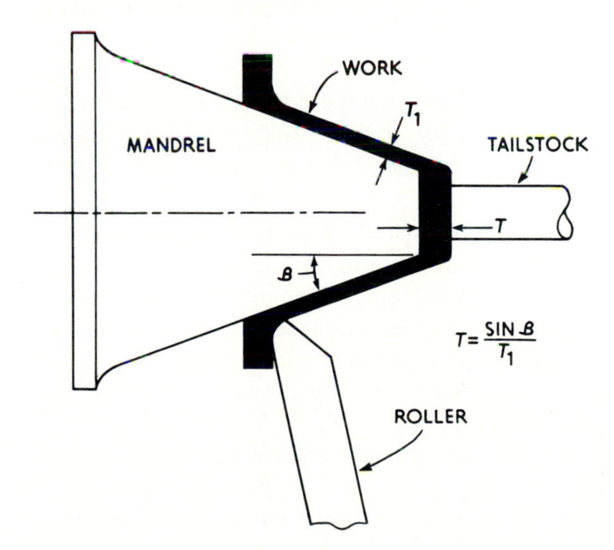

FIGURE 13.61
Roll flowing a conical part.

the original blank so as to provide predetermined final size for the part, allowing a specified amount of material for trimming.

Roller pressures are approximately 200,000 lb/in.^{2} This cold working will, of course, result in considerable work hardening. Tensile strengths are doubled frequently.

This process is being used when production requirements are usually not large, such as in duct work or dairy equipment (cream separators, etc.).

Limitations of Press Techniques

The engineer, in order to reduce costs, may become overenthusiastic about using stampings and press techniques to make his parts. The successful production-design engineer must always consider the limitations of every process. For example, a punched hole does not make a good bearing. An inside diameter, to be used as a bearing, should be drilled, reamed, and, in some cases, bushed. It is difficult to have holes in alignment when a blank is formed after piercing the holes. Holes required to be in alignment should be punched, drilled, or line-reamed after forming. The punched hole is not always suitable for tapping. Sheet metal tolerances are usually $\pm\frac{1}{32}$ in., which is large compared to machining.

Die Designing

Die designing is a field in itself. The types of dies and presses run the full range of man's ingenuity and abilities. Problems of stripping and use of springs, rubber, air cushions, movable pads, automatic stops, feeders, and ejectors should be understood, but space is not available to cover these items.

Some practical hints for the die designer, as developed by the Royal Typewriter Company, are:

1. Always check die drawing with part prints to make certain that the die will produce the part correctly with reference to burr side, relation of grain direction to forms, shear forms, embossments, blanking pressure required, and strength of fragile projections.
2. Dimension the detail drawing with the information in the form required by the diemaker, so that he has a minimum amount of translating to do.
3. As soon as the design of a die is decided upon, the material for the die should be ordered so that there are no unnecessary delays. Tryout material for the part should be ordered at the same time.
4. Always conform to customer or company specifications for type of die construction, kind of steel, die sets, and press data.
5. Whenever possible, specify standard die parts from catalogs or technical literature of reputable manufacturers.
6. All die parts should be designed to be easily removable with minimum disassembly.

7. The safety and the relation of efficiency to fatigue of the press operator should be considered in the planning of every die desgin.
8. In designing a die it is essential to know the type of press in which it is to be used. In selecting the proper size and type of press for a specific job, the designer should consider the following factors:
   a. The size and type of die required
   b. The amount of stroke necessary
   c. The press pressure required for doing the work
   d. The distance above the bottom of the stroke where the pressure first occurs
   e. Any additional pressure required owing to attachments, such as the blank holder, ironing wrinkles, or stretching the materials in drawn work
   f. The method of feeding, the direction of the feed, and the size of the sheet, blank, or workpiece
9. Remember that the die opening sizes the outside contour of the blank and that the perforators size the openings in the blank.
10. Split the punch pad if there are a great number of punches; it simplifies the lining-up process.
11. Pilots and punches must be correct in the lineup with back gage and blanking center of die.
12. Whenever possible removable pilots should be used to facilitate grinding of the punches.
13. Punches should always be as short as practical. Long production runs require punches long enough to grind from $\frac{1}{2}$ to 1 in.
14. Pilots must be long enough to locate the strip accurately before punches start cutting or forming.
15. On complicated forms always use a template to check the parts and the layout of punch and die.
16. The die thickness should be determined by the requirements: production run, degree of precision, and budget of both time and cost of building.
17. All delicate projections or weak parts in a die should be designed as inserts if possible.
18. If stock guides are indicated, their length should be made twice the width of the stock before the first operation.
19. The spring stripper should be of sufficient length and width to provide for additional springs if necessary. It should be thick enough to avoid damage against the stripper bolts, which sometimes loosen.
20. When punches are guided in a spring stripper, the stripper must be supported on guide posts for accurate alignment. The stripper should be bushed, with dowels in the die placed to engage the stripper before the punches enter the die. When punches are not guided in the stripper, ample clearance must be allowed around small punches to assure that they will not be thrown out of line. The stripper should be fitted close to one or two stronger, heavier punches (0.001- to 0.005-in. clearance) and allow 0.010- to 0.015-in. clearance on the smaller punches.

21. The solid or bridge stripper should always be well doweled to prevent it from shifting. This is of special importance when the stripper does heavy guiding.
22. The punch holder must be thick enough to allow proper travel on draw and form pads, strippers, spring pins, spring pilots, stripper bolts, and moving punches.
23. The shoulders of draw pads or shedders should never strike against the retaining sections. Make knockout pins the correct length.
24. Never form too close to the edge of any opening in the blank without proper support along the edge.
25. Use a standard commercial bolster plate whenever possible. These are usually torch-cut to any size or shape. The design of the bolster plate must be the most effective compromise possible between a strong, rigid support for the die and a means of locking the die assembly in the press.
26. Use a stationary stripper whenever possible. The stationary stripper screwed and doweled in the die is more positive and accurate for lining up punches going through bushing guides. It is also more economical to make than a spring stripper.
27. Spring strippers are used where visibility is an important factor and where flatness of parts is required.
28. Spring strippers are also needed when lugs, draws, or embossments necessitate the placing of the stripper at some distance from the die to allow space for the strip to feed through.
29. If possible, use the insert and split-die-section construction for small and difficult die openings.
30. Always determine whether a more satisfactory design can be obtained by the use of an inverted die.

Practical hints for the tool and die maker are:

1. When dies for rectangular draws give excessive trouble in the corners, it is well to check the draw radii and fit between punch and die.
2. Tap holes in unhardened tool steel parts with tap 0.003 in. oversize to allow for shrinkage and distortion.
3. Punches should be polished in their longitudinal direction (since stripping of the work takes place in this direction). In the case of slender perforating punches working in heavy stocks, annular marks from turning or polishing tend to weaken such punches.
4. The punch plate must be of adequate thickness to support its punch or punches. This is of particular importance when punches are not guided in the stripper.
5. Taper-ream all slug holes in the die shoe.
6. Normalize dies or sections after rough machining, particularly for intricate

shapes or when extreme accuracy is required; then finish-machine and harden.

7. A punch or shedder with a base only on one side, or with a split base, will usually warp in hardening; therefore, allow grinding stock to compensate for the warpage.

It must be remembered that the tool designer has only one chance to make his tool. He can modify and improve within the limits of his tool budget, but he cannot come back to his manager and say that he has a better way of doing the operation and request an appropriation for a new tool. He must be right the first time. This is in contrast to the designer, who can make his models and test them and, if necessary, radically change them. The experienced tool designer usually obtains close to the ultimate results with the first set of tools. Of course, when major savings can be shown by a new method, money should be appropriated to make the saving.

High-Speed Operations

With the advent of cemented carbide and tough cast alloy tools, the speed of punching has been increased to such an extent that punching of material is not a shearing or tearing action, but a breaking action. Thick materials are being punched with cleaner holes and fewer burrs. Presses are now available that operate at 2300 strokes per minute. Material is fed 5000 to 10,000 in./min. Hexagonal nut blanks have been pierced, countersunk, blanked, and crowned at a rate of 1150 nuts per minute. Ball-bearing and roller-bearing guide pins have been developed to withstand the high speeds of the press. There are presses that may operate rams in vertical and two horizontal side positions, enabling the forming of complex parts in progressive operations. The field of wire and strip forming is a speciality in itself.

Presses

Presses for sheet-metal work range from the massive machines used to make automobile body parts to the small bench types. They may be driven hydraulically or mechanically. Table 13.7 provides a comparison of the performance characteristics of these two types of drive. The reader should understand that mechanical presses develop their rated tonnage at a specific distance above the bottom of the stroke. Consequently, it is important to identify the point on the stroke where work begins. Forming and drawing begin at a point above the bottom of the stroke equal to or greater than the depth of the draw. Of course, at this point, the force exerted at the press is less than the rated tonnage.

Blanking and piercing by mechanical presses is done near the bottom of the stroke, where developed tonnage is the greatest. Hydraulic presses are

TABLE 13.7
Comparison of performance characteristics of hydraulic and mechanical drive presses

Characteristic	Hydraulically driven	Mechanically driven
Variation in force	Force is constant throughout press stroke	Force varies with position of the crank
Length of stroke	Easily adjusted and controlled	Fixed by the throw of the crank
Speed adjustment	Adjustable over wide range	Limited by the speed of drive
Overloading	Can deliver only a preset force; slide motion stops upon reaching that force	Can be overloaded. Damage to press can take place unless the press is equipped with overload protection
Speed	Slower cycle time	Fast cycle time; suited for high production
Motor size	Larger motor is required. May be up to 2.5 times as large as mechanical press	Smaller motor required since energy is stored in the press flywheel
Ram velocity	Low	High, thus this style useful in high-impact type work such as blanking and piercing

better suited for making deep and/or intricate draws, when the length of stroke can be adjusted easily to accommodate the operation and the force is constant throughout the press stroke. A number of styles of stamping and drawing presses are compared in Table 13.8.

In addition to tonnage, the engineer should consider the width through the back of the press, shut height, stroke, bolster size, and flywheel speed before selecting a given press to perform an operation. The shut height is important from the standpoint of placing work in the dies. The stroke is important when drawing operations are performed. Obviously deep-draw presses are those that are capable of having greater strokes than a press used for blanking and piercing. The flywheel speed determines the rate of production; the width through the back is important if the completed part or scrap is to be ejected through the back. The bolster size places a limitation on the die size and consequently the size of the work being processed.

Presses can be equipped with such auxiliary equipment as automatic feeding mechanisms, scrap outfits, and stock straighteners in one integrated automatic setup. Figure 13.22 illustrates an inclinable punch press equipped with a roll feed, scrap cutter, and a straightening machine for pulling the coiled stock from a brake reel.

Presses are usually classified by power-transmitting mechanisms, frame

TABLE 13.8
Stamping and drawing presses

Machine	Size	Operations	Features	Uses
Foot press	Small sizes only	Small work—single small holes, light stampings, embossing	Hand fed; may have adjustable bed or horn to accommodate large work	Jewelry, buttons, silverware, radio parts, etc.
Bench press	1,000 pounds to 12 tons	Embossing, stamping, etc., upon light material	May have roll feed, ratchet dials, magazines, etc.	Watch parts, novelties, jewelry, etc.
Inclinable press—open-back (OBI)	4 to 90 tons	Blanking, bending, stamping, forming, assembling	Inclined from vertical to 45° backward, may have drawing attachments	For light sheet metal
Open-back gap press (OBG)	1 to 225 tons	Punching, shearing, cutting-out, trimming, forming, etc.	Ram driven by cam or eccentric	Automobile parts, etc., for large or irregular work
End-wheel gap press	Small to 50 tons	Blanking, forming, notching, piercing, cutting	Flywheel at rear and crankshaft at right angles to bed	For work with long, narrow strips of metal
Deep-gap punch press	Medium sizes	Punching	Deep clearance in back frame	For wide sheets
Horning press	10 to 100 tons	Forming, stamping, blanking, wiring, punching, riveting	Horn bolted to frame; may have swinging table also	For hollow cylindrical work, as steel drums, etc.
Double-crank overhanging press	Small	Blanking, cutting, piercing	Usually automatic or semiautomatic feed-overhanging frame gives large die space—flywheel—or gear-driven	For large, light sheet metal
Notching press	Small to medium	Cutting slots in edges of usually circular work	Usually short stroke machine, with work driven by rocker arm—450–650 strokes per min	Notching motor laminations
Single- and double-action presses	Heavy	Shaping, blanking, forming, etc.	Frame consisting of a base, a crown, and two uprights tied together with steel tie rods	Heavy stock

TABLE 13.8
(*Continued*)

Machine	Size	Operations	Features	Uses
Arch press	Light	Cutting, trimming, shaping, blanking	Offers large bed area	For light, large-area sheet metal
Double-crank straight-side press	To 2000 tons	Punching, cutting, bending, blanking, shaping	Slide often counter-balanced by air cylinder—flywheel or geared	Wide variety of uses in many sizes
Four-point suspension press	100 to 1500 tons	For large-scale deep drawing, etc.	Four corners of slide suspended, giving even pressure on work	Automobile body tops, tec.
Knuckle-joint press	25 to 250 tons	Coining, upsetting, swaging, embossing, extrusion	Knuckle-joint operating slide makes short stroke necessary	Coining money, light to medium thickness metals
Straight-side high-speed press	10 to 400 tons	Blanking, stamping, etc.	Heavy construction to eliminate vibration—high-speed roll feed, and scrap cutter, variable speed motor—about 400 rpm	For high production with comparatively light metal
Dieing machine	Small	Progressive die operation, etc.	Ram operated with a pulling stroke rather than a thrust, by vertical rods passing down through bed—about 350 strokes per min	Electrical appliance parts, small automobile parts, etc.

Oscillating-die	small	Blanking, cutting, etc.	Die plate moves horizontally back and forth; strip fed through continuously. About 1000 strokes per min	Small parts of light-gage material
Multislide machine	Small	Combination of operations, as blanking, forming, bending, etc.	As many as 8 operations may be performed in succession by the different slides. Very accurate feeds. To 300 pieces per min	Operations upon light-gage metal
Double-action press	Small, medium, of large	Blanking, drawing, etc.	Stock is successively blanked and worked by an outer and an inner ram	Applications requiring two related operations
Pillar press	70 to 250 tons	Blanking, shaping, trimming, etc.	Subpress dies may be attached to the face of the ram for close-tolerance work	Light- to medium-gage work
Hydraulic press	To 1000 tons	High-speed pumping units can give this type of press operating speeds comparable to other large presses		Largest sizes of work
Horizontal draw press	Small to medium	Shell may be pushed through die for redrawing, or knockouts and stripper plates provided if shouldered		Cylindrical shells
Multiple-plunger eyelet machine	Small	Successive draws made by a row of plungers, the work transferred from station to station by finger conveyors		Light-gage shells
Rack-and-pinion deep-drawing press	10 to 30 tons	For long uniform redrawing operations with accurate length of stroke		Cylindrical shells

design, and purpose. Thus, under power transmitting mechanisms there are:

1. Crank, eccentric
2. Cam, cam and crank, cam and eccentric
3. Knuckle joint
4. Toggle
5. Rack and pinion
6. Hydraulic
7. Pneumatic

Frame design further classifies the style of press as:

1. C frame
2. Gap frame
3. Arbor
4. Horning

The purpose of the press will help identify special-purpose presses as:

1. Flat-edge trimming
2. Stretching
3. Quenching

Drawing Presses

The terms single-, double-, and triple-action are often referred to in drawing operations. These terms refer to the number of separate movements in the tools. Thus, in a single-action press, work is done in one movement without any auxiliary movement to hold the blank. In drawing operations this can be done, without wrinkling, when the material is relatively thick in comparison to its blank diameter.

In a double-action setup, a separate movement of the press gives a blank-holding action on the periphery of the blank, while it is being drawn, so as to avoid the formation of wrinkles.

Where three successive and separate pressures are required, such as holding the blank, drawing, and redrawing, a triple-action press may be employed. The triple-action press is used for such work as automobile fenders, bathtubs, and refrigerators. In the triple-action setup, the metal must be able to withstand the second draw without annealing.

Turret-type Punch Press

In the turret-type press, an inventory of differently sized and shaped punches is maintained. The turret may contain 50 or more different punches and make from 75 to 175 hits per minute. Thus, each tool can be considered as an individual station. A circular plate above the material opening holds the punches and a similar plate below the material gap holds the dies. The turrets automatically rotate to select the correct punch and die and firmly lock into position. The time for this is typically two seconds per station.

Material is mounted on a movable table that can accurately locate the sheet in the *x* and *y* planes in preparation for punching.

Through the use of numerical control operations, sheet metal may be processed automatically on this equipment. Complicated patterns of openings can be made by using single stations and combinations of different sizes and shapes.

SUMMARY

The range of application of cold metal-working processes has widened considerably in the last few years in view of added technology. Today, many parts that were formerly produced by other methods have become press-department products. The power press has always been a production machine and where quantities are high and price is an important criterion, the engineer should always consider the feasibility of making the design using the cold-working operations explained in this chapter.

Today, three- and four-cylinder presses are being built that exceed the 1:2 draw ratio standard by adding a third and or fourth cylinder, which allows three or four operations to be performed on one machine. For example, in one operation sequence a four-cylinder press can draw, reverse draw, and redraw twice. The advantage in production is apparent. One company producing closed-end cylinders typically produced parts on three different presses with three different operators at 300 parts per hour. After installing a three-cylinder press, they were able to produce 450 parts per hour with only one operator operating the single press.

Currently, we find press work being used to form large parts such as quarter-panels for automobile bodies as well as small brass eyelets. In the production of small items such as eyelets, as many as a quarter of a million pieces can be produced from one machine in a day. To take advantage of cold metal-working operations, the product-design function should be coordinated with the process-design function, since many of the answers concerning the possibilities of using a particular process to advantage are based on experience and empirical calculations.

QUESTIONS

13.1 What is the effect of cold-working operations on the strength and hardness of a metallic material?

13.2. Explain what occurs when a piece of hot-rolled steel passes through the yield point.

13.3 What is meant by bulging?

13.4 What type of force occurs when the components of the force are parallel to a surface?

13.5. If the application of a force tends to move atoms closer together, the direct stress is termed______stress.

13.6. What is meant by ductility?

13.7. What is the "yield-point elongation?" Why is the yield-point elongation an important factor in the plastic working of metals?

13.8. How do the work-hardening characteristics of a metal affect its deep-drawing capabilities?

13.9. How much tonnage would be required to pierce and blank a 10-gage mild steel washer of 5 in. OD and 3 in. ID?

13.10. What are the desirable characteristics of cold-drawn bar stock?

13.11. What are the two types of lubrication?

13.12. What type lubricant would you recommend in the cold extrusion of steel?

13.13. Make a sketch illustrating the impact-extrusion process?

13.14. Explain when and why you would advocate double-shear as opposed to single-shear on a punch.

13.15. Explain the difference between punching and blanking.

13.16. Make a sketch of a die that produces $2\frac{1}{2}$ in. round blanks that cannot have fractured edge characteristics.

13.17. When would you advocate using the warm-forming process?

13.18. What is the Hooker process?

13.19. Describe the knuckle-joint press.

13.20. How does coining differ from embossing?

13.21. Describe the swaging process.

13.22. What advantages does thread rolling have over thread cutting?

13.23. Define pitch diameter.

13.24. Upon what two principles are forming processes based?

13.25. What would be the developed length of a bend in $\frac{3}{32}$ in. aluminum alloy steel with a $\frac{3}{4}$ in. inside radius?

13.26. Make a simple sketch showing a single bend being made in a punch-press die. Show the punch, pad, stop or gage, inside radius, die, and angle relieved for overbending.

13.27. Illustrate by a sketch how lock-type joints may be made on a pan brake.

13.28. Give several applications of rubber forming.

13.29. What is a box-forming brake?

13.30. How can a tube be bent without its walls buckling?

13.31. Indicate by a sketch a roll-forming operation. Show the positions of the rolls.

13.32. Explain why no springback occurs during the stretch-forming process.

13.33. What would be the important criteria in determining whether it would be more advantageous to spin or to deep-draw a part?

13.34. What are the principal causes of buckling or wrinkling in the deep-drawing process?

13.35. In producing a deep-drawn part, identify where the wall thickness would be the thinnest, and where it would be the thickest.

13.36. What is nibbling?

13.37. What determines the size of a hole in a piercing die?

13.38. Why is it advisable to taper the working end of the punch? How much taper is practical?

13.39. How much clearance should there be between the punch and the die?

13.40. What is the "grain" in sheet stock? Why should material be bent across the grain?

13.41. What is the continental method of tool construction?

13.42. Outline the important practical hints for the die designer.

13.43. If material continually tears when being deep-drawn, outline the possible courses of action.

13.44. What lubricant would you recommend for deep-drawing aluminum? Why?

13.45. In mechanical presses, at what position of the stroke is the developed tonnage the greatest?

13.46. When would you advocate the purchase of a hydraulic press as opposed to a mechanical press?

PROBLEMS

13.1. One kilobar is equal to how many atmospheres?

13.2. A power shear cuts a length of 5 ft from a sheet of $\frac{1}{8}$ in. steel, SAE 1020 (annealed). The shear blade has a slope of $\frac{1}{8}$ in./ft. Calculate the shearing load. Assume penetration is 40 percent.

13.3. Find the maximum press capacity in tons for cutting a blank of 10 in. diameter from annealed aluminum of $\frac{1}{16}$ in. thickness. The shear strength of the metal is 8000 lb/in.^{2}

13.4. How much energy will be required in the operation given in Problem 2? Assume 60 percent penetration.

13.5. The given material of a washer is cold-worked brass; 20 percent penetration, ID = 6 in., OD = 9 in., six holes each $\frac{3}{4}$ in. diameter, thickness of sheet 16 gage. With reference to Fig. P13.5.

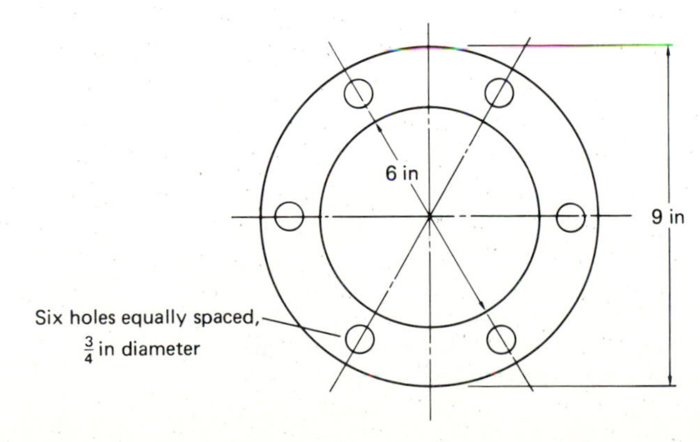

FIGURE P13.5

a. Determine the characteristics of the tooling detail as to the shears, etc. Disregard the detail of the accessories such as springs, etc.
b. Calculate all pressures and work per stroke.
c. Specify the size of equipment—using a safety factor of 2.
d. Specify the stroke for one washer at a time.

13.6. The sheet-metal department of a plant has been requested to make the part shown in Fig. P13.6. The department has a punch with a $\frac{3}{8}$-in.-thick blade and an included angle of 89° as shown, however the lower die must be designed to meet the requirements of this particular part. The chief industrial engineer has asked you to answer the following question.
a. Show the dimensions of the lower die.
b. What is the developed width of the sheet?
c. Determine the minimum number of forming operations required on a press brake. Sketch each one in sequence.
d. Show the gaging points and position of the tools at each operation. Give appropriate dimensions for proper location of stops and the workpiece.
e. Show one additional method to reduce costs by changing tools or using another process.
f. What standard width of sheet should be used (nearest $\frac{1}{2}$ in.)?

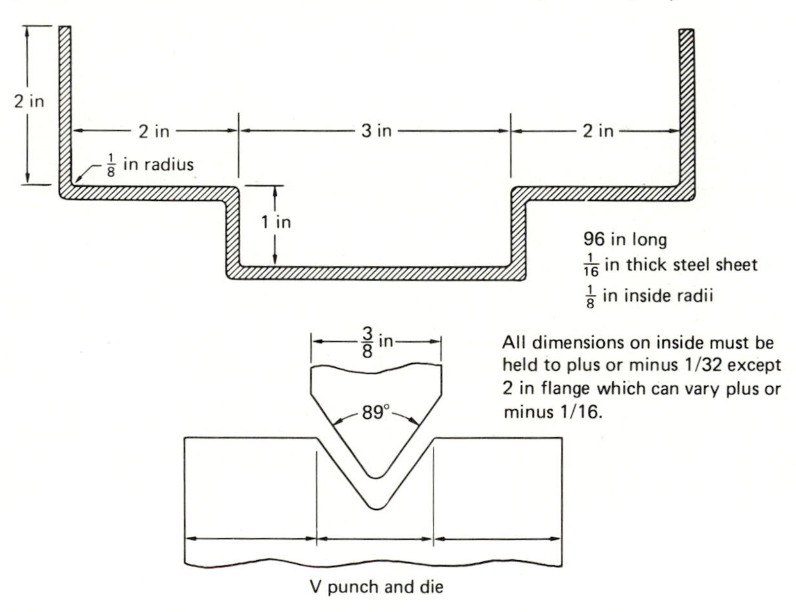

FIGURE P13.6

REFERENCES

1. Alting, Manufacturing Engineering Processes, Marcel Dekker, New York, 1982.
2. American Society for Metals: *Metals Handbook,* Metals Park, Ohio, 1985.
3. Avitzur, B.: *Handbook of Metal-Forming Processes,* Wiley, New York, 1983.

4. Avitzur, B.: *Metal Forming: Processes and Analysis,* McGraw-Hill, New York, 1968.
5. Crane E. V.: *Plastic Working of Metals and Nonmetallic Materials in Presses,* 3d Ed., Wiley, New York, 1944.
6. Doyle, L. E., Leach, James L., Keyser, Carl A., Schrader, George F., and Singer, Morse B.: *Manufacturing Processes and Materials for Engineers,* 3d Ed., Prentice-Hall, Englewood Cliffs, N.J., 1985.
7. Eary, D. F., and Reed, E. A.: *Techniques of Pressworking Sheet Metal,* Prentice-Hall, Englewood Cliffs, N. J., 1958.
8. Krauss, G.: *Deformation, Processing, and Structure,* American Society for Metals, Metals Park, Ohio, 1984.
9. Sachs, George: *Principles and Methods of Sheet Metal Fabrication,* Reinhold, New York, 1951.
10. Shaw, M. C.: Metal Cuting Principles, Oxford University Press, Oxford, 1984.

CHAPTER 14

JOINING PROCESSES

INTRODUCTION

Early man attached handles to his hunting clubs and cooking vessels, heads and feathers to his arrow shafts. His descendants forge-welded Damascus swords, chain links, and devices needed by armorers and shipbuilders. Spinning, knitting, weaving, felting, and matting—of textiles and paper—are not new processes. But only in the twentieth century has man advanced in threaded fasteners, welding, rubber cements, synthetic resins, and other organic cements.

The major industrial joining processes are presented in Table 14.1. This chapter will concentrate on the welding, brazing and soldering processes. Welding is one of the most effective methods for joining metals. However, it is also one of the most complex processes, since a variety of metallurgical phenomena occur in a brief time interval while the weld is being made. The engineer, when specifying a welding operation, is concerned with gas–metal reactions, slag–metal reactions, surface phenomena, solid-state reactions, heat flow, solidification, and of course, reliability and cost. Thus, the engineer is not just concerned with the area of coalescence produced by the welding process, but is concerned with the entire welded joint, which includes the region around the coalesced zone as well as the zone itself. This chapter introduces the broad

TABLE 14.1
Production joining processes

Arc welding	Resistance welding
Shielded metal-arc	Spot
Gas metal-arc (MIG)	Projection
Submerged arc	Seam
Gas tungsten-arc (TIG)	Flash
Flux cored-arc	Upset
Plasma-arc	Electroslag
Arc-spot	
Stud	
Pulsed arc	
Gas welding	**Solid state welding**
Oxyfuel gas	Friction
	Ultrasonic
	Diffusion bonding
Brazing	**Other welding processes**
Furnace brazing	Electron beam
Gas brazing	Thermit
Induction brazing	Laser
Vacuum brazing	Explosive
Related processes	**Soldering**
Arc cutting	Hand soldering
Oxyacetylene cutting	Dip soldering
Hard-facing	Wave soldering
Plasma-arc cutting	
Adhesive bonding	**Mechanical fastening**
Synthetic resins	Rivets
Elastomeric adhesives	Screws, nuts, and bolts
Natural adhesives	Special fasteners
	Staking
	Lock seam, etc.
	Stitching

subject of joining of materials from an engineering-design standpoint. It is far from complete, as this area of technology warrants several texts.

WELDING: GENERAL CONDITIONS

Welding is the joining of two polycrystalline workpieces—usually of metal—by bringing their fitted surfaces into such intimate contact that crystal-to-crystal

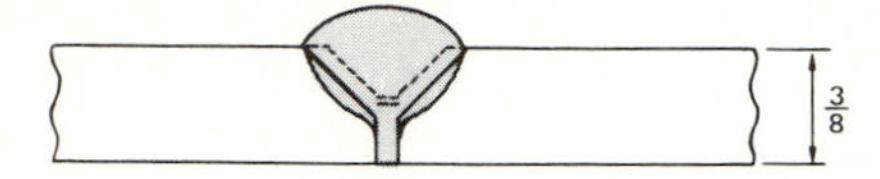

FIGURE 14.1
Typical single-V-butt welded joint showing edge preparation and filler metal.

bonding occurs. Industrial welding usually entails local heat from a burning gas or an electric arc, or heat generated by electrical resistance. The fitted surfaces may melt together, or a filler rod may melt between them to form a connecting bridge. The only nonthermal welding process is cold welding. In this, joining is accomplished through controlled plastic deformation of the members being joined.

Oxides impede welding. A small disk of indium and another of silver will bond at room temperature when pinched between thumb and forefinger—but only if the surfaces are first abraded. Equally, there is the phenomenon of welding that is *not* wanted. A weld deposit builds up on the edge of a cutting tool, causing chatter or poor finish on the workpiece. Bearings gall when overloaded or underlubricated. The parts of an instrument that rub together while unfolding from a satellite in space bond together despite the cold. One remedy is to pair a metal with a nonmetal.

Some welding involves further metal in addition to the workpiece, as in brazing or soldering. More important is the joining of steel plates with a consumable electrode that penetrates the joint and deposits a weld bead as shown in Fig. 14.1. Welding equipment can be simple, such as that used in a farm shop or in automotive repair shop or complex, as in an establishment which offers production joining as a business. Such establishments have positioners, shears, annealing furnaces, and booths for sand blasting and painting.

The Effect of Heat on Welding

A designer must anticipate two problems inherent in the welding process: (1) the effect of localized heating and cooling on the microstructure and properties of the base metal, and (2) the effect of the residual stresses that are locked in the weldment as a result of the uneven cooling of the weld deposit. In general, weldments have poorer fatigue and impact resistance than correctly designed castings or forgings. For example, critically stressed joints in hopper cars for railroads are riveted rather than welded to avoid fatigue failure in service.

The heat-affected zone is the region of the base metal, adjacent to the weld zone, where the temperature has caused the microstructure to change (Fig. 14.2). Note the difference in microstructure between the hot-worked metal near the weld joint and the cold-worked structure in more remote areas. The weld is a flash butt weld in which there was no weld metal added. In the case of fusion-welded steel, the weld deposit is usually stronger than the base metal because the grain growth in the heat-affected zone makes that metal somewhat weaker than the fine-grained dendritic structure that is typical of the weld deposit. Consequently, a tensile test specimen taken from the weld area

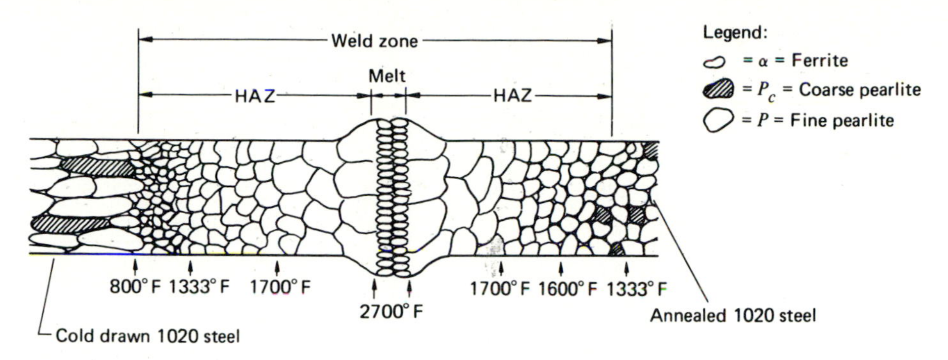

FIGURE 14.2
Microstructure of heat-affected zone (HAZ) of flash butt weld. (*Redrawn from Joseph Datsko,* Material Properties and Manufacturing Processes, *Wiley, New York, 1966.*)

in low-carbon steel will break adjacent to the weld deposit rather than in the weld deposit itself. The reader should understand that the influence of welding on some materials can act simultaneously as a hardening and an annealing cycle, depending on the part of the heat-affected zone.

In carbon and alloy steels the heat-affected zone is particularly important because of the metallurgical changes that occur in steels when subjected to heating to a high temperature and then to rapid cooling. In Fig. 14.2, the steel has been heated above 1200°F throughout the heat-affected zone. A large part of the metal within this zone has transformed to austenite because of the intense heat from the welding arc. Upon subsequent cooling, the properties of the metal within the heat-affected zone are determined by the cooling rates and consequent decomposition of the austenite in relation to the CCT continuous-cooling temperature diagram. This curve lies below and to the right of the corresponding isothermal diagram curves and the iron–iron carbide phase diagram for the steel that was welded.

A profile curve of the maximum temperatures reached at various locations within the heat-affected zone of a weldment from 0.3 percent carbon steel may be correlated with the iron–iron carbide phase diagram (see Fig. 14.3). The structures found in the various regions of the heat-affected zone may be analyzed as follows.

Point 1. The steel in that zone has been heated to excess of 2400°F, so the austenite grains have grown to a large size.

Point 2. The alloy in this zone has been heated to 1800°F and is fully austenitized but there has been little grain growth.

Point 3. The steel in this region has been heated far enough above the austenite transformation temperature so that the austenite has achieved a homogeneous structure.

Point 4. The metal in this region has been heated to 1400°F. Only part of the alloy has changed to austenite, so the resultant mixture of structures has poor mechanical properties.

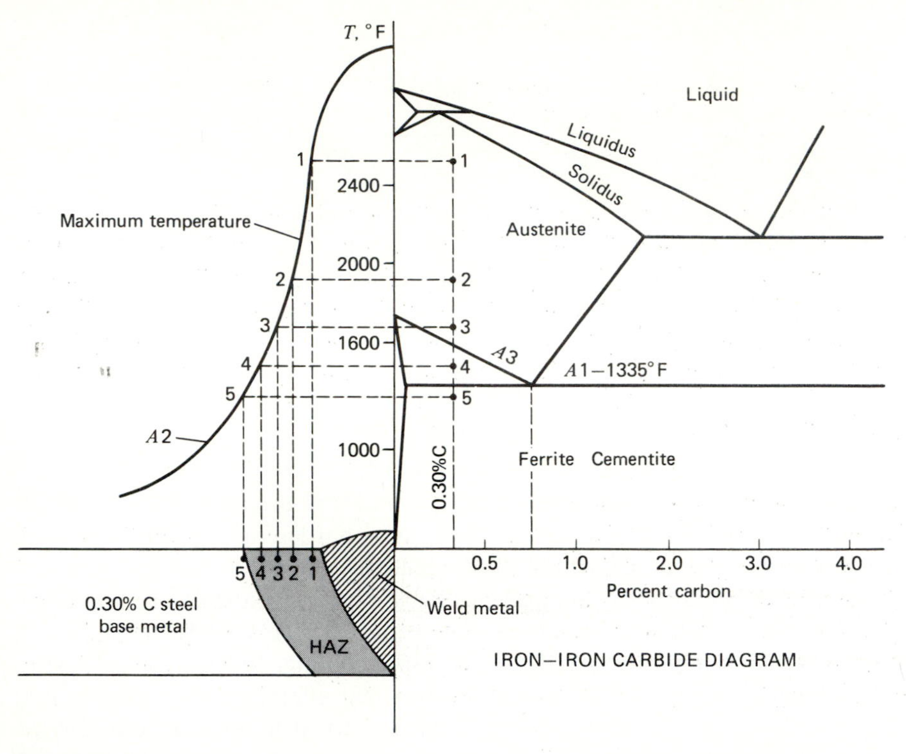

FIGURE 14.3
Relation between peak temperature in weld and the iron–carbon phase diagram.

Point 5. No change has occurred at this point, which represents the base metal outside the heat-affected zone.

In the study of heat flow in welding, the engineer is concerned with the "arc energy input," or the quantity of energy that is introduced per length of weld. The energy input may be computed as the ratio of the total input power to its travel velocity. Thus

$$H = \frac{P}{V}$$

where H is the energy input (joules per millimetre), P is the total input of heat source (watts) and V is the velocity of heat source (millimetres/second). If the heat source is an arc,

$$H_{\text{net}} = \frac{fEI}{V}$$

where f is the heat transfer efficiency, E is the voltage, in volts and I is the current in amperes.

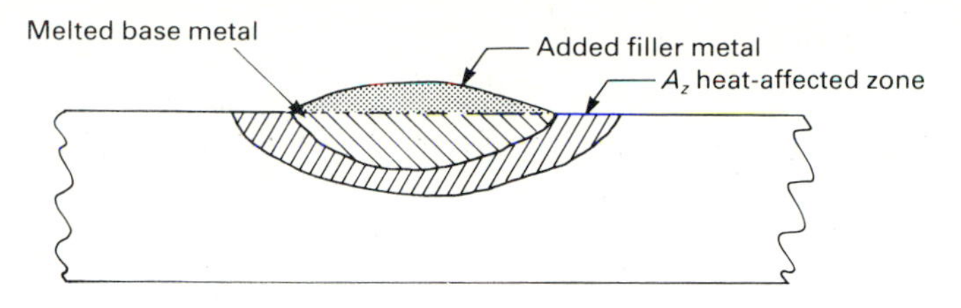

FIGURE 14.4
Cross section of typical weld showing added filler metal, melted base metal, and the heat-affected zone.

Referring to Fig. 14.4, the cross section of weld metal A_w is equal to the sum of the cross-sectional area of that portion of the weld bonded by the weld fusion line and the original top surface of the work, A_m, plus the cross section of the added filler metal, sometimes referred to as the overfill A_r. The heat-affected zone, which is that portion of the surrounding solid metal that has undergone a change in structure or properties, has been identified as A_z. Since the base metal and the filler metal are both molten, those portions identified as A_m and A_r (Fig. 14.4) are in reality a mixture of the base metal and the filler metal.

An approximation of heat Q required to melt a given volume of metal is[1]

$$Q = \frac{(T_m + 273)^2}{300,000} \quad \mathrm{J/mm^3}$$

where T_m is the melting temperature of the metal in degrees Celsius.

The reader should understand the effect of thermal conductivity on melting efficiency. In view of the more rapid conduction of heat away from the weld region, the higher the thermal conductivity of the metal being welded, the lower the melting efficiency. For example, in the oxyacetylene welding of aluminum, only about 5 percent of the heat is used for melting the metal—the majority is dissipated by conduction.

Residual Stresses

Residual stresses result from the restrained expansion and contraction that occur during the localized heating and cooling in the region of weld deposit. The magnitude of such stresses depends upon the design of the weldment, the support and clamping of the components, their material, and the welding process used. The relationship between the thermal and shrinkage stresses is highly complex in both the stress direction and time phase because of the steep

---

[1] *Welding Handbook,* vol. 1, 7th Edn., American Welding Society, 1976.

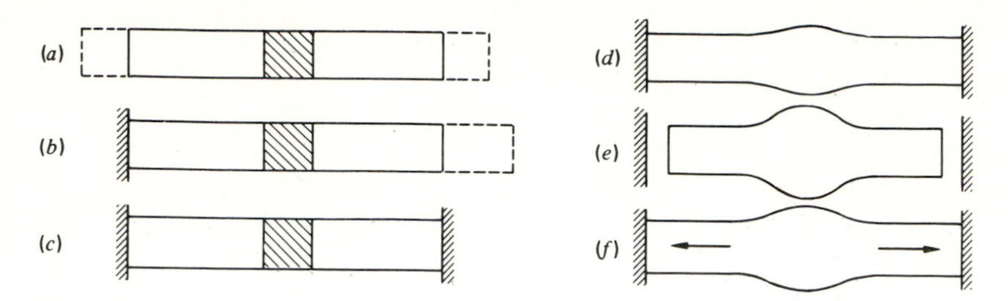

FIGURE 14.5
Comparison of effects on a heated bar with various constraints.

thermal gradients and the movement of the heat source in welding. Part of a weld bead is solidifying while a short distance away the molten pool is forming and the base metal is still gaining heat. The sequence can be analyzed by studying a simple bar heated in the center. In Fig. 14.5*a* the heated bar expands at either end. However, if one end is fixed (Fig. 14.5*b*), the other end moves twice as far relative to the original centerline. No lengthwise movement can occur if each end is restrained (Fig. 14.5*c*), so the center must bulge (Fig. 14.5*d*). When the bar cools again, it is shorter than before (Fig. 14.5*e*). However, if the bar is fully restrained, when it is heated a bulge occurs at the center and residual tensile stresses build up in the bar as it tries to contract, but the walls will not permit it to return to its shortened length (Fig. 14.5*f*).

In fusion welding, the molten pool solidifies like a casting poured into a metal mold. It is restrained from contracting by an amount that varies with the welding process. The cooling rate has a great influence on the amount and nature of the residual stress. In general, there are residual tensile stresses in the weld deposit that counterbalance compressive stresses in the base metal (Fig. 14.6). They are proportional in intensity to the weld size, and the maximum stress occurs in the direction of welding, with the transverse stress the next highest in intensity. The stress in the thickness direction is least, because there is the least hindrance to contraction in that direction.

The stress distribution around the weld deposit can be represented

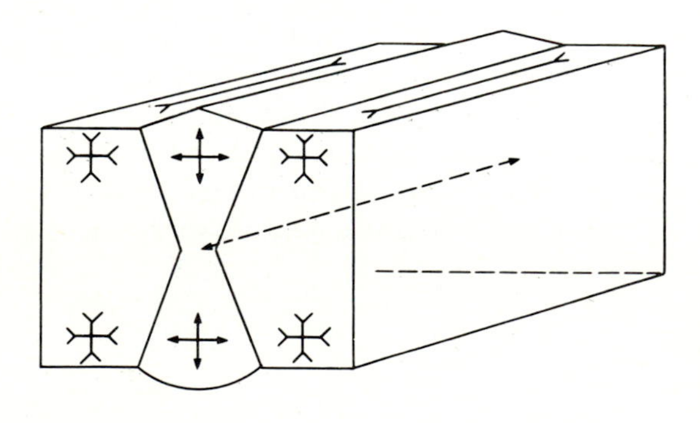

FIGURE 14.6
Residual stresses in the region of a welded joint.

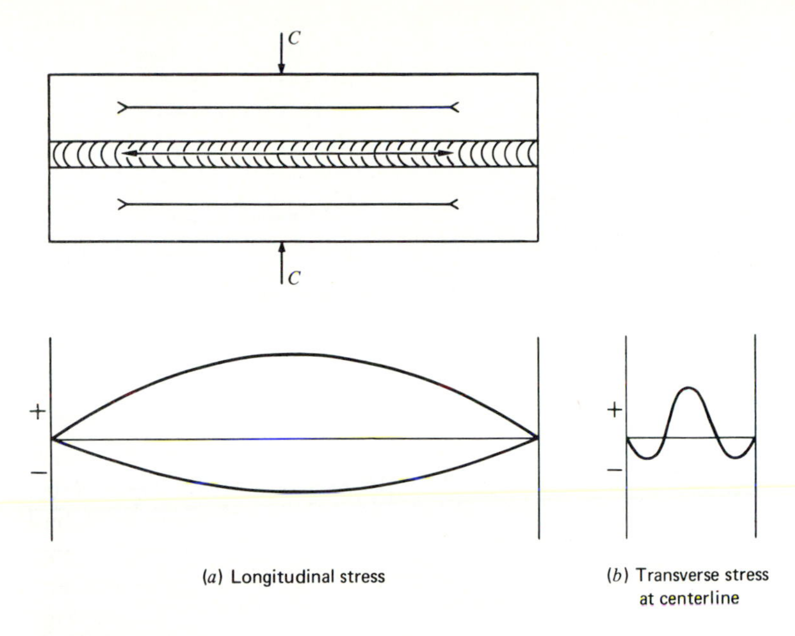

FIGURE 14.7
Typical stress distribution for a single-pass weld.

schematically as in Fig. 14.7. Note that the center of the weld is the point of maximum stress.

Distortion

The welding sequence is important in minimizing distortion and residual stresses. For example, in ship construction welding is started at the keel at midship and then continues simultaneously fore and aft and upward. It is wise in any complex weldment to balance heat input and distortion by welding first on one side and then the other. In fact one of the greatest cost items in a custom-welded frame is the straightening operation. In production-welded structures such as railroad cars, balanced welding sequences and proper fixtures reduce distortion to a minimum.

Welding fixtures can be used to reduce distortion, especially on thin pieces. The thermal input to the workpiece can be reduced and the mechanical restraint permits movement in only one direction (Fig. 14.8). Proper design can also reduce residual stresses and their resultant distortion.

The transverse shrinkage for a butt weld may be estimated from the following empirical relationship:

$$S_t = 0.18\frac{A_w}{t} + 0.05d$$

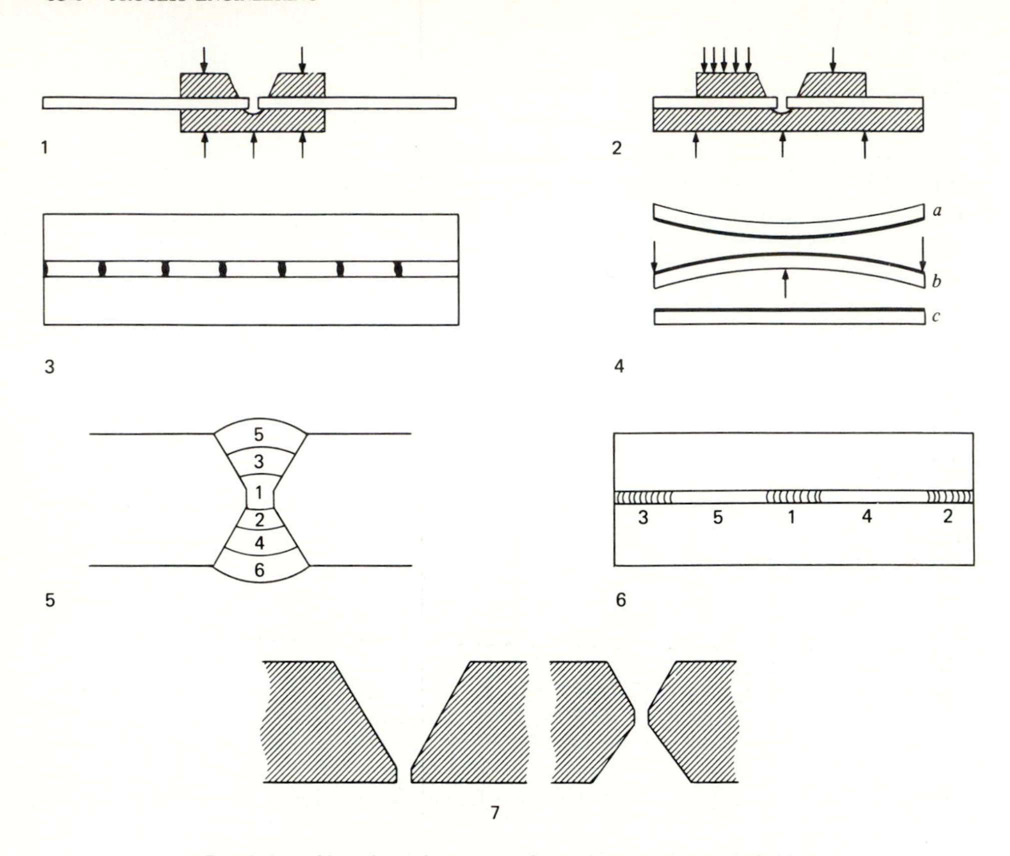

1 Restriction of heat input by means of massive or water-cooled chill blocks.
2 Controlled restraint to permit some movement during heating and cooling.
3 Tack welding to provide restraint during welding.
4 Prestraining to expected distortion level (*a*) distortion to be expected without prestrain, (*b*) welding in prestrained condition, (*c*) and release of prestrain to obtain undistorted product.
5 Bead sequence as shown, to allow opposing stresses to correct distortion obtained due to prior bead.
6 Skipping sequence of welding to prevent stress accumulation of continuous weld and allow opposing stresses to correct previous distortion.
7 Root gap to permit the weld metal to shrink transversely and reduce rotational distortion, the double V and sequence (no. 5) produce minimum distortion.

FIGURE 14.8
Methods for reducing the distortion that occurs in welding.

where S_t is the transverse shrinkage, in., A_w is the cross-sectional area of weld, in.^{2}, t is the thickness of material, in. and d is the root opening, in.

The lengthwise contraction for a V-butt weld (in./in.) may be estimated as

$$S_1 = 0.025 \frac{A_w}{A_p}$$

where A_w is the cross-sectional area of weld, in.^{2} and A_p is the cross-sectional area of weldment, in.^{2}

Methods for reducing distortion are given in Fig. 14.8.

Heat changes the mechanical and physical properties of a metal, which in turn affect the heat flow and the uniformity of heat distribution. As the temperature increases, the yield strength, Young's modulus, and thermal conductivity decrease, but the coefficient of thermal expansion increases (Fig. 14.9). Finding suitable values for those properties of an alloy at elevated temperature, one should be able to estimate the severity of the distortion that results from welding operations. A high coefficient of thermal expansion leads to high shrinkage on cooling and the possibility of distortion, especially if the alloy has a low yield strength, as would be the case for aluminum alloy, whereas high thermal conductivity would serve to counteract the effect of the high coefficient of thermal expansion. The higher the yield strength of an alloy in the weld area, the greater the residual stress; conversely, the lower the yield strength the less the severity of the distortion.

Longitudinal bending or cambering of a rolled section such as an *I* beam or angle results from contraction that has occurred at some distance from the neutral axis. The amount of distortion is a function of the shrinkage moment and the resistance of the member to bending as given by its moment of inertia. The following formula can be used to estimate the amount of distortion in such

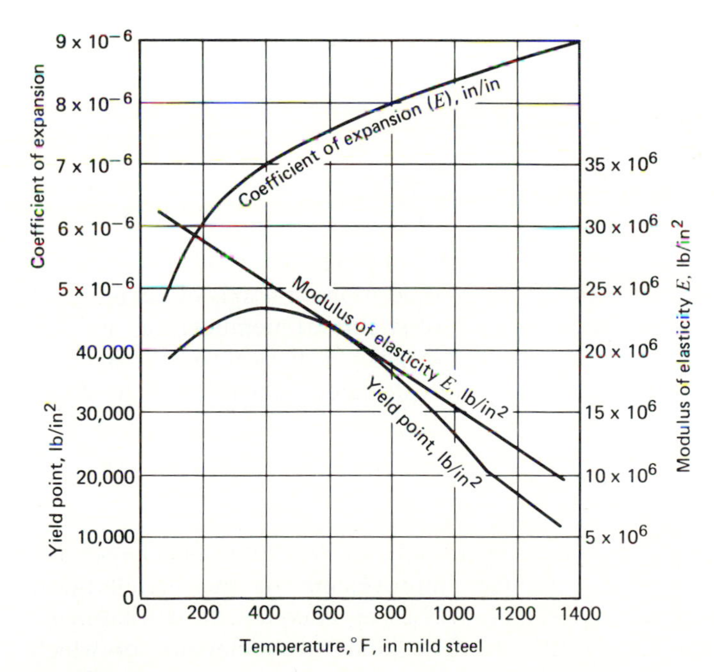

FIGURE 14.9
Change in the properties of mild steel when subjected to heat.

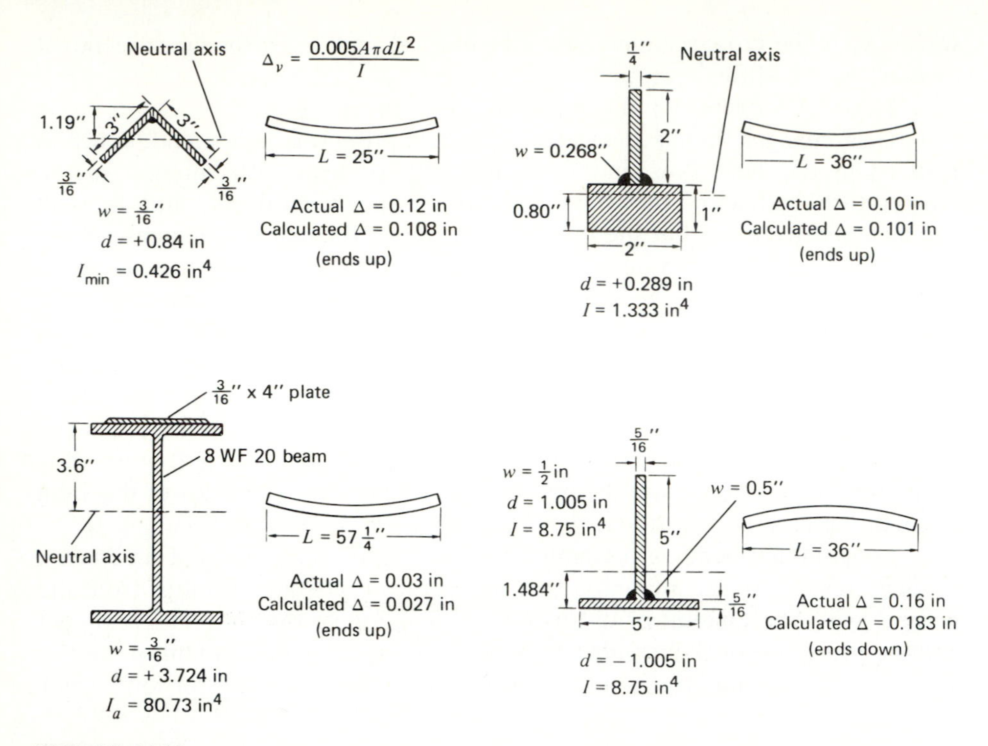

FIGURE 14.10
Longitudinal distortion in weldments. (*From Omar Blodgett,* Welding Engineer, *Feb. 1972.*)

cases:

$$\Delta_{\text{long}} = 0.005 \frac{A_w l^2 d}{I}$$

where A_w is the total cross-sectional area within fusion lines of all welds, in.^{2}, d is the distance between center of gravity of weld to neutral axis of beam, in., l is the full length of weld (should be length of structural member), in. and I is the moment of inertia of the structural member about its neutral axis, in.^{4}

Some typical experimental values and calculated values for such beams are given in Fig. 14.10.

Relieving Stress

The type of material, weld, and process (usually arc welding), and the service requirements of the part may require improvement of the metallurgical properties and the removal of residual stresses by adequate heat treatments such as stress relieving, tempering, etc. The kinds of weldments for which stress relief is advisable and the kinds that need no stress relief are listed below.

1. In general, statically loaded structures composed of light sections, nonrigid connections, and members containing only a small amount of welding need not be stress-relieved. Buildings, bed plates, housings, tanks, certain classes of pressure vessels, motor frames, etc., are typical structures of this class.
2. Work that requires accurate machining must not become distorted during machining or warp later as a result of the redistributing of the streses. Jigs, fixtures, lathes, planer beds, etc., are structures of this class, and require stress relief.
3. Parts subjected to dynamic loads should, in general, be stress-relieved because experience has shown residual stresses to be more serious for loadings of this type than for static loadings. Rotors for electrical machinery, diesel engine crankcases, gears, etc., are in this class. There are some outstanding exceptions in this class of structures, such as bridges, freight cars, locomotive cabs, etc. The latter are generally composed of thin plates or long members containing small amounts of welding; consequently they can be considered as belonging to the first classification.
4. Structures composed of heavy or large parts and containing large amounts of welding should preferably be annealed, because they generally contain large residual stresses. The Boulder Dam cylinder gates for the intake towers are a typical example of this class of structure.
5. Materials that undergo metallurgical changes in the base or weld metal and produce hard zones and high residual stresses should be strain-relieved by heating. These materials include high-carbon and alloy steels.

From the previous classifications, it might appear at first sight that the major portion of all welded structures should be stress-relieved or annealed. This, however, is not the case, because most of the structures fall within the first classification. The others actually comprise only a small percentage of all welded structures.

The process of stress relieving consists of slowly heating parts to a suitable temperature, allowing them to soak at this temperature for a definite period of time, and then slowly cooling them in the furnace. For steel the annealing temperatures generally used range from 1100 to 1200°F. (This temperature range has been established experimentally and is recommended by all welding codes.) These temperatures are not high enough to produce grain refinement, but they do have a tempering effect, which tends to increase the ductility of the weld and slightly lower the tensile strength. The tempering effect that results from strain annealing is often found to be beneficial in softening hard areas adjacent to the weld, thereby facilitating machining operations.

The theory of strain annealing to relieve residual stresses depends upon three things. First, within certain limits of a strain, a ductile metal will not develop a stress greater than its yield point. Second, the yield point of metals decreases at high temperatures. Third, the phenomenon of creep, or plastic flow, takes place at high temperatures after an extended period of time. The function of these factors during the annealing operation is best explained by an example.

Assume that a weldment containing residual stresses close to the yield point (30,000 lb/in.^{2}) of the base metal is to be annealed at 1100°F. When the temperature of the structure increases above 600°F, the yield point of the material decreases. The residual stresses also decrease (provided the amount of plastic strain is small) because the metal cannot develop a stress greater than the yield point. When the annealing temperature (1100°F) is reached, the remaining residual stress will be approximately 12,000 lb/in.^{2}, which is the yield point of the base metal and the weld deposit at this temperature.

If the structure is cooled uniformly immediately after 1100°F is reached, the final residual stress will be 12,000 lb/in.^{2} If, however, instead of cooling the structure immediately after it has reached the annealing temperature, it is allowed to soak at this temperature, after a period of time the phenomenon of creep occurs, and the metal flows plastically, thus further reducing the residual stresses. The longer the soaking time, within limits, the lower the final residual stresses will be. Note that subcritical stress relief does not wipe out residual stress completely but can reduce it to a minimum level of 5000 lb/in.^{2}

Data obtained from welded test plates show that residual stresses, regardless of their initial values, can be reduced to a negligible amount, and that their final value decreases as the annealing temperature increases and the soaking time increases.

Annealing temperatures above 1200°F can be used, but they introduce excessive distortion and consequently are seldom used unless required to produce metallurgical changes in the weld or the heat-affected areas.

Weldability

Weldability denotes the relative ease of producing a weld that is free from defects such as cracks, hard spots, porosity, or nonmetallic inclusions and is able to perform satisfactorily in its intended service. Weldability depends on one or more of five major factors.

Melting point. When welding low-melting-point alloys such as aluminum in thin sections, care must be taken to avoid melting too much base metal.

Thermal conductivity. Alloys with higher thermal conductivity, such as aluminum, are difficult to bring to the fusion point. Welds in certain alloys may cool quickly and crack. A high-intensity heat source is very important. For example, for a given size of weld, aluminum alloy requires up to three times as much heat per unit volume as does steel.

Thermal expansion. Rapid cooling for alloys with high thermal coefficients of expansion results in large residual stresses and excessive distortion.

Surface condition. Surfaces coated with oils, oxides, dirt, or paint hinder fusion and result in excessive porosity.

Change in microstructure. Not only are steels above 0.4 percent carbon subject to grain growth in the HAZ (heat-affected zone) but martensite is also formed wherever the temperature exceeds 1330°F for a sufficient time.

The reader can appreciate that in order to produce sound welded joints, the engineer should have an understanding of the metallurgical factors involved. Typical welding problems are cracking of the weld metal, cracking of the base metal, porosity, and inclusions. Table 14.2 identifies the cause of these problems and remedies that will generally solve the difficulty.

Steels. Steels vary widely in their weldability, especially according to their carbon content. Plain low-carbon steels have excellent weldability, but for steels above 0.3 percent carbon, special operations such as preheating and postheating are needed. The amount of preheating and postheating increases with carbon content. Alloy steels and quenched and tempered steels also require special preheat interpass temperature control and/or postheat cycles.

Alloying elements in steel may be added to increase the hardenability. This is done by reducing the critical cooling rate and the temperature of the austenite-to-martensite transformation. The elements having the greatest effect on the hardenability of steel are carbon, manganese, molybdenum, chromium, vanadium, nickel, and silicon. The effect of these elements on controlling the tendency to form heat-affected zone martensite, and thus cold cracking, is often expressed as carbon equivalent (CE). One computation of CE that is used frequently is

$$CE = \%C + \%Mn/4 + \%Ni/20 + \%Cr/10 + \%Cu/40 - \%Mo/50 - \%V/10$$

The hardenability of weld metal and base metal in carbon and low-alloy steels affects the weldability in two ways:

1. The chances for hydrogen underbead cracking increase with hardenability.
2. The depth at which hard brittle martensite is formed increases.

When CE calculated by the above formula exceeds a value of 40, underbead cracking can occur. Cracking is also influenced by welding preheat.

Since steels are particularly susceptible to hydrogen embrittlement, care must be exercised to avoid introducing moisture into the weld zone. Low-hydrogen electrodes should be used; moreover, they should be stored in special warming ovens to prevent absorption of moisture from the atmosphere.

Stainless steels. Stainless steel is a special case. The molten pool must be as free as possible from oxides if the stainless properties are to be retained in the weldment. The gas tungsten-arc process is best for high-quality work. Only

TABLE 14.2
Welding problems, their causes and remedies

Problem	Cause	Remedies
Cracking of weld metal	High joint rigidity	Preheat parts, thus reducing the cooling rate. Relieve residual stresses by peening. Increase strength of weld by building heavier cross section
	Excessive alloy pickup from the base metal	Change the current level and the travel speed; weld with straight polarity (emissive coating on wires and electrodes). Overlay the base metal at low amperages prior to welding the joint.
	Defective electrodes. This can be from moisture, eccentricity, poor striking end, poor core wire.	Change electrode; grind striking ends to proper dimensions; bake electrodes.
	Poor fit-up	Reduce root gap; clad edges
	Small bead	Use larger electrode; increase cross-sectional area of weld
	High sulphur in the base metal (carbon and low-alloy steels)	Use process with high level of sulphur–fixing elements. Use lime-rich fluxes for example the covering of EXX18 electrodes
	Angular distortion (weld root in tension)	Change to balanced welding on both sides; preheat; peen to relieve residual stresses
	Crater cracks	Fill in crater prior to withdrawing electrodes
Cracking of base metal	Hydrogen in the welding atmosphere	Use hydrogen-free process—gas metal-arc welding, gas tungsten-arc welding, submerged arc welding, etc. Use postweld aging or annealing
	Hot short cracking	Use low-heat-input, high-speed welding; use thin beads
	High-strength low-ductility material	Use annealed or stress-relieved material
	Hot tensile cracking in heat-affected zone (copper alloys)	Weld using an electrode whose melting melting point matches the base metal
	Excessive stresses	Redesign the joint. Use intermediate stress relief. Change welding sequence (cladding technique)

TABLE 14.2 (*Continued*)

Problem	Cause	Remedies
	High hardenability	Preheat; change conditions to slow cool the weld beads; weld with austenitic electrode
	Brittle phases	Solution heat treat prior to welding
	High lead content	Weld with minimum heat input. Change material
Porosity	Excessive hydrogen, oxygen, nitrogen in the welding atmosphere	Use low-hydrogen process; use filler with high deoxidizers. Check gas flow rate and travel speed
	High rate of weld freezing	Preheat; increase heat input; use lower-melting filler
	Oil, paint, or rust on base metal	Thoroughly clean all joint surfaces
	Dirty surface on gas metal-arc welding electrode	Use cleaned wire
	Improper arc length and/or current	Control these parameters
	Zinc volatilization in copper	Use ECuSi electrodes; reduce heat input
	Galvanized coating on steel	Use E6010 electrode and manipulate so as to volatilize zinc ahead of the molten pool
	Excessive moisture in electrode or joint	Use dry materials
	High-sulphur base metal	Use electrode with basic slagging reactions
Inclusions	Failure to remove slag from previous deposit	Clean surfaces and previous beads
	Entrapment of refractory oxides	Wire-brush previous beads
	Tungsten in weld metal	Use high frequency to initiate arc; improve manipulation
	Improper joint design	Increase included angle of joint
	Oxide inclusions	Provide proper shielding and coverage

Source: Welding Handbook, 7th Edn., Vol. 1, Fundamentals of Welding, American Welding Society, Miami, Fl., 1976.

low-carbon grades or grades with special carbide stabilizers should be specified if service in corrosive atmospheres is expected. Standard grades of stainless steel may fail by precipitation of carbides to the grain boundaries of the heat-affected zone where intergranular corrosion occurs. To minimize carbide precipitation:

1. Use a low-carbon (less than 0.03 percent C) base metal such as 304L stainless steel.
2. Use a columbium stabilized filler rod with stabilized base metal such as 316.
3. Heat above 1800°F and quench immediately after welding.

Generally, the 400 series of martensitic stainless steels are more weldable than the 400 series of ferritic stainless steels because the latter have grain growth that leads to lower ductility, toughness, and corrosion resistance. Austenitic types of stainless steel (the 300 series) make up the bulk of stainless steel weldments.

Stainless steels can be spot-welded easily because of their low thermal conductivity, which is about 17 percent that of steel. The low conductivity and absence of a high-resistance surface coating make the fabrication of steel by

TABLE 14.3
Weldability of aluminum in various methods

Type of aluminum	Gas shielded-arc	Shielded metal-arc	Resistance	Gas
1060	A[a]	A	B	A
1100	A	A	A	A
3003	A	A	A	A
3004	A	A	A	A
5005	A	A	A	A
5050	A	A	A	A
5052	A	A	A	A
2014	C	C	A	X
2017	C	C	A	X
2024	C	C	A	X
6061	A	A	A	A
6063	A	A	A	A
6070	A	B	A	C
6071	A	A	A	A
7070	A	X	A	X
7072	A	X	A	X
7075	C	X	A	X

[a] A—Readily weldable
B—Weldable in most applications; may require special techniques
C—Limited weldability
X—Not recommended

resistance spot-welding possible at much lower current settings than needed for aluminum alloys.

Aluminum alloys. Many aluminum alloys are easily welded by the gas tungsten-arc (TIG) and gas metal-arc processes (MIG), but their weldability varies considerably according to the alloying element and its amount. Their low melting point, high thermal conductivity (about 10 times that of steel), high chemical activity, and high thermal expansion have made aluminum alloys rather difficult to weld in the past. But today, the TIG process provides high unit energy and good protection from oxidation. See Table 14.3 for a comparison of the weldability of aluminum alloys.

Other alloys. Many other metals and alloys can be welded, but most of them are of limited use in production. Zirconium, beryllium, and titanium can all be welded if the gas tungsten-arc process is used. For welding titanium, the root side of the weld must also be protected. Nickel, and most magnesium- and copper-based alloys, can be welded successfully. Cast irons can be welded provided preheat and postheat treatments are used, but high-nickel electrodes are required, especially if the weldment is to be machined. Cast iron can also be braze-welded, but then service temperatures should not exceed 500°F. Most cast-iron welds are repair welds rather than welds for production joining.

Welding Versus Other Processes

Since 1920, welding has competed with riveting, bolting, and casting. Without it, neither high-pressure boilers nor nuclear power plants would be economically possible. A welded joint permits no play; nor, if continuous, will it allow leakage of fluid. The demand for sheet-metal products, including items too thin in cross section to be cast, has increased markedly with the development of resistance welding. The cost of fixtures and of cleaning surfaces in preparation for finishing is generally less for welding than for casting, especially if the production quantity is low. Yet welding a given product can cost more in labor than casting it. Sometimes the expense of strain relief precludes a welded design.

In a number of cases, steel weldments can replace gray iron castings on an economic basis, but good engineering-design practice would suggest that a valid decision could only be made if a casting redesign were carried out at the same time. A good example of superiority of weldments would be motor generator bases for dc power supplies or air-conditioning units. Such welded steel bases are lighter and less expensive for about the same rigidity. On the other hand, most machine-tool bases are cast if they are complex and if there are more than a few to be built.

The strength of cast components is equivalent to that of weldments provided equivalent designs and alloys are compared. In tension, gray iron is weaker and less stiff than steel; but in compression, the opposite is true;

therefore, when there are combined stresses, as there are in most engineering applications, the strength advantage of steel is considerably reduced. Thus, it is wise to analyze each design application carefully before deciding which production method is superior.

On balance, the designer of a product must consider both the properties of materials and the characteristics of available equipment. While welding is more costly than casting, bending, or cold forging in some cases, it is often the most useful—particularly if the material in question is easily welded and if suitable welding equipment has already been installed.

WELDING PROCESSES

Since the heat of the electric arc may be concentrated and effectively controlled for fusion, several welding processes use this method for joining metal. The electric arc consists of a high-current discharge through a thermally ionized gaseous column referred to as a *plasma*. This gas is composed of similar numbers of electrons and ions. The ions flow out of a negative terminal (cathode) and move toward the positive terminal (anode). In addition to the plasma, there are other materials such as molten metals, slags, vapors, and neutral and excited gaseous atoms that are mixed together.

Arc Welding—Consumable Electrodes

SHIELDED METAL-ARC WELDING. Manual arc welding is widely used in the construction and fabrication of metal sheets, plates, and roll-formed products. The equipment includes a source of direct or alternating electric current, a ground, an electrode holder, and proper safety equipment. The latter consists of a helmet with dark eye protection, long sleeves, and a leather apron. The ultraviolet light from the welding arc can cause the equivalent of sunburn or snow blindness.

A conventional electrode forms a molten pool in the joint area. A gaseous shield and slag protect the weld deposit from oxidation and rapid loss of heat (Fig. 14.11*a*). Unskilled operators find the drag-type electrode, with large amounts of iron powder in the electrode coating, much easier to use (Fig. 14.11*b*). The iron powder increases the rate of deposition, but reduces the penetration and permits the core to burn away so that the coating can drag along the surface and the arc length stays constant. Thereby a good deposit can be made by an operator with relatively little skill.

In shielded-metal-arc welding the arc is started by momentarily striking the electrode against the base metal and quickly withdrawing to form an arc. The arc must not be too long, as this gives an opportunity for contamination by the atmosphere, and it is more difficult to control its application to the joint. The current and voltage must be under close control; they are governed by the quality of equipment and its inherent regulating characteristics.

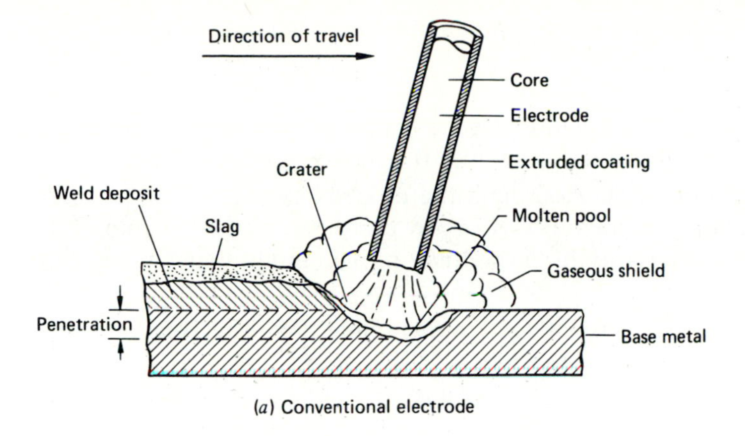

(*a*) Conventional electrode

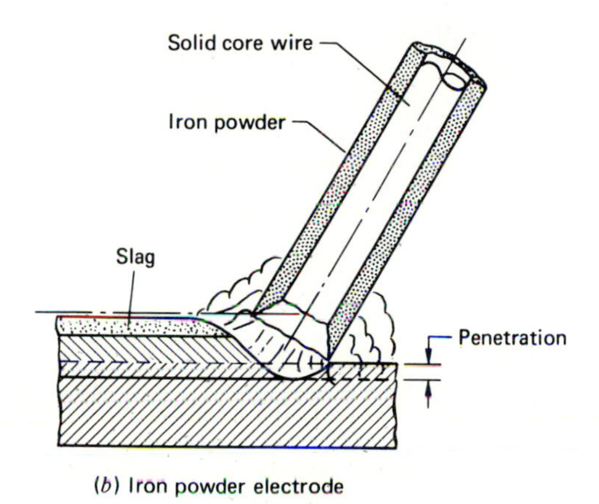

(*b*) Iron powder electrode

FIGURE 14.11
Shielded metal-arc welding with conventional and iron-powder electrodes.

When the arc forms between the base metal and the electrode, the immediate surface is melted, and, with the use of an electrode that cannot conduct the heat away rapidly, some of the metal is vaporized. These droplets and the vaporized metal flow along the stream of the arc path to the base metal where they condense, build up, and solidify. (Motion-picture studies of this action have been made and are available from leading welding equipment suppliers.) Therefore, the arc process is primarily a localized casting process that is influenced by the action of the electrode, current, flux, and operator. In the liquid and gaseous state, it is essential that no harmful chemical action (such as oxidation and forming of nitrides) occur, that gas occlusions escape, that flux inclusions be avoided, and that the material cool without tearing or cracking.

SHIELDING. The molten metal is protected from contamination by four general methods.

1. The electrode is coated with a flux that melts and forms a gaseous envelope around the arc and a liquid covering over the molten metal.
2. An inert gas envelope is blown around the electrode as it melts. The inert gas is argon, helium, or mixtures of argon and helium. This method has proved especially satisfactory in the welding of aluminum, magnesium, and copper alloys.
3. The end of the electrode is submerged in a granular flux. This surrounds the arc and protects it and the molten metal.
4. Carbon dioxide is used as a shielding gas to weld low-carbon and alloy ferritic steels. Carbon dioxide is less expensive than inert gases and gives deeper penetration into the base metal. When carbon dioxide is used with filler metals of proper chemistry, it produces welds of high quality and soundness. A mixture of carbon dioxide and argon increases the stability of the arc over either a pure carbon dioxide or pure argon gas. Pure carbon dioxide causes splatter in and about the weld. Pure argon prevents oxidation of the metal with resulting surface tension and uneven surfaces. With a combination of carbon dioxide and argon, slight oxidation takes place and the result is a smooth weld surface. Argon plus 25 percent carbon dioxide is used for welding steel. Especially for stainless steel, argon plus 1 to 5 percent oxygen may be used. Shielding gases are used at rates of 10 to 35 ft^3/h.

There are two basic types of coated stick electrodes: low-hydrogen types and cellulosic-coating types.

Low-hydrogen coating. This is an all-mineral coating, containing no material that will form hydrogen. The coating is baked at high temperature during manufacture to remove moisture. Electrodes are sealed in containers by the manufacturer and stored under elevated temperatures by the user to prevent moisture absorption.

Cellulosic coating. This is not a low-hydrogen coating. It can contain moisture. It essentially is a mixture of binders with cellulose and carbon as well as the necessary minerals for alloying.

Only coated electrodes with low hydrogen content that are kept in a drying oven can be used for welding low-alloy and high-tensile-strength steels. This type of electrode depends upon the transfer of the molten mineral slag coating for shielding. Therefore, it is important that an extremely close or short arc be used when welding with low-hydrogen electrodes.

For welding low-carbon mild steels, electrodes that contain cellulose in the coating can be used. The burning of the cellulose furnishes a gas shield that is primarily carbon dioxide. Therefore, maintenance of a short arc is not critical with this type of electrode.

Gas shielding. In the gas-metal-arc welding process, various inert gas combinations can be used to protect the weld metal and molten pool. The process was originally called MIG welding for metal-inert-gas welding. But carbon dioxide and other gases are not inert; the process has changed but the name remains. Carbon dioxide is a slightly oxidizing gas and therefore electrodes should be selected with sufficient deoxidizer content to produce sound welds. Deoxidizers include silicon, aluminum, manganese, and others. The following combinations are used:

CO_2 + solid wires with sufficient deoxidizers in the wire chemistry

CO_2 + cored wires having sufficient deoxidizers as a flux in the core with a low-carbon steel shell (Fig. 14.12) and self-shielded fluxed cored wire welding (Fig. 14.13)

CO_2 + argon with solid wires having sufficient deoxidizers in wire chemistry

The CO_2 with solid wire is more economical because of no slag removal and CO_2 is cheaper than argon.

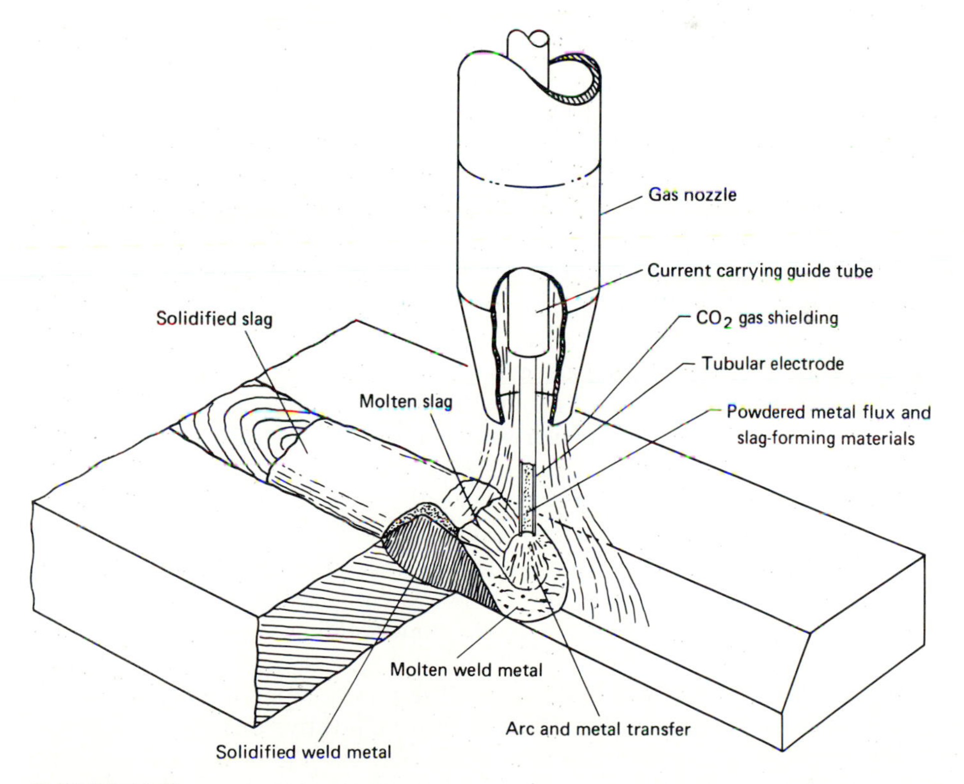

FIGURE 14.12
Schematic flux core electrode and gas shield.

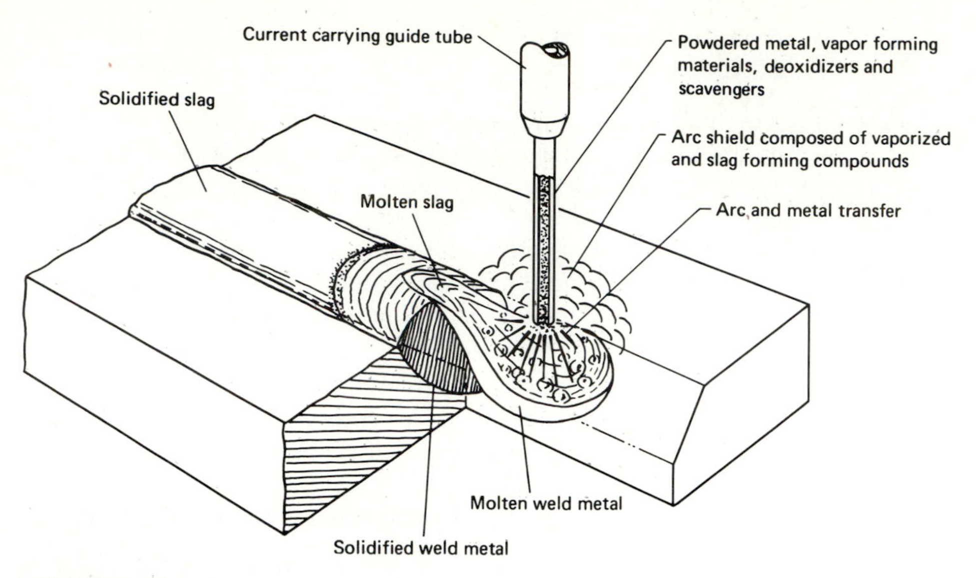

FIGURE 14.13
Schematic self-shielded flux core electrode.

More effective welds can be obtained by proper selection of alloy electrodes, proper flux or gas protections, and automatic equipment to reduce variations caused by manual operation.

Flux cored electrodes. Many gas shielded-metal-arc applications are being replaced by semiautomatic welding facilities that use flux cored wire electrodes. The operation can be fully shielded by the flux in the core of the continuously fed electrode, or auxiliary carbon dioxide gas shielding can be applied through the electrode holder. The flux materials are sealed into the center of the electrode during its manufacture. The use of an internal flux makes the continuous feeding of the electrode simple and ensures good electrical contact between the power source and the electrode.

Compared with an equivalent stick-electrode operation, the flux-cored wire facility gives deeper penetration, faster deposition rates, and continuous operation; and, because of the lightness of the equipment, the operator can make out-of-position welds with considerably less effort than he can with the conventional setup. In addition, the process is easily learned and can use the same power source as is used for stick-electrode welding.

POWER SUPPLIES FOR ARC WELDING. The power supply to the arc may be dc or ac, and the polarity of the electrode may be positive or negative when dc is used. The type of electrode, the base material, and the position of the joint determine the polarity that should be used for the electrode. With electrode-negative polarity, more heat is concentrated at the tip of the electrode and more electrode is usually melted off per minute than with electrode-positive polarity. There is usually less tendency for burn-through, and better fill-in on poor fit-up when a negative electrode is used. With electrode positive and work negative, the base metal is hotter than the electrode.

Current control. Too much current tends to burn and scatter the metal. Too little current does not fuse the base material to the welding material. Welding equipment should include voltage- and current-regulating controls to compensate for variations in the supply voltage caused by sudden overloads or reduction in loads on the power circuit to which the welding equipment is connected. An experienced operator can determine by observation whether he is getting proper penetration and fusion without burning the material. It is helpful experience for the designer to learn to weld, so that he can better visualize the problems of the welder and detect improper welds.

The operator is blind for the interval between the lowering of the helmet and the striking or the arc. After he strikes the arc, he can see the welding processes through his dark glasses. Precautions should always be taken to shield the eyes and exposed portions of the skin from the arc flash to avoid a severe burn.

Arc blow. The arc itself is a flexible gaseous conductor of current and is subject to deflection by outside magnetic forces. The passing of current creates magnetic lines of force that pass through the base material. When the base material is magnetic (steel, for instance), the phenomenon of arc blow[1] occurs, and it is often difficult to get an arc to penetrate into a corner or fillet. Alternating current has less arc blow than direct current. The following factors are related to arc blow.

1. If the direction of current changes, the arc will change its position.
2. Magnetic material around the arc will cause a change in direction of the arc when its position and volume change.
3. Eddy currents of ac or pulsating dc welding affect the arc according to the position of hot-welded material and magnetic shunts near the arc.

SEMIAUTOMATIC WELDING. In semiautomatic welding, flexible rubber hose, metal tubes, and wires carry the electrode wire, flux, gas, and electric current to a portable nozzle that can be manipulated by the operator about as easily as a stick-electrode holder (Fig. 14.14). Thus, the parts to be joined need not be

---

[1] The term *arc blow* is used when the arc is unstable and cannot be directed properly by the operator. Sometimes it is impossible to deposit metal in a joint because of arc blow. Arc blow is not limited to magnetic base materials.

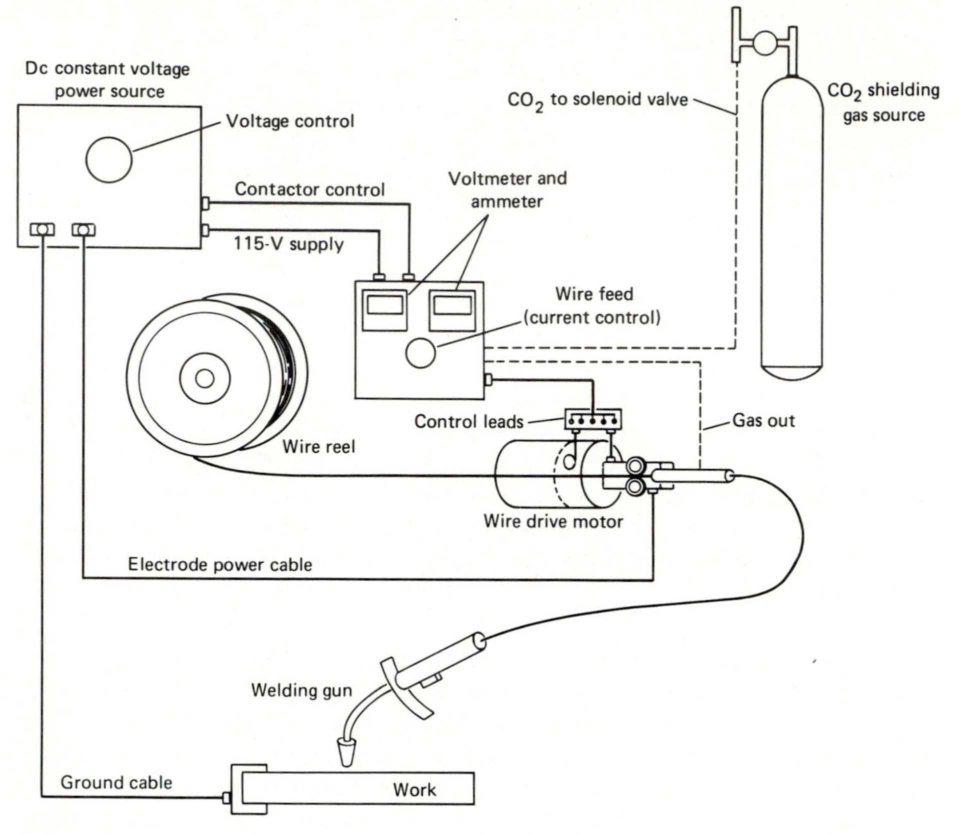

FIGURE 14.14
Schematic layout of typical flux-cored electrode welding facility.

positioned on the machine, but can be welded in any position provided that the proper small diameter wire and machine settings are used. In semiautomatic welding the wire is fed through the the hand-guided gun automatically and the arc length is always maintained.

The distinguishing feature of gas metal-arc welding is the mode of metal transfer, which can be (1) globular, (2) spray, or (3) short circuit. Which type occurs depends on the diameter of the electrode wire, the arc voltage, the welding current, the type of shielding gas, and travel speed.

1. *Globular transfer* occurs at currents that are lower than those used for spray transfer. The molten drop on the end of the electrode wire grows to several times the wire diameter before it passes to the molten pool. Globular transfer is not a practical welding process because it results in too little pentration, poor arc stability and severe spatter.
2. *Spray transfer* occurs in the presence of argon gas, which permits high current density and a large electrode diameter. The resulting deep

penetration is good for heavy-gage metals but it causes burn-through on thin-gage alloys. The bead appearance is excellent with very little spatter.

3. *Short-circuiting transfer* occurs because as a molten droplet forms at the end of the wire, it bridges the arc when it contacts the molten pool and shorts it out from 20 to 200 times a second depending on the control setting. A number of gas mixtures can be used from 25 percent CO_2 balance argon to all CO_2. Short-circuit transfer results in a smaller, cooler molten pool which is particularly adaptable to welding out of position or for joining thin sheets.

Usually dc power sources with constant potential and slope control are used for MIG welding. Reversed polarity (wire positive) and current settings of about 100 to 400 A for thicknesses of 0.050 to $\frac{1}{2}$ in. provide for good melting, deep penetration, and proper cleaning action during the spray-transfer process. In the short-circuit arc process, only shallow penetration can be achieved.

AUTOMATIC WELDING USING CONSUMABLE ELECTRODES. In automatic welding the machine provides the wire feed, the correct arc length, and the weld travel speed (Fig. 14.15). The bare-wire electrode is rarely used without a flux or gas envelope. Cored wire contains flux and alloys within a central core. The wire is used alone or with a gas shield. Advantages of automatic welding over hand welding include the following.

1. A much higher rate of welding speed can be maintained.
2. Welding is continuous from the beginning to end of a seam, and intermittent craters are eliminated.
3. A more steady and uniform arc can be maintained.
4. Better fusion is possible because higher welding currents can be used.
5. No welding wire is lost in the stub ends. This means higher deposition efficiency and no time lost for electrode changes and slag removal.
6. A welding operator of limited experience can handle the welding machine and produce satisfactory results.
7. Slag removal is eliminated when solid wire is used or much reduced when using fluxed-cored wire.

SUBMERGED ARC WELDING. Automatic welding in conjunction with proper tooling has three broad types of applications: circular welds, linear welds, and mass production of identical parts. Automatic welding has been made possible by the development of welding heads that strike the arc, feed the electrode, and maintain an arc of proper length and current. The metal electrode wire is coiled on reels and fed continuously. These heads are mounted on adjustable supports that may move along the welded joint, or the part to be welded may move under the head (Fig. 14.15). Various kinds of equipment are available

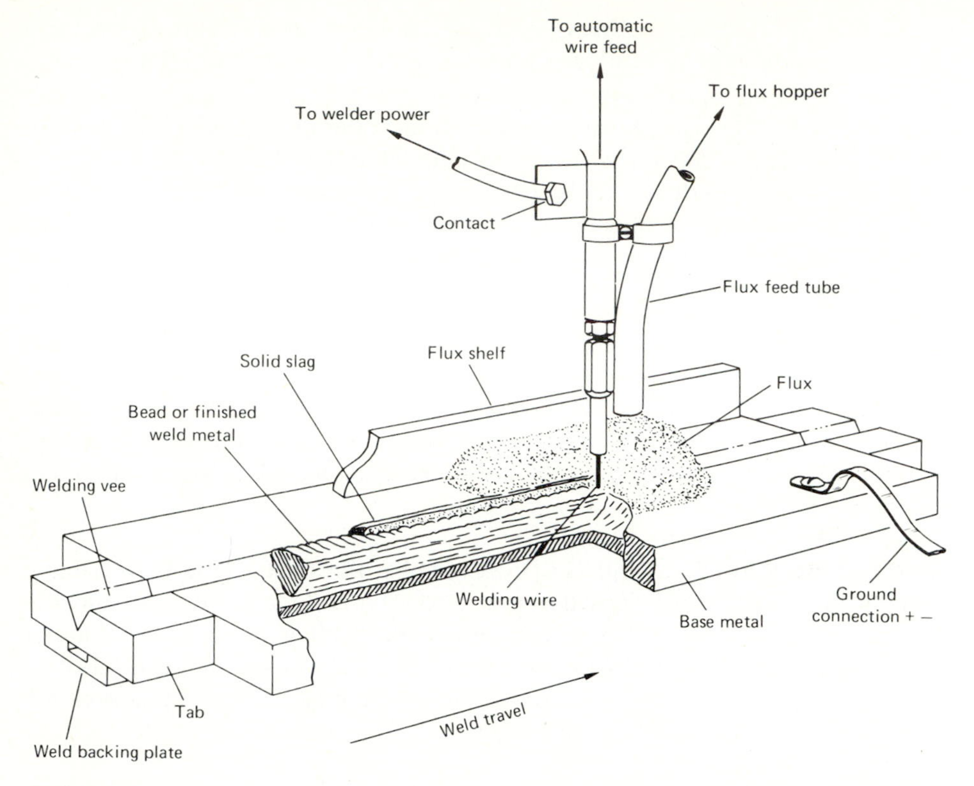

FIGURE 14.15
Submerged arc welding operation. (*Courtesy Hobart Bros., Troy, Ohio.*)

for clamping the parts into position, for positioning the part, and for feeding the wire electrode.

The submerged arc is an important process used with automatic welding equipment. The automatic submerged arc processes include:

1. Single ac or dc—straight or reverse polarity (this is the most common method)
2. Series arc—ac (used for cladding)
3. Three-phase, two-wire—ac, dc, or combination ac and dc
4. Multiwire—ac for high deposition rates

Submerged-arc welding is used in connection with automatic and semiautomatic welding equipment. The submerged arc process uses a mineral powdered flux that surrounds the electrode and arc. The electrode can melt rapidly and fuse with the parent metal under a protective atmosphere. The flux is easily removed. No arc flash appears. The operator judges the proper location by observing the general direction of the wire and the welded material. High current densities (up to 40,000 A/in.^{2}) and high rates of metal deposition are possible with high quality, deep penetration, and high welding speed.

Direct-current automatic welding seldom uses currents above 600 to 1000 A, and is used usually for alloy and stainless steels. Alternating-current welding, which predominates, uses currents up to 2000 A for standard equipment, and more for special equipment; ac welding is usually used for low-carbon steels.

A manually operated submerged arc welder has been developed that has the flexibility of hand operation and the advantages of automatic welding. The electrode wire ($\frac{1}{8}$ in.) is fed through a flexible tube up to 55 in./min, and with 450 A of current applied at the nozzle. The powdered flux is fed around the electrode at the nozzle by compressed gas which carries the powder through a

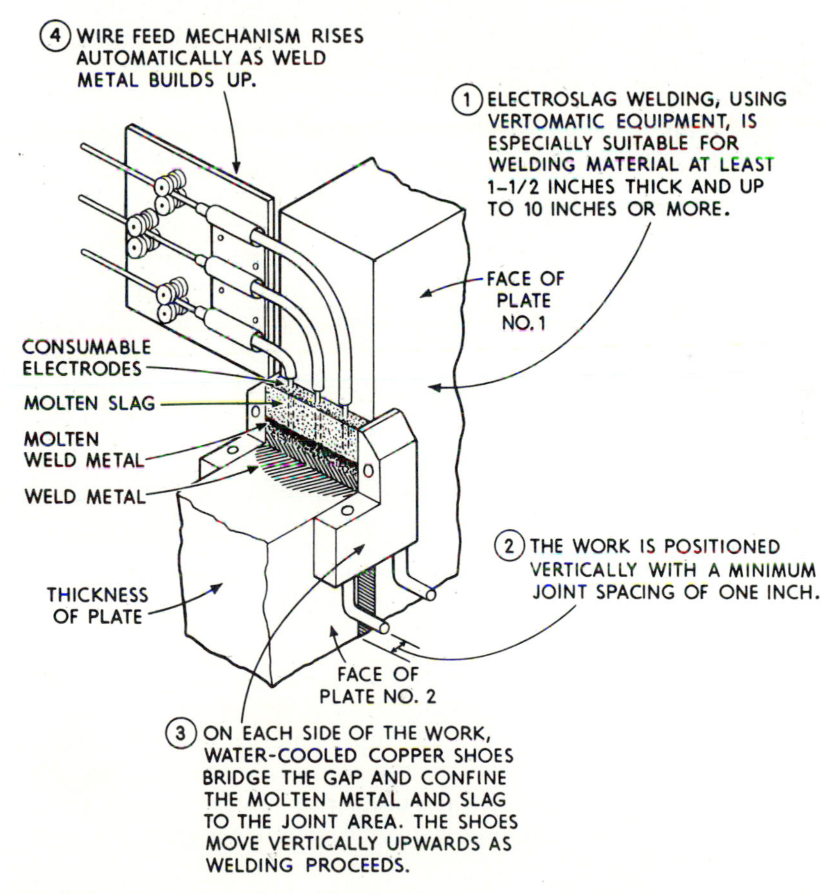

FIGURE 14.16
Electroslag welder. (*Courtesy Arcos. Corporation.*)

tube connected to the nozzle. The operator is much more comfortable because of the absence of smoke, spatter, and visible arc rays.

ELECTROSLAG WELDING. The electroslag welding process is unique in that it joins plates by casting the filler metal between the butted edges of the parent metal. The joint is made in one pass. The filler metal can be alloyed as desired through the introduction of alloy granules on top of the molten flux or through the cored consumable electrode. The process is explained in Fig. 14.16 and is applied to vertical or nearly vertical joints. Run on taps are used to start the arc. Special fixtures and procedures can produce circumferential welds to join thick-walled pipe of large diameter.

The process uses conductive slags, both to protect the weld and to melt the filler metal and the plate edges. An arc is used to begin the process by melting the slag and preheating the parent materials. Upon the formation of a layer of molten slag, the arc is extinguished and the heat conducted by the slag's resistance to current flow ($EIT = H$) provides the energy for welding.

The welding current will vary depending on the plate thickness and the number of electrodes used. When welding heavy sections with multiple electrodes, thousands of amperes are required. Deposition rates for multiple-electrode machines approximate a hundred pounds per hour.

Arc Welding—Nonconsumable Electrodes

GAS TUNGSTEN-ARC WELDING (TIG). In this process, the arc usually passes between a tungsten electrode and the metal joint. High temperature (up to 10,000°F) is concentrated at the end of the arc, where a small pool of molten metal is formed. The arc passes from the electrode to the work and is shielded by an inert gas such as helium or argon or a mixture of the two. Formerly a carbon electrode was also used, but in the 1950s the tungsten-inert-gas (TIG) process proved to be economically superior. The TIG name is still used in the shop but the AWS designation is technically gas tungsten-arc welding (Fig. 14.17).

TIG welding was originally developed for joining magnesium alloys, but it is now used for all alloys. It is particularly adapted to welding dissimilar metals and for hard-facing worn or damaged dies. It can also be adapted to welding light-gage sheet.

In general, an ac power source is best for TIG welding nonferrous alloys except deoxidized copper. For ferrous alloys, the dc power source with straight polarity (electrode negative) is better for gas tungsten-arc welding because it greatly reduces the volumetric loss from the tungsten electrode. For example, a TIG torch that has a rating of 250 A when used with straight polarity (dcsp) must be de-rated to 15 to 25 A when used with reverse polarity (dcrp-electrode positive).

Arc spot welds can also be made with a TIG torch fitted with special adaptors. In this case a $\frac{1}{8}$-in. electrode is often used for a total cycle time of $\frac{1}{2}$ to 3 s depending on the alloy and its thickness.

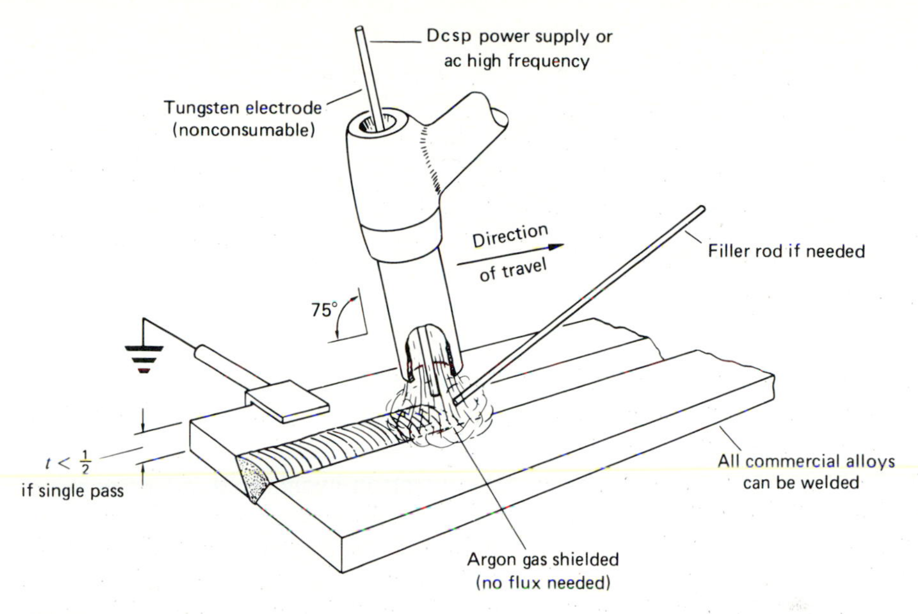

FIGURE 14.17
Gas tungsten-arc (TIG) welding operation. (*Courtesy of Linde Division, Union Carbide Corp., New York.*)

PLASMA-ARC WELDING. Plasma welding provides temperatures of 28,000°F or higher, as compared with 10,000°F for an electric arc. Therefore, the high-temperature flame can be used (1) to melt any metal or ceramic powder for surfacing metals or other materials, (2) to cut any material that will melt, and (3) to weld metals by heating the joint with or without filler metal.

A nonconsumable tungsten electrode within a water-cooled nozzle is enveloped by a gas (see Fig. 14.18). The gas is forced past an electric arc through a constricted opening at the end of the water-cooled nozzle. As the

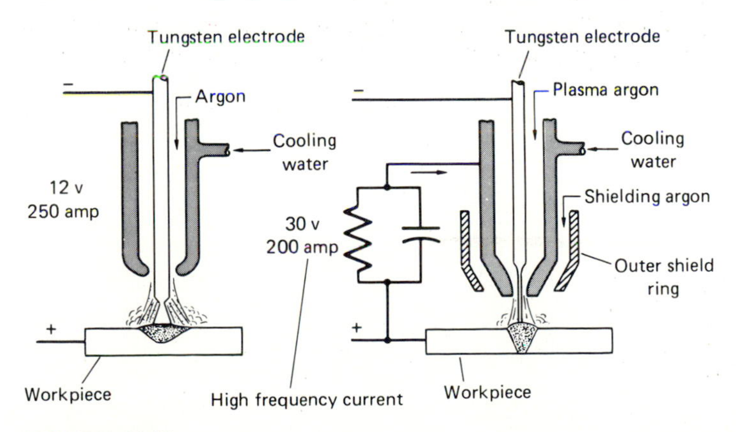

FIGURE 14.18
Schematic comparison of TIG and plasma arc. (*Courtesy of AWS.*)

gas passes through the arc, it is dispersed. This releases more energy and raises the temperature of the gases, leaving the nozzle to 28,000°F or more. The arc may pass between the electrode and the workpiece, in which case it is called a transferred arc; or it may pass between the electrode and the nozzle, in which case it is called a nontransferred arc. An example of the use of a transferred arc would be the cutting of a 1-in. thick aluminum plate at a rate of 50 in./min.

The types of arc welding are similar to gas tungsten-arc welding in that the heat is applied at the joint to obtain a weld and, when required, a filler rod can introduce material into the joint.

ELECTRON-BEAM WELDING. In this process the heat for coalescence is derived from the impingement of a stream of high-velocity electrons on the area to be joined. The process requires a focused electron beam, control circuits, and a high vacuum or at least low pressure to avoid collisions between the electrons and molecules of air. Most electron-beam welders are powered by a triode gun and accessory circuits (Fig. 14.19) and are available up to 1 A at 60,000 V. The trend is toward machines with even greater power and larger vacuum chambers.

After the electron beam is aligned with the joint seam, its power, spot size, and time of application can be easily varied to control the width and depth of the weld. The great concentration of energy in the beam (up to 0.5 to 10 kW/mm^2) permits the width of the heat-affected zone to be extremely

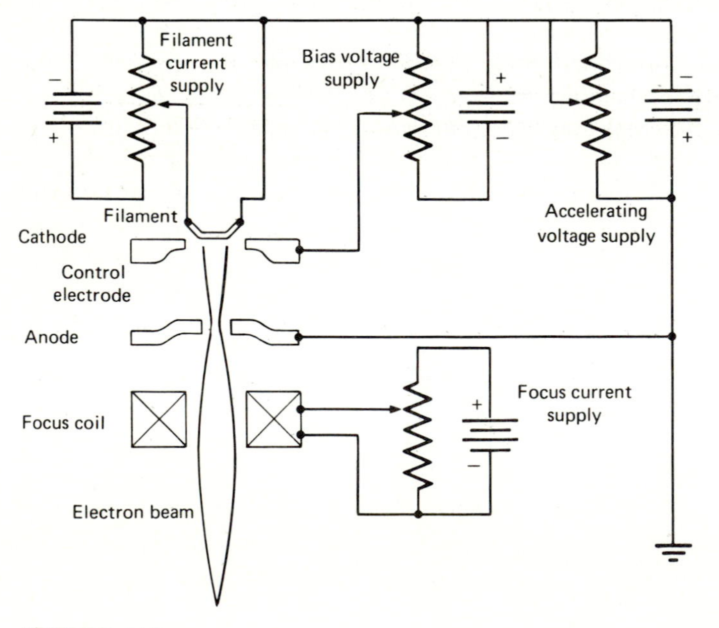

FIGURE 14.9
Triode electron-beam gun. (*Courtesy Sciaky Bros., Chicago, Ill.*)

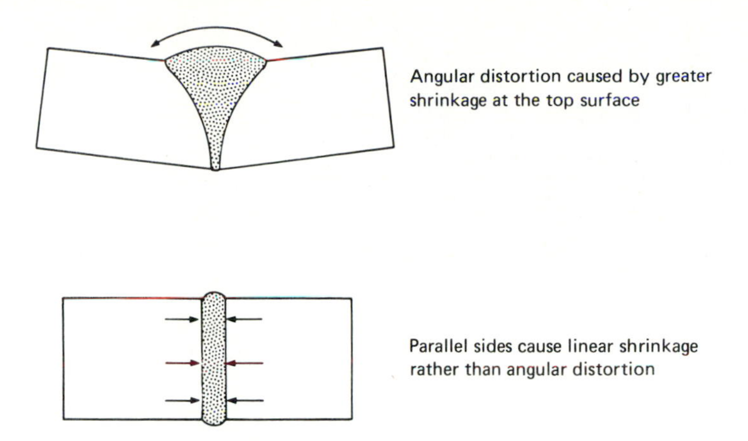

FIGURE 14.20
Effect of weld geometry on weldment distortion in electron-beam welding. (*Courtesy Sciaky Bros., Chicago, Ill.*)

narrow when compared with any other welding process (Fig. 14.20). Thus heat dissipation is minimized so that even pieces up to 5 in. thick can be welded from one side. However, when specifying electron-beam welding, it is important that the mating parts fit together accurately to ensure ease of proper alignment of the mating parts. It is common practice to specify that faying surfaces be machined to close tolerances prior to assembly and joining by electron-beam welding.

The gun-transport system and the work-holding device must be designed and built with the accuracy of a machine tool. In fact, optical aligning devices and closed-circuit television are integral parts of a large machine. Modern welders have specially designed electronic scanning devices that continuously locate the center of the weld within ±0.0005 in. on a routine basis. Feedback circuits correct for any misalignment and recenter the beam if there is any tendency to drift. The same device can be used to follow broken surfaces or circular or irregular contours.

The process usually requires a vacuum of about 10^{-4} Torr (1 Torr = $\frac{1}{760}$ standard atmosphere) for optimum quality. However, commercial welding can be successful in vacuums of as little as 0.1 Torr. Such vacuums can be achieved in only a few seconds with roughing pumps alone.

Commercial applications. Electron-beam welding has found wide application in aerospace and defense applications, but it is also useful under low-vacuum conditions for high-volume production in the automotive and appliance industries. Electron-beam welding has been successfully applied to the production of automotive distributor cams (Fig. 14.21). In this case, an automated dual-sliding-seal, 12-station, dial-feed welder equipped with a soft vacuum produced 1400 welded assemblies per hour (Fig. 14.22). No produc-

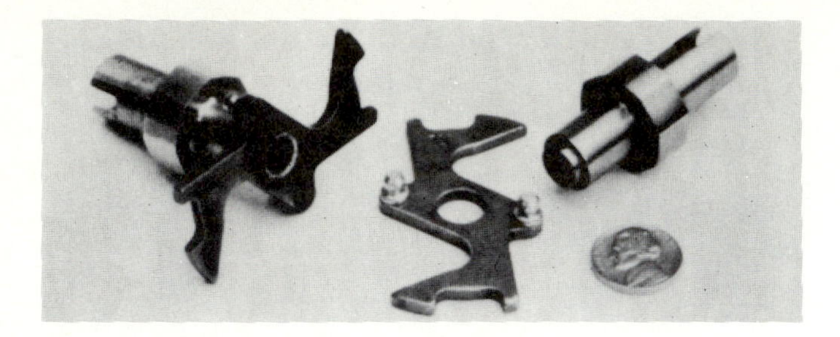

FIGURE 14.21
Closeup view of the distributor cam piece parts and the welded assembly. (*Courtesy Sciaky Bros., Chicago, Ill.*)

tion time was needed for the pump-down operation because while one assembly was being welded in the inner chamber, the next fixture was loaded in the outer chamber and evacuation was begun there. Although the complete installation cost more than several hundred thousand dollars, it paid for itself in less than a year because not only was operation time saved but valuable space was made available when the former furnace brazing equipment was retired.

Another production application is a retainer ring for a roller bearing assembly for nuclear reactors welded at 200 pieces per hour with no damage to the rolls and fuel elements, and assembled without danger to the operator.

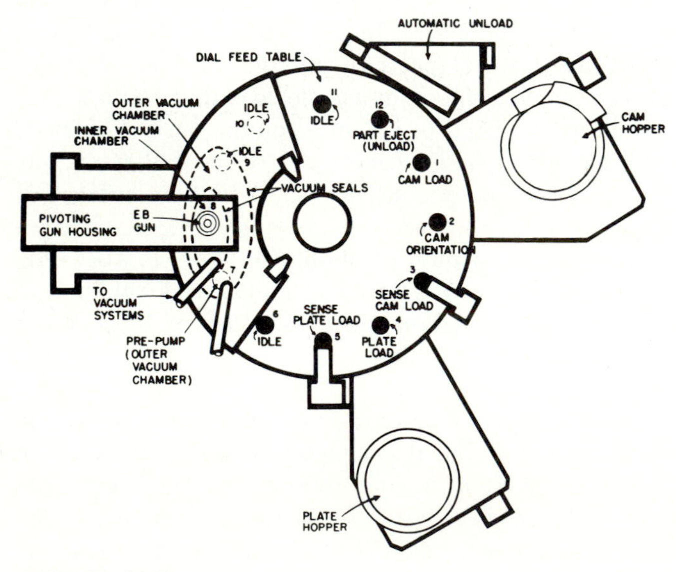

FIGURE 14.22
Schematic of high-speed dial-feed unit.

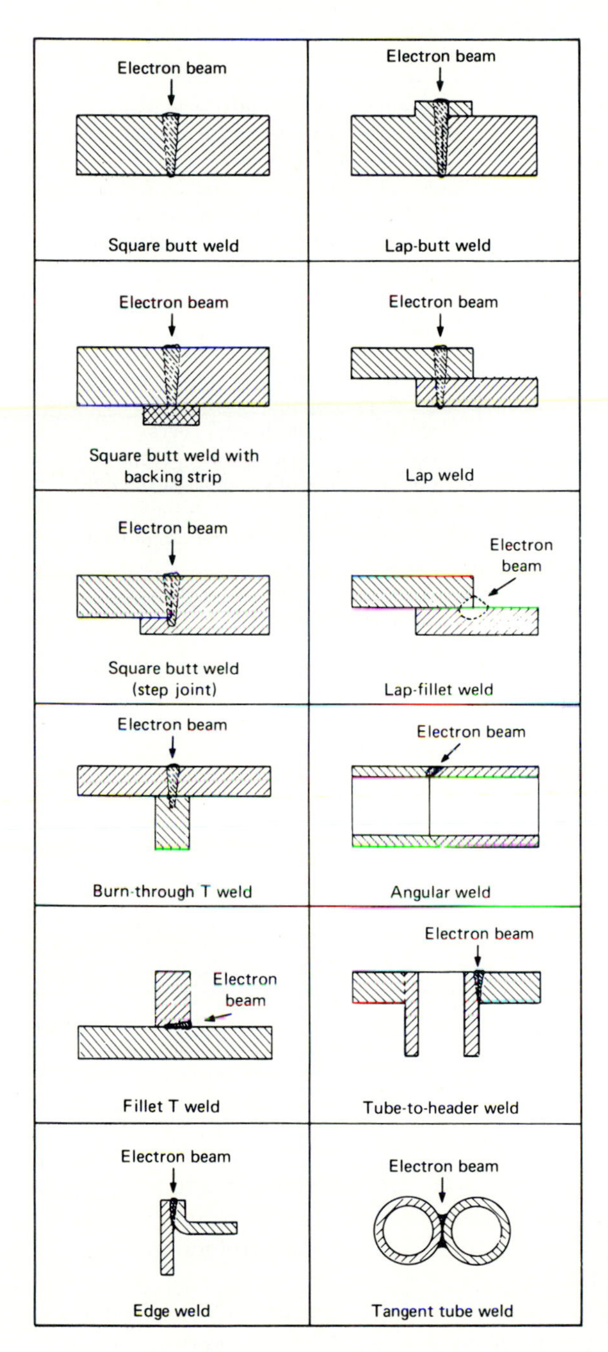

FIGURE 14.23
Typical joint configuration for electron-beam welding. (*Courtesy Sciaky Bros., Chicago, Ill.*)

TABLE 14.4
Joint design

Do's	Don't's
1. Linear joints should be machined square within 0.005 in.[a]	1. Avoid poorly machined weld joints
2. Gapping should held to a maximum of 0.005 in.[a]	2. Avoid partial penetration annular welds in areas of high restraint
3. For annular welds, a 0.0005 to 0.0001 in. press fit should be considered.	3. Avoid loose fitting assemblies for annular welds in areas of close final tolerances
4. For partial penetration annular welds, a relief groove in the root area should be considered to equalize shrinkage.	4. Avoid partial penetration annular or girth welds in areas of high stress
5. For girth welds a self-locating step should be used whenever possible in low-stress areas.	5. Avoid sharp corners when using a step joint
6. Self-locating step joints can be used to eliminate underbead on the ID in low stress.	6. Avoid welding parts having internal cavities which are not vented properly
7. For linear or girth welds that employ a step joint, 0.005 in. gap should be allowed in the lower portion of the step to accommodate lateral shrinkage.	7. Avoid making square butt welds on sheet-metal parts with sheared edges.
8. Step joint corners should be chamfered.	8. Avoid using backing bars or strips that are not metallurgically compatible with the base material
9. For combination of annular and girth welds on the same assembly, the girth welds should be made first	9. Avoid lap weld joints in areas of high stress and fatigue
10. Parts having internal cavities must be vented properly	10. Avoid gapping between sheets of material when using lap weld joint
11. If backing bars or strips are used, they should be of the same alloy as base material	
12. Utilize lap-weld joints for use with sheet metal and thin-gage materials wherever possible	
13. Consider seal pass techniques for annular welds to eliminate the possibility of gapping when other than interference fits are employed	

[a] Larger values are feasible. Usable values of joint squareness and fit depend on the joint thickness, the material being welded, and the weld quality that is acceptable. In joints of more than 0.25 in. thickness, gaps of 0.010 to 0.015 in. may be acceptable in some instances; in 1-in.-thick joints, gaps of 0.030 in. can be tolerated if filler metal is wire-fed into the joint during its welding.

Typical joint designs. Joint design—material, preparation, and fixturing—are important factors in the economic application of electron-beam welding of quality assemblies. It also has a major influence on the production rate that can be achieved.

Typical joint designs are given in Fig. 14.23, and checklists of do's and don't's have been borrowed from the Sciaky Bros., Chicago, and are given in Tables 14.4 to 14.7.

TABLE 14.5
Material weldability

Do's	Don't's
1. High-quality wrought or forged materials should be considered whenever possible 2. Aluminum-killed steel should be considered as a first-shoice material 3. Silicon-killed steel material should be considered as a second-choice material 4. Rimmed steel can be used as long as maximum amounts of "rim" and minimum amount of "core" are mixed into the weld zone 5. Minimum amounts of case should be mixed into the weld zone when welding case-hardened materials 6. Utilize slow welding speeds when welding cast materials 7. Utilize aluminum spray, shims, or wire when welding nonkilled steels 8. Consider preheating and/or postheating when welding materials containing more than 0.35% carbon	1. Castings of poor density should be avoided 2. Avoid rimmed steel unless maximum rim cannot be mixed into the weld zone 3. Avoid case-hardened materials unless minimum amounts of case can be mixed into weld zone 4. Do not use parts that are case-hardened with any process using ammonia (this leaves a high nitrogen content in the surface-hardened area) 5. Avoid cast materials unless slow welding speeds can be used 6. Avoid free-machining steels that contain high percentages of sulfur or phosphorus 7. Avoid welding materials such as brass, bronze, zinc, magnesium, and powdered metals containing high percentages of non-metallic binders

TABLE 14.6
Material preparation

Do's	Don't's
Minimum requirements	
1. Materials should be free of contaminants such as oil, grit, and heavy scale 2. For material having irregular surfaces, such as corrugations, pickling can be used 3. Clean with acetone or equivalent solvent 4. Demagnetize ferrous materials before welding	1. Avoid grinding of nonferrous materials 2. Avoid wire-brushing of nonferrous materials 3. Avoid solvents of oil when machining nonferrous materials 4. Avoid use of solvents containing chlorides 5. Avoid use of solvents or oils when machining case materials
Treatment of the joint area for optimum weld quality	
1. Remove all traces of scale and surface oxides 2. Materials should be machined to approximately 32 rms finish 3. Steel alloys can be machined or surface-ground 4. Nonferrous materials should be machined dry (no solvents or oils) 5. Nonferrous materials should be hand scraped or chemically cleaned to remove surface oxides 6. Cast materials should be machined dry if possible	

TABLE 14.7
Weldment tooling

Do's	Don't's
1. Tooling should be made of nonmagnetic materials in areas of close proximity to the beam	1. Avoid use of magnetic materials in close proximity to the beam
2. Use forged-wrought or densely cast materials	2. Avoid use of porous cast materials
3. Blind holes or grooves should be vented	3. Avoid materials that contain lead or zinc plating
4. Tooling fixture and clamps must have a good groud	4. Avoid use of nonmetallics, such as ceramics, rubber, and plastic that are porous
5. Parts that are insulated for preheating purposes must have good ground return	5. Avoid severe restraint tooling when making annular welds
6. Tooling should provide sufficient lateral pressure to minimize gapping during welding	6. Avoid unsupported edges when welding sheet metal or thin-gage material
7. When making girth welds on light-gage materials, spring-load for follow-up pressure	7. Avoid use of plastics and other electrically insulating or nonconductive materials in close proximity to the beam
8. For girth welds in heavy thicknesses of material use rigid clamping	
9. Maximum support and clamping close to the weld joint should be considered when welding sheet metal or light-gage material	

Resistance Welding

Resistance welding is the heating of material at the junction to be welded by local resistance to passage of electric current. Spot, projection, seam, flash, percussion, and butt welding are forms of resistance welding. The material is raised to a temperature that causes it actually to melt and, under pressure, it is fused or forged together. The principle is the same as that used in any blacksmith-forged joint.

The amount of heat depends upon the amount of current and the length of time it is applied ($H = I^2RT$). The amount of current depends upon the voltage applied and the total resistance or impedance of the circuit; therefore, voltage must be consistent regardless of variations in the power required. Some welding equipment, especially that used for welding aluminum, places heavy demands on power lines and often requires special feeders and transformers to maintain suitable electrical capacity and voltage. The total resistance or impedance of the welding or welding equipment's circuit depends upon the following factors.

1. The impedance of the welding circuit varies as the position of the part within the welder changes. If the part is magnetic, the lines of force will pass through the material and reduce the current. Therefore, if a resistance weld is made when a small portion of the part is near the electric circuit, it will receive more current than it does when the part is moved, so that a large portion is included in the electric circuit.

2. The resistance or impedance of the electrical equipment producing the current influences the amount of current. These parts can be designed with suitable electronic control, so that variations in current can be compensated for to a great-extent—even variations in position of part, line voltage, and resistance of the joint.
3. Resistance of the joint is composed of:
   a. contact resistance between electrodes or clamps and material
   b. contact resistance at joint of mating parts
   c. base resistance of mating material
   d. resistance of the electrodes

Contact resistance (3.a and 3.b above) are significantly affected by the surface conditions such as cleanliness, uniformity, and freedom from oxides and other compounds. Also, contact resistances are directly related to the resistivity of the materials being joined and resistivity force.

The base material resistance is proportional to the resistivity of the metal and the length of the current path.

The reader should understand that in resistance welding high-resistivity materials necessitate greater consideration of base material resistances, while low resistivity materials such as aluminum call for more consideration of contact resistances.

Successful application of resistance welding to designs depends upon consistency of material composition, surface, pressure and current applied, and time of their application. Low-carbon steel is the most common material welded by resistance welding and is assumed in all data given in this chapter.

Resistance welding is usually performed with alternating current. The greatest advance in the use of various forms of resistance welding came in the early 1920s when engineers realized that consistency in each of the related factors would assure good welds. First, mechanical devices were developed to apply proper pressure and control the length of time the current was applied. Recently, with air and hydraulic systems for applying pressure at the correct time and in the right amount through electronic controls, resistance welding has been advanced. The length of time for applying current can be controlled from one-half cycle to as many cycles as desired. A shot of high current for a short time produces the best weld, since the heat is concentrated at the joint and does not have time to spread, cause distortion, or affect the material adjacent to the joint. Since the magnitudes of the resistances in resistance welding are low (on the order of 100 $\mu\Omega$) currents are large—running into the tens of thousands of amperes. When capacitor-discharge supplies are used, the current may be as high as 200,000 A, but of course for only a very short duration. Resistance welds in automobiles, airplanes, passenger cars, and all forms of sheet-metal equipment are accepted without question on the part of the public and in many cases have replaced arc welding and riveting. Equipment designed for resistance welding (such as electrical switch gear) has reduced scrap, weight, and labor, and has increased the use of standard parts.

Spot welding. Spot welding, the most common form of resistance welding, consists of joining two pieces of material by placing them between two electrodes (Fig. 14.24) and passing a current to heat the material sufficiently at the joint to cause plastic flow and the union of the two parts. The parts are held together while they cool sufficiently to regain mechanical strength. As outlined under the general principles of resistance welding, the greatest resistance should be between the two parts to be joined. The initial pressure should be great enough to obtain contact and then provide a forging action.

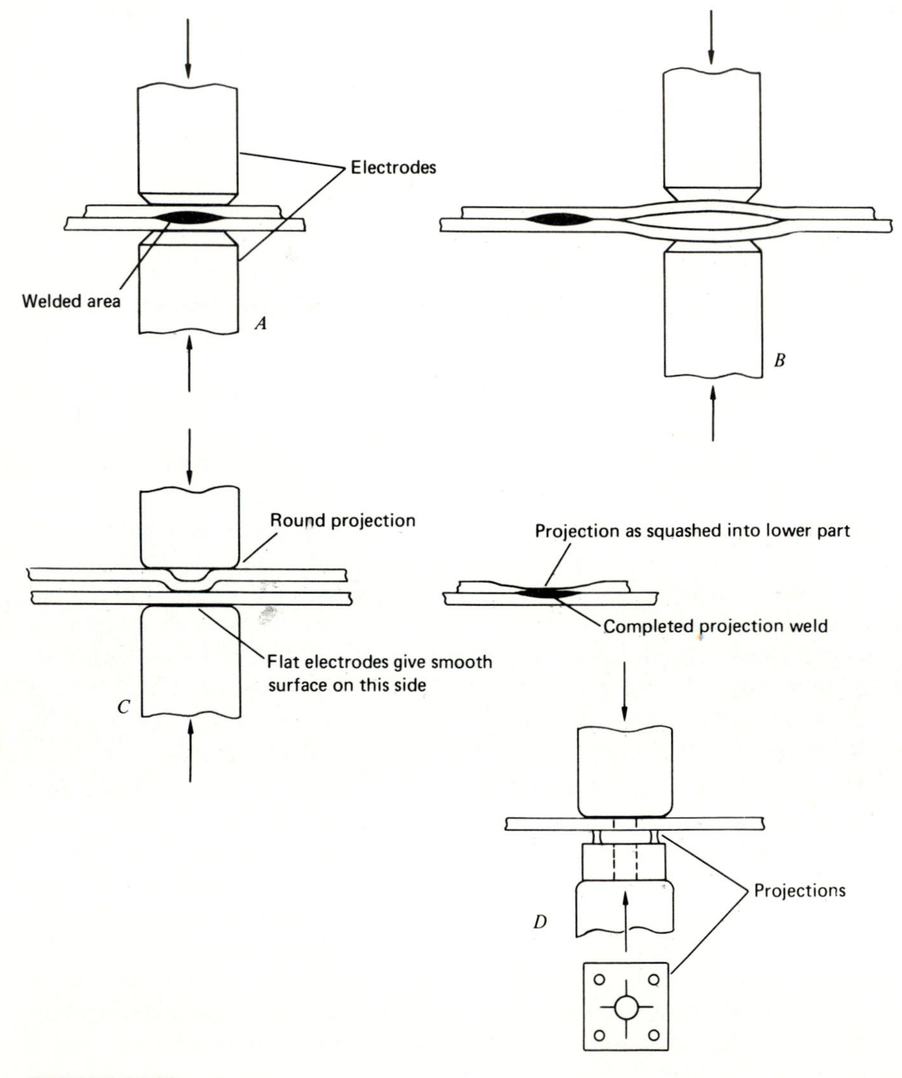

FIGURE 14.24
Resistance welding. *A*, spot welding; *B*, pressure is sometimes necessary to force parts together before welding; *C*, projection welding; *D*, projection welding of a special nut to a sheet.

The current must be controlled to give sufficient heat yet not melt or burn the material.

The diameter of the electrode end must remain the same and not mushroom and increase the area of contact. When this happens, the current density is reduced. The area increases in proportion to the square of the diameter, and the heat generated is reduced in proportion to the square of the current (I^2R); therefore, the electrodes should be machined to size and should not be filed by an operator. The contact surface should meet the material surface evenly. The electrodes are usually water-cooled to prevent softening. Electrodes of special alloys are stronger, but have more electrical resistance. Copper is the best electrode material for general application.

Spot welds, unlike rivets, require no holes or riveting, heading, or squeezing operations. The spot can be placed easily because the operator can see the work as the weld is made. The time for assembling and positioning the parts in the spot welder is greater than the few seconds required for making a spot weld; therefore, it is common practice to make multiple spot welds. Strength equivalent to that of riveted joints can be obtained by an equivalent spot-welded structure. Since there is no movement of joints, squeaks are avoided. Shear and tensile strength are close to that of the material welded. The strength of spot welds should be determined by experiment, using material and equipment suppliers' data as a guide. Portable spot welders are used to join members in large structures. They permit the same flexibility that is obtained with riveting air hammers.

Sometimes the parts do not have proper contact at the joint of the weld and considerable pressure is required to force the two surfaces together and make a weld at the place desired (Fig. 14.24*A*). This condition may cause a poor weld. Multiple welds made with electrodes in series or parallel are difficult to make unless all factors are controlled. For this reason, projection welding has come into general use.

The weldability of metals for spot and projection welding is shown in Table 14.8.

Projection welding. Projection welding is the use of a projection on one or both pieces to be welded, which forms a spot weld. More than one projection weld can be made at a time, as the area and location of the spot weld and contact pressures can be controlled. The current will be distributed uniformly between the multiple spots and good welds will result (Fig. 14.24*C*). For example, in Fig. 14.24*D* a special nut is shown projection-welded to a sheet. Spot- and projection-welded joints, like riverted joints, are not liquid- or gas-tight; therefore, seam welding was developed to make a continuous joint.

In all projection welding, it is necessary to have point or line contact in order to start a weld. When joining dissimilar sections, the projection should be placed on the heavier part. Heavy sections lend to joining by projection welding. The American Welding Society's "Recommended Practices for Resistance Welding" include recommended projections for sheet and plate

TABLE 14.8

Spot welding—weldability of materials

	Aluminum	Brass	Bronze		Copper	Cop. sil.	Iron-steel (low[b])	Iron-galv.[a] (hot dip)	Kovar	Lead	Magnesium	Molybdenum	Monel	Nickel
Aluminum	+													
Brass	0	ø												
Bronze	0	ø	+											
Copper		ø	ø		ø									
Copper silicon	0	ø	+		ø	+								
Iron, steel (low carbon)[b]	0	ø	ø		ø	ø	+							
Iron, galv. (hot dip)[a]	0	#	#		#	#	#	#						
Kovar							ø		+					
Lead		#	#			ø		0						
Magnesium	+		0								+			
Molybdenum	0	#	#		#		×					×		
Monel	0	ø			ø	ø	+	#	ø	0			+	

Blank—Combination not tried
+—Good weld
ø—Good weld with limitations
×—Completely miscible but brittle weld
#—Poor weld
0—No weld

															Nickel silver	Silver	Steel, cobalt	Steel, copper plated[a]	Steel, nickel plated[a]	Steel, stainless (austenitic)	Steel, tin plated[a]	Tin	Zinc
Nickel	0	Ø	Ø			Ø	+	#	Ø			×	+	+									
Nickel silver	0	Ø	Ø			Ø	+	#					+	+	+								
Silver		Ø	Ø				Ø	#	Ø					+		Ø							
Steel, cobalt	#	#	#		#	#	+	#	#			#	#	#	#	#	#						
Steel, copper plated[a]	0	Ø	Ø		Ø		+	#	Ø				+	+	+	Ø	+	+					
Steel, nickel plated[a]	0	Ø	Ø		Ø	Ø	+	#	Ø	#			+	+	+		+	Ø	+				
Steel, stainless (austenitic)	0	Ø	Ø		Ø	Ø	×	#									#	×	×	+			
Steel, tin plated[a]	#	Ø	Ø		Ø	Ø	Ø	#	Ø	Ø			+	+	+		#	+	+	×	Ø		
Tin	#	Ø	Ø		Ø	Ø	#	#		+			+	0	+					#		+	
Zinc	+	×	×		×	×	0	0		0			×	0	×				0	0	0	0	+

[a] In the course of spot welding coated materials the coating frequently dissolves in the other metals present or burns away.

[b] Limited to 0.15 percent carbon content for single thicknesses of less than $\frac{1}{8}$ in.; limited to 0.25 percent carbon content for single thicknessof $\frac{1}{8}$ to $\frac{1}{2}$ in.

This table should be taken only as a general guide, bearing in mind that some of the pairs marked "Ø" call for special manipulation or equipment and that others so marked may produce unsatisfactory welds in certain varieties, as for instance, high-carbon stainless steel. Some of the pairs marked "0" can be welded by means of special techniques. Welding copper to copper becomes very difficult when heavy sections are involved.

In many cases, the success of a weld between these pairs depends upon the rating, timing control, and other characteristics of the available welding equipment.

It is probable that in time to come, somc of the "0" signs will be changed to " + ". It will be noted that most combinations with aluminum are marked "0" This means that, as a regular shop process, aluminum cannot be welded to these materials.

By special techniques, such as interposing thin pieces of high-resistance material, it is sometimes possible to accomplish some of these welds.

In case of doubt the designer should consult the specialist of the process or material company.

Limitations refer to equipment (size and type of control equipment), thickness of metal, and application of special technique. The welding section of the division involved should recommend use of limited applications before starting production.

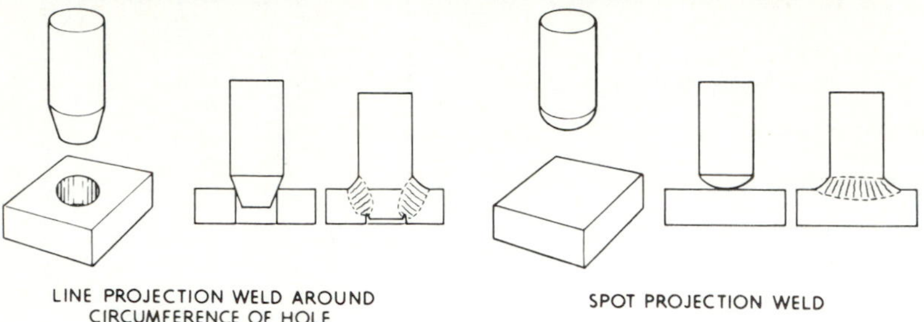

FIGURE 14.25
Line-type projection welds.

sections up to 0.50 in. It should be remembered that as sections increase in thickness, the diameter and height of projections are increased to develop greater strength.

Only clean, scale-free surfaces should be used in projection welding. A dirty surface will cause considerable variation in the resistance between the parts being joined, with resulting variation in current flow and weld strength.

Line projection welds (see Fig. 14.25) are recommended over point welds when sections are subject to heavy static or dynamic loads.

Seam welding. A seam weld is a joint being continuously welded by the resistance-welding process. The electrodes are disks which are driven as the two pieces to be welded pass between them. Pressure and current are applied to the joint, as in spot welding. When seam or continuous welding was first developed, the current was continuous, but the heat was difficult to control. It was soon discovered that overlapping spot welds was more successful. By making a series of spot welds in rapid succession, the operator finds that slight variations in contact pressure, surface conditions, and electrode contact resistance result in a better weld. Representative seam welds are 12 spots per inch on stock 0.01 in. thick at a speed of 100 in./min, and 5 spots per inch for $\frac{1}{8}$ in. thickness at a speed of 25 in./min. Water is sprayed on the electrodes to cool them and the weld material. Sometimes only one roller is used and a bar is substituted for the lower roller. More than one seam weld can be made at a time on special machines. An example of this is welding parallel seams in refrigerator radiator shells. Most seam welding is limited to sheet metals 0.01 to 0.125 in. thick (Fig. 14.26). Intermittent spots are made rapidly on seam-welding equipment—600 spots per minute, $\frac{1}{2}$ in. apart.

Seam welding is applied to lap-joint seams of cylinders and cabinets and to circular seams for welding bottoms in ends of cylindrical tanks.

Flash-butt welding. Flash welding is a resistance-welding process where joining is produced simultaneously over the entire area of abutting surfaces.

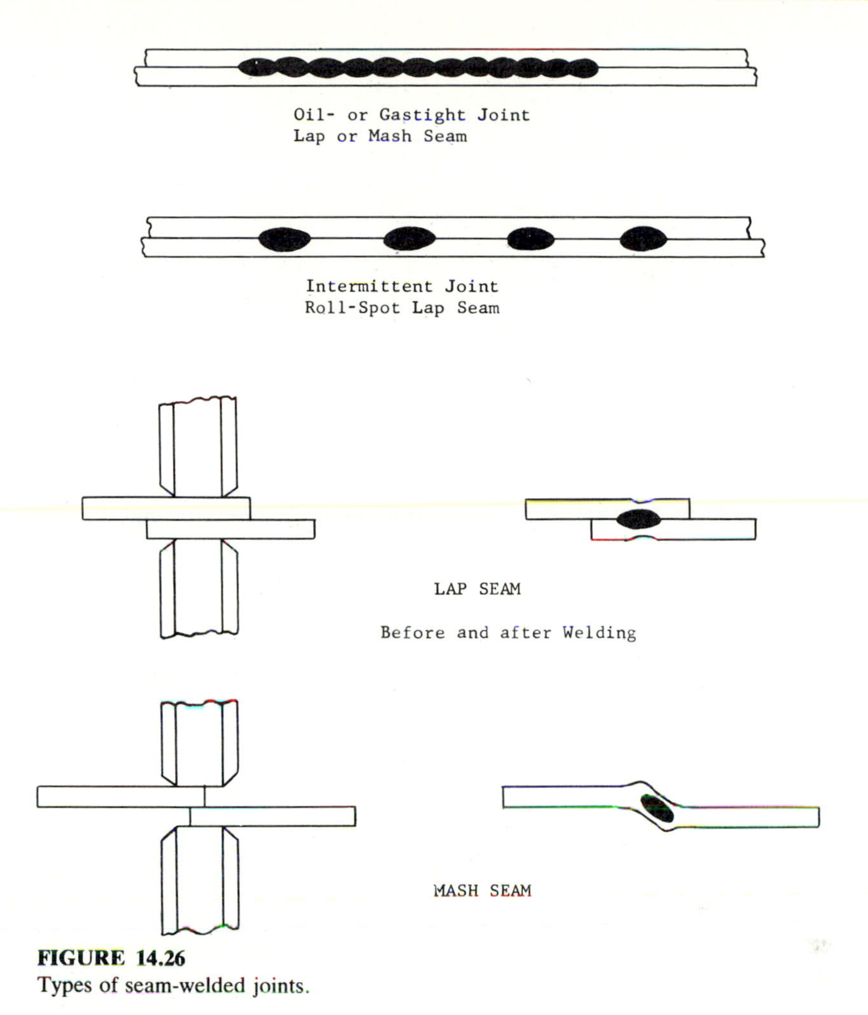

FIGURE 14.26
Types of seam-welded joints.

The flash-butt welding process produces a homogeneous weld between two sheets, wires, or bars without overlapping and without the addition of any materials (Fig. 14.27). Dissimilar materials may be flash- and butt-welded. Flash welding is limited only by the amount of current available to heat the surfaces and the pressure available for forging the parts together. Flash welding joins sheets together in fabricating the typical automobile body. Some companies find it economical to weld small sheets into a large sheet from which the car top is made.

Percussion or stored-energy welding. Percussion welding is similar to flash, stud, and spot welding in that a very high current is passed instantaneously through the surfaces to be joined, and the parts are joined immediately thereafter. The parts are moved toward one another rapidly. Just prior to

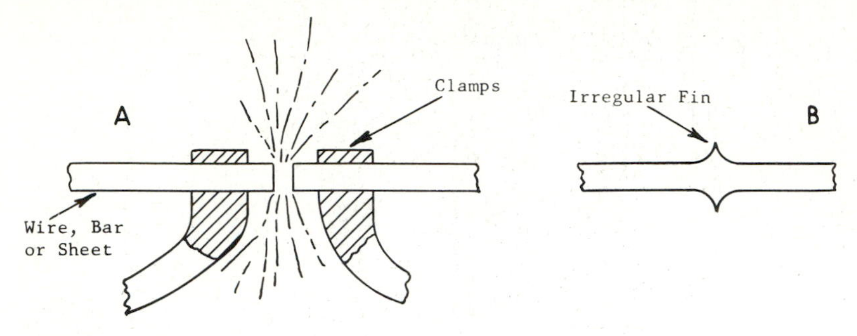

FIGURE 14.27
Flash-welding operation.

contact the current is passed through the two conductors being joined. This is usually accomplished by the discharge of a condenser. Dissimilar metals are welded with very little penetration of the heat within the parts.

Stored-energy welding involves an electrical means of storing and releasing large amounts of energy. Spot or percussion welding utilizes a condenser or a transformer circuit, or a combination of the two. Thus, less capacity is required in feeder lines and transformers. Stored-energy systems are used frequently for welding aluminum.

Stud welding. In the past it was difficult to flash-weld a stud to a flat plate because the stud would burn away and the plate would not heat sufficiently to permit a forged weld. A stud welder has been developed that is a combination arc welder and flash welder. The stud welder holds the stud as an electrode within an insulated tube. The end of the tube is placed over the location of the stud, and the stud is momentarily allowed to strike the plate. An arc is struck and passes between the stud and plate. The plate is heated by the arc, the stud is melted, and molten material is built up on the plate as in arc welding. After sufficient time, the stud welder cuts off the current, releases the stud, and drives the stud and built-up molten material together to make a forged weld similar to a flash weld. A strong weld results. This is a great saving over fastening studs to plates or flanges by drilling and tapping, welding threaded lugs to the plate, or arc welding studs to plates.

Friction Welding

Friction welding has been used successfully to join the ends of two components. One of the two components is rotated rapidly about an axis of symmetry and against the other component. Friction heat is developed at the interface. After the correct temperature is reached, rotation is stopped and the two components are upset forged together (see Fig. 14.28). The average interface temperature is always below the melting point of both members being

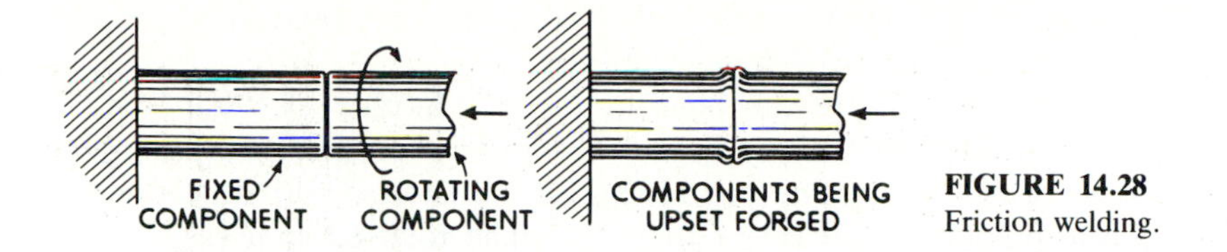

FIGURE 14.28
Friction welding.

joined, resulting in a bond achieved by diffusion as opposed to fusion. For this reason, the process is suitable for joining dissimilar metals.

This process is applicable to steels, brass, aluminum, titanium, and practically all other metals, as well as ceramics and thermoplastics. Carbon steel can be joined to stainless steel or aluminum. It can be used to join tubes, solid bars, studs, etc., either to each other or to plates. Today, it can be considered as a substitute for both resistance and gas-pressure butt welding. This method eliminates the need for flux and special gaseous atmospheres.

The American method uses high speeds and low pressures (12,000 r/min and 2000 lb/in.^{2}) as compared to the Russian method (1200 r/min and 14,000 lb/in.^{2}). The higher rotational speeds ensure better cleansing of the abutting surfaces. The lower pressures involve less deformation, less chance for cracking, and less likelihood of premature distortion of the joined parts.

Ultrasonic Welding

Ultrasonic welding is used principally for joining two members of which one or both is of foil or sheet thickness. It can produce spot welds, both straight and circular, between the parent members.

The process uses vibratory energy, which is transferred from the clamping jaws to the work in a plane that is parallel to the weld surface. The process produces a metallurgical bond between the parent metals, which may be dissimilar. The temperatures reached are below the melting temperatures of the metals being joined.

The acoustic energy required to weld a given material increases with the material's hardness and its thickness. An estimate of the energy required to produce a given spot weld can be made from the following equation:

$$E_a = 63H^{3/2}t^{3/2}$$

where E_a is the acoustical energy in joules, H is the Vickers microindentation hardness number and t is the thickness of material adjacent to the ultrasonically active clamp, in.

Laser-Beam Welding

The focused high-power coherent monochromatic light beam, or laser beam, is being used to join a variety of metals and alloys today. Laser beam welds can be made porosity-free with tensile strengths equivalent to those of the base metal.

The laser beam will cause the base metal to vaporize at the point of focus, resulting in a deep penetration column of vapor surrounded by a liquid pool of the base metal, which is moved along the path to be joined producing welds with depth-to-width ratios larger than 4:1. The focused spot size d of a laser beam is given by $d = f\theta$, where f is the focal length of the lens and θ is the full-angle beam divergence. The typical CO_2 laser-beam welder with a power level of 5 kW will weld 0.1-in. thick carbon or stainless steel at the rate of approximately 200 in./min. When the material to be joined is 0.2 in. thick, the speed is about 100 in./min at 5 kW. Thus, for material in this range of thickness, the welding speed is approximately proportional to power.

Laser-beam welding offers some distinct advantages over electron beam welding.

1. A vacuum environment is not required for the workpiece.
2. X-rays are not generated by the laser beam.
3. The laser beam may be directed, shaped, and focused with reflective optics, thus permitting mechanization of the process.
4. Since the laser beam has lower power density, there is less underbead spatter, incomplete fusion, and root bead porosity.

The electrical efficiency of the process is typically between 10 and 20 percent. Laser-beam welding today is readily automated, although the cost of the equipment is high, necessitating volume production if total mechanization is introduced.

BRAZING, SOLDERING, AND COLD WELDING

Brazing

Brazing is the joining of metallic parts by filler metals (such as copper and silver alloys) that have a melting point greater than 800°F and below the solidus of the base metal. The metals are usually at red heat and a flux is needed. The filler material adheres intimately to the surface of the metal and, with copper filler, forms an alloy with iron that has strength very near to that of the iron itself. It should be recognized that this alloying action is at the surface only; therefore, it is necessary to keep the thickness of the filler metal to a minimum—a few thousandths of an inch thickness is adequate. The strength of a brazed joint depends upon the shear area, the shear strength of the filler material, and the triaxiality of the applied stress. Therefore, it is desirable that the filler metal cover the entire area without blowholes or gaps. The ability of the filler material to travel between the mating parts depends upon a very clean surface, proper fluxes, and the capillary attraction provided in the joint. The fluid material cannot travel within a joint if the separation is excessive. The parts must be machined accurately, prepositioned, and held firmly in place without distortion while being brazed. Typical brazed joints are shown in Fig. 14.29.

The principal advantage of brazing lies in the fact that a strong joint can be made without undue distortion of parts. In continuous or batch furnaces, more than one joint can be brazed at the same time. The parts are heated and

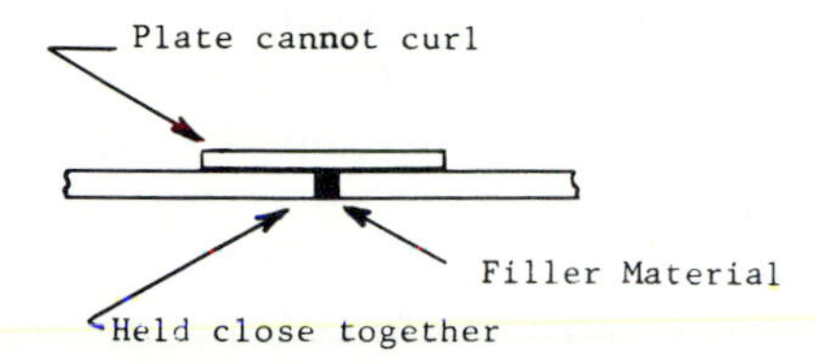

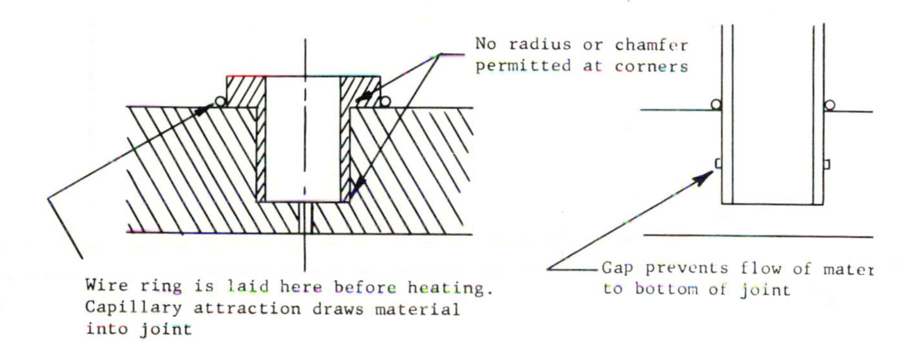

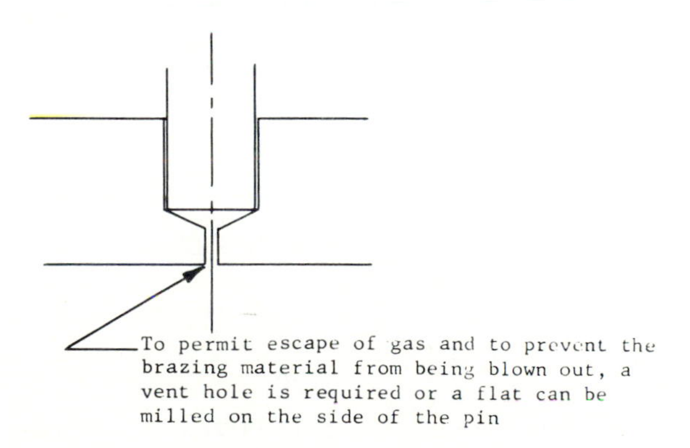

FIGURE 14.29
Typical brazed joints.

allowed to cool gradually, usually in a nonoxidizing atmosphere. Gas heat, oxyacetylene heat, and induction heating can be used to advantage in brazing operations. In all of these methods, even heating, clamping the parts to be brazed between jaws that are heated by the current passing through them enables the brazing of copper bars and leads on electrical equipment.

Induction brazing. Induction brazing refers to the process in which the parent metals are brought to the fusion temperature of the filler metal by placing them in an electromagnetic field that exists when an alternating current passes through an induction coil. The current flowing through the inductor sets up magnetic lines of force that pass through the surface of the parent materials and induce a flow of energy. Heat is developed in the parent materials owing to hysteresis and eddy currents—and this heat in turn melts the filler metal.

It should be recognized, since induction brazing depends upon heat being applied by way of magnetic lines of force, that ferrous materials that are magnetic are well suited to the process. However, when ferrous materials lose their magnetic properties, such as when heated above the magnetic transfor-

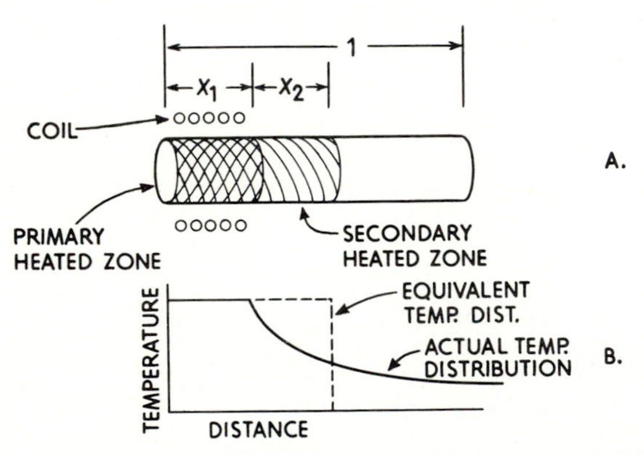

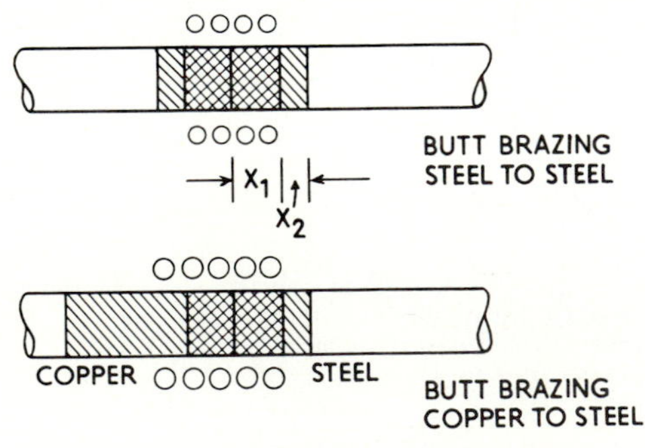

FIGURE 14.30 Brazing of similar and dissimilar alloys.

mation point, they become increasingly difficult to braze. Copper and aluminum are unfavorably suited to induction welding in view of their magnetic properties.

In induction brazing, there are two considerations concerning the thermal aspects of the process that the production-design engineer should be cognizant of. These are the power necessary to heat the work to the required temperature and the generator rating necessary to get this power. The power in kilowatts required to raise the temperature of a given amount of material to a given temperature may be calculated from the following relationship:

$$P = \frac{1.06WCT}{t}$$

where P is the power, kW, W is the weight of material, lb, C is the specific heat of material, Btu/(lb · °F), T is the temperature rise, °F and t is the time, s

After computing the thermal power needed, giving due consideration to the difficulties of bringing the power into the work, the production-design engineer will be able to determine the generator rating required. Figure 14.30 illustrates the temperature distribution within an induction-heated rod and Fig. 14.31 shows the extension of heating, x_2, for copper, brass, and steel.

The configuration of the work should be such that the inductor coil can be located to give close coupling to the section to be joined. Figure 14.32 illustrates some typical induction-coil configurations.

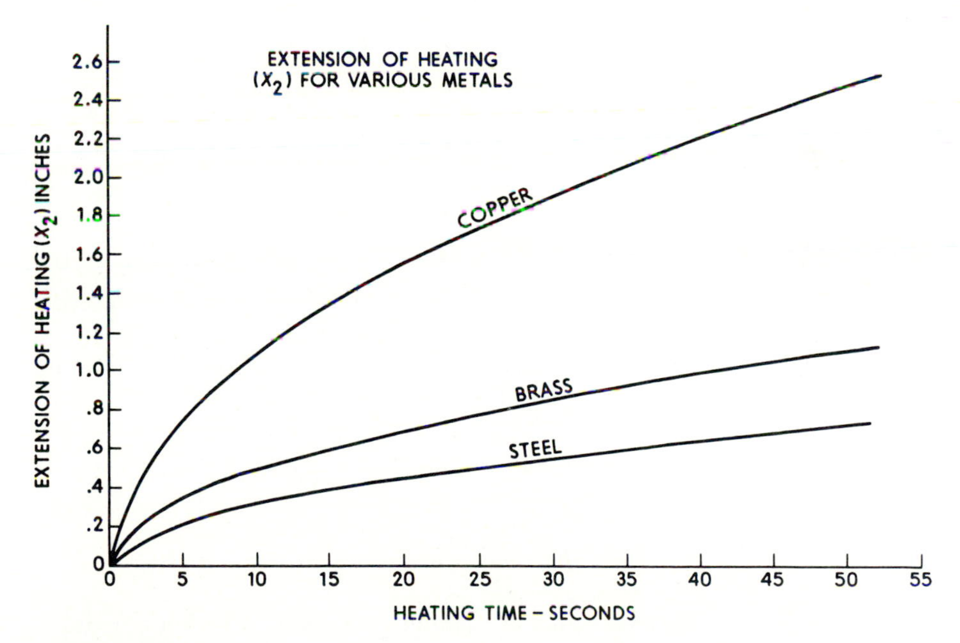

FIGURE 14.31
Extension of heating for various alloys.

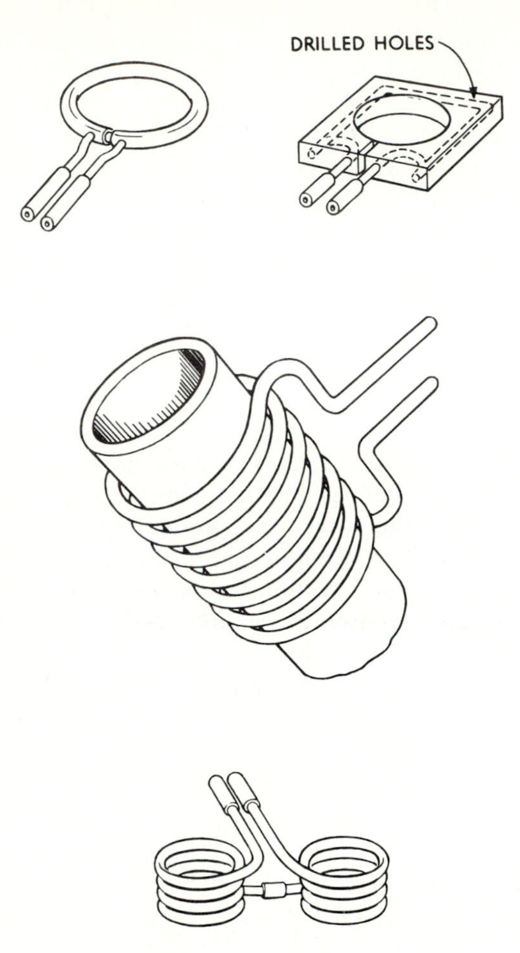

FIGURE 14.32
Typical coils for induction heating.

Braze Welding

Braze welding should be distinguished from brazing as a process used to repair a part by adding a nonferrous filler metal, such as naval brass. Such items as gears and machine frames made of steel or cast iron may be joined or built up by preparing the surface to be repaired and braze welded.

Nonferrous filler materials offer the advantage of accomplishing the work at lower heat input and consequent development of less thermal stress than if a ferrous filler material is used.

Soldering

Soldering is one of the most familiar joining processes. The filler material (solder) has low strength and melting temperature below 840°F, and results in less alloying between the base metal and the filler. The strength of a soldered joint depends upon adhesion of the solder to the metals being joined and the type of mechanical joint. The design of the joint should permit properly fitted surfaces so that the solder, through capillary attraction, will be distributed completely.

In order to insure reliable soldered joints, mating parts need to fit closely together, surfaces need to be clean, a flux needs to be applied on the metal surface that permits the molten solder to wet and flow into the joint. Finally, heat and the solder need to be applied correctly.

Soldering processes are identified according to the method of heating. Today the principal soldering processes used in industry include:

DS Dip soldering
INS Iron soldering
RS Resistance soldering
TS Torch soldering
IS Induction soldering
FS Furnace soldering

In addition to the above, infrared and ultrasonic soldering are also used. In all soldering processes, the thickness of the filler material should be kept to a minimum. In general, the same factors apply to soldering as to brazing, except that the soldering operation is performed at lower temperatures. Noncorrosive fluxes should be used wherever possible, since it is difficult to clean the flux thoroughly from the parts when the joint is cooled.

Cold Welding

Cold welding is the welding of metals by pressure at room temperature. Nonferrous metals, such as aluminum, cadmium, lead, copper, nickel, zinc, and silver, can be welded by this process. The surface must be clean of any contamination, including oxidation. The cleaning is best done by the use of rotating wire brushes before pressure is applied. Owing to the high pressure, the metal is squeezed together and the resulting joint is thinner than the sum of the two original thicknesses.

DESIGN OF WELDMENTS

About 70 percent of all fabricated steel is joined by arc welding—in production most often by the MIG process. But a design engineer is most interested in the major design factors that influence the economy and service life of typical weldments. We therefore review weld design fundamentals.

Welding Symbols

Until recently there were no standards regarding how information should be conveyed from the designer to the weld shop. The American Welding Society

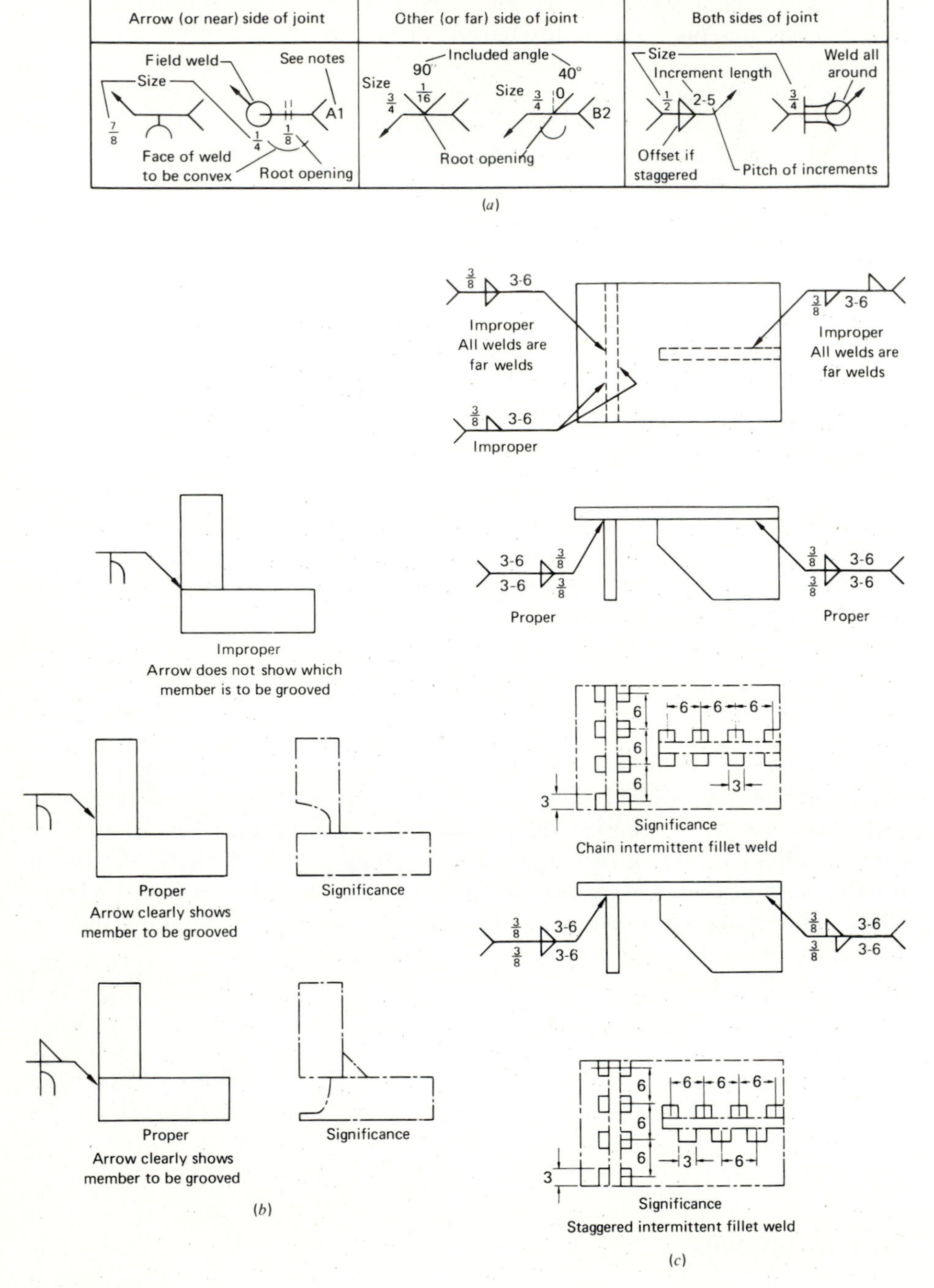

FIGURE 14.33
AWS standard joint symbols.

has gained the cooperation of many of the leading industrial organizations who specialize in welding fabrication, and a set of conventional symbols has been developed and is now widely adopted. The AWS drawing nomenclature and standard symbols are given in the five-volume set of the American Welding Society *Handbook,* which covers the full range of joints, weld sizes, and edge preparation for each major welding process. The basic welding symbol is an arrow connected to a reference line. All pertinent data for the desired weld are designated by symbols around the reference line. In the case of fillet, groove, flange, and flash welding, the arrow is drawn to touch the weld joint (Fig. 14.33). For plug, spot, seam, and projection welding the arrow should contact the centerline of the weld. The location of the weld can be represented by the location of the arrow with respect to the weld. If the weld is to be on the arrow side of the joint, the weld symbol is on the bottom of the reference line. If the joint is to be on the other side, the weld symbol is above the reference line. If both sides are welded, the symbol is located at both the top and bottom of the reference line. Further refinements and more detailed information can be obtained from the AWS *Handbook.*

Joint Design

The American Welding Society has developed standard symbols for most types of welds including four basic weld types (butt, fillet, flange, and plug), seven kinds of edge preparation, and the major resistance welds (Fig. 14.34). A weld bead is used for joining two pieces in the same plane end to end, which results in a butt weld (Fig. 14.35).

Three types of edge preparation are commonly used in butt welding: (1) the *square edge,* the simplest and most economical, used for plates up to $\frac{1}{8}$ to $\frac{1}{4}$ in. thick, depending on the process; (2) bevel or J joints for thicker plates

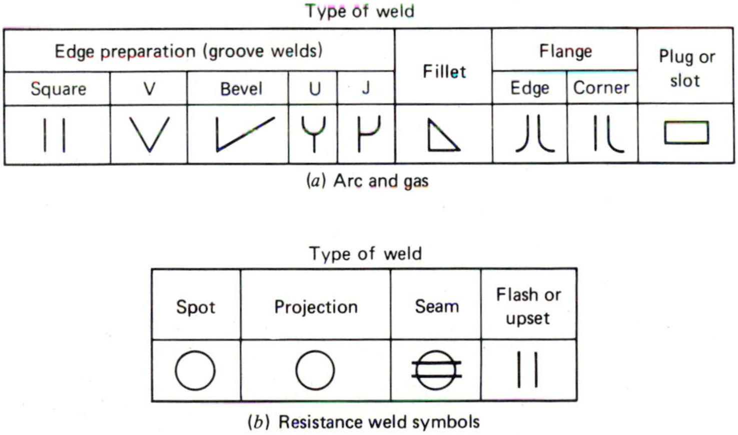

Type of weld

Edge preparation (groove welds)					Fillet	Flange		Plug or slot
Square	V	Bevel	U	J		Edge	Corner	

(*a*) Arc and gas

Type of weld

Spot	Projection	Seam	Flash or upset

(*b*) Resistance weld symbols

FIGURE 14.34
AWS standard symbols for welded joints.

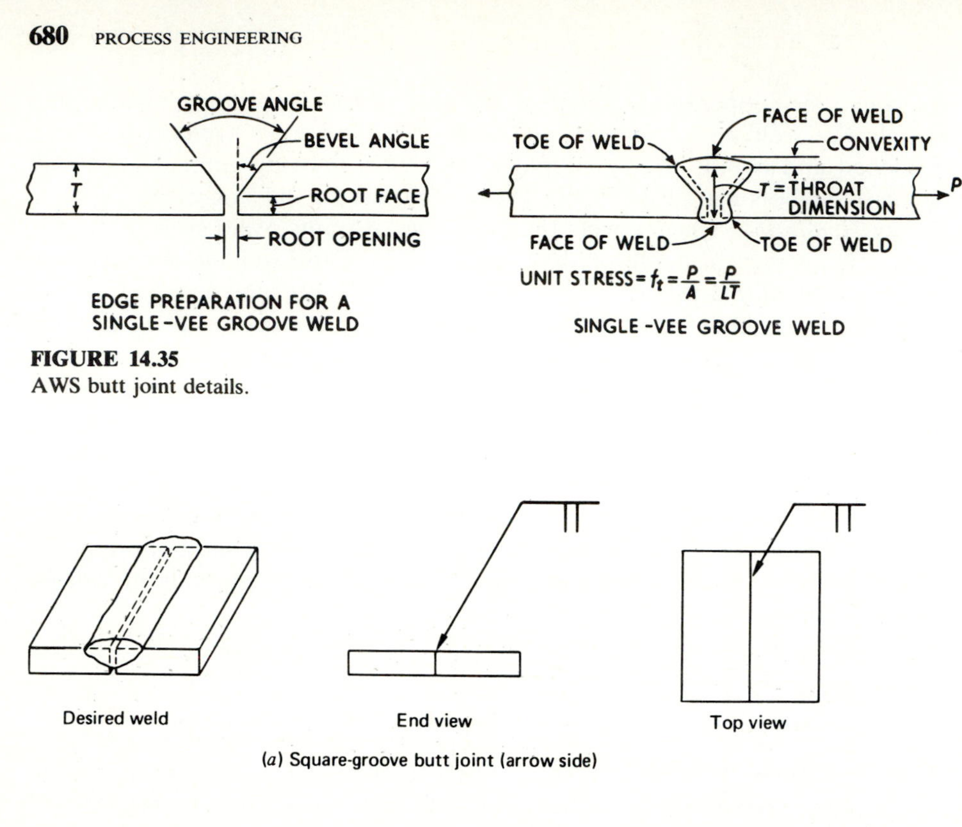

FIGURE 14.35
AWS butt joint details.

(*a*) Square-groove butt joint (arrow side)

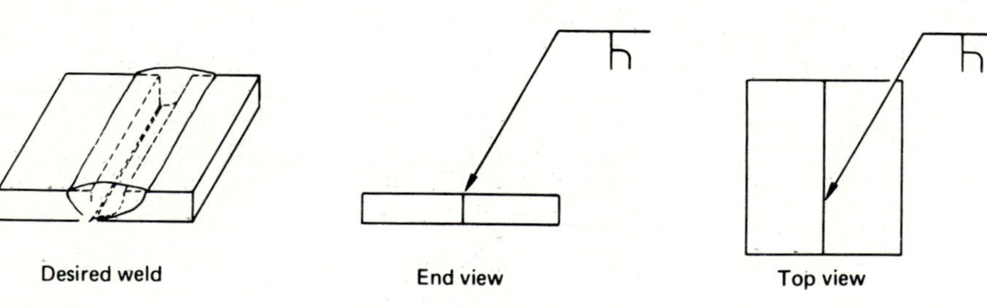

(*b*) Single-J-groove butt joint (arrow side)

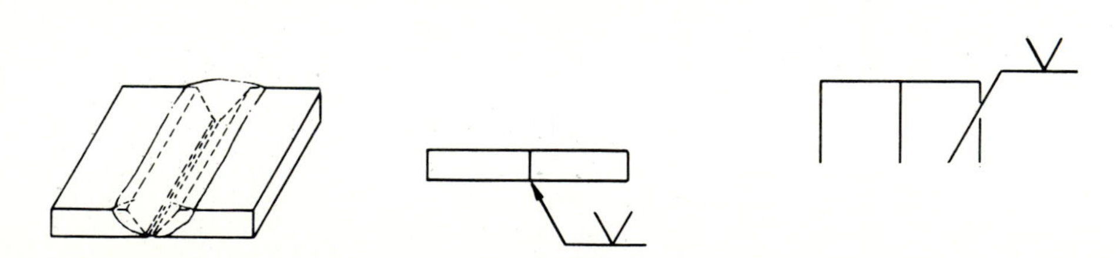

(*c*) Single-V-groove butt joint (other side)

FIGURE 14.36
Typical joints for butt welds. (*Redrawn from Joseph Datsko,* Material Properties and Manufacturing Processes, *Wiley, New York, 1966.*)

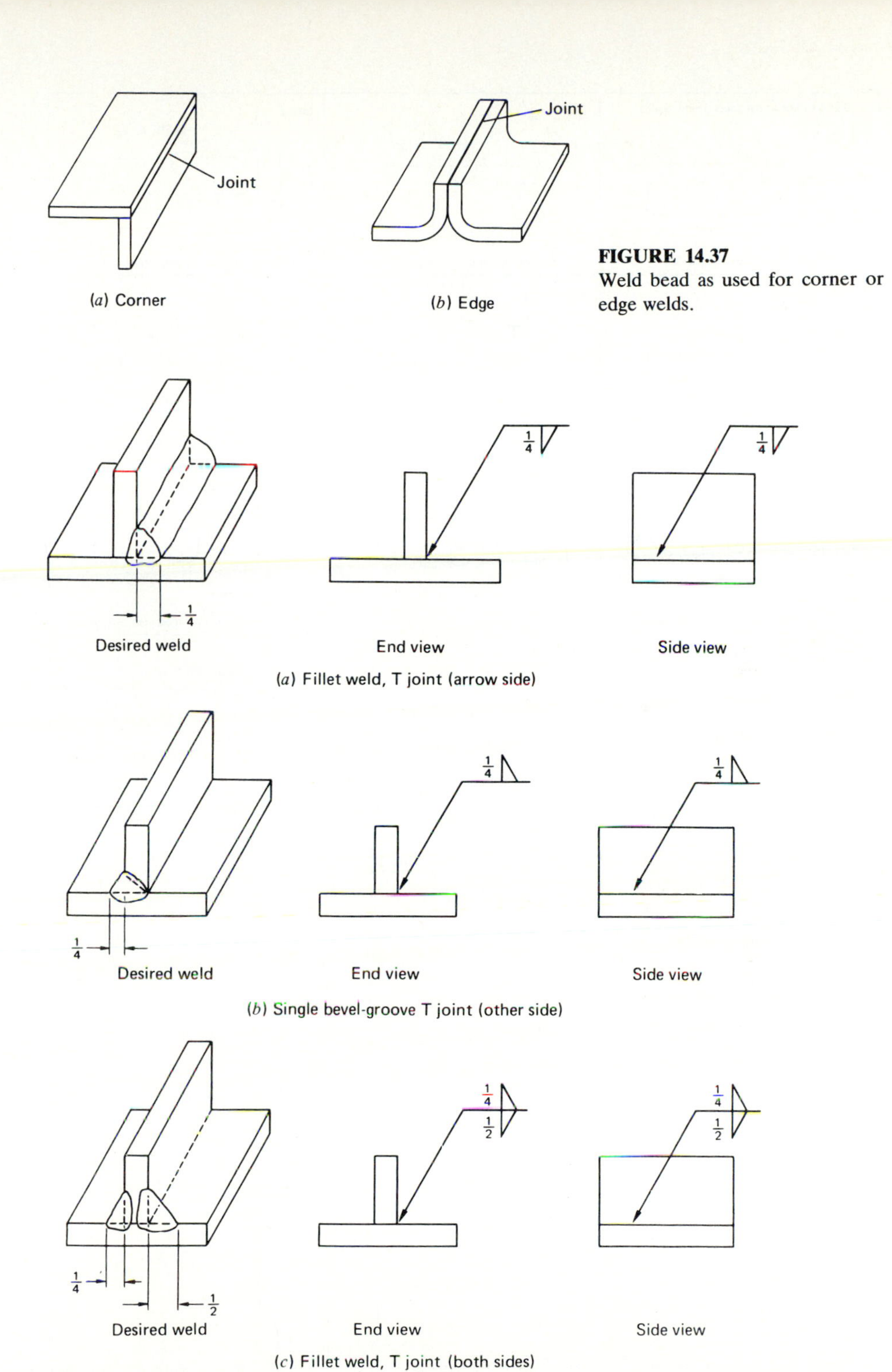

FIGURE 14.37
Weld bead as used for corner or edge welds.

FIGURE 14.38
Typical fillet welds. (*Redrawn from Joseph Datsko,* Material Properties and Manufacturing Processes, *Wiley, New York, 1966.*)

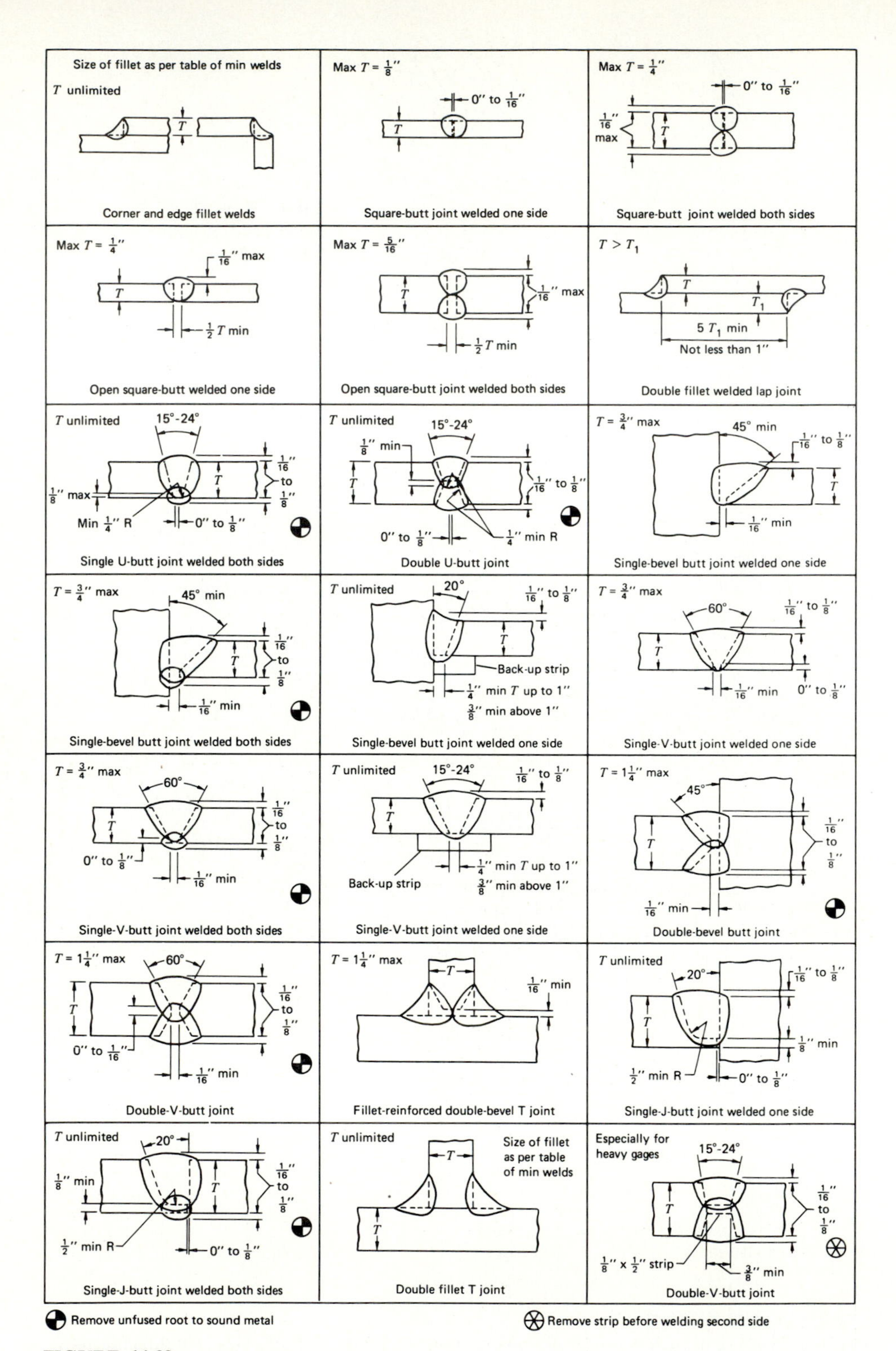

FIGURE 14.39
Edge preparation and typical joint details.

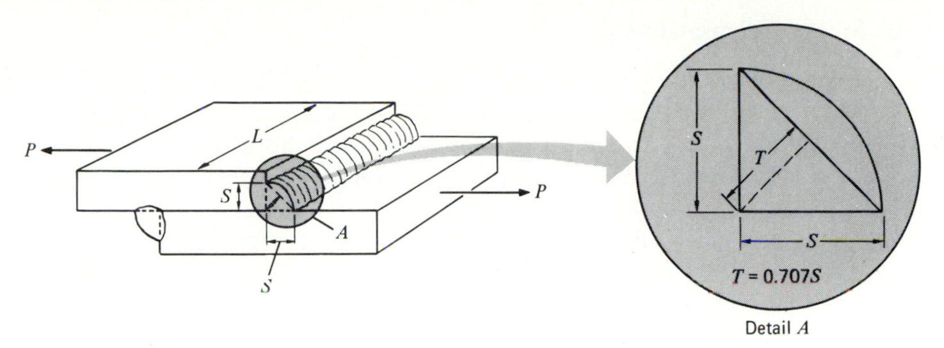

FIGURE 14.40
Double fillet weld.

where only one plate must be prepared; and (3) V or U joints, which require edge preparation on both plates (Fig. 14.36). It is frequently advantageous to use double-edge preparation so that welding can be carried out on both sides. For joints over $\frac{1}{2}$ in. thick designers should carefully investigate the economics of welding on one or both sides of the joint. Weld beads may also be used for corner and edge welds (Fig. 14.37).

In the T joint, two members are joined by depositing two welds of triangular cross section, called fillet welds, on either side of the vertical member of the T (Fig. 14.38).

Edge preparation and typical joint details for a number of types of welds are given in Fig. 14.39.

The lap joint, as the name implies, refers to those joints where one member laps the other and a weld is made along the exposed ends of one or both members. Figure 14.40 illustrates a double fillet-welded lap joint. Fillet welds are usually stressed in shear at the throat area. The throat T is defined as the perpendicular to the hypotenuse of the largest isosceles right triangle which can be inscribed within the weld bead. For example, in Fig. 14.40 the weld stress would be

$$\sigma = \frac{P}{2LT} = \frac{P}{2(0.707)SL} = \frac{0.707P}{SL}$$

Corner joints are usually made with either a fillet or groove weld. Typical corner joints are illustrated in Fig. 14.41.

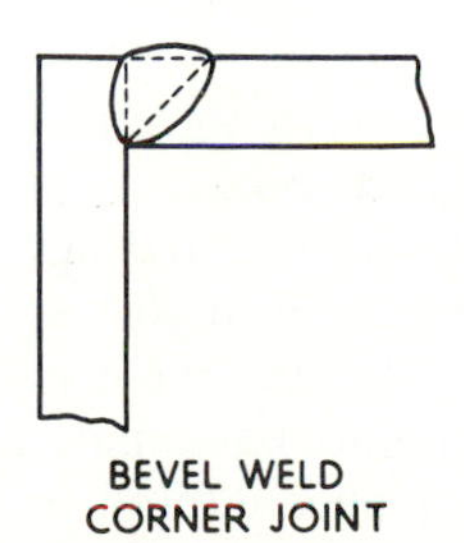

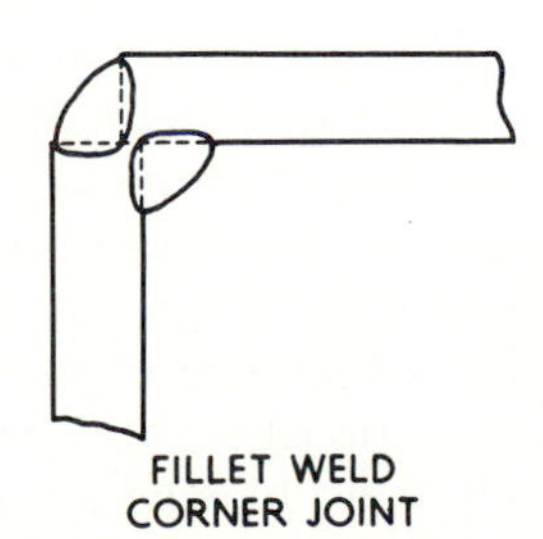

FIGURE 14.41
Typical corner joints.

Weld Stress Calculations

The stresses in typical welded joints may be calculated by using the relationships given in Fig. 14.42. The direction of the applied load is shown by the arrow *P*. The size of a weld can be calculated if the load *P* is known or reliably estimated and the code requirement or factor of safety (usually 3, but up to 5 in some cases) is known.

Ever since welding has been used, it has had to meet the opposition of conservative engineers who knew the reliability and uniformity of the material used in riveted and bolted joints; therefore, welded joints have always been calculated carefully and made under proper controls. Many experiments and extensive studies have been made to prove their reliability and agreement with calculated values. The mathematical derivation of formulas and the tests results made to prove their authenticity are shown in Fig. 14.43. An explanation of the figure is as follows.

The stress-concentration factor for welded joints on mild steel is

$$\text{Actual stress} = K \times \text{calculated stress}$$

For reinforced butt welds, $K = 1.2$; for toe of transverse fillet weld, $K = 1.5$; for end of parallel fillet weld, $K = 2.7$; for T butt with sharp corners, $K = 2.0$.

Stress-concentration factors are only important under conditions of dynamic loading and in cases where the joint is subjected to a static load with a superimposed dynamic load. The following formula will be helpful:

$$\frac{K(S_{max} - S_{min})}{2S_{wd}} + \frac{(S_{max} + S_{min})}{2S_{ws}} = 1$$

where S_{ws} is the static design working stress, S_{wd} is the dynamic design working stress.

A welded joint can be made of strength, ductility, uniformity, and reliability equal to or greater than the base material. In butt joints where no X-ray of the joint is specified, it is common practice to use 80 percent of the strength of the base material for the strength of the joint. If X-ray inspection is specified, then the strength of the base material can be used.

Arc-welded steel structures are three to six times as strong as ordinary gray cast iron, and the material is more uniform and reliable. They are 2 to $2\frac{1}{2}$ times as stiff as cast iron and 5 times as great in impact strength. Material can be added where it is most effective in meeting the functional requirements of the part. High-strength materials (alloy steels) can be incorporated where additional strength is desired. Steel-plate or sheet-metal walls are thinner and weigh less than cast walls. The cost of a pattern is eliminated; however, in repetitive parts or parts that are difficult to fabricate, welding fixtures and jigs may be necessary. Fixture costs should be compared with the pattern cost as well as comparative labor costs in making the choice between welding and casting (Fig. 14.44). Welding construction permits homogeneous base materials and the absence of sand inclusions; blowholes are not encountered in

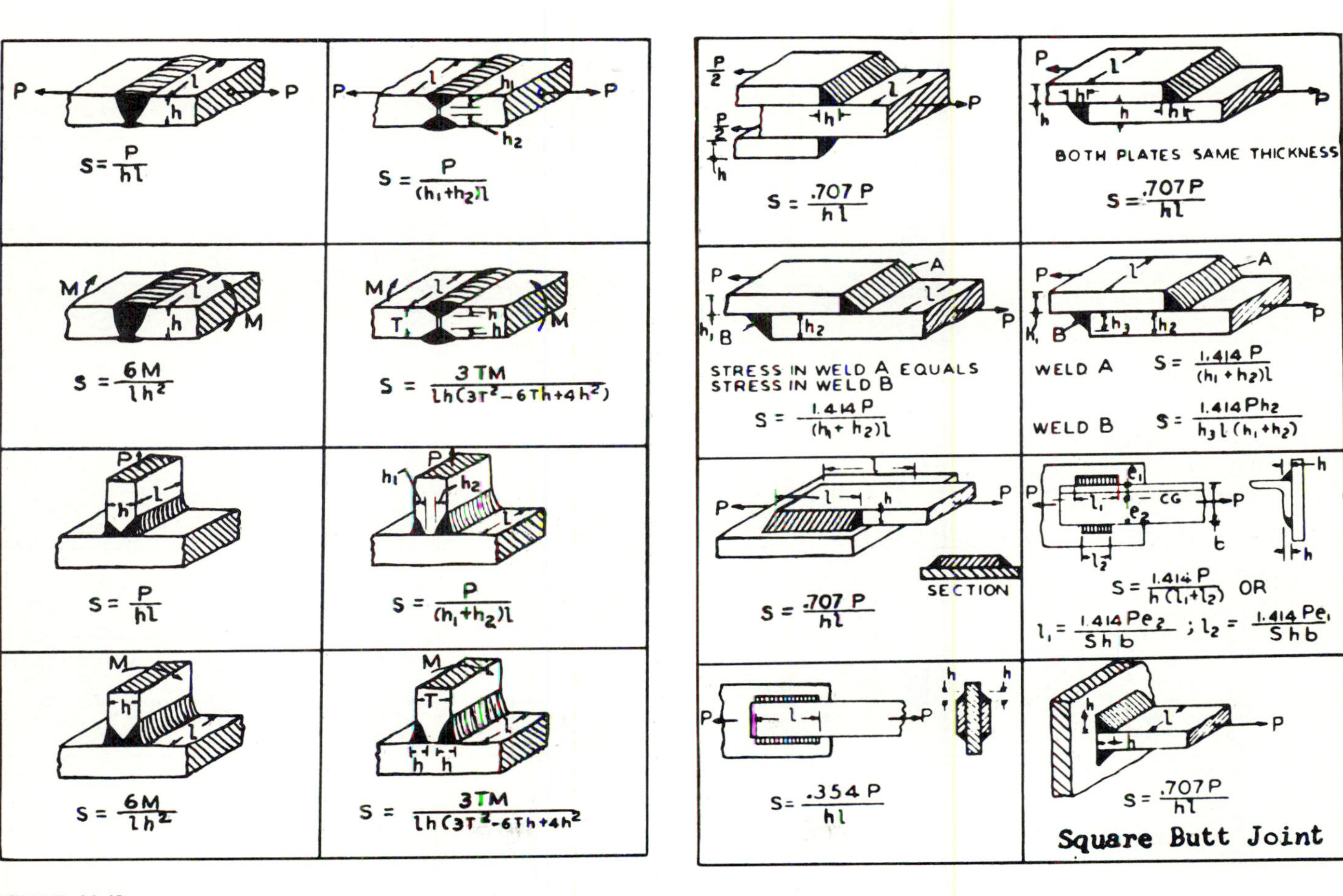

FIGURE 14.42

Stress formulas for typical welded joints. S = normal stress, lb/in.^{2}; I = moment of inertia, inch units; D = external load, lb; h = size of weld; S_s = unit stress, lb/in.^{2}; M = bending moment, in. · lb; L = linear distance; l = length of weld, in. (*Courtesy Charles Jennings Texas Laboratory, Inc., Dallas, Texas.*)

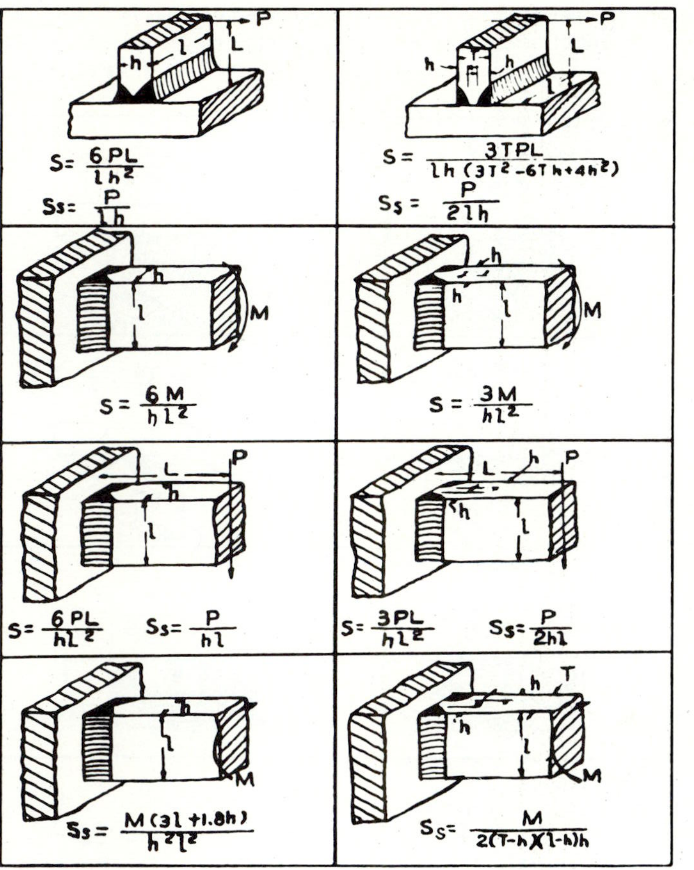

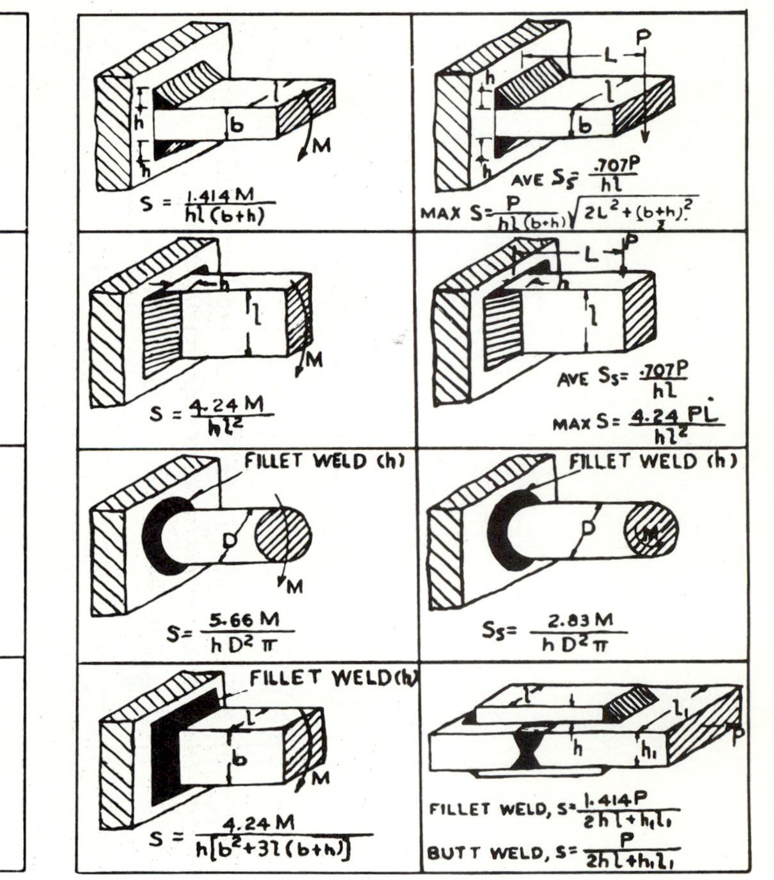

FIGURE 14.42 *(continued)*

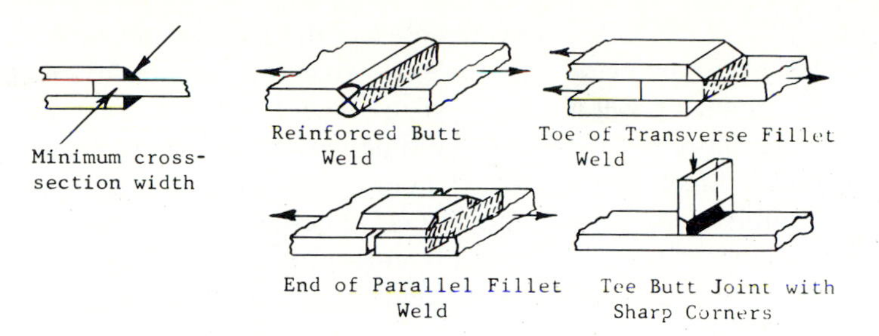

FIGURE 14.43
Working stresses for mild steel welds.

machining. The General Motors locomotive diesel-engine crankcases are welded in the production line. Cast construction was discarded because of the above factors. Hard surfacing materials can be added where wear due to abrasion may be encountered. Parts are stronger and consequently can be subjected to heavier machine cuts. Frequently, welded construction allows holding parts to closer tolerances, and less allowance for material to be removed by machining is necessary. Because of the smooth surface of

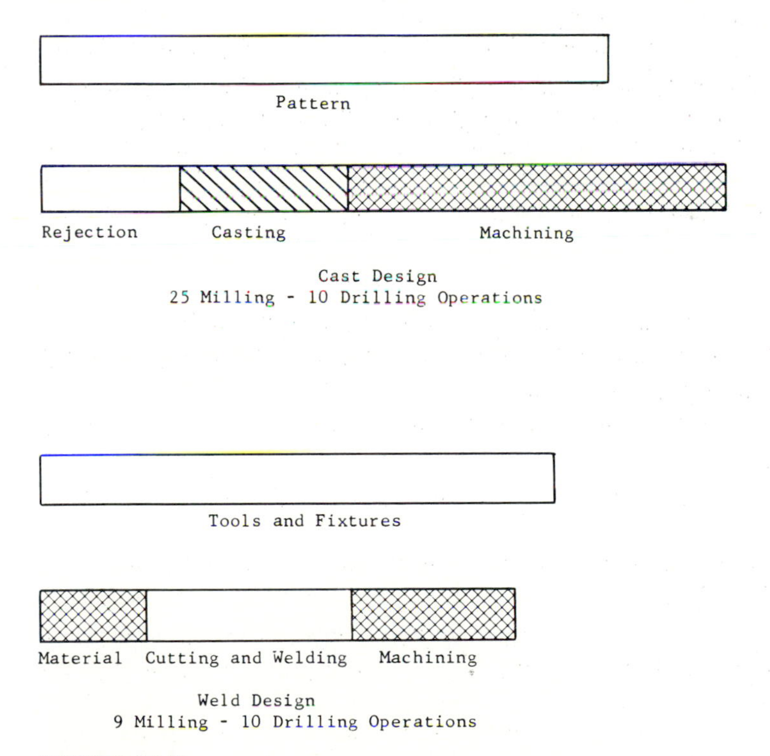

FIGURE 14.44
Cost comparisons of cast and welded designs for a part.

steel-mill plates and bars, it is often not necessary to specify machined surfaces. Changes in design can be made without expensive pattern alterations. The production-design engineer should recognize that, as the structure becomes complicated and more forming and welding are required, the cost may exceed casting costs. Steel castings have about the same strength as welded structures of equivalent section thickness. Therefore, it is often profitable to cast such items as complex brackets and housings and combine them with structural steel members by welding them into the completed part. A part may be advantageously designed for welded construction when the quantity requirements are small, but, when the quantity increases, other processes and materials may be more economical. Thus, quantity of manufacture is an important factor in determining whether welding will be economical.

Design of a Weld Size

Before the proper sizes and length of a weld bead can be specified for a given weldment, considerable time must be devoted to analyzing the forces on the welded joints in terms of their magnitude and direction. Low-cost welding can only be achieved by specifying and depositing the correct size weld bead. For applications that require maximum strength, groove welds must be used to assure 100 percent penetration. There is no need to calculate the strength of a groove weld, because it is more than equal to the strength of the plates being joined. Fillet weld sizes must be determined, because they can be either too large or too small.

As a rule of thumb, the leg size of a fillet weld should be three-quarters of the plate thickness to develop the full strength of the plate. This rule of thumb is based on the assumption that:

1. The fillet weld is on both sides of the plate.
2. The fillet weld is the full length of the plate.
3. The thinner plate is the controlling factor if the plates are of unequal thickness.

If the weld is designed for rigidity only, then the weld size may be $\frac{3}{8}t$. For thick plates, i.e., those $\frac{3}{8}$ in. and over, the fillet weld sizes recommended by AWS are as given in Table 14.9.

Estimation of weld size by treating the weld as a line. In this method, as given by O. W. Blodgett of the Lincoln Electric Co.,[1] the weld is considered to be a line with no cross-sectional area, having a definite length and outline as shown in Fig. 14.45. Then use the information in Table 14.10 to find the properties of

[1] Used by permission of the James F. Lincoln Arc Welding Foundation, Cleveland, Ohio.

TABLE 14.9
Minimum weld sizes for thick plates (AWS)

Thickness of thicker plate joined t, in.	Minimum of leg size of filler weld,[a] in.
To $\frac{1}{2}$ incl.	$\frac{3}{16}$
$\frac{1}{2}$–$\frac{3}{4}$	$\frac{1}{4}$
$\frac{3}{4}$–$1\frac{1}{2}$	$\frac{5}{16}$
$1\frac{1}{2}$–$2\frac{1}{4}$	$\frac{3}{8}$
$2\frac{1}{4}$–6	$\frac{1}{2}$
Over 6	$\frac{5}{8}$

[a] Minimum leg size need not exceed thickness of the thinner plate.

the weld when treated as a line. Insert the appropriate property of the weld connection into standard design formulas from strength of materials to find the forces on the weld joint when subjected to the application of a particular load (Table 14.11). Thus, the force on the welded connection is found in terms of pounds per lineal inch of weld.

For example, for tension or compression we have the standard design formula, which is the same formula as used for a weld, treating the weld as a line. That is,

$$\sigma = \frac{P}{A} \text{ lb/in.}^2 \qquad \text{or} \qquad f = \frac{P}{A_w} \text{ lb/in.}$$

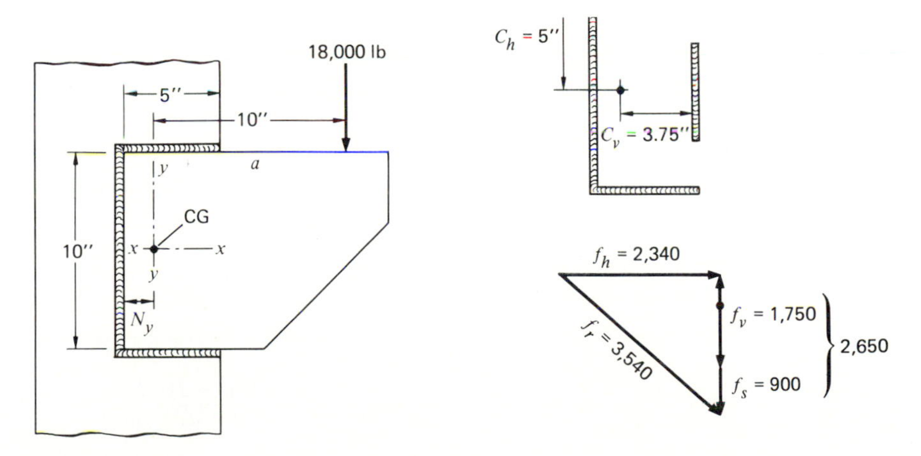

FIGURE 14.45
Determination of weld size by treating the weld as a line.

TABLE 14.10
Properties of weld treated as line

Outline of welded joint *b*-width *d*-depth	Bending (about horizontal axis *xx*)	Twisting
	$S_w = \frac{d^2}{6}$ in.^{2}	$J_w = \frac{d^3}{12}$ in.^{3}
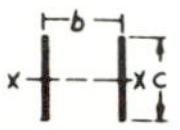	$S_w = \frac{d^2}{3}$	$J_w = \frac{d(3b^2 + d^2)}{6}$
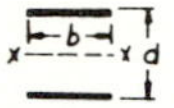	$S_w = bd$	$J_w = \frac{b^3 + 3bd^2}{6}$
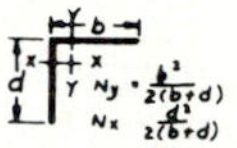 $N_y = \frac{b^2}{2(b+d)}$ $N_x = \frac{d^2}{2(b+d)}$	$S_w = \frac{4bd + d^2}{6} = \frac{d^2(4b + d)}{6(2b + d)}$ top bottom	$J_w = \frac{(b + d)^4 - 6b^2d^2}{12(b + d)}$
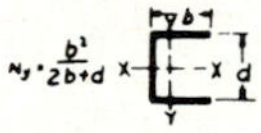 $N_y = \frac{b^2}{2b+d}$	$S_w = bd + \frac{d^2}{6}$	$J_w = \frac{(2b + d)^3}{12} - \frac{b^2(b + d)^2}{(2b + d)}$
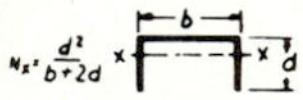 $N_x = \frac{d^2}{b+2d}$	$S_w = \frac{2bd + d^2}{3} = \frac{d^2(2b + d)}{3(b + d)}$ top bottom	$J_w = \frac{(b + 2d)^3}{12} - \frac{d^2(b + d)^2}{(b + 2d)}$
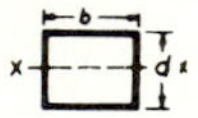	$s_w = bd + \frac{d^2}{3}$	$J_w = \frac{(b + d)^3}{6}$
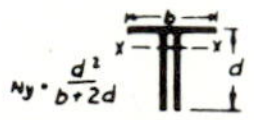 $N_y = \frac{d^2}{b+2d}$	$S_w = \frac{2bd + d^2}{3} = \frac{d^2(2b + d)}{3(b + d)}$ top bottom	$J_w = \frac{(b + 2d)^3}{12} - \frac{d^2(b + d)^2}{(b + 2d)}$
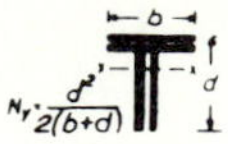 $N_y = \frac{d^2}{2(b+d)}$	$S_w = \frac{4bd + d^2}{3} = \frac{4bd^2 + d^3}{6b + 3d}$ top bottom	$J_w = \frac{d^3(4b + d)}{6(b + d)} + \frac{b^3}{6}$

TABLE 14.10 *(continued)*

Outline of welded joint *b*-width *d*-depth	Bending (about horizontal axis *xx*)	Twisting
	$S_w = bd + \frac{d^2}{3}$	$J_w = \frac{b^3 + 3bd^2 + d^3}{6}$
	$S_w = 2bd + \frac{d^2}{3}$	$J_w = \frac{2b^3 + 6bd^2 + d^3}{6}$
	$S_w = \frac{\pi d^2}{4}$	$J_w = \frac{\pi d^3}{4}$
	$S_w = \frac{\pi d^2}{2} + \pi D^2$	

where σ is the normal stress in standard design formulas, lb/in.^{2}, f is the load per inch of weld bead, lb/in., P is the tensile or compressive load, lb, A is the area of flange material held by welds in horizontal shear, in.^{2} and A_w is the length of the weld measured along the contour of the weld, in.

After determining the force, the weld size is determined by dividing the resultant force by the allowable strength of the appropriate weld (groove or fillet) as found in Table 14.12 for constant loads or in Table 14.13 for fatigue loads. Consider the following example, which illustrates the use of the procedure for determining the weld size required for the joint in Fig. 14.45, which is to be subjected to a load of 18,000 lb.

Example

Step 1. Find the properties of the weld, treating it as a line (use Table 14.10).

The following definitions are needed to estimate the weld size using the concept of defining the weld as a line:

- n Number of welds—shear stress in standard design formulas
- y Distance between the center of gravity of the flange material and the neutral axis of the whole section, in.
- I Moment of inertia of the whole section, in.^{4}
- c Distance to the outermost fiber from center of gravity, in.
- t Thickness of plate, in.
- S Section modulus of the section, in.^{3}
- J Polar moment of inertia of the section, in.^{4}

TABLE 14.11
Determining force on weld

Type of loading		Standard design formula stress, lb/in.^{2}	Treating the weld as a line, force, lb/in.
		Primary welds (transmit entire load at this point)	
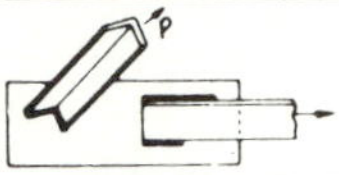	Tension or compression	$\sigma = \frac{P}{A}$	$f = \frac{P}{A_w}$
	Vertical shear	$\sigma = \frac{V}{A}$	$f = \frac{V}{A_w}$
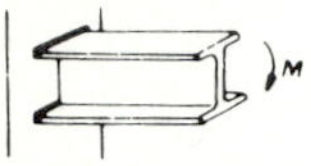	Bending	$\sigma = \frac{M}{S}$	$f = \frac{M}{S_w}$
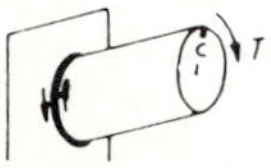	Twisting	$\sigma = \frac{Tc}{J}$	$f = \frac{Tc}{J_w}$
		Secondary welds (hold section together—low stress)	
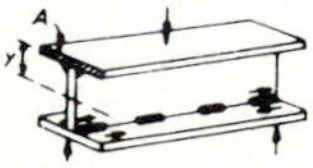	Horizontal shear	$\tau = \frac{VAy}{It}$	$f = \frac{VAy}{In}$
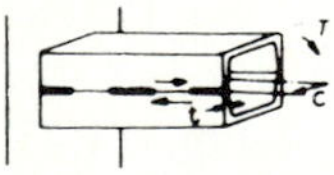	Torsional horizontal shear[a]	$\tau = \frac{T}{2A_t}$	$f = \frac{T}{2A}$

A = area contained within median line.

[a] Applies to closed tubular section only.

b	Width of the connection, in.
d	Depth of the connection, in.
V	Vertical shear load, lb.
M	Bending moment, in. · lb
T	Twisting moment in. · lb
S_w	Section modulus of the weld, in.^{2}
J_w	Polar moment of inertia of the weld, in.^{3}
N_x	Distance x axis to face, in.
N_y	Distance y axis to face, in.
σ_y	Yield strength, lb/in.^{2}
τ	Horizontal shear stress, lb/in.^{2}

TABLE 14.12
Allowable steady loads (lb/linear in of weld)

Fillet weld (for 1-in. weld leg)	Groove weld (for 1-in. weld thickess)	Partial-penetration[a] groove weld[b] (for 1-in. weld thickness)
	Parallel load	
E60 or SAW—1 weld 9,6000 (AWS)	$\tau = 0.40\ \sigma_y$ of base metal (shear) (AWS)	E60 or SAW—1 weld 13,600 (AISC)
E70 or SAW—2 weld 11,200 (AWS)		E70 or SAW—2 weld 15,800 (AISC)
	Transverse load	
E60 or SAW—1 weld 11,200	$\tau = 0.60\ \sigma_y$ of base metal (tension) (AWS)	E60 or SAW—1 weld 13,600 (AISC)
E70 or SAW—2 weld 13,100		E70 or SAW—2 weld 15,800 (AISC)

[a] For tension transverse to axis of weld or shear-use table; for tension parallel to axis of weld or compression, weld same as plate.

[b] For bevel joint, deduct first $\frac{1}{8}$ in. for effective throat if done by manual electrode.

For the weld in this example the following calculations are needed:

$$N_y = \frac{b^2}{2b + d} = \frac{5^2}{2(5) + 10} = 1.25 \text{ in.}$$

$$J_w = \frac{(2b + d)^3}{12} - \frac{(b)^2(b + d)^2}{(2b + d)}$$

$$= \frac{(2 \times 5 + 10)^3}{2(5 + 10)} - \frac{(5)^2(5 + 10)^2}{2(5) + 10}$$

$$= 385.9 \text{ in.}^3$$

$$A_w = 5 + 10 + 5 = 20\ in.$$

Step 2. Find the various forces on the weld, inserting the properties of the weld found above.

The forces are a maximum at point *a*. The twisting force per inch of weld is broken into horizontal and vertical components by proper values of *c*.

Twisting (horizontal component):

$$f_{t_h} = \frac{Tc_h}{J_w} = \frac{(18{,}000 \times 10)(5)}{385.9} = 2340 \text{ lb/in.}$$

TABLE 14.13

Allowable fatigue stress for A7, A373, and A36 steels and their welds

	2,000,000 cycles	600,000 cycles	100,000 cycles	But not to exceed
Base metal in tension connected by fillet welds, but not to exceed →	$\sigma = \frac{7{,}500}{1 - \frac{2}{3}K}$ lb/in.^{2} P_t	$\sigma = \frac{10{,}500}{1 - \frac{2}{3}K}$ lb/in.^{2} P_t	$\sigma = \frac{15{,}000}{1 - \frac{2}{3}K}$ lb/in.^{2} P_t	$\frac{2P}{3K}$ lb/in.^{2}
Base metal compression connected by fillet welds	$\sigma = \frac{7{,}500}{1 - \frac{2}{3}K}$ lb/in.^{2}	$\sigma = \frac{10{,}500}{1 - \frac{2}{3}K}$ lb/in.^{2}	$\sigma = \frac{15{,}000}{1 - \frac{2}{3}K}$ lb/in.^{2}	P_c lb/in.^{2} $\frac{P_c}{1 - K/2}$ lb/in.^{2}
Butt weld in tension	$\sigma = \frac{16{,}000}{1 - \frac{8}{10}K}$ lb/in.^{2}	$\sigma = \frac{17{,}000}{1 - \frac{7}{10}K}$ lb/in.^{2}	$\sigma = \frac{18{,}000}{1 - K/2}$ lb/in.^{2}	P_t lb/in.^{2}
Butt weld compression	$\sigma = \frac{18{,}000}{1 - K}$ lb/in.^{2}	$\sigma = \frac{18{,}000}{1 - 0.8K}$ lb/in.^{2}	$\sigma = \frac{18{,}000}{1 - K/2}$ lb/in.^{2}	P_c lb/in.^{2}
Butt weld in shear	$\tau = \frac{9{,}000}{1 - K/2}$ lb/in.^{2}	$\tau = \frac{10{,}000}{1 - K/2}$ lb/in.^{2}	$\tau = \frac{13{,}000}{1 - K/2}$ lb/in.^{2}	13,000 lb/in.^{2}
Fillet welds, ω = leg size	$f = \frac{51{,}000\,\omega}{1 - K/2}$ lb/in.	$f = \frac{7{,}100\,\omega}{1 - K/2}$ lb/in.	$f = \frac{8{,}800\,\omega}{1 - K/2 \text{ lb/in.}}$	8,800 ω lb/in.

K = min/max.
P_c = allowable unit compressive stress for member.
P_t = allowable unit tensile stress for member.
Source: Adapted from AWS *Bridge Specifications*.

Twisting (vertical component):

$$f_{t_v} = \frac{Tc_v}{J_w} = \frac{18{,}000 \times 10(5.00 - N_y)}{385.9} = 1{,}750 \text{ lb/in.}$$

Vertical shear:

$$f_{s_v} = \frac{P}{A_w} = \frac{18{,}000}{20} = 900 \text{ lb/in.}$$

Step 3. Determine the resultant force on the weld.

$$f_r = \sqrt{f_{t_h}^2 + (f_{t_v} + f_{s_v})^2}$$
$$= \sqrt{(2{,}340)^2 + (1{,}750 + 900)^2} = 3{,}540 \text{ lb/in.}$$

Step 4. Now find the required leg size of the fillet weld connecting the bracket, using Table 14.12:

$$\omega = \frac{\text{actual force}}{\text{allowable force}}$$
$$= \frac{3{,}540}{9{,}600}$$
$$= 0.368 \quad (\text{approx. weld size} = \tfrac{3}{8} \text{ in.})$$

Therefore make the leg $\frac{3}{8}$ in.

FASTENERS

"Fastners" is a general term including such widely separated and varied materials as nails, screws, nuts and bolts, locknuts and washers, retaining rings, rivets, and adhesives, to mention but a few.

Fasteners are of many different types. Some fasteners, like nuts, bolts, and washers, have been in use for years. Other fasteners, such as Rivnuts and retaining rings, are fairly new in the field. A large number of different fasteners are used in most products. As an example, 80 fasteners are used in a telephone receiver and 136 different kinds of fasteners are needed throughout the country to erect, brace, and guy telephone poles.

Fasteners can be divided into two classifications: those that do not permanently join the pieces, and those that do. Included in the first category are such fasteners as nuts and bolts, lock washers, locknuts, and retaining rings. The second classification includes such fasteners as rivets, metal stitching, and adhesives.

Nuts, Bolts, and Screws

Nuts, bolts, and screws are undoubtedly the commonest means of joining materials. Since they are so widely used, it is essential that these fasteners attain maximum effectiveness at the lowest possible cost. Bolts are, in reality, carefully engineered products with a practically infinite use over a wide range of services.

An ordinary nut loosens when the forces of vibration overcome those of friction. In a nut and lock washer combination, the lock washer supplies an independent locking feature preventing the nut from loosening. The lock washer is useful only when the bolt might loosen because of a relative change between the length of the bolt and the parts assembled by it. This change in the length of the bolt can be caused by a number of factors—creep in the bolt, loss of resilience, difference in thermal expansion between the bolt and the bolted members, or wear. In the above static cases, the expanding lock washer holds the nut under axial load and keeps the assembly tight. When relative changes are caused by vibration forces, the lock washer is not nearly as effective.

The slotted nut and cotter pin provides an assembly that locks. Since the diameter of the pin is smaller than the hole in the bolt and the slot in the nut, the nut can back off under vibration until these clearances are reduced and the cotter pin jams the assembly. Another method of locking is the use of a self-locking nut for which the top threads are manufactured at a reduced pitch diameter. During assembly, they clamp the threads of the bolt and produce greater frictional forces than an ordinary nut does. Another type of locknut that has received considerable attention is the nonmetallic-insert type. The elastic locking medium in this nut is independent of bolt loading, and seals the bolt threads against external moisture or leakage of internal pressure or liquids. The primary limitation in the use of this type nut is the inability of the insert to maintain its elasticity and locking characteristics at elevated temperatures.

When dissimilar materials, such as sheet metal and plastics or wood are to be joined, or when one side of the work is blind and the application of nuts to bolts becomes practically impossible, self-tapping and drive screws are available. They are used to fasten nearly everything from sheet metal and castings to plastics, fabrics, and leather.

Lead holes are necessary for the application of self-tapping or drive screws. Self-tapping screws have specially designed threads suitable to their applications. There are cutting threads and forming threads. For varied applications, they are available in all the standard head forms, either slotted or with a recess. Drive screws are usually self-tapping and are driven with a hammer. The holding power is greater than that of a nail and in many cases they may be backed out with a screw driver.

A special type of screw is the self-piercing screw. This is often used for attaching wood panels to light structural parts without having to drill lead holes. The screw is hammered through the wood to the metal and is then turned down with a screw driver.

For fastening metallic elements, machine screws, setscrews, and cap screws frequently are used. Machine screws are usually threaded the full length of the shaft, and are applied with a screwdriver. Cap screws have square, hexagonal, or knurled heads, and are threaded for only part of the shank.

Setscrews are threaded for the length of the shaft and may be either headed or headless. The headless setscrew is usually fluted or provided with a slot or hexagonal socket for driving. Setscrews are used to secure machine tools or some machine element in a precise setting. Many types of points are provided on these screws for various applications.

Rivets

Rivets are permanent fasteners. They depend on deformation of their structure for their holding action. Rivets are usually stronger than the thread-type fastener and are more economical on a first-cost basis. Rivets are driven either hot or cold, depending upon the mechanical properties of the rivet material. Aluminum rivets, for instance, are cold-driven, since cold working improves the strength of aluminum. Most large rivets are hot-driven, however.

A hammer and bucking bar are used for heading rivets. The bar is held against the head of the inserted rivet, while the hammer heads the other end. Squeeze heading usually replaces hammer and bucking-bar methods. In this method, the rivet is inserted and brought between the jaws of a compression tool, which does the setting by mechanical, hydraulic, or pneumatic pressure. Where production runs are larger, riveting machines are used, exerting pressure on the rivet to head it rather than heading it by hammering. Improperly formed heads fail quickly when placed under stress.

Riveting is the commonest method of assembling aircraft. A medium bomber requires 160,000 rivets and a heavy bomber requires 400,000 rivets. Some of the forms of rivets are solid rivets with chamfered shanks, tubular or hollow rivets, semitubular rivets, swaged rivets, split rivets, and blind rivets. Solid rivets are used where great strength is required. Tubular rivets are used in the fastening of leather braces. Split rivets are used frequently in the making of suitcases. The materials used in making rivets are aluminum alloys, Monel metal, brass, and steel. Aluminum rivets make possible the maximum saving in weight, and are also quite resistant to corrosion. Anodic coating of aluminum improves the resistance of the rivets to corrosion and also provides a better surface for painting. Steel rivets are stronger than aluminum rivets and offer certain advantages in ease of driving from the standpoint of equipment required; however, their use is limited to those applications in which the structure can be protected adequately against corrosion by painting. Some of the types of rivet heads are button, mushroom, brazier, universal, flat, tinners, and oval. Brazier and mushroom heads are used for interior work. The flat head is used for streamlined pieces. Aluminum-alloy rivets can be distinguished by their physical appearance: 2017 aluminum has a small dimple and is the most-used of the aluminum alloys, while 2024 has two small raised portions on either side of the head. It is poor practice to use a hard rivet in a soft alloy. For example, if a 2017 rivet should be driven into soft plate, such as 3003, the plate will be distorted and the resulting appearance will be poor. If the joint is not to be highly stressed, a soft rivet may sometimes be used in a hard alloy plate. In general, it is most advantageous and practical to use a rivet having similar properties to the material in which it is driven.

In determining the length, $1\frac{1}{2}$ times the diameter should extend below the piece being riveted. In addition to this, for a $\frac{1}{2}$- to 1-in. rivet add $\frac{1}{16}$ in.; for a 1- to $1\frac{1}{2}$-in. rivet add $\frac{1}{8}$ in.; and for a $1\frac{1}{2}$- to 2-in. rivet add $\frac{3}{16}$ in. Rivet holes may be punched, drilled, or subpunched and reamed to size. The clearance between rivets and rivet holes should be 0.004 in. on an average. Blind rivets are often specified. For this purpose, cherry rivets and explosive rivets are frequently used. The cherry rivet is a hollow rivet that is fastened by pulling a pin partly through it.

The explosive rivet contains a small charge that is set off as the temperature is raised by an electrical contact. Also included in the category of rivets is the Rivnut. It resembles a hollow rivet, but has threads on its inner diameter. A screw is turned into the rivet and then pulled, while the head is held stationary. The sides give way, thus holding the Rivnut.

Metal Stitching

Metal stitching is used to join thin-section metals and nonmetals at high production rates. A big advantage is that this can be done without precleaning, drilling, punching, or hole alignment. In comparison with nut-and-bolt and riveting methods of assembly, metal stitching has increased production as much as 700 percent and effected material savings up to 50 percent. These stitches are formed in as little as $\frac{1}{5}$ s, and have high strength and durability. Wire cost is low, being less than $0.06 per hundred stitches in some instances. High material savings are possible, because flange widths need be only $\frac{1}{4}$ in. for stitches, compared to $\frac{1}{2}$ in. for rivets. To the cost-conscious production engineer, metal stitching is an important process worthy of consideration in many designs.

Adhesives

Adhesive bonding is becoming more and more important, particularly in manufacturing and construction. There are two types of adhesive action: (1) specific adhesion that results from interatomic or intermolecular action between the adhesive and the assembly, and (2) mechanical adhesion produced by the penetration of the adhesive into the assembly, after which the adhesive hardens and is anchored. Since many adhesives are proprietary chemical products, their effective use depends upon consultation with the manufacturer.

In many applications, adhesive bonding is replacing rivets, bolts, screws, nails, and other types of mechanical fasteners. The principal advantages of adhesive bonding are the capability of joining dissimilar materials, bonding very thin sections to heavy sections without distortion, joining heat-sensitive alloys, producing bonds with unbroken surfaces, and the fact that bonding can be performed inexpensively. The major disadvantage of the process is the inability of the joint to sustain substantial impact and/or shear loads, although some of the epoxy-based adhesives can produce joint strengths up to 10,000 lb/in.^{2} in shear or tension if cured at 350°F for a few hours under pressures of approximately 150 lb/in.^{2} Adhesive bonded joints cannot sustain temperatures exceeding 500°F.

It is important for the production-design engineer to be familiar with the nomenclature of adhesives. The following definitions have appeared in the *Materials and Methods* magazine:

Tack. The characteristic that causes one surface, coated with an adhesive, to adhere to another on contact. It might be regarded as the essential stickiness of the adhesive.

Wet-strength. The bond strength that is realized immediately after adhesive-coated surfaces are joined, and before curing occurs.

Ultimate bond strength. The strength of the bond after the cure has been completed, or substantially completed.

Nonlocking properties. This is opposed to tack, and indicates the freedom of the adhesive-coated materials from sticking to unwanted material, such as the hands of the operator.

Can stability. The length of time the adhesive can be held in storage without deterioration. Pot life is the length of time the adhesive remains usable after being put into serviceable condition.

Specific service conditions. Such items as color, nonstaining properties, thermal range, resistance to solvents, and cleanup.

Adhesives can be used to join practically all industrial materials. They can be obtained in solid, liquid, powder, or tape form. The elimination of rivets and other fasteners results in smoother surfaces and considerable savings in weight and space—features that are very important in airplane construction. In addition, adhesive joints prevent fluid leakage and bimetallic corrosion at joints as well as providing some degree of thermal and electrical insulation. Adhesive bonding is suited to mass-production techniques. There are many types of high-speed hand and mechanical applicators, including brushes, sprays, rollers, scrapers, extruding devices such as pressure guns, and tapes. Some machines automatically apply the correct bonding pressure and cause adhesion quickly with radiofrequency heating.

Most adhesives have a limited temperature range of service. The surfaces to be bonded must be very clean and most glues require clamping and heat for proper setting, because the bond is not instantaneous. Another disadvantage is that most adhesives are good for only one specific set of service conditions.

Elastometric adhesives. Rubber, polyurethane, and silicones are the essential elements in adhesives used to bind such materials as felt to metals. There are two main types of rubber adhesives: rubber dissolved in an organic solvent, and water dispersions of finely divided rubber. Organic solvent rubber adhesives are easy to handle and apply and give strong bonds, but they release inflammable vapors and attack certain materials. Rubber adhesives of the water-dispersion type possess immediate tack and are therefore good for such applications as cloth backing. They have the disadvantages of being subject to destruction by freezing and of being limited in use to porous materials, because the water must either be absorbed or be removed.

Synthetic-resin adhesives. These are excellent for bonding plastics of the same composition. Thermoplastic-resin adhesives include cellulosics, vinyls, acrylics, and styrenes. They are usually of formaldehyde composition. These have a limited field of application, having been used most successfully for wood, rubber, thermosetting resins, and ceramics. Their cost is moderate and they produce a high-strength, waterproof bond that is resistant to attack by fungi. Their shelf life is limited, and since they have little initial tack, they must be clamped until the glue is set.

Design suggestions. Poor engineering design causes most of the failures in adhesive applications. Important points to remember in designing for adhesive bonding include the following.

1. The assembly should be designed so that a sufficiently large bonded area is obtained, thus preventing failures.
2. The bond should be stressed in shear rather than tension.
3. Joints should be tapered for greater strength.
4. The adhesive should be applied in uniform thin layers so as to avoid irregularities that may create points of stress concentration.
5. Combinations of materials should be selected such that stresses arising from differences in coefficients of thermal expansion are kept to a minimum.

Joining Plastics

Plastics are joined in one of four ways: (1) adhesive bonding, (2) mechanical fastening, (3) solvent cementing, and (4) thermal welding.

Solvent cementing is applicable only to certain thermoplastic resins (acrylics, polystyrenes, cellulosics, and some vinyls). The production-design engineer should observe the following pointers when specifying the joining of thermoplastic materials by solvent cementing.

1. The surfaces to be joined should be smooth, clean, and dry.
2. After application of a suitable solvent, constant pressure must be applied until the joint has become dry and hard.
3. Additional processing on the joint parts should be delayed until the joint has had ample time to set.

If these three pointers are followed and a suitable solvent is selected and applied, the joint properties will be similar to those of the parent material.

All thermoplastics may be joined by thermal methods. Heat is applied to the area to be joined either by burning gases (the most common method), induction means, heated tools, or friction. Upon joining, the welded section is cooled, thus hardening the materials and completing the joining process.

In gas welding, the procedure is similar to gas welding of metals. A welding rod of the same composition as the parent plastic is brought to the fusion temperature with the parent material in order to provide filler material. Joint preparation in plastics is as important as in metals. Butt welds, for example, should be shaped so as to have an included angle of 60° at the weld. Welding torches for welding thermoplastics convey compressed gas through electrically heated coils that raise the gas temperature to between 350 and 600°F. Power requirements are approximately 500 W for the heating element in the torch.

Heated tool joining is done without a filler material. Heat and pressure are applied directly with a tool until fusion takes place. The heating tool may take the form of a simple soldering iron or hot plate or it may involve a resistance-heated form. In this category falls the heat-sealing of plastic film that is characteristic of many packaging applications.

In induction-heating joining methods, a metallic conductor is placed adjacent to the areas where the parent plastics are to be joined. Heat from the conductor brings the plastics to the fusion temperature.

Friction welding is also a viable method of joining thermoplastic components that have circular cross section. The process requires the rotating member to be symmetrical about the axis of rotation, while the other member can be of any geometry within the constraints of the clamping machine.

The reader should understand that thermosetting plastics are not weldable. However, they can be joined readily by adhesive bonding.

SUMMARY

There is a large number of joining processes from which the production-design engineer may choose. Most products include several distinct varieties of joining. For each set of conditions, a particular process is superior to competing techniques. Some general rules that the production-design engineer should consider before deciding on a specific process include the following.

1. Do not overjoin. For example, use intermittent instead of continuous welds; however, do not sacrifice quality for simplicity.
2. Consider relative costs of supply materials in addition to time involved when comparing joining methods, for example, cost of silver brazing alloys versus cost of explosive rivets per assembly.
3. Consider overhead costs before deciding to standardize on a specific joining process.

4. Fabricated construction is not always the most economical. Perhaps a casting or forging would be a better solution.
5. Factors of safety at joined sections should be higher than at other portions, since joined areas can readily lead to points of stress concentration.
6. Keep appearance in mind when specifying a joining process. A projection-welded nut gives a much more pleasing appearance than one that has been arc-welded to a surface.
7. Consider adhesive bonding for sheet-metal assemblies of aluminum or stainless steel.
8. Contact cement is finding many uses such as the application of veneers to table tops or desks.
9. Specify joining methods that will be satisfactory for all conditions to which the end product will be subjected.
10. Keep maintenance in mind when specifying joining operations. Some parts may be interchangeable and may have to be replaced at times.

QUESTIONS

14.1. What parameters are of concern to the engineer in specifying a welding operation?

14.2. Explain why a tensile test specimen taken from the weld area in low-carbon steel will fracture adjacent to the weld deposit rather than in the weld itself.

14.3. Explain what is meant by overfill or reinforcement in connection with the weld.

14.4. What amount of heat would be required to melt 200 mm^3 of metal whose melting temperature is 1050°C?

14.5. What are the principal causes of cracking of weld metal? Of cracking of base metal?

14.6. What would you recommend should be done to eliminate tungsten inclusions in the weld metal?

14.7. What are the typical magnitudes of currents used in resistance welding?

14.8. How much heat would be generated in the spot welding of two sheets of 1.00-mm (0.04-in.) thick steel that required a current of 10,000 A for 0.1 second? An effective resistance of 100 $\mu\Omega$ is assumed.

14.9. What are the principal advantages of laser-beam welding when compared to electron-beam welding?

14.10. Describe the friction welding process as applied to thermoplastics.

14.11. What problems are presented when aluminum alloys are welded?

14.12. Give the spot-welding symbol; the brazing symbol; the seam-welding symbol.

14.13. When should welded assemblies be stress-relieved?

14.14. What methods are used to protect the molten metal from contamination in arc-welding processes?

14.15. What gases are used usually in the gas shielded-arc-welding processes?

14.16. What is the danger of a long arc in electric welding?

14.17. What is the effect of too much current in electric welding?

14.18. What determines the polarity that should be used with electric welding?

14.19. Why are long seams more difficult to weld than short seams?

14.20. What effect does peening have on welding beads?

14.21. What nine points should be observed in the design of joints?

14.22. How is the stress in a square butt joint computed?

14.23. What is the relationship in strength of arc-welded steel structures to similar structures composed of gray cast iron?

14.24. What are some of the advantages of automatic welding? Semiautomatic welding?

14.25. What is meant by submerged arc welding?

14.26. List the arc-welding processes and give major features of each.

14.27. Why is stud welding usually thought of as an arc-welding process?

14.28. In resistance welding, what is the composition of the resistance of the joint?

14.29. When would you recommend projection welding?

14.30. Would you recommend the spot welding of Monel sheet to copper plate? Why?

14.31. What is flash welding?

14.32. Upon what does the strength of brazing depend?

14.33. What is cold welding?

14.34. What is the advantage of the explosive rivet?

14.35. When is metal stitching practical?

14.36. What are two types of adhesives? When would you recommend the use of each type?

PROBLEMS

14.1. A welded butt joint in an SAE 1020 steel plate is to carry a total force of 30,000 lb. The width of the plate is 4 in. and the working stress is specified to be 13,000 lb/in.^{2} What standard plate thickness should be used? What process would you select for the welding operation?

14.2. A rectangular bar of machine steel that is $\frac{3}{8}$ in. thick is to be welded to a machine frame by side fillet welds. The estimated tensile force that will be applied to the bar is 35,000 lb. Calculate the length of each $\frac{1}{4}$-in. side fillet weld, if the allowable shear stress in the weld is 13,600 lb/in.^{2}

14.3. The tensile force P applied to a double-fillet welded lap joint is 120,000 lb and the length of the joint is 10 in. The plate thickness is sufficient to cause failure to occur in the weld rather than in the plate. If the allowable shear stress on the nominal throat section of a fillet weld is 13,600 lb/in.^{2}, find the required size of the fillet weld to the nearest $\frac{1}{16}$ in.

14.4. Calculate the amount of transverse shrinkage for the butt welded assembly shown in Fig. P14.4.

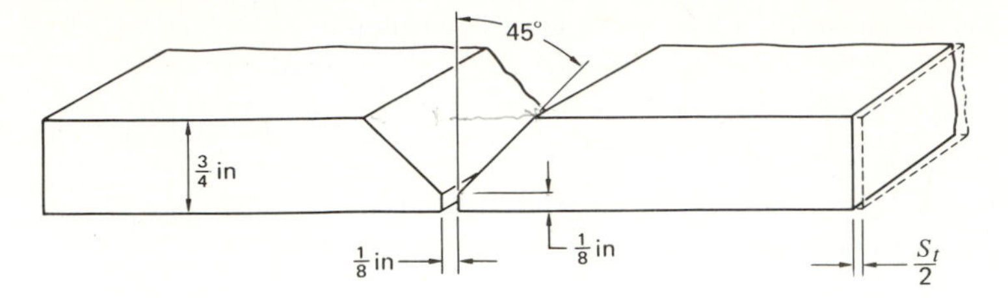

FIGURE P14.4

14.5. Two structural steel plates 2 in. wide by $\frac{1}{2}$ in. thick are joined by a single rivet in a lap joint. The rivet is 1 in. diameter. If the joint is subjected to a tensile load of 20,000 lb, determine the factors of safety with respect to tension, shear, and bearing. The ultimate strengths for tension, shear, and bearing are respectively: 55,000 lb/in.; 44,000 lb/in.; and 95,000 lb/in. Specify a competitive welded joint design and welding process. Under what conditions would a riveted joint be used rather than a weldment?

14.6. A steel pipeline is being field-welded. A number of short pieces of random lengths have accumulated during the construction period. Now the field engineer must either use the pieces by welding them into longer lengths or scrap them. He has called on you to help him to determine the minimum length of pipe which can be economically salvaged based on the following data: Pipe diameter, 8 in.; wall thickness, $\frac{1}{8}$ in.; specific weight of steel, 0.287 lb/in.^{3}; welding rod cost, \$0.08 in. of weld; average welding time, 1.7 min/in. deposit; cost of pipe, \$0.72/lb; scrap value, \$0.06/lb; welder's pay, \$10.52/h; overhead, 240 percent.

14.7. A firm is considering welding drive pulleys to automotive generator shafts. Specify the size of weld needed between the pulley and the shaft. The shear strength of the weld is 38,000 lb/in.^{2} maximum, acceleration 0 to 9000 rev/min in 2 s with 5.0 in. · lb electrical torque load under the most severe conditions. Use a factor of safety of 2 and neglect friction.

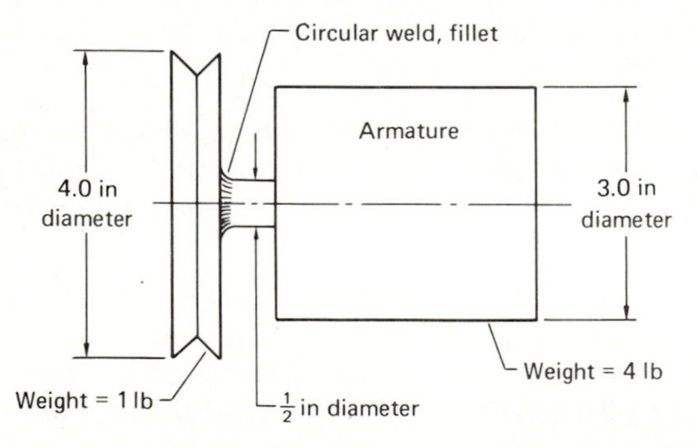

FIGURE P14.7

14.8. An elliptical flange with a major diameter of 100 in. and a minor diameter of 80 in. on its outer surface is to be cut from a plate $\frac{3}{8}$ in. thick. The inner major and minor diameters are to be 80 and 60 in., respectively. Determine the scrap weight if the flange is flame-cut as a unit from a rectangular plate. Calculate the saving if the flange is cut as four pieces and welded into a unit after flame-cutting. Cut the sections along axes at 45° to the major axis of the ellipse. Determine the best plate size graphically and use current prices for labor and material. A lot size of 100 pieces is needed.

SELECTED REFERENCES

1. American Welding Society: *Welding Handbook,* 7th ed., Miami, Fla., 1976.
2. American Welding Society: *Standard Welding Terms and Definitions,* Miami, Fla., 1985.
3. Blodgett, O. M.: *Design of Weldments,* James F. Lincoln Arc Welding Foundation, Cleveland, Ohio, 1963.
4. Bruckner, W. H.: *Metallurgy of Welding,* Pitman, New York, 1954.
5. Funk, Edward R., and Rieber, Lloyd, J.: *Handbook of Welding,* Breton Publishers, Boston, Mass, 1985.
6. Giachino, J. W., Weeks, W., and Johnson, G. G.: *Welding Technology,* American Technical Society, Chicago, 1968.
7. Lancaster, J. F.: *The Metallurgy of Welding, Brazing, and Soldering,* Allen and Unwin, Boston, Mass, 1965.
8. Lincoln Electric Co.: *Procedure Handbook of Arc Welding Design and Practice,* Lincoln Electric Co., Cleveland, Ohio, 1957.
9. Linnert, G. E.: *Welding Metallurgy,* 3d ed., vols. 1 and 2, American Welding Society, Miami, Fla., 1965.
10. Patton, W. J.: *The Science and Practice of Welding,* Prenctice-Hall, Englewood Cliffs, N.J., 1967.
11. Rossi, B. E.: *Welding and Its Application,* McGraw-Hill, New York, 1954.
12. West E. G.: *The Welding of Nonferrous Metals,* Wiley, New York, 1952.

CHAPTER 15

DECORATIVE AND PROTECTIVE COATINGS

In most merchandised products, the cost of finish is an apppreciable percentage, amounting in some products to as high as 35 percent or more of total factory cost. Finishes include the coverage type, such as paint, lacquer, and varnish; the chemical-changing-of-the-surface type such as anodizing, sheradizing, or bonderizing; the plating type, which may be polished and buffed after the plating process; and the temporary or permanent type that covers materials in order to protect them against corrosion during storage or shipment.

The key to successful finishes is a clean surface to assure good adhesion. The control of temperature, humidity, and constituents of the finish is also important. There is considerable difference in cost of finishing materials and processes for applying the various finish types. There are competitive means of cleaning, such as water-based solvents versus chemical solvents; competitive means of drying, such as infrared heat versus steam, oil, or gas heat; and competitive means of application, such as dip, spray, flow, and centrifuge. Cementation, metal cladding, cathode sputtering, and vapor coating in a vacuum are specialized processes for applying finishes.

Finishes are used for the purposes of decoration, surface protection, corrosion resistance, and the providing of a hard surface. The colors should not fade. The covering should be uniform and free from runs, checks, or peelings. It should be pliable so that it will expand or contract under weather or operating conditions. The surface should be hard so that it can be cleaned and will not permit imbedding of dust or grit or staining by oil. The time taken for drying often determines the choice of finish. The development of quick-drying finishes has eliminated the high cost of storing for several days automobile parts and bodies finished with slow-drying enamels and varnishes.

CLEANING

Cleaning operations are performed both preparatory to and after finishing operations. These rather common operations are worthy of careful analysis because they frequently consume considerable time and cost. They are used to remove dirt, oil, oxides, scale, and other harmful ingredients that affect the operation of the equipment and the life of protective finishes. Cleaning is not a simple matter. It is often expensive and it requires considerable ingenuity and knowledge of cleaning processes to remove unwanted surface contamination effectively.

The engineer should consider the following prior to the selection of the cleaning process: identification of the contaminant that is to be removed, identification of the substrate to be cleaned, degree of cleanliness required, availability and capability of cleaning equipment, estimated cost of competitive processes. With reference to the type of contaminant, there are five classes that are prominent: unpigmented oil and grease, chips and cutting fluid, polishing and buffing compounds, pigmented drawing compounds, and rust and scale. Typical shop oils and greases such as drawing lubricants, rust-preventive oils, lubricating oils, quenching oils, etc. can be removed efficiently by several cleaning processes. Most cutting and/or grinding fluids are readily removed by plain boiling water or steam and the chips fall away during the cleaning. Of course, this would not be true if the chips or part being cleaned becomes magnetic. Chemical processes, such as pickling; solvent, alkaline, electrolytic, and emulsion cleaning are used to insure clean parts and surfaces before the desired finish is applied. Table 15.1 provides a guide to the contaminant-removal process to be used in preparation for plating, phosphating and painting.

There are several fundamental methods for removing scale and rust from ferrous mill products, forgings, castings, and fabricated components. These are abrasive blasting (dry or wet), tumbling (dry or wet), brushing, acid pickling, salt bath descaling, alkaline descaling, and acid cleaning. The engineer planning the removal of rust and/or scale need consider the thickness of the contaminant, composition of the base metal, the allowable metal loss, the surface finish requirements of the part, the shape and size of the work, the production requirements, the availabilty of rust removing equipment, and the cost involved.

At times, it may be advantageous to utilize two or more of the methods enumerated in order to satisfy requirements at the least cost.

TABLE 15.1
Preparative removal of contaminants

Contaminant to be removed	Preparation for plating	Preparation for phosphating	Preparation for painting
Pigmented drawing compounds	Hot emulsion or alkaline soak, hot rinse	Alkaline or acid soak, hot rinse	Alkaline soak, hot rinse
Unpigmented oils and grease	Automatic vapor degrease Electrolytic alkaline rinse Emulsion soak and barrel rinse	Emulsion spray, rinse Vapor degrease	Vapor degrease Phosphoric acid clean
Chips and cutting fluids	Alkaline soak, rinse	Alkaline dip and emulsion surfactant	Alkaline dip and emulsion surfactant
Polishing and buffing compounds	Surfactant alkaline soak and spray rinse	Surfactant alkaline soak and spray rinse	Surfactant alkaline soak and spray rinse

Source: Adapted from a table by George A. Shepard, Republic Steel Research Center.

Acid Pickling

The most common method of removing unwanted oxides (scale) on large-tonnage products such as merchant bar, blooms, billets, sheet, strip, wire, and tubing is by acid pickling. Either sulfuric acid (the most common pickling liquor) or hydrochloric acid is sprayed on the part or the part is dipped into a tank, agitated, and then washed and rinsed thoroughly. Acid solutions are difficult to maintain because of carryover into rinsing tanks or dilution from previous cleaning operations. Splash and vapors from the acid solution corrode equipment and tanks. The maintenance cost is high and working conditions are often disagreeable. Acid cleaning of steel parts creates hydrogen which is absorbed by the steel and causes "hydrogen embrittlement." This is often the cause of excessive spring breakage. The hydrogen in the steel can be reduced by heating the parts after pickling. Inhibitors are usually added to acid pickling solutions to reduce the extent of hydrogen embrittlement and minimize the material loss.

Solvent Cleaning

Solvent cleaning is designed to remove oil and grease. It can be accomplished in room-temperature baths and by the use of vapor degreasing methods. Some processes involve soaking the part in a liquid solvent, and others use a spray, but most processes use a vapor. In the vapor type, a boiling liquid gives off vapors into which the cold pieces are dipped. The vapor condenses on the part and penetrates and dilutes the oil or grease, which runs off the part. The part is then removed, with a very light coat of oil remaining. This acts as a rust preventive for a brief period of time until the next operation. The vapors are condensed by cooling coils at the top of the tank. The solvent can be reused by removing the oil and dirt. In order to obtain chemically clean surfaces, other cleaning processes, such as electrolytic cleaning, are necessary.

Room-temperature solvent cleaning is known as cold cleaning. The solvents are typically aliphatic petroleums, chlorinated hydrocarbons, chlorofluorocarbons. The parts to be cleaned are soaked and may be agitated to help remove all soil contaminants.

Alkaline Cleaning

The most prevalent type of cleaning for removing soils from the surface of metals is with alkali. It is used when there are no soluble compounds present. Its cost is low because it uses water and inexpensive cleaning compounds. Parts must be rinsed thoroughly after exposure to the alkaline solution to prevent residue on the metal and lack of adhesion of organic finishes. It should not be used on aluminum, zinc, tin, or brass. The principal soils removed by this cleaning include metallic, carbon and silica particles, and oils and greases.

Alkaline cleaners include three principal components: builders (the alkaline salts, including phosphates to soften the water and silicates to provide alkalinity), organic or inorganic additives (to provide additional cleaning or surface modification), and surfactants (to provide detergency, emulsification, and wetting).

The cleaners are applied either by immersion or by spray. Cleaning takes place by emulsification, dispersion, saponification, or a combination of these processes.

Electrolytic Cleaning

In electrolytic cleaning, an alkaline cleaning solution is used with electric current passing through the bath, causing the generation of oxygen at the positive pole and hydrogen at the negative pole. This action breaks up the oil film holding dirt to the metal surface and results in chemically clean surfaces suitable for plating. The material from which the part is made and the cleaning action desired determine whether the parts should be made the anode or the cathode.

Typically, 3 to 12 V dc is applied, resulting in current densities of 10 to 150 A/ft^2 of work area. An alkaline soak to remove major soils typically precedes the electrolytic cleaning operation.

Abrasive Blast Cleaning

The applications of abrasive blast cleaning include removal of such imperfections as scale, burrs, and flash; removal of unwanted materials from a previous operation such as paint, foundry sand, etc.; surface preparation, such as providing a matt surface or roughened surface in preparation for painting or bonding.

The principal abrasives used include iron grit, steel shot, malleable iron shot, glass beads, sand, and walnut shells. The size, shape, and hardness of the material being cleaned determine the best abrasive size and type. The abrasives may be propelled either by a pressure or suction blast nozzle system or an airless blast blade or vane-type wheel.

Emulsion Cleaning

Emulsion cleaning combines the action of a solvent and an emulsifier. It is the least expensive cleaning process and can be applied on most materials (metals and nonmetals) at room temperature by either spraying or dipping for removing soils from the surfaces. The parts should be constructed so that they can be rinsed and dried easily—the process is not particularly suitable to parts having deep pockets that will trap the liquid. The solvent is generally of petroleum origin, while the emulsifier (soap) includes nonionic polyethers, and high-molecular-weight sodium or amine soaps of hydrocarbon sulfonates, amine salts of alkyl aryl sulfonates, fatty acid esters of polyglycerides, glycerols and polyalcohols.

Ultrasonic Cleaning

When ultrasonic vibrations of sufficient power level are transmitted in a liquid, cavitation occurs. This action in liquids has two effects: *bulk,* which refers to those effects within the liquid, and *surface,* which occurs between the liquid and the solid. Many organic compounds are broken down by cavitation; thus the dirt and grease clinging to solid articles placed in an ultrasonic cleaning tank are ripped apart and emulsified.

Frequencies of approximately 30,000 Hz are characteristic of ultrasonic cleaning. Typical fluids used are water to which has been added a detergent or solvents such as cyclohexane and trichloroethylene.

Ultrasonic cleaning is adaptable to complete assemblies, since cavitation tends to occur in crevices and corners that under other cleaning methods would be difficult if not impossible to reach. Thus, one of the main advantages of ultrasonic cleaning is the ability to reach inaccessible areas. Ultrasonic cleaning is sometimes used in conjunction with solvent cleaning to loosen and remove contaminants from deep recesses.

A typical ultrasonic cleaning facility is composed of a generator, a transducer, and a cleaning tank. The generator, of course, produces the ac electric energy. The transducer converts the electric impulses into high-frequency sound waves, and the tank holds the cleaning fluid into which the transducer transmits its sound energy

FINISH-TYPES

The principal types of finishes applied to metal products are:

1. Organic finishes
2. Inorganic finishes
3. Metallic coatings
4. Conversion coatings

Types of Organic Finishes

If an opaque, organic coating is applied so that the metallic coating or base metal is not discernible, the finish is then classified as an organic finish.

Organic coatings may be applied directly to the parent metal, although they are frequently applied after a metal and/or phosphate coating has been given to the base metal so as to increase the resistance to corrosion of the part. Organic finishes usually include two coats: first, a priming coat, then the second coat. Thickness of organic coats varies from 0.0002 to 0.002 in.

In order for the organic coating to be bonded firmly to the substrate, it is important that surface contaminants such as oils, greases, dirt, rust, mill scale, water and salts are removed. Frequently, both mechanical and chemical methods should be used in order to insure a clean surface.

Oil paint. Exterior surfaces are often finished with oil paint. This completely hides the surface to which it is applied. Paint requires a relatively long drying time, and painted articles cannot be stacked, because they will stick together and spoil the finish when separated. The drying time, hardness, and elasticity of paint films depend principally on the drying oil or combination of drying oils used. Slow-drying oils are soybean and linseed; the faster-drying oils are tung and castor.

Enamels. Whereas paints may be classified as organic finishes in which a pigment is dispersed in a drying-oil vehicle, enamels represent those organic finishes in which the pigment is dispersed in a varnish or a resin or a combination of both. Thus, enamels may dry by either or both oxidation and polymerization. Both air-drying and baking-type enamels are available.

Because of their availability in all colors, ease of application, and ability to resist corrosive atmospheres and attack of most of the chemical agents usually encountered, enamels are the most widely used organic coating in the metal-processing industry.

Baked enamels (baking time may vary from a few minutes to half an hour) usually provide a finish that is harder and more abrasion resistant than typical air-drying enamels. The automotive and electrical-appliance industries are large users of baked enamels.

Varnishes. Varnishes can be clear or may contain dye, but they do not hide a surface when applied to it. The addition of pigment to a varnish results in an enamel, and this type of finish does hide the surface to which it is applied. The oleoresinous-type varnishes and enamels are fast-drying, produce glossy, hard films, and have various degrees of elasticity, toughness, and durability. On the whole, the chemical resistance of this group can be as good as that of the alkyds.

An alkyd coating is one of the most popular finishes for metal products. There are several types of alkyd coatings. The oil-modified alkyds are resistant to heat and solvents; however, the phenolic coatings are more resistant to water and alkalies. The alkyd-amine baking enamels are quite resistant to alkali, and give a hard-wearing, glossy surface. These enamels are used for refrigerators, stoves, and other indoor appliances. The alkyd-phenolic coatings have greater resistance to water and alkali; consequently, they are used on tank liners for such things as washing-machine bowls. Alkyd-silicone finishes have better heat resistance, but they are more costly. High-silicone coatings are resistant to temperatures up to 700°F. Alkyd-styrene enamels are fast-drying and so are in demand to provide a good, one-coat, fast-drying finish.

The alkyd types of varnish are quick-drying, especially under the application of heat, and have good adherence to smooth surfaces. Their exterior durability is very satisfactory, making them suitable for vehicle finishes.

Lacquers. Lacquers are noted primarily for their property of drying in a few minutes. The main drawback of this group is the small coverage obtainable from one unit as compared to that obtained from an equal unit of paint, varnish, or enamel. At least two coats of lacquer are required to give the protection that one coat of varnish or enamel affords. A unit of lacquer will cover only 50 to 60 percent as much area as an equal unit of the other finishes. Their poor adherence to metal surfaces requires the use of a priming coat for best results. Poor exterior durability is another drawback of this group, although, when well pigmented, lacquers have better exterior durability than oleoresinous varnishes. They are still inferior to the alkyd group, but the extremely fast-drying property of this group sometimes outweighs its drawbacks. Clear lacquers are used for protection against indoor atmospheres, while pigmented lacquers are suitable for outdoor protection as well.

Vinyl lacquers have properties that make them useful for lining food and beverage containers. They are impermeable to water, chemically resistant, and free from odor, taste, and toxicity.

Shellac. Shellac dries quickly and does not penetrate wooden surfaces deeply. It is often used as a sealing coat on wood, since it gives a durable film. Since shellac is soluble only in alcohol, lacquers and varnishes can be applied over it without the two running together.

Stain. Stain is not a paint, but is nevertheless very important in finishing wood. Its primary purpose is to color, not to cover; hence, dyes and not pigments are used to supply the color.

Luminescent paints. Luminescent paints are coming into more prominence, and many unusual effects can be produced by them for both daylight and night applications. They are suitable for both indoor and outdoor exposure.

Metal-flake paints. Metal-flake paints are made by adding very small polished flakes of a metal such as brass, copper, or aluminum to a so-called "bronzing" liquid. The hiding and sealing properties of this type of paint are very good.

Pearl essence. Pearl-essence finishes originally used pigment obtained from fish scales, but are now made synthetically. They have little hiding value, but leave a beautiful pearl-like irridescent finish such as that found on jewelry.

Crystal finish. By dissolving in clear lacquer chemicals that crystallize upon drying, finishes known as crystal finishes are obtained. These form beautiful, regularly shaped crystal-type patterns. Special varnishes dried in atmospheres containing products of combustion form frosted surfaces in regular patterns that are also very decorative.

Wrinkle finish. Another of these special finishes is the wrinkle finish. Ordinarily wrinkling is not wanted, but, if controlled, it can produce a very effective finish. The surface to which this type of finish is to be applied need not be smooth, which is sometimes a definite advantage in hiding surface defects. Only one coat is required to produce a good covering coat. The film is usually applied by spraying and is then baked. Some types of wrinkle finishes available can be applied by dipping. These are air-dried. This type of finish requires special materials and techniques for good results.

Crackle finish. Closely akin to the wrinkle finish is the crackle finish. However, a crackle finish usually employs two different colors of lacquer enamel. The surface is first sprayed with a color, such as yellow, and then sprayed with a crackle lacquer containing a high percentage of pigment, which in this case might well be brown. When this second coat dries, it shrinks and leaves cracks, the sizes of which are determined by the thickness of the crackle finish applied. Usually this process is followed by a coat of gloss or clear lacquer for protection.

Hammered finish. A finish similar to the crackle finish is the hammered finish. Two coats of the finish are applied in the same manner as the crackle finish; but the second coat, which is called a spatter thinner, hits the surface in the form of droplets. This finish is customarily baked.

Flock finishing. Flock finishing is used when it is desired to give the article the feel of fabric. The flock adhesive, which is either a clear varnish or a lacquer, is applied and either the flock is sprayed on or the article is dropped into a pile of it. Flock is actually fibers of fabric about 0.02 in. long.

Organic Finishes—Method of Application

Organic finishes are applied by brushing, spraying, flow coating, dipping, tumbling, centrifugally and electrostatically.

Brushing. Brushing requires the greatest amount of labor and the least amount of material.

Spraying. Spraying is rapid and gives a smooth coat, but is wasteful of material. As much as 70 percent of the paint sprayed might be lost because of overspray. See Fig. 15.1. Booths are required to collect the overspray, to prevent contamination of the air, and to remove fire hazard. Some material can be recovered by water screens in the booths. Electrostatically charging the part or the sprayed paint or lacquer will cause the spray to collect on the part and in this way reduces waste. Using the electrostatic method can result in as little as 15 percent of the paint being lost. Heating the sprayed materials increases their fluidity and so decreases the amount of solvent required. These methods have reduced the cost of material sprayed per part and have increased productivity per man hour.

Flow coating. Flow coating paint, enamel, or varnish involves the pumping of the finishing material from a storage tank through strategically positioned nozzles onto all surfaces of the parts being conveyed through the coating area. This process has limited applications on large tanks and frames that permit the paint to run off evenly (Fig. 15.2). The finishing material sets and some solvent escapes as the paint drains from the collecting racks to the accumulating tank for recirculation.

The coating material applied by flow coating tends to be thinner at the top of the part and thicker at the bottom. The flow coating process utilizes approximately 95 percent of the paint.

Roller coating. Roller coating can be applied to sheets only.

Powder coating. In powder coating a suitable formulation in the form of the powder is spread over the surface to be coated, which is subsequently subjected to heat thus melting the powder allowing it to flow and fuse into a uniform continuous coating. Before baking, powder coatings are pulverized plastic compositions. Both thermoplastic and thermosetting powders are used today. Thermoplastic resin systems are more applicable as thicker coatings (three to six mils) and are applied usually by the *fluidized* bed technique.

FIGURE 15.1
Spraying enamel on refrigerator shells.

FIGURE 15.2
Flow coating transformer frames.

Popular thermoplastic coatings include polyethylene, polypropylene, nylon, polyvinyl chloride and thermoplastic polyester. Thermosetting resins are used more where thin paint-like surface coatings are desired (one to three mils). Resins used here include: epoxy, polyester and acrylic.

The most used powder coating for interior use is probably epoxy which may be used for both thick film functional end uses and for thin film decorative applications. Epoxies are typically baked at temperatures between 250°F. (121°C.) and 275°F. (135°C.) for a period of 20 to 30 minutes.

Epoxy coatings give products excellent mechanical and resistance properties but poor exterior weatherability. Chalking will take place in a period of a few months when exposed to the elements giving a poor appearance. However, the coating will continue to maintain its mechanical and resistance properties.

In the fluidized bed process, a preheated part is immersed into a fluidized

FIGURE 15.3
Modern-type tumble barrel suitable for tumble finishing.

powder bed. Thickness of coat can be controlled to some extent by the temperature of the part and the length of time the part is immersed in the powder bed. Figure 15.4 illustrates the principle of the fluidized bed.

Since different powders have different melting points, part temperature for optimum coating at the point of immersion in the bed will depend upon the plastic powder being used. Then too, part temperature for a given powder will depend upon the material being coated and its size and shape.

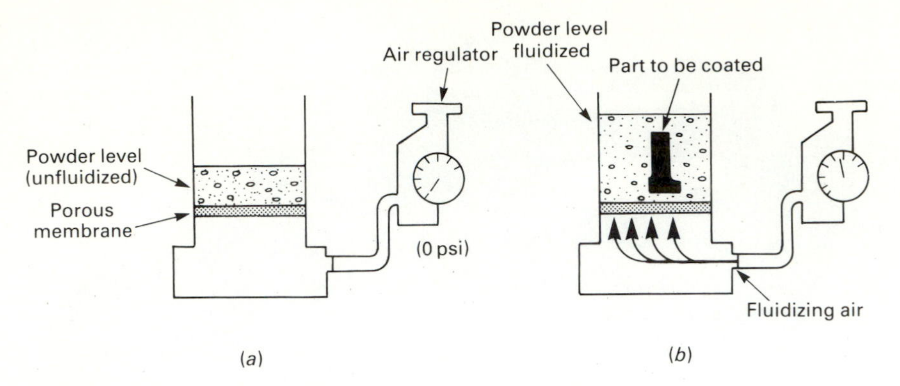

FIGURE 15.4
Principle of powder fluidized bed: (*a*) shows powder level during unfluidized condition; (*b*) illustrates immersed part in fluidized bed.

The electrostatic fluidized bed process is based upon creating a cloud of charged powder particles above the powder bed. Grounded parts are immersed or can be conveyed into this cloud. The powder particles and the grounded parts produce an electrical field of attraction. Figure 15.5 illustrates this process. The reader should understand that in this process, preheating of the part is not necessary. The electrical forces (+ to −) results in the deposition of the powder no matter if the part surface is hot or cold. Those plastic powders usually used in this process include: epoxy, polyester, acrylic, polyethylene and polypropylene.

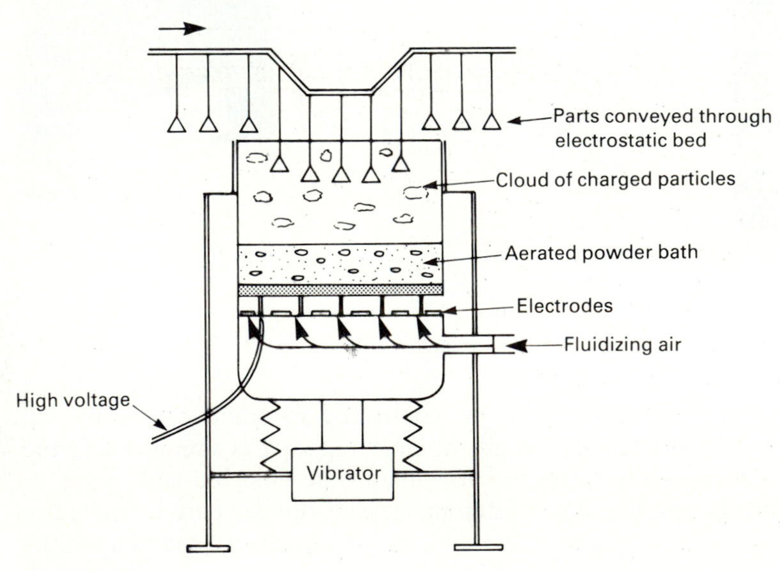

FIGURE 15.5
Coating of parts by passing through automated electrostatic fluidized bed.

The deposition rate is less when the part is more remote from the charged powder. Consequently parts having large vertical heights cannot be coated uniformly. The process lends itself to coating flats where coating of but one side can take place while immersion will coat the entire part.

The most widely used method of applying powder coating materials is by electrostatic spraying. The equipment required here is: powder feed unit, powder spray gun, spray booth, electrostatic voltage source, powder recovery unit.

The overspray by this process is recovered readily and reapplied. The shrinkage, due to overspray, is typically one to two percent. Usually, one coat coverage without a primer is adequate. Coverage takes place without sags, runs, tears, etc. In the majority of cases, the cost of application is lower than liquid coatings. Figure 15.6 illustrates patient walker legs being conveyed through an electrostatic spraying unit.

Tumble finishing. Tumble or rotary barrel finishing is used to finish many small parts (Fig. 15.3). The parts are put into the tumbler along with the finishing substance and tumbled until completely covered. They can be removed for drying or can remain in the tumbler after draining. The rotary

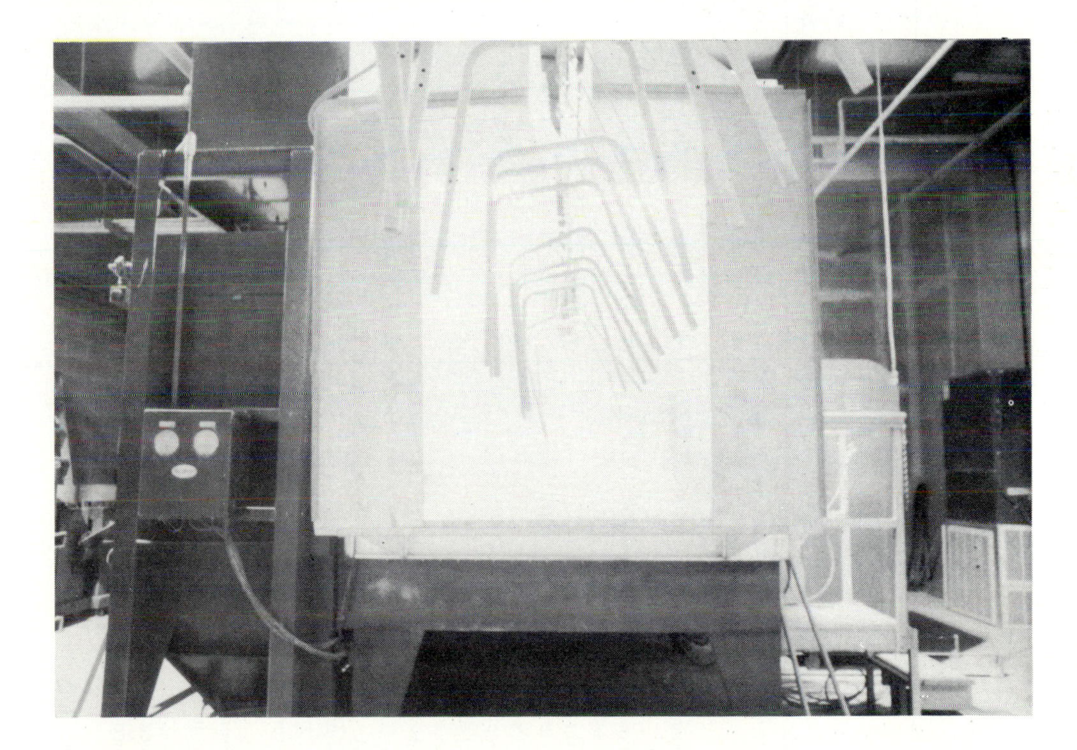

FIGURE 15.6
Coating of steel tubing parts with epoxy powder by electrostatic spraying.

TABLE 15.2
Organic coatings

	Alkyd				Cellulose		Epoxy		
	Alkyd	Alkydamine	Styrenated alkyd	Acrylic	Nitro-cellulose	Butyrate	Epoxy	Epoxy-urea	Epoxy-ester
Cost, c/(ft^2)(mil), dry	Inexpensive (3.0–6.0)	Inexpensive (3.5)	Inexpensive (3.5)	Moderate (5.5)	Moderate (5.0–9.0)	Moderate (5.5–14.0)	Moderate (10.0–14.0)	Inexpensive (4.00)	Inexpensive (3.50)
Appearance (choice of color)	Unlimited	Unlimited	Unlimited	Unlimited	Unlimited	Unlimited	Somewhat limited	Somewhat limited	Somewhat limited
Maintenance of dimensional tolerance	0.0005	0.0005	0.0005	0.0005	0.0005	0.0005	0.0005	0.005	0.005
Typical thickness, mils	1.5	1.5	1.5	1.0	1.0	1.0	1.8	1.8	1.5
Resistance to atmosphere (salt spray)	Good	Good	Good	Excellent	Excellent	Excellent	Excellent	Very good	Excellent
Resistance to elevated temperature (maximum service temp., °F)	200	250	200	180	180	180	400	400	300
Abrasion resistance Tabor GS-10 wheel cycles)	Fair 3,500	Good 5,000	Good 5,000	Fair 2,500	Fair 2,500	Fair 2,500	Good 5,000	Good 5,000	Good 5,000
Surface finish	Smooth or wrinkle	Smooth or wrinkle	Smooth or wrinkle	Smooth or wrinkle	Smooth or wrinkle	Smooth or wrinkle	Smooth or wrinkle	Smooth or wrinkle	Smooth or wrinkle
Color retention	Good	Good	Good	Excellent	Very good	Excellent	Good	Good	Good
Gloss	Excellent	Excellent	Excellent	Excellent	Excellent	Excellent	Very good	Very good	Excellent
Gloss retention	Excellent	Good	Good	Excellent	Very good	Excellent	Fair	Fair	Good
Exterior durability	Excellent	Excellent	Fair-good	Excellent	Good-excellent	Good-excellent	Good	Good	Good-excellent
Resistance to alcohols	Fair	Good	Good	Poor	Good	Good	Excellent	Excellent	Fair
Resistance to gasoline	Good	Excellent	Fair-excellent	Good	Good	Good	Excellent	Excellent	Excellent
Resistance to hydrocarbons	Good	Excellent	Excellent	Fair	Fair	Fair	Excellent	Excellent	Very good

Resistance to esters, ketones	Poor	Poor	Poor	Poor	Poor	Poor	Fair	Very good	Fair
Resistance to chlorinated solvents	Poor	Poor	Poor	Poor	Poor	Poor	Excellent	Excellent	Fair
Resistance to salts	Very good	Excellent	Excellent	Very good	Good	Very good	Excellent	Excellent	Excellent
Resistance to alkalies, concentrated	Poor-good	Good	Very good	Fair	Poor	Poor	Excellent	Excellent	Excellent
Resistance to acids (mineral), concentrated	Poor	Poor	Poor	Poor	Poor-fair	Poor	Good	Fair	Poor
Resistance to acids (oxidizing), concentrated	Poor	Poor	Poor	Poor	Poor	Poor	Poor	Poor	Poor
Resistance to acids (organic as acetic, formic), concentrated	Poor	Poor	Poor	Poor	Poor	Poor	Poor	Poor	Poor
Resistance to acids (organic, as oleic, stearic), concentrated	Fair	Good	Fair	Fair	Fair	Fair	Excellent	Excellent	Excellent
Resistance to acids (phosphoric)	Poor	Poor	Poor	Poor	Poor	Poor	Poor	Poor-good	Poor
Resistance to H_2O (salt and fresh)	Fair	Good	Good	Excellent	Fair-good	Excellent	Very good	Good	Very good
Rockwell R	24	30	28	24	26–30	26	36	34	30
Flexibility	Excellent	Very good	Good	Excellent	Excellent	Excellent	Excellent	Very good	Good-excellent
Toxicity	None	Slight	Slight	None	None	None	Slight	Slight-none	Slight-none
Impact resistance	Very good	Excellent	Good	Excellent	Excellent	Excellent	Good	Good	Excellent
Dielectric properties	Good	Good	Good	Very good	Poor	Good	Very good	Very good	Very good
Adhesion to ferrous metals	Excellent	Excellent	Excellent-fair	Very good	Excellent	Very good	Excellent	Excellent	Excellent
Adhesion to nonferrous metals	Fair	Excellent	Excellent-fair	Very good	Good	Good	Excellent	Excellent	Very good
Adhesion to old paints	Very good	Good	Very good	Poor	Poor	Poor	Poor	Poor	Fair-very good
Ease of application	Excellent	Bake required	Excellent	Very good	Very good	Very good	Catalyst required	Bake required	Excellent

TABLE 15.2 *(continued)*

	Alkyd				Cellulose		Epoxy		
	Alkyd	Alkydamine	Styrenated alkyd	Acrylic	Nitro-cellulose	Butyrate	Epoxy	Epoxy-urea	Epoxy-ester
Surface preparation	Primer	No primer	No primer	Primer	Primer	Primer	No primer	No primer	Primer
Bake-drying time	30 min (275°F)	20 min (320°F)	15 min (300°F)	. . . . . .	. . . . . .	. . . . . .	. . . . . .	30 min (350°F)	30 min (320°F)
Air-drying time	2 h	. . . . . .	10 min	5 min	5 min	5 min	45 min	. . . . . .	1 h
Coverage	450	450	400	350	200	200	450	500	450

				Rubber						
	Fluoro-carbon (air-dried)	Phenolic	Polyamide	Chlorinated rubber	Neoprene	Hypalon	Silicone	Urethane	Vinyl	Linseed and tung oils
Cost, $c/(ft^2)(mil)$, dry	High (30.0)	Inexpensive (3.5-8.0)	Moderate	Inexpensive (3.0–10.0)	Inexpensive	Moderate	Moderate (12.0)	Moderate-high (18.0)	Moderate	Inexpensive (1.0–1.5)
Appearance (choice of color)	Unlimited	Limited	Limited	Limited	Limited	Unlimited	Unlimited	Unlimited	Unlimited	Unlimited
Maintenance of dimensional tolerance	0.0005	0.0005	0.0006	0.0006	0.0007	0.0006	0.0006	0.0006	0.0005	
Typical thickness, mils	1.0	1.5	2–30	1.5	2–10	2	1.0	1–2	1.0	1.5–2.5
Resistance to atmosphere (salt spray)	Excellent	Excellent	Fair	Excellent	Excellent	Excellent	Excellent	Excellent	Excellent	Good
Resistance to elevated temperature (maximum service temp., °F)	220–550	350	300	200–250	200	250	550–1,200	300	150–180	
Abrasion resistance (Tabor GS-10 wheel cycles)	Poor 1,000	Good 5,000	Good 5,000	Good 5,000	Good 5,000	Good 5,000	Fair 2,500	Good 5,000	Good 5,000	Fair
Surface finish	Smooth or wrinkle	Smooth or wrinkle	Smooth	Smooth	Smooth	Smooth	Smooth or wrinkle	Smooth or wrinkle	Smooth or wrinkle	Smooth

Color retention	Good	Poor	Very good	Good	Good	Excellent	Excellent	Good	Very good
Gloss	Excellent	Very good	Good	Fair	Poor	Poor	Excellent	Excellent	Good
Gloss retention	Fair	Fair	Fair	Fair	Fair	. . . . . .	Excellent	Fair	Good
Exterior durability	Excellent	Excellent	Poor	Excellent	Excellent	Excellent	Excellent	Excellent	Excellent
Resistance to alcohols	Fair-good	Excellent	Good	Fair-excellent	Excellent	. . . . . .	Poor-fair	Fair-excellent	Fair-excellent
Resistance to gasoline	Fair-excellent	Excellent	Good	Good	Fair	Poor	Fair-good	Fair-good	Excellent
Resistance to hydrocarbons	Poor-excellent	Excellent	Excellent	Poor	Poor	. . . . . .	Very good	Fair-excellent	Good
Resistance to esters, ketones	Poor	Fair-excellent	Good	Poor	Good	Poor	Poor	Fair	Poor
Resistance to chlorinated solvents	Poor-excellent	Fair-good	Excellent	Poor	Poor	Poor	Poor	Poor excellent	Poor
Resistance to salts	Excellent	Excellent	Very good	Excellent	Excellent	Excellent	Good	Excellent	Excellent
Resistance to alkalies concentrated	Excellent	Poor	Good	Fair-excellent	Excellent	Fair	Fair	Fair	Excellent
Resistance to acids (mineral), concentrated	Excellent	Excellent-poor	Poor	Poor–good	Poor	Excellent	Poor	Poor	Good
Resistance to acids (oxidizing), concentrated	Excellent	Good–poor	Poor	Poor-fair	Poor	Poor	Poor	Poor	Good
Resistance to acids (organic as acetic, formic), concentrated	Excellent	Poor	Poor	Poor	Poor	Poor	Poor	Poor	Poor
Resistance to acids (organic, as oleic, stearic), concentrated	Excellent	Excellent	Very good	Excellent-fair	Fair–poor	Fair	Poor–good	Fair–excellent	Excellent
Resistance to acids (phosphoric)	Excellent	Fair	. . . . . .	Good	Very good	Excellent	Poor–fair	Fair–excellent	Excellent
Resistance to H_2O (salt and fresh)	Excellent	Excellent	Fair	Very good	Excellent	Excellent	Excellent	Excellent	Excellent
Rockwell R	20	38	. . . . . .	24	10	10	16	35–65	20
Flexibility	Good	Good	Good	Very good	Excellent	Excellent	Fair	Excellent	Excellent
Toxicity	Slight	None	. . . . . .	Slight	None	. . . . . .	Slight–none	Slight	None

TABLE 15.2 (*continued*)

	Alkyd				Cellulose		Epoxy			
	Alkyd	Alkydamine	Styrenated alkyd	Acrylic	Nitro-cellulose	Butyrate	Epoxy	Epoxy-urea	Epoxy-ester	
Impact resistance	Excellent	Good	Very good	Good	Excellent	Excellent	Good–fair	Excellent	Excellent	Very good
Dielectric properties	Excellent	Excellent	Good	Excellent	Fair	Very good	Excellent	Excellent	Excellent	
Adhesion to ferrous metals	Very good	Excellent	Very good	Fair	Very good	Very good	Good–fair	Excellent	Good	Fair
Adhesion to nonferrous metals	Very good	Excellent	Very good	Very good	Very good	Very good	Fair–excellent	Excellent	Very good	Fair
Adhesion to old paints	Poor	Good	......	Excellent–fair	......	Fair	Excellent	Fair–good	Good–fair	
Ease of application	Very good	Excellent	Good	Excellent	Very good	Very good	Excellent	Excellent	Very good–fair	
Surface preparation	Primer	No primer	No primer	Primer	No primer	Primer	Primer	primer	Primer	Primer
Bake-drying time	15 min (300°F)	30 min (350°F)	......	15 min (300°F)	15 min (300°F)	15 min (300°F)	1 h (400°F)	30 min (325°F)	15 min (300°F)	
Air-drying time	5 min	10 min	......	45 min	15 min	15 min	45 min	45 min	15 min	
Coverage	200	350	......	450	350	300	350	300–750	250	

barrel is typically loaded with a mixture of parts to be finished, medium compound, and water and then rotated so that the force of gravity causes the parts to slide over one another (the barrel is loaded only to about 60 percent of capacity). This action will deburr, generate corner radii, remove scale and rust and relieve surface stress.

Centrifugal finishing. Centrifugal finishing is also used for small parts. Such parts are placed in a strongly made wire basket, dipped, and then put in the centrifuge and whirled. This throws the excess finish material from the surface of the parts. The parts are finally removed and hung to dry. Parts treated by this method cannot have pockets that will accumulate the paint.

Air drying and baking. Some of the finishes are air-dried, but this process cannot be successfully accomplished under 70°F, and is usually done between 80 and 100°F. Some finishes are force-dried in ovens between 100 and 200°F; most synthetic primers and enamels are baked between 200 and 450°F. Baking is done only with nonporous articles and is to be preferred (except with lacquers) as it ordinarily gives a harder and more durable finish than air drying.

All enamels have a "smooth" look, such as that found on automobiles, produced by rubbing the finish with fine abrasives, applying wax, and polishing.

Preliminary treatment. The above finishes should be applied on a sound foundation. The surface for the finish should be pretreated as carefully as the finish is applied. It should be free of dirt, rust, scale, grease, oil, or flux residue, and should have a large surface area of a texture that will provide mechanical adhesion. The chromate-phosphate finishes, known in the trade as *parkerizing* or *bonderizing* finishes, are good foundations.

SUMMARY. Organic coatings constitute a family of coatings made up of a vehicle that consists of either a drying oil or a resin, and a pigment. This family includes paints, enamels, varnishes, and shellacs, with vehicles of synthetic resins, rubber, linseed, and tung oils. As a family, these finishes are relatively inexpensive, provide the opportunity for a variety of pleasing colors, and give good resistance to corrosion. They do not allow the holding of close dimensional tolerance and they have only moderate resistance to abrasion. Their resistance to elevated temperature is poor in comparison with that of other coating families. Table 15.2 provides details of the more important organic coatings.

Inorganic Coatings

Inorganic coatings are made up of refractory compounds. As a class, they are harder, more rigid, and have greater resistance to elevated temperatures than the organic coatings. The principal characteristics that lead the production-design engineer to specify an inorganic coating are eye-appealing finish, resistance to corrosion, a protection against elevated-temperature oxidation, and a surface that provides thermal insulation.

Porcelain enamels. Porcelain enamels equal or surpass the organic finishes in beauty and permanence. Porcelain-enameled surfaces combine the strength and stability of steel with the beauty and utility of glass. They are easily cleaned, have color stability, and are durable. The cost has been reduced by using one coat for some applications, by using a less expensive steel, and by using conveyorized equipment for spraying, dipping, drying, and firing at 1500 to 1600°F. Porcelain enamels will resist temperatures up to 1000°F. Where appearance is not critical, a single coat of 0.003 to 0.004-in. thickness has sufficient opacity for eye appeal and for protection.

Design considerations that will provide the best conditions for a durable porcelain finish include the following.

1. Avoid sharp edges; make radii as large as possible; avoid re-entrant corners.
2. Strengthen edges by flanging and stiffen large flat areas by embossing. This applies especially to porcelain-enameled parts. The firing temperatures cause expansion and contraction, which require that the metal be distributed uniformly and not be restricted.
3. If there must be enclosed spaces that can trap and hold finish or cleaning materials, provide holes for adequate drainage. They will also prevent dragover of the various solutions used.

Porcelain enamel may be applied by dipping, flow coating, manual spray, electrostatic spray, and dry powder. The engineer will decide which method to use by consideration of base material, quantity, quality requirements, and cost. The firing of porcelain enamel may be done in either continuous or batch furnaces. The firing time and temperature will vary inversely. Thus, a part that is fired at 1475°F for $2\frac{3}{4}$ min may provide similar properties as if the same part were fired at 1440°F for $5\frac{1}{2}$ min.

Ceramic coatings. Ceramic coatings are vitreous and metallic oxide coatings that are more refractory than the porcelain enamels. Typical ceramic coatings have a higher alumina content than porcelain enamels and, in addition to protecting metal surfaces from oxidation and corrosion, increase their strength and rigidity. This latter characteristic is especially important when the part is subjected to elevated temperatures. Thus, these coatings may be applied to metals to protect them from wear, corrosion, and oxidation at both room and elevated temperatures. The greatest industrial usage of ceramic coatings is of those prepared from silicate powders. However, coatings based on oxide

TABLE 15.3
Inorganic coatings

Property	Porcelain enamel	Ceramic coating		
		Enamel refractory oxide	Refractory oxide	Cermet
Cost	High	High	High	High
Appearance (choice of colors)	Wide range of colors	Dull	Dull	Dull
Maintenance of dimensional tolerance	0.0002	0.0002	0.0002	0.0002
Typical thickness, mils	3 to 5 per coat; 3 coats typical	2–3	2–50	2–10
Resistance to atmosphere	Excellent	Good	Good	Good
Resistance to elevated temperature, °F	700–2000°F	......	2500°	4000°
Abrasion resistance	Good	Good	Excellent	Excellent
Surface finish	Smooth, glassy appearance	Coarse–smooth	Coarse–smooth	Coarse–smooth
Base matrials	Sheet iron or steel or cast iron with low carbon content; few imputities; aluminum; copper; gold	......	......	Refractory and high-temp. materials, columbium, molybdenum, tantalum, tungsten
Number of coats	1–3	1	1	1
Method of application	Dipping or spraying	Dipping or spraying	Spraying	Spraying
Chemical resistance	Good	......	Very good	
Impact resistance	Fair to excellent			
Torsion resistance	Fair to excellent			
Dielectric strength, V/mil	500–1,000			
Hardness (Vickers)	4–7 mohs	......	1,400	1,350

materials, carbides, silicides, and phosphates are also used. These coatings may be applied by spraying, dipping, flow coating, etc.

SUMMARY. Inorganic coatings constitute a family of protective coatings that have a glass like finish. They include the porcelain enamels and ceramic coatings composed of inorganic mineral materials that are fused to base metals. They may be readily applied to both ferrous and nonferrous surfaces. These coatings provide excellent resistance to corrosion and elevated temperatures. They also provide good appearance and resistance to abrasion. As a family, their cost is high and they afford limited ability to maintain close tolerances. Table 15.3 summarizes the properties of the most important inorganic coatings.

Metallic Coatings

Metallic coatings may be applied by electroplating, hot immersion (galvanizing), chemical deposition, or spraying of molten metal (metallizing). They are used to provide a decorative finish, protection against corrosion, and resistance to wear; they serve as a base for painting to provide a reflectant surface and to provide a thermally or electrically conductive surface.

ELECTROPLATING. Electroplating is the process of passing current through the material to be deposited, the solution or bath, and the part on which the material is to be plated.

The base metal is made the cathode in an aqueous solution of a salt of the coating metal. Anodes of the coating material are used to complete the circuit and replenish the solution. Also, in order to increase its conductivity, other chemicals that will ionize strongly are added (e.g. sulfuric acid is added to an acid copperplating bath). Plating solutions attack metals and containers, and plating equipment is expensive to maintain. The cost of plating is small compared to cost of surface preparation, cleaning, and handling of the parts. Polishing also adds considerably to the cost of plating; therefore, the advantages of improving appearance and corrosion resistance must be balanced against the increased cost.

The properties of the coating will vary with the composition of the plating solution, current density, agitation, solution pH, and solution temperature. Generally, the three properties usually sought when specifying an electroplate are hardness, resistance to corrosion, and appearance. Brinell hardness values of electrodeposited metals are shown in Table 15.4. Resistance-to-corrosion characteristics will vary with the thickness of the plate and, of course, the plating material. Appearance is usually thought of in terms of brightness of the finished coat. Bright plating can be realized by minimizing the buffing coat. The usual procedure is to buff the softer underlying metal (such as copper or nickel) and to follow with a hard bright coat (such as chromium) that requires little or no buffing.

TABLE 15.4
Hardness of electrodeposited metals

Metal	Brinell hardness
Cadmium	35–50
Chromium	700–1000
Copper	60–150
Gold	5
Iron	150–500
Lead	5
Nickel	150–500
Rhodium	400–800
Silver	50–150
Tin	5
Zinc	40–50

The development of "periodic reverse current electroplating" has reduced the cost of polishing after plating. At the same time, the plated deposit shows superior qualities of strength, elasticity, density, and freedom from flaws like porosity. It involves a novel plating cycle in which plating cycle current is reversed briefly at short periodic intervals to deplate what may be unsound and inferior metal deposited in the previous plating period. Many microscopically thin increments of sound metal are built up to make a deposit more dense and of greater homogeneity than is possible with conventional, continuous-current methods. Work has been done with silver, copper, brass, zinc, cadmium, gold, nickel, and iron. Equipment for timing the cycles is available in mechanical and electronic types.

The protective value of electroplated surfaces depends on the thickness and porosity of the plate and its uniformity. The uniformity of thickness of plate on a given part depends on the shape of the piece and the "throwing power" of the bath—the ability to deposit material in remote areas, such as on thread roots and in deep holes.

The throwing power of cyanide and alkaline baths is good; thus, parts including complex geometry should be plated in this type of bath rather than an acid bath. The throwing power of the chromium bath is poor.

An electroplate may be applied directly to the base metal or it may be applied over another electroplate. Thus, a nickel plate may be applied over a copper plate and chromium over a nickel plate. Lamination of plates allows an accumulation of specific desirable characteristics. For example, a base metal may be copper-plated and so acquire a greater polishability. The copper plate may be covered with nickel, which is less porous and thus has greater resistance to corrosion. A final lamination of chromium may then be applied for appearance.

Prevention of hydrogen embrittlement is an important precaution in connection with the plating of case or through-hardened parts that have not been stress-relieved. In order to avoid delayed fracture of stress parts owing to hydrogen embrittlement, the plated parts should be baked for about 4 hours at 200°C (392°F) within a few hours after the plating operation.

Electroplating theory. Electroplating theory is based on the work done by Michael Faraday in 1833. The laws developed by Faraday state that the quantity of material liberated at either the anode or cathode during electrolysis is proportional to the quantity of electricity that passes through the solution. Also, the amount of a given element liberated during electrolysis is proportional to its equivalent weight. Thus in silver plating if a coulomb (C) deposits 1.118 mg of silver on the cathode, 5 C will deposit 5.590 mg.

Knowing the quantity of one metal that is liberated by a given amount of electricity, we can determine the amount of other metals deposited. For example, if silver, having an atomic weight of 107.88, is deposited at the rate of 1.118 mg/C, cadmium, having an atomic weight of 112.41 and a valence of 2, will be deposited at a rate of

$$\frac{112.41/2}{107.88} \times 1.118 = 0.586\ \text{mg/C}$$

Of course, the actual amount of material deposited will never quite equal the theoretical value because of such factors as solution variations and current leakage. The amount actually deposited divided by the theoretical amount is known as the *cathode efficiency*. Average cathode-current efficiencies of common plating solutions are shown in Table 15.5.

Zinc plating. Zinc plating is one of the commonest and most widely used metallic coating methods. Zinc is the lowest-priced metal used for electrode-

TABLE 15.5
Cathode efficiencies in electroplating

Metal	Type of bath	Usual cathode efficiency
Cadmium	Cyanide	88–95
Chromium	Chromic acid–sulfate	12–16
Copper	Acid sulfate	97–100
Copper	Cyanide	30–60
Copper	Rochelle–cyanide	40–70
Lead	Fluoborate	100
Lead	Fluosilicate	100
Nickel	Acid sulfate	94–98
Silver	Cyanide	100
Zinc	Acid sulfate	99
Zinc	Cyanide	85–90

posited coatings. Since zinc is anodic to iron and steel, it provides greater protection when applied in thin films (0.3 to 0.5 mil) than do the same thicknesses of nickel and other cathodic coatings (with the exception of cadmium in salt-water environments). It is readily applied in tank, barrel, or continuous plating equipment. Zinc tarnishes readily and its vapors are toxic; therefore, when welding zinc-plated parts, adequate ventilation should be provided. Representative applications of zinc-plated commodities are switch boxes, hardware, screw machine parts, and conduits. The appearance of electrodeposited zinc on smooth steel varies from a dull, light-gray surface to a mirror-bright, gray appearance.

Cadmium plating. Cadmium plates are more expensive than zinc, as cadmium is the highest-priced of the metals commonly used in electrodepositing protective coatings on metals; however, cadmium plates provide greater protection to corrosion in saline atmospheres. Furthermore, cadmium is usually applied from 0.0002 to 0.0005 in. thick for ordinary commercial parts.

Thickness of plate is readily controlled, since both the covering power and throwing power of the cyanide bath, from which the cadmium deposit is obtained, is good. Cadmium-plated parts have a lower coefficient of friction than zinc-plated parts; they are solderable, although they do make spot welding erratic. Since a cadmium-plated surface has a dull, silverlike appearance that is readily tarnished by many chemicals, it is not used as a decorative plate. Cadmium plating is specified on many aircraft and marine parts and is used on electronic instruments and miscellaneous hardware.

Nickel plating. Nickel plates have a wide range of characteristics depending upon the plating that is used. Three classifications of baths comprise the finishes available utilizing nickel: general purpose, black, and bright baths. The anodes employed are of high purity, containing a minimum of 99 percent nickel. A general-purpose nickel-plating bath contains nickel sulfate, nickel chloride, and boric acid. In general, nickel plates provide excellent wear and corrosion resistance, good uniformity, and ability to be joined by both brazing and soldering.

Electroless nickel plating results in a coating without the use of an electric current. Here, the deposition is a result of an autocatalytic chemical reduction of nickel ions by hypophosphite, aminoborane or borohydride compounds.

Tin plating. Tin plates are quite good but are more expensive. Consequently, tin plating of ferrous parts is seldom resorted to strictly for the prevention of atmospheric corrosion; however, since tin plates are especially resistant to tarnishing, and their oxides are not toxic and do not have an objectionable taste, they are used on "tin cans," kitchenware, and food containers. Also, since tin-plated parts solder quite easily, they are used in radio, television, and electronic equipment.

Copper plating. Copper coatings on steel have resistance to corrosion, but once the base metal is laid bare by a pinhole or scratch, strong voltaic action oxidizes the base metal at the break and in a fanning-out development beneath the coating. This results in a peeling away of the protective plate. For this reason, copper plate should be rather heavy (1 to 2 mils) and antipitting additives should be employed in order to promote pore-free deposits when subsequent coats of other metals are applied. Since copper-plated parts polish well, this plate is used frequently as a base in preparation for subsequent coats of nickel or chromium. Copper plates have decorative value if protected from oxidation. This can be accomplished by coating the plated part with clear lacquer.

The electrolytes used under most conditions have high throwing power and consequently produce adequate coverage in recessed areas. Although copper can be electrodeposited from several electrolytes, sulfate and fluoborate acid baths and cyanide and pyrophosphate alkaline baths are the principal ones. Copper plating is typically applied to a thickness of between 2.0 and 0.3 mils.

Brass plating. Brass plates are frequently used as a base for bonding rubber and rubberlike materials to metal. Since brass tarnishes, it must be protected with lacquer when it is being used for decorative purposes. Brass plates are used on steel, zinc, aluminum, and can be applied on top of a copper plate.

Brass is deposited from a cyanide bath and, consequently, thickness of plate can be controlled on intricate sections.

Lead plating. Lead plates provide good atmospheric protection and also give an excellent base to which paints can be applied. The plate holds well to the base metal and can be severely deformed without stripping off. Typical applications include cable sheathing, roofing, and paint cans.

Silver plating. Silver plates are used principally for decorative purposes; they give a pleasing effect when polished to a bright luster. They also have high electical conductivity, good bearing qualities, and resistance to oxidation at elevated temperatures, which may result in their selection for a certain process. The principal disadvantages of silver plate are its high cost and susceptibility to tarnishing. Typical silver-plated items are cutlery, jewelry, musical and surgical instruments (such as electrocardiogram probes), and electronic components in which tarnishing or destabilization will result in no functional problem.

Gold plating. Gold plates, similarly to silver plates, find their greatest use in decorative parts and in electronic related components. Gold is extremely expensive; consequently, the thickness of plate used for appearance is very thin. In addition to eye appeal, gold is resistant to most chemicals, is a good electrical conductor, and resists tarnishing. Electronic components require thicker, low porosity, and harder coatings.

Chromium plating. Chromium plating is performed primarily for two reasons: for its decorative and protective properties, and for its high wear-resistant properties. Chromium plates have high reflectivity, giving an attractive, bright appearance.

Thick or "hard" chromium plates are applied to ferrous materials as well as aluminum and zinc, with the predominant purpose of providing a hard wear-resistant surface. Typical applications are molds for producing plastic parts, plug gages, taps, reamers, rebuilding worn parts, valve stems, and piston rings. The electrodeposition solution contains chromic acid (CrO_3) and a catalytic anion in the correct proportion.

Since the chromium bath has poor throwing power, it is difficult to maintain uniformity of plating thickness on complex shapes. To remedy this, specially shaped anodes can be designed to surround or insert in the object to be plated so as to reduce the irregularity of "throwing" distance. The anodes are usually of lead alloy or insoluble lead and chromium is supplied by the chromic acid in the electrolyte.

Since the ratio of cost for equal thickness of plate of chromium compared with nickel is about 20:1, thin thicknesses of chromium are used when the object is for decorative purposes. If high resistance to corrosion is necessitated in addition to eye appeal, as in automobile hardware, then the chromium is applied over coats of nickel and/or copper. The sequence of operations performed on a typical automobile bumper might be:

1. Clean
2. Nickel plate (at least 0.0008 in. thick)
3. Wash
4. Copper plate (at least 0.001 in. thick)
5. Wash
6. Nickel plate (at least 0.008 in. thick)
7. Wash
8. Dry
9. Polish and buff
10. Chromium plate

Chromium plate runs from 0.000,02 to 0.000,03 in. for standard chrome plate. Duplex chrome deposits are about 0.000,06 in. total, evenly divided between two layers.

ELECTROLESS PLATING. Electroless plating is a chemical plating process. Nickel and copper are the two metals that are most often used in this process, most other metal-plating systems are not available as electroless plating. This technique necessitates strict control of both the temperature and composition of the chemical bath, resulting in costs that frequently exceed electroplating.

An important advantage of electroless plating is that it does not produce hydrogen embrittlement. Therefore, it is often used to plate case-hardened hardware for which embrittlement in service would be intolerable.

PLATING OF PLASTICS. Plated plastics—knobs, handles, buttons, and other parts made to look like metal parts—has been one of the fastest-growing businesses in recent years. The biggest consumer of plated plastic components is the automobile industry. Also growing in importance are such nonauto products as TV, radio, and appliance knobs and trim. In the plating of plastic parts (acrylonitrile-butadiene-styrene and polypropylene are the plastics most used), it is necessary to first deglaze or etch the surface of the parts. Then onto this conditioned surface a very thin layer of precious metal (silver or gold) is applied. This forms the base for a thin layer of either nickel or copper, that provides a conductive surface for the final layer, which usually is chromium.

THERMAL SPRAY COATINGS. This method of depositing metallic and sometimes non-metallic coatings is frequently referred to as metallizing. The coatings can be sprayed from rod or wire stock or from powdered material. There are three systems that are used: plasma-arc spray, electric-arc spray, and the flame spray. In the flame spray process, a spray gun is used that feeds the wire through a nozzle surrounded by an oxyacetylene flame. A blast of air breaks the melted metal into globules and sprays it onto the surface. The wire is fed by a turbine driven by air pressure. The globules are oxidized on their surfaces and are molten on the inside. They strike the surface and flatten out. The oxidized surface opens and the molten material adheres to the oxidized surface of the flattened globules previously deposited. There are voids between the surfaces that serve to absorb oil. The surface acts as an oil reservoir similar to powdered metal materials and forms an excellent bearing material. The surface on which the metal is deposited should be rough like that obtained from a very rough turn on the lathe, and it must be clean and free of moisture or oil. This provides a mechanical lock for a deposited material. A metallized coating adheres to steels, stainless steels, Monel, nickel, iron manganese, and most aluminum alloys. It saves machining and can be used as a base for other less-expensive materials. The properties of porosity and ability to hold oil have led to spraying babbitt in bearing sleeves. This sprayed-babbitt material is superior to cast-babbitt bearings because of its ability to absorb oil.

A technique known as spray welding is used to hard-surface materials. The alloy is in powdered form held by a binder that is extruded into a wire form. The metallizing process vaporizes and consumes the binder so that the pure molten material is deposited without any trace of the binder. An even and uniform coating can be applied, contrasting with the rough and thick coating of the arc-welding process of hard surfacing. The coating is then heated with a torch or in a furnace, which fuses it to the base material and forms a hard, dense, homogeneous surface. The surface is free from porosity and the uniformity of the deposit reduces machine-finishing time.

The electric-arc spray process uses no external heat source, as the heating takes place when two electrically opposed wires are fed together in such a manner that a controlled arc results at the intersection. The wires themselves are the spray material. The molten metal is atomized and subsequently is propelled by compressed air to the substrate.

In the plasma-arc spray process, the external heat source is electrically induced plasma.

Thermal spray coatings are used extensively for building up worn parts, for protection from corrosion, and for improving wearing surfaces. Cloth and paper are coated for use in electrical condensers.

It can be seen that metallizing is a production tool as well as a repair tool. The engineer needs to consider the possibility of corrosion resulting from strong galvanic couples between the coating metal and the substrate.

SOLDER SEALING. Solder sealing of metal parts to glass or ceramic parts presupposes that the metal and nonmetal parts will have approximately the same coefficient of expansion during the wide variation of temperature necessary in the sealing process or in the operation of the apparatus. The procedure followed is to chemically deposit a metal on the glass or ceramic. The metallic compounds make the surface of the ceramic or glass chemically pure and deposit the metal at the same time. This simultaneous action enables the metal to adhere firmly to the glass or ceramic. A soft solder is added to this metal deposit. The metal container receiving the glass or ceramic is also prepared for soldering and the two are joined by the usual soldering operations.

HOT-DIP COATING. The process of hot-dip coatings is quite common. Galvanizing is a hot-dip process in which the base material is immersed in a tank of molten zinc. Adhesion results from the tendency of the molten zinc to diffuse into the base metal. The protective coating is made up of several layers. The layer that is closest to the base metal is made up of iron–zinc compounds while the outermost layer is primarily zinc.

Hot dipping is a rapid, inexpensive process that allows the coating of corrosion-resistant metals onto base metals at less cost than by electroplating. However, the process is limited to shapes that will not trap the molten metal upon extraction from the dip tank.

Tin, lead, and aluminum coatings may be applied to the base metals in addition to zinc. Base metals are restricted to the materials with higher melting temperatures such as cast iron, steel, and copper. In order to avoid brittleness of the iron–zinc layer, the substrate cast iron should be low in phosphorus (<0.02 percent) and low in silicon (<0.15 percent). Similarly, steel should contain less than 0.25 percent carbon, less than 0.05 percent phosphorous, less than 1.35 percent manganese and less than 0.05 percent silicon in order to allow satisfactory galvanizing.

Application of these coatings involve elevated temperatures (from 850°F for zinc coatings to 1200°F for aluminum). Thus, galvanizing and aluminizing can have significant effects on the chemical properties of the substrate material.

IMMERSION COATINGS. Immersion coatings are applied by dipping a base metal having a higher solution potential into an aqueous solution containing ions of the coating metal. No electric current is used. Deposition occurs while the base metal is in the solution. The coat is usually quite thin and can be controlled quite closely. Nickel, tin, zinc, gold, and silver are used as immersion-coating materials.

Nickel immersion coatings are used on steel parts that are so constructed that it would be difficult to maintain a uniform electrodeposited coating, such as valves, gears, threaded parts, and other items having deep recesses.

Tin immersion coatings provide a bright, decorative, and protective coating on such items as paper clips, pins, and needles.

VAPOR-DEPOSITED COATINGS. As the name implies, vapor-deposited coatings result from the condensation of a metal film on the base metal. Aluminum is the most widely applied vapor film. The aluminum is heated and vaporized in a vacuum. The aluminum vapor then condenses on the surfaces of the base metal. As would be expected, vapor deposited films are very thin. Consequently, they have had little appliction where extreme resistance to corrosion is required. The principal use of this method of deposition has been for decorative purposes such as on trim for television and radio sets, for other household furniture, and on costume jewelry.

MECHANICAL PLATING. This is a proprietary procedure where metal powders are mechanically welded to a substrate. The parts to be plated are tumbled in a medium of metal powder and glass beads. During the tumbling, the surface of the parts is activated. This process is used to plate soft materials such as tin, cadmium, and zinc on steel.

SUMMARY. Metallic coatings as a family are those coatings in which a metal is deposited on a surface by electroplating, hot dipping, immersion, metallizing, flame spraying, or electroless methods. Since most plating, dipping, and immersion tanks are limited in size, metallic coatings are usually applied to smaller parts. An upper constraint is approximately 2000 cubic inches in most plants.

This family of coatings provides for bright metallic surfaces with close tolerance control and good resistance to abrasion. The cost of metallic coatings as a family is somewhat higher than that of organic coatings and less than that of inorganic coatings. Metallic coatings as a family provide adequate resistance to corrosion and elevated temperatures. Table 15.6 provides detailed information on design characteristics of the principal electroplated coatings.

Conversion Coatings

Conversion coatings are those produced when a film is deposited on the base material as a result of a chemical reaction. The most widely used conversion coatings are:

1. Phosphate coatings
2. Chromate coatings
3. Anodic coatings

Phosphate coatings. Phosphate coatings are used principally as a base for the application of paint or enamel. Also, they are used to aid in the forming of sheet and cold heading of bar and rod. The process itself does tend to rust-proof the base material, which is usually iron, steel, aluminum, or zinc. The metal surface is treated with a dilute solution of phosphoric acid and other elements so that a mildly protective layer of insoluble crystalline phosphate is obtained. A typical phosphate coating process consists of spray cleaning, phosphatizing, water-spray rinse, and chromic acid-spray rinse. The phosphate coating is usually 0.0001 to 0.003 in. in thickness. The process is very rapid because the capacity of the equipment rather than the cycle time limits the rate of production. This process is widely used in the automotive and electrical-appliance industries for preparing automobiles, washing machines, refrigerators, and similar products to receive an organic finish.

Chromate coatings. Chromate dip coating may be applied at the mill in order to provide for corrosion protection of galvanized sheet. A chromate dip-coated galvanized product is said to be "passivated" or "stabilized." Chromate coatings are also used on nonferrous materials including aluminum, magnesium, zinc-coated materials, and cadmium-coated materials. Chromate coatings are generally quite thin, being less than 0.00002 in. thick, and are used chiefly for added resistance to corrosion and as a base for paint. Chromate conversion coatings result when the base metal is either sprayed or immersed in a solution of chromic acid, chromium salts, together with hydrofluoric acid or hydrofluoric acid salts, phosphoric acid, or other mineral acid. The resulting chemical attack produces a protective film containing chromium compounds. The process can give either of two types of films: a yellow iridescent film or a clear film. Through the use of acidified organic dyes, the process can be used to impart a variety of popular colors to the treated part, including red, blue, lemon, violet, orange, green, and black.

Chromate films are less expensive to apply than anodized coatings because the process is faster and less overhead is involved. They also give greater resistance to corrosion than do anodic films; however, anodic films offer superior wear-resistant qualities.

A representative breakdown of the operations of a chromate treatment

TABLE 15.6
Metallic coatings (electroplated)

Characteristic	Cadmium	Chromium	Copper	Gold	Iron
Cost, \$/ft^2	0.005–0.075/ ft^2 moderate (3)	Moderate (4)	Inexpensive (2)	Very high (10)	Fairly expensive (6)
Appearance	Bright white	White, mirror-like, highly decorative	Bright or semi-bright pink, red	Bright yellow, highly decorative	Matte gray
Maintenance of dimensional tolerance, decimal in.	0.0003	0.0003	0.0003	0.0002	0.001
Typical thickness, mils	0.1–0.2 (indoor), 0.3–0.7 (outdoor)	0.01–0.05 (decorative), 0.05–2.00-up (hard)	0.3–0.5, (undercoat), 2.0–3.0 (functional topcoat)	0.002–0.01 (decorative); 0.01–2.0 (functional); 0.004–0.15 (electroforming)	2.0–10.0
Resistance to atmospheric corrosion	Very good (3)	Good (when over copper and nickel) (5)	Fair (6)	Excellent (1)	Poor (10)
Resistance to elevated temperature; metling point, °F	610	2939	1931	1944	2795
Abrasion resistance	Fair (8)	Excellent (1)	Poor (9)	Poor (8)	Good (5)
Surface finish	Smooth (bright to dull)	Very smooth	Very smooth (bright to dull)	Very smooth (bright to dull)	Moderately smooth
Reflectivity at 6000 Å	Fair	Very good (66%)	Fair (44%)	good (47%)	Poor (60%)
Throwing power	(1)	(10)	(2)		(3)
Base metals	Steel, stainless steel, wrought iron, gray iron, copper and its alloys	Ferrous, nonferrous metals	Most ferrous and nonferrous metals	Copper, brass, nickel, silver	Ferrous metals, copper and alloys
Specific gravity	8.65	7.10	8.93	19.30	7.85
Resistivity, $\mu\Omega \cdot$ cm	7.54 at 18°C	2.6 at 20°C	1.72 at 20°C	2.44 at 20°C	0.6 at 20°C
Thermal conductance, cal/(cm · °C · s)	0.217	0.65	0.023	0.707	0.190
Brinnell hardness	2–22	400–1,000	40–130	(65–325 Knoop)	140–350
Adhesion	Good	Excellent	Excellent	Excellent	Good
Type bath and cathode efficiency (%)	Cyanide 88 to 95	Chromic and sulfate 12 to 16	Acid sulfate 97 to 100, cyanide 30 to 60, rochell-cyanide 40 to 70	Cyanide 70 to 90	Acid chloride 90 to 98, acid sulfate 95 to 98
Grams deposited per A · h	2.0968	0.3233	2.371 (ous) 1.186 (ic)	7.356 (ous) 2.450 (ic)	1.042
Solubility	Soluble in acids; ammonium nitrate	Soluble in HCl, dilute H_2SO_4; insoluble HNO_3	Soluble in HNO_3; hot H_2SO_4; slightly soluble in HCl and ammonium hydroxide	Soluble in potassium cyanide aqua regia; hot selenic-acid, insoluble most acids	

(*Continued on p. 740*)

TABLE 15.6 (*continued*)

Lead	Nickel	Rhodium	Silver	Tin	Zinc
Fairly expensive (5)	Moderate (3)	Very high (10)	High (9)	Moderate (4)	0.015–0.02 ft^2 inexpensive (1)
Matte gray	White, either dull or highly decorative bright	Mirror-bright white, highly decorative	Bright white, highly decorative	Dull white, highly decorative	Matte gray to attractive bright
0.0005	0.0003	0.0002	0.0002	0.0003	0.0003
0.5–8	0.1–1.5 (decorative), 5–20 (industrial)	0.001–1	0.1 (with undercoat) to 1.0	0.015–0.5	0.1–0.2 (indoor), 0.3–0.7 (outdoor)
Good (5)	Very good (4)	Very good (4)	Good (5)	Good (5)	Very good (4)
621	2651	3553	1760	448	786
Poor (9)	Good (5)	Good (4)	Good (5)	Poor (8)	Poor (8)
Smooth	Very smooth (bright to dull)	Smooth	Very smooth	Smooth	Smooth (bright to dull)
Poor (3)	Good (60%) (5)	Excellent (76%)	Excellent (90%)	Fair (1)	Fair (55%) (8)
Ferrous metals, copper and alloys	Most ferrous, nonferrous metals	Most ferrous, nonferrous metals	Most ferrous, nonferrous metals	Most ferrous metals and nonferrous metals	Ferrous metals excepting cast iron, copper and its alloys
11.35	8.90	12.50	10.50	6.75	7.14
22 at 20°C	7.8 at 20°C	5.11 at 0°c	1.63 at 20°c	11.5 at 20°C	5.75 at 0°C
0.083	0.210	0.210	0.974	0.157	0.268
5	125–550	594–641	60–79	8–9	40–50
Good	Very good	Good	Good	Good	Excellent
Fluoborate 98, fluosilicate 98	Acid sulfate 94 to 98	Acid sulfate 10 to 18, acid phosphate 10 to 18	Cyanide 99	Acid sulfate 90 to 95, stannate 70 to 90	Acid sulfate 98, cyanide 85 to 90
8.865	1.095	1.280	4.025	2.214 (ous) 1.107 (ic)	1.102
Soluble in HNO_3 and hot concentrated H_2SO_4	Soluble in dilute HNO_3, slightly soluble in HCl and H_2SO_4, insoluble in ammonia	Soluble in sulfuric-hydrochloric acid; hot concentrated H_2SO_4, slightly soluble in acids	Soluble in HNO_3; hot H_2SO_4, potassium cyanide, insoluble in alkalies	Decomposes in HCl, H_2SO_4, dilute HNO_3, aqua regia, hot potassium hydroxide	Soluble in acids, alkalies, acetic acid

(*Continued on p. 741*)

TABLE 15.6 *(continued)*

Characteristic	Cadmium	Chromium	Copper	Gold	Iron
Preparation for	Grit blast or polish, clean chemically with organic solvent, alkali cleaner, or electrolytic alkali cleaner, prepare chemically by acid nickel or dip				
Remarks	Price of cadmium is 8–10 times cost of zinc. A 0.0005-in. coat will withstand 96 h of salt spray without showing iron rust or white salts	Usually applied from 0.00001 to 0.00005 in. over copper and nickel. Composite thickness runs from 0.00041 min for noncorrosive applications to 0.0020 for outdoor service. Hard chromium deposits from 0.0001 to 0.010 in without undercoating	......	......	......

Note: Numbers in parantheses indicate relative ranking on a scale from 1 to 10: 1 represent most attractive and 10 the least.

might be:

1. Degrease
2. Rinse
3. Clean in alkali
4. Rinse
5. Soak in hydrofluoric acid
6. Rinse
7. Place in dichromate bath (45-min soak)
8. Rinse
9. Dry

Chromate coatings must not be used on galvanized steel parts that are to be resistance-welded or phosphated and painted. Chromate will cause problems in welding and will interfere with proper phosphating, thus resulting in poor paint adhesion.

Anodic coatings. Anodic coatings result when an oxide is applied to aluminum and magnesium and their alloys. Here the base metal is connected as an anode

TABLE 15.6 (*continued*)

Lead	Nickel	Rhodium	Silver	Tin	Zinc
Cathodic to iron and steel and must be used in thick deposits to prevent galvanic corrosion	Cathodic to steel and therefore used over copper	......	Tarnishes in the vapors of sulfur compounds	Cathodic to iron	

in an electrolytic immersion, and an oxide film is deposited on the base metal that increases its resistance to atmospheric as well as galvanic corrosion and offers a good foundation for painting.

A typical breakdown of the operations in applying an anodic coating to aluminum is as follows.

1. Degrease
2. Rinse
3. Clean in alkali
4. Rinse
5. Apply anodic coating (3 to 10 percent solution of chromic acid)
6. Rinse
7. Dry

The process outlined, using chromic acid, will give a coating of aluminum oxide and chromium salts varying in thickness from 0.00003 to 0.0002 in. The color will range from a light gray to black. It provides excellent corrosion protection and has been used extensively by the aircraft industry.

Anodic coatings are also applied by sulfuric acid, oxalic acid, and boric acid bath processes.

It is possible to impart excellent colored coatings by immersing the parts

TABLE 15.7
Conversion coatings

Property	Phosphate	Chromate	Oxide
Cost, $/ft^2	Low (0.004–0.02/ft^2)	Medium (0.01–0.03/ft^2)	High (0.02–0.40/ft^2)
Appearance (choice of colors)	Poor (preparatory to painting)	Wide choice: clear, golden iridescent, olive drab	Unlimited in anodizing process
Maintenance of dimensional tolerance	No appreciable change	No appreciable change	0.0001
Typical thickness, mils	0.01–0.02	0.01–0.02	0.04–0.5 (2.0–4.0 for wear, abrasion resistance)
Resistance to atmosphere	Fair	Good	Good
Resistance to elevated temperature, °F	Good	Good	Good
Abrasion resistance	Poor	Poor	Good
Surface finish	Thin amorphous film	Thin amorphous film	Thin crystalline structure
Base materials	Iron, steel, zinc, aluminum, cadmium, and tin	Zinc, copper, tin, magnesium, silver, aluminum	Nonferrous (aluminum and magnesium)
Number of coats	1	1	1
Method of application	Dipping, brushing, or spraying	Dipping or spraying	Dipping, electrolytic
Procedure	Pretreatment: phosphating, rinsing, drying	Degrease and etch-rinse; chromate; rinse warm; rinse dry	Degrease; rinse; anodize; rinse; dry
Types	Zinc, iron, and magnesium	Chemical polishing; non-polishing colored coating applied	Sulfuric, chromic
Purpose	Paint adhesion and oil retention, corrosion resistance		Decoration, corrosion resistance
Coefficient of friction	Medium	Medium	Low

in warm dye solutions, and then sealing the dye in the porous oxide coatings by dipping in dilute nickel acetate.

SUMMARY. Conversion coatings constitute those coatings used on ferrous or nonferrous metals in which either a phosphate, chromate, or oxide salt represents the protective coating. As a family, these coatings are inexpensive and allow the maintenance of close tolerances. However, in view of the fact that they are very thin, they do not in themselves provide a great amount of resistance to corrosion or abrasion. Table 15.7 provides the principal properties of phosphate, chromate, and oxide conversion coatings.

SUMMARY

The production-design engineer has a vast number of finishes from which to choose. For a given application, one finish may be superior to another from the standpoint of adaptability, service, eye appeal, and cost.

The following characteristics represent the principal criteria for selection of the decorative and protective coating to be applied.

Cost. Cost considerations enter into the selection process of all designs. The analyst is interested in the total cost resulting from a given coating. Thus, cost must include cost of surface preparation prior to receiving the coating, cost of coating the surface, and cost of any recommended treatment after coating.

Appearance (color). The characteristic appearance (color) has to do with the clarity of colors that the coating is capable of achieving. Not only is the range of colorability significant but also the durability of the color in the coating. Thus color clarity, range, and durability should be considered in this characteristic.

Appearance (bright-metallic). This characteristic relates to the ability of the coating to provide and retain a bright, reflective, metallic appearance.

Resistance to corrosion. This characteristic is based on the ability of the coating to protect the covered surface from typical corrosive agents including salt spray, ozone, oxygen, and SO_2.

Ability to maintain close tolerance. This characteristic refers to the tolerance capability, under normal conditions, of the coating process.

Abrasion resistance. This characteristic is related to the ability of the coating to resist galling, wear, and scraping when a like or disimilar material is repeatedly rubbed against the coated surface.

Resistance to elevated temperature. This characteristic represents the ability of a coating to permit the part to provide service at temperatures significantly above normal room temperature.

By identifying the pertinent characteristics desired of the decorative and protective coating, the well-informed production-design engineer, with the help of the guide sheets provided, should be able to select the most appropriate finish for his company's products.

QUESTIONS

15.1. What are the principal ways in which a metallic surface can be cleaned?

15.2. What major considerations should be made by the engineer in the selection of the most effective cleaning process?

15.3. To prepare a SAE 1112 steel for plating, what procedure would you recommend in order to remove a buffing compound from an earlier operation?

15.4. What four types of finish may be applied to metal products?

15.5. When would you advocate solvent cleaning? Alkaline cleaning? Electrolytic cleaning? Emulsion cleaning?

15.6. What advantages do oleoresinous varnishes have over oil paint?

15.7. What is the main drawback of lacquers?

15.8. What products have made use of vinyl lacquers?

15.9. For what type of products would you advocate the use of porcelain enamels?

15.10. Explain how "hydrogen embrittlement" is caused. How can parts that have hydrogen embrittlement be treated so as to avoid delayed fracture?

15.11. What does "cold cleaning" involve?

15.12. What cleaning process would you recommend for small assemblies that have deep inaccessible crevices?

15.13. How may metallic coatings be applied?

15.14. From an engineering point of view, why is cadmium plating popular?

15.15. Why is a copper plate not particularly suitable to parts subject to atmospheric corrosion?

15.16. Explain why flow coating usually does not provide a uniform thickness of coat.

15.17. Describe the electroless plating process.

15.18. Where is metallizing used?

15.19. How are conversion coatings classified?

15.20. What is the future of anodic and chromate coatings?

15.21. With a cathode efficiency of 94 percent, what would be the rate of deposition of zinc?

15.22. What seven criteria are usually considered in the selection of a decorative and protective coating?

15.23. Explain the procedure used in the chrome plating of plastic components used on the dashboard of the modern automobile.

15.24. What type of plating would you recommend for a design having deep recesses and subject to atmospheric corrosion? Why?

PROBLEMS

15.1. A Unified National Coarse (UNC) thread is to be plated to a specific tolerance on the pitch diameter after plating. Determine the limits for the pitch diameter after

machining, if the plating thickness upper and lower limits are 0.0005 in. and 0.0002 in., respectively. The final limits on the pitch diameter are: $\frac{0.2854\text{ in.}}{0.2830\text{ in.}}$.
How much of the tolerance does the plating tolerance use?

SELECTED REFERENCES

1. American Society for Metals: *Metals Handbook* edited by Howard E. Boyer and Timothy L. Gall, Metals Park, Ohio, 1985.
2. Blum, W., and Hogaboom, G.B.: *Principle of Electroplating and Electroforming,* 3d ed., McGraw-Hill, New York, 1949.
3. Burns, Robert M., and Bradley, W. W.: *Protective Coatings for Metals,* 3d ed., Reinhold, New York, 1967.
4. Lowenheim, Frederick A.: *Modern Electroplating,* 2d ed., Wiley, New York, 1963.
5. Society of Manufacturing Engineers: *User's Guide to Powder Coating,* edited by Emery Miller, Dearborn, Michigan, 1987.

CHAPTER

16

PROCESS ENGINEERING

In most metal-cutting or metal-removing facilities, the task of planning piece/part operations and sequences is the responsibility of the process planner. The process planner is charged to implement virtually all of the knowledge contained in this book, as well as elsewhere, in order to plan for efficient production. This individual holds a key to the profitability of a specific product, but little has been done to aid the process planner in the performance of his job. With the cost of machinery sky-rocketing, and as the degree of automation increases, greater emphasis has been placed on the importance of process planning and engineering.

This chapter describes process plans and outlines the responsibilities and functions carried out by the process planner. The decisions required of the process planner or process-planning system are considered. Two approaches to *automated* processing planning, called *variant* and *generative* planning are presented. A discussion of group technology is also presented.

INTRODUCTION

Today one of the major concerns in the United States is to continue to improve manufacturing productivity, thereby improving our standard of living. The productivity increases that have occurred recently are largely due to new technological developments in manufacturing. The 1950s saw the introduction of numerical control (NC) and the APT programming language.[1] In the 1960s, the concept and first application of direct numerical control (DNC) was demonstrated. The 1970s and 1980s have brought even more applications of the computer to manufacturing with the advent of computer numerical control (CNC) and computer integrated manufacturing systems (CIMS).

As computer applications continue to emerge in the manufacturing environment, planning for totally computer-integrated manufacturing systems has begun. In these systems, parts for production will be drawn using a computer-aided design (CAD) system. Once the design has been finalized, economic lots will be selected; materials will be ordered; production plans will be created; and the machinery to produce the part(s) will be selected and scheduled: all via either a single computer or an integrated network of computers. Today, the technology exists to automatically perform each of the above activities, attempting to make total computer control a reality.

PROCESS PLANNING

As manufacturing becomes more automated, the cost, as well as the level of control, of a manufacturing system also increases. This then introduces the problem of economic effectiveness. When and what design is specified in the selection of a manufacturing system? Also, how does one plan the operations for such a system? It appears somewhat obvious that the "best guess" method of selecting and planning an automated or semiautomated manufacturing system may cause the system to function poorly, even though the system employs the most modern technology.

NC machine tools can be controlled more fully and in many cases more accurately than conventional machines. The repeatability of these machines has eliminated some classical tools. (An example of this is the piloted counterboring tool.) In general, planning for production on these machines has undergone an evolution. Machining speeds and feeds can be controlled accurately and changed during the most complex operations, and for some machines five separate axes are controlled simultaneously. As the level of detail required to control a machine increases, the complexity of planning its operation also increases. Planning for production for many parts has become the most time-consuming operation required in manufacturing [1, 2].

Delays, errors and higher than necessary production costs are likely to result from parts planned solely by trained or experienced planners. More importantly, this type of planning often precludes the thorough analysis of alternative processes and of the optimization of these processes. The manual

[1] APT or automatically programmed tool is a set of computer programs used to create machine control information for NC machines.

planning of operations virtually always results in the nonstandardization of process plans and increases the likelihood of nonstandard tooling [2, 3].

Automated process planning and selection is not a new idea; it was mentioned as early as 1965 [4]. Today, several computer-aided process-planning systems are available for use for a variety of manufacturing operations. Reports have appeared that describe process-planning programs available for sheet-metal fabrication [11]. The results obtained from these process planning-systems are encouraging. One report claims that a unit of quality control equipment was designed, planned, built, and delivered in two weeks using a process-planning program; this is opposed to a traditional lead time of six to twelve weeks for similar equipment planned manually [11]. In addition, there is an increasing awareness that productivity goals in the metal-working industry are incompatible with the present labor-intensive methods of process planning. In a U.S. Air Force study, it was estimated that although between 200,000 and 300,000 process planners were needed in 1980, only 150,000–200,000 were available. Because of the time required for process planning and the impact that process planning has on manufacturing, a significant amount of attention has been given to this important activity. Before continuing, process planning will be more precisely defined.

What is Process Planning?

Process planning has been defined as "The subsystem responsible for the conversion of design data to work instruction" [12]. A more specific definition of process planning is "that function within a manufacturing facility that establishes the processes and process parameters to be used (as well as those machines capable of performing these processes) in order to convert a piece-part from its initial form to a final form that is predetermined (usually by a design engineer) on a detailed engineering drawing." The input (raw) material to a process may take a number of forms (in machining, these materials normally result from a metal-forming process; the most common of which are bar stock, castings, forgings, or perhaps just a slab of metal, other processes have other input materials). Another machining material might be a burn-out (a part produced by a flame-cutting operation) cut to some rough dimension, or just a rectangular block of material. This input material might have almost any shape and physical property. Some processes may alter the size or surface texture of a part. Other processes, like heat treating, change the physical properties of materials. More specifically, annealing would tend to lessen the material hardness and decrease the workpiece tensile strength. Figure 16.1 contains a classification of some basic manufacturing process employed by the process planner.

With these types of raw materials as a base, the process planner must prepare a list of those (machining) processes needed to convert this normally predetermined material into its specified final geometry. The commonest metal-removal processes that a process planner has at his disposal are turning,

Process	Operation
Metal forming	Hot and cold rolling, forging, casting, drawing, extruding, etc.
Machining	Drilling, boring, reaming, turning, milling, planing, broaching, etc.
Heat treating	Normalizing, stress relieving, carburizing, nitriding, flame hardening, etc.
Inspection	Gaging, measurement apparatus, quality control, etc.
Assembly	Fixturing, accumulation, assembly.

FIGURE 16.1
Process-planning activities.

OPERATION SHEET

Part No. S563-1 Material Cast Iron 15 lbs/pc (approx.)
Part Name Eccentric
Orig.________ Changes________
Checked________
Approved________

Operation		Machine	Set-up		Operate
No.	Description		Description	Hr	Hr/Unit
5	Rough & Fin. Turn & Face: $6\frac{3}{4}''$ D $6\frac{1}{8}''$ D, one side reverse piece, rechuck and repeat for other side	16″ Monarch Engine Lathe (Kender #122)	Univ. 4 jaw chuck on $6\frac{1}{8}''$ cast. diam.	—	0.124
10	Rough & Fin. Turn, Face $3\frac{1}{2}''$ D eccentric boss, one side reverse piece, repeat other side. Rough & Fin. bore $2\frac{1}{2}''$ D hole Ream $2\frac{1}{2}''$ D hole	16″ Monarch Engine Lathe (Kender #138)	Four jaw chuck	—	0.28
15	Broach $\frac{1}{8}'' \times \frac{1}{16}''$ Keyway	La Pointe Broach (Kender #310)	Gang six castings in fixture	—	0.052
20	Drill $\frac{22}{64}''$ and tap $\frac{1}{16}''$ Set screw holes	L. G. Multi-Spindle Drill Press (Kender #110)	—		0.041

FIGURE 16.2
Operations sheet for eccentric cap.

facing, milling, drilling, boring, broaching, shaping, gundrilling, reaming, planing, sawing, trepanning, burnishing, punching, and grinding. Some manufacturing people may consider some of the operations as subsets of a major category. Reaming is often considered a subset of drilling. Others may define further major categories. Some less-familiar processes such as electric discharge machining (EDM), electrochemical machining (ECM), and laser machining are also used for material removal. All applicable processes that are available for production should be considered by the process planner.

An example of a process plan, taken from McKinney and Rosenbloom [25], and referred to as an operation sheet, appears in Fig. 16.2. These plans are typical of what one will find in contemporary machine shops. Note that the headings include columns for operation number, operation description, machine to be used, set-up description, and operation time information.

Elements of Process Planning

Doyle [13] separates the activities performed in process planning into seven general categories.

1. Interpret the specification requirements
2. Position the part on the machine.
3. Determine the intermediate product requirements at each stage of processing.
4. Select the major pieces of equipment to handle the processing.
5. Select the tooling and sequence of processing steps within each operation.
6. Compute the process time requirements.
7. Document the process plan.

Some of these activities can be divided into smaller units; however, the activities described provide a convenient categorization for analysis.

Manual Process Planning

A process planner normally operates under the following constraints.

1. He plans for a given set of machines.
2. The machines are capable of a limited number of manufacturing operations.
3. The machines have a specific burden/workload.

Given these machine constraints, and a set of engineering drawings containing specific part-geometry requirements, the process planner relies on his experience to develop a set of processes capable of producing a part. The selection of the processes is neither entirely random nor totally predictable. There is usually more than one process capable of producing a specific surface:

the process planner must choose what he believes is the best process. This is normally done by recalling similar parts, or at least similar surfaces, and the means utilized in manufacturing that part. In this manner, the planner compares specific processes used to obtain some set of final specifications and chooses what he believes to be the best alternative. This type of planning is known as *man-variant* process planning and is the commonest type of planning used for production today.

Planning the operations to be used to produce a part requires knowledge of two groups of variables: the part requirements (as indicated by an engineering drawing); and the available machines and processes, and the capabilities of each process. Given these variables, the planner selects the *combination* of processes required to produce a finished part. In selecting this combination of processes, a number of criteria are employed. Production cost or time are usually the dominant criteria in process selection; however, machine utilization and routing often affect the plans chosen. In general, the process planner tries to select the best set of processes and machines to produce a whole family of parts rather than just a single part.

As one might imagine, a large part of process planning, as it currently exists, is an *art* rather than a science. Process tolerance information is frequently recorded in a "black book." This information, may or may not be reliable. Similarly, the planner must also select operational information (such as machine speed, feed, and depth of cut). Again, a black book is employed to select these operating characteristics—this time the black book is somewhat more sophisticated (a machining data handbook); however, it does not guarantee operating efficiency, only feasibility.

Automated Process Planning

A practical process-planning system with good decision rules for each activity will seldom generate bad plans. Such a practical system could be structured by reducing the decision-making process to a series of mechanical steps. However, even if a system of good decision rules were to be developed, the human interaction of man-variant process planning could still create problems. Process planning can become a boring and tedious job. Accordingly, man-variant planning often produces erroneous process plans. This, coupled with the labor intensity of man-variant planning, has led many industries to investigate the automation of process planning. Whether a planning system is to be automated or manual, the seven general requirements of process planning as defined by Doyle must be performed. In the following section, the framework for automated process planning will be described.

Spur and Optiz [14, 15] were among the first to write on the automation of manufacturing systems and the role that process planning should play in these systems. Spur was perhaps the first to define variant and generative methods of process planning and the mechanization and implementation of such planning systems. The *variant* method of process planning is based on the principle of group technology and essentially consists of two steps.

1. Build a catalog (or "menu" as it is often called) of process plans to produce a gamut of parts, given a set of machine tools.
2. Create the software necessary to examine the part that is being planned and find the closest facsimile in the catalogue, then retrieve the associated process plans.

The *generative* method of process planning essentially consists of four steps.

1. Describe a part in detail.
2. Describe a catalog of processes available to produce parts.
3. Describe the machine tool(s) that can perform these processes.
4. Create the software to inspect the part, process, and available machinery to determine whether all three are compatible.

In general, planning using generative principles requires a detailed description of the part as well as a detailed understanding of manufacturing processes and their accuracies. Manufacturing plans based on the variant principle are determined by activating several standard solutions for individual operations and adapting or adjusting them where necessary.

In addition to the different types of process planning, different degrees of automation have also been proposed [5]. A completely automated process planning system would eliminate all human effort between the preparation of an engineering drawing and a complete process plan for every manufacturing operation. Schematically, the system would correspond to Fig. 16.3*a*. The block labeled LOGIC would include the capability to scan and interpret the drawing, to convert this information into process requirements and to select machines, tools and operations to yield an economically acceptable product.

Such a system would be truly automated. Self-contained logic would check for contradictory requirements on the engineering drawing. These requirements would be checked for compatibility with available processes. Selection of the processes (such as turning, milling or stamping) would be based on product requirements, quantities ordered, and process capabilities. Whenever contradictory or incompatable requirements were detected, a printed message would indicate the source of the problem and recommend remedial action.

A less automated system is illustrated in Fig. 16.3*b*. This system needs human assistance to code the engineering drawing data. Thereafter, the system is fully automated. In Fig. 16.3*c*, a still less automated system is shown. For this arrangement, a man must select the process as well as interpret and code the drawing information.

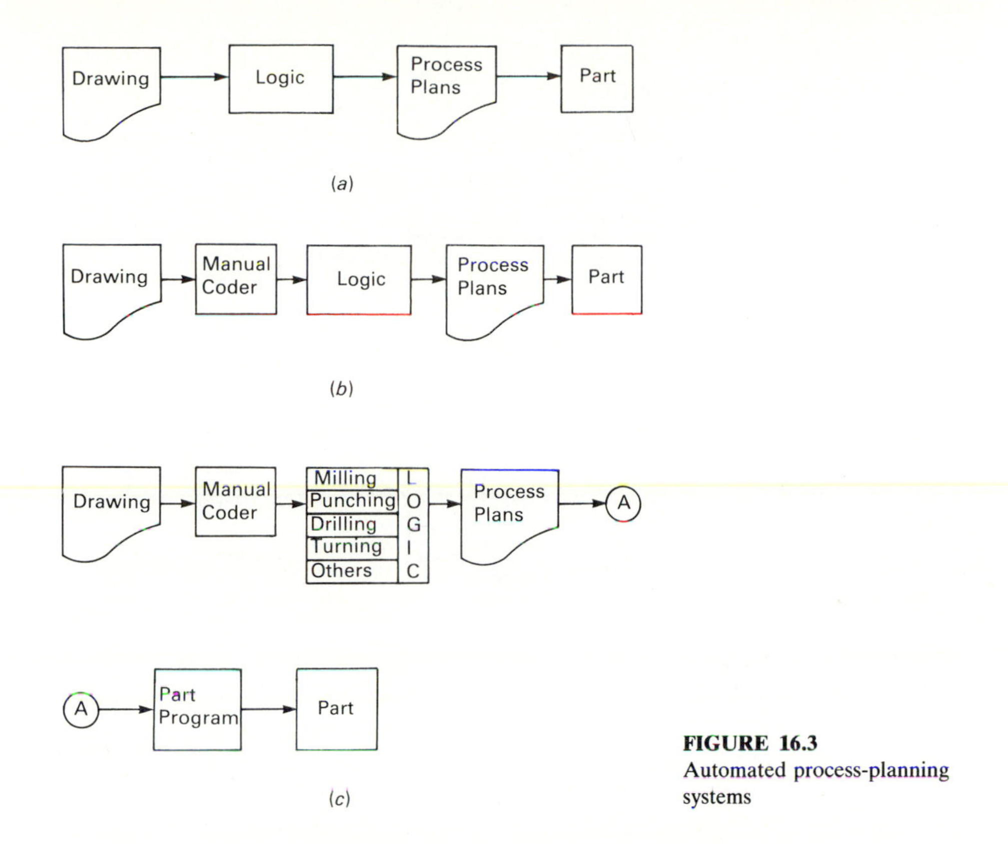

FIGURE 16.3
Automated process-planning systems

Figure 16.3*c* also contains a block for manually producing a part program. For turning operations, the part geometry describes the motions required by the machine tool. For this case, creating part programs is quite easy. For milling operations, symmetry is not required as it was for turning. Because symmetry need not be present, milling adds a new dimension to the creation of part programs. Because of the added complexity, part programming is contained in a separate block that may or may not be automated.

Many developments in computer-aided process planning have focused on eliminating the process planner from the entire planning function. Computer-aided process planning can reduce some of the decision making required during a planning process. Advantages of computer-aided process planning include the following.

1. It can reduce the skill required of a planner.
2. It can reduce the process planning time.
3. It can reduce both process planning and manufacturing cost.
4. It can create more consistent plans.
5. It can produce more accurate plans.
6. It can increase productivity.

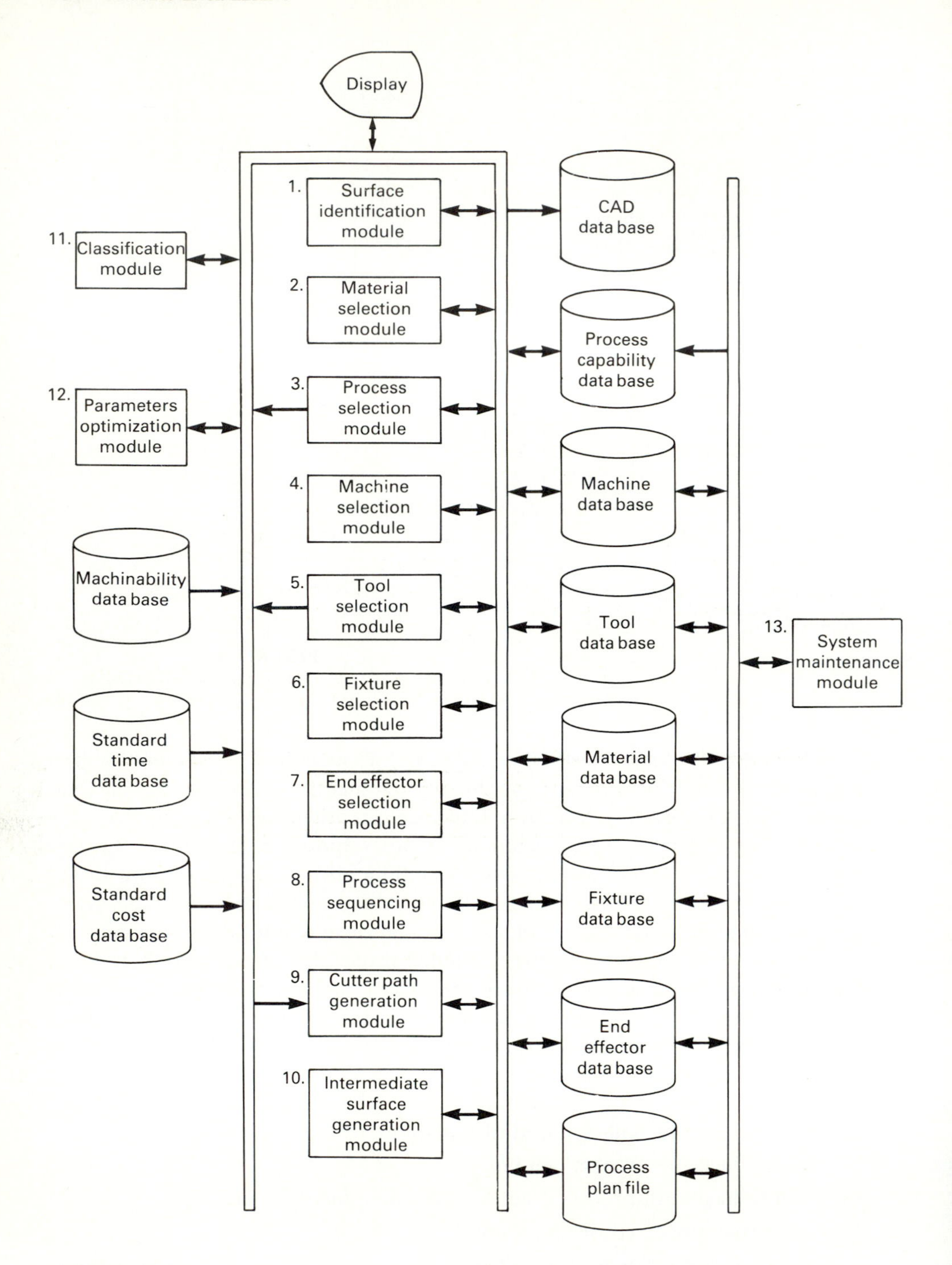

FIGURE 16.4
Process-planning modules and databases. (*From [36].*)

Benefits from computer-aided process planning have been documented by several industries [28–31]. Such systems can reduce planning time from days to hours or from hours to minutes.

Figure 16.4 represents the structure of a complete computer-aided process planning system. Although no existing turnkey system integrates all of the functions shown in Fig. 16.4 (or even a goodly portion of them), the figure illustrates the functional dependences of a complete process planning system. It also helps to illustrate some of the constraints imposed on a process-planning system (e.g., available machines, tooling, jigs, etc.) as well as to illustrate that process planning is critical for computer-integrated manufacturing.

In Fig. 16.4, the modules are not necessarily arranged on the basis of importance or decision sequence. The system monitor controls the execution sequence on the individual modules. Each module may require execution several times in order to obtain the optimum process plan. Iterations are required to reach feasibility as well as a good economic balance. Planning, whether manual or computer-assisted, consists of each of the noted functions and is constrained by available tooling and machines.

In a completely computer-integrated manufacturing system, input to the system will come from a CAD database. The model will contain not only the shape and dimensioning information, but also the tolerances and special features. Process plans will be routed directly to the production planning system and production control system. Time estimates and resource requirements can be sent to the production planning system for scheduling. Part programs and material-handling control programs will also be sent to the control system.

Process planning is the critical bridge between design and manufacturing. Design information can be translated into manufacturing language only through process planning. Today, both automated design (CAD) and manufacturing (CAM) have been implemented. Integrating or bridging these functions requires automated process planning.

GROUP TECHNOLOGY

Since the beginning of human culture, people have tried to apply reason to their actions. One important way to apply reason is to relate similar things. Biologists classify items into genus and species. We relate to such things as mammals, marsupials, batrachians, amphibians, fish, mollusks, crustaceans, birds, reptiles, worms, insects, and so on. A chicken is a bird with degenerated wings. A tiger, jaguar, and domestic cat are all members of a single family.

The same concept applied to natural phenomena can also be applied to fabricated products. When a vast amount of information needs to be kept and ordered, a *taxonomy* is normally employed. Librarians use taxonomies to classify books in stacks. Similarly, in manufacturing, thousands of items are produced yearly. When one looks at the parts that construct the product, the number is exceptionally large. Each part has a different shape, size, and

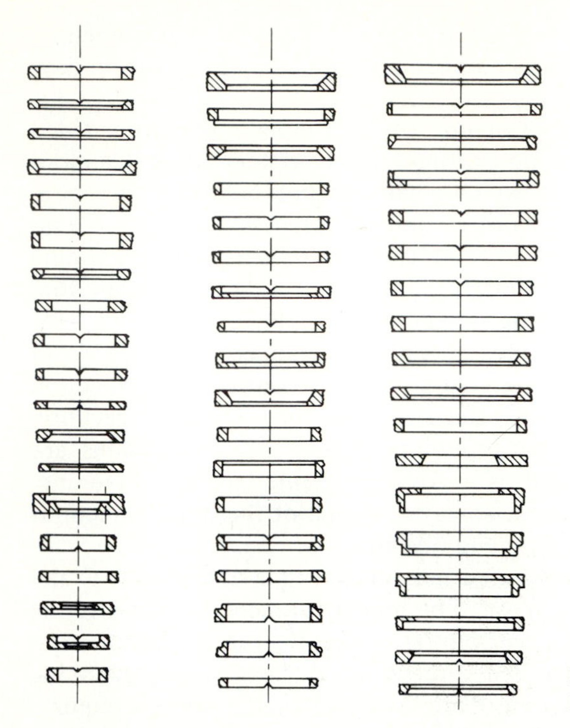

FIGURE 16.5
Design family.

function. However, when one looks closely, one may again find similarities among components (Fig. 16.5); a dowel and a small shaft may be very similar in appearance but different in function. Spur gears of different sizes require similar manufacturing processes. Manufactured components can be classified into families similar to biological families or library taxonomies. Parts classified and grouped into families produce a much more tractable database for management.

Although this simple concept has been in existence for a long time, it was not until 1958 that S. P. Mitrofanov, a Russian engineer, formalized the concept and put together a book entitled *The Scientific Principles of Group Technology*. Group technology (GT) has been defined [32] as follows:

> Group technology is the realization that many problems are similar, and that by grouping similar problems, a single solution can be found to a set of problems, thus saving time and effort.

Although the definition is quite broad, one usually relates group technology only to production applications. In production systems, group technology can be applied in different areas. For component design, it is clear that many components have a similar shape (Fig. 16.5). Similar components therefore can be grouped into design families. A new design can be created by modifying an existing component design from the same family. Using this

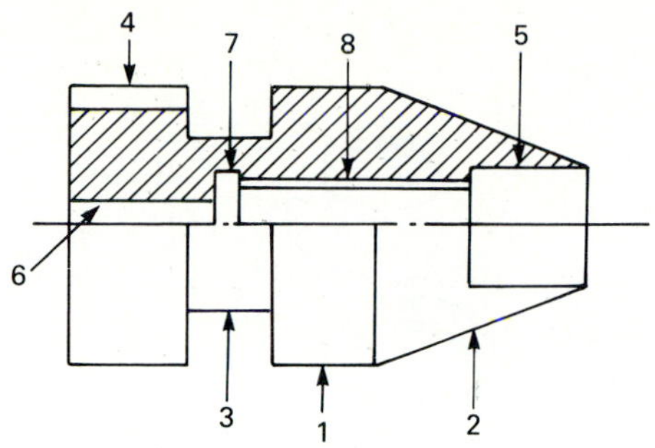

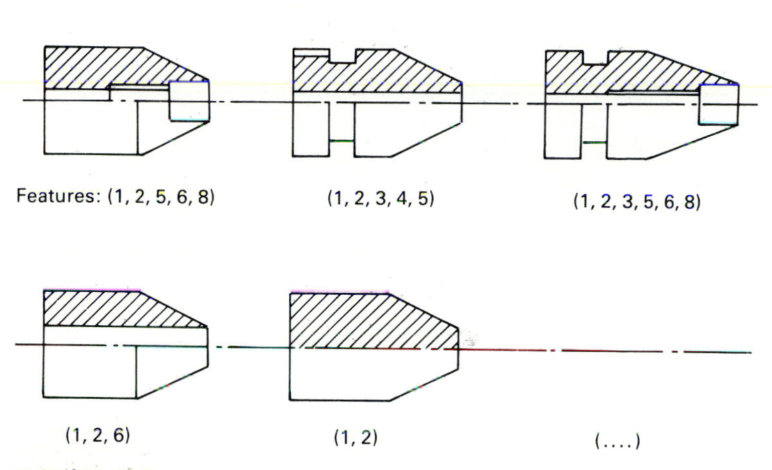

FIGURE 16.6
Composite component.

concept, composite components can be identified. Composite components are parts that embody all the design features of a design family or design subfamily. An example is illustrated in Fig. 16.6. Components in the family can be identified from features of the composite components.

For manufacturing purposes, GT represents a greater importance than simply a design philosophy. Components that are not similar in shape may still require similar manufacturing processes. For example, in Fig. 16.7 most components have different shapes and functions; but all of them require internal boring, face milling, hole drilling, and so on. Therefore, it can be concluded that the components in the figure are similar. The set of similar components can be called a production family. From this, process planning work can be facilitated. Since similar processes are required for all family members, a machine cell can be built to manufacture the family. This makes production planning and control much easier, since only similar components are considered for each cell. Such a cell-oriented layout is called a group technology layout.

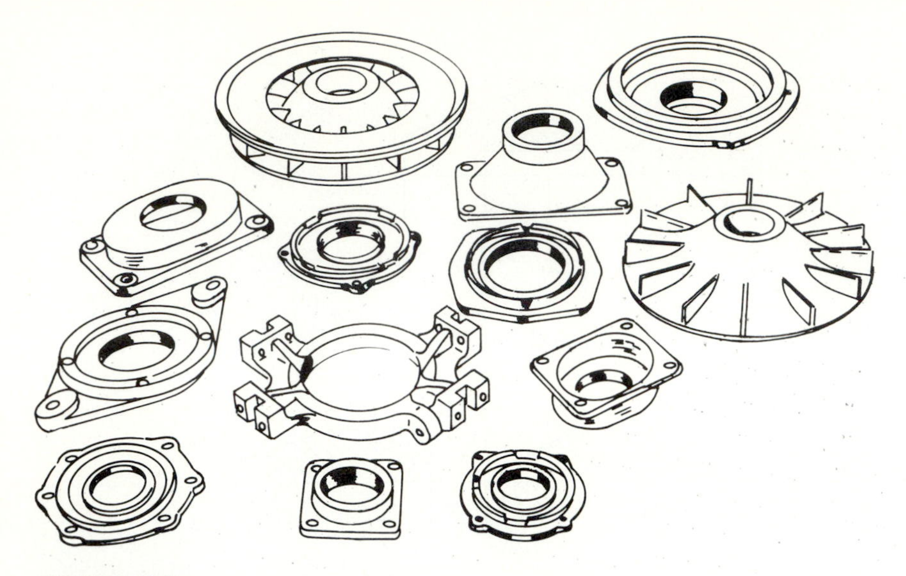

FIGURE 16.7
Production family.

The following techniques are employed in GT:

1. Coding and classification
2. Production flow analysis
3. Group layout

Although both production flow analysis and group layout are usually based on coding and classification methods, they can still be distinguished as different activities. Since group layout is not "directly" related to process planning, it will not be discussed further in detail.

CODING AND CLASSIFICATION

Coding is a process of establishing symbols to be used for meaningful communication. Classification is a separate process in which items are separated into groups based on the existence or absence of characteristic attributes. Coding can be used for classification purposes, and classification requirements must be considered during the construction of a coding scheme. The two topics are hence closely related.

Before a coding scheme can be constructed, a survey of all component features must be completed; then code values can be assigned to the features. The selection of relevant features depends on the application of the coding scheme. For example, tolerance is not important for design retrieval; therefore, it is not a feature in a design-oriented coding system. However, in a

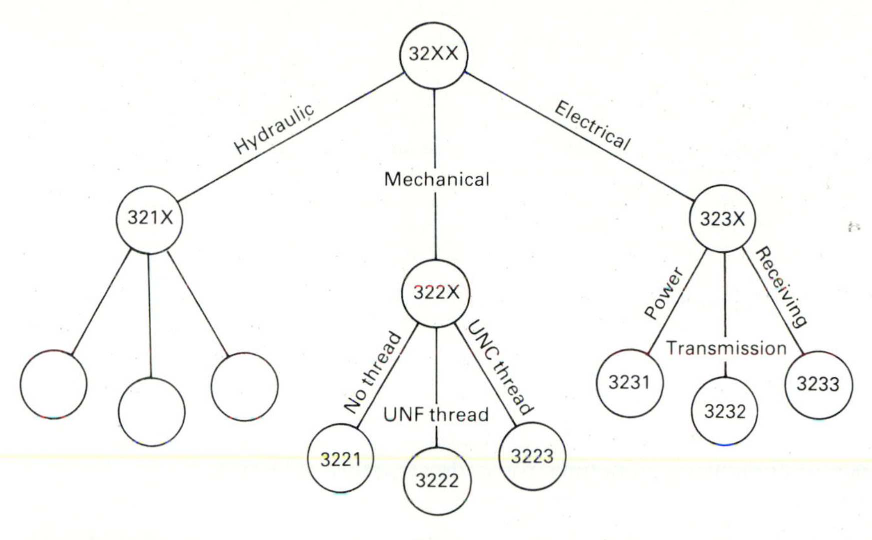

FIGURE 16.8
Hierarchical structure.

manufacturing-oriented coding system, tolerance is indeed an important feature.

Because the code structure affects its length, the accessibility and the expandability of a code (and the related database) is of importance. There are three different types of code structures in GT coding systems: (1) hierarchical, (2) chain (matrix), and (3) hybrid.

A hierarchical structure is also called a monocode. In a monocode, each code number is qualified by the preceding characters. For example, in Fig. 16.8, the fourth digit indicates threaded or not threaded for a 322x family. One advantage of a hierarchical structure is that it can represent a large amount of information with very few code positions. A drawback is potential complexity of the coding system. Hierarchical codes are difficult to develop because of all the branches in the hierarchy that must be defined.

A chain structure is called a polycode. Every digit in the code position represents a distinct bit of information, regardless of the previous digit. A chain-structured coding scheme is presented in Fig. 16.9. A "2"in the third position always means a cross hole, no matter what numbers are given to positions one and two. Chain codes are compact and are much easier to construct and use. The major drawback is that they cannot be as detailed as hierarchical structures with the same number of coding digits.

The third type of structure, the hybrid structure, is a mixture of the hierarchical and chain structures (Figure 16.10). Most existing coding systems use a hybrid structure to obtain the advantages of both structures. A good example is the widely used Opitz code (Figure 16.11).

There are more than a hundred GT coding systems used in industry

	Digit position			
	1	2	3	4
Class of feature	**External shape**	**Internal shape**	**Holes**	**Other**
Possible value 1	**Shape 1**	**Shape 1**	**axial**	**features**
2	Shape 2	Shape 2	Cross	
3	Shape 3	Shape 3	Axial and cross	
4				

FIGURE 16.9
Chain structure.

today. The structure selected is based primarily on the application. Table 16.1 provides comprehensive guidelines for code structure selection.

The physical coding of a component can be shown best by example. In Fig. 16.12, a rotational component is coded using the Opitz system. Going through each code position, the resulting code becomes 01112. This code represents this component and all others with similar shape and diameter.

Computer-variant process planning normally evolves through two stages—a preparatory and a production stage.

The Preparatory Stage

During the preparatory stage, existing components are coded, classified, and subsequently grouped into families. A family matrix is also constructed. The

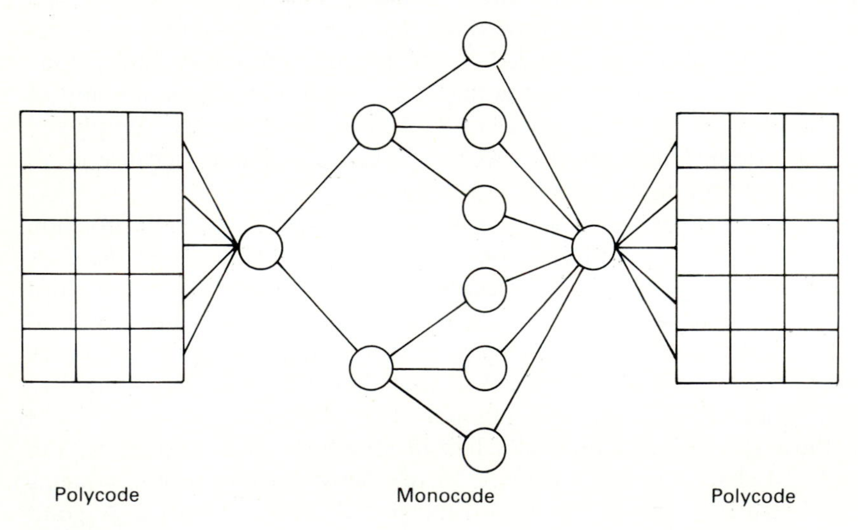

FIGURE 16.10
Hybrid structure.

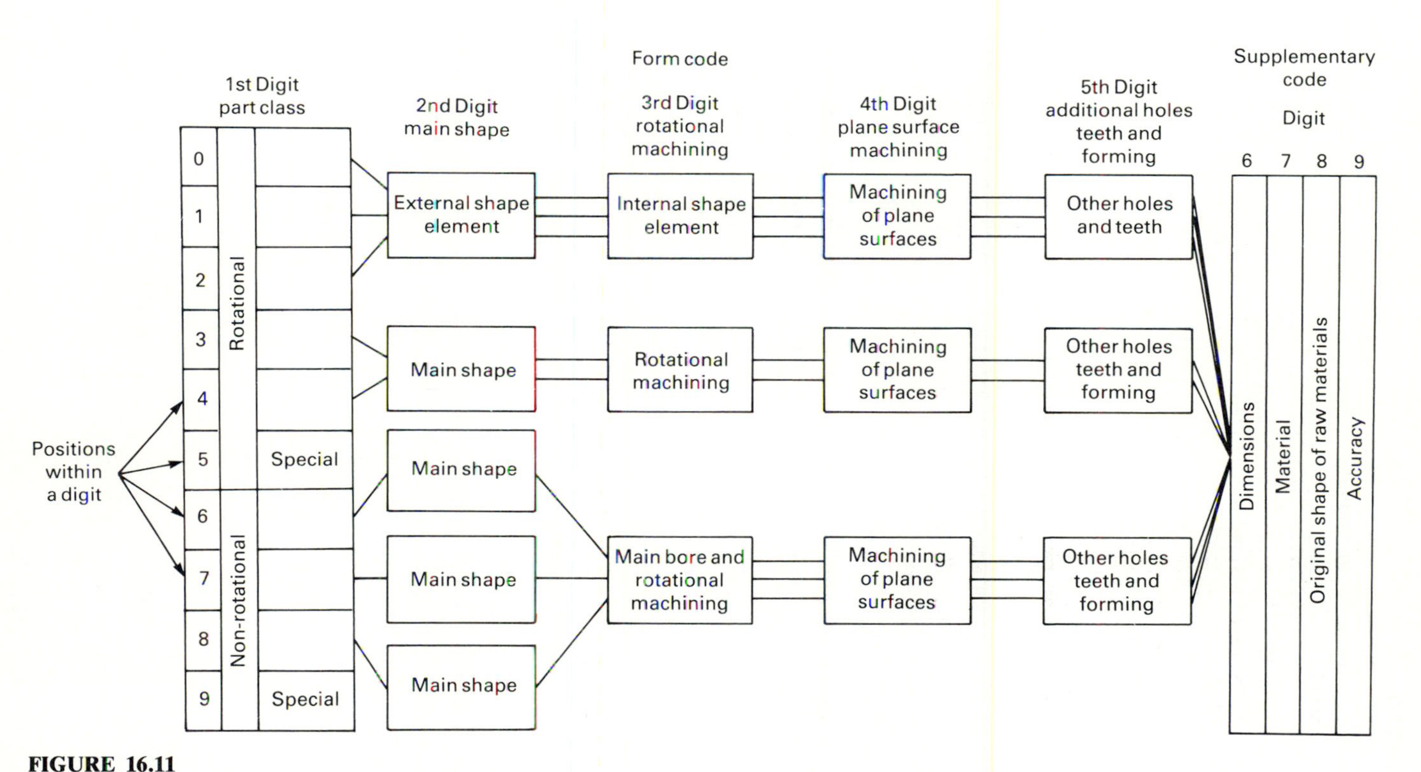

FIGURE 16.11
Opitz coding and classification system. (*Reprinted with permission from [37].*)

TABLE 16.1
Code structure selection

Major item class	Resolution required	Flexibility needed	Code system type[a]
Raw materials	Moderate	Low	Hybrid (H/C)
Commercial items	High	Low	H
Designed piece parts	Moderate	High	C
Assemblies models	Moderate	Moderate	Hybrid (H/C)
Machinery	Moderate	Moderate	Hybrid (H/C)
Technical information	Moderate	Low	H
Tools	Moderate	Low	H
Commercial			
Proprietary	Moderate	High	C
Gages/fixtures	Moderate	Low	H
Supplies	High	Low	H

[a] H, hierarchical; C, chain.

Source: Reference 33; courtesy of Numerical Control Society.

process begins by summarizing process plans that already exist for components in the family. Standard plans are then stored in a database and indexed by family matrices (Fig. 16.13). The preparatory stage is a labor-intensive process. Some reports indicate that it can take 18 to 24 person-years to complete preparation [34]. The efficiency of a variant process planning system depends

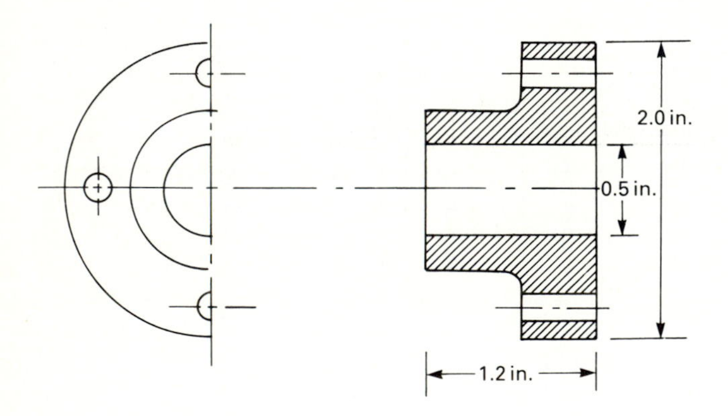

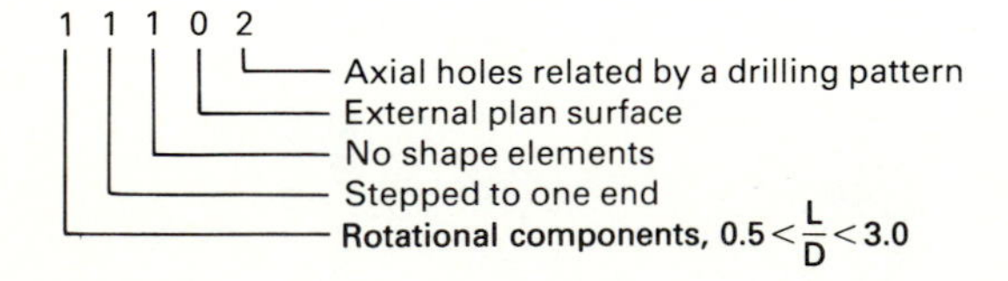

FIGURE 16.12
Rotational component. (*Reprinted with permission from [37].*)

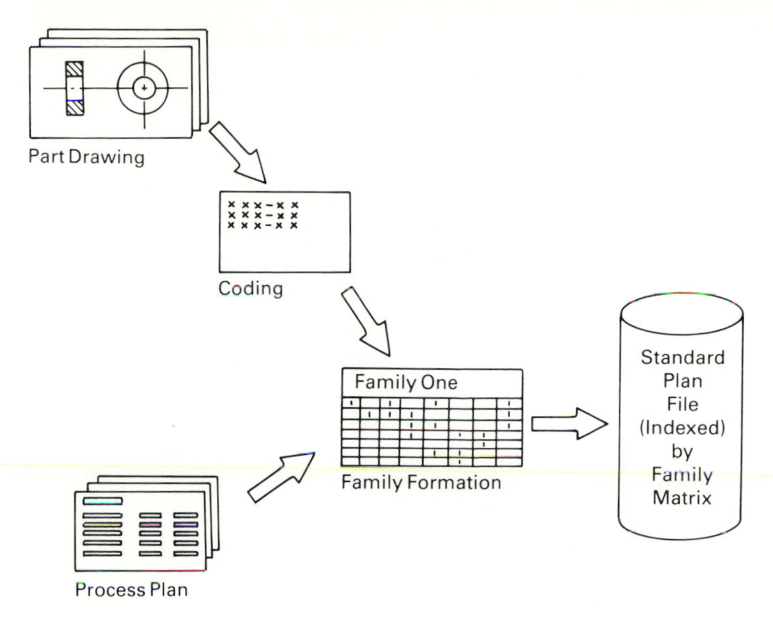

FIGURE 16.13
Preparatory stage.

directly on the standard plans stored for each family. Much human judgement is used here.

The Production Stage

The operation stage occurs when the system is ready for production. New components can now be planned. An incoming component is first coded. The code is then input to a part family search routine to find the family to which the component belongs. The family number is then used to retrieve a standard plan. The human planner may modify the standard plan to satisfy the component design. Figure 16.14 illustrates the flow of the production stage.

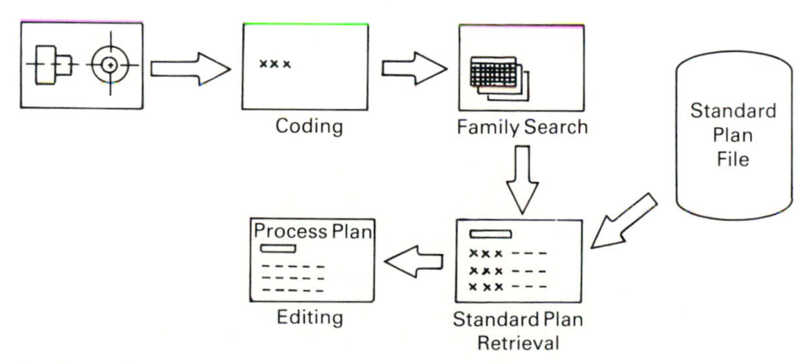

FIGURE 16.14
Production stage.

Some other functions, such as parameter selection and standard time calculations, can also be added to make the system more complete.

GENERATIVE PROCESS PLANNING

Generative process planning can be concisely defined as a system that synthesizes process information in order to create a process plan for a new component automatically. In a generative planning system, process plans are created from information available in a manufacturing database with little or no human intervention. Upon receiving the design model, the system can generate the required operations and operations sequence for the component. Knowledge of manufacturing must be captured and encoded into efficient software. By applying decision logic, a process planner's decision-making process can be imitated. Other planning functions, such as machine selection, tool selection, process optimization, and so on, can also be automated using generative planning techniques.

The generative planning approach has the following advantages.

1. It can generate consistent process plans rapidly.
2. New components can be planned as easily as existing components.
3. It can potentially be interfaced with an automated manufacturing facility to provide detailed and up-to-date control information.

Decisions on process selection, process sequencing, and so on, are all made by the system. However, transforming component data and decision rules into a computer-readable format is still a major obstacle to be overcome before generative planning systems become operational. Successful implementation of this approach requires the following key developments.

1. The logic of process planning must be identified and captured.
2. The part to be produced must be clearly and precisely defined in computer-compatible format (e.g., three-dimensional model, GT code, etc.).
3. The captured logic of process planning and the part description data must be incorporated into a unified manufacturing database.

Today, the term "generative process planning" is usually relaxed from the definition given above to a less complete system. Systems with built-in decision logic are often called generative process planning systems. The decision logic consists of the unusual ability to check some conditional requirements of the component and select a process. Some systems have decision logic to select several "canned" process-plan fragments and combine them into a single process plan. However, independent of the decision logic used and how extensively it is used, the system is usually categorized as a generative system.

Ideally, a generative process-planning system is a turnkey system with all the decision logic contained in the software. The system possesses all the necessary information for process planning; therefore, no preparatory stage is required. This is not always the case, however. In order to generate a more universal process-planning system, variables such as process limitations and capabilities, process costs, and so on, must be defined prior to the production stage. Systems such as CPPP [29] require user-supplied decision logic (process models) for each component family. A wide range of methods have been and can be used for generative process planning.

CAD Models

Since a design can be modeled effectively in a CAD system, using a CAD model as input to a process-planning system can eliminate the human effort of translating a design into a code or other descriptive form. The increased use of CAD in industry further points to the benefits of using CAD models as process-planning data input.

A CAD model contains all the detailed information about a design. It can provide information for all planning functions. However, an algorithm to identify a general machined surface from a CAD model does not currently exist. Additional research is needed to specify the machined surface shape from raw material shape. In the future, any codification required for part description will come directly from a CAD system. This, however, is impossible with today's pattern-recognition technology except for a very small set of components.

Decision Logic

In a generative process-planning system, the system decision logic is the focus of the software and directs the flow of program control. The decision logic determines how a process or processes are selected. The major function of the decision logic is to match the process capabilities with the design specification. Process capabilities can typically be described by "IF . . . THEN . . ." expressions. Such expressions can be translated into logical statements in a computer program. Perhaps the most efficient way to translate these expressions is to code process capability expressions directly into a computer language. Information in handbooks or process boundary tables can easily be translated using a high-level computer language. However, such programs can be very long and inefficient. Even more disadvantageous is the inflexibility (difficulty of modification) of such software—this inflexibility leaves customized codes of this type virtually useless in process planning.

Several methods can be used to describe the decision structure of process planning. Knowledge-representation methods are related directly to the decision logic in these systems. The static data are the representation and the dynamic use of the data becomes the decision logic. The following decision procedures are applied to process planning systems:

1. Decision trees
2. Decision tables
3. Artificial intelligence

Decision trees. A decision tree is a natural way to represent process information. Conditions (IF) are set on branches of the tree and predetermined actions can be found at the junction of each branch.

A decision tree can be implemented as either (1) computer code, or (2) presented as data. When a decision tree is implemented in computer code, the tree can be directly translated into a program flowchart. The root is the start node (Fig. 16.15), and each branch is a decision node. Each branch has a decision statement (a true condition, and a false condition). At each junction, an action block is included for the true condition. For a false condition, another branch might be taken or the process might be directed to the end of

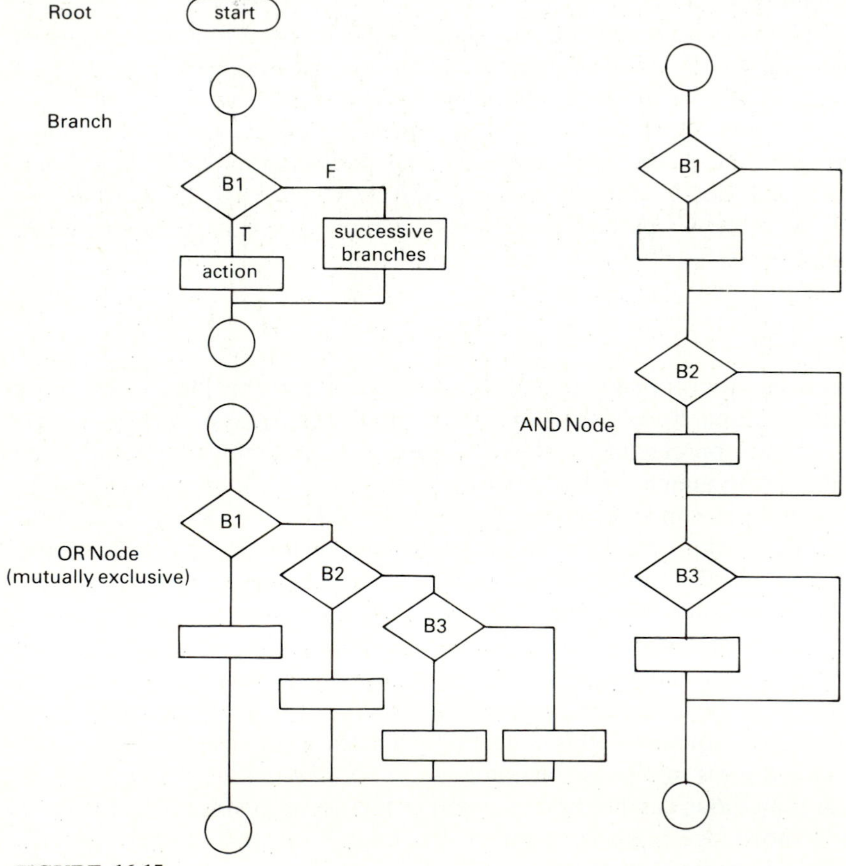

FIGURE 16.15
Structured flowchart corresponds to a decision tree.

the logic block. When the false conditions includes another branch, these two branches are said to branch from an OR node. When the false condition goes directly to the end of an action block (which is rooted from the same decision statement), the current branch and the following branch are part of the same AND node. A decision statement can be a predicate or a mathematical expression.

Decision tables. Decision tables have long been a popular method of presenting complex engineering data. Decision tables can also be easily implemented on a computer. Using decision tables for process planning,

		SELECTION						
		1	2	3	4	5	6	7
Conditions	Hole	×	×	×	×	×		
	Diameter = 0.0	×	×	×	×	×		
	Slot						×	
	Internal thread							×
	T.P. ≤ 0.002	×						
	0.002 < T.P. ≤ 0.01		×					
	0.01 < T.P.			×	×	×		
	Tol ≤ 0.002			×				
	0.002 < Tol ≤ 0.01				×			
	0.01 < Tol					×		
Conclusions	Rapid travel out	×						
	Finish bore		×	×				
	Semifinish bore				×			
	Drill					×		
	Mill						×	
	Tap							×
	T.P. = 0.01	×						
	T.P. = 0.02		×					
	Tol = 0.01			×				
	Tol = 0.02				×			
	Diameter = 0					×		

FIGURE 16.16
Decision table.

however, normally requires a special preprocessor program or computer language to implement the table and control the operation of the table.

The decision table is the most essential part of a decision table language program. It is represented in its original table format. For example, the decision tree in Fig. 16.15 can be represented by the decision table in Fig. 16.16.

Artificial Intelligence

Artificial intelligence (AI) has become one of the major topics of research and application in computer science. AI can be defined as the ability of a device to perform functions that are normally assocated with human intelligence. These functions include reasoning, planning, and problem solving. Applications for AI have been included in the areas of natural language processing, intelligent database retrieval, expert consulting systems, theorem proving, robotics, scheduling, and perception problems. Process-planning applications have been considered as part of an expert consulting system.

In an expert system, the knowledge of human experts is represented in an appropriate format. The most common approach is to represent knowledge by using rules. Rule-based deduction is frequently used to find an action. Since AI is too large a subject to discuss in this text, potential applications in process planning will be illustrated.

There are two types of knowledge involved in process-planning systems: component knowledge and process knowledge. The component knowledge defines the current state of the problem to be solved (it is also called *declarative knowledge*). On the other hand, the knowledge of processes defines how the component can be changed by processes (it is also called *procedural knowledge*). Applying the process knowledge to a component in a logical manner is called *control knowledge.*

There are several methods available for representing declarative knowledge. First-order predicate calculus (FOPC), frames, and semantic networks [35] are popular methods. FOPC is a formal language in which a wide variety of statements can be expressed. In predicate calculus, a statement is expressed by predicate symbols, variable symbols, function symbols, and constant symbols. For example:

$$\text{Depth}\,(\text{Hole}(X), 2.5)$$

represents the depth of hole(X) as 2.5 units. Such a representation is called an atomic formula. In the atomic formula, depth is a predicate symbol. Hole is a function symbol, X is a variable symbol, and 2.5 is a constant. A legitimate atomic formula is called the well-formed formula (wff); the above atomic formula is a wff. A wff can be either T (true) or F (false). For example, the depth of hole (1) is 2.0; therefore, depth (HOLE(1), 2.5) is false. When we use FOPC to describe a component, all wff must be true. The example shown in Fig. 16.15 can also be considered descriptive knowledge for the hole design.

Procedural knowledge can be represented by

IF (Condition) THEN (action)

statements that are similar to decision trees or decision tables. In AI, such statements can be called production rules. A system using production rules to describe a procedural knowledge is called a production system. In a production rule, a condition can be a conjunction of predicates (wff).

Process Planning and Selection Efficiency—General Program Structures

In order to more fully describe the basic mechanisms of the two types of process planning systems, an example of each system and the mechanism by which it works is useful. Although there are several *variant* process planning systems being used, the *CAM-I CAPP* system [12, 24] is probably the best-known system. The CAPP program, as well as most variant planning programs, is basically a data management and retrieval system. A menu or catalog of process plans and machining parameters (speed, feed, and depth of cut) are stored and accessed, on the basis of a "coded" part description. Process plans are stored and retrieved on the basis of part similarities. The flow representation of the CAPP system is given in Fig. 16.17. The CAPP system, as it is configured, is modular in structure and can handle a variety of coding systems. However, as also can be seen in Fig. 16.17, CAPP is a "pure retrieval system for user supplied, stored, standard information" [24].

In CAPP or any other variant process-planning system, stored planning information is simply retrieved. If a new machine tool were to be developed, a

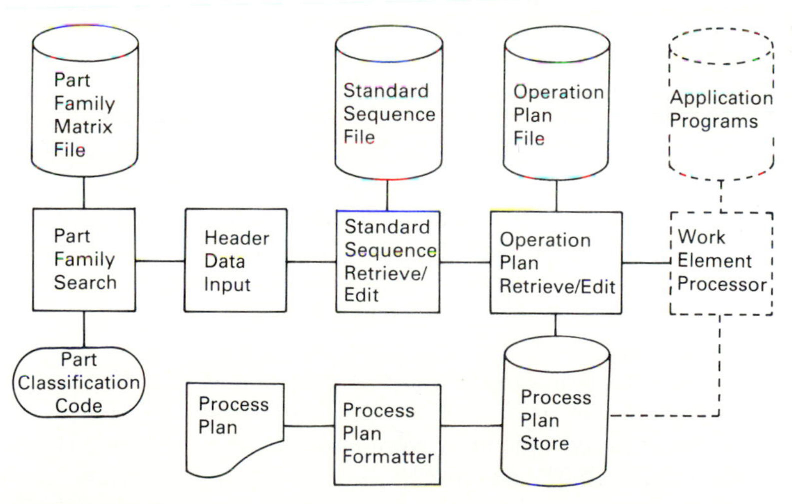

FIGURE 16.17
CAPP flow diagram.

set of parts and plans for that machine would have to be manually developed, and tested by a process planner. In a generative scheme, process planning is accomplished by simply describing the processes that the machine can perform (drill, ream, bore, mill, etc.), the accuracies of the process and the machine, and finally the machine capacities (horsepower, maximum feed and speed, X, Y, and Z axis travel, etc.). The planning of parts is then performed by the system logic. This logic consists primarily of five steps.

1. Examine the geometric form of a machined surface to determine which types of machines can produce it.
2. Examine the geometric tolerances to determine which processes can produce it.
3. From the tolerance, determine the maximum machining constraints (maximum feed, maximum speed, maximum depth of cut, etc.).
4. Develop the optimum machining conditions (feed, speed, and depth of cut) for minimum machining cost or time for all feasible tool materials.
5. Select the most appropriate process and tooling to produce a specific machined surface.

As can be seen from the two basic planning mechanisms, the efficiencies of the two types of automated process-planning schemes are quite different. Variant process planning should produce good results for a consistent manufacturing system (one of which few technological developments occur). However, if machine tools, tool materials, and new process developments continue as they have, all of this planning information will have to be developed by a process planner and catalogued on the retrieval system; otherwise, the company will be planning by yesterday's standards. Variant process planning does not necessarily imply that the optimal process plans will be produced. Unfortunately, no operational purely generative planning system currently exists. Some machine-specific systems have been developed.

An Illustrative Example

It will be helpful to illustrate some of the major points that have been made in this chapter by planning a specific part. Figures 16.18 and 16.19 contain the drawings of the part to be planned, Fig. 16.18 contains the drawing of the forging used as the raw material for a bevel gear. Figure 16.19 contains the drawing of the finished gear.

It was mentioned earlier that many different methods exist to produce a part. To prove this point as well as to illustrate the differences in efficiencies of these plans, two different process planners were asked to plan the part. These plans appear in Figs. 16.20, 16.21, and 16.22.

The quantity and lot size of the part being planned play a major role in process planning. This point was not mentioned earlier. An example of this

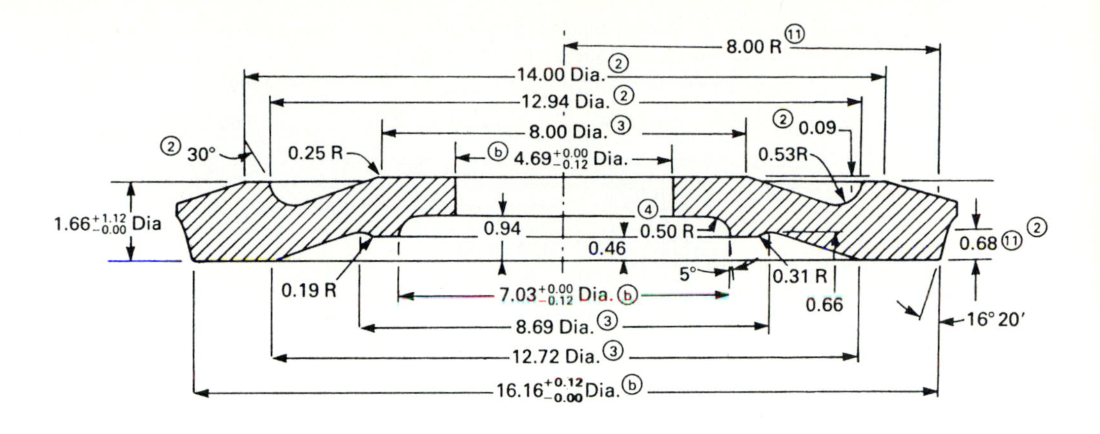

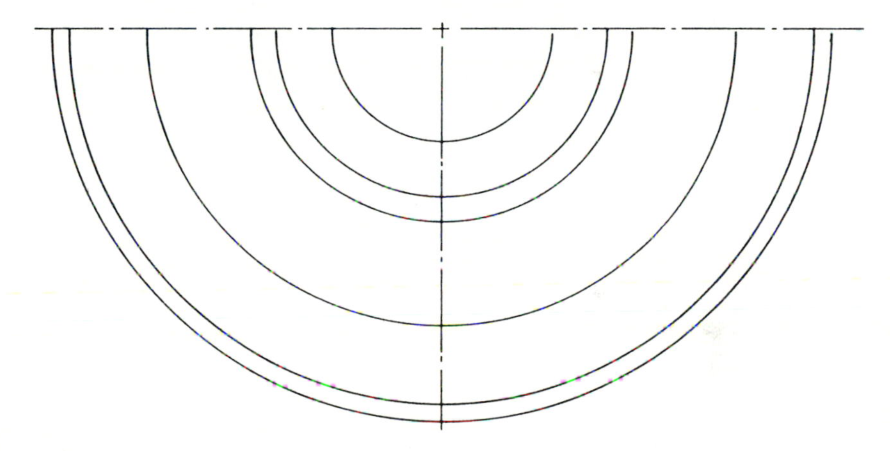

FIGURE 16.18
Forging for bevel gear.

aspect of process planning is that if only a few parts in total were going to be produced, a planner would typically avoid the use of specialized tooling, fixtures, and automated equipment because the initial investment required for the tooling and fixturing and the initial setup costs for automated equipment cannot be absorbed by the few parts that will be produced.

To facilitate the planning of our bevel gear, as well as to attempt to create as general a set of plans as possible, the planners were told that large production volumes of parts would be planned in a flow-shop configuration. This implies that automated equipment if so selected could be justified, and that if special tooling were appropriate, this tooling could also be justified.

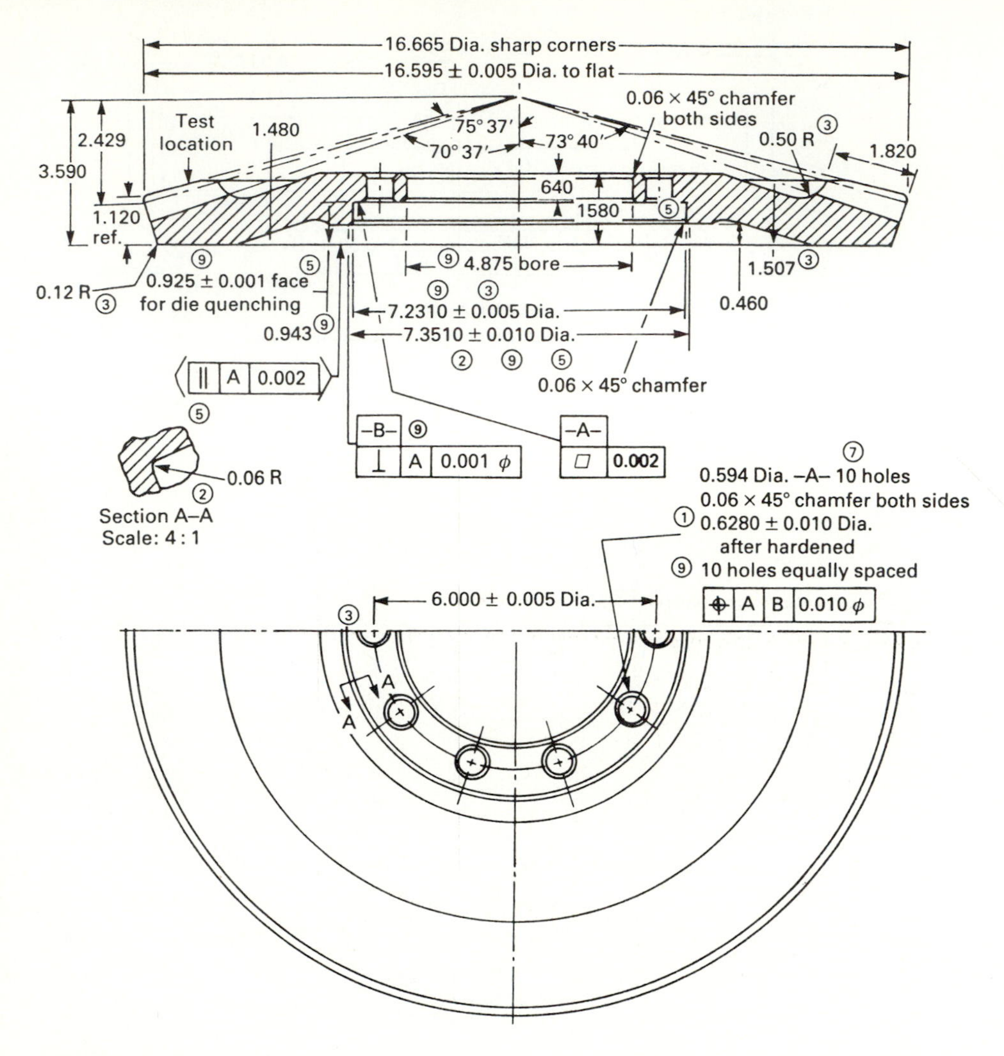

FIGURE 16.19
Bevel gear machine dimensions.

Another point previously unmentioned is that the detailed output from a process planner may vary. For instance, a planner for NC equipment is typically charged with producing an NC tape or tape image. However, a process planner in a job shop might simply route a part to a specific machine and note the operations required. The machinist, in turn, is given the responsibility for the detailed planning of the part. Figures 16.20 and 16.21 are plans from the same planner; however, Fig. 16.21 contains the detailed machining information, while Fig. 16.20 is more aggregately stated.

In looking at the production routing summaries in Figs. 16.20 and 16.21, one can note several discrepancies in the process plan. It was stated earlier that

Operation	Machine used	Speed (ft/min)	Feed (in./rev)
1. Remove slag and rough edges	Hand grinder		
2. Machine inside face of blank	24″ engine lathe (C-7 insert)	300	0.010
3. Machine back face bore and chamfer 7.230 dim. bore and chamfer 4.875 dim. holding 1.580 and 0.640 thickness	Engine lathe (C-7 insert)	300	0.005
4. Drill holes 19/32	Drill press and drill jig (carbide tipped drill)	250	0.005
5. Chamfer holes	Drill press 45° countersink (HSS tool)	85	Hand fed
6. Machine gear tooth Face and outside diam.	Engine lathe (C-7 carbide)	300	0.010
7. Machine gear teeth	Gear generator		(rate—6 pcs/h)
8. Wash and heat treat	Heat treat		

FIGURE 16.20
Production routing summary #1.

Operation	Machine	Speed	Feed	Depth of cut
1. Face back of gear, bore and c'bore Straddle face flange, Form radii & chamfer	Warner and Swasey Auto Lathe (C-7 tooling)	400 fpm	0.015 in./rev	0.2 in.
2. Rough and finish turn—face angle Form and finish turn back angle, Form OD flat, tooth Clearance and radius	Gisholt Automatic Lathe (C-7 tooling)	400 fpm	0.010 in./rev	0.17–0.18 in.
3. Cut gear teeth	Gleason Gear Generator		(Rate—6 pcs/h)	
4. Remove breakout burrs from bevel gear teeth	Deburring machine w/high-speed grinders		(Rate—30/h)	
5. Carburize	Heat treat			
6. Drill (10) flange holes	Multiple drill (HSS drills)	45 fpm	0.005 in./rev	
7. Chamfer flange hole	Chamfer machine		(Rate—13/h)	
8. Wash and harden	Heat treat			

FIGURE 16.21
Production routing summary #2.

Operations sheet for bevel gear

1. Machine—hand grinder
   A. Remove slag and rough edges to provide smooth clamping surface.
2. Machine—29" engine lathe
   A. Using 4-jaw chuck clamp gear blank with face No. 9 against chuck.
   B. Indicate gear bland to ±0.010
   C. Machine 1/64 material from face of gear using spindle speed of 300 rpm 0.010 feed.
   D. Remove gear blank from the chuck, reverse the gear blank and chuck the blank with the machined surface against the chuck.
   E. Indicate the blank.
   F. Face surface No. 9—Bore and chamfer 7.230 dim. and chamfer 4.875 dimension holding 1.580 and 0.640 thickness dimension.
3. Machine—drill press
   A. Place blank in drill fixture locating on 7.2300 dim.
   B. Using 19/32 drill proceed to drill 10 holes on 6.000 dia. bolt circle.
   C. Remove blank from fixture and chamfer both sides of the holes, using a 45° countersink.
4. Machine—engine lathe
   A. Mount fixture in 4-jaw chuck and indicate.
   B. Mount gear blank on fixture locating on 7.2300 dimension.
   C. Machine gear tooth face.
   D. Machine gear outside dimension.
5. Machine—gear generator
   A. Place gear fixture in workhead and indicate.
   B. Place gear on fixture and advance to cutting position. Gear blank is fed into.
   C. Two reciprocating tools ground as the shape of the tooth, the workpiece is then fed into the reciprocating tool to full depth of tooth.
   D. Workpiece indexed to next position to begin cut on next tooth.
   E. Repeat cycle until all teeth are cut.
   F. Remove gear from machine.

FIGURE 16.22
Production routing summary #3.

man-variant process planning is quite prone to variation and that the efficiency of manufacture is a function of this variation. Take, for instance, the lathe operation performed on the part. The speed and feed stated for the operations are different. The cutting velocity (v) for plan #1 is specified as 300 ft/min, while plan number 2 specified 400 ft/min. The feed rate (f) similarly varies from 0.010 in./rev to 0.015 in./rev and from 0.005 in./rev to 0.010 in./rev.

Recall that total machining time (T_m) can be calculated by

$$T_m = \frac{l\pi d}{12Vf}$$

where l is the length of cut (in.) and d is the diameter (in.)

Knowing that the length and diameters of part do not vary from planner to planner, the percent variation for the two plans can be calculated.

$$T_{m1} = \frac{C\ (\text{in.}^2/\text{rev})}{(12\ \text{in./ft})(300\ \text{ft/min})(0.010\ \text{in./rev})} = \frac{C}{36}\quad \text{min}$$

$$T_{m2} = \frac{C\ (\text{in.}^2/\text{rev})}{(12\ \text{in./ft})(400\ \text{ft/min})(0.015\ \text{in./rev})} = \frac{C}{72}\quad \text{min}$$

Therefore, the machine time required in plan #2 will be 50 percent of that of plan #1. Similar differences can be noted throughout the planning summary. One of particular interest is the drilling operation. In one case, a multiple drill is used where all holes are drilled simultaneously, while in the other case a standard drill press is used. One might again calculate the percentage differences in the two processes to determine their efficiency.

An important note here is that during the next few decades, computer-aided process planning will most likely simply automate human planning activities. If the "standard plans" for a variant planning system or the "knowledge base" for a generative system are inefficient then *all* process plans generated by the system will be inefficient.

SUMMARY

In this chapter, process planning was defined as "converting design information into detailed work instruction." In turn, different types of process planning were also defined. The most frequently used type of process planning was identified as "man-variant planning" and some of the shortcomings of this type of planning were identified.

Two frameworks for automated process planning (computer variant and generative) were described. The advantages and disadvantages of these two methods were briefly described. Finally, two examples of man-variant planning for the same part were given. In these examples, different planning efficiencies were noted thereby reinforcing the need for standardizing the process planning activity.

QUESTIONS

16.1. Automated process planning in order to be implemented will have to prove cost-effective. List the economic benefits, item by item, that can be derived from automated planning.

16.2. Variant process planning is billed as nothing more than a pure retrieval system. Create a flow graph of how a variant process-planning system could be configured.

16.3. What are some of the advantages of a generative as opposed to a variant planning system?

16.4. Many experts on automated process planning believe that the use of process-planning programs will evolve from the predominantly variant planning systems used today through a combination of variant and generative phases to completely generative planning systems. Describe how these two types of planning might be combined.

16.5. Calculate the times required for the hole drilling operations in both plan #1 (Fig. 16.20) and plan #2 (Fig. 16.21).

PROBLEMS

16.1. Create the process plans required to produce the part shown in Fig. P16.1.

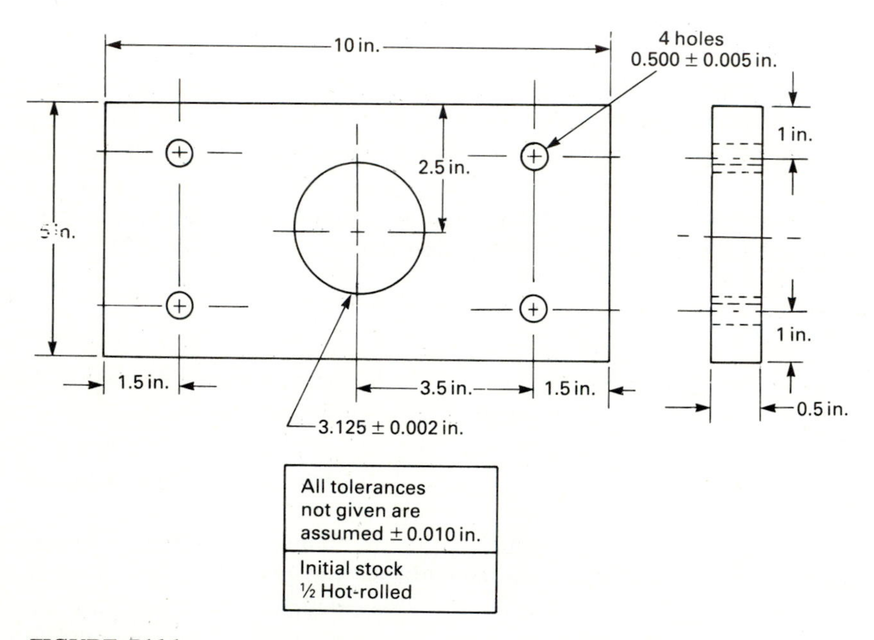

FIGURE P16.1

16.2. Calculate the processing times required to produce the holes given in the part shown in Fig. 16.19. Calculate the times for both sets of plans.

16.3. Create the decision tree logic for a process planning system to make cabinets. A rip saw, a cut-off saw, a planer, a sander, and a drill press are the processes that should be considered.

16.4. Choose a reasonably simple part and make a sketch of the part along with the process plans required to produce the part.

REFERENCES

1. Leslie, W. H. P.: *Numerical Control User's Handbook,* McGraw-Hill, New York, 1970.
2. Wandmacher, Richard R.: *CAM-I and the Process Planning Project,* Second NSF/RANN grantees' Conference, January, 1975.

3. Mann, Wilbur S.: "Commline Interview," *N/C COMMLINE,* July, 1976.
4. Niebel, B. W.: "Mechanized Process Selection for Planning New Designs," *ASME Paper No. 737,* 1965.
5. Sheck, Donald E.: "Feasibility of Automated Process Planning," Ph.D. Thesis, Purdue University, W. Lafayette, Ind., 1966.
6. Berra, P. B., and Barash, M. M.: "Investigation of Automated Planning and Optimization of Metal Working Processes," *Report No. 14,* Purdue Laboratory for Applied Industrial Control, July, 1968.
7. Warner and Swasey: *CUTS,* A brochure on Computer Utilized Turning System, 1973.
8. Cincinnati Milicron Company, *CINTURN II,* A brochure on Macro-Programming Systems for NC Turning Centers, Publication O.L-550, 1975.
9. "The *MITURN* Programming System for Lathes," Metal/Institute TNO, December, 1974, The Netherlands.
10. Giddings and Lewis: "Programming Instruction Manual," Vertical Turret Lathe with Numeri Path 80L, 1974.
11. Bennett, Keith W.: "Smoothly Blending Design and Production," *Iron Age,* July, 1976.
12. Link, C. H.: "CAM-I Process Planning Program, 1976," *CAM-I Special Projects, 1976, PR-75-ASPP-01.*
13. Doyle, Lawrence E.: *Tool Engineering,* Prentice-Hall, Englewood Cliffs, N.J. 1950.
14. Optiz, Herwart: "Planning Flexible Manufacturing Systems," Unpublished Research Paper, 1974.
15. Spur, G.: "Automation of Manufacturing Planning," A paper presented at CIRP conference in Chicago, 1974.
16. Rose, D. W.: "Coding For Machining," *Masters Thesis,* Purdue University, W. Lafayette, Ind., 1977.
17. *Dimensioning and Tolerancing for Engineering Drawing,* USAS, ANSI Y14.5, 1973.
18. Fortini, Earlwood, T.: *Dimensioning for Inter-changeable Manufacturing,* Industrial Press, New York, 1967.
19. Wysk, Richard A.: "An Automated Process Planning and Selection Program: APPAS," Ph.D. Thesis, Purdue University, W. Lafayette, Ind., 1977.
20. Wick, Charles H.: "Gundrilling Precise Shallow Holes," *Manufacturing Engineering,* September, 1975.
21. Trucks, H. E.: *Designing for Economic Production,* SME, 1974.
22. Eary, Donald F., and Johnson, Gerald E.: *Process Planning for Manufacturing,* Prentice-Hall, Englewood Cliffs, N.J., 1962.
23. Scarr, A. J. T.: *Metrology and Precision Engineering,* McGraw-Hill, London, 1967.
24. Link, C. H., "CAPP-CAM-I Automated Process Planning System," Proceeding of the 1976 N/C Conference.
25. McKinney, J. L., and Rosenbloom, R. S.: *Cases in Operations Management,* pp. 66–68, Wiley, New York, 1969.
26. Phillips, R. H., and ElGomayel, J.: *Group Technology Applied to Product Design,* MAPEC Module, 1977.
27. Carrie, A.: *Group Technology: Part Family Formation and Machine Grouping Techniques,* MAPEC Module, 1978.
28. Dunn, M. S., and Mann, W. S.: "Computerized Production Process Planning," Proc. 15th Numerical Control Society Annual Meeting and Technical Conference, Chicago, April, 1978.
29. Kotler, R. A.: "Computerized Process Planning—Part 2," *Army Man. Tech. J.,* vol. 4, no. 5, pp. 20–29, 1980.
30. Tulkoff, J.: "Lockheed's GENPLAN," Proc. 18th Numerical Control Society Annual Meeting and Technical Conference, Dallas, Tex., May, 1981, pp. 417–421.
31. Vogel, S. A.: "Integrated Process Planning at General Electric's Aircraft Engine Group," Proc. AUTOFACT WEST, Society of Manufacturing Engineers, Anaheim, Calif., November, 1980, pp. 729–742.

32. Solaja, V. B., and Urosevic, S. M.: "The Method of Hypothetical Group Technology Production Lines," *CIRP Annals, vol.* 22, *no.* 1, 1973.
32. Krag, W. B.: "Toward Generative Manufacturing Technology," Proc. 15th Numerical Control Society Annual Meeting and Technical Conference, Chicago, April 9–12, 1978.
34. Planning Institute, Inc., "PII Press release—4-15-80," PII, Arlington, Tex., 1980.
35. Nau, D. S.: "Expert Computer Systems, and Their Applicability to Automated Manufacturing," Technical Report, Industrial Systems Division, National Bureau of Standards, Washington, D.C., 1981.
36. Chang, T. C., and Wysk, R. A.,: "An Introduction to Automated Process Planning Systems," Prentice Hall, Englewood Cliffs, N.J., 1984.
37. Opitz, H.: *A Classification System to Describe Workpieces,* Pergamon Press,

CHAPTER

17

COMPUTER-INTEGRATED MANUFACTURING

INTRODUCTION

Throughout the history of our industrial society, many inventions have been patented and whole new technologies have evolved. Whitney's concept of interchangeable parts, Watt's steam engine, and Ford's assembly line are but a few developments that are most noteworthy during our industrial period. Each of these developments has impacted manufacturing as we know it, and has earned these individuals deserved recognition in our history books. Perhaps the single development that has impacted manufacturing more quickly and significantly than any previous technology is the ditigal computer.

Today, only 35 years after its commercial development, the computer is used for a variety of manufacturing control functions. Manufacturing facilites without a computer are the exception rather than the rule today. Facilities employing as few as a hundred people may have half that number of computers or computer-based controllers performing a variety of activities. Today's modern factory environment has become an example of computer controlled manufacturing. Facilities currently exist in which parts are created on a computer (on a computer-aided design (CAD) system). Production plans are created from the CAD data using an automated process planning system. Numerically controlled part programs are created using the tool path requirements on the CAD system, and the parts are manufactured under the control of a computer.

In this chapter, the basics of CAD and CAM will be discussed. The chapter is intended to provide an introduction to CAD and CAM and create a vocabulary for future chapters. A limited discussion of CAD is provided in this book, but a reasonably detailed presentation of CAM is provided in Chapters 18 through 21.

COMPUTER-AIDED DESIGN

Computer-aided design, or CAD as it is more commonly known, has grown from a narrow activity and concept to a methodology of design activities that include a computer or group of computers used to assist in the analysis, development, and drawing of product components. The original CAD systems developed and used in industry could more realistically be classified as computer-aided drafting systems. However, the benefits of using basic geometric information for structural analysis and planning for manufacturing were quickly recognized and included in many CAD systems. Today, as in the past, the basis for CAD is still the drafting features or interactive computer graphics (ICG) that these systems were originally designed to perform. However, the scope of these systems has taken on a new meaning. A typical CAD system is illustrated in Fig. 17.1. As can be seen, the graphics display portion of the system is the most noticeable feature. Computational processes have, however, been added to the system for increased capabilities.

In general, there are four basic reasons for implementing CAD systems.

1. *A reduction in design time.* The total time required from inception of an idea to its complete specification can be reduced by an order of magnitude by using easily alterable geometric models. Design perturbations/changes can be completed in minimal time. Whole scenarios of design possibilities can be constructed quickly.
2. *Improved product design.* Because CAD systems allow the designer to alter the product without major redrawing with considerable time commitment, many final designs can be constructed in a reasonable period of time. Similarly, these designs can be automatically analyzed for structural characteristics by using computer-aided engineering (CAE) software such as finite-element modeling (FEM).
3. *Improved information access.* Because CAD drawings are stored in a large computer database, they can be accessed quickly and easily. Parts can be coded on the basis of geometric shape, and similar parts can be called up to assist in the design and specification of new parts. "Standard parts" can be employed whenever possible, rather than having to re-invent the wheel over and over again.
4. *Manufacturing data creation.* With the advent of numerical control (NC) came the need to automatically generate the tool path required for

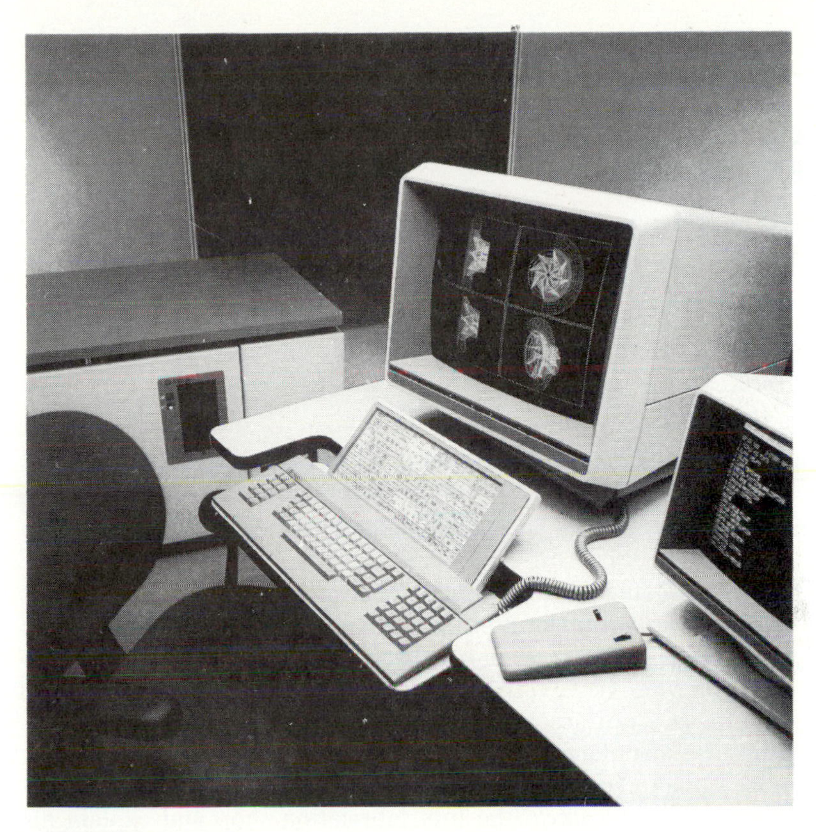

FIGURE 17.1
Some of the important components in a computer-aided design system. (*Courtesy of Computervision Corp.*)

machining. Since the part geometry dictates the machining required, knowing the part shape can allow for (semi-)automatic part-program preparation. CAD data can also be used for automated process planning.

It is interesting to note that twenty years ago if a part of reasonable geometric and manufacturing sophistication was created, hundreds of design and drafting hours would be required. After the part was specified, manufacture would begin. This planning would normally require some minor design changes (back to the designer and draftsman), and might take as long as the original design process. Special tooling, fixturing, etc., might also be specified during the planning for manufacture. In all, the entire process of product and process design could take several weeks or months. With today's CAD systems, designing (again a reasonably sophisticated component) and generating manufacturing plans, preparing part programs and producing the part is possible in days rather than weeks. In general, the total engineering and manufacturing time has been cut markedly using integrated CAD/CAM methodologies.

USING CAD FOR MANUFACTURED COMPONENT DESIGN

Although CAD represents the integrated use of a computer in the design process, the use of CAD does not change the design process requirements. As in the past, the designer is still responsible for

1. Developing the geometric detail required for the product (shape, dimension, tolerance, etc.)
2. Performing the necessary analysis on the design
3. Reviewing and altering the design on the basis of functional and economic characteristics
4. Communicating the design via an understandable drafting system

Independent of whether a CAD system was employed in the design or whether the designer used a standard drafting table, each of these design steps must be addressed either formally or informally. An example of these activities would be an engineer who needs a special wrench to adjust a positioning nut that is located in a place not accessible using a standard straight-shank tool. Several bends must be manufactured into a standard open-end wrench.

In order to obtain the wrench in the most expeditious manner, the engineer would most likely walk to the tool fabrication shop and explain his needs to one of the machinists. If the number of bends required in the wrench are few and simple, the engineer may simply explain his needs verbally to the machinist. Using this description, the machinist may be able to understand the geometric requirements and select the material based on the physical requirements (maximum torque conditions, length of the shaft, etc.). Without a drawing ever being made, the part may be designed and conveyed in the heads of the engineer and machinist and never take form on paper. As the part is being bent, the machinist may alter his initial conception of the part to ease the manufacturing requirements.

As the complexity of the part increases, the likelihood of the process requiring a sketch or drawing also increases. If a very complex tool was necessary, then a detailed engineering drawing with specific tolerances would be required.

DESIGN USING CAD SYSTEMS

The basic tasks that are performed automatically or semiautomatically using CAD systems include

1. Geometric modeling
2. Engineering analysis

3. Data storage and retrieval
4. Automated drafting

The functions performed by the CAD system are those functions that are normally employed in the design of any product, independently of whether a CAD system is used. Again, these design functions might be employed formally or might be employed as part of an informal design and manufacturing system.

GEOMETRIC MODELING

The geometric model that a designer creates represents the basic geometry of the object being modeled. When drawn by hand, this model can be represented as a traditional multiview drawing, as shown in Fig. 17.2. This drawing usually consists of two or three views (top, front, and right side). The same drawing, however, can be represented as a three-dimensional isometric representation: see Fig. 17.3.

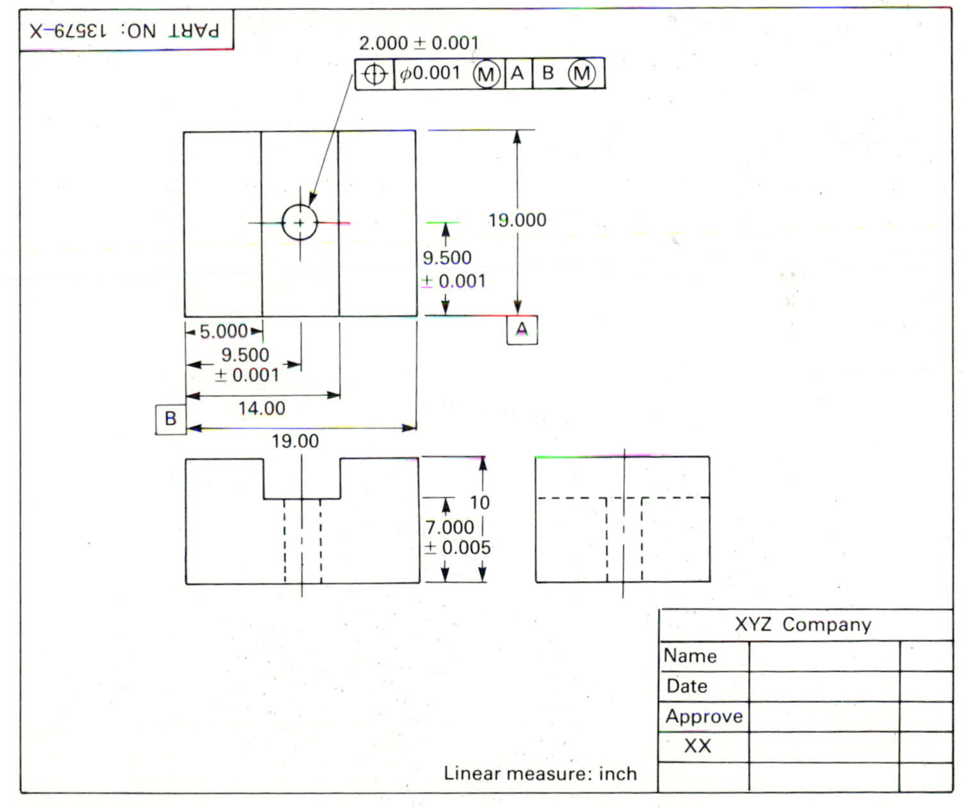

FIGURE 17.2
Multiview drawing of a bracket.

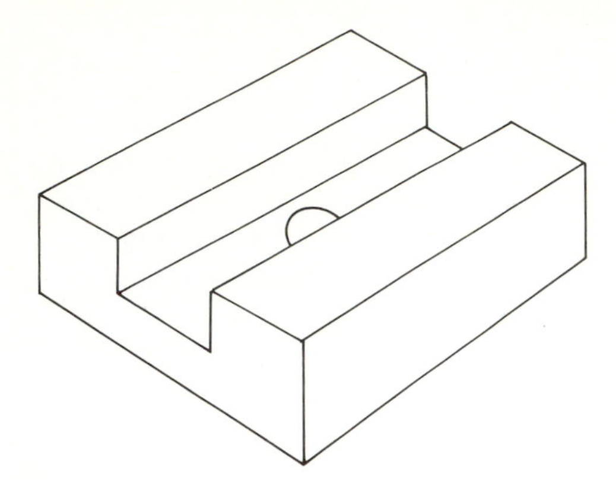

FIGURE 17.3
Isometric view of bracket.

In a CAD system, the geometry of the object is not stored in the same manner as it is represented on paper. Instead, the object is represented using computer-compatible mathematical symbology. The mathematical description allows the object not only to be stored in the computer efficiently (the part could also have been digitized as a series of zeros or ones representing locations on the paper), but also allows the user to display an image of the object on his graphics display terminal in different ways. The traditional multiview representation as well as three-dimensional perspective represenation can normally be constructed with little difficulty on a CAD system. In CAD, the computer's internal representation of the system is usually referred to as the geometric model.

Designers normally create their geometric model at a terminal using three types of constructs. The first construct is used to create the basic geometric elements such as points, lines, circles, etc. The second type of construct is used to scale, rotate, or transform the basic constructs in some way. The third type of construct allows the designer to combine two or more shape elements into one item. During the process of creating the part, the designer normally interacts with the CAD system using a menu keypad, and a mouse or light pen. The items called out onto the screen are converted to the internal geometric model within the CAD system.

The object that the designer creates can be represented internally to the computer in several methods. The most basic of these methods is called a wire-frame model. Wire-frame models are normally classified as one of three types: 2D (two-dimensional models), $2\frac{1}{2}$D (two-and-a-half-dimensional models, and 3D (three-dimensional models). Figure 17.4 shows examples of a wireframe boundary and solid modelling. Note that in a 2D isometric projection, hidden lines are not removed. In a $2\frac{1}{2}$D model, the hidden lines can be automatically "clipped" from the drawing. 3D models normally employ a different internal geometric model from 2D systems.

In a 2D drawing system, a set of drawing primitives are normally used to create a drawing. These drawing primitives normally consist of entities like

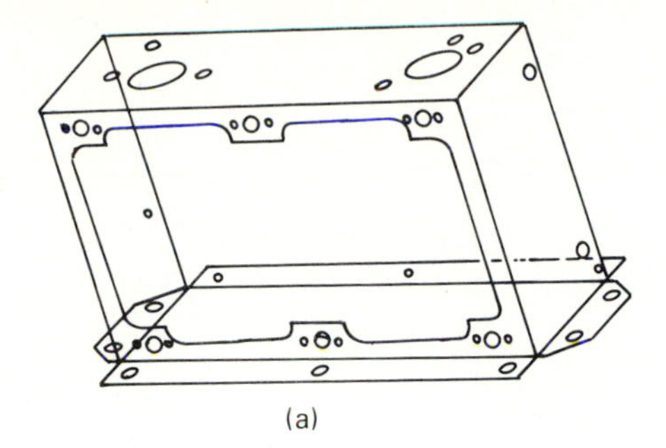

(a)

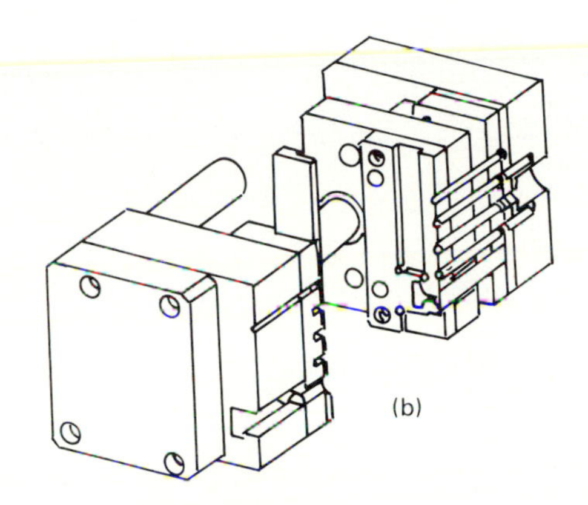

(b)

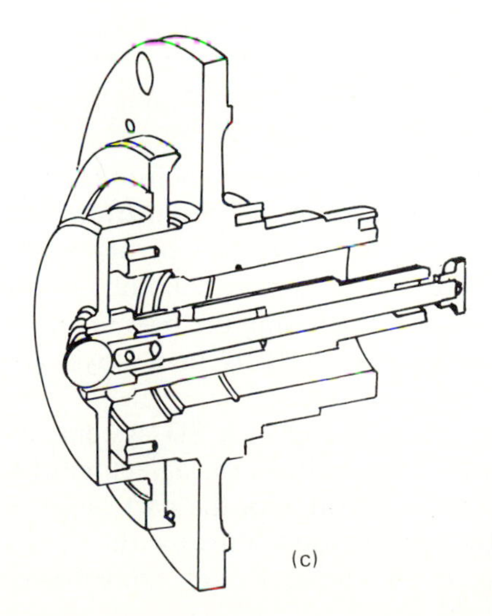

(c)

FIGURE 17.4
Wireframe boundary and solid drawing of a part.

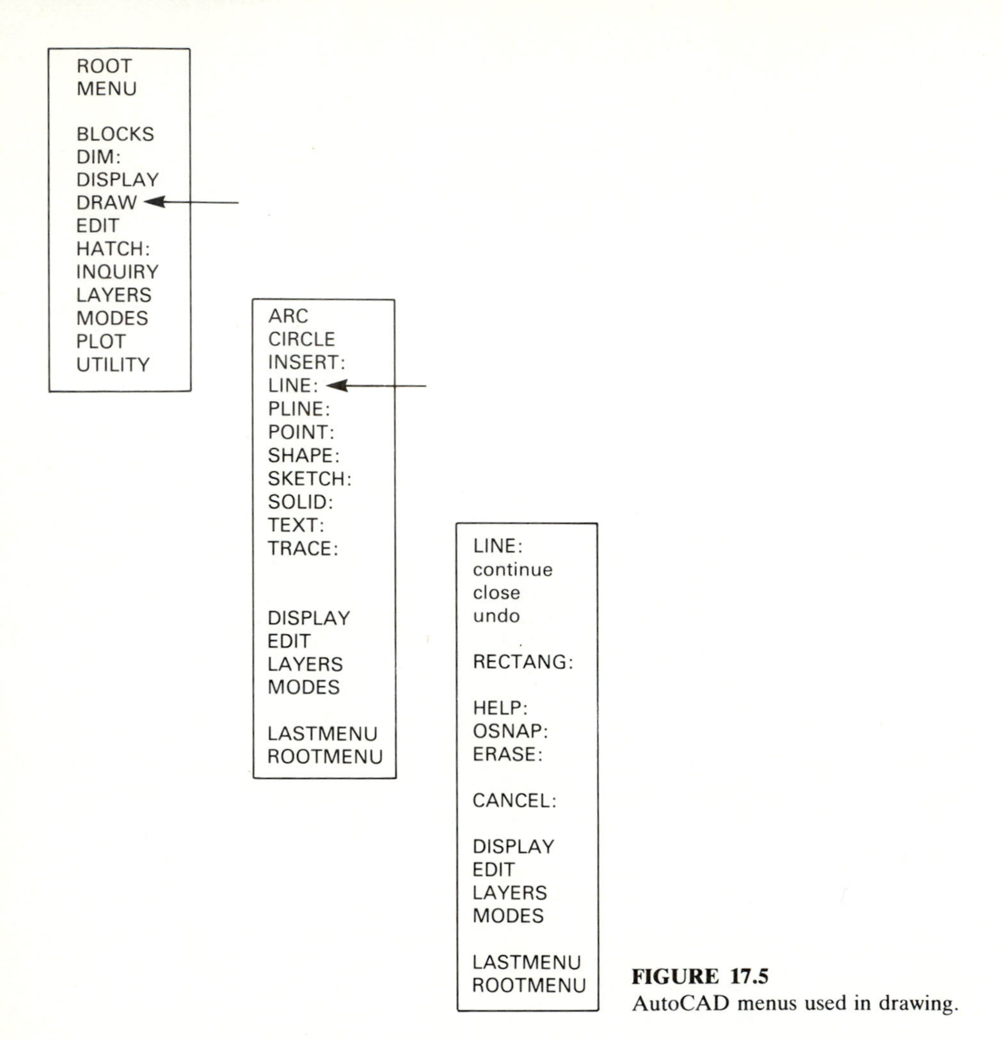

FIGURE 17.5
AutoCAD menus used in drawing.

line, circle, arc, point, etc. These one- and two-dimensional features are combined to form complex 2D drawings. Figure 17.5 illustrates the drawing menus used for AutoCAD. Since AutoCAD (version 2.6x or lower) is a $2\frac{1}{2}$D drawing system, features like solid are used to denote solid objects that can be "extruded" to take on a third dimension. Extruding the part in a CAD system is limited to linear elongation of the part, just as in a physical extrusion process.

The same drawing constructs can be used in a 3D system to construct polygonal faces on a part. The inside or outside (surface vector) of the part is then defined by the user, so that the solid portion of the part can be identified, i.e., a hole and cylinder are identical geometric features with opposite surface vectors. This type of drawing is called a boundary model. Drawings can also be constructed using building blocks much like toy blocks that can be scaled. Solid components such as: box, sphere, cylinder, cone, etc., are used to assemble a complex part. This type of 3D system is called constructive solid geometry (CSG).

ENGINEERING ANALYSIS

Virtually all products and parts produced today are specified for a particular application, and must meet some minimum functional specification. The product must undergo some type of engineering analysis, ranging from simple stress calculations to dynamic system behavior calculations. In order to perform the majority of these analysis calculations, the part geometry is required. Furthermore, the calculations are usually cumbersome enough to necessitate the use of a computer to perform the computations. Since the part geometry is available along with a computer on the CAD system, it makes engineering analysis as part of a CAD system an ideal marriage. Many CAD systems either have their own software or can be interfaced to engineering analysis software. The most common analysis employed on these systems includes analysis of mass properties and finite-element modeling. Mass calculations are important to determine the surface area, mass, weight, and center of gravity of various parts. When performed by hand, these calculations can take several hours, compared to a few seconds on a CAD system. Finite-element analysis modeling is normally employed to decompose a part into a finite set of locations at which the structural and dynamic characteristics of the part can be determined. Figure 17.6 contains a view of a typical component decomposed into finite components.

DATA STORAGE AND RETRIEVAL

Any drawing, created either on a manual drafting system or on a CAD system, must be stored so that it can be retrieved when needed. Conventional drawing rooms for reasonably sized industries may have contained 2 to 10 people to simply file, retrieve, and copy drawings. In addition, thousands of square feet of floor space were required to physically store the drawings. The retrieval time needed to obtain a drawing to review is not insignificant. If a designer felt compelled to review several similar drawings to compare his concept to existing components, hours could be consumed simply trying to locate existing drawings. Retrieving these drawings could then take several additional hours.

Today's CAD systems employ sophisticated database-management systems that allow the designer and planner access not only to the part in question, but also to a variety of similar components that can be identified by type or part code. Drawings can be reviewed at a CAD drafting station without ever having to leave the room. Since many alternative components can be reviewed, the final part drawing on a CAD system is normally more fully developed than if the part were created on a manual drawing system.

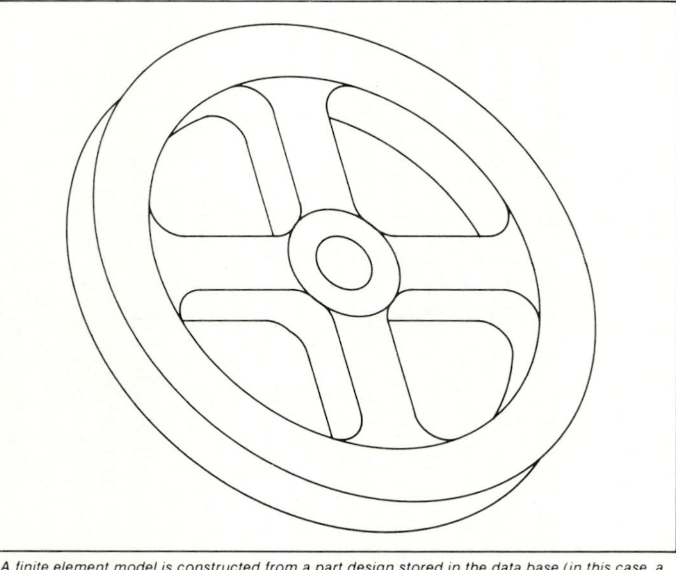

A finite element model is constructed from a part design stored in the data base (in this case, a wheel).

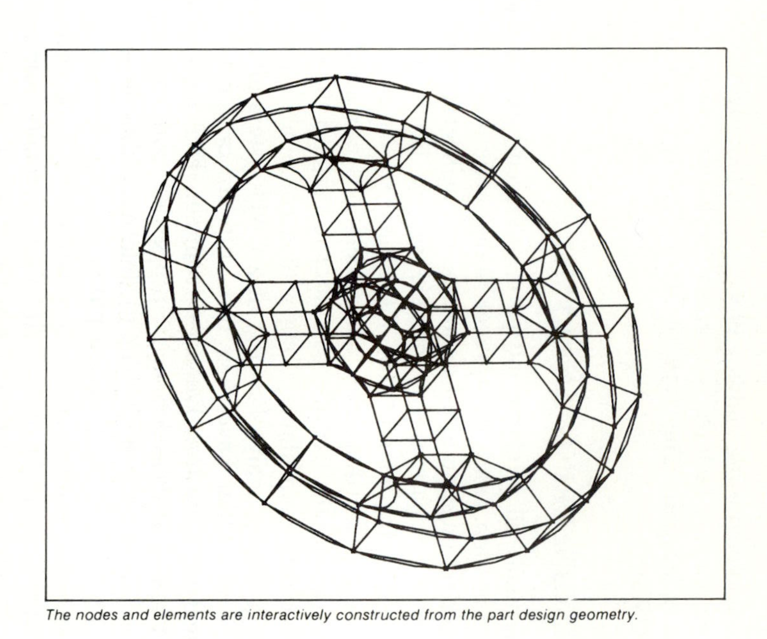

The nodes and elements are interactively constructed from the part design geometry.

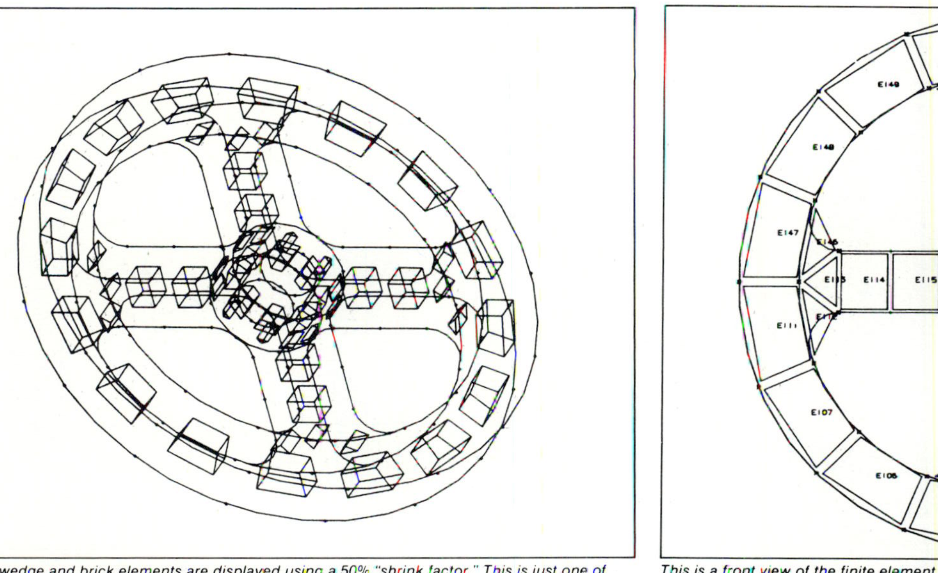

The 3-D wedge and brick elements are displayed using a 50% "shrink factor." This is just one of the numerous features which contribute to productive model building.

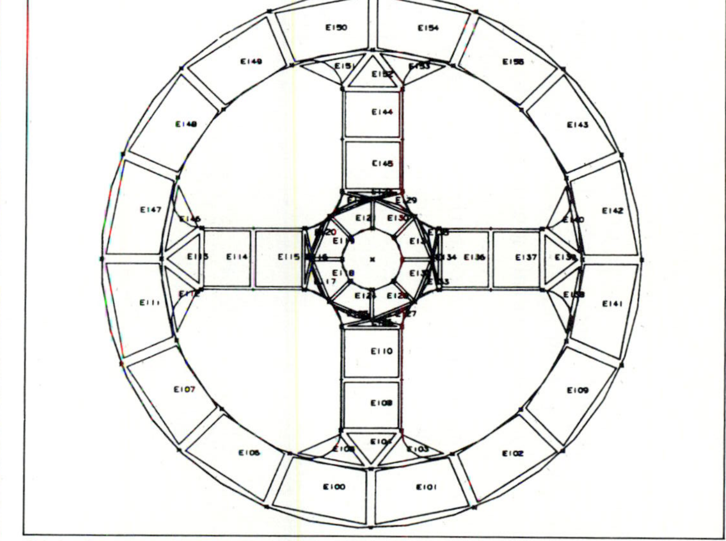

This is a front view of the finite element model with nodes, elements and element numbers.

FIGURE 17.6
Finite-element modelling for stress–strain analysis. Graphics display shows strained part superimposed on unstrained part for comparison. (*Courtesy of Applicon Inc.*)

AUTOMATED DRAFTING

Although there are many forms in which product information can be transmitted (recall the example of the engineer in need of a special wrench), the standard communication medium today is still the engineering drawing. Today's CAD systems are capable of depicting a part drawing in many different ways. Once a part has been created on a CAD system, the internal geometric model can be used to generate several alternative drawings. Virtually every existing commercial CAD system is capable of producing multiview drawings of a part. In addition, oblique, isometric, and perspective views can also be created; see Figs 17.7 and 17.8. Many systems are capable of generating drawings from as many as six different views. Scaling, dimensioning, and rotating the part are also possible on many CAD systems.

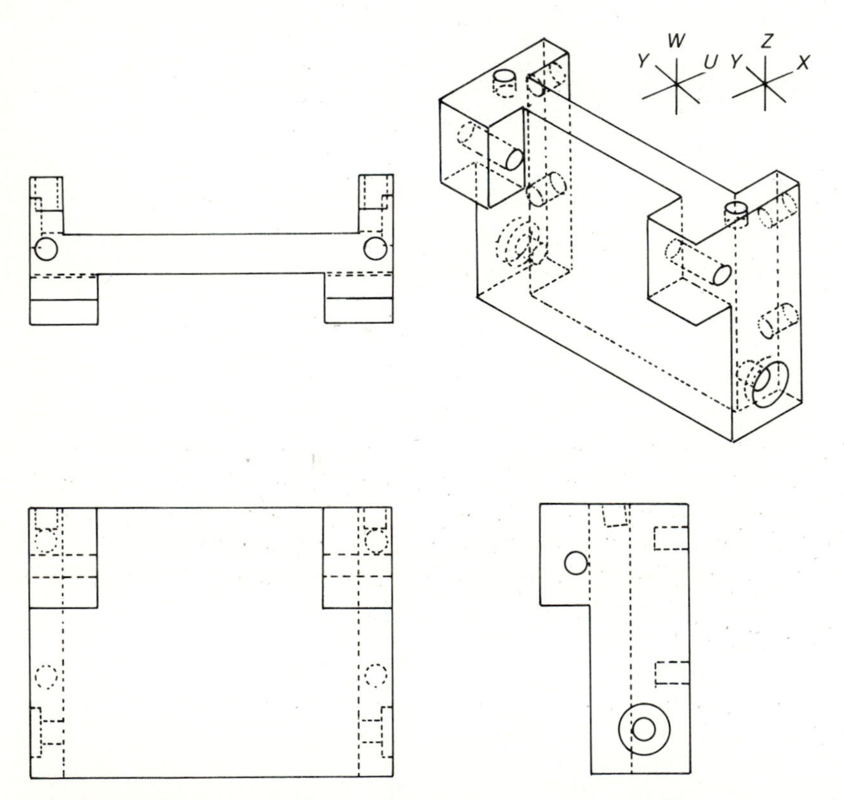

FIGURE 17.7
Engineering drawing with four views generated automatically by a CAD system. (*From W. Fitzgerald, F. Gracer, and R. Wolfe, "GRIN: Interactive Graphics for Modeling Solids,"* IBM Journal of Research and Development, vol. 25, no. 4, July, 1981. Copyright 1981 by International Business Machines Corporation; reprinted with permission.)

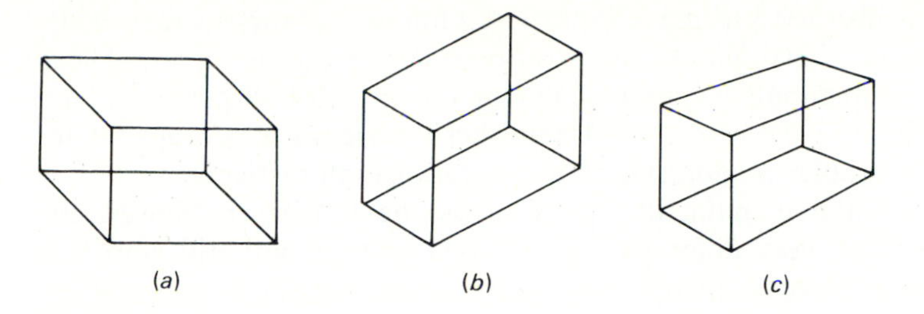

FIGURE 17.8
Three views of a wire frame block: (a) oblique, (b) isometric, (c) perspective.

CAD HARDWARE

Today's CAD systems consist of various hardware components. Figure 17.9 depicts some typical components employed in a CAD system. Input devices are generally used to transfer information from the designer to a computer on which CAD functions are conducted. A keyboard is still the standard input device used to transmit alphanumeric data to the system. In many systems, a

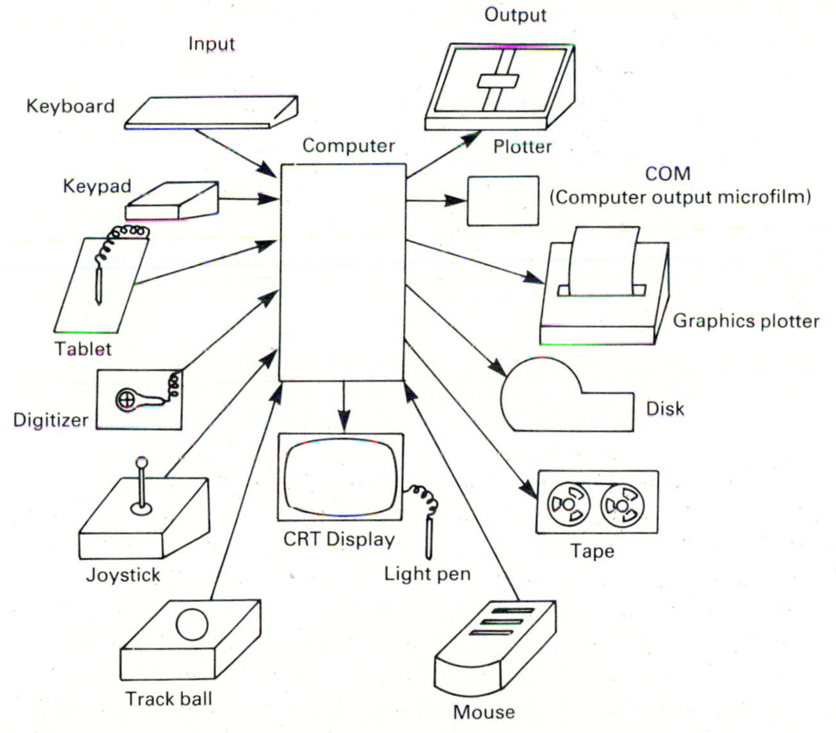

FIGURE 17.9
I/O peripherals of a CAD system.

function keypad is also used to make input less tedious. Joysticks, track balls, and mouses are used to manipulate and position a cursor on the screen. These devices are used to manipulate a graphic cursor on the CRT display and feed back this information to the computer. Using these devices (or perhaps a light pen), the designer is able to address terminal locations interactively.

Depending on the manufacturer as well as the CAD capabilities, the CAD system can look very different. Some CAD systems are still stored in large air-conditioned rooms filled with large storage peripherals. However, systems are also available today that are self-contained workstations containing all or most of the hardware shown in Figure 17.9.

COMPUTER-AIDED MANUFACTURING

The scientific study of metal-cutting and automation techniques are products of the twentieth century. Two pioneers of these techniques were Frederick Taylor and Henry Ford. During the early 1900s, the improving U.S. standard of living brought a new high in personal wealth. The major result was the increased demand for durable goods. This increased demand meant that manufacturing could no longer be treated as a blacksmith trade, and the use of scientific study was employed in manufacturing analysis. Taylor pioneered studies in "scientific management" in which methods for production by both men and machines were studied. Taylor also conducted metal-cutting experiments at the Midvale Steel Company that lasted 26 years and produced 400 tons of metal chips. The result of Taylor's metal-cutting experiments was the development of the Taylor tool-life equation that is still used in industry today. This tool-life equation is still the basis of determining economic metal cutting and has been used in adaptive controlled machining.

Henry Ford's contributions took a different turn from Taylor's. Ford refined and developed the use of assembly lines for the major component manufacturer of his automobile. Ford felt that every American family should have an automobile, and if they could be manufactured inexpensively enough then every family would buy one. Several mechanisms were developed at Ford to accommodate assembly lines. The automation that Ford developed was built into the hardware, and Ford realized that significant demand was necessary to offset the initial development and production costs of such systems.

Although manufacturing industries continued to evolve, it was not until the 1950s that the next major development occurred. For some time, strides to reduce human involvement in manufacturing were being taken. Speciality machines using cams and other "hardwired" logic controllers had been developed. The U.S. Air Force recognized the development time required to produce this special equipment and that the time required to make only small sequence changes was excessive. As a result, the Air Force commissioned the Massachusetts Institute of Technology to demonstrate programmable or numerically controlled (NC) machines (also known as "softwired" machines). With this first demonstration in 1952 came the beginning of a new era in manufacturing. Since then, digital computers have been used to produce input either in a directed manner to many NC machines, direct numerical control (DNC), or in a more dedicated control sense, computer numerical control (CNC). Today, machine control languages such as APT (Automatic Programming Tool) have become the standard for creating tool control for NC machines.

It is interesting to note that much of the evolution in manufacturing has come as a response to particular changes during different periods. For instance, the technology that evolved in the nineteenth century brought with it the need for higher-precision machining. (This resulted in the creation of many new machine tools, a more refined machine design, and new production processes.) The early twentieth century became an era of prosperity and industrialization that created the demand necessary for mass-production techniques. In the 1950s it was estimated that as the speed of an aircraft increased, the cost of manufacturing the aircraft (because of geometric complexity) increased proportionately with the speed. The result of this was the development of NC technology.

A few tangential notes on this history include the following. As the volume of parts manufactured increases, the production cost for the parts decrease (this is generally known as "economy of scale"). Some of the change in production cost is due to fixed versus variable costs. For instance, if only a single part is to be produced (such as a space vehicle), all of the fixed costs for planning and design (both product and process) must be absorbed by the single item. If, however, several parts are produced, the fixed charges can be distributed over several parts. Changes in production cost, not reflected in this simple fixed- versus variable-cost relationship, are usually the result of different manufacturing procedures—transfer-line techniques for high-volume items versus job-shop procedures for low-volume items. Figure 17.10 illustrates the fundamental relationship of volume versus production system, and Fig. 17.11 depicts cost versus volume for different types of systems.

The U.S. Department of Commerce has pointed out that in the United States, 95% of all products are produced in lots of size 50 or fewer. This indicates that although high-volume techniques are desirable from a consumer standpoint (lower cost), these techniques are not appropriate from a manufacturing standpoint (lower cost); the reason being the volume will not offset the setup expenses. The manufacturing alternative to produce those parts is through the use of flexible manufacturing systems (FMSs). These systems are nothing more than programmable job shops. However, a major economic expense still exists before one can begin employing such systems more fully. This obstacle is that a considerable setup (planning) expense is still required. The alternative to eliminate this expensive setup is through integration of computer-aided design and computer-aided manufacturing (CAD/CAM). In an integrated CAD/CAM system, parts will be detailed using a CAD system. This system will store the geometric information necessary to create process plans and generate the machine instructions necessary to control the machine

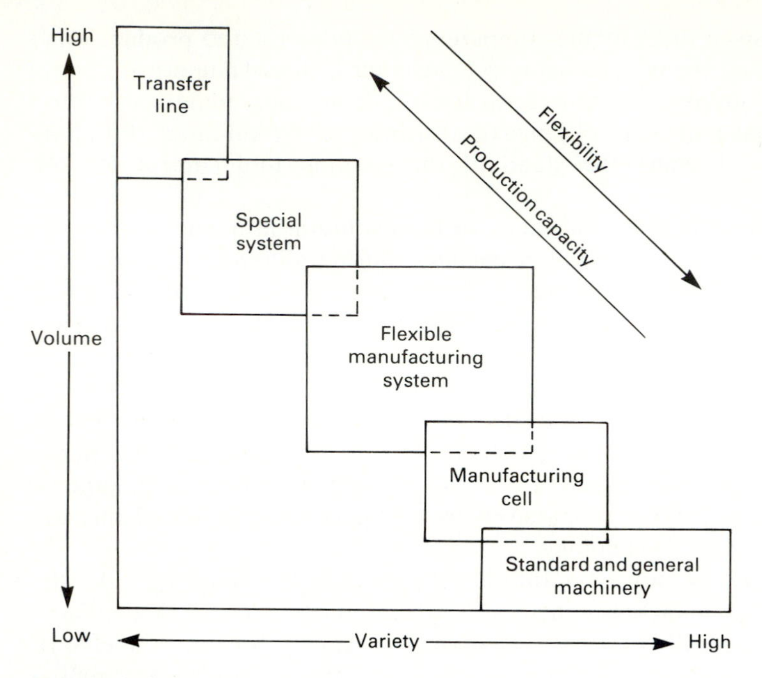

FIGURE 17.10
System selection versus volume and variety of parts.

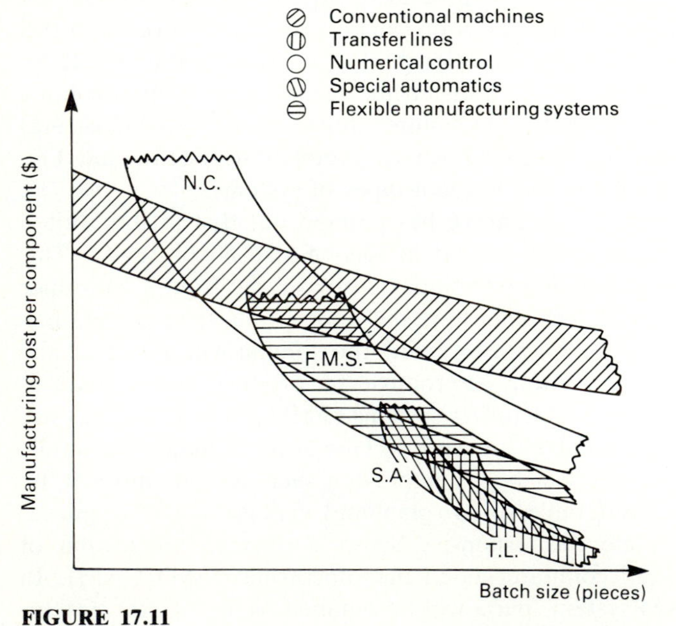

FIGURE 17.11
Comparison of the average cost of various machining systems as a function of the number of pieces per production run.

tools. Some estimates suggest that this approach will reduce planning time for FMS parts by more than 95%.

AUTOMATED MANUFACTURING SYSTEMS

An automated manufacturing system consists of a collection of automatic or semiautomatic machines linked together by a "intrasystem" material-handling system. These systems have been around since before Henry Ford began to manufacture his Model T on his moving assembly line. These automated systems have been used to produce machined components, assemblies, electrical components, food products, chemical products, etc. The total number of products produced on a single system varies with the production method. However, the principles of designing the production systems are the same independently of the product being manufactured. Figure 17.12 contains a schematic of the symbology for production flow diagrams along with a typical production system. The workstations in a production system can be manual, semiautomatic or fully automatic. The automatic stations can be programmable or hardwired.

The purpose of any production system is to produce a product or family of products in the most economical manner. Automated production systems are no different from any other type of manufacturing system. In order to

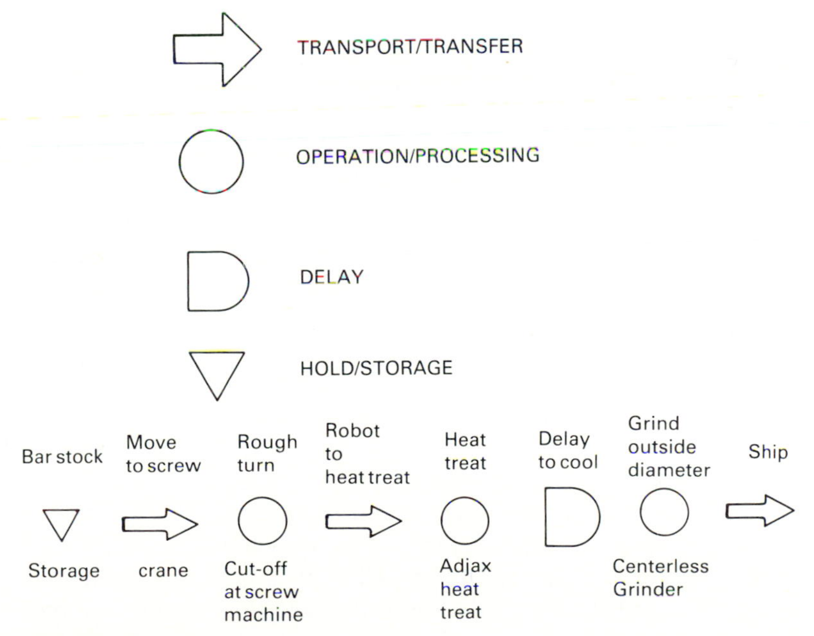

FIGURE 17.12
Schematic diagram of a production system.

employ any form of automation, the implementation must be economically justifiable. Automation has traditionally been most appropriate for high-volume products. However, flexible automation equipment has brought automation to some relatively low-volume products. In general, automated manufacturing is employed to:

1. Reduce labor cost
2. Improve product quality
3. Increase production rates
4. Reduce in-process inventory
5. Reduce material handling cost and time
6. Increase manufacturing control

WORKPIECE HANDLING

In an automated manufacturing system that consists of more than one machine or processing station, a transfer mechanism between stations is required. The transfer mechanism must move a part from one machine to the next and orient and locate the part accurately enough so that the necessary processing can be performed with the required accuracy. Although many mechanisms can be used for the transfer, these mechanisms fall into the following three categories:

1. Continuous transfer
2. Intermittent or synchronous transfer
3. Nonsynchronous or power-and-free transfer

The type of transfer used normally dictates many of the other system characteristics. The control of the system for these basic transfer procedures also changes significantly, and many different mechanisms can be used to accommodate this transfer.

Continuous Transfer

As the name implies, in a continuous-transfer system the parts are moved at a constant speed through the system. Because the parts are constantly in motion, the processing stations must be able to follow the part. In order to accommodate this movement, stations workheads are frequently set in rotary shells that are timed to rotate at the same speed as the transfer system. An example of this type of system is a beer-bottling facility, where the bottles flow at constant speed through the system and are filled by a head that rotates to the speed of the line. Some automotive assembly plants currently employ robots on continuous transfer systems. On these systems, timing marks are normally employed to signal the robot with a reference position, and the program for the robot's activities accounts for the additional movement of the product.

Continuous-transfer systems are usually employed on relative high-volume products with few options or changes occurring over time.

Intermittent Transfer

Intermittent transfer moves the parts through a system in an intermittent or timed discrete motion. All parts move at the same time, as in the case of the continuous transfer systems; however, the parts dwell motionless at the processing stations. This allows the processing stations to work on parts in a fixed position. However, it also requires that unbalanced processing stations will wait for the transfer to occur.

Nonsynchronous Transfer

The use of continuous and intermittent transfer mechanisms requires a good "line balance" where all station workloads are approximately equal. If this is not the case, parallel workstations or additional shifts must be employed. Because of the imbalance, in-process queues are required and the transfer of parts through the system must be conducted independently of other processing stations. Examples of nonsynchronous systems include "power-and-free conveyors" used to cycle parts through a production facility, or a robot that moves parts between several processing stations.

HARDWARE FOR AUTOMATION

Automated production systems have been around for many years and take on many different appearances. In general, an automated manufacturing system can be described as a collection of mechanical, electric, and electronic components coupled together to perform one or more manufacturing tasks. In early automated systems, most of the functions to be automated were automated using a variety of mechanical devices. These mechanisms were integrated into a large system using cams, timing chains, and a variety of other mechanical integrating devices. The early screw machine, as shown in Fig. 17.13, was a typical example of mechanical automation. One general characteristic of these systems was that they could normally only be used to produce one type of part or a very few types, and the set-up time required to change over from one part to another was usually orders of magnitude greater than the cycle time. This type of automation could be used effectively to produce large production runs. The second generation of automation equipment integrated more flexible electric controls such as relays, timers, and counters into the system. These systems could be set-up for different parts production far more quickly than the first-generation systems. Finally, today's manufacturing systems employ programmable electronic controls that in some cases can

FIGURE 17.13
Automatic screw machine.

be set up for random parts sequences without incurring any set-up time between different runs of parts. Second- and third-generation systems will be covered in detail in Chapters 18 and 19.

It would be impractical within the scope of this book to present a treatise of mechanisms used for automation. Only two basic mechanisms will be covered in this chapter—one a rotary mechanism and the other a linear mechanism. Depending on the size of the components being manufactured, the number of operations required, the accuracy requirements, and the weight of the parts, either a rotary or linear system may be more appropriate. In

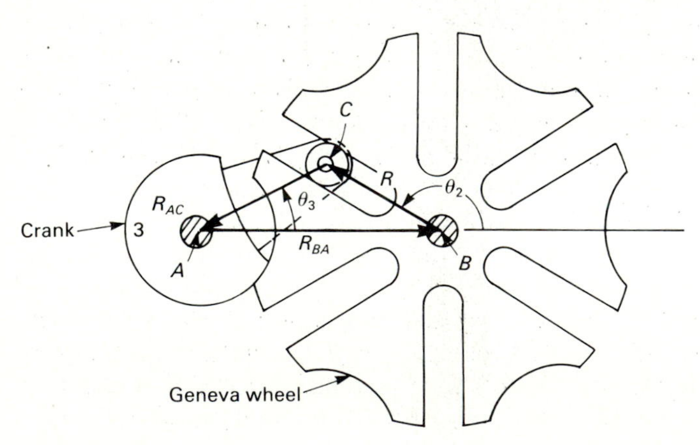

FIGURE 17.14
The Geneva mechanism.

general, if few operations are required on relatively small parts, then rotary transfer mechanisms are usually employed. In this case, parts are fixtured on a rotary table and move intermittently through the system. A common device employed in these systems is the Geneva Mechanism shown in Fig. 17.14. The Geneva Mechanism has several characteristics that make it ideal for use in automation.

THE GENEVA MECHANISM

The Geneva mechanism is a cam-like mechanism that is widely used in both low- and high-speed machinery. It has widespread use where a spindle, turret, or worktable must be indexed. Although the Geneva mechanism shown in Fig. 17.14 has six slots, these mechanisms can have between three and a large number of slots or indexing positions. (A more realistic upper limit for the number of indexing positions is 12.) A feature that makes these mechanisms most valuable is that the centerlines of the drive crank and indexer are perpendicular at initial engagement and disengagement. This feature allows the use of constant-velocity motors without acceleration and deceleration problems.

In designing these mechanisms, a known drive radius or indexing positioner radius, and the number of indexing positions required is usually the logical starting point. As shown in Fig. 17.15, the rotation required for an n-position table can be calculated as

$$Q_i = \frac{360°}{n} \tag{17.1}$$

The indexer radius can then be expressed as

$$r_i = d \cos\left(\frac{Q_i}{2}\right) \tag{17.2}$$

The drive radius can also be expressed as

$$r_d = d \cos\left(\frac{Q_d}{2}\right) \tag{17.3}$$

The ratio of r_i to r_d can then be expressed as

$$\frac{r_i}{r_d} = \frac{\sin(Q_i/2)}{\sin(Q_d/2)} \tag{17.4}$$

This ratio is important both for the design as well as for the operation of the Geneva mechanism. The ratio provides the ratio of drive to dwell time, and a given dwell requirement dictates the drive crank speed. Figure 17.16 illustrates the position of the roller in the slot during the drive cycle. From the

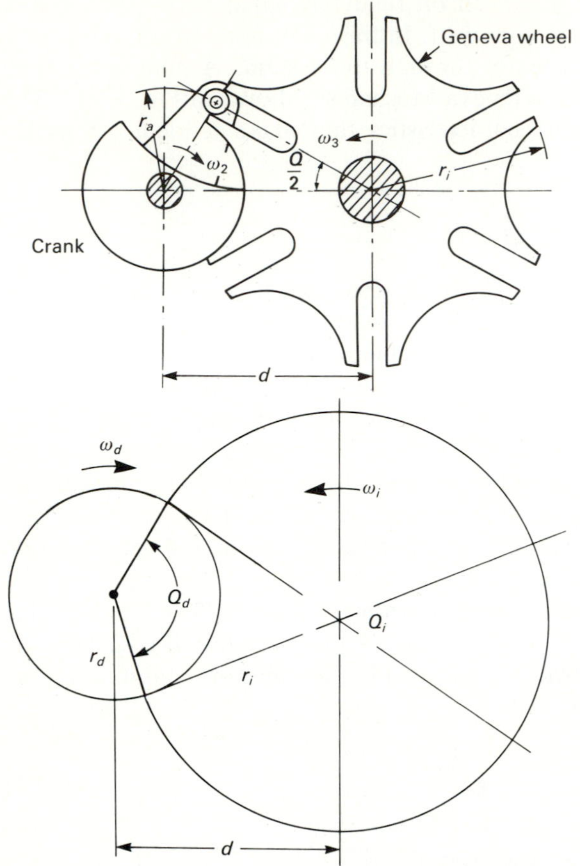

FIGURE 17.15
A Geneva mechanism in the drive cycle.

figure, the indexing angle can be calculated as

$$\tan Q_i = \frac{\sin(Q_d/2)}{(d/r_d) - \cos(Q_d/2)} \tag{17.5}$$

The angular velocity of the wheel can be determined for any value of Q_d by

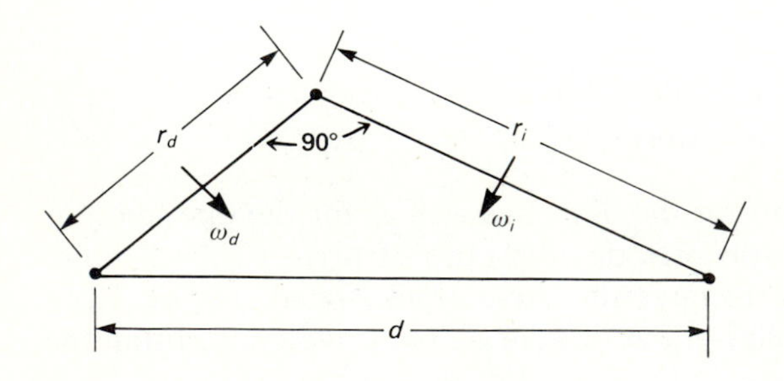

FIGURE 17.16
A six-slot Geneva mechanism.

differentiating Eq. (17.5) with respect to time. This produces

$$\omega_i = \omega_d \left[\frac{(d/r_d)\cos(Q_d/2) - 1}{1 + (d^2/r_i^2) - 2(d/r_d)\cos(Q_d)} \right] \tag{17.6}$$

The maximum wheel velocity occurs when the crank drive angle is zero, or when

$$\omega_i^{\max} = \omega_d \frac{r_d}{d - r_d} \tag{17.7}$$

If a system is being planned that requires six stations and a station cycle time of $\frac{1}{2}$ minute, for an indexer of radius 12 in., the indexing angle can be computed as

$$Q_i = \frac{360°}{6} = 60°$$

The distance between drive and index centers is

$$d = \frac{r_i}{\cos(Q_i/2)} = \frac{12}{\cos 30°} = 13.86 \text{ in.} \tag{17.8}$$

The drive diameter is

$$r_d = (d^2 - r_i^2)^{1/2} = 6.93 \text{ in.} \tag{17.9}$$

The drive angle is

$$Q_d = 2\left(90 - \frac{Q_i}{2}\right) = 120°$$

Since the nondrive time must correspond to the cycle time, the rotation of the drive motor can be calculated as

$$\text{RPM}_d = \left[\frac{360}{360 - Q_d} t_s \times \tfrac{1}{60} \text{ min/sec} \right]^{-1}$$

where t_s is the cycle time. In our case, the drive motor would rotate one revolution every 45 seconds. The ratio of time spent processing to time spent indexing would be 30:15 if a constant speed motor were used.

THE WALKING BEAM

Conveyors undoubtedly comprise the majority of linear-transfer devices used today. However, today's conveyor systems can take on a variety of control characteristics, ranging from simple to programmable control. Chapter 18 will discuss some of the merit of programmable applications. As a result, the walking beam transfer device will be illustrated as the linear device of choice. The walking beam is a reasonably simple device used for the intermittent transfer of parts to stations laid out in a linear manner. The power drive for

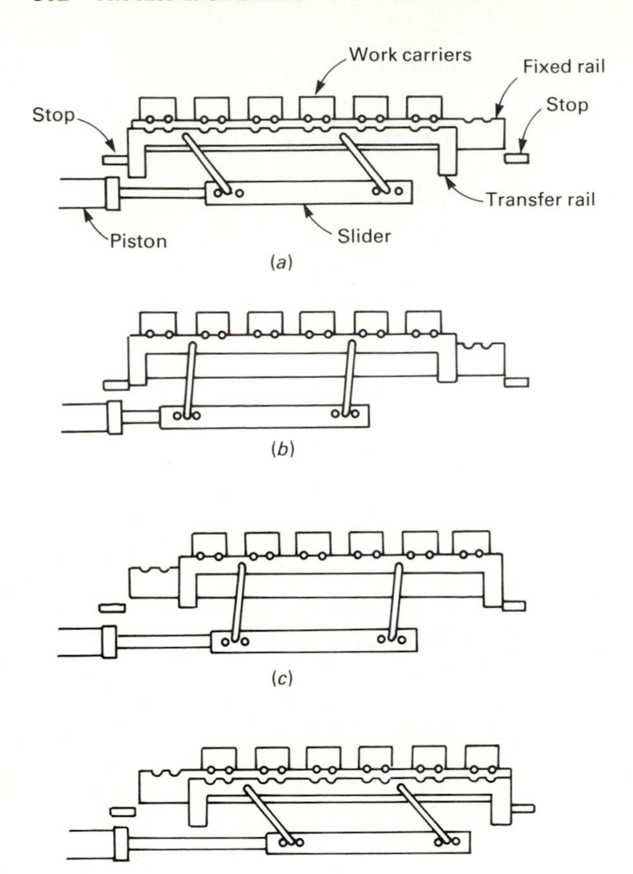

FIGURE 17.17
Walking beam transfer system, showing various stages during parts transfer cycle. (*Reprinted from [5].*)

the walking beam is typically a pneumatic or hydraulic driven two-stroke piston.

The walking beam is illustrated in Fig. 17.17. As can be seen in the figure, the piston provides both the horizontal as well as vertical drive for the system. As such, the drive stroke of the piston must be greater than the distance that the transfer rail and parts will travel. The action, as seen in the figure, begins with withdrawal of the piston. The transfer rail moves backward until the rail makes contact with the back stop. The rail is then driven up by the walking beams until the rail falls to rest on the forward rest. The piston then drives the transfer rail forward until the forward stop is impacted. The forward stop inhibits the forward movement of the transfer rail, driving it back off the beams.

Figure 17.18 shows the geometry of the walking beam. To design a walking beam, the designer normally begins with the distance between workstations, D_x, and the required lifting height, h_z. A walking beam size, b, is then selected (usually about 10 times the lifting height). From the geometry

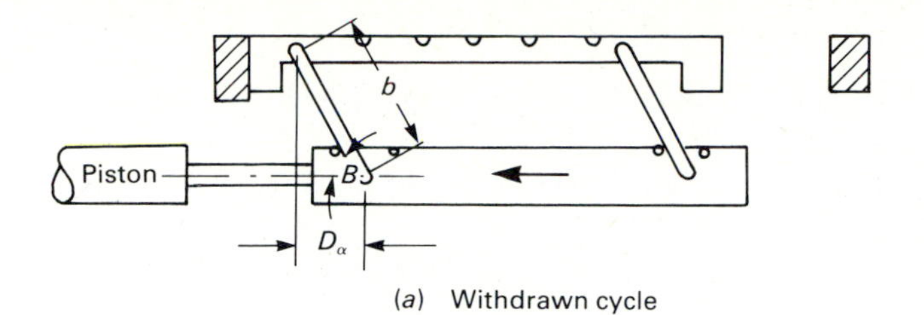

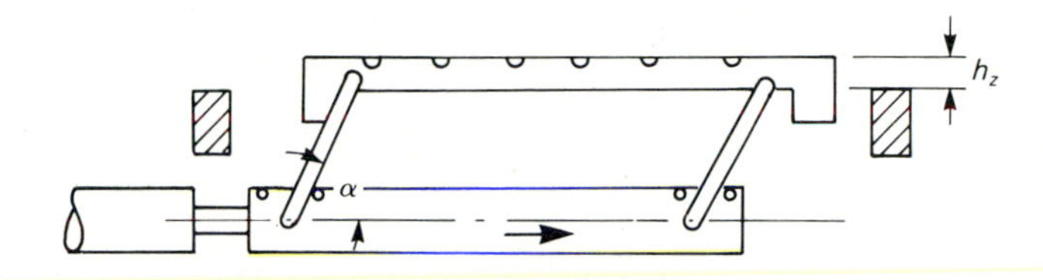

(b) Advance cycle

FIGURE 17.18
A walking beam, with geometric detail.

in the figure, the following can be determined

$$D_x = b \sin(90° - B) \tag{17.10}$$

$$h_z = b - b \cos(90° - B) \tag{17.11}$$

The minimum and maximum piston stroke can then be computed. The minimum piston stroke corresponds to the stroke required to lift the beams to a vertical position. This can be expressed as

$$S_{\min} = D_x + b \sin(90° - B) + b \sin(90° - a) \tag{17.12}$$

The maximum piston stroke corresponds to that driven from stop to stop and can be expressed as

$$S_{\max} = D_x + 2[b \sin(90° - B) + b \sin(90° - a)] \tag{17.13}$$

A piston whose stroke falls somewhere between these values is then specified.

The type of hardware selected for any manufacturing system usually depends on many parameters. The mechanisms described here are simple mechanisms that are frequently used in transfer devices. These mechanisms can be embellished in many ways to improve their performance. For instance, a Geneva mechanism will frequently employ a motor and timer. By doing so, the efficiency of the transfer device can normally be improved. This control adds to the complexity of the system and may or may not be justifiable. This type of control will be discussed in the next chapter.

QUESTIONS

17.1. Briefly describe the benefits of using CAD, of using CAM, and of using CAD/CAM.

17.2. What are the basic functions that a designer must perform? How can these functions be assisted using a CAD system?

17.3. What is a Geometric model? What is its role in CAD? How does engineering analysis interface with the geometric model?

17.4. Create a schematic diagram of an automated assembly system for your version of a lawn-mower production line. The major components to consider are: the motor, the chassis, the wheels, the handle, and the fasteners. Assume whatever order you feel is logical.

PROBLEMS

17.1. Create a Geneva mechanism for an eight-station automated table. The indexer diameter is set to 22 in. The station cycle time (time required for each operation) is 35 s. Show both your engineering design and your calculations.

17.2. Create a walking beam mechanism for the system described in Problem 17.1 that will set the workstations "in-line" rather than in a circle. Space the workstations 2 feet apart.

17.3. Many "fast food" restaurants use a variety of automated devices in the preparation of their hamburgers. Design a rotary type of system that might be used to make hamburgers. Try to make some reasonable time estimates of the activities at each station. Your concept should include approximate dimensions for the workstations.

17.4. Many "service-related" activities involve repeated processing of checks, bills, licences, etc. Describe a service system that you are familiar with and develop an automation conception for the system. Do you think that there is an opportunity to apply your design to the system? Why?

REFERENCES

1. Groover, M. P.: *Automation, Production Systems, and Computer-Integrated Manufacturing,* Prentice-Hall, Englewood Cliffs, N.J. 1987.
2. *AutoCAD Reference Manual,* Autodesk, Inc. Sausalito, Calif., 1986.
3. Groover, M. P., and Zimmers, E. W.: *CAD/CAM: Computer-Aided Design and Manufacturing,* Prentice-Hall, Englewood Cliffs, N.J., 1984.
4. Chang, T. C., and Wysk, R. A.: *An Introduction to Automated Process Planning Systems,* Prentice-Hall, Englewood Cliffs, N.J., 1985.
5. Boothroyd, G., and Redford, A. H.: *Mechanized Assembly,* McGraw-Hill, London, 1968.

CHAPTER 18

PROGRAMMABLE CONTROLLERS

BACKGROUND

The automation of any manufacturing process or system is a complicated task that requires mechanical components, electrical components, and a significant amount of instrumentation to integrate these components. The automation of the early twentieth century was primarily built of mechanical systems coupled by other mechanical components. Electrical switches were integrated into the equipment to supplement end-of-travel fixed stops and provide direction control and increased flexibility to the equipment. Soon these machines evolved into large complexes of wiring, switches, and relays. Unfortunately, the most unreliable link in many of these systems became the switches and relay panels. In this chapter, we will provide the reader with an overview of these basic switching devices and introduce programmable controllers as a means of process control.

Programmable controllers (PCs or PLCs) were first introduced in 1968 as a substitute for hardwired relay panels. The original purpose of the PC was to replace mechanical switching devices (relay panels) that were known to wear out with an electronic board that had no moving parts and could last forever. Bedford Associates (now a division of Gould, Inc.) first coined the term *programmable controller* and patented the invention. Since 1968, the capabilities of the PC have been enhanced significantly. Although PCs are capable of many manufacturing functions, their main purpose is still to replace relay panels. Their use, however, extends from simple process control applications to manufacturing system control and monitoring. Today's PC's are capable of high-speed digital processing, high-speed digital communication, high-level computer language support, and of course basic process control.

The National Electrical Manufacturing Association (NEMA) defines the programmable controller as "A digitally operating electronic apparatus that uses a programmable memory for the internal storage of instructions by implementing specific functions such as logic sequencing, timing, counting, and arithmetic to control, through digital or analog input/output modules, various types of machines or processes." The digital computer that is used to perform the functions of a programmable controller is considered to be within this scope; excluded are drum and other similar mechanical sequencing controllers. This definition from NEMA implies that PC is an electronic interface device used to perform logical operations on input signals or responses. Input for a PC typically comes from discrete signal devices such as push-buttons, micro-switches, photocells, limit switches, proximity switches, machines, etc.; or from analog devices such as thermocouples, voltmeters, potentiometers, etc. Output from a PC is normally directed to switching closures for motors, valves, motor starters, etc.

In this chapter, we will present an introduction to the world of programmable controllers. Discussion will begin with an introduction to basic switching and relay devices and proceed to the application and programming of PCs.

SWITCHING DEVICES

Since programmable controllers are primarily intended to replace relay panels, switching devices, counters, and timers, it is appropriate to begin with a discussion of some basic switching or key functions. Perhaps the most familiar switching device is the light-switch that controls the lights in our homes and in other buildings. This device usually consists of two metal conductors called springs and a mechanical device to open and close the springs. When the springs touch, a closed circuit results and the switch provides a closed path for an electric current to pass. When the springs are open, an open circuit exists.

Although there are numerous types and styles of switches or keys, they can be classified into two general categories: nonlocking and locking (see Fig. 18.1). A nonlocking switch returns to its initial state when released; a locking switch would retain its changed state even after release. A light-switch would be an example of a locking switch; a door-bell switch would be an example of a nonlocking switch. Switches can further be classified into two other classes: normally open or normally closed. An example of a normally open switch would be the simple door bell. Contact is made by physically depressing the switch ("MAKE" contact). Normally closed switches would operate in the

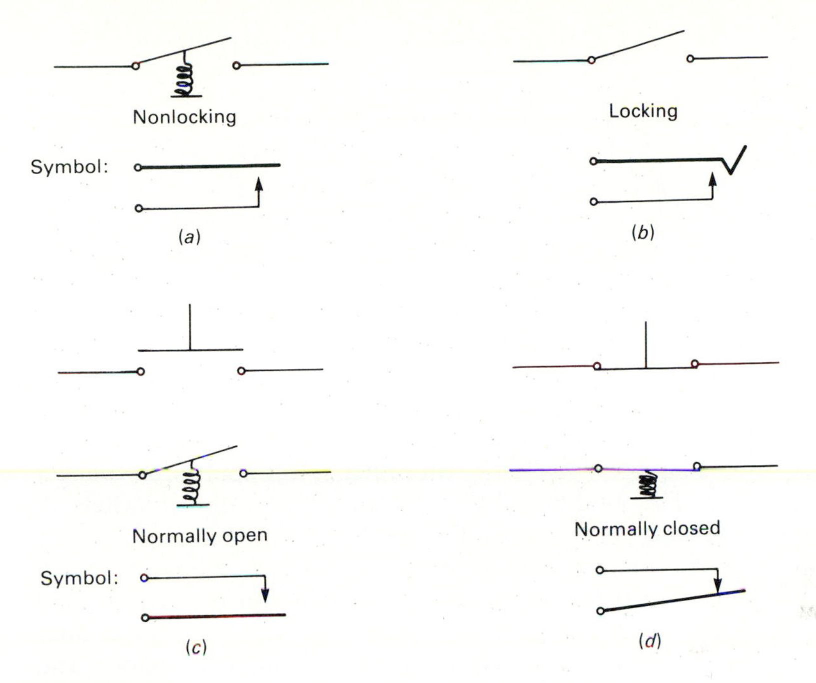

FIGURE 18.1
Types of switches. (*a*) Non-locking, break (normally open) contact; (*b*) Locking, break contact; (*c*) break (normally open), nonlocking contact; (*d*) make (normally closed), nonlocking contact.

opposite manner, in which contact would interrupt (BREAK) the operation. An example would be the switch on the door of a microwave, where contact is broken upon activation.

Thus far, the description has addressed two-state switches—open or closed. These switches certainly constitute the majority of applications and are called single-throw switches. Pictures of typical single-throw switches are shown in Fig. 18.2. These switches can be found in many household appliances

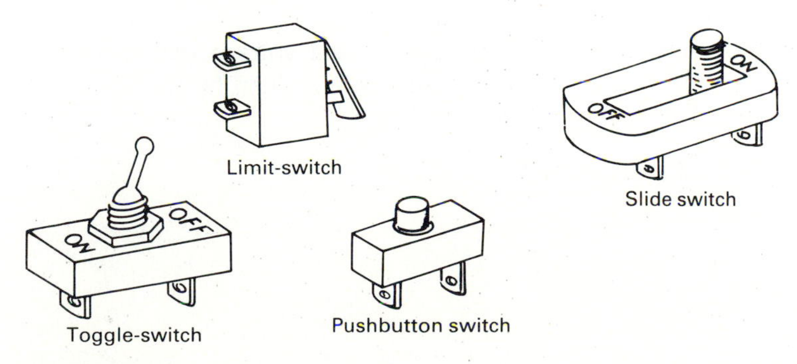

FIGURE 18.2
Single-throw (ST) switches.

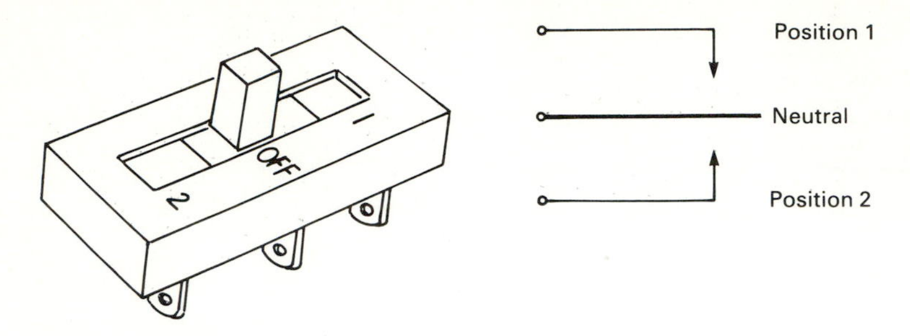

FIGURE 18.3
Double-throw (DT) switch.

as well as machine controls. There are some switches that have three states—a release and two operating positions. One can either select neutral or connect to one of two circuits. This kind of switch is called a double-throw switch. A typical double-throw switch is depicted in Fig. 18.3.

For some circuits, contact can be made or broken several times in succession. There are two types of transfer contact in which "makes" and "breaks" can be combined. A "break-before-make," or transfer, contact does as the name suggests—breaks one contact before another is made. The minimum hardware required for such a transfer would consist of three springs (two separately operated and a common spring). When the switch is operated, there is a certain amount of time when the common spring is not in contact. Thus, a break-before-make results. A "make-before-break" or continuity transfer provides the same function as a transfer contact. However, continuity always exists for one or the other contact. Figure 18.4 illustrates the symbols used to represent these switches.

The switches discussed above all have a single pole (moving part), and subsequently are called single-pole switches. In order to close (or break) two or more contacts at the same time, multiple-pole switches become necessary. The most widely seen multiple-pole switch is the double-pole switch. A double-pole switch has two moving parts (poles), and both poles move simultaneously.

Switches are not only rated by their current and voltage (for example, 10A, 125 Vdc or 5A, 500 Vac, etc.), but can also be classified by the way they operate. There are the following types of switches.

1. Basic switch, operated mechanically
2. Pushbotton switch
3. Slide switch
4. Thumbwheel switch
5. Limit switch
6. Proximity switch
7. Photoelectric switch

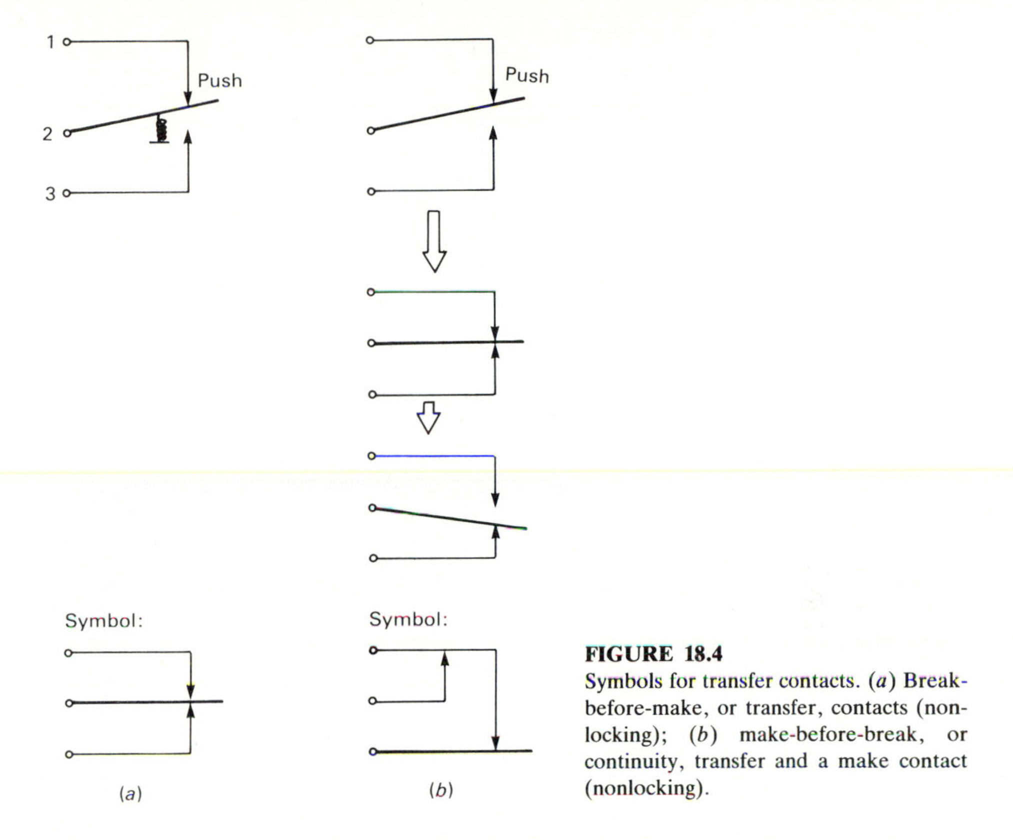

FIGURE 18.4
Symbols for transfer contacts. (*a*) Break-before-make, or transfer, contacts (nonlocking); (*b*) make-before-break, or continuity, transfer and a make contact (nonlocking).

The last two types of switches are noncontact switches. A proximity switch operates when an object is moved into close proximity of the switch. Many elevators use proximity switches. They are also used widely in industrial control. Photoelectric switches are operated by a light beam. They can be used to detect the presence of an object on a conveyor, or of a human in a dangerous zone.

In designing complex switch circuits, one usually employs a diagram of detached contact symbols. Although there are a variety of different types of switches, the contact symbol for all switches is reasonably standard. The symbol used for a normally open switch is ⊣ ⊢. The symbol used for a normally closed switch is ⊣/⊢. Normally no special feature is used to denote the type of switch or whether the switch is a latched or nonlatched device. Detached contact symbol diagrams with these switches will be used through this chapter.

RELAYS

A switch whose operation is activated by an electromagnet is called a "relay." The contact and symbology for relays is usually the same as for switches. To

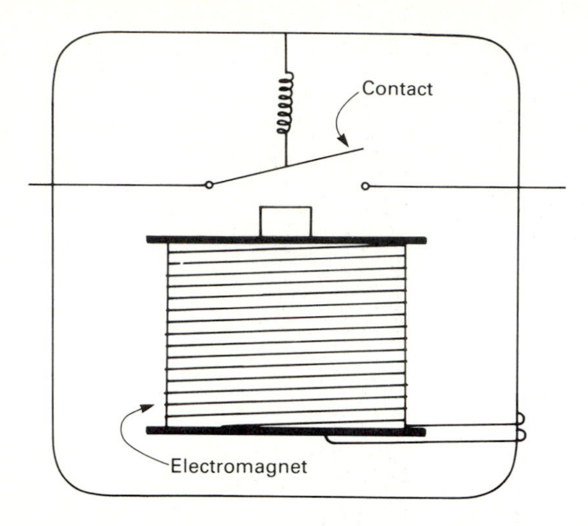

FIGURE 18.5
A single-pole magnetic relay.

activate current-carrying conductors, a magnetic field acts upon a nearby magnetic material and draws it to the most dense part of the magnetic field (typically iron to a magnet). These magnetic forces exist at every air gap in a magnetic circuit. Since the forces must balance each other (an equilibrium condition), a variety of "movable iron" devices that convert magnetic attraction to mechanical work have been developed for switching applications.

A simple relay mechanism is shown in Fig. 18.5. A small current passes through the magnet, causing the contact to close. Usually the magnet is rated between 3 and 100 volts and at a few hundred milliamperes. Therefore, it is operated at a very low power (current and voltage). A circuit carrying a much heavier rating can be switched using a relay; with the two circuits totally separated.

A make contact relay is usually noted by —○—; a break relay is usually noted by —∅—. A number of different graphic symbols are used to denote relays. Figure 18.6 illustrates some typical circuits containing relays. In the figure, a coil symbol is used to denote a relay.

Although several types of relays exist, (e.g., polarized relays, AC relays, sensitive relays, etc.), the most common relay is probably the general-purpose dc relay. This type of relay operates when the current through its coil exceeds a certain value (called the operate current) and remains operational until the current diminishes below another value (called the release current). Typical values for these currents range from a few milliamperes to 1 A. The time it takes for the contacts to close after the application of the operate current ranges from a few milliseconds to several hundred milliseconds. An important characteristic of the operation of a relay is contact stagger. When a relay operates, the contacts do not all open or close instantaneously, and there may be a delay of several milliseconds between the operation of two contacts of the same relay. In the design of a relay circuit, this delay should always be taken

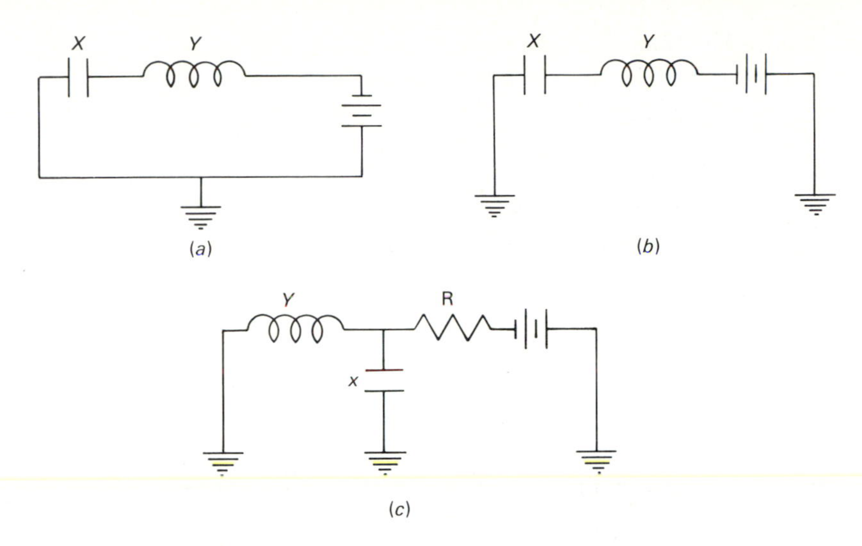

FIGURE 18.6
Methods of controlling relays. (*a*) Circuit for series control of relay *Y* by contact *x*; (*b*) form of (*a*) to be used in this book (*c*) shunt control of relay *Y* by *x*.

into account. Voltages used to activate relays are usually between a few volts and 100 V.

The commonest method for controlling the operation of a relay is shown in Fig. 18.6*a* and 18.6*b*. Here, a contact is placed in series with the relay coil and a battery. When the contact is open, no current flows and the relay does not operate. When the contact is closed, current flows through the relay coil and the relay operates. Occasionally, a relay is controlled by a shunt contact as in Fig. 18.6*c*. When this contact is closed, it places a very low-impedance path in parallel with the relay coil, and the current from the battery flows through

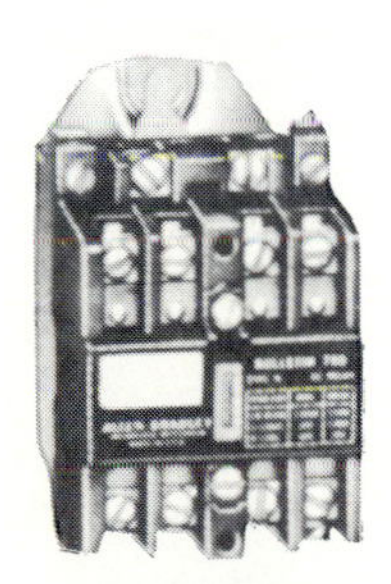

Type N Relay
4 Poles

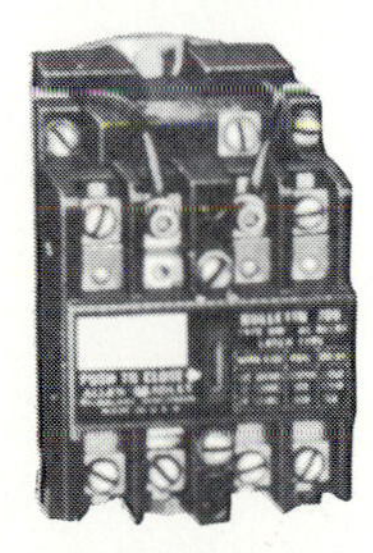

Type NM Relay
2 Poles

Type N Relay
8 Poles

FIGURE 18.7
Typical relays. These are characterized by features, number of poles, rated voltage, power consumption, contacts rating, service life.

this shunt path rather than through the relay coil, preventing the relay from operating. When the contacts are open, the current flows through the relay and it operates (provided that the value of R is chosen properly).

From this discussion, one can see that a relay is really a magnet-operated contact switch. The contact switch inside a relay can also be classified by the number of poles and throws. Although most relays are single-throw, it is not uncommon to have multiple-pole relays. Figure 18.7 depicts typical multiple pole relays.

COUNTERS

On the basis of their structure, counters can be classified as mechanical or digital. Mechanical counters are devices that the programmable controller typically replaces. A digital counter is used to count the number of input impulses. When the count value reaches a preassigned value, it gives an output. The output is usually in the form of a relay contact closure. A digital counter is illustrated in Fig. 18.8. Whenever there is a pulse coming from the input, the count value is updated (either incremented or decremented). The count value is then compared with the value in the count register (which has a preassigned value). When the count value is equal to the count-register value, an output contact changes state.

The operation of a counter can best be shown by a timing diagram. Figure 18.9 depicts a timing diagram for a counter. The preassigned count register value is 4. There are up counters and down counters. An up counter counts starting from zero and increments the value when there is an input. A down counter, on the other hand, counts down from an initial value. They both serve the same purpose—they count a certain number of inputs then output to a relay contact closure. A typical counter is characterized by the number of counting digits, input electrical rating, reset system, output contact rating, and power source. Counters can be cascaded to form a larger counter. For example, an eight-digit counter can be made by cascading two four-digit counters. Input to a counter is normally from a contact. To initialize the

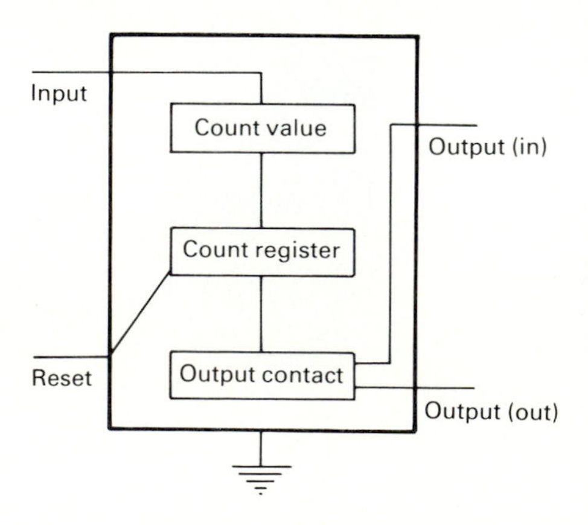

FIGURE 18.8
A counter.

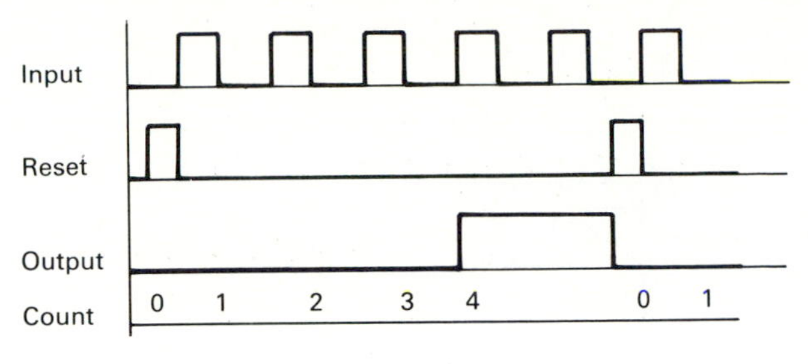

FIGURE 18.9
Counter operation time diagram.

counting, a reset input is used. The output contact is rated in the same way as a regular contact switch.

Setting the count register value (the preassigned value) is usually done directly on the counter. Figure 18.10 shows a typical digital counter. The display on the counter shows the current count value. The lower panel portion of the figure shows the count register value setting thumbwheel.

One example of a counter application is the loading of boxes with parts. Assume that a conveyor carries parts to a box-packaging area. Each box is designed to hold 10 parts. A photoelectric switch is installed at the end of the conveyor. Whenever a part drops into the box, the photoelectric switch outputs a pulse. The layout of the station is shown in Fig. 18.11.

In the system, the output of the counter is a normally closed contact. Therefore, the output of the counter is opposite to the one shown in Fig. 18.9. The output is connected to the conveyor drive motor. Before the system is turned on, an operator must put an empty box at the packaging station. After the power switch is turned on, parts start to arrive at the station. Each part dropped into the box triggers the photoelectric switch. When a count of 10 is reached, the output is opened, which in turn switches the converyor off. The

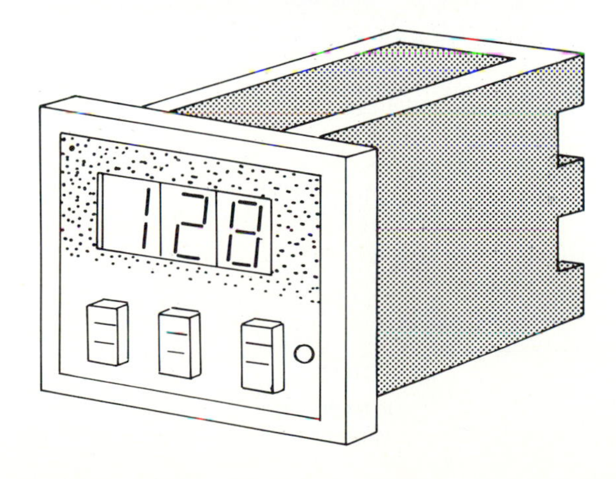

FIGURE 18.10
A digital counter.

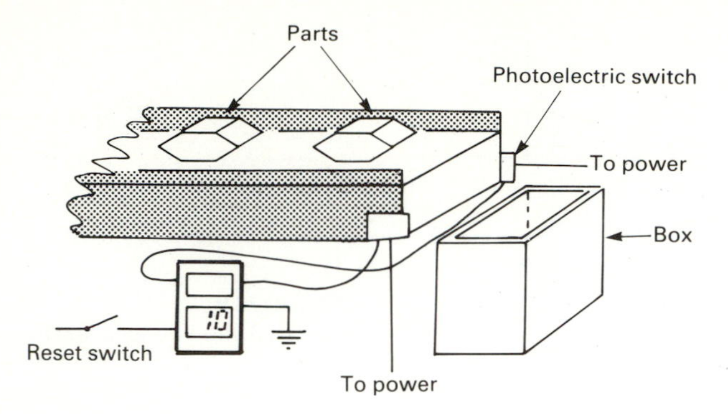

FIGURE 18.11
Example of using a counter.

operator replaces the box, then pushes the reset switch, and the cycle begins again.

TIMERS

A timer, as its name implies, is used for some timing purpose. In process control, a significant portion of operations must be timed. For example, in a chemical process, curing of certain products, mixing several chemicals, etc., all require a certain length of time for completion. In process control, synchronization of operations is also essential. There are two ways to synchronize operations, namely, event-triggered synch and time-controlled synch. Event-triggered synch can be achieved by using sensors and switches to detect the triggering event. For time-controlled synch, each operation is given a fixed time period to finish; therefore, a clock or timer becomes necessary.

A timer starts timing after receiving a start signal. When a preassigned timing value is reached, it outputs a signal. The basic structure is illustrated by the diagram shown in Fig. 18.12. The operation of a timer can be shown by a timing diagram (Fig. 18.13). It is very similar to that for a counter (Fig. 18.9). In the diagram, the preassigned timing value is 4 seconds. Each clock pulse represents 1 second.

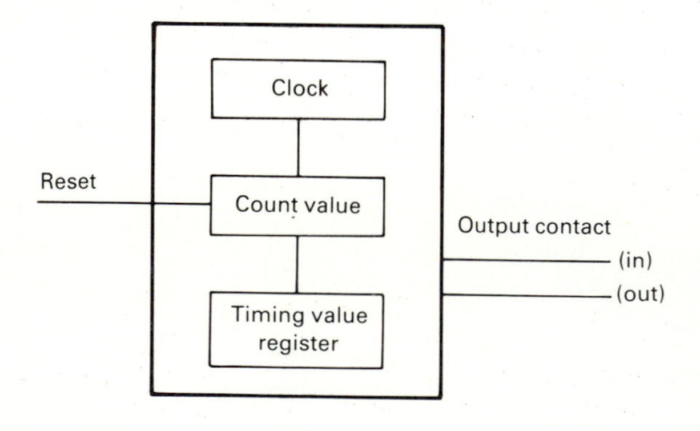

FIGURE 18.12
Timer structure.

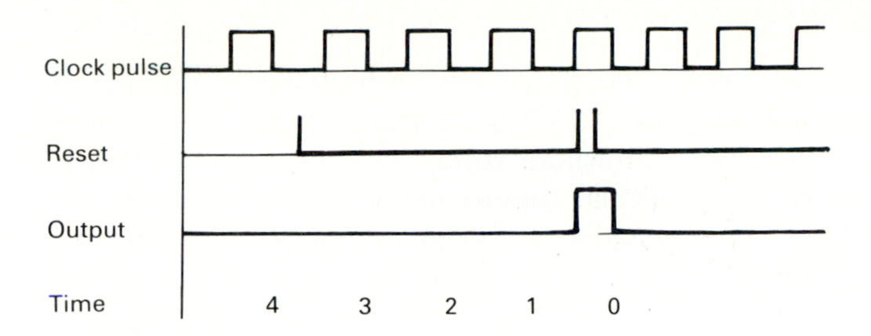

FIGURE 18.13
Timer timing diagram.

Depending on the specific application, the timing diagram for other timers may have some small deviation. However, the basic principle is the same. Timers are usually characterized by the following factors:

1. Special features: size, mounting, display type, etc.
2. Time range: e.g., 0.1 second to 30 hours
3. Rated voltage: operating voltage, e.g., 120 V, 60 Hz
4. Accuracy: e.g., 1 percent
5. Contact rating: e.g., 250V 3A, 250V 5A, etc.
6. Output contact: e.g., SPDT, DPDT, 4PDT, etc.

The following is an example to show how a timer can be used for a delayed motor start. It is desirable to have a motor start running 5 seconds after a switch is closed. A timer is used in the circuit to generate the 5-second time delay. Figure 18.14 depicts the wiring of the circuit.

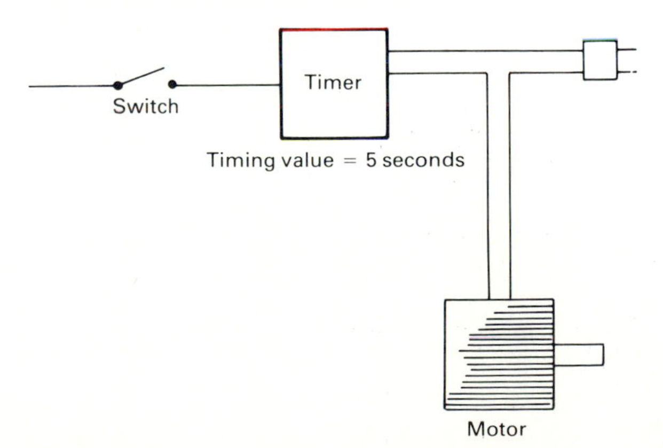

FIGURE 18.14
A delayed motor control.

PROGRAMMABLE CONTROLLERS

From the preceding sections, the reader should have a basic understanding of what relay logic systems are. In order to design these systems, an engineer must know the characteristics (time, threshold voltage, etc.) of standard relays or must design his own relay for special requirements. It is not difficult to envision the timing and logic problems that are encountered in breadboarding a relay logic system. Furthermore, the time required to construct these systems is nontrivial. If these problems are not sufficiently difficult, one can also note that relays (because they are mechanical devices) wear out with some regularity. It has been noted that the mechanical relay is the most unreliable component in a control circuit. It was for these reasons that the PC was spawned.

Basic PC Functions

The basic function of a PC is to examine the status of a set of input devices and, depending on their status, regulate a set of output responses, thus controlling a process. The combinations of the input and output data reflect the logic of the system. The combination of the input and related output responses reflects the control plan or program resident in the PC. The control plan is normally programmed into the PC using a "program loader."

The logic of a PC can usually be embellished to include: (1) timing delays of output responses, (2) counting functions for cumulative switching of output responses, and (3) arithmetic functions for basic algorithmic processes.

Input and Output

The input and output (I/O) for a PC is normally a set of modular plug-in peripherals. The I/O modules allow the PC to accept signals from a variety of external devices, e.g., limit switches, optimal sensors, proximity switches, etc. The signals (two-state signals for the devices mentioned—open or closed) are converted from an external voltage (115 Vac, 230 Vac, 24 Vdc, etc.) to a TTL (transistor to-transistor logic) signal of 0 to 5 Vdc. The PC processor then uses these signals to determine the appropriate output response. A 0 or 5 Vdc signal is transmitted to the appropriate output module, which converts the signal to the appropriate response domain (115 Vac, 230 Vac, 10 V, 24 Vdc, etc.).

I/O modules are typically housed in a rack separate from the PC. Lights are usually included on the I/O module to indicate the current state (this is ideal for trouble-shooting). In addition, each module is normally fused and isolated from the processor. Some typical I/O modules include:

1. AC voltage input and output
2. DC voltage input and output
3. Low-level analog input

4. High-level analog input and output
5. Special-purpose modules, e.g., high-speed timers, stepping-motor controllers, etc.

The Processor

The brain of a PC is a microprocessor or the electronic logic circuit analog. Although early PCs used special-purpose logic circuits, most current PCs are microprocessor-based systems. The processor (central processing unit, or CPU as it is often called) scans the status of the input peripherals examines the control logic to see what action to take, and then executes the appropriate output responses.

Normally, a peripheral interface adapter (PIA) is used to transfer the status of the input peripherals to some prespecified memory location. The user defines thc location of the peripheral on the I/O housing in the program. Each I/O location is assigned to a specific memory location. This makes accessing input by the CPU a task of loading the contents of a specific memory into a storage register. Output changes are equally easy for the CPU to alter. The contents of a particular memory location are altered. Owing to the electrical differences between the CPU and external I/O peripheral, I/O points and internal memory are actually electrically isolated. In a more advanced design, a separate I/O processor is used to bring the external I/O status to an internal memory location.

The microprocessor-based PC has significantly increased the logical and control capacities of programmable controllers. High-end PC systems allow the user to use arithmetic operations, logic operations, block memory moves, computer interface, local area network functions, and so on. Some of them even support high-level programming languages. PCs are becoming more and more powerful.

Memory

The memory of the PC is important because the control program along with the peripheral status is stored in the PC's memory. Whether the PC is microprocessor-based or not, it will operate using digital electronic memory. Because they operate digitally, the fundamentals of microcomputers relate directly to PCs. The fundamental unit of the PC memory is a bit. A bit is a two-state entity that can be analogized to a switch (on or off). Most data in a microprocessor or PC is usually transferred (and often operated upon) in 8-bit chunks. These 8-bit chunks are called bytes and in an 8-bit processor are also called words. In a PC, a word usually refers to the size of the storage registers.

Memory size in a PC is measured in bits, bytes, or words. Because many words of memory are required, it is usually measured in "K" increments (where $1\text{ K} = 1024$). An 8 K byte memory would contain 8192 bytes (8×1024 bytes) or 65536 bits.

Although several types of memory are used in a PC, memory can be classified into two basic categories: volatile, and nonvolatile. Volatile memory is memory that loses state (the stored information) when the power is removed. This may seem perfectly appropriate. However, you must remember that the program is stored in memory, and if the power fails (if, say, the plug is pulled), the program must be retyped or reread into memory—a potentially time-consuming activity. Nonvolatile memory, on the other hand, maintains the information in memory even if the power is removed. Some of the types of memory used in a PC include:

1. Core memory
2. ROM (read-only memory)
3. RAM (random-access memory)
4. PROM (programmable read-only memory)
5. EPROM (erasable programmable read-only memory)
6. EAPROM (electronically alterable programmable read-only memory)

Of these memory types, the only volatile memory is RAM (oddly enough, it is probably the most commonly used memory). The other memory types maintain their status even after power is lost. Many RAM-based memory systems use battery back-up to preserve the contents of memory in case of a power failure. These RAMs are built using CMOS technology. CMOS devices consume minimal amounts of power.

Memory can be further classified as "read-only" or "read–write." Of the memory listed, only core and RAM memory are of the "read–write" variety. RAM and core memory can be easily changed by the processor. The other types of memory requires additional hardware to alter or program.

Power Supply

The power supply operates on ac power to provide dc power required for the controller's internal operation. It is designed to take either 115 Vac or 220 Vac. Some power supplies can take either voltage with a jumper switch for selection. The internal operations of I/O modules are also supported by the PC's power supply. However, separate power sources are required in order to close the circuits of external devices such as switches, motors, etc.

Peripherals

A number of peripheral devices are available. They are used to program the PC, prepare the program listing, record the program, and display the system status. The following is a partial list of peripherals.

1. Hand-held programmer (loader)
2. CRT programmer

3. Operator console
4. Printer
5. Simulator
6. EPROM loader
7. Cassette loader
8. Network communication interface

Usually, programs are entered into a PC either by a hand-held programmer or a CRT programmer. A hand-held programmer is a low-cost programming device that can display only one or a few program statements. A CRT programmer is a computer-based special CRT terminal. A special keyboard that has keys representing the ladder-diagram elements of a PC program makes programming easier. Entire pages of program can be displayed on a CRT. Programs can also be stored on either built-in cassette subsystems or floppy disk drives. Most CRT programmers also allow users to track the operation of the PC by highlighting the status of ladder-diagram elements in real time. Programmers are used only during the programming and debugging period.

An operator console or operator display unit is used for operator data input or system monitoring. It can be a small numeric keypad with an LED display or a full-sized CRT display with a typewriter keyboard. A machine operator can enter machine-control parameters on the console. However, a PC program can not be changed through it.

A printer is used to print either PC programs or operation messages. When a hard copy of the operating conditions or production statistics is needed, a printer can be valuable. A simulator usually consists of some lights and switches. It is used for the debugging of a program. It can be connected to the PC I/O module. PC program logic can be tested by flipping switches and observing lights.

An EPROM loader can load a program from an EPROM (erasable programmable read-only memory) to a PC's memory. Some models also allow programs in the PC memory to be dumped to an EPROM. A cassette loader serves the same purpose.

More advanced PCs can also communicate with other computers or devices through a communications network. Most major PC manufacturers support their own local-area network (LAN). A unified LAN (MAP, proposed by General Motors) is being developed by nearly all control-device manufacturers. Upon its completion, all PCs and control devices will be able to be linked to the same network and share the same resources.

The basic components of a PC have been discussed. Most PCs employ computer communication standards, and can, therefore, be expanded easily or can communicate with a computer. Figure 18.15 illustrates a typical PC installation.

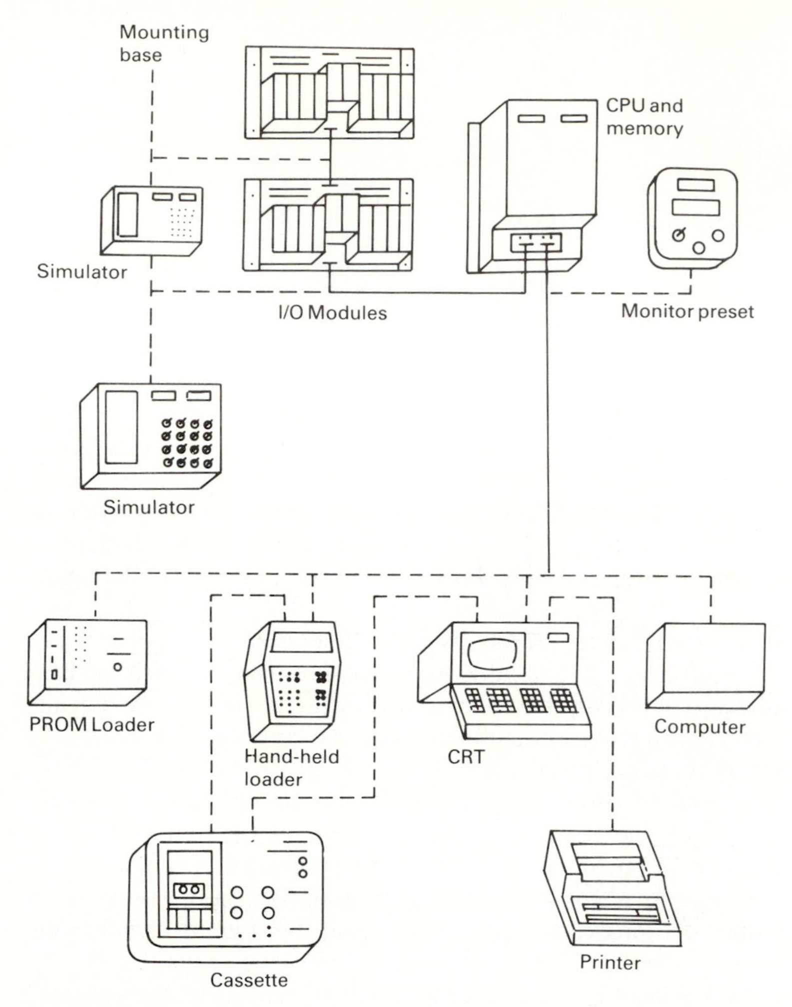

FIGURE 18.15
A programmable control system.

INTERNAL OPERATION OF A PROGRAMMABLE CONTROLLER

The programmable controller basically acts as a programmable sequencer that runs through cycles. In each cycle, it first senses the input conditions from the input module, then solves the programmed logic. The programmed logic takes the input conditions then sets the output condition. At the end of the cycle, all the output modules are set. In PCs that contain a separate I/O processor, the I/O operation runs concurrent to the program logic. It runs in a loop and repeats the cycle until the power is turned off or a stop-running switch is turned on. The time it takes to finish one cycle is called the scan time. The

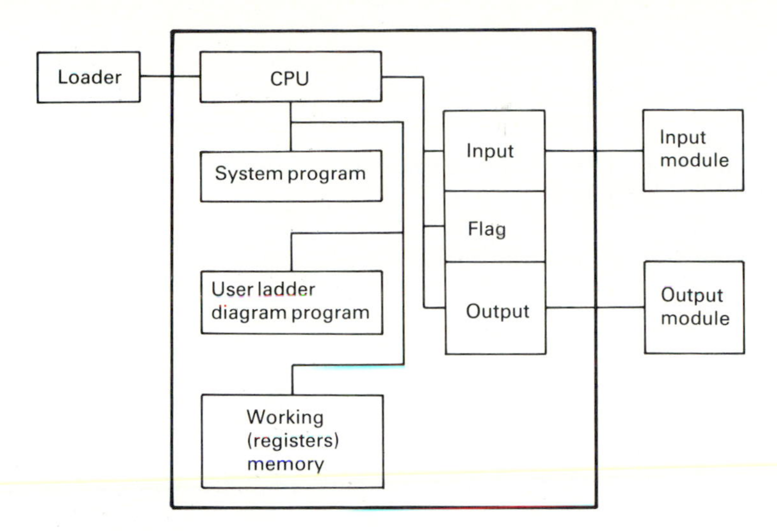

FIGURE 18.16
PC structure block diagram.

length of the scan time is based on the number of I/O points and the length of the program.

The block diagram in Fig. 18.16 shows the basic architecture of a PC. In a PC, the system program is the operating system. It is stored in nonvolatile memory, such as ROM or PROM. The operating system controls the entire operation of a PC. It consists of three major portions:

1. An editor
2. I/O drivers
3. A ladder logic interpreter

The editor provides the capability of (1) entering a user program (ladder-diagram logic), (2) editing an existing program in memory, and (3) saving a program on an external storage device (such as a tape recorder, disk, etc.). The I/O drivers take care of the communication protocol of the I/O devices. Each I/O device has different characteristics, which the I/O driver is designed to handle. The function of the ladder logic interpreter is similar to that of the BASIC interpreter in a microcomputer. Since the CPU can only operate on machine code, a ladder diagram program cannot be executed directly. An interpreter must first interpret the user's ladder diagram program line by line (rung by rung), and perform the appropriate action to assemble the ladder logic into machine-executable code.

The user ladder-diagram program is usually stored in either RAM or EPROM. During the program development stage, a program is written directly to RAM. RAM allows the user to modify the program directly without going through special devices and lengthy editing procedures. When a program is entirely debugged, it is more desirable to save the program in nonvolatile

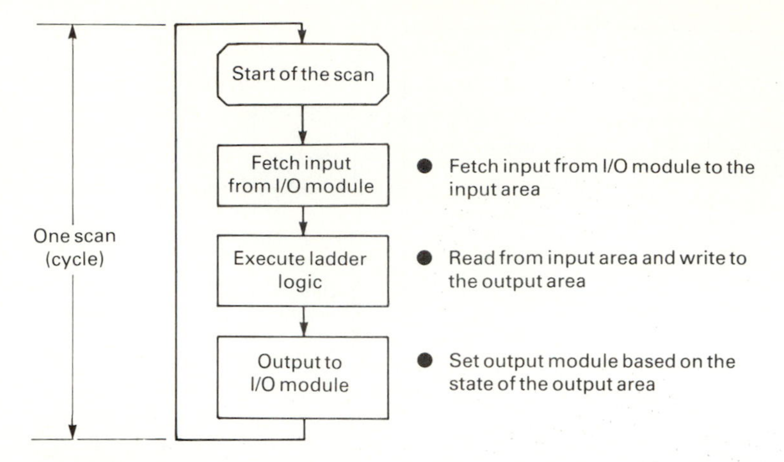

FIGURE 18.17
A complete scan.

memory (RAM is volatile memory). The program can be "burned" into EPROM using a simple device called an *EPROM burner*. A program can also be stored on a cassette tape or floppy diskette.

Working memory is used to create the stack logic for the PC. It is used as temporary storage, and must be read–write memory, either RAM or core (core is seldom used anymore). The intermediate result of ladder logic operation is saved in the working stack until the operations on the entire rung are finished. The final result is sent to the output area.

The input and output areas are also memory locations in the controller. Each I/O point is assigned a location. The ladder logic takes the input value directly from the input area and writes the output to the output area. There are also interfaces between the input/output area to the I/O modules. Internally, PCs use low voltage (5 V) and low current (a few milliamperes) transistor to transistor logic (TTL). In order to deal with industrial equipment, the PC must take signals at different voltage levels and send signals out at a greater current. I/O modules are used to connect directly to external devices. The specification of I/O modules will be discussed in the following section.

The operation of a PC is shown in Fig. 18.17. When the required sample rate is very high, the scan time becomes critical. If the scan time is too long, the PC may not be able to catch some of the data. Notice that the PC fetches data from the I/O module only once in each scan.

PROGRAMMABLE CONTROLLER CONFIGURATIONS

1. Basic controller
2. I/O modules
3. Programmable hardware

The basic controller hardware normally consists of a processor, memory module, power supply, and rack (See Fig. 18.18). Figure 18.19 illustrates a typical processor. The processor is usually enclosed in a metal housing with a power supply (either in the same housing or in a separate housing). The power supply provides power for the operation of the processor and sometimes for the programming equipment. Separate power supplies are usually required for driving the I/O modules.

The memory module typically plugs into the controller housing. Memory can consist of any of the memory types discussed earlier. The commonest type

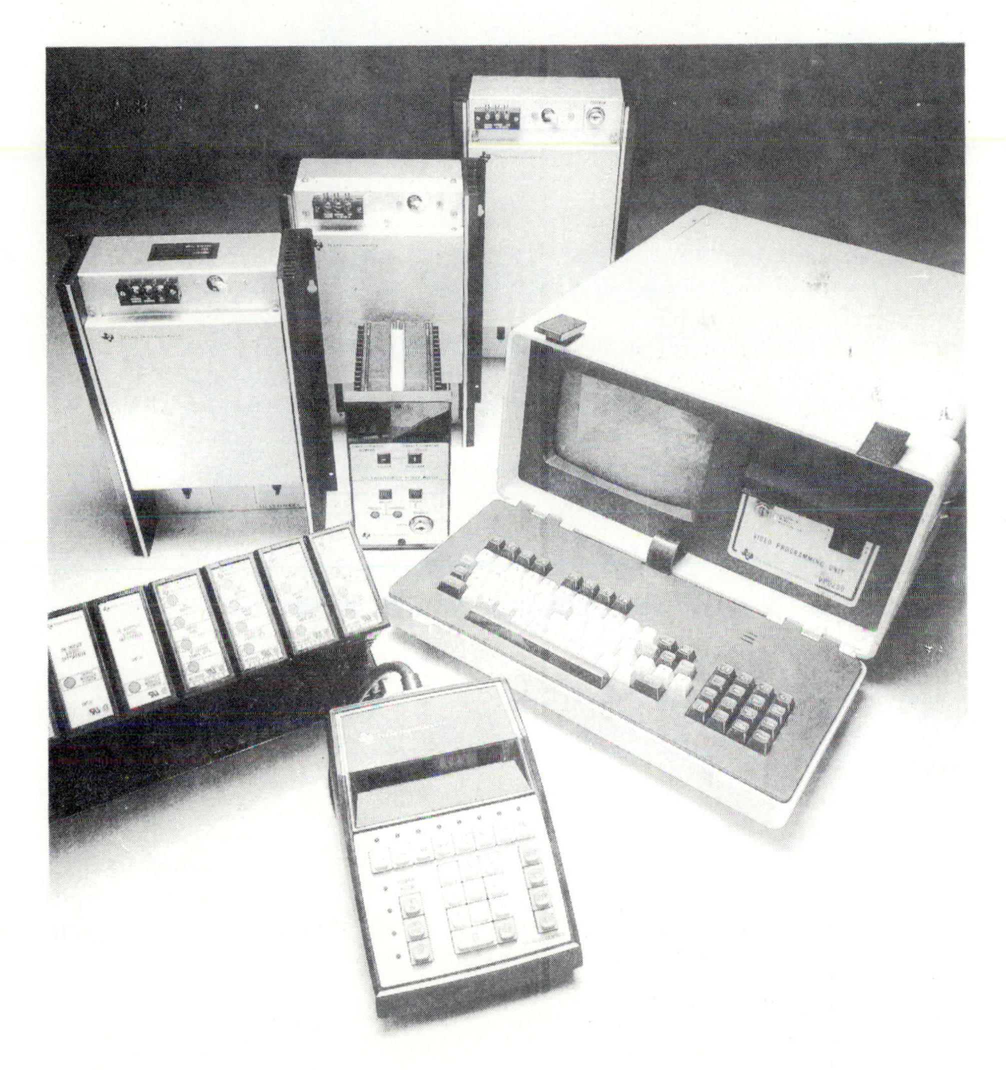

FIGURE 18.18
Basic programmable controller components.

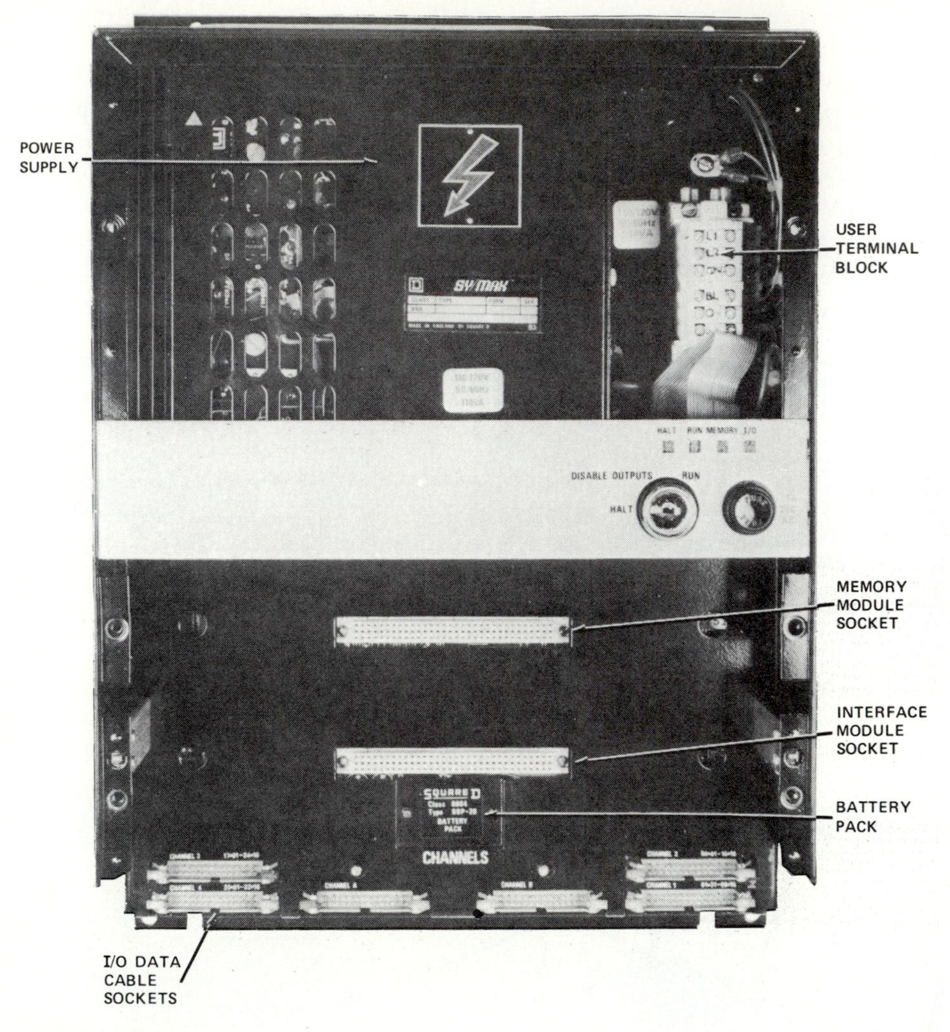

FIGURE 18.19
A typical programmable controller.

of memory employed in PCs is RAM with battery backup. Reprogramming and programming changes are possible as often as desired. Memory type, however, is usually inconsequential to the user since they all perform the same function and are somewhat "invisible" to the user.

I/O MODULES

Although the basic decision making is done by the processor, I/O status must come to and leave the processor in the appropriate form. The processor

(a)

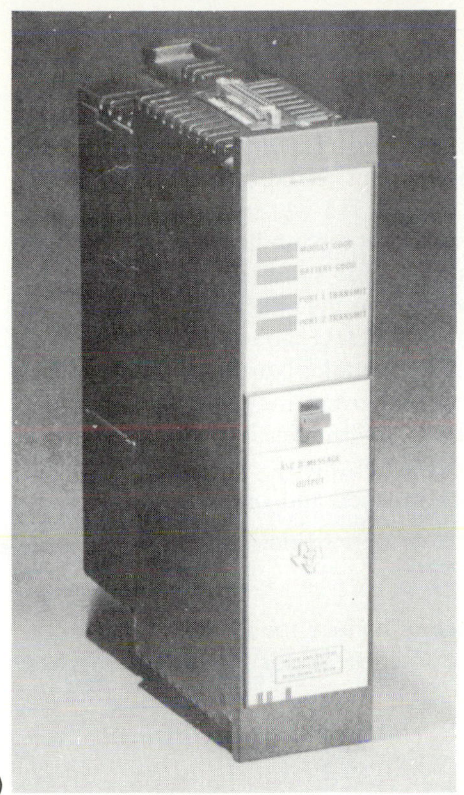

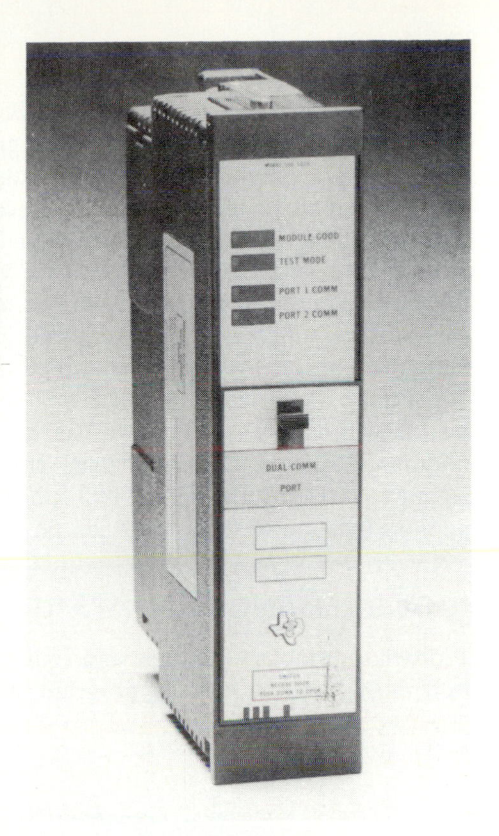

(b)

FIGURE 18.20
(*a*) I/O modules in rack assembly. (*b*) I/O modules.

operates on TTL level signals (a +5 Vdc signal is the active state and a 0 Vdc signal is inactive). These 0 or 5 Vdc signals are the only means by which the processor can communicate with the outside world. The I/O bus either carries 0 or 5 V signals to the PC (input), or carries 0 or 5 Vdc signals from the PC (output). If the PC is to receive a signal from a 115 Vac switch, then the line voltage must be transformed to 0 or 5 Vdc for compatibility. Similarly, if the PC were required to turn on a 24-Vdc motor, solid state relays and/or power transistor amplifiers would be required external to the processor. The external electronic control circuitry is contained in the PC I/O modules.

Figure 18.20 shows a typical I/O control module. Normally, a cable from the PC will carry the TTL level signals to and from the processor. The I/O modules can normally be "daisy-chained" (with serial connections) for expanded I/O needs. Each I/O position is normally "wired" to specific memory location so that status can be both checked and/or changed.

PROGRAMMING HARDWARE

In order to establish the logic and sequence of control for any process, the PC user must enter a control program into the processor's memory. This is

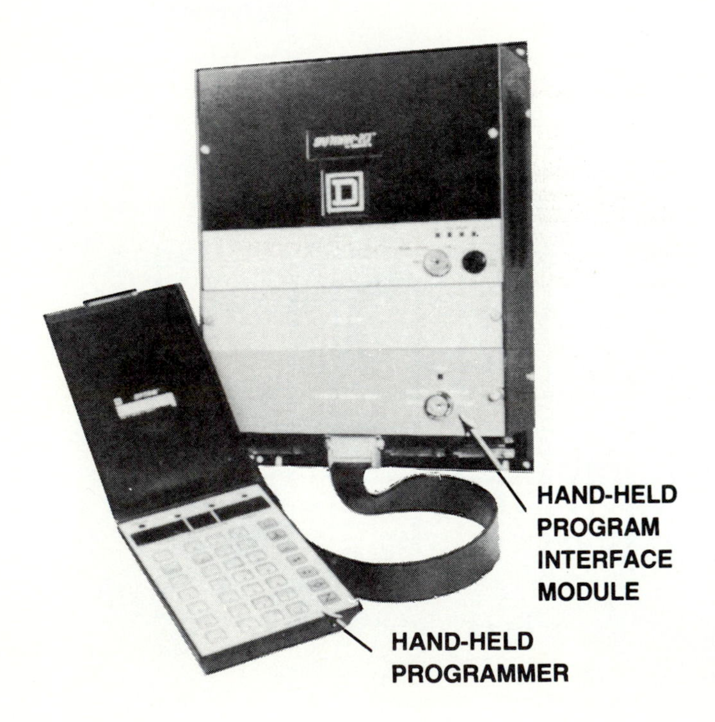

FIGURE 18.21
A hand-held programmer.

normally done using one of two types of programming devices:

1. A CRT program loader
2. A hand-held loader (Fig. 18.21)

Both devices pass the same information to the processor and look similar except for the display. Although the keystrokes might be slightly different, transferring from one device to the other does not usually present the programmer with any major difficulty. Specifics of each PC will be addressed separately.

PROGRAMMING A PROGRAMMABLE CONTROLLER

The instructions that specify input, output, and operational rules are called a *Program.* Although PCs may be used to control many different processes, the control components of these processes are usually quite similar. Physical devices are usually turned on or off to accomplish some particular activity; timers are usually set and timed out; counters are usually initialized and incremented; and so forth. Although only discrete logic will be addressed here, most PC systems are capable of far more sophisticated logic. The sophistication and program requirements, however, vary from PC to PC.

CONTACT SYMBOLOGY DIAGRAMS

A "contact symbology diagram" (commonly referred to as a ladder diagram) is a means of graphically representing the logic required in a relay logic system. Ladder diagrams precede the PC, but still represent the basic logic required by a relay panel or PC. The fundamental ladder diagram consists of a series of inputs, timers, and counters. Most simply, the ladder diagram represents the actions required (relay closure) as a function of a series of inputs that are either on or off.

A ladder diagram consists of two rails of the ladder and various control circuits—rungs (see Fig. 18.22). Each rung starts from the left rail and ends at

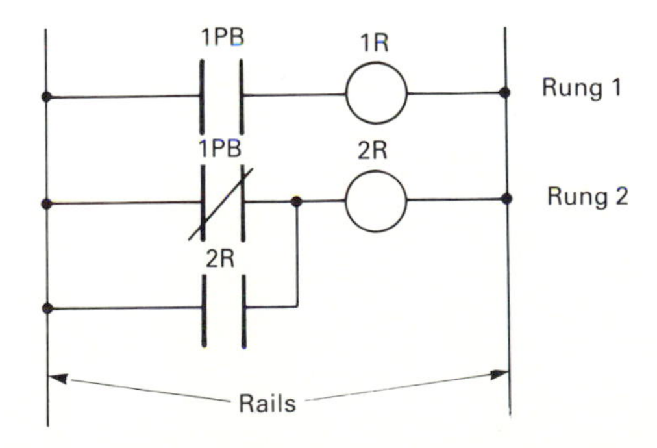

FIGURE 18.22
A typical ladder diagram.

the right rail. We can consider that the left rail is the power wire and the right rail is the ground wire. Power flows from the left rail to the right rail, and each rung must have an output to prevent a short. The output is connected to physical devices such as motors, lights, solenoids, etc. To control the output, some switches are used on the rung to form the AND and OR logic. Different rungs are not connected except through the rails. Each rung can contain only one output.

As mentioned above, a rung consists of a group of inputs and an output. In addition to input and output, there are also some internal components such as timers and counters. Inputs and outputs are actually connected to the physical switches and devices; however, internal components are not. Because of the operating voltage differences and the logic circuit requirements, the output is usually not connected to the motor or other devices directly. Instead, a relay is normally used. In Fig. 18.22, 1PB is the input pushbutton switch and 1R is the output relay. The physical wiring diagram is illustrated in Fig. 18.23.

In the circuit, when the pushbutton switch 1PB is pushed, relay 1R will energize and turn on the output circuit. In the diagram, the relay has a SPST (single-pole–single-throw) contact. If a multiple-pole contact is used, some of the internal contacts can be used in other rungs as inputs.

The basic devices that commonly appear in a ladder diagram include switches, relays, timers, and counters. Their symbology representations are shown in Fig. 18.24. In the following sections, we will discuss how to construct a ladder diagram. It is also worth noting that different PC manufacturers may use different symbols in their ladder diagrams. Relay and output symbols are usually very similar; however, timers, counters, and other register manipulation functions do not have a standard symbol. Some PC manufacturers may also provide some special function components. In this book, one set of symbols is used. We will try to make them consistent throughout instead of trying to incorporate all of the existing symbols.

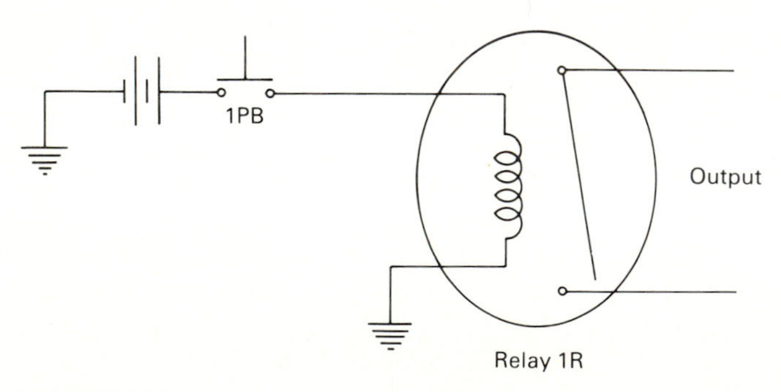

FIGURE 18.23
Circuitry represented by rung #1 in Figure 18.22.

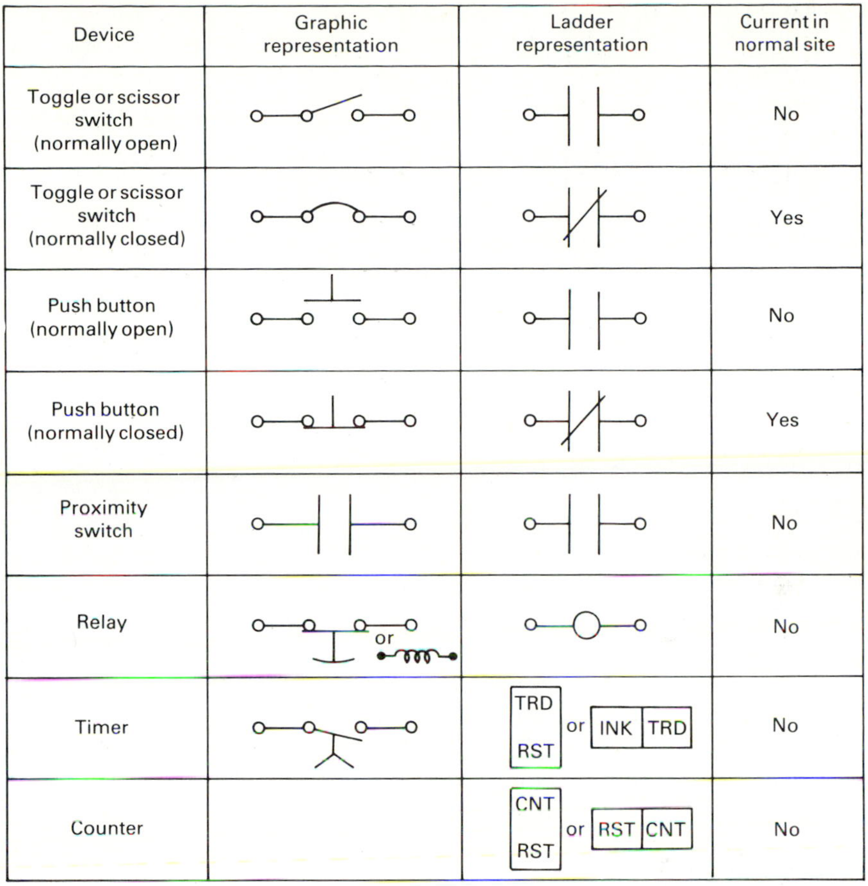

Device	Graphic representation	Ladder representation	Current in normal site
Toggle or scissor switch (normally open)			No
Toggle or scissor switch (normally closed)			Yes
Push button (normally open)			No
Push button (normally closed)			Yes
Proximity switch			No
Relay	or		No
Timer		TRD RST or INK TRD	No
Counter		CNT RST or RST CNT	No

FIGURE 18.24
Contact symbology representation.

LOGIC

By using serial and parallel connections, various types of logic can be represented in a ladder diagram. The logic states of a component are either ON (TRUE, contact closure, energize, etc.) or OFF (FALSE, contact open de-energize, etc.). In Fig. 18.22, on rung one, 1PB and 1R are connected in series. On rung two, 1PB and 2PB are in parallel then they are connected with 2R in series. Complex control logic can be represented using this simple graphical logic. In this section, the fundamentals of logic construction will be discussed. Ladder diagrams, equivalent logic representations, and truth tables are presented.

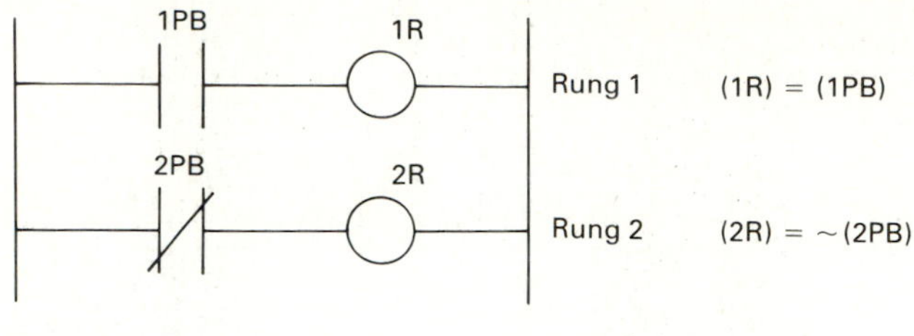

Basic logic (rung 1)

1PB	1R
ON	ON
OFF	OFF

Negation (rung 2)

2R	2PB
ON	OFF
OFF	ON

FIGURE 18.25
Simple ladder logic.

Basic Logic

The ladder diagram in Fig. 18.25 depicts the simplest circuit logic. The output is solely determined by the input. Rung one uses a normally open contact, where the state of the output is the same as the state of the input. On rung two, the output (2R) has the opposite state of the input (2PB). In the control diagram, when switch (2PB) is pushed, the output (2R) is turned OFF.

In the following sections, the symbol "( )" is used to represent the logic state of a component. The symbol "~" represents negation. AND and OR are logic operators. These symbols will be used throughout the text.

AND Logic

As mentioned, AND logic is achieved by connecting two components in series. The two rungs shown in Fig. 18.26 consist of AND logic. AND logic is the

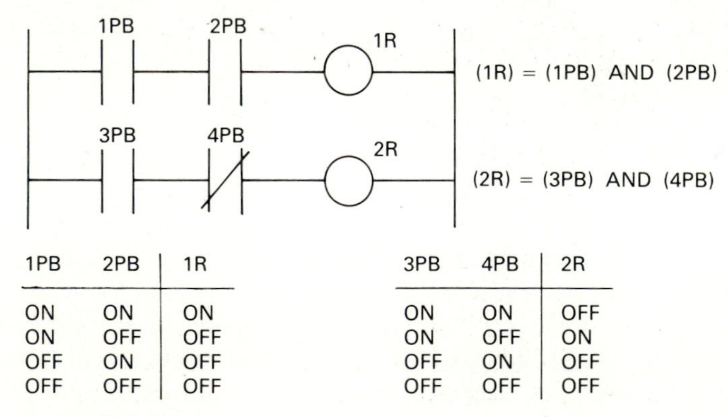

1PB	2PB	1R
ON	ON	ON
ON	OFF	OFF
OFF	ON	OFF
OFF	OFF	OFF

3PB	4PB	2R
ON	ON	OFF
ON	OFF	ON
OFF	ON	OFF
OFF	OFF	OFF

FIGURE 18.26
AND logic.

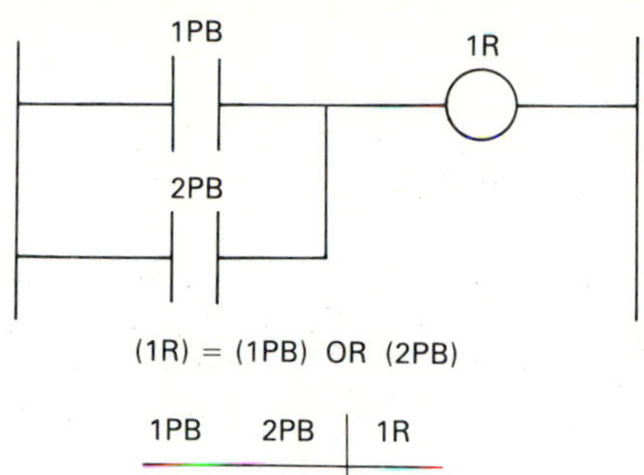

1PB	2PB	1R
ON	ON	ON
ON	OFF	ON
OFF	ON	ON
OFF	OFF	OFF

FIGURE 18.27
OR logic.

most commonly used logic. For example, in a punch-press control the operator has a foot pedal to initiate the action of the punch. A safety guarding device, such as photodiode is also used in order to prevent the accidental triggering of the punch while part of the operator's body is under the punch. The circuit on rung two can be used for this process. Let 3PB be the pedal switch, 4PB be the photodiode detector, and 2R be the control of the clutch.

OR Logic

OR is used when either one of two switches (or one or more switches out of several) is pushed and the logic output needs to be TRUE. In Fig. 18.27, either switch 1PB or 2PB can turn the output 1R ON.

Combined AND and OR Logic

A combination of AND and OR logic can also be included in one rung. For example, if we want to turn 1R on when either (1) 1PB is pushed, or (2) 2PB and 3PB are pushed, the logic can be represented by an equation.

$$(1R) = (1PB) \quad OR \quad ((2PB) \quad AND \quad (3PB))$$

The corresponding rung is:

1PB 1R

2PB 3PB

Latches

In the basic logic circuit shown in Fig. 18.25, the output 1R is controlled by the input 1PB. If a regular push button switch is used in 1PB, when 1PB is released, the output 1R is turned OFF. Sometimes it is desired to keep the output ON, and a special device must be added. One way to satisfy this requirement is to use a push button switch that has a locking device. When the switch is pushed, the output state will stay ON until it is pushed again. The same function can be accomplished by using a latched relay. Using the internal switch of a relay, one can design a latch as:

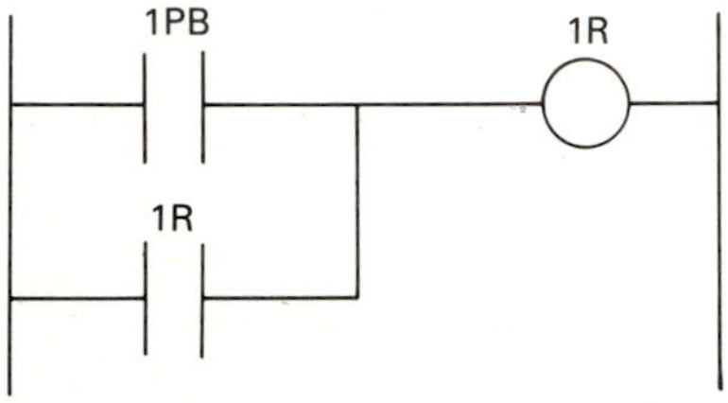

When 1PB is pushed, the relay 1R is active. The internal switch in 1R becomes closed. Even after 1PB is released, current still can flow through the switch 1R, thus keeping the output active.

In order to break the output, an additional switch is required. In the following diagram, the pushbutton switch 2PB is used to break the output:

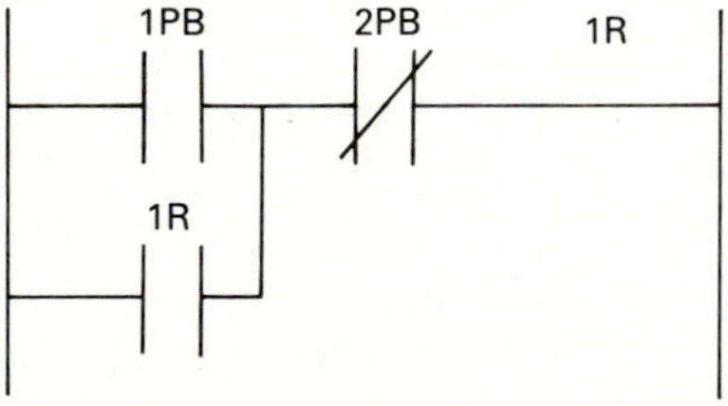

2PB is normally closed, therefore, current can flow through 2PB if it is not pushed. As soon as it is pushed, all the inputs are blocked from the output.

It is worth noting that a logic truth table alone cannot represent the logic in the above latch circuit. A timing diagram is necessary to represent the state changes over time. The timing diagram for the above diagram is shown in Fig. 18.28.

Internal Contacts

As mentioned before, an internal contact in a relay, such as one of the contacts in a double-pole relay contact, can be used in the circuit as input. In a

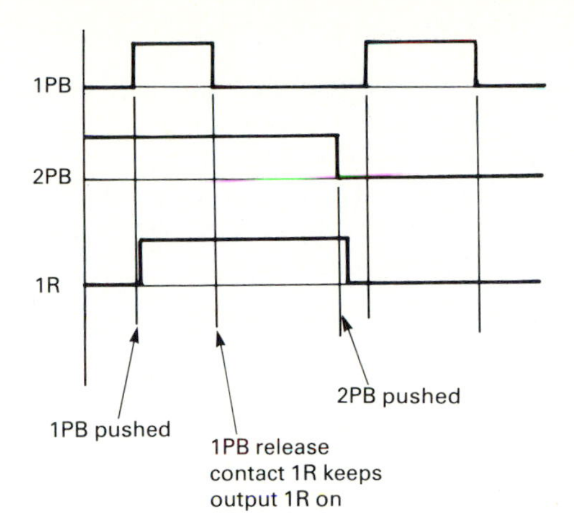

FIGURE 18.28
The timing of a latch relay.

mechanical relay panel, each internal contact must physically exist in the relay. However, a programmable controller is more flexible. Since there is no physical relay inside the PC, software internal contacts can easily be constructed. Normally a regular contact or a relay contact is represented in a PC by a memory location. The state of the memory location is set or reset by either (1) an external switch in case of a regular contact, or (2) a logical operation resulting in the closure of a relay contact. The state of a memory location can be referenced as many times as one wishes. Therefore, any contact can be used several times throughout the ladder diagram. The example in Fig. 18.29 shows that 1PB is used in two rungs, and 1R is used as input to the second rung.

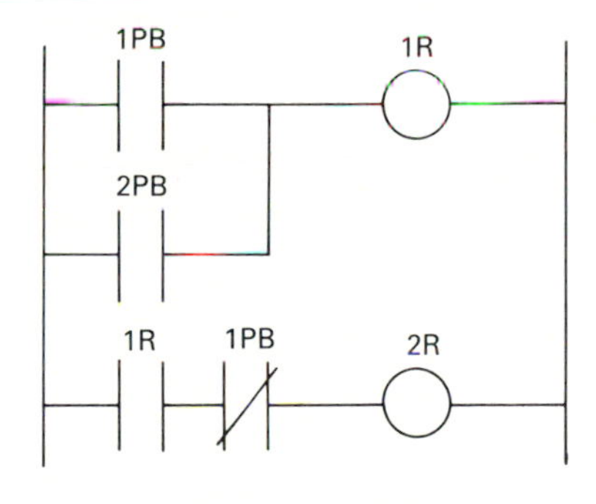

1PB	2PB	1R	2R
ON	ON	ON	OFF
ON	OFF	ON	OFF
OFF	ON	ON	ON
OFF	OFF	OFF	OFF

FIGURE 18.29
An example of a contact used in more than one rung.

Transitional Contacts

So far, all the contacts used in the discussion are conventional normally open and normally closed contacts. Most PCs also provide transitional contacts, illustrated below. Transitional contacts remain in the ON state for exactly one scan whenever the signal to which they are referenced (either an internal relay contact or switch) changes from either ON to OFF or OFF to ON. When used with reference to an output, they are referred to as "one-shot."

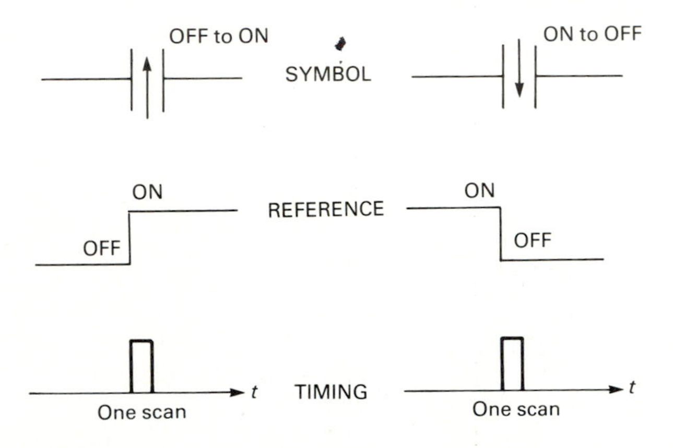

An Example Problem

Although the relay contact logic discussed so far is fairly simple, basic machine control can be accomplished using these simple circuit components. A simple application will be presented to illustrate these components.

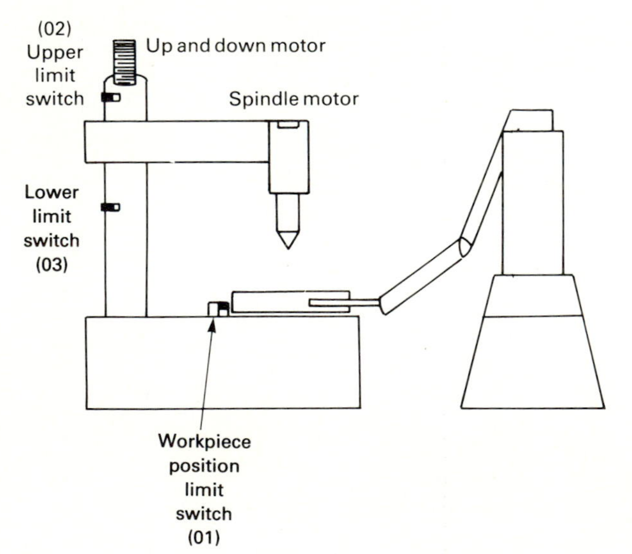

FIGURE 18.30
Drilling system for our example problem.

TABLE 18.1
I/O used for drilling example

I/O number	Usage	
1	Input	Limit switch for the workpiece position
2	Input	Spindle upper limit switch
3	Input	Spindle lower limit switch
41	Output	Spindle motor start
42	Output	Spindle up
43	Output	Spindle down

A simple drilling operation is to be automated. Four holes are to be drilled using a gang drill. The workstation consists of a drill press with a gang drill, a material handling robot, and a conveyor line. A workpiece is picked up by the robot from a conveyor, and is then placed on the drill table. When the precise location is reached, a drill cycle starts. After the drilling is complete, the robot puts the workpiece back on the conveyor and starts the next cycle. The layout of the system is shown in Fig. 18.30.

In the system, the robot is preprogrammed to perform the necessary operations. The robot is interfaced with the controller using a PC output. Output (40) triggers the robot to place the workpiece back to the conveyor and pickup a new workpiece. When the robot senses that the output (40) is on, it begins executing its cycle. Other inputs/outputs used in the drill control are listed in Table 18.1. The flowchart for the example is given in Fig. 18.31.

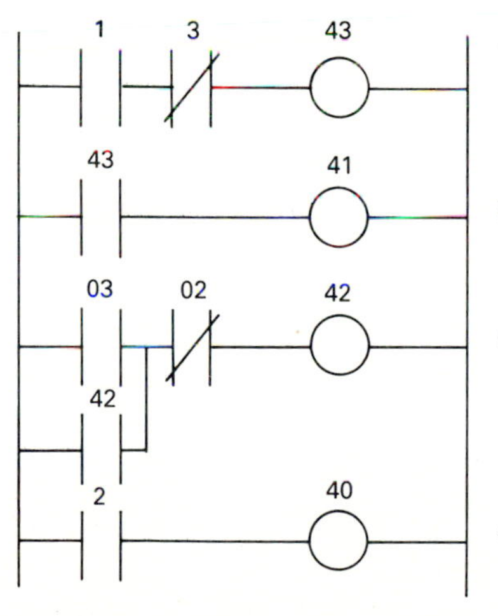

FIGURE 18.31
Ladder diagram for drilling system.

TIMERS

A timer is another ladder-diagram component that is widely used for machine control. There are several different ways to represent a timer. Basically, a timer consists of a time value, an enable/reset input, and an output contact. A simple timer rung is shown below.

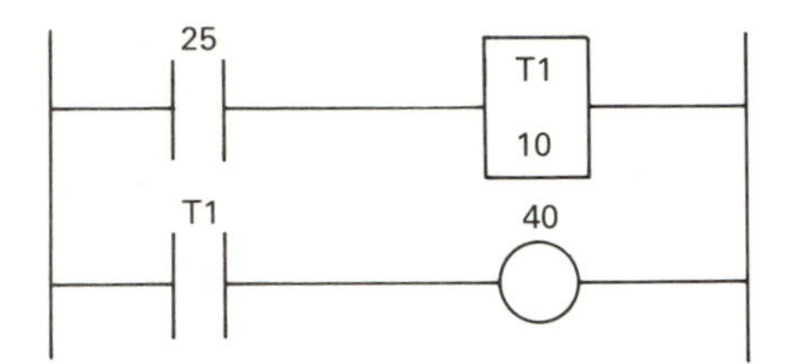

Time T1 is located in rung one. Switch 25 is connected to the enable/reset input to the timer. The timing value is set at 10 seconds. When switch 25 is closed, timer T1 begins timing. Ten seconds later, the timer generates an output. The output can be used in other rungs. For example, the input contact on rung two uses the timer T1 internal contact. Several different kinds of timing logic can be generated using a timer. The most widely used ones are discussed in the following section.

One-shot Timers

A one-shot timer is a timing device that generates a single-shot pulse output. The diagram shown in Fig. 18.32 represents a one-shot circuit. When there is

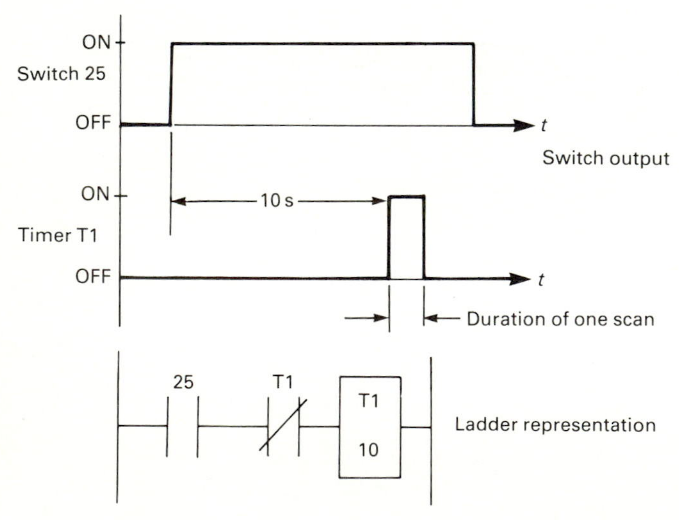

FIGURE 18.32
A one-shot timer.

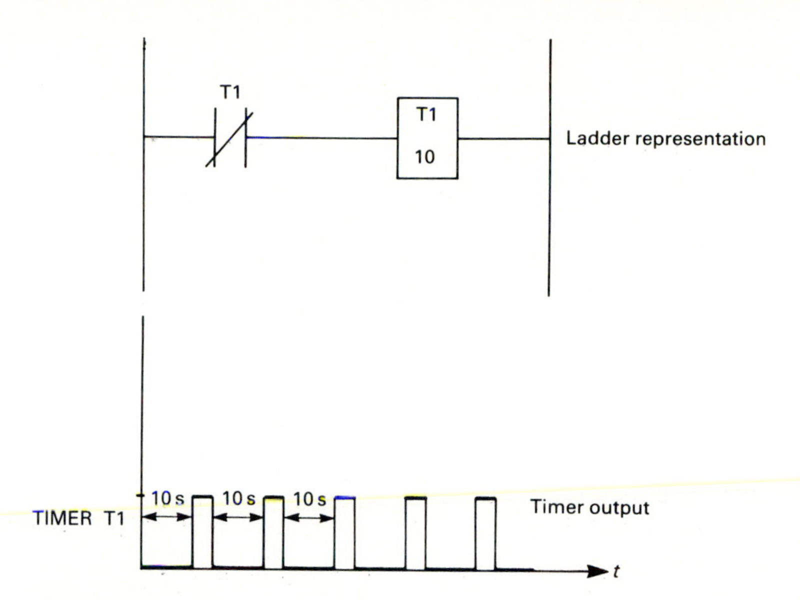

FIGURE 18.33
A repeat-cycle timer.

an input at switch 25, ten seconds later the timer outputs a shot pulse. The pulse lasts for one program scan. This output is transitional, it can be used to trigger another part of the circuit.

Repeat-cycle Timers (Self-resetting)

When cyclic output is required, a repeat-cycle timer is best suited. A repeat-cycle timer circuit outputs a pulse every several time units. It can be used in a process that has fixed cycle time; see Fig. 18.33.

Time-out Timers

A time-out timer generates an output as soon as a start button is pushed. After a certain time interval, the output is turned off. In Fig. 18.34, the output (40) is the output of the time-out timer. When #1 is pushed, the output is turned on for 10 seconds. A time-out timer can be used to control process time, such as cooling time.

Interval Timers

An interval timer turns things OFF/ON for a specific amount of time after a certain time has elapsed and then turns them ON/OFF after another specified time has passed. An interval timer is shown in Fig. 18.35. The output of T1 is

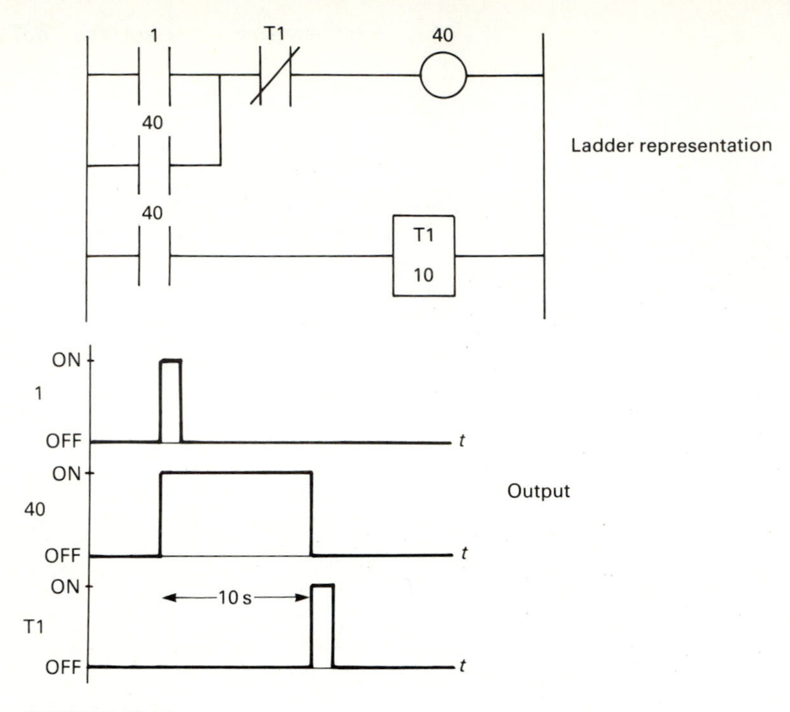

FIGURE 18.34
A time-out timer.

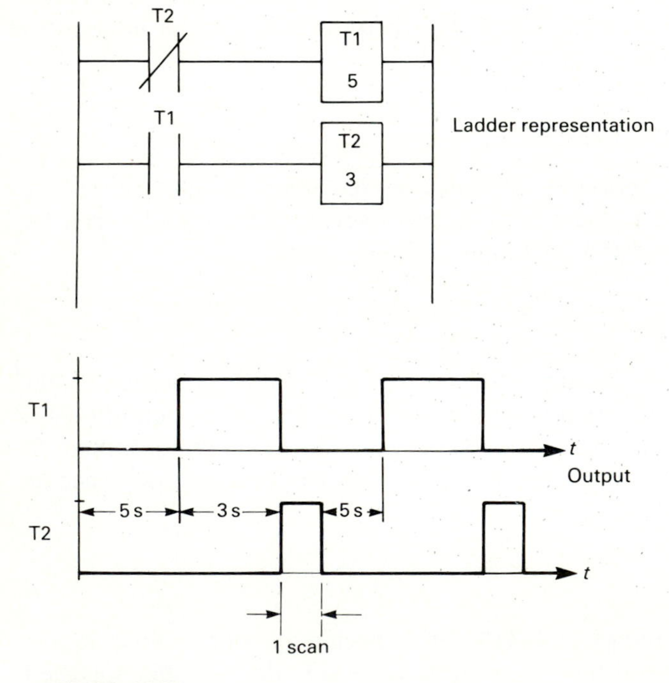

FIGURE 18.35
An interval timer.

turned on for 3 seconds and then off for 5 seconds. The intervals are adjustable by changing the timer values.

Illustrative Examples

Example 1. For our example, we will try to control the robot system shown in Fig. 18.36. A simple robot arm is used to handle the material for a machine. Three switches are used to detect the workpiece. When the three switches are on, clamp 1 is activated. Half a second later, the robot arm moves back and clamp 2 is activated. When the robot arm is moved outside the operation area (detected by switch 04), a go signal is sent to the machine (relay output 44). When the machine has finished its task, a signal is sent to switch 05. Clamp 2 is released (turn off 41), and the robot arm moves forward to pickup the part. It takes 1 second for the robot to finish the grasp operation. At the end of the 1 second, clamp 1 is released, and the robot begins moving back. The ladder logic used to control this system is shown in Fig. 18.37.

Example 2. The example is that of an automated conveyor sorting system. The conveyor system is shown in Fig. 18.38. Basically, the system operates as follows. A master switch activates the entire system. The master switch turns the conveyor on, activates the photocells, etc. When operating, the conveyor travels at a velocity of 50 ft/min. The photocells are adjusted so that: (1) parts greater than 6 in. are sorted into Bin #1; (2) parts greater than 4 in. and less than 6 in. are sorted into Bin #2; (3) the remainder of parts travel to Bin #3. The cam-piston requires a single ON signal for a duration of at least 0.1 second.

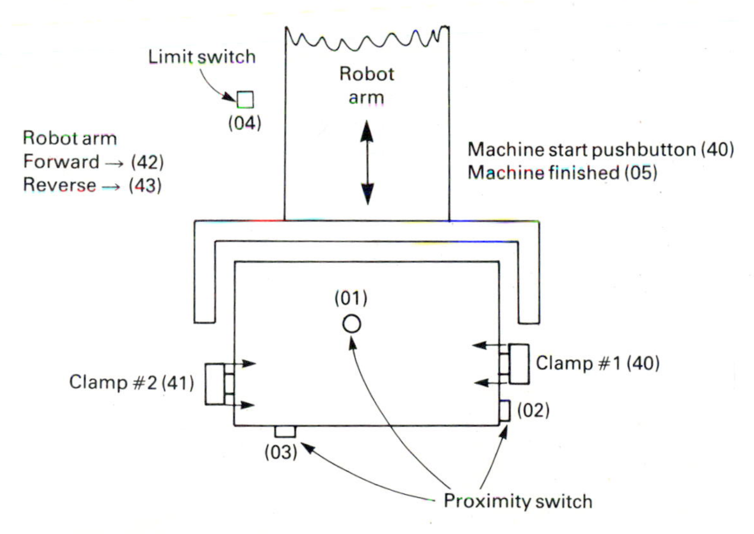

FIGURE 18.36
Robot system.

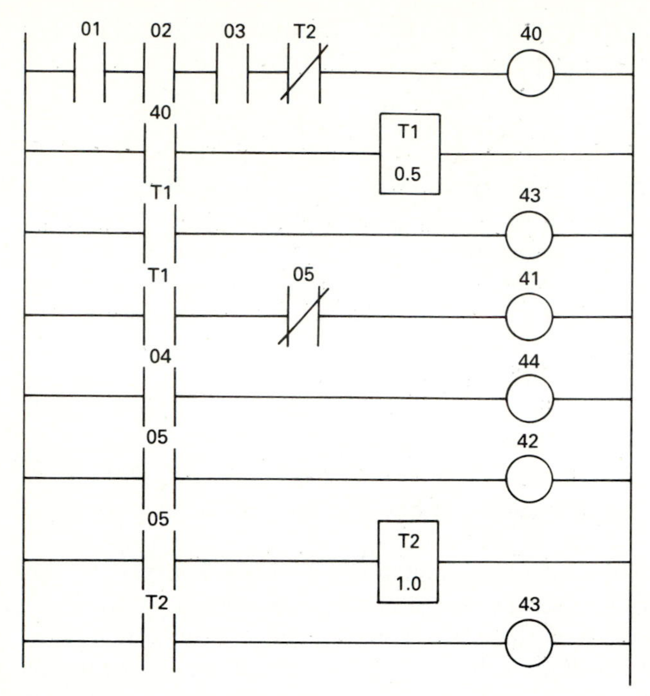

FIGURE 18.37
Ladder logic to control robot systems.

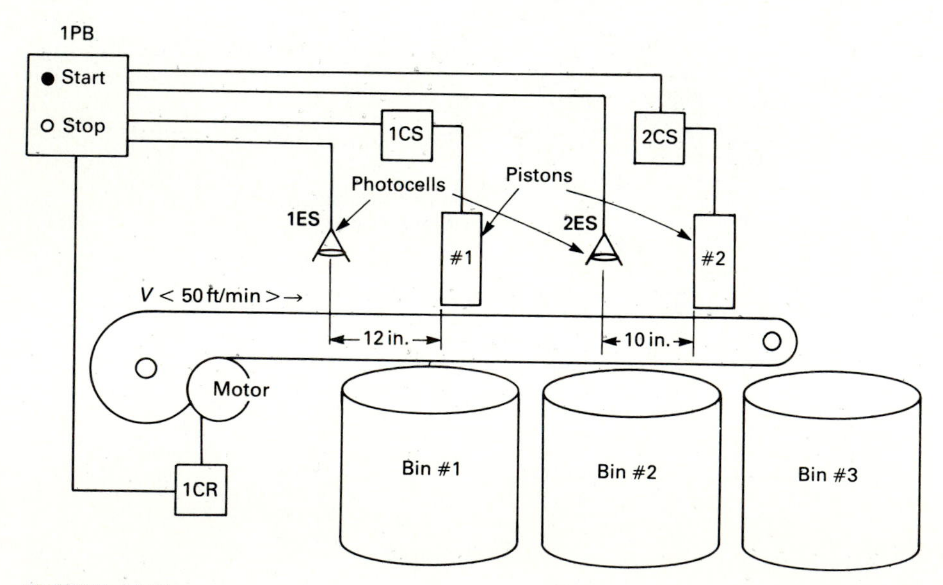

FIGURE 18.38
Conveyor sortation system.

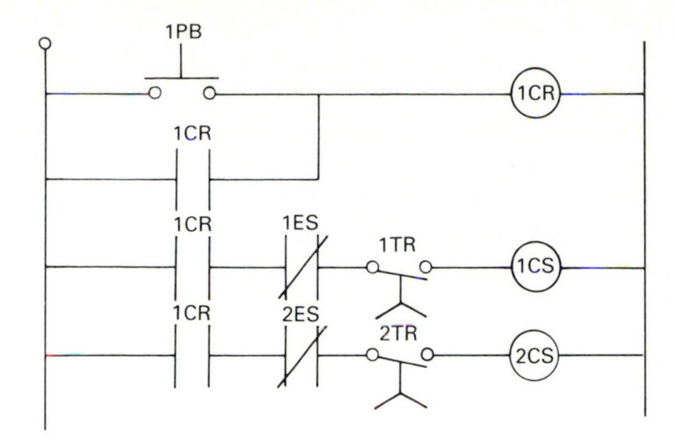

FIGURE 18.39
Conveyor sortation relay logic.

Figure 18.39 contains the relay control logic for the system. The following calculations are also necessary to calculate timer values.

1. Timing delay between the first photo cell and the first piston:

$$(50\ \text{ft/min})(12\ \text{in./ft})(1/60\ \text{min/s}) = 10\ \text{in./s}$$

$$T = \frac{S}{V} = \frac{12\ \text{in.}}{10\ \text{in./s}} = 1.2\ \text{s}$$

2. Timing delay between the second photocell and the second piston:

$$T = \frac{S}{V} = \frac{10\ \text{in.}}{10\ \text{in./s}} = 1\ \text{s}$$

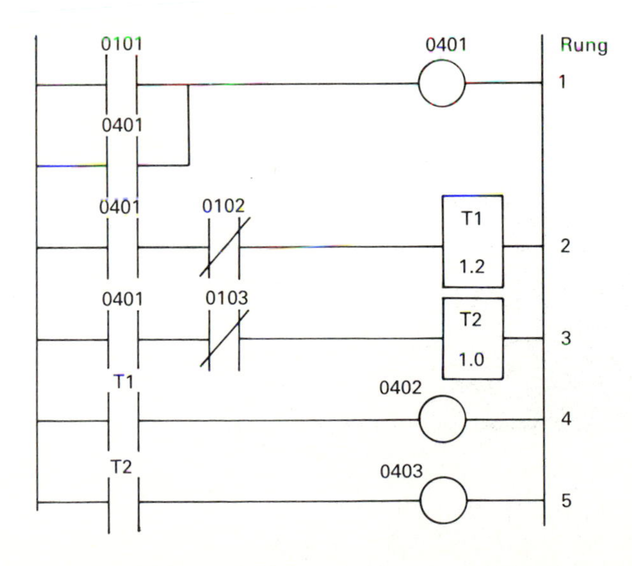

FIGURE 18.40
Ladder logic diagram for conveyor sortation.

The first rung of the relay logic indicates that, once the switch has been activated, a control relay (1 CR) is energized and latched. The second rung of the relay logic indicates that, when active, a photocell makes continuous contact until a part passes between it and its reflector. When this occurs, a timer is activated and times until a time when the piston is energized in order to push the part into Bin #1. The procedure is the same for Bin #2 and, unless otherwise indicated, all remaining parts move on to Bin #3.

Figure 18.40 illustrates how this control is translated into ladder logic. As can be seen, the translation from relay graphic symbology to a ladder diagram is a one-to-one translation.

COUNTERS

As mentioned earlier, a counter consists of an input, a reset, a count register, and an output. To use a counter, one must assign a count value to the counter. An input switch provides the input and a reset switch initializes the counting. The output is in the form of an internal relay. A typical counter with its response is shown in Fig. 18.41. The counter in the figure will count five inputs from switch #1, then output to output #40.

Example. A packaging operation was shown in Fig. 18.11. A microswitch and a counter are used to count the number of parts placed in each of the boxes.

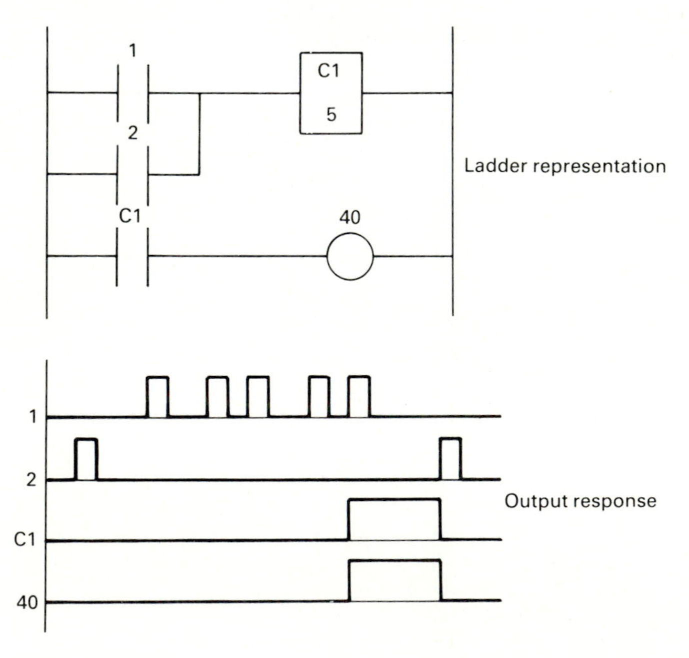

FIGURE 18.41
A typical counter.

The following is the ladder diagram for the control:

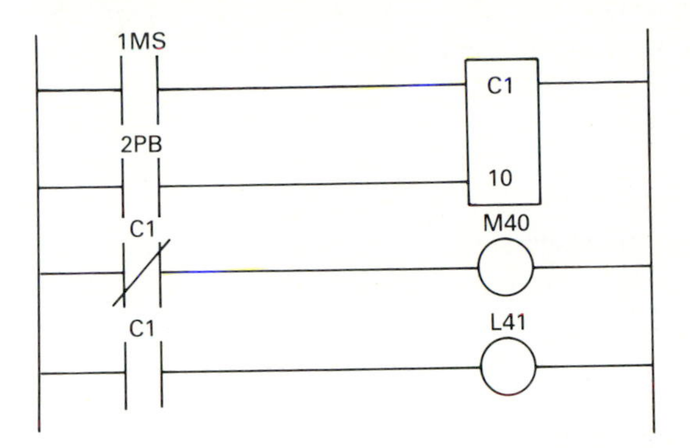

1MS is the microswitch; 2PB is a reset switch; C1 is the counter; M40 is the motor; and L41 represents a "box-full" output. The conveyor is initially running. Whenever a part is loaded into the box, the counter increments its count. When there are 10 parts in the box, the motor is turned off and the box-full output is turned on. When the reset switch is pushed, the cycle starts again.

MASTER CONTROL RELAYS

It is frequently desirable to skip a network of ladder rungs for certain scans. A master relay start and master relay stop can be placed in the program to contain several rungs. When there is an open contact on the start relay, all the rungs between the start and the stop are frozen. They are not scanned until the start relay is closed.

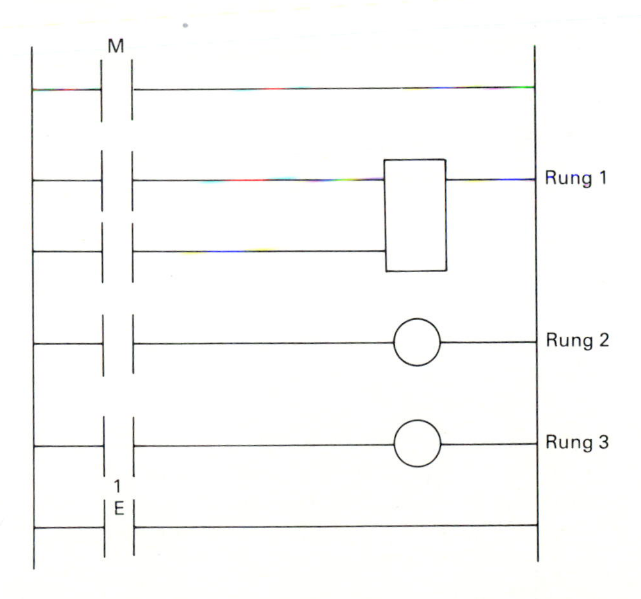

When M1 is open Rung 1 to Rung 3 will not be scanned. By using master control relays, one can save scan time. Those I/Os associated with a master relay will not be checked and can be skipped completely.

SEQUENCERS

Sequencers are used with machines or processes involving repeating operating cycles that can be segmented into steps. For a machine that requires four steps of sequences and three ON–OFF operations, a sequencer pattern is as illustrated below:

	Output			
Step	**A**	**B**	**C**	**Dwell time, s**
1	ON	OFF	OFF	5
2	ON	ON	OFF	10
3	OFF	OFF	ON	3
4	OFF	ON	OFF	9

The ONs and OFFs are repeated until the power of the machine is turned off. The dwell time defines the length of time a step requires.

There are time-driven sequencers and event-driven sequencers. Time-driven sequencers are like the one described above. A dwell time is defined for each step. Each step is up for exactly the amount of time defined.

For an event-driven sequencer, each time the sequencer is triggered, it advances one step. Since sequencer programming varies greatly, no example is given here.

SUMMARY

By now you should have become acquainted with the basic functions of a programmable controller. Since the expanded functions of a PC vary in symbology and capability from vendor to vendor, they have not been directly addressed in this text. Instead, a hypothetical PC symbology was used. After you have learned the basic concepts, you should have little difficulty in using any PC and expanding your PC command library.

The PC is an inexpensive, flexible control device that is quickly becoming the standard control device for individual processes as well as for process integration. However, it is worthwhile to note that the PC is not the only means of control. The conventional relay and computer control are two other alternatives. Although, for most control applications, the PC is both attractive and economic, in some applications, relay logic and/or computer control may be more appropriate. We should not overlook any feasible tool.

The major advantages of the PC over other devices are ease of programming the interface and robustness. When an application requires only ten or fewer relays, hardwired relay logic is probably more economic. When the applications requires more complex control, then the PC becomes a more attractive alternative.

The PC is not as flexible as computer control. However, since the PC is microcomputer-based and specifically designed for industrial control, it is much easier to use in the harsh shop environment in which a computer will not survive long. A PC can operate without problems but, if the application requires some functions that a PC does not support or a sampling rate that is greater than the shortest scan time can provide, then a computer is more desirable. If you have experience with computer interfacing and assembly programming, you will appreciate how easy it is to use a PC for control.

Actually, a PC can be seen as a computer dedicated to process control. It has some standardized interface modules, and runs a special interpretive control language. Most of the languages are ladder logic, although some PCs may support procedural languages. Following the advance of microcomputer technology and the development of interfaces and languages, the PC has become more like a computer. More and more mathematical functions have been added to PCs. Color graphics display, network capability, etc., have totally changed the image of a PC as a relay-panel substitute. In the future, we will see a greater use of PCs for all kinds of manufacturing systems control and device control. The failure to recognize and utilize these devices will undoubtedly handicap manufacturing, electrical, and mechanical engineers.

REVIEW QUESTIONS

18.1. Using the classification presented, classify the following switches by type:
  1. Simple light-switch
  2. Automobile ignition switch
  3. Automobile door light switch
  4. Step pad for a door opener.

18.2. What is nonvolatile memory? Give some examples of nonvolatile memory.

18.3. How many bits are there in 48 Kbytes?

18.4. What are some advantages of the PC over a relay logic panel?

18.5. What does a processor do?

18.6. If a PC has a 16-bit word and 8 K words of RAM, how many bits of information can it store?

18.7. If a PC were to drive a 5 Vdc motor, would an additional I/O module be required? Defend your answer.

18.8. What are the advantages of RAM versus nonvolatile memory?

18.9. Design a ladder diagram to turn light #40 on 1 second for every 3 seconds:

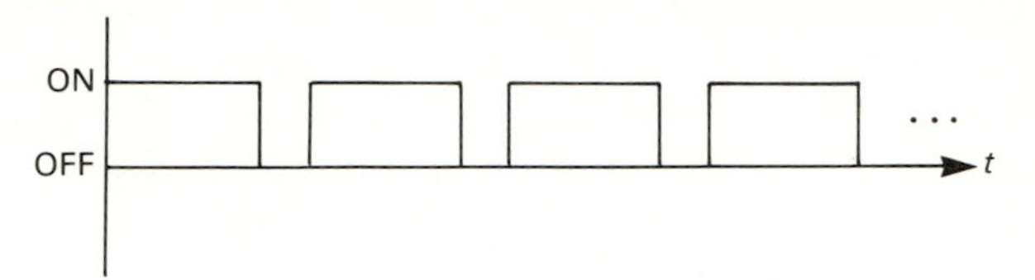

18.10. Design an XOR logic with a ladder diagram. Use switches #1, #2 as inputs, and light #40 as the output.

REVIEW PROBLEMS

18.1. Using the symbols provided, draw the circuits for each of the switches in Review Question 18.1.

18.2. Create the wiring diagram for the design shown in Fig. R18.2. Boxes are traveling on the main conveyor. When the photosensor detects the presence of a box, and there is nothing on the lower level conveyor, the push rod is activated. The push rod requires 110 Vdc, all the switches are rated at 5 Vdc. Select appropriate relays for your design. Include in your design the appropriate power sources.

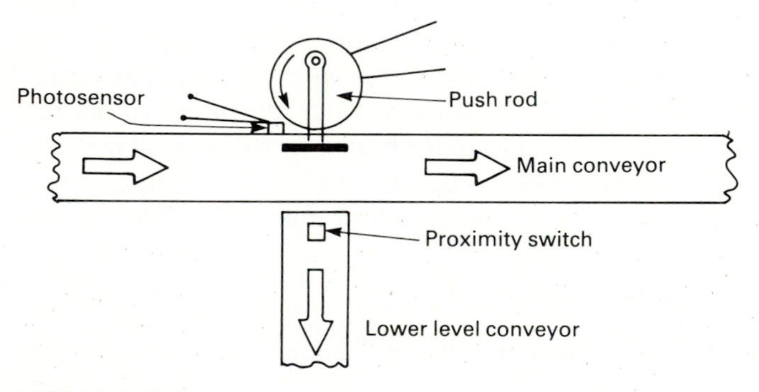

FIGURE R18.2

18.3. Create a simple control system for an automatic washing machine using a PC.

18.4. Create a simple control system for a clothes dryer.

18.5. The example illustrated in Figs 18.38 and 18.39 is to be embellished for the following. The conveyor segment shown in Fig. 18.38 is only a small segment of a much larger conveyor system. In order to reduce energy costs, the segment is turned OFF whenever no parts are on the conveyor. This is accomplished by placing another photocell on the conveyor segment immediately proceeding the one shown in Fig. 18.39. Create the control logic required to control the conveyor.

18.6. Design a ladder diagram that will turn on two lights #40, #41 when switch #1 is closed. When switch #1 is released, light #40 goes off, but light #41 will still be on.

18.7. Design a ladder diagram to control the drilling machine shown in Fig. R18.7. *Operation procedure.* A part is on the table (proximity switch closes), spindle motor starts. After 3 seconds, the Z axis moves down until the lower limit switch is touched. When the lower limit switch is touched, the spindle motor stops and moves up. One second later, the part is ejected off the table by a solenoid ejector. When the spindle reaches the upper limit, the Z axis motor stops. For every 10 parts drilled, a tool-check light comes on to warn the operator to check the tool life.

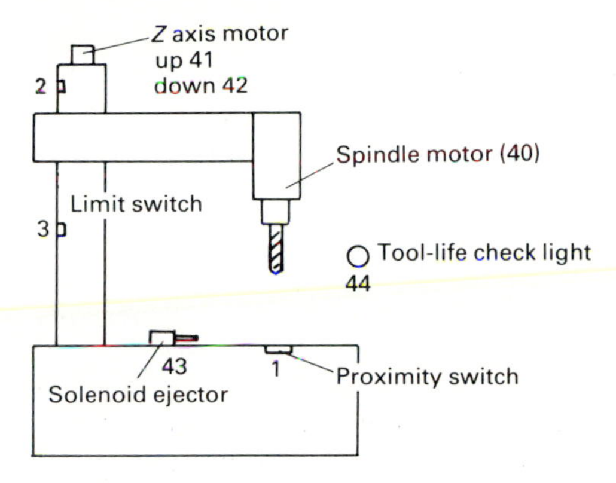

FIGURE R18.7

REFERENCES

1. McCluskey, E. J.: *Electrical and Electronic Engineering*, McGraw-Hill, New York, 1965.
2. Carlson, A. Bruce, and Gisser, David G.: *Electrical Engineering Concepts and Applications*, Addison-Wesley, Reading, Mass., 1981.
3. Keister, W., Ritchie, A. E., and Washburn, S. H.: *The Design of Switching Circuits*, Van Nostrand Princeton, N.J., 1957.
4. Caldwell, S. H.: *Switching Circuits and Logical Design*, Wiley, New York, 1958.
5. Morley, R. E., and Libbey, A. H.: "The PC from A to Z—Part 1," *Assembly Engineering*, pp. 34–39, May 1982.
6. Bennett, W. L.: "The PC from A to Z—Part 2," *Assembly Engineering*, pp. 32–38, June 1982.
7. Rusch, B. R.: "The PC from A to Z—Part 3," *Assembly Engineering*, pp. 44–47, July 1982.
8. Hickey, Jack: "Programmable Controller Roundup," *Control Systems*, pp. 57–64, July 1981.
9. Inglesby, Tom: "The PC and Automation," *Assembly Engineering*, pp. 28–31, August 1982.
10. Schaldach, Matt K.: "Management Information and Graphics from your Programmable Controller," *Man Machine Interference for Industrial Control*, Kopess and Williams, Control Engineering, pp. 77–86.
11. Ellis, K.: "Color Graphics CRT's Provide 'Window' into Factor Operations," *Instruments and Control Systems*, pp. 29–32, Feb. 1983.
12. PC course—A ten part article, *Instruments and Control Systems*, March–November, 1980.
13. Whitehouse, Robert A.: "The Role of Programmable Controllers in Industrial Data Highways," *Autofact Europe Conference Proceedings*, pp. 7-31 to 7-48, 1983.

CHAPTER 19

NUMERICAL CONTROL AND ROBOTICS

NUMERICAL CONTROL

BACKGROUND

Two generations of automated manufacturing have already evolved through U.S. industry. The first generation of mechanization evolved from simple mechanisms to complex mechanical machine systems that are still being developed today. The second generation of relays and switching logic, which brought automation to a variety of mid- to high-volume production systems, also continues. Flexible manufacturing marks the third generation of modern manufacturing and has come upon us with a bang (as a revolution rather than evolution). Today's flexible automation is driven by increased international competition from Japan, West Germany and other advanced countries.

The basis of third-generation manufacturing is numerical control, or NC as it is more commonly known, and robotics. In the 25 years since its commercial development, numerical control has impacted on manufacturing as perhaps no previous invention. Today, the operation and appearance of production systems has changed, propelling us into the world of computer-aided manufacture (CAM).

Following World War II, the U.S. and the U.S.S.R. were involved in the "Cold War" in which development of aviation supremacy was paramount. U.S. Air Force studies showed that in the development of new aircraft, the manufacturing cost was directly proportional to the maximum flight speed. This, of course, was due to the sophisticated contours required by these aircraft. For one of the few times in our history, our military readiness was being constrained by our manufacturing ability.

John Parsons, of the Parsons Corporation, experimented with the idea of generating three-axis curve data automatically, and using the data to control a machine tool. In 1947, Parsons developed a jig borer that was coupled to a computer. Using punched cards, Parsons was able to control the machine's position. In 1949 the U.S. Air Force, encouraged by Parson's success, commissioned the Massachusetts Institute of Technology to develop a prototype of a "programmable" milling machine. In 1952, a modified three-axis Cincinnati Hydrotel milling machine was demonstrated, and the term "numerical control" was coined (Fig. 19.1). Since then, NC has taken on a spectrum of activities, including tool changing, process monitoring, computer integration, etc. By 1955, the Air Force began awarding some $35 million for the manufacture of approximately 100 NC machines.

By 1960, NC machines appeared on a commercial basis. Some 100 NC machines manufactured by a variety of companies were demonstrated at the 1960 Machine Tool Show in Chicago (*American Machinist,* 1964). These original machines were quickly sold, usually at a price of around $50,000 or less. Today, a wide variety of NC equipment is designed and sold throughout

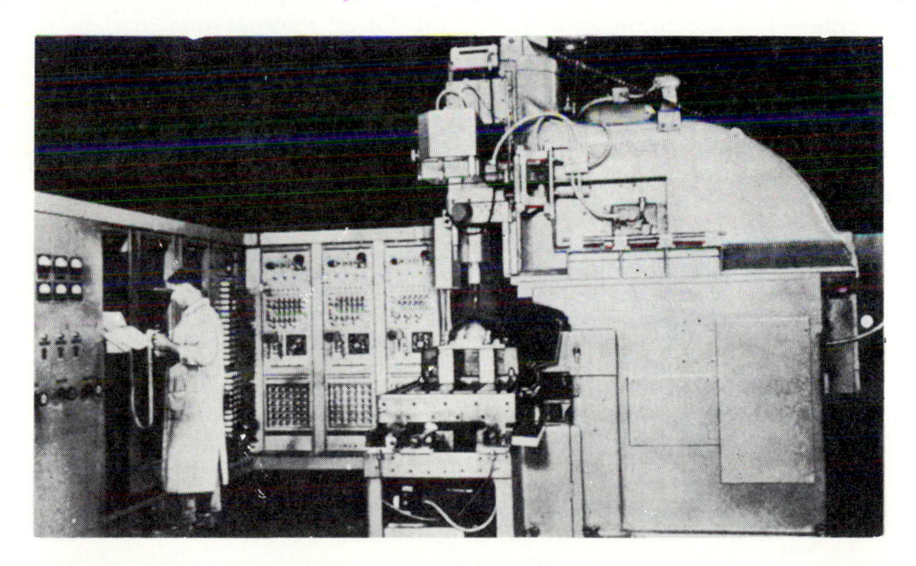

FIGURE 19.1
Believed to be the first successful numerical control application, this milling machine was demonstrated in 1952 at the Servo Mechanism Laboratory of the Massachusetts Institute of Technology.

the world. Applications including lathes, milling machines, drilling machines, punch presses, bending machines, flame cutters, wire wrapping, etc., have appeared. NC equipment is delivered ranging in size from table-top applications to skin mills and flame cutters more than 100 feet long. In this chapter, we will explore the basics of numerical control, numerical control programming, and robotics.

BASICS OF NUMERICAL CONTROL

Controlling a machine tool using a punched tape or stored program is known as numerical control (NC). NC has been defined by the Electronic Industries Association (EIA) as "a system in which actions are controlled by the direct insertion of numerical data at some point. The system must automatically interpret at least some portion of this data." The numerical data required to produce a part is known as a part program.

A numerical control machine tool system contains a machine control unit (MCU) and the machine tool itself (see Fig. 19.2). The MCU is further divided into two elements: the data processing unit (DPU) and the control loops unit (CLU) [10]. The DPU processes the coded data read from the tape or other media and passes information on the position of each axis, required direction of motion, feed rate, and auxiliary function control signals to the CLU. The CLU operates the drive mechanisms of the machine, receives feedback signals concerning the actual position and velocity of each of the axes, and signals the completion of operation. The DPU sequentially reads the data. When each line has completed execution as noted by the CLU, another line of data is read.

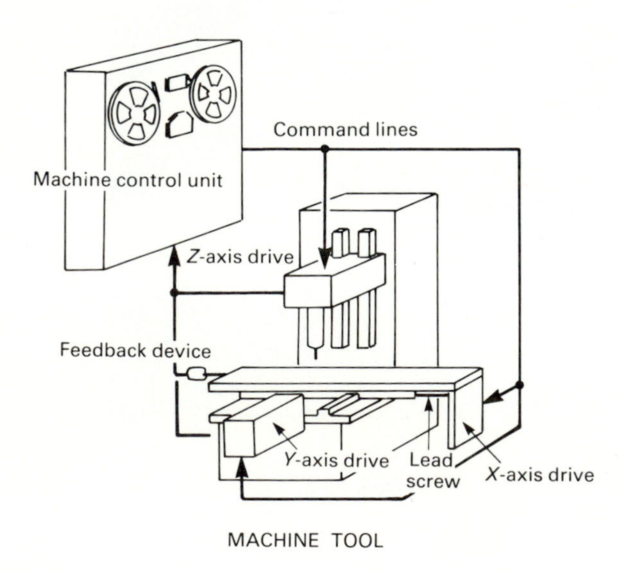

FIGURE 19.2
Numerical control system. (*Reproduced from [5].*)

A data processing unit consists of some or all of the following parts:

1. Data input device such as a paper tape reader, magnetic tape reader, RS232-C port, etc.
2. Data-reading circuits and parity-checking logic
3. Decoding circuits for distributing data among the controlled axes
4. An interpolator, which supplies machine-motion commands between data points for tool motion

A control loops unit, on the other hand, consists of the following:

1. Position control loops for all the axes of motion, where each axis has a separate control loop
2. Velocity control loops, where feed control is required
3. Deceleration and backlash takeup circuits
4. Auxiliary functions control, such as coolant on/off, gear changes, spindle on/off control

Geometric and kinematic data are typically fed from the DPU to the CLU. The CLU then governs the physical system based on the data from the DPU.

CLASSIFICATION OF NUMERICAL CONTROL MACHINES

Numerical control machines can be generally classified using the following categories [11]:

1. Power drives: hydraulic, pneumatic and electric
2. Machine tool control: point-to-point and contouring (or continuous path)
3. Positioning systems: incremental positioning and absolute positioning
4. Control loops: open-loop and closed-loop
5. Coordinate definition: right-hand and left-hand coordinate system

Power Drives

One of the most notable distinguishing features of an industrial-quality NC machine is its power source. The power source usually determines the range of the machine's performance capabilities and in turn its feasibility for various applications. The three principal power sources are hydraulic, pneumatic, and electric.

Most high-power machines are generally driven by hydraulic power. Hydraulics can deliver large forces, so that the machine slides move with more uniform speed. Offsetting this advantage is cost, which is usually higher for hydraulic machines than for electric or pneumatic models of equivalent rating. Hydraulic power also requires additional peripherals, such as a reservoir, valves, etc. The major disadvantages of hydraulic power are the noise normally associated with these units, and hydraulic contamination from leaking fluid.

Pneumatic machines are often the least expensive power alternative. The availability of shop air at about 90 to 100 psi that can be tapped to power a pneumatic machine is an added advantage. Each axis on a pneumatic machine is normally controlled only at the end points. However, by varying the timing and sequence, an infinite variation of programmed setups is possible. The motion in a pneumatic machine is usually nonuniform in nature (typically called "bang-bang," with high acceleration and deceleration).

Electric drives are the most applicable for precision jobs or when close precision control is desired. Sophisticated motion-control features are typical of electrical machines. There are two major groups among electric drives. One type uses stepper motors, which are driven a precise angular rotation for every voltage pulse issued by the controller. Stepper motor movements can be very precise provided the torque load does not exceed the motor's design limits [1]. Because of this inherent accuracy, stepper motor systems are usually of the open-loop type. The other kind of electric drive is the servodrive. These motors invariably incorporate feedback loops for the driven components back from the driver to a controller. There is continuous monitoring of positions and error conditions are promptly corrected by issuing appropriate voltage or current response changes until the position and velocity error goes to zero. The servodrive is a continuous-position device whose position is measured by an encoded transducer, inductosyn, resolver, or other similar feedback device. Servomotors typically provide a smoother and more continuously controllable movement.

Motion Control: Point-to-Point and Contouring

Point-to-point (PTP) control systems position the tool from one point to another within a coordinate system. Each tool axis is controlled separately, and the motion of the axes is either one-axis-at-a-time (Fig. 19.3*a*) or multiple-axes motion (Fig. 19.3*b*) with constant velocity on each axis. The control of motion is always defined by the programmed points, not by the path between them. The simplest example of such a system is a drilling machine, in which the workpiece or tool is moved along two axes until the center of the tool is positioned over the desired hole location. The drill path and its feed while traveling from one point to the next point are assumed to be unimportant. The path from the starting point to the final position is not controlled. The data for the desired position is given by coordinate values. Rapid traverse is usually a point-to-point operation, even on contouring systems.

A contouring control system independently controls the speed of each of the driving motors i.e., $\frac{dx}{dt}$, $\frac{dy}{dt}$, and $\frac{dz}{dt}$ are controlled at all times. It therefore

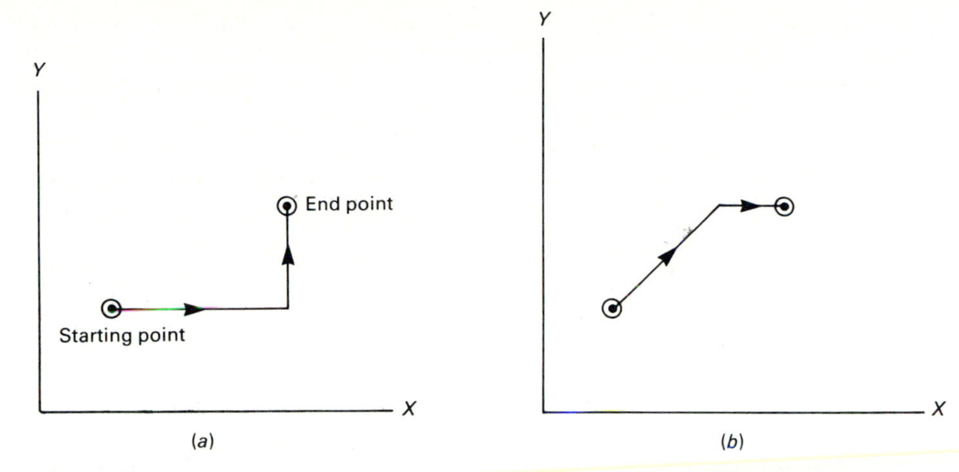

FIGURE 19.3
Point-to-point motion control. (*a*) One axis at a time; (*b*) multiple-axes motion.

is able to regulate the rate of travel of the table (or spindle) for at least two axes simultaneously. In the example illustrated in Fig. 19.4, note that as the cutter moves 9.397 in. to the right on the *X*-axis, it also moves 3.420 in. up in the *Y*-axis. Note that 0.9397 and 0.3420 are, respectively, the cosine and sine of 20 degrees. This type of cut requires that the motion of the two axes be translated by motors running at unequal speeds. This control requires an elaborate control and drive system. Also, contouring control normally requires a great deal of computation work by the programmer. It is often desirable to utilize the services of a computer as an interpolator to make calculations more quickly and reliably.

An interpolator is an electronic device that automatically computes the feed rate and path computations for a large number of standard curves such as circles and parabolas. Given only the mathematical description of a circle or other standard curve, an interpolator can calculate the feedrates and all the intermediate points necessary to produce a continuous path resulting in a

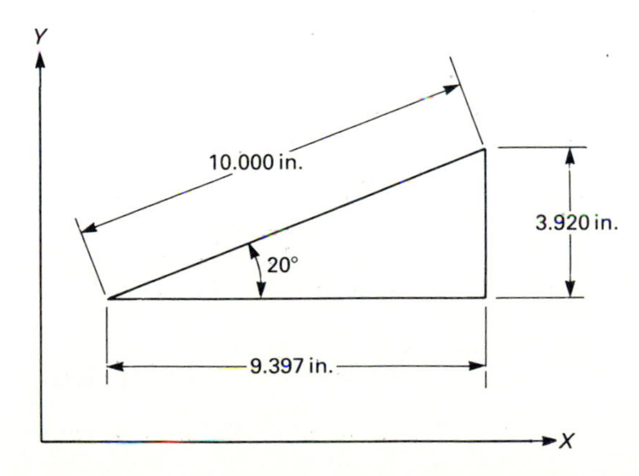

FIGURE 19.4
Contouring motion control using linear interpolation.

smooth curve. This feature saves a great deal of programming time, cost, and errors. The interpolator is usually a part of the DPU. Three types of interpolations are standard with a variety of NC systems: linear, circular and parabolic.

LINEAR INTERPOLATION. Control of the rate of travel in two directions, proportionally to the distance moved, is called linear interpolation [6]. The ratio between the pulse frequencies in two axes is equal to the ratio between the required incremental distance.

The functional concepts of this simplest and commonest interpolator are shown in Fig. 19.5. In this example, the linear interpolator inputs are the X- and Y-distance increments and the path velocity to be maintained. The output from the interpolator is a stream of command pulses to the X- and Y-axis position control systems. As each pulse represents a position increment, the total number of pulses fed to an axis determines the total axis displacement, while the pulse rate determines the axis velocity [7].

Example. If the example path given in Fig. 19.4 is to be followed by a machine

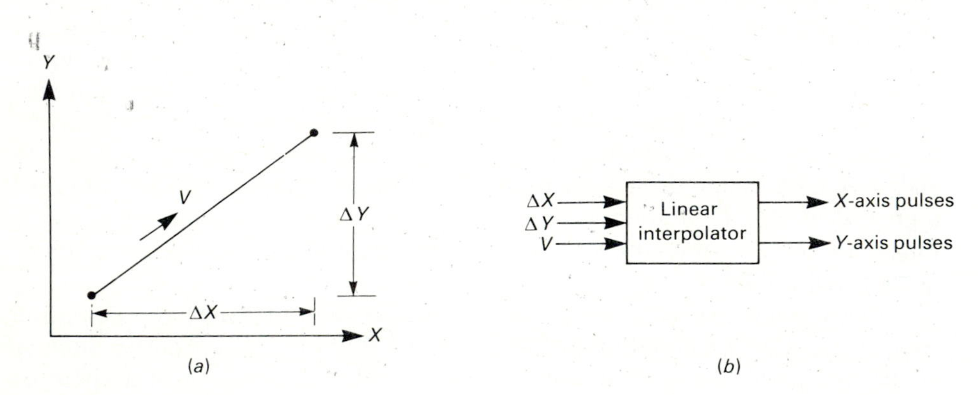

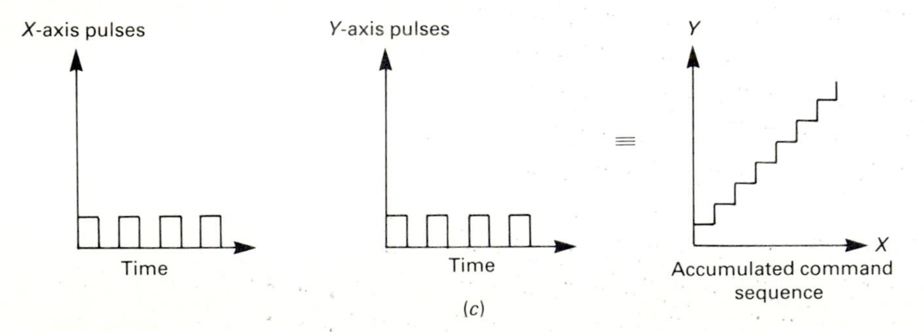

FIGURE 19.5
Linear interpolation—a functional description. (a) The desired path; (b) the functional characteristics; (c) the output characteristics. (*Adapted from [7].*)

at a velocity of 60 in./min and a basic length unit (BLU) of 0.0001 in./pulse is used, the interpolator must perform the following calculations and communicate the results to the CLU:

1. 9.397 in./(0.0001 in./pulse) = 93970 pulses for the X-axis motor controller
2. 3.420 in./(0.0001 in./pulse) = 34200 pulses for the Y-axis motor controller
3. The path length, $S = \sqrt{(9.397)^2 + (3.420)^2}$
   $= \sqrt{100} = 10$ in.
4. The path velocity = 60 in./min (given)
5. Total cutting time = 10 in./(60 in./min) = 10 s
6. Pulse rate in X-axis = 93,970 pulses/10 s
   = 9370 pulses/s
7. Pulse rate in Y-axis = 34,200 pulses/10 s
   = 3420 pulses/s

CIRCULAR INTERPOLATION. Circular interpolation is the process of generating commands that create circular machine movements as shown in Fig. 19.6. In the case of circular interpolation, the interpolated points lie on a specified arc between a pair of end points. In many cases, the circular interpolation is limited to one quadrant in the machine-tool system, i.e., the given end points are located in the same quadrant of the circle.

The inputs required are the X- and Y-axis distances of the center of the circle with respect to the starting points (i and j respectively in Fig. 19.6), X- and Y-axis distances (X and Y-axis distances (x and y) of the end point with respect to the starting point, and the direction of cutter travel (clockwise CW in Fig. 19.6). Circular arcs are normally generated in one of two manners. The first method uses two digital differential analyzers (DDA) to regulate the velocity of the tool through the cutting space. The cutter moves along a circular path in a smooth motion.

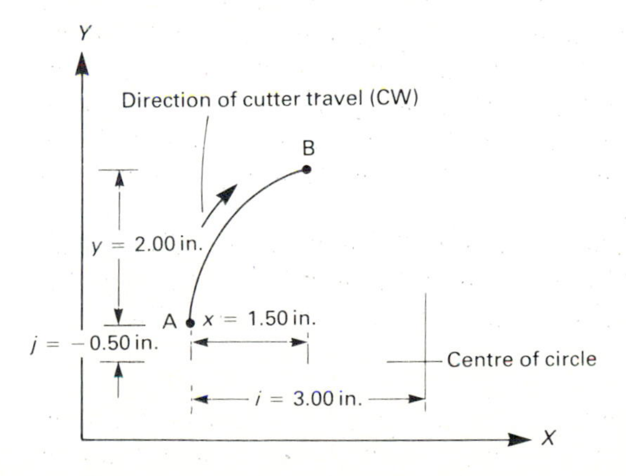

FIGURE 19.6
Clockwise circular interpolation.

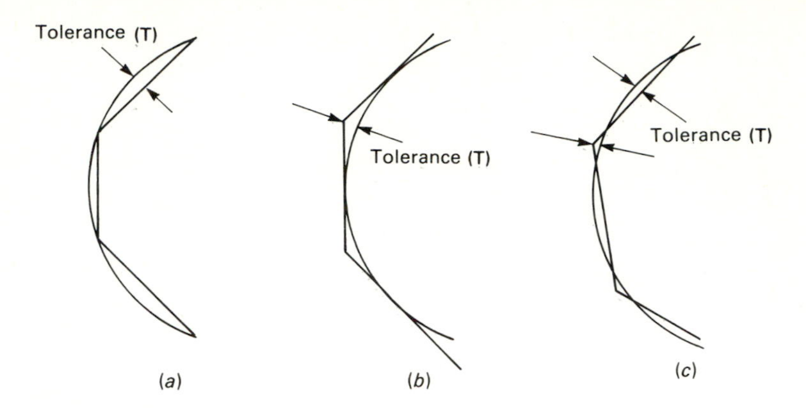

FIGURE 19.7
Straight-line approximations of a circle. (*a*) Chords; (*b*) tangents; (*c*) secants.

In the second method, a linear interpolator is used to cut curves to very close tolerances by cutting a series of straight lines—chords, tangents, or secants (Fig. 19.7). It is not difficult to compute mathematically the x- and y-coordinates of the end of each straight line around a circular path, however, it can be time-consuming. Hence, computer assistance is frequently sought to calculate the x- and y-coordinate values, which are then used in the program. Alternatively, higher level NC programming languages like APT can be used to generate these coordinates and other machine parameters automatically.

Chords for circular approximation can be defined using a prescribed

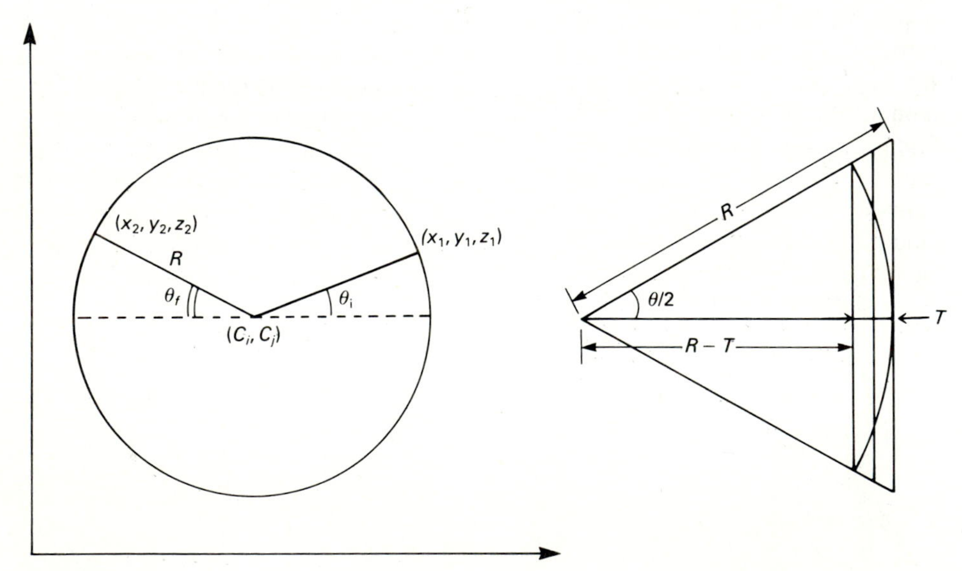

FIGURE 19.8
Circular interpolation.

inner, outer, or total tolerance. Once the tolerance has been defined, the maximum chord angle, θ, can be computed (for inner tolerance) as:

$$\cos\frac{\theta}{2}=\frac{R-T}{R} \tag{19.1}$$

$$\theta=2\cos^{-1}\left(\frac{R-T}{R}\right) \tag{19.2}$$

for the chord shown in Fig. 19.8. The points corresponding to the ends of these chords can also be defined using simple geometry.

PARABOLIC INTERPOLATION. Many workpieces contain curves that can be approximations of polynomials of degree higher than two. The parabola (and the ellipse and the hyperbola) are second-order polynomials but can fit most any curve within the required tolerance provided a sufficiently small interval is taken between each set of three data points. The parabolic or elliptic shapes can be quite helpful for free-form contours where the path accuracy does not require a complex combination of linear and circular segments.

A parabolic arc is defined by two end points M and N and the intersection T of the tangents to the arc at those end points (Fig. 19.9). The output from a circular interpolator can be modified to produce either parabolic or elliptic path shapes. This can be done by using a scaler multiplication on one axis of a circular path (see Fig. 19.10). The multiplicant for an ellipse is a number less than 1. For a parabola the number is greater than 1.

CONTOURING CONTROL. Contouring or continuous-path systems are the most flexible and versatile of the NC controllers. They are capable of performing both point-to-point (PTP) and straight-cut operations. The path of the tool (or workpiece) is continuously controlled to generate the desired geometry of the workpiece in these systems (rapid traverse is generally PTP). Circular interpolation as well as linear interpolation or parabolic interpolation features can also be employed in such systems. Milling machines are good examples of contouring systems, in which all axes of motion may move simultaneously, each at a diferent velocity. When a nonlinear path is required, the axial velocity changes at a different rate for each axis to produce the

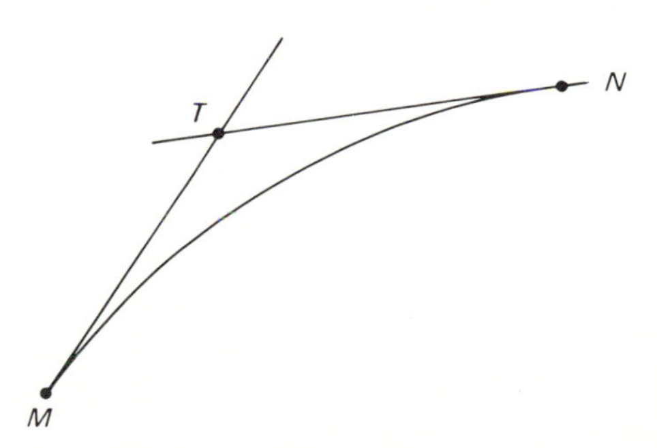

FIGURE 19.9
Defining a parabolic arc.

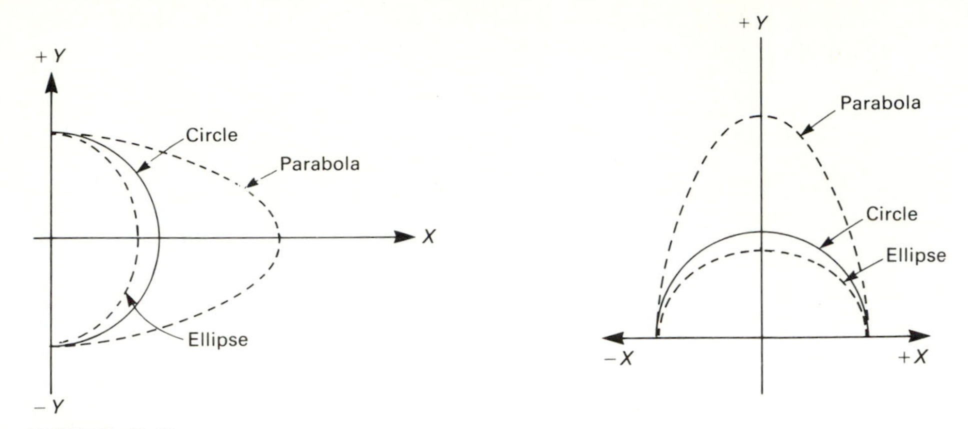

FIGURE 19.10
Generation of parabolic/elliptic paths.

required feed rate along a specified tool path. For example, cutting a circular contour requires a sinusoidal rate of change of speed in one axis, with a corresponding cosinusoidal rate of change of speed in the other.

Positioning Systems: Absolute and Incremental

In incremental NC systems, the distance for the current position is measured from the preceding position of the table. For example, consider Fig. 19.11, where $P1$, $P2$, and $P3$ are the successive points of the cutter. The X- and

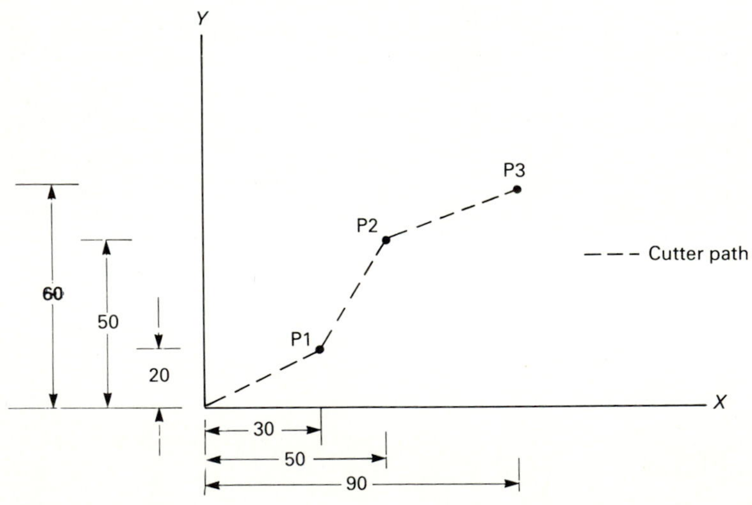

FIGURE 19.11
Absolute and incremental systems of NC programming.

Y-axis dimensions will be expressed as follows:

Point 1	$x + 30$	$y + 20$
Point 2	$x + 20$	$y + 30$
Point 3	$x + 40$	$y + 10$

Both the programming and the feedback elements are referenced with respect to the last point in an incremental system.

In absolute NC systems, all positions are referenced relative to the same zero or reference point. All positional movements come from the same zero point at all times. For example in Fig. 19.11, the X- and Y-axis dimensions will be expressed as follows:

Point 1	$x + 30$	$y + 20$
Point 2	$x + 50$	$y + 50$
Point 3	$x + 90$	$y + 60$

Most point-to-point NC machines are of an absolute type. It is estimated that more than 90 percent of the point-to-point NC machines presently installed use absolute programming [11]. However, incremental controls are generally cheaper to build.

Absolute systems have advantages over incremental systems [11].

1. In case of an interruption caused by cutting tool breakage or unprogrammed checking, the machine table might be moved manually to fix the cause of interruption. With the absolute system, the cutting tool will automatically return to the next position from where the interruption occurred. Since it always moves to the absolute coordinate called for, the machining proceeds from the block where it was interrupted. With an incremental system, it is difficult to bring the tool precisely to the beginning of the segment in which the interruption occurred.
2. With an absolute system, changing the dimensional data in the part program is not difficult. Since, distances are taken from one reference point, a modification or addition of a position instruction does not affect the rest of the part program. In the case of an incremental system, the part must be reprogrammed from the point at which the original program has been modified.

However, incremental systems have several advantages over absolute systems [11].

1. During part program preparation, it is easier to check geometric errors if an incremental system is used, since in using incremental dimensions, the sum of the position commands for each axis must equal zero.
2. Mirror-image programming, which can simplify the part programming of

symmetrical geometry of a part about one or two axes, is facilitated with an incremental system.

3. With incremental systems, part dimensions on each axis can be easily recognized simply by checking the part programs without any other calculations.

Both absolute and incremental systems have their logical areas of application. Many modern CNC systems permit both programming methods.

Control Loops: Open and Closed-loop Control

Every control system, including CNC systems, may be designed as either an open or a closed-loop control system. Open-loop systems provide no check or measurement to verify that a specific position has actually been reached. No feedback information is passed from the machine tool back to the controller. Closed-loop systems used in NC are characterized by the presence of feedback that is used to return positional information on the machine slides back to the controller.

Stepping motor-driven systems are typical examples of open-loop NC control. The stepping motor is an electromechanical device driven by an electrical pulse train to produce a sequence of angular (rotational) movements corresponding to the number of pulses [4]. Since there is no feedback from the slide position, the system's accuracy is solely a function of the motor's ability to step through the exact number of steps provided as its input [11]. Whether open-loop or closed-loop control is used, it does not affect the part programming.

Two different feedback principles are used in closed-loop NC systems. These are *indirect feedback* and *direct feedback*. The indirect feedback system compares the command position signal with the drive signal of the servomotor. This system is unable to sense backlash or leadscrew windup due to varying loads. Direct feedback, with its drive signal originated by the table, is the preferred system, because it monitors the actual position of the table on which the part is mounted [8].

The feedback system measures the machine's actual position and compares it with the desired position. This feedback comparison provides the ability to correct any position and velocity errors. Standard feedback devices include:

1. Digital transducers—linear and rotary
2. Analog transducers—linear and rotary

The rotary transducer, commonly known as rotary encoder, is a shaft-driven device delivering electrical pulses with frequency directly proportional to the shaft speed. The encoder consists of a glass disk marked with a

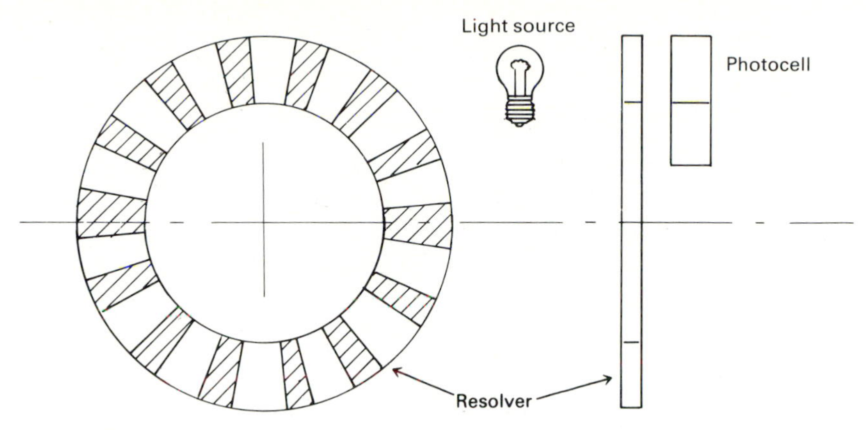

FIGURE 19.12
Optical encoder to determine incremental shaft position.

precise circular pattern of alternate clear and opaque segments on its periphery, as shown in Fig. 19.12. As the disk rotates, light from a fixed source on one side of the disk falls periodically on a photocell on the other side of the disk. A square wave output signal is produced when the motor rotates. This signal is fed through an amplifier to the associated logic circuits to determine the speed and direction of rotation and hence the position of the shaft [5].

Open-loop control provides a relatively cheap solution to NC control systems, while closed-loop control is especially suited for varying load conditions and contouring control systems.

Cartesian Coordinate Conventions: Left-handed and Right-handed

Cartesian coordinates are mutually perpendicular x, y, and z coordinates. Machine-tool motions are generally described in $x-y-z$ Cartesian space, making the X-, Y- and Z-axis coordinates the basis for NC programming. Figure 19.13 shows a vertical milling machine with the axes labelled [9]. The X- and Y-axes correspond to the motions of the table that holds the workpiece. The Z-axis represents the vertical motions of the spindle. For example, moving the table to the right causes the tool to make a cut to the left. This creates an X-motion in the negative direction (with respect to the tool).

Usually, some consistent rationale in naming the axis of a machine is followed [1]. The X-axis and the Y-axis describe a plane that is orthogonal to the penetration axis of the tool. The tool penetration axis is usually the Z-axis, with negative Z representing penetration and positive Z representing withdrawal. In the $X-Y$ plane, the axis with the greatest range of travel is usually taken as the X-axis, while the minor axis is the Y-axis.

Two major coordinate conventions used are the right-hand rule and the left-hand rule. The right-hand rule (Fig. 19.14) specifies that if the right hand is

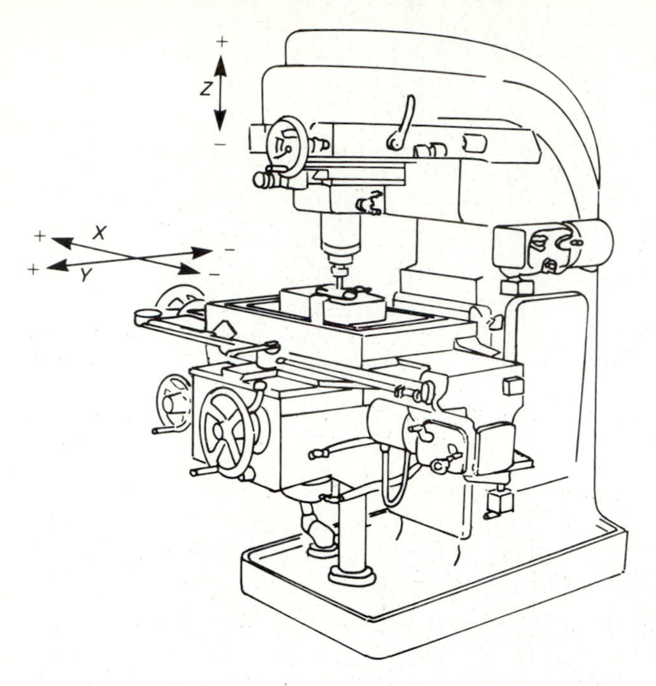

FIGURE 19.13
A vertical milling machine with the axes labeled.

placed at the origin on the $X-Y$ plane, with the fingers wrapped around the Z-axis and the forefinger extended, then the forefinger would point in the positive direction of the y-axis and the thumb toward positive Z. The vertical milling machine in Fig. 19.13 conforms to the right hand convention.

In the left-hand rule, the left hand is placed at the origin on the $X-Y$ plane and the left forefinger points to the $+Y$ direction and the thumb in the $+Z$-axis.

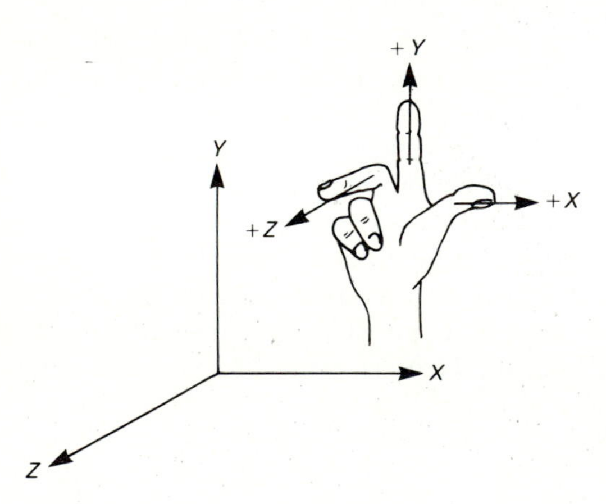

FIGURE 19.14
A right-handed coordinate system.

PROGRAMMING NUMERICALLY CONTROLLED MACHINES

Introduction

A numerical control machine must be programmed for each different part it will produce. The collection of instructions used to produce a part on an NC machine is commonly known as a part program. Part programming is defined as, "the planning and specification of the sequence of processing steps to be performed on the NC machine" [4]. It is composed of a set of computer procedures and a group of special words used to specify the required operations. There are two primary ways to produce part programs:

1. Manual part programming
2. Computer-assisted part programming

In manual part programming, the machining instructions are recorded on a document, called a part-program manuscript, by the part programmer. The manuscript is essentially an ordered list of the positions of the cutting tool and the workpiece that must be followed to machine the workpiece. The manuscript is input via a variety of hardware to produce a program. The standard medium for storing a part program is punched-tape. Each symbol on the manuscript—alphanumeric or special characters—corresponds to perforations on the tape and is referred to as a character. Each line of the manuscript is equivalent to a block on the punched tape, followed by an end of block (EOB) character.

In computer-assisted part programming, one uses a general-purpose computer as an aid in programming. Special-purpose high-level programming languages are used for performing the various tedious calculations necessary to prepare the punched tape. The computer allows economical programming for the machining of complex parts that could not be manually programmed. The part programmer's job is divided into two tasks. First, the programmer has to define the configuration of the workpiece in terms of basic geometric elements, namely, points, line, surfaces, circles, etc. Second, the programmer must direct the cutting tool to perform the machining steps along these geometric elements. Several programming languages that run on general-purpose computers have been developed. These languages are based on common English words and mathematical notations that are relatively simple to use.

NC Words

In the numerical control literature, a *word* refers to an ordered set of characters that are used to create a specific action of a machine tool. Each line of code in an NC manuscript is equivalent to one block. A block of code in an NC drilling operation might contain information on the coordinates of the hole location, the feed and speed at which the cut should take place and also perhaps a specification of the drill. A block consists of one or more of the following NC words:

1. *Sequence number* (*N*) is a block identifier. It usually consists of three digits, e.g., N001, N099, etc.
2. *Preparatory function* (*G*) prepares the machine control unit to perform a specific operation. For example, the word G03 prepares the NC controller for circular interpolation along an arc in a counter clockwise manner. Some typical examples are shown in Table 19.1.
3. *Dimension words* (*X, Y,* and *Z*) specify the coordinate position of the cutting tool. For example, X − 12.453 Y3.234, etc. In some systems the decimal point is omitted.
4. *Feed function* (*F*) specifies the feed rate of the cutting tool in a machining operation. Its use is generally restricted to contouring or straight-cut machines, whereas with point-to-point machines a constant maximum feed is generally used [5]. Feed can be expressed in terms of a four-digit integer feed rate number (FRN). There are several methods of expressing the FRN.

   Inverse Time Code. In a linear motion,

$$\text{FRN} = 10 \times \left(\frac{\text{velocity along path of motion, in./min}}{\text{length of tool path, in.}}\right)$$

   For example, for a velocity of 20 in./min on a tool-path of 5 in. the FRN will be $10 \times (20/5) = 40$. The feed word will be F0040.

   In circular cutting motions,

$$\text{FRN} = 10 \times (\text{velocity along arc, in./min/arc radius, in.})$$

   For example, for a velocity of 10 in./min on an arc of radius 2 in., the FRN will be $10 \times (10/2) = 50$. The feed word will be F0050.

   Magic Three Code (FRN) is expressed as a three-digit integer number. The last two digits are the two most significant digits of the feed expressed in inches per minute. The first digit is 3 greater than the number of digits to the left of the decimal point in the feed. For example, for a feed rate of 214.3 in./min, the FRN will be 621. The feed word will be F621.

   If the feed is less than 1.0, then the first digit in the FRN is 3 minus the number of zeros immediately following the decimal point. For example, for a feed rate of 0.053 inches per min, the FRN will be 253. The feed word will be, F253.

   EIA RS-236 feed code is a two-digit code number defined by EIA standard RS-326. Each successive feed number is 12 percent greater than the previous one. For example, FRN = 20 is 0.100 in./min; FRN = 21 is 0.112 in./min, etc.

TABLE 19.1
Examples of G-codes

G-code	Function
G00	Positioning
G01	Linear interpolation
G02	Circular interpolation, CW
G03	Circular interpolation, CCW
G04	Dwell
G17	X–Y plane selection
G18	Z–X plane selection
G20	Inch data input
G21	Metric data input
G27	Zero return check
G28	Automatic zero return
G45	Tool offset (increase)
G46	Tool offset (decrease)
G47	Tool offset (double increase)
G48	tool offset (double decrease)
G73	Canned cycle #10
G76	Canned cycle #11
G80	Canned cycle cancel
G81	Canned cycle #1
G82	Canned cycle #2
G83	Canned cycle #3
G84	Canned cycle #4
G85	Canned cycle #5
G86	Canned cycle #6
G87	Canned cycle #7
G88	Canned cycle #8
G89	Canned cycle #9
G90	Absolute programming
G91	Incremental programming
G92	Programming of absolute zero point
G98	Return to initial point level in canned cycle
G99	Return to R point level in canned cycle

5. *Cutting speed* (S). The cutting speed of the process is specified by the rate of spindle rotation in revolutions per minute. Again it can be expressed by a three-digit number. The usual method of expressing the speed word is by using the Magic Three Coding scheme, as described earlier.
6. *Tool selection* (T) is a code necessary to access a particular tool from an automatic tool changer or tool turret on a machining center. Each cutting tool has a different code number, by which it is selected.
7. *Miscellaneous functions* (M) are used to specify certain miscellaneous or auxiliary functions available on a particular machine tool. These are two-digit words used for miscellaneous tool commands such as spindle on/off, coolant on/off, executive tool change, end of tape, etc. Some typical examples are shown in Table 19.2.

TABLE 19.2
Examples of M-code

M-code	Function
M00	Stop the machine
M03	Start spindle clockwise
M04	Start spindle counter clockwise
M05	Stop spindle
M06	Execute tool change
M07	Turn coolant on
M09	Turn coolant off
M13	Start spindle clockwise
M14	Start spindle counter clockwise
M30	End-of-tape. Rewind the tape

NC Data Formats

NC data format has been defined as the "physical arrangement of the input media as possible locations of holes on a punched tape or as magnetized areas on a magnetic tape. Also, the general order in which information appears on the input media" [3]. A tape format means "an agreed order in which the various types of words will appear within a block" [5]. There are three primary types of formats in use at present: fixed-address, tab-sequential, and word-address. Within each format, however, there are some variations owing to differences in the type of machines, and their features.

FIXED-ADDRESS FORMAT. In fixed-address format, all information is inserted in a prescribed order. Also, each word must be of the same format and length. Both plus and minus signs and zero distance of movement have to be specified. This is the least flexible of the formats. For example:

N	G	X	Y	S	M
216	03	+ 7500	+ 2500	716	04

The data contained in the above example are: block sequence number 216; circular interpolation, counterclockwise direction; 7.5 in positive X direction 2.5 in. in positive Y direction; 16×10^4 r/min; and start spindle in counterclockwise direction.

TAB-SEQUENTIAL FORMAT. In a tab-sequential format, words are inserted in a fixed and prescribed sequence and separated from each other by a TAB. The TAB is omitted from the end of the last word in the block and the EOB code that denotes the end of the block is punched instead. This format is quite convenient to the part programmer as the data is laid out in the form of a table with words spaced horizontally. The previous example in a tab-sequential format would be,

N	G	X	Y	S	M
216 TAB	03 TAB	+7500 TAB	+2500 TAB	716 TAB	04 (EOB)
217 TAB	TAB	+5000 TAB	TAB	TAB	(EOB)

WORD-ADDRESS. A word-address system is the most commonly used and most flexible format in NC. It is a variable-block format in that the words can be of different length, can be arranged in any order and the number of words in an individual block can vary. Each word is preceded by an alphabetic character denoting the word address letter, identifying the word. The above example in a word address format would be,

N216 GO3 X + 7500 Y2500 S716MO4 (EOB)

The plus sign before any coordinate is optional. Also, there may or may not be spaces between each word. Again, if a word remains unchanged from the preceding block or is not required, it can be omitted from the block. This type of format is more commonly used in contouring systems.

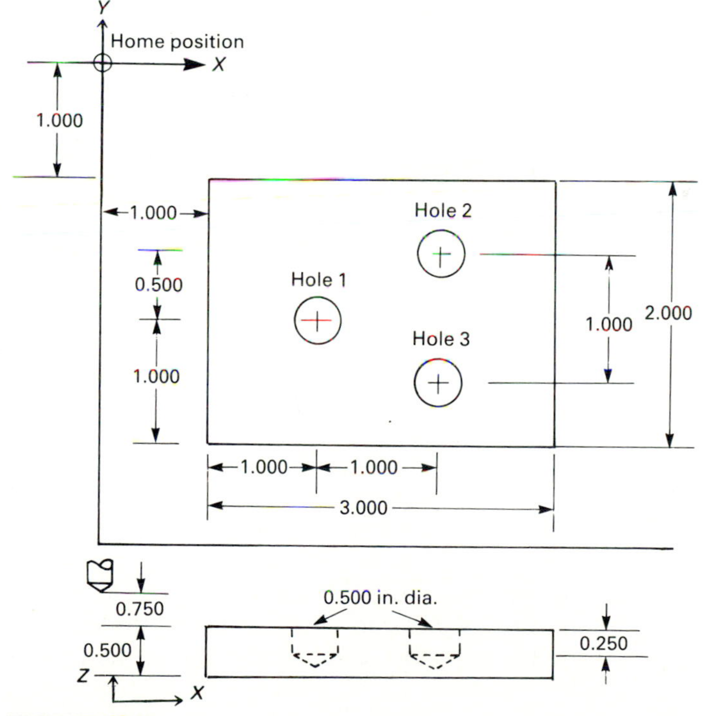

FIGURE 19.15
Example of manual programming.

Examples

Case I. An NC machine is at the home position shown in Fig. 19.15 and is commanded to drill three holes in the sequence shown. The origin for the Z-axis is 0.5 in. above the workpieces. The holes are to be drilled to a depth of 0.250 in. with a drill of diameter 0.500 in. An illustration of NC programming using an incremental system of positioning and a word address format is given below.

$$\text{Spindle speed, } S = \frac{\text{recommended cutting speed}}{\pi \times \text{dia. of drill}}$$

$$= \frac{100\ \text{ft/min} \times 12\ \text{in./ft}}{\pi \times 0.5\ \text{in./rev}}$$

$$= 764\ \text{rev/min}$$

$$\text{Feed rate, } F = (\text{recommended feed per rev}) \times (\text{spindle speed})$$

$$= 0.0010\ \text{in./rev} \times 764\ \text{rev/min}$$

$$= 0.764\ \text{in./min}$$

Word-address format

```
N01  G00  G17  G90  X2.00  Y-2.00  F676  S376
N02  G01  Z-1.00  M03  M07
N03  G01  Z-1.00
N04  G01  X-1.00  Y0.50
N05  G01  Z-1.00
N06  G01  Z-1.00
N07  G01  Y-1.00
N08  G01  Z-1.00
N09  G01  Z-1.00
N10  G01  X-3.00  Y2.5  M05  M09  M30
```

COMPUTER-ASSISTED PART PROGRAMMING: INTRODUCTION TO PROGRAMMING LANGUAGES

There are two major classes of part-programming languages [7];

1. *Machine oriented.* These create tool paths by doing all the necessary calculations in one computer-processing stage by computing directly the special coordinate-data format and the coding for speed and feed requirements.
2. *General purpose.* These break the computer processing into two stages. A processing stage and a postprocessing stage (Figure 19.16). The processing

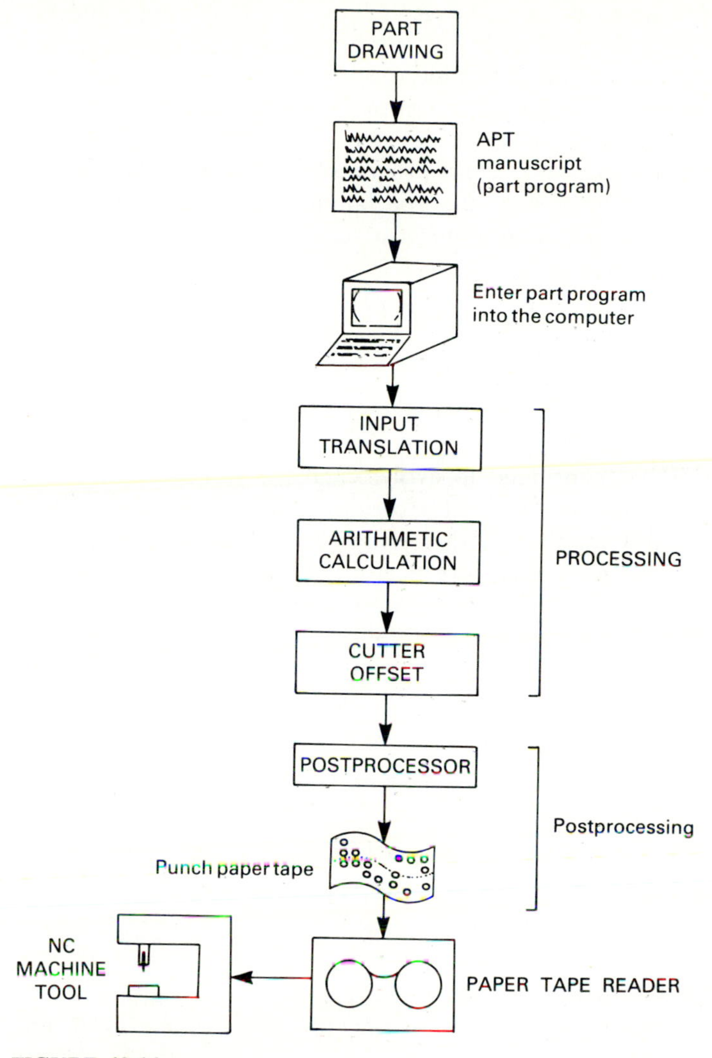

FIGURE 19.16
Steps in general-purpose computer-assisted part programming. (*Adapted from [4].*)

stage creates an intermediate set of data points called *cutter location data* (CL DATA). There are three steps in the processing stage:

a. *Translate input symbols.* This function translates symbolic inputs contained in the part program into computer-usable form. It also establishes the link between the human and computer.
b. *Arithmetic calculation.* The arithmetic calculation unit performs geometric and trigonometric calculations required to generate the part surface.
c. *Cutter offset calculations.* The cutter offset unit calculates the path of the center line of the cutter based on the part-outline information.

At the postprocessing stage, the CL DATA is converted by the computer to translate the data to the format and work configuration corresponding to a particular machine tool, and a punched tape (a tape image) is prepared. The output of the postprocessor is the NC tape written in an appropriate format for the machine on which it is to be used.

Since 1956, about a hundred NC part programming languages have been developed. Some of the languages are special-purpose machine-oriented languages. However, most are general-purpose languages. Some languages have stood the test of time; many have not.

APT

APT is an acronym for Automatically Programmed Tool. Initially developed in 1956 at M.I.T., it is the most popular part-programming language in the United States [3]. APT has continued to evolve, and the Illinois Institute of Technology Research Institute (IITRI) has continued the development and administrative responsibility for the fourth version of APT developed in 1970. APT is the most powerful general-purpose part-programming system, against which other systems are commonly compared and evaluated. Summary characteristics of APT are described below [IITRI, 1967].

1. Three-dimensional unbounded surfaces and points are defined to represent the part to be made.
2. Surfaces are defined in an X–Y–Z coordinate system chosen by the part programmer.
3. In programming, the tool does all the moving; the part is stationary.
4. The tool path is controlled by pairs of three-dimensional surfaces; other motions, not controlled by surfaces, are also possible.
5. A series of short straight-line motions are calculated to represent curved tool paths (linear interpolation).
6. The tool path is calculated so as to be within specified tolerances of the controlling surfaces.
7. The X-, Y-, and Z-coordinates of successive tool-end positions along the desired tool path are recorded as the general solution to the programming problem.
8. Additional processing (postprocessing) of the tool end-point coordinates generates the exact tape codes and format for a particular machine.

There are four types of statements in the APT language.

1. *Geometry statements*. These define a scaler or geometric quantity.
2. *Motion statements*. These describe a cutter path.

3. *Postprocessor statements.* These define machining parameters like feed, speed, coolant on/off, etc.
4. *Auxiliary statements.* These describe auxiliary machine tool functions to identify tool, part, tolerances, etc.

There are two other important features provided in the APT language.

1. *Macros.* Individual macros similar to FORTRAN subroutines can be created for adding to the APT program routines. A library of frequently used routines and definitions can be created as special macros.
2. *Loops.* An individual section of an APT program can repeat itself until a specified result is obtained.

Example 1. The workpiece that was manually programmed and shown in Fig. 19.15 will be programmed using APT in order to illustrate the programming procedure step by step.

The part program:

```
PARTNO, DRILL 3 HOLES          ; defines part name and work to be done
MACHIN/P & W HORIZON V         ; defines machine part
CLPRNT
SETPT = POINT/0, 0, 0          ; defines home position of tool
CUTTER/.500                    ; defines the outer diameter
P = POINT/2.0, −2.0            ; defines location of hole 1
P2 = POINT/3.0, −1.5           ; defines location of hole 2
P3 = POINT/3.0, −2.5           ; defines location of hole 3
SPINDL/764 CLW                 ; spindle speed 764/r/min clockwise
FEDRAT/7640                    ; feed rate 764 in./min
COOLNT/ON                      ; turn coolant on
FROM/SETPT                     ; defines the initial location of the
                                 cutter
GOTO/P1                        ; move cutter to hole 1
GODLTA/0, 0, −1.0              ; move tool down to drill hole 7
GODLTA/0, 0, 1.0               ; move tool up
GOTO/P2                        ; move cutter to hole 2
GODLTA/0, 0, −1.0              ; move tool down to drill hole 2
GODLTA/0, 0, 1.0               ; move tool up
GOTO/P3                        ; move cutter to hole 3
GODLTA/0, 0, −1.0              ; move tool down to drill hole 3
GODLTA 0, 0, 1.0               ; move tool up
RAPID
GOTO/SETPT                     ; go back to the home position
COOLNT/OFF                     ; turn coolant off
FINI                           ; termination of part program
```

Example 2. A vertical NC milling machine is in the home position shown in Fig. 19.17 and is commanded to scribe 5 circles of diameter 1 in. in the form of an olympic crest. The circles are to be milled to a depth of 0.500 in. A complete APT program to achieve the above is given below.

```
PARTNO, OLYMPIC CREST                ; defines part name
MACHIN/P & W HORIZON V               ; defines machine type
SETPT = POINT/0, 0.5, 0.5            ; defines the home position of tool
C1 = CIRCLE/1.5, 3.0, 0.5            ; defines the location of circle 1
C2 = CIRCLE/2.75, 3.0, 0.5           ; defines the location of circle 2
C3 = CIRCLE/4.0, 3.0, 0.5            ; defines the location of circle 3
C4 = CIRCLE/2.125, 2.25, 0.5         ; defines the location of circle 4
C5 = CIRCLE/3.375, 2.25, 0.5         ; defines the location of circle 5
L1 = LINE/LEFT, TANTO, C1,
LEFT, TANTO, C2                      ; defines location of line 1
L2 = LINE/RIGHT, TANTO, C4,
RIGHT, TANTO, C5                     ; defines location of line 2
$$ PART DESCRIPTION HAS NOW
BEEN COMPLETED
CUTTER/0.250                         ; cutter diameter
INTOL/.001                           ; inner tolerance
OUTTOL/.001                          ; outer tolerance
SPINDL/1740, CLW                     ; spindle RPM
FEDRAT/2500                          ; feedrate
FROM/SETPT                           ; defines the initial location of outer
GO/TO, L1                            ; moves tool to line 1
GORGT/L1, TANTO, C1                  ; go right on line 1, tangent to circle
GODLTA/0, 0, −0.75                   ; go down
GOFWD/C1, TANTO, L1                  ; go forward on circle 1, tangent to line 1
GODLTA/0, 0, +0.75                   ; go up
GORGT/L1, TANTO, C2                  ; go right on line 1, tangent to circle 2
GODLTA/0, 0, −0.75                   ; go down
GOFWD/C2, TANTO, L1                  ; go forward on circle 2, tangent to line 7
GODLTA/0, 0, +0.75
GORGT/L1, TANTO, C3
CODLTA/0, 0, −0.75
GOFWD/C3, TANTO, L1
GODLTA/0, 0, +0.75
GOTO/L2
GOLFT/L2, TANTO, C4
GODLTA/0, 0, 0.75
GOFWD/C4, TANTO, L2
GODLTA/0, 0, +0.75
GORGT/L2, TANTO, C5
GODLTA/0, 0, −0.75
GOFWD/C5, TANTO, L2
GODLTA/0, 0, +0.75
RAPID                                ; change feed rate to rapid
GOTO/SETPT                           ; go back to home position of tool
COOLNT/OFF                           ; turn coolant off
FINI                                 ; termination of part program
```

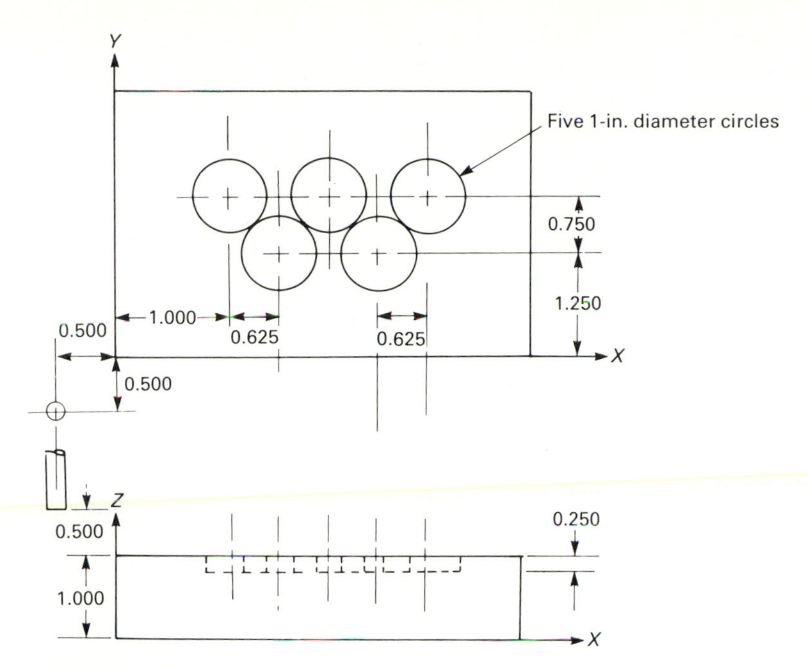

FIGURE 19.17
APT programming Example 2.

Other Programming Languages

ADAPT (Adaptation of APT) was the first attempt to adapt more commonly used APT routines for smaller computers. It was developed by IBM under a U.S. Air Force contract and was constructed in a modular manner, to provide greater flexibility for the user to add and delete routines. It has full two-dimensional and some limited three-axis capabilities and has routines for curve fitting, inclined planes, polygonal pockets, and macro definitions.

UNIAPT was developed by the United Computing Corp. of Carson, Calif., and was the first successful attempt to handle the full power of APT on a small computer. Externally completely compatible with APT, it differs only in the internal design of the processor.

EXAPT (Extended Subset of APT) was developed in Germany around 1964 jointly by several German technical universities, to make APT more appropriate to European conditions.

AUTOSPOT (Automatic System for Positioning Tools) was developed by IBM for three-axis point-to-point motion control around 1962. It was subsequently combined with ADAPT to provide an effective language for both point-to-point and continuous-path applications.

COMPACT was developed by by the Manufacturing Data Systems, Inc.

(MDSI), Ann Arbor, Mich. for simultaneous servicing of multiple users from a remote computer over telephone lines. The COMPACT system converts its language statements to machine control codes in a single computer iteration, thus eliminating the postprocessing stage completely [7]. COMPACT II, the latest version, is available to the user on a time-sharing basis.

SPLIT (Sundstrand Processing Language Internally Translated) was developed by the Sundstrand Corporation, intended for their machine tools. It can handle up to five-axis positioning and possesses contouring capability. The postprocessor is built right into the program. Each machine tool uses its own version of the SPLIT package.

It is clear from this discussion that computer-aided part programming has received a great deal of attention over the past twenty years. The languages developed can be of tremendous advantage to the part programmer. With the advent of computer-aided design (CAD), the geometric information needed for part programming is already resident in the computer. The new generation of computer-aided part-programming systems is capable of automatically generating some limited part programs. Future systems should be able to do this more effectively.

ROBOTICS

INTRODUCTION

Industrial robots are relatively new electromechanical devices that are beginning to change the appearance of modern industry. Industrial robots are not like the science fiction devices that possess human-like abilities and provide companionship to space travelers. Research to enable robots to "see," "hear," "touch," and "listen" has been underway for two decades and is beginning to bear fruit. However, the current technology of industrial robots is such that most robots contain only an arm rather than all the anatomy a human possesses. Current control only allows these devices to move from point to point in space, performing relatively simple tasks.

Basic Concepts

The Robotics Institute of America defines a robot as "a reprogrammable multifunction manipulator designed to move material, parts, tools, or other specialized devices through variable programmed motions for the performance of a variety of tasks." An NC machining center would qualify as a robot if one were to interpret different types of machining as different functions. Most manufacturing engineers do not consider a NC machining center a robot, even though these machines have a number of similarities. The power drive and controllers for both NC machines and robots can be quite similar. Robots, like NC machines can be powered by electrical motors, hydraulic systems, or pneumatic systems. Control for either device can be either open-loop or closed-loop. In fact, many of the developments used in robotics have evolved from the NC industry, and many of the manufacturers of robots also manufacture NC machines or NC controllers.

A physical robot is normally composed of a main frame (or arm) with a wrist and some tooling (usually some type of gripper) at the end of the frame. An auxiliary power system may also be included with the robot. A controller with some type of teach pendant, joy-stick, or key-pad is also part of the system. A typical robotic system is shown in Fig. 19.18.

Robots are usually characterized by the design of the mechanical system. A robot whose main frame consists of three linear axes is called a Cartesian robot. The Cartesian robot derives its name from the coordinate system. Travel normally takes place linearly in three-space. Some Cartesian robots are constructed like a gantry to minimize deflection along each of the axes. These robots are referred to as gantry robots. Figure 19.19 shows examples of Cartesian robots. These robots behave and can be controlled similarly to conventional three-axis NC machines. Gantry robots are generally the most accurate physical structure for robots. Gantry robots are commonly used for assembly where tight tolerance and exact location are required.

A cylindrical robot is composed of two linear axes and one rotary axis.

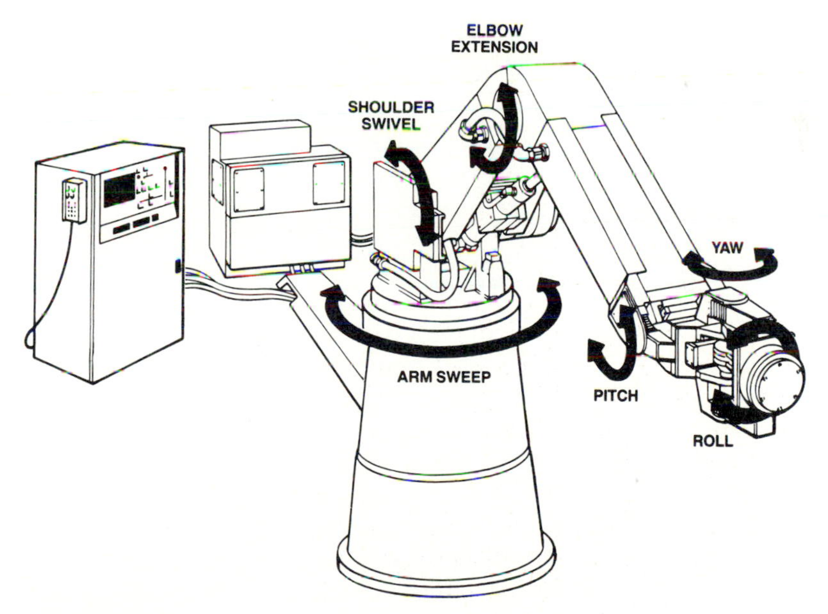

FIGURE 19.18
A typical robot system. (*Courtesy of Cincinnati Milacron.*)

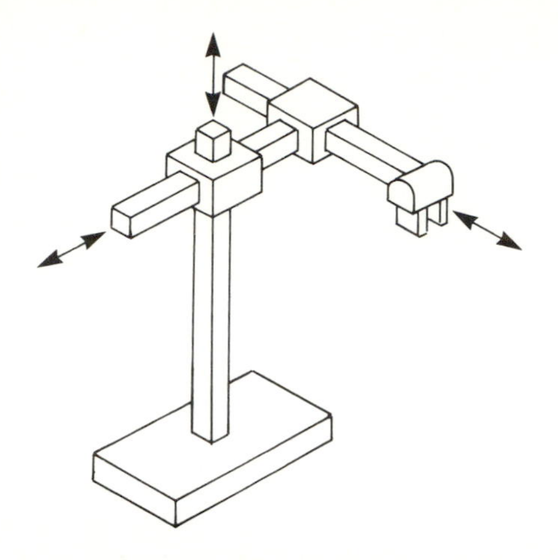

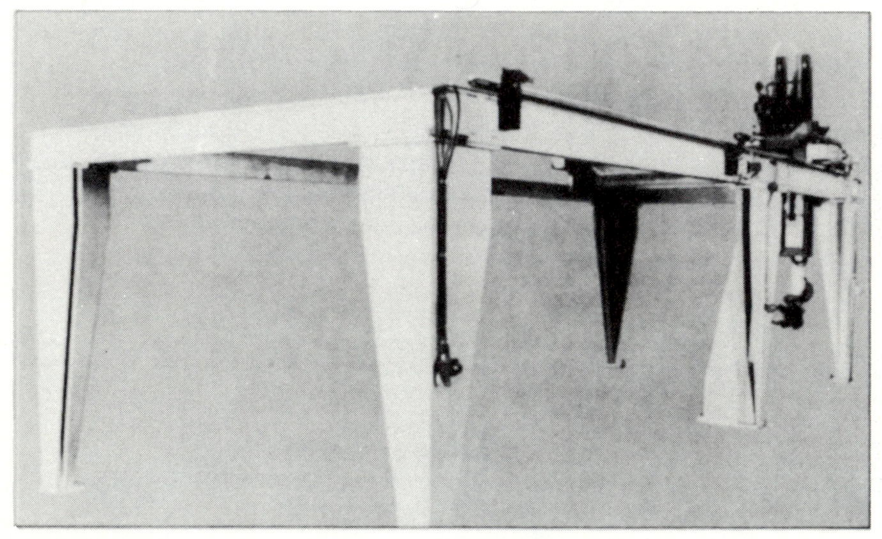

FIGURE 19.19
Cartesian robots. (*Photo courtesy of Cincinnati Milacron.*)

This robot derives its name from the work envelope (the space in which it operates), which is created by moving the axes from limit to limit. Figure 19.20 shows typical cylindrical robots. Cylindrical robots are used for a variety of applications, but most frequently for material-handling operations.

A spherical robot consists of one linear axis and two rotational axes. Again, the spherical robot derives its name from the work envelope that is accessible to the robot. Figure 19.21 illustrates a typical spherical robot. Spherical robots are used in a variety of industrial activities. Spot-welding and material handling are two activities commonly performed by spherical robots.

A jointed, articulated or anthropomorphic robot consists of three

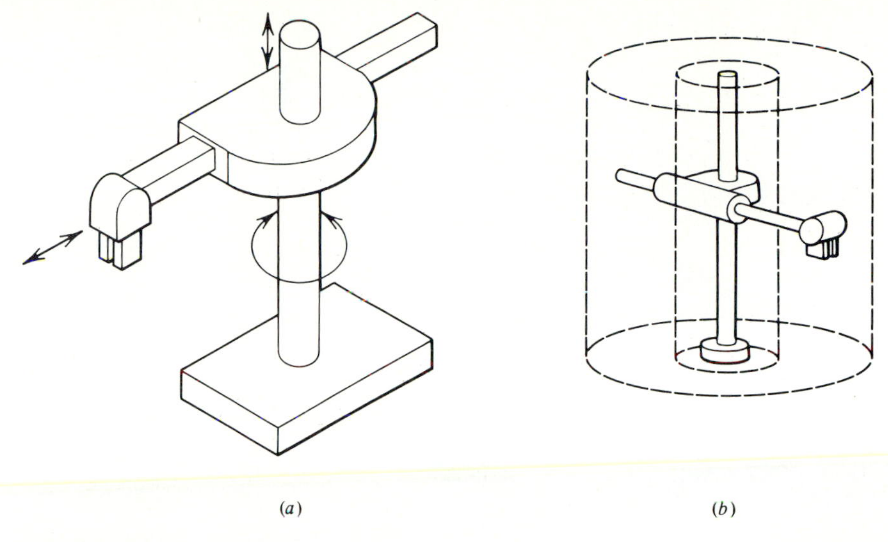

FIGURE 19.20
Cylindrical robots. (*a*) Cylindrical coordinate manipulator; (*b*) work volume shape of cylindrical manipulator. (*Courtesy of U.S. Air Force.*)

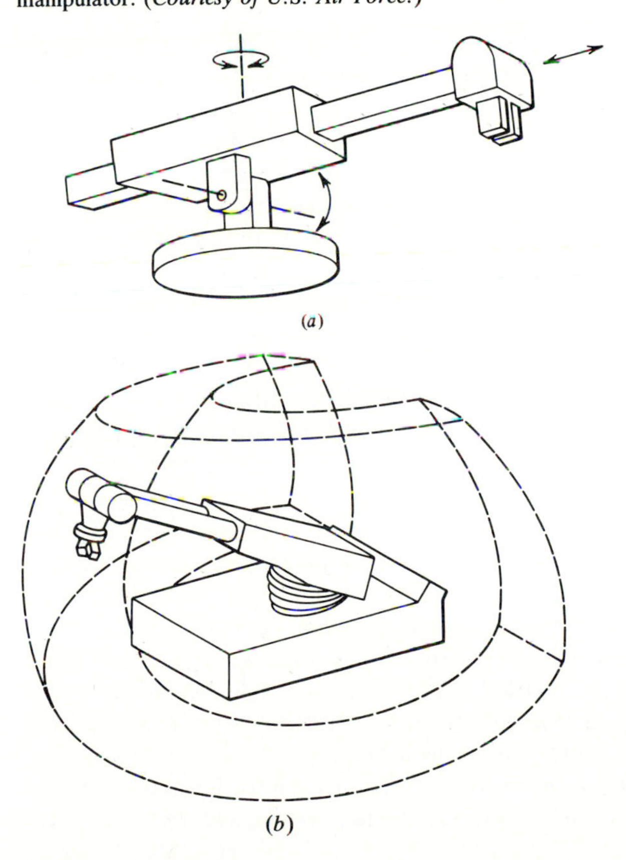

FIGURE 19.21
Spherical robots. (*a*) Spherical coordinate manipulator; (*b*) work volume shape of spherical manipulator. (*Courtesy of U.S. Air Force.*)

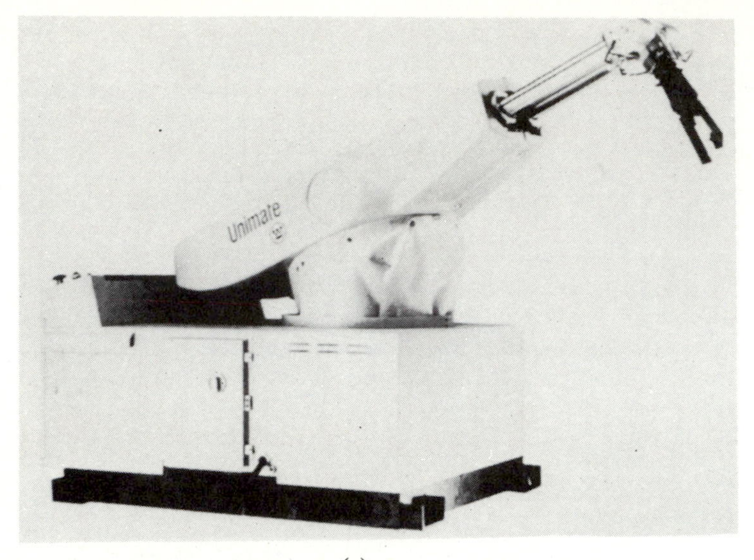

(c)

FIGURE 19.21, continued

rotational axes connecting three rigid links and a base, as shown in Fig. 19.22. The articulated robot is frequently called an anthropomorphic arm because it closely resembles a human arm. The notation used to describe the articulated arm also parallels that of the human arm. The base of the robot provides rotation similarly to that the body would provide for the human arm. The first joint above the base is referred to as the shoulder. The shoulder joint is connected to the upper robot arm, which is connected to the elbow joint. The elbow joint is connected to the forearm, which connects to the wrist. Articulated arms are used for a wide range of activities ranging from welding to assembly. Normally these robots can move quickly throughout the work envelope. Because of the cantilever effect of each of the links, loads can cause significant deflection of the end-effector (loaded versus unloaded).

One style of robot that has recently become quite popular is a combination of the articulated arm and the cylindrical robot. The robot is a four-axis rather than a three-axis device and looks like an articulated arm mounted to the staff of a cylindrical robot. The robot is called the Scara robot and is shown in Fig. 19.23. The Scara robot has some major advantages over the articulated arm and the cylindrical robot. You can note from the figure that the rotary axes are mounted vertically rather than horizontally. This means that the weight of the joint links is carried by the bearings rather than the drive motors. Deflection of the robot loaded versus unloaded will be reduced over the articulated arm. Many Scara robots are accurate enough to perform insertion of components on printed circuit boards.

The end-effector or end-of-arm tooling is connected to the main frame of the robot by a wrist. Most typical robots (Cartesian, cylindrical, spherical, and jointed) contain three major axes used to move the end of arm tooling throughout the work envelope. The end-of-arm tooling may not be oriented in the proper manner to perform the required operation. The orientation of the end-or-arm tooling is referred to as the *attitude*. For most machining operations, the tool is kept orthogonal to the part surface. However, since a robot performs a variety of operations, the orientation of the tooling may require a variety of attitudes. For instance, in a butt welding operation of two

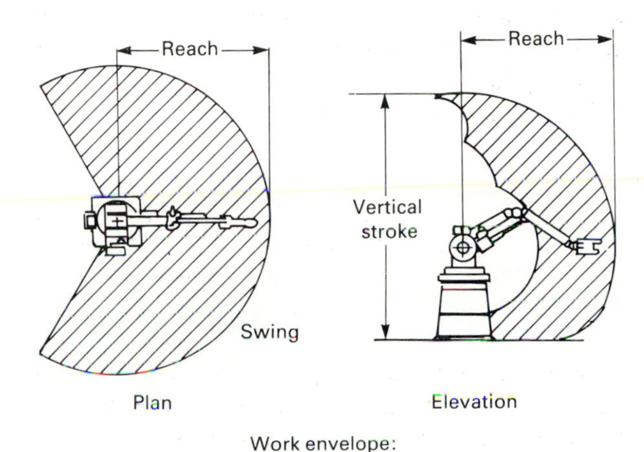

FIGURE 19.22
Jointed, or articulated, robots.

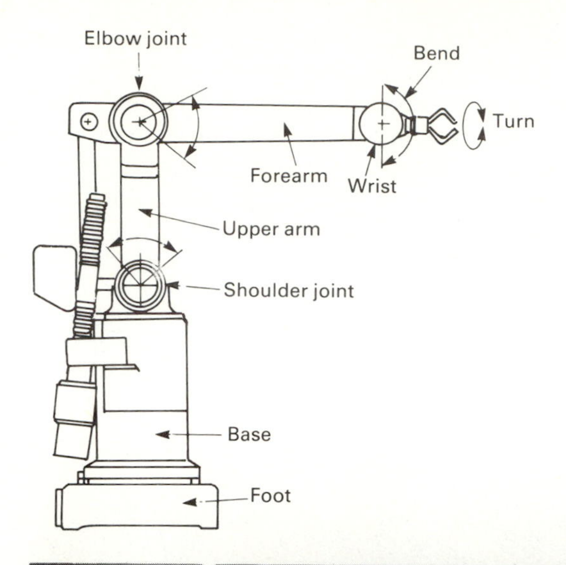

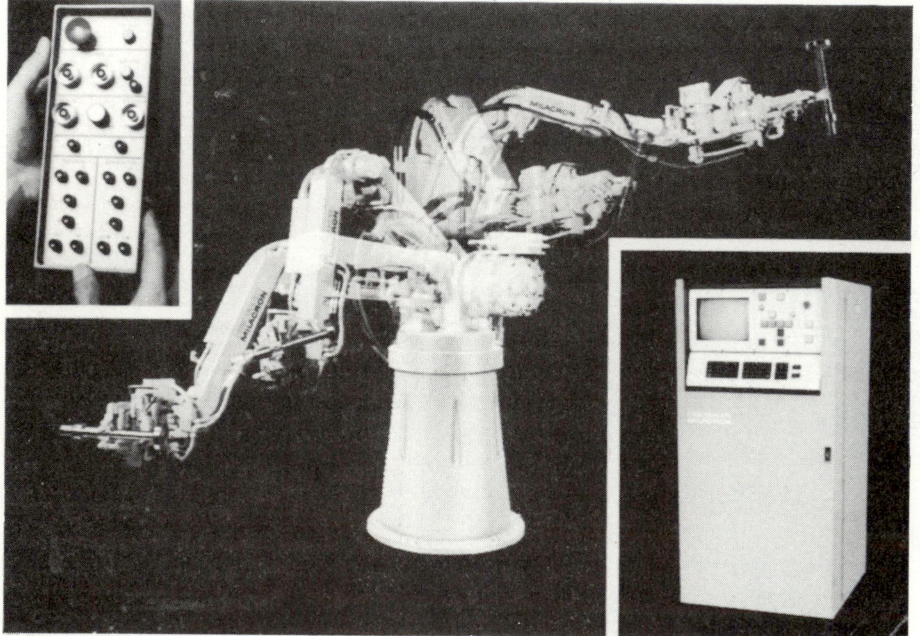

FIGURE 19.22, continued

flat workpieces, the tool would be kept perpendicular to the base but at 45 degrees to the weld, as shown in Fig. 19.24.

Most wrists consists of either two or three joints/axes. The axes are defined like most kinematic systems. The axis that causes the end-effector to rotate about the major tool face is called wrist roll. The other axes are defined like those of a ship, in that the axis that alters the horizontal presentation of

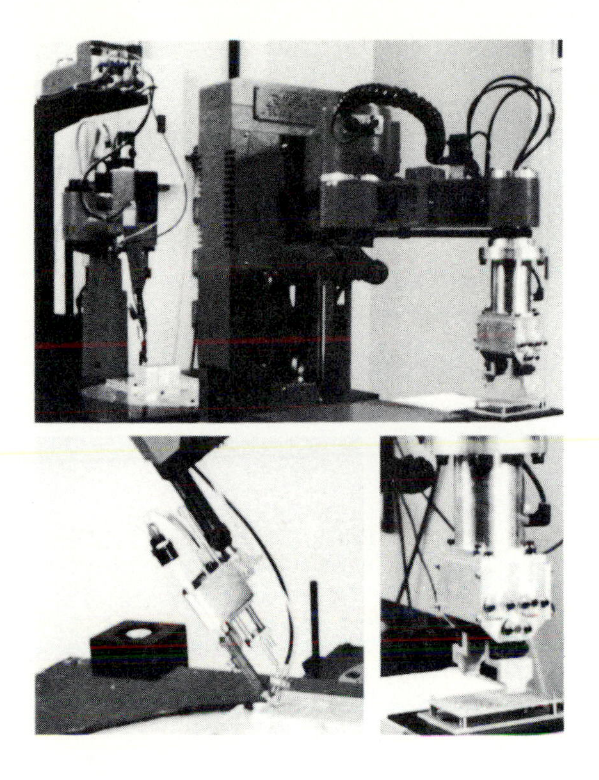

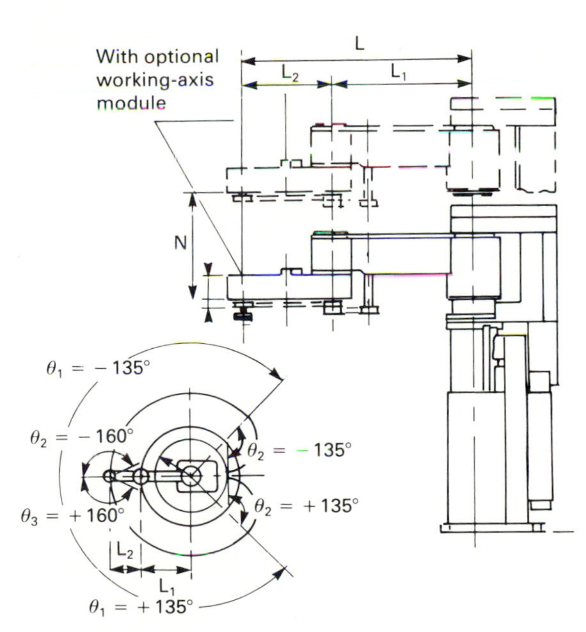

FIGURE 19.23
Scara robots. (*Courtesy Rhino Robots.*)

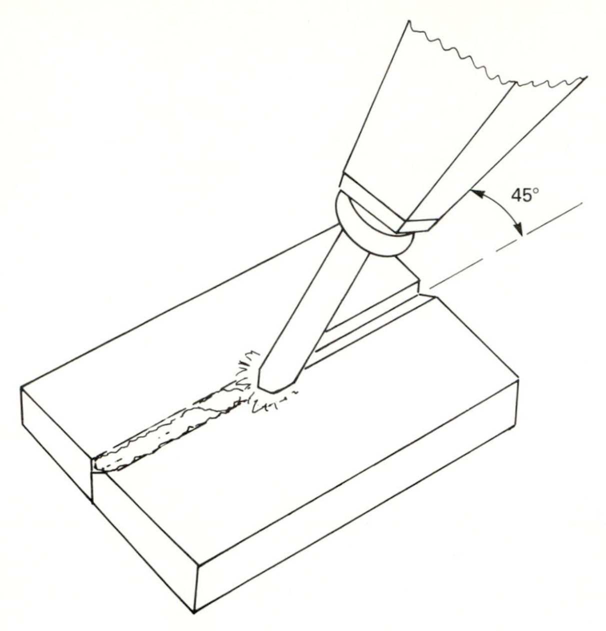

FIGURE 19.24
Robot attitude for arc welding.

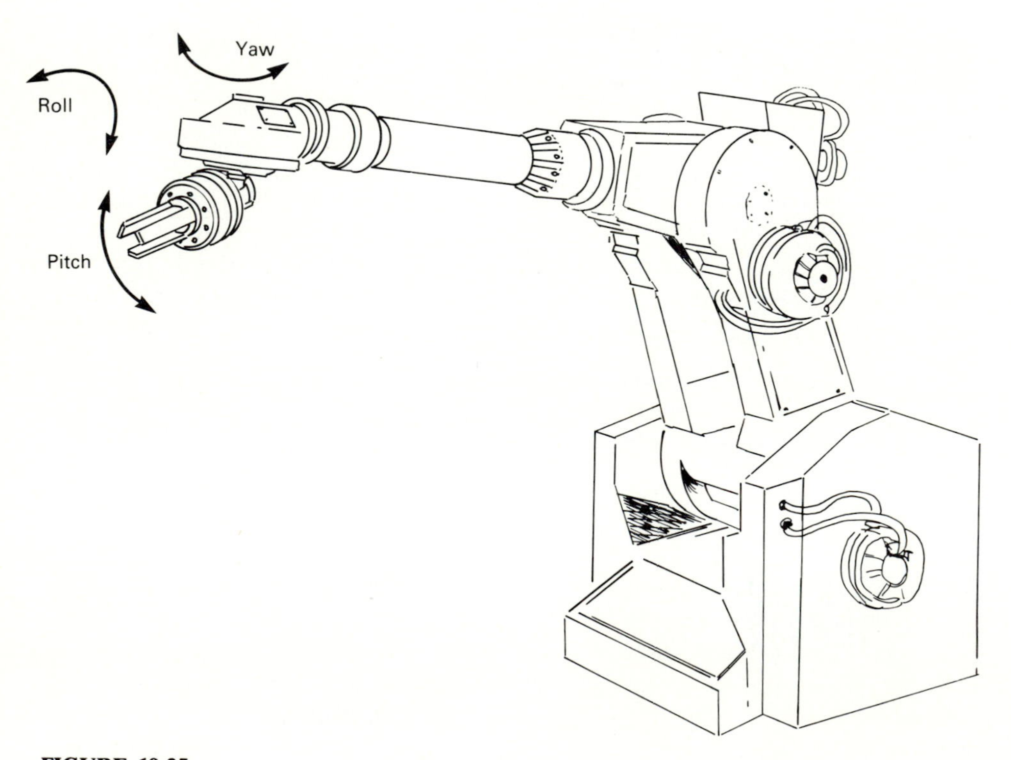

FIGURE 19.25
Wrist configuration for a robot.

the end of arm is called the pitch. The axis affecting the directional orientation of the end of arm is called yaw. Normally, only two of these axes are required, because by rotating the first axis by 90 degrees the presentation of the next two axes is exchanged (pitch for yaw or yaw for roll). Figure 19.25 shows a typical wrist configuration for a robot.

PROGRAMMING A ROBOT

In order for a device to qualify as a robot, it must be easily reprogrammable. Nonprogrammable mechanisms, regardless of their potential flexibility by reassembly or rewiring, do not qualify as robots. A class of devices that fits this category are fixed- or variable-sequence robots. Many of these robots are pneumatically driven. Rather than controlling the robot path, the device is driven to fixed stops or switches via some form of ladder logic. Although the ladder programming qualifies for the definition of a robot, the switches or stops must normally be physically moved in order to alter the tasks being performed. Since programmable controllers were covered in detail in Chapter 18, this type of programming will not be addressed here. The inherent logic and procedure for programming these robots is the same as for any programmably controlled mechanism. Drive actuators or motors are turned "on" or "off" depending on the desired sequence of tasks and switch states. Robot operations for this type of system are normally limited to rather simple applications.

Programming of more conventional robots normally takes one of three forms: (1) walk-through or pendant teaching, (2) lead-through teaching, or (3) offline programming. Each robot normally comes with one or more of these types of programming systems. Each has advantages and disadvantages depending on the application being considered.

Walk-through or pendant teaching or programming is the most commonly used robot programming procedure. In this type of programming, a pendant that normally contains one or more joy-sticks is used to move the robot throughout its work envelope. At the end of each teach point, the current robot position is saved. As was the case with NC machines, some robots allow the programmer the option of defining the path between points. Again, these robots are called continuous-path systems. Systems that do not allow the user to specify the path taken are called point-to-point systems. Many continuous-path robots allow the user to define the path to take between successive points. That is, the user may define a straight-line, circular, or joint-interpolated path. In a straight-line path, the robots move between successive points in a straight-line in Cartesian space. Circular moves, as the name implies take place in circles along one of the major planes. The path that the robot takes using a joint-interpolation scheme is not always easy to determine. In joint interpolation, each of the robot joints are moved at a constant rate so that all the axes start and stop at the same time. For Cartesian robots, straight-line and joint-interpolation schemes produce the same path. For the other types of robot systems, this is not true.

Pendant programming systems normally have supplemental commands that allow the programmer to perform auxiliary operations such as close the end-effector, wait, pause, check the status of a switch or several switches, return a complete status to a machine, etc. The programmer walks the robot through the necessary steps required to perform a task, saving each intermediate step along with the auxiliary information. The teach pendant used to program the Fanuc M1 robot is shown in Fig. 19.26.

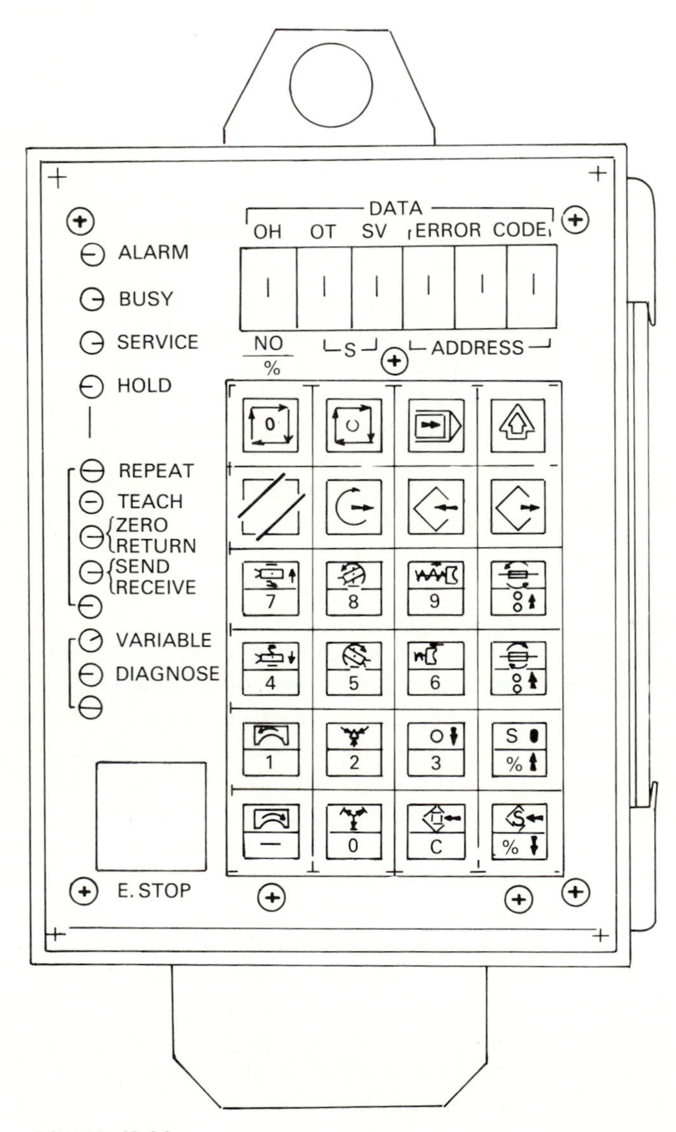

FIGURE 19.26
Fanuc M1 teach pendant.

Lead-through programming is one of the simplest programming procedures used to program a robot. As the name implies, the programmer simply physically moves the robot through the required sequence of motions. The robot controller records the position and speed as the programmer leads the robot through the operation. The power is normally shut down while the programmer is leading the robot through the necessary moves, so that the robot will not generate any "gliches" that might injure the operator. Although lead-through programming is the easiest programming method to learn, it does introduce some severe limitations to the robot's application. For instance, when the robot is being led through the operation, the operator carries the weight of the robot. The gears, motor and lead screw may introduce a false resolver reading, so that when the weight of the robot, and perhaps a part, must be supported by the system the actual end-effector position may be significantly different from the position taught to the robot. Another problem with this method is that since the position and speed are recorded as the robot is being led through the desired path, a significant amount of data is generated. This data must be stored and later recalled. Storage and retrieval space and time can cause the programmer problems. Perhaps the major problem associated with lead-through programming is that the human who leads the robot through the process is capable of only finite accuracy and may introduce inconsistencies into the process. Human-induced errors and inaccuracies eliminate some of the major advantages of using robots.

Offline programming for robots is a relatively new technology that provides several advantages over both lead-through and teach programming. The principles of offline programming are similar to using an offline language such as APT for NC programming. Several languages have been developed at major universities as well in industry throughout the U.S. Examples of these languages include VAL created by Unimation, AR-Basic by American Robot Corp., ARMBASIC by Microbot, Inc. and AML by IBM. To illustate offline programming, AR-Basic will be used. AR-Basic allows the user to:

1. Define the position of the robot
2. Control the motion of the robot
3. Input and output control data

AR-Basic is an interpretive BASIC system that employs many of the same functions as the familiar BASIC programming language. In AR-Basic, points and tools are defined as a set of primitive data. Points are defined using the convention:

$$X, Y, Z, R, P, Y$$

where X, Y, and Z are the Cartesian space occupied by the end-effector, and R, P, and Y are the roll, pitch and yaw of the tool. Each of the point definitions can be specified as either absolute or relative points (again defined in a similar manner as for NC machines).

TABLE 19.3
Point and tool definition statements

```
TEACH_ABS POINT_1                        ! Define an Absolute Point based on
                                         !    the position of the robot's tool tip.
POINT_2 = DEF_REL(30,-5,4,,45,180)       ! Define a Relative Point by specifying
                                         !    the six components.
REL_POINT_1 = CONVERT_REL(POINT_1)       ! Convert an Absolute Point definition
                                         !    to a Relative Point definition
                                         !    (based on the current frame).
DRILL_BIT = DEF_TOOL(8)                  ! Define a simple 8-in. straight Tool
                                         !    (all unspecified parameters
                                         !    default to 0).
PRINT POINT_2.PITCH                      ! Print on the screen the Pitch
                                              component of POINT_2.
POINT_1.Y = SQR(ABS(-144))               ! Set the Y component of POINT_1
                                         !    to 12.
```

Tool-definition commands are used to define the location of any tooling that might be required for an operation. The tool definition specifies the midpoint of the robot's faceplate, and consists of the same six data used to describe a point.

The robot is set into motion using a set of motion control commands. The motion commands allow the programmer to:

1. Define the type of path to take (straight-line, circular, or joint-coordinated)
2. Define the end of tooling speed
3. Define the frame of reference
4. Describe the current tool tip

AR-Basic also allows the programmer to input and output data to any

TABLE 19.4
Motion-control command statements

```
SET_SPEED TO 30                          ! Set tool tip speed to 30 in./s.
SET_MOTION TO STRAIGHT                   ! Specify straight line motion.
SET_TOOL TO DRILL_BIT                    ! Tool defined in last example.
SET_FRAME TO POINT_1                     ! Set current frame using an
                                         !    Absolute Point definition.
MOVE TO REL_POINT_1                      ! Move to position specified by
                                         !    relative Point definition
                                         !    (relative to frame specified by
                                         !    Absolute Point POINT_1).
MOVE TO POINT_2, AT SPEED 20, USING JOINT MOTION
                                         ! Move to POINT_2, temporarily
                                         !    overriding the current default
                                         !    speed and motion style.
MOVE $XYZ_TABLE TO POINT_1               ! Move an independently defined
                                         !    XYZ table.
```

device to which the robot is interfaced. Analog and digital data can be sent to A/D converters, D/A converters, parallel, or serial I/O ports. Table 19.3 contains Point and tool definition examples. Table 19.4 contains AR-Basic examples for motion control.

ECONOMICS

Although industrial robots have been around since 1961, the use of these devices did not become commonplace until recently. Their use, like that of NC machines, has been driven by the economics of their application. Robots and NC machines must be justified like any other piece of capital equipment. Savings and expenses must be estimated and documented. In the following section, an example of a robot application will be discussed along with the justification process.

ROBOT SELECTION EXAMPLE

Owing to the hot and heavy workparts used in a hot-forging operation at the ACME Wrench Company, the IE staff has considered using a robot for handling the parts. Another factor involved in this decision is that the Hotpart-Handler Union's latest contract stipulates that the current part handlers must be paid a 200 percent wage differential owing to the uncomfortable and hazardous work conditions. This amounts to $20.00 per hour for labor, so to reduce costs of labor and eliminate an unpleasant job, a robot is being considered. The question then arises as to which robot to use for the job. But first, a description of the job is in order.

Job Description

1. Part blanks weighing 20 lb are preheated in a furnace. Parts travel through the furnace on a conveyer at a rate of 3 parts/minute.
2. A part blank is removed from the accumulating conveyor end and is loaded into the forging press.
3. The forging press forges the hot blank with a cycle time of 15 seconds, including load and unload time.

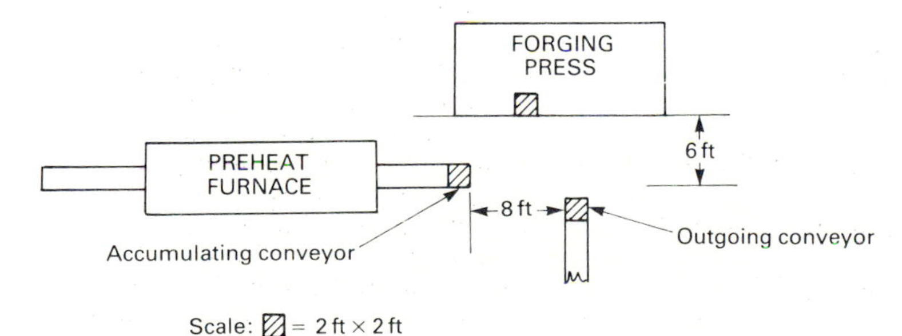

4. The finished forging is removed from the forging press and is placed on the outgoing conveyor.

Details of Job

1. Part
   a. Size: 4 × 2 × 18 in.
   b. Weight: 20 lb
2. Workspace information
   a. Maximum distance between work locations: 10.77 ft (in-conveyor point to out-conveyor)
   b. Overhead clearance: 14 ft
   c. Floor load capacity: 800 lb/ft^2
   d. Electrical power buss overhead
   e. Compressed air line overhead
   f. Pillars or columns: none
   g. Conveyor heights
      (1) In: 42 in. off floor
      (2) Out: 36 in. off floor
   h. Forge data
      (1) Bed height: 40 in. off floor
      (2) Die separation: 12 in. open
      (3) Tolerance of blank placement in dies = ±0.5 in.

Initial Robot Specification

1. From the job description, we can surmise that more costly contour path robots are not necessary for the task. Hence, a PTP robot is required.
2. A minimum reach of 5.385 ft ($10.77 \times \frac{1}{2}$ ft) is required.
3. A minimum load capacity of 20 lb is required.
4. The work envelope of the robot should contain the three points of P_1 = in-conveyor, P_2 = forge dies, and P_3 = out conveyor. If we let P_1 = (0, 0, 42) (in.) then P_2 = (72, 48, 40) and P_3 = (120, −48, 36).

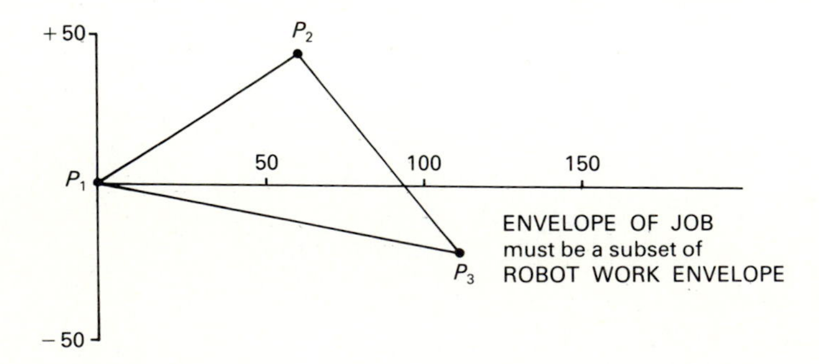

5. Accuracy required is ±0.50 in.
   Repeatability required is ±0.25 in.
   (based on placement of part blanks in the die)
6. Vertical stroke required is 8 in. = (42–36).
7. Based on a minimum reach of 5.385 ft, a minimum swing of 180° is required. (This may be reduced if the reach exceeds 5.385 ft.)

Other Considerations

1. Reliability should exceed 0.98.
2. Maintenance to be performed by plant personnel.
3. May be electric, pneumatic, or hydraulic, since noise is not a consideration.
4. Gripper should withstand heat.
5. Easy programming desired.
6. Robot should be able to control the forging press.
7. Open-loop system is acceptable.

The Feasible Set of Robots

Once the requirements/specifications of the robot(s) have been determined a finite set of feasible or acceptable units has been defined. This may be expressed as a mixed integer program (MIP) with an objective of minimum cost (similar to Tanchoco lift-truck selection model). However, such a formal development is not truly necessary for solution of practical selection problems.

There are currently over 23 manufacturers of industrial robots, and many more individual models of robots. Development of a robot specification data manual on available robots should be undertaken if robotics are seriously considered. From such a specification database, a set of feasible units for an application can readily be identified.

Perusal of these robot specification data manuals indicates two robots as feasible for the forge loading/unloading application. What follows here is simply an engineering economic analysis of the alternatives. A five-year planning horizon is specified.

Cashflows

Robot A		Robot B	
0	$−45,000	0	$−1,000
1	−1,500	1	−1,000
2	−2,000	2	−1,200
3	−3,000	3	−1,800
4	−3,000	4	−2,000
5	−3,000	5	−2,000
5	+10,000	5	+32,000

A present worth comparison with a MARR of 20 percent is specified:

$$\text{Present worth of A} = -\$48{,}008$$

$$\text{Present worth of B} = -\$66{,}616$$

Economic analysis would indicate the selection of Robot A. However, a strict cash flow approach ignores several important factors, such as improved quality and consistency, the elimination of undesirable or hazardous jobs, etc. These non-economic factors need be considered also.

SUMMARY

NC machines and robots have quickly changed the appearance of today's factory. These automated marvels have increased productivity throughout the world and have changed manufacturing forever. They have, however, not come to use without a penalty. That is, applications of flexible automation must be well thought-out and justified. The potential gains from these systems are matched only by the potential loss that such expensive capital equipment can bring.

QUESTIONS

19.1. Briefly describe some of the similarities as well as differences between a robot and an NC machine.

19.2. What are some of the advantages of utilizing NC machines?

19.3. What are some of the advantages of using robots? Are the advantages for NC different than for robots? Why?

19.4. Name at least six different processes that NC is currently being employed to perform. Can you think of other processes that NC could be beneficial for? Briefly describe one or two.

19.5. What are the five different types of robot systems? Describe the advantages and disadvantages of each.

19.6. Programming an NC machine is a fairly standard practice; while programming a robot off-line is not very common. Why?

19.7. What are the advantages of off-line robot programming? What are the disadvantages? Why is so little off-line robot programming performed?

PROBLEMS

19.1. For the part geometry shown below, write an NC program using a word-address format to machine the part. All operations will be performed with a 0.5-in. end-mill at a feed of 0.20 in./rev and at a velocity of 80 sfpm.

19.2. Write an APT program to machine the part in Problem 19.1.

19.3. A part is currently being produced on a conventional milling machine. It takes 4.2 minutes on the average to produce a single part. An NC machine is being considered as a replacement for the manual operation. On the NC machine, it

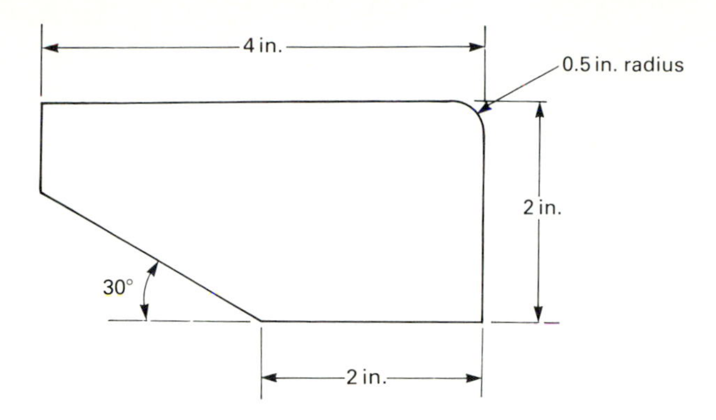

FIGURE P19.1

would take 3.2 minutes to produce a part, and a single operator could tend two machines. The current milling machine has long since been paid for by the company, and requires on average $1500 of maintenance a year. The company has gotten a bid of $60,000 for a new NC machine, which they would finance at 10 percent interest a year (100 percent financing of the machine has already been approved). A maintenance contract for $1100 per year would also be required if the NC machine was purchased. The operator of the current machine would also operate the NC machine. He is paid $35 per hour including overhead and fringe benefits. Should the NC machine be purchased?

19.4. A circle is to be painted using a robot that has only a straight-line interpolation system. The circle is 12 in. in diameter and should have a total tolerance of no more than 0.150 in. How may cords are required to make the circle? The circle's center is to be at point (0, 0, 0) in world coordinate space, and the circle will be in the X–Y plane. If the circle will start at point (0, 6, 0) and will be created clockwise, what are the world coordinates of the next three points?

REFERENCES

1. Asfahl, C. R.: *Robots and Manufacturing Automation,* Wiley, New York, 1985.
2. Bezier, P.: *Numerical Control: Mathematics and Application,* Wiley, New York, 1972.
3. Childs, J. J.: *Principles of Numerical Control,* Industrial Press, Inc., New York, 1982.
4. Groover, M. P.: *Automation, Production Systems and Computer Aided Manufacturing,* Prentice-Hall, Englewood Cliffs, N.J., 1980.
5. Koren, Y., and Ben-Uri, J.: *Numerical Control of Machine Tools,* Khanna Publishers, Delhi, India, 1978.
6. Roberts, A. D., and Prentice, R. C.: *Programming for Numerical Control Machines,* McGraw-Hill, New York, 1978.
7. Smith, D. N., and Evans, L.: *Management Standards for Computer and Numerical Control,* The University of Michigan, Ann Arbor, Mich., 1977.
8. Pousztai and Sava, 1983.
9. Pressman, R. S., and Williams, J. E.: *Numerical Control and Computer Aided Manufacturing,* John Wiley & Sons, New York, 1977.
10. Koren, Y.: *Computer Control of Manufacturing Systems,* McGraw-Hill, New York, 1983.
11. IITRI, 1967.

CHAPTER 20

RELIABILITY AND QUALITY CONTROL

As the public escalates its demand for product performance and safety, the production-design engineer is faced with the task of incorporating the most effective product-assurance procedures with due regard for cost. The fact that foreign competition is not only equaling United States goods from the standpoint of quality but exceeding it in some instances is placing the burden of "proof" on the production-design engineer. Japan has been referred to as the "quality miracle of the century" and they may well lead the world market from the standpoint of product performance in the 1980s and 1990s. Today, many consumers feel that the workmanship in Europe at least equals that in the United States.

In this chapter, we will briefly discuss reliability and quality control from the standpoint of the production-design engineer. By reliability, we mean the probability that a system (e.g., component, part, equipment, etc.) will operate for a period of time without failure under specified usage conditions. For example, a part-failure rate of 0.0001 percent per 1000 hours of service is about as good a reliability performance as can be expected in electronic equipment today. The 2000-part 1974 automobile had an average part-failure rate of 1.5 percent per 1000 hours of service while todays automobile has about half that failure rate.

"Quality control" may be defined in its broadest sense as all those activities within an organization that positively affect the quality of the products produced. Thus, quality control includes those processes or operations of testing, measuring, and comparing the manufactured part or apparatus with a standard, and then determining whether it should be accepted, rejected, adjusted, or reworked. Other quality control activities include the initial specifications applying to allowable tolerances and the extent of inspections considered essential or feasible. Every successful product is a reflection of the effectiveness of quality control. Consequently, the production-design engineer is vitally interested in this activity. The high reliability of ball and roller bearings is taken for granted today because of many years of careful control of materials, processes, and workmanship.

Controls are exercised by federal, state, and local governments through codes and inspectors. The armed services and insurance companies place their own inspectors in product source plants to control the product quality.

In most organizations, the quality control and reliability department managers reports to the manager of manufacturing; in other cases they report directly to the plant manager. They usually have equal status with the manufacturing and engineering managers.

Members of the quality control and reliability department contact suppliers of material in order to explain the requirements of purchase specifications and to assist in eliminating substandard material. They also contact customers who report failures or have difficulty with the equipment. They obtain firsthand information on the failure and endeavor to determine its cause. The quality control and reliability department usually allocates charges to the department responsible for that failure.

As the representative of the customer, the inspector or tester must ask himself, "Would I be satisfied with this equipment if I were purchasing it?" As a representative of the engineer, he must understand the standards set for performance and appearance and should be able to exercise good judgment when variations from these standards occur. As a representative of the accounting department, he is responsible for recording the number of pieces produced by each operation and for determining whether the work is acceptable or whether the piece should be rejected or salvaged. This same responsibility applies to purchased material. The action of the quality-control department affects every operation from the first to the last.

All the activities of the quality-control and reliability department are promoted through others. If the operators, material handlers, production clerks, supervisors, and engineers are quality conscious, the work of the department is less difficult and can be directed toward more constructive and creative efforts. A newly designed line of equipment can be ruined if proper controls are not established and maintained in the shop. The production-design engineer relies on the quality-control and reliability person to assist in the education of the shop personnel as to the importance of various features of the design.

Supervisors and engineers may be consulted, but the decision as to acceptance, rejection, adjustment, reoperation, or repair is the responsibility of the quality-control and reliability department. Its decisions should be such that delivery of the product on time is assured, product quality is maintained, and costs are minimized.

QUALITY-CONTROL PROCEDURES—GENERAL

The inspector for the quality-control and reliability department may inspect in any of the following ways.

1. Inspect each piece on each operation. This is called "100 percent inspection" and is quite expensive.
2. Inspect the first piece, and other pieces selected at random. This is called random inspection.
3. Check the final part and complete assembly, using random or 100 percent inspection.
4. Inspect the partial assembly or the final assembly only, using random or 100 percent inspection.
5. Inspect a part or complete apparatus taken from the production line to a laboratory or inspection station, where critical dimensions and functions are carefully checked. A refrigerator is carefully checked for color, placed in a tumbling machine to determine its ability to stand shipment, and given an accelerated life test under extreme conditions. The door may be opened and closed until failure occurs. Critical parts are removed and carefully measured. Such control checks insure a product with practically no hidden defects.

"One hundred percent inspection" is practically never 100 percent. If there are 100 known defectives in a batch of 5000 pieces, the ordinary inspector will miss 5, 10, 20, or more of them. At the very best, 100 percent inspection is seldom more than 98 percent efficient. Monotony, fatigue, and ineptitude take their inevitable toll.

In one study made under particularly good conditions, it was found that batches containing not more than 2 percent defective parts could be screened so that about 0.1 percent of the parts shipped were defective, but as the work coming into inspection increased to 5, 8, and 10 percent or more defective, the proportion of substandard pieces slipping by inspectors increased to 2 percent and even 5 percent.

Quality and Wage Payment

Under incentive wage payment plans, the operator is usually paid for the good pieces he makes. In some organizations, the operator's pay is determined according to the inspector's records at the end of the day or at the completion of the order. Later, if the part proves defective owing to improper workmanship, the operator usually cannot be penalized, because more than one operator may have performed the operation. Therefore, the responsibility of the quality-control man in approving an operator's pay is very important.

Rejections and Inspection Orders

When a part is rejected, it can be reworked by the operator (usually at his expense), salvaged, repaired in another department, or used as a substandard or special part. Usually, the part or assembly is removed from its original production order and is made a part of a separate order. For convenience in describing this procedure, the term "IO order" will be used, meaning an inspection order.

The IO order can be used: (1) to deliver the material or parts to the salvage department where the assembly may be dismantled, the good parts salvaged, and the material sold for scrap; (2) to return to supplier; (3) to reoperate and return to the original order or deliver to its proper section.

The IO order must carry the necessary charges and time values for performing the additional work. Tags to identify the parts are provided and serve to notify supervisors, operators, and production control people that trouble has occurred. These tags can be removed only by an inspector. The production-design engineer is often involved in the IO order procedure, since newly developed items on first production runs are usually substandard and are often chargeable to the accounts for which the engineer is responsible. Copies of IO orders are distributed to: (1) the person receiving the charge, (2) the production control department, to notify them of trouble that may delay their orders, and (3) the accounting department. The original copy of the IO order travels with the work to notify the operators and supervisors of the action to be taken.

Dimensional Measurement

A product is normally qualified in a variety of ways. A part can simply be tested after assembly to determine whether it performs its prescribed function. This, however, can be exceptionally costly if many components constitute a single assembly, since any of the pieces may cause the part to fail. Typically, most components are inspected to determine whether they comply with engineering specifications called out on an engineering drawing. These specifications can include geometric dimension, geometric tolerance, hardness, tensile strength, electrical resistance, etc. Each of these features must reside within the specified limits if the part was designed properly and is to perform correctly. Over the past three decades several drafting standards designed to improve drafting practice and improving the clarity of dimension specification have evolved. In the following sections a brief discussion of these standards is given.

Features, Datums and Dimensions

A datum is any point, line, plane, cylinder, or shape element specified on a component drawing in order to provide a basis for measure and interpretation. Datums are used to establish reference points, lines, and surfaces for tolerance specification and measurement. Datums can be explicitly called out or implicitly defined on a drawing. For instance, in Fig. 20.1, no datums have been explicitly specified. However, the drawing implies that the hole position is referenced from surfaces 1 and 2 of the drawing. An implicit datum referring to these surfaces for tolerance measurement is assumed.

Basic dimensions on engineering drawings are normally designated by appending the letters BSC to the measurement. A basic dimension on an engineering drawing is a statement indicating that the measure is exact less any other form call out. For instance, in Fig. 20.1, the hole location measure is given as a basic measure; however, the position is qualified by the true position datum.

There are three types of geometric tolerance used to specify a drawing: position or location tolerance, form tolerance, and runout tolerance. Position

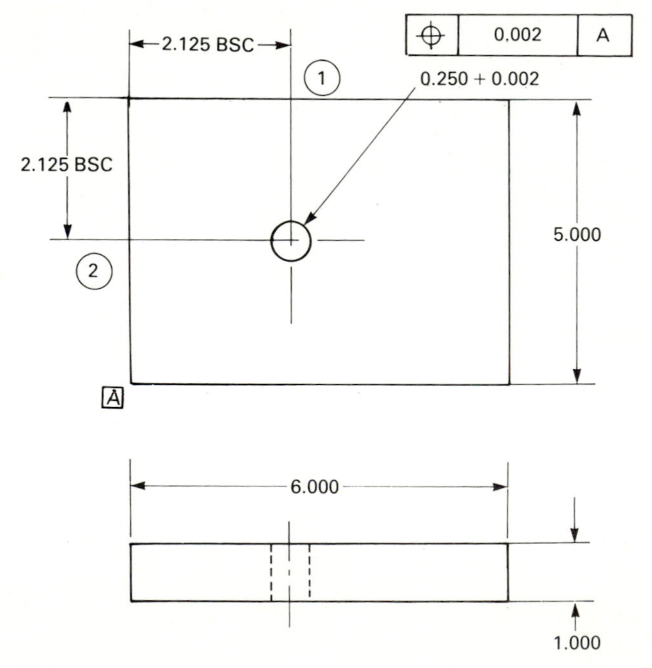

FIGURE 20.1
Positional tolerancing with a datum.

tolerance specifies the range of variation in the location of a feature. The specification may again be implicit or explicit. Position tolerance involves at least two features—one noting the size of one feature and the other noting the relationship between the two features. The positional tolerance of a feature is assumed to be taken from the axis noted by the dimension line otherwise specified.

Form tolerance specifies the range of allowable variation from a perfect form called out on a drawing. Form tolerances can be used with one- or

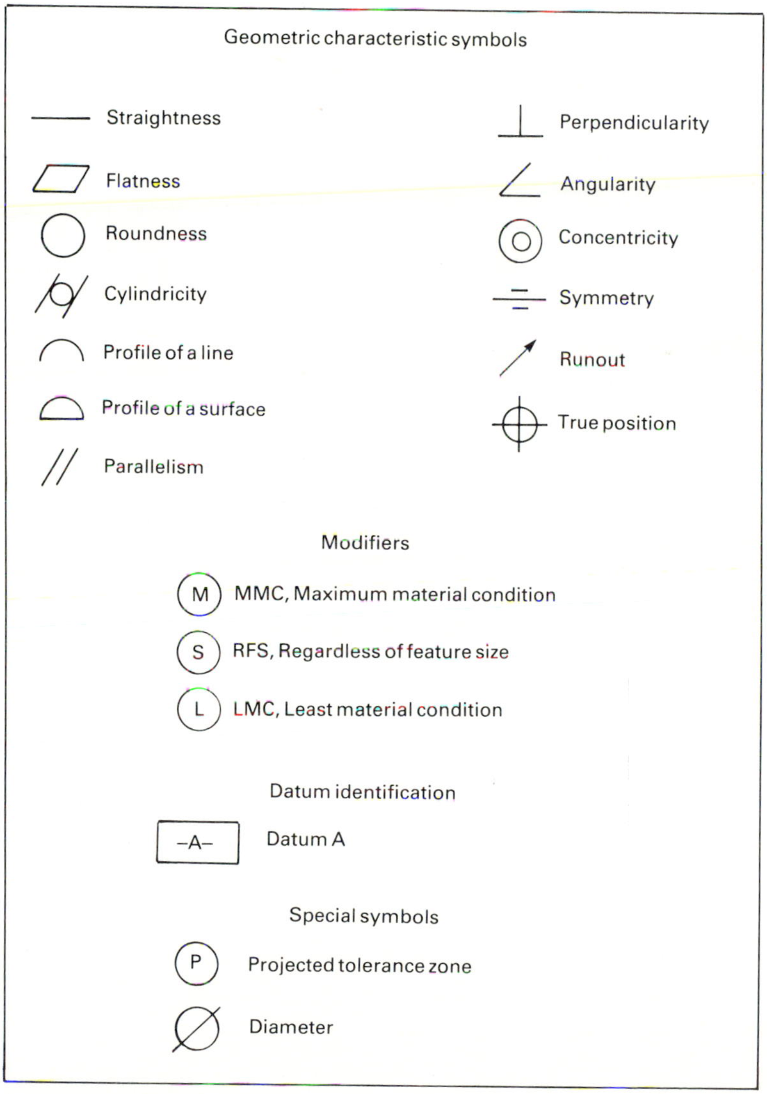

Geometric characteristic symbols

Straightness	Perpendicularity
Flatness	Angularity
Roundness	Concentricity
Cylindricity	Symmetry
Profile of a line	Runout
Profile of a surface	True position
Parallelism	

Modifiers

(M) MMC, Maximum material condition

(S) RFS, Regardless of feature size

(L) LMC, Least material condition

Datum identification

-A- Datum A

Special symbols

(P) Projected tolerance zone

Diameter

FIGURE 20.2(*a*)
Geometric tolerancing symbols.

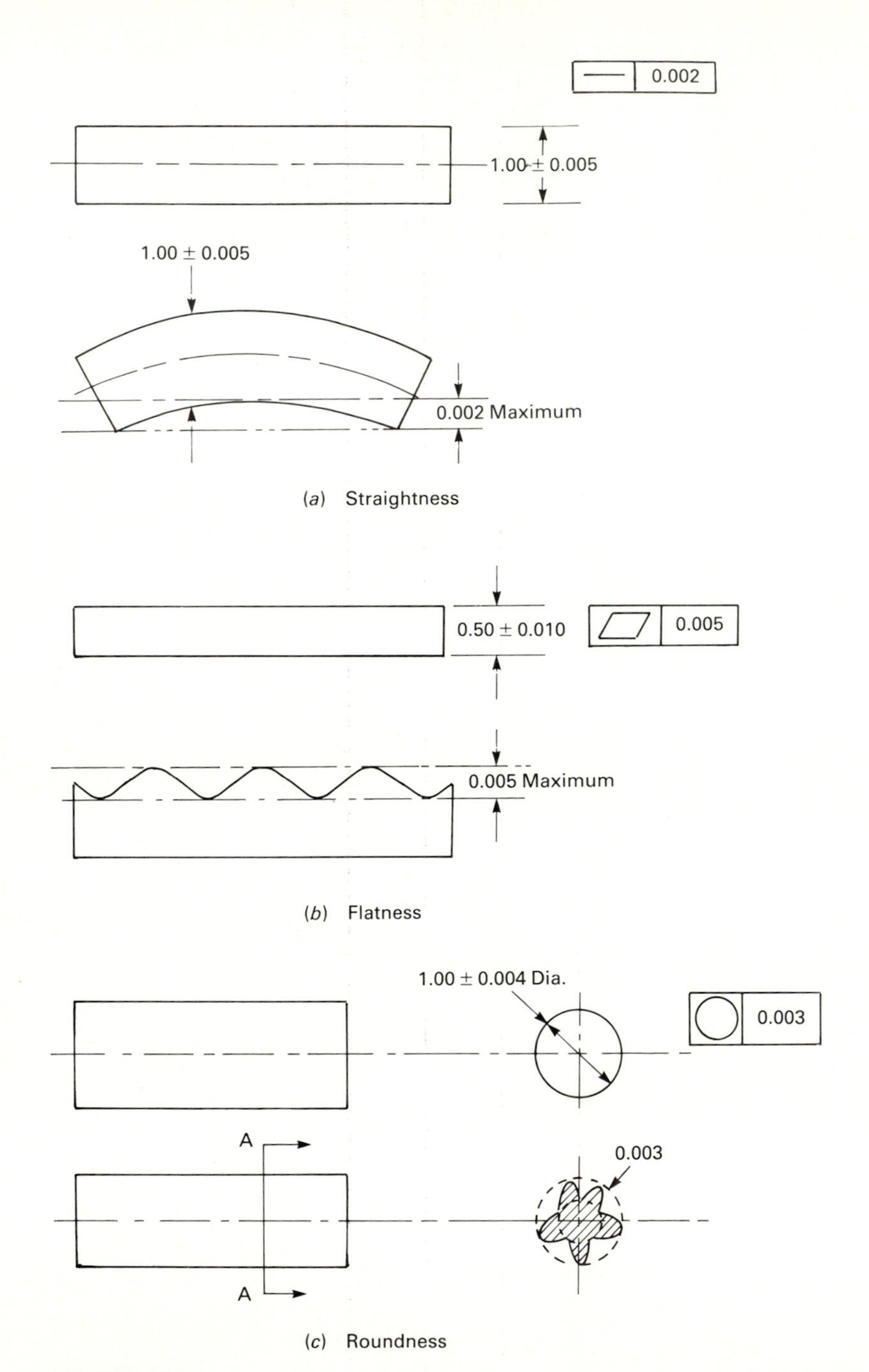

(*a*) Straightness

(*b*) Flatness

(*c*) Roundness

FIGURE 20.2(*b*)
Illustration of the form geometry symbols.

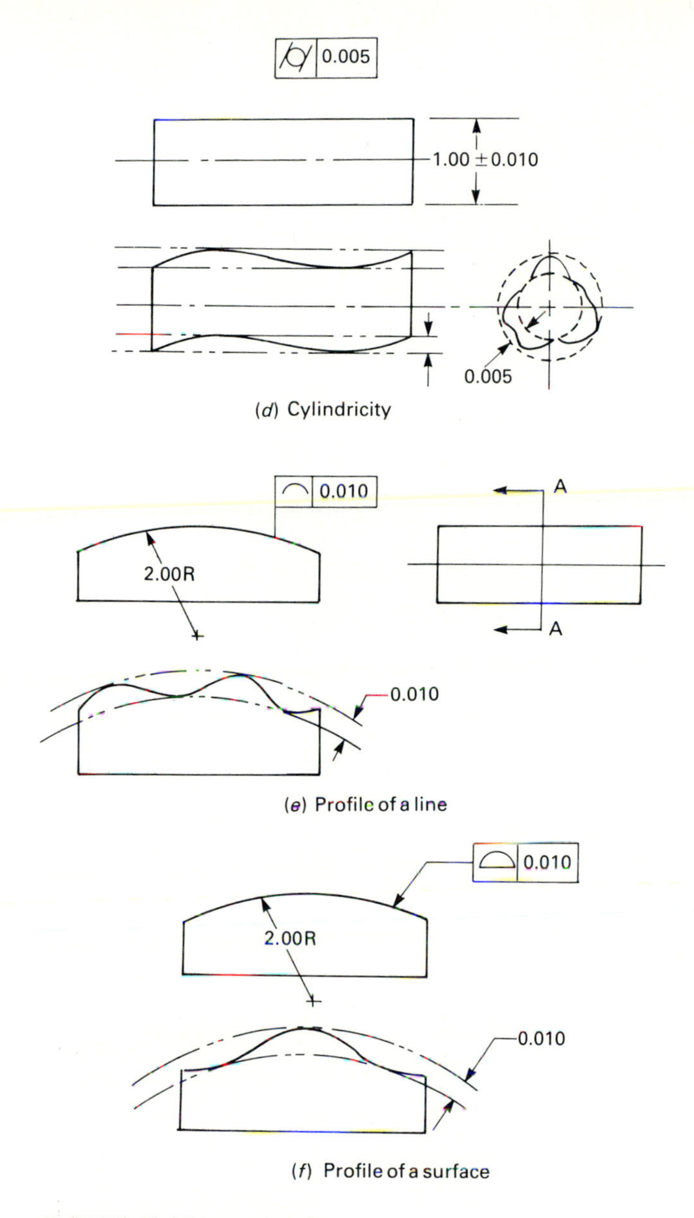

(*d*) Cylindricity

(*e*) Profile of a line

(*f*) Profile of a surface

FIGURE 20.2(*b*), continued.

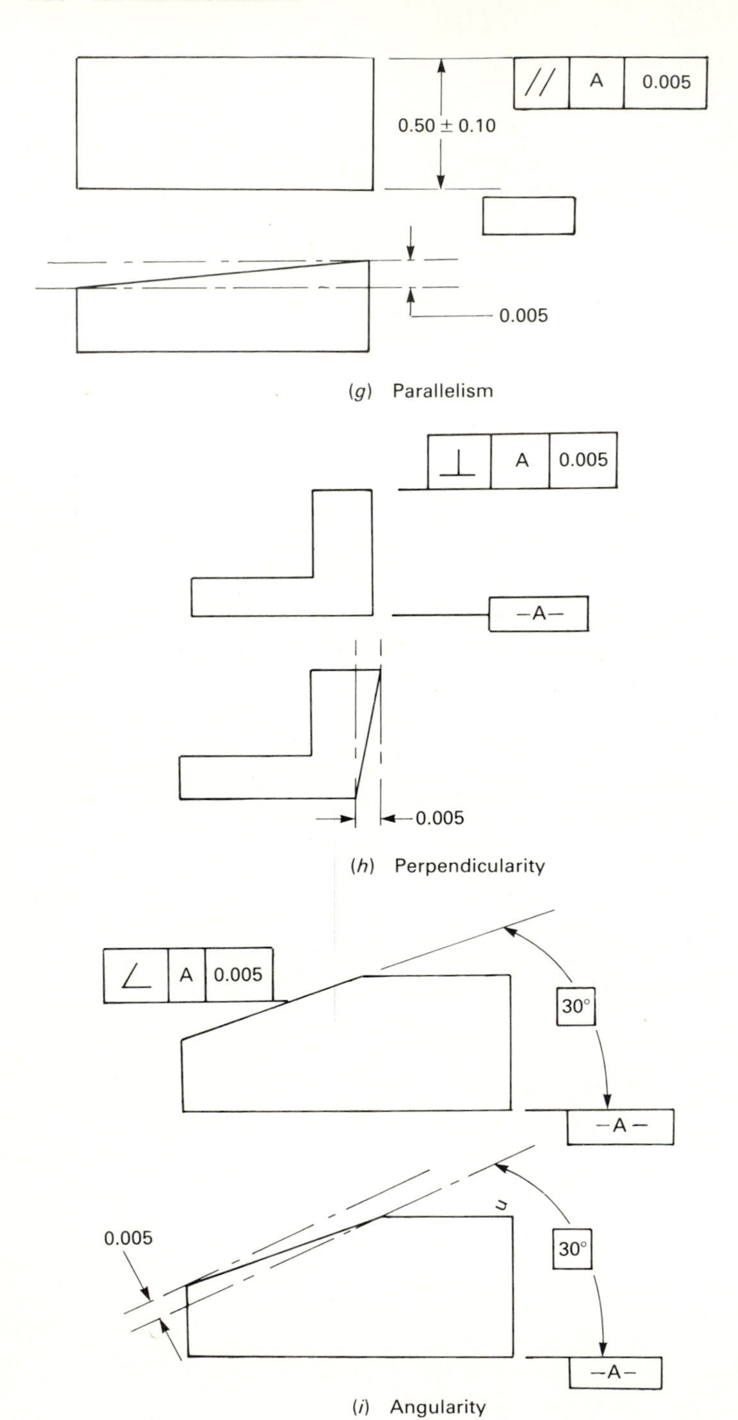

FIGURE 20.2(*b*), continued.

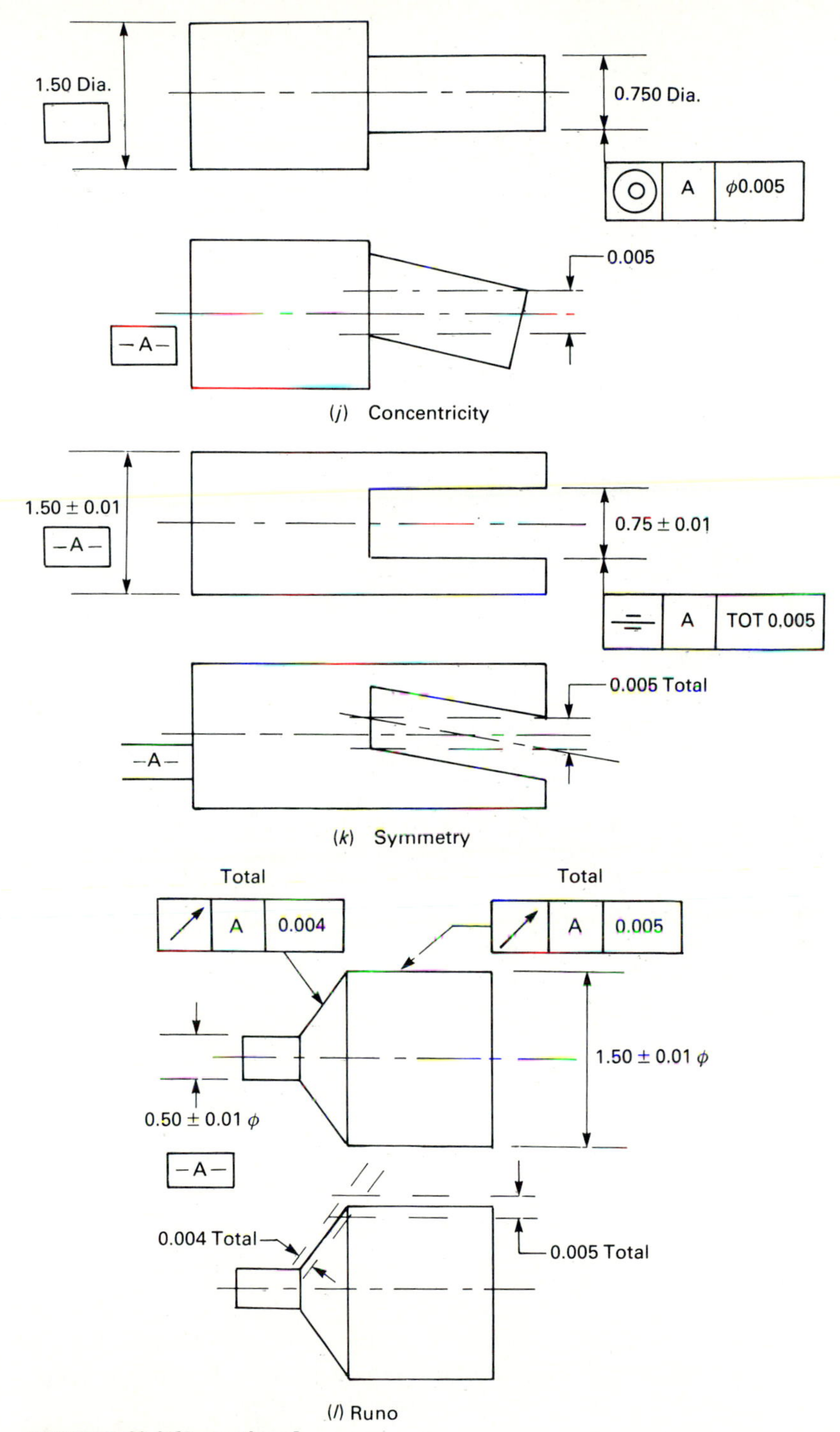

FIGURE 20.2(*b*), continued.

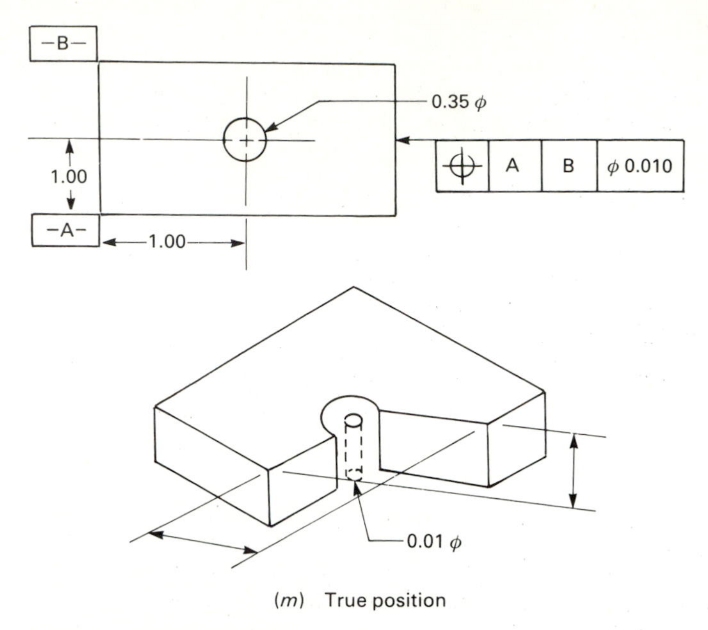

(*m*) True position

FIGURE 20.2(*b*), continued.

two-part features. For instance if circularity of a hole is called out on a drawing, this would relate to only the hole—a singular feature. However, perpendicularity, angularity, or parallelism all relate to the relationship of two features. Form tolerance features are illustated in Fig. 20.2.

Runout toletance specifies the range of variation allowed when a rotational feature is rotated about its major axis. Runout is normally used to specify the variation in axial alignment of a cylindrical part of more than two diameters. This feature is also illustated in Fig. 20.2.

Material Conditions

Tolerance specification are normally made under a set of material specifics. The three material conditions used in standard drafting systems are:

1. Maximum material conditions (MMC)
2. Least material conditions (LMC)
3. Regardless of feature size (RFS)

Maximum material conditions correspond to a part occupying the maximum allowable volume (or material). Maximum material for a hole specification corresponds to the smallest allowable hole diameter. A shaft call out under MMC would correspond to the largest allowable shaft diameter.

Least material conditions represent the opposite of maximum material conditions. They correspond to the least amount of allowable volume. A shaft call out under LMC would correspond to the smallest possible diameter. A hole call out would correspond to the maximum allowable diameter.

Regardless of feature size conditions indicate that some relationship between two features must hold independently of the initial feature size. To determine the adequacy of a part under RFS conditions, one must first qualify the size of the feature noted as the datum.

LIMIT AND FIT

In Chapter 18, we discussed the concept of group technology, where we tried to take advantage of "sameness' in order to avoid reinventing the wheel over and over again. The same concept can also be applied to dimensioning and tolerancing. In fact, a tolerancing system for mating components was developed as part of U.S. Standards—USAS B4.1—1967. This standard provides a table of recommended specifications for hole/shaft fit for various functionality.

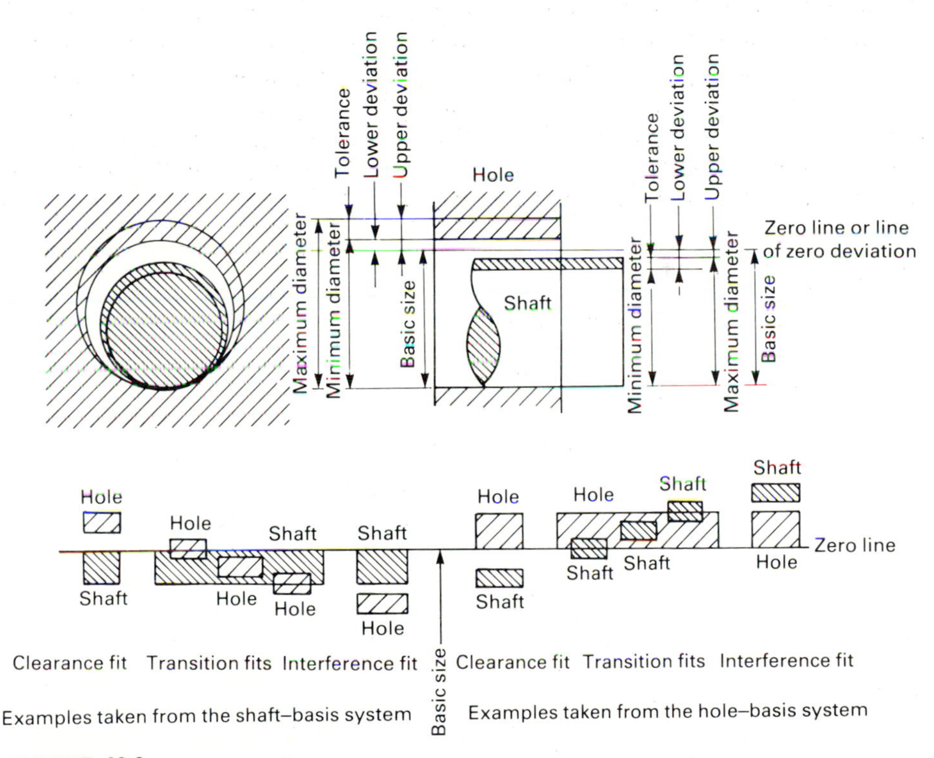

FIGURE 20.3
"Shaft-basis" and "hole-basis" systems for specifying fits in the ISO system. (*By permission from ISO Recommendation R286-1962,* System of Limits and Fits, *copyright 1962, American National Standards Institute, New York.*)

TABLE 20.1

American standard clearance locational fits

Limits are in thousands of an inch.
Limits for hole and shaft are applied algebraically to the basic size to obtain the limits of size for the parts.
Data in bold face are in accordance with ABC agreements.
Symbols H9, f8, etc., are Hole and Shaft designations used in ABC System.

Nominal size range, in.		Class LC 1			Class LC 2			Class LC 3			Class LC 4			Class LC 5		
		Limits of clearance	Standard limits		Limits of clearance	Standard limits		Limits of clearance	Standard limits		Limits of clearance	Standard limits		Limits of clearance	Standard limits	
Over	To		Hole H6	Shaft h5		Hole H7	Shaft h6		Hole H8	Shaft h7		Hole H10	Shaft h9		Hole H7	Shaft g6
0	0.12	0 0.45	+0.25 −0	+0 −0.2	0 0.65	+0.4 −0	+ −0.25	0 1	+0.6 −1	+0 −0.4	0 2.6	+1.6 −0	+0 −1.0	0.1 0.75	+0.4 −0	−0.1 −0.35
0.12	0.24	0 0.5	+3 −0	+0 −0.2	0 0.8	+0.5 −0	+0 −0.3	0 1.2	+0.7 −0	+0 −0.5	0 3.0	+1.8 −0	+0 −1.2	0.15 0.95	+0.5 −0	−0.15 −0.45
0.24	0.40	0 0.65	+0.4 −0	+0 −0.25	0 1.0	+0.6 −0	+0 −0.4	0 1.5	+0.9 −0	+0 −0.6	0 0.6	+2.2 −0	+0 −1.4	0.2 1.2	+0.6 −0	−0.2 −0.6
0.40	0.71	0 0.7	+0.4 −0	+0 −0.3	0 1.1	+0.7 −0	+0 −0.4	0 1.7	+1.0 −0	+0 −0.7	0 4.4	+2.8 −0	+0 −1.6	0.25 1.35	+0.7 −0	−0.25 −0.65
0.71	1.19	0 0.9	+0.5 −0	+0 −0.4	0 1.3	+0.8 −0	+0 −0.5	0 2	+1.2 −0	+0 −0.8	0 5.5	+3.5 −0	+0 −2.0	0.3 1.6	+0.8 −0	−0.3 −0.8
1.19	1.97	0 1.0	+0.6 −0	+0 −0.4	0 1.6	+1.0 −0	+0 −0.6	0 2.6	+1.6 −0	+0 −1	0 6.5	+4.0 −0	+0 −2.5	0.4 2.0	+1.0 −0	−0.4 −1.0
1.97	3.15	0 1.2	+0.7 −0	+0 −0.5	0 1.9	+1.2 −0	+0 −0.7	0 3	+1.8 −0	+0 −1.2	0 7.5	+4.5 −0	+0 −3	0.4 2.3	+1.2 −0	−0.4 −1.1
3.15	4.73	0 1.5	+0.9 −0	+0 −0.6	0 2.3	+1.4 −0	+0 −0.9	0 3.6	+2.2 −0	+0 −1.4	0 8.5	+5.0 −0	+0 −3.5	0.5 2.8	+1.4 −0	−0.5 −1.4

4.73	7.09	0 1.7	+1.0 −0	+0 −0.7	0 2.6	+1.6 −0	+0 −1.0	0 4.1	+2.5 −0	+0 −1.6	0 10	+6.0 −0	+0 −4	0.6 3.2	+1.6 −0	−0.6 −1.6
7.09	9.85	0 2.0	+1.2 −0	+0 −0.8	0 3.0	+1.8 −0	+0 −1.2	0 4.6	+2.8 −0	+0 −1.8	0 11.5	+7.0 −0	+0 −4.5	0.6 3.6	+1.8 −0	−0.6 −1.8
9.85	12.41	0 2.1	+1.2 −0	+0 −0.9	0 3.2	+2.0 −0	+0 −1.2	0 5	+3.0 −0	+0 −2.0	0 13	+8.0 −0	+0 −5	0.7 3.9	+2.0 −0	−0.7 −1.9
12.41	15.75	0 2.4	+1.4 −0	+0 −1.0	0 3.6	+2.2 −0	+0 −1.4	0 5.7	+3.5 −0	+0 −2.2	0 15	+9.0 −0	+0 −6	0.7 4.3	+2.2 −0	−0.7 −2.1
15.75	19.69	0 2.6	+1.6 −0	+0 −1.0	0 4.1	+2.5 −0	+0 −1.6	0 6.5	+4 −0	+0 −2.5	0 16	+10.0 −0	+0 −6	0.8 4.9	+2.5 −0	−0.8 −2.4
19.69	30.09	0 3.2	+2.0 −0	+0 −1.2	0 5.0	+3 −0	+0 −2	0 8	+5 −0	+0 −3	0 20	+12 −0	+0 −8	0.9 5.9	+3.0 −0	−0.9 −2.9
30.09	41.49	0 4.1	+2.5 −0	+0 −1.6	0 6.5	+4 −0	+0 −2.5	0 10	+6 −0	+0 −4	0 26	+16.0 −0	+0 −10	1.0 7.5	+4.0 −0	−1.0 −3.5
41.49	56.19	0 5.0	+3.0 −0	+0 −2.0	0 8.0	+5 −0	+0 −3	0 13	+8 −0	+0 −5	0 32	+20.0 −0	+0 −12	1.2 9.2	+5.0 −0	−1.2 −4.2
56.19	76.39	0 6.5	+4.0 −0	+0 −2.5	0 10	+6 −0	+0 −4	0 16	+10 −0	+0 −6	0 11	+25.0 −0	+0 −16	1.2 11.2	+6.0 −0	−1.2 −5.2
76.39	100.9	0 8.0	+5.0 −0	+0 −3.0	0 13	+8 −0	+0 −5	0 20	+12 −0	+0 −8	0 50	+30.0 −0	+0 −20	1.4 14.4	+8.0 −0	−1.4 −6.4
100.9	131.9	0 10.0	+6.0 −0	+0 −4.0	0 16	+10 −0	+0 −6	0 26	+16 −0	+0 −10	0 65	+40.0 −0	+0 −25	1.6 17.6	+10.0 −0	−1.6 −7.6
131.9	171.9	0 13.0	+8.0 −0	+0 −5.0	0 20	+12 −0	+0 −8	0 32	+20 −0	+0 −12	0 8	+50.0 −0	+0 −30	1.8 21.8	+12.0 −0	−1.8 −9.8
171.9	200	0 16.0	+10.0 −0	+0 −6.0	0 26	+16 −0	+0 −10	0 41	+25 −0	+0 −16	0 100	+60.0 −0	+0 −40	1.8 27.8	+16.0 −0	−1.8 −11.8

TABLE 20.2
Abbreviated table of running and sliding fits

Extracted from American Standard *Limits and Fits from Cylindrical Parts,* USAS B4.1–1967, with the permission of the publisher, The American Society of Mechanical Engineers, United Engineering Center, 345 East 47th Street, New York, 10017.
Limits are in thousandths of an inch.
Limits for hole and shaft are applied algebraically to the basic size to obtain the limits of size for the parts.

Nominal size range, in.		Class RC 1			Class RC 2			Class RC 3			Class RC 4		
			Standard limits			Standard limits			Standard limits			Standard limits	
Over	To	Limits of clearance	Hole H5	Shaft g4	Limits of clearance	Hole H6	Shaft g5	Limits of clearance	Hole H7	Shaft f6	Limits of clearance	Hole H8	Shaft f7
0	0.12	0.1 0.45	+0.2 0	−0.1 −0.25	0.1 0.55	+0.25 0	−0.1 −0.3	0.3 0.95	+0.4 0	−0.3 −0.55	0.3 1.3	+0.6 0	−0.3 −0.7
0.12	0.24	0.15 0.5	+0.2 0	−0.15 −0.3	0.15 0.65	+0.3 0	−0.15 −0.35	0.4 1.12	+0.5 0	−0.4 −0.7	0.4 1.6	+0.7 0	−0.4 −0.9

0.24	0.40	**0.2 0.6**	**+0.25 0**	**−0.2 −0.35**	**0.2 0.85**	**+0.4 0**	**−0.2 −0.45**	**0.5 1.5**	**+0.6 0**	**−0.5 −0.9**	**0.5 2.0**	**+0.9 0**	**−0.5 −1.1**
0.40	0.71	**0.25 0.75**	**+0.3 0**	**−0.25 −0.45**	**0.25 0.95**	**+0.4 0**	**−0.25 −0.55**	**0.6 1.7**	**+0.7 0**	**−0.6 −1.0**	**0.6 2.3**	**+1.0 0**	**−0.6 −1.3**
0.71	1.19	**0.3 0.95**	**+0.4 0**	**−0.3 −0.55**	**0.3 1.2**	**+0.5 0**	**−0.3 −0.7**	**0.8 2.1**	**+0.8 0**	**−0.8 −1.3**	**0.8 2.8**	**+1.2 0**	**−0.8 −1.6**
1.19	1.97	**0.4 1.1**	**+0.4 0**	**−0.4 −0.7**	**0.4 1.4**	**+0.6 0**	**−0.4 −0.8**	**1.0 2.6**	**+1.0 0**	**−1.0 −1.6**	**1.0 3.6**	**+1.6 0**	**−1.0 −2.0**
1.97	3.15	**0.4 1.2**	**+0.5 0**	**−0.4 −0.7**	**0.4 1.6**	**+0.7 0**	**−0.4 −0.9**	**1.2 3.1**	**+1.2 0**	**−1.2 −1.9**	**1.2 4.2**	**+1.8 0**	**−1.2 −2.4**
3.15	4.73	**0.5 1.5**	**+0.6 0**	**−0.5 −0.9**	**0.5 2.0**	**+0.9 0**	**−0.5 −1.1**	**1.4 3.7**	**+1.4 0**	**−1.4 −2.3**	**1.4 5.0**	**+2.2 0**	**−1.4 −2.8**
4.73	7.09	**0.6 1.8**	**+0.7 0**	**−0.6 −1.1**	**0.6 2.3**	**+1.0 0**	**−0.6 −1.3**	**1.6 4.2**	**+1.6 0**	**−1.6 −2.6**	**1.6 5.7**	**+2.5 0**	**−1.6 −3.2**
7.09	9.85	**0.6 2.0**	**+0.8 0**	**−0.6 −1.2**	**0.6 2.6**	**+1.2 0**	**−0.6 −1.4**	**2.0 5.0**	**+1.8 0**	**−2.0 −3.2**	**2.0 6.6**	**+2.8 0**	**−2.0 −3.8**

Three classes of fit are used to specify mating interaction. These classes of fit are: (1) clearance fits, (2) transitional fits, and (3) interference fits. Figure 20.3 illustrates this basic concept. As their names imply, a clearance fit indicates that a clearance remains between the shaft and hole after they have been assembled. This allows the shaft to rotate or move about the major hole axis. The USAS standard uses nine subclasses of fit to describe hole/shaft fit. These subclasses range from RC1, a close sliding fit where no perceivable play can be observed, to RC9, a loose running fit where the shaft fits more loosely into the hole. The designer simply selects from the subclasses the one that best fits his needs knowing that the higher the specification subclass, the less expensive the manufacturing.

Transition fits are normally used to specify tolerance for parts that are stationary. Location clearance fits (LC1 to LC11) are used for parts that are assembled together and can be disassembled for service. The accuracy for these components is not exact. Transition location fits (LN1 to LN6) are specified when the location accuracy is of importance but a small clearance or interference is acceptable.

Locational interference fits (LN1 to LN3) are specified when both rigidity and accuracy are required. Location interference fit parts can be assembled and disassembled but not without special tooling (usually a shaft or wheel puller) and considerable time. Other interference fit parts normally require special operations for assembly. Tight drive fits (FN1) are used on parts requiring nominal assembly pressure. Force fits (FN5) are used for drive applications where the hole element is normally heated to expand the diameter prior to assembly. Tables 20.1 and 20.2 contain the specification for various classes of fit.

Inspection Equipment

In order to detect errors, measure performance, and check materials and parts with standards, inspection and testing facilities must be available. Tensile-strength test machines, hardness-testing machines, and other equipment is set up in laboratories that check materials received, made, or treated in the shop. Elaborate checking gages and fixtures are designed to inspect complicated parts and assemblies. Complete sets of gages and tools for measuring critical dimensions should be available to the operator, as well as the inspector. Where controls are extensive, it is necessary to have a gage section that checks, repairs, and adjusts gages by using master gages of the highest accuracy. Thus duplicate and triplicate sets of gages are sometimes necessary to maintain control.

Errors in measurement depend upon the accuracy of the measuring equipment, which is subject to errors. These errors are the result of one or more of the following.

1. Inherent errors in the measuring instrument

2. Errors in the "master gage" used to set the instrument
3. Errors resulting from temperature variation and different coefficients of linear expansion of instrument and part being gaged
4. Errors due to the "human element" of the inspector

The engineer should not specify a dimension, characteristic, or function of a part or apparatus that cannot be measured. Fortunately, owing to the extensive development of inspection and testing apparatus, it is now possible to measure to a very high degree of accuracy. Instruments for measuring roughness of surfaces are available, and through the use of oscilloscopes, oscillographs, X-rays, and other types of sensitive measuring equipment, the quality of products can be controlled. Some of these instruments, such as air gages for measuring close dimensions, are rugged enough to use in production operations and thus the operator has a means of controlling the quality of the part he is making.

Inspection operations such as detection of surface defects and blowholes can be a part of the operator's responsibility. Checking gages can be built into jigs, fixtures, and equipment that can be used by the operator and inspector. This is especially valuable on medium-activity and complicated parts.

Inspection and testing equipment may be stationary or portable and destructive or nondestructive of the material or part.

The ability to measure and control dimensions has progressed. A few years ago it was said, "We can work to 0.001 in. and talk about holding to 0.0001 in.; now it can be said, we can work to 0.0001 in. and talk about holding to 0.00001-in. tolerance." The basic equipment and standards for measurement are measuring blocks, known as Jo blocks, invented by Johansson. Interchangeable manufacture would not be possible without these carefully maintained standards.

Inspection equipment can be broadly classified as general-purpose or special-purpose equipment. To measure reasonably simple parts or very low-volume items, general-purpose inspection equipment is normally used. To inspect very intricate or high-volume parts, special gages are normally designed in order to reduce the amount of time required for the inspection process. A general characteristic of these special gages is that they are designed to fit over some aspect of the part when the part is within dimension (A GO GAGE). An alternate gage is used to determine whether a part is too small or too large (A NO-GO GAGE).

GENERAL MEASUREMENT EQUIPMENT

Scales

A simple scale is perhaps the most widely known and used measuring instrument. Scales come in a variety of sizes, graduations, and accuracies. Scales are normally used for measuring part accuracies greater than $\frac{1}{64}$ in. or

0.5 mm. When accuracies greater than this are noted, more precise measurement equipment is used.

Vernier Caliper

A vernier caliper is a general-purpose measuring device capable of making both outside and inside measurements. A typical vernier caliper is illustrated in Fig. 20.4. The scale on the vernier caliper looks similar to a standard scale and resolution directly from the scale results in about the same accuracy as that attained from a standard scale. However, an auxiliary scale, as shown in the figure, can be used to obtain a finer resolution. The lower scale normally provides 0.025 in. of dimension in 0.001 in. increments, or some other similar resolution. In order to obtain the reading from the vernier caliper, a base measure is first attained from the scale istelf. The vernier scale shown below it is then used to obtain the finer graduations between the scale readings. For instance in Fig. 20.4, a base measure of 1.900 is read directly from the scale. By aligning the scale lines on the vernier, a reading of 0.006 is obtained. The vernier reading is then added to the scale reading to obtain a reading of 1.906 (1.900 + 0.006). Although the scales on verniers may vary, the procedure used to obtain the reading is usually quite similar.

Micrometer

A micrometer is frequently used for measurements requiring somewhat greater precision than the vernier caliper can achieve. Micrometers come in a variety of shapes and styles required to measure different geometric features. Figure 20.5 shows some typical micrometers. As can be seen in the figure, a micrometer is a screw-actuated instrument. The accuracy of a micrometer may vary; however, one can normally expect 0.0005 in. of precision with a micrometer. Some micrometers contain ratchet drives to produce a consistent pressure on the part. A vernier may also be included to obtain a finer

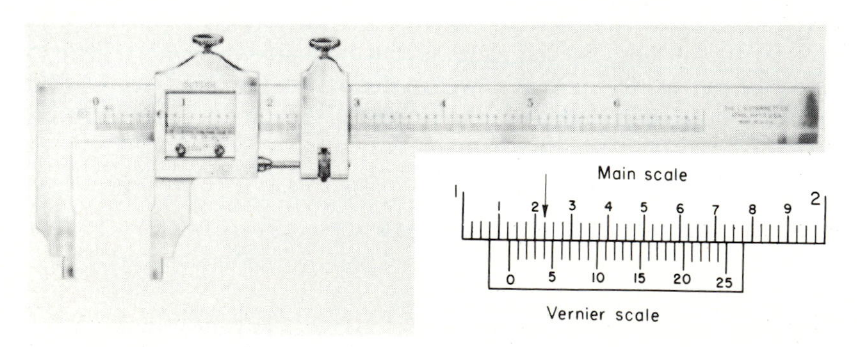

FIGURE 20.4
A vernier caliper. (*Courtesy The L. S. Starrett Co.*)

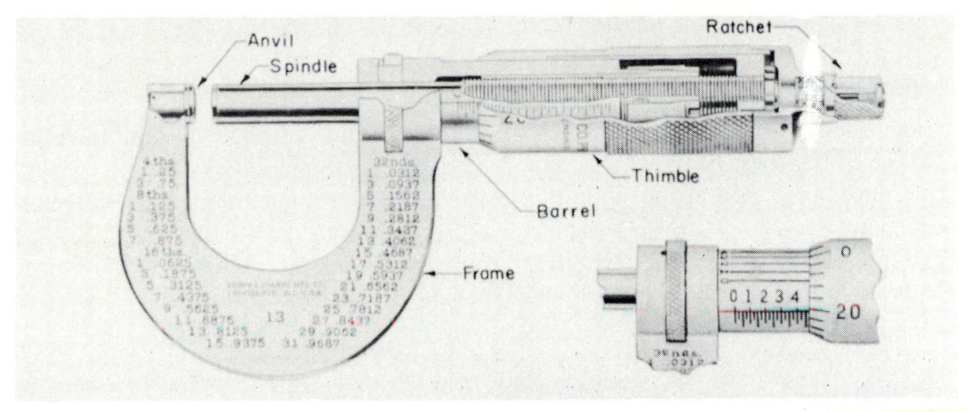

(*a*) A sectional view of a micrometer caliper. (*Courtesy Brown and Sharpe Mfg. Co.*)

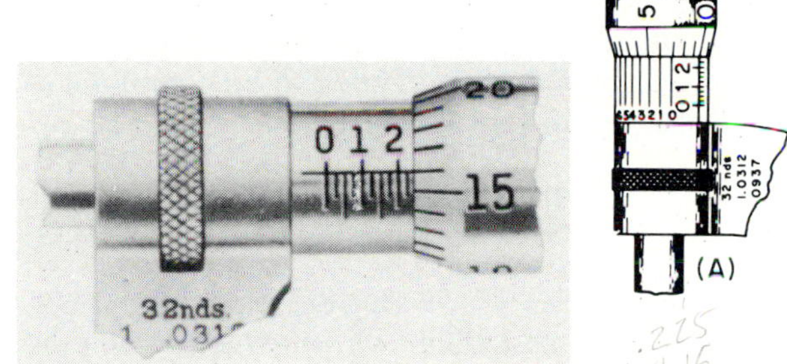

(*b*)(i) A micrometer reading of 0.241 in. (*Courtesy Brown and Sharpe Mfg. Co.*)

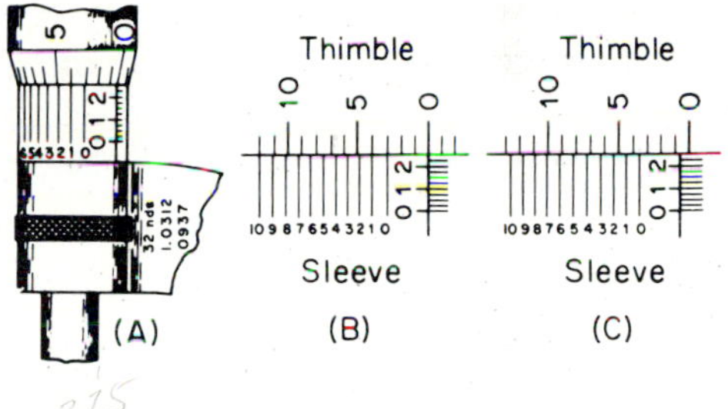

(*b*)(ii) The scales on a vernier micrometer caliper.

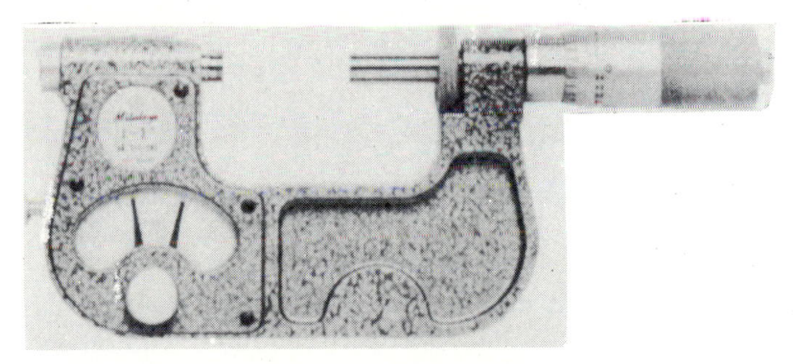

(*c*) An indicating micrometer. (*Courtesy Federal Products Corp.*)

FIGURE 20.5
Some micrometers.

resolution. Micrometers are general-purpose gages used to obtain a reasonable accuracy without requiring an exceptionally large amount of inspection time.

Gage Blocks

Gage blocks are accurate in height, flatness, and parallelism and have a Rockwell hardness of over C65. The following set of gage blocks with accuracy tolerances indicated are available today.

	Grade .	Tolerance, in.
Laboratory set	AA	±0.000001
Inspection set	A	±0.000004
Working set	B	±0.000008
Working set	C	±0.000010

Dimension can be obtained by combining various blocks. These blocks are so accurate that they cling together because of the surface tension of the adsorbed water film and must be slid or pulled apart. They can be combined to check snap gages, height gages, micrometers, and verniers.

Optical Flats

Optical flats made of fused quartz are available in two grades: AA Grade, ±0.000001 in.; A Grade, ±0.000002 in. One millionth of an inch can be measured with these flats by means of interference of light waves.

Air and Electric Comparator

Air and electric comparator gages can be set by Jo blocks or other standards so that deviations from standard can be measured quickly by the operator. Large scales that can be read easily are provided. The range of variations measured is between 0.000001 and 0.001 in.

Optical comparators throw a profile of the part on a screen so that it can be compared to a master. The profile of the part is blown up many times so that deviations can be seen and measured easily.

A variety of other measuring equipment is also available for dimensional measurement. Figures 20.6 and 20.7 contain a cross section of equipment normally used for manual measurement. The accuracy, workpiece geometry, and annual production rate, along with a variety of other factors will dictate the type of equipment used for inspection. As is the case in manufacturing, general-purpose equipment is usually used for inspection when only a few parts are involved. However, as the number of pieces that require inspection increases, the use of special gages becomes more economic. High-volume

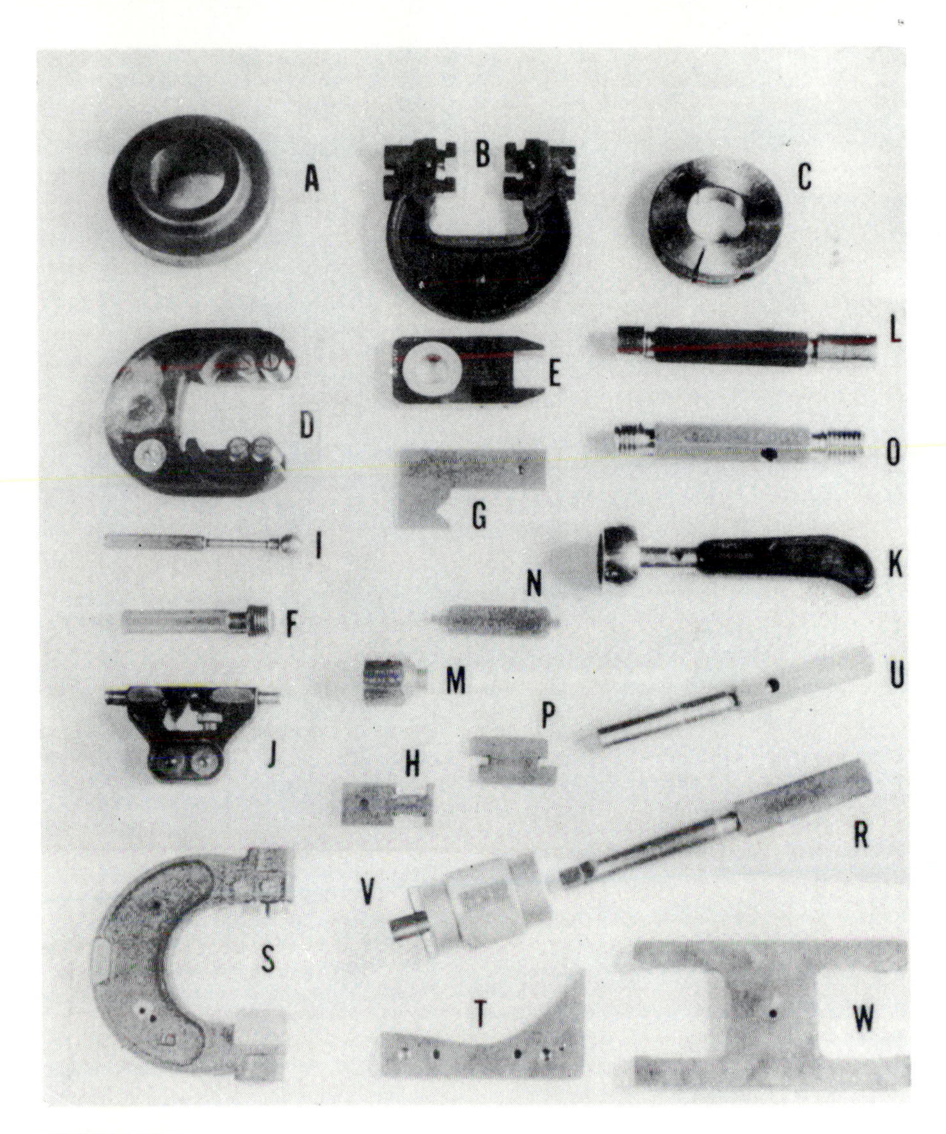

FIGURE 20.6
Typical gages. (A) Plain ring gage. (B) Adjustable limit progressive thread snap gage. (C) Adjustable thread ring gage. (D) Fixed limit progressive thread snap gage. (E) Fixed limit ring-snap gage. (F) Tapered thread plug gage. (G) Form gage. (H) Fixed limit snap gage. (I) Single end Blass plug gage. (J) Adjustable limit inside diameter or length gage. (K) Single end spherical plug gage. (L) Cylindrical double end plug gage. (M) Flush pin gage. (N) Double end slot width gage. (O) Double end thread plug gage. (P) Double end step gage. (R) Taper plug gage. (S) Adjustable limit progressive snap gage. (T) Form gage. (U) Cylindrical progressive plug gage. (V) Tapered ring gage. (W) Fixed limit double end snap gage.

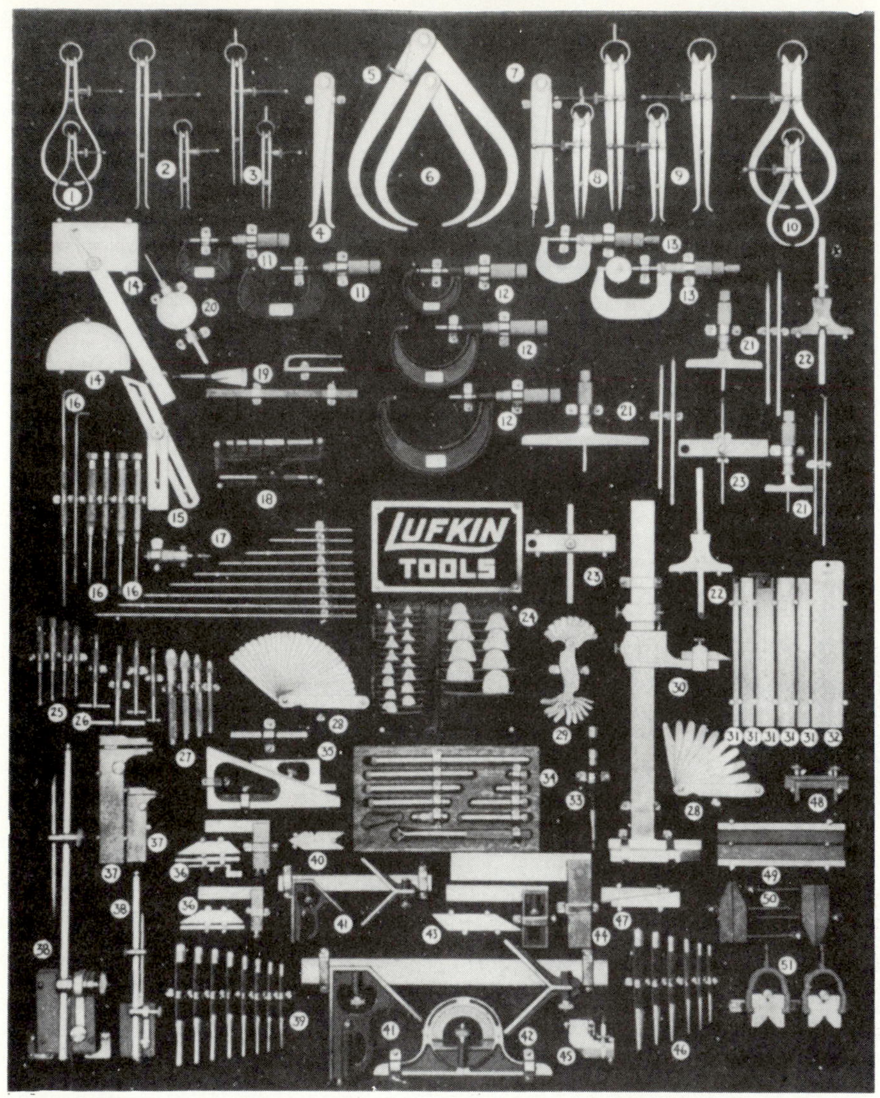

FIGURE 20.7
Standard measuring instruments. (1) Outside spring calipers, round leg. (2) Inside spring calipers. (3) Spring dividers. (4) Inside firm joint caliper. (5) Outside firm joint caliper with adjusting screw. (6) Outside firm joint caliper. (7) Firm joint hermaphrodite caliper. (8) Spring divider. (9) Inside spring caliper. (10) Outside spring caliper. (11) Outside micrometer calipers. medium weight. (12) Outside micrometer calipers, heavy duty. (13) Outside micrometer calipers with lock nut and ratchet stop (14) Steel protractors. (15) Universal bevel. (16) Scribers. (17) Inside micrometer set, solid rods. (18) Steel rule set and holder. (19) Universal indicator. (20) Dial test indicator. (21) Micrometer depth gages. (22) Depth rules. (23) Depth gage. (24) Radius gage set. (25) Small hole gages. (26) Telescoping gages. (27) Pin vises. (28) Thickness gages. (29) Screw pitch gage. (30) Vernier height gage. (31) Steel rules. (32) Mechanics' reference table. (33) Automatic center punch. (34) Inside micrometer set, tubular. (35) Planer and shaper gage. (36) Diemakers' squares. (37) Slide calipers. (38) Universal surface gages. (39) Drive pin punches. (40) Center gage. (41) Combination square. (42) Bevel protractor. (43) Double squares. (44) Steel square. (45) Right-angle rule clamps. (46) Center punches. (47) Tapered parallels. (48) Rule clamps. (49) Hold down parallels. (50) Toolmakers' parallel clamps. (51) V-blocks and clamps.

production systems usually employ a variety of special gages, many of which are built into the production system itself.

Flexible inspection equipment, like flexible production equipment, has also emerged during the past three decades. The use of coordinate measurement machines (CMM) has become an effective alternative to manual

FIGURE 20.8
Coordinate measurement machine.

dimensional inspection. These machines are used and programmed in much the same way as numerically controlled machines. Figure 20.8 contains an example of one such machine. A tactile sensor/strain gage is used to locate surfaces on the workpiece. The location information is used to determine the geometric characteristics of the workpiece. The characteristics are then compared to the specification on the part in order to determine whether the part is acceptable. Programs to inspect individual workpieces are prepared and executed as required. Automatic inspection systems can be economically effective for batch production of even small to medium lots (30 to 500 parts).

Dimensional inspection is not limited to inspection machines requiring physical contact. Several optical inspection systems have also evolved during the past decade. In these systems, a camera or laser is used to detect the edge of a workpiece. This information is used in much the same manner as the surface location from a coordinate measuring machine. The edges of the workpiece are used to construct a dimensional representation of the part. These measurements are again compared to those called out on the drawing in order to determine whether the part is good or not.

A variety of inspection equipment is used to qualify the dimensionality of products manufactured both here and abroad. The use of the equipment is

TABLE 20.3
Major features of typical measuring instruments and gages (dimensions in inches)

Name	Type[a]	Common range[b]	Discrimination[c]	Sensitivity[d]	Practical use[e]
Steel scale	1	6–24	0.001–0.02	0.01	±0.03
Vernier caliper	1	6–24	0.025	0.001	±0.003
Plain micrometer	1	1	0.001	0.0005	±0.002
Vernier micrometer	1	1	0.001	0.0001	±0.0005
Dial indicator (mechanical)[f]	2	0.010–0.250	0.00005–0.001	0.00005–0.0005	±0.0005–±0.003
Electronic comparator[f]	2	0.0005–0.005	0.00001–0.0001	0.000005–0.00005	±0.00001–±0.0001
Air gage[f]	2	0.0001–0.06	0.000002–0.001	0.000002–0.0005	±0.00001–±0.001
Optical comparator (screen type)	1	10–30	0.001	0.005	±0.00032–±0.001
Gage blocks (Class A)	3	0.10–24	0.0001	0.000008	±0.000004–±0.0001

[a] 1, direct reading; 2, comparison or transfer; 3, end standard.
[b] The range is the distance over which readings can be made.
[c] Discrimination is the usual size shown by the smallest graduation.
[d] Sensitivity is the smallest increment of dimension that has meaning.
[e] Practical use designates the smallest increment for which the instrument is normally reliable. Some models are used for smaller tolerances than those shown.
[f] Various models available.

predicated not only on the geometry and accuracy called out but also on the economics of the inspection process. Inspection does not transform a product, but it does add to the value of the end item by improving the reliability of the end product.

Strain Gages

Strain gages give a picture of the stresses existing in parts under static or dynamic load. There are mechanical, magnetic, electrical-resistance, and electrical-capacity types of strain gages that record strains where the gage can be fastened to the surface of the part. This equipment is valuable in designing apparatus.

Photoelasticity.

Photoelasticity is the process of using special transparent material having the profile of the shape to be studied. Polarized light is passed through the part. When it is strained, colors indicate where stresses occur.

Stress Paints

Stress paints are another means of obtaining a picture of strains existing in parts under stress. Stress paints cling to the surface of the part and crack when stressed because they are brittle. The pattern of cracks indicates the direction and magnitude of the strain. Some quantitative values can be obtained by measuring the photograph of the part.

Hardness Testers

The Brinell, Rockwell, and Scleroscope type of hardness testers are universally used to control hardness and measure strength of materials. Hardness and strength tend to have a direct relation to each other in the same materials. Conversion tables are available. The Rockwell and Scleroscope hardness testers may be portable, and can measure hardness on castings, billets, and parts that would be difficult to transport to a machine. Hardness testers leave a mark on the surface, which limits their use.

NONDESTRUCTIVE INSPECTION

X-ray

Powerful X-ray equipment is now available for medical, industrial, and military use. Parts and apparatus can be radiographed by exposing objects to X-rays and recording on a photographic film. Irregularities in the material, such as blowholes, slag, cracks, and nonuniform section, are revealed and then identified by the expert. Mines and ammunition may be disarmed by observing the device by X-rays and determining its design and method for disarming. X-rays can be focused so that any plane passing through the object can be observed with the other parts in the background. X-rays are used to control continuously the thickness of sheets rolled in steel, brass, or aluminum mills. The control is sensitive enough to measure variations of 0.001 in.

Ultrasonic Equipment

Ultrasonic equipment with electronic transmitters and receivers is available for detecting flaws in material. Fine cracks within material that cannot be detected by X-ray or magnetic means are picked up using ultrasonic methods.

Fluorescent Penetrants

A penetrating fluid that carries highly fluorescent dyes and is able to enter any minute crack shows up under ultraviolet light. This process, and magnaflux, are often used to inspect each part in a production line, such as a piston rod in an airplane, a gas engine housing, or any other vital part.

TEST EQUIPMENT

Test equipment is usually designed to check, adjust, and determine the extent of the functional characteristics of the apparatus, such as strength, capacity, and size. The equipment may include proving grounds where actual service conditions can be simulated. It may be individual equipment designed to check a specific characteristic of material or operation. For example, the automobile manufacturer has engineering laboratories equipped with testing equipment that can test and measure the performance of each part of a car or truck—rear-axle housings and differentials, transmissions, carburetors, generators, suspension springs, shock absorbers, and engines. These testing laboratories are sometimes supervised by the engineering department, which uses them to check current production assemblies and parts and to determine the characteristics of new equipment.

Such test equipment, including electrical test panels for transformers, circuit breakers, motors, radio, and radar, is usually so complicated and specialized that the quality-control department designs and builds its own. Fortunately, standard test equipment and components are available, such as dial gages, meters, relays, tensile and torsional strength-testing machines, hardness testers, and electronic units. These can be built into special test equipment.

Test operations are expensive and the equipment is costly. To govern test practices, manuals are written for the guidance of testers and engineers, and are useful for instructing engineering graduates who serve their apprenticeship on the test floor.

Testing operations can be included in the production line and so enable the operator to adjust, reject, or accept the part or assembly before it is built into the product. The testing of samples of steel or alloys in a steel mill before pouring is an example of controls by testing. The production-design engineer can assist in determining these points of test in order to save disassembly and repair.

The errors that are discovered on test can be reduced by removing the possibility of error by the operator. For example, pegboards for wiring electrical panels can be equipped with guides and markings so that the wires can be bent to shape, laid into position, and tied together as a unit. They are then placed in the radio or on the back of a control panel and can be easily connected to the proper terminal. Each wire can be identified by a tag or color code or by position. The use of wiring pegboards reduces the skill required and the errors of connection, and gives a neat appearance.

The supervisor can determine his best operators by observing inspection and test records. By studying test records, the engineer can determine components that can be improved in design or manufacture. One manufacturer of small motors has the following facilities to maintain quality.

Electrical Laboratory

1. Curve test equipment—to test all forms of motors and generators in order to determine their characteristics
2. Strobo scope—to observe objects rotating at high speed as though they were standing still
3. Cathode-ray oscillograph—to observe electric energy in forms of curves that can be photographed and measured
4. Weather-duplicating equipment (sandstorm box, humidity cabinet, salt-spray cabinet)
5. Radiofrequency test equipment
6. Dynamometer
7. Motor-life test for aircraft—to duplicate high altitude, temperature, and humidity
8. Sound-level room

Physics Laboratory

1. Vibration-testing equipment.
2. High-speed camera.
3. Equipment to test:
    a. Tensile, compressive, shear, and torsional strength
    b. Ductility
    c. Resistance to abrasion

   d. Hardness and depth of case-carburizing
   e. Depth of decarburization
   f. Response to heat treatment.
4. Identometer—to identify steels by checking against known master specimens. Operation depends on the thermoelectric properties of metals that conduct electricity. An identity can be established in 1 or 2 min.
5. Taber abrader—to evaluate the resistance of a plated, painted, or plastic finish to abrasion. Samples are rotated on the spindle, while an abrasive wheel is allowed to roll on the surface in question. The number of cycles the surface will withstand before a breakdown is measured with an automatic counter.
6. Durometer (used principally on resilient materials such as rubber).
7. Magnaflux—to determine quickly whether subsurface imperfections exist in a magnetic metal. The equipment employs the wet process under which very fine iron particles suspended in kerosene are poured over a specimen held in a powerful magnetic field. Any break in the continuity of the metal causes the iron filings to assume a definite pattern at the surface.

Chemical Laboratory

1. Portable gas analyzer
2. Electroanalyzer
3. Saybolt viscosimeter
4. pH meter
5. Carbon combustion train
6. Spectrophotometer
7. Complete facilities for chemical analysis of inorganic compounds
8. Paint-spray booth and complete spraying equipment
9. Centrifuges for various purposes
10. Facilities for determining properties of greases, oils, and solvents
11. Salt-spray cabinet—to determine the resistance to corrosion of various finishes
12. Humidity cabinet—to study the effect of excess humidity on materials or parts
13. Cold box—to conduct tests at −70°F

Metallurgical Laboratory

1. Marco etching
2. Bend testing
3. Fracture testing
4. Polishing
5. Profilometer

6. X-ray
7. Metallograph
8. Binocular microscope

Product-performance Test Laboratory

Equipment for testing apparatus for service life and performance during operation.

Test Equipment Facilities

Sufficient equipment is available to test and inspect mechanical and electrical apparatus on the production line. The manufacture of this equipment and its maintenance and calibration require machine tools and other equipment as well as master gages and electrical calibrating equipment.

QUALITY CONTROL—SURFACE FINISHES

Surface finishes have been discussed under the subjects of machining, honing, grinding, superfinishing, lapping, polishing, buffing, plating, and painting. In the control of surface finish, methods of measuring and comparing with samples are used. This discussion will be confined to the controlling of surfaces that can be measured and designed on drawings. Some of the instruments for measuring are the profilometer and the Brush surface analyzer. They have a tracer or stylus that moves over the surface. The up-and-down movement is recorded on an electric meter or on paper or film that moves at a uniform rate as the measurement is taken (Fig. 20.9).

In the past it has been difficult to duplicate, measure, and designate desired surfaces. The degree of surface roughness must be controlled and maintained because it affects:

1. The life of the product by controlling friction, abrasive wear, corrosion, and galling.

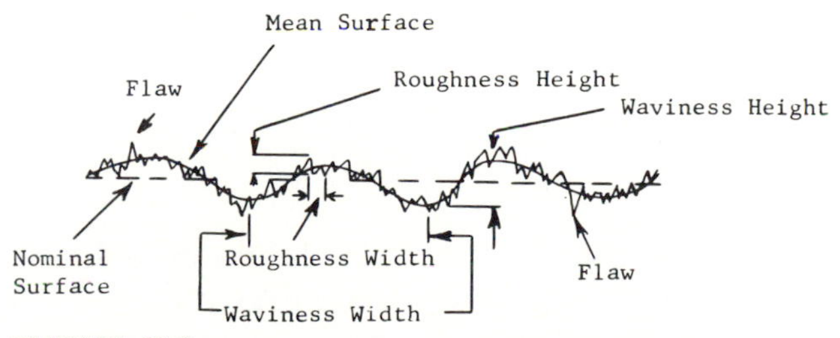

FIGURE 20.9
Enlarged surface profile with some surface properties indicated schematically.

2. The function of the product by permitting the smooth surfaces of the parts to slide freely, fit properly, serve as bearings, reduce leakage, and rub against packings.
3. The appearance of plated and other decorative surfaces.
4. The safety of the part by preventing stress concentration, fatigue, and notch sensitivity. For these reasons, airplane engine manufacturers are especially conscious of the value of controlling surface finish.
5. The heat transfer, because smooth surfaces offer better heat conductivity.

Since it costs more to produce smooth surfaces, they should be specified only when desired. Progress has been made in advancing the methods of controlling surface finishes through the adoption of national standards.

Measurement or Evaluation

For compliance with specified ratings, surfaces may be evaluated by comparison with specified reference surfaces or observational standards, or by direct instrumental measurements.

In many applications, these comparisons can be made by sight, feel, or instrument. In making comparisons, care should be exercised to avoid errors due to differences in material, contour, and type of operation represented by the reference surface and the work.

In using instruments for comparison, or for direct measurement, care should be exercised to insure that the specified quality or characteristic of the surface is measured.

Roughness measurements, unless otherwise specified, are taken across the lay of the surface or in the direction that gives the maximum value of the reading. The physical measurement of the roughness height value shall be the maximum *sustained* reading of a series of readings. It shall be the minimum *sustained* reading in case a minimum permissible value is specified also.

The physical measurement of the waviness height shall be the algebraic difference between the maximum and minimum readings of a dial gage over a distance not exceeding 1 in., if no other definite waviness width is specified. The waviness height can also be determined by means of a straight edge. The recommended values for roughness classification, in microinches, are:

1	16	125
2	32	250
4	63	500
8		1000

The use of only one number or class to specify the height or width of irregularities shall indicate the maximum value. Any lesser degree or class on the actual surface of the part shall be satisfactory. When two numbers are used on the drawing or specification, they shall specify the maximum and minimum permissible values.

Surface Symbol

The symbol used to designate surface irregularities is the check mark and a horizontal extension. The point of the symbol may be on the line indicating the surface (on the witness line) on the drawing, or on an arrow pointing to the surface. The long leg and extension shall preferably be to the right, as the drawing is read.

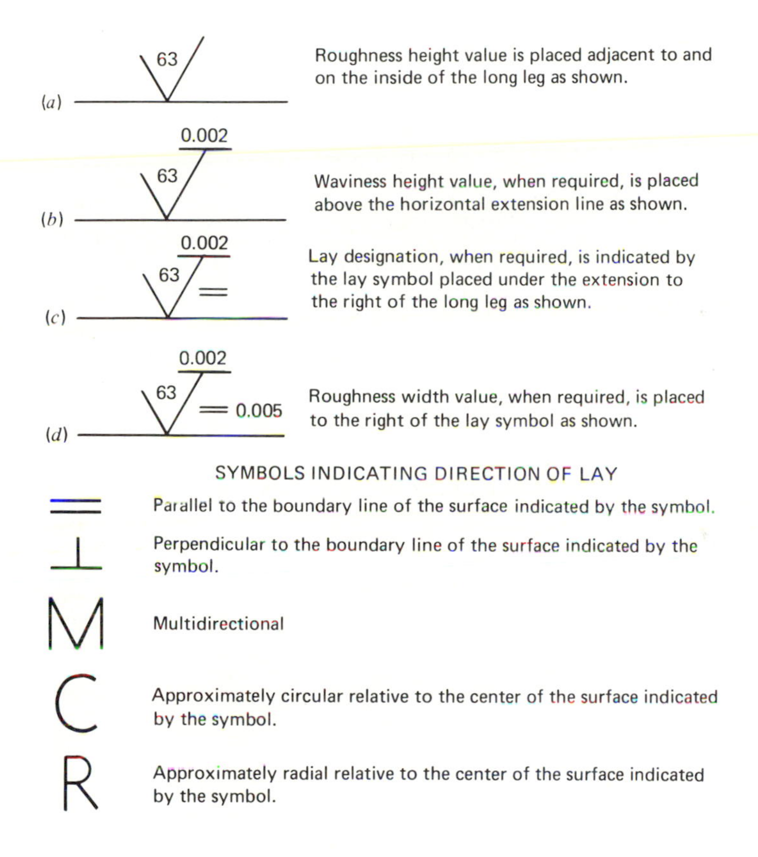

Application

Figures 20.10 to 20.13 show the proper method of designating surface qualities on the drawings, with examples for different ratings. The new symbol will replace the old finish mark.

The roughness of natural surfaces is shown in Fig. 20.14 and characteristic maximum surface roughness in common machine parts is illustrated in Fig. 20.15.

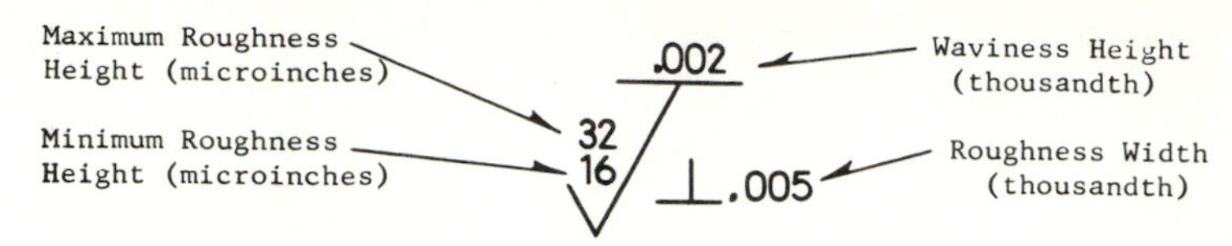

FIGURE 20.10
The surface-quality symbols conform to the "Aeronautical Standard" published by the Society of Automotive Engineers and they meet the ASA standard on "Surface Rougness, Waviness, and Lay," part I, B46.1.

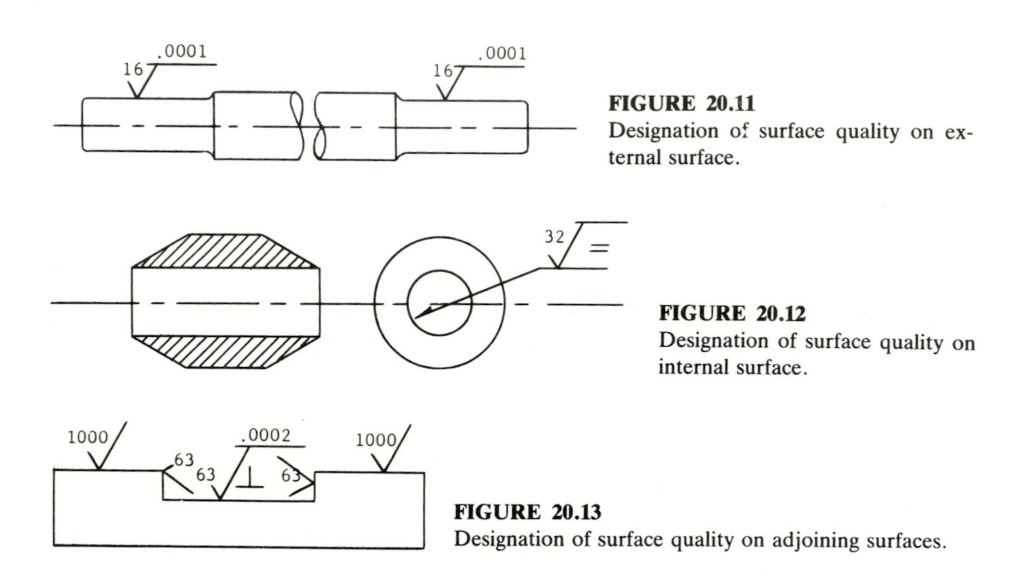

FIGURE 20.11
Designation of surface quality on external surface.

FIGURE 20.12
Designation of surface quality on internal surface.

FIGURE 20.13
Designation of surface quality on adjoining surfaces.

STATISTICAL CONTROL

Statistical quality control is an analytical tool that can be used to evaluate machines, materials, and processes by observing capabilities and trends in variation so that comparisons and predictions may be made to control the desired quality level. This tool makes possible:

1. Decreased inspection costs
2. Reduced rejects, scrap, and rework
3. More uniform product
4. Greater quality assurance
5. Anticipation of production trouble
6. More efficient use of materials
7. Rational setting of tolerances
8. Better purchaser–vendor relationships
9. Improved and concise reports to top management on the quality picture

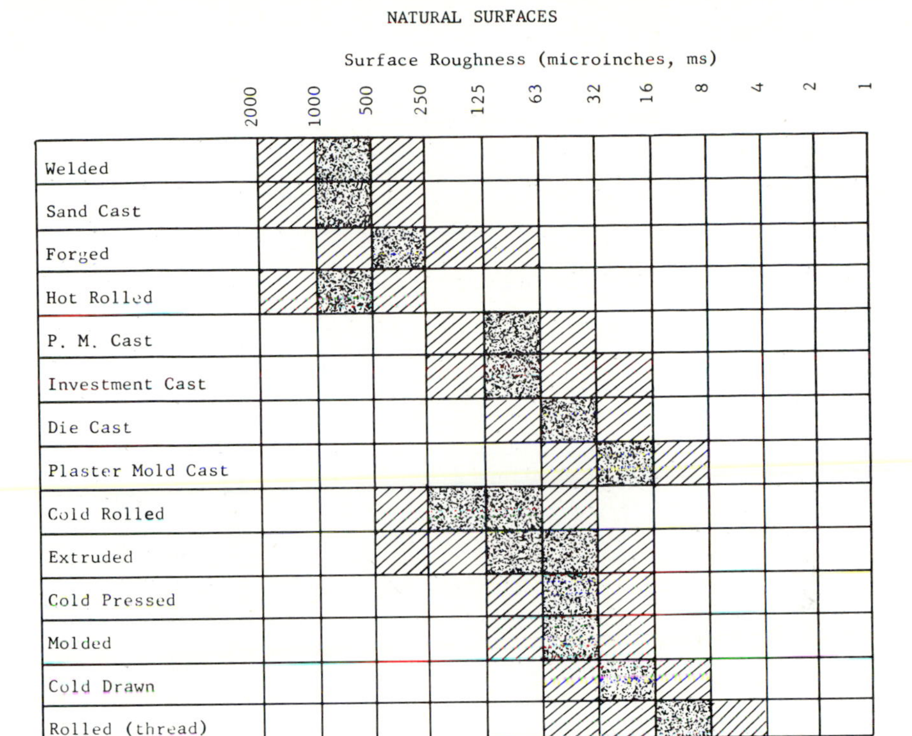

FIGURE 20.14
Chart showing overall and average range of natural surface-roughness characteristics with the nonmachining processes.

In manufacturing, accuracy and precision are often used interchangeably. Unfortunately, this is incorrect and produces misleading results. Figure 20.16 illustrates the relationship of accuracy and precision. As can be seen from the figure, accuracy generally refers to the centroid or target dimension or call out. Precision generally refers to the scatter resulting from the process. In order to quantify these variables, statisticians have developed two measures commonly used in quality control to describe these features. They are the sample mean and standard deviation. In the following sections, some basic statistical quality control terms will be defined for manufacturing application.

Statisticians who had studied the laws of variation in nature became interested in the possibility of controlling quality by applying the laws of probability and dispersion. For example, they kenw that in firing artillery weapons the shots fall in a pattern around the target in spite of the greatest care in obtaining uniform powder, projectile, gun, and accuracy of sighting. This pattern of shots was true to mathematical laws that had been worked out in other probability cases. That is, the statisticians could predict the number of

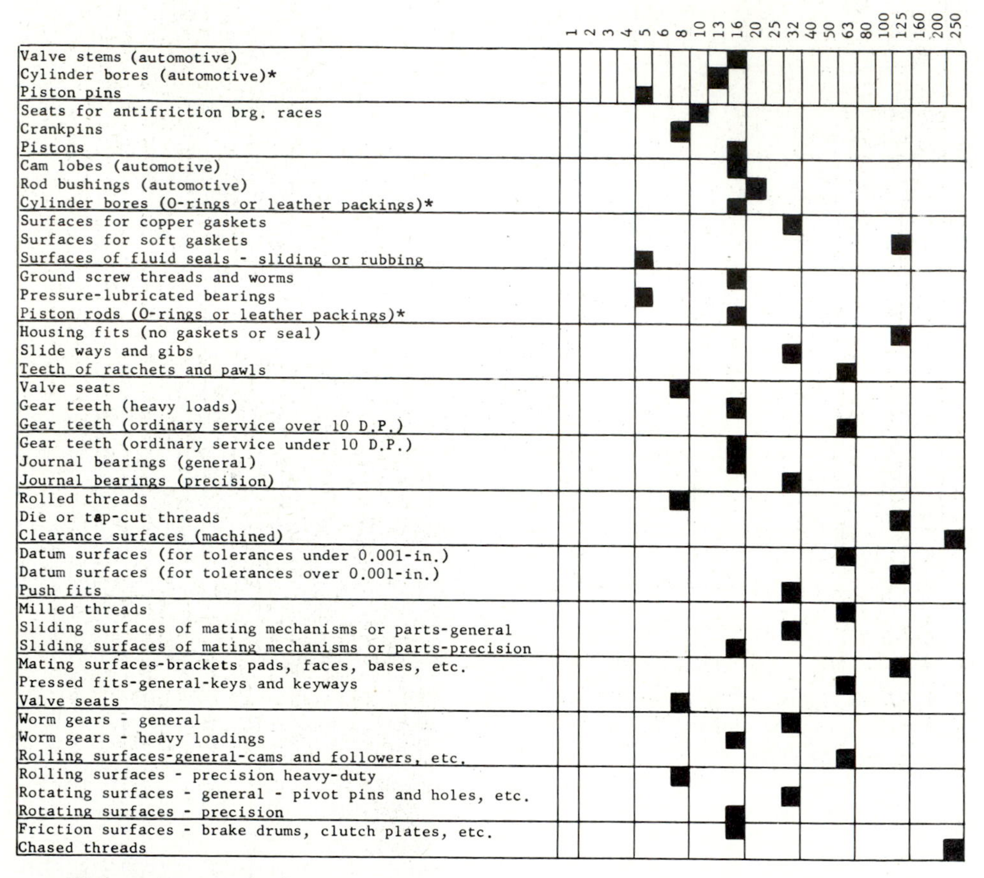

FIGURE 20.15
Characteristic maximum surface roughness for common machine parts. Charted from various manufacturers' practices on such parts, these data will assist in judging practical finishes.

shots that would hit the target, the number that would fall 5 ft away, 10 ft away, and 20 ft away. In studying a screw machine, they found similar data. The machine capable of producing uniform parts would make a certain quantity that would fall on the dimensions required, so many would be 0.001 above or below, so many 0.002 above and below, and so on. As the tool wore, the number below size decreased and the number above size increased. By studying this pattern, the observer could make the amount of correction needed to bring the machine back to best performance (Figs. 20.17 and 20.18).

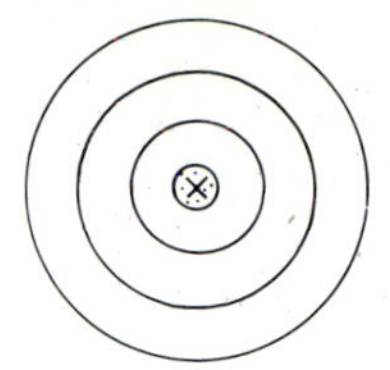

(*a*) Accurate and precise

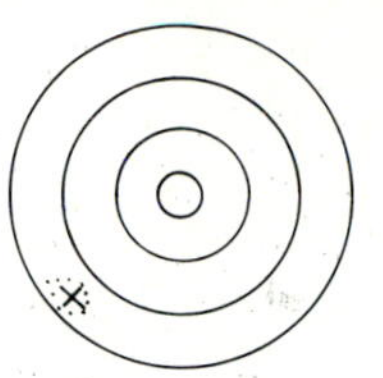

(*b*) Precise, not accurate

X - denotes sample average

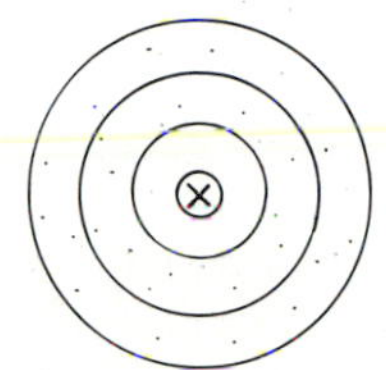

(*c*) Accurate, not precise

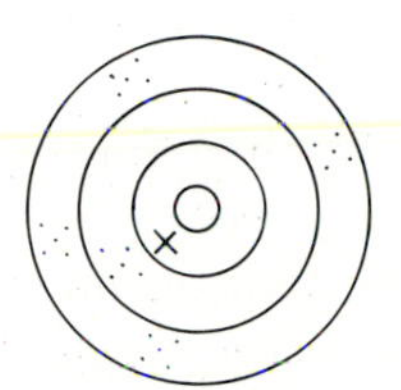

(*d*) Precise within sample. Not precise between samples. Not accurate over all or within sample.

FIGURE 20.16
Accuracy versus precision in process. Dots in targets represent location of shots. Cross represent location of the average position of all shots.

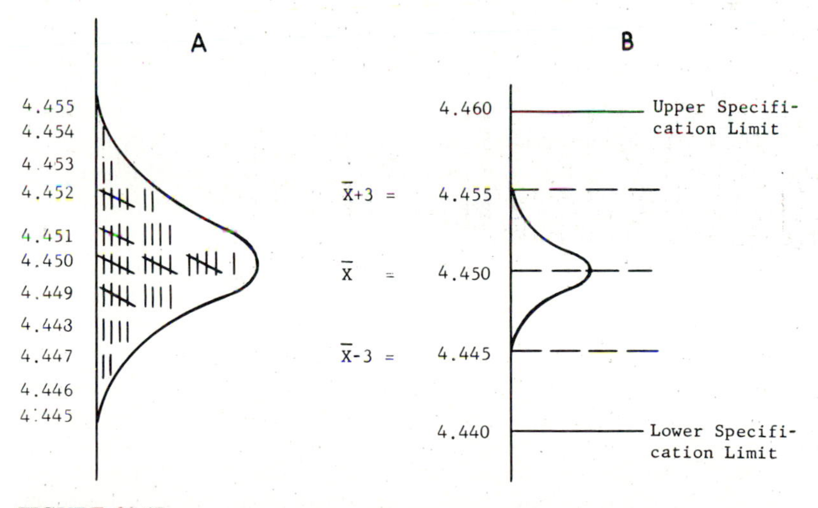

FIGURE 20.17
(A) A frequency curve of 50 measurements on screw machine part. (B) Relative position of frequency curve of the process relative to the specification limits shows good statistical control.

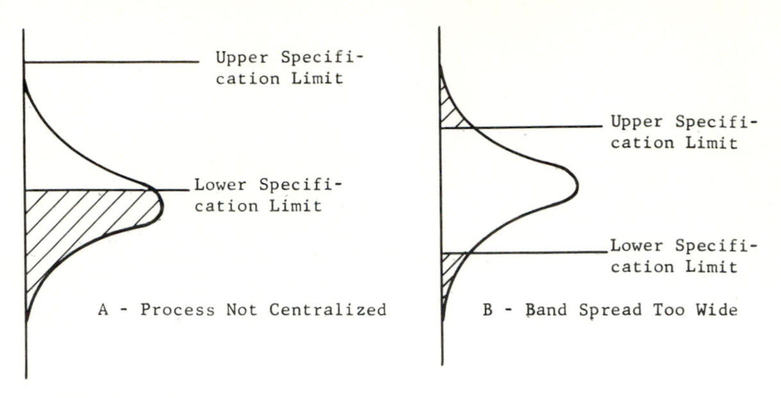

FIGURE 20.18
(A) Frequency curve of a process that is not centralized. (B) Frequency curve of a process that has too large a spread of variation.

In statistical quality control, observable features of a product or process are measured. A study of these measurements usually determines the ability of the operator or equipment to produce within the desired specifications. The percentage gives the same story as if the product were inspected or tested 100 percent. The size of sample is determined by the nature of the product and the standard of quality required.

The two statistics that the production-design engineer will use most frequently are the mean and the standard deviation. We define the mean, $\bar{x}$ of the set of numbers x_i, as

$$\bar{x} = \frac{1}{n} \sum_{i=1}^{n} x_i$$

The standard deviation is a most useful measure of the extent to which an individual item may deviate from the expected mean. To employ the standard deviation, we must also adopt confidence limits; that is the percentage of probability that the product will fall between the "upper and lower" limits. No result can be absolutely assured and nothing is perfect, at least in production or anticipated sales. The standard deviation can be defined as the root-mean-square deviation in that it is the square root of the mean of the squares of the deviations. The standard deviation of a sample of data is usually symbolized by s. The value of s is given by the equation:

$$s = \sqrt{\frac{1}{n-1} \sum_{i=1}^{n} (x_i - \bar{x})^2}$$

The analysis of random errors of observation in gathering statistics is based on probability theory. The production-design engineer should be cognizant of the fact that the mean of a sample of data represents only an estimate of the mean of the parent distribution. Similarly, the standard

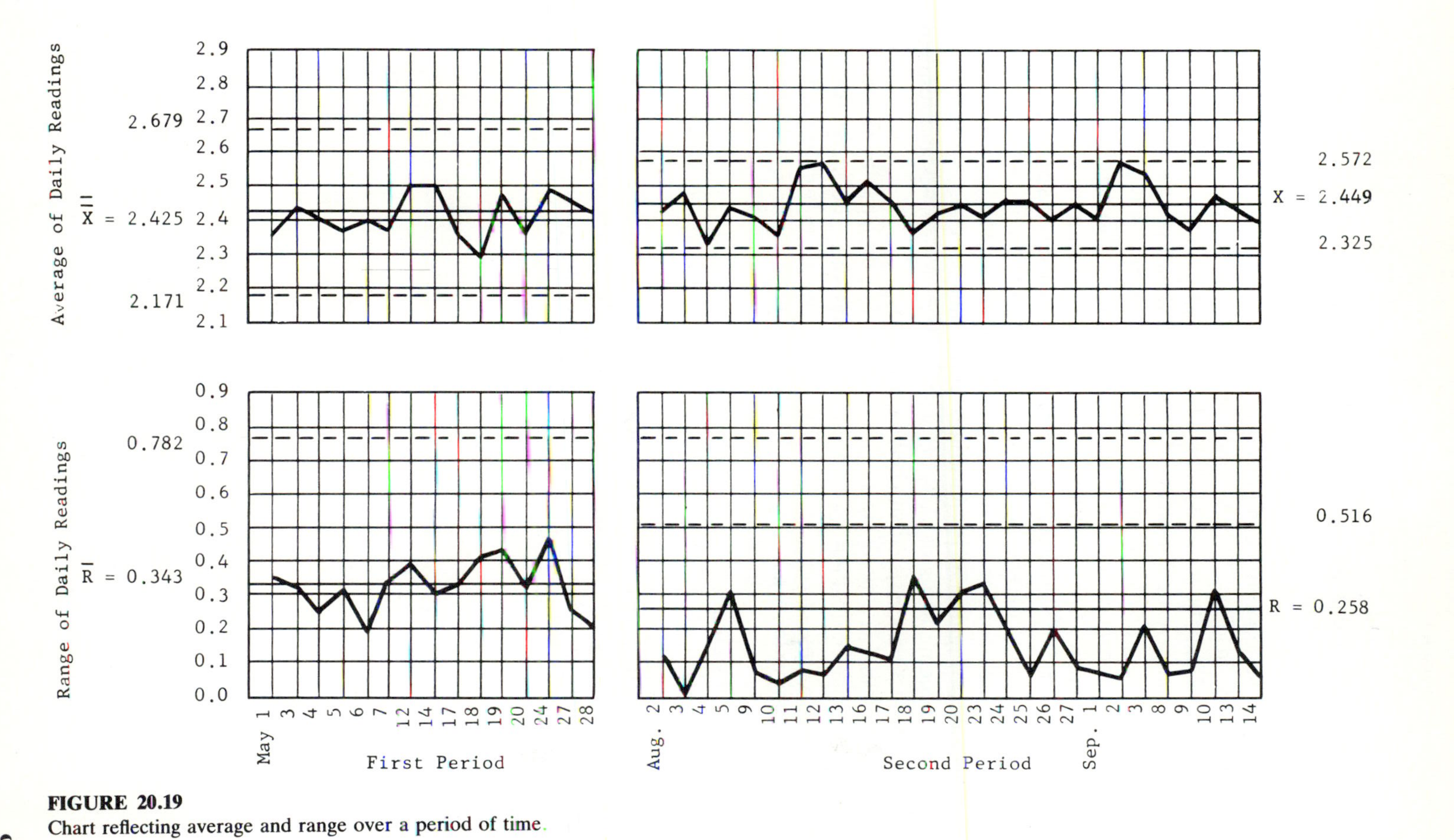

FIGURE 20.19
Chart reflecting average and range over a period of time.

deviation of the sample is an estimate of the standard deviation of the population. For most distributions, the precision of the estimate of both the mean and the standard deviation increases with the size of the sample of data taken.

Therefore the operator can select at random parts to check, plot the results as shown in Fig. 20.17A, and determine the performance of his machine. The tool may be wearing gradually. As it approaches the limit, it can be adjusted to compensate for wear. If it is chipped and producing defective parts, the sample immediately shows this and the entire lot will be rejected. Then 100 percent inspection of the lot will be necessary to salvage as many good parts as possible. When statistical control is established for an operation, the average size of part is plotted on one chart for each set of samples and the range of size is plotted on another chart for each set of samples. By observing the two charts, the operator can see the trends and know when the machine is exceeding limits. Figure 20.19 illustrates how such a chart revealed the possibility of reducing limits of variation without increasing costs and resulted in a better product that assisted succeeding operations.

The patternmaker, mold maker, toolmaker, and designer have recently realized the value of making patterns, molds, and dies the same size for multiple patterns and cavities. For example, an electric-motor bracket hub varied to such an extent that a special machining chuck was required to hold the part. A study of deviations revealed that the patterns differed from each other. When they were made alike, the normal casting deviation permitted the use of a standard chuck and the reduction of material allowed for machining (Fig. 20.20).

The production-design engineer will benefit by studying the current articles and textbooks on statistical control. The subject has been reduced to "everyday" understandable terms that can be applied by quality-control men, shop supervisors, and engineers in any organization.

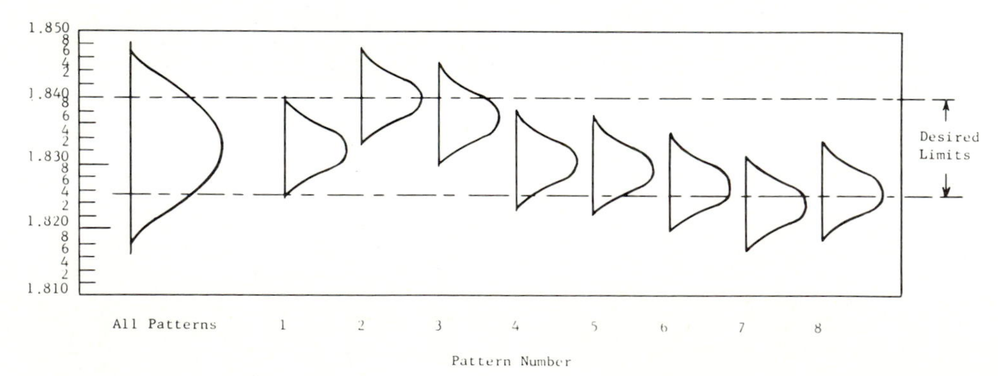

FIGURE 20.20
Frequency distribution of OD of hub diameter.

MANUFACTURING RELIABILITY

The concepts of reliability need to be incorporated in the functional design of the product and into all manufacturing operations in view of the long life requirements for many of today's products.

When we design for reliability, the product must withstand actual service conditions for a given period of time. We can prescribe the time period, but we can only estimate the service conditions. Generally speaking, there are four types of failure while in service.

1. *Infant mortality.* These are the early failures due to faulty material, or manufacturing errors such as sharp corners, scratches, heavy undercuts, etc., that cause points of stress concentration. Sound quality-control procedures should minimize infant mortality.
2. *Chance failures.* Chance failures may be referred to as random or constant hazard failures. These are failures that take place during the product's life due to chance. A specific cause could be inadequate lubrication due to a clogged wick, leaking due to the development of excessive porosity, a burnout due to a surge of power, etc. Chance failures are often mistaken for wear-outs.
3. *Abuse or misuse.* These failures are due to using the design in installations beyond the intended purpose of the design. Known excessive loads are typical examples.
4. *Wear-out.* These failures are due to aging, fatigue, corrosion, etc. Wear-out failures are usually progressive until the product capability is such that it is retired because of inefficiency. Wear-out can usually be postponed by proper maintenance and preventive maintenance.

Figure 20.21 illustrates a theoretical curve, which is seldom fulfilled in practice, of the relation in time and the magnitude of the failure rate of these four types of failure. Through a well-conceived reliability and quality control program, a reliability performance similar to that shown in Fig. 20.22 can be achieved. Here, it can be seen that both the failure rate and the period of infant mortality has been reduced and the start of wear-out failures has been deferred.

Probability, the Tool of Reliability

As has already been explained, reliability is expressed as a probability, and this probability usually must be expressed as a function of time.

Typical industrial practice is to specify a safety factor that is the average design parameter (strength, for example) minus the maximum operating parameter (stress, for example) divided by the standard deviation of the design parameter (strength). Another typical method is to specify the mean time to failure. As has been shown (Fig. 20.21) wear-out failures of most designs take

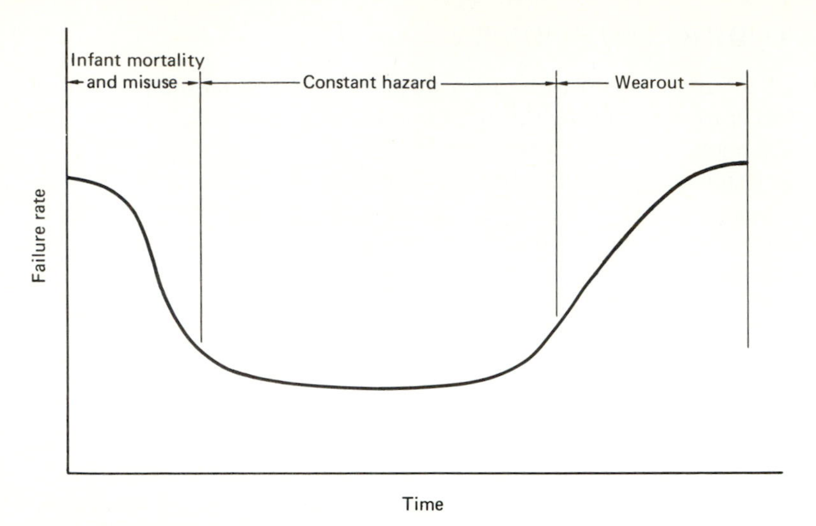

FIGURE 20.21
Mortality curve based on Robert Lusser's concepts.

a normal probability distribution with a mean failure time equal to G and a standard deviation equal to σ. Figure 20.23 shows this distribution and the resulting reliability R that provides the expected percentage of designs operating after t hours of operation. Figure 20.24 illustrates the possible failure zone on a design that is subjected to an environment that is normally distributed. Figure 20.25 illustrates how the failure zone can be reduced by narrowing the tolerance of the design.

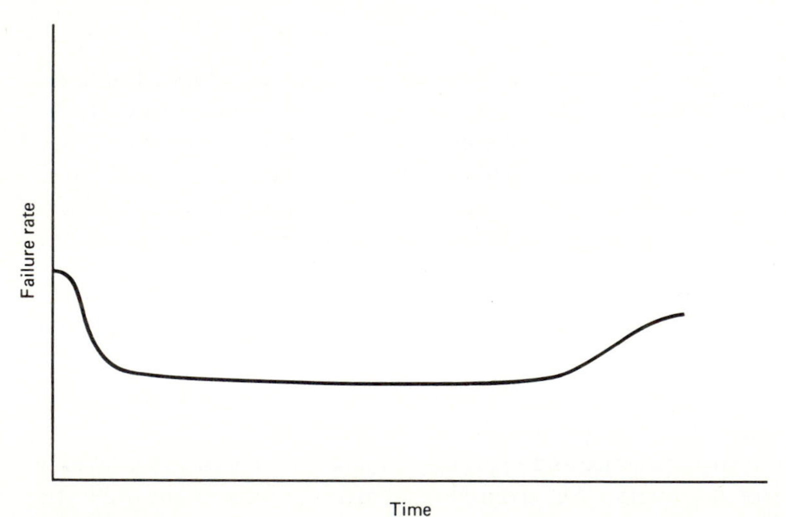

FIGURE 20.22
Mortality curve on product designed and built under a well-conceived quality-control and reliability program.

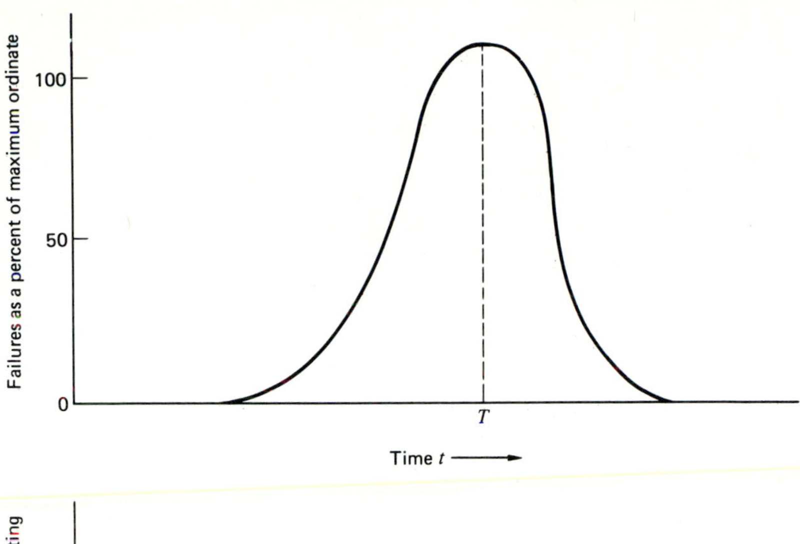

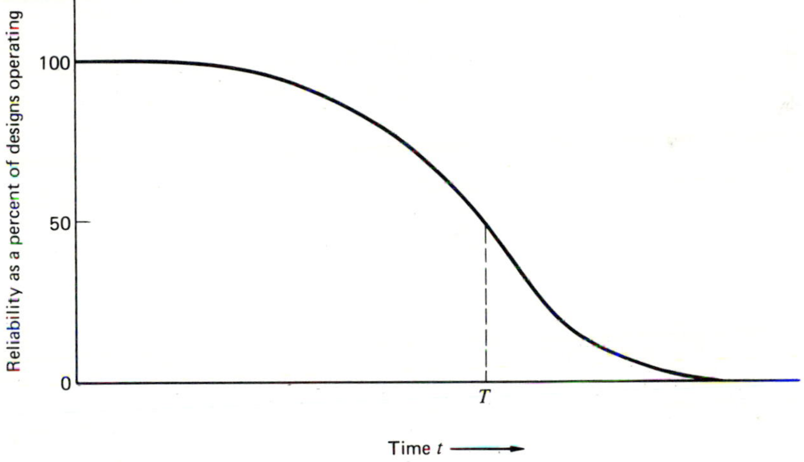

FIGURE 20.23
Typical wear-out failure with mean failure time t.

Where constant-hazard-rate failures take place, the failures when plotted as a percentage of the maximum ordinate will plot as an exponential curve. Here

$$y = \frac{1}{T} e^{-t/T}$$

where y = the number of failures at time t.

If a product represents the assembly of several components and each component has its own pattern of failure, then obviously the probability distribution of failure will be different for each component. The reliability of

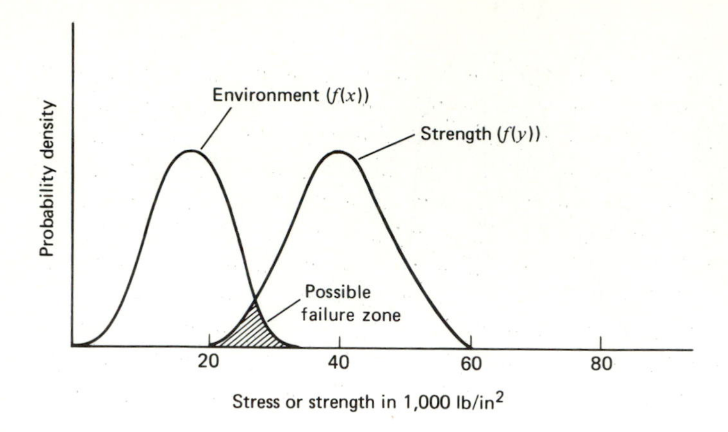

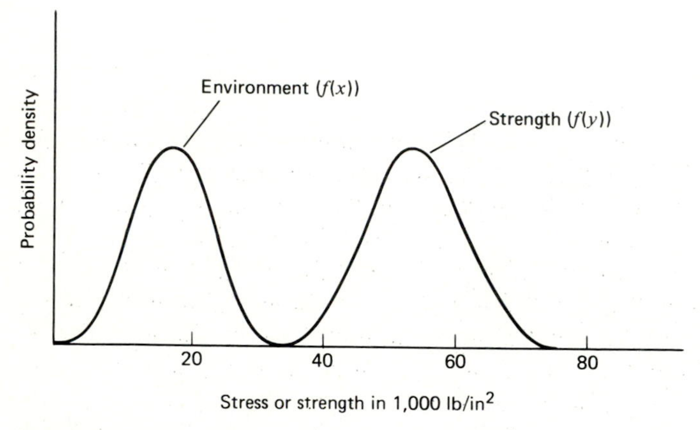

FIGURE 20.24
The possible failure zone can be removed by increasing the strength of the material.

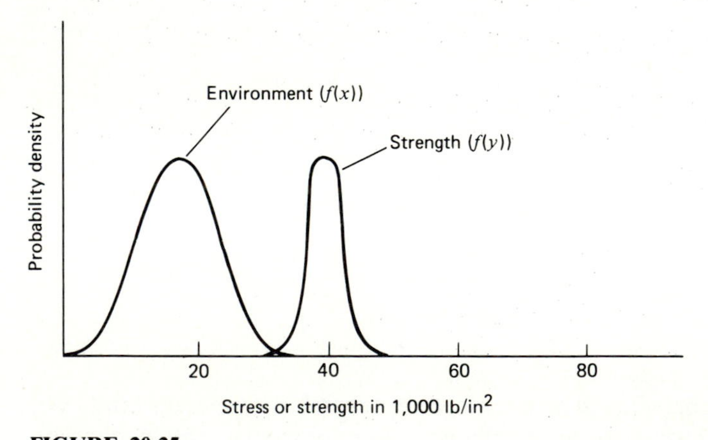

FIGURE 20.25
The possible failure zone can be reduced or removed by narrowing the tolerance of the material.

the assembly will be equal to the product of the separate reliabilities of the components going into the assembly. For example, if an assembly is made up of four components with the following reliabilites: $R_1 = 0.95$; $R_2 = 0.98$; $R_3 = 0.97$; $R_4 = 0.90$, the reliability of the assembly will be:

$$\begin{aligned} R_a &= R_1 \times R_2 \times R_3 \times R_4 \\ &= (0.95)(0.98)(0.97)(0.90) \\ &= 0.813 \end{aligned}$$

Reliability Operations

In order to maintain a successful reliability program, a "reliability mindedness" attitude must be developed in design, manufacturing, purchasing, and management.

It will be necessary to review or develop standards for procured parts and material, manufactured parts, and assemblies. Test and inspection procedures for maintaining these standards will need to be developed and implemented.

Reliability data, including operational life, failures, consumption, etc., will need to be gathered both from the user field and within the plant and be analyzed in such a way that significant feedback can be realized.

The work of the reliability personnel will continually involve the review, study, and analysis of equipment, specifications, control drawings, and manufacturing and inspection procedures, so as to assure reasonable product life. Product life can be short for cheap noncritical products; it must be long for items like costly machines (e.g., generators). Infant mortality must be practically zero for aircraft or space vehicles. Reliability personnel will make improvement recommendations, and follow up such recommendations to verify that corrective action was taken resulting in improved product reliability.

SUMMARY

The cost of quality control and reliability is proportional to the level of reliability and accuracy desired in materials, processes, and functions of the apparatus. The accuracy desired is based upon the judgment of management, which depends upon the advice of the sales, engineering, and manufacturing departments. The level of reliability established usually emanates from the customer. The ultimate decision as to acceptance or rejection rests upon the quality control and reliability department. In general, the company that is willing to maintain quality at considerable expense survives competition that has lower quality standards. Over the long pull, the high-quality, reliable product wins because customers are satisfied. Through good engineering, a good-quality product can be obtained at less cost than the former poor-quality product produced under uncontrolled conditions. By going to the source of trouble and eliminating the cause, by providing the shop operators and members of the quality control department with adequate equipment, by improving maintenance of tools and machines, and inspecting in the critical places, quality control has become an asset to competitive industry.

The quality control and reliability departments of industry are becoming effective in building quality into the product with minimum overall costs of production, scrap, rework operation, and service adjustments. Goodwill is maintained by verifying the quality and reliability before the product is shipped. Years of control have brought about uniform materials and parts that are produced in enormous quantities. These consistent materials make possible our mass-production lines producing reliable products today.

The steps involved in developing a sound quality-control and reliability program include the following.

1. Initiate, plan, and direct preprototype manufacturing and assembly work.
2. Determine and delineate manufacturing and assembly problem areas that affect product quality and product life.
3. Provide critical analysis of manufacturing operations to determine the relative merits of "make or buy" not only from a price standpoint, but also from a quality control and reliability standpoint.
4. Identify the areas where process control should take place. Determine the "how" of this process control.
5. Determine the degrees of environmental control required for the various manufacturing operations and specify the specific conditions required in order to maintain the required quality and assure the desired reliability. The conditions specified may include:
   a. Humidity level
   b. Dust count and particle size
   c. Temperature
   d. Differential pressure
   e. Wearing apparel
   f. Use of cosmetics
6. Review and analyze critical storage and handling requirements for materials, parts, and components.
7. Specify functional and maintenance characteristics of tools, fixtures, and handling equipment to ensure precision and repeatability.
8. Review manufacturing plans and procedures and check all development hardware for compliance to precision.
9. During the manufacturing cycle, audit manufacturing operations and initiate corrective action where needed so as to help assure meeting long-life requirements.
10. Evaluate vendor facilities, methods, and capabilities. Provide assistance where necessary.

QUESTIONS

20.1. Why is it that 100 percent inspection seldom ensures that a shipment of parts will be 100 percent to specifications?

20.2. In what way does incentive wage payment improve the quality of a product?

20.3. What is the purpose of Jo blocks?

20.4. For what type of work is the optical comparator used?

20.5. What advantage does the Brinell hardness tester have over the Rockwell tester?

20.6. What is the purpose of Magnaflux inspection?

20.7. What does the durometer measure?

20.8. What equipment would you recommend to test the ability of a part to resist corrosion?

20.9. For what reasons is it advisable to control the surface roughness of a part?

20.10. What is meant by the root-mean-square deviation?

20.11. What is the value of 1 μin?

20.12. Give the sequence in ascending order based on surface roughness of the following processes: cold rolled, die cast, plaster cast, hot rolled, and cold drawn.

20.13. What do we mean by statistical quality control?

20.14. Define standard deviation.

20.15. What is a normal distribution?

20.16. Why is it desirable to plot the "range" of successive samples?

20.17. Why do chance failures plot as an approximate straight line when failure rate is plotted against time?

20.18. Show graphically the effect of a strong maintenance and preventive maintenance program on the relation between failure rate and time.

PROBLEMS

20.1. A motor has two brushes, two ball bearings, and a small fan on the shaft. An analysis of electrical failures is known to be of the constant-hazard type, with a mean time to failure of 20,000 h. The brushes have a peaked wear-out failure centering on 1000 h and a standard deviation of 200 h. The bearings have a more spread-out failure centering on 1800 h with a standard deviation of 600 h. Mechanical failures are of the constant-hazard variety and have a mean time to failure of 10,000 h. What percentage of motors will still be running at the end of 500 h?

20.2. The weights of ceramic compacts were taken at random and recorded as follows:

Specimen weights, kg				
1.24	1.31	1.29	1.34	1.26
1.32	1.41	1.37	1.28	1.31
1.40	1.34	1.29	1.32	1.30
131	1.26	1.28	1.34	1.31
1.25	1.29	1.34	1.31	1.32

Calculate the mean and standard deviation of the samples. Assume that the mean and standard deviation of the samples is representative of the population. Find the upper and lower control limits and plot an $\bar{x}$ control chart. Is the process in statistical control? Explain why.

20.3. Given a process in statistical control with a population mean of 5.000 in. and a population standard deviation of 0.001 in., determine (*a*) the natural tolerance limits of the process and (*b*) if the specification limits are 5.001 ± 0.002 in., what percentage of the product is defective, assuming that the process output deviations are normally distributed.

20.4. Given a process with a population mean of 10.00 and a population standard deviation of 1, what can be said about the interval 10 ± 2 if the distribution is normal.

20.5. Determine the surface roughness values of the partial surface trace given in Fig. P20.5 in terms of the maximum peak to valley roughness, the arithemetic average, AA, and the root-mean-square, rms.

Point	Deviations from the mean at Δx intervals, μin.	Point	Deviations from the mean at Δx intervals, μin.
1	4	10	13
2	19	11	23
3	23	12	15
4	16	13	6
5	31	14	12
6	20	15	18
7	27	16	21
8	20	17	17
9	31	18	9

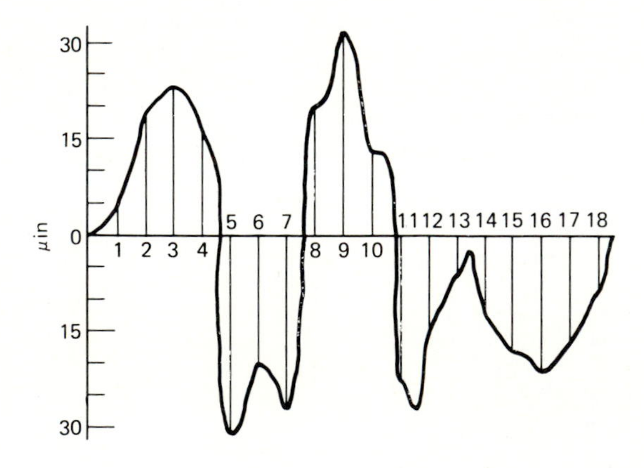

Partial trace of a surface

FIGURE P20.5

20.6. On an interference fit the basic size of a hole is 3.5000 in. The interference between the shaft and the hole must be at least 0.0015 in. The tolerance on the shaft and the hole is 0.0009 in. using the basic hole system and unilateral tolerances. Divide the tolerances on each item into three groups so that the small shafts mate with the small holes, the medium shafts mate with the medium holes, etc. Thereby there will be as nearly uniform as possible interference between the mating parts. This procedure is called selective assembly and is used when more precise metal fits are needed than can be obtained by conventional interchangeable manufacture.

20.7. A wrist pin $\frac{7}{16}$ in. diameter is designed to have a medium force fit in the cast iron piston bosses of a small internal engine and a snug fit in the steel connecting rod.

a. Using ASA standards, determine the limits for the pin and the hole in the connecting rod.

b. Calculate the limits for the pin and the holes in the piston bosses.

c. Devise an assembly method that would double the machining tolerance but still maintain the same average interference.

d. Specify the sizes of the parts and how they will be prepared for assembly using the improved method of assembly.

e. Using the same components, develop a functional design that will provide better wear characteristics than the present design and that will require only minor changes in tooling.

REFERENCES

1. Groover, M. P.: *Automation, Production Systems, and Computer-Integrated Manufacturing,* Prentice-Hall, Englewood Cliffs, N.J. 1987.
2. Groover, M. P., and Zimmers, E. W.: *CAD/CAM: Computer-Aided Design and Manufacturing,* Prentice-Hall, Englewood Cliffs, N.J., 1984.
3. Chang, T. C., and Wysk, R. A.: *An Introduction to Automated Process Planning Systems,* Prentice-Hall, Englewood Cliffs, N.J., 1985.

RECOMMENDED READING

Walpole, R. and Myers, R.: *Probability and Statistics for Engineers and Scientists,* Macmillan Publishing Inc., 1986.

Hanson, B. L.: *Quality Control Theory and Applications,* Prentice-Hall Inc., Englewood Cliffs, N.J., 1974.

CHAPTER 21

FLEXIBLE MANUFACTURING SYSTEMS

INTRODUCTION

Today, the industrial world has become a true international marketplace. Our transportation networks have created a "world market" that we participate in on a daily basis. For any industrial country to compete in this market, it is necessary for a company to provide an economic, high-quality product to the market place in a timely manner. The importance of integrating product design and process design cannot be over emphasized. However, even once a design has been finalized, today's manufacturing industries must be willing to accommodate their customers by allowing for last-minute engineering design changes without affecting their shipping schedule or altering their product quality.

Most U.S.-based manufacturing companies look toward CAD/CAM and CIM to provide this flexibility to their system. Today, the use of computers in manufacturing is as common as the use of computers in education. Manufacturing systems are being designed that not only process parts automatically, but also move the parts from machine to machine and sequence the ordering of operations in the system. Today, manufacturing systems used in industry employ a variety of automation features. Numerical control machines and flexible manufacturing systems (FMSs) have become a part of how durable goods are produced not only in the United States but throughout the world.

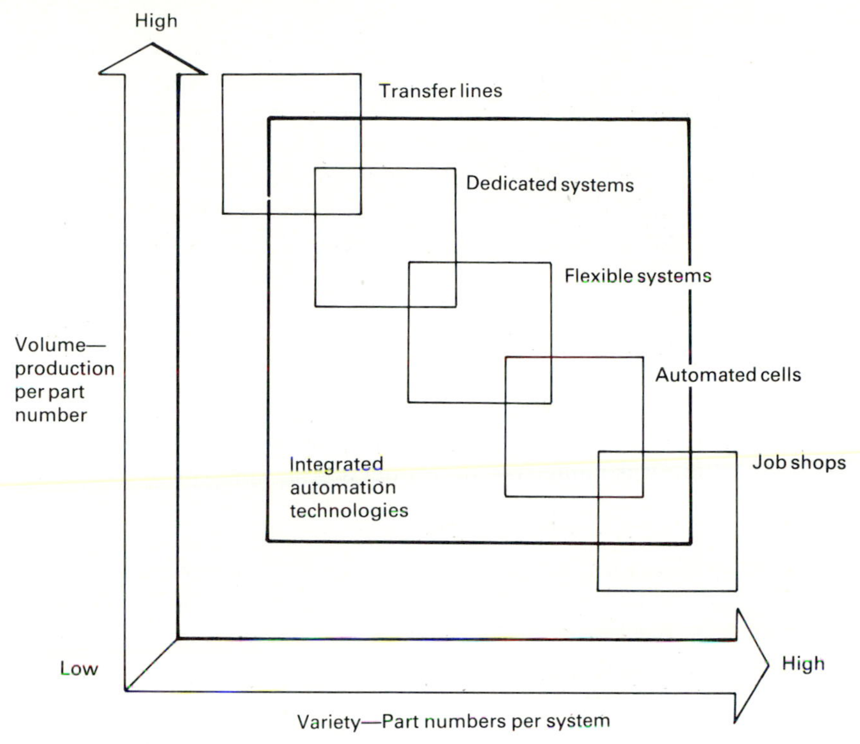

FIGURE 21.1
Volume versus variety regions for economic manufacturing. (*Courtesy of Cincinnati Milacron, Inc.*)

There is no single solution to all manufacturing needs, and the set of solutions seem to change constantly.

Figure 21.1 contains a plot of the economic regions of manufacturing. It should be noted that manual handcrafted goods will always have a market in the U.S. as well as abroad. This is also true of industrial products—there will continue to be a need for special one-of-a-kind items. The spectrum from unique one-of-a-kind goods through high volume goods will dictate that a variety of manufacturing methods be used to serve our various industrial needs. Some of these systems will look like the factories that our grandparents labored in; others will take on a futuristic look, more like our space-program control room. In the following sections, a discussion of flexible manufacturing systems along with the economics of manufacturing systems will be presented.

FLEXIBLE MANUFACTURING SYSTEMS

A flexible manufacturing system, or FMS as such a system is more commonly known, is a reprogrammable manufacturing system capable of producing a variety of products automatically. Since Henry Ford first introduced and modernized the transfer line, we have been able to perform a variety of manufacturing operations automatically. However, altering these systems to accommodate even minor changes in the product was at times impossible. Whole machines might have to be introduced to the system while other machines or components are retired to accommodate small changes in a product. In today's competitive market place, it is necessary to accommodate customer changes or the customer will find someone else who will accommodate his changes. Conventional manufacturing systems have been marked by one of two distinct features:

1. Job-shop types of systems were capable of producing a variety of product but at a large cost.
2. Transfer equipment could produce large volumes of a product at a reasonable cost, but was limited to the production of one, two, or very few different parts.

The advent of numerical control and robotics has provided us with the basic processing capabilities to reprogram a machine's operation with minimum set-up time. NC machines and robots provide the basic physical building blocks for reprogrammable manufacturing systems.

EQUIPMENT

Machines

In order to meet the requirements of the definition of a FMS, the basic processing in the system must be automated. Since the automation must be programmable to order, so as to accommodate a variety of product-processing requirements, easily alterable as well as versatile machines must perform the basic processing. For this reason, CNC machining centers, CNC turning centers, and robotic workstations comprise the majority of equipment in these systems. These machines are not only capable of being easily reprogrammed, but are also capable of accommodating a variety of tooling via a tool-changer and tool-storage system. It is not unusual for a CNC machining center to contain 60 or more tools (mills, drills, boring tools, etc.), and for a CNC turning center to contain 12 or more tools (right-hand turning tools, left-hand turning tools, boring bars, drills, etc.). The automatic tool-changer and storage capabilities of NC machines make them a natural choice for the material processing equipment (see Figs. 21.2 and 21.3).

Parts must also be moved between processing stations automatically. Several different types of material-handling systems are employed to move these parts from station to station. The selection of the type of material-handling system is a function of several system features. The material-handling

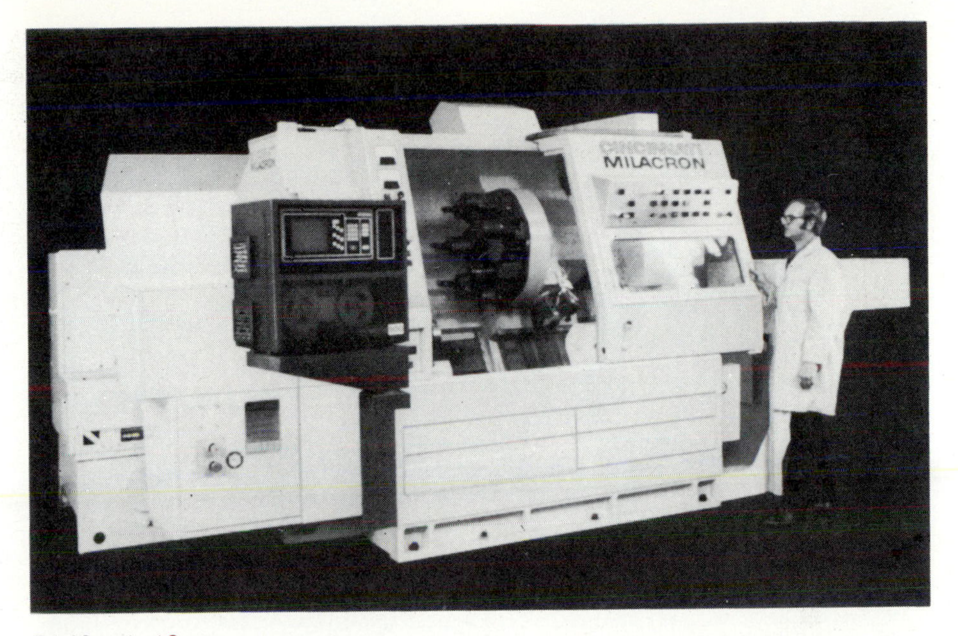

Brief Standard Specs

Models	12C	15C	18C	21C	24C	28C
Chuck sizes	12″	15″	18″	21″	24″	28″
Spdl. rpm	23/2240	18/1800	16/1600	16/1600	12/1200	12/1200
Thru Hole	3.04″	3.04″	4.56″	4.56″	6.5″	6.5″
Spdl. Type	A1-8	A1-8	A1-11	A1-11	A1-15	A1-15
HP	50	50	50	60	60	60
Z-Axis Travel	42.5″	42.5″	42.5″	42.5″	42.5″	42.5″
No. Turrets	2	2	2	2	2	2
No. Tools	7 OD-7 ID	7 OD-7 ID	6 OD-6 ID	7 OD-7 ID	6 OD-6 ID	6 OD-6 ID
Tailstock Opt.	Swing-up	Swing-up	Swing-up	N/A	N/A	N/A

FIGURE 21.2
Cincinnati Milacron CINTURN 2-axis CNC Turning Center. (*Courtesy Cincinnati Milacron, Inc.*)

system must first be able to accommodate the load of the part and perhaps the part fixture. Large, heavy parts require large, powerful handling systems such as roller conveyors, guided vehicles, or track-driven vehicle systems. The number of machines and the layout of the machines also present another design consideration. If a single material handler is to move parts to all the machines in the system, then the work envelope of the handler must be at least as large as the physical system. A robot is normally only capable of addressing one or two machines and a load and unload station. A conveyor or automatic guided vehicle (AGV) system can be expanded to include miles of factory floor. The material-handling system must also be capable of moving parts from one machine to another in a timely manner. The machines in the system will

Brief Standard Specs	
Longitudinal travel	26″
Vertical travel	26″
Cross travel	26″
Index table size	18″ sq.
Table workload	1500 lbs. each pallet
Spindle HP	10
Spindle rpm - std. - opt.	20/4000 30/6000
Spindle taper	#50 ANSI V-flange
No. Tools ATC - std. - opt.	30 45

FIGURE 21.3
Cincinnati Milacron T-10 CNC Machining Center and specifications. (*Courtesy Cincinnati Milacron, Inc.*)

be unproductive if they spend much of their time waiting for parts to be delivered by the material handler. If many parts are included in the system and they require frequent visits to machines, then the material-handling system must be capable of supporting these activities. This can usually be accommodated either by using a very fast handling device or by using several devices in parallel, e.g., instead of using a single robot to move parts to all the machines in the system, a robot would only support service to a single machine. (See Figs. 21.4 and 21.5.)

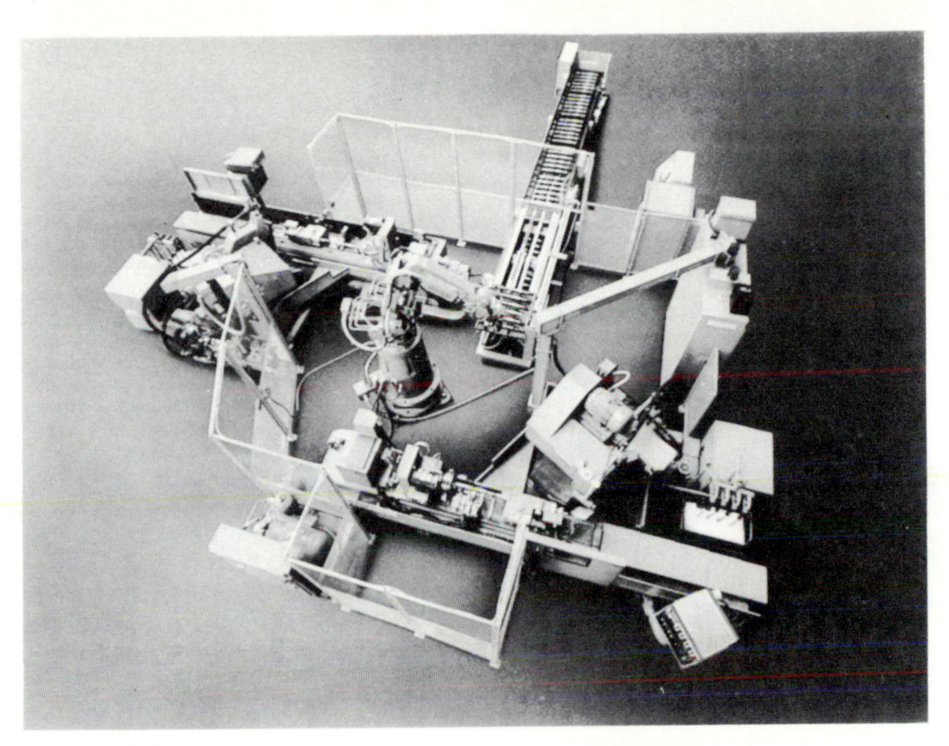

FIGURE 21.4
A robot-tended machining system. (*Courtesy Cincinnati Milacron Inc.*)

Tooling and Fixtures

Versatility is the key to most FMSs, and as such, the tooling used in the system must be able of supporting a variety of products or parts. The use of special forming tools in a FMS is generally not employed in practice. The contours obtained by using forming tools can usually be obtained through a contour-control NC system and a standard mill. The standard mill can then be used for a variety of parts rather than to produce a single special contour. An economic analysis of the costs and benefit of any special tooling is necessary to determine the best tooling combination. However, since NC machines have a small number of tools that are accessible, very few special tools should be used.

One of the commonly neglected aspects of a FMS is the fixturing used in these systems. Since fixtures are part of the tooling of the system, one could argue that they should also be standard for the system. Work on creating "flexible fixtures" that could be used to support a variety of components has only begun recently. One unique aspect of many FMSs is that the part is also moved about the system in the fixture. Fixtures are made to the same dimension so that the material-handling system can be specialized to handle a single geometry. Parts are located precisely on the fixture and moved from one station to another on the fixture. Fixtures of this type are usually called pallet

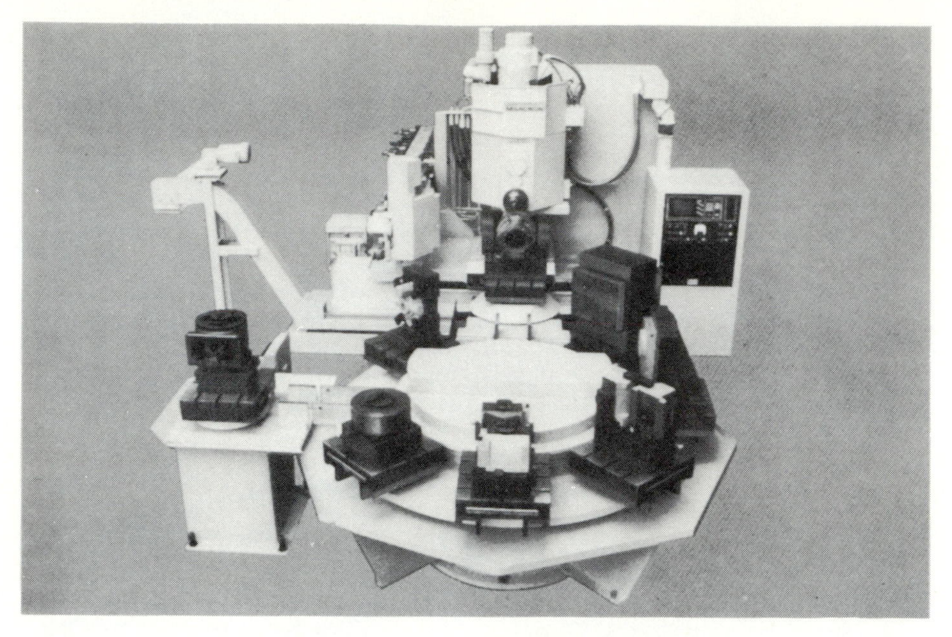

Brief Standard Specs	
Longitudinal travel	26″
Vertical travel	26″
Cross travel	26″
8-pallet Automatic Workchanger	Std.
Index pallet size	19.7″ sq., 360 pos.
Table workload	1500 lbs. ea. pallet
Spindle HP	10
Spindle rpm - std. - opt.	20/4000 30/6000
Spindle taper	No. 50 ANSI V-flange
No. Tools ATC - std. - opt.	45 60, 90

FIGURE 21.5
A CNC machining center with an eight-pallet workchanger.

fixtures or pallets. Many of the pallet fixtures employed today have standard "T-slots" cut in them, and use standard fixture kits to create the part-locating and holding environment needed for machining.

COMPUTER CONTROL OF FLEXIBLE MANUFACTURING SYSTEMS

FMS Architecture

An FMS is a complex network of equipment and activities that must be controlled via a computer or network of computers. In order to make the task

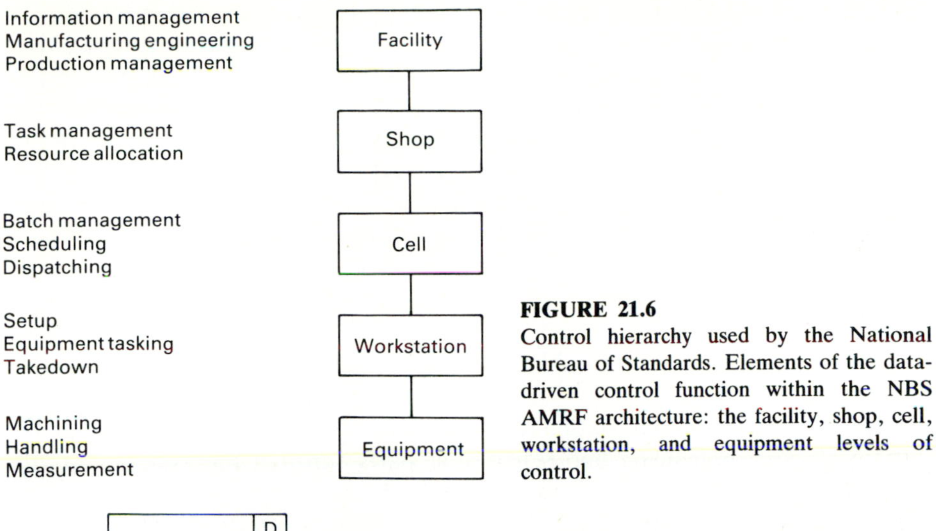

FIGURE 21.6
Control hierarchy used by the National Bureau of Standards. Elements of the data-driven control function within the NBS AMRF architecture: the facility, shop, cell, workstation, and equipment levels of control.

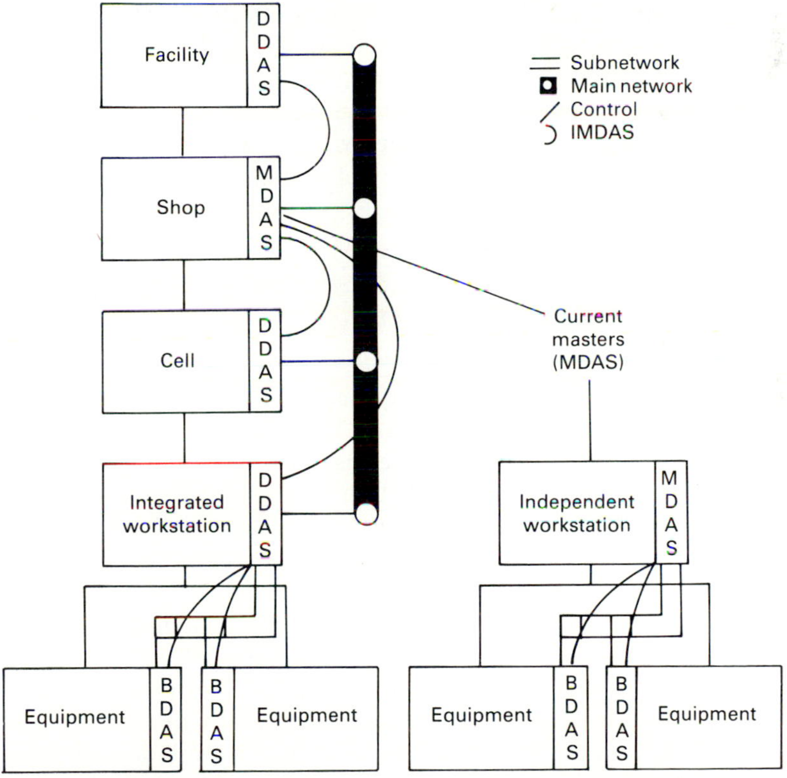

FIGURE 21.7
The relationship between the data administration systems (DAS) in the NBS architecture. The topologies of the IMDAS data administration system, the network data communication network, and the hierarchical system of data-driven control; data preparation is implied in the facility level of control.

of controlling an FMS more tractable, the system is usually decomposed in a task-based hierarchy. One of the standard hierarchies that have evolved is the National Bureau of Standards factory control hierarchy. This hierarchy consists of five levels and is illustrated in Figs. 21.6 and 21.7. The system consists of the physical machining equipment at the lowest level of the system. Workstation equipment resides just above the process level and provides integration and interface functions for the equipment. For instance, pallet fixtures and programming elements are part of the workstation. The workstation typically provides both man–machine interface as well as machine–part interface. Off-line programming, such as APT for NC, or AML for a robot, would reside at the workstation level.

The cell is the unit in the hierarchy where interaction between machines becomes part of the system. The cell controller provides the interface between the machines and the material-handling system. As such, the cell controller is responsible for sequencing and scheduling parts through the system. At the shop level, integration of multiple cells occurs, as well as the planning and management of inventory. The facility level is the place in the hierarchy where the master production schedule is constructed and manufacturing resource planning is conducted. Ordering of materials, planning inventories, and analyzing business plans are part of the activities that affect the production system. Poor business and manufacturing plans will incapacitate the manufacturing system just as surely as the unavailability of a machine.

EXAMPLES OF FLEXIBLE MANUFACTURING SYSTEMS

A Cart-driven System at Caterpillar Tractor, Inc.

One of the first systems installed in the United States is located in East Peoria, Illinois, at the Caterpillar Tractor Company. The system is still somewhat unique in that it produces parts that are quite large by most standards; these are transmission housings for earth-moving equipment. The parts produced on the system are not only large but also quite detailed and need to be very accurate. These requirements produce a unique set of constraints for the system, which are reflected in the FMS design.

To accommodate the part volume (approximately a 40-inch cube), a special set of machines were necessary. The milling and some of the hole-producing operations are performed on a set of Sunstrand Omnimil machining centers. These five-axis NC machines have a 60-tool carousel and tool changer capable of storing the necessary tooling for the milling and many of the drilling operations. The system also contains three Sundstrand Omnidrils to perform the remainder of the hole-making operations. These machines are four-axis NC machines with 100-tool capacity tooling system. The remaining two NC machines in the system are Giddings and Lewis vertical turret lathes that are used to machine the faces on the housings.

The material handling in the system is conducted with two programmable carts that move along a set of tracks between the rows of machines. The physical system is shown in Fig. 21.8. The figure also shows sixteen load and unload stations and a DEA automatic inspection station. Parts are manually loaded and unloaded onto pallet fixtures at the load and unload stations. This loading and off-loading is the only manual activity in the system. From these stations, the carts pick up the pallets and move them from machine to machine as required. Once all of the required metal-cutting operations on a part have been completed, the part is taken to the inspection station to be automatically inspected. The carts used in the system are capable of easily lifting and moving the 600-lb parts.

The entire system is controlled with a DEC PDP 11 minicomputer. Part programs are down-loaded to the NC machines as required from the PDP 11. The computer keeps track of which parts are at which machines along with the part status. The nine NC machines along with the two carts and inspection machine are run and maintained by as few as three people. Tool-room personnel also replace and qualify the system tooling off-line.

The Pennsylvania State University FMS

Several universities have been developing FMSs in order to demonstrate flexible concepts to students. One such system is located at The Pennsylvania State University in State College, Pennsylvania. The system consists of a Cartrac material-handling system surrounded by several robots and NC machines (see Fig. 21.9). The system is designed specifically for flexibility, i.e., producing a wide variety of parts. In the system, parts are manually placed on the Cartrac system. The parts then cycle through the system, stopping at several control points. At each of these points, the system controller checks to see whether the part is to be processed at the related station. If the part is to be processed at the station, the status of the machine and buffer is checked. If room exist at the machine, the part is moved to the station. If room does not exist, or if the part is not processed at this station, it is routed to the next control point.

The system contains a variety of computers and programmable controllers. A PDP 11 is used as the system controller. The PDP 11 down-loads part programs to the NC machines and robots as needed. Each machine is front-ended with a personal computer that can validate part programs and serve as a network communication interface with the PDP 11. Part flow through the system is controlled with a programmable controller using only standard ladder logic.

The system is capable of producing parts requiring turning, milling/drilling, and assembly operations. The Cartrac material-handling system is capable of moving parts weighing up to 200 lb; however, the off-loading robots are restricted to a 40-lb load. The system, as configured, provides economic flexibility for typical metal-removal operations.

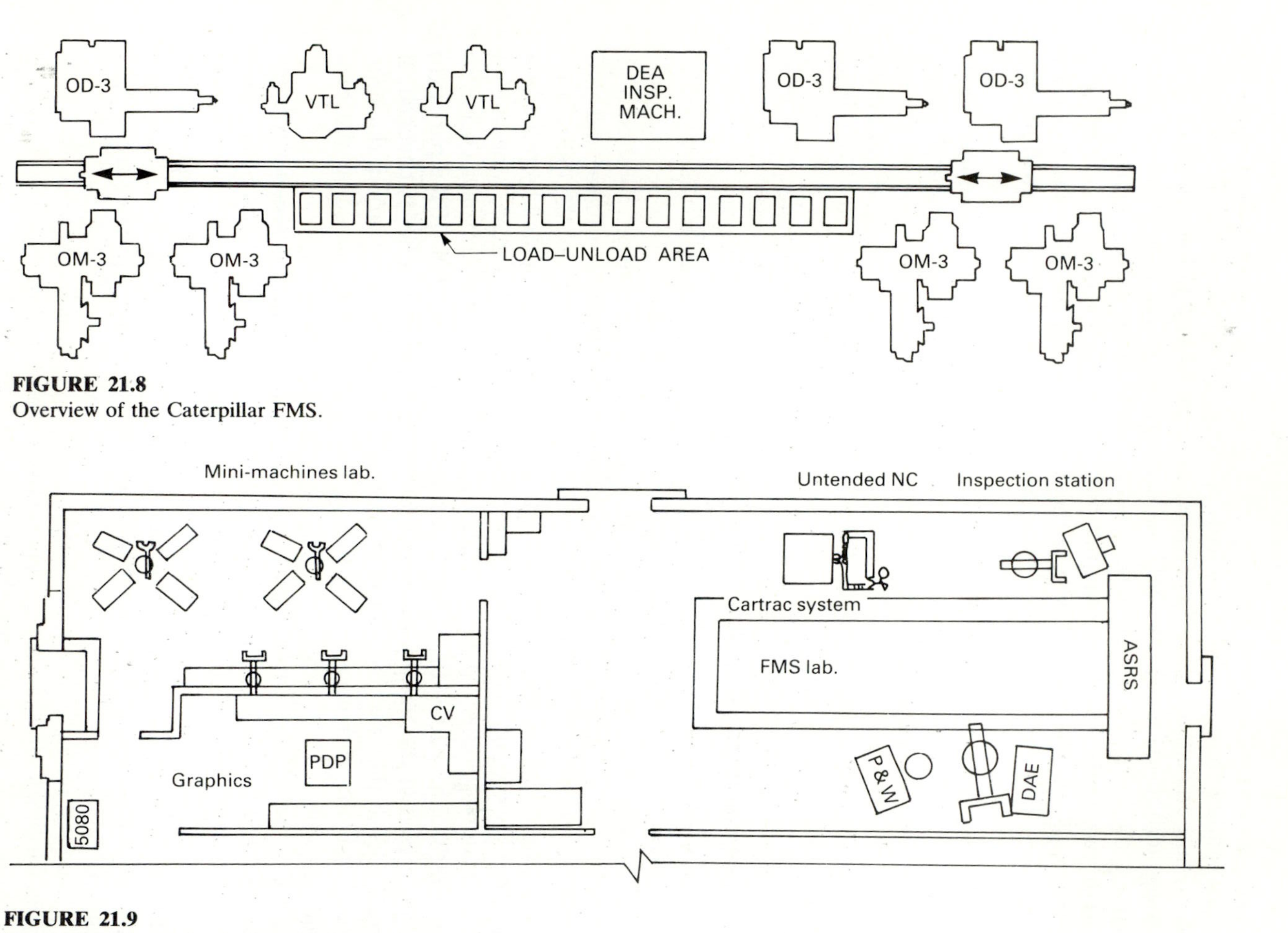

FIGURE 21.8
Overview of the Caterpillar FMS.

FIGURE 21.9
The FMS at Pennsylvania State University.

SUMMARY

CAD/CAM and flexible manufacturing systems have arrived. These systems are capable of producing a wide range of parts. As we move closer and closer to computer-integrated manufacturing, a greater range of engineering and computer knowledge will be required by manufacturing, mechanical, and industrial engineers. If we in the United States are to continue to be a world leader in industrial productivity, future engineers must understand manufacturing science, mechanical design principles, industrial engineering principles, and computer science. Integration of our manufacturing environment can only be achieved through an integration of our education. Flexible manufacturing systems promise to provide the logical solution for the automatic production of low to medium batch size manufacturing. However, the design and control of these systems will provide manufacturing engineers with a wealth of design and control problems during their construction.

QUESTIONS

21.1. Briefly explain what the "flexibility" in a flexible manufacturing system means.

21.2. An FMS is integrated by an automatic material-handling system. What type of system would be appropriate for small parts requiring many short operations? Defend your answer.

21.3. Keeping in mind your answer to Question 21.2, what type of considerations should be made for large parts requiring few, long processing operations?

21.4. Referring to Fig. 21.9 (Penn State's FMS), position the decision-control points on the Cartrac system and create the necessary ladder logic to control the flow of parts through the system.

21.5. How does the process plan affect the selection of FMS equipment and control computers? Explain your answer.

21.6. Many different machine features are offered when selecting either an NC machine or a robot. The NC machine could have a tool changer and pallet changer. What are the advantages and disadvantages of selelcting machines with expanded features?

21.7. One of the problems with converting to large automated manufacturing systems is that they must prove economic. Describe how these systems would be justified? List the savings as well as the additional costs that these systems would produce.

REFERENCES

1. Simpson, J. A., Hocken, R. J., and Albus, J. S.: "The Automated Manufacturing Research Facility of the National Bureau of Standards," *Journal of Manufacturing Systems,* vol. 1, no. 1, 1985.
2. Sandford, J. E.: "DNC lines link cutting to a new future," *Iron Age,* Oct. 26, 1972.
3. Williams, V. A.: "DNC for flexibility," *Production Engineering,* 1974.
4. Swyt, D. A.: "CIM, Data and Standardization within the NBS AMRF," *NBS Internal Report,* 1986.

5. Dupont-Gatelmand, C.: "A Survey of Flexible Manufacturing Systems," *Journal of Manufacturing Systems,* vol. 1, No. 1, 1983.
6. Sutton, G. P.: "Introduction To Flexible Manufacturing Systems: Their Applications, Classification Opportunities," *Proceedings of FMS-West,* Society of Manufacturing Engineers, 1983.
7. Johnson, I. S. C.: "FMS Operations Management Decision Aids," *Proceedings of FMS-West,* Society of Manufacturing Engineers, 1983.
8. Mehlhope, K. D.: "Integrated Sheet Metal Center (ISMC); An Integrated System for Management Control," *Proceedings of FMS-West,* Society of Manufacturing Engineers, 1983.
9. Seaman, F. D.: "Extending Flexible Manufacturing Systems with High-Power Lasers," *Proceedings of FMS-West,* Society of Manufacturing Engineers, 1983.
10. Knabb, W. F.: "Implementing a FMS at Hughes Aircraft Company," Proceedings of FMS-West, Society of Manufacturing Engineers, 1983.
11. Lee, Y. C., and Fu, K. S.: "A Relational Approach to the Integrated Database Management System for CAM," *CIDMAC,* Purdue University, 1983.
12. Groover, M. P.: *Automation, Production Systems, and Computer-Integrated Manufacturing,* Prentice-Hall, Englewood Cliffs, N.J., 1987.
13. *AutoCAD Reference Manual,* Autodesk, Inc. Sausalito, Calif., 1986.
14. Groover, M. P., and Zimmers, E. W.: *CAD/CAM: Computer-Aided Design and Manufacturing,* Prentice-Hall, Englewood Cliffs, N.J., 1984.
15. Chang, T. C., and Wysk, R. A.: *An Introduction to Automated Process Planning Systems,* Prentice-Hall, Englewood Cliffs, N.J., 1985.

APPENDIX A

SPECIAL TABLES

CONVERSION FACTORS FOR ENGINEERING DATA—SI UNITS

Base SI Units

In the SI system, there are seven base units. No numerical factor other than unity is used in forming derived units. For example, when unit length (one metre) is multiplied by unit length (one metre), unit area is obtained (one square metre). When unit length (one metre) is divided by unit time (one second), unit velocity is developed (one metre per second). Unit mass (one kilogram) multiplied by unit acceleration (one metre per second squared) gives unit force (one kilogram-metre per second squared), referred to as one newton.

Multiples

In the SI system, prefixes are used to indicate multiples and submultiples. These prefix names along with their corresponding symbols follow:

Multiple	Prefix name	Prefix symbol
10^{12}	tera	T
10^{9}	giga	G
10^{6}	mega	M
10^{3}	kilo	k
10^{2}	hecto	h
10	deca	da
10^{-1}	deci	d
10^{-2}	centi	c
10^{-3}	milli	m
10^{-6}	micro	μ
10^{-9}	nano	n
10^{-12}	pico	p

For example, one million newtons per square metre would be expressed as one meganewton per square metre, or $1\ MN/m^2$, and one millionth of a metre would be shown as $1\ \mu m$.

Stress Conversions

The unit of stress in the SI system is the newton per square metre (N/m^2) referred to as pascal (Pa).

Impact Values

The joule (J) is the unit of energy in the SI system. Impact values are given in this unit. One joule is equivalent to one newton (force) × one metre (distance) or $1\ N \cdot m$.

Temperature

The SI unit of temperature is the kelvin (K). The kelvin is used for the expression of temperature when the datum is absolute zero. When temperature differences are being considered, 1°C is exactly equal to 1 K, and wherever a temperature interval is denoted in kelvin (K), it can also be in degrees Celsius (°C).

Thermal Conductivity

In the SI system, thermal conductivity is measured in watts per metre-kelvin ($W/(m \cdot K)$).

Specific Heat Capacity

In the SI system, specific heat capacity is measured in joules per kilogram per kelvin (J/(kg K).

Density

The unit of density in the SI system is the kilogram per cubic metre (kg/m^3).

Magnetic Permeability

In the SI system, the unit for magnetic permeability is the microhenry per metre (μH/m).

Remanent Magnetism

The tesla, T, is the unit of remanent magnetism in the SI system. In the CGS system, the unit is the gauss.

Coercive Force

In the SI system, the unit for coercive force is the ampere per metre, A/m.

Hysteresis Loss

Hysteresis loss may be related to unit volume or unit mass. In the SI system, the unit of hysteresis loss is either the joule per cubic metre (J/m^3) or the watt per kilogram (W/kg).

Electrical Resistivity

The unit of electrical resistivity is the ohm-metre; the microohm-metre ($\mu\Omega \cdot m$) is frequently used for metals.

Variable quantity	SI unit name	SI symbol
Length	metre	m
Mass	kilogram	kg
Time	second	s
Electric current	ampere	A
Thermodynamic temperature	kelvin	K
Luminous intensity	candela	cd
Amount of substance	mole	mol

Conversion factors: given unit × conversion factor = equivalent unit

Given unit	Conversion factor	Equivalent unit
Atmosphere (normal)	76	centimetre of Hg (0°C)
Atmosphere (normal)	29.9213	inches of Hg
British Thermal Unit (mean)	1050	joule
BTU/foot square/hour	0.316998	watt/metre squared
BTU/hour	0.293071	watt
Calorie (20°C)	4.1819	joule
Centimetre	0.0328084	foot
Centimetre	0.3937008	inch
Centimetre cubed	0.03381402	ounce (fluid)
Centimetre cubed	0.00211338	pint
Centimetre of Hg (0°C)	0.0131579	atmosphere (normal)
Centimetre of Hg (0°C)	1333.22	newton/square metre
Circular mill	0.0000051	centimetre squared
circular mill	0.0005067	millimetre squared
Degree (angle)	0.0174533	radian
Fathom	1.8288	metre
Foot	30.48	centimetre
Foot	0.3048	metres
Foot-candle	10.76391	lumen/square metre
Foot cubed	1728	inch cubed
Foot cubed	28.32	liter
Foot-pound force	1.355818	joule
Foot-pound force/minute	0.0225957	watt
Foot-pound force/hour	0.0003766	watt
Foot-pound force/second	1.355818	watt
Gallon	3.785	liter
Horsepower	746	watt
Inch	2.54	centimetre
Inch	0.08333	foot
Inch	0.0254	metre
Inches of Hg	0.0334	Atmosphere (normal)
Joule	0.00095	British Thermal Unit
Joule	0.0239	calorie (20°C)
Joule	0.7376	foot-pound force
Joule	0.00000028	kilowatt-hour
Joule	0.0002778	watt-hour
Kilogram	35.274	ounce mass (avoir)
Kilogram	2.20462	pound mass
Kilogram force	9.80665	newton
Kilonewton	224.809	pound force
Kilowatt-hour	3,600,000	joule
Kip	4,448.22	newton
Liter	0.0353147	foot cubed
Liter	0.2642	gallon
Liter	1.057	quart
Lumen per square metre	0.092903	foot-candle
Metre	0.546807	fathom
Metre	3.28084	foot
Metre	39.37008	inch

Conversion factors, continued

Given unit	Conversion factor	Equivalent unit
Mile/hour	1.609344	kilometre/hour
Mile/hour	0.86898	knot
Millimetre	0.03937	inch
Newton	0.10197	kilogram force
Newton	0.0002248	kip
Newton	0.224809	pound force
Ounce (fluid)	29.57353	centimetre cubed
Pint	473.1765	centimetre cubed
Pound Force	0.004448	kilonewton
Pound Force	4.448	newton
Pound Mass	0.45359	kilogram
Quart	0.9463	liter
Radian	57.29578	degree
Watt	341.2142	BTU/hour
Watt	4.425372	foot-pound force/minute
Watt	13.40483	horsepower (electric)
Watt-hour	0.003600	joule

APPENDIX B

Glossary of Controller Terms

Glossary of programmable controller terms

This glossary was abstracted from, "Glossary of Programmable Controller Terms" compiled by Allen Bradley Company.

ac input module I/O rack module that converts various ac signals originating in user switches to the appropriate logic level for use within the processor

ac output module I/O rack module that converts the logic levels of the processor to a usable output signal to control a user's ac load.

address A location in the processor's memory; usually used in reference to the data table.

analog input module An I/O rack module that converts an analog signal from a user device to a digital signal that may be processed by the processor.

analog output module An I/O rack module that converts a digital signal from the processor into an analog output signal for use by a user device.

arithmetic capability The ability to do addition, subtraction, and in some cases multiplication and division, with the PC processor.

ASCII An eight-bit (7 bits + a parity bit) code that forms the American Standard Code for Information Interchange.

asynchronous Not related through repeating time patterns.

asynchronous shift register A shift register which does not require a clock. Register sigments are loaded and shifted only at date entry.

baud (1) A unit of data transmission speed equal to the number of code elements (bits) per second. (2) A unit of signaling speed equal to the number of discrete conditions or signal events per second.

BCD (binary coded decimal) The 4-bit binary notion in which individual decimal digits (0 through 9) are represented by 4-bit binary numerals; e.g., the number 23 is represented by 0010 0011 in the BCD notation.

binary A numbering system using only the digits 0 and 1. Also called "base 2."

binary word A related grouping on ones and zeros having meaning assigned by position, or numerical value in the binary system of numbers.

bit (1) An acronym for binary digit; the smallest unit of information in the binary numbering system. Represented by the digits 0 and 1. (2) The smallest division of a PC word.

bit rate The rate at which binary digits, or pulses representing them, pass a given point in a communication line.

Boolean algebra Shorthand notation for expressing logic functions.

buss An electrical channel along which data can be sent or received.

byte A sequence of binary digits usually operated upon as a unit. (The exact number depends on the system, but often 8 bits.)

card reader A device for reading information from punched cards.

cassette recorder A magnetic tape recording and playback device for entering or storing programs.

central processing unit (CPU) Another term for processor. It includes the circuits controlling the interpretation and execution of the user-inserted program instructions stored in the PC memory.

character One symbol of a set of elementary symbols, such as a letter of the alphabet or a decimal numeral. Characters may be expressed in many binary codes. For example, an ASCII character is a group of 7 bits.

chip A tiny piece of semiconductor material on which microscopic electronic components are photoetched to form one or more circuits. After connection leads and a case are added to the chip, it is called an integrated circuit.

circuit card A printed circuit board containing electronic components.

clear To return a memory to a nonprogrammed state, usually represented as "O" or OFF (empty).

clock A pulse generator that synchronizes the timing of various logic circuits and memory in the processor.

clock rate The speed (frequency) at which the processor operates, as determined by the rate at which words or bits are transferred through internal logic sequences.

CMOS Complementary metal-oxide semiconductor circuitry. An integrated circuit family that has low power consumption.

coding The preparation of a set of instructions or symbols that, when used by a programmable controller, have a special external meaning.

computer interface A device designed for data communication between a central computer and another unit such as a PC processor.

contact symbology diagram Commonly referred to as a ladder diagram, it expresses the user-programmed logic of the controller in relay-equivalent symbology.

core memory A device used to store information in ferrite cores. Each may be magnetized in either polarity, which are represented by a logical "1" or "0". This memory is nonvolatile: the contents of the memory are retained while power is off.

counter In relay-panel hardware, an electromechanical device that can be wired and preset to control other devices according to the total cycles of one ON and OFF function. In a PC, a counter is internal to the processor, i.e., it is controlled by a user-programmed instruction. A counter instruction has greater capability than any hardware counter. Therefore, PC applications do not require hardware counters.

CRT terminal A terminal containing a cathode ray tube to display programs as ladder diagrams that use instruction symbols similar to relay characters. A CRT terminal can also display data lists and application reports.

cursor A visual movable pointer used on a CRT by the PC programmer to indicate where an instruction is to be added to the PC program. The cursor is also used during editing functions.

data link Equipment, especially transmission cables and interface modules, that permits the transmission of information.

debugging The process of detecting, locating, and correcting mistakes in hardware (system wiring) or software (program).

diagnostic program A test program to help isolate hardware malfunctions in the programmable controller and application equipment.

digital The representation of numerical quantities by means of discrete numbers. It is possible to express in binary digital form all information stored, transferred, or processed by dual-state conditions; e.g., on/off, open/closed, octal, and BCD values.

documentation An orderly collection of recorded hardware and software data such as tables, listing, diagrams, etc., to provide reference information for PC application, operation, and maintenance.

downtime The time when a system is not available for production owing to required maintenance.

duplex Two-way data transmission. Full-duplex describes two data paths that allow simultaneous data transmission in both directions. Half-duplex describes one data path that allows data transmission in either of two directions, but only in one direction at a time.

edit To deliberately modify the user program in the PC memory.

execution The performance of a specific operation, such as would be accomplished through processing one instruction, a series of instructions, or a complete program.

execution time The total time required for the execution of one specific operation.

feedback The signal or data fed back to the PC from a controlled machine or process to denote its response to the command signal.

filter Electrical device used to suppress undesirable electrical noise.

firmware A series of instructions in ROM (read-only memory). These instructions are for internal processor functions only, and are transparent to the user.

flow chart A graphical representation for the definition, analysis, or solution of a problem. Symbols are used to represent a process or sequence of decisions and events.

full-duplex A mode of data transmission that is the equivalent of two paths—one in each direction simultaneously.

half-duplex A model of data transmission capable of communicating in one of two directions, but in only one direction at a time.

hard copy Any form of printed document, such as ladder diagram program listing, paper tape, or punched cards.

hardware The mechanical, electrical, and electronic devices that compose a programmable controller and its applications.

input devices Devices such as limit switches, pressure switches, push buttons, etc. that supply data to a programmable controller. These discrete inputs are of two types: those with common return, and those with individual returns (referred to as isolated inputs). Other inputs include analog devices and digital encoders.

instruction A command or order that will cause a PC to perform one certain prescribed operation.

interfacing Interconnecting a PC with its application devices, and data terminals through various modules and cables, interface modules convert PC logic levels into external signal levels, and vice versa.

I/O Abbreviation for input/output.

I/O electrical isolation: Separation of the field/wiring circuits from the logic circuits of the PC, typically done with optimal isolation.

I/O module The printed board that is the termination for field wiring of I/O devices.

I/O rack A chassis that contains I/O modules.

I/O scan The time required for the PC processor to monitor all inputs and control all outputs. The I/O scan repeats continuously.

isolated I/O module A module that has each input or output electrically isolated from every other input or output on that module. That is to say, each input or output has a separate return wire.

ladder diagram An industry standard for representing control logic relay systems.

language A set of symbols and rules for representing and communicating information (data) among people, or between people and machines.

large-scale integration (LSI) Any integrated circuit that has more than 100 equivalent gates manufactured simultaneously on a single slide of semiconductor material.

latching relay A relay with two separate coils, one of which must be engaged to change the state of the relay; it will remain in either state without power.

line In communications, describes cables, telephone lines, etc., over which data is transmitted to and received from the terminal.

line driver An integrated circuit specifically designed to transmit digital information over long lines—that is, over extended distances.

line printer A high-speed printer device that prints an entire line at one time.

location A storage position in memory.

logic A means of solving complex problems through the repeated use of simple functions that define basic concepts. Three basic logic functions are AND, OR, and NOT.

logic diagram A drawing that represents the logic functions AND, OR, NOT, etc.

logic level The voltage magnitude associated with signal pulses representing ones and zeros ("1" and "0") in binary computation.

matrix A two-dimensional array of circuit elements such as wires, diodes, etc., that can transform one type to another.

memory A grouping of circuit elements that has data storage retrieval capability.

memory module A processor module consisting of memory storage and capable of storing a finite number of words (e.g. 4096 words in a 4K memory module). Storage capacity is usually rounded off and abbreviated, with K representing each 1024 words.

microelectronic Refers to circuits built from miniaturized components and includes integrated circuits.

microprocessor An electronic computer processor position implemented in relatively few IC chips (typically LSI) which contains arithmetic, logic, register, control, and memory functions.

microsecond (μs) One millionth of a second: 1×10^{-6} or 0.000001 second.

millisecond (ms) One thousandth of a second: 10^{-3} or 0.001 second.

modem Acronym for data set (MOdulator/DEModulator).

module An interchangeable "plug-in" item containing electronic components that may be combined with other interchangeable items to form a complete unit.

MOS Metal–oxide semiconductor. It is a technology for manufacturing low-current-drain transistors and integrated circuits.

multiplexing The time-shared scanning of a number of data lines into a single channel. Only one data line is enabled at any instant.

noise Extraneous signals; any disturbance that causes interference with the desired signal or operation.

nonvolatile memory A memory that does not lose its information while its power supply is turned off.

offline Describes equipment or devices that are not connected to the communications line.

online Describes equipment or devices that are connected to the communications line.

online operation Operation in which the programmable controller is directly controlling the machine or process.

output Information transferred from PC through output modules to control output devices.

output devices Devices such as solenoids, motor starters, etc., that receive data from the programmable controller.

parallel operation Type of information transfer whereby all digits of a word are handled simultaneously.

parallel output Simultaneous availability of two or more bits, channels, or digits.

parity A method of verifying the accuracy of recorded data.

parity bit An additional bit added to a memory word to make the sum of the number of "1's" in a word always even or odd "even parity" or "odd parity."

parity check A check that tests whether the number of "1's" in an array of binary digits is odd or even.

PC Abbreviation for programmable controller.

peripheral equipment Units that may communicate with the programmable controller, but are not part of the programmable controller, e.g., Teletype, cassette recorder, CRT terminal, tape reader, etc.

printed circuits A board on which a predetermined pattern of printed connections has been formed.

processor A unit in the programmable controller that scans all the inputs and outputs in a predetermined order. The processor monitors the status of the inputs and outputs in response to the user-programmed instructions in memory, and it energizes or de-entergizes outputs as a result of the logical comparisons made through these instructions.

program A sequence of instructions to be executed by the PC processor to control a machine or process.

program panel A device for inserting, monitoring, and editing a program in a PC.

program scan The time required for the PC processor to execute all instructions in the program once. The program scan repeats continuously. The program monitors inputs and controls outputs through the input and output image tables.

programmable controller (PC) A solid-state control system that has a user-programmable memory for storage of instructions to implement specific functions such as I/O control logic, timing, counting, arithmetic, and data manipulation. A PC consists of central processor input/output interface, memory, and a programming device that typically uses relay-equivalent symbols. A PC is purposely designed as an industrial control system that can perform functions equivalent to a relay panel or a wired solid-state logic control system.

PROM Abbreviation for programmable read-only memory. A digital storage device that can be written into only once but read any number of times.

protocol A defined means of establishing criteria for receiving and transmitting data through communication channels.

RAM A random-access memory is an addressable LSI device used to store information in microscopic flip-flops or capacitors. Each may be set to an ON or OFF state, representing logical "1" or "0". This type of memory is volatile. That is to say, memory is lost while power is off, unless battery back-up is used.

read To sense the presence of information in some type of storage, which includes RAM memory, magnetic tape, punched tape, etc.

resolution A measure of the smallest possible increment of change in the variable output of a device.

ROM A read-only memory is a digital storage device specified for a single function. Data is loaded permanently into the ROM when it is manufactured. This data is available whenever the ROM address lines are scanned.

rung A grouping of PC instructions that controls one output. This is represented as one section of a logic-ladder diagram.

scan time The time necessary to completely execute the entire PC program one time.

self-diagnostic Describes the hardware and firmware within a controller which allows it to continuously monitor its own status and indicate any fault which might occur within it.

serial operation Type of information transfer within a programmable controller whereby the bits are handled sequentially rather than simultaneously as they are in parallel operation. Serial operation is slower than parallel operation for an equivalent clock rate. However, only one channel is required for serial operation.

significant digit A digit that contributes to the precision of a number. The number of significant digits is counted beginning with the digit contributing the highest value, called the most-significant digit, and ending with the one contributing the lowest value, called the least-significant digit.

software The user program that controls the operation of a programmable controller.

solid-state devices (semiconductors) Electronic components that control electron flow through solid materials such as crystals; e.g., transistors, diodes, integrated circuits.

special-purpose logic Proprietary features of a programmable controller that allow it to perform logic not normally found in relay ladder logic.

start-up The time between equipment installation and the full operation of the system.

state The logic "0" or "l" condition in PC memory or at a circuit's input or output.
storage Synonymous with memory.
strip printer A peripheral device used with a PC to provide a hardcopy of process numbers, status, and functions.
surge A transient variation in the current and/or potential at a point in the circuit.
synchronous shift register Shift register that uses a clock for timing of a system operation and where only one state change per clock pulse occurs.
system A collection of units combined to work as a larger integrated unit having the capabilities of all the separate units.

tape reader A unit that is capable of sensing data from punched tape.
Teletype A peripheral electromechanical device for inserting or recording a program into or from a PC memory in either a punched paper tape or printed ladder diagram format.
termination (1) The load connected to the output end of a transmission line. (2) The provisions for ending a transmission line connecting to a bus bar or other terminating device.
thumbwheel switch A rotating numeric switch used to input numerical information to a controller.
timer In relay-panel hardware, an electromechanical device that can be wired and preset to control the operating interval of other devices. In a PC, a timer is internal to the processor, which is to say it is controlled by a user-programmed instruction. A timer instruction has greater capability than any hardware timer. Therefore, PC applications do not require hardware timers.
transducer A device used to convert physical parameters, such as temperature, pressure, or weight, into electrical signals.
translator package A computer program that allows a user program (in binary) to be converted into a usable form for computer manipulation.
truth table A matrix that describes a logic function by listing all possible combinations of inputs and indicating the outputs for each combinaton.
TTL Abbreviation for transistor–transistor logic. A family of integrated circuit logic (usually 5 volts is high or "1", and 0 volts is low or "0", $5\ V = 1$, $0\ V = 0$).
TTY An abbreviation of Teletype.

UV-erasable PROM An ultraviolet-erasable PROM is a programmable read-only memory that can be cleared (set to "0") by exposure to intense ultraviolet light. After being cleared, it may be reprogrammed.

volatile memory A memory that loses its information if the power is removed from it.

weighted value The numerical value assigned to any single bit as a function of its position in the code word.
word A grouping or a number of bits in a sequence that is treated as a unit and is stored in one memory location.
word length The number of bits in a word. In PC literature these are generally only data bits. One PC word = 16 data bits.
write The process of loading information into a memory. Writing to a printer is obtained by first writing to special designated memory. Special hardware manages the transference of data to the printer.

INDEX